MW01258860

ASAE STANDARDS 1988

Standards, Engineering Practices and Data developed and adopted by the American Society of Agricultural Engineers

EDITORS:
Russell H. Hahn, PE
Director of Standards
Evelyn E. Rosentreter
Administrative Assistant for Standards

35th Edition
Published June, 1988

2950 Niles Road, St. Joseph, MI 49085-9659 USA
Phone 616-429-0300, Eastern Time Zone
FAX: 616-429-3852

The American Society of Agricultural Engineers is a professional and technical organization of members worldwide interested in engineering knowledge and technology for agriculture, associated industries, and related resources.

ISBN 0-916150-93-3
ISSN 8755-1187
Library of Congress Card Number: 54-14360

PREFACE

This 35th edition includes all current Standards, Engineering Practices and Data developed and adopted through the Cooperative Standards Program of the American Society of Agricultural Engineers (ASAE). The technical knowledge and experience of individuals from manufacturing industries, agribusiness, private practice and public service is brought together in a consensus process for expression as published ASAE Standards to benefit agriculture.

ASAE continues to grow in its mission to **be a world center for engineering standards and practices for agriculture.** This book includes 186 Standards, Engineering Practices and Data of which 55 have been newly adopted, technically revised or reconfirmed within the past year. In addition over 100 standards projects (see Table of Contents) are active within ASAE committees and subcommittees. ASAE STANDARDS is published annually and supersedes the title, Agricultural Engineers Yearbook of Standards, published 1954 - 1983.

ASAE invites and encourages you to:

- Provide feedback for use in improving current standards.
- Identify standards that must be developed now to meet the needs of tomorrow's agriculture.
- Design products and systems to conform to consensus standards.
- Use standards in testing, evaluating and comparing products.
- Educate tomorrow's engineers in the development, use and importance of standards.
- Help assure adequate funding of the Cooperative Standards Program.
- Participate on ASAE committees and subcommittees to develop needed standards.

Russell H. Hahn, PE
Director of Standards
ASAE

The ASAE COOPERATIVE STANDARDS PROGRAM

STANDARDS—ESSENTIAL TO AGRIBUSINESS AROUND THE WORLD

ASAE Standards, Engineering Practices, and Data, developed and maintained through the Cooperative Standards Program, provide:

- **INTERCHANGEABILITY** among interfunctional products manufactured by two or more organizations.
- **REDUCTION IN VARIETY OF COMPONENTS** required to serve an industry, thus improving availability and economy.
- **IMPROVED DEGREE OF HUMAN SAFETY** in operation and application of products, materials and equipment.
- **PERFORMANCE CRITERIA** for products, materials, or systems.
- **A COMMON BASIS FOR TESTING, DESCRIBING, OR INFORMING** regarding performance and characteristics of products, methods, materials or systems.
- **INCREASED EFFICIENCY** of engineering effort in design and production.
- **DESIGN DATA** in readily available form.
- **A SOUND BASIS FOR CODES, LEGISLATION, EDUCATION** related to the agricultural industry. Promote uniformity of practice.
- **INPUT FOR ISO DRAFT PROPOSALS** (International Organization for Standardization).

CONTENTS

USER INFORMATION

These questions and answers may be of assistance in using ASAE Standards 1988.

- *What are ASAE Standards?*

ASAE Standards are consensus documents developed and adopted by the American Society of Agricultural Engineers to meet standardization needs within the scope of the Society; principally agricultural equipment, processes and systems for producing, storing, handling, processing, packaging, and distributing agricultural products—including food, feed, fiber and other products of renewable resources.

- *How are ASAE Standards developed?*

Documents are developed by task groups and committees (see Table of Contents for Technical Committee Roster) in accordance with due process, consensus procedures adopted by the ASAE Board of Directors (see Table of Contents for Standardization Procedures). ASAE's procedures are accredited by the American National Standards Institute, ANSI.

- *How are documents classified?*

ASAE documents are classified as Standards, Tentative Standards, Engineering Practices or Data. Classifications are defined in Standardization Procedures published herein.

- *How are documents organized within this book?*

Documents are organized in numerical sequence within seven broad subject matter classifications as shown in the Table of Contents and tabbed for convenience. The sections are:

General Engineering for Agriculture
Agricultural Equipment
Powered Lawn and Garden Equipment
Electrical and Electronic Systems
Food and Process Engineering
Structures, Livestock, and Environment
Soil and Water Resource Management

Each section is preceded by its own table of contents listing all of the documents in numerical order that are included in that section. Complete keyword and numerical indexes are located toward the end of this book.

- *How should standards improvements or needs be reported?*

A form is provided inside the front cover for convenience in reporting errors or in suggesting specific improvements in a standard. The form may also be used to identify areas where new standards are needed. Your feedback is welcome and important. Forms or correspondence should be sent to ASAE Headquarters, 2950 Niles Road, St. Joseph, MI 49085, Attention: Director of Standards.

- *What do the alpha-numeric designations mean?*

The capital letters S, EP and D designate the document classifications of Standard, Engineering Practice and Data, respectively. The capital letter, T, after the numeric designation indicates Tentative Standard classification; such as S404T. The primary 3-digit number is a number assigned when development of the standard began as an ASAE project. The number has no significance. The digit following the decimal identifies a specific technical revision of the document. For example, S201.4 has been technically revised 4 times since the document was first adopted by ASAE.

- *How are changes within a document identified?*

To assist the reader, symbols are placed in the margins of documents that have been changed since the document was last revised. The symbol, T, preceding or in the margin adjacent to section headings, paragraph numbers, figure captions, or table headings indicates a **technical** change has been made in that area of the document. The symbol, T, preceding the title of a document indicates significant changes have been made in the entire document. The symbol, E, used similarly indicates **editorial** changes or corrections have been made with no intended change in the technical meaning of the document.

- *How can the history of a document be tracked?*

The brief paragraph beneath the title of each document provides a chronological history of the document's development, adoption, revision and reconfirmation. All ASAE documents must be reviewed every five years unless technically revised in the interim. Tentative Standards must be reviewed annually.

- *Are there related ISO Standards?*

A listing of international standards developed by ISO Technical Committee, ISO/TC 23, Tractors and Machinery for Agriculture and Forestry, is included (see Table of Contents).

- *What standards are being developed by ASAE?*

A list of current ASAE standardization projects is included (see Table of Contents).

- *Is related technical information available from ASAE?*

Approximately 900 technical papers are presented annually at ASAE meetings. This information is keyword indexed by the Society. Papers may be available relating directly or indirectly to the scope of specific Standards, Engineering Practices, and Data. Contact ASAE Headquarters for more information.

STANDARDS, ENGINEERING PRACTICES AND DATA

GENERAL ENGINEERING FOR AGRICULTURE

ASAE Notation:

The letter S preceding numerical designation indicates ASAE Standard; EP indicates Engineering Practice; D indicates Data. A decimal and numeral following the file number indicate the number of times a document has been revised. Thus ASAE S201.4 indicates Standard number 201, four times revised. The letter T after the designation indicates tentative status. Always refer to ASAE documents by complete designation to avoid confusion with standards of other organizations. For example: ASAE S201.4.

The symbol **T** preceding or in the margin adjacent to section headings, paragraph numbers, figure captions, or table headings indicates a technical change was incorporated in that area when this document was last revised. The symbol **T** preceding the title of a document indicates essentially the entire document has been revised. The symbol **E** used similarly indicates editorial changes or corrections have been made with no intended change in the technical meaning of the document.

ASAE Data: ASAE D271.2

PSYCHROMETRIC DATA

Reviewed by ASAE's Structures and Environment Division and the Food Engineering Division Standards Committees; approved by the Electric Power and Processing Division Standards Committee; adopted by ASAE December 1963; reconfirmed December 1968; revised April 1974, April 1979; reconfirmed December 1983.

SECTION 1—PURPOSE AND SCOPE

1.1 The purpose of this Data is to assemble psychrometric data in chart and equation form in both SI and English units.

1.2 Psychrometric charts are presented that give data for dry bulb temperature ranges of -35 to 600 °F in English units and -10 to 120 °C in SI units.

1.3 Many analyses of psychrometric data are made on computers. The equations given in Sections 2 and 3 enable the calculation of all psychrometric data if any two independent psychrometric properties of an air-water vapor mixture are known in addition to the atmospheric pressure. In some cases, iteration procedures are necessary. In some instances, the range of data covered by the equation has been extended beyond that given in the original source. The equations yield results that agree closely with values given by Keenan and Keyes (1936) and existing psychrometric charts.

TABLE 1—SYMBOLS

Symbol	Definition
h	Enthalpy of air-vapor mixture, J/kg dry air or Btu/lb dry air
h_{fg}	Latent heat of vaporization of water at saturation, J/kg or Btu/lb
h'_{fg}	Latent heat of vaporization of water at T_{wb}, J/kg or Btu/lb
h''_{fg}	Latent heat of vaporization of water at T_{dp}, J/kg or Btu/lb
h_{ig}	Heat of sublimation of ice, J/kg or Btu/lb
h'_{ig}	Heat of sublimation of ice at T_{wb}, J/kg or Btu/lb
h''_{ig}	Heat of sublimation of ice at T_{dp}, J/kg or Btu/lb
H	Humidity ratio, kg water/kg dry air or lb water/lb dry air
ln	Natural logarithm (base e)
P_{atm}	Atmospheric pressure, Pa or psi
P_s	Saturation vapor pressure at T, Pa or psi
P_{swb}	Saturation vapor pressure at T_{wb}, Pa or psi
P_v	Vapor pressure, Pa or psi
rh	Relative humidity, decimal
T	Dry-bulb temperature, kelvin or rankine
T_{dp}	Dew-point temperature, kelvin or rankine
T_{wb}	Wet-bulb temperature, kelvin or rankine
V_{sa}	Air specific volume, m^3/kg dry air or ft^3/lb dry air

SECTION 2—PSYCHROMETRIC DATA IN SI UNITS

2.1 Psychrometric charts; two presented. One for a temperature range of -10 to 55 °C and one for a temperature range of 20 to 120 °C.

2.2 Psychrometric equations, SI units. Symbols are defined in Table 1.

2.2.1 Saturation line. P_s as a function of T

$$\ln P_s = 31.9602 - \frac{6270.3605}{T} - 0.46057 \ln T$$

Brooker (1967)

$255.38 \leq T \leq 273.16$

and

$$\ln(P_s/R) = \frac{A + BT + CT^2 + DT^3 + ET^4}{FT - GT^2}$$

Adapted from Keenan and Keyes (1936)

$273.16 \leq T \leq 533.16$

where

R =	22,105,649.25	D =	0.12558×10^{-3}
A =	-27,405.526	E =	-0.48502×10^{-7}
B =	97.5413	F =	4.34903
C =	-0.146244	G =	0.39381×10^{-2}

2.2.2 Saturation line. T as a function of P_s

$$T - 255.38 = \sum_{i=0}^{i=8} A_i \left[\ln (0.00145 P_s)\right]^i$$

$620.52 < P_s < 4,688,396.00$ Steltz and Silvestri (1958)

A_0 = 19.5322
A_1 = 13.6626
A_2 = 1.17678
A_3 = -0.189693
A_4 = 0.087453
A_5 = -0.0174053
A_6 = 0.00214768
A_7 = -0.138343×10^{-3}
A_8 = 0.38×10^{-5}

2.2.3 Latent heat of sublimation at saturation

$$h_{ig} = 2,839,683.144 - 212.56384 (T - 255.38)$$

$255.38 \leq T \leq 273.16$ Brooker (1967)

2.2.4 Latent heat of vaporization at saturation

$$h_{fg} = 2,502,535.259 - 2,385.76424 (T - 273.16)$$

$273.16 \leq T \leq 338.72$ Brooker (1967)

$$h_{fg} = (7,329,155,978,000 - 15,995,964.08\, T^2)^{1/2}$$

$338.72 \leq T \leq 533.16$ Brooker (Unpublished)

2.2.5 Wet bulb line

$$P_{swb} - P_v = B'(T_{wb} - T)$$

Brunt (1941)

where

$$B' = \frac{1006.9254(P_{swb} - P_{atm})\left(1 + 0.15577 \frac{P_v}{P_{atm}}\right)}{0.62194\, h'_{fg}}$$

Substitute h'_{ig} for h'_{fg} where $T_{wb} \leq 273.16$

$255.38 \leq T \leq 533.16$

2.2.6 **Humidity ratio**

$$H = \frac{0.6219\ P_v}{P_{atm} - P_v}$$

$255.38 \leq T \leq 533.16$

$P_v < P_{atm}$

2.2.7 **Specific volume**

$$V_{sa} = \frac{287\ T}{P_{atm} - P_v}$$

$255.38 \leq T \leq 533.16$

$P_v < P_{atm}$

2.2.8 **Enthalpy**

Enthalpy = enthalpy of air + enthalpy of water (or ice) at dew-point temperature + enthalpy of evaporation (or sublimation) at dew-point temperature + enthalpy added to the water vapor (super-heat) after vaporization.

$$h = 1006.92540\ (T - 273.16) - H[333{,}432.1 + 2030.5980(273.16 - T_{dp})] + h''_{ig}\ H + 1875.6864\ H(T - T_{dp})$$

$255.38 \leq T_{dp} \leq 273.16$

and

$$h = 1006.92540(T - 273.16) + 4186.8\ H\ (T_{dp} - 273.16) + h''_{fg}\ H + 1875.6864\ H\ (T - T_{dp})$$

$273.16 \leq T_{dp} \leq 373.16$

2.2.9 **Relative humidity**

$$rh = P_v/P_s$$

SECTION 3—PSYCHROMETRIC DATA IN ENGLISH UNITS

3.1 Three psychrometric charts, are presented with temperature ranges of -35 to 50 °F, 32 to 120 °F and 32 to 600 °F, respectively.

3.2 Psychrometric equations, English Units. Symbols are defined in Table 1.

3.2.1 **Saturation line. P_s as a function of T**

$$\ln P_s = 23.3924 - \frac{11286.6489}{T} - 0.46057 \ln T$$

Brooker (1967)

$459.69 \leq T \leq 491.69$

$$\ln(P_s/R) = \frac{A + BT + CT^2 + DT^3 + ET^4}{FT - GT^2}$$

Adapted from Keenan and Keyes (1936)

$491.69 \leq T \leq 959.69$

where

$R = 3206.18$
$A = -27405.5$
$B = 54.1896$
$C = -0.045137$
$D = 0.215321 \times 10^{-4}$
$E = -0.462027 \times 10^{-8}$
$F = 2.41613$
$G = 0.00121547$

3.2.2 **Saturation line. T as a function of P_s**

$$T - 459.69 = \sum_{i=0}^{i=8} A_i\ [\ln(10P_s)]^i$$

Steltz and Silvestri (1958)

$0.09 \leq P_s \leq 680$

where

$A_0 = 35.1579$
$A_1 = 24.5926$
$A_2 = 2.11821$
$A_3 = -0.341447$
$A_4 = 0.157416$
$A_5 = -0.0313296$
$A_6 = 0.00386583$
$A_7 = -0.249018 \times 10^{-3}$
$A_8 = 0.684016 \times 10^{-5}$

3.2.3 **Latent heat of sublimation at saturation**

$$h_{ig} = 1220.844 - 0.05077\ (T - 459.69)$$

$459.69 \leq T \leq 491.69$ Brooker (1967)

3.2.4 **Latent heat of vaporization at saturation**

$$h_{fg} = 1075.8965 - 0.56983\ (T - 491.69)$$

Brooker (1967)

$491.69 \leq T \leq 609.69$

$$h_{fg} = (1354673.214 - 0.9125275587\ T^2)^{1/2}$$

Brooker (Unpublished)

$609.69 \leq T \leq 959.69$

3.2.5 **Wet bulb line**

$$P_{swb} - P_v = B'(T_{wb} - T)$$

Brunt (1941)

where

$$B' = \frac{0.2405\ (P_{swb} - P_{atm})\ (1 + 0.15577 P_v/P_{atm})}{0.62194\ h'_{fg}}$$

Substitute h'_{ig} for h'_{fg} when $T_{wb} \leq 491.69$

$459.69 \leq T \leq 959.69$

3.2.6 **Absolute humidity (humidity ratio)**

$$H = \frac{0.6219\ P_v}{P_{atm} - P_v}$$

$459.69 \leq T \leq 959.69$

$P_v < P_{atm}$

3.2.7 **Specific volume**

$$V_{sa} = \frac{53.35 \times T}{144\ (P_{atm} - P_v)}$$

$459.69 \leq T \leq 959.69$

$P_v < P_{atm}$

3.2.8 Enthalpy

Enthalpy = enthalpy of air + enthalpy of water (or ice) at dew-point temperature + enthalpy of evaporation (or sublimation) at dew-point temperature + enthalpy added to the water vapor (super-heat) after vaporization.

$$h = 0.2405\,(T - 459.69) - H\,[143.35 + 0.485(491.69 - T_{dp})] + h''_{ig}\,H + 0.448\,H\,(T - T_{dp})$$

$$459.69 \leqslant T_{dp} \leqslant 491.69$$

$$h = 0.2405\,(T - 459.69) + H\,(T_{dp} - 491.69) + h''_{fg}\,H + 0.448H\,(T - T_{dp})$$

$$491.69 \leqslant T_{dp} \leqslant 671.69$$

3.2.9 Relative humidity

$$rh = P_v/P_s$$

Note: Psychrometric charts are printed with permission from the American Society of Heating, Refrigerating and Airconditioning Engineers, Inc., 345 E. 47th St., New York, NY; Proctor & Schwartz, Inc., 7th St. and Tabor Rd., Philadelphia, PA; and Carrier Corp., Carrier Parkway, Syracuse, NY.

References: Last printed in 1981 AGRICULTURAL ENGINEERS YEARBOOK; list available from ASAE Headquarters.

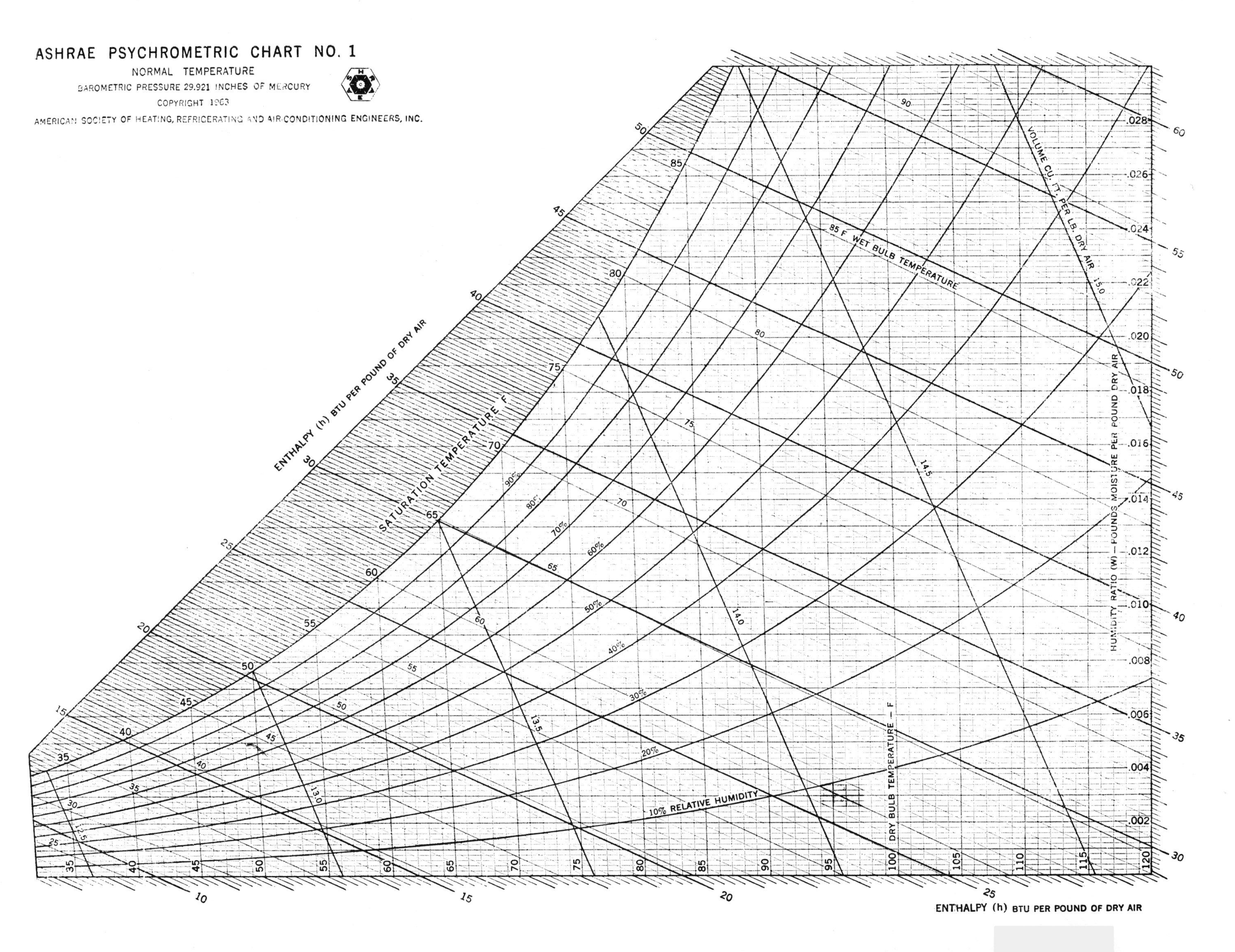
ASHRAE PSYCHROMETRIC CHART NO. 1
NORMAL TEMPERATURE
BAROMETRIC PRESSURE 29.921 INCHES OF MERCURY
COPYRIGHT 1963
AMERICAN SOCIETY OF HEATING, REFRIGERATING AND AIR-CONDITIONING ENGINEERS, INC.
ENTHALPY (h) BTU PER POUND OF DRY AIR
SATURATION TEMPERATURE — F
85 F WET BULB TEMPERATURE
VOLUME CU. FT. PER LB. DRY AIR
HUMIDITY RATIO (W) — POUNDS MOISTURE PER POUND DRY AIR
DRY BULB TEMPERATURE — F
10% RELATIVE HUMIDITY
ENTHALPY (h) BTU PER POUND OF DRY AIR

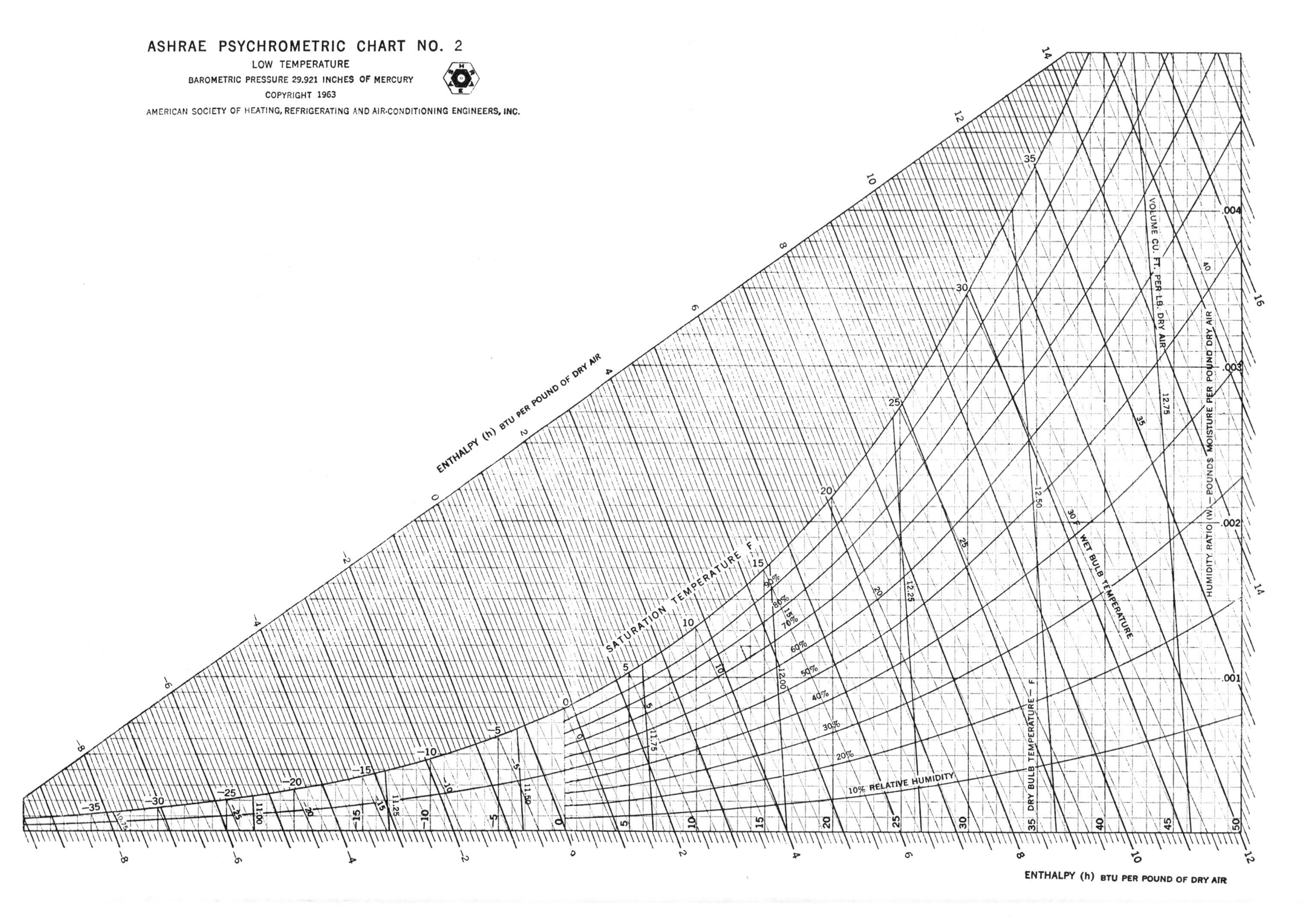
ASHRAE PSYCHROMETRIC CHART NO. 2
LOW TEMPERATURE
BAROMETRIC PRESSURE 29.921 INCHES OF MERCURY
COPYRIGHT 1963
AMERICAN SOCIETY OF HEATING, REFRIGERATING AND AIR-CONDITIONING ENGINEERS, INC.
ENTHALPY (h) BTU PER POUND OF DRY AIR
SATURATION TEMPERATURE F
VOLUME CU. FT. PER LB. DRY AIR
HUMIDITY RATIO (W)—POUNDS MOISTURE PER POUND DRY AIR
30 F WET BULB TEMPERATURE
DRY BULB TEMPERATURE — F
10% RELATIVE HUMIDITY
ENTHALPY (h) BTU PER POUND OF DRY AIR

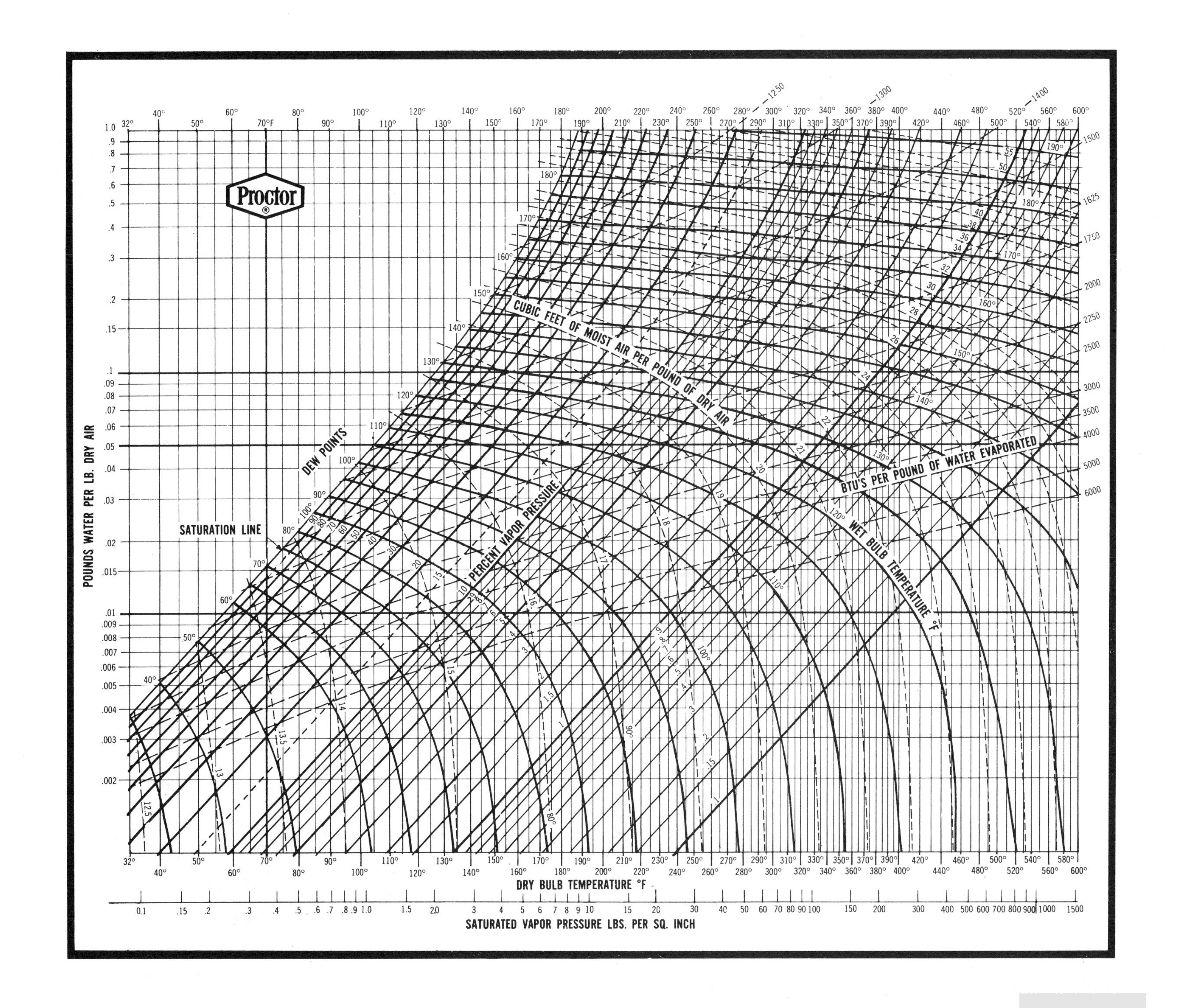

Proctor
SATURATION LINE
DEW POINTS
PERCENT VAPOR PRESSURE
CUBIC FEET OF MOIST AIR PER POUND OF DRY AIR
BTU'S PER POUND OF WATER EVAPORATED
WET BULB TEMPERATURE °F
DRY BULB TEMPERATURE °F
SATURATED VAPOR PRESSURE LBS. PER SQ. INCH
POUNDS WATER PER LB. DRY AIR

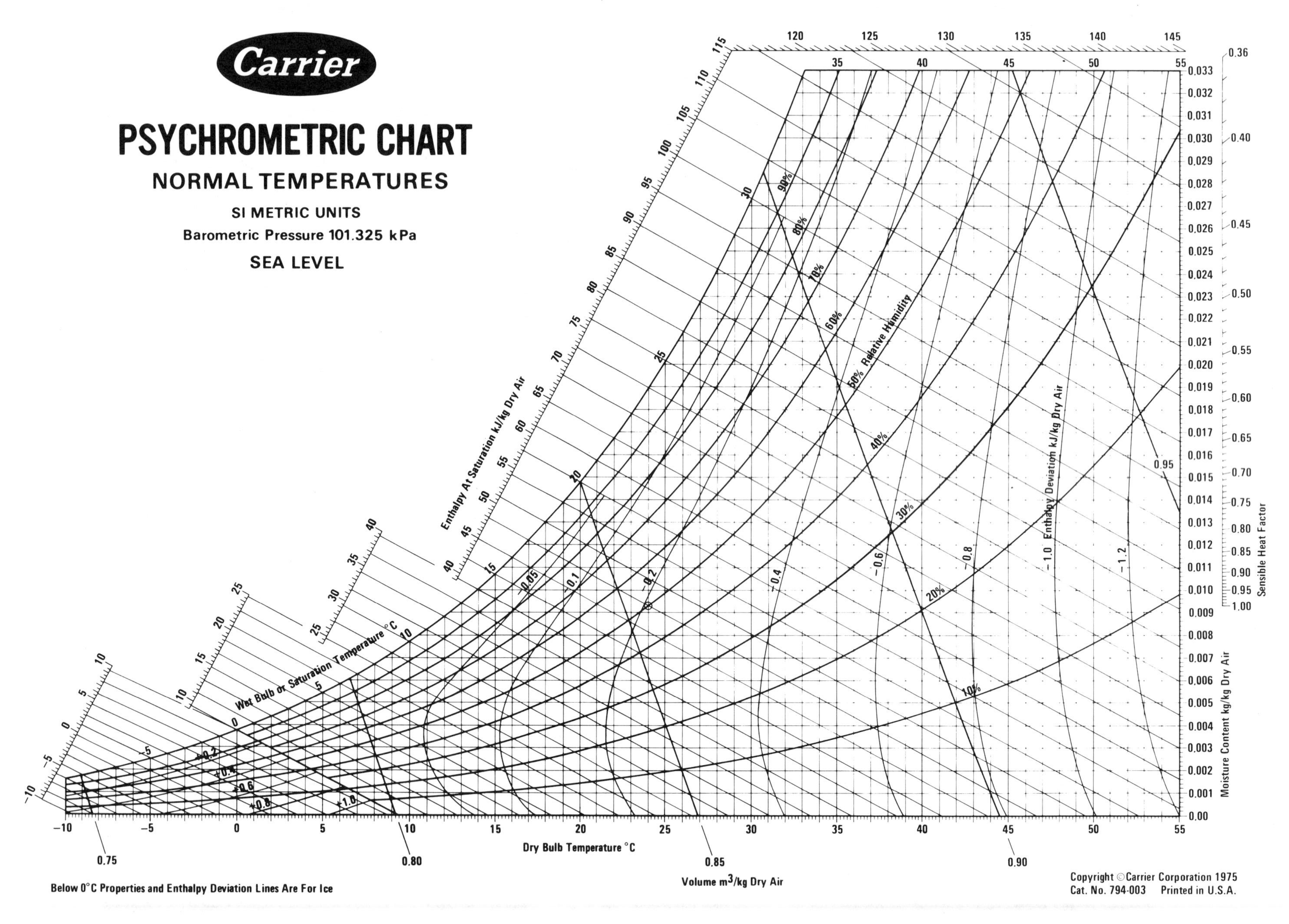
Carrier
PSYCHROMETRIC CHART
NORMAL TEMPERATURES
SI METRIC UNITS
Barometric Pressure 101.325 kPa
SEA LEVEL
Enthalpy At Saturation kJ/kg Dry Air
Wet Bulb or Saturation Temperature °C
Relative Humidity
Enthalpy Deviation kJ/kg Dry Air
Dry Bulb Temperature °C
Volume m3/kg Dry Air
Moisture Content kg/kg Dry Air
Sensible Heat Factor
Below 0°C Properties and Enthalpy Deviation Lines Are For Ice
Copyright © Carrier Corporation 1975
Cat. No. 794-003 Printed in U.S.A.

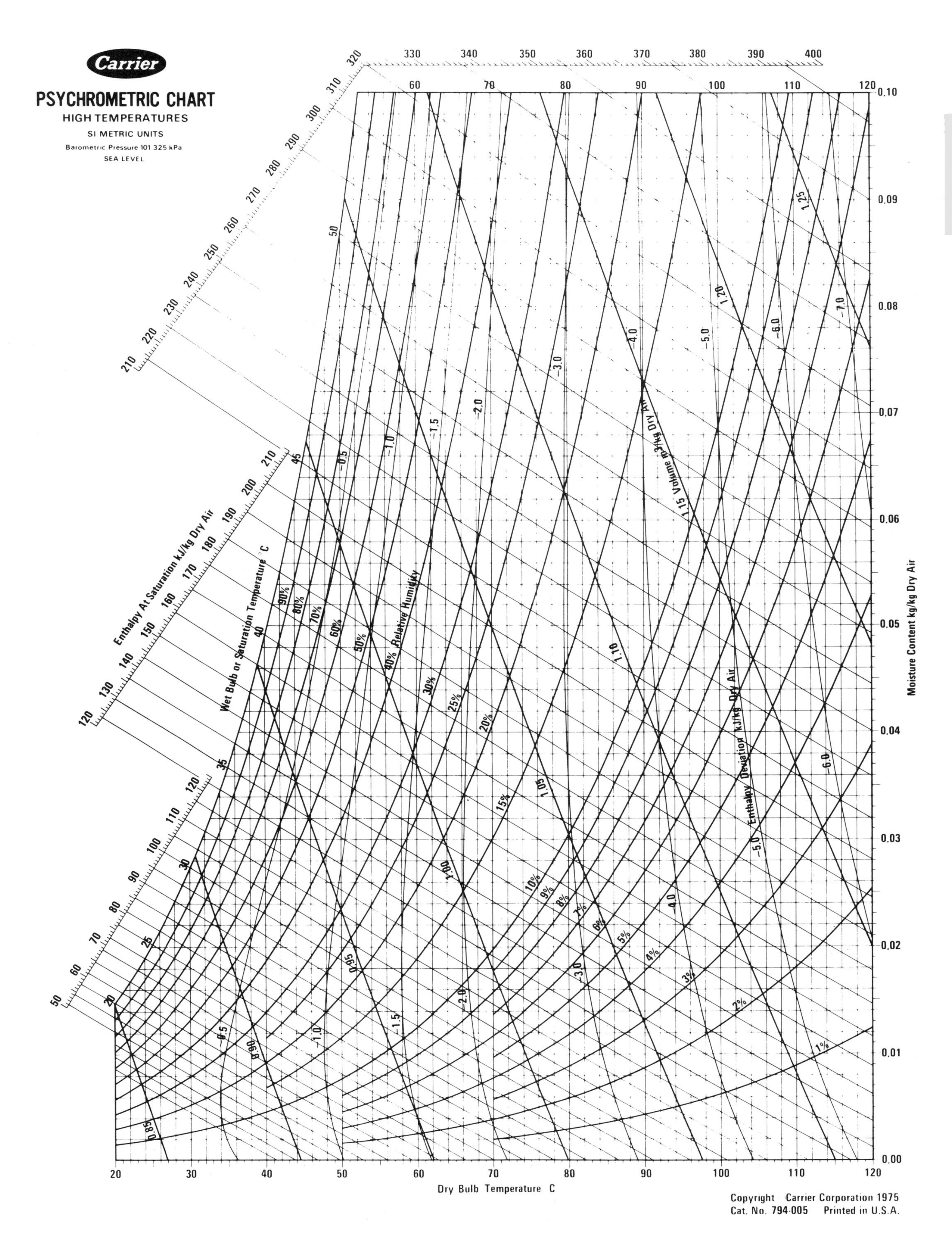
Carrier
PSYCHROMETRIC CHART
HIGH TEMPERATURES
SI METRIC UNITS
Barometric Pressure 101 325 kPa
SEA LEVEL
Enthalpy At Saturation kJ/kg Dry Air
Wet Bulb or Saturation Temperature C
Relative Humidity
Volume m3/kg Dry Air
Enthalpy Deviation kJ/kg Dry Air
Moisture Content kg/kg Dry Air
Dry Bulb Temperature C
Copyright Carrier Corporation 1975
Cat. No. 794-005 Printed in U.S.A.

USE OF SI (METRIC) UNITS

Prepared under the general direction of ASAE Committee on Standards (T-1); reviewed and approved by ASAE division standardizing committees: Power and Machinery Division Technical Committee (PM-03), Structures and Environment Division Technical Committee (SE-03), Electric Power and Processing Division Technical Committee (EPP-03), Soil and Water Division Standards Committee (SW-03), and Education and Research Division Steering Committee; adopted by ASAE as a Recommendation December 1964; reconfirmed December 1969, December 1970, December 1971, December 1972; revised by the Metric Policy Subcommittee December 1973; revised March 1976; revised and reclassified as an Engineering Practice April 1977; revised April 1979; revised editorially December 1979; revised September 1980, February 1982; revised editorially January 1985; reconfirmed December 1985; revised March 1988.

SECTION 1—PURPOSE AND SCOPE

1.1 This Engineering Practice is intended as a guide for uniformly incorporating the International System of Units (SI). It is intended for use in implementing ASAE policy, "Use of SI Units in ASAE Standards, Engineering Practices, and Data." This Engineering Practice includes a list of preferred units and conversion factors.

SECTION 2—SI UNITS OF MEASURE

2.1 SI consists of seven base units, two supplementary units, a series of derived units consistent with the base and supplementary units. There is also a series of approved prefixes for the formation of multiples and submultiples of the various units. A number of derived units are listed in paragraph 2.1.3 including those with special names. Additional derived units without special names are formed as needed from base units or other derived units, or both.

2.1.1 Base and supplementary units. For definitions refer to International Organization for Standardization ISO 1000, SI Units and Recommendations for the Use of Their Multiples and of Certain Other Units.

Base Units

meter (m) — unit of length
second (s) — unit of time
kilogram (kg) — unit of mass
kelvin (K) — unit of thermodynamic temperature
ampere (A) — unit of electric current
candela (cd) — luminous intensity
mole (mol) — the amount of a substance

Supplementary units

radian (rad) — plane angle
steradian (sr) — solid angle

2.1.2 SI unit prefixes

Multiples & Submultiples	Prefix	SI Symbol
10^{18}	exa	E
10^{15}	peta	P
10^{12}	tera	T
10^{9}	giga	G
10^{6}	mega	M
10^{3}	kilo	k
10^{2}	hecto	h
10^{1}	deka	da
10^{-1}	deci	d
10^{-2}	centi	c
10^{-3}	milli	m
10^{-6}	micro	μ
10^{-9}	nano	n
10^{-12}	pico	p
10^{-15}	femto	f
10^{-18}	atto	a

2.1.3 Derived units are combinations of basic units or other derived units as needed to describe physical properties, for example, acceleration. Some derived units are given special names; others are expressed in the appropriate combination of SI units. Some currently defined derived units are tabulated in Table 1.

SECTION 3—RULES FOR SI USAGE

3.1 General. The established SI units (base, supplementary, derived, and combinations thereof with appropriate multiple or submultiple prefixes) should be used as indicated in this section.

3.2 Application of prefixes. The prefixes given in paragraph 2.1.2 should be used to indicate orders of magnitude, thus eliminating insignificant digits and decimals, and providing a convenient substitute for writing powers of 10 as generally preferred in computation. For example:

12 300 m or 12.3×10^{3} m becomes 12.3 km, and
0.0123 mA or 12.3×10^{-6} A becomes 12.3 μA

It is preferable to apply prefixes to the numerator of compound units, except when using kilogram (kg) in the denominator, since it is a base unit of SI and should be used in preference to the gram. For example:

Use 200 J/kg, not 2 dJ/g

With SI units of higher order such as m^2 or m^3, the prefix is also raised to the same order.
For example:

mm^2 is $(10^{-3}\ m)^2$ or $10^{-6}\ m^2$

3.3 Selection of prefix. When expressing a quantity by a numerical value and a unit, a prefix should be chosen so that the numerical value preferably lies between 0.1 and 1000, except where certain multiples and submultiples have been agreed for particular use. The same unit, multiple, or submultiple should be used in tables even though the series may exceed the preferred range of 0.1 to 1000. Double prefixes and hyphenated prefixes should not be used.
For example:

use GW (gigawatt) not kMW

3.4 Capitalization. Symbols for SI units are only capitalized when the unit is derived from a proper name; for example, N for Isaac Newton (except liter, L). Unabbreviated units are not capitalized; for example kelvin and newton. Numerical prefixes given in paragraph 2.1.2 and their symbols are not capitalized; except for the symbols M (mega), G (giga), T (tera), P (peta) and E (exa).

3.5 Plurals. Unabbreviated SI units form their plurals in the usual manner. SI symbols are always written in singular form.
For example:

50 newtons or 50 N
25 millimeters or 25 mm

3.6 Punctuation. Whenever a numerical value is less than one, a zero should precede the decimal point. Periods are not used after any SI unit symbol, except at the end of a sentence. English speaking countries use a dot for the decimal point, others use a comma. Use spaces instead of commas for grouping numbers into threes (thousands).
For example:

6 357 831.376 88
not 6,357,831.376,88

TABLE 1—DERIVED UNITS

Quantity	Unit	SI Symbol	Formula
acceleration	meter per second squared	___	m/s^2
activity (of a radioactive source)	disintegration per second	___	(disintegration)/s
angular acceleration	radian per second squared	___	rad/s^2
angular velocity	radian per second	___	rad/s
area	square meter	___	m^2
density	kilogram per cubic meter	___	kg/m^3
electrical capacitance	farad	F	A·s/V
electrical conductance	siemens	S	A/V
electrical field strength	volt per meter	___	V/m
electrical inductance	henry	H	V·s/A
electrical potential difference	volt	V	W/A
electrical resistance	ohm	Ω	V/A
electromotive force	volt	V	W/A
energy	joule	J	N·m
entropy	joule per kelvin	___	J/K
force	newton	N	$kg{\cdot}m/s^2$
frequency	hertz	Hz	(cycle)/s
illuminance	lux	lx	lm/m^2
luminance	candela per square meter	___	cd/m^2
luminous flux	lumen	lm	cd·sr
magnetic field strength	ampere per meter	___	A/m
magnetic flux	weber	Wb	V·s
magnetic flux density	tesla	T	Wb/m^2
magnetomotive force	ampere	A	___
power	watt	W	J/s
pressure	pascal	Pa	N/m^2
quantity of electricity	coulomb	C	A·s
quantity of heat	joule	J	N·m
radiant intensity	watt per steradian	___	W/sr
specific heat	joule per kilogram-kelvin	___	J/kg·K
stress	pascal	Pa	N/m^2
thermal conductivity	watt per meter-kelvin	___	W/m·K
velocity	meter per second	___	m/s
viscosity, dynamic	pascal-second	___	Pa·s
viscosity, kinematic	square meter per second	___	m^2/s
voltage	volt	V	W/A
volume	cubic meter	___	m^3
wavenumber	reciprocal meter	___	(wave)/m
work	joule	J	N·m

3.7 Derived units. The product of two or more units in symbolic form is preferably indicated by a dot midway in relation to unit symbol height. The dot may be dispensed with when there is no risk of confusion with another unit symbol.
For example:

Use N·m or N m, but not mN

A solidus (oblique stroke, /), a horizontal line, or negative powers may be used to express a derived unit formed from two others by division. For example:

$$m/s \quad \frac{m}{s} \quad \text{or} \quad m{\cdot}s^{-1}$$

Only one solidus should be used in a combination of units unless parentheses are used to avoid ambiguity.

3.8 Representation of SI units in systems with limited character sets. For computer printers and other systems which do not have the characters available to print SI units correctly, the methods shown in ISO 2955, Information Processing — Representation of SI and Other Units for Use in Symbols with Limited Character Sets, is recommended.

SECTION 4—NON-SI UNITS

4.1 Certain units outside the SI are recognized by ISO because of their practical importance in specialized fields. These include units for temperature, time and angle. Also included are names for some multiples of units such as "liter" (L)* for volume, "hectare" (ha) for land measure and "metric ton" (t) for mass.

*The International symbol for liter is either the lowercase "l" or the uppercase "L". ASAE recommends the use of uppercase "L" to avoid confusion with the numeral "1".

4.2 Temperature. The SI base unit for thermodynamic temperature is kelvin (K). Because of the wide usage of the degree Celsius, particularly in engineering and nonscientific areas, the Celsius scale (formerly called the centigrade scale) may be used when expressing temperature. The Celsius scale is related directly to the kelvin scale as follows:

one degree Celsius (1 °C) equals one kelvin (1 K), exactly

A Celsius temperature (t) is related to a kelvin temperature (T), as follows:

$$t = T - 273.15$$

4.3 Time. The SI unit for time is the second. This unit is preferred and should be used when technical calculations are involved. In other cases use of the minute (min), hour (h), day (d), etc., is permissible.

4.4 Angles. The SI unit for plane angle is the radian. The use of arc degrees (°) and its decimal or minute (′), second (″) submultiples is permissible when the radian is not a convenient unit. Solid angles should be expressed in steradians.

SECTION 5—PREFERRED UNITS AND CONVERSION FACTORS

5.1 Preferred units for expressing physical quantities commonly encountered in agricultural engineering work are listed in Table 2. These are presented as an aid to selecting proper units for given applications and to promote consistency where interpretation of the general rules of SI may not produce consistent results. Factors for conversion from old units to SI units are included in Table 2.

TABLE 2—PREFERRED UNITS FOR EXPRESSING PHYSICAL QUANTITIES

1. **Quantities are arranged in alphabetical order by principal nouns. For example, surface tension is listed as tension, surface.**
2. **All possible applications are not listed, but others such as rates can be readily derived. For example, from the preferred units for energy and volume the units for heat energy per unit volume, kJ/m^3, may be derived.**
3. **Conversion factors are shown to seven significant digits, unless the precision with which the factor is known does not warrant seven digits.**

Quantity	Application	From: Old Units	To: SI Units	Multiply By:
Acceleration, angular	General	rad/s^2	rad/s^2	
Acceleration, linear	Vehicle	(mile/h)/s	(km/h)/s	1.609 344*
	General (includes acceleration of gravity)#	ft/s^2	m/s^2	0.304 8*
Angle, plane	Rotational calculations	r (revolution)	r (revolution)	
		rad	rad	
	Geometric and general	° (deg)	°	
		′ (min)	° (decimalized)	1/60*
		′ (min)	′	
		″ (sec)	° (decimalized)	1/3600*
		″ (sec)	″	
Angle, solid	Illumination calculations	sr	sr	
Area	Cargo platforms, roof and floor area, frontal areas,	in.^{2}	m^2	0.000 645 16*
	fabrics, general	ft^2	m^2	0.092 903 04*
	Pipe, conduit	in.^{2}	mm^2	645.16*
		in.^{2}	cm^2	6.451 6*
		ft^2	m^2	0.092 903 04*
	Small areas, orifices, cross section area of structural shapes	in.^{2}	mm^2	645.16*
	Brake & clutch contact area, glass, radiators, feed opening	in.^{2}	cm^2	6.451 6*
	Land, pond, lake, reservoir, open water channel (Small)	ft^2	m^2	0.092 903 04*
	(Large)	acre	ha	0.404 687 3(‖) E
	(Very Large)	mile^2	km^2	2.589 998
Area per time	Field operations	acre/h	ha/h	0.404 687 3
	Auger sweeps, silo unloader	ft^2/s	m^2/s	0.092 903 04*
Bending Moment	(See Moment of Force)			
Capacitance, electric	Capacitors	μF	μF	
Capacity, electric	Battery rating	A·h	A·h	
Capacity, heat	General	Btu/°F	kJ/K†	1.899 101
Capacity, heat, specific	General	Btu/(lb·°F)	kJ/(kg·K)†	4.186 8*
Capacity, volume	(See Volume)			
Coefficient of Heat Transfer	General	Btu/(h·ft^2·°F)	W/(m^2·K)†	5.678 263 E
Coefficient of Linear Expansion	Shrink fit, general	°F^{-1}, (1/°F)	K^{-1}, (1/K)†	1.8*
Conductance, electric	General	mho	S	1*
Conductance, thermal	(See Coefficient of Heat Transfer)			
Conductivity, electric	Material property	mho/ft	S/m	3.280 840
Conductivity, thermal	General	Btu·ft/(h·ft^2·°F)	W/(m·K)†	1.730 735
Consumption, fuel	Off highway vehicles (See also Efficiency, fuel)	gal/h	L/h	3.785 412
Consumption, oil	Vehicle performance testing	qt/(1000 miles)	L/(1000 km)	0.588 036 4
Consumption, specific, oil	Engine testing	lb/(hp·h)	g/(kW·h)	608.277 4
		lb/(hp·h)	g/MJ	168.965 9
Current, electric	General	A	A	
Density, current	General	A/in.^{2}	kA/m^2	1.550 003
		A/ft^2	A/m^2	10.763 91
Density, magnetic flux	General	kilogauss	T	0.1*
Density, (mass)	Solid, general; agricultural products, soil, building materials	lb/yd^3	kg/m^3	0.593 276 3
		lb/in.^{3}	kg/m^3	27 679.90
		lb/ft^3	kg/m^3	16.018 46
	Liquid	lb/gal	kg/L	0.119 826 4
	Gas	lb/ft^3	kg/m^3	16.018 46
	Solution concentration	—	g/m^3, mg/L	—
Density of heat flow rate	Irradiance, general	Btu/(h·ft^2)	W/m^2	3.154 591††
Consumption, fuel	(See Flow, volume)			
Consumption, specific fuel	(See Efficiency, fuel)			
Drag	(See Force)			
Economy, fuel	(See Efficiency, fuel)			

TABLE 2—PREFERRED UNITS FOR EXPRESSING PHYSICAL QUANTITIES (cont'd)

1. Quantities are arranged in alphabetical order by principal nouns. For example, surface tension is listed as tension, surface.
2. All possible applications are not listed, but others such as rates can be readily derived. For example, from the preferred units for energy and volume the units for heat energy per unit volume, kJ/m^3, may be derived.
3. Conversion factors are shown to seven significant digits, unless the precision with which the factor is known does not warrant seven digits.

Quantity	Application	From: Old Units	To: SI Units	Multiply By:
Efficiency, fuel	Highway vehicles			
	economy	mile/gal	km/L	0.415 143 7
	consumption	——	L/(100 km)	§
	specific fuel consumption	lb/(hp·h)	g/MJ	168.965 9
	Off-highway vehicles			
	economy	hp·h/gal	kW·h/L	0.196 993 1
	specific fuel consumption	lb/(hp·h)	g/MJ	168.965 9
	specific fuel consumption	lb/(hp·h)	kg/(kW·h)§§	0.608 277 4
Energy, work, enthalpy, quantity of heat	Impact strength	ft·lbf	J	1.355 818
	Heat	Btu	kJ	1.055 056
		kcal	kJ	4.186 8*
	Energy usage, electrical	kW·h	kW·h	
		kW·h	MJ	3.6
	Mechanical, hydraulic, general	ft·lbf	J	1.355 818
		ft·pdl	J	0.042 140 11
		hp·h	MJ	2.684 520
		hp·h	kW·h	0.745 699 9
Energy per area	Solar radiation	Btu/ft^2	MJ/m^2	0.011 356 528
Energy, specific	General	cal/g‡	J/g	4.186 8*
		Btu/lb	kJ/kg	2.326*
Enthalpy	(See Energy)			
Entropy	(See Capacity, heat)			
Entropy, specific	(See Capacity, heat, specific)			
Floor loading	(See Mass per area)			
Flow, heat, rate	(See Power)			
Flow, mass, rate	Gas, liquid	lb/min	kg/min	0.453 592 4
		lb/s	kg/s	0.453 592 4
	Dust flow	g/min	g/min	
	Machine work capacity, harvesting, materials handling	ton (short)/h	t/h, Mg/h**	0.907 184 7
Flow, volume	Air, gas, general	ft^3/s	m^3/s	0.028 316 85
		ft^3/s	m^3/min	1.699 011
	Liquid flow, general	gal/s (gps)	L/s	3.785 412
		gal/s (gps)	m^3/s	0.003 785 412
		gal/min (gpm)	L/min	3.785 412
	Seal and packing leakage, sprayer flow	oz/s	mL/s	29.573 53
		oz/min	mL/min	29.573 53
	Fuel consumption, micro irrigation emitter flow	gal/h	L/h	3.785 412
	Pump capacity, coolant flow, oil flow, irrigation sprinkler capacity	gal/min (gpm)	L/min	3.785 412
	Irrigation pump capacity, pipe flow	gal/min (gpm)	L/s	0.063 090 20
	River and channel flow	ft^3/s	m^3/s	0.028 316 85
Flux, luminous	Light bulbs	lm	lm	
Flux, magnetic	Coil rating	maxwell	Wb	0.000 000 01*
Force, thrust, drag	Pedal, spring, belt, hand lever, general	lbf	N	4.448 222
		ozf	N	0.278 013 9
		pdl	N	0.138 255 0
		kgf	N	9.806 650
		dyne	N	0.000 01*
	Drawbar, breakout, rim pull, winch line pull[(h)], general	lbf	kN	0.004 448 222
Force per length	Beam loading	lbf/ft	N/m	14.593 90
	Spring rate	lbf/in.	N/mm	0.175 126 8
Frequency	System, sound and electrical	Mc/s	MHz	1*
		kc/s	kHz	1*
		Hz, c/s	Hz	1*
	Mechanical events, rotational	r/s (rps)	s^{-1}, r/s	1*
		r/min (rpm)	min^{-1}, r/min	1*
	Engine, power-take-off shaft, gear speed	r/min (rpm)	min^{-1}, r/min	1*
	Rotational dynamics	rad/s	rad/s	
Hardness	Conventional hardness numbers, BHN, R, etc., not affected by change to SI.			

TABLE 2—PREFERRED UNITS FOR EXPRESSING PHYSICAL QUANTITIES (cont'd)

1. **Quantities are arranged in alphabetical order by principal nouns. For example, surface tension is listed as tension, surface.**
2. **All possible applications are not listed, but others such as rates can be readily derived. For example, from the preferred units for energy and volume the units for heat energy per unit volume, kJ/m³, may be derived.**
3. **Conversion factors are shown to seven significant digits, unless the precision with which the factor is known does not warrant seven digits.**

Quantity	Application	From: Old Units	To: SI Units	Multiply By:
Heat	(See Energy)			
Heat capacity	(See Capacity, heat)			
Heat capacity, specific	(See Capacity, heat, specific)			
Heat flow rate	(See Power)			
Heat flow - density of	(See Density of heat flow)			
Heat, specific	General	cal/g·°C	kJ/kg·K	4.186 8*
		Btu/lb·°F	kJ/kg·K	4.186 8*
Heat transfer coefficient	(See Coefficient of heat transfer)			
Illuminance, illumination	General	fc	lx	10.763 91
Impact strength	(See Strength, impact)			
Impedance, mechanical	Damping coefficient	lbf·s/ft	N·s/m	14.593 90
Inductance, electric	Filters and chokes, permeance	H	H	
Intensity, luminous	Light bulbs	candlepower	cd	1*
Intensity, radiant	General	W/sr	W/sr	
Leakage	(See Flow, volume)			
Length	Land distances, maps, odometers	mile	km	1.609 344*‖
	Field size, turning circle, braking distance, cargo platforms, rolling circumference, water depth, land leveling (cut and fill)	rod	m	5.029 210 ‖
		yd	m	0.914 4
		ft	m	0.304 8*
	Row spacing	in.	cm	2.54*
	Engineering drawings, product specifications, vehicle dimensions, width of cut, shipping dimensions, digging depth, cross section of lumber, radius of gyration, deflection	in.	mm	25.4*
	Precipitation, liquid, daily and seasonal, field drainage (runoff), evaporation and irrigation depth	in.	mm	25.4*
	Precipitation, snow depth	in.	cm	2.54*
	Coating thickness, filter particle size	mil	μm	25.4*
		μin.	μm	0.025 4*
		micron	μm	1*
	Surface texture Roughness, average	μin.	μm	0.025 4*
	Roughness sampling length, waviness height and spacing	in.	mm	25.4*
	Radiation wavelengths, optical measurements (interference)	μin.	nm	25.4*
Length per time	Precipitation, liquid per hour	in./h	mm/h	25.4*
	Precipitation, snow depth per hour	in.h	cm/h	2.54*
Load	(See Mass)			
Luminance	Brightness	footlambert	cd/m²	3.426 259
Magnetization	Coil field strength	A/in.	A/m	39.370 08
Mass	Vehicle mass, axle rating, rated load, tire load, lifting capacity‡‡, tipping load, load, quantity of crop, counter mass, body mass general	ton (long)	t, Mg**	1.016 047
		ton (short)	t, Mg**	0.907 184 7
		lb	kg	0.453 592 4
		slug	kg	14.593 90
	Small mass	oz	g	28.349 52
Mass per area	Fabric, surface coatings	oz/yd²	g/m²	33.905 75
		lb/ft²	kg/m²	4.882 428
		oz/ft²	g/m²	305.151 7
	Floor loading	lb/ft²	kg/m²	4.882 428
	Application rate, fertilizer, pesticide	lb/acre	kg/ha	1.120 851
	Crop yield, soil erosion	ton (short)/acre	t/ha**	2.241 702
Mass per length	General, structural members	lb/ft	kg/m	1.488 164
		lb/yd	kg/m	0.496 054 7
Mass per time	Machine work capacity, harvesting, materials handling	ton (short)/h	t/h, Mg/h**	0.907 184 7
Modulus of elasticity	General	lbf/in.²	MPa	0.006 894 757
Modulus of rigidity	(See Modulus of elasticity)			
Modulus, section	General	in.³	mm³	16 387.06
		in.³	cm³	16.387 06
Modulus, bulk	System fluid compression	psi	kPa	6.894 757
Moment, bending	(See Moment of force)			
Moment of area, second	General	in.⁴	mm⁴	416 231.4
		in.⁴	cm⁴	41.623 14

TABLE 2—PREFERRED UNITS FOR EXPRESSING PHYSICAL QUANTITIES (cont'd)

1. **Quantities are arranged in alphabetical order by principal nouns. For example, surface tension is listed as tension, surface.**
2. **All possible applications are not listed, but others such as rates can be readily derived. For example, from the preferred units for energy and volume the units for heat energy per unit volume, kJ/m³, may be derived.**
3. **Conversion factors are shown to seven significant digits, unless the precision with which the factor is known does not warrant seven digits.**

Quantity	Application	From: Old Units	To: SI Units	Multiply By:
Moment of force, torque, bending moment	General, engine torque, fasteners, steering torque, gear torque, shaft torque	lbf·in.	N·m	0.112 984 8
		lbf·ft	N·m	1.355 818
		kgf·cm	N·m	0.098 066 5*
	Locks, light torque	ozf·in.	mN·m	7.061 552
Moment of inertia, mass	Flywheel, general	lb·ft²	kg·m²	0.042 140 11
Moment of mass	Unbalance	oz·in.	g·m	0.720 077 8
Moment of momentum	(See Momentum, angular)			
Moment of section	(See Moment of area, second)			
Momentum, linear	General	lb·ft/s	kg·m/s	0.138 255 0
Momentum, angular	Torsional vibration	lb·ft²/s	kg·m²/s	0.042 140 11
Permeability	Magnetic core properties	H/ft	H/m	3.280 840
Permeance	(See Inductance)			
Potential, electric	General	V	V	
Power	General, light bulbs	W	W	
	Air conditioning, heating	Btu/min	W	17.584 27
		Btu/h	W	0.293 071 1
	Engine, alternator, drawbar, power take-off, hydraulic and pneumatic systems, heat rejection, heat exchanger capacity, water power, electrical power, body heat loss	hp (550 ft·lbf/s)	kW	0.745 699 9
Power per area	solar radiation	Btu/ft²h	W/m²	3.154 591
Pressure	All pressures except very small	lbf/in.² (psi)	kPa	6.894 757
		in.Hg (60 °F)	kPa	3.376 85
		in.H_2O (60 °F)	kPa	0.248 84
		mmHg (0 °C)	kPa	0.133 322
		kgf/cm²	kPa	98.066 5
		bar	kPa	100.0*
		lbf/ft²	kPa	0.047 880 26
		atm (normal = 760 torr)	kPa	101.325*
	Very small pressures (high vacuum)	lbf/in.² (psi)	Pa	6 894.757
Pressure, sound level	Acoustical measurement-When weighting is specified show weighting level in parenthesis following the symbol, for example dB(A).	dB	dB	
Quantity of electricity	General	C	C	
Radiant intensity	(See Intensity, radiant)			
Resistance, electric	General	Ω	Ω	
Resistivity, electric	General	Ω·ft	Ω·m	0.304 8*
		Ω·ft	Ω·cm	30.48*
Sound pressure level	(See Pressure, sound, level)			
Speed	(See Velocity)			
Spring rate, linear	(See Force per length)			
Spring rate, torsional	General	lbf·ft/deg	N·m/deg	1.355 818
Strength, field, electric	General	V/ft	V/m	3.280 840
Strength, field, magnetic	General	oersted	A/m	79.577 47
Strength, impact	Materials testing	ft·lbf	J	1.355 818
Stress	General	lbf/in.²	MPa	0.006 894 757
Surface tension	(See Tension, surface)			
Temperature	General use	°F	°C	$t_{°C}=(t_{°F}-32)/1.8$*
	Absolute temperature, thermodynamics, gas cycles	°R	K	$T_K=T_{°R}/1.8$*
Temperature interval	General use	°F	K†	1 K=1 °C=1.8 °F*
Tension, surface	General	lbf/in.	mN/m	175 126.8
		dyne/cm	mN/m	1*
Thermal diffusivity	Heat transfer	ft²/h	m²/h	0.092 903 04
Thrust	(See Force)			
Time	General	s	s	
		h	h	
		min	min	
	Hydraulic cycle time	s	s	
	Hauling cycle time	min	min	

TABLE 2—PREFERRED UNITS FOR EXPRESSING PHYSICAL QUANTITIES (cont'd)

1. Quantities are arranged in alphabetical order by principal nouns. For example, surface tension is listed as tension, surface.
2. All possible applications are not listed, but others such as rates can be readily derived. For example, from the preferred units for energy and volume the units for heat energy per unit volume, kJ/m³, may be derived.
3. Conversion factors are shown to seven significant digits, unless the precision with which the factor is known does not warrant seven digits.

Quantity	Application	From: Old Units	To: SI Units	Multiply By:
Torque	(See Moment of force)			
Toughness, fracture	Metal properties	ksi·in.$^{0.5}$	MPa·m$^{0.5}$	1.098 843
Vacuum	(See Pressure)			
Velocity, angular	(See Velocity, rotational)			
Velocity, linear	Vehicle	mile/h	km/h	1.609 344*
	Fluid flow, conveyor speed, lift speed, air speed	ft/s	m/s	0.304 8*
	Cylinder actuator speed	in./s	mm/s	25.4*
	General	ft/s	m/s	0.304 8*
		ft/min	m/min	0.304 8*
		in./s	mm/s	25.4*
Velocity, rotational	(See Frequency)			
Viscosity, dynamic	General liquids	centipoise	mPa·s	1*
Viscosity, kinematic	General liquids	centistokes	mm²/s	1*
Volume	Truck body, shipping or freight, bucket capacity, earth, gas, lumber, building, general	yd³	m³	0.764 554 9
		ft³	m³	0.028 316 85
	Combine harvester grain tank capacity	bushel	L	35.239 07
	Automobile luggage capacity	ft³	L	28.316 85
	Gas pump displacement, air compressor, air reservoir, engine displacement			
	Large	in.³	L	0.016 387 06
	Small	in.³	cm³	16.387 06
	Liquid - fuel, lubricant, coolant, liquid wheel ballast	gal	L	3.785 412
		qt	L	0.946 352 9
		pt	L	0.473 176 5
	Small quantity liquid	oz	mL	29.573 53
	Irrigation, reservoir	acre·ft	m³	1 233.489‖
			dam³	1.233 489‖
	Grain bins	bushel (U.S.)	m³	0.035 239 07
Volume per area	Application rate, pesticide	gal/acre	L/ha	9.353 958
Volume per time	Fuel consumption (Also see Flow)	gal/h	L/h	3.785 412
Weight	May mean either mass or force—avoid use of weight			
Work	(See Energy)			
Young's modulus	(See Modulus of elasticity)			

*Indicates exact conversion factor.
†In these expressions K indicates temperature intervals. Therefore K may be replaced with °C if desired without changing the value or affecting the conversion factor. kJ/(kg·K) = kJ/(kg·°C).
‡Not to be confused with kcal/g. kcal often called calorie.
§Convenient conversion: 235.215 ÷ (mile per gal) = L/(100 km).
‖Official use in surveys and cartography involves the U.S. survey mile based on the U.S. survey foot, which is longer than the international foot by two parts per million. The factors used in this standard for acre, acre foot, rod are based on the U.S. survey foot. Factors for all other old length units are based on the international foot. (See ANSI/ASTM Standard E380-76, Metric Practice).
#Standard acceleration of gravity is 9.806 650 m/s² exactly (Adopted by the General Conference on Weights and Measures).
**The symbol t is used to designate metric ton. The unit metric ton (exactly 1 Mg) is in wide use but should be limited to commercial description of vehicle mass, freight mass, and agricultural commodities. No prefix is permitted.
††Conversions of Btu are based on the International Table Btu.
‡‡Lift capacity ratings for cranes, hoists, and related components such as ropes, cable chains, etc., should be rated in mass units. Those items such as winches, which can be used for pulling as well as lifting, shall be rated in both force and mass units for safety reasons.
§§ASAE S209 and SAE J708, Agricultural Tractor Test Code, specify kg/(kW·h). It should be noted that there is a trend toward use of g/MJ as specified for highway vehicles.

SECTION 6—CONVERSION TECHNIQUES

6.1 Conversion of quantities between systems of units involves careful determination of the number of significant digits to be retained. To convert "1 quart of oil" to "0.946 352 9 liter of oil" is, of course, unrealistic because the intended accuracy of the value does not warrant expressing the conversion in this fashion.

6.2 All conversions, to be logically established, must depend upon an intended precision of the original quantity — either implied by a specific tolerance, or by the nature of the quantity. The first step in conversion is to establish this precision.

6.3 The implied precision of a value should relate to the number of significant digits shown. The implied precision is plus or minus one-half unit of the last significant digit in which the value is stated. This is true because it may be assumed to have been rounded from a greater number of digits, and one-half unit of the last significant digit retained is the limit of error resulting from rounding. For example, the number 2.14 may have been rounded from any number between 2.135 and 2.145. Whether rounded or not, a quantity should always be expressed with this implication of precision in mind. For instance, 2.14 in. implies a precision of ±0.005 in., since the last significant digit is in units of 0.01 in.

6.4 Quantities should be expressed in digits which are intended to be significant. The dimension 1.1875 in. may be a very accurate one in which the digit in the fourth place is significant, or it may in some cases be an exact decimalization of a fractional dimension, 1-3/16 in., in which case the dimension is given with too many decimal places relative to its intended precision.

6.5 Quantities should not be expressed with significant zeros omitted. The dimension 2 in. may mean "about 2 in.," or it may, in fact, mean a very accurate expression which should be written 2.0000 in. In the latter case, while the added zeros are not significant in establishing the value, they are very significant in expressing the proper intended precision.

SECTION 7—RULES FOR ROUNDING

7.1 Where feasible, the rounding of SI equivalents should be in reasonable, convenient, whole units.

7.2 Interchangeability of parts, functionally, physically, or both, is dependent upon the degree of round-off accuracy used in the conversion of the U.S. customary to SI value. American National Standards Institute ANSI/ASTM E380-76, Metric Practice, outlines methods to assure interchangeability.

7.3 Rounding numbers. When a number is to be rounded to fewer decimal places the procedure shall be as follows:

7.3.1 When the first digit discarded is less than 5, the last digit retained shall not be changed. For example, 3.463 25, if rounded to three decimal places, would be 3.463; if rounded to two decimal places, would be 3.46.

7.3.2 When the first digit discarded is greater than 5, or if it is a 5 followed by at least one digit other than 0, the last figure retained shall be increased by one unit. For example, 8.376 52, if rounded to three decimal places, would be 8.377; if rounded to two decimal places, would be 8.38.

7.3.3 Round to closest even number when first digit discarded is 5, followed only by zeros.

7.3.4 Numbers are rounded directly to the nearest value having the desired number of decimal places. Rounding must not be done in successive steps to less places.

For example:

27.46 rounded to a whole number = 27. This is correct because the ".46" is less than one-half. 27.46 rounded to one decimal place is 27.5. This is a correct value. But, if the 27.5 is in turn rounded to a whole number, this is successive rounding and the result, 28, is incorrect.

7.4 Inch-millimeter linear dimensioning conversion. 1 inch (in.) = 25.4 millimeters (mm) exactly. The term "exactly" has been used with all exact conversion factors. Conversion factors not so labeled have been rounded in accordance with these rounding procedures. To maintain intended precision during conversion without retaining an unnecessary number of digits, the millimeter equivalent shall be carried to one decimal place more than the inch value being converted and then rounded to the appropriate significant figure in the last decimal place.

Cited Standards:

ASAE S209, Agricultural Tractor Test Code

ANSI/ASTM E380-76, Metric Practice

ISO 1000, SI Units and Recommendations for the Use of Their Multiples and of Certain Other Units

ISO 2955, Information Processing — Representation of SI and Other Units for Use in Systems with Limited Character Sets

WET-BULB TEMPERATURES AND WET-BULB DEPRESSIONS

Approved by ASAE Electric Power and Processing Division Technical Committee (EPP-03); adopted by ASAE June 1967; reconfirmed December 1971, December 1976, December 1981; revised June 1987.

SECTION 1—PURPOSE AND SCOPE

1.1 These data allow estimation of weather parameters for locations in the continental United States and southern Canada. The accompanying sets of maps show mean wet-bulb temperatures, mean wet-bulb depressions, and their standard deviations, in °C and in °F.

SECTION 2—WET-BULB TEMPERATURES AND WET-BULB DEPRESSIONS

2.1 Graphical data. Celsius and Fahrenheit maps are based upon different weather records. Celsius maps are based upon weather records from 127 U.S. and 13 Canadian sites. United States records averaged 21 yr while Canadian records averaged 12 yr. Fahrenheit maps are based upon weather records averaging 10 yr from 82 U.S. sites.

2.2 Use of standard deviation values. Standard deviation isolines are shown on both the Fahrenheit and Celsius maps. However, the Fahrenheit isolines have a different meaning from the Celsius isolines.

2.2.1 Standard deviation lines on the Fahrenheit maps estimate standard deviations of monthly mean values about the average of monthly means for the years of record. These values can be used to estimate year-to-year variations in monthly mean values. For example, for October in Central Iowa, the mean wet-bulb temperature is 47 °F with a standard deviation of 4 °F. Assuming variations are normally distributed, about 68% of the years the mean wet-bulb temperature for the month will be within one standard deviation of 47 °F (between 43 and 51 °F). About 95% of the years, the mean monthly wet-bulb temperature for the month will be within two standard deviations of 47 °F (between 39 and 55 °F). Estimation of year-to-year variation in mean monthly temperature is of interest in an application like natural-air grain drying which requires several weeks to complete and which is not influenced very much by daily temperature fluctuations.

2.2.2 Standard deviation isolines on the Celsius maps estimate standard deviations of hourly readings about the monthly mean. These standard deviation values are two to four times as large as the Fahrenheit standard deviations and can be used to estimate short-term (hourly) temperature variations. For example, for October in Central Iowa, the mean wet-bulb temperature is 8 °C with a standard deviation of 5.5 °C. Assuming variations are normally distributed, about 68% of the hourly temperature readings during October will be within one standard deviation of 8 °C (between 2.5 and 13.5 °C). About 95% of the hourly temperature readings will be within 2 standard deviations of 8 ° (between −3 and 19 °C). Estimation of the range of hourly temperatures during the month is of value in applications like environmental control of buildings where the systems are responsive to short-term temperature fluctuations.

References:

1. Waite, P. J., C. J. Bern. 1986. Wet-bulb temperatures and wet-bulb depressions in the United States and southern Canada. TRANSACTIONS of the ASAE 30(6):1827-1832.

2. Schmidt, J. L., P. J. Waite. 1962. Summaries of wet-bulb temperature and wet-bulb depression for grain drier design. TRANSACTIONS of the ASAE 5(2):186-189.

WET-BULB TEMPERATURES °C

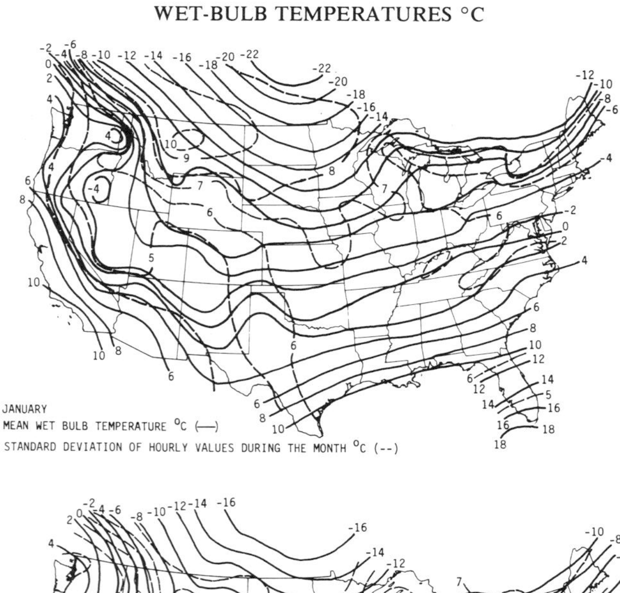

JANUARY
MEAN WET BULB TEMPERATURE °C (—)
STANDARD DEVIATION OF HOURLY VALUES DURING THE MONTH °C (--)

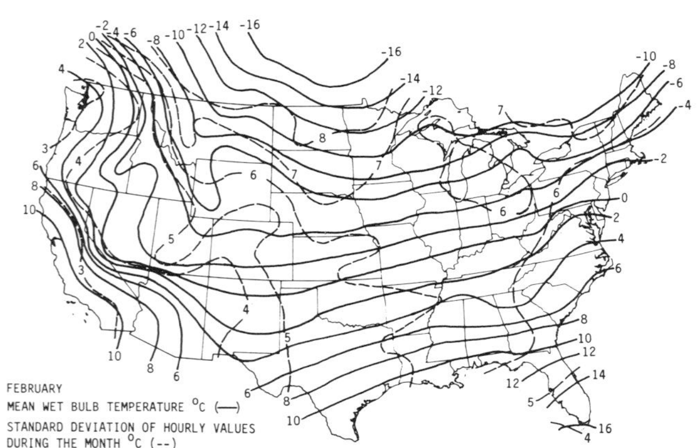

FEBRUARY
MEAN WET BULB TEMPERATURE °C (—)
STANDARD DEVIATION OF HOURLY VALUES DURING THE MONTH °C (--)

WET-BULB DEPRESSIONS °C

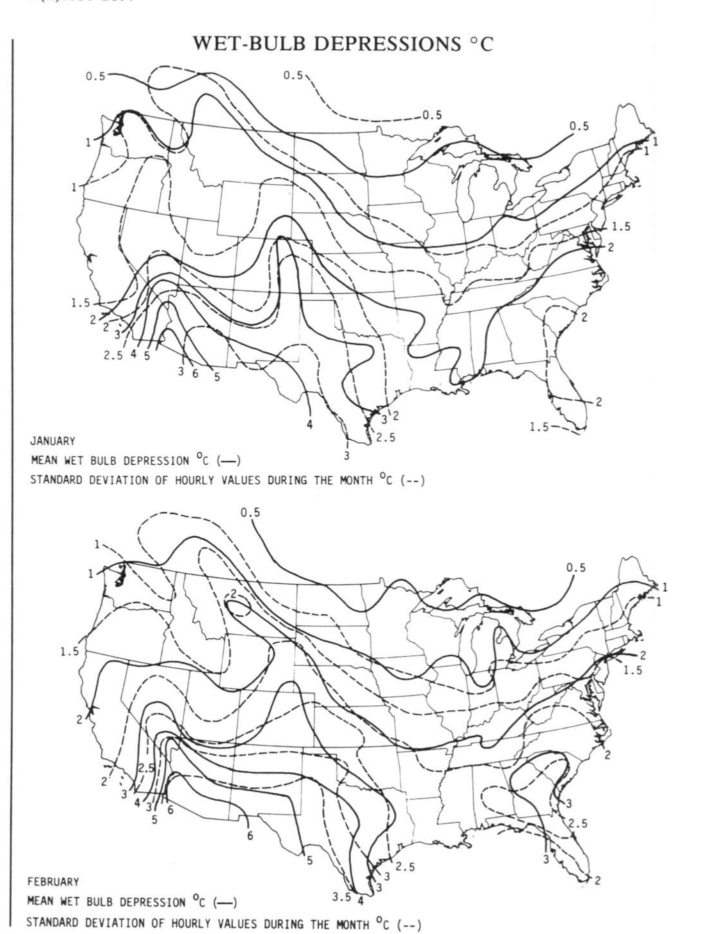

JANUARY
MEAN WET BULB DEPRESSION °C (—)
STANDARD DEVIATION OF HOURLY VALUES DURING THE MONTH °C (--)

FEBRUARY
MEAN WET BULB DEPRESSION °C (—)
STANDARD DEVIATION OF HOURLY VALUES DURING THE MONTH °C (--)

WET-BULB TEMPERATURES °C

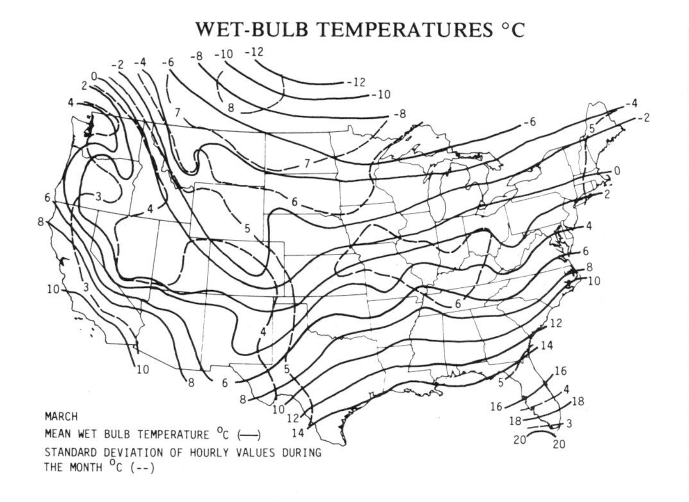

MARCH
MEAN WET BULB TEMPERATURE °C (—)
STANDARD DEVIATION OF HOURLY VALUES DURING THE MONTH °C (--)

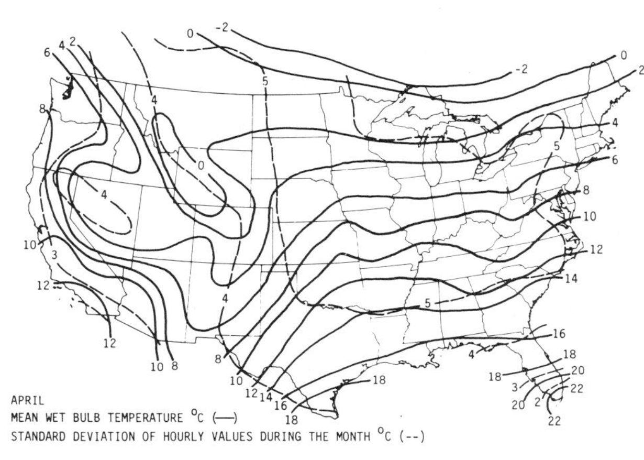

APRIL
MEAN WET BULB TEMPERATURE °C (—)
STANDARD DEVIATION OF HOURLY VALUES DURING THE MONTH °C (--)

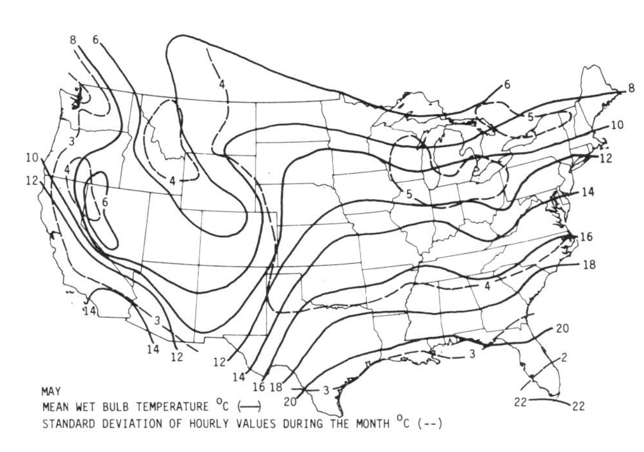

MAY
MEAN WET BULB TEMPERATURE °C (—)
STANDARD DEVIATION OF HOURLY VALUES DURING THE MONTH °C (--)

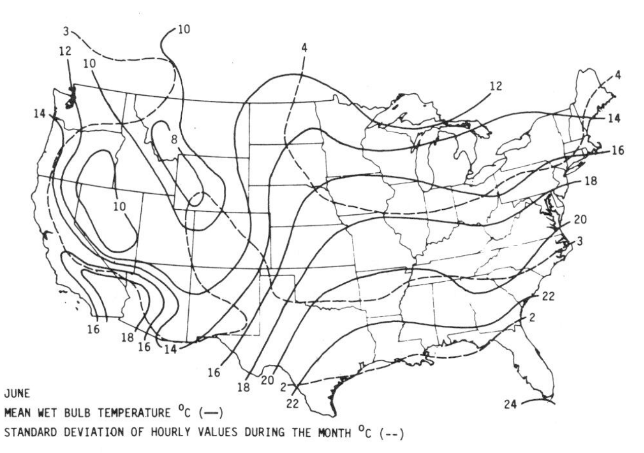

JUNE
MEAN WET BULB TEMPERATURE °C (—)
STANDARD DEVIATION OF HOURLY VALUES DURING THE MONTH °C (--)

WET-BULB DEPRESSIONS °C

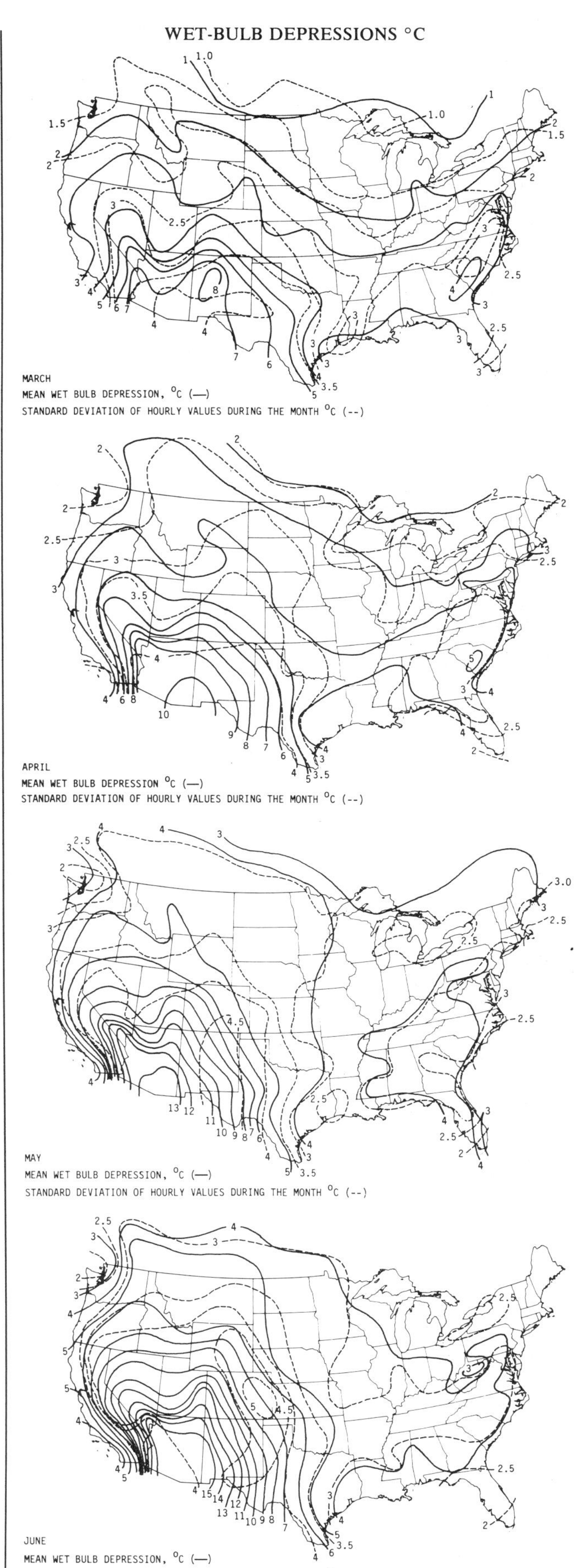

MARCH
MEAN WET BULB DEPRESSION, °C (—)
STANDARD DEVIATION OF HOURLY VALUES DURING THE MONTH °C (--)

APRIL
MEAN WET BULB DEPRESSION °C (—)
STANDARD DEVIATION OF HOURLY VALUES DURING THE MONTH °C (--)

MAY
MEAN WET BULB DEPRESSION, °C (—)
STANDARD DEVIATION OF HOURLY VALUES DURING THE MONTH °C (--)

JUNE
MEAN WET BULB DEPRESSION, °C (—)
STANDARD DEVIATION OF HOURLY VALUES DURING THE MONTH °C (--)

WET-BULB TEMPERATURES °C

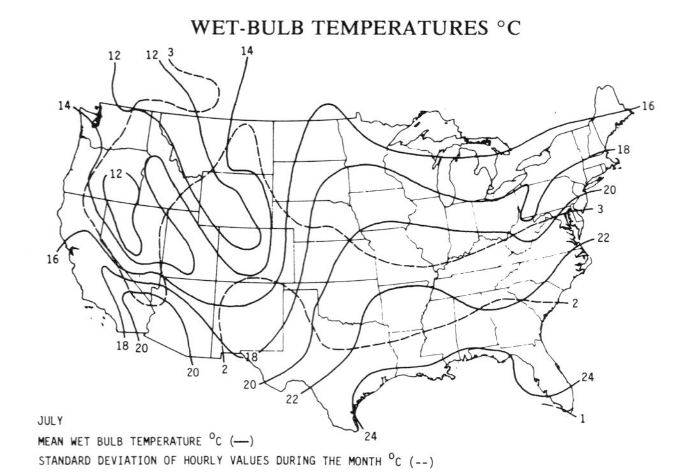

JULY
MEAN WET BULB TEMPERATURE °C (—)
STANDARD DEVIATION OF HOURLY VALUES DURING THE MONTH °C (--)

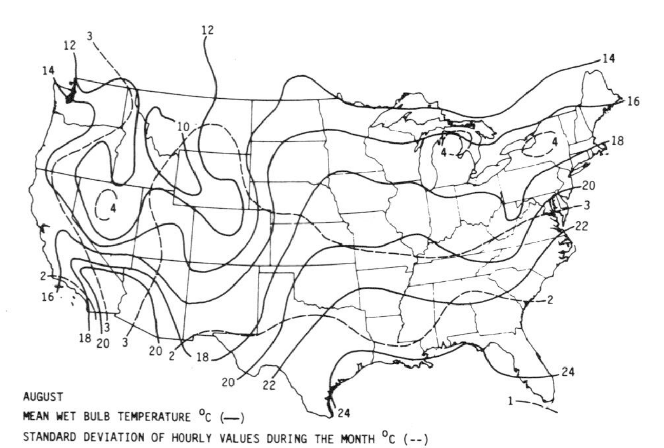

AUGUST
MEAN WET BULB TEMPERATURE °C (—)
STANDARD DEVIATION OF HOURLY VALUES DURING THE MONTH °C (--)

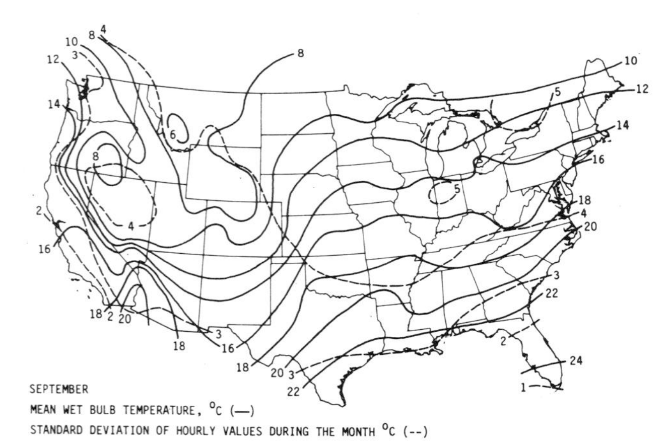

SEPTEMBER
MEAN WET BULB TEMPERATURE, °C (—)
STANDARD DEVIATION OF HOURLY VALUES DURING THE MONTH °C (--)

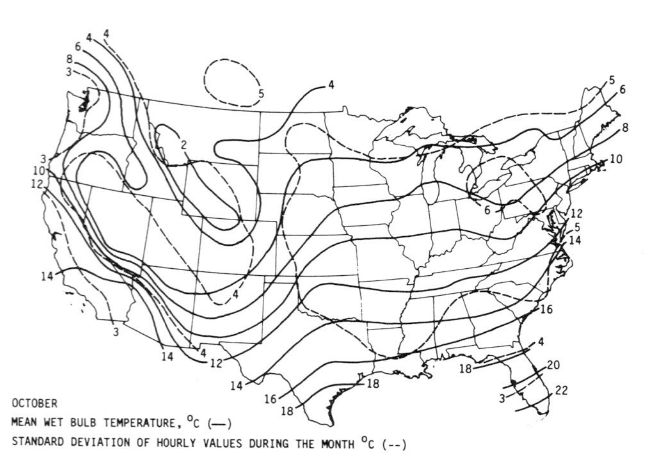

OCTOBER
MEAN WET BULB TEMPERATURE, °C (—)
STANDARD DEVIATION OF HOURLY VALUES DURING THE MONTH °C (--)

WET-BULB DEPRESSIONS °C

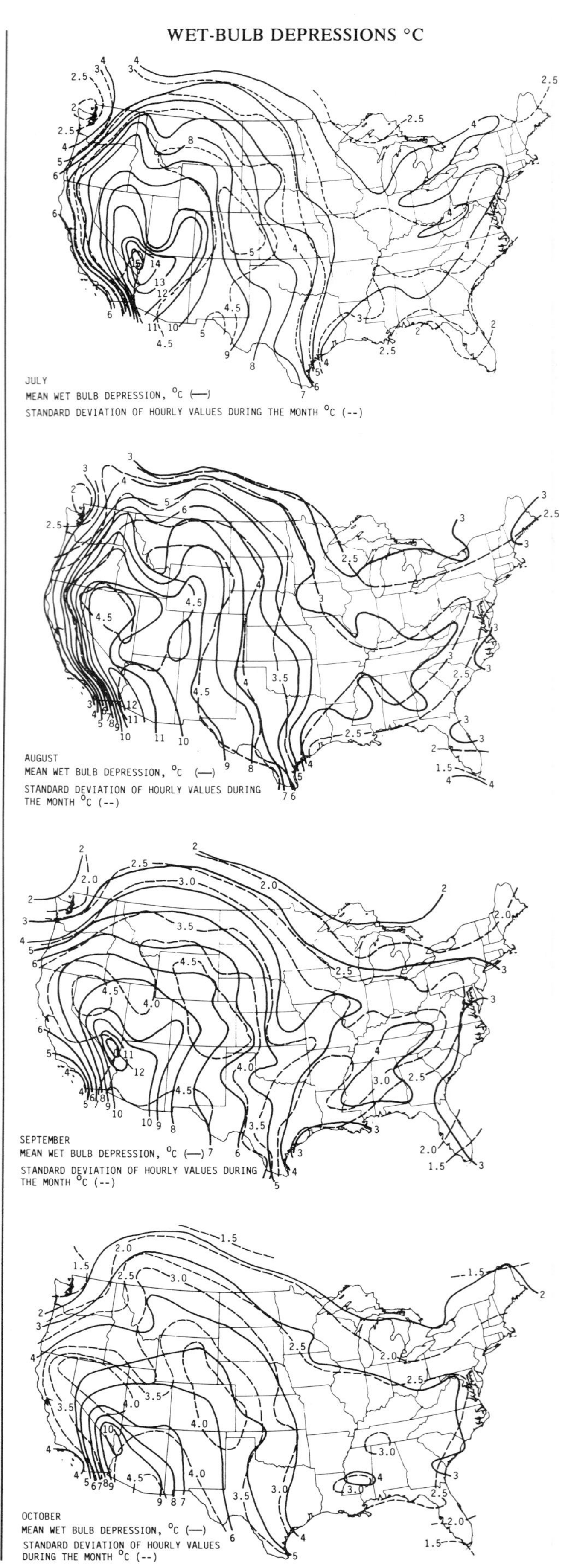

JULY
MEAN WET BULB DEPRESSION, °C (—)
STANDARD DEVIATION OF HOURLY VALUES DURING THE MONTH °C (--)

AUGUST
MEAN WET BULB DEPRESSION, °C (—)
STANDARD DEVIATION OF HOURLY VALUES DURING THE MONTH °C (--)

SEPTEMBER
MEAN WET BULB DEPRESSION, °C (—)
STANDARD DEVIATION OF HOURLY VALUES DURING THE MONTH °C (--)

OCTOBER
MEAN WET BULB DEPRESSION, °C (—)
STANDARD DEVIATION OF HOURLY VALUES DURING THE MONTH °C (--)

WET-BULB TEMPERATURES °C

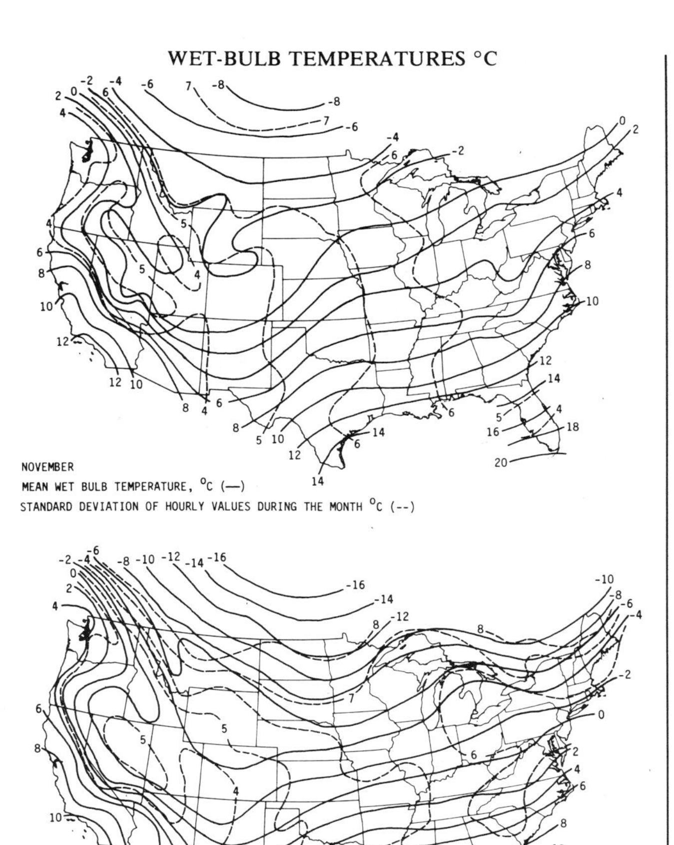

NOVEMBER
MEAN WET BULB TEMPERATURE, °C (—)
STANDARD DEVIATION OF HOURLY VALUES DURING THE MONTH °C (--)

DECEMBER
MEAN WET BULB TEMPERATURE, °C (—)
STANDARD DEVIATION OF HOURLY VALUES DURING THE MONTH °C (--)

WET-BULB DEPRESSIONS °C

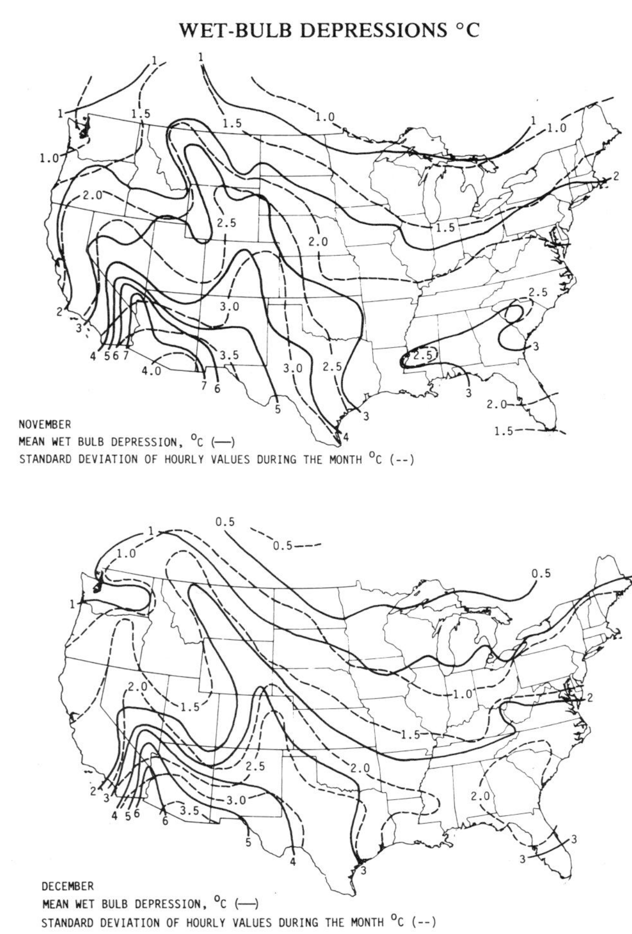

NOVEMBER
MEAN WET BULB DEPRESSION, °C (—)
STANDARD DEVIATION OF HOURLY VALUES DURING THE MONTH °C (--)

DECEMBER
MEAN WET BULB DEPRESSION, °C (—)
STANDARD DEVIATION OF HOURLY VALUES DURING THE MONTH °C (--)

WET-BULB TEMPERATURES °F

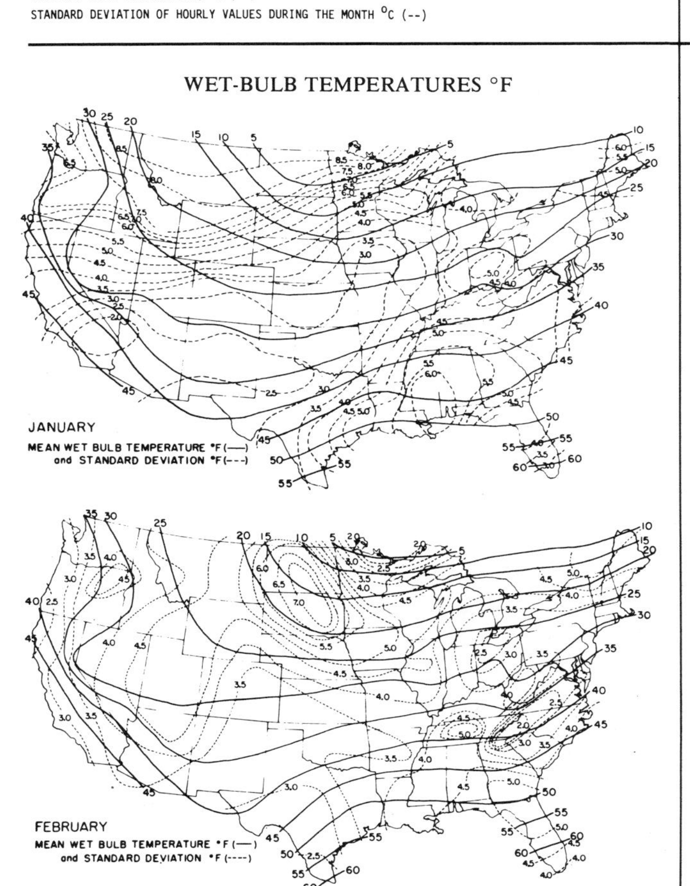

JANUARY
MEAN WET BULB TEMPERATURE °F (—)
and STANDARD DEVIATION °F (---)

FEBRUARY
MEAN WET BULB TEMPERATURE °F (—)
and STANDARD DEVIATION °F (----)

WET-BULB DEPRESSIONS °F

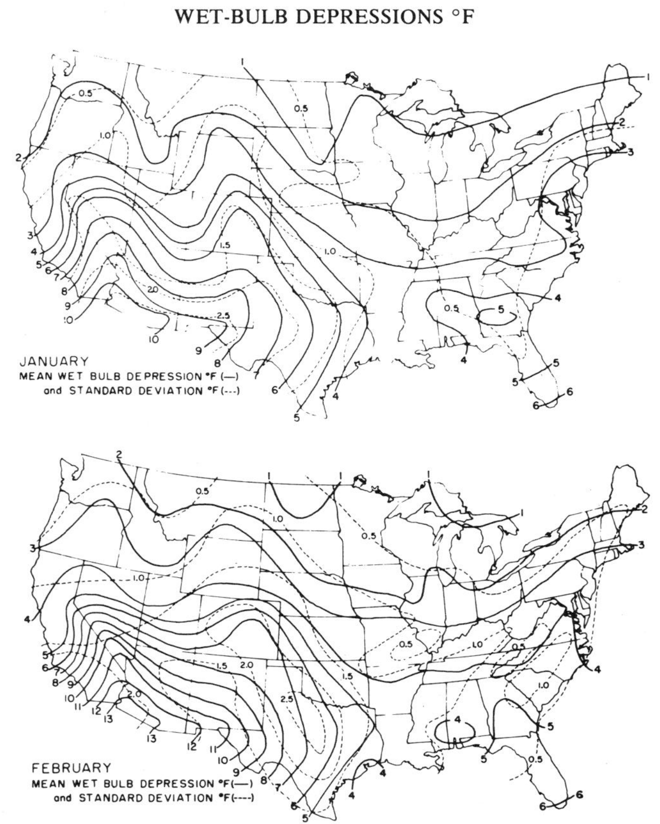

JANUARY
MEAN WET BULB DEPRESSION °F (—)
and STANDARD DEVIATION °F (---)

FEBRUARY
MEAN WET BULB DEPRESSION °F (—)
and STANDARD DEVIATION °F (----)

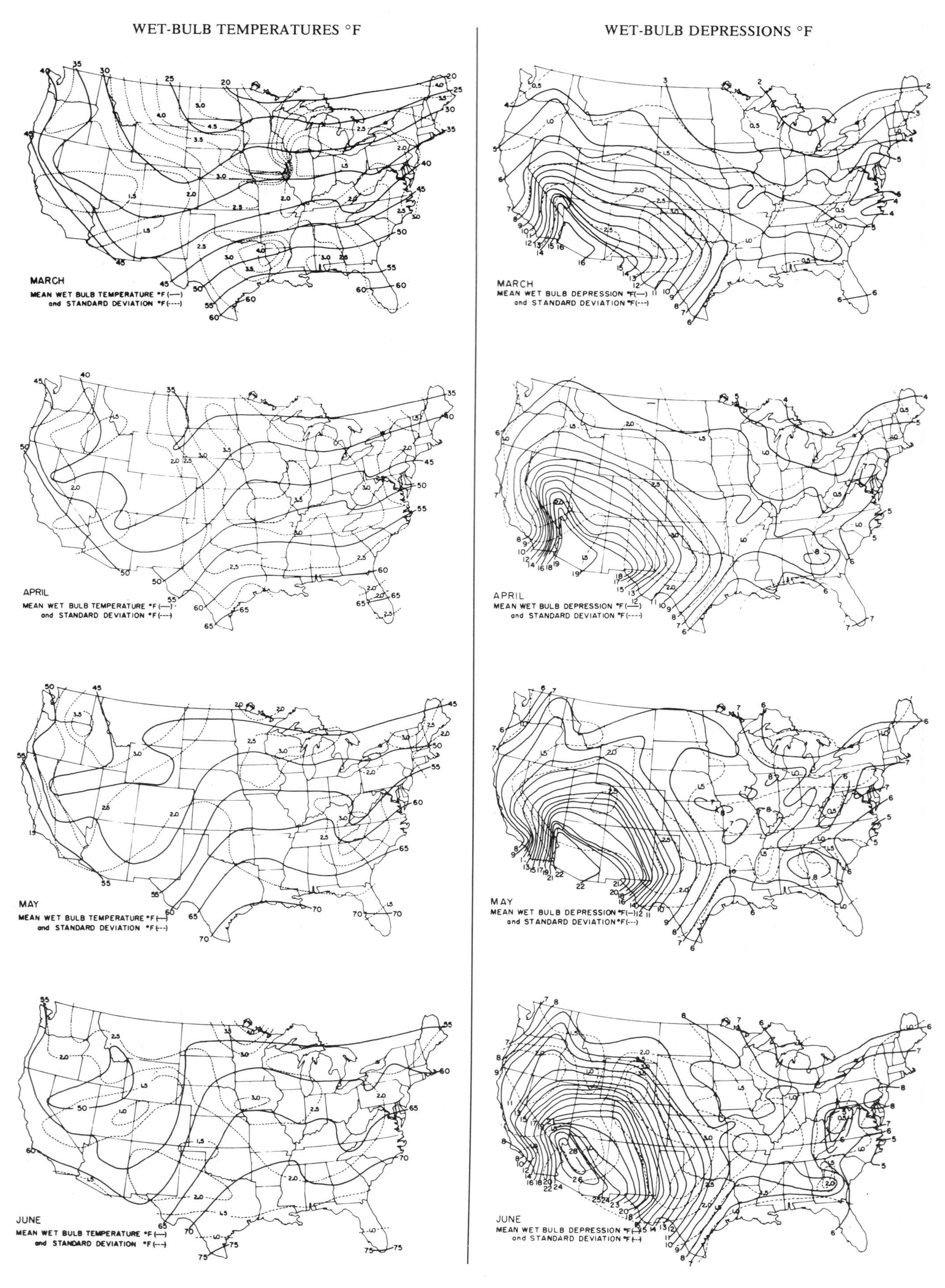
WET-BULB TEMPERATURES °F
WET-BULB DEPRESSIONS °F
MARCH
MEAN WET BULB TEMPERATURE °F (—)
and STANDARD DEVIATION °F (----)
MARCH
MEAN WET BULB DEPRESSION °F (—)
and STANDARD DEVIATION °F (---)
APRIL
MEAN WET BULB TEMPERATURE °F (—)
and STANDARD DEVIATION °F (---)
APRIL
MEAN WET BULB DEPRESSION °F (—)
and STANDARD DEVIATION °F (----)
MAY
MEAN WET BULB TEMPERATURE °F (—)
and STANDARD DEVIATION °F (--)
MAY
MEAN WET BULB DEPRESSION °F (—)
and STANDARD DEVIATION °F (---)
JUNE
MEAN WET BULB TEMPERATURE °F (—)
and STANDARD DEVIATION °F (---)
JUNE
MEAN WET BULB DEPRESSION °F (—)
and STANDARD DEVIATION °F (--)

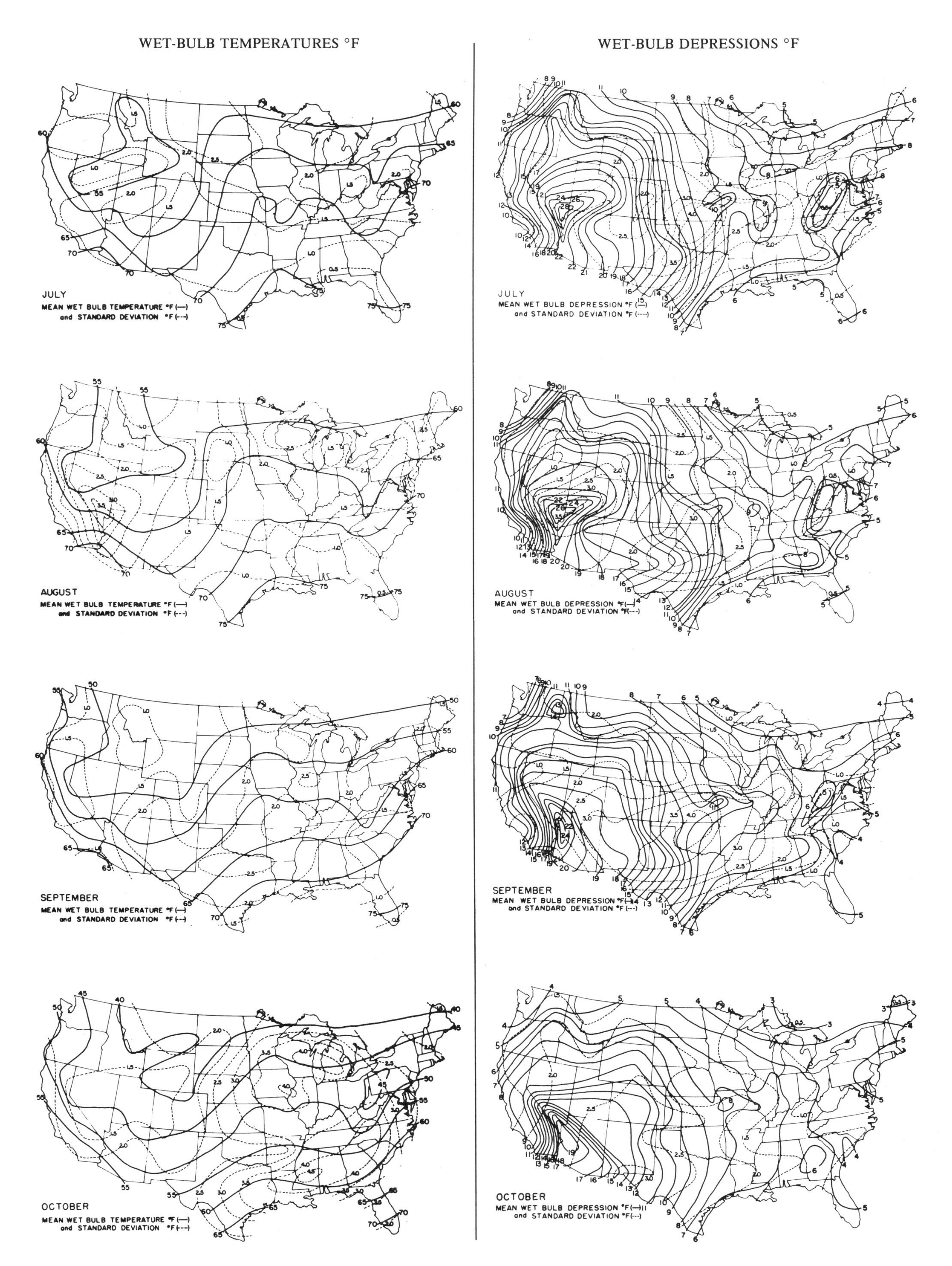
WET-BULB TEMPERATURES °F
WET-BULB DEPRESSIONS °F
JULY
MEAN WET BULB TEMPERATURE °F (—)
and STANDARD DEVIATION °F (---)
JULY
MEAN WET BULB DEPRESSION °F (—)
and STANDARD DEVIATION °F (---)
AUGUST
MEAN WET BULB TEMPERATURE °F (—)
and STANDARD DEVIATION °F (---)
AUGUST
MEAN WET BULB DEPRESSION °F (—)
and STANDARD DEVIATION °F (---)
SEPTEMBER
MEAN WET BULB TEMPERATURE °F (—)
and STANDARD DEVIATION °F (---)
SEPTEMBER
MEAN WET BULB DEPRESSION °F (—)
and STANDARD DEVIATION °F (---)
OCTOBER
MEAN WET BULB TEMPERATURE °F (—)
and STANDARD DEVIATION °F (---)
OCTOBER
MEAN WET BULB DEPRESSION °F (—)
and STANDARD DEVIATION °F (---)

WET-BULB TEMPERATURES °F

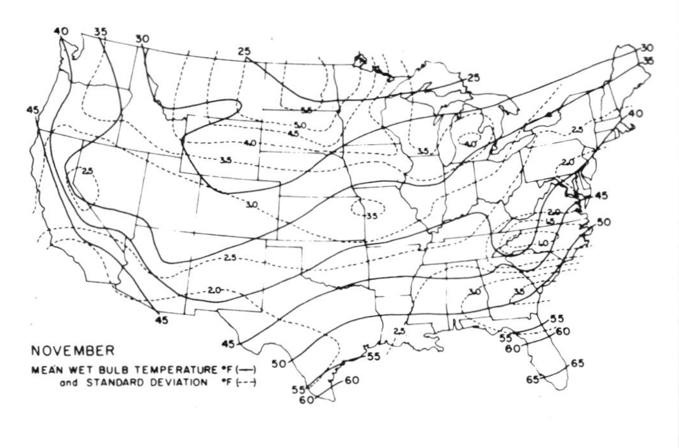

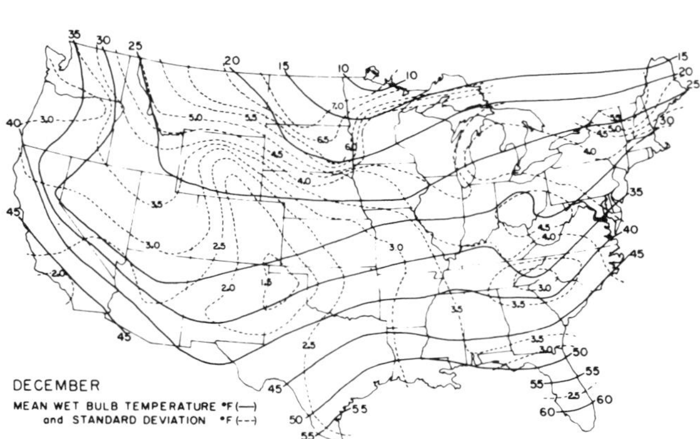

WET-BULB DEPRESSIONS °F

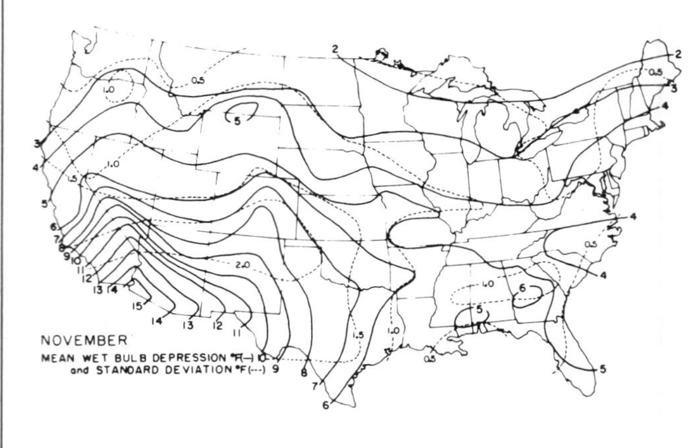

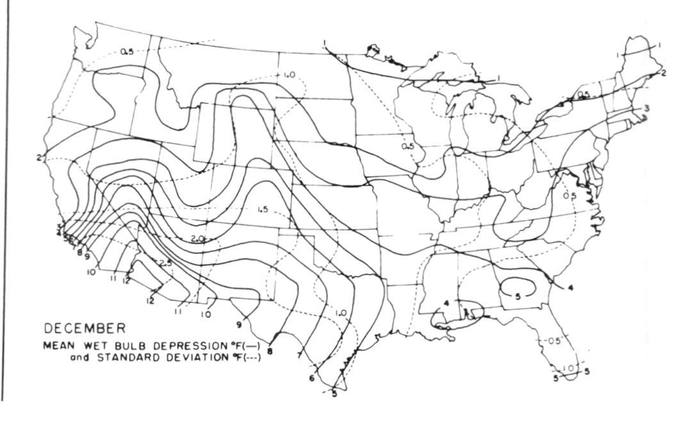

ASAE Standard: ASAE S351 (ANSI/ASAE S351)

HAND SIGNALS FOR USE IN AGRICULTURE

Proposed by the Tractor and Machinery Committee of the National Institute for Farm Safety; approved by ASAE Power and Machinery Division Standards Committee; adopted by ASAE as a Recommendation February 1972; reconfirmed December 1976; reclassified as a Standard December 1978; reconfirmed December 1981; approved by ANSI as an American National Standard December 1982; reconfirmed December 1986.

SECTION 1—PURPOSE AND SCOPE

1.1 This Standard provides for hand signals to be used in agricultural operations especially when noise or distance precludes the use of normal voice communication.

1.2 The purpose of the hand signals is to provide an easy means of communication, particularly in the interest of safety.

SECTION 2—GENERAL

2.1 These hand signals are in general agreement with U.S. Army Field Manual, FM21-60, Section II, Standard Arm and Hand Signals. Many of them are also used in construction and other industries where noise or distance precludes the use of voice communications.

SECTION 3—HAND SIGNALS (Figs. 1-11)

FIG. 1—THIS FAR TO GO—Place palms at ear level facing head and move laterally inward to indicate remaining distance to go.

FIG. 2—COME TO ME—Raise the arm vertically overhead, palm to the front, and rotate in large horizontal circles.

FIG. 3—MOVE TOWARD ME—FOLLOW ME—Point toward person(s), vehicle(s), or unit(s), beckon by holding the arm horizontally to the front, palm up, and motioning toward the body.

FIG. 4—MOVE-OUT—TAKE OFF—Face the desired direction of movement; hold the arm extended to the rear; then swing it overhead and forward in the direction of desired movement until it is horizontal, palm down.

FIG. 5—STOP—Raise hand upward to the full extent of the arm, palm to the front. Hold that position until the signal is understood.

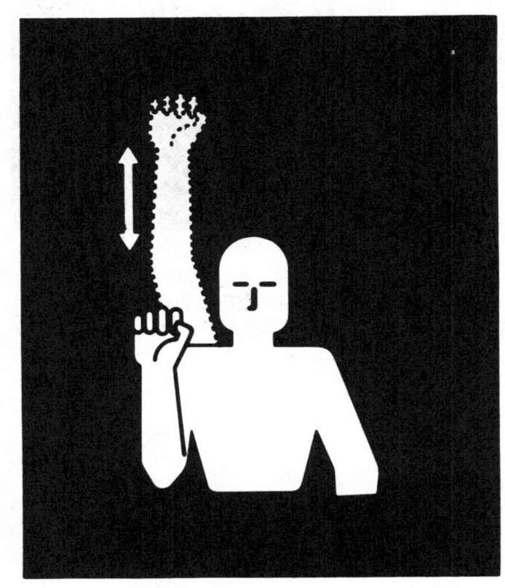

FIG. 6—SPEED IT UP—INCREASE SPEEDP—Raise the hand to the shoulder, fist closed; thrust the fist upward to the full extent of the arm and back to the shoulder rapidly several times.

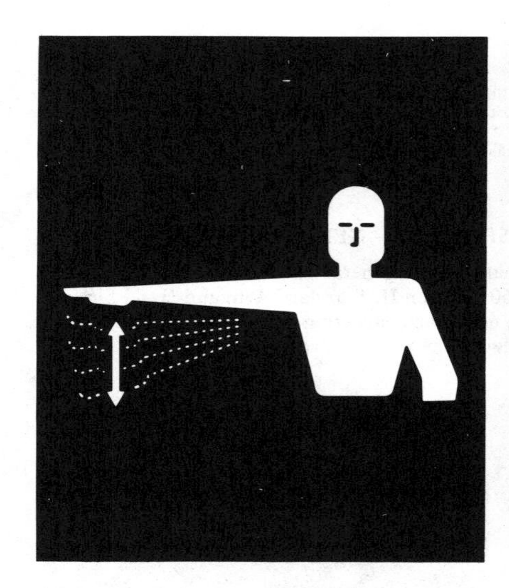

FIG. 7—SLOW IT DOWN—DECREASE SPEED—Extend the arm horizontally sideward, palm down, and wave arm downward 45 deg minimum several times, keeping the arm straight. Do not move arm above horizontal.

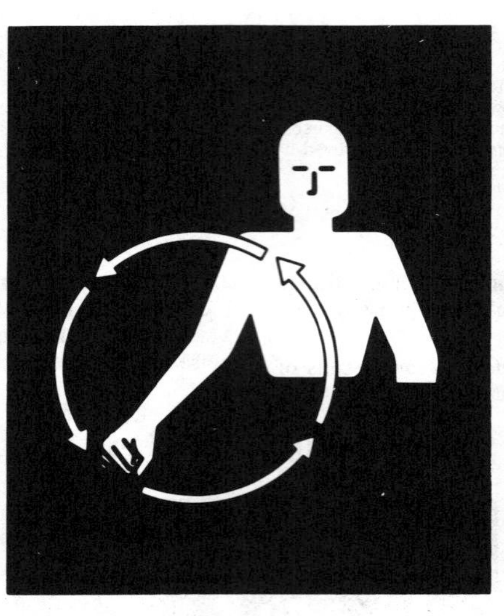

FIG. 8—START THE ENGINE—Simulate cranking of vehicles by moving arm in a circular motion at waist level.

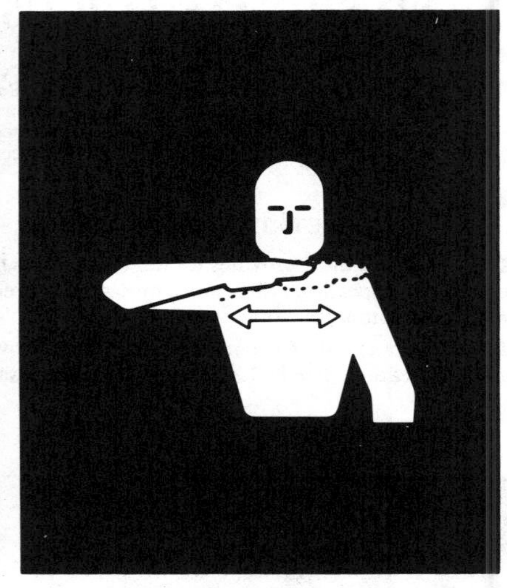

FIG. 9—STOP THE ENGINE—Draw right hand, palm down, across the neck in a "throat cutting" motion from left to right.

FIG. 10—LOWER EQUIPMENT—Make circular motion with either hand pointing to the ground.

FIG. 11—RAISE EQUIPMENT—Make circular motion with either hand at head level.

RADIATION QUANTITIES AND UNITS

Developed by the Radiation Committee; approved by the ASAE Electric Power and Processing Division Standards Committee; adopted by ASAE December 1980; revised editorially November 1981; reconfirmed December 1985.

SECTION 1—PURPOSE AND SCOPE

1.1 This Engineering Practice provides the SI units of measurement associated with quantities useful in describing all types of radiation for agricultural engineering applications. As an aid in establishing uniformity in the agricultural engineering literature, it also provides the generally accepted symbols which represent those quantities and the approved symbols for the units of the SI system applicable to those radiation quantities. The principal customary units that have been used for radiation quantities are also listed, and factors are provided for conversion of those customary units to SI units.

1.2 Table 1 provides the preferred names for radiation quantities, quantity symbols, names of customary units and SI units, symbols for the SI units, and identification of authoritative sources of the information relating to conversion factors and the quantities and units. These references may be consulted for more detailed information in many cases.

References: Last printed in 1981 AGRICULTURAL ENGINEERS YEARBOOK; list available from ASAE Headquarters.

TABLE 1—RADIATION QUANTITIES AND UNITS

Quantity	Quantity Symbol	Units		SI Unit Symbol	Conversion Factor‡	Authoritative Reference
		Customary†	SI			
Electromagnetic Radiation (General)						
Frequency	f,ν	cycles/second	hertz	Hz	1.000 000 E+00*	(1)
Wavelength	λ	angstrom	meter	m	1.000 000 E-10*	(1)
		micron			1.000 000 E-06*	
Energy	E,W,U	watt-hour	joule	J	3.600 000 E+03*	(1)
		erg			1.000 000 E-07*	(1)
		electronvolt			1.602 19 E-19	(1)
Electromagnetic Fields and Circuits						
Electric field strength	E,K	volt/inch	volt/meter	V/m	3.937 008 E+01	(1,2)
Magnetic field strength	H	oersted	ampere/meter	A/m	7.957 747 E+01	(1,2)
Electric charge	Q		coulomb	C		(2)
Electric current	I		ampere	A		(2)
Potential difference	V,φ		volt	V		(2)
Electromotive force	V,E,U		volt	V		(2)
Electric flux density	D		coulomb/square meter	C/m^2		(2)
Magnetic flux density	B	gauss	tesla (weber/square meter)	T	1.000 000 E-04*	(1,2)
Magnetic flux	φ	maxwell	weber	Wb	1.000 000 E-08*	(1,2)
Electric flux	Ψ		coulomb	C		(2)
Electric current density	J,S		ampere/square meter	A/m^2		(2)
Resistance	R		ohm	Ω		(2)
Capacitance	C		farad	F		(2)
Inductance	L		henry	H		(2)
Impedance	Z		ohm	Ω		(2)
Conductance	G	mho	siemens	S	1.000 000 E+00*	(1,2)
Susceptance	B	mho	siemens	S	1.000 000 E+00*	(1,2)
Admittance	Y	mho	siemens	S	1.000 000 E+00*	(1,2)
Conductivity	σ	mho/cm	siemens/meter	S/m	1.000 000 E+02*	(2)
Permittivity	ε		farad/meter	F/m		(2)
Complex relative permittivity	ϵ_r^*, κ^*		numeric			(2)
Permeability	μ		henry/meter	H/m		(2)
Relative permeability	μ_r		numeric			(2)
Current density	J		ampere/square meter	A/m^2		(2)
Power	P		watt	W		(2)
Radiometric Quantities						
Radiant energy	Q, Q_e		joule	J		(3)
Radiant exposure	H, H_e		joule/square meter	J/m^2		(3)
Radiant energy density	w, w_e		joule/cubic meter	J/m^3		(3)
Radiant power or flux	Φ, Φ_e		watt	W		(3)
Radiant intensity	I, I_e		watt/steradian	W/sr		(3)
Irradiance	E, E_e		watt/square meter	W/m^2		(3)
Radiant excitance	M, M_e		watt/square meter	W/m^2		(3)

TABLE 1—RADIATION QUANTITIES AND UNITS (cont'd)

Quantity	Quantity Symbol	Units			Conversion Factor‡	Authoritative Reference
		Customary†	SI	SI Unit Symbol		
Radiance	L, L_e		watt/square meter, steradian	$W/(m^2 \cdot sr)$		(3)
Spectral§ radiance	$L_\lambda, L_{e\lambda}$		watt/square meter, steradian, nanometer	$W/(m^2 \cdot sr \cdot nm)$		(3)
Emissivity	ϵ		numeric			(4)
Absorptance	α		numeric			(4)
Reflectance	ρ		numeric			(4)
Transmittance	τ		numeric			(4)
Scattering coefficient	S		reciprocal meter	m^{-1}		(6)
Absorption coefficient	K		reciprocal meter	m^{-1}		(6)
Photon energy	Q, Q_p	einstein‖	quantum#	q	6.022 045 E+23	(3)
Photon flux	Φ, Φ_p		quantum/second	q/s		(3)
Photon-flux intensity	I, I_p		quantum/second, steradian	$q/(s \cdot sr)$		(3)
Incident photon-flux density	E, E_p		quantum/second, square meter	$q/(s \cdot m^2)$		(3)
Photon-flux exitance	M, M_p		quantum/second, square meter	$q/(s \cdot m^2)$		(3)
Photon-flux sterance (radiance)	L, L_p		quantum/second, square meter, steradian	$q/(s \cdot m^2 \cdot sr)$		(3)
			Photometric Quantities			
Luminous energy	Q, Q_v		lumen-second (talbot)	$lm \cdot s$		(3,4)
Luminous exposure	H, H_v		talbot/square meter	$lm \cdot s/m^2$		(3)
Luminous flux or power	Φ, Φ_v		lumen	lm		(3,4)
Luminous intensity	I, I_v		candela	cd		(3,4)
Illuminance (Illumination)	E, E_v	phot (lm/cm^2)	lux lumen/square meter	lx lm/m^2	1.000 000 E+04*	(1)
		footcandle (lm/ft^2)			1.076 391 E+01	(1)
Luminous exitance	M, M_v		lumen/square meter	lm/m^2		
Luminance (photometric brightness)	L, L_v	stilb (cd/cm^2)	candela/square meter (nit)	cd/m^2	1.000 000 E+04*	(4)
		apostilb (π^{-1} cd/m^2)			$1/\pi$ E+00*	(4)
		lambert (π^{-1} cd/cm^2)			$1/\pi$ E+04*	(1)
		footlambert (π^{-1} cd/ft^2)			3.426 259 E+00	(1)
Luminous efficacy	K		lumen/watt	lm/W		(4)
Luminous efficiency	V		numeric			(4)
			Solar Radiation			
Solar radiant energy	Q_s	Btu (thermochemical)	joule	J	1.054 350 E+03	(1)
		calorie (gram, thermochemical)			4.184 000 E+00*	(1)
Solar radiant exposure	H_s	langley (gram calorie/cm^2)	joule/square meter	J/m^2	4.184 000 E+04*	(1)
		Btu/sq. foot			1.134 893 E+04	(1)
Solar irradiance (insolation)	E_s	langley/minute	watt/square meter	W/m^2	6.973 333 E+02	(1)
		Btu/sq. foot/minute			1.891 489 E+02	(1)
			Ionizing Radiation			
Activity		curie	becquerel	Bq, s^{-1}	3.700 000 E+10*	(1)
Absorbed dose		rad	gray	Gy, J/kg	1.000 000 E-02*	(1,5)
			Acoustics			
Sound pressure	p	$dyne/cm^2$	pascal	Pa	1.000 000 E-01*	(7,8)
Sound particle velocity	u, v	cm/sec	meter/second	m/s	1.000 000 E-02*	(7)
Sound particle acceleration	α		meter/second-squared	m/s^2		(7)
Sound intensity	I(J)	$erg/sec \cdot cm^2$	watt/square meter	W/m^2	1.000 000 E-03*	(7,8)
Sound power	P(K)	erg/second	watt	W	1.000 000 E-07*	(7)
Sound energy	E, W	erg	joule	J	1.000 000 E-07*	(7,8)
Sound energy density	w(e)	erg/cm^3	joule/cubic meter	J/m^3	1.000 000 E-01*	(7,8)
Linear displacement	x		meter	m		(7)
Rotational displacement	ϕ		radian	rad		(7)
Sound particle displacement	ξ	cm	meter	m	1.000 000 E-02*	
Acoustic impedance	Z_a	$\frac{dyne\ sec}{cm^5}$	pascal second/cubic meter	$Pa \cdot s/m^3$	1.000 000 E+05	(7,8)
Acoustic resistance	R_a	$\frac{dyne\ sec}{cm^5}$	pascal second/cubic meter	$Pa \cdot s/m^3$	1.000 000 E+05*	(7)
Acoustic reactance	χ_a	$\frac{dyne\ sec}{cm^5}$	pascal second/cubic meter	$Pa \cdot s/m^3$	1.000 000 E+05*	(7)
Acoustic admittance	γ_a	$\frac{cm^5}{dyne\ sec}$	cubic meter/pascal second	$m^3/(Pa \cdot s)$	1.000 000 E-05*	(7)
Acoustic conductance	G_a	$\frac{cm^5}{dyne\ sec}$	cubic meter/pascal second	$m^3/(Pa \cdot s)$	1.000 000 E-05*	(7)
Acoustic susceptance	B_a	$\frac{cm^5}{dyne\ sec}$	cubic meter/pascal second	$m^3/(Pa \cdot s)$	1.000 000 E-05*	(7)

†Where no customary units are listed, the SI unit has been the unit of customary use.
‡Factor by which customary units must be multiplied to obtain SI units. An asterisk indicates that the conversion is exact.
§Other radiometric quantities become spectral quantities when taken per nanometer.
‖Usage varies with respect to wavelength range.
#Energy dependent upon wavelength, photon and photon flux units listed here are SI compatible rather than approved SI units.

ASAE Engineering Practice: ASAE EP415

SAFETY COLOR CODE FOR THE TRAINING AND EDUCATIONAL SHOP

Developed by the ASAE Instruction in Agricultural Mechanization Committee; approved by ASAE Electric Power and Processing, Food Engineering, and Power and Machinery Divisions Standards Committees; adopted by ASAE April 1982; reconfirmed December 1986.

SECTION 1—PURPOSE AND SCOPE

1.1 The purpose of this Engineering Practice is to establish a safety color code in training and educational shops to alert and inform persons to take precautionary or other appropriate action in the presence of hazards.

1.2 This Engineering Practice is intended to improve personal safety in training and educational shops during the operation and application of tools, equipment, machinery and materials.

1.3 This Engineering Practice is intended to help promote uniform identification of potential hazards in training and educational shop activities.

1.4 This Engineering Practice specifies the safety colors and their applications to specific purposes in training and educational shops.

1.5 This Engineering Practice shall apply to the use of safety color coding for the identification of and marking of potential physical hazards, safety equipment, other protective equipment, stationary shop equipment and machines, and safety signs and markers in training and educational shops.

1.6 This Engineering Practice shall apply to existing structures, facilities, equipment and machinery, to subsequent modifications and additions, and to new structures.

1.7 This Engineering Practice specifies color codes for safety only. The primary purpose of safety color coding in training and educational shop operations is to promote safety. The use of color is not a substitute for engineering and administrative controls, such as safety training and the elimination of identifiable hazards. In addition, hazard control and accident prevention are dependent upon the awareness, concern, and prudence of personnel involved in the operation, transport, maintenance and storage of equipment or in the use and maintenance of facilities.

1.8 It is the intent of this Engineering Practice to be in agreement with all existing standards and regulations. However, due to some discrepancies between organizations, this Engineering Practice may vary slightly, specifically with the American National Standards Institute. Steps are being taken to resolve these minor differences.

SECTION 2—GENERAL PRINCIPLES

2.1 The color safety treatment includes painting the equipment, safety signs, walls, ceiling, and lanes on the floor of the training and educational shop where desirable. Adequate illumination should be provided so that objects, safety signs, and colors can be readily and positively identified.

2.2 Color contrast shall be made between the color-coded object, and/or the safety sign, and its background.

2.3 Bright colors should not be placed in the line of vision of the operator because they are disturbing to the operator, causing the operator to change the focus of vision from one object to another, resulting in fatigue.

2.4 The colors of the walls, ceiling, floor cabinets and base equipment should be coordinated.

2.5 Color coding should be such as to distinguish the dangerous from the less dangerous parts of a machine. The operator must focus on the work being done, and at the same time must readily observe the moving parts, adjusting levers, and controls of the machine.

2.6 Discretion must be used in whether a total surface should be painted or the outline of the surface be painted. Frequently, too much paint will defeat the purpose of color coding. For example, a band of stripes on a container or an area may be more effective than painting the entire area.

2.7 In addition to safety color coding and correct lighting, proper training in the use of safety equipment and in accident prevention is necessary for the safe operation of shops.

SECTION 3—SAFETY COLORS FOR SHOP EQUIPMENT AND TYPICAL USES*

3.1 **Safety Red** represents danger. It shall identify areas or items of danger and emergency. The following applications are examples of Safety Red:

3.1.1 Danger signs. The word DANGER is written in white letters on a red background.

3.1.2 Emergency stop bars on hazardous machines. Stop buttons or electrical switches used for *emergency stopping* of machinery.

3.1.3 Fire equipment and apparatus. If it is not practical to paint the fire equipment and apparatus, then the background should be painted Safety Red†.

3.1.4 Containers of flammable or combustible materials having a flash point at or below 27 °C (80 °F). Identify further by a yellow band at least 1/4 the height of the container, or with yellow labeling.

3.2 **Safety Orange** represents a warning. It shall designate machine hazards. The following applications are examples of Safety Orange:

3.2.1 Guards, covering a specific hazard, that *must be in place before operation is begun.*

3.2.2 The electrical box containing the stop button, lever, or toggle of an electrical switch.

3.2.3 Parts of a machine which might cut, crush, shock, or otherwise injure a worker.

3.2.4 Hazards exposed when guards or shields for gears, pulleys, chains, etc., are removed (such as gears, pulleys, chains, etc.) or inspection doors are open (such as the inside of doors of fuse boxes and switch boxes).

3.3 **Safety Yellow** represents caution. Solid yellow, or diagonal yellow and black stripes, can be used in accordance with the color contrast between the safety sign, color-coded object and its background. The following applications are examples of Safety Yellow:

3.3.1 Critical parts of machines, adjusting wheels, levers, and knobs, or *any part that may need adjustment before or during use of the machine.*

3.3.2 Diagonal black and yellow stripes on nonmoving hazards that may cause stumbling, falling, or strike-against, such as low beams, machinery which projects into traffic lanes, and moving hazardous parts which extend beyond the machine when operated.

3.3.3 Lettering, such as "Flammable, Keep Fire Away", on storage cabinets containing flammable materials. Lettering shall be conspicuous with a high visibility contrast.

3.3.4 Containers for explosives, corrosives, or unstable materials. These shall be yellow or identified by a yellow band at least 1/4 of the container height. The contents of the container shall be identified thereon.

3.3.5 Caution signs.

---

*This Engineering Practice is intended to be compatible with Occupational Safety and Health Administration Regulation 1910.144, Safety Color Code for Marking Physical Hazards, and American National Safety Institute Standard Z53.1, Safety Color Code for Marking Physical Hazards.

†There are questions being raised whether both danger areas and fire equipment should be safety red. Steps are being taken toward resolution.

3.4 Safety Blue represents informational signs if warning or caution is implied. The following applications are examples of Safety Blue:

3.4.1 To indicate caution against using defective equipment or machines under repair or out of order. Large signs painted blue, with white letters and the words, "Out of Order", "Do Not Operate", or "Do Not Move", should be hung on a machine to warn potential users of unsafe conditions.

3.5 Safety Green represents safety and designates the location of first aid equipment. The following applications are examples of Safety Green:

3.5.1 To show the location of medical first-aid kits, gas masks, stretchers, safety deluge showers, or safety bulletin boards. On first-aid kits, it is used either with a white cross on a green background or vice versa.

3.6 Safety Black, Safety White, or combination of Safety Black with Safety White shall be for designating traffic and housekeeping markings. The following applications are examples of Safety Black, Safety White, or combination of Safety Black and Safety White:

3.6.1 To identify barricades.

3.6.2 Combination of Safety Black and Safety White paint bands (not entire object) to identify movable objects, such as trash containers, etc.

3.7 Ivory (not a safety color) is used as a spotlight color to improve visibility. The following applications are examples of Ivory:

3.7.1 To form a background for viewing work being done.

3.7.2 To outline the vertical edges of tables, other work surfaces or as a background for instruments; to draw attention to work areas that may have safety hazards.

3.8 Miscellaneous applications of color

3.8.1 Safety Purple or Magenta is used to designate radiation hazards. The following applications are examples of the radiation hazard symbol colors:

3.8.1.1 Rooms and areas (outside and inside buildings) where radioactive materials are stored or handled, or which have been contaminated with radioactive materials. Used in combination with Safety Yellow for markers such as tags, labels, signs and floor markers.

3.8.2 Safety Traffic Gray, Safety White or Safety Yellow may be used for floor lanes or work areas around machines if machines are stationary.

3.8.3 Safety NonGlare Traffic Gray may be used for metal work bench tops and work bench tops with natural finish wood to provide a restful color contrast. (An ivory band around the vertical edge of all tables makes the edges much more visible.) It is not intended that all table tops be painted—only at users' discretion.

3.8.4 Aluminum (not a safety color) may be used to paint waste containers. Except for containers for oily rags and other flammable materials, a black band about 1/5 the height of the container should be painted around the top, to indicate contents. An orange band about 1/5 the height of the container should be used around the top of waste containers for oily rags or other flammable materials. The band should be labeled to identify the material.

3.9 Electrical receptacles (See National Fire Protection Association Standard NFPA No. 70, National Electrical Code)

3.9.1 On all 120-V receptacle and toggle switch cover plates, use a color which contrasts with the background—preferably the original color of the cover (ivory, brown, zinc plated, stainless steel). In many instances, it is not considered desirable to paint cover plates. This is left up to the discretion of the individual.

3.9.2 All 240-V, single- and three-phase receptacle outlets shall be provided with special connections which are not interchangeable among them or 120-V outlets so that a special color is unnecessary.

3.10 Luminescent pigments and reflective tape

3.10.1 Luminescent pigments and reflective tape are useful to mark danger spots which are not visible at night, or location of signs and switches. The following applications are examples of luminescent pigments and reflective tape:

3.10.1.1 Stripes on stair railing and risers to make them visible at night.

3.10.1.2 Signs or directional arrows for fire escape or exit signs which are not lighted or readily seen when lights are out.

3.10.1.3 Switch plates of the luminescent type so that they may be readily located at night.

3.11 Piping systems‡. The following applications are examples of color coding of piping. (All should have labels at point of use to identify materials carried by the pipe, such as compressed air, natural gas, etc.)

3.11.1 It is not necessary to paint all pipes; however, all pipes should identified. If painted, use the same color as the background walls or ceiling. Do not paint rubber and flexible synthetic hoses or pipes. Identify pipes and hoses with decals or stencils (and optional bands) and with the following color combinations:

3.11.1.1 Safety Red with Safety Black lettering—Fuel gas pipes, such as gasoline, acetylene, etc.

3.11.1.2 Safety Green with Safety Black lettering—Liquid gas pipes, such as oxygen and inert liquids.

3.11.1.3 Safety Yellow with Safety Red lettering—Natural gas, steam at 100 °C (212 °F) or higher, and other hazardous and flammable materials and high pressure fluids.

3.11.1.4 Safety Blue with Safety White lettering—Compressed air (inert gases at low pressure).

3.11.1.5 Safety Black bands—Vent lines.

3.11.1.6 Safety Gray bands—Service water and sprinkler pipes.

3.11.2 Other application of bands of color

3.11.2.1 To further identify pipes and hoses, at room entrances and exits, color bands may be painted around the pipe with intermediate bands 2.5 cm (1 in.) wide every 3 m (10 ft).

3.11.2.2 Valve bodies.

‡There is a slight difference of opinion between OSHA and ANSI as to the usage of some of the colors for identifying piping. This is because of accidents to color-blind people who were not able to correctly identify the contents of the piping. Therefore, changes may be forthcoming, but both groups agree that all piping should be appropriately identified by signs and tags.

Cited Standards:

ANSI Z53.1, Safety Color Code for Marking Physical Hazards
NFPA No. 70, National Electrical Code
OSHA 1910.144, Safety Color Code for Marking Physical Hazards

AGRICULTURAL EQUIPMENT

ASAE Notation:

The letter S preceding numerical designation indicates ASAE Standard; EP indicates Engineering Practice; D indicates Data. A decimal and numeral following the file number indicate the number of times a document has been revised. Thus ASAE S201.4 indicates Standard number 201, four times revised. The letter T after the designation indicates tentative status. Always refer to ASAE documents by complete designation to avoid confusion with standards of other organizations. For example: ASAE S201.4.

The symbol **T** preceding or in the margin adjacent to section headings, paragraph numbers, figure captions, or table headings indicates a technical change was incorporated in that area when this document was last revised. The symbol **T** preceding the title of a document indicates essentially the entire document has been revised. The symbol **E** used similarly indicates editorial changes or corrections have been made with no intended change in the technical meaning of the document.

ASAE Standard: *ASAE S201.4 (ANSI/ASAE S201.4/SAE J716 DEC84)

APPLICATION OF HYDRAULIC REMOTE CONTROL CYLINDERS TO AGRICULTURAL TRACTORS AND TRAILING-TYPE AGRICULTURAL IMPLEMENTS

*Corresponds in substance to previous revision SAE 716a.

Proposed by the Engineering Committee of the Farm and Industrial Equipment Institute; approved by the ASAE Power and Machinery Division Technical Committee; adopted by ASAE March 1949; revised July 1951; reconfirmed 1961; revised December 1964, December 1965, June 1966; reconfirmed December 1971, December 1976, December 1981; approved by ANSI as an American National Standard December 1982; reconfirmed December 1986.

SECTION 1—PURPOSE AND SCOPE

1.1 The purpose of this Standard is to establish common mounting and clearance dimensions for hydraulic remote control cylinders and trailing-type agricultural implements with such other specifications as are necessary to accomplish the following objectives:

1.1.1 To permit use of any make or model of trailing-type agricultural implement adapted for control by a hydraulic remote cylinder with any remote cylinder furnished as part of any make or model of agricultural tractor in a drawbar horsepower size suitable for operating that implement.

1.1.2 To facilitate changing the hydraulic cylinder from one implement to another and decrease the possibility of introducing dirt or other foreign material into the hydraulic system by reducing the necessity for supplemental hose lengths or piping with certain types of implements.

SECTION 2—DEFINITIONS

2.1 Anchor end: The closed end of the cylinder.

2.2 Attaching pins: Removable pins provided in the yokes for attaching the cylinder to the implement.

2.3 Bottom of cylinder: The side of the cylinder opposite that to which the oil lines are attached.

SECTION 3—CLASSIFICATION AND RATING

3.1 For the purpose of application of remote hydraulic cylinders, tractors are divided into categories as shown in Table 1.

3.2 A hydraulic cylinder with minimum extending stroke thrust capacity as shown in Table 1 will be considered as part of the tractor remote control hydraulic system.

3.3 Length of cylinder stroke and attaching pin size will be as shown in Table 1.

3.4 Minimum thrust capacity will be based on the calculated area of the cylinder and 80 percent of the minimum relief valving setting.

3.5 Implements requiring cylinder thrust over 18,000 lb (8165 kg) should provide for use of 16-in. (406 mm) stroke cylinders.

SECTION 4—CLEARANCE DIMENSIONS

4.1 Implements will provide clearance for the maximum cylinder size as shown in Fig. 1.

4.2 Hydraulic remote cylinder with hose is considered part of the tractor. The cylinder falls within the clearance dimensions given in Fig. 1 and provides anchor and rod yoke clearances as shown in Figs. 2 and 3.

SECTION 5—STANDARD HOSE LENGTHS FOR REMOTE CYLINDERS

5.1 The standard hose length shall be sufficient so the remote cylinder is operable when the front attaching pin is located in accordance with dimensions in Table 1.

5.1.1 On tractors for which the ASAE-SAE standard power take-off shaft is available, the ASAE-SAE agricultural tractor drawbar hitch point shall be used, shown in Fig. 4.

5.1.2 On all other track-type tractors without ASAE-SAE standard power take-off shaft, the SAE standard track-type drawbar hitch point, per Society of Automotive Engineers SAE J749, Industrial (Track-Type) Tractor Drawbars, shall apply, shown in Fig. 5.

5.1.3 On an implement attached to the tractor by means of an ASAE S217 Three-Point Free Link Hitch and using a tractor hydraulic remote cylinder, the maximum spherical radius for locating the cylinder front anchor pin on the implement, shall be measured from a point 7 in. (178 mm) horizontally ahead of the midpoint between the lower hitch points with lower links horizontal, shown in Fig. 6.

5.2 When a three-point hitch quick-attaching coupler (ASAE S278, Attachment of Implements to Agricultural Wheel Tractors Equipped with Quick-Attaching Coupler for Three-Point Free-Link Hitch) is made available by the tractor manufacturer, remote hydraulic cylinder hose length must provide for moving the implement 4.06 in. (103 mm) rearward.

SECTION 6—OTHER SPECIFICATIONS

6.1 Both single- and double-acting cylinders will operate to raise the implement (or deangle disk harrows) on their extending strokes.

6.2 Variable stroke control necessary in the application of hydraulic control to some implements will be incorporated in the cylinder or hydraulic system and applied to the retracting stroke. Provisions will be made on the implement to accommodate the fully extended cylinder.

6.3 Operating time at maximum full load engine speed (ASAE S209, Agricultural Tractor Test Code, paragraph 2.2), cylinders extending, will be 1.5 to 2.0 seconds for 8-in. (203 mm) stroke cylinders, and 3 to 4 seconds for 16-in. (406 mm) stroke cylinders.

6.4 Hose support as required for remote cylinder hose, will be considered a part of the implement.

6.5 Hose connections to cylinders will be such that the hose will not interfere with bars extending through the yoke on either end of the cylinder.

6.6 Attaching pins will be considered a part of the cylinder. They will be easily removed and attached.

Cited Standards:

ASAE S209, Agricultural Tractor Test Code
ASAE S217, Three-Point Free-Link Attachment for Hitching Implements to Agricultural Wheel Tractors
ASAE S278, Attachment of Implements to Agricultural Wheel Tractors Equipped with Quick-Attaching Coupler for Three-Point Free-Link Hitch
SAE J749, Industrial (Track-Type) Tractor Drawbars

TABLE 1—HYDRAULIC REMOTE CONTROL CYLINDERS FOR TRAILING-TYPE AGRICULTURAL IMPLEMENTS

Category	Max Drawbar hp*	Length of Stroke in. Min	Length of Stroke in. Max	Length of Stroke mm Min	Length of Stroke mm Max	Attaching Pin Max Dia in.	Attaching Pin Max Dia mm	Min Thrust Per Drawbar hp† lb	Min Thrust Per Drawbar hp† kg	Spherical Radius to Front Attaching Pin § in.	Spherical Radius to Front Attaching Pin § mm
I	Up to 45	8.0	8.125	203.0	206.2	1.005	25.53	150	68	60	1524
II	40 to 100	8.0	8.125	203.0	206.2	1.005	25.53	150	68	84	2134
III	Over 80	8.0	8.125	203.0	206.2	1.005	25.53	150	68	96	2438
		Or									
		16.0	16.125	406.0	409.2	1.255	31.88	150	68	96	2438

*Based on ASAE S209, Agricultural Tractor Test Code, paragraph 2.5.
†It is intended that implements be designed to require thrust per drawbar horsepower below this figure.
§Refer to Fig. 4, 5 and 6.

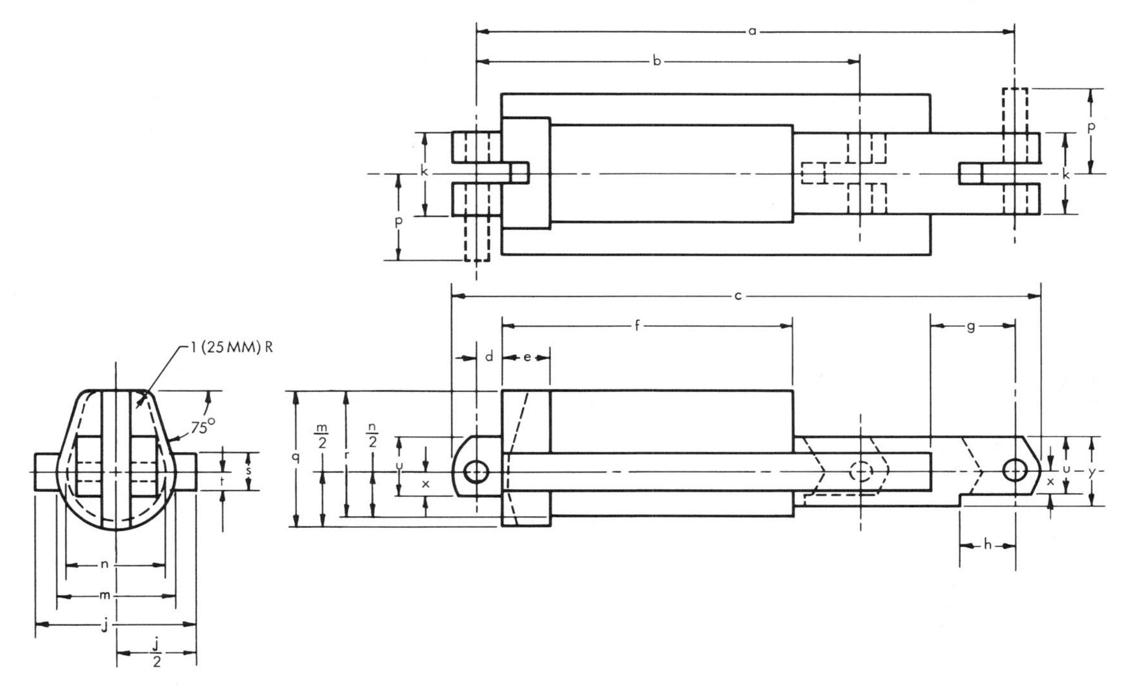

FIG. 1—CLEARANCE MODEL OF AGRICULTURAL HYDRAULIC CYLINDER

CLEARANCE DIMENSIONS FOR AGRICULTURAL CYLINDER FOR FIG. 1

Code	Description	8-in. (203 mm) Stroke in.	8-in. (203 mm) Stroke mm	16-in. (406 mm) Stroke in.	16-in. (406 mm) Stroke mm
	Length				
a	Between pin centerlines, extended, max	28.38	721	47.62	1210
b	Between pin centerlines, retracted, min	20.25	514	31.50	800
c	Overall, extended	30.88	785	50.38	1280
d	Anchor pin centerlines to cylinder body	1.25	32	1.25	32
e	Cylinder end flange	2.50	64	3.50	89
f	Cylinder body	15.50	394	26.38	670
g	Rod pin centerlines to stop mechanism	4.50	114	10.62	270
h	Rod pin centerlines to stop collar	3.00	76	3.00	76
	Width				
j	Overall, stop mechanism	8.56	217	9.50	241
k	Yoke	4.50	114	4.50	114
m	Cylinder end flange dia	6.00	152	7.00	178
n	Cylinder outside dia	5.00	127	6.00	152
p	For pin removal only	4.50	114	4.50	114
	Height				
q	Overall, cylinder end flange	7.50	190	8.50	216
r	Overall, cylinder body	7.00	178	8.00	203
s	Stop mechanism	2.38	60	2.38	60
t	Rod centerline to bottom stop mechanism	1.19	30	1.19	30
u	Yoke	3.50	89	3.50	89
x	Rod centerline to bottom of yoke	1.50	38	1.50	38
y	Stop collar, dia	4.00	102	4.00	102

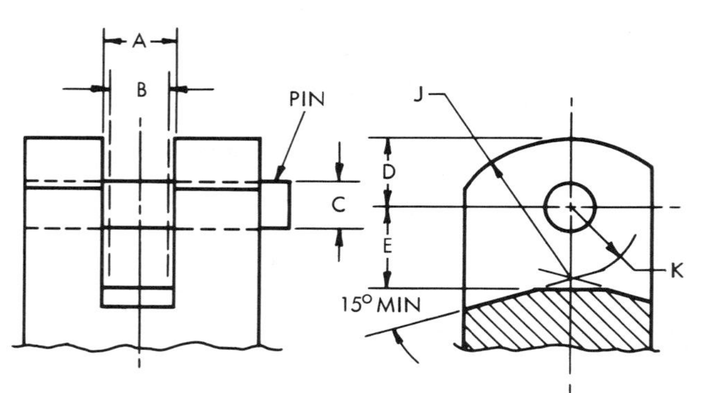

FIG. 2—YOKE CLEARANCES (ANCHOR END)

FIG. 3—YOKE CLEARANCES (ROD END)

YOKE CLEARANCE (FOR FIGS. 2 AND 3)

Code	Description	8-in. (203 mm) Stroke		16-in. (406 mm) Stroke	
		in.	mm	in.	mm
A	Yoke throat clearance, min	1.06	26.9	1.06	26.9
	max	1.12	28.4	1.12	28.4
B	Thickness of bar cleared, max	1.02	25.9	1.02	25.9
	min recommended	0.86	21.8	0.86	21.8
C	Pin dia, nom	1.000	25.40	1.250	31.75
	max	1.005	25.53	1.255	31.88
D	Length, pin centerline to end of yoke, max	1.25	31.8	1.38	35.0
E	Length, pin centerline to bottom of throat, min (anchor end)	1.62	41.1	1.75	44.4
F	Radius of yoke end (rod end)	1.25	31.8	1.38	35.0
G	Radius of throat clearance (rod end)	1.38	35	1.62	41.1
H	Length, pin centerline to bottom of throat, min (rod end)	1.62	41.1	2.25	57.2
J	Radius of yoke end (anchor end)	2.62	66.5	2.62	66.5
K	Radius of throat clearance (anchor end)	1.38	35.0	1.50	38.1
L	Clearance angle (rod end)	30*		35*	

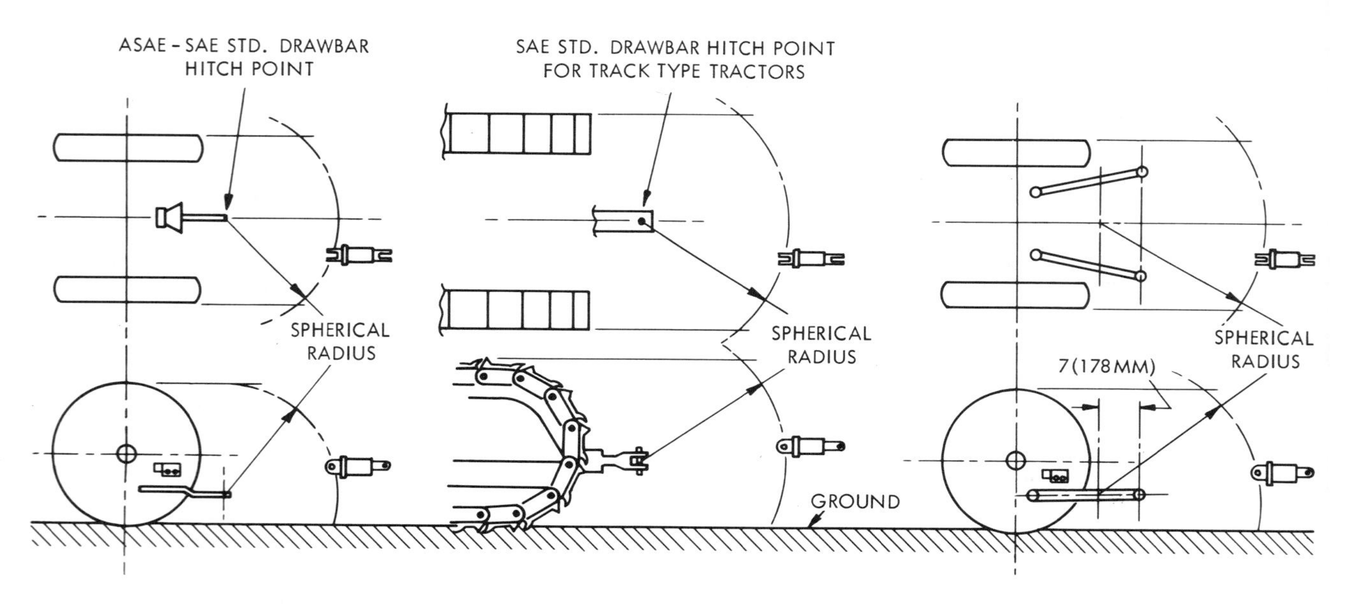

FIG. 4—HOSE LENGTH DIAGRAM FOR WHEEL-TYPE TRACTOR DRAWBAR

FIG. 5—HOSE LENGTH DIAGRAM FOR TRACK-TYPE TRACTOR DRAWBAR

FIG. 6—HOSE LENGTH DIAGRAM FOR HITCH-MOUNTED IMPLEMENTS

ASAE Standard: ASAE S203.10 (SAE J1170 NOV84)

REAR POWER TAKE-OFF FOR AGRICULTURAL TRACTORS

Originally developed in 1926 by a conference of engineers representing tractor manufacturers; adopted by ASAE April 1927; revised July 1928, March 1931, August 1941, June 1952; revisions submitted by Farm and Industrial Equipment Institute; approved by ASAE Power and Machinery Division Standards Committee 1958, 1961, June 1964, June 1966, December 1966, December 1967, December 1968, December 1969; revised editorially March 1973; reconfirmed December 1974; revised and combined with ASAE S204.6, 1000-RPM Power Take-Off for Agricultural Tractors, March 1976; revised March 1978, January 1982; reconfirmed December 1986.

SECTION 1—PURPOSE

1.1 This Standard specifies a suitable means of mechanical power transmission from the tractor to the implement and satisfactory hitching of the implement to the tractor, and promotes dimensional interchangeability of tractors and towed implements with the same Type power take-off (see Table 1).

SECTION 2—SCOPE

2.1 This Standard provides dimensions relating to the tractor power take-off shaft, master shield and drawbar.

2.2 This Standard provides specifications for the splined power take-off-shaft and the mating connector.

2.3 This Standard establishes and defines Type 1, 2 and 3 power take-off (see Table 1).

2.4 The successful performance of all tractor and implement combinations likely to be met in field service requires consideration of factors other than the dimensional relationship provided in this Standard.

SECTION 3—DEFINITIONS

3.1 Power take-off (PTO): An external shaft on the rear of the tractor to provide rotational power to implements.

SECTION 4—SPECIFICATIONS

4.1 The tractor PTO shaft shall conform to the specifications shown in Figs. 1, 2, 3 and 4 and Table 1.

4.2 The location of the tractor PTO shaft shall be within the limits of 25 mm (1 in.) to the right or left of the tractor centerline, the centerline being recommended location.

4.3 The direction of PTO shaft rotation shall be clockwise when facing in the direction of forward travel.

4.4 A means to indicate when the PTO shaft is operating at standard speed shall be provided on tractors capable of driving the 540 r/min shaft in excess of 600 r/min and the 1000 r/min shaft in excess of 1100 r/min.

4.5 Tractors capable of driving the 540 r/min shaft in excess of 630 r/min and the 1000 r/min shaft in excess of 1170 r/min shall also include a suitable warning of operation in excess of those speeds.

4.6 The tractor master shield shall conform to ASAE Standard ASAE S318, Safety for Agricultural Equipment. Tractor master shield dimensions shall conform to Figs. 6 and 7.

4.7 Dimensions associated with the drawbar shall conform to Fig. 5 and Table 2.

4.8 The drawbar hitch point shall be directly in line with the centerline of the tractor PTO shaft, and provisions shall be made on the tractor for locking the drawbar in this position.

4.9 Through a 1.57 rad (90 deg) turn, right or left, the tractor drawbar shall clear an implement clevis having the following dimensions:

	mm	in.
With Type 1 and 2 PTO shafts:		
Vertical opening	76	3.0
Throat depth from hitch pin center	76	3.0
With Type 3 PTO shaft:		
Vertical opening	102	4.0
Throat depth from hitch pin center	102	4.0

Cited Standard:

ASAE S318, Safety for Agricultural Equipment

TABLE 1—POWER TAKE-OFF SHAFT DIMENSIONS* (SEE FIG. 1)

	Type 1	Type 2	Type 3
Nominal Diameter	35 mm (1 3/8 in.)	35 mm (1 3/8 in.)	45 mm (1 3/4 in.)
Standard Operating Speed-r/min	540 ± 10	1000 ± 25	1000 ± 25
A Groove to end of shaft	38.1 (1.50) ± 0.8 (0.03)	25.4 (1.00) ± 0.8 (0.03)	38.1 (1.50) ± 0.8 (0.03)
B Effective spline length with relation gage, min	76.2 (3.00)	63.5 (2.50)	88.9 (3.50)
C Chamfer	7.1 (0.28) ± 0.8 (0.03)	4.8 (0.19) ± 0.8 (0.03)	7.6 (0.30) ± 0.8 (0.03)
D Chamfer angle	0.5 rad (30 deg) ± 0.05 (3.0)	0.5 rad (30 deg) ± 0.05 (3.0)	0.5 rad (30 deg) ± 0.05 (3.0)
E ID of groove	29.46 (1.160) / 29.26 (1.152)	29.46 (1.160) / 29.26 (1.152)	37.34 (1.470) / 37.13 (1.462)
F Radius of groove	6.86 ± 0.25 (0.270 ± 0.010)	6.86 ± 0.25 (0.270 ± 0.010)	8.38 ± 0.25 (0.330 ± 0.010)
G Spherical clearance radius on tractor, min	82.6 (3.25)	82.6 (3.25)	101.6 (4.00)
H Location of center of clearance radius	0.0	12.7 (0.50)	0.0
J Break sharp corner of chamfer	Yes	Optional	Optional

*Dimensions are in mm (in.) except where indicated otherwise.

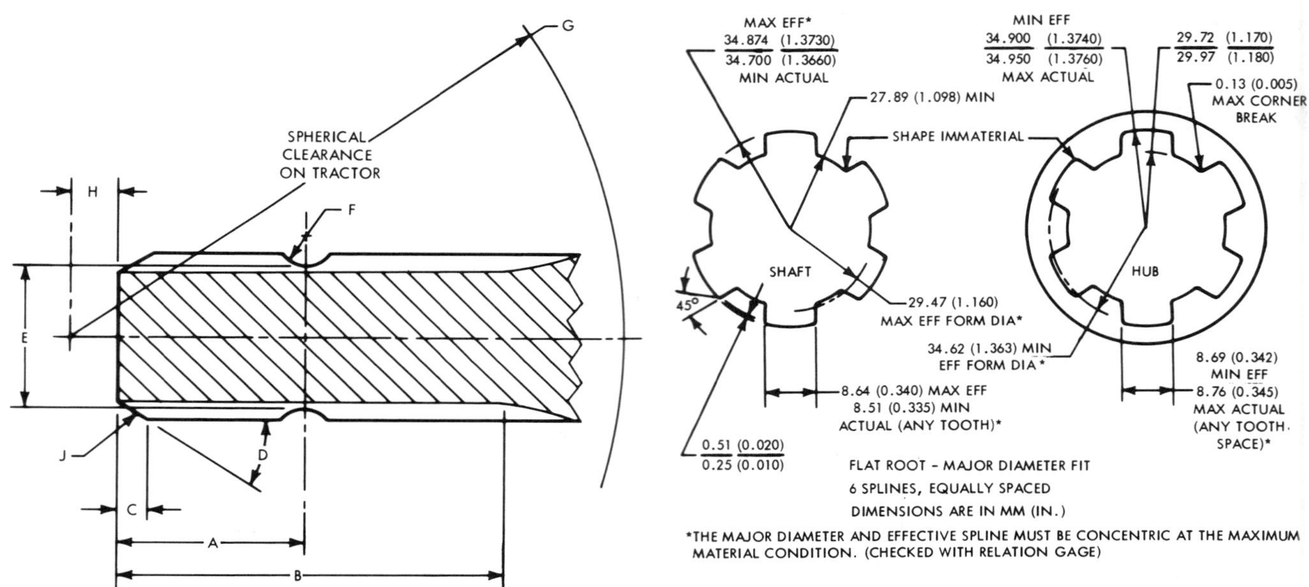

FIG. 1—POWER TAKE-OFF SHAFT (See Table 1)

The circumferential groove is provided for a locking means in the implement hub.

Effective spline length, B, to be heat treated for surface durability (within Rockwell C 48-56).

FIG. 2—TYPE 1 POWER TAKE-OFF. 540 r/min—35 mm (1 3/8 in.) diameter—straight side spline dimensions.

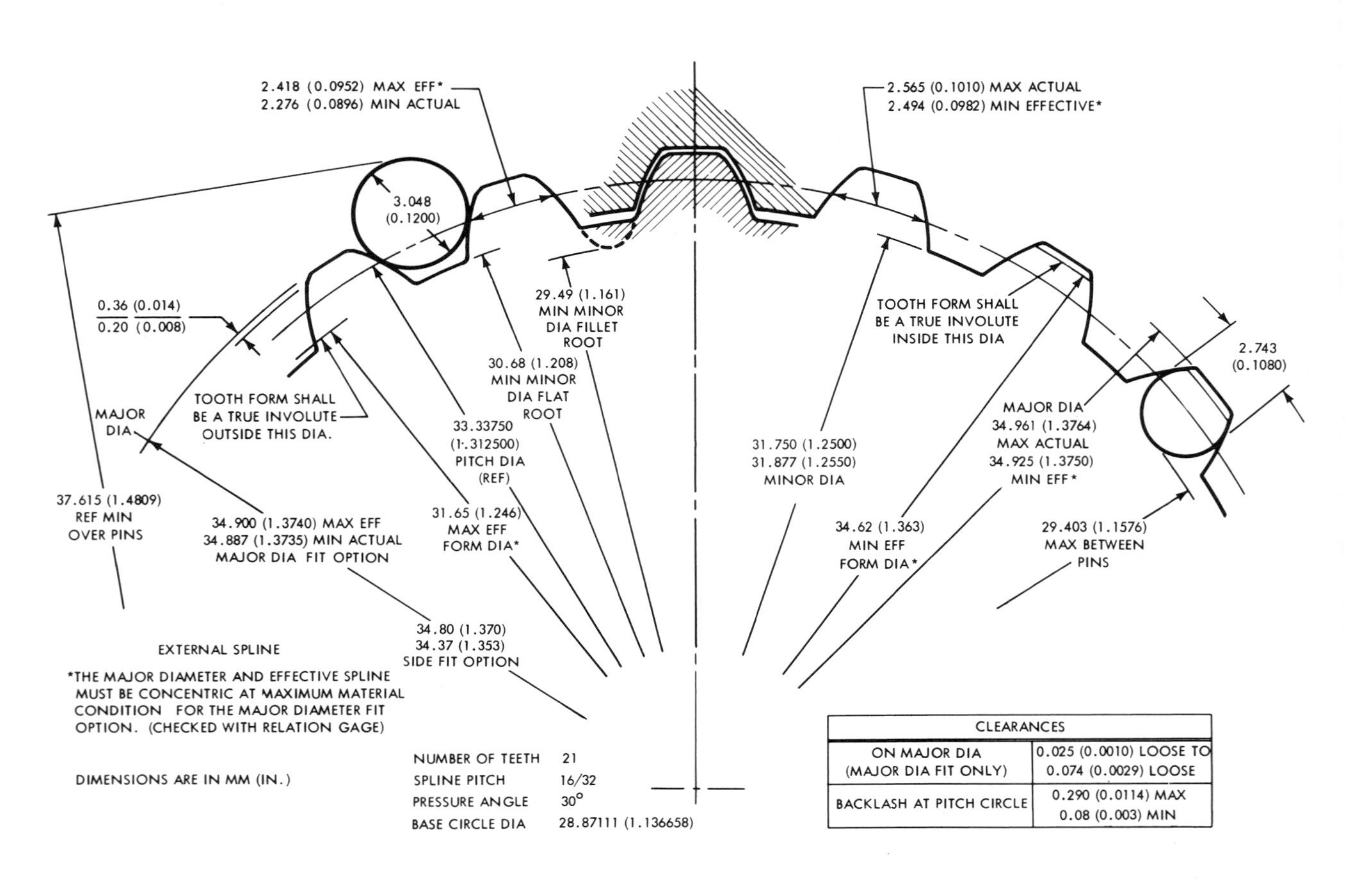

CLEARANCES	
ON MAJOR DIA (MAJOR DIA FIT ONLY)	0.025 (0.0010) LOOSE TO 0.074 (0.0029) LOOSE
BACKLASH AT PITCH CIRCLE	0.290 (0.0114) MAX 0.08 (0.003) MIN

FIG. 3—TYPE 2 POWER TAKE-OFF. 1000 r/min—35 mm (1 3/8 in.) diameter—involute spline dimensions.

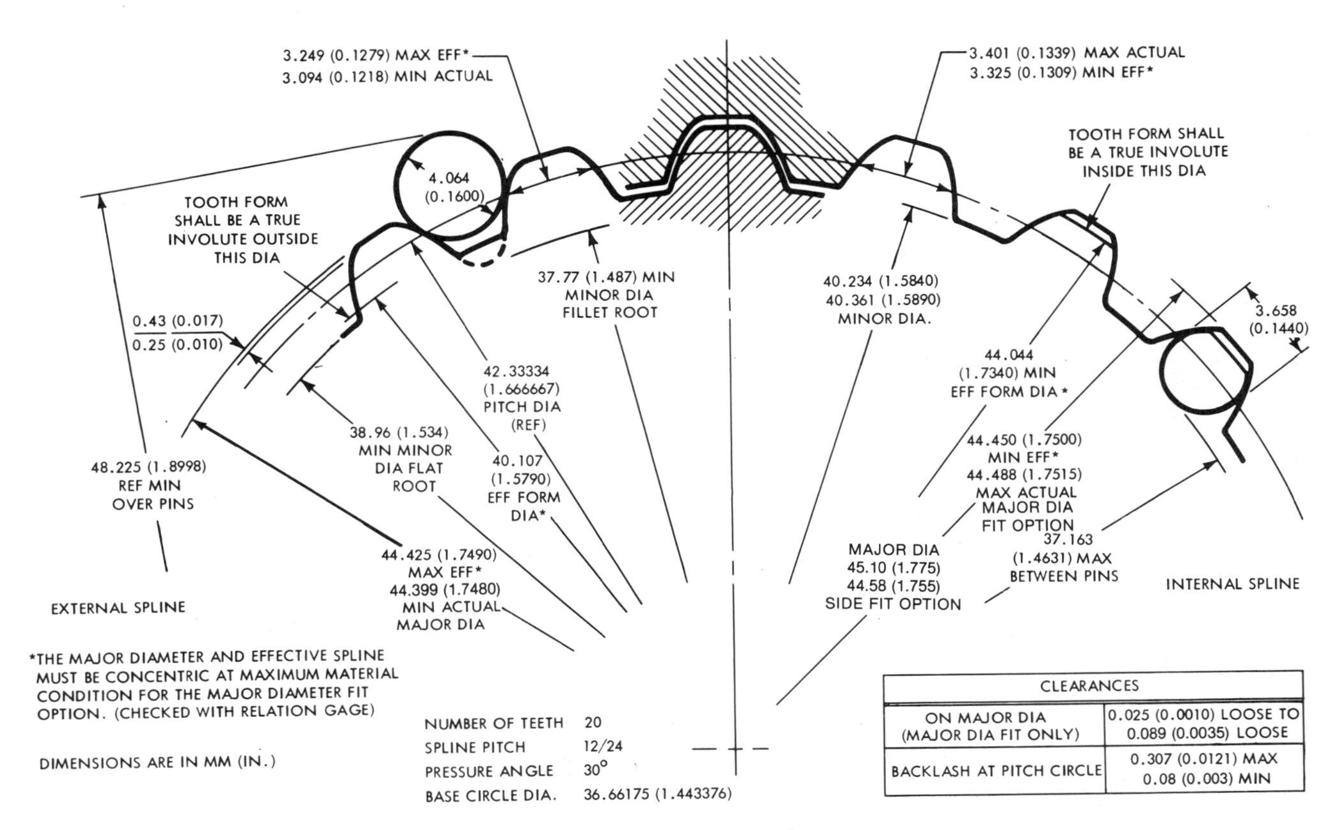

FIG. 4—TYPE 3 POWER TAKE-OFF. 1000 r/min—45 mm (1 3/4 in.) diameter—involute spline dimensions.

TABLE 2—DIMENSIONS ASSOCIATED WITH DRAWBAR AND POWER TAKE-OFF* (SEE FIG. 5)

		Type 1	Type 2	Type 3
	Nominal Diameter	35 mm (1 3/8 in.)	35 mm (1 3/8 in.)	45 mm (1 3/4 in.)
	Standard Operating Speed-r/min	540 ± 10	1000 ± 25	1000 ± 25
K	Hitch pin hole dia., min	20.6 (0.81)	20.6 (0.81)	33.1 (1.31)
L	Auxiliary hole dia.	17.3 (0.68)	17.3 (0.68)	17.3 (0.68) min
M	Auxiliary hole spacing	102 (4.0) ± 0.8 (0.03)	102 (4.0) ± 0.8 (0.03)	102 (4.0) ± 0.8 (0.03)
N	Hitch pin and clevis clearance plane below PTO shaft centerline, min	122 (4.8)	122 (4.8)	165 (6.5)
P	Hitch pin and clevis clearance plane from hitch pin centerline, min	203 (8.0)	203 (8.0)	203 (8.0)
R	†Horizontal distance from hitch pin to tire OD:			
	Preferred Min.	25 (1.0)	25 (1.0)	25 (1.0)
	Permissible forward of tire OD	25 (1.0)	25 (1.0)	25 (1.0)
S	Height of drawbar with popular tires Max. PTO power			
	up to 119 kW (160 hp)	330 (13.0) to 508 (20.0)	330 (13.0) to 508 (20.0)	330 (13.0) to 508 (20.0)
	over 93 kW (125 hp)	381 (15.0) to 559 (22.0)	381 (15.0) to 559 (22.0)	381 (15.0) to 559 (22.0)
T	End of PTO shaft to hitch pin hole	356 (14.0)	406 (16.0)	508 (20.0)
U	Top drawbar to PTO centerline:			
	Preferred	203 (8.0)	203 (8.0)	254 (10.0)
	Min	152 (6.0)	152 (6.0)	203 (8.0)
	Max	305 (12.0)	305 (12.0)	305 (12.0)

*Dimensions are in mm (in.)
†Largest code R1 tire specified for use by the tractor manufacturer.

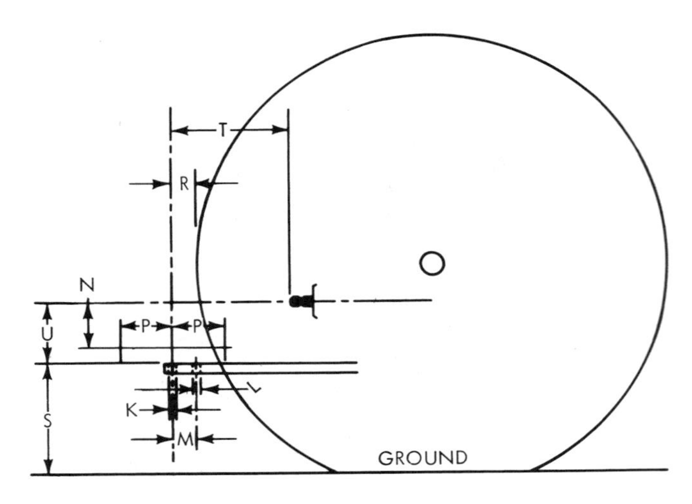

FIG. 5—TRACTOR POWER TAKE-OFF AND DRAWBAR (See Table 2)

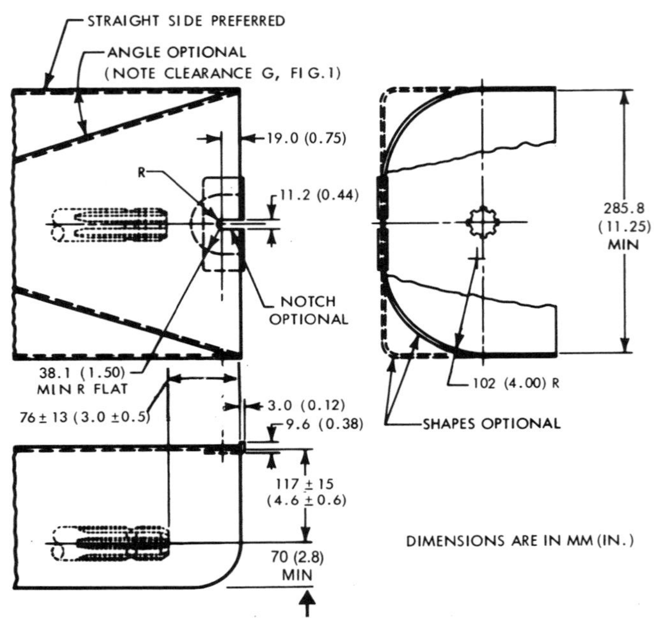

FIG. 6—POWER TAKE-OFF MASTER SHIELD FOR TRACTOR WITH TYPES 1 AND 2 PTO.

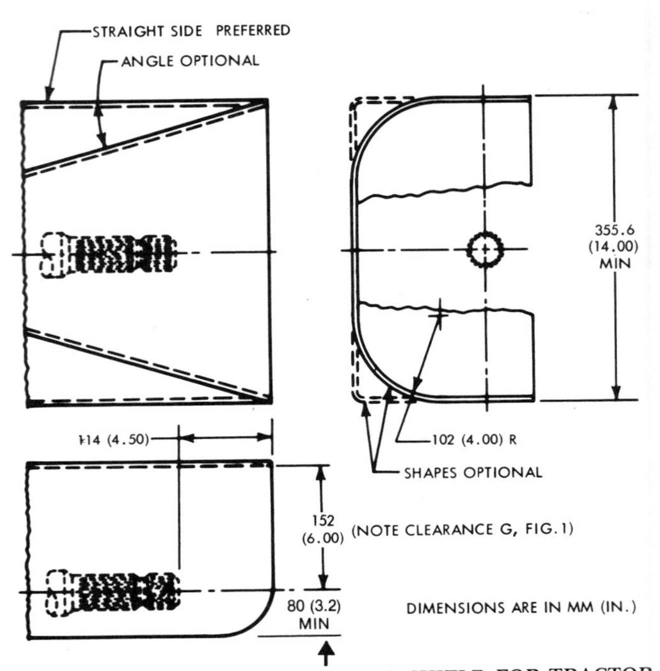

FIG. 7—POWER TAKE-OFF MASTER SHIELD FOR TRACTOR WITH TYPE 3 PTO.

ASAE Standard: ASAE S205.2 (SAE J722 NOV84)

POWER TAKE-OFF DEFINITIONS AND TERMINOLOGY FOR AGRICULTURAL TRACTORS

Proposed by Advisory Engineering Committee of Farm and Industrial Equipment Institute; adopted by ASAE as a Recommendation March 1955; reconfirmed December 1961, December 1966; revised May 1967; reconfirmed December 1971; title revised June 1972; reconfirmed December 1976; reclassified as a Standard December 1978; reconfirmed December 1983.

SECTION 1—PURPOSE AND SCOPE

1.1 This Standard deals only with the definitions and terminology pertaining to the power shafts of farm tractors in which the power take-off rotational speed is proportional to the engine speed. This is the type that prevails in the United States, Canada, England, and generally throughout the world. The following terminology will facilitate a clear understanding for engineering discussions, comparisons and the preparation of technical papers.

SECTION 2—TERMINOLOGY

2.1 Clutch, master: The term "master clutch" is generally used to describe a clutch which transmits all power from the engine and controls both travel and the power take-off. Likewise, when disengaged, both stop.

2.2 Power take-off, continuous-running: Power to operate both the transmission and the power-take-off is transmitted through a master clutch. Both operate only when the master clutch is engaged. Auxiliary means are provided for stopping the travel of the tractor without stopping the power-take-off. The continuous-running power take-off ceases to operate at any time when the master clutch is disengaged.

2.3 Power take-off, independent: Power to operate the transmission and power take-off is transmitted through independent transmission and power-take-off clutches. Travel of the tractor may be started or stopped by operation of the transmission clutch without affecting operation of the independent power take-off. Likewise, the power take-off may be started or stopped by the power take-off clutch without affecting tractor travel.

2.4 Power take-off, transmission-driven: Power to operate both the transmission and the power take-off is transmitted through a master clutch, which serves primarily as a traction clutch. The power take-off operates only when the master clutch is engaged. The transmission-driven power take-off ceases to operate at any time the master clutch is disengaged.

ASAE Standard: ASAE S207.11 (*SAE J721 NOV84)

OPERATING REQUIREMENTS FOR TRACTORS AND POWER TAKE-OFF DRIVEN IMPLEMENTS

*Corresponds to previous revision S207.10.

Proposed by the Engineering Committee of Farm and Industrial Equipment Institute; adopted by ASAE as a Recommendation March 1953; revised January 1958, December 1962, June 1964, June 1966, December 1966, January 1968, December 1969, December 1970, May 1975; reclassified as a Standard March 1978; revised March 1980; revised editorially December 1980; reconfirmed December 1984; revised March 1987; revised editorially April 1988.

SECTION 1—PURPOSE AND SCOPE

1.1 This Standard was prepared to assist manufacturers of tractors and implements in providing suitable means of transmitting power from the tractor power take-off to the implement, and satisfactory hitching of the implement to the tractor.

1.2 ASAE Standard S203, Rear Power Take-Off for Agricultural Tractors, provides dimensions relating to the tractor power take-off shaft, master shield and drawbar; provides specifications for the splined power take-off shaft and the mating connector; and establishes and defines power take-off types 1, 2, and 3.

1.3 ASAE Standard S217, Three-Point Free-Link Attachment for Hitching Implements to Agricultural Wheel Tractors, sets forth requirements for the attachment of three-point hitch implements or equipment to the rear of agricultural wheel tractors.

1.4 The successful performance of all tractor and implement combinations likely to be met in field service require consideration of factors other than the dimensional relationship provided in the aforementioned ASAE Standards.

SECTION 2—DEFINITIONS

2.1 Power take-off (PTO): An external shaft on the rear of a tractor to provide rotational power to implements.

2.2 Implement input driveline (IID): Two universal joints and their connecting member(s) and fastening means for transmitting rotational power from the tractor PTO to the implement input connection. A double Cardan, constant velocity is considered a single joint. The IID also includes integral shielding (guarding) where provided.

2.3 Implement input connection (IIC): The shaft or other connecting means to which the rear joint of the IID is attached on the implement.

SECTION 3—INSTRUCTIONS FOR THE OPERATOR

3.1 The implement manufacturer shall provide a sign in a prominent place on the implement specifying the required tractor drawbar hitch point location and/or implement hitch adjustments.

3.2 The operator's manual for the implement shall also include the above information.

3.3 If a conversion assembly is made available for changing tractors or implements from 540 to 1000 rpm or from 1000 to 540 rpm, these conversion assemblies shall include a sign specifying the power take-off speed and the corresponding drawbar adjustments.

3.4 For recommended safety instructions, refer to ASAE Standard S318, Safety for Agricultural Equipment.

SECTION 4—IMPLEMENT INPUT DRIVELINE AND HITCH REQUIREMENTS

4.1 Provisions should be made in the IID, IIC and hitch of the implement to prevent any of the following during normal operation when attached to any tractor which conforms to ASAE Standards S203, Rear Power Take-Off for Agricultural Tractors, and S217, Three-Point Free-Link Attachment for Hitching Implements to Agricultural Wheel Tractors, and operated according to the instructions of the implement manufacturer:

4.1.1 The universal joints in the IID from reaching a locking angle.

4.1.2 The telescoping section of the implement driveline from separating beyond the point where there is sufficient bearing to provide for proper operation.

4.1.3 The IID from sustaining damage from telescoping to a solid position.

4.1.4 The IID or its shields from sustaining damage due to contacting the implement hitch or hitch pin, or any tractor parts such as master shield or three-point hitch linkage.

4.2 Vertical loads on drawbars

4.2.1 The minimum vertical static loads which the tractor drawbar must withstand are shown in Table 1.

4.2.2 The maximum vertical static loads which the implement shall impose upon the tractor drawbar are shown in Table 1. The dynamic loads imposed upon the tractor drawbar and implement hitch will be considerably higher than static load ratings.

4.2.3 The use of a hitch extender will require a reduction in the vertical static load to limit the maximum bending moment in the tractor drawbar to the bending moments corresponding to the allowable vertical static loads in paragraph 4.2.2.

SECTION 5—MAXIMUM BENDING LOAD LIMITATIONS FOR POWER TAKE-OFF SHAFT DRIVES EMPLOYING V-BELTS OR CHAINS

5.1 The PTO shaft of tractors is designed primarily to transmit torsional loads. The total bending load imposed on the tractor PTO shaft by V-belt or chain drives should not be in excess of values shown in the following table:

Position of Load Application	1-3/8 in. Diameter Power Take-Off		1-3/4 in. Diameter Power Take-Off	
	kN	(lbf)	kN	(lbf)
At the end of the PTO shaft	2.22	(500)	3.56	(800)
Between the PTO shaft rear bearing and/or at the groove in the outside diameter of the PTO shaft splines	2.67	(600)	4.45	(1000)

The tractor PTO shaft and bearing mountings should successfully withstand this magnitude of bending loads (above).

TABLE 1—VERTICAL STATIC LOADS ON DRAWBARS

Max Power†	Drawbar Load	
	kN	lb
14.9-74.6 kW (20-100 hp)	3.34+0.15 per kW for excess over 14.9 kW	750+25 per hp for excess over 20 hp
74.6-186.4 kW (100-250 hp)	12.23+0.06 per kW for excess over 74.6 kW	2750+10 per hp for excess over 100 hp
186.4-372.8 kW (250-500 hp)	18.90+0.03 per kW for excess over 186.4 kW	4250+5 per hp for excess over 250 hp

†Maximum drawbar power established per ASAE Standard S209, Agricultural Tractor Test Code.

SECTION 6—TORSIONAL LOAD CONSIDERATIONS

6.1 Because of the large amount of kinetic energy available at the PTO shaft, instantaneous torsional loads and fluctuating operating loads in excess of the average rated power of the tractor may be transmitted.

TABLE 2—PTO THRUST FORCES

	PTO Power		Thrust	
	kW	(hp)	kN	(lbf)
1-3/8 in. PTO				
	15 - 25	(20.1 - 33.5)	7.00	(1575)
	Over 25 - 40	(33.5 - 53.6)	9.00	(2025)
	Over 40 - 60	(53.6 - 80.5)	11.00	(2475)
	Over 60 - 110	(80.5 - 147.5)	13.00	(2925)
	Over 110	(147.5)	14.00	(3150)
1-3/4 in. PTO				
	Over 110	(147.5)	18.00	(4050)

SECTION 7—PTO SHAFT AND IMPLEMENT INPUT DRIVELINE THRUST LOAD LIMITATIONS

7.1 The tractor PTO shall be designed to accept IID telescoping thrust force values in Table 2 based on PTO power at rated engine speed as established per ASAE Standard S209, Agricultural Tractor Test Code. A properly maintained implement at its designed power shall not impose IID telescoping thrust forces upon the tractor PTO in excess of the values in Table 2, recognizing that instantaneous thrust forces may exceed these values.

Cited Standards:

ASAE S203, Rear Power Take-Off for Agricultural Tractors
ASAE S209, Agricultural Tractor Test Code
ASAE S217, Three-Point Free-Link Attachment for Hitching Implements to Agricultural Wheel Tractors
ASAE S318, Safety for Agricultural Equipment

ASAE Standard: ASAE S209.5 (SAE J708 DEC84)

AGRICULTURAL TRACTOR TEST CODE

Original Agricultural Wheel-Type Tractor Test Code approved in 1937; revised in 1948, 1956, 1957; rewritten and adopted as Agricultural Tractor Test Code in 1959; revised March 1962; ASAE-SAE liaison responsibility established June 1964; SAE revisions adopted December 1964, December 1967, December 1972; reconfirmed December 1977; SAE revisions adopted March 1981; reconfirmed December 1985.

Purpose. The purpose of this Standard is to define test conditions, give a description of the tests to be made, specify data to be obtained, show formulas and calculations, define terms, and establish a uniform method of reporting so that performance data obtained on various makes and models of tractors, tested in accordance with this Standard, will be comparable regardless of where the test is made. It is obvious, because of the many present day tractor models available in a number of types with numerous items of special or optional equipment, that the scope of this Standard must be limited to obtaining and reporting only the most significant of widely used performance data.

Outline of Code

SECTION 1—TEST CONDITIONS

1.1 The tractor tested shall represent a production model in all respects. The tractor manufacturer shall supply the tractor when a certified test report is desired. The manufacturer shall supply a list of technical specifications of the tractor's construction giving detailed information regarding the powerplant, transmission, final drive, miscellaneous special and optional equipment, center of gravity of tractor with operator, machine clearance circles, and turning radii with and without brakes. The manufacturer shall also supply printed information covering all operating and servicing instructions necessary for the satisfactory operation of the tractor. All specifications shall be subject to verification.

1.1.1 The tractor manufacturer shall appoint an official representative to be present during certification runs. It shall be the duty of the manufacturer's representative to make all decisions where permissible choices or company policy are concerned. He shall also prepare the tractor for test, operate it during limber-up, make any adjustments required, and at conclusion of the test, prepare tractor for final inspection and reassemble after inspection.

1.1.2 The tractor shall be equipped with the most frequently installed items of optional equipment; however, the manufacturer should ascertain the optional equipment requirements of the specific test station, when tests are to be made for certification. Power consuming accessories shall be disconnected only if it is practical for the operator to do so as a normal farming practice. Any equipment on the tractor shall be complete and operable. The equipment shall be such that it does not interfere with the conduct of the test.

1.1.3 The tractor shall be operated as recommended by the manufacturer. Tracks or wheel and tire equipment, adjustments, servicing operations, and selection of fuel and types of lubricants shall conform to instructions printed in the published information delivered with the tractor.

1.1.4 Commercially available fuel shall be used providing it meets the manufacturer's specifications.

1.1.5 Commercially available lubricants shall be used providing they meet the manufacturer's specifications.

1.1.6 Where choices of adjustments or operating conditions are made by the manufacturer, the guide in making such choices should be the suitability for general operation.

1.1.7 Unless otherwise specified, controls that are easily manipulated from tractor seat may be used to secure optimum performance during runs.

1.1.8 The maximum drawbar hitch-point height shall be established in accordance with paragraph 4.3 as defined under Section 4.

1.1.9 Drive wheel or track slippage shall be calculated as shown under Section 4, and shall not exceed 15 percent for tractors equipped with pneumatic tires or 7 percent for tractors equipped with steel lugs or tracks.

1.2 All measurements shall be obtained with instruments and test equipment having an accuracy representative of good laboratory practice. Power at the mechanical power outlet shall be measured by means of a dynamometer. The laboratory fuel supply shall be arranged so that the fuel pressure at the carburetor or the fuel transfer pump is equivalent to that which exists when the tractor fuel tank is half full. The equipment should be arranged so that the fuel temperature is comparable to that which exists in normal operation of the tractor when fuel is taken from the tractor fuel tank.

1.2.1 Laboratory air temperature shall be 23 ± 7 C (73 ± 13 F) dry bulb, and readings shall be taken at a sufficient distance from the tractor to record actual ambient temperature. The test area shall be well ventilated. Engine exhaust gas shall be discharged from the test area. If an auxiliary laboratory exhaust stack is used, it shall be of such design that it does not change the engine performance. Laboratory atmospheric pressure shall not be less than 96.6 kPa (28.6 in. Hg).

1.2.2 The test tractor shall be equipped with track or tire and wheel equipment regularly supplied to the trade. Additional mass may be added as ballast if the manufacturer regularly supplies it for sale. When liquid ballast is used in tires, the inflation pressure shall be determined at the same height as the valve with the valve in the lowest position. Individual tire inflation pressures shall be in accordance with the manufacturer's instructions, and at these pressures, the tire load including the mass of a 75 kg (165 lb) operator on the tractor seat shall not exceed the limitation of ASAE Standard ASAE S295, Agricultural Tractor Tire Loadings, Torque Factors and Inflation Pressures. Because the traction coefficient of a tire changes with wear, the tread bar height of the test tire prior to the start of run 2.3.1 shall be not less than 65 percent of that of a new tire. To establish tread bar height, the tire shall be mounted and inflated as for the runs covered in paragraph 2.3.1. The height of the tire tread bars shall be measured by use of a 3-point gage. The gage shall be placed astride of the tread bar and perpendicular to the direction of the tread bar as close to the tire center line as possible. Two legs of the gage shall be positioned at the base of the tread bar (at the point of tangency between the tire carcass and the radius joining the tread bar to the carcass). The third point of the gage shall be in the center of the tread bar. The tread bar height shall be the difference in elevation between the two outside legs of the gage and the center point. The tread bar height measured in this manner shall be taken for a minimum of four equally spaced locations around the periphery of the tire. The results of these measurements shall be averaged and compared to similar data on a new tire of the same make, size, and type.

1.3 The drawbar test course shall be a hard surface on which data can readily be reproduced. It shall be constructed according to modern highway construction standards and of sufficient strength to withstand the heaviest tractor. The drawbar test runway or runways shall be straight, level, and not less than 91.4 m (300 ft) long with the approach of such length that speed and pull can be stabilized before entering the runway. For tractors equipped with rubber tires, the recommended surface materials of the drawbar test course are, in the order of preference:

1.3.1 Concrete. The runway or runways of the drawbar test course shall have a minimum of expansion joints. The surface shall have a uniform gritty texture with a corrugated appearance. This type of finish is also known as a "belted" finish.

1.3.2 Bituminous. These materials are generally known as tar-macadam or asphaltic concrete.

1.3.3 Earth. Test courses having earthen surfaces shall be well-packed and substantially free of loose material. This requires a soil that will adhere together when properly prepared and maintained. Suitable maintenance equipment shall be provided for grading, applying water, and packing both the subsurface and the surface. The use of this type of surface is discouraged for testing tractors equipped with pneumatic tires. For tractors having steel lugs or tracks, the test course should be earth as described in paragraph 1.3.3.

1.4 All information published in the test report shall represent the performance of a complete tractor. Power measurements shall be taken as delivered to the test equipment from a mechanical power outlet, if available, and from the drawbar.

1.4.1 The test report shall accurately define the tractor type and list all items of special or optional equipment used during the run.

1.5 For official certification, the Test Station shall provide facilities and personnel to conduct the performance runs, record all data, prepare, certify, and publish the report.

1.6 Until satisfactory correction formulas are developed for all tractors, only observed data will be published. However, the necessary wet- and dry-bulb air temperatures and barometric pressure are recorded and published so that correction formulas may be applied.

SECTION 2—DETAILED DESCRIPTION OF TEST PROCEDURE

2.1 Preparation of tractor for performance runs

2.1.1 The purpose of these preparatory runs is to stabilize the tractor performance for the later runs by operating the tractor to remove stiffness, check its condition, and make permissible adjustments to assure normal operation.

2.1.2 The tractor shall be limbered up in accordance with the manufacturer's recommendations. Limber-up shall be accomplished with approximately the same mass as will be used for the maximum drawbar power runs. Minor adjustments are permissible during and at the end of this run. Adjustments shall be limited to those which conform to the published instructions supplied with the tractor. The recording of hours of operation for the entire test shall begin with the start of this run. Prior to the start of this run, the engine crankcase shall be drained and refilled with new oil of the type and viscosity recommended by the manufacturer as stated in the published information delivered with the tractor. The oil used to fill the crankcase, any oil added or withdrawn during the test, and the oil drained from the crankcase at the conclusion of the test shall be weighed and specific gravity taken, in order that the total volume of oil used during the entire test can be determined. Specific gravity data are to be obtained at or converted to 15/15 C (60/60 F). Prior to the start of this run, the transmission and other oil reservoirs on the tractor shall be filled with lubricants of the type and viscosity recommended by the manufacturer.

2.2 Mechanical power outlet performance

2.2.1 Maximum power—fuel consumption

2.2.1.1 The purpose of this run is to determine the maximum power as delivered through a mechanical power outlet to a dynamometer at the manufacturer's specified engine or mechanical power outlet speed; and to record the corresponding fuel consumption.

NOTE: This power can be measured through a belt pulley, power take-off shaft, or any other mechanical power outlet depending upon limitations of test equipment.

2.2.1.2 During the preparation for this run, the manufacturer shall establish fuel settings and ignition or injection timing, which shall remain unchanged throughout the test. The governor and the position of the manually operated governor control shall be adjusted to provide the high idle engine or power outlet speed specified by the manufacturer for maximum power operation.

2.2.1.3 Data recorded at intervals of no more than 10 min shall include engine crankshaft revolutions per minute, dynamometer revolutions per minute, mechanical power outlet shaft revolutions per minute, coolant temperature, wet- and dry-bulb air temperatures, fuel consumed, and dynamometer torque. Speeds of engine, mechanical power outlet, and dynamometer shall be taken simultaneously. The coolant temperature shall be taken in the radiator top tank. The barometric pressure shall be recorded at the beginning of the run and at 1 h intervals thereafter. The duration of the run shall be a minimum of 2 h continuous operation.

NOTE: In order to determine belt slippage, simultaneous determinations of the revolutions of both drive and driven pulleys shall be taken at no-load for a minimum of 1000 revolutions of the drive pulley with the belt tension used for this run. Belt slippage shall be calculated as shown under Section 4. Belt tension shall be adjusted for optimum power and remain unchanged throughout run. Usually optimum power is obtained with approximately 1 percent slippage.

2.2.2 Varying power—fuel consumption

2.2.2.1 The purpose of this run is to determine fuel consumption and speed when power is varied.

2.2.2.2 All adjustments shall be the same as in paragraph 2.2.1.2.

2.2.2.3 Data recorded shall be the same as in paragraph 2.2.1.3. The duration of the run shall be for 2 h of continuous operation.

2.2.2.4 The run shall consist of six power settings, each to be run for a period of 20 min in the following order:

(a) 85 percent of dynamometer torque obtained at maximum power, run 2.2.1.

(b) Zero dynamometer torque.

(c) One-half of 85 percent of dynamometer torque obtained at maximum power, run 2.2.1.

(d) Dynamometer torque at maximum power.

(e) One-quarter of 85 percent of dynamometer torque obtained at maximum power, run 2.2.1.

(f) Three-quarters of 85 percent of dynamometer torque obtained at maximum power, run 2.2.1.

NOTE: These percentages represent long and continuous past practice and are necessary to maintain continuity in procedure and meaning of the results.

2.2.3 Power at standard power take-off speeds (This run is made only when the engine speed at maximum power does not correspond to the engine speed at SAE and ASAE standard power take-off speeds, as specified in ASAE Standard S203, Rear Power Take-Off for Agricultural Tractors.)

2.2.3.1 The purpose of this run is to determine power at the standard power take-off speed or speeds, and to record the corresponding fuel consumption.

2.2.3.2 All adjustments shall be the same as in paragraph 2.2.1.2.

2.2.3.3 Data recorded shall be the same as in paragraph 2.2.1.3. The duration of each run shall be a minimum of 1 h of continuous operation.

2.3 Drawbar performance

2.3.1 Maximum drawbar power

2.3.1.1 The purpose of this run is to determine the maximum power in not more than 12 forward gears or 12 travel speeds as selected by the manufacturer. The maximum travel speed shall not exceed the safety limitations of track or test equipment.

2.3.1.2 All engine adjustments shall be the same as in paragraph 2.2.1.2 unless the manufacturer specifies a different engine revolutions per minute for drawbar operation than for mechanical power outlet operation, run 2.2.1. In this case, the position of the manually operated governor control shall be adjusted to provide the maximum high idle engine or power outlet revolutions per minute specified by the manufacturer for drawbar operation. The tractor tire and wheel or track equipment shall conform to the manufacturer's recommendations. Ballast added shall meet requirements specified under paragraph 1.2.2 of Section 1. The tractor shall be weighed with operator and ballast (if ballast is used) after tractor has been properly serviced, fuel tank and radiator filled, and all test equipment in place. For wheel type tractors, the total weight of the tractor, the weight on the front wheels, and the weight on the rear wheels shall be recorded and reported with and without ballast, but including a 75 k, (165 lb) operator. In the interests of steering control and to obtain more uniform results, the maximum drawbar hitch point height shall be established in accordance with paragraph 4.3 as defined under Section 4. Preliminary drawbar runs at maximum drawbar pull shall be made to establish maximum hitch point height. The height as related to the tractor shall remain unchanged throughout all drawbar tests. The effective circumference of the drive wheels or tracks shall be determined by driving the tractor over the drawbar test runway or runways and counting the revolutions of each drive wheel or track. The tractor shall be driven at low speed, without drawbar pull, and with all ballast in place (if ballast is used). Drive wheel or track slippage is calculated as shown under Section 4. If drive member slippage is excessive, the drawbar pull shall be reduced until slippage does not exceed 15 percent for tractors equipped with pneumatic tires and 7 percent for tractors equipped with steel lugs or tracks. Where there are two or more travel speeds in which maximum power shall be limited by drive member slippage, only one travel speed shall be checked.

2.3.1.3 Data recorded shall include: Average drawbar pull and engine crankshaft revolutions per minute maintained over test runway or runways, drive member revolutions to traverse the test runway or runways, time to traverse the test runway or runways, engine coolant temperature, wet- and dry-bulb air temperatures, and barometric pressure. Test results reported are recommended to be an average of two or more suitable runs over the drawbar test runway or runways. Two suitable runs would be those in which test conditions are stabilized, and during which the average travel speed for the two runs is within 1 percent of each other.

2.3.2 Varying drawbar power, fuel consumption, and sound level at operator station

2.3.2.1 The purpose is threefold: First, to determine fuel consumption with the tractor operating at maximum available power, at 75 percent of the drawbar pull obtained at maximum power in run 2.3.1, and at 50 percent of the drawbar pull obtained at maximum power in run 2.3.1. Second, to determine the sound level at the operator's station under these load conditions. Third, to determine whether the tractor will maintain a preselected drawbar power output for 10 h and to measure fuel consumed during the run.

2.3.2.2 Except for the run specified under paragraph 2.3.2.4d, all engine adjustments and other items shall be the same as in paragraph 2.3.1.2. The gear or travel speed shall be selected by the manufacturer within the speed range normally used in agricultural operations. All runs shall be of sufficient duration to assure accuracy of data.

2.3.2.3 Data recorded shall be the same as obtained in paragraph 2.3.1.3 with two exceptions as follows:

(a) The drawbar pull shall be the average determined from draft data taken over the complete circuit of the drawbar test course.

(b) Sound level measurements at the operator's station are added.

The sound level measurements at the operator's station shall be conducted according to the instrumentation and procedure specified in Society of Automotive Engineers Standard J919, Operator Sound Level Measurement Procedure for Powered Mobile Construction Machinery—Singular Type Test. The load conditions for these measurements will be those described in paragraph 2.3.2.4 of this code instead of the load conditions described in SAE Standard J919, Operator Sound Level Measurement Procedure for Powered Mobile Construction Machinery—Singular Type Test, paragraph 3.3.

2.3.2.4 Fuel consumption runs and measurement of sound level at the operator's station shall be made with the tractor operating as follows:

(a) At maximum available power, which is the maximum sustained power the tractor is capable of delivering at the drawbar for a predetermined length of time.

(b) At 75 percent of the drawbar pull obtained in run 2.3.1.

(c) At 50 percent of the drawbar pull obtained in run 2.3.1.

(d) At reduced engine speed, where both engine and travel speed ratio are adjusted to produce the pull and travel speed recorded under 2.3.2.4c.

NOTE: For tractors equipped with torque multiplier with "lockout," the four runs shall be made in both torque-multiplier drive and with the torque multiplier in "lockout".

2.3.3 Drawbar pull versus travel speed

2.3.3.1 The purpose of this run is to determine the drawbar pull-travel speed characteristic, or "lugging" ability, of the tractor in the gear or transmission setting selected in run 2.3.2.

2.3.3.2 All adjustments and other items shall be the same as in paragraph 2.3.1.2. (It would be desirable to make this run immediately after the maximum drawbar power run has been made in run 2.3.1 in the gear or at the travel speed selected for run 2.3.2.) Drive member slippage shall not exceed 15 percent for tractors equipped with pneumatic tires or 7 percent for tractors equipped with steel lugs or tracks.

2.3.3.3 Data recorded shall be the same as paragraph 2.3.1.3.

2.3.3.4 A series of runs shall be made starting at maximum power. In each succeeding run, the necessary drawbar pull shall be applied to the tractor to reduce the drive member speed in increments of approximately 10 percent, using drive member speed at maximum power as 100 percent. A sufficient number of runs shall be made to establish speed at which maximum drawbar pull is obtained, or speed at which maximum drawbar pull is limited by drive member slippage or cooling capacity of torque multiplier when tractor is equipped with torque multiplier.

NOTES:

A. For tractors equipped with torque multiplier and "lockout," runs shall be made in both torque multiplier drive and with torque multiplier in "lockout."

B. For tractors equipped with an automatic powershifting fixed ratio transmission, the same procedure shall be followed except that the test shall be terminated at the first automatic shift.

C. For tractors equipped with transmissions having infinitely variable engine-to-final drive ratios under the control of the operator, no runs are required. Actual drawbar pull, drawbar power, and speed obtained in run 2.3.1 shall be published.

2.3.4 Exterior sound level. The exterior sound level measurements will be conducted according to the instrumentation and procedures specified in SAE J1008, Exterior Sound Level Measurement Procedure for Self-Propelled Agricultural Field Equipment.

SECTION 3—FINAL INSPECTION

3.1 The purpose of the final inspection is to check some of the most significant items of the manufacturer's tractor specifications against the tractor tested, and to inspect the condition of some of the most critical tractor parts. It is suggested that the check of specifications include bore and stroke, valve lift, valve sizes, compression ratio and carburetor size.

SECTION 4—CALCULATIONS AND FORMULAS

4.1 Drive wheel or track slippage

$$\text{Percent slip} = 100 \left(\frac{R-r}{R}\right)$$

R = total drive wheel revolution count to traverse the drawbar runway under load

r = total drive wheel revolution count to traverse the drawbar test runway under no-load

4.2 Belt Slippage

$$\text{Percent belt slip} = 100 \left(\frac{RN-n}{RN}\right)$$

$$R = \frac{\text{Driven pulley revolutions with no-load}}{\text{Drive pulley revolutions with no-load}}$$

N = drive pulley rpm with load
n = driven pulley rpm with load

4.3 Stability Factor

$$\text{Stability factor } K = \frac{Fw \times Wb}{P \times h}$$

Fw = static front end weight
Wb = wheelbase
P = maximum drawbar pull parallel to ground
h = height of static line of pull perpendicular to ground
K = 1.25 (minimum)—may be more

SECTION 5—DEFINITION OF TERMS

5.1 Agricultural tractor is a vehicle designed and advertised to pull, propel, and supply power to operate machinery used in agricultural operations.

5.2 Ballast is any substance that is added to or removed from the tractor for the purpose of changing traction or stability. It may be added or removed at the will of the operator, and its presence is not always essential for operation of the complete tractor. Mounted equipment may be used as ballast.

5.3 Spark ignition engine is an internal combustion engine in which the ignition of the air-fuel mixture is accomplished by a spark inside the combustion chamber but generated from a source outside the combustion chamber. The mixing of the fuel and air in conventional spark ignition engines occurs before compression, the fuel metering being accomplished by a carburetor or similar device. The control on the amount of air-fuel mixture admitted to the cylinders is usually secured by means of a throttle valve in accordance with power requirements.

5.3.1 Compression ignition engine is an internal combustion engine in which the ignition of the air-fuel mixture is accomplished by the heat of compression. The initial mixing of the fuel and air in the conventional compression ignition engine occurs near the end of the compression stroke, the fuel being metered into the combustion chamber by means of a suitable fuel metering mechanism, with the quantity varied in accordance with power requirements.

5.3.2 Rated speed of the engine is the speed in revolutions per minute specified by the tractor manufacturer for continuous operation at maximum power for a particular operation. There may be one or more speeds specified. For example, there may be different rated speeds for operation at the mechanical power outlet, at the power take-off shaft, and at the drawbar.

5.3.3 High idle engine or power outlet speed is the high idle speed in revolutions per minute, sometimes called maximum no-load speed, necessary to provide the required full load speed, sometimes called rated speed, for maximum power at the mechanical power outlet, power take-off shaft, or drawbar.

5.4 Selective gear fixed ratio transmission is a transmission of such design that the ratio of engine speed to final drive member speed can be changed but with interruption of power to the final drive members.

5.4.1 Torque multiplier is a mechanism capable of automatically multiplying the engine torque. In this case, the increased torque produces a corresponding increase in drawbar pull.

NOTE: The torque converter falls into this classification.

5.4.2 Torque multiplier lockout is a means by which the torque multiplier can be made inoperative.

5.4.3 Automatic power shifting fixed ratio transmission is of such a design that it automatically shifts from one fixed ratio to another.

5.4.4 Operator controlled power shifting fixed ratio transmission

5.4.4.1 Full range. This type of transmission is of such design that the operator can change from any fixed ratio to any other fixed ratio with the tractor under power.

5.4.4.2 Partial range. This type of transmission is of such a design that the operator can change from certain fixed ratios to other fixed ratios with the tractor under power.

5.4.5 Infinitely variable transmission. This type of transmission is of such design that the operator can infinitely vary the ratio between engine and final drive members throughout part or all of the speed range of the tractor. This type of transmission includes electrical, hydrostatic, friction, and any other devices.

5.5 Mechanical power outlet is any outlet through which the tractor engine power can be delivered to a dynamometer, such as a belt pulley, power take-off shaft, or any other shaft or outlet.

5.5.1 Maximum power at the mechanical power outlet is the average maximum sustained power the tractor delivers to a dynamometer through a mechanical power outlet with adjustments as specified by the manufacturer.

5.5.2 Power at standard power take-off speeds is the power the tractor delivers to a dynamometer through a mechanical power outlet at SAE and ASAE standard power take-off speed or speeds.

5.6 Maximum drawbar power is the maximum power the tractor is capable of delivering at the drawbar on the test runway or runways.

5.7 Maximum available drawbar power is the average maximum sustained power the tractor delivers at the drawbar during several complete circuits of the drawbar test course. Power shall be determined from the average drawbar pull obtained over two or more complete circuits of the drawbar test course and the average travel speed concurrently obtained on the test runway or runways.

5.8 Specific fuel consumption is the ratio between the mass of fuel consumed per unit of time and the corresponding measured power. Distinction should be made between specific fuel consumption based on the power obtained at a mechanical power outlet and that based on the power obtained at the drawbar.

5.9 Specific volumetric consumption is the ratio between the measured power and the corresponding volume of fuel consumed per unit of time. Distinction should be made between specific volumetric consumption based on the power obtained at mechanical power outlet and that based on the power obtained at the drawbar.

5.10 Drawbar pull versus travel speed is sometimes referred to as the "lugging" ability of the tractor. This shows the ability of the tractor to produce increased drawbar pull with the corresponding reduction in travel speed.

SECTION 6—UNIFORM METHOD OF PUBLISHING RESULTS

All information shall be published in either the SI units [Systeme Internationale, See ASAE Standard EP285, Use of SI (Metric) Units] and U.S. customary units, or SI units only. The following shows that information shall be published in the certified report, how it is obtained, and how it shall be published.

6.1 Mechanical power outlet performance

6.1.1 Maximum power-fuel consumption at the mechanical power outlet. Power, engine and/or power outlet revolutions per minute, coolant temperature, wet- and dry-bulb air temperatures, and barometric pressure shall be the average for the run. The rate of fuel consumption shall be reported in L/h (gal/h). Specific fuel consumption shall be determined on the basis of average power developed and fuel consumed during the run, and shall be reported as kg/kW·h (lb/hp·h). Specific volumetric consumption shall be reported as kW·h/L (hp·h/gal).

6.1.2 Varying power-fuel consumption at the mechanical power outlet. Power, engine and/or power outlet revolutions per minute, coolant temperature, and wet- and dry-bulb air temperatures shall be averaged for each run. The rate of fuel consumption shall be reported in L/h (gal/h) for each run. Specific fuel consumption shall be determined on the basis of the average power developed and fuel consumed during each run, and shall be reported as kg/kW·h (lb/hp·h). Specific volumetric consumption shall be reported as kW·h/L (hp·h/gal). The last, or seventh, line in the tabulation shall be a summary of the six runs determined as follows: Power, engine and/or mechanical power outlet revolutions per minute, coolant temperature, wet- and dry-bulb temperatures, and barometric pressure shall be the average of the six runs. The rate of fuel consumption shall be reported in L/h (gal/h) for the run. Specific fuel consumption shall be determined on the basis of the average power developed and fuel consumed during the run, and shall be reported as kg/kW·h (lb/hp·h). Specific volumetric consumption shall be reported as kW·h/L (hp·h/gal).

6.1.3 Power at standard power take-off speeds. Power engine and/or power outlet revolutions per minute, coolant temperature, wet- and dry-bulb air temperatures, and barometric pressure shall be the average for the run. The rate of fuel consumption shall be reported in L/h (gal/h). Specific fuel consumption shall be determined on the basis of average power developed and fuel consumed during the run, and shall be reported as kg/kW·h (lb/hp·h).

Specific volumetric consumption shall be reported as kW·h/L (hp·h/gal).

6.2 Drawbar performance

6.2.1 Maximum drawbar power. For each forward gear or travel speed tested, power, drawbar pull, travel speed, engine revolutions per minute, percent drive member slippage, coolant temperature, wet- and dry-bulb air temperatures, and barometric pressure shall be reported as an average from two or more suitable runs as described in paragraph 2.3.1.3.

6.2.2 Varying drawbar power-fuel consumption. Power, drawbar pull, travel speed, engine revolutions per minute, percent drive member slippage, coolant temperature, wet- and dry-bulb air temperatures, and barometric pressure shall be reported as an average from data obtained during each run. The rate of fuel consumption shall be reported in L/h (gal/h). Specific fuel consumption shall be determined on the basis of the average power developed and fuel consumed during each run, and shall be reported as kg/kW·h (lb/hp·h). Specific volumetric consumption shall be reported as kW·h/L (hp·h/gal). Sound level at the operator's station shall be reported in dB(A) as specified in SAE Standard J919, Operator Sound Level Measurement Procedure for Powered Mobile Construction Machinery—Singular Type Test.

6.2.3 Drawbar pull versus travel speed. Information published shall be drawbar pull, drawbar power, travel speed, engine revolutions per minute, and percent drive member slippage.

6.2.4 Exterior sound level shall be reported in dB(A) as specified in SAE J1008, Exterior Sound Level Measurement Procedure for Self-Propelled Agricultural Field Equipment.

6.3 Fuel

Type .... Grade ....
Specific Gravity at, or Converted to, 15/15 C (60/60F).
Octane No. Motor .... Research ....
Cetane No. ....

6.4 Engine oil

SAE Viscosity No. .... API Service Classification ....
Transmission and Final Drive Lubricant: SAE No. ....
Type .... Oil to Engine .... gal(L). Oil from Engine .... gal(L)

6.5 Total hours of operation ....

6.6 Any parts replaced or repaired during the entire test shall be listed in the published report.

Cited Standards:

ASAE S295, Agricultural Tractor Tire Loadings, Torque Factors, and Inflation Pressures
ASAE EP285, Use of SI (Metric) Units
SAE J919, Operator Sound Level Measurement Procedure for Powered Mobile Construction Machinery—Singular Type Test
SAE J1008, Exterior Sound Level Measurement Procedure for Self-Propelled Agricultural Field Equipment

ASAE Standard: ASAE S210.2

TRACTOR BELT SPEED AND PULLEY WIDTH

Revised January 1944; reconfirmed December 1965; revised December 1966; reconfirmed December 1971, December 1976, December 1981, December 1986.

1.1 The standard belt speed for agricultural tractors shall be 3100 ft (944.9 m) per minute, plus or minus 100 ft (30.5 m) per minute.

1.2 The minimum width of agricultural tractor belt pulleys shall be such as to provide for the use of a belt 6 in. (152 mm) in width.

V-BELT AND V-RIBBED BELT DRIVES FOR AGRICULTURAL MACHINES

Adopted by ASAE June 1950; revised 1960, 1962; revision proposed by a joint committee representing the Rubber Manufacturers Association and the Farm and Industrial Equipment Institute; approved by ASAE Power and Machinery Division Technical Committee December 1968; reconfirmed December 1973, December 1978, December 1983; revised May 1986; revised editorially February 1987.

SECTION 1—PURPOSE

1.1 The purpose of this Standard is to provide sufficient technical data for the uniform physical application of belt drives to farm machines and mobile industrial equipment. Use of this Standard will contribute to the design of simple and economical drives.

SECTION 2—SCOPE

2.1 This Standard establishes acceptable manufacturing tolerances, methods of measuring, and proper application for drives using V-belts or V-ribbed belts. They may be used individually or in matched sets.

2.2 This Standard is unique to agricultural belt drives and should be used in lieu of standards for industrial drives published by The Rubber Manufacturers Association.

2.3 This Standard does not specify the load-life characteristics of belts.

2.4 This Standard does not include belts for automotive accessory drives, flat conveyor belting, flat power transmission belts, or synchronous belts.

2.5 The term "belt(s)" used throughout this Standard means V-belt(s) and V-ribbed belt(s).

2.6 In the interest of international standardization, metric-SI units, consistent with International Organization for Standardization Standard 1000, SI Units and Recommendations for the Use of Their Multiples and of Certain Other Units, are included in Tables 1A through 16A.

SECTION 3—DEFINITIONS

3.1 Effective width of V-belt groove: A groove width characterizing the groove profile. It is a defined value not subject to tolerance and is usually located at the outermost extremities of the straight side walls of the groove. For all V-belt measuring sheaves and for most machined-type sheaves, it coincides with the actual top width of the groove within reasonable tolerances.

3.2 Effective diameter of V-belt sheave: The diameter of the sheave at the effective width of the sheave groove. (Formerly designated "Effective outside diameter".)

3.3 Effective diameter of V-ribbed sheave: The outside diameter of the sheave as a defined value at the specified sheave groove dimensions (without tolerances).

3.4 Effective length: The length of a line circumscribing a belt at the level of the effective diameter of the measuring sheaves with the belt at a prescribed tension.

3.5 Pitch width: The width of the belt at its neutral zone.

3.6 Pitch width of groove: That width of the sheave groove which has the same dimension as the pitch width of the belt used with this sheave.

3.7 Pitch diameter of sheave: The diameter of the sheave at the pitch width of sheave groove.

3.7.1 For V-belts, the pitch diameter is the effective diameter less the 2a value shown in Tables 7, 8, and 9.

3.7.2 For V-ribbed belts, the pitch diameter is the effective diameter plus the 2a value shown in Table 10.

3.7.3 For double-V belts and joined V-belts, the pitch diameter coincides with the effective diameter of the sheave so that the 2a value is zero.

TABLE 1—NOMINAL DIMENSIONS OF CROSS SECTIONS, INCHES

Belt type	Cross section	b_b	h_b	h_{bb}*	s_g†
Classical V-belts	HA	0.50	0.31	0.41	0.625
	HB	0.66	0.41	0.50	0.750
	HC	0.88	0.53	0.66	1.000
	HD	1.25	0.75	0.84	1.438
Narrow V-belts	H3V	0.38	0.31	0.38	0.406
	H5V	0.62	0.53	0.62	0.688
	H8V	1.00	0.91	1.00	1.125
Double-V belts	HAA	0.50	0.41		
	HBB	0.66	0.53		
	HCC	0.88	0.69		
Adjustable speed V-belts	HI	1.00	0.50		
	HJ	1.25	0.59		
	HK	1.50	0.69		
	HL	1.75	0.78		
	HM	2.00	0.88		
	HN	2.25	0.94		
	HO	2.50	1.00		
V-ribbed belts	J	See	0.16		0.092
	L	Fig.	0.38		0.185
	M	1	0.66		0.370

*Classical and narrow V-belts are also available in the joined belt configuration as illustrated in Fig. 1.
†S_g is specified sheave groove spacing (see Tables 7 or 10).

TABLE 1A—NOMINAL DIMENSIONS OF CROSS SECTIONS, MILLIMETERS

Belt type	Cross section	b_b	h_b	h_{bb}*	s_g†
Classical V-belts	13F	13	8	10	15
	16F	16	10	13	19
	22F	22	13	17	25.5
	32F	32	19	21	36.5
Narrow V-belts	9FN	9	8	10	10.3
	15FN	15	13	16	17.5
	25FN	25	23	25	28.6
Double-V belts	13FD	13	10		
	16FD	16	13		
	22FD	22	18		
Adjustable speed V-belts	25FV	25	13		
	32FV	32	15		
	38FV	38	18		
	44FV	44	20		
	51FV	51	22		
	57FV	57	24		
	63FV	63	26		
V-ribbed belts	FPJ	See	4		2.34
	FPL	Fig.	10		4.70
	FPM	1	17		9.40

*Classical and narrow V-belts are also available in the joined belt configuration as illustrated in Fig. 1.
†S_g is specified sheave groove spacing (see Tables 7A or 10A).

3.8 Speed ratio and belt speed: Speed ratio is the ratio of the pitch diameter of the sheaves; generally expressed as a number equal to or greater than unity. Belt speed is the linear speed of the belt calculated using the pitch diameter of the driver sheave.

3.9 Installation allowance: A design length factor permitting the unforced installation of a belt (see Tables 13, 14, 15, and 16).

3.10 Take-up allowance: A design length factor to permit sufficient tensioning over the life of the drive (see Tables 13, 14, 15 and 16).

3.11 Measuring sheaves: Sheaves used for determining the effective length of a belt (see Tables 5 and 6 for dimensions).

3.12 "Y" center distance: The center distance between measuring sheaves used to determine the effective length of a belt (see paragraph 6.1 for procedure).

3.13 Clutching allowance: A design length factor to facilitate the belt drive systems operation as a clutch (see paragraph 8.5).

SECTION 4—CROSS SECTIONS

4.1 Nominal dimensions of belt cross sections for agricultural machines are shown in Tables 1 and 2. Because of different constructions and methods of manufacture, the cross-sectional shape, dimensions, and included angle between the sidewalls may differ among manufacturers. However, all belts of a given cross section shall operate interchangeably in standard grooves of the same cross section, but belts of different manufacturers should never be mixed on the same drive (see Tables 7, 8, 9 and 10).

SECTION 5—AVAILABLE LENGTHS

5.1 The length ranges for agricultural belts are shown in Tables 2 and 2A.

SECTION 6—METHOD OF MEASURING BELTS

6.1 The effective length of an agricultural belt is determined using a measuring fixture (see Fig. 3), consisting of two sheaves of equal diameter having standard groove dimensions (see Tables 5, 5A, 6, and 6A). One of the sheaves is fixed in position while the other is movable along a graduated scale with a specified force applied to it. The belt is rotated around the sheaves at least twice to properly seat it in the sheave grooves and to determine the midpoint of the center distance range. Effective length of the belt is determined by adding twice the average center distance measured on the fixture to the effective circumference of the measuring sheave specified in Tables 5, 5A, 6 or 6A.

6.2 The belt ride dimension is checked by measuring the distance from the top of the belt to the top of the measuring sheave groove (see Figs. 4, 5, and 6). Belt ride shall be within the maximum limit given in Tables 5, 5A, 6, and 6A. For V-belts the belt ride dimension is the only method of determining proper belt fit in the groove.

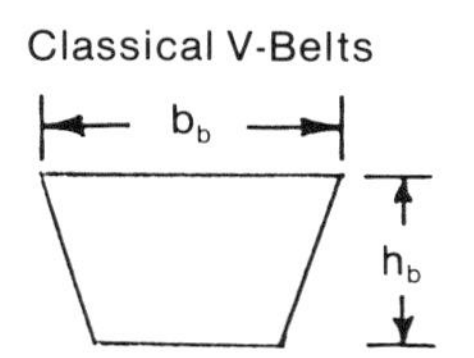

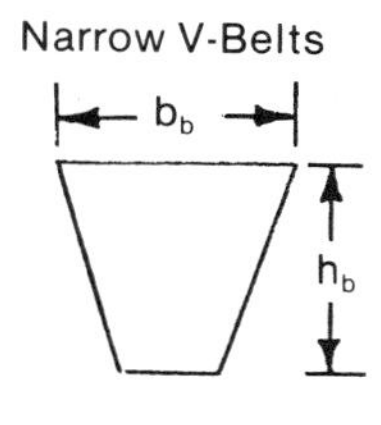

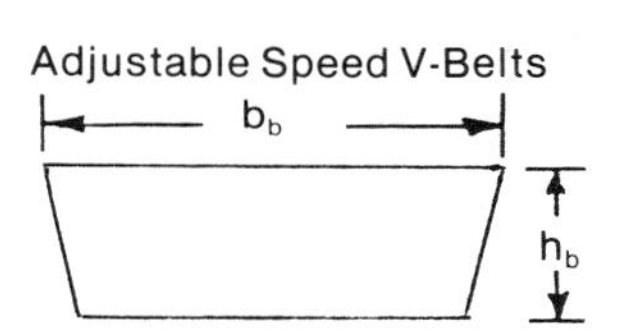

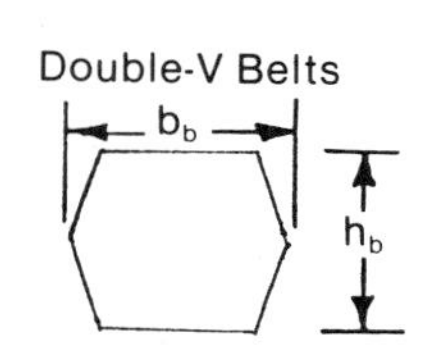

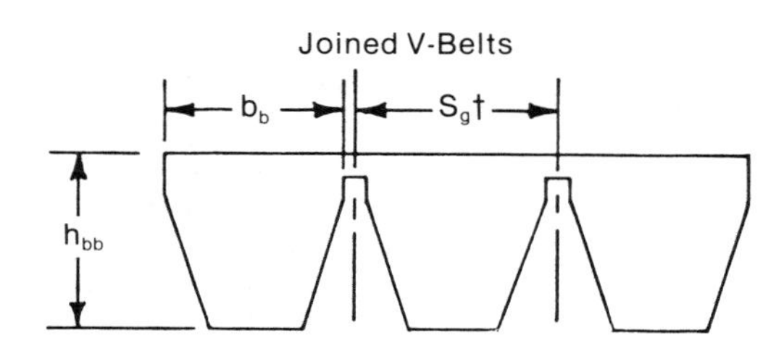

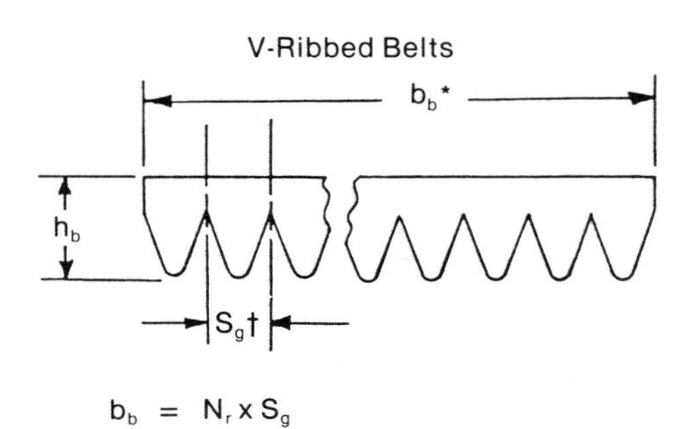

$b_b = N_r \times S_g$

where: N_r = number of ribs

†S_g is specified sheave groove spacing (see Tables 7, 7A, 10 or 10A).

FIG. 1—BELT TYPES

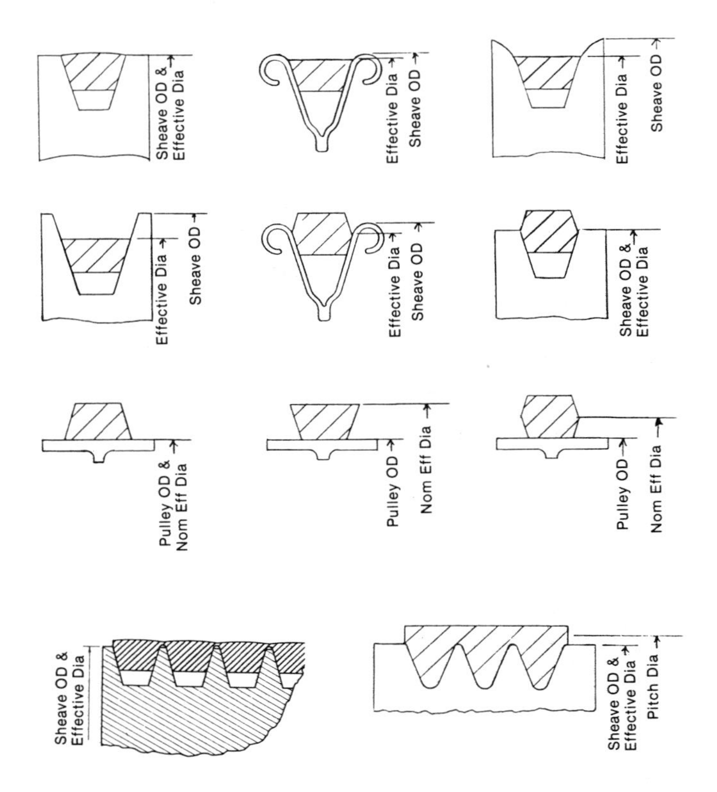

FIG. 2—RELATIONSHIP BETWEEN SHEAVE OR PULLEY OUTSIDE DIAMETER AND THE CORRESPONDING EFFECTIVE DIAMETER

TABLE 2—EFFECTIVE LENGTH RANGES, INCHES

V-Belts				V-Ribbed Belts
Classical*	Narrow*	Adjustable speed	Double-V	
HA 25.0-130.0	H3V 25.0-140.0	HI 40.0-125.0	HAA50.0-130.0	J 18.0-100.0
HB 30.0-300.0	H5V 50.0-355.0	HJ 50.0-160.0	HBB 50.0-300.0	L 50.0-145.0
HC 55.0-365.0	H8V 100.0-600.0	HK60.0-180.0	HCC 85.0-365.0	M90.0-365.0
HD120.0-365.0		HL70.0-200.0		
		HM80.0-200.0		
		HN85.0-200.0		
		HO90.0-200.0		

*Includes joined belts.

TABLE 2A—EFFECTIVE LENGTH RANGES, MILLIMETERS

V-Belts				V-Ribbed Belts
Classical*	Narrow*	Adjustable speed	Double-V	
13F 635-3300	9FN 635-3560	25FV 1020-3175	13FC 1270-3300	FPJ 455-2540
16F 760-7620	15FN 1270-9020	32FV 1270-4065		
22F 1400-9270	25FN 2540-15240	38FV 1525-4570	16FD 1270-7620	FPL 1270-3685
32F 3050-9270		44FV 1780-5080		
		51FV 2030-5080	22FD 2160-9270	FPM 2285-9270
		57FV 2160-5080		
		63FV 2285-5080		

*Includes joined belts.

TABLE 3—EFFECTIVE LENGTH TOLERANCE, INCHES

Effective length range	Effective length tolerance
Up through 51	±0.40
Over 51 to and incl. 98	±0.50
Over 98 to and incl. 124	±0.60
Over 124 to and incl. 157	±0.80
Over 157 to and incl. 197	±1.00
Over 197 to and incl. 248	±1.25
Over 248 to and incl. 315	±1.60
Over 315 to and incl. 390	±2.00

TABLE 3A—EFFECTIVE LENGTH TOLERANCE, MILLIMETERS

Effective length range	Effective length tolerance
Up through 1300	±10
Over 1300 to and incl. 2500	±13
Over 2500 to and incl. 3150	±16
Over 3150 to and incl. 4000	±20
Over 4000 to and incl. 5000	±25
Over 5000 to and incl. 6300	±32
Over 6300 to and incl. 8000	±40
Over 8000 to and incl. 10000	±50

TABLE 4—LIMITS OF DIFFERENCE IN EFFECTIVE LENGTH FOR MATCHING SETS, INCHES

Effective length range	Matching limits for one set	
	Normal tensile modulus	High tensile modulus*
Up through 54	0.16	0.08
Over 54 to and incl. 111	0.24	0.12
Over 111 to and incl. 236	0.39	0.20
Over 236 to and incl. 390	0.63	0.24

*Examples of high tensile modulus belts are those containing aramid, fiberglass, or steel cable reinforcement.

TABLE 4A—LIMITS OF DIFFERENCE IN EFFECTIVE LENGTH FOR MATCHING SETS, MILLIMETERS

Effective length range	Matching limits for one set	
	Normal tensile modulus	High tensile modulus*
Up through 1375	4	2
Over 1375 to and incl. 2820	6	3
Over 2820 to and incl. 6000	10	5
Over 6000 to and incl. 10,000	16	6

*Examples of high tensile modulus belts are those containing aramid, fiberglass, or steel cable reinforcement.

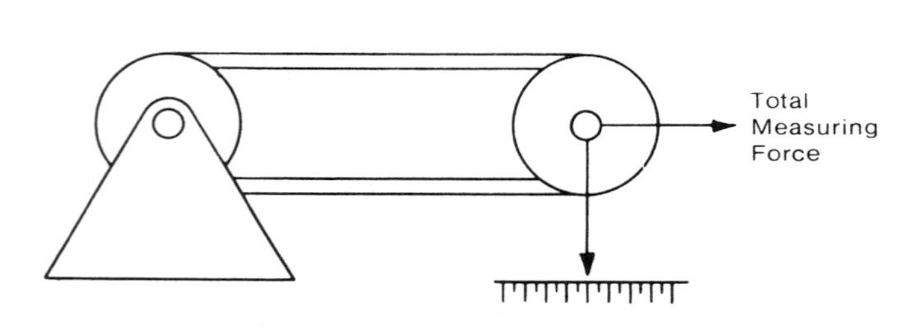

FIG. 3—DIAGRAM OF A FIXTURE FOR MEASURING BELTS

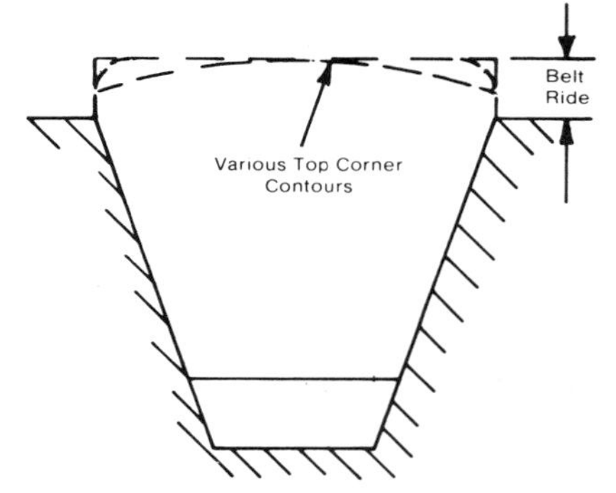

FIG. 4—MEASURING BELT RIDE, V-BELT

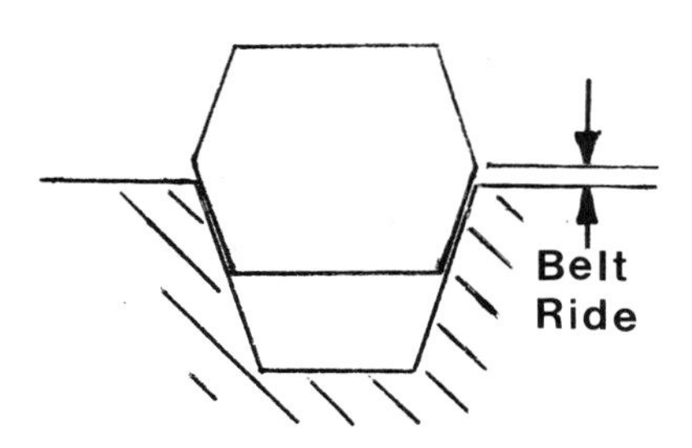

FIG. 5—MEASURING BELT RIDE, DOUBLE-V BELT

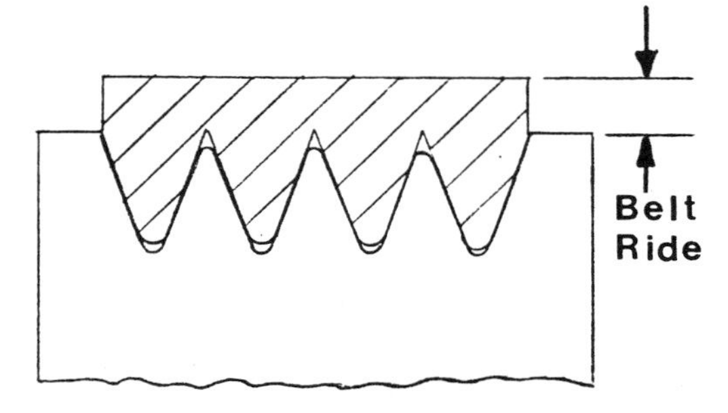

FIG. 6—MEASURING BELT RIDE, V-RIBBED BELT

TABLE 5—DATA FOR USE IN MEASURING BELT EFFECTIVE LENGTH AND BELT RIDE (SEE FIG. 7)

Belt cross section	Sheave outside diameter ±0.005 in.	Sheave effective circum-ference, in.	Sheave groove angle α ±0.25 deg	Sheave groove top width b_g, reference in.	Diameter ball or rod d_B ±0.0005 in.	Diameter over balls or rods ±0.005 in.	Groove depth h_g, min. in.	Total measuring force per belt, lb	Maximum ride position of belt with respect to top of groove, inches: Not joined	Joined
HA*	3.183	10.000	32	0.490	0.4375	3.499	0.490	65	0.09	0.18
HB*	4.775	15.000	32	0.630	0.5625	5.181	0.580	100	0.09	0.20
HC	7.958	25.000	34	0.879	0.7812	8.536	0.780	190	0.09	0.25
HD	11.141	35.000	34	1.259	1.1250	11.996	1.050	405	0.12	0.28
HAA	3.183	10.000	32	0.490	0.4375	3.499	0.490	65	0.03	
HBB	4.775	15.000	32	0.630	0.5625	5.181	0.580	100	0.03	
HCC	7.958	25.000	34	0.879	0.7812	8.536	0.780	190	0.03	
H3V	3.820	12.000	38	0.350	0.3438	4.203	0.340	100	0.10	0.20
H5V	7.958	25.000	38	0.600	0.5938	8.633	0.590	225	0.12	0.25
H8V	15.916	50.000	38	1.000	1.0000	17.083	0.990	500	0.16	0.30
HI	6.366	20.000	26	1.000	0.9531	7.225	0.813	180	0.16	
HJ	9.549	30.000	26	1.250	1.1875	10.601	0.938	290	0.16	
HK	9.549	30.000	26	1.500	1.4375	10.879	1.000	405	0.18	
HL	9.549	30.000	26	1.750	1.6875	11.158	1.125	560	0.20	
HM	9.549	30.000	26	2.000	1.9062	11.266	1.188	740	0.20	
HN	12.732	40.000	26	2.250	2.1481	14.683	1.339	740	0.22	
HO	12.732	40.000	26	2.500	2.3881	14.907	1.456	740	0.22	

*Measuring sheave dimensions for HA and HB grooves are different than those recommended for production sheaves. The dimensions in this table reflect previous recommendations so that precision measuring sheaves will not need to be replaced and to assure correlation of length measurement.

TABLE 5A—DATA FOR USE IN MEASURING BELT EFFECTIVE LENGTH AND BELT RIDE (SEE FIG. 7)

Cross section	Sheave outside diameter ±0.10 mm	Sheave effective circum-ference mm	Sheave groove angle α ±0.25 deg	Sheave groove top width b_g, reference mm	Diameter ball or rod d_B ±0.01 mm	Diameter over balls or rods ±0.10 mm	Groove depth h_g, min. mm	Total measuring force per belt, newtons	Maximum ride position of belt with respect to top of groove, mm: Not joined	Joined
13F	95.5	300	34	13	12.5 ±0.01	108.2	12	300	1.5	4.5
16F	143.2	450	34	16	15.5 ±0.02	159.4	14	450	1.5	5.0
22F	222.8	700	36	22	21.0 ±0.02	244.0	19	850	1.5	6.5
32F	318.3	1000	34	32	30.5 ±0.02	348.5	26	1800	1.5	7.0
13FD	95.5	300	34	13	12.5 ±0.01	108.2	12	300	0.8	
16FD	143.2	450	34	16	15.5 ±0.02	159.4	14	450	0.8	
22FD	222.8	700	36	22	21.0 ±0.02	244.0	19	850	0.8	
9FN	95.5	300	38	8.89	8.50±0.01	104.3	8.6	445	2.5	5.1
15FN	191.0	600	38	15.24	15.00±0.02	207.8	15.0	1000	3.0	6.4
25FN	318.3	1000	38	25.40	25.00±0.02	346.3	25.1	2225	4.1	7.6
25FV	127.3	400	26	25.40	24.50±0.01	150.7	20	800	4.1	
32FV	159.2	500	26	31.75	30.50±0.01	187.8	23	1300	4.1	
38FV	191.0	600	26	38.10	36.50±0.01	224.8	26	1800	4.6	
44FV	222.8	700	26	44.45	42.50±0.01	261.7	29	2500	5.1	
51FV	254.6	800	26	50.80	48.50±0.01	298.7	32	3300	5.1	
57FV	286.5	900	26	57.00	54.50±0.01	336.4	34	3300	5.6	
63FV	318.3	1000	26	63.00	60.00±0.01	372.1	37	3300	5.6	

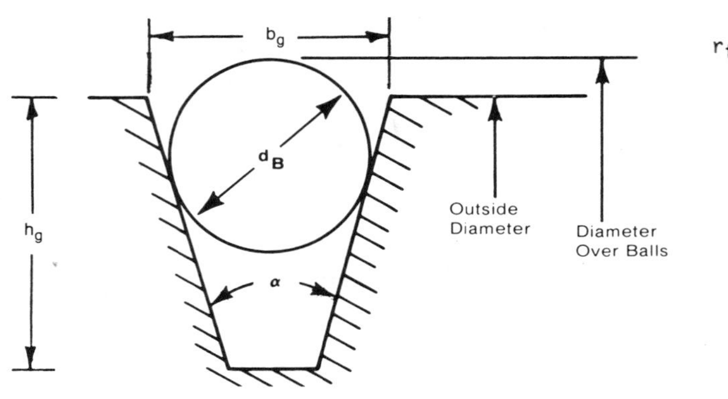

FIG. 7—V-BELT MEASURING SHEAVE GROOVE

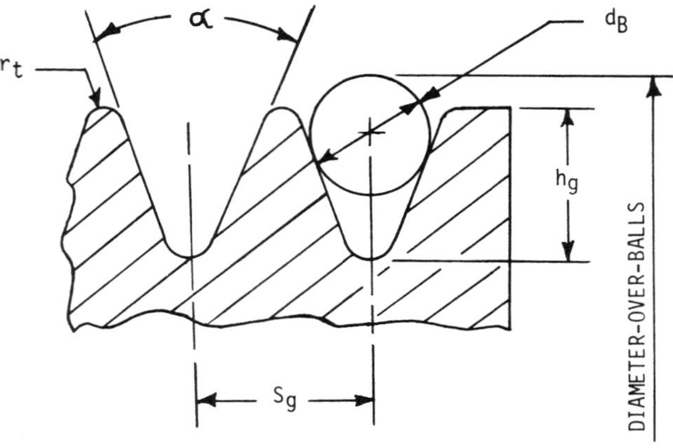

FIG. 8—V-RIBBED BELT MEASURING SHEAVE GROOVE

TABLE 6—DATA FOR USE IN MEASURING EFFECTIVE LENGTHS AND RIDE OF V-RIBBED BELTS (SEE FIG. 8)

Cross section	Sheave outside diameter, reference in.	Sheave effective circumference, in.	Sheave groove angle α ±0.25 deg	Sheave groove spacing S_g, in.	Diameter ball or rod d_B ±0.0005 in.	Groove depth h_g, min. in.	Top radius r_t +0.005 −0.000 in.	Maximum ride position of belt with respect to top of groove, in.	Total measuring force per rib, lb
J	3.183	10.000	40	0.092 ±0.001	0.0625	0.082	0.008	0.10	11
L	6.366	20.000	40	0.185 ±0.002	0.1406	0.196	0.015	0.22	45
M	9.549	30.000	40	0.370 ±0.0003	0.2812	0.393	0.030	0.30	100

TABLE 6A—DATA FOR USE IN MEASURING EFFECTIVE LENGTHS AND RIDE OF V-RIBBED BELTS (SEE FIG. 8)

Cross section	Sheave outside diameter, reference mm	Sheave effective circumference, mm	Sheave groove angle α ±0.25 deg	Sheave groove spacing S_g, mm	Diameter ball or rod d_B ±0.01 mm	Diameter over ball or rod ±0.1 mm	Groove depth h_g, min. mm	Top radius r_t +0.15 −0.00 mm	Maximum ride position of belt with respect to top of groove mm	Total measuring force per rib, newtons
FPJ	95.5	300	40	2.34 (±0.03)	1.50	97.5	2.06	0.20	2.50	50
FPL	159.2	500	40	4.70 (±0.05)	4.00	163.5	4.92	0.40	5.60	200
FPM	254.6	800	40	9.40 (±0.08)	7.00	259.2	10.03	0.75	7.60	450

TABLE 7–DIMENSIONS FOR SHEAVES USING V-BELTS AND DOUBLE-V BELTS, INCHES (SEE FIGS 9 AND 10)

Cross section	Effective diameter range	Groove angle, degrees	Regular Groove Dimensions							Minimum recommended effective diameter
			b_g	h_g min.	2a	R_B min.	d_B ±0.0005	S_g	S_e	
HA	Up through 5.65	34	0.494	0.460	HA=0.25	0.148	0.4375	0.625	0.375	3.0
HAA	Over 5.65	38	0 504		HAA=0.00	0.149			+0.090	
		(±0.33)	(±0.005)					(±0.025)	−0.062	
									0.500	
HB	Up through 7.35	34	0.637	0.550	HB=0.35	0.189	0.5625	0.750	+0.120	5.4
HBB	Over 7.35	38	0.650		HBB=0.00	0.190			−0.065	
		(±0.33)	(±0.006)					(±0.025)		
HC	Up through 8.39	34	0.879		HC=0.40	0.274			0.688	
HCC	Over 8.39 to & incl. 12.40	36	0.887	0.750	HCC=0.00	0.276	0.7812	1.000	+0.160	9.0
	Over 12.40	38	0.895			0.277			−0.070	
		(±0.33)	(±0.007)					(±0.025)		
	Up through 13.59	34	1.259			0.410			0.875	
HD	Over 13.59 to & incl. 17.60	36	1.271	1.020	0.60	0.410	1.1250	1.438	+0.220	13.0
	Over 17.60	38	1.283			0.411			−0.080	
		(±0.33)	(±0.008)					(±0.025)		
	Up through 3.49	36				0.181			0.344	
H3V	Over 3.49 to & incl. 6.00	38	0.350	0.340	0.05	0.183	0.3438	0.406	+0.094	2.65
	Over 6.00 to & incl. 12.00	40				0.186			−0.031	
	Over 12.00	42				0.188				
		(±0.25)	(±0.005)					(±0.015)		
	Up through 9.99	38				0.329			0.500	
H5V	Over 9.99 to & incl. 16.00	40	0.600	0.590	0.10	0.332	0.5938	0.688	+0.125	7.10
	Over 16.00	42				0.336			−0.047	
		(±0.25)	(±0.005)					(±0.015)		
	Up through 15.99	38				0.575			0.750	
H8V	Over 15.99 to & incl. 22.40	40	1.000	0.990	0.20	0.580	1.0000	1.125	+0.250	12.50
	Over 22.40	42				0.585			−0.062	
		(±0.25)	(±0.005)					(±0.015)		
HA	Up through 5.96	34	0.589	0.615	HA=0.56	−0.009	0.4375	0.750	0.438	
HAA	Over 5.96	38	0.611		HAA=0.31	−0.008			+0.090	
		(±0.33)	(±0.005)					(±0.025)	−0.062	
									0.562	
HB	Up through 7.71	34	0.747	0.730	HB=0.71	−0.007	0.5625	0.875	+0.120	
HBB	Over 7.71	38	0.774		HBB=0.36	−0.008			−0.065	
		(±0.33)	(±0.006)					(±0.025)		
HC	Up through 9.00	34	1.066			−0.035			0.812	
HCC	Over 9.00 to & incl. 13.01	36	1.085	1.055	HC=1.01	−0.032	0.7812	1.250	+0.160	
	Over 13.01	38	1.105		HCC=0.61	−0.031			−0.070	
		(±0.33)	(±0.007)					(±0.025)		
	Up through 14.42	34	1.513			−0.010			1.062	
HD	Over 14.42 to & incl. 18.43	36	1.541	1.435	1.430	−0.009	1.1250	1.750	+0.220	
	Over 18.43	38	1.105			−0.008			−0.080	
		(±0.33)	(±0.008)					(±0.025)		
	Up through 3.71	36	0.421			0.070			0.375	
H3V	Over 3.71 to & incl. 6.22	38	0.425	0.449	0.268	0.073	0.3438	0.500	+0.094	
	Over 6.22 to & incl. 12.22	40	0.429			0.076			−0.031	
	Over 12.22	42	0.434			0.078				
		(±0.25)	(±0.005)					(±0.015)		
	Up through 10.31	38	0.710			0.168			0.562	
H5V	Over 10.31 to & incl. 16.32	40	0.716	0.750	0.420	0.172	0.5938	0.812	+0.125	
	Over 16.32	42	0.723			0.175			−0.031	
		(±0.25)	(±0.005)					(±0.015)		
	Up through 16.51	38	1.180			0.312			0.844	
H8V	Over 16.41 to & incl. 22.92	40	1.190	1.252	0.724	0.316	1.000	1.312	+0.250	
	Over 22.92	42	1.201			0.321			−0.062	
		(±0.25)	(±0.005)					(±0.015)		

NOTES:

1. Summation of the deviations from S_g for all grooves in any one sheave shall not exceed ± 0.031 in.
2. The variation in pitch diameter between the grooves in any one sheave must be within the following limits:
   Up through 19.9 in. outside diameter and up through 6 grooves, 0.010 in. Add 0.0005 in. for each additional groove.
   20.0 in. and over an outside diameter and up through 10 grooves, 0.0015 in. Add 0.0005 in. for each additional groove.
   This variation can be obtained easily by measuring the distance across two measuring balls or rods placed diametrically opposite each other in a groove. Comparing this diameter over balls or rods measurement between grooves will give the variation in pitch diameter.
3. Deep groove sheaves are intended for drives with belt offset such as quarter-turn or vertical shaft drives (see Rubber Manufacturers Association Power Transmission Belt Technical Information Bulletin No. IP-3-10).
4. Joined belts will not operate in deep groove sheaves.
5. Other sheave tolerances:
   Outside Diameter—Up thru 8.0 in. outside diameter, ± 0.020 in. For each additional inch of outside diameter, add ± 0.0025 in.
   Radial Runout—Up thru 10.0 in. outside diameter, 0.010 in. total indicator reading. For each additional inch of outside diameter, add 0.0005 in.
   Axial Runout—Up thru 5.0 in. outside diameter, 0.005 in. total indicator reading. For each additional inch of outside diameter, add 0.001 in.

TABLE 7A—DIMENSIONS FOR SHEAVES USING V-BELTS AND DOUBLE-V BELTS, MILLIMETERS (SEE FIGS. 9 AND 10)

			Regular Groove Dimensions										
Cross section	Effective diameter range	Groove angle, degrees	b_g	h_g min.	2_a	R_B min.	d_B	S_g	S_e	b_e Ref.	Minimum recommended effective diameter		
13F	Up through 140	34	13	12	13F=3.0	6.11	12.5	15.88	10	13	80		
13FD	Over 140	38	(±0.13)		13FD=0.0	6.34	(±0.01)	(±0.6)	+2				
		(±0.33)							−1				
									12.5				
16F	Up through 180	34	16	14	16F=4.0	7.78	15.5	19	+3	16	140		
16FD	Over 180	38	(±0.15)		16FD=0.0	8.04	(±0.02)	(±0.6)	−1				
		(±0.33)											
22F	Up through 200	34	22	19	22F=6.0	10.06	21.0	25.5	18		224		
22FD	Over 200 to & incl. 315	36	(±0.18)		22FD=6.0	10.27	(±0.02)	(±0.6)	+4	22			
	Over 315	38				10.47			−2				
		(±0.33)											
	Up through 355	34	32	26	9.0	14.65	30.5	36.53	26				
32F	Over 355 to & incl. 450	36	(±0.20)			14.96	(±0.02)	(±0.6)	+6		355		
	Over 450	38				15.25			−3				
		(±0.33)											
	Up through 90	36	8.89	8.6	1.27	4.09	8.5	10.3	9				
9FN	Over 90 to & incl. 150	38	(±0.13)			4.18	(±0.01)	(±0.40)	+2	899	67		
	Over 150 to & incl. 305	40				4.26			−1				
	Over 305	42				4.33							
		(±0.25)											
	Up through 255	38	15.24	15.0	2.54	8.16	15.0	17.5	13				
15FN	Over 255 to & incl. 405	40	(±0.13)			8.26	(±0.02)	(±0.40)	+3	1524	180		
	Over 405	42				8.36			−1				
		(±0.25)											
	Up through 405	38	25.40	25.1	5.08	13.76	25.0	28.6	19				
25FN	Over 405 to & incl. 570	40	(±0.13)			13.92	(±0.02)	(±0.40)	+6	2549	315		
	Over 570	42				14.07			−2				
		(±0.25)											
13F	Up through 148	34	15.45	16	13F=11.0	2.06	12.50		11			8	84
13FD	Over 148	38	15.75		13FD=8.0	2.31	(±0.01)	17.5	+2				
		(±0.33)	(±0.13)					(±0.6)	−1				
									14				145
16F	Up through 190	34	19.06	19	16F=14.0	2.72	15.50		+3			10	
16FD	Over 190	38	19.44		16FD=10.0	3.00	(±0.02)	22	−4				
		(±0.33)	(±0.15)					(±0.6)					
22F	Up through 214	34	26.28			2.99			20			14	231
22FD	Over 214 to & incl. 329	36	26.55	26	22F=20.0	3.20	21.0	30	+4				
	Over 329	38	26.82		22FD=14.0	3.41	(±0.02)	(±0.6)	−2				
		(±0.33)	(±0.18)										
	Up through 364	34	37.50			5.56			29				364
32F	Over 364 to & incl. 459	36	37.85	35	27.0	5.87	30.5	43	+6				
	Over 459	38	38.20			6.16	(±0.02)	(±0.6)	−3				
		(±0.33)	(±0.20)										
	Up through 96	36	10.69			1.30			10	8.89	1.27	5.54	73
9FN	Over 96 to & incl. 156	38	10.80	11.4	6.8	1.38	8.50	12.7	+2				
	Over 156 to & incl. 311	40	10.90			1.48	(±0.01)	(±0.4)	−1				
	Over 311	42	11.02			1.54							
		(±0.25)	(±0.13)										
	Up through 263	38	18.03			4.08			14	15.24	2.54	8.11	188
15FN	Over 263 to & incl. 413	40	18.19	19.0	10.6	4.18	15.00	20.6	+3				
	Over 413	42	18.36			4.27	(±0.02)	(±0.4)	−1				
		(±0.25)	(±0.13)										
	Up through 418	38	29.97			7.08			21	25.40	5.08	13.30	328
25FN	Over 418 to & incl. 583	40	30.23	31.8	18.4	7.24	25.00	33.3	+6				
	Over 583	42	30.51			7.37	(±0.02)	(±0.4)	−2				
		(±0.25)	(±0.13)										

NOTES:

1. Summation of the deviations from S_g for all grooves in any one sheave shall not exceed ± 0.79 mm.
2. The variation in pitch diameter between the grooves in any one sheave must be within the following limits:
   Up through 505 mm outside diameter and up through 6 grooves, 0.25 mm. Add 0.013 mm for each additional groove.
   506 mm and over on outside diameter and up through 10 grooves, 0.38 mm. Add 0.013 mm for each additional groove.
   This variation can be obtained easily by measuring the distance across two measuring balls or rods placed diametrically opposite each other in a groove. Comparing this diameter over balls or rods measurement between grooves will give the variation in pitch diameter.
3. Deep groove sheaves are intended for drives with belt offset such as quarter-turn or vertical shaft drives (see Rubber Manufacturers Association Power Transmission Belt Technical Information Bulletin No. IP-3-10).
4. Joined belts will not operate in deep groove sheaves.
5. Other sheave tolerances:
   Outside Diameter—Up thru 203 mm outside diameter, ± 0.5 mm. For each additional 25 mm of outside diameter, add ± 0.064 mm.
   Radial Runout—Up thru 254 mm outside diameter, 0.25 mm total indicator reading. For each additional 25 mm of outside diameter, add 0.13 mm.
   Axial Runout—Up thru 127 mm outside diameter, 0.13 mm total indicator reading. For each additional 25 mm of outside diameter, add 0.025 mm.

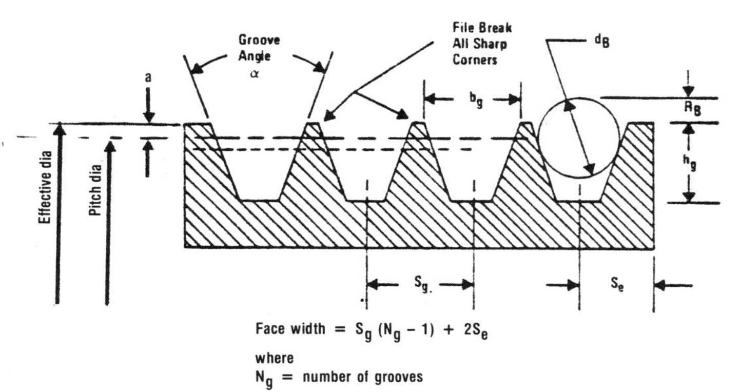

FIG. 9—STANDARD GROOVE DIMENSIONS

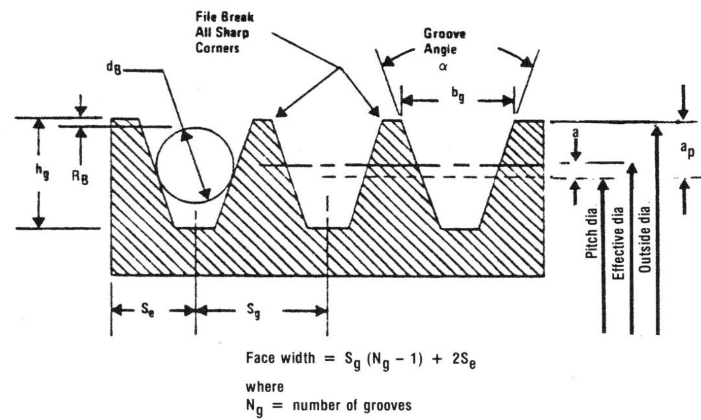

FIG. 10—DEEP GROOVE DIMENSIONS

TABLE 8—DIMENSIONS FOR SHEAVES USING ADJUSTABLE SPEED BELTS, INCHES

Cross section	Recommended minimum outside diameter	Groove angle α ±1 deg	b_g	b_{go}*	2a	2av
HI	7.50	26	1.00	1.65	0.30	2.84
HJ	9.75	26	1.25	2.11	0.37	3.73
HK	11.50	26	1.50	2.45	0.45	4.62
HL	14.00	26	1.75	3.02	0.52	5.52
HM	16.00	26	2.00	3.48	0.60	6.41
HN	17.75	26	2.25	3.94	0.68	7.36
HO	19.25	26	2.50	4.41	0.75	8.27

*b_{go} is calculated to provide for a clearance of 0.250 in. as shown in Fig. 11 (dimension C_L).

TABLE 8A—DIMENSIONS FOR SHEAVES USING ADJUSTABLE SPEED BELTS, MILLIMETERS

Cross section	Recommended minimum outside diameter	Groove angle α ±1 deg	b_g	b_{go}*	2a	2av
25FV	175	26	25.40	42.0	7.6	72.0
32FV	225	26	31.75	54.7	9.4	95.5
38FV	265	26	38.10	65.1	11.4	117.0
44FV	310	26	44.45	76.9	13.2	140.5
51FV	355	26	50.80	88.7	15.2	164.0
57FV	400	26	57.00	100.2	17.2	186.9
63FV	440	26	63.00	112.0	19.0	210.0

*b_{go} is calculated to provide for a clearance of 6.35 mm as shown in Fig. 11 (dimension C_L).

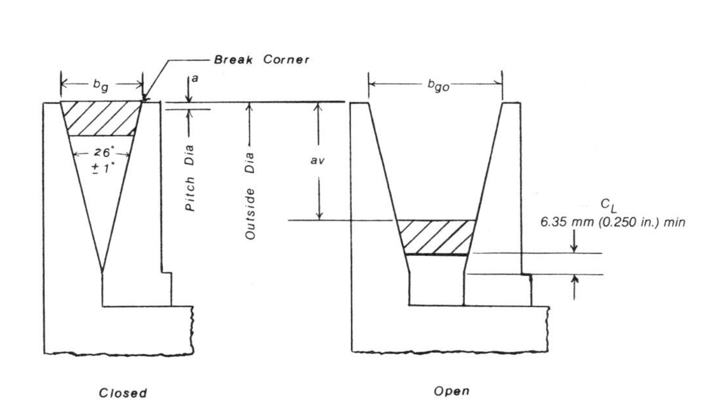

FIG. 11—ADJUSTABLE SHEAVE DIMENSIONS

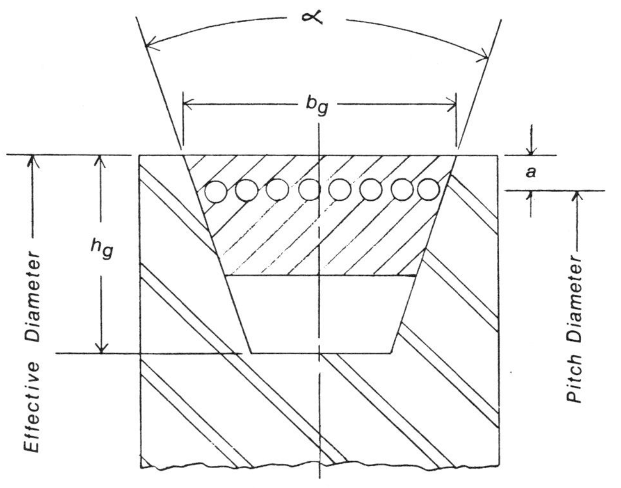

This is a fixed sheave operating in conjunction with an adjustable sheave where relatively low speed range change is required.

FIG. 12—ADJUSTABLE SPEED COMPANION SHEAVE DIMENSIONS

TABLE 9—DIMENSIONS FOR ADJUSTABLE SPEED COMPANION OR IDLER SHEAVES, INCHES (SEE FIG. 12)

Cross section	Minimum recommended effective diameter	Groove angle α ±0.5 deg	b_g ±0.001	h_g min.	2a
HI	4.75	26	1.00	0.78	0.30
HJ	6.10	26	1.25	0.94	0.37
HK	7.25	26	1.50	1.05	0.45
HL	8.50	26	1.75	1.25	0.52
HM	9.50	26	2.00	1.40	0.60
HN	10.50	26	2.25	1.56	0.68
HO	11.00	26	2.50	1.69	0.75

TABLE 9A—DIMENSIONS FOR ADJUSTABLE SPEED COMPANION OR IDLER SHEAVES, MILLIMETERS (SEE FIG. 12)

Cross section	Minimum recommended effective diameter	Groove angle α ±0.5 deg	b_g ±0.25	h_g min.	2a
25FV	120	26	25.4	20	7.6
32FV	155	26	31.75	24	9.4
38FV	185	26	38.10	27	11.4
44FV	215	26	44.45	32	13.2
51FV	240	26	50.80	36	15.2
57FV	270	26	57.00	40	17.3
63FV	280	26	63.00	43	19.1

TABLE 10—DIMENSIONS FOR SHEAVES USING V-RIBBED BELTS, INCHES (SEE FIG. 13)

Cross section	Minimum recommended effective diameter	Groove angle α ±0.25 deg	S_g*	r_t +0.005 −0.000	r_b	h_g min.	S_e	d_B ±0.0005	2a
J	0.80	40	0.092 ±0.001	0.008	0.015 +0.000 −0.005	0.071	0.125 +0.030 −0.015	0.625	0.030
L	3.00	40	0.185 ±0.002	0.015	0.015 +0.000 −0.005	0.183	0.375 +0.075 −0.030	0.1406	0.058
M	7.00	40	0.370 +0.003	0.030	0.030 +0.000 −0.010	0.377	0.500 +0.100 −0.040	0.2812	0.116

*Summation of the deviations from S_g for all grooves in any one sheave shall not exceed ±0.010 in.

TABLE 10A—DIMENSIONS FOR SHEAVES USING V-RIBBED BELTS, MILLIMETERS (SEE FIG. 13)

Cross section	Minimum recommended effective diameter	Groove angle α ±0.25 deg	S_g*	r_t +0.15 −0.00	r_b	h_g min.	S_e	d_B ±0.001	2a
FPJ	20	40	2.34 ±0.03	0.20	0.40 +0.00 −0.15	1.77	3.0 + 0.8 − 0.4	1.50	0.76
FPL	75	40	4.70 ±0.05	0.40	0.40 +0.00 −0.15	4.63	10.0 + 2.0 − 1.0	4.00	1.54
FPM	100	40	9.40 +0.08	0.75	0.75 +0.00 −0.25	9.74	13.0 + 3.0 − 1.0	7.00	2.88

*Summation of the deviations from S_g for all grooves in any one sheave shall not exceed ±0.25 mm.

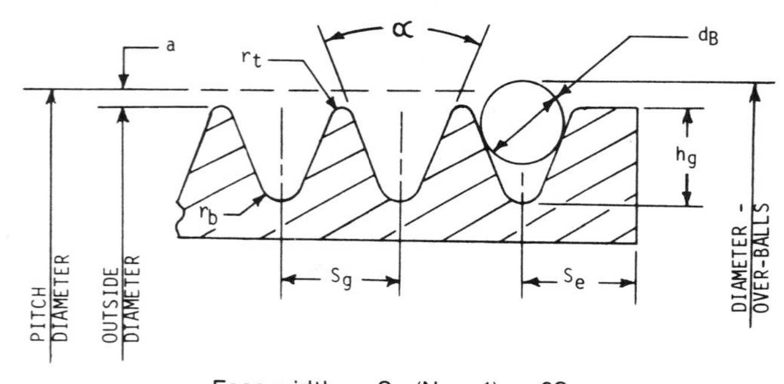

Face width = $S_g (N_g - 1) + 2S_e$

where

N_g = number of grooves

FIG. 13—GROOVE DIMENSIONS FOR SHEAVES USING V-RIBBED BELTS

SECTION 7—SPECIFICATIONS FOR SHEAVES USED WITH V-BELTS AND V-RIBBED BELTS

7.1 Construction

7.1.1 Sheaves used with agricultural V-belts or V-ribbed belts shall be made of a material which is resistant to abrasion between the groove wall and the belt. The material should be sufficiently close-grained to allow the machining or forming of a smooth groove sidewall.

7.1.2 Machined sheaves shall have surface finishes equal to or smoother than the following values:

Machined surface area	Maximum surface roughness height, (arithmetic average)* µin.	µm
Sheave groove sidewall	125	3.2
Adjustable sheave sidewall	63	1.6
Flat pulley rim O.D.	250	6.3
Rim edges, rim I.D.	500	12.7

*NOTE: The measuring methods defined in American National Standard ANSI/ASME B46.1 Surface Texture, shall be used to determine these values.

7.1.3 Sheaves formed from sheet metal shall be made so that the groove width and angle are uniform throughout the circumference of the sheave. The gage of the sheet metal used should be such that the groove will not deflect under the load imposed by the belt.

7.1.4 Adjustable-speed sheaves should be so designed that the movable disk is perpendicular to the axis of rotation at all times without appreciable runout or wobble. Failure to accomplish this results in a non-uniform groove width which materially reduces belt life and may set up undesirable vibration of the machine on which is is used (see Tables 8 and 8A).

7.2 Groove dimensions and tolerances for agricultural belts are shown in Tables 7, 7A, 8, 8A, 9, 9A, 10, and 10A. Deep groove sheaves are used on quarter turn drives or other drives where the belt must enter the groove at an angle.

SECTION 8—RECOMMENDED DESIGN PRACTICES

8.1 Sheave diameters. In designing belt drives, it should be recognized that the use of larger sheave diameters will result in lower bearing loads and can result in the use of smaller and less expensive belt cross sections.

8.2 Idlers

8.2.1 Idlers may be necessary on agricultural belt drives to provide take-up or to increase the arc of contact to obtain the required drive capacity. If an idler is needed, it should be located on the slack side of the drive. Other factors that affect the location of the idler are its effectiveness in belt take-up and its effect on arcs of contact.

8.2.2 An idler should have its axis of rotation perpendicular to the plane of the belt strand on which it runs. The idler mounting should be strong enough to maintain this relationship at all times.

8.2.3 If grooved idlers are used, the groove dimensions should be as shown in Tables 7, 7A, 9, 9A, 10, or 10A.

8.2.4 Minimum diameters recommended for idlers are shown in Tables 11 and 11A.

8.3 Length calculations

8.3.1 The approximate belt length for a two-sheave drive may be calculated using the formula

$$L_e = 2C + 1.57\,(D_e + d_e) + \frac{(D_e - d_e)2}{4C} \quad \ldots\ldots\ldots\ldots [1]$$

where

L_e = effective length of belt
C = distance between centers of sheaves
D_e = effective diameter of large sheave
d_e = effective diameter of small sheave
(See Fig. 14)

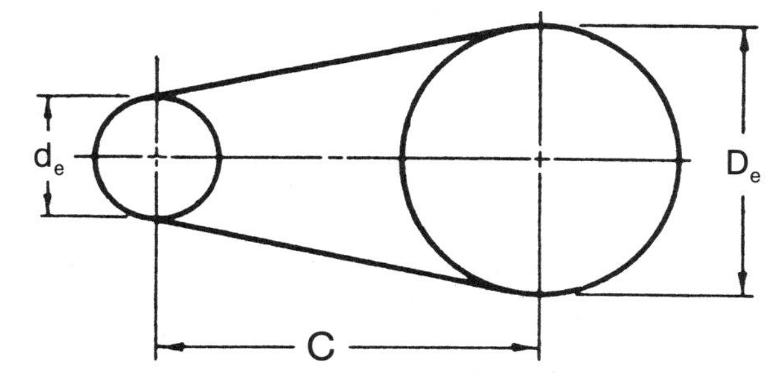

FIG. 14—RELATION BETWEEN CENTER DISTANCE AND BELT LENGTH FOR DRIVES WITH TWO SHEAVES

If sheave effective diameters (D_e and d_e) and belt effective length (L_e) are known, the approximate center distance between sheaves may be calculated by means of formula [2] as follows

$$C = a + \sqrt{a^2 - b} \quad \ldots\ldots\ldots\ldots [2]$$

where

$a = L_e/4 - 0.393\,(D_e + d_e)$
$b = 0.125\,(D_e - d_e)^2$

8.3.2 To determine belt length when more than two sheaves are used on a drive (see Fig. 15), lay out the sheaves in terms of their effective diameters to scale in the position desired when a new belt is applied and first brought to driving tension. The length of belt shall be the sum of the tangents and the connecting arcs around the effective diameters of the sheaves. The length of the connecting arcs can be calculated by the formula

Length of arc = $D_eA/115$

where

D_e = the effective outside diameter of the sheave
A = the angle in degrees subtended by the arc of belt contact on the sheave

8.3.3 Belt manufacturers have computer programs for calculating belt length and will provide assistance in solving complex drive geometries.

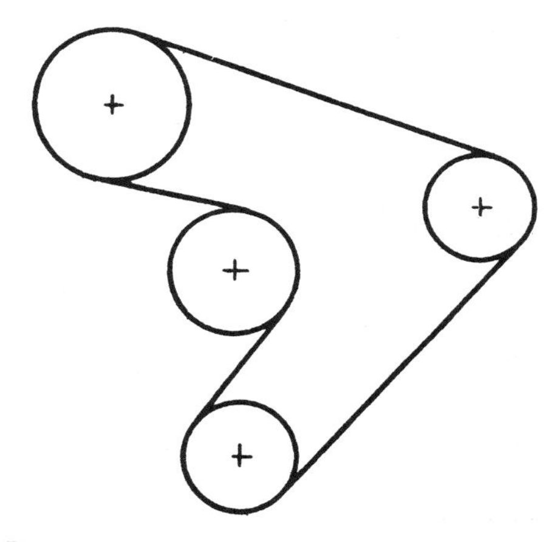

FIG. 15—BELT DRIVE WITH MORE THAN TWO SHEAVES

TABLE 11—MINIMUM RECOMMENDED DIAMETERS FOR IDLERS, INCHES

Cross section	Minimum O.D. of grooved inside idler	Minimum O.D. of flat inside idler	Minimum O.D. of outside idler	Minimum face width of flat idler‡
HA, HAA	2.75	2.25	4.25	1.00
HB, HBB	4.00	3.75	6.00	1.25
HC, HCC	6.75	5.75	8.50	1.50
HD	9.00	7.50	13.50	2.00
H3V	2.65	*	4.25	1.13
H5V	7.10	*	10.00	1.38
H8V	12.50	*	17.50	1.75
J	0.80	0.65	1.25	†
L	3.00	2.63	4.50	†
M	7.00	6.25	10.50	†
HI	5.50	4.50	*	1.75
HJ	6.75	5.63	*	2.00
HK	8.00	6.75	*	2.25
HL	9.25	7.75	*	2.50
HM	10.50	8.75	*	2.75
HN	11.75	9.88	*	3.00
HO	13.00	11.00	*	3.25

*Not recommended.
†Belt width + 0.75 in.
‡For both inside and outside idlers.

TABLE 11A—MINIMUM RECOMMENDED DIAMETERS FOR IDLERS, MILLIMETERS

Cross section	Minimum O.D. of grooved inside idler	Minimum O.D. of flat inside idler	Minimum O.D. of outside idler	Minimum face width of flat idler‡
13F, 13FD	70	57	108	25
16F, 16FD	102	95	152	32
22F, 22FD	172	146	216	38
32F, 32FD	229	190	343	51
9FN	67	*	108	29
15FN	180	*	254	35
25FN	318	*	444	44
FPJ	20	16	32	†
FPL	76	67	114	†
FPM	178	159	267	†
25FV	140	114	*	44
32FV	172	143	*	51
38FV	203	172	*	57
44FV	235	197	*	64
51FV	267	222	*	70
57FV	299	251	*	76
63FV	330	279	*	83

*Not recommended.
†Belt width + 19 mm.
‡For both inside and outside idlers.

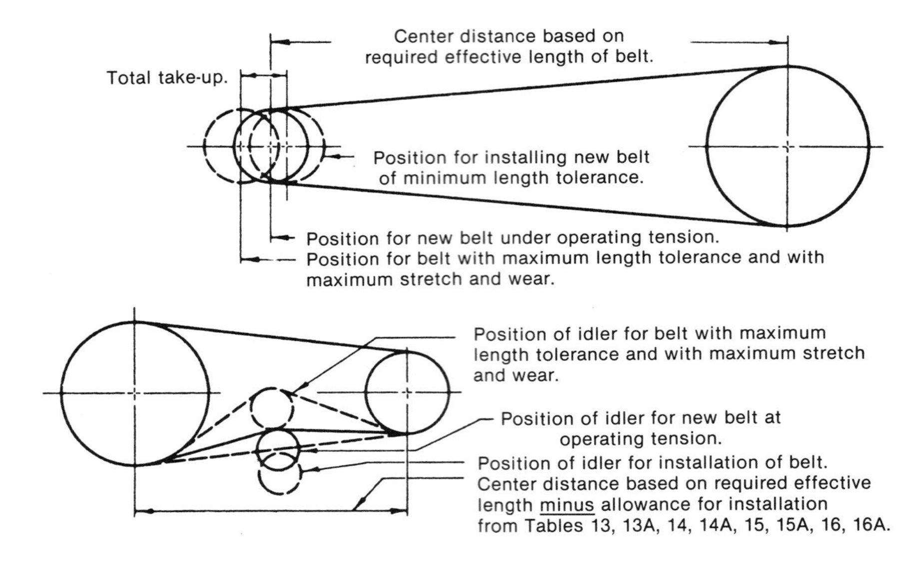

FIG. 16—INSTALLATION AND TAKE-UP OF AGRICULTURAL BELTS

8.4 Installation and take-up (see Fig. 16)

8.4.1 The calculated belt length (paragraph 8.3) shall be the effective length of an ideal belt under operating tension. A belt drive shall be arranged so that any belt within the length tolerances given in Tables 4 and 4A can be placed in the sheave groove without forcing. In addition, provision shall be made to compensate for the change in effective length caused by the seating of the belt in the sheave groove and by the stretch and wear of the belt during its life.

8.4.2 Installation and take-up allowance specified in Tables 13, 13A, 14, 14A, 15, 15A, 16, and 16A shall be provided on every belt drive to insure satisfactory operation.

8.5 Clutching belt drives. Properly designed belt drive systems can be used as a clutching mechanism. The required belt effective length is determined by adding the clutching allowance to the calculated declutched length. The clutching allowance is calculated from the formula

$$\text{Clutching allowance (min.)} = 3.14\, h_b \left(\frac{0}{360 \text{ deg}}\right) + (\text{Minus belt length tol.})$$

where

h_b = belt thickness (see Tables 1 and 1A)

0 = arc of contact on clutching sheave, deg

Minus belt length tolerance = values from Tables 3 and 3A

8.5.1 The calculated clutching allowance should be equal to or exceed the allowance for installation given in Tables 13 through 16 and 13A through 16A.

8.5.2 The drive design should provide proper belt guide(s) to permit the belt to disengage from the driving sheave.

8.6 Cross drives, mule drives, and other twisted-belt drives

8.6.1 The minimum tangent length for a 180 deg twist in a belt is shown in Tables 12 and 12A. The minimum tangent length for any amount of twist other than 180 deg can be obtained by multiplying the minimum tangent length by the fraction

$$\frac{\text{degrees of twist required}}{180}$$

Adjustable-speed belts are not recommended for these drives.

8.6.2 Quarter-turn drives. On quarter-turn drives, the angle of entry of the belt into the plane of the sheave grooves should not exceed 5 deg. A center distance at least 5 1/2 times the diameter of the large sheave is necessary to insure this condition where one belt is used.

8.7 Specification of belt drives. In submitting a drive design problem to engineering departments of the different belt manufacturers, it is strongly recommended that complete information be given. Appendix A provides examples of the data needed.

TABLE 12—MINIMUM TANGENT LENGTHS FOR 180 DEG TWIST, INCH

Cross section	Minimum tangent length, in.
HA	18
HB	22
HC	28
HD	37
H3V	18
H5V	28

NOTE: For all other cross sections, consult belt manufacturer.

TABLE 12A—MINIMUM TANGENT LENGTHS FOR 180 DEG TWIST, MILLIMETER

Cross section	Minimum tangent length, mm
13F	460
16F	560
22F	710
32F	940
9FN	460
15FN	710

NOTE: For all other cross sections, consult belt manufacturer.

TABLE 13—INSTALLATION AND TAKE-UP ALLOWANCES FOR CLASSICAL, JOINED CLASSICAL AND DOUBLE-V BELTS, INCHES*

Effective length range	Allowance for installation*								Allowance for stretch and wear†	
	HA HAA	HA joined	HB HBB	HB joined	HC HCC	HC joined	HD	HD joined	Normal tensile modulus	High tensile modulus
Up through 51.2	1.09	1.25	1.25	1.39	1.44	1.64			1.54	1.28
Over 51.2 to & incl. 98.4	1.38	1.54	1.54	1.68	1.73	1.93			2.95	2.46
Over 98.4 to & incl. 124.0	1.58	1.74	1.74	1.88	1.93	2.13			3.72	3.10
Over 124.0 to & incl. 157.5	1.92	2.07	2.07	2.22	2.26	2.47	2.61	2.75	4.73	3.94
Over 157.5 to & incl. 196.9			2.43	2.57	2.62	2.83	2.97	3.11	5.91	4.92
Over 196.9 to & incl. 248.0			2.89	3.03	3.08	3.28	3.42	3.56	7.44	6.20
Over 248.0 to & incl. 315.0			3.50	3.65	3.69	3.90	4.04	4.18	9.45	7.88
Over 315.0 to & incl. 393.7					4.41	4.61	4.75	4.89	11.81	9.84

TABLE 13A—INSTALLATION AND TAKE-UP ALLOWANCES FOR CLASSICAL, JOINED CLASSICAL AND DOUBLE-V BELTS, MILLIMETERS*

Effective length range	Allowance for installation*								Allowance for stretch and wear†	
	13F 13FD	13F joined	16F 16FD	16F joined	22F 22FD	22F joined	32F	32F joined	Normal tensile modulus	High tensile modulus
Up through 1300	28	32	32	35	37	42			39	33
Over 1300 to & incl. 2500	35	39	39	43	44	49			75	62
Over 2500 to & incl. 3150	40	44	44	48	49	54			95	79
Over 3150 to & incl. 4000	49	53	53	56	57	63	66	70	120	100
Over 4000 to & incl. 5000			62	65	67	72	75	79	150	125
Over 5000 to & incl. 6300			73	77	78	83	87	90	189	157
Over 6300 to & incl. 8000			89	93	94	99	103	106	240	200
Over 8000 to & incl. 10,000					112	117	121	124	300	250

TABLE 14—INSTALLATION AND TAKE-UP ALLOWANCES FOR NARROW AND JOINED NARROW BELTS, INCHES*

Effective length range	Allowance for installation*						Allowance for stretch and wear†	
	3V	3V joined	5V	5V joined	8V	8V joined	Normal tensile modulus	High tensile modulus
Up through 51.2	1.09	1.20	1.44	1.58			1.54	1.28
Over 51.2 to & incl. 98.4	1.38	1.49	1.73	1.87	2.32	2.47	2.95	2.46
Over 98.4 to & incl. 124.0	1.58	1.69	1.93	2.07	2.53	2.67	3.72	3.10
Over 124.0 to & incl. 157.5			2.26	2.40	2.86	3.00	4.73	3.94
Over 157.5 to & incl. 196.9			2.62	2.76	3.22	3.36	5.91	4.92
Over 196.9 to & incl. 248.0			3.08	3.22	3.67	3.81	7.44	6.20
Over 248.0 to & incl. 315.0			3.69	3.83	4.29	4.43	9.45	7.88
Over 315.0 to & incl. 393.7					5.00	5.15	11.81	9.84

TABLE 14A—INSTALLATION AND TAKE-UP ALLOWANCES FOR NARROW AND JOINED NARROW BELTS, MILLIMETERS*

Effective length range	Allowance for installation*						Allowance for stretch and wear†	
	9FN	9FN joined	15FN	15FN joined	25FN	25FN joined	Normal tensile modulus	High tensile modulus
Up through 1300	28	30	37	40			39	33
Over 1300 to & incl. 2500	35	38	44	47	59	63	75	62
Over 2500 to & incl. 3150	40	43	49	53	64	68	95	79
Over 3150 to & incl. 4000			57	61	73	76	120	100
Over 4000 to & incl. 5000			67	70	82	85	150	125
Over 5000 to & incl. 6300			78	82	93	97	189	157
Over 6300 to & incl. 8000			94	97	109	113	240	200
Over 8000 to & incl. 10,000					127	131	300	250

TABLE 15—INSTALLATION AND TAKE-UP ALLOWANCES FOR V-RIBBED BELTS, INCHES

Effective length range	Allowance for installation* J	L	M	Allowance for stretch and wear† Normal tensile modulus	High tensile modulus
Up through 51.2	0.86	1.20		1.54	1.28
Over 51.2 to & incl. 98.4	1.15	1.49	1.93	2.95	2.46
Over 98.4 to & incl. 124.0		1.69	2.13	3.72	3.10
Over 124.0 to & incl. 157.5		2.03	2.47	4.73	3.94
Over 157.5 to & incl. 196.9			2.83	5.91	4.92
Over 196.9 to & incl. 248.0			3.28	7.44	6.20
Over 248.0 to & incl. 315.0			3.90	9.45	7.88
Over 315.0 to & incl. 393.7			4.61	11.81	9.84

*Allowance for installation includes the minus manufacturing length tolerance from Table 3, the difference between the length of belt under no tension and the length under installation tension, and an amount for installing the belts over the sheave flanges without injury.

†Allowance for stretch and wear includes the plus manufacturing tolerance from Table 3 as well as an allowance for the stretch and wear of the belt resulting from service on the drive (see Tables 13, 14, 15, & 16).

Installation and take-up methods are shown in Fig. 13. In the first sketch, the center distance of the drive can be adjusted to furnish the necessary installation and take-up allowances. In the second sketch, the center distance is fixed, and the allowance for installation and take-up is provided by the idler pulley.

Examples of the calculation of center distance, effective length, and installation and take-up allowances are shown in Appendix A.

TABLE 15A—INSTALLATION AND TAKE-UP ALLOWANCES FOR V-RIBBED BELTS, MILLIMETERS

Effective length range	Allowance for installation* FPJ	FPL	FPM	Allowance for stretch and wear† Normal tensile modulus	High tensile modulus
Up through 1300	22	30		39	33
Over 1300 to & incl. 2500	29	38	49	75	62
Over 2500 to & incl. 3150		43	54	95	79
Over 3150 to & incl. 4000		52	63	120	100
Over 4000 to & incl. 5000			72	150	125
Over 5000 to & incl. 6300			83	189	157
Over 6300 to & incl. 8000			99	240	200
Over 8000 to & incl. 10,000			117	300	250

*Allowance for installation includes the minus manufacturing length tolerance from Table 3A, the difference between the length of belt under no tension and the length under installation tension, and an amount for installing the belts over the sheave flanges without injury.

†Allowance for stretch and wear includes the plus manufacturing tolerance from Table 3A as well as an allowance for the stretch and wear of the belt resulting from service on the drive (see Tables 13A, 14A, 15A, & 16A).

Installation and take-up methods are shown in Fig. 13. In the first stretch, the center distance of the drive can be adjusted to furnish the necessary installation and take-up allowances. In the second sketch, the center distance is fixed, and the allowance for installation and take-up is provided by the idler pulley.

TABLE 16—INSTALLATION AND TAKE-UP ALLOWANCES FOR ADJUSTABLE SPEED BELTS, INCHES*

Effective length range	Allowance for installation* HI	HJ	HK	HL	HM	HN	HO	Allowance for stretch and wear† Normal tensile modulus	High tensile modulus
Up through 51.2	1.39	1.53	1.69	1.83	1.99	2.08	2.18	1.54	1.28
Over 51.2 to & incl. 98.4	1.68	1.82	1.98	2.12	2.28	2.37	2.47	2.95	2.46
Over 98.4 to & incl. 124.0	1.88	2.02	2.18	2.32	2.48	2.57	2.67	3.72	3.10
Over 124.0 to & incl. 157.5	2.22	2.36	2.51	2.66	2.81	2.91	3.00	4.73	3.94
Over 157.5 to & incl. 196.9			2.87	3.01	3.17	3.27	3.36	5.91	4.92

*Installation allowance on a drive using two adjustable sheaves can be neglected.

†Allowance for stretch and wear includes the plus manufacturing tolerance from Table 3 as well as an allowance for the stretch and wear of the belt resulting from service on the drive (see Tables 13, 14, 15, & 16).

Installation and take-up methods are shown in Fig. 13. In the first sketch, the center distance of the drive can be adjusted to furnish the necessary installation and take-up allowances. In the second sketch, the center distance is fixed, and the allowance for installation and take-up is provided by the idler pulley.

Examples of the calculation of center distance, effective length, and installation and take-up allowances are shown in Appendix A.

TABLE 16A—INSTALLATION AND TAKE-UP ALLOWANCES FOR ADJUSTABLE SPEED BELTS, MILLIMETERS*

Effective length range	Allowance for installation* 25FV	32FV	38FV	44FV	51FV	57FV	63FV	Allowance for stretch and wear† Normal tensile modulus	High tensile modulus
Up through 1300	35	39	43	46	51	53	55	39	33
Over 1300 to & incl. 2500	43	46	50	54	58	60	63	75	62
Over 2500 to & incl. 3150	48	51	55	59	63	65	68	95	79
Over 3150 to & incl. 4000	56	60	64	68	71	74	76	120	100
Over 4000 to & incl. 5000			73	76	81	83	85	150	125

*Installation allowance on a drive using two adjustable sheaves can be neglected.

†Allowance for stretch and wear includes the plus manufacturing tolerance from Table 3A as well as an allowance for the stretch and wear of the belt resulting from service on the drive (see Tables 13A, 14A, 15A, & 16A).

Installation and take-up methods are shown in Fig. 13. In the first sketch, the center distance of the drive can be adjusted to furnish the necessary installation and take-up allowances. In the second sketch, the center distance is fixed, and the allowance for installation and take-up is provided by the idler pulley.

APPENDIX A
EXAMPLES OF THE CALCULATION OF BELT LENGTH, CENTER DISTANCE, INSTALLATION AND TAKE-UP ALLOWANCES, AND INSPECTION REQUIREMENTS

EXAMPLE 1 (Refer to Fig. 17)

The drive consists of two sheaves; one of the shafts may be moved for installation and take-up. Effective diameters have been determined. The preferred center distance is about 20 in.

Belt length and center distance

1. Substitute the effective diameters and preferred center distance in formula [1]. The effective belt length required is 79.02 in.
2. Round to the nearest tenth of an inch, or 79.0 in.
3. This length substituted in formula [2] will give a center distance of 19.97 in.

Installation allowance

1. From Table 13 the installation allowance will be 1.54 in.
2. Subtract this amount from the effective belt length of 79.0 in. to get a length of 77.46 in.
3. This length substituted in formula [2] will give a center distance of 19.16 in., the minimum center distance needed for installation of the belt.

Take-up allowance

1. From Table 13 the allowance needed for take-up is 2.95 in.
2. Add this amount to the effective belt length of 79.0 in. to get a maximum length of 81.95 in.
3. This length substituted in formula [2] will give the maximum required center distance of 21.51 in.

Inspection requirements: Fill in the inspection requirements for the belt required in Example 1.

1. Fill in values from Table 5 as follows

   Tension = 100 lb
   h_g = 0.580 in.
   b_g = 0.630 in.
   α = 32 deg
   OD = 4.775 in.

   Maximum ride position of belt with respect to top of groove is 0.09 in. (Note: From Table 5, the effective circumference of the measuring sheave is 15.000 in.)
2. From the effective length of 79.0 in., subtract 15.000 in. and divide the remainder by 2 to find Y, or

$$Y = (79.0 - 15.000)/2 = 32.0 \text{ in.}$$

3. From Table 3, the length tolerance is ± 0.50 in. The tolerance on dimension Y will be equal to these length tolerances divided by 2, or

$$\text{Tolerance on } Y = \pm 0.25 \text{ in.}$$

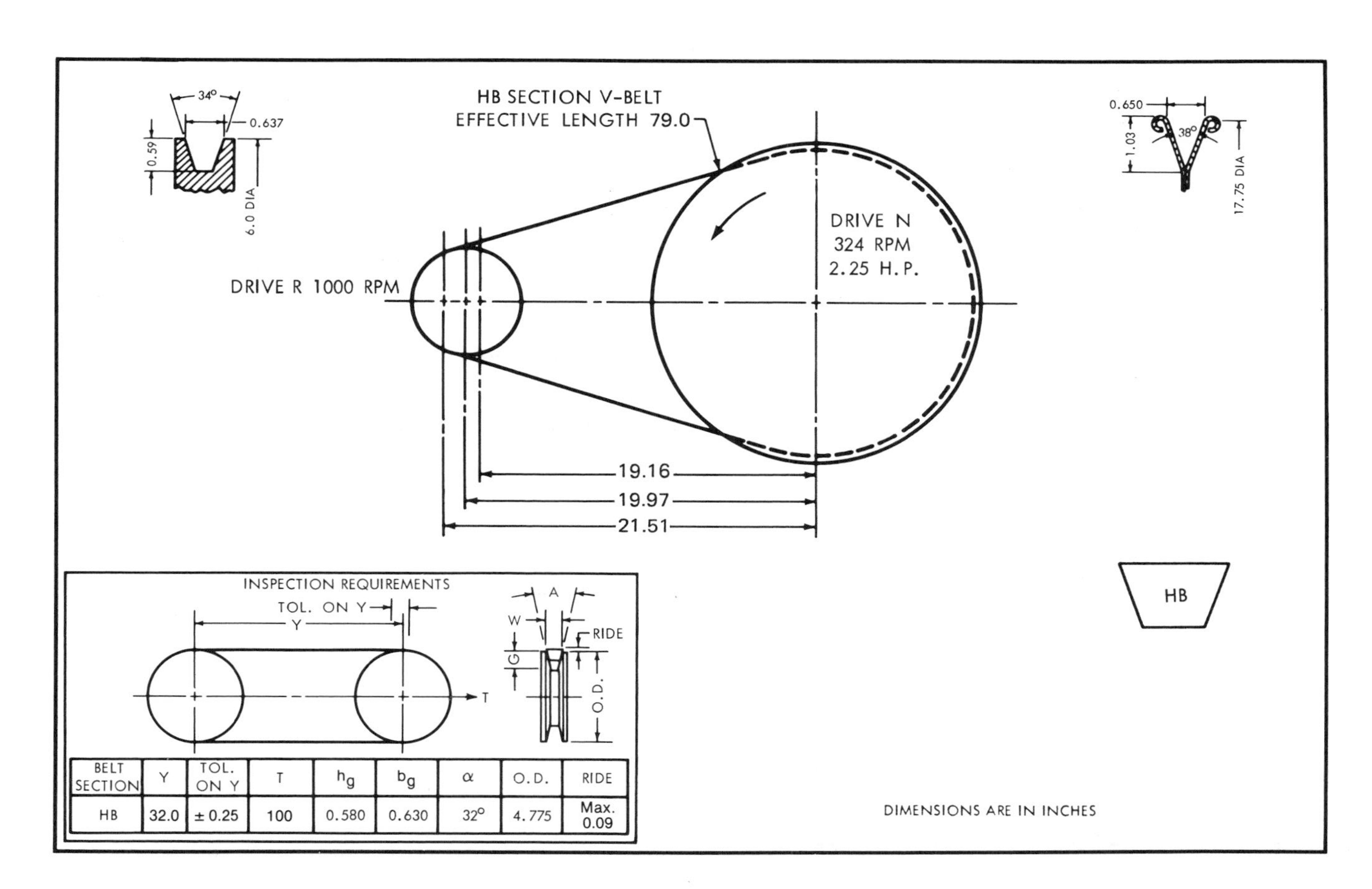

BELT SECTION	Y	TOL. ON Y	T	h_g	b_g	α	O.D.	RIDE
HB	32.0	± 0.25	100	0.580	0.630	32°	4.775	Max. 0.09

FIG. 17—TYPICAL TWO-SHEAVE DRIVE WITH ONE SHAFT MOVABLE FOR TAKE-UP

EXAMPLE 2 (Refer to Fig. 18)

The effective diameters have been determined. Both shafts are fixed in position and the center distance is 26.66 in. An 8.5 in. outside diameter flat idler will be used for take-up on the drive.

Belt length

1. Substitute the effective outside diameters of the sheaves and the fixed center distance of 26.66 in. in formula [1]. The resulting belt length is 80.01 in.
2. Since the centers cannot be moved for installation, the shortest possible belt must go on the drive with the idler out of the way. Consequently, the installation allowance must be added to the belt length obtained above. The installation allowance from Table 13 is 1.73 in. This added to the length of 80.01 in. gives a required effective belt length of 81.74 in.

Take-up allowances

From Table 13, the take-up allowance needed for this belt is 2.95 in. This amount added to the effective belt length of 81.74 in. gives a maximum length of 84.69 in. By one of the methods outlined above for determining belt length when more than two sheaves are used on a drive, locate the position of the idler so that it will provide take-up for this length of belt.

Inspection requirements:

Fill in the inspection requirements for the belt required in Example 2.

1. Fill in values from Table 5 as follows

   Tension = 190 lb
   h_g = 0.780 in.
   b_g = 0.879 in.
   α = 34 deg
   OD = 7.958 in.

   Maximum ride position of belt with respect to top of groove is 0.09 in. (Note: From Table 5, the effective circumference of the measuring sheave is 25.000 in.)
2. From the effective length of 81.74 in. subtract 25.000 in. and divide the remainder by 2 to find Y, or

$$Y = (81.74 - 25.000)/2 = 28.37 \text{ in.}$$

3. From Table 3, the length tolerance is ± 0.50 in. The tolerance on dimension Y will be equal to these length tolerances divided by 2, or

$$\text{Tolerance on Y} = \pm 0.25 \text{ in.}$$

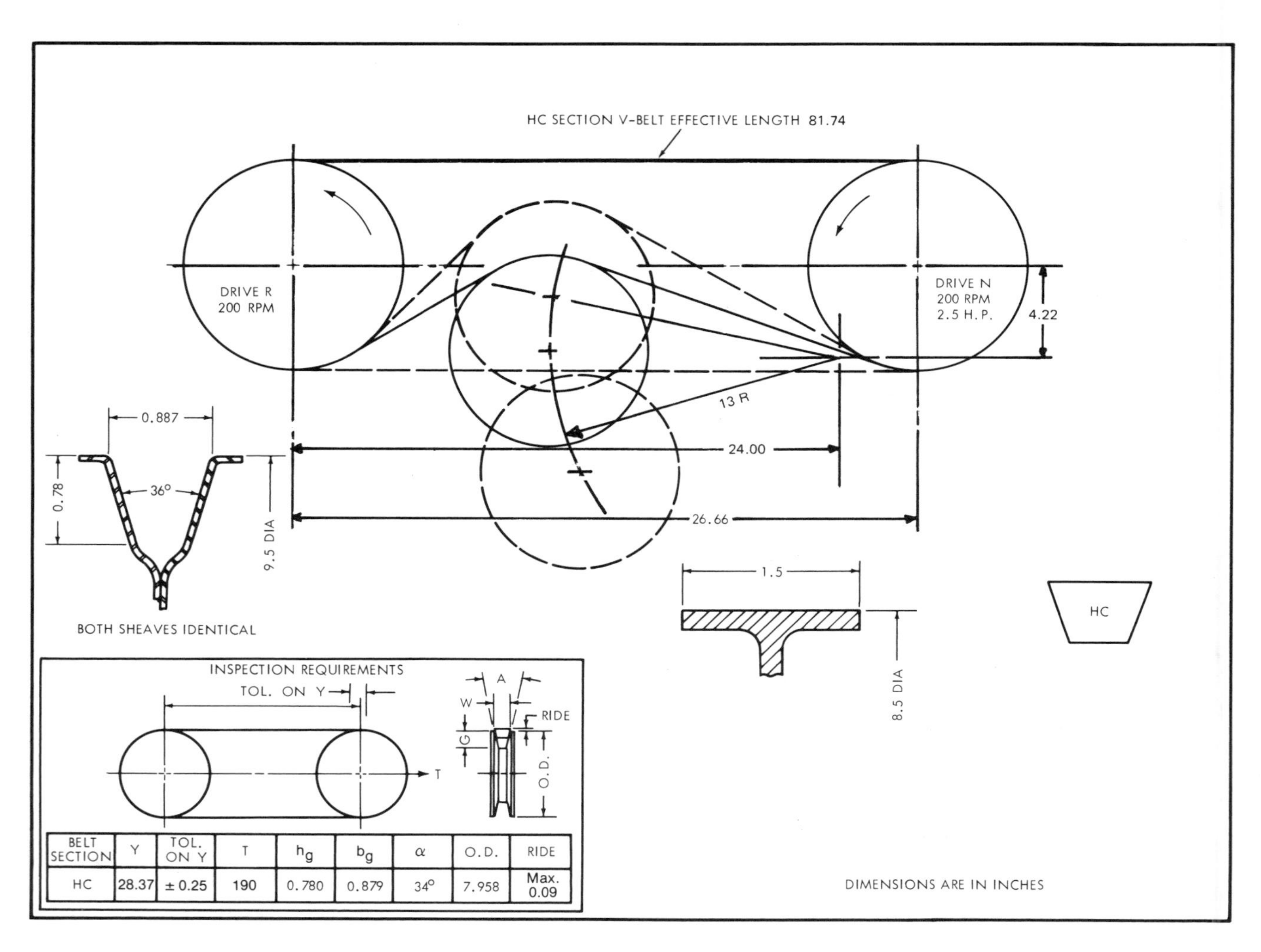

BELT SECTION	Y	TOL. ON Y	T	h_g	b_g	α	O.D.	RIDE
HC	28.37	± 0.25	190	0.780	0.879	34°	7.958	Max. 0.09

FIG. 18—TWO-SHEAVE DRIVE WITH CENTERS FIXED AND IDLER USED FOR TAKE-UP

EXAMPLE 3 (Refer to Fig. 19)

The effective diameters have been selected and shaft centers have been located approximately. All shafts will be fixed in position and belt take-up will be accomplished by means of a grooved idler pulley.

Belt length

1. With the idler in its "installation position", use one of the methods outlined above for determining belt length when more than two sheaves are used on a drive.
2. To find the length of belt for the drive, add to the length obtained in step 1 the allowance for installation from Table 13.

Take-up allowance

To the length of belt for the drive, add the allowance for take-up from Table 13. Check the drive with the idler in its maximum take-up position to see that this length of belt can be accommodated.

Inspection requirements:

Fill in the inspection requirements for the belt required in Example 3.

1. Fill in values from Table 5 as follows

   Tension = 190 lb.
   h_g = 0.780 in.
   b_g = 0.630 in.
   α = 34 deg
   OD = 7.958 in.

   Maximum ride position of belt with respect to top of groove is 0.03 in. (Note: From Table 5, the effective circumference of the measuring sheave is 25.000 in.)

2. From the effective length of 176.7 in., subtract 25.000 in. and divide the remainder by 2 to find Y, or

   $$Y = (177.2 - 25.000)/2 = 75.85 \text{ in.}$$

3. From Table 3, the length tolerance is ± 1.00 in. The tolerance on dimension Y will be equal to these length tolerances divided by 2, or

   $$\text{Tolerance on } Y = \pm 0.50 \text{ in.}$$

Cited Standards:

ANSI/ASME B46.1, Surface Texture

ISO 1000, SI Units and Recommendations for the Use of Their Multiples and of Certain Other Units

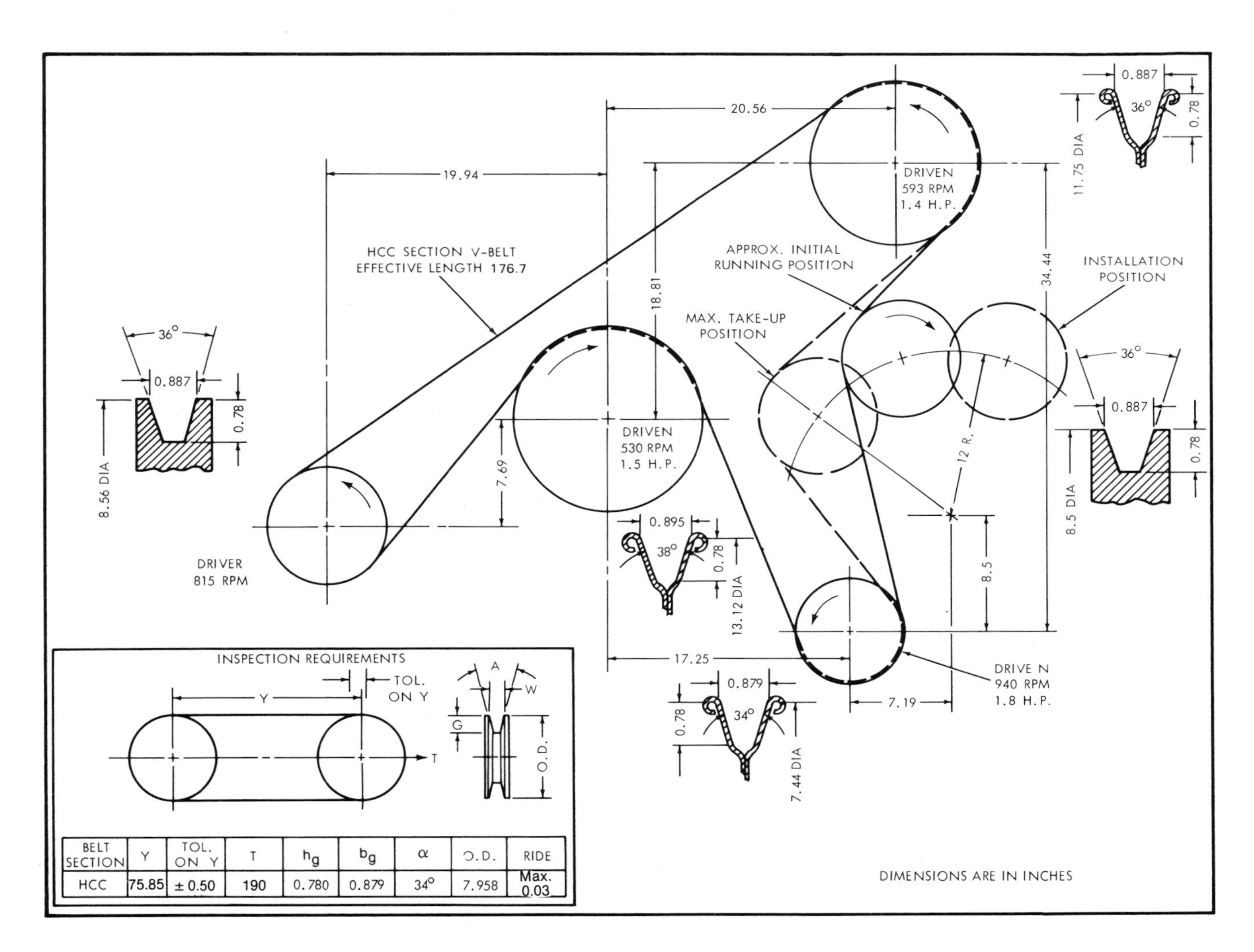

BELT SECTION	Y	TOL. ON Y	T	h_g	b_g	α	O.D.	RIDE
HCC	75.85	± 0.50	190	0.780	0.879	34°	7.958	Max. 0.03

FIG. 19—DOUBLE-V BELT DRIVE WITH FOUR SHEAVES ON FIXED CENTERS

ASAE Standard: ASAE S212.1

LABORATORY PROCEDURE FOR TESTING V-BELTS

Recommendation prepared by V-Belt Standardization Subcommittee of ASAE Power and Machinery Technical Committee; adopted by ASAE as a Recommendation 1962; reconfirmed December 1966; revised December 1968; reconfirmed December 1973; reconfirmed and reclassified as a Standard December 1978; reconfirmed December 1983.

SECTION 1—PURPOSE

1.1 This Standard covers a procedure for accelerated laboratory testing designed to establish a quality control standard for acceptance or rejection of belts against a proven field-tested application.

1.2 It is difficult to satisfactorily duplicate field conditions in a laboratory test set up. It is, therefore, desirable to establish a universal testing system for fast, uniform results which can be duplicated. Standardization of test procedures will permit efficient use of test equipment.

SECTION 2—SCOPE

2.1 This procedure establishes an accelerated method for comparing the quality of belts. The stress fatigue (endurance) limit and belt elongation information can be obtained under controlled conditions of speed, load, and tension. The test does not evaluate wear characteristics in service.

SECTION 3—TEST EQUIPMENT

3.1 The equipment consists of a motor and drive sheave, driving a dynamometer and driven sheave, through the belt to be tested, and a tensioning device together with the necessary instrumentation. If an engine is used, it should have a sufficiently large flywheel to smooth out the power strokes. Either the driver or driven member can be mounted on a track which permits horizontal movement so that constant tension can be applied to the belt drive. All parts of the assembly should be accurately machined and mounted in suitable anti-friction bearings to minimize frictional forces (see Fig. 1).

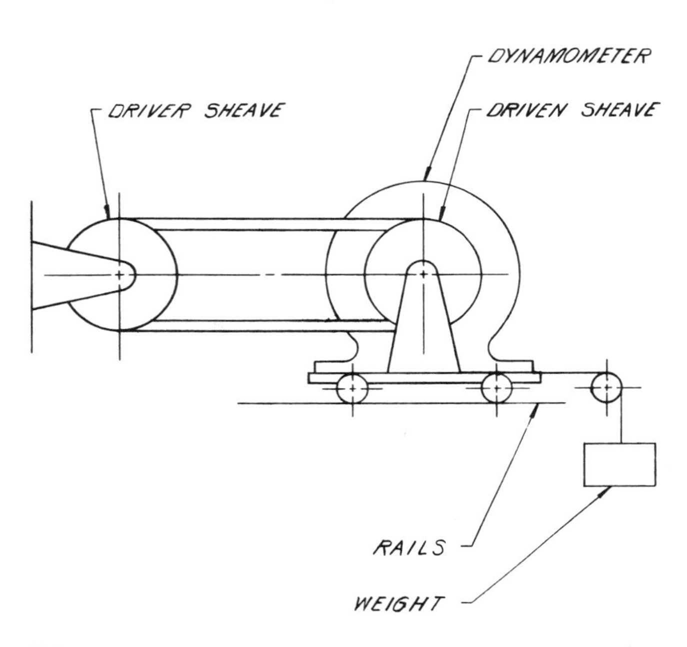

FIG. 1—V-BELT TEST EQUIPMENT

SECTION 4—TEST SHEAVES

4.1 Test sheave grooves shall be surface hardened to a minimum hardness of 450 Brinell and shall have a surface finish of 32 microinches maximum.

4.2 Eccentricity on the groove shall not exceed 0.010 inch total indicator reading. Sheaves shall not have side wobble, as measured at the pitch diameter, exceeding 0.001 inch per inch of P.D. measured as total indicator reading.

SECTION 5—TEST CONDITIONS

5.1 Table 2 gives the essential items to be controlled and measured in running the test. The output horsepower and total belt tension shall be controlled within ±2 percent of the assigned values. Total belt tension is the sum of the two strand tensions which is also the total force exerted by the belt on each test sheave. The driver and driven sheaves (Table 1) shall have equal diameters for each of the belt sections listed in Table 2.

SECTION 6—CONDITIONS FOR TEST WITH REVERSE BENDING OF V- AND DOUBLE V-BELTS

6.1 As explained in Sections 1 and 2 laboratory testing described herein is for the purpose of quality control of a belt quality level which has been proven by field testing or by other experience. Despite this limiting purpose, it is sometimes considered to be desirable to test belts with reverse bending over idlers when their actual usage involves this condition. Fig. 3 and Table 3 provide the recommended factors and conditions for such tests.

SECTION 7—TEST PROCEDURE

7.1 The ambient temperature shall be 90 deg F ±5 degrees.

7.2 Condition the belt by running for five minutes at specified conditions of full speed, load and constant tension.

7.3 Shut off the test machine and allow the belt to coast to a stop. Measure the center distance. This is the initial center distance for determination of effective stretch.

7.4 Measure the outside length with a thin tape. This is the initial outside length value. For the purpose described in paragraph 7.8.2 it shall be considered sufficiently accurate in this measurement to take the tape over the top of idler No. 1 in Fig. 3 for the reverse bend test.

7.5 Run the test under the specified test conditions.

7.6 Stop the test machine for observation at least every 24 hours, and record additional measurements of center distance and outside length.

7.7 Continue testing until belt breaks or until it no longer transmits specified horsepower under the specified test condition.

7.8 Record the following:

7.8.1 **Effective stretch.** This is obtained by doubling the difference between the initial center distance, Step 3, and the final center distance before failure, Step 6, and dividing this doubled amount by the outside length value obtained in Step 4. This figure is usually expressed as a percentage, so the above figure should be multiplied by 100.

7.8.2 **Percentage change in outside length.** This is obtained by dividing the difference between the initial Step 4 and final measurements of the outside length by the initial outside length, Step 6.

Comment: The difference between the two values in paragraphs 7.8.1 and 7.8.2 reflects a combination of a belt side wear and seating.

7.8.3 **Hours to failure**

SECTION 8—INTERPRETATION OF TEST RESULTS

8.1 The quality requirements shall be measured primarily by the number of hours to failure, in accordance with agreement between the user and the manufacturer, when tested under the conditions specified.

8.2 Requirements as to elongation of a belt shall be as agreed upon between the user and the manufacturer.

8.3 As a minimum for qualification testing, four belts of a sample size are recommended. The average hours to failure of the belts tested is to be equal to or larger than the number of hours agreed upon by user and manufacturer.

8.4 If the minimum hours to failure of any one of the four belts tested is less than ½ the agreed required average, it may be necessary to test four additional belts. In such an event, not more than one of the total of eight belts tested should have a total running time to failure of less than ½ of the agreed required average.

TABLE 1—V-BELT SHEAVE DIMENSIONS
(Refer to Fig. 2)

Belt Section	Outside Dia ±0.005 in. ±0.13 mm		A ±0° 20'	W* (ref)		G (min)		Ball or Rod Dia d		Dia Over Balls or Rod ±0.005 in. ±0.13 mm	
	in.	mm	deg	in.	mm	in.	mm	in.	mm	in.	mm
HA	3.250	82.55	34	0.494	12.55	0.490	12.45	0.4375	11.112	3.568	90.63
HB	4.500	114.30	32	0.630	16.00	0.580	14.73	0.5625	14.288	4.906	124.61
HC	6.800	172.72	34	0.879	22.33	0.780	19.81	0.7812	19.844	7.378	187.40
HD	10.600	269.24	34	1.259	31.98	1.050	26.67	1.1250	28.575	11.455	290.95
HI	5.550	140.97	26	1.058	26.85	0.906	23.01	0.9531	24.209	6.157	156.39
HJ	6.620	168.16	26	1.308	33.22	1.125	28.58	1.1875	30.162	7.421	188.49
HK	7.700	195.58	26	1.558	39.57	1.375	34.92	1.4375	36.512	8.779	222.99
HL	8.520	216.41	26	1.808	45.92	1.500	38.10	1.6875	42.862	9.878	250.90
HM	9.350	237.49	26	2.058	52.27	1.750	44.45	1.9036	48.419	10.816	274.72

*The sheave groove dimensions have been determined so that an adjustable speed belt of nominal dimensions will ride down in the sheave 0.12 in. (3.2 mm), and conventional cross section belts of nominal dimensions will ride flush with O.D.

FIG. 2—SHEAVE DIMENSIONS

TABLE 2—FACTORS TO BE CONTROLLED IN V-BELT TESTING

Belt Section	HP Load at DriveN	Total Belt Tension		Approx. Belt Velocity		DriveR
		lb	kg	ft per min	m per min	rpm
HA	2.75	100	45	1375	419	1750
HB	7.00	180	82	1950	594	1750
HC	16.5	280	127	2930	893	1750
HD	40.0	430	195	4580	1396	1750
HI	4.6	100	45	2290	698	1750
HJ	13.2	240	109	2750	838	1750
HK	23.3	360	118	3210	978	1750
HL	32.2	460	209	3550	1082	1750
HM	41.0	520	236	3895	1187	1750

Note:
The values of horsepower and total belt tension shown in Table 2 have been selected as approximate guides for practical laboratory operation in the range of 100 to 200 hours duration at tight-side/slack-side tension ratio of 5 for average length belts. Experience with successfully field tested belts may show that changes in the horsepower and tension values (keeping the same tight-side/slack-side ratio of 5) may be necessary to achieve this practical laboratory operation.

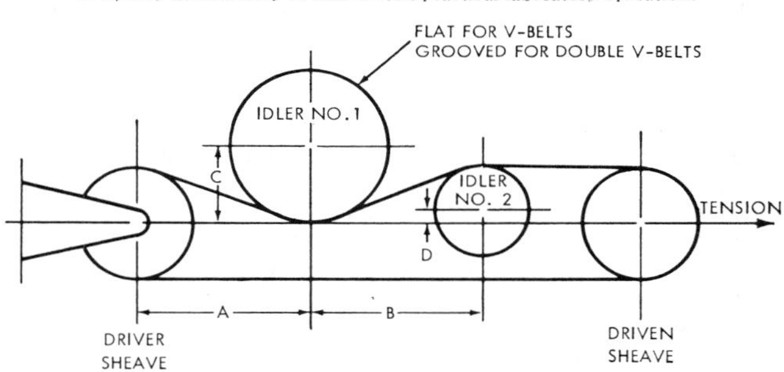

FIG. 3—REVERSE BEND TEST EQUIPMENT

TABLE 3—FACTORS FOR TESTS WITH REVERSE BEND FOR V- AND DOUBLE V-BELTS. (See Fig. 3)

Belt Section	DriveR and DriveN		Idler #1 Flat Idler (V-belts)		Idler #1 Ground Idler (double V-belts)		Idler #2 Grooved		Approx. Belt Velocity		DriveR
	Effective Outside Diameters*										
	in.	mm	in.	mm	in.	mm	in.	mm	ft per min	m per min	rpm
HA HAA	3.250	82.55	4.25	108.0	2.75	69.8	2.75	69.8	1375	419	1750
HB HBB	4.500	114.30	6.00	152.4	4.00	101.6	4.00	101.6	1950	594	1750
HC HCC	6.800	172.72	8.50	215.9	5.75	146.0	5.75	146.0	2930	893	1750
HD HDD	10.600	269.24	13.50	342.9	9.00	228.6	9.00	228.6	4580	1396	1750

Fixed Center Distance For Idlers

Belt Section	A		B		Flat Idlers (V-Belts) C		Grooved Idler (Double V-Belts) C		D	
	in.	mm	in.	mm	in.	mm	in.	mm	in.	mm
HA HAA	3.88	98.4	3.88	98.4	1.75	44.4	1.00	25.4	0.25	6.4
HB HBB	5.50	139.7	5.50	139.7	2.38	60.3	1.38	34.9	0.25	6.4
HC HCC	8.75	222.2	8.75	222.2	3.00	76.2	1.62	41.3	0.50	12.7
HD	14.00	355.6	14.00	355.6	4.50	114.3	2.25	57.2	0.75	19.0

*Groove dimensions and tolerances are to be in accordance with Table 1.
Note: Values of horsepower and total belt tension are to be determined by agreement between user and supplier to result in 100 to 200 hours test duration and are to be in such relationship to provide a tension ratio of 5 between the parallel tight and slack strand tensions at the driven sheave.

ASAE Standard: ASAE S216

SELF-POWERED ELECTRIC WARNING LIGHTS

Proposed by ASAE Farm Safety Committee; approved by ASAE Power and Machinery Division Technical Committee; reconfirmed without technical change and with editorial revision December 1965, December 1970; reconfirmed December 1975, December 1978, December 1979, December 1984.

SECTION 1—DEFINITION

1.1 A self-powered electric warning light is a self-powered device capable of providing and displaying a warning light, either flashed or steady burning, indicating to the driver of an approaching motor vehicle that a hazard is present during other than daylight hours.

SECTION 2—GENERAL REQUIREMENTS

2.1 A self-powered electric warning light shall show red to the rear and at a 90-deg angle to the axis at the lamp. If visible from the front, it shall show amber. Preferably, it shall be not less than 4 in. in diameter.

2.2 The light mounting shall be adaptable to the bracket as specified in ASAE Standard ASAE S277, Mounting Brackets and Socket for Warning Lamp and Slow-Moving Vehicle (SMV) Identification Emblem.

2.3 The device shall be furnished for test with the size and type of battery as specified by the manufacturer, and the battery shall be representative of standard batteries in regular production.

SECTION 3—TEST PROCEDURES AND PERFORMANCE REQUIREMENTS

3.1 Samples for test. Sample lamps submitted for laboratory test shall be representative of the devices as regularly manufactured and marketed. Where necessary each sample shall include all accessory equipment peculiar to the device and necessary to operate it in normal manner. Each sample shall also be mounted in its normal operating position on a supporting bracket designed to be bolted rigidly to the vibration rack. Dust and photometric test may be made on a second set of unmounted samples, if desired, to expedite completion of tests.

3.2 Lamp bulbs. Lamp bulbs shall be supplied by the device manufacturer to provide a complete unit, and shall be representative of standard bulbs in regular production. They shall be selected for accuracy in accordance with specifications approved by the National Bureau of Standards and shall be operated at their rated mean spherical candlepower except as otherwise specified during the tests.

3.3 Laboratory facilities. The laboratory shall be equipped with all facilities necessary to make accurate physical and optical tests in accordance with established laboratory practices.

3.4 Vibration test. A sample unit, as mounted on the support supplied shall be bolted to the anvil end of the table of the vibration test machine and vibrated approximately 750 cpm through a distance of 1/8 in. The table shall be spring mounted at one end and fitted with steel calks on the under side of the other end. These calks are to make contact with the steel anvil once during each cycle at the completion of the fall. The rack shall be operated under a spring tension of 60 to 70 lb. This test shall be continued for 1 hr. The unit shall then be examined. Any unit showing evidence of material physical weakness, lens or reflector rotation, displacement or rupture of parts shall be considered to have failed, provided that rotation of lens or reflector shall not be considered as a failure when tests show compliance with specifications despite such rotation. (See Society of Automotive Engineers Standard, SAE J577, Vibration Test Machine.)

3.5 Moisture test. A sample unit shall be mounted in its normal operation position with all drain holes open and subjected to a precipitation of 0.1 in. of water per minute, delivered at an angle of 45 deg from a nozzle with a solid cone spray. During the moisture test the lamp shall revolve about its vertical axis at a rate of 4 rpm. This test shall be continued for 12 hr. The water shall then be turned off and the unit permitted to drain for 1 hr. The unit shall then be examined. Moisture accumulation in excess of 2 cc shall constitute a failure.

3.6 Dust Test. A sample unit with any drain hole closed shall be mounted in its normal operating position, at least 6 in. from the wall in a box measuring 3 ft in all directions, containing 10 lb of fine powdered cement in accordance with American Society for Testing and Materials Standard, ASTM C 150, Specification for Portland Cement. At intervals of 15 minutes, this dust shall be agitated by compressed air or fan blower by projecting blasts of air for a 2-sec period in a downward direction into the dust in such a way that the dust is completely and uniformly diffused throughout the entire cube. The dust is then allowed to settle. This test shall be continued for 5 hr. After the dust test the exterior surface shall be cleaned; and if the maximum candlepower is within 10 percent of the maximum as compared with the condition after the unit is cleaned inside and out, it shall be considered adequately dusttight. Where sealed units are used, the dust test shall not be required.

3.7 Corrosion test. A sample unit shall be subjected to a salt spray (fog) test in accordance with ASTM Standard ASTM B117, Salt Spray Testing, for a period of 50 hr, consisting of two periods of 24 hr exposure and 1 hr drying time each. There shall be no evidence of excessive corrosion immediately after the above test has been completed, which would affect the proper functions of the device.

3.8 Color test

3.8.1 Red—Transmitting media which are to be classed as red must conform to the specifications herein and may be examined for compliance with the requirements by comparison with a red limit glass. This glass has been prepared so that it represents the closest permissible approach to amber (yellow) and has at the same time the maximum purity of color. A red transmitting medium shall not be acceptable if it is paler or yellower than the light limit standard glass when the two are illuminated by incadescent lamp light. In case of doubt or samples close to the limit, resort must be had to spectrophotometric determinations of the color of the sample.

3.8.2 Amber (Yellow)—Transmitting media which are to be classed as amber (yellow) must conform to the specifications herein and may be examined for compliance with the requirements by comparison with amber (yellow) limit glasses. These glasses have been prepared so that one represents the closest permissible approach to red and the other the closest permissible approach to yellow-green, both having at the same time the maximum purity of color. An amber (yellow) transmitting medium shall not be acceptable if it is paler or greener than the light limit standard or redder than the dark limit standard when the lens and the standards are illuminated by incandescent lamp light. In case of doubt or samples close to the limit, resort must be had to spectrophotometric determinations of the color of the sample.

3.9 Warpage test. A sample unit shall be mounted in its normal position and operated at rated voltage in an oven for 1 hr at 120 deg F ambient temperature. The device should be operating in the test in the same manner as it will be operated in service. The lens color shall be identical to that intended for use in the device. After this warpage test has been completed, there shall be no evidence of warpage of lenses which would affect the proper functioning of the device.

3.10 Reliability test. In the case of lanterns which can be turned on or off at will, a sample unit shall be set up in complete form and operated for 1,000 cycles, using the operating unit or switch submitted with the device as a part thereof. This test shall be made at a rate not to exceed 50 times per minute. In the case of flashing units, the rate shall be slow enough to permit the unit to flash at least twice for each operation of the switch. When this test is completed, the operating unit shall show no evidence of material physical weakness, excessive wear, or high resistance. The lantern shall be turned on for a period of 100 hr. During this test, the "on" period for the flasher, if one is provided, shall be long enough at all times to permit the source to come up to required visibility or full brightness. The rate of flashing during the test shall not be less than 60 nor more than 150 cpm. If during these tests the flasher (if one is provided) continues to function as specified, and if after the tests the lantern with the battery used in the tests meets all optical requirements, the lantern shall be considered satisfactory from the reliability standpoint.

3.11 Photometric test. All beam candlepower measurements shall be made with a bar photometer or equivalent with the center of the light at a distance of at least 4 ft and preferably, 10 ft from the photometer screen. In measuring distances and angles the incandescent filaments shall be taken as the center of light. The lamp axis shall be taken as the horizontal line through the light source and parallel what would be the longitudial axis of the vehicle if the lamp was mounted in its normal position on the vehicle.

3.12 Candlepower requirements. The photometric minimum candlepower requirements for self-powered electric warning lights are given in Table 1. The minimum photometric requirements for an amber lens shall be the same as specified for the red lens.

Cited Standards:

ASAE S277, Mounting Brackets and Socket for Warning Lamp and Slow-Moving Vehicle (SMV) Identification Emblem
ASTM B117, Salt Spray Testing
ASTM C150, Specification for Portland Cement
SAE J577, Vibration Test Machine

TABLE 1—CANDLEPOWER

Test points, degrees		Red and Amber
10U	5 L	0.20
	V	0.50
	5 R	0.20
5U	20 L	0.20
	10 L	0.50
	5 L	1.00
	V	1.50
	5 R	1.00
	10 R	0.50
	20 R	0.20
H	20 L	0.30
	10 L	0.70
	5 L	1.5
	V	5.00
	5 R	1.5
	10 R	0.70
	20 R	0.30
5D	20 L	0.20
	10 L	0.50
	V	1.50
	10 R	0.50
	20 R	0.20
10D	5 L	0.20
	V	0.50
	5 R	0.20

ASAE Standard: ASAE S217.10 (ANSI/SAE J715 SEP83/ASAE S217.10)

THREE-POINT FREE-LINK ATTACHMENT FOR HITCHING IMPLEMENTS TO AGRICULTURAL WHEEL TRACTORS

Proposed by Advisory Engineering Committee of Farm and Industrial Equipment Institute; adopted by ASAE March 1959; revised 1961, 1962, 1963, June 1964, December 1964, December 1966, March 1971, February 1972, April 1974, February 1975; corrected editorially April 1977; reconfirmed December 1979, December 1984.

SECTION 1—SCOPE

1.1 This Standard sets forth requirements for the attachment of three-point hitch implements or equipment to the rear of agricultural wheel tractors by means of a three-point free-link in association with a power lift.

1.2 In order to assure proper performance of certain implements, standard dimensions for mast height, mast pitch adjustment and implement leveling adjustment are included. Location of link-attachment points is not restricted and is, therefore, left to the discretion of the tractor designer.

1.3 If draft links are used for trailing power take-off implements, a means shall be included for locking the draft links in a fixed position, and a drawbar hitch point shall be positioned in conformance with power take-off standards.

1.4 Dimensions comprising the standard specifications are divided into four categories:

Category	Maximum Drawbar Power* kW
I	15 to 35 (20 to 45 hp)
II	30 to 75 (40 to 100 hp)
III and III-N†	60 to 168 (80 to 225 hp)
IV and IV-N†	135 to 300 (180 to 400 hp)

*Based on ASAE Standard ASAE S209, Agricultural Tractor Test Code

†Refer to Section 3, Special Hitch Categories.

SECTION 2—DEFINITION OF TERMS

2.1 **Linkage:** The combination of one upper link and two lower links, each articulated to the tractor and the implement at opposite ends in order to connect the implement to the tractor.

2.2 **Upper link, lower link:** Elements in the linkage.

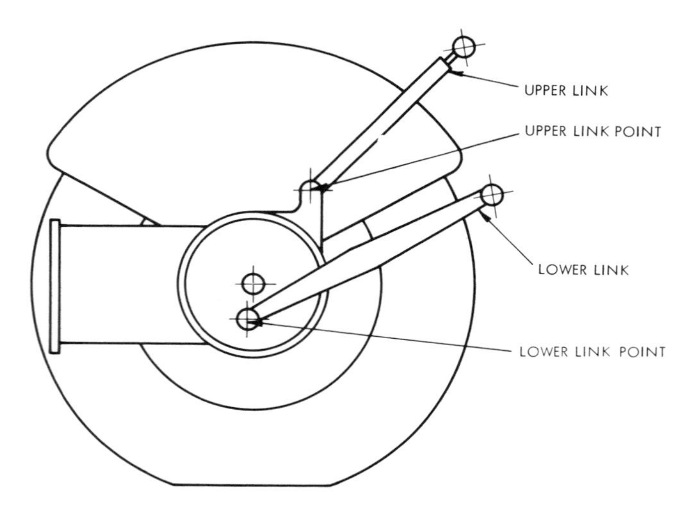

FIG. 1—TRACTOR LINKAGE

2.3 **Hitch point:** The articulated connection between a link and the implement. For geometrical analysis, the hitch point is established as the center of the articulated connection between a link and the implement.

2.4 **Link point:** The articulated connection between a link and the tractor. For geometrical analysis, the link point is established as the center of the articulated connection between a link and the tractor.

2.5 **Upper hitch point:** The articulated connection between the upper link and the implement.

2.6 **Upper link point:** The articulated connection between the upper link and the tractor.

2.7 **Lower hitch point:** The articulated connection between a lower link and the implement.

2.8 **Lower link point:** The articulated connection between a lower link and the tractor.

2.9 **Upper hitch pin:** The pin that connects the upper link to the implement.

2.10 **Upper link pin:** The pin that connects the upper link to the tractor.

2.11 **Lower hitch stud or pin:** The stud or pin, attached to the implement, on which a lower link is secured.

2.12 **Linchpin:** The retaining pin used in the hitch pins or studs.

2.13 **Mast:** The member that provides attachment of the upper link to the implement.

2.14 **Mast height:** The perpendicular distance between the upper hitch point and common axis of the lower hitch points.

2.15 **Mast adjustment:** The usable range of movement of the mast in a vertical plane. It is measured as the maximum and minimum heights of the lower hitch points above the ground between which a mast of standard height can be adjusted to any inclination between the vertical and five degrees from the vertical towards the rear. Adjustment of the mast controls the pitch of the implement. Specifying the mast adjustment to be provided enables the tractor designer to determine the minimum acceptable adjustment of the length of the top link in relation to the point of attachment of the linkage; it also permits the implement designer to determine the range of operating depths of the implement over which pitch adjustment can be obtained.

2.16 **Leveling adjustment:** The adjustment of the lower links so that one lower hitch point may be moved vertically with respect to the other lower hitch point to provide an inclination of the implement.

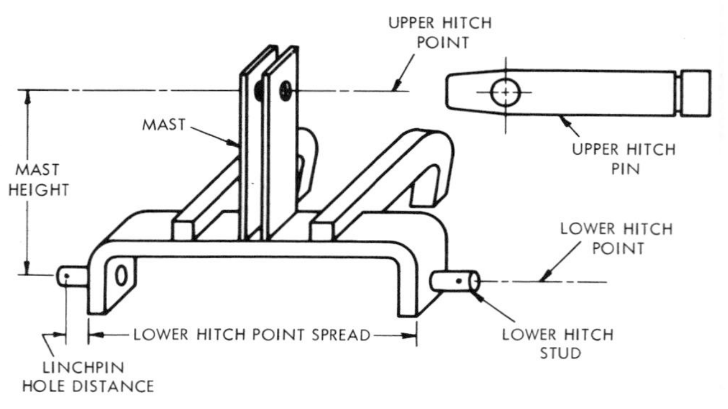

FIG. 2—DIMENSIONS ASSOCIATED WITH IMPLEMENT

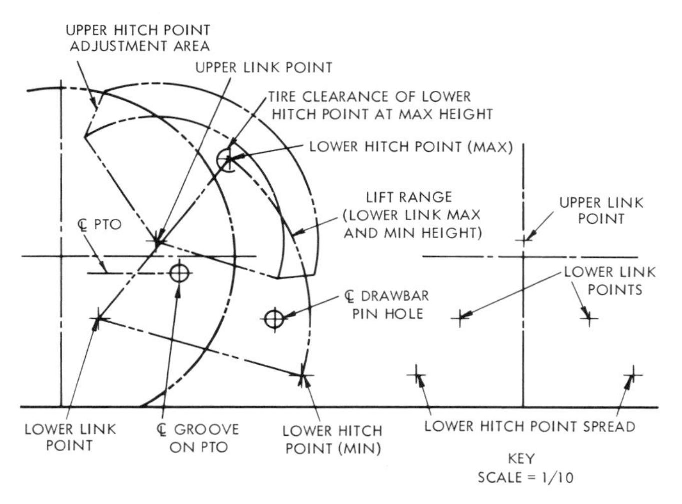

FIG. 3—DIMENSIONS ASSOCIATED WITH TRACTOR

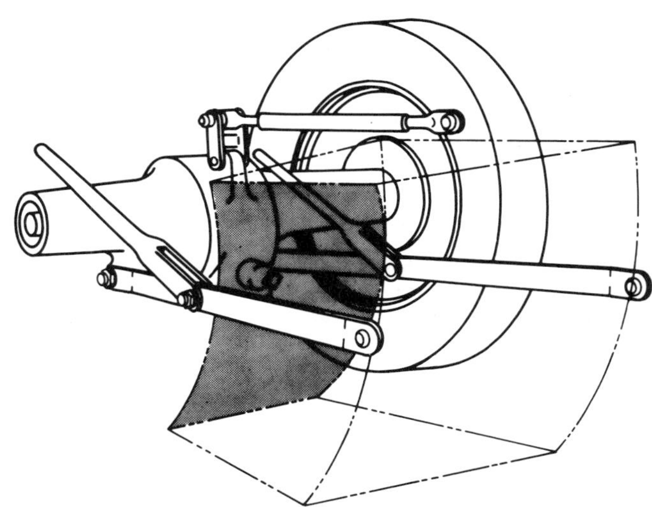

FIG. 4—LOWER HITCH-POINT TRACTOR CLEARANCE

2.17 Lower hitch-point spread: The distance between lower hitch points measured at the base of the lower hitch stud, or the distance between the inner most restraining means provided on the implement.

2.18 Linchpin hole distance: The distance between the linchpin hole centerline and the lower link stud base.

2.19 Lift linkage: The connecting linkage that transmits force to the lower links for raising and lowering.

2.20 Lift range: The range of movement of the lower hitch points utilizing the extent of manual adjustment provided in the lift linkage in conjunction with the power range, expressed as the maximum and minimum possible heights of the lower hitch points above ground level, the lower hitch point axis being maintained horizontal to the ground.

2.21 Power range: The total vertical movement of the lower hitch point excluding any adjustment in the linkage or lift linkage.

2.22 Lower hitch-point tire clearance: Clearance expressed as a radial dimension from the lower hitch point to the outside diameter of the tire with the implement in raised position and all side sway removed from the links.

2.23 Lower hitch-point tractor clearance: The horizontal dimension between the rearmost parts of the tractor in the area between the two draft links and the horizontal line through the two lower hitch points throughout the range of vertical movement of the hitch points (See Figure 4). The power take-off master shield may be removed, if necessary to meet this dimension.

SECTION 3—SPECIAL HITCH CATEGORIES

3.1 Certain farming operations require dual rear wheels and narrow spacing. Experience has shown that some large implements used in these circumstances require a special narrow hitch.

3.1.1 Special Category III narrow hitch (III-N). This special hitch differs from the Category III hitch in only one dimension; namely, the lower hitch point spread is 822.5 mm (32.38 in.)/825.5 mm (32.50 in.) (same as Category II). All other dimensions for the Category III-N three point free-link hitch are the same as Category III in Table 1 and Table 2.

TABLE 1 – DIMENSIONS ASSOCIATED WITH IMPLEMENT

	Category I				Category II				Category III‖				Category IV‖			
	Millimeter		Inches		Millimeter		Inches		Millimeter		Inches		Millimeter		Inches	
	Min	Max	Min	Max	Min	Max	Min	Max	Min	Max	Min	Max	Min	Max	Min	Max
UPPER HITCH POINT																
Width inside	44.5	–	1.75	–	52.3	–	2.06	–	52.3	–	2.06	–	65	–	2.56	–
Width outside	–	85.9	–	3.38	–	95.3	–	3.75	–	95.3	–	3.75	–	132	–	5.20
Clearance radius for upper link* (†)	57.2	–	2.25	–	57.2	–	2.25	–	57.2	–	2.25	–	76.2	–	3.00	–
Hitch pin hole diameter	19.3	19.56	0.76	0.77	26.65	25.91	1.01	1.02	32.0	32.26	1.26	1.27	45.2	45.5	1.78	1.79
LOWER HITCH POINT																
Stud diameter	21.84	22.10	0.86	0.87	28.19	28.45	1.11	1.12	36.32	36.58	1.43	1.44	49.7	50.8	1.96	2.00
Linchpin hole distance†	38.86	–	1.53	–	48.52	–	1.91	–	48.52	–	1.91	–	68	–	2.68	–
Linchpin hole diameter	11.68	12.19	0.46	0.48	11.68	12.19	0.46	0.48	11.68	12.19	0.46	0.48	17.5	18	0.69	0.71
Lower hitch point spread	681.0	684.3	26.81	26.94	822.5	825.5	32.38	32.50	963.7	966.7	37.94	38.06	1165	1168	45.87	45.99
Clearance radius for lower link* (†)	63.5	–	2.50	–	73.2	–	2.88	–	82.6	–	3.25	–	82.6	–	3.25	–
Implement encroachment in front of lower hitch point if implement extends laterally behind tire	–	12.7	–	0.5	–	12.7	–	0.5	–	12.7	–	0.5	–	12.7	–	0.5
IMPLEMENT MAST HEIGHT ‡ §	457		18		483		19		559		22		686		27	

* Some tractors with quick-attachable connectors require 140 mm (5.50 in.) space for clearance above the upper hitch point and below the lower hitch points.

† Refer to standard for attachment of implements to agricultural wheel tractors equipped with quick-attaching coupler for three-point free link hitch.

Tractors equipped with a standard quick-attaching coupler for the three-point free-link hitch require an auxiliary attaching pin on the implement mast located 76 mm (3.0 in.) for categories I and III, and 102 mm (4.0 in.) for Category II, below the standard upper hitch point. To facilitate attachment and detachment of the implement, a clearance zone must be maintained 76 mm (3.0 in.) rearward from and extending 104 mm (4.10 in.) above and 216 mm (8.50 in.) below this pin. In addition, a clearance zone must be maintained 94 mm (3.70 in.) rearward from and extending 25 mm (1.0 in.) above and 211 mm (8.30 in.) below each lower hitch point.

To facilitate the attachment and detachment of the implement with tractors equipped with a standard Category IV-N or IV quick-attaching coupler for the three-point free-link hitch, a clearance zone must be maintained 85 mm (3.35 in.) rearward from and extending 120 mm (4.72 in.) above and 252 mm (9.92 in.) below the standard upper hitch point on the implement mast. In addition, a clearance zone must be maintained 94 mm (3.7 in.) rearward from and extending 32 mm (1.26 in.) above and 272 mm (10.71 in.) below each lower hitch point.

‡ The mast height is not necessarily a mechanical dimension on the implement itself. It is a figure used in design and if properly used for design of both implement and tractor, a well performing interchangeable implement and tractor combination will be achieved. This standard makes it possible to produce tractors and implements that will give good performance in any combination; therefore, consideration to hitch geometry is essential. This makes it desirable to establish a standard mast height and a standard mast adjustment within a working range, because these items influence the position of hitch points that are common to both the implement and the tractor.

Mast height is one of the essential factors in establishing the virtual hitch point of the free-link system, draft signal for the draft-responsive system, loads on the linkage and hitch points, changes in implement pitch corresponding to changes in working depth, implement pitch when the implement is in transport position, clearance of the implement with the tractor, especially in transport position and clearance of the hitch links with the implement or with the tractor, especially in the transport position.

When an implement mast height is made different than standard to accomplish some specific performance feature, care should be exercised to insure that the desired performance is secured with tractors likely to operate the implement.

§ Some Category II tractors are designed to accommodate a 559 mm (22 in.) mast height for optimum performance. In the design of implements for use on these tractors care should be taken to investigate the need for providing a 559 mm (22 in.) mast height. Tractors that are designed only for use with 559 mm (22 in.) mast height must be properly identified.

‖ See Section 3 for Category IV-N or III-N, Narrow Hitch, dimensions.

3.1.2 Special Category IV narrow hitch (IV-N). This special hitch differs from the Category IV hitch in the following dimensions: Lower hitch point spread is 919 mm (36.18 in.)/922 mm (36.30 in.). Leveling adjustment is same as Category III. All other dimensions for the Category IV-N three-point free-link hitch are the same as Category IV in Table 1 and Table 2.

SECTION 4—TRACTOR LIFT FORCE CAPACITY

4.1 Tractors shall have the following minimum lift force available throughout the power range, at a distance of 610 mm (24 in.) to the rear of the lower hitch points, when tested in accordance with ASAE Standard ASAE S349, Test Procedure for Measuring Hydraulic Lift Force Capacity on Agricultural Tractors Equipped with Three-Point Hitch.

4.1.1 Through 65 kW maximum drawbar power: 310 N per kW drawbar power.

4.1.2 Above 65 kW maximum drawbar power: 20.15 kN plus 155 N per kW drawbar power above 65 kW drawbar power.

4.2 In customary units, the expressions for 4.1.1 and 4.1.2 are:

4.2.2 Through 85 maximum drawbar horsepower: 52 lb per drawbar horsepower.

4.2.3 Above 85 maximum drawbar horsepower: 4420 lb plus 26 lb per drawbar horsepower above 85 drawbar horsepower.

Cited Standard:

ASAE S349, Test Procedure for Measuring Hydraulic Lift Force Capacity on Agricultural Tractors Equipped with Three-Point Hitch

TABLE 2 – DIMENSIONS ASSOCIATED WITH TRACTOR

	Category I				Category II				Category III‖				Category IV‖			
	Millimeter		Inches		Millimeter		Inches		Millimeter		Inches		Millimeter		Inches	
	Min	Max	Min	Max	Min	Max	Min	Max	Min	Max	Min	Max	Min	Max	Min	Max
UPPER LINK																
Width at hitch point	–	43.9	–	1.73	–	51.1	–	2.01	–	51.1	–	2.01	–	64	–	2.52
Radius at hitch point	–	50.8	–	2.00	–	50.8	–	2.00	–	50.8	–	2.00	–	64	–	2.52
Hitch pin hole diameter	19.3	19.56	0.76	0.77	25.6	25.91	1.01	1.02	32.0	32.26	1.26	1.27	45.2	45.5	1.78	1.79
Side sway at hitch point	(See footnote*)				(See footnote*)				(See footnote*)				(See footnote*)			
UPPER HITCH PIN																
Diameter	18.8	19.05	0.74	0.75	25.15	25.40	0.99	1.00	31.50	31.75	1.24	1.25	44.2	45	1.74	1.77
Distance from head to centerline of linchpin hole	92.5	–	3.64	–	101.6	–	4.00	–	101.6	–	4.00	–	140	–	5.51	–
Linchpin hole diameter	11.68	12.19	0.46	0.48	11.68	12.19	0.46	0.48	11.68	12.19	0.46	0.48	17.5	18	0.69	0.71
LOWER LINK																
Width at hitch point	34.80	35.05	1.37	1.38	34.80	44.7	1.37	1.76	44.2	44.7	1.74	1.76	57	57.5	2.24	2.26
Radius at hitch point	–	44.5	–	1.75	–	66.5	–	2.62	–	76.2	–	3.00	–	76.2	–	3.00
Stud hole diameter	22.35	22.61	0.88	0.89	28.70	28.96	1.13	1.14	36.83	37.08	1.45	1.46	51	51.5	2.01	2.03
Lower hitch point tire clearance with largest Code R-1 tire offered†	76.2	–	3	–	76.2	–	3	–	76.2	–	3	–	76.2	–	3	–
Lower hitch point tractor clearance	457	–	18	–	457	–	18	–	457	–	18	–	457	–	18	–
Side sway at hitch point each side of center position with draft links horizontal ‡	102	–	4	–	127	–	5	–	127	–	5	–	127	–	5	–
Horizontal distance from end of PTO shaft to lower hitch points with draft links horizontal	508	559	20	22	508	559	20	22	508	559	20	22	508	559	20	22
	(See footnote §)				(See footnote §)				(See footnote §)				(See footnote §)			
LIFT RANGE, POWER RANGE, ADJUSTMENTS																
Lift range																
Max height for lower position	–	203	–	8	–	203	–	8	–	203	–	8	–	203	–	8
Min height for highest position	813	–	32	–	914	–	36	–	1016	–	40	–	1120	–	44	–
Power range	559	–	22	–	610	–	24	–	660	–	26	–	762	–	30	–
Leveling adjustment																
Higher	102	–	4.0	–	114	–	4.5	–	127	–	5.0	–	150	–	5.90	–
Lower	102	–	4.0	–	114	–	4.5	–	127	–	5.0	–	150	–	5.90	–
Mast adjustment																
Min height for highest position	508	–	20	–	610	–	24	–	660	–	26	–	710	–	27.95	–
Max height for lowest position	–	254	–	10	–	254	–	10	–	254	–	10	–	254	–	10
TRACTOR LIFT FORCE CAPACITY (See Section 4)																

* Side sway of the upper link must be compatible with that provided at the lower links plus necessary additional allowance for lateral leveling adjustment.

† Code R-1 (regular agricultural) tire as specified in ASAE Standard ASAE S220, Tire Selection Tables for Agricultural Machines of Future Design.

‡ Means should be provided to lock the draft links in a rigid lateral position for power take-off, for other operations where side sway cannot be tolerated and when the hitch is raised to the transport position. No maximum dimensions for side sway are specified: this must be limited in each individual application so that hitch or implement components will not come in contact with the tractor tires.

§ Dimensions pertain to 1-3/8 and 1-3/4 in. diameter, 540 RPM power take-off shafts and 1-3/8 in. diameter, 1000 RPM power take-off shafts. On some previously designed models of tractors the minimum horizontal distance for Category I and Category II is 457 mm (18 in.). Dimensions shown should be increased by 102 mm (4.00 in.) on tractors equipped with 1-3/4 in. diameter 1000 RPM power take-off shaft.

‖ See Section 3 for Category III-N or IV-N, Narrow Hitch, dimensions.

ASAE Standard: ASAE S218.2 (SAE J714b)

WHEEL MOUNTING ELEMENTS FOR AGRICULTURAL EQUIPMENT DISC WHEELS

Developed by the Society of Automotive Engineers; adopted by ASAE as a Recommendation March 1955; reconfirmed December 1961; revised December 1965; reconfirmed December 1970; reconfirmed December 1975; revised and reclassified as a Standard March 1978; reconfirmed December 1982, December 1987.

SECTION 1—PURPOSE AND SCOPE

1.1 This Standard includes all wheel mounting elements subject to standardization in a series of agricultural equipment disc wheels and a variety of offsets.

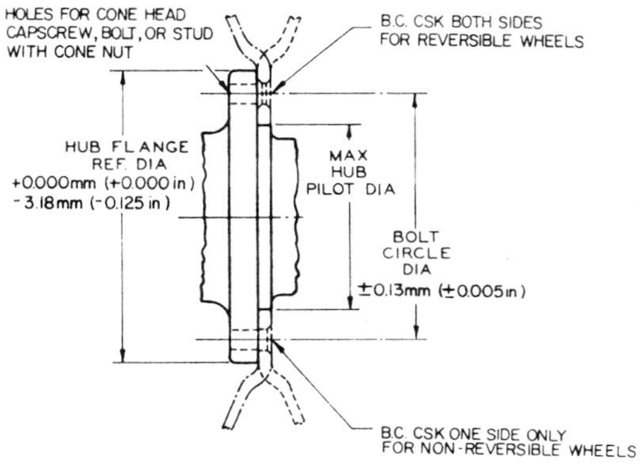

HOLES FOR CONE HEAD CAPSCREW, BOLT, OR STUD WITH CONE NUT

B.C. CSK BOTH SIDES FOR REVERSIBLE WHEELS

HUB FLANGE REF. DIA +0.000mm (+0.000 in) -3.18mm (-0.125 in)

MAX HUB PILOT DIA

BOLT CIRCLE DIA ±0.13mm (±0.005in)

B.C. CSK ONE SIDE ONLY FOR NON-REVERSIBLE WHEELS

FIG. 1—HUB FLANGE FOR CIRCLE FARM EQUIPMENT DISC WHEELS

1.2 The disc may be reversible or nonreversible and concave or convex offset.

TABLE 1—DIMENSIONS, mm (in.)

No. Bolts	Bolt Circle	Hub Pilot Max. Dia.	Attaching Hub Bolt Dia.	B.C. Csk.	Hub Flange Dia. Ref.
4	127.0 (5.000)	92.1 (3.625)	12.00 (0.50)	90°	160.3 (6.31)
5	114.3 (4.500)	79.3 (3.125)	11.00 (0.44)	75°	144.3 (5.68)
5	139.7 (5.500)	101.6 (4.000)	12.00* (0.50)*	90°	177.8 (7.00)
6	152.4 (6.000)	117.4 (4.625)	12.00* (0.50)*	90°	185.7 (7.31)
8	203.2 (8.000)	152.4 (6.00)	14.00 or 16.00 (0.56) or (0.62)	90°	242.8 (9.56)
8	275.0 (10.826)	220.0 (8.70)	18.00 (0.75)	90°	317.5 (12.50)
10	335.0 (13.187)	280.0 (11.02)	18.00 or 22.00 (0.75) or (0.88)	90°	385.0 (15.16)
12	422.3 (16.625)	368.3 (14.50)	18.00 or 22.00 (0.75) or (0.88)	90°	476.25 (18.75)

*Bolt size may be increased to 14.00 (0.56) only if cone head bolt is used.

AGRICULTURAL TRACTOR AND EQUIPMENT DISC WHEELS

Developed by the Society of Automotive Engineers; adopted by ASAE as a Recommendation December 1952; revised 1961; reconfirmed December 1965, December 1970, December 1975; revised and reclassified as a Standard March 1978; reconfirmed December 1982, December 1987.

The purpose of this Standard is to provide a selection of disc wheels for agricultural tractor and equipment use with a maximum of interchangeability.

This is accomplished by establishing 5 groups of disc wheels, in each of which the hub mounting elements are common. These groups are designated 4 bolt, 5 in. bolt circle; 5 bolt, 4.5 in. bolt circle; 5 bolt, 5.5 in. bolt circle; 6 bolt, 6 in. bolt circle; and 8 bolt, 8 in. bolt circle.

Further, this Standard establishes an SAE part number and the maximum rated radial load for each standard wheel. In addition, the Standard requires the wheel manufacturer's name or trademark to be impression stamped on the wheel with location at the discretion of the manufacturer.

4 BOLT, 5 IN. BOLT CIRCLE GROUP—This group is provided for light duty implement service. There are 4 mounting holes on a 5 in. (127 mm) diameter bolt circle.

There are 4 wheels in this group in 2 offsets and 4 rims of 14 in. and 15 in. diameter.

See Fig. 1 for SAE part numbers, dimensions and rated loads.

5 BOLT, 4.5 IN. BOLT CIRCLE GROUP—This group is provided for light duty implement, lawn and garden tractor, or front tractor wheel service.

There are 5 mounting holes on a 4.5 in. (114.3 mm) diameter bolt circle.

There are 6 wheels in this group in 1 offset and 5 rims of 12 in. diameter.

See Fig. 2 for SAE part numbers, dimensions and rated loads.

5 BOLT, 5.5 IN. BOLT CIRCLE GROUP—This group is provided for medium duty implement service. There are 5 mounting holes on a 5.5 in. (139.7 mm) diameter bolt circle.

There are 6 wheels in this group in 1 offset and 6 rims of 14 in. and 15 in. diameters.

See Fig. 3 for SAE part numbers, dimensions and rated loads.

6 BOLT, 6 IN. BOLT CIRCLE GROUP—This group is provided for general tractor and implement service. There are 6 mounting holes on a 6 in. (152.4 mm) diameter bolt circle.

There are 40 wheels in this group in 1 offset and 22 rims of 14, 15, 16, 16.1, 18 and 20 in. diameter.

See Fig. 4 for SAE part numbers, dimensions and rated loads.

8 BOLT, 8 IN. BOLT CIRCLE GROUP—This group is provided for heavy duty tractor and implement service. There are 8 mounting holes on an 8 in. (203.2 mm) diameter bolt circle.

There are 7 wheels in this group in 1 offset and 7 rims in 15, 16 and 16.1 in. diameter.

See Fig. 5 for SAE part numbers, dimensions and rated loads.

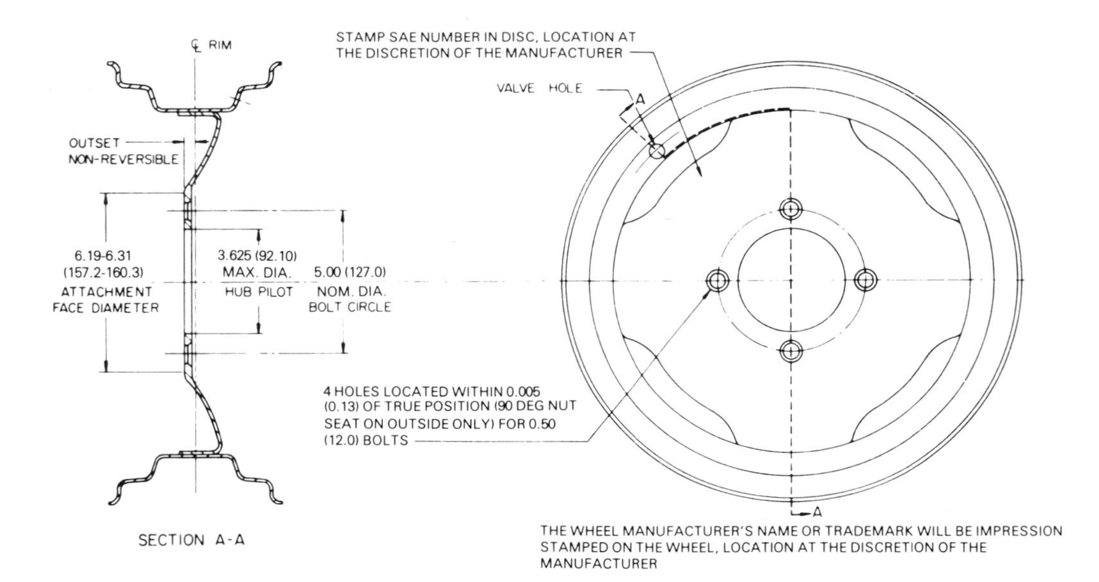

FIG. 1—4 BOLT, 5 IN. BOLT CIRCLE GROUP

SAE Wheel No.	Rim Size	Outset, mm (in.)	Max. Rated Radial Wheel Load*, kg (lb)
401	14 x 5 KB	10.0 (0.38)	570 (1250)
402	14 x 6 KB	10.0 (0.38)	570 (1250)
403	15 x 4 J	11.0 (0.44)	570 (1250)
404	15 x 4-1/2 KB or K	10.0 (0.38)	570 (1250)

*Determine wheel loads with machine at rest. This rating applies up to 20 mph (32 km/h) maximum travel speed.

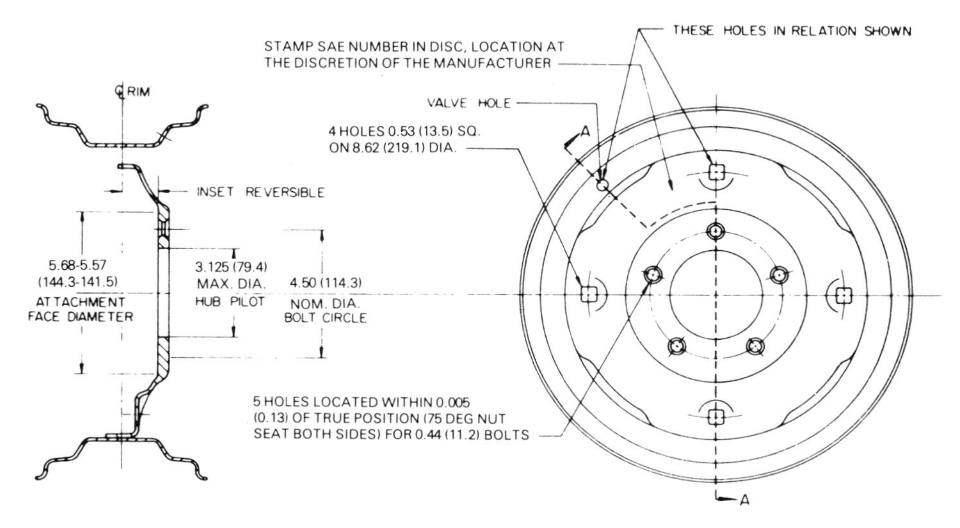

FIG. 2—5 BOLT, 4.5 IN. BOLT CIRCLE GROUP

SAE Wheel No.	Rim Size	Inset, mm (in.)	Max. Rated Radial Wheel Load*, kg (lb)
501	12 x 3.00 D	32.0 (1.25)	410 (900)
502	12 x 4 JA	32.0 (1.25)	410 (900)
503	12 x 5 JA	32.0 (1.25)	410 (900)
504	12 x 7 JA	32.0 (1.25)	230 (500)
505	12 x 7 JA	32.0 (1.25)	410 (900)
506	12 x 8-1/2 JA	32.0 (1.25)	410 (900)

*Determine wheel loads with machine at rest. This rating applies up to 20 mph (32 km/h) maximum travel speed.

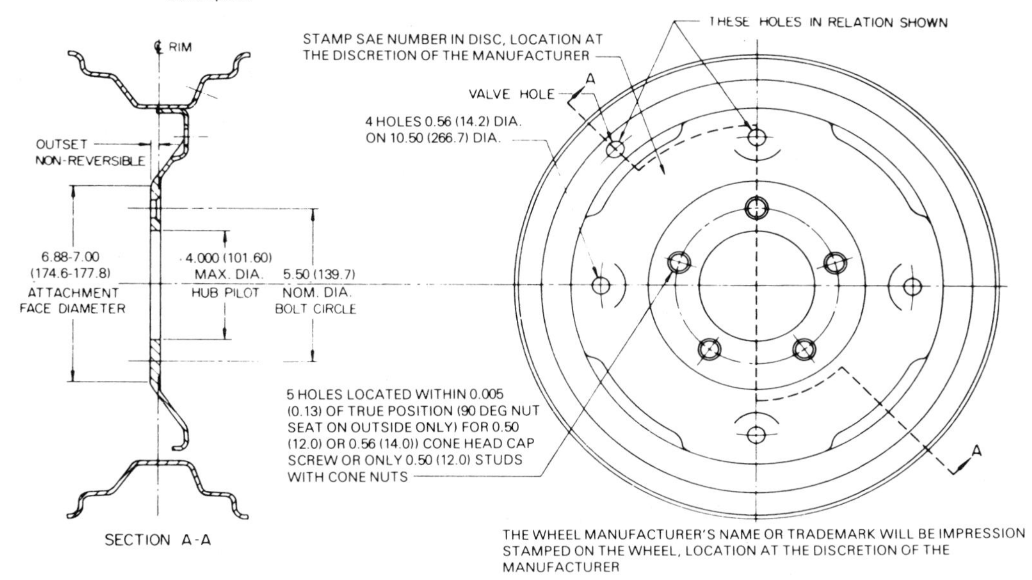

FIG. 3—5 BOLT, 5.5 IN. BOLT CIRCLE GROUP

SAE Wheel No.	Rim Size	Outset, mm (in.)	Max. Rated Radial Wheel Load*, kg (lb)
551	14 x 5 KB	6.0 (0.25)	820 (1800)
552	14 x 6 KB	6.0 (0.25)	820 (1800)
553	14 x 8 KB	6.0 (0.25)	820 (1800)
554	15 x 5 KB or K	6.0 (0.25)	820 (1800)
555	15 x 6 LB or L	6.0 (0.25)	820 (1800)
556	15 x 8 LB	6.0 (0.25)	820 (1800)

*Determine wheel loads with machine at rest. This rating applies up to 20 mph (32 km/h) maximum travel speed.

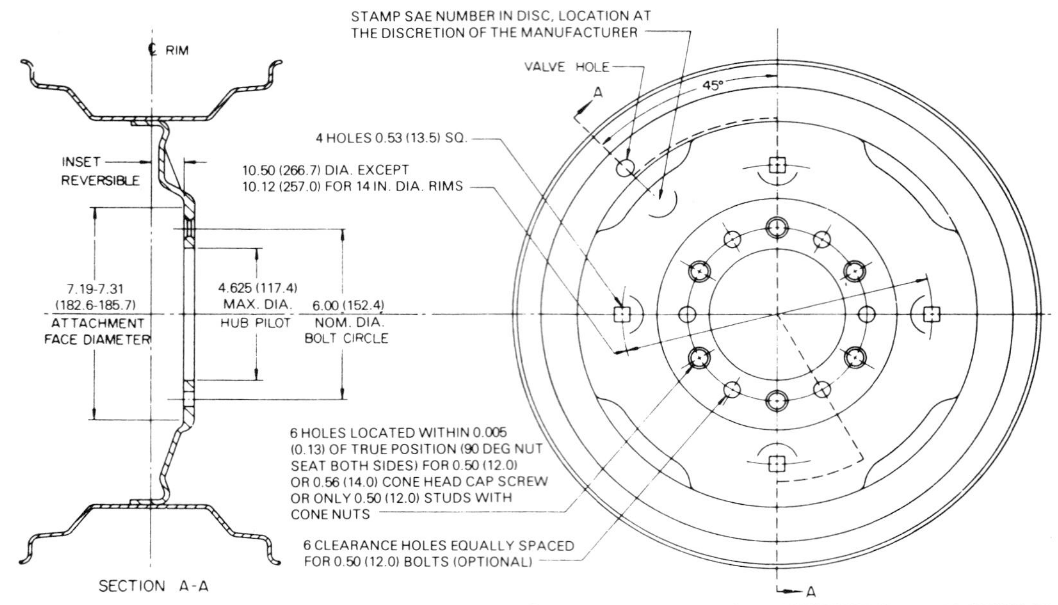

FIG. 4—6 BOLT, 6 IN. BOLT CIRCLE GROUP

SAE Wheel No.	Rim Size	Inset, mm (in.)	"Max. Rated" Radial Wheel Load*, kg (lb)
601	14 x 5 KB	28.0 (1.12)	820 (1800)
603	14 x 5 KB	28.0 (1.12)	1045 (2300)
605	14 x 6 KB	28.0 (1.12)	820 (1800)
607	14 x 6 KB	28.0 (1.12)	1270 (2800)
632	14 x 8 KB	28.0 (1.12)	1270 (2800)
609	15 x 3.00 D	28.0 (1.12)	455 (1000)
610	15 x 5 K or KB	28.0 (1.12)	820 (1800)
612	15 x 6 L or LB	28.0 (1.12)	820 (1800)
614	15 x 6 L or LB	28.0 (1.12)	1270 (2800)
616	15 x 6 L or LB	28.0 (1.12)	1500 (3300)
634	15 x 8 LB	28.0 (1.12)	1270 (2800)
635	15 x 8 LB	28.0 (1.12)	1590 (3500)
636	15 x W8L	28.0 (1.12)	1270 (2800)
637	15 x W8L	28.0 (1.12)	1590 (3500)
638	15 x 10 LB	28.0 (1.12)	1270 (2800)
639	15 x 10 LB	28.0 (1.12)	1590 (3500)
617	16 x 4.00 E	28.0 (1.12)	820 (1800)
619	16 x 4.50 E	28.0 (1.12)	820 (1800)
621	16 x 4.50 E	28.0 (1.12)	1045 (2300)
623	16 x 5.50 F	28.0 (1.12)	1270 (2800)
624	16 x 5.50 F	28.0 (1.12)	1500 (3300)
625	16 x 6 LB	28.0 (1.12)	1270 (2800)
627	16 x 8 LB	28.0 (1.12)	1270 (2800)
640	16 x W8L	28.0 (1.12)	820 (1800)
641	16 x W8L	28.0 (1.12)	1270 (2800)
642	16 x W8L	28.0 (1.12)	1590 (3500)
643	16 x 10 LB	28.0 (1.12)	1270 (2800)
644	16 x 10 LB	28.0 (1.12)	1590 (3500)
645	16 x W10L	28.0 (1.12)	1270 (2800)
646	16 x W10L	28.0 (1.12)	1590 (3500)
647	16.1 x W11C	28.0 (1.12)	1270 (2800)
648	16.1 x W11C	28.0 (1.12)	1590 (3500)
628	18 x 5.50 F	28.0 (1.12)	1180 (2600)
649	18 x 5.50 F	28.0 (1.12)	1590 (3500)
630	20 x 5.50 F	28.0 (1.12)	1270 (2800)
651	20 x 5.50 F	28.0 (1.12)	1590 (3500)
650	20 x W7A or W7B	28.0 (1.12)	1270 (2800)
652	20 x W7A or W7B	28.0 (1.12)	1590 (3500)
653	20 x W8B	28.0 (1.12)	1270 (2800)
654	20 x W8B	28.0 (1.12)	1590 (3500)

*Determine wheel loads with machine at rest. This rating applies up to 20 mph (32 km/h) maximum travel speed.

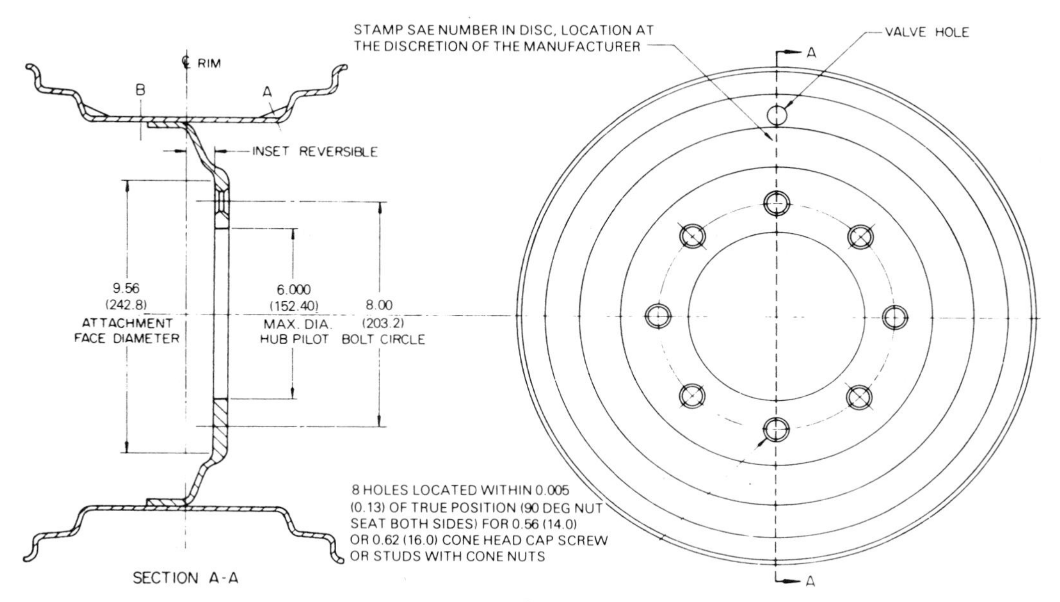

FIG. 5—8 BOLT, 8 IN. BOLT CIRCLE GROUP

SAE Wheel No.	Rim Size	Inset, mm (in.)	"Max. Rated" Radial Wheel Load*, kg (lb)	V/H
801	15 x 8 LB	28.0 (1.09)	2045 (4500)	A
802	15 x 10 LB	28.0 (1.09)	2045 (4500)	A
803	16 x W8L	28.0 (1.09)	2270 (5000)	A
804	16 x W10L	28.0 (1.09)	2270 (5000)	A
805	16.1 x W11C	28.0 (1.09)	2270 (5000)	A
806	16.1 x 14 LB	28.0 (1.09)	2270 (5000)	B
807	16.1 x 16 LB	28.0 (1.09)	2270 (5000)	B

*Determine wheel loads with machine at rest. This rating applies up to 20 mph (32 km/h) maximum travel speed.

ASAE Standard: ASAE S220.4 (SAE J711 DEC84)

TIRE SELECTION TABLES FOR AGRICULTURAL MACHINES OF FUTURE DESIGN

Developed by the Society of Automotive Engineers; administered through official liaison with SAE Tire Subcommittee; adopted by ASAE as a Recommendation March 1961; revised December 1964, March 1970, February 1972; revised and reclassified as a Standard March 1978; reconfirmed December 1982, December 1987.

This Standard is established for the purpose of providing selection tables of tires preferred for application to machines of future design.

The objective of the tables is to minimize the number of tire sizes.

TABLE 1—PREFERRED DRIVE TIRE SIZES FOR USE ON AGRICULTURAL MACHINES OF FUTURE DESIGN

Grouped by Rim Dia. in	Tire Size	Ply Rating	Std Rim Width in	Design Section Width in	Static Loaded Radius, in	Design Overall Dia. in
CODE R-1 (Regular Agricultural)						
16	9.5–16	4	W8, W8L, 8LB	9.5	15.1	33.26
	18.4–16.1	6	16.1–16LB	18.4	19.5	44.76
24	8.3–24	4	W-7	8.3	18.1	39.10
	9.5–24	4	W-8	9.5	19.0	41.26
	11.2–24	4	W-10	11.2	19.8	43.44
	12.4–24	4, 6, 8	W-11	12.4	20.8	45.62
	13.6–24	4	W-12	13.6	21.5	47.62
	14.9–24	6	W-13	14.9	22.4	49.82
	16.9–24	6, 8	W-15L	16.9	23.4	52.48
26	14.9–26	6, 8	W-13	14.9	23.5	51.82
	16.9–26	6, 8	W-15L	16.9	24.4	54.48
	18.4–26	6, 8, 10	W-16L, DW-16	18.4	25.5	57.10
	23.1–26	8, 10	DW-20	23.1	28.0	63.20
	28L–26	10	DW-25	28.1	28.2	63.60
28	11.2–28	4	W-10	11.2	21.8	47.44
	12.4–28	4	W-11	12.4	22.6	49.62
	13.6–28	4, 6	W-12	13.6	23.5	51.62
	14.9–28	6	W-13	14.9	24.4	53.82
	16.9–28	6, 8	W-15L	16.9	25.4	56.48
	18.4–28	6	W-16L, DW-16	18.4	26.6	59.10
30	14.9–30	6	W-13	14.9	25.5	55.82
	16.9–30	6	W-15L	16.9	26.5	58.48
	18.4–30	6, 8, 10	W-16L, DW-16	18.4	27.4	61.10
	23.1–30	8	DW-20	23.1	29.8	67.20
32	24.5–32	10	DW-21	24.5	31.5	71.00
34	16.9–34	6	W-15L	16.9	28.5	62.48
	18.4–34	6, 8, 10	W-16L, DW-16	18.4	29.5	65.10
	20.8–34	8	W-18L	20.8	30.3	68.20
	23.1–34	8	DW-20	23.1	31.8	71.20
36	13.9–36	6	W-12, DW-12	13.9	26.8	58.20
38	13.6–38	4, 6	W-12, DW-12	13.6	28.5	61.62
	15.5–38	6, 8	W-14L, DW-14	15.5	28.5	61.76
	16.9–38	6, 8	W-15L	16.9	30.4	66.48
	18.4–38	6, 8, 10, 12	W-16L	18.4	31.5	69.10
	20.8–38	8, 10	W-18L	20.8	32.6	72.20
CODE R-2 (Cane and Rice)						
26	18.4–26	6, 8, 10	W-16L, DW-16	18.4	26.2	58.96
	23.1–26	8, 10	DW-20	23.1	29.3	65.44
	28L–26	10	DW-25	28.1	29.4	65.86
30	23.1–30	8	DW-20	23.1	31.2	69.44
32	24.5–32	10	DW-21	24.5	32.8	73.34
34	18.4–34	6, 8	W-16L, DW-16	18.4	30.1	66.96
	20.8–34	8	W-18L	20.8	31.4	70.25
	23.1–34	8	DW-20	23.1	33.1	73.44
38	15.5–38	6	W-14L	15.5	30.0	63.18
	18.4–38	6, 8, 10	W-16L	18.4	32.0	70.96
	20.8–38	8, 10	W-18L	20.8	33.4	74.26
CODE R-3 (Industrial and Sand)						
16	9.5–16	4	W-8, W8L, 8LB	9.5	14.7	32.30
	12.4–16	6, 8	W-11	12.4	16.3	36.66
	18.4–16.1	6, 8	16.1–16LB	18.4	18.9	43.80
24	8.3–24	4	W-7	8.3	17.8	38.14
	9.5–24	4	W-8	9.5	18.4	40.30
	14.9–24	6	W-13	14.9	22.4	48.86
	16.9–24	6	W-15L	16.9	22.8	51.52
26	16.9–26	6	W-15L	16.9	23.8	53.52
	18.4–26	6, 10	W-16L, DW-16	18.4	25.3	56.14
	23.1–26	8	DW-20	23.1	26.9	62.24
CODE R-4 (Industrial Tractor, Intermediate Tread)						
24	14.9–24	6, 8	W-13	14.9	22.2	48.86
	16.9–24	6, 8, 10	W-15L	16.9	23.1	51.52
	17.5L–24	6	W-15L	17.5	22.0	48.84
28	14.9–28	6, 8	W-13	14.9	24.2	52.86
	16.9–28	6, 8	W-15L	16.9	25.1	55.52

TABLE 2—PREFERRED STEERING TIRE SIZES FOR USE ON AGRICULTURAL MACHINES OF FUTURE DESIGN

Grouped by Rim Dia. in	Tire Size	Ply Rating	Std Rim Width in	Design Section Width in	Static Loaded Radius, in	Design Overall Dia. in
CODE F-1 (Single Rib)						
16	10.00-16SL	6	W8L, W8, 8LB	10.8	16.3	35.66
	11.00-16SL	8	W10L, 10LB	12.4	17.8	38.54
18	7.50-18SL	6	5.50F	8.0	16.1	34.26
20	7.50-20SL	6	5.50F	8.0	17.2	36.26
CODE F-2 (Regular Agricultural)						
12	4.00-12SL	4	3.00D	4.4	10.0	21.24
14	6.00-14SL	4	5KB	6.6	12.6	27.10
15	5.00-15SL	4	3.00D	5.1	12.0	26.14
	7.5L-15SL	6, 8	6LB	8.2	13.5	29.36
	9.5L-15SL	6, 8	W8L, 8LB	9.5	13.9	30.80
	11L-15SL	6, 8	W8L, 8LB	11.0	14.4	32.00
16	5.50-16SL	4	4.00E	5.9	13.2	28.06
	6.00-16SL	4, 6	4.00E	6.2	13.5	29.10
	6.50-16SL	4, 6	4.50E	6.8	13.9	29.96
	7.50-16SL	4, 6, 8	5.50F	8.0	14.8	31.80
	10.00-16SL	6, 8	W81, W8, 8LB	10.8	16.0	35.20
	11.00-16SL	6, 8	W10L, 10LB	12.4	17.3	38.08
	14L-16.1SL	6	W11C-16.1	14.0	17.4	38.80
18	7.50-18SL	4, 6	5.50F	8.0	15.9	33.80
20	7.50-20SL	6	5.50F	8.0	17.0	35.80
	9.50-20SL	6, 8	W7L	10.0	17.8	38.50
CODE F-3 (Industrial Multiple Rib)						
10	9.00-10SL	4	6.00F	9.2	12.2	26.54
16	7.50-16SL	6	5.50F	8.0	14.7	30.78
	11.00-16SL	6, 8	W10L, 10LB	12.4	16.8	36.80
CODE I-1 (Rib Implement)						
9	4.00-9SL	4	3.00D	4.4	8.1	17.44
10	9.00-10SL	4	6.00F	9.2	11.1	26.60
12	4.00-12SL	4	3.00D	4.4	9.5	20.44
	8.5L-14SL	6, 8	6KB	8.5	12.8	28.38
14	9.5L-14SL	6, 8	8KB	9.5	13.1	29.18
	11L-14SL	6	8KB	11.0	13.2	29.60
15	5.00-15SL	4	3.00D	5.1	11.4	25.16
	5.90-15SL	4	4½KB	5.9	12.0	26.20
	6.40-15SL	4, 6	4½KB	6.4	12.4	26.92
	6.70-15SL	4, 6	4½KB	6.7	12.7	27.70
	7.60-15SL	4, 6, 8	5½K	7.6	13.1	28.90
	9.5L-15SL	8	W8L, 8LB	9.5	13.4	30.18
	10.00-15SL	8	8LB, W8L	10.8	14.8	33.60
	11L-15SL	6, 8, 10	8LB, W8L	11.0	13.8	30.60
	12.5L-15SL	6, 8	10LB	12.5	14.3	32.40
16	5.50-16SL	4	4.00E	5.9	12.3	27.04
	6.00-16SL	4, 6	4.00E	6.2	12.6	28.04
	6.50-16SL	4, 6, 8	4.50E	6.8	13.0	28.92
	7.50-16SL	4, 6, 8, 10	5.50F	8.0	14.0	30.92
	9.00-16SL	8, 10	6.00F	9.2	15.0	33.40
	11L-16SL	6, 8, 10	W8, W8L, 8LB	11.0	14.3	31.60
	12.5L-16SL	8	W10L, 10LB	12.5	14.8	33.40
18	4.00-18SL	2, 4	3.00D	4.4	12.5	26.44
	7.50-18SL	4, 6	5.50F	8.0	15.0	32.92
20	7.50-20SL	4, 6	5.50F	8.0	16.0	34.92
24	7.50-24SL	4	W7	8.8	18.0	39.26
	9.00-24SL	6, 8	W8	10.7	19.4	43.06
	11.25-24SL	8	W10H	12.8	21.0	46.10
28	11.25-28SL	10	W10H	12.8	23.0	50.10
CODE I-2 (Moderate Traction)						
16	13.50-16.1SL	6, 8	W11C-16.1	13.9	17.4	40.70
	16.5L-16.1SL	6, 8, 10	16.1-14LB	16.5	17.3	40.80
CODE I-3 (Traction Implement)						
15	5.00-15SL	4	3.00D	5.1	11.6	25.78
	5.90-15SL	4	4½KB	5.9	12.2	26.82
	6.70-15SL	4	4½KB	6.7	12.8	28.35
	12.5L-15SL	6, 8	10LB	12.5	14.4	33.15
16	7.50-16SL	4	5.50F	8.0	14.1	31.54
	13.50-16.1SL	6	W11C-16.1	13.9	17.5	41.08
18	7.50-18SL	4	5.50F	8.0	15.2	33.54
20	7.50-20SL	4	5.50F	8.0	16.2	35.54
24	7.50-24SL	4	W7	8.8	18.3	39.88
CODE I-6 (Smooth Implement)						
8	4.00-8SL	2, 4	3.00D	4.4	7.4	16.00

NOTES: For vehicle design clearance purposes:

1. Maximum grown tire widths may exceed the above design new tire section widths by 8 percent for drive tires and 9 percent for steering wheel and implement tires.
2. Maximum grown tire outside diameters may exceed the above design new tire outside diameters for drive tires by 6 percent and for steering wheel and implement tires by 8 percent of the difference between the design new tire outside diameter and the nominal rim diameter.

INTERCHANGEABILITY OF DISK HALVES FOR AGRICULTURAL EQUIPMENT PRESS AND GAGE WHEELS

Approved by ASAE as a Recommendation March 1955; reconfirmed December 1965, December 1970, December 1975; reclassified as a Standard March 1978; reconfirmed December 1980, December 1985.

Disk wheels for mounting semipneumatic press and gage wheel tires are made up of two identical disks provided with tire-driving lugs and clamped together by a series of bolts near the rim. A second series of bolts mounts the disks on the hub flange. A pilot hole at the center of the wheel is of sufficient diameter to clear the hub.

Rim contours and numbers and dimension of tire-driving lugs are those which have been approved by the Tire and Rim Assoication.

The purpose of this Standard is to establish dimensions and relationships to insure interchangeability of disk halves as supplied by different manufacturers. These disk halves fall into three categories (See Figs. 1, 2, and 3). In each of the three categories and for each wheel size, the following items are established as recommended practice:

Diameter of clamp-bolt circle with ±0.005 tolerance.

Diameter of mounting-bolt circle with ±0.005 tolerance.

Diameter of pilot hole with ±0.015 tolerance.

Number and size of clamp-bolt holes.

Number and size of mounting-bolt holes.

Number of drive lugs.

In every case, the clamp-bolt holes, mounting-bolt holes, and drive lugs are equally spaced.

In the case of hub mountings other than shown, it is recommended that one drive lug, one clamp-bolt hole and one mounting-bolt hole be maintained on a common center line.

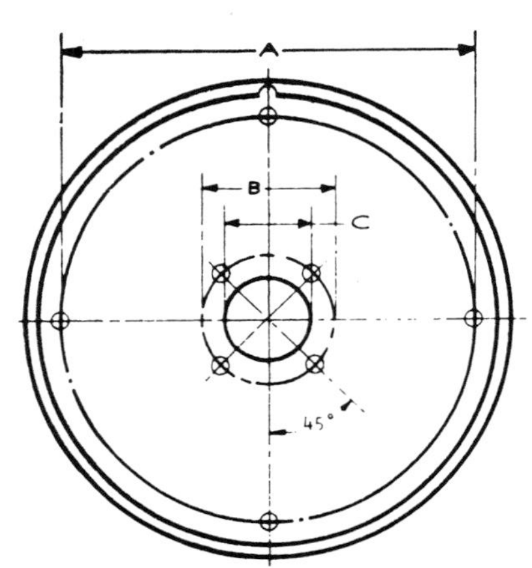

CENTERLINE OF ONE DRIVE LUG AND ONE CLAMP BOLT HOLE AT 45° TO CENTERLINES OF MOUNTING BOLT HOLES

	A	B	C	CLAMP BOLT HOLES	MOUNTING BOLT HOLES	DRIVE LUGS
2X13	9-1/2	4	2-1/2	4-13/32	4-13/32	4
4X16	10-3/4	4	2-1/2	4-13/32	4-13/32	6

FIG. 1

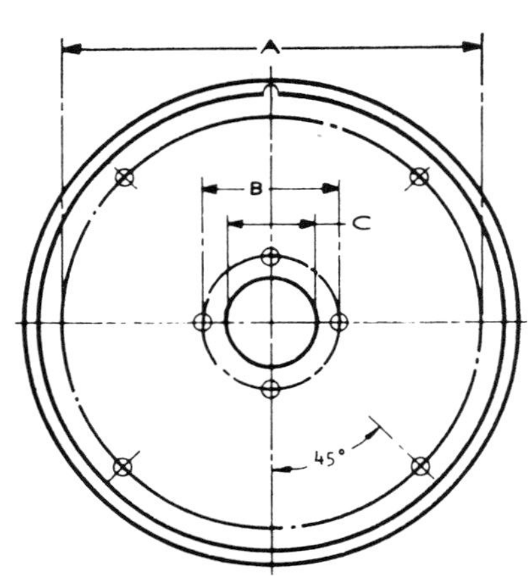

CENTERLINE OF ONE DRIVE LUG AND ONE MOUNTING BOLT HOLE AT 45° TO CENTERLINES OF CLAMP BOLT HOLES

	A	B	C	CLAMP BOLT HOLES	MOUNTING BOLT HOLES	DRIVE LUGS
3-1/2X12	6-7/8	4	2-1/2	4-13/32	4-13/32	4
4X12	6-7/8	4	2-1/2	4-13/32	4-13/32	4

FIG. 2

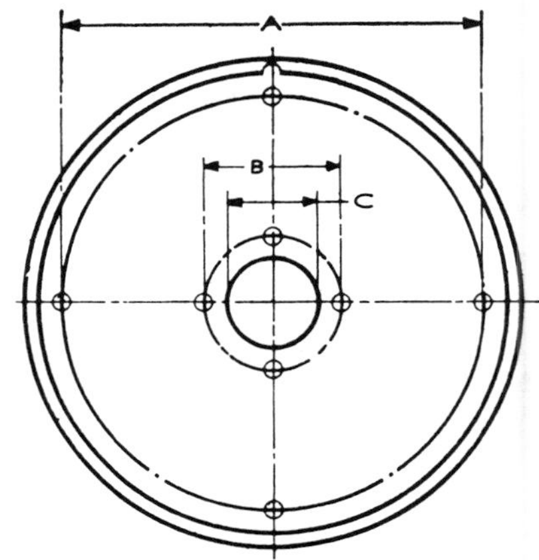

ONE DRIVE LUG, ONE CLAMP BOLT HOLE AND ONE MOUNTING BOLT HOLE ON COMMON CENTERLINE

	A	B	C	CLAMP BOLT HOLES	MOUNTING BOLT HOLES	DRIVE LUGS
2X20	16-1/4	4	2-1/2	4-13/32	4-13/32	6
4X18	12-3/4	—	—	6-13/32	—13/32MIN	6
4X20	14-3/4	—	—	6-13/32	—13/32MIN	6

FIG. 3

GENERAL NOTES - APPLY TO ALL SIZES.

CLAMP BOLT HOLES, MOUNTING BOLT HOLES, AND DRIVE LUGS ARE EQUALLY SPACED.

IN CASE OF ANY SPECIAL HUB MOUNTING, MAINTAIN ONE DRIVE LUG, ONE CLAMP BOLT HOLE, AND ONE MOUNTING BOLT HOLE ON COMMON CENTERLINE.

RIM CONTOURS FOR AGRICULTURAL PRESS AND GAGE WHEELS

This Standard includes specifications from PG-1 established by the Tire and Rim Association; approved by ASAE Power and Machinery Division Technical Committee; adopted by ASAE March 1955; reconfirmed March 1962, December 1966, December 1971, December 1976, December 1981, December 1986.

SECTION 1—PURPOSE AND SCOPE

1.1 This Standard specifies rim contours of agricultural press and gage wheels for tire sizes included in ASAE Standard ASAE S223, Agricultural Press and Gage Wheel Tires.

SECTION 2—SPECIFICATIONS

2.1 Fig. 1 and Table 1 cover rim dimensions for two sizes of press wheels, and Fig. 2 and Table 2 cover rim dimensions for three sizes of gage wheels. All dimensions, unless otherwise noted, are in inches.

Cited Standard:

ASAE S223, Agricultural Press and Gage Wheel Tires

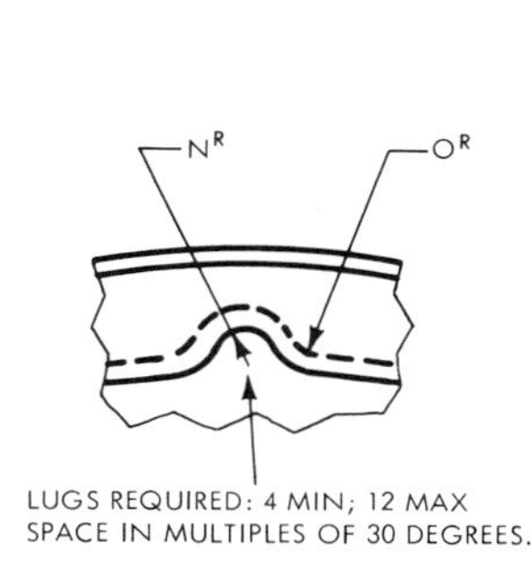

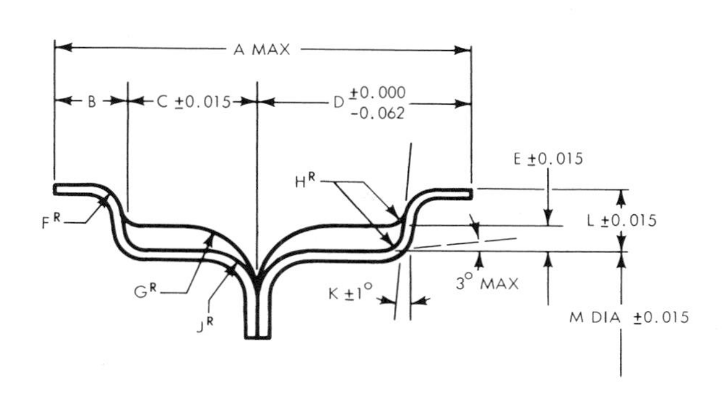

FIG. 1—PRESS WHEEL RIM DIMENSIONS

TABLE 1—RIM DIMENSIONS FOR PRESS WHEELS*

Size	A	B	C	D	E	F	G	H	J	K	L	M	N	O
2 x 13	2	0.31	0.69	1	0.125	0.156	0.375	0.094	0.25	3°	0.31	10.63	0.156	0.078
2 x 20	2	0.31	0.69	1	0.125	0.156	0.375	0.094	0.25	3°	0.31	17.63	0.156	0.078

*Dimensions in inches unless otherwise indicated.

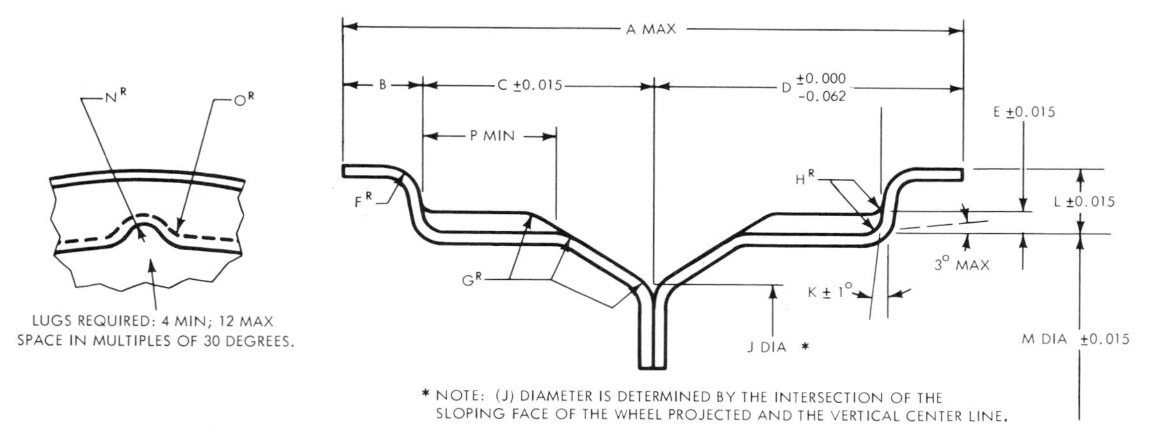

FIG. 2—GAGE WHEEL RIM DIMENSIONS

TABLE 2—RIM DIMENSIONS FOR GAGE WHEELS*

	A	B	C	D	E	F	G	H	J	K	L	M	N	O	P
3½ x 12	3.50	0.44	1.31	1.75	0.140	0.156	0.25	0.094	8	3°	0.38	8.63	0.187	0.125	0.75
4 x 12	4.00	0.44	1.56	2.00	0.140	0.156	0.25	0.094	8	3°	0.38	8.63	0.187	0.125	1.00
4 x 16	4.00	0.44	1.56	2.00	0.140	0.156	0.25	0.094	12	3°	0.38	12.63	0.187	0.125	1.00

*Dimensions in inches unless otherwise indicated.

ASAE Standard: ASAE S223

AGRICULTURAL PRESS AND GAGE WHEEL TIRES

This Standard includes specifications from PG-1 established by the Tire and Rim Association; approved by ASAE Power and Machinery Technical Committee; adopted by ASAE March 1955; reconfirmed March 1962, December 1966, December 1971, December 1976, December 1981, December 1986.

SECTION 1—PURPOSE AND SCOPE

1.1 This Standard applies to tires for agricultural press and gage wheels, *not intended for transport.*

1.2 Tires built in accord with this Standard are intended for speeds not exceeding a maximum of 8 mph.

SECTION 2—SPECIFICATIONS

2.1 These tires shall have twelve (12) transverse grooves (Fig. 1) across the base spaced 30 deg apart, to prevent creep or slip on the rims.

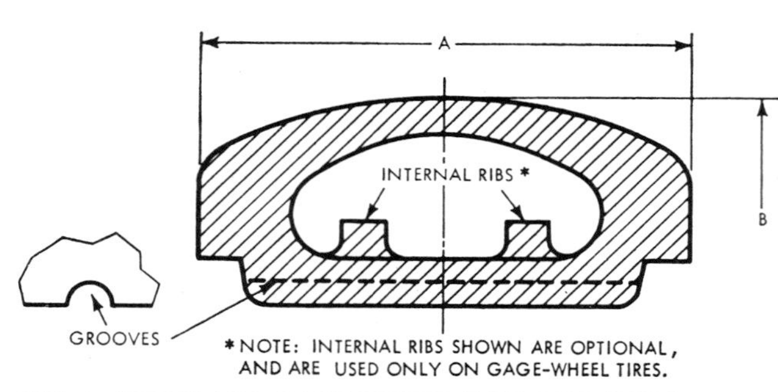

FIG. 1—PRESS AND GAGE WHEEL TIRE CONTOURS

2.2 Table 1 gives tire size, maximum load, rim size, and maximum tire dimensions (mounted) for the two grain-drill press-wheel sizes and the three gage-wheel sizes included in this Standard.

TABLE 1—TIRE DIMENSIONS FOR AGRICULTURAL PRESS AND GAGE WHEEL TIRES

Tire Size	Maximum Load, lb Light Duty	Maximum Load, lb Heavy Duty	Rim Sizes, in.	Maximum Tire Dimensions (Mounted), in. Width A*, in.	Maximum Tire Dimensions (Mounted), in. OD B, in.
			Grain Drill Press Wheel Sizes		
2 x 13	60	—	2 x 13	2.00	13.00
2 x 20	100	—	2 x 20	2.00	20.00
			Gage Wheel Sizes		
3½ x 12	100	200	3½ x 12	3.50	12.00
4 x 12	150	300	4 x 12	4.00	12.00
4 x 16	175	350	4 x 16	4.00	16.00

*Tire width is that of a new tire including normal sidewalls, but does not include protective ribs, bars, or decorations.

SECTION 3—RIMS FOR USE WITH TIRES

3.1 Standard rim contours for wheels for the above tire sizes are covered in ASAE Standard: ASAE S222, Rim Contours for Agricultural Press and Gage Wheels.

Cited Standard:

ASAE S222, Rim Contours for Agricultural Press and Gage Wheels

ASAE Standard: ASAE S224.1

AGRICULTURAL PLANTER PRESS WHEEL TIRES

This Standard includes specifications from PW-1 established by the Tire and Rim Association; approved by ASAE Power and Machinery Technical Committee; adopted by ASAE March 1955; reconfirmed March 1962; revised December 1966; reconfirmed December 1971, December 1976, December 1981, December 1986.

SECTION 1—PURPOSE AND SCOPE

1.1 This Standard applies to tire and rim dimensions for agricultural planter press wheel sizes, *not intended for transport.*

1.2 Tires built in accord with this Standard are intended for speeds not exceeding a maximum of 8 mph.

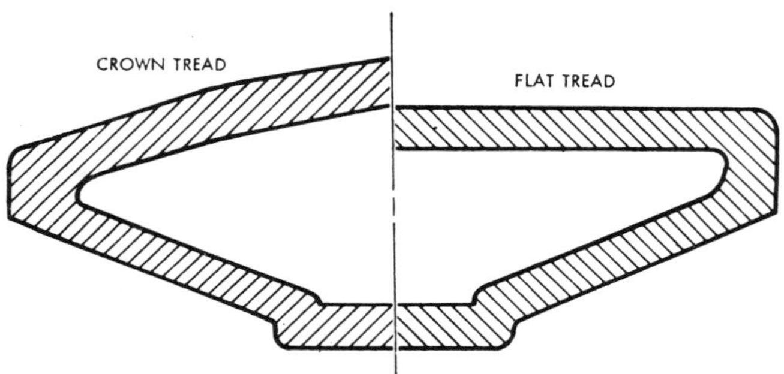

FIG. 1—TIRE CONTOURS

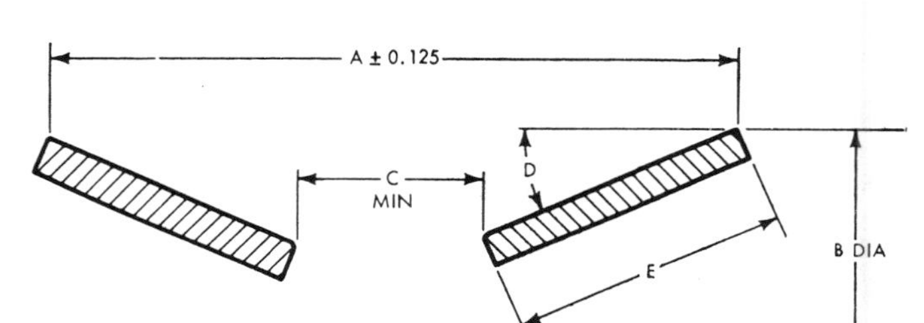

FIG. 2—RIM SECTION

TABLE 1—TIRE DIMENSIONS FOR PLANTER PRESS WHEELS*

Tire Size†	Width max, in.	OD, max, in.	Loaded Radius, approx, in.	Load max, lb
6½x12(C)	6.6	14.5	6.5	200
6½x16(C)	6.6	18.8	8.5	200
6½x20(C)	6.6	23.2	10.5	200
7x16(F)	7.1	18.0	9.0	200
7x16(C)	7.0	18.8	8.5	200
7x18(F)	7.1	20.0	10.0	200
7x18(C)	7.1	21.2	9.5	200
7x22(F)	7.1	24.0	12.0	200
7x24(F)	7.1	26.0	13.0	200
8x20(C)	8.5	23.2	10.5	200

*All dimensions are with tires mounted.
†C = crown tread; F = flat tread.

TABLE 2—RIM DIMENSIONS FOR PLANTER PRESS WHEELS

Tire Size	A in.	B in.	C in.	D deg.	E in.
6½x12	6.5	12	1.5	22.5	2.7
6½x16	6.5	16	2.2	22.5	2.2
6½x20	6.5	20	2.2	22.5	2.2
7x16	6.8	16	2.2	22.5	2.5
7x18	7.0	18	2.2	22.5	2.5
7x22	7.0	22	2.2	22.5	2.5
7x24	7.0	24	2.2	22.5	2.5
8x20	8.0	20	2.3	22.5	3.1

ASAE Standard: ASAE S225.1

CHISEL PLOW, FIELD AND ROW CROP CULTIVATOR SHANKS AND GROUND TOOL MOUNTINGS

Adopted by ASAE March 1955; reconfirmed by Power and Machinery Division Standards Committee December 1961, December 1966, December 1971; revised by ASAE Cultural Practices Equipment Committee February 1973; reconfirmed December 1977, December 1982, December 1987.

SECTION 1—PURPOSE AND SCOPE

1.1 The purpose of this Standard is to provide greater interchangeability of cultivator sweeps and shovels used on various types of footpieces for row crop cultivators, field cultivators, and chisel plots; to provide a good fit and thereby better mechanical support for sweeps and shovels on such footpieces; and to provide ground clearance between the lowest portion of footpieces or attaching bolts and the cutting edge of sweeps to accommodate wear.

1.2 This Standard includes the following specifications for row crop cultivators, field cultivators and chisel plows:

1.2.1 Mounting specifications for the ground tool attaching portion of the shank as applied to spring trip and friction trip shanks, stiff shanks and square coil spring tool shanks, and flat spring tooth shanks.

1.2.2 Attaching specifications for sweeps with curved stems; double pointed shovels, teeth and spikes; and spear point and single end shovels.

SECTION 2—SPECIFICATIONS FOR GROUND TOOL PORTION OF SHANK

2.1 Dimensions related to the following specifications are shown in Fig. 1 and Table 1.

2.1.1 Width of footpiece, dimension A. Dimensions are nominal or maximum as specified.

2.1.2 Width of flat surface on footpiece, dimension B. The width of the flat surface on the shank footpiece shall not be less than the minimum specified unless the footpiece is recessed at the bolt holes. All shank footpieces with less than the specified flat surface width shall be recessed at the bolt holes to accommodate the extrusion on the sweep stem.

2.1.3 Thickness of footpiece, dimension C. This dimension applies only to trip shanks of row crop cultivators.

2.1.4 Radius of curvature, dimension D. The arc of curvature extends through dimensions E, H, and F.

2.1.5 Length of curvature extension above bolt hole, dimension E. A good fit requires radius of curvature of the shank to extend a mimimum distance above the upper bolt hole. This dimension is measured on the chord.

2.1.6 Length of curvature extension below bolt hole, dimension F. A good fit requires radius of curvature of the shank to extend a minimum distance below the lower bolt hole, but not more than the maximum specified. This dimension is measured on the chord.

2.1.7 Angle of attachment, dimension G. A chord between the attaching bolt centers, at the attaching surface of the footpiece, shall make an angle with the ground level which provides the desired working angle for standard sweeps.

2.1.8 Attaching bolt hole spacing, dimension H. Bolt hole spacing as specified shall be measured on the chord at the attaching surface. The upper hole may be slotted to increase bolt hole spacing by 0.25 in. (6.4 mm).

2.1.9 Bolt holes size, dimension I. Bolt holes shall provide clearance for number 3 plow bolts as specified in American National Standard B18.9, Plow Bolts.

2.1.10 Height of lower bolt hole, dimension J. The center of the lower bolt hole at the attaching surface shall be within the maximum distances above the lowest point of the footpiece. This is to provide adequate support without protruding lower than the sweep.

SECTION 3—SPECIFICATIONS FOR SWEEPS

3.1 Dimensions related to the following specifications are shown in Fig. 2 and Table 2.

3.1.1 Radius of curvature at shank attaching surface, dimension A. The arc of curvature with the specified radius extends through dimensions B, C and D.

3.1.2 Stem extension above top bolt hole, dimension B. The stem of the sweep shall extend above the upper bolt hole the specified distance.

3.1.3 Length of curvature extension below lower bolt hole, dimension C. Good fit requires the radius of curvature to extend below the lower bolt hole the specified distance. This dimension is measured on the chord.

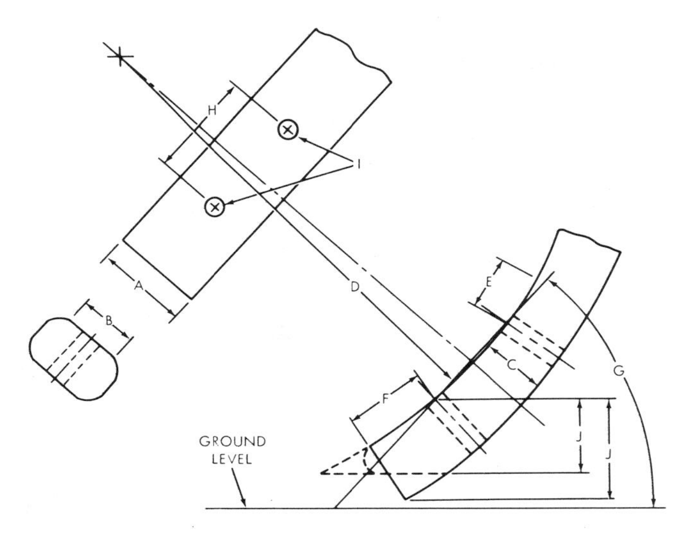

FIG. 1—CHISEL AND CULTIVATOR SHANKS

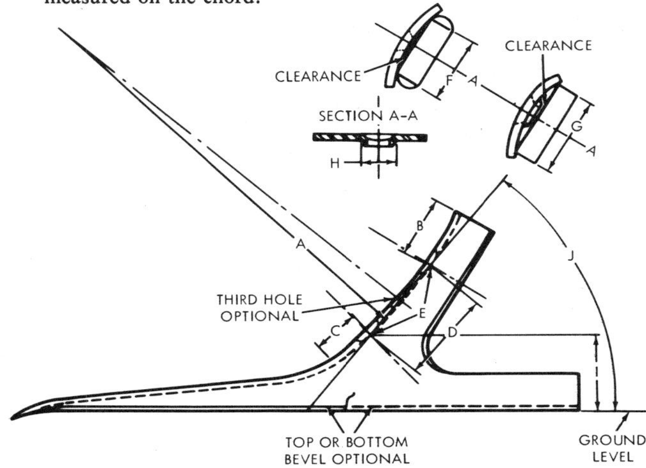

FIG. 2—CHISEL AND CULTIVATOR SWEEPS

3.1.4 Attaching bolt hole spacing, dimension D. Bolt hole spacing as specified shall be measured on the chord at the attaching surface.

3.1.5 Bolt hole size, dimension E. Bolt holes shall provide clearance for No. 3 plow bolts as specified in American National Standard B18.9, Plow Bolts.

3.1.6 Stem scross section, dimensions F and G. The shank attaching surface of the sweep stem shall be so constructed that it will provide clearance over a specified chord, Dimension F, and be sufficient width to extend beyond a specified chord, Dimension G, for the full contact length.

3.1.7 Extrusion diameter, dimension H. If the holes are extruded, the maximum diameter of the extrusion shall not exceed that specified.

3.1.8 Height of lower bolt hole, dimension I. The height of the center of the lower bolt hole at the attaching surface shall be a minimum distance above the base of the sweep as specified.

3.1.9 Angle of attachment, dimension J. A chord between the attaching bolt centers on the attaching surface of the sweep shall make an angle with the ground level as specified. This assumes adjusting to working depth with no load.

SECTION 4—SPECIFICATIONS FOR DOUBLE POINTED AND SPEAR POINT SHOVELS

4.1 Dimensions related to the following specifications are shown in Table 3 and Figs. 3 and 4.

4.1.1 Radius of curvature at shank attaching surface, dimension A. The arc of curvature with the speficied radius extends through dimensions B, C and D.

4.1.2 Length of curvature extension above upper bolt, dimension B. Good fit requires that the radius of curvature of the attaching surface extend above the upper bolt hole the specified distance. This dimension is measured on the chord.

4.1.3 Length of curvature extension below lower bolt, dimension C. Good fit requires that the radius of curvature of the attaching surface extend below the lower bolt hole the specified distance. This dimension is measured on the chord.

4.1.4 Attaching bolt hole spacing, dimension D. Bolt hole spacing as specified shall be measured at the attaching surface on the chord.

4.1.5 Bolt hole size, dimension E. Bolt holes shall provide clearance for No. 3 plot bolts as specified in American National Standard B18.9, Plow Bolts.

4.1.6 Shovel cross section, dimension E. The shank attaching surface of the shovel shall be so constructed that it will provide clearance over a specified chord.

4.1.7 Extrusion diameter, dimension G. If the bolt holes are extuded, the maximum diameter of the extrusion shall not exceed that specified.

Cited Standard:

ANSI B18.9, Plow Bolts

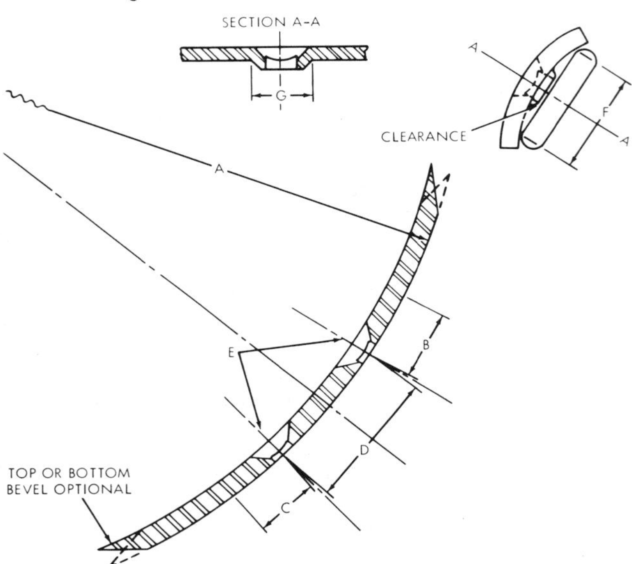

FIG. 3—DOUBLE POINT SHOVEL

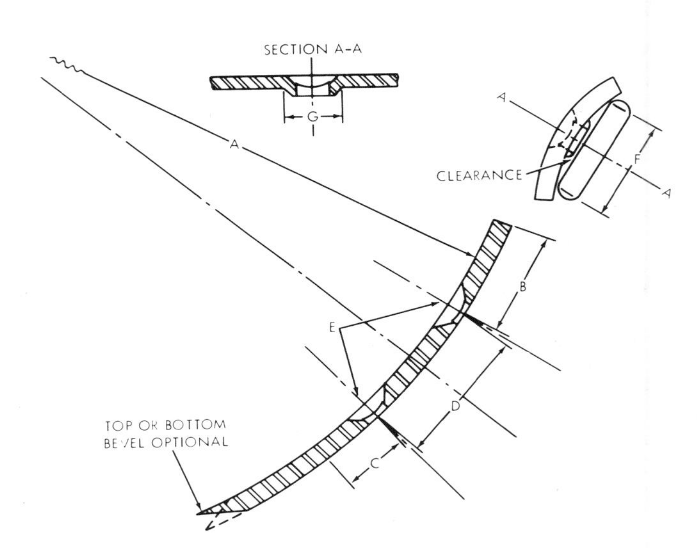

FIG. 4—SPEAR POINT SHOVEL

TABLE 1–DIMENSIONS FOR GROUND TOOL ATTACHING PORTION OF SHANK (SEE FIG. 1)

		CHISEL PLOW				FIELD CULTIVATOR				SPRING TRIP SHANKS*				FLAT SPRING SHANKS			
		Inches		Millimeters		Inches		Millimeters		Inches		Millimeters		Inches		Millimeters	
Dimension	Description	Min	Max	Min	Max	Min	Max	Min	Max	Min	Max	Min	Max	Min	Max	Min	Max
A.	Width of footpiece	2.0 Nominal		51 Nominal		1.75 Nominal		44.5 Nominal		–	1.5	–	38	1.75 Nominal		44.5 Nominal	
B.	Width of footpiece flat surface	1.6	–	41	–	1.6	–	41	–	–	–	–	–	1.6	–	41	
C.	Thickness of footpiece	–	–	–	–	–	–	–	–	–	1.38	–	35	–	–	–	–
D.	Radius of curvature in tool attaching area	11.0 Nominal		279 Nominal		9.0 Nominal		229 Nominal		8.0 Nominal		203 Nominal		8.0 Nominal		203 Nominal	
E.	Length of curvature ext. above top bolt hole	0.75	–	19.1	–	0.75	–	19.1	–	0.75	–	19.1	–	0.75	–	19.1	–
F.	Length of curvature ext. below lower bolt hole	1.0	–	25	–	1.0	–	25	–	1.0	–	25	–	1.0	–	25	
G.	Angle of attachment	†	†	†	†	†	†	†	†	†	†	†	†	†	†	†	†
H.	Attaching bolt hole spacing	2.25 Nominal‡		57.2 Nominal‡		1.75 Nominal‡		44.5 Nominal‡		2.0 Nominal		51 Nominal		2.0 Nominal		51 Nominal	
I.	Holes to provide clearance for specified No. 3 plow bolt	0.44 Std. 0.50 Opt.		11.2 Std. 12.7 Opt.		0.38 Std. 0.44 Opt.		9.5 Std. 11.2 Opt.		0.44 Nominal		11.2 Nominal		0.44 Nominal		11.2 Nominal	
J.	Height of lower bolt hole	–	1.5	–	38	–	1.38		35	–	1.38		35.4	–	1.38		35

*Includes square coil spring, friction trip and stiff shanks.
†Shanks shall provide desired working angle for standard sweeps.
‡The upper hole may be slotted to increase the center to center distance by 0.25 in. (6.4 mm).

TABLE 2—SWEEP DIMENSIONS (SEE FIG. 2)

		CHISEL PLOW				FIELD CULTIVATOR				ROW CROP CULTIVATORS CURVED STEM			
		Inches		Millimeters		Inches		Millimeters		Inches		Millimeters	
Dimension	Description	Min	Max	Min	Max	Min	Max	Min	Max	Min	Max	Min	Max
A.	Radius of curvature at shank attaching surface	11.0 Nominal		279 Nominal		9.0 Nominal		229 Nominal		8.0 Nominal		203 Nominal	
B.	Stem extension above top bolt hole	0.75	–	19.1	–	0.75	–	19.1	-	0.75	-	19.1	-
C.	Length of curvature ext. below lower bolt hole	0.3	–	8	–	0.3	–	8	–	0.3	-	8	-
D.	Attaching bolt hole spacing	2.25 Nominal		57.2 Nominal		1.75 Nominal		44.5 Nominal		2.00 Nominal		50.8 Nominal	
E.	Bolt holes to provide clearance for specified No. 3 plow bolt	0.44 Std. 0.50 Opt.		11.2 Std. 12.7 Opt.		0.38 Std. 0.44 Opt.		9.5 Std. 11.2 Opt.		0.44 Nominal		11.2 Nominal	
F.	Stem to provide clearance over specified chord	1.6	–	41	–	1.6	-	41	–	1.6	-	41	–
G.	Stem to extend beyond specified width	2.0	–	51	–	1.75	–	44.5	–	1.75		44.5	-
H.	Extrusion diameter	–	–	–	–	–	0.88	-	22.4	–	0.88	-	22.4
I.	Height of lower bolt hole	1.75	–	44.5	-	1.62	-	41.1	-	1.75	–	44.5	-
J.	Angle of attachment	49 deg	51 deg	49 deg	51 deg	47 deg	49 deg	47 deg	49 deg	54 deg	56 deg	54 deg	56 deg

TABLE 3—DOUBLE POINTED SPIKE AND SINGLE SPEAR POINT SHOVELS (SEE FIGS. 3 AND 4)

		CHISEL PLOWS DOUBLE POINT SHOVELS				FIELD CULTIVATORS DOUBLE POINT SHOVELS				DOUBLE POINT SHOVELS				ROW CROP CULTIVATORS SPEAR POINT SHOVELS			
		Inches		Millimeters		Inches		Millimeters		Inches		Millimeters		Inches		Millimeters	
Dimension	Description	Min	Max	Min	Max	Min	Max	Min	Max	Min	Max	Min	Max	Min	Max	Min	Max
A.	Radius of curvature at shank attaching surface	11.0 Nominal		279 Nominal		9.0 Nominal		229 Nominal		8.0 Nominal		203 Nominal		8.0 Nominal		203 Nominal	
B.	Lengths of curvature ext. above top bolt hole	1.5	–	38	–	1.5	-	38	-	1.5		38		To Top	-	To Top	-
C.	Length of curvature ext. below lower bolt hole	1.5	–	38	–	1.5	-	38	-	1.5		38		1.5		38	-
D.	Attaching bolt hole spacing	2.25 Nominal		57.2 Nominal		1.75 Nominal		44.5 Nominal		2.0 Nominal		51 Nominal		2.0 Nominal		51 Nominal	
E.	Bolt holes to provide clearance for specified No. 3 plow bolt	0.44 Std. 0.50 Opt.		11.2 Std. 12.7 Opt.		0.38 Std. 0.44 Std.		9.5 Opt. 11.2 Opt.		0.44 Nominal		11.2 Nominal		0.44 Nominal		11.2 Nominal	
F.	Shovel to provide clearance over specified chord	1.6	–	41	-	1.6	–	41	-	1.6	-	41	-	1.6	–	41	
G.	Extrusion diameter	–	-	-	–	–	0.88	-	22.4		0.88		22.4		0.88		22.4

ASAE Standard: ASAE S226.2

HEADED DRILLED PINS

Proposed by the Farm and Industrial Equipment Institute; adopted by ASAE August 1952; revised by Power and Machinery Division Standards Committee March 1962; reconfirmed December 1966, December 1971; revised March 1973; reconfirmed December 1976, December 1981, December 1986.

SECTION 1—PURPOSE

1.1 The purpose of this Standard is to establish a line of headed drilled pins for use by agricultural equipment manufacturers.

SECTION 2—SCOPE

2.1 This Standard provides the dimensional specifications for headed drilled pins.

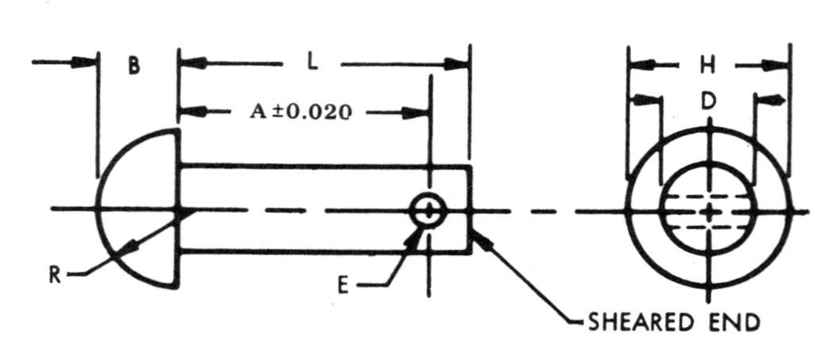

FIG. 1—DIMENSIONS OF HEADED DRILLED PINS (See Table 1)

SECTION 3—SPECIFICATIONS

3.1 Dimensions of headed drilled pins are shown in Fig. 1 and Tables 1 and 2.

3.2 Material. Headed drilled pins shall be produced from rivets manufactured in accordance with American National Standards B18.1.1, Small Solid Rivets (7/16 Inch Nominal Diameter and Smaller) and B18.1.2, Large Rivets (1/2 Inch Nominal Diameter and Larger). Plating and hardening are optional.

Cited Standards:

ANSI B18.1.1, Small Solid Rivets (7/16 Inch Nominal Diameter and Smaller)
ANSI B18.1.2, Large Rivets (1/2 Inch Nominal Diameter and Larger)

TABLE 2—LENGTH TOLERANCE

Nominal Pin Diameter	Length Tolerance	
	Pins Up To and Including 6 in.	Pins Over 6 in.
3/16 and smaller	± 0.016	
1/2 and 5/8	± 0.03	
3/4 and 7/8	± 0.06	± 0.12
1 and larger	± 0.09	± 0.19

TABLE 1—DIMENSIONS OF HEADED DRILLED PINS, INCHES (SEE FIG. 1)

Nominal Length, L		Nominal Diameter	1/4	5/16	3/8	7/16	1/2	5/8	3/4	7/8	1
	D	Body Diameter Max.	0.253	0.316	0.380	0.443	0.520	0.655	0.780	0.905	1.030
		Body Diameter Min.	0.244	0.304	0.365	0.428	0.478	0.600	0.725	0.850	0.975
	H	Head Diameter	0.460	0.572	0.684	0.798	0.938	1.157	1.390	1.609	1.828
			0.430	0.538	0.646	0.754	0.844	1.063	1.281	1.500	1.719
	B	Head Thickness	0.196	0.243	0.291	0.339	0.406	0.500	0.593	0.687	0.781
			0.180	0.225	0.271	0.317	0.375	0.469	0.562	0.656	0.750
	R	Radius of Head	0.221	0.276	0.332	0.387	0.443	0.553	0.664	0.775	0.885
	E	Hole Size	0.141	0.141	0.141	0.203	0.203	0.203	0.266	0.266	0.266
1/2	A	Distance to Hole	0.38	0.34	0.34	0.34					
5/8	A		0.50	0.47	0.47	0.47	0.41				
3/4	A		0.63	0.59	0.59	0.59	0.53	0.53			
7/8	A		0.75	0.72	0.72	0.72	0.66	0.66			
1	A		0.88	0.84	0.84	0.84	0.78	0.78			
1-1/4	A		1.13	1.09	1.09	1.09	1.03	1.03	1.00		
1-1/2	A		1.38	1.34	1.34	1.34	1.28	1.28	1.25		
1-3/4	A			1.59	1.59	1.59	1.53	1.53	1.50	1.47	1.47
2	A			1.84	1.84	1.84	1.78	1.78	1.75	1.72	1.72
2-1/4	A				2.09	2.09	2.03	2.03	2.00	1.97	1.97
2-1/2	A				2.34	2.34	2.28	2.28	2.25	2.22	2.22
2-3/4	A				2.59	2.59	2.53	2.53	2.50	2.47	2.47
3	A				2.84	2.84	2.78	2.78	2.75	2.72	2.72
3-1/4	A					3.09	3.03	3.03	3.00	2.97	2.97
3-1/2	A					3.34	3.28	3.28	3.25	3.22	3.22
3-3/4	A					3.59	3.53	3.53	3.50	3.47	3.47
4	A					3.84	3.78	3.78	3.75	3.72	3.72
4-1/4	A					4.09	4.03	4.03	4.00	3.97	3.97
4-1/2	A					4.34	4.28	4.28	4.25	4.22	4.22
4-3/4	A					4.59	4.53	4.53	4.50	4.47	4.47
5	A					4.84	4.78	4.78	4.75	4.72	4.72
5-1/2	A							5.28	5.25	5.22	5.22
6	A							5.78	5.75	5.72	5.72
6-1/2	A									6.16	6.12
7	A	Distance to Hole								6.66	6.62

ASAE Standard: ASAE S229.6 (ANSI/ASAE S229.6-1982)

BALING WIRE FOR AUTOMATIC BALERS

Proposed by Advisory Engineering Committee of Farm and Industrial Equipment Institute; approved by ASAE Power and Machinery Division; adopted by ASAE May 1952; revised January 1958, February 1962; reconfirmed with new format December 1966; revised June 1967, December 1969, December 1970; reconfirmed December 1975; revised December 1976; revised editorially March 1980; reconfirmed December 1981; approved by ANSI as an American National Standard December 1982; reconfirmed December 1986.

SECTION 1—SCOPE AND DESIGNATION

1.1 This specification shall cover annealed baling wire for automatic balers.

1.2 The wire shall be furnished in two sizes of coils: 960 m (3150 ft) minimum and 1981 m (6500 ft) minimum.

SECTION 2—STANDARD PRACTICE

2.1 Unless otherwise specified, standard practices as covered by the American Iron and Steel Institute in the latest revision of the AISI Steel Products Manual, Wire and Rods, Carbon Steel, shall govern on all points covered by this specification.

SECTION 3—REQUIREMENTS

3.1 Physical Properties:

345 to 483 MPa (50,000 to 70,000 psi) tensile strength

12 percent minimum permanent elongation in 254 mm (10 in.) length

3.2 Rewinding Practices

3.2.1 Wire shall be wound with uniform tension and shall be furnished with a protective coating that will provide adequate lubricity and corrosion resistance but will not harden or produce a gummy condition in the baler tying mechanism.

3.2.2 Wire shall be free from injurious heavy scale and surface imperfections.

3.2.3 Wire shall be free of kinks and shall be of continuous length; therefore, it shall contain no twisted splices. Welded joints shall be dressed down to wire diameter.

3.2.4 Wire shall be capable of being uncoiled from both the inside and the outside of the coil.

3.2.5 Outside end of wire shall bear a tag containing only the manufacturer's identification; inside end of coil shall be bent over the tie at the inside diameter.

3.3 Dimensions

Wire shall be furnished in rewound coils, conforming to the following requirements:

Description	960 m coil	3150 ft coil
Wire gage no.	14 1/2	14 1/2
Wire diameter	1.93 mm ± 0.05 mm	0.076 in. ± 0.002 in.
Inside diameter	76.2 mm ± 12.5 mm	3 in. ± 0.5 in.
Outside diameter	251 mm ± 12.5 mm	9.875 in. ± 0.5 in.
Coil width	95 mm ± 12.5 mm	3.75 in. ± 0.5 in.
Length	960 m minimum	3150 ft minimum

Description	1981 m coil	6500 ft coil
Wire gage no.	14 1/2	14 1/2
Wire diameter	1.93 mm ± 0.05 mm	0.076 in. ± 0.002 in.
Inside diameter	206 mm + 6, − 3 mm	8.125 in. + 0.25, − 0.125 in.
Outside diameter	336 mm ± 6 mm	13.25 in. ± 0.25 in.
Coil width	152 mm + 6, − 0 mm	6 in. + 0.25, − 0 in.
Length	1981 m minimum	6500 ft minimum

SECTION 4—PACKAGING

4.1 Coils shall be banded securely with a minimum of three ties on the 1981 m (6500 ft) coil and the 960 m (3150 ft) coil, evenly spaced.

4.2 Coils of wire shall be shipped in suitable containers providing adequate protection for shipping, storage, and distribution.

4.3 The 960 m (3150 ft) coil shall be shipped two in a carton which shall be marked as follows:

Package No. 3150 ASAE Standard Baling Wire (2 coils),
14½ gage, 960 m (3150 ft) minimum per coil

4.4 The container shall also carry, in a different location, the name and/or brand name of the wire manufacturer, and necessary instructions for storing and handling.

4.5 The 1981 m (6500 ft) coil shall be shipped one in a carton which shall be marked as follows:

Package No. 6500 ASAE Standard Baling Wire (1 coil),
14½ gage, 1981 m (6500 ft) minimum

4.5.1 The maximum outside dimensions of the empty carton shall be 349.3 x 355.6 x 165.1 mm (13.75 x 14.00 x 6.50 in.). A knockout hole with a maximum diameter of 190.5 mm (7.50 in.) shall be provided on each side of the carton. It is intended that this carton fit balers designed subsequent to the December 1969 revision of this standard. The 355.6 mm (14.00 in.) dimension may cause some interference with balers currently in the field.

4.6 The container shall also carry, in a different location, the name and/or brand name of the wire manufacturer, and necessary instructions for storing and handling.

SECTION 5—INSPECTION AND REJECTION

5.1 Any coil not conforming to the foregoing specifications may be rejected and the manufacturer or distributor notified.

Cited Standard:

AISI, Steel Products Manual

AGRICULTURAL MACHINERY MANAGEMENT DATA

Proposed by ASAE Farm Machinery Management Committee; approved by ASAE Power and Machinery Division Standards Committee; adopted by ASAE February 1963; revised December 1965, February 1971; revised editorially December 1974; revised December 1977; revised editorially September 1979; Table I corrected December 1981; reconfirmed December 1982; revised December 1983.

SECTION 1—PURPOSE

1.1 These Data include representative values of farm machinery operation parameters as an aid to managers, planners, and designers in estimating the performance of field machines.

1.2 These Data are intended for use with ASAE Engineering Practice EP391, Agricultural Machinery Management.

1.3 Some Data are reported in equation form to permit use in computer and calculator mathematical models.

SECTION 2—SCOPE

2.1 These Data report typical values for tractor performance, implement power requirements, repair and maintenance costs, depreciation, fuel and oil use, reliability for field operation, probable working days, and timeliness coefficients as measured by experiment, modeling, or survey.

2.2 Where possible, variation in sampled data is reported using the range, a standard deviation (S.D.), or a coefficient of variation (C.V.) defined as S.D./mean. In a normal distribution 68% of the population should be contained in a range of one S.D. about the mean 95% will be contained in a ± 2 S.D.

SECTION 3—TRACTOR PERFORMANCE

3.1 Drawbar performance of tractors depends primarily on engine power, weight distribution on drive wheels, type of hitch, and soil surface. Drive wheel slip (sl) is a power loss but has an optimum value for the tractive efficiency ratio (TE) to be a maximum. Fig. 1 is a graphical presentation of performance factors as they apply to two drive wheel, single tire, rear axle drive tractors with selective gear transmissions. Four surface conditions and three types of hitch are included variables. The drive tire size is that just large enough to carry the expected dynamic loading.

3.2 Equations of pneumatic-tired, single-wheel performance are useful for design specifications, prediction of vehicle performance, and computer simulation of vehicle productivity. The following relationships apply to most agricultural, earthmoving, and forestry prime movers.

3.2.1 Rolling resistance (RR) (as defined in ASAE Standard S296, Uniform Terminology for Traction of Agricultural Tractors, Self-Propelled Implements, and Other Traction and Transport Devices)

$$RR = W\left(\frac{1.2}{Cn} + 0.04\right)$$

where W is the dynamic wheel load in force units normal to the soil surface and Cn is a dimensionless ratio which is a function of the cone index (Ci) for the soil (see ASAE Standard S313, Soil Cone Penetrometer), the unloaded tire section width (b), the unloaded overall tire diameter (d) and W.

$$Cn = \frac{Ci\ b\ d}{W}$$

3.2.1.1 Values of Ci and Cn for agricultural drive-wheel tires ($bd/W \cong 0.25$) on typical soil surfaces are:

Soil	Ci	Cn
Hard	200	50
Firm	120	30
Tilled	80	20
Soft, Sandy	60	15

These values are applicable to soils which are not highly compactible and to tires operated at pressures which produce tire deflections of approximately 20% radially of the undeflected tire section height.

3.2.1.2 The coefficient of rolling resistance (Crr) is a ratio of the rolling resistance to dynamic wheel load.

$$Crr = \frac{\text{Rolling resistance}}{W} = \left(\frac{1.2}{Cn} + 0.04\right)$$

3.2.2 Net pull or traction (P) (as defined in ASAE Standard S296, Uniform Terminology for Traction of Agricultural Tractors, Self-Propelled Implements, and Other Traction and Transport Devices)

$$P = 0.75W\left(1 - e^{-0.3Cn\ sl}\right) - W\left(\frac{1.2}{Cn} + 0.04\right)$$

where the added terms are slip (sl) (decimal) (as defined in ASAE Standard S296, Uniform Terminology for Traction of Agricultural Tractors, Self-Propelled Implements, and Other Traction and Transport Devices) and e is the base of natural logarithms.

3.2.3 Predicted slip (sl) (decimal)

$$sl = \frac{1}{0.3\ Cn}\ln\left(\frac{0.75}{0.75 - \left(\frac{P}{W} + \frac{1.2}{Cn} + 0.04\right)}\right)$$

3.2.4 Tractive efficiency (TE)

$$TE = (1 - sl)\left(1 - \frac{\frac{1.2}{Cn} + 0.04}{0.75\left(1 - e^{-0.3Cn\ sl}\right)}\right)$$

3.3 Fuel efficiency varies by type of fuel and by percent load on the engine. Typical farm tractor and combine engines above 20% load are modeled by the equations below. Typical fuel consumption is given in L/kW·h [gal/hp-h] where X is the ratio of equivalent PTO power required by an operation to that maximum available from the PTO. These equations model fuel consumption 15% higher than typical Nebraska Tractor Test performance to reflect used equipment.

Gasoline . . . . $2.74X + 3.15 - 0.203\sqrt{697X}$

$\left[0.54X + 0.62 - 0.04\sqrt{697X}\right]$

Diesel . . . . $2.64X + 3.91 - 0.203\sqrt{738X + 173}$

$\left[0.52X + 0.77 - 0.04\sqrt{738X + 173}\right]$

LPG (liquefied petroleum gas) . $2.69X + 3.14 - 0.203\sqrt{646X}$

$\left[0.53X + 0.62 - 0.04\sqrt{646X}\right]$

3.4 Oil consumption is defined as the volume per hour of engine crankcase oil replaced at the manufacturer's recommended change interval. Consumption is in L/h [gal/h] where P is the rated engine power in kW (hp).

Gasoline . . . . 0.00056P + 0.02487

[0.00011P + 0.00657]

Diesel . . . . 0.00059P + 0.02169

[0.00021P + 0.00573]

LPG (liquefied petroleum gas) . . . . 0.00041P + 0.02

[0.00008P + 0.00755]

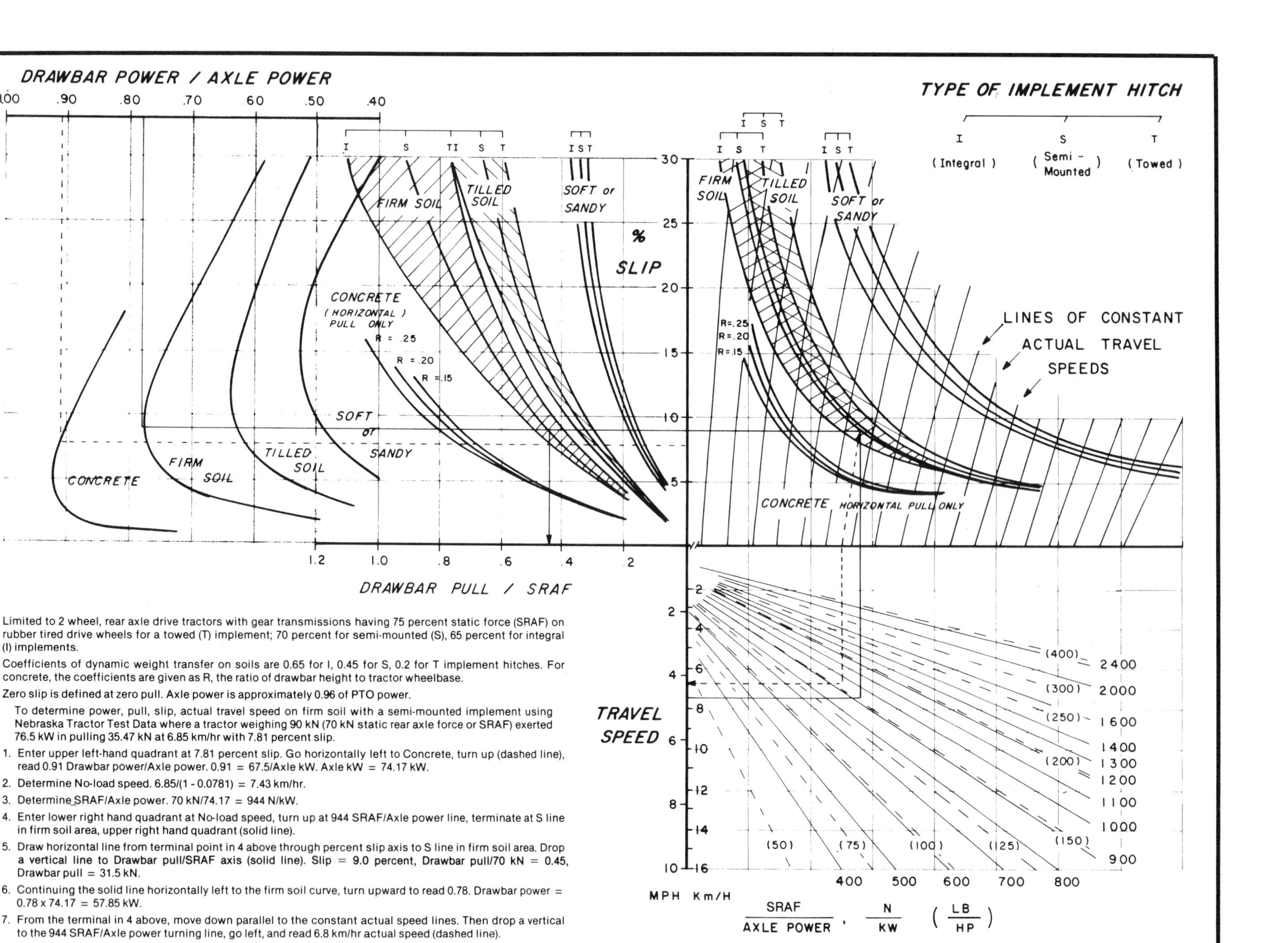

Limited to 2 wheel, rear axle drive tractors with gear transmissions having 75 percent static force (SRAF) on rubber tired drive wheels for a towed (T) implement; 70 percent for semi-mounted (S), 65 percent for integral (I) implements.

Coefficients of dynamic weight transfer on soils are 0.65 for I, 0.45 for S, 0.2 for T implement hitches. For concrete, the coefficients are given as R, the ratio of drawbar height to tractor wheelbase.

Zero slip is defined at zero pull. Axle power is approximately 0.96 of PTO power.

To determine power, pull, slip, actual travel speed on firm soil with a semi-mounted implement using Nebraska Tractor Test Data where a tractor weighing 90 kN (70 kN static rear axle force or SRAF) exerted 76.5 kW in pulling 35.47 kN at 6.85 km/hr with 7.81 percent slip.

1. Enter upper left-hand quadrant at 7.81 percent slip. Go horizontally left to Concrete, turn up (dashed line), read 0.91 Drawbar power/Axle power. 0.91 = 67.5/Axle kW. Axle kW = 74.17 kW.
2. Determine No-load speed. 6.85/(1 - 0.0781) = 7.43 km/hr.
3. Determine SRAF/Axle power. 70 kN/74.17 = 944 N/kW.
4. Enter lower right hand quadrant at No-load speed, turn up at 944 SRAF/Axle power line, terminate at S line in firm soil area, upper right hand quadrant (solid line).
5. Draw horizontal line from terminal point in 4 above through percent slip axis to S line in firm soil area. Drop **a vertical line to Drawbar pull/SRAF axis (solid line). Slip = 9.0 percent, Drawbar pull/70 kN = 0.45,** Drawbar pull = 31.5 kN.
6. Continuing the solid line horizontally left to the firm soil curve, turn upward to read 0.78. Drawbar power = 0.78 x 74.17 = 57.85 kW.
7. From the terminal in 4 above, move down parallel to the constant actual speed lines. Then drop a vertical to the 944 SRAF/Axle power turning line, go left, and read 6.8 km/hr actual speed (dashed line).

FIG. 1—TRACTOR DRAWBAR PERFORMANCE

SECTION 4—DRAFT AND POWER REQUIREMENTS

4.1 Draft data are reported as the force required in the horizontal direction of travel. Only functional draft (soil and crop resistance) is reported. Rolling resistance of transport wheels may have to be added to get total implement draft.

4.1.1 Tillage

4.1.1.1 Moldboard plows. Draft per unit cross-section of furrow slice is for bottoms equipped with high-speed moldboards, coulters, and landslide. Draft in N/cm² [lb/in.²], speed (S) in km/h [mph].

Silty clay (South Texas)	$7 + 0.049\ S^2$ [$10.24 + 0.185\ S^2$]
Decatur clay loam	$6 + 0.053\ S^2$ [$8.77 + 0.2\ S^2$]
Silty clay (N. Illinois)	$4.8 + 0.024\ S^2$ [$7 + 0.09\ S^2$]
Davidson loam	$3 + 0.020\ S^2$ [$4.5 + 0.08\ S^2$]
Sandy silt	$3 + 0.032\ S^2$ [$4.4 + 0.21 S^2$]
Sandy loam	$2.8 + 0.013\ S^2$ [$4 + 0.05\ S^2$]
Sand	$2 + 0.013\ S^2$ [$3 + 0.05\ S^2$]

Multiply equations by 1.07 for an added jointer or coverboard. An increase of 1 soil moisture percentage point can decrease draft 10%. An increase in apparent specific gravity of 0.1 can increase draft 10%. A C.V. of 0.13 is common in plow unit draft measurements.

4.1.1.2 Disk plows. Draft per unit cross-section of furrow slice for 66 cm [26 in.] diameter disk, 0.38 rad [22 deg] tilt, 0.785 rad [45 deg] angle. Draft in N/cm² [lb/in.²], speed (S) in km/h [mph].

Decatur clay	$5.2 + 0.039\ S^2$ [$7.6 + 0.15\ S^2$]
Davidson loam	$2.4 + 0.045\ S^2$ [$3.4 + 0.17\ S^2$]

4.1.1.3 Listers. Draft per 36 cm [14 in.] bottom at 6.76 km/h [4.2 mph]. Draft in N/bottom [lb/bottom], depth (d) in cm [in.].

Silty clay loam	$21.5\ d^2$ [$31.2\ d^2$]

4.1.1.4 Disk harrows. Draft per mass [weight] at any speed, typical working depth. Draft in N [lb], Mass (M) in kg [lb].

Clay	14.7 M [1.5 M]
Silt loam	11.7 M [1.2 M]
Sandy loams	7.8 M [0.8 M]

4.1.1.5 Chisel plows and field cultivators. Draft in firm soil per tool spaced at 30 cm [1 ft]. Includes wheel rolling resistance. Depth (d) = 8.26 cm [3.25 in.]. Draft (D) in N [lb] per tool, speed (S) in km/h [mph].

Loam (Saskatchewan)	520 + 49.2 S [117 + 17 S]
Clay loam (Saskatchewan)	480 + 48.1 S [108 + 16 S]
Clay (Saskatchewan)	527 + 36.1 S [118 + 12 S]

Draft at depth x (d_x) follows the relationship:

$$D_x = D_{8.26\ cm} \left(\frac{d_x}{8.26}\right)^2$$

$$\left[D_x = D_{3.25\ in.} \left(\frac{d_x}{3.25}\right)^2\right]$$

Variations in draft of 10% about the mean are common.

4.1.1.6 Rotary tillers. Effective draft per unit of cross-section of furrow slice, 45 cm [18 in.] diameter rotor, 10 cm [4 in.] depth, 6.7-11.7 r/s [400-700 rpm]. Unit draft in N/cm² [lb/in.²], bite length (b) in cm [in.].

Dry silt loam	$43.9\ b^{-0.46}$ [$41.8\ b^{-0.46}$]
Negative unit drawbar force for forward turning rotor	0.14 b [0.5 b]

4.1.1.7 One-way disk plow with seeder attachment. Draft per unit width, 7.5 cm [3 in.] depth of tillage. Includes rolling resistance. Draft in kN/m [lb/ft], speed (S) in km/h [mph].

Loam (Saskatchewan)	1.6 + 0.13 S [110 + 14 S]
Clay loam (Saskatchewan)	1.7 + 0.13 S [120 + 14 S]
Clay (Saskatchewan)	2 + 0.17 S [140 + 18 S]

Variations in draft of 10% about the mean are common. Increase draft 3.5% in loam, 7% in clay loam and 20% in clay for each cm increase in depth.

4.1.1.8 Subsoiler. Draft per shank per unit depth. Draft in N/shank [lb/shank], depth (d) in cm [in.].

Sandy loam	120-190 d [70-110 d]
Medium or clay loam	175-280 d [100-160 d]

4.1.1.9 Minor tillage tools. Draft per unit width, average for all soils. Draft in N/m [lb/ft].

Land plane	4400-11 600 [300-800]
Spike tooth harrow	440-730 [30-50]
Spring tooth harrow	1460-2190 [100-150]
Rod weeder	880-1830 [60-125]
Roller or packer	440-880 [30-60]

4.1.2 **Seeding**

4.1.2.1 **Row planters.** Draft per row. Includes rolling resistance. Draft in N/row [lb/row], loam soils, good seedbed.

Seeding only	450-800/row [100-180/row]
Seed, fertilizer herbicides	1100-2000/row [250-450/row]

4.1.2.2 **Grain drills.** Draft per furrow opener. Includes rolling resistance. Draft in N/opener [lb/opener].

Regular	130-450 [30-100]
Deep furrow	335-670 [75-150]

4.1.3 **Cultivation.** All draft per unit effective width at typical field speeds. Draft in N/m [lb/ft], speed (S) km/h [mph], depth (d) in cm [in.].

Row cultivator	115-230 d [20-40 d]
Lister cultivator	730-2200 d [50-150 d]
Rotary hoe	440 + 21.7 S [30 + 2.4 S]

4.1.4 **Fertilizer and chemical application**

Anhydrous ammonia applicator	1800 N per knife [400 lb per knife]
Fertilizer, pesticide distributors	Rolling resistance only

4.2 Rolling resistance is an additional draft force that must be included in computing implement power requirements. Values of rolling resistance depend on transport wheel dimensions, tire pressure, soil type, and soil moisture. Soil moistures are assumed to be less than field capacity for implement operations. Coefficients of rolling resistance are defined in ASAE Standard S296, Uniform Terminology for Traction of Agricultural Tractors, Self-Propelled Implements, and Other Traction and Transport Devices, and predicted by paragraph 3.2.1.2.

4.2.1 The values given in paragraph 3.2.1 are for single wheels in undisturbed soil. For loose, tilled soils and for sands, the coefficient for a rear wheel operating in the track of a front wheel is about 0.5 of the given value. For stubble ground the value is 0.9. For firm surfaces there is no reduction.

4.2.2 Extra width or flotation tires will reduce the coefficient appreciably on soft soils but will increase it for hard soils and concrete.

4.2.3 Coefficients of rolling resistance increase with increased tire pressure in soft soils. Doubling the tire pressure to 200 kPa [29 psi] causes the coefficient to increase to -0.0135 + 1.27 x coefficient at 100 kPa.

4.2.4 An effective coefficient of rolling resistance (C_e) can be computed for use on slopes.

$$C_e = C \cos \alpha \pm \sin \alpha$$

where C is the coefficient on level land (see paragraph 3.2.1.2) and α is the slope. The minus sign is to be used for going down slopes.

4.3 Rotary power data are reported as functional power required at the implement engine or tractor PTO shaft. Rolling resistance must be added to obtain total power requirements. Power in kW [hp] per unit given, S in km/h [mph]. Feed rate (F) typical material, wet basis (wb), in kg/s [lb/s].

Cutterbar mower, alfalfa	1.2/m [0.5/ft]
Cutterbar mower-conditioner, alfalfa	3.7-4.9/m [1.5-2/ft]
Flail mower-conditioner, alfalfa	8.2 + 2.13 F [11.0 + 1.3 F]
Conditioner only, alfalfa	2.45/m [1/ft]
Side delivery rake 2.44 m width [8 ft width]	-0.186 + 0.052 S [-0.25 + 0.25 S]
Baler field, rectangular, normal hay or straw	2.95 F [1.8 F]

Multiply by 1.5 for tough crop, high density.

Forage harvester, flail, green forage. . . (see flail-mower conditioner)
Multiply by 2.0 for other forages.

Forage harvester, shear bar

Corn	1.5 + 3.3 F [2 + 2 F]

Multiply by 1.33 for green alfalfa.
Multiply by 2.0 for low moisture forage and hay.
A recutter screen may double the above.

Combines, SP F, typical feed rate; based on 20% wb strawrack material.

Soybeans & small grain	7.5 + 7.5 F [10 + 4.6 F]

For corn multiply by 3 (estimated).
Power peaks may be 100% greater than above.
For windrowed material, multiply by 0.9.

Cotton pickers	7.5-11.0/row [10-15/row]
Cotton strippers	1.5-2.2/row [2-3/row]
Beet topper	3.7-5.2/row [5-7/row]
Beet harvester (PTO)	1.5-3.0/row [2-4/row]
plus draft, kN [lb]	2.0-4.0/row [450-900/row]
Potato digger (PTO)	0.75-1.5/row [1-2/row]
plus draft, kN [lb]	2.2-3.5/row [500-800/row]

SECTION 5—MACHINE PERFORMANCE

5.1 Performance rates for field machines depend upon achievable field speeds and upon the efficient use of time. Field speeds may be limited by heavy yields, rough ground, and adequacy of operator control. Small or irregularly shaped fields, heavy yields and high capacity machines may cause a substantial reduction in field efficiency. Typical speeds and field efficiencies are given in Table 1.

5.2 Slippage of drive wheel, decimal, for ground driven implements (see paragraph 3.2.3)

$$\text{Slippage (decimal)} = \frac{1}{0.3\ Cn} \ln \left(\frac{0.75}{\frac{T}{rW} + \frac{1.2}{Cn} + 0.079} \right)$$

where T is the torque due to mechanism operation on the drive wheel of rolling radius r.

SECTION 6—COSTS OF USE

6.1 Ownership costs for field machines are those fixed costs that are not associated with the amount of use. Nationwide averages are shown below for on-farm remaining values of machines after depreciation. Formulas for the highly variable costs of repair and maintenance are shown based on both the 1970 Midwest values and on earlier nationwide values. Multiply data by $(1+i)^n$ for inflation effects.

6.1.1 Remaining values as a percentage of the list price at the end of year n.

Tractors	$68(0.920)^n$
All combines, cotton pickers, SP windrowers	$64(0.885)^n$
Balers, forage harvesters, blowers, and SP sprayers	$56(0.885)^n$
All other field machines	$60(0.885)^n$

6.1.2 Repair and maintenance costs are highly variable and unpredictable as to time of occurrence. Surveys of accumulated repair and maintenance costs (no charge for farmer labor) related to accumulated use do show consistent trends; however, a standard deviation equal to the mean is a typical variation in these data. Repair and maintenance factors based upon the accumulated use of the machine are given in Table 1. Values listed are for machines used under typical field conditions and speeds.

SECTION 7—RELIABILITY

7.1 Operational reliability is a probability of satisfactory machine function over any given time period. It is computed as one minus the probability of a failure.

TABLE 1—FIELD EFFICIENCY, FIELD SPEED AND REPAIR AND MAINTENANCE COST PARAMETERS

Machine	Field efficiency		Field speed				Estimated life	Total life repairs	Repair factors	
	Range %	Typical %	Range mph	Typical mph	Range km/h	Typical km/h	h	Percent of list price	RF1	RF2
TRACTORS										
2 wheel drive & stationary							10 000	120	0.012	2.0
4 wheel drive & crawler							10 000	100	0.010	2.0
TILLAGE										
Moldboard plow	70-90	80	3.0 - 6.0	4.5	5.0 - 10.0	7.0	2 000	150	0.43	1.8
Heavy-duty disk	70-90	85	3.5 - 6.0	4.5	5.5 - 10.0	7.0	2 000	60	0.18	1.7
Tandem disk harrow	70-90	80	3.0 - 6.0	4.0	5.0 - 10.0	6.5	2 000	60	0.18	1.7
Chisel plow	70-90	85	4.0 - 6.5	4.5	6.5 - 10.5	7.0	2 000	100	0.38	1.4
Field cultivator	70-90	85	3.0 - 8.0	5.5	5.0 - 13.0	9.0	2 000	80	0.30	1.4
Spring tooth harrow	70-90	85	3.0 - 6.0	5.0	5.0 - 10.0	9.0	2 000	80	0.30	1.4
Roller-packer	70-90	85	4.5 - 7.5	6.0	7.0 - 12.0	10.0	2 000	40	0.16	1.3
Mulcher-packer	70-90	80	4.0 - 6.0	5.0	6.5 - 10.0	8.0	2 000	40	0.16	1.3
Rotary hoe	70-85	80	5.0 - 10.0	7.0	8.0 - 16.0	11.0	2 000	60	0.23	1.4
Row crop cultivator	70-90	80	2.5 - 5.0	3.5	4.0 - 8.0	5.5	2 000	100	0.22	2.2
Rotary tiller	70-90	85	1.0 - 4.5	3.0	2.0 - 7.0	5.0	1 500	80	0.36	2.0
PLANTING										
Row crop planter:										
No-till tillage	50-75	65	2.0 - 4.0	3.0	3.2 - 6.4	4.8	1 200	80	0.54	2.1
Conventional tillage	50-75	60	3.0 - 7.0	4.5	4.8 - 9.7	6.4	1 200	80	0.54	2.1
Grain drill	65-85	70	2.5 - 6.0	4.0	4.0 - 9.7	6.4	1 200	80	0.54	2.1
HARVESTING										
Corn picker sheller	60-75	65	2.0 - 4.0	2.5	3.0 - 6.5	4.0	2 000	70	0.14	2.3
Combine:										
Pull-type	60-75	65	2.0 - 5.0	3.0	3.0 - 6.5	5.0	2 000	90	0.18	2.3
Self-propelled	65-80	70	2.0 - 5.0	3.0	3.0 - 6.5	5.0	2 000	50	0.12	2.1
Mower	75-65	80	4.0 - 7.0	5.0	6.5 - 11.0	8.0	2 000	150	0.46	1.7
Mower-conditioner	55-80	75	3.0 - 6.0	4.5	5.0 - 10.0	7.0	2 000	80	0.26	1.6
Side delivery rake	70-85	80	4.0 - 5.0	4.5	6.5 - 8.0	7.0	2 000	100	0.38	1.4
Baler	60-85	75	2.5 - 5.0	3.5	4.0 - 8.0	5.5	2 000	80	0.23	1.8
Big bale baler	55-75	65	3.0 - 5.0	3.5	5.0 - 8.0	5.5	2 000	80	0.23	1.8
Long hay stacker	55-75	60	2.5 - 4.5	3.5	4.0 - 7.0	5.5	2 000	80	0.23	1.8
Forage harvester:										
Pull-type	50-75	65	1.5 - 5.0	2.5	2.5 - 8.0	4.0	2 000	80	0.23	1.8
Self-propelled	60-85	70	1.5 - 6.0	3.0	2.5 - 10.0	5.0	2 500	60	0.12	1.8
Sugar beet harvester	60-80	70	2.5 - 5.0	3.0	4.0 - 8.0	5.0	2 500	70	0.19	1.4
Potato harvester	55-70	60	1.5 - 4.0	2.0	2.5 - 6.5	3.0	2 500	70	0.19	1.4
Cotton picker or stripper	60-75	70	2.0 - 4.0	3.0	3.0 - 6.0	4.5	2 000	60	0.17	1.8
MISCELLANEOUS										
Fertilizer spreader	60-70	70	3.0 - 5.0	4.5	5.0 - 8.0	7.0	1 200	120	0.95	1.3
Boom-type sprayer	50-80	65	3.0 - 7.0	6.5	5.0 - 11.5	10.5	1 500	70	0.41	1.3
Air-carrier sprayer	55-70	60	2.0 - 5.0	3.0	3.0 - 8.0	5.0	2 000	60	0.20	1.6
Bean puller-windrower	70-90	80	2.0 - 5.0	3.5	3.0 - 8.0	5.5	2 000	60	0.20	1.6
Beet topper stalk chopper	60-80	70	2.0 - 3.0	2.5	3.0 - 5.0	4.0	2 000	60	0.23	1.4
Forage blower							2 000	50	0.14	1.8
Wagon							3 000	80	0.19	1.3

7.1.1 Midwestern U.S. reports by farmers (1970) of field failures show the probability of failure (tractors and implements combined) per 40 ha [100 acres] of use and the average S.D. of the total downtime per year for farms of over 200 ha [500 acres].

	Breakdown time h/yr	S.D.	Breakdown probability per 40 ha [100 acres]	Reliability per 40 ha [100 acres]
Tillage	13.6	24.1	0.109	0.89
Planting corn	5.3	5.4	0.133	0.87
Planting soybeans	3.7	2.4	0.102	0.90
Row cultivation	5.6	6.3	0.045	0.96
Harvest soybeans (SP)	8.2	9.6	0.363	0.64
Harvest corn (SP)	12.3	12.6	0.323	0.68

7.1.2 Breakdown probabilities for machine systems increase with an increase in the size of the farm.

Crop area, ha [acres]	Probability of at least one failure per year	Reliability of tractor-machine system per year
0-80 [0-200]	0.435	0.56
80-160 [200-400]	0.632	0.37
160-240 [400-600]	0.713	0.29
240+ [600+]	0.780	0.22

7.1.3 Downtime and reliability appear to be independent of use for some machines while others have shown an increase with accumulated use. Midwestern U.S. data show: Moldboard plows average 1 h of downtime for each 400 ha [1000 acres] of use; row planters average 1 h of downtime for each 250 ha [600 acres] of use; S.P. combines had little downtime for the first 365 ha [900 acres] of use. Downtime was a constant 1 h for each 30 ha [70 acres] afterward; and tractors had a constantly increasing downtime rate with use. The accumulated hours of downtime depend upon the accumulated hours of use (X).

Spark ignition $0.0000021\, X^{1.9946}$

Diesel $0.0003234\, X^{1.4173}$

SECTION 8—WORKING DAYS, TIMELINESS

8.1 Freezing temperatures, precipitation, excessive deficient soil moistures, and other weather related factors may limit field machine operations. As weather variability is great, any prediction of the number of future working days can only be made probabilistically.

8.2 The number of working days in any time period is a function of: climatic region, slope of soil surface, soil type, drainage characteristics, operation to be performed, and traction and flotation devices.

8.3 Probabilities for a working day (pwd) are given in Table 2 for both 50 and 90% confidence levels. The probabilities obtained from the table are averages for biweekly periods. That is, a probability of 0.4 implies that 0.4 x 14 or 5.6 working days could be expected in that 2 wk period. If the probability were taken at the 50% level, the 5.6 day figure would be exceeded in 5 yrs out of 10. If at the 90% level, the 5.6 day figure would be exceeded in 9 yrs out of 10.

TABLE 2—PROBABILITIES FOR A WORKING DAY

	Region	Central Illinois		State of Iowa		South-eastern Michigan		State of South Carolina		Southern Ontario Canada		Mississippi Delta	
	Soil	Prairie Soils		State Average		Clay Loam		Clay Loam		Clay Loam		Clay	
	Notes	18 yrs. data. In early spring and late fall, pwd in Iowa and Illinois may be 0.07 greater in North and West and 0.07 less in South and East		17 yrs. data		Simulation (tillage only)		Simulation (tillage only) Sandy soils can be worked all months and have higher pwd		Simulation (tillage only) Start 7-10 days earlier on sandy soils, 0.15 greater pwd		Simulation (tillage only) Non-tillage field work pwd and pwd for sandy soils some greater in winter and early spring	
		Probability Level, Percent											
Ave. Date	Biweekly Period	50	90	50	90	50	90	50	90	50	90	50	90
Jan. and Feb.	—	0.0	0.0	0.0	0.0	0.0	0.0	0.01	0.0	0.0	0.0	0.07	0.0
Mar. 7	1	0.0	0.0	0.0	0.0	0.0	0.0	—	—	0.0	0.0	—	—
Mar. 21	2	0.29	0.0	0.0	0.0	0.0	0.0	0.03	0.0	0.0	0.0	0.18	0.0
Apr. 4	3	0.42	0.13	0.39	0.16	0.0	0.0	—	—	0.01	0.0	—	—
Apr. 18	4	0.47	0.19	0.57	0.38	0.20	0.0	0.29	0.06	0.07	0.0	0.35	0.08
May 2	5	0.54	0.31	0.66	0.48	—	—	—	—	0.62	0.02	—	—
May 16	6	0.61	0.34	0.68	0.47	0.61	0.32	0.64	0.37	0.60	0.02	0.58	0.28
May 30	7	0.63	0.40	0.66	0.47	—	—	—	—	0.79	0.16	—	—
June 13	8	0.66	0.41	0.69	0.52	0.69	0.42	0.72	0.46	0.77	0.22	0.69	0.39
June 27	9	0.72	0.53	0.74	0.57	—	—	—	—	0.80	0.23	—	—
July 11	10	0.72	0.52	0.77	0.64	0.75	0.52	0.67	0.43	—	—	0.63	0.25
July 25	11	0.72	0.54	0.80	0.67	—	—	—	—	—	—	—	—
Aug. 8	12	0.78	0.64	0.80	0.68	0.74	0.53	0.73	0.51	—	—	0.72	0.45
Aug. 22	13	0.86	0.74	0.86	0.79	—	—	—	—	—	—	—	—
Sept. 5	14	0.81	0.66	0.79	0.64	0.70	0.35	—	—	—	—	—	—
Sept. 19	15	0.65	0.42	0.69	0.46	—	—	0.72	0.46	—	—	0.80	0.58
Oct. 3	16	0.72	0.52	0.71	0.48	0.59	0.26	—	—	—	—	—	—
Oct. 17	17	0.76	0.58	0.79	0.64	—	—	0.61	0.23	—	—	0.76	0.42
Nov. 1	18	0.72	0.50	0.75	0.55	0.42	0.06	—	—	—	—	—	—
Nov. 15	19	0.67	0.47	0.73	0.54	—	—	0.33	0.02	—	—	0.43	0.0
Nov. 29	20	0.54	0.43	0.82	0.70	0.07	0.0	—	—	—	—	—	—
Dec. 13	21	—	—	—	—	—	—	0.02	0.0	—	—	0.10	0.0

Adjust for Sundays and holidays by multiplying pwd's above by 0.86, 0.82, 0.78 and 0.75 for months 0, 1, 2 and 3 holidays.

8.3.1 Two types of field operations are identified—soil working operation such as tillage and seeding and traffic operations where a crop is processed and the soil needs to be dry enough only to provide machine support. The Illinois and Iowa data in Table 2 are reports of actual observed operations and include both types of operations. The other data are simulations for tillage operations only.

8.3.2 Dry western farms and farms under irrigation are likely to have a pwd approaching 1.0.

8.4 Persistence is recognized in weather data. Given that a particular day is a working day, the succeeding day has about a 0.8 (Midwest) probability of being a working day also. The probability of 5 consecutive working dates is the pwd for day 1 multiplied by $(0.8)^4$.

8.5 Timeliness considerations (see ASAE Standard S322, Uniform Terminology for Agricultural Machinery Management, Section 2—Terms Used Primarily for System Analysis) are important to efficient selection of farm machinery. An economic value for timeliness is required to include the penalty for both quantity and quality reductions in the crop return from prolonged field machinery operations. Timeliness costs vary widely. Variation is expected among regions, crop varieties, time of the season, and machine operations. Timeliness costs are essentially zero for those tillage and other operations where there is little need to finish quickly.

8.6 A timeliness coefficient (K) (see ASAE Standard S322, Uniform Terminology for Agricultural Machinery Management, Section 2—Terms Used Primarily for System Analysis) is a factor that permits computation of timeliness costs, (see ASAE Engineering Practice EP391, Agricultural Machinery Management, Section 8—Selection of Field Machine Capacity). This factor assumes linear timeliness costs with calendar days and is expressed as a decimal of maximum value of the crop per unit area per day either before or after the optimum day. These coefficients can be calculated from measured crop returns as they vary with the timing of machine operations. For example, if 10 days delay in an operation reduces the eventual return from the crop by 5%, K is calculated as 0.05/10 or 0.005 per unit area per day. The cost of operating on 6 ha of $100/ha crop 7 days after the optimum would be 0.005x6x7x100 = $21. (For the timeliness costs for harvesting a total field, see ASAE Engineering Practice EP391, Agricultural Machinery Management, Section 8—Selection of Field Machine Capacity).

8.7 Values of K have been determined for several operations. The following list was derived from crop research reports.

Operation	K		
Tillage (depends on whether planting is delayed by prior tillage)		0.000-0.010	
Seeding			
Corn (Indiana, Illinois, Iowa, Eastern Nebraska, Eastern Kansas)			
Available moisture in root zone at planting, cm	April	May	June
10	0.010	0.000	-0.002
20	0.006	0.001	-0.003
30	0.003	0.004	-0.007
Wheat, Utah		0.008	
North Dakota		0.007	
Soybeans, Wisconsin, May & June		0.005	
Missouri, Illinois, June		0.006	
Double crop after wheat, Illinois		0.010	
Cotton, Lubbock, Texas			
April		0.004	
May		0.020	
Mississippi, April & May		0.007	
Barley, Utah		0.008	
North Dakota		0.007	
Oats, Illinois and Michigan		0.010	
Wisconsin after May 6		0.012	
Alabama, Fall		0.000	
Utah		0.008	
Rape, Manitoba		0.003	
Rice, California, May		0.010	
Row Cultivation, Illinois, soybeans		0.011	
Rotary hoeing, Iowa, soybeans		0.028	
Harvest			
Haymaking, Michigan, June		0.018	
Shelled corn, Iowa		0.003	
Ear corn, Illinois, after Oct. 26		0.007	
Soybeans, Illinois (depends on variety)		0.006-0.010	
Wheat, Ohio		0.005	
Cotton, Alabama		0.002	
Rice, California		0.009	
Sugar cane, Queensland, Australia			
pre-optimum		0.002	
post-optimum		0.003	

Cited Standards:

ASAE S296, Uniform Terminology for Traction of Agricultural Tractors, Self-Propelled Implements, and Other Traction and Transport Devices
ASAE S313, Soil Cone Penetrometer
ASAE S322, Uniform Terminology for Agricultural Machinery Management
ASAE EP391, Agricultural Machinery Management

ASAE Engineering Practice: ASAE EP236

GUIDE FOR PLANNING AND REPORTING TILLAGE EXPERIMENTS

Recommendation prepared by ASAE Cultural Practices Committee of the Power and Machinery Division; approved by the ASAE Power and Machinery Division Technical Committee; adopted by ASAE as a Recommendation 1962; reconfirmed December 1966, December 1971, December 1976; reclassified as an Engineering Practice December 1977; reconfirmed December 1981, December 1986.

The suggested guide for reporting tillage experiments has been written in an effort to establish a uniform method of recording data. It has been recognized that all researchers will not wish to record and report all of the data which might be taken from a given tillage research project. The complete outline represents the optimum information which might be taken. The minimum outline varies only in the record section (Section 6).

It is believed that every tillage project, regardless of extent, should include at least those items which are marked with an asterisk.*

SECTION 1—ABSTRACT

1.1 Objective of project

1.2 Year work initiated

1.3 Results of research in abstract form. Give only the important findings. Include soil type, plot size, crop variety, row spacing, short description of equipment, and number of years of results being reported.

SECTION 2—DESCRIPTION OF EXPERIMENT

2.1 Design of experiment

2.1.1 Plot layout

2.1.2 Statistical methods

2.2 Treatments

2.2.1 Conventional practice for area (check plots)

2.2.2 Variations from conventional

SECTION 3—DESCRIPTION OF IMPLEMENTS USED

3.1 Primary tillage equipment

3.2 Secondary tillage equipment

3.3 Fertilizer applicators

3.4 Planting equipment

3.5 Pesticide equipment

3.6 Cultivating equipment

SECTION 4—PROCEDURE

4.1 Pertinent steps required to perform experiment

SECTION 5—RESULTS

5.1 Crop yields

5.2 Weed yields

5.3 Speed and power requirements (note item 6.19)

5.3.1 Tillage tools

5.3.2 Planter

5.3.3 Fertilizer applicator

5.3.4 Cultivator

5.3.5 Harvesting equipment

5.4 Costs and labor

5.4.1 Requirements for practical application

SECTION 6—RECORDS, OBSERVATIONS, AND MEASUREMENTS

(All Listed Here Are Desirable—
Items Marked With An Asterisk Are Essential)

6.1 Location of plots

6.1.1 State, county, section, township, nearby city

6.2 History of field

6.2.1 Cropping history (previous two years)

6.2.1.1 Crops

6.2.1.2 Yields

6.2.1.3 Residues returned or removed

6.2.1.4 Commercial fertilizer

6.2.1.5 Manure

6.2.1.6 Lime

6.2.1.7 Green manure or cover crop

6.2.2 Tillage practices (previous year)

6.2.2.1 Description

***6.3** Topography of test plots

***6.3.1** Degree of slope

6.3.2 Length of slope

6.3.3 Uniformity of slope

6.3.4 Row direction with respect to slope

6.3.5 Row direction with respect to compass bearings

***6.4** Soil

***6.4.1** Soil type and phase

***6.4.1.1** Restrictive layer if any

***6.4.1.2** Degree of erosion

***6.4.2** Soil test (pH-P & K)

***6.4.2.1** Soil test results-specify method used

***6.4.2.2** Fertility level-high, medium or low

***6.4.3** Organic matter content

***6.4.4** Soil moisture

***6.4.4.1** Optimum-wet or dry at tillage time

6.4.4.2 Moisture characteristic curve by horizon

6.4.4.2.1 Determine before tillage

6.4.4.2.2 Determine at tillage

6.4.4.2.3 Determine at harvest

***6.4.5** Drainage condition (good, fair or poor)

***6.4.5.1** Tile spacing and depth-size, length of run

***6.4.5.2** Tile depth to plot area

6.4.6 Plastic limit

6.4.6.1 Determination of lower plastic limit by horizons

6.4.6.2 Bulk density

6.4.6.2.1 Before treatments

6.4.6.2.2 After treatments

***6.5** Rainfall record during growing season

***6.5.1** Inches of water and dates fell

6.5.2 Estimate of duration of oxygen deficiency

6.5.3 Runoff determination

***6.6** Irrigation

***6.6.1** Used or not used

6.6.2 Type

***6.6.3** System used to determine time of application

6.6.4 Total water applied

***6.6.5** Number and dates of application

***6.7** Seed

***6.7.1** Germination test

***6.7.2** Laboratory test

6.7.3 From seed tag

***6.7.4** Variety

*6.7.5 Description
*6.7.6 Treatment of seed
6.7.7 Planting rate
*6.8 Row space
*6.9 Plant population
6.9.1 At time of emergence
*6.9.2 At time of harvest
*6.10 Commercial fertilizer
*6.10.1 Type
*6.10.2 Grade
*6.10.3 Rate
6.10.4 Placement
*6.10.5 Method of application
6.10.6 Date of application
6.11 Manure
6.11.1 Kind
6.11.2 Rate
6.11.3 Date of application
6.12 Lime
6.12.1 Kind
6.12.2 Screen size
6.12.3 Rate
6.12.4 Date of application
*6.13 Primary tillage
*6.13.1 Name of tool used
*6.13.2 Depth
*6.13.3 Spacing
6.13.4 Date performed
6.13.5 Speed
6.13.6 Draft
6.13.7 Special adjustments
6.14 Secondary tillage
*6.14.1 Name of tool used
*6.14.2 Degree of refinement
6.14.3 Date performed
6.14.4 Speed
6.14.5 Draft
6.14.6 Special adjustments
*6.15 Planting equipment
*6.15.1 Description of planter
*6.15.1.1 Mounted or pull type
*6.15.1.2 Type of furrow opener
*6.15.1.3 Press wheels (initial seed and final)
*6.15.1.3.1 Type
*6.15.1.3.2 Pressure on soil (compaction in psi plus description)
*6.15.1.3.3 Technique used to determine pressure
*6.15.1.4 Seed covering devices
*6.15.1.5 Packing or smoothing devices other than press wheel
*6.15.2 Seed placement
*6.15.2.1 Depth
*6.15.2.2 Uniformity
6.15.2.3 Soil condition in area about seed
*6.16 Weed control
6.16.1 General statement of weed problem
6.16.2 Varieties of weeds
*6.16.3 Mechanical control methods used
6.16.3.1 Description of equipment used
6.16.3.2 Dates of cultivation
6.16.3.3 Estimate of effectiveness
*6.16.4 Chemical control methods used
6.16.4.1 Date of application
6.16.4.2 Description of application equipment
6.16.4.3 Chemicals used
*6.17 Miscellaneous crop information
*6.17.1 Planting date
*6.17.2 Notes on growth characteristics
*6.17.3 Harvest date
*6.17.4 Degree of lodging
*6.17.5 Insect or other damage
*6.17.6 Barren stalks
*6.18 Special treatments
*6.18.1 Fumigation
6.18.2 Mulching
6.18.3 Other
6.19 Cost study should be done on practical field scale
6.19.1 Time or field performance rates
6.19.2 Power requirements
6.19.2.1 General description in instrumentation, soil conditions, etc.
6.19.2.2 Horsepower used per hour per acre
6.19.2.3 Fuel consumption per acre
6.19.3 Power, labor, time and cost comparisons between experimental and conventional
*6.20 Illustrations
*6.20.1 Photographs
6.20.1.1 Equipment
6.20.1.2 Plots (before and after each treatment)
6.20.2 Drawings
6.20.2.1 Schematic or perspective drawings of equipment
6.20.2.2 Plot layouts

ASAE Standard: ASAE S238.1

VOLUMETRIC CAPACITY OF FORAGE WAGONS, WAGON BOXES, AND FORAGE HANDLING ADAPTATIONS OF MANURE SPREADERS

Adopted by ASAE 1961; editorial revision and reconfirmation without technical change by the Power and Machinery Division Technical Committee December 1965; reconfirmed December 1970, December 1975, December 1980; revised April 1986.

SECTION 1—PURPOSE

1.1 This Standard is intended to provide a uniform method for calculating and expressing the volumetric capacity of forage wagons, wagon boxes, and forage handling adaptations of manure spreaders.

SECTION 2—CAPACITY DETERMINATION

2.1 To conform to this Standard, wagons must be rated in cubic meters (cubic feet), struck level, using the inside dimensions of the box. In the case of forage wagons with mixing or agitating mechanisms at the front or rear of the box, the length shall be limited by a vertical plane tangent to the restraining member of the beater, or to any restricting surface should the surfaces fall inside the restraining member of the beater. If the beaters are not in place in a common vertical plane, the dimensions shall be calculated in accordance with a rectilinear system of vertical and horizontal tangential planes to the restraining member (see Fig. 1).

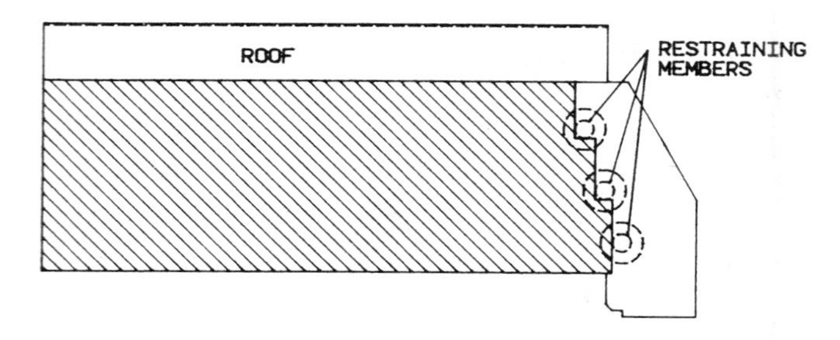

FIG. 1—CONFIGURATION FOR CALCULATING CAPACITY OF FORAGE BOXES

HITCH AND BOX DIMENSIONS FOR AGRICULTURAL GRAIN WAGONS

Adopted by ASAE February 1954; reconfirmed by the Power and Machinery Division Technical Committee December 1962, December 1967, December 1972, December 1977, December 1978, December 1979, December 1980; revised April 1986.

SECTION 1—PURPOSE

1.1 This ASAE Standard was developed to assist manufacturers of wagons and grain-harvesting machinery in providing the farmer with units that will permit ready coupling between wagons and grain-harvesting machines, or between wagons and tractors, if the grain-harvesting unit is mounted on the tractor.

1.2 The specifications cover the dimensions of a farm wagon designed to permit satisfactory use when the wagon is hitched to a grain-harvesting unit. Only essential dimensions are specified to permit the designers of wagons and grain-harvesting machines the greatest possible latitude in their work.

SECTION 2—SPECIFICATIONS

2.1 The rcommended wagon-tongue length is 1956 mm (77 in.) from the centerline of the front wheels to the centerline of the hitch pin. When an adjustable-length tongue is provided, one adjustment should be the recommended length of 1956 mm (77 in.). It is further recommended that this position be marked for ready identification.

2.2 The wagon box should not extend more than 559 mm (22 in.) from the centerline of the front wheels. A 45-deg line with the horizontal, sloping toward the rear and passing through a point 1626 mm (64 in.) above the ground line and 559 mm (22 in.) forward of the centerline of the front wheels, establishes the clearance line for the forward end of the wagon box. The wagon box may have a maximum height of 1778 mm (70 in.) above the ground, provided the forward end does not project beyond the 45-deg clearance line.

2.3 The overall width of the wagon and/or wagon box should not exceed 2438 mm (96 in.).

2.4 The recommended bolster heights from the ground line are as follows

711 mm (28 in.) maximum for an autosteer wagon
813 mm (32 in.) for a fifth-wheel wagon

2.5 The recommended stake spacing for the bolster is adjustable between 965 mm (38 in.) and 1067 mm (42 in.).

2.6 If a towplate is used on the wagon-gear rear axle for tandem hitching, the hitch hole should be 20.6 mm (0.81 in.) diameter and the dimensions and mounting of the plate shall conform to the appropriate drawbar and hitch specifications of ASAE Standard S203, Rear Power Take-Off for Agricultural Tractors. The towplate is to be mounted on the centerline of the wagon with the hitch-hole centerline not less than 50 mm (2 in.) behind the axle.

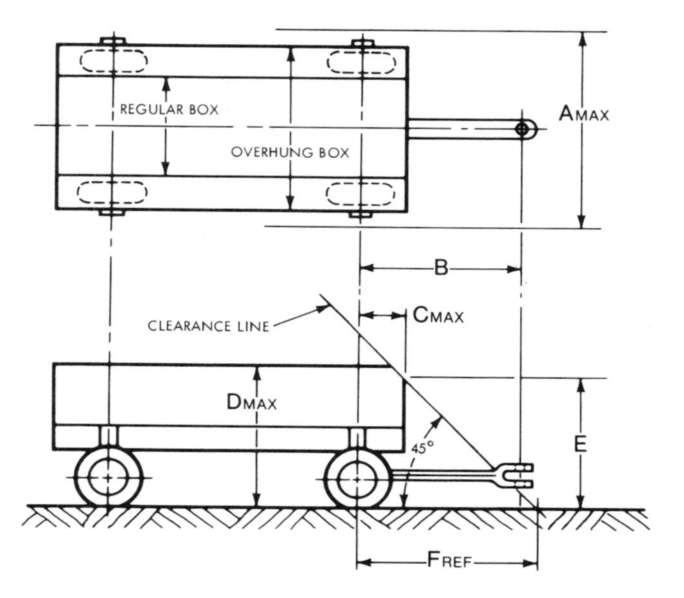

FIG. 1—HITCH AND BOX DIMENSIONS FOR AGRICULTURAL GRAIN WAGONS

TABLE 1—HITCH AND BOX DIMENSIONS FOR AGRICULTURAL GRAIN WAGONS

	mm	in.
A	2438	96
B	1956	77
C	559	22
D	1778	70
E	1626	64
F	2184	86

Cited Standard:

ASAE S203, Rear Power Take-Off for Agricultural Tractors

ASAE Standard: ASAE S276.3 (ANSI/ASAE S276.3/SAE J943 SEP83)

SLOW-MOVING VEHICLE IDENTIFICATION EMBLEM

Developed cooperatively by the ASAE Farm Safety Committee and the National Safety Council's Farm Conference Studies and Research Committee; approved by the ASAE Power and Machinery Division Technical Committee and adopted by ASAE as a Recommendation in December 1964; revised and adopted as a Standard December 1966; revised March 1968; approved by ANSI as an American National Standard January 1971; reconfirmed December 1973; revised December 1975; error corrected in percent peak reflectance, Table 1, June 1976; reconfirmed December 1980; revision approved by ANSI as an American National Standard July 1984; reconfirmed December 1985.

SECTION 1—PURPOSE

1.1 The purpose of this Standard is to establish specifications which define a unique identification emblem to be used only for slow-moving vehicles when operated or traveling on highways.

SECTION 2—SCOPE

2.1 This Standard establishes emblem dimensional specifications, performance requirements, related test procedures and mounting requirements.

2.2 This unique identification emblem shall be used only on slow-moving vehicles which are designed for and travel at rates of speed less than 40 km/h (25 mph).

2.3 The identification emblem shall supplement but not replace warning devices such as tail lamps, reflectors, or flashing lights and shall not be used to identify stationary objects or stopped vehicles.

2.4 The dimensions and color patterns of the emblem have been established as a unique identification and shall not be altered to permit advertising or other markings on the face of the emblem, except as permitted in paragraph 4.2.

SECTION 3—DEFINITIONS

3.1 Highway: The entire width between the boundary lines of every way publicly maintained when any part thereof is open to the use of the public for purposes of vehicular travel.

3.2 Permanent-mounted emblem: A yellow-orange triangle with a dark red border as illustrated in Fig. 1 and securely affixed to a slow-moving vehicle.

3.3 Portable emblem: A yellow-orange triangle with a dark red border as illustrated in Fig. 1 securely affixed to a backing material as illustrated in Fig. 2 and displayed on a slow-moving vehicle.

SECTION 4-DESCRIPTION

4.1 The identification emblem, Fig. 1, consists of a fluorescent, yellow-orange equilateral triangle with a dark red retro-reflective border positioned with a point of the triangle up. The yellow-orange fluorescent triangle provides for daylight identification. The reflective border defines the shape of the fluorescent color in daylight and appears as a hollow red triangle in the path of motor vehicle headlights at night. The emblem may be permanently mounted or portable as defined in paragraphs 3.2 and 3.3.

4.2 The emblem manufacturer shall place his name and address on the emblem, and may state that the emblem meets the requirements of this Standard. This information shall be clearly and permanently marked on the face of the emblem. It shall appear only in the lower center or lower right-hand corner of the emblem. On portable emblems, the information may be located on the reverse side of the backing material. When the information is located on the face of the emblem, it shall not include trademarks, symbols, or other types of promotional communications, and the total area used for such information on the face of the emblem shall not exceed 6.5 cm² (2 in.²).

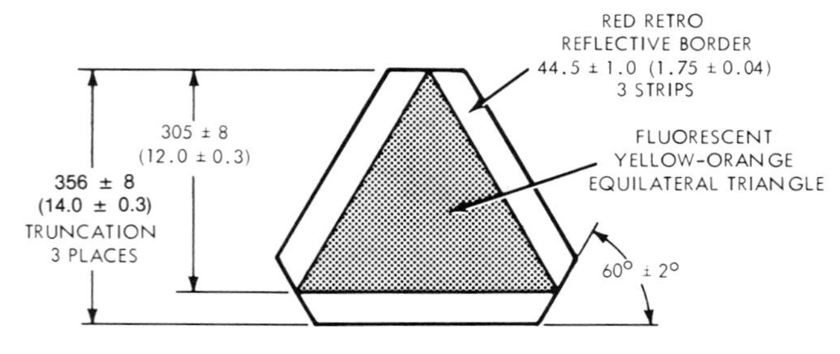

FIG. 1—SLOW-MOVING VEHICLE IDENTIFICATION EMBLEM
NOTE: Emblem must be mounted with the point upward.

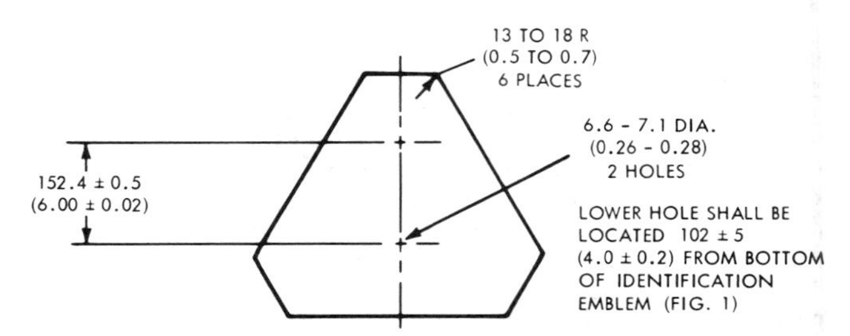

FIG. 2—BACKING MATERIAL FOR PORTABLE SLOW-MOVING VEHICLE IDENTIFICATION EMBLEM

SECTION 5—PERFORMANCE REQUIREMENTS

5.1 Visibility. The emblem shall be entirely visible in daylight and at night from all distances between 183 m and 30 m (600 ft and 100 ft) from the rear when directly in front of the lawful upper beam of headlamps.

5.2 Emblem dimensional requirements. The size of the emblem shall be as shown in Fig. 1.

5.3 Backing material dimensional requirements. The size of the backing material for portable emblems shall be as shown in Fig. 2.

5.4 Emblem material. The reflective and fluorescent materials shall be tough, flexible and of sufficient thickness and strength to meet the requirements of Sections 5 and 6. After the durability test, paragraph 6.2, the fluorescent and reflective materials shall show no appreciable cracking, crazing, blistering, loss of durable bond or dimensional change, and reflective material shall show no appreciable discoloration.

5.4.1 Fluorescent material The yellow-orange color, purity, luminance and peak reflectance of the fluorescent material shall be within the values shown in Table 1 before and after durability tests specified in paragraph 6.2. The test procedure for measuring fluorescent material is specified in paragraph 6.6.

TABLE 1—FLUORESCENT VALUES

	Before Exposure Test	After Outdoor Exposure Test
Dominant wavelength, nm	602-610	585 min
Purity, percent	84 min	77 min
Luminance, percent	28 min	50 max
Peak reflectance observable at wavelength nearest dominant, percent	over 100	75 min

TABLE 2—MINIMUM REFLECTIVE LUMINANCE VALUES, R

Observation Angle, deg	Entrance Angle, deg (±)	Before Exposure Test	After Exposure Test
0.2	4	12.0	9.5
0.2	15	9.0	7.0
0.2	30	5.0	4.0
0.5	4	6.0	4.5
0.5	15	4.0	3.0
0.5	30	2.0	1.5

5.4.2 Reflective material. The dark red reflective material shall have minimum intensity values at each of the angles listed in Table 2 before and after durability tests specified in paragraph 6.2. The test procedure for measuring the reflective intensity values is specified in paragraph 6.5.

5.5 Backing material for portable identification emblems

5.5.1 Backing material for portable identification emblems shall be equivalent to 1.0 mm (0.04 in.) minimum thickness aluminum; 22-gage, 0.8 mm (0.03 in.) minimum thickness mill-galvanized or coated sheet steel; or 2.0 mm (0.08 in.) minimum thickness ABS plastic as specified in American Society for Testing and Materials Standard D1788, Specifications for Rigid Acrylonitrile-Butadiene-Styrene (ABS) Plastics.

5.5.2 The backing material shall be weatherable, semi-rigid and have a surface receptive to a durable bond. The edges of the backing material shall be shaped to minimize personal injury during handling and when mounted on a slow-moving vehicle. These backing materials shall withstand a minimum impact of 14 J (10 ft lbf) using the falling dart procedure as described in paragraph 6.4.

5.6 All of these requirements are minimal and do not preclude the use of materials having superior performance.

SECTION 6—TEST PROCEDURE

6.1 The emblem shall be tested in conformance with the following sections from Society of Automotive Engineers Standard SAE J575f, Tests for Motor Vehicle Lighting Devices and Components:

Paragraph 2 — Samples for Tests
Paragraph 3 — Laboratory Facilities
Paragraph 4.1 — Vibration Test
Paragraph 4.4 — Corrosion Test (pertains to face of emblem only)

6.2 Durability test. Samples mounted on backing material specified in paragraph 5.5.1 shall be exposed to the sun at an angle of 45 deg to horizontal, facing upward and south, in non-metallic racks, per ASTM Standard D1014, Conducting Exterior Exposure Tests of Paints on Steel. After the durability test (Table 3), the emblem material shall show no appreciable discoloration, cracking, crazing, blistering, loss of adhesion, or dimensional change, and shall meet the requirements set forth in paragraph 5.4.1 for fluorescent material, 5.4.2 for reflective materials and the visibility requirements of 5.1.

6.3 Drop test. A portable emblem shall be dropped from a height of 1.8 m (6 ft) to a smooth hard surface equivalent to rigid metal or concrete. Each portable emblem shall be submitted to three drop tests; corner drop, edge drop, and flat face surface drop. Failure shall be considered to have occurred when the emblem or the backing material will no longer meet the requirements of this Standard. The drop tests shall be conducted at both 24 °C (75 °F) and -23 °C (-10 °F).

6.4 Impact resistance of backing material. This test procedure provides the means of determining the force required to fracture backing materials by a free falling metal cylinder dropped vertically. The impact hammer shall be 15.88 mm (0.625 in.) diameter and have a 15.88 mm (0.625 in.) nose radius. The base shall be 31.8 mm (1.25 in.) diameter, and the test specimen shall be a minimum of 102 mm (4 in.) square. Test conditions shall be at room temperature of 24 ± 2 °C (75 ± 3 °F), and failure will be any evidence of fracture or rupture of the backing material [see ASTM Standard D2794, Test for Resistance of Organic Coatings to the Effects of Rapid Deformation (Impact)].

TABLE 3—DURABILITY TEST

Location	Minimum Test Periods, Months	
	Fluorescent Material	Reflective Material
Outside in South Florida	12	12

6.5 Minimum reflective luminance values, R. Measurements shall be conducted in accordance with photometric testing procedures for reflex-reflectors as specified in SAE Standard SAE J594, Reflex Reflectors, except reference to area limitations need not be followed as long as area is greater than 6452 mm² (10 in.²). The maximum dimension of the test surface shall not be greater than 1.5 times the minimum dimension. The reflective luminance is computed from the equation:

$$R = \frac{(Lr)\ (d^2)}{(Ls)\ (A)}$$

where

R = reflective luminance, candelas per square meter per incident lux (candle power per square foot per incident foot-candle).
Lr = illumination incident upon the receiver at observation point, lux (foot-candles).
Ls = illuminance incident upon a plane perpendicular to the incident ray at the test specimen position, lux (foot-candles).
d = distance from test specimen to observation point, meters (feet).
A = area of test surface, square meters (square feet).

6.6 Fluorescent color and peak reflectance. The spectrophotometric color values of the fluorescent material shall be determined by using a Signature Model D-1, Color-Eye Spectrophotometer (Instrument Development Laboratories Division of Killmorgan Corporation) per Method "C" of instruction manual #4001-A or an equivalent spectrophotometer. Luminance shall be compared to that of barium sulfate under the International Commission of Illumination CIE standard source C illuminant.

NOTE: "Vitrolite" standard plate calibrated by the National Bureau of Standards may be substituted for the barium sulfate standard.

If Signature Model D-1 is used, this procedure shall be followed. Barium sulfate standard and the specular insert shall be used. X, X′, Y and Z shall be determined, and the values for x and y shall be calculated as shown in paragaph 6.7. The dominant wave length and purity shall be determined using x and y from CIE diagrams. The values for Y shall be the luminance factor recorded as percent. The peak reflectance obtained shall be recorded as the peak reflectance. If the peak reflectance is too high to be measured with the microdial, reverse the microdial, reverse the positions of the sample and standard so standard is at the rear and sample in front. Then take a new set of readings. The new reciprocal readings may be converted to peak reflectance by using the Reciprocal Table on page F247-1 270 in the "Handbook of Chemistry and Physics" 51st edition. The peak reflectance must be no lower than the values shown in paragraph 5.4.1. For other instruments, use the manufacturer's recommended procedures.

6.7 Calculation procedure for fluorescent material

Reflectance measurements: Illuminant C—Barium sulphate as a standard

$^{X}CIE = 0.783\ (^{X}Color\text{-}Eye) + 0.197\ (^{X'}Color\text{-}Eye)$
$^{Y}CIE = {}^{Y}Color\text{-}Eye$
$^{Z}CIE = 1.180\ (^{Z}Color\text{-}Eye)$

Transmittance measurements: Illuminant C—Either white Vitrolite or barium sulphate as a standard

$^{X}CIE = 0.783\ (^{X}Color\text{-}Eye) + 0.197\ (^{X'}Color\text{-}Eye)$
$^{Y}CIE = {}^{Y}Color\text{-}Eye$
$^{Z}CIE = 1.180\ (^{Z}Color\text{-}Eye)$

The coefficients in the reflectance formulae have been integrated from spectrophotometric curves on a General Electric recording spectrophotometer. In the transmittance formulae, the actual values of the glass coefficients do not enter the formula. Having obtained X, Y, Z (CIE), the coordinated x, y and luminosity ^{Y}CIE may be obtained as follows:

$$x = \frac{X}{X + Y + Z\ \ CIE}$$

$$y = \frac{Y}{X + Y + Z\ \ CIE}$$

$$Y_{CIE} = Y_{CIE}$$

When making measurements on CIE basis, the light source must be operated at calibrated voltage and the sphere must be clean. One method of comparison on the CIE basis is to compute the factors x, y, Y in the CIE system for sample and standard. The values for the standard will give the approximate location in the CIE diagram to determine the size of unit tolerance ellipses.*

SECTION 7—MOUNTING

7.1 Both the permanently mounted emblem and the portable emblem shall be mounted with a point of the triangle upward (see Fig. 1).

7.2 Emblems shall be mounted in a plane perpendicular, ± 10 deg, to the direction of travel and visible from the rear of a slow-moving vehicle in accordance with Section 5.

7.3 The emblem shall be displayed as near to the rear and centered, or as near to the left of center of the vehicle or equipment as practicable. It shall be located 0.6 to 1.8 m (2 to 6 ft) above the ground measured from lower edge of the emblem.

* For CIE color difference work, reference is made to "Handbook of Colorimetry" by A.C. Hardy for descriptive information and CIE graphs, and "Colorimetry" by D.B. Judd for ellipse and other information.

7.4 The emblem shall be securely and rigidly affixed to the equipment. Portable emblems may be mounted by using the socket and identification emblem bracket specified in ASAE Standard ASAE S277, Mounting Brackets and Socket for Warning Lamp and Slow-Moving Vehicle (SMV) Identification Emblem, or by other means that provide secure and rigid attachment.

7.5 The effective reflectivity and fluorescence of the emblem shall be unobscured to the extent that the triangular shape is readily identifiable both day and night.

7.6 When more than one vehicle is being operated or transported in a train or series, the emblem may be mounted on either the first vehicle or attached implement as long as the emblem is visible and identifiable from the rear of vehicles involved as described in paragraph 7.5.

Cited Standards:

ASAE S277, Mounting Brackets and Socket for Warning Lamp and Slow-Moving Vehicle (SMV) Identification Emblem
ASTM D1014, Conducting Exterior Exposure Tests of Paints on Steel
ASTM D1788, Specifications for Rigid Acrylonitrile-Butadiene-Styrene (ABS) Plastics
ASTM D2794, Test for Resistance of Organic Coatings to the Effects of Rapid Deformation Impact
SAE J575f, Tests for Motor Vehicle Lighting Devices and Components
SAE J594, Reflex Reflectors

ASAE Standard: ASAE S277.2 (ANSI/ASAE S277.2/SAE J725 NOV84)

MOUNTING BRACKETS AND SOCKET FOR WARNING LAMP AND SLOW-MOVING VEHICLE (SMV) IDENTIFICATION EMBLEM

Proposed by the Engineering Committee of the Farm and Industrial Equipment Institute; approved by the ASAE Power and Machinery Division Technical Committee; adopted by ASAE December 1964; revised June 1967, February 1972; approved by ANSI as an American National Standard November 1976; reconfirmed December 1976, December 1981; reaffirmed and redesignated by ANSI June 1984; metric dimensions corrected editorially February 1985; reconfirmed December 1986.

SECTION 1—PURPOSE

1.1 This standard defines mounting devices for use with warning lamps and SMV emblems.

SECTION 2—SCOPE

2.1 The bracket shown in Fig. 1 is intended for fixed attachment to SMV emblem and to provide, through the tapered portion, a means of detachably mounting the emblem assembly in cooperation with the socket shown in Fig. 3.

2.2 The bracket shown in Fig. 2 is intended for fixed attachment to the warning lamp and to provide, through the tapered portion, a means of detachably mounting the warning lamp assembly in cooperation with the socket shown in Fig. 3.

2.3 The socket shown in Fig. 3 is intended for fixed attachment to accommodate the tapered portion of brackets shown in Figures 1 and 2.

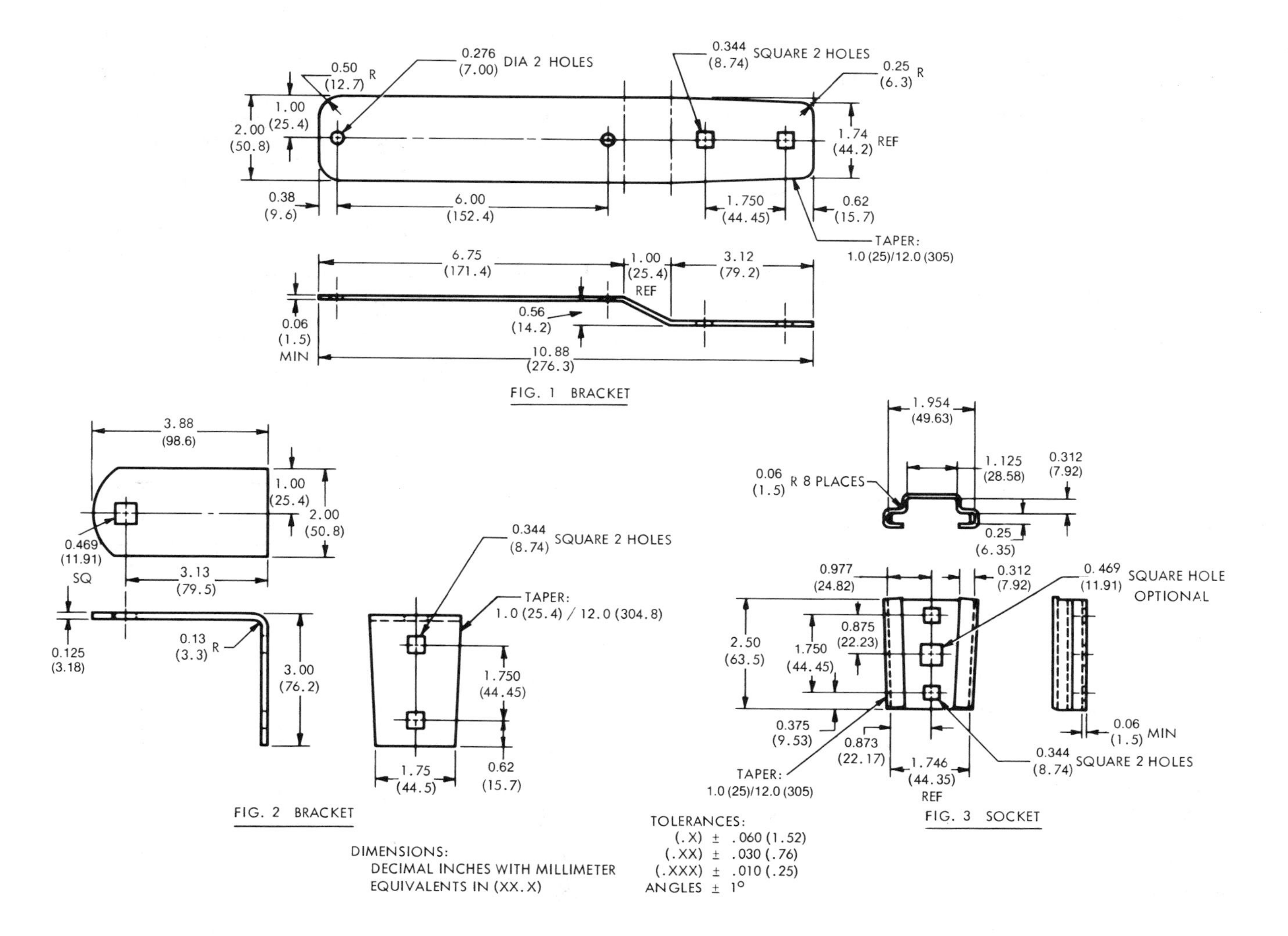

FIG. 1 BRACKET

FIG. 2 BRACKET

FIG. 3 SOCKET

DIMENSIONS:
DECIMAL INCHES WITH MILLIMETER EQUIVALENTS IN (XX.X)

TOLERANCES:
(.X) ± .060 (1.52)
(.XX) ± .030 (.76)
(.XXX) ± .010 (.25)
ANGLES ± 1°

ASAE Standard: ASAE S278.6 (*SAE J909 NOV84)

ATTACHMENT OF IMPLEMENTS TO AGRICULTURAL WHEEL TRACTORS EQUIPPED WITH QUICK-ATTACHING COUPLER

*Corresponds in substance to previous revision S278.5.

Proposed by Advisory Engineering Committee of Farm and Industrial Equipment Institute; approved by ASAE Power and Machinery Division Technical Committee; adopted by ASAE December 1964; revised December 1965, June 1967; reconfirmed December 1971; revised April 1974, February 1975; corrected editorially April 1977; revised December 1978, March 1980; reconfirmed December 1984.

SECTION 1—PURPOSE AND SCOPE

1.1 This Standard sets forth the requirements for the attachment of three-point hitch implements or equipment to the rear of agricultural wheel tractors equipped with quick-attaching couplers.

1.2 It is intended to establish those dimensions which are necessary to assure adequate clearance between components and to assure proper functioning of the tractor-implement combination when the implement is attached to the tractor by means of quick-attaching coupler.

1.3 Design of the latching mechanism and individual components of the coupler not restricted by this Standard are left to the discretion of the manufacturer.

SECTION 2—QUICK-ATTACHING COUPLER FOR TRACTORS EQUIPPED WITH A THREE-POINT FREE-LINK HITCH

2.1 In general, the dimensions associated with tractor and implement for use with the quick-attaching coupler are the same as those for the three-point hitch specified in ASAE Standard S217, Three-Point Free-Link Attachment for Hitching Implements to Agricultural Wheel Tractors. This Standard sets forth those additional requirements that are necessary to provide for use with the coupler. It does not provide for interchangeability between tractors and couplers.

SECTION 3—QUICK-ATTACHING COUPLER AS INTEGRAL PART OF TRACTOR HITCH

3.1 In some instances it may be desirable to provide linkage between the tractor and coupler of a configuration and attachment per the manufacturers discretion where the coupler is not intended to be removed and the capability to attach implements directly to the tractor linkage shall not exist. This section sets forth the requirements for the coupler as an integral part of the tractor linkage for hitch categories, III, III-N, IV, and IV-N.

3.2 Integral quick attaching coupler. All dimensions related to the linkage between the tractor and quick-attaching coupler shall be at the discretion of the manufacturer.

3.2.1 Coupler lower socket tire clearance (Fig. 3) with largest Code R-1 tire offered shall be 76.2 mm (3.00 in.).

3.2.2 Lift range, power range, mast adjustment, leveling adjustment, tractor lift force capacity and side sway provisions shall be in accordance with ASAE Standard S217, Three-Point Free-Link Attachment for Hitching Implements to Agricultural Wheel Tractors.

3.2.3 Dimension C, Lower Socket Offset, and Dimension J, Upper Hook Offset, in Table 1 and Fig. 1 shall be at the discretion of the manufacturer. All other dimensions shall be in accordance with Tables 1 and 2 and Figs. 1 and 2, as set forth for quick-attaching coupler for tractors equipped with a three-point free-link hitch.

3.2.4 Horizontal distance from end of power take-off shaft to coupler lower socket with coupler positioned in conformance with dimension for drawbar hitch point per ASAE Standard S203, Rear Power Take-Off for Agricultural Tractors, shall be as follows:

1-3/8 in. diameter PTO Shaft, 540 rpm =
508 mm (20 in.)/559 mm (22 in.)

1-3/8 in. diameter PTO Shaft, 1000 rpm =
508 mm (20 in.)/559 mm (22 in.)

1-3/4 in. diameter PTO Shaft, 1000 rpm =
610 mm (24 in.)/661 mm (26 in.)

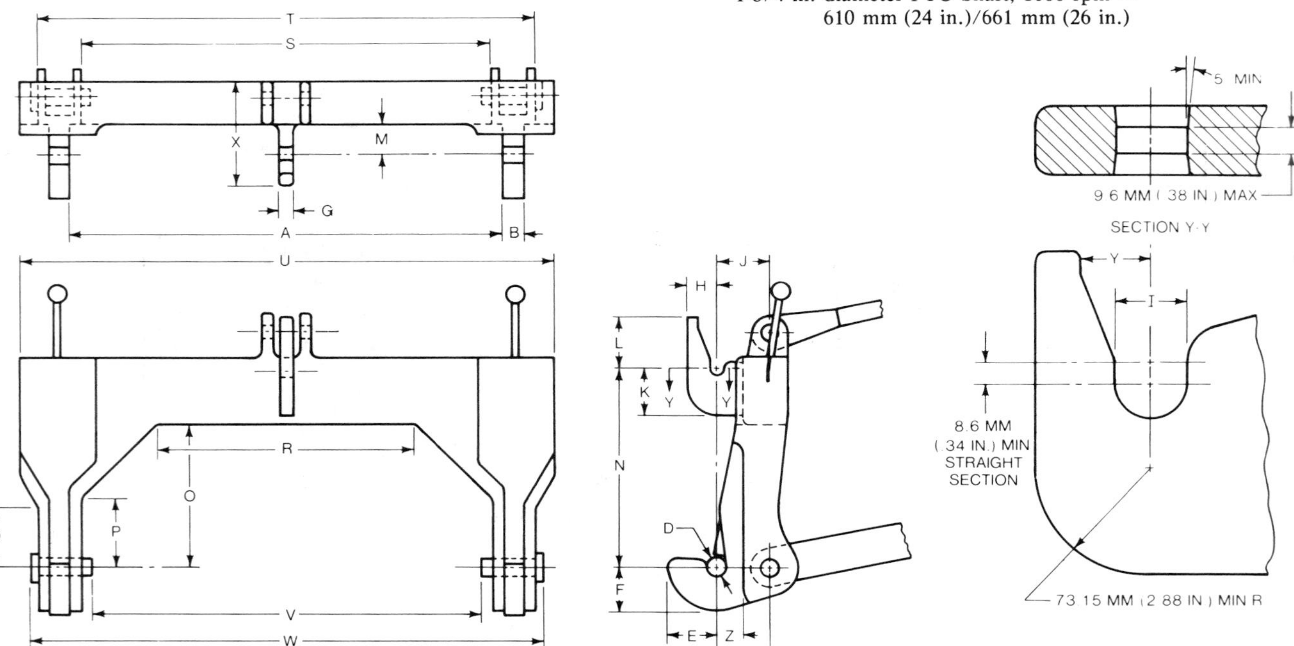

FIG. 1—DIMENSIONS ASSOCIATED WITH QUICK-ATTACHING COUPLER (Letter designations A, B, C, etc., correspond to dimensions in Table 1.)

TABLE 1 — DIMENSIONS ASSOCIATED WITH QUICK-ATTACHING COUPLER (SEE FIG. 1)

Dimensions designated in Fig. 1. (A, B, etc.)	Category I				Category II				Category III §				Category IV §			
	Millimeters		Inches		Millimeters		Inches		Millimeters		Inches		Millimeters		Inches	
	Min	Max	Min	Max	Min	Max	Min	Max	Min	Max	Min	Max	Min	Max	Min	Max
A Lower socket inside span	686.56	689.61	27.03	27.15	828.56	834.13	32.62	32.84	969.78	975.36	38.18	38.40	1171	1174	46.10	46.22
B Lower socket width	27.94	29.46	1.10	1.16	64.26	66.55	2.53	2.62	64.26	66.55	2.53	2.62	87	89	3.42	3.50
C Lower socket offset*		103.13		4.06		127		5.00	—	127		5.00	—	165	—	6.50
D Lower socket diameter	38.10	38.61	1.50	1.52	38.10	38.61	1.50	1.52	38.10	38.61	1.50	1.52	52	52.5	2.05	2.07
E Lower socket overhang	—	88.9	—	3.50	—	88.9	—	3.50	—	88.9	—	3.50	—	89	—	3.50
F Lower socket depth	—	88.9	—	3.50	—	88.9	—	3.50	—	88.9	—	3.50	—	127	—	5.00
G Upper hook width †	—	31.75	—	1.25	—	31.75	—	1.25	—	31.75	—	1.25	—	45	—	1.77
H Upper hook overhang		73.15		2.88	—	73.15		2.88	—	73.15	—	2.88	—	82	—	3.23
I Upper hook opening	32.51	33.27	1.28	1.31	32.51	33.27	1.28	1.31	32.51	33.27	1.28	1.31	45.7	46.5	1.80	1.83
J Upper hook offset*		103.13		4.06	—	127		5.00	—	127		5.00	—	165	—	6.50
K Upper hook depth	—	91.44		3.60		91.44	—	3.60	—	91.44	—	3.60	—	102	—	4.02
L Upper hook height		98.30	—	3.87		98.30		3.87	—	103.13	—	4.06	—	117	—	4.61
M Implement mast clearance ‡	41.15		1.62	—	41.15		1.62		41.15	—	1.62	—	54	—	2.13	—
N Upper hook vertical spacing	375.41	377.96	14.78	14.88	375.41	377.96	14.78	14.88	477.01	479.56	18.78	18.88	680	683	26.77	26.89
O Coupler frame height	283.21		11.15	—	283.21		11.15		365.76	—	14.40	—	508	—	20.00	—
P Coupler leg clearance height inside	203		8		203		8		203	—	8	—	330	—	13.00	—
Q Coupler leg clearance height outside	178	—	7	—	178	—	7	—	178	—	7	—	305	—	12.00	
R Coupler frame clearance width	559	—	22	—	559	—	22	—	559	—	22	—	660	—	25.98	
S Coupler frame inside width	657.10	—	25.87	—	796.81	—	31.37	—	931.68	—	36.68	—	1104	—	43.46	—
														(See footnote ‖)		
T Coupler frame lower outside width	—	806.46		31.75	—	1005.08	—	39.57	—	1130.30	—	44.50	—	1420	—	55.91
U Coupler frame upper outside width		879.61	—	34.63		1065.53		41.95	—	1158.75	—	45.62	—	1420	—	55.91
V Lower link attaching pin inside	620.8		24.44	—	762.0		30.00	—	903.2	—	35.56	—	1104	—	43.46	—
W Lower link attaching pin outside	—	828.12	—	32.60		1019.05		40.12	—	1174.75	—	46.25	—	1420	—	55.91
														(See footnote ‖)		
X Coupler frame overall span	—	222.25	—	8.75	—	222.25	—	8.75	—	222.25	—	8.75	—	305	—	12.01
Y Upper hook reach	47.75	—	1.88	—	47.75	—	1.88	—	47.75	—	1.88	—	63	—	2.48	—
Z Implement lower frame clearance ‡	41.15		1.62	—	41.15	—	1.62	—	41.15	—	1.62	—	54	—	2.13	

*The upper hook offset (Dimension J) is to be no more than 15.75 mm (0.62 in.) greater or 6.35 mm (0.25 in.) less than the lower socket offset (Dimension C).

†The implement must provide clearance when attached to the coupler to permit lowering all elements of the coupler 120.9 mm (4.75 in.) for Category I, II and III and 146 mm (5.75 in.) for Category IV-N and IV minimum for satisfactory attachment and detachment of the implement from the coupler.

The upper hook on the quick coupler shall be on center of the lower sockets within 3.0 mm (0.12 in.) and the upper hook opening on the implement shall be on center of the lower hitch pin shoulders within 3.0 mm (0.12 in.).

Provisions shall be made for adequate upper hook clearance on those implements which require landing or leveling.

‡The lower implement attaching point on the quick-attaching coupler shall be located in the vertical position such that the lift range, power range, and leveling adjustment, as specified in Table 2 of ASAE Standard S217, Three-Point Free-Link Attachment for Hitching and Implements to Agricultural Wheel Tractors, are fulfilled.

Implement components, other than the hitch pins, that are in alignment with the Lower Socket Width (Dimension B) shall not extend forward of the centerline of the lower socket for a distance of 203 mm (8 in.) above the lower socket.

Components above this height and extending laterally more than 381 mm (15 in.) from the coupler centerline shall not extend more than 25 mm (1.0 in.) forward of the vertical centerline through the upper hook opening.

The lower socket width (B) is to be maintained within the area defined by Dimensions P, Q, and Z.

§See Section 4 for Category IV-N or III-N, narrow hitch, dimensions.

‖Lower link to coupler pins may be recessed to provide design freedom to obtain structural integrity in coupler frame, resulting in a common dimension of S and V, T and W, respectively.

TABLE 2 — DIMENSIONS ASSOCIATED WITH IMPLEMENT (SEE FIG. 2)

	Category I				Category II				Category III §				Category IV §			
	Millimeters		Inches		Millimeters		Inches		Millimeters		Inches		Millimeters		Inches	
	Min	Max	Min	Max	Min	Max	Min	Max	Min	Max	Min	Max	Min	Max	Min	Max
A_1 Lower hitch pin or adapter outside diameter*	36.33	36.58	1.43	1.44	36.33	36.58	1.43	1.44	36.33	36.58	1.43	1.44	49.7	50.8	1.96	2.00
B_1 Lower hitch pin inner shoulder spread*	681.0	684.3	26.81	26.94	822.5	825.5	32.38	32.50	963.7	966.7	37.94	38.06	1162	1165	45.75	45.87
C_1 Lower hitch pin outer shoulder spread*	750.83	753.88	29.56	29.68	970.28	973.33	38.20	38.32	1111.50	1114.55	43.76	43.88	1358	1361	53.46	53.58
D_1 Lower hitch pin inner and outer shoulder dia.*	50.55	63.75	1.99	2.51	50.55	63.75	1.99	2.51	50.55	63.75	1.99	2.51	63	101.6	2.48	4.00
E_1 Coupler mast pin diameter	31.50	31.75	1.24	1.25	31.50	31.75	1.24	1.25	31.50	31.75	1.24	1.25	44.2	45	1.74	1.77
F_1 Coupler mast pin vertical spacing	379.48	382.53	14.94	15.06	379.48	382.53	14.94	15.06	481.08	484.13	18.94	19.06	684.5	687.5	26.95	27.07
G_1 Coupler mast pin horizontal spacing†	—	38.1	—	1.50	—	38.1	—	1.50	—	38.1	—	1.50	—	50.8	—	2.00
H_1 Lower socket clearance‡																
J_1 Upper hook clearance‡																
K_1 Lower hitch pin or adapter alignment	To be in line within 0.015 mm (0.015 in.) per 1 mm (1 in.) of pin length.															

*For implements with cantilever mounted lower hitch pins, special quick coupler hitch pins must be supplied to Dimensions A_1, B_1, C_1, and D_1. Adapter bushings may be supplied which convert existing three-point hitch pins to Dimensions A_1, B_1, C_1, and D_1 in lieu of special quick coupler hitch pins.

For three-point hitch implements on which the lower hitch points consist of straddle mounted pins, no additional thrust surfaces are required, providing the pin diameter and support dimensions conform to Dimensions A_1, B_1, C_1, and G_1.

The method used and dimensions related to attaching the pins or adapter bushings to the implement shall be at the discretion of the implement manufacturer.

†The lower implement attaching point on the quick-attaching coupler shall be located in the vertical position such that the lift range, power range, and leveling adjustment, as specified in Table 2 of ASAE Standard S217, Three-Point Free-Link Attachment for Hitching Implements to Agricultural Wheel Tractors, are fulfilled.

Implement components, other than the hitch pins, that are in alignment with the Lower Socket Width (Dimension B, See Fig. 1 and Table 1) shall not extend forward of the centerline of the lower socket for a distance of 203 mm (8 in.) above the lower socket.

Components above this height and extending laterally more than 381 mm (15 in.) from the coupler centerline shall not extend more than 25 mm (1.0 in.) forward of the vertical centerline through the upper hook opening.

‡The implement must provide clearance when attached to the coupler to permit lowering all elements of the coupler 120.9 mm (4.75 in.) for Category I, II, and III and 146 mm (5.75 in.) for Category IV-N and IV minimum for satisfactory attachment and detachment of the implement from the coupler.

The upper hook on the quick coupler shall be on center of the lower sockets within 3.0 mm (0.12 in.) and the upper hook opening on the implement shall be on center of the lower hitch pin shoulders within 3.0 mm (0.12 in.).

Provision shall be made for adequate upper hook clearance on those implements which require landing or leveling.

§See Section 4 for category IV-N or III-N, narrow hitch, dimensions.

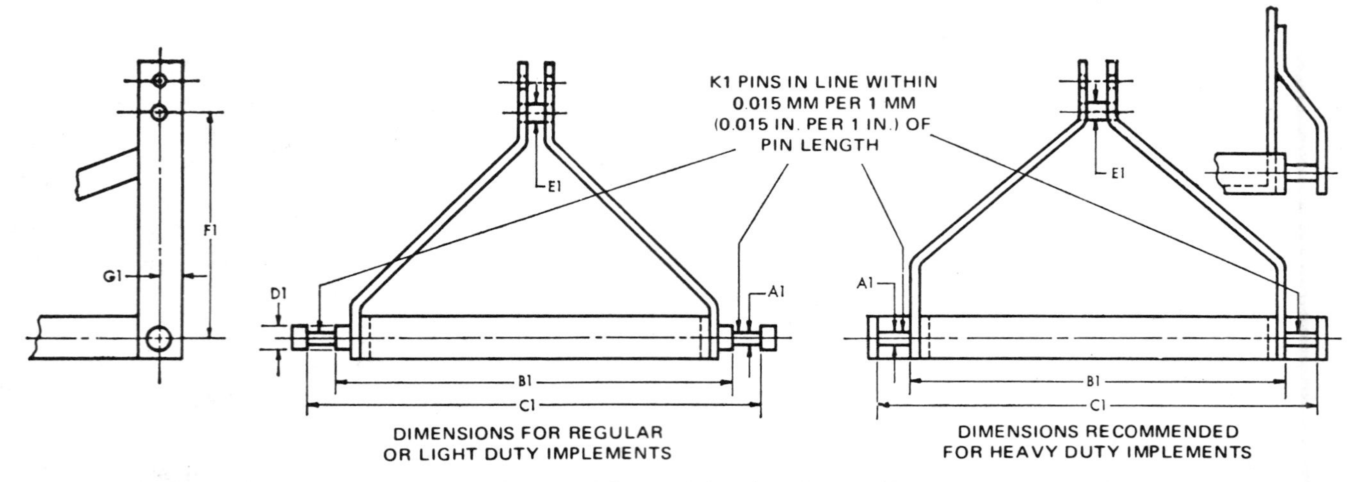

FIG. 2—DIMENSIONS ASSOCIATED WITH IMPLEMENT (Letter designations A1, B1, C1, etc., correspond to dimensions in Table 2.)

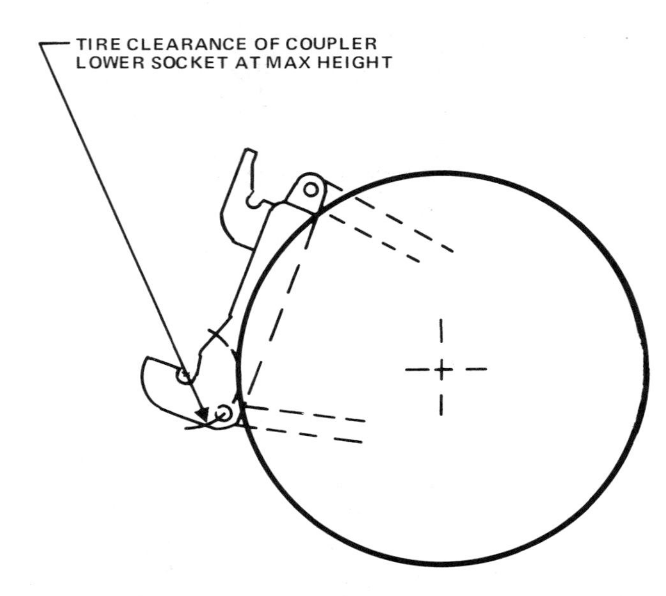

FIG. 3—DIMENSIONS ASSOCIATED WITH TRACTOR

SECTION 4—SPECIAL HITCH COUPLER CATEGORIES

4.1 It has been found from experience that certain large implements used with dual rear tires and narrow crop row spacings require a special category narrow hitch.

4.2 Special category III-N narrow hitch coupler. The dimensions of special category III-N shall be as follows:

4.2.1 Dimensions A, S, T, U, V, and W in Table 1 shall be in accordance with Category II. All other dimensions shall be in accordance with Category III.

4.2.2 Dimensions B_1 and C_1 in Table 2 shall be in accordance with Category II. All other dimensions shall be in accordance with Category III.

4.3 Special category IV-N narrow hitch coupler. The dimensions of special category IV-N shall be as follows:

4.3.1 Dimension A, S, T, V and W in Table 1 shall be:
A = 925 mm (36.42 in.)/930 mm (36.61 in.)
S = 858 mm (33.78 in.)
T = 1174 mm (46.22 in.)
V = 858 mm (33.78 in.)
W = 1174 mm (46.22 in.)

All other dimensions shall be in accordance with Category IV.

4.3.2 Dimension B_1 in Table 2 shall be 919 mm (36.18 in.)/922 mm (36.30 in.). Dimension C_1 in Table 2 shall be in accordance with Category III. All other dimensions shall be in accordance with Category IV.

Cited Standard:

ASAE S217, Three-Point Free-Link Attachment for Hitching to Agricultural Wheel Tractors

ASAE Standard: ASAE S279.9 (*ANSI/ASAE S279.8/SAE J137 MAR83)

T LIGHTING AND MARKING OF AGRICULTURAL FIELD EQUIPMENT ON HIGHWAYS

*Corresponds in substance to previous document ASAE S279.8.

Proposed by the Engineering Committee of the Farm and Industrial Equipment Institute; approved by ASAE Power and Machinery Division Technical Committee; adopted by ASAE December 1964; revised December 1965, June 1966, June 1967, December 1969, June 1971, March 1974; approved by ANSI as an American National Standard November 1976; revised April 1977, March 1982; revision approved by ANSI July 1984; reconfirmed December 1986; revised April 1988.

SECTION 1—PURPOSE

1.1 This Standard provides specifications for lighting and marking of agricultural field equipment whenever such equipment is operated or traveling on a highway.

SECTION 2—DEFINITIONS

2.1 Agricultural equipment: Refer to ASAE Standard S390, Classifications and Definitions of Agricultural Equipment, for definitions of terms.

2.2 Highway: The entire width between the boundary lines of every way publicly maintained, when any part thereof is open to the use of the public for purposes of vehicular travel (Uniform Vehicle Code).

2.3 Reflectors: Reflex reflectors as described in Society of Automotive Engineers Standard J594, Reflex Reflectors, or reflective material that shall be visible at night from all distances within 183 to 31 m (600 to 100 ft) when directly in front of lawful lower beams of headlamps. Reflective material shall meet the durability requirements for such materials as specified in ASAE Standard S276, Slow-Moving Vehicle Identification Emblem.

2.4. Lamp location: Dimensions in this Standard, unless specified otherwise, are based on measurements to the lamp filament.

SECTION 3—LIGHTING AND MARKING REQUIREMENTS

3.1 Lighting and marking of tractors and self-propelled machines shall be as follows:

3.1.1 At least two headlamps generally conforming to SAE Standard J975, Headlamps for Agricultural Equipment, mounted at the same height and spaced laterally as widely as practicable. Headlamps or the low beams of headlamps, if so equipped, shall be aligned such that measured at 7.5 m (25 ft) from the lamp, the horizontal line separating the upper edge of the lighted zone (line at which the intensity is decreased to 10% or less of the peak intensity) is 0.1 x H minimum below the center of the lamp, where H is the height of the lamp from the ground. The headlamp beams shall be centered laterally (see Fig. 1). Flood lamps or general service lamps shall be aimed downward to provide illumination close to the machine and shall not project rearward.

3.1.2 At least one red tail lamp conforming to SAE Standard J585, Tail Lamps, mounted to the rear of the machine and positioned less than 1.5 m (5 ft) to the left of the machine center. If two tail lamps are used, the second shall be placed to the right of the machine center and should be symmetrical with the left lamp location.

3.1.3 At least two amber flashing warning lamps conforming to SAE Standard J974, Flashing Warning Lamp for Agricultural

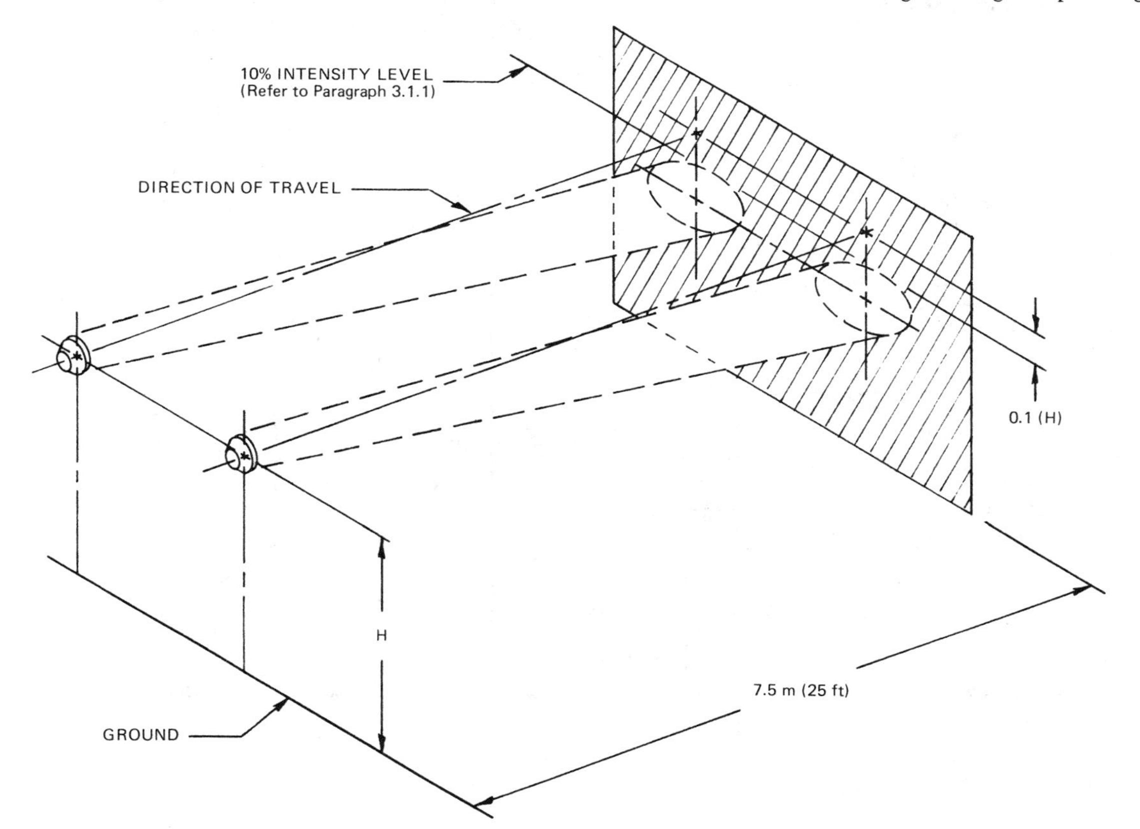

FIG. 1—ILLUSTRATION OF HEADLAMP AIMING PROCEDURE

Equipment, as symmetrically mounted and as widely spaced laterally as practicable, visible from both front and rear, mounted at least 1 m (39 in.) high. Lamps shall flash in unison at a rate of 60 to 85 flashes per minute.

3.1.3.1 On machines over 4 m (13 ft) wide, at least two amber flashing warning lamps conforming to SAE Standard J974, Flashing Warning Lamp for Agricultural Equipment, visible from front and rear shall be provided. The lamps shall be placed a minimum of 1 m (39 in.) high and within 400 mm (16 in.) of the lateral extremities of the machine, and shall flash in unison with warning lamps described in paragraph 3.1.3. The extremity dimension includes such items as dual wheels, wide axles, headers, etc. These lamps may be used in addition to, or in place of, the lamps prescribed in paragraph 3.1.3.

3.1.4 When turn indicators are provided, the amber flashing warning lamps shall be used as the turn indicators. When a turn is to be signaled, the flashing lamp(s) opposite the direction of turn shall become steady burning until the turn signal is cancelled. The flashing lamp(s) in the direction of turn shall increase in flashing rate a minimum of 20 flashes per minute, but shall not exceed 110 flashes per minute.

3.1.5 At least two red reflectors visible to the rear and mounted to indicate, as nearly as practicable, the extreme left and extreme right projections. Reflectors may be incorporated as part of lensing in tail lamps described in paragraph 3.1.2.

3.1.6 One SMV (slow-moving vehicle) identification emblem as described in ASAE Standard S276, Slow-Moving Vehicle Identification Emblem, or means for mounting the SMV emblem such as a mounting socket as described in ASAE Standard S277, Mounting Brackets and Socket for Warning Lamp and Slow-Moving Vehicle (SMV) Identification Emblem.

3.1.7 One seven-terminal receptacle conforming to SAE Standard J560, Seven-Conductor Electrical Connector for Truck-Trailer Jumper Cable, mounted on the machine and located as shown in Fig. 2. Tractors and self-propelled machines not primarily used with agricultural implements described in paragraphs 3.3.1 and 3.3.2 are excluded. (Examples are small garden and compact utility tractors, self-propelled windrowers, and high clearance sprayers.)

3.1.7.1 As a minimum the receptacle terminal numbers 1, 3, 5 and 6 (ground, flashing and turn signals, and taillights), shall be wired for service.

3.1.7.2 The circuit designations for the breakaway connector defined in paragraph 3.1.7 are:

Tractor Receptacle

Conductor identification	Wire color	Terminal number	Circuit
Wht	White	1	Ground
Blk	Black	2	Work lights
Yel	Yellow	3	Left-hand flashing and turn signals
Red	Red	4	Auxiliary
Grn	Green	5	Right-hand flashing and turn signals
Brn	Brown	6	Tail lamp
Blu	Blue	7	Auxiliary

3.2 Marking of agricultural implements shall be as follows:

3.2.1 Implements extending more than 1.2 m (4 ft) to the left of the center of the propelling machine shall have at least one amber reflector visible to the front and positioned to indicate, as nearly as practicable, the extreme left projection of the implement.

3.2.2 Implements extending more than 10 m (33 ft) behind the hitch point shall have amber reflectors visible from the left and right sides. The reflectors shall be spaced at intervals of 5 m (16.4 ft) maximum on both sides measuring from the transport hitch point. The rear most reflector shall be positioned as far rearward as practicable.

3.2.3 Implements extending more than 1.2 m (4 ft) to the rear of the hitch point of the propelling machine or more than 1.2 m (4 ft) to the right or left of the centerline of the propelling machine shall have at least two red reflectors visible to the rear and mounted to indicate, as nearly as practicable, the extreme left and extreme right projections.

3.2.4 Implements that obscure the SMV emblem on the propelling machine or extend more than 10 m (33 ft) to the rear of the hitch point shall be equipped with one SMV emblem as described in ASAE Standard S276, Slow-Moving Vehicle Identification Emblem, or means for mounting the SMV emblem, such as a mounting socket as described in ASAE Standard S277, Mounting Brackets and Socket for Warning Lamp and Slow Moving Vehicle (SMV) Identification Emblem.

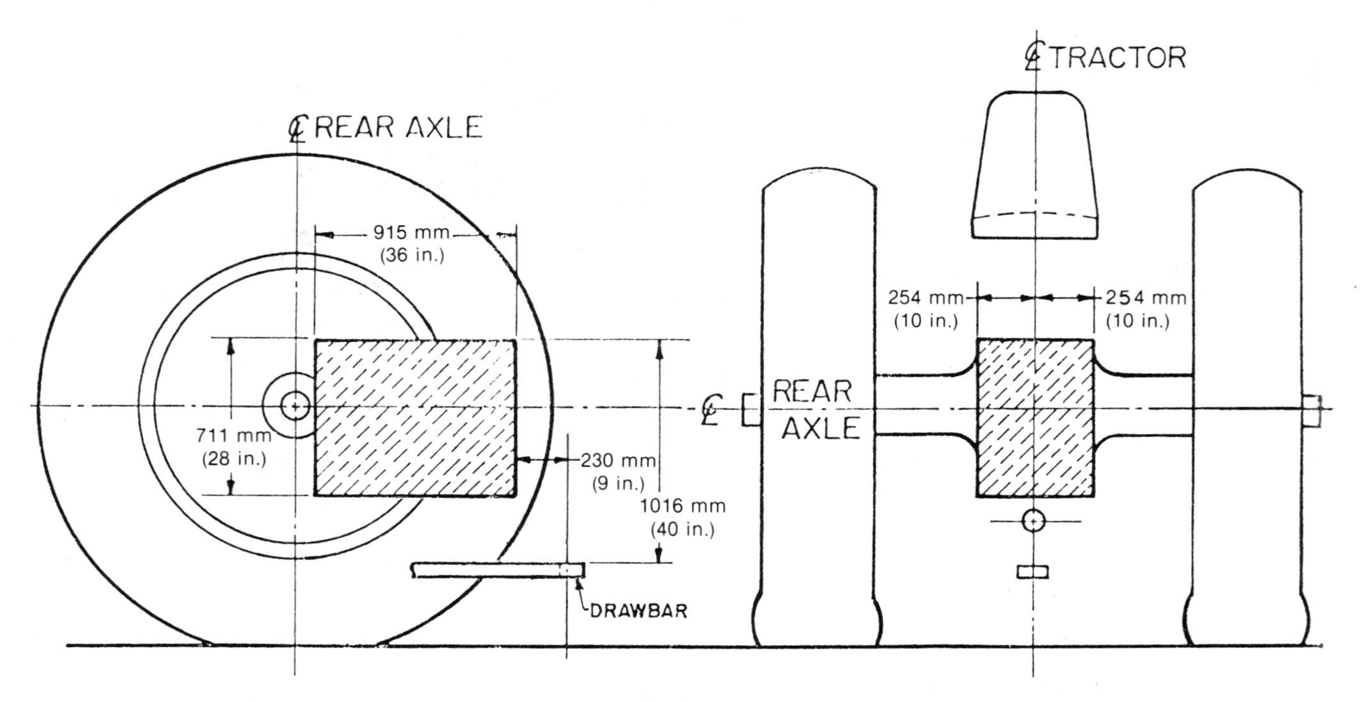

FIG. 2—LOCATION ZONE FOR SEVEN-TERMINAL CONNECTOR

3.3 Lighting of agricultural implements shall be as follows:

3.3.1 Implements which obscure the effective illumination of any flashing warning lamp or extremity lamp on the propelling machine shall have lighting as described in paragraphs 3.3.3 and 3.3.4. If the tail lamps on the propelling machine are obscured, at least one tail lamp conforming to SAE Standard J585, Tail Lamps, shall be mounted to the rear of the implement and positioned to the left of the implement center. If two tail lamps are used, the second shall be placed to the right of the implement center and should be symmetrical with the left tail lamp location.

3.3.2 Implements which (1) are more than 4 m (13 ft) wide or extend over 2 m (79 in.) to the left or right of the centerline and beyond the left or right extremity of the propelling machine, or (2) extend more than 10 m (33 ft) to the rear of the hitch point shall have lighting as described in paragraphs 3.3.3 and 3.3.4.

3.3.3 At least two amber flashing warning lamps conforming to SAE Standard J974, Flashing Warning Lamp for Agricultural Equipment, visible from front and rear shall be provided. The lamps shall be spaced to within 400 mm (16 in.) of the lateral extremities of the machine, preferably mounted at least 1 m (39 in.) but not over 3 m (10 ft) in height, and shall flash in unison with warning lamps described in paragraph 3.1.3. On non-symmetrical implements extending only to the left or right, such as moldboard plows or windrowers, one flashing warning lamp shall be provided spaced laterally to within 400 mm (16 in.) of the left or right extremity.

3.3.4 When turn signals are provided on the propelling machine, the amber flashing warning lamps of the implement shall be used as turn indicators as described in paragraph 3.1.4.

3.3.5 A seven-terminal plug conforming to SAE Standard J560, Seven-Conductor Electrical Connector for Truck-Trailer Jump Cable, shall be provided for operating remote flashing warning lamps, turn indicators, and tail lamp(s). The plug location and cable length shall be compatible with the location of the seven-terminal receptacle on the tractor or self-propelled machine (see paragraph 3.1.7) as shown in Fig. 2.

Cited Standards:

ASAE S276 Slow-Moving Vehicle Identification Emblem
ASAE S277, Mounting Brackets and Socket for Warning Lamp and Slow-Moving Vehicle (SMV) Identification Emblem
ASAE S390, Classifications and Definitions of Agricultural Equipment
SAE J560, Seven-Conductor Electrical Connector for Truck-Trailer Jumper Cable
SAE J585, Tail Lamps (Rear Position Lamps)
SAE J594, Reflex Reflectors
SAE J974, Flashing Warning Lamp for Agricultural Equipment
SAE J975, Headlamps for Agricultural Equipment

CAPACITY DESIGNATION FOR FERTILIZER AND PESTICIDE HOPPERS AND CONTAINERS

Proposed December 1962 by the Fertilizer Application Subcommittee of the ASAE Power and Machinery Division Chemical Application Committee; approved by the ASAE Power and Machinery Division Technical Committee; adopted by ASAE December 1964; revised December 1967; reconfirmed December 1972; revised March 1976; reconfirmed December 1980, December 1985; revised April 1987.

SECTION 1—PURPOSE AND SCOPE

1.1 This Standard is intended to provide a uniform method for calculating and expressing the capacity of fertilizer and pesticide hoppers and containers.

SECTION 2—DRY HOPPERS

2.1 Hoppers for dry materials shall be rated volumetrically and gravimetrically, both based on the inside measurements of the hopper, struck level, with the level low enough to permit normal closing of covers.

2.2 The volumetric rating shall be expressed in cubic meters (cubic feet).

2.3 The gravimetric rating shall be expressed in kilograms (pounds). The gravimetric rating shall be calculated by multiplying the volumetric rating by the nominal bulk density of materials for which the hopper is designed. A bulk density value of 1000 kg/m^3 (62 lb/ft^3) shall be used in rating all general-purpose dry fertilizer hoppers and 700 kg/m^3 (43 lb/ft^3) for granular pesticide hoppers.

SECTION 3—LIQUID AND FERTILIZER CONTAINERS

3.1 Liquid fertilizer containers (tanks or hoppers) shall be rated volumetrically and gravimetrically.

3.2 The volumetric rating shall be expressed in liters (U.S. gallons), measured with the container in the same position as occupied during filling by the user.

3.3 The gravimetric rating shall be expressed in kilograms (pounds). The gravimetric rating shall be calculated by multiplying the volumetric rating by the nominal density of the fertilizers for which the container is designed. A density value of 1.2 kg/L (10 lb/U.S. gal) shall be used in rating all general-purpose, liquid fertilizer containers.

SECTION 4—LIQUID PESTICIDE TANKS

4.1 Liquid pesticide tanks need to be rated volumetrically only.

4.2 The volumetric rating shall be expressed in liters (U.S. gallons), measured with the container in the same position as occupied during filling by user.

SECTION 5—LABELING HOPPERS AND CONTAINERS

5.1 Pesticide and fertilizer hoppers and containers used on farm equipment should have affixed to the hopper or container a statement of the capacity rating specified in Section 2—Dry Hoppers, Section 3—Liquid Fertilizer Containers and Section 4—Liquid Pesticide Tanks. The statement should include the density value used in establishing the gravimetric rating.

SECTION 6—FERTILIZER DENSITY

6.1 Labels for both liquid and dry fertilizer products should include an average bulk density value. The average bulk density for dry fertilizer shall be expressed in kilograms per cubic meters (pounds per cubic feet) and for liquid fertilizer, kilogram per liter (pounds per U.S. gallons) at 15 °C (59 °F). Dry fertilizer densities shall be based on the loose fill of a volumetric box struck level. Minimum dimensions for a volumetric box shall be 200 × 200 × 200 mm (8 × 8 × 8 in.). To calculate the actual weight of a particular fertilizer required to fill a hopper or container, the volumetric rating is multiplied by the density of the particular fertilizer.

DETERMINING CUTTING WIDTH AND DESIGNATED MASS OF DISK HARROWS

Proposed by the Disk Harrow Standardization Subcommittee of the ASAE Cultural Practices Equipment Committee; approved by ASAE Power and Machinery Division Technical Committee; adopted by ASAE June 1965; reconfirmed December 1970, December 1975; revised June 1976; reconfirmed December 1981, December 1986.

SECTION 1—PURPOSE AND SCOPE

1.1 The purpose of Section 2 of this Standard is to establish a standard method for determining the cutting width of:

1.1.1 Single disk harrows

1.1.2 Tandem disk harrows

1.1.3 Double offset disk harrows

1.1.4 Offset disk harrows

1.2 The purpose of Section 3 of this Standard is to establish a standard method for designating the mass of disk harrows.

SECTION 2—CUTTING WIDTH

2.1 Cutting width is to be expressed in meters (feet).

2.1.1 The cutting width for single disk harrows shall be the transverse distance between the top or bottom cutting edges of the end blades when the harrow gangs are set at an 18-degree working angle. The cutting width shall be determined by the formula:

$$\text{Cutting width} = 0.95NS + 0.3D$$

where

N = number of spaces between disk blades
S = blade spacing
D = diameter of disk blades

2.1.2 The cutting width for tandem disk harrows shall be the transverse distance between the top or bottom cutting edges of the end blades of the harrow rear gangs set at an 18-degree working angle. The cutting width shall be determined by the formula:

$$\text{Cutting width} = 0.95NS + 1.2D$$

where

N = number of spaces between disk blades
S = blade spacing
D = diameter of disk blades

2.1.3 The cutting width for double offset disk harrows shall be the transverse distance between the top or bottom cutting edges of the end blades of the harrow rear gangs set at an 18-degree working angle. The cutting width shall be determined by the formula:

$$\text{Cutting width} = 0.95NS + 0.85D$$

where

N = number of spaces between disk blades
S = blade spacing
D = diameter of disk blades

2.1.4 The cutting width for offset disk harrows shall be the transverse distance between top or bottom cutting edge of the end blade on the front gang and the opposite end blade on the rear gang, when the gangs are set at an 18-degree working angle (equivalent to 36-degree included angle). The cutting width shall be determined by the formula:

$$\text{Cutting width} = 0.95NS + 0.6D$$

where

N = number of spaces between disk blades on rear gang
S = blade spacing
D = diameter of disk blades

SECTION 3—MASS DESIGNATION

3.1 The standard mass of disk harrows shall be expressed in kilograms (pounds).

3.2 The standard mass of disk harrows shall be the mass of a field functional machine which includes equipment such as scrapers, hydraulic components, and tires.

ASAE Engineering Practice: ASAE EP291.1

TERMINOLOGY AND DEFINITIONS FOR SOIL TILLAGE AND SOIL-TOOL RELATIONSHIPS

Proposed and developed by ASAE Cultural Practices Equipment Committee, with assistance by ASAE Conservation Tillage Methods Committee, and by the Soil Effects Subcommittee of the Society of Automotive Engineers' Construction and Industrial Machinery Technical Committee and the Tillage Terminology Committee of the Soil Science Society of America; approved by ASAE Power and Machinery Division Technical Committee; adopted by ASAE as a Recommendation June 1965; reconfirmed December 1970; revised June 1972; reconfirmed December 1976; reclassified as an Engineering Practice June 1978; reconfirmed December 1981, December 1986.

SECTION 1—BASIC TILLAGE GOALS

1.1 Tillage action. The action of a tillage tool in executing a specific form of soil manipulation such as soil cutting, shattering, or inversion.

1.2 Tillage objective. A desired soil condition that is to be produced by one or more tillage actions.

1.3 Tillage requirement. The soil physical conditions, which after a complete evaluation of basic utilitarian and economic requirements, are deemed necessary and can be feasibly produced by tillage.

SECTION 2—GENERAL TILLAGE TERMS AND KINDS OF TILLAGE

2.1 Broadcast (overall) tillage. Coverage of an entire area as contrasted to a partial coverage as in bands or strips.

2.2 Cultivation, soil. Shallow tillage operations performed to create soil conditions conducive to improved seration, infiltration, and moisture conservation or to control weeds to promote the growth of crop plants.

2.3 Earth moving. Tillage action and transport operations utilized to loosen, load, carry, and unload soil.

2.4 Harvesting depth (raking, picking, lifting). The depth below the soil surface to which harvesting tools perform their intended function.

2.5 Land forming. Tillage operations which move soil to create desired soil configurations. Forming may be done on a large scale such as contouring or terracing, or on a small scale such as ridging or pitting.

2.5.1 Land grading. Tillage operations which move soil to establish a desired soil elevation and slope. Examples: Leveling, contouring, cutting, and filling.

2.5.2 Land planing. A tillage operation that cuts and moves small layers of soil to provide a smooth, refined surface condition.

2.6 Oriented tillage. Tillage operations which are oriented in specific paths or directions with respect to the sun, prevailing winds, previous tillage actions, or field base lines.

2.7 Rotary tillage. A tillage operation employing power-driven rotary action to cut, break up, and mix soil.

2.8 Tillage. The mechanical manipulation of soil for any purpose; but in agriculture the term is usually restricted to the changing of soil conditions for crop production.

2.9 Tillage depth. Vertical distance from the initial soil surface to a specified point of penetration of the tool.

2.10 Tillage, deep. A primary tillage operation which manipulates soil to a greater depth than normal plowing. It may be accomplished with a very heavy-duty moldboard or disk plow which inverts the soil, or with a chisel plow or subsoiler which shatters soil.

2.11 Tillage, primary. That tillage which constitutes the initial major soil-working operation. It is normally designed to reduce soil strength, cover plant materials, and rearrange aggregates.

2.12 Tillage, secondary. Any of a group of different tillage operations, following primary tillage, which are designed to create refined soil conditions before the seed is planted.

2.13 Tool depth. Vertical distance from the initial soil surface to the lowest point of the tool.

SECTION 3—TILLAGE SYSTEMS

3.1 Conventional tillage. The combined primary and secondary tillage operations normally performed in preparing a seedbed for a given crop and area.

3.2 Minimum tillage. The minimum soil manipulation necessary for crop production or for meeting tillage requirements under the existing soil conditions.

3.3 Mulch tillage. Tillage or preparation of the soil in such a way that plant residues or other mulching materials are specifically left on or near the surface.

3.4 No-tillage planting. A procedure whereby a planting is made directly into an essentially unprepared seedbed.

3.5 Optimum tillage. An idealized system which permits a maximized net return for a given crop under given conditions.

3.6 Precision tillage. By common usage, subsoiling under the plant row prior to planting.

3.7 Reduced tillage. A system in which the primary tillage operation is performed in conjunction with special planting procedures in order to reduce or eliminate secondary tillage operations.

3.8 Strip tillage. A system in which only isolated bands of soil are tilled.

SECTION 4—SPECIFIC TILLAGE OPERATION

4.1 Anchoring. A tillage operation used to partially bury and thereby prevent movement of foreign materials, such as plant residue or paper mulches, to prevent movement.

4.2 Bed planting. A method of planting in which several rows are planted on an elevated level bed with beds being separated by furrows or ditches.

4.3 Bedding (ridging). A tillage operation which mounds soil into a specific configuration. (See Fig. 1)

4.4 Bulldozing. The pushing or rolling of soil by a steeply inclined blade.

4.5 Chiseling (subsoiling). A tillage operation in which a narrow tool is used to break up hard soil. It may be performed at other than the normal plowing depth. Chiseling at depths greater than 16 in. (406 mm) is termed subsoiling. (See paragraph 4.16)

4.6 Combined tillage operations. Operations simultaneously utilizing two or more different types of tillage tools or implements (subsoil-lister, lister-planter, or plow-planter combinations) to simplify, control, or reduce the number of trips over a field.

4.7 Harrowing. A secondary tillage operation which pulverizes, smooths, and packs the soil in seedbed preparation, or controls weeds.

4.8 Incorporating (mixing). Tillage operations which mix or disperse foreign materials, such as pesticides, fertilizers, or plant residues, into the soil.

4.9 Lister planting. A method of planting in which the seed is planted in the bottom of lister furrows.

4.10 Listing. A tillage and land-forming operation using a tool which splits the soil and turns two furrows laterally in opposite directions, thereby providing a ridge-and-furrow soil configuration.

4.11 Middlebreaking. The use of a lister in a manner that opens the furrow midway between two previous rows of plants.

4.12 Plowing. A primary tillage operation which is performed to shatter soil with partial or complete soil inversion.

4.13 Residue processing. Operations that cut, crush, anchor or otherwise handle residues in conjunction with soil manipuation.

4.14 Ridge planting. A method of planting in which the seed is planted on ridges which are formed with a lister or similar tillage tool.

4.15 Ridging (bedding). A tillage operation which mounts soil into a specific configuration. (See Fig. 1)

4.16 Subsoiling. Deep chiseling, below 16 in. (406 mm) for the purpose of loosening soil for root growth and/or water movement (See paragraph 4.5)

4.17 Vertical mulching. A subsoiling operation in which a vertical band of mulching material is injected into the slit immediately behind the chisel.

SECTION 5—NOMENCLATURE FOR TILLAGE TOOLS AND IMPLEMENTS

5.1 Applicator, soil-additive. A machine utilized to apply, or to apply and incorporate by means of tillage soil additives. Examples: Granular herbicide applicator, lime or manure spreader, fumigation and fertilizer distributor, or chemical incorporator.

5.2 Bed shaper. A soil handling implement which forms uniform ridges of soil to predetermined shapes.

5.3 Blade. A soil working tool, consisting of an edge and a surface, which is primarily designed to cut through soil (e.g., rotary tiller blades, anhydrous ammonia blades).

5.4 Coulter. A tool which cuts plant material and soil ahead of another tool.

5.5 Draft. The force to propel an implement in the direction of travel. Equal and opposite to drawbar pull.

5.6 Effective operating width. Operating width minus overlap. (See paragraph 5.19)

5.7 Edge-clearance angle. The effective angle which is included between the line of travel and a line drawn through the back or nonsoil-working surface of the tool at its immediate edge.

5.8 Ground clearance. Minimum vertical distance between the soil surface and a potentially obstructing implement or machine element.

5.9 Hitch. The portion of an implement designed to connect the implement to a draft source.

5.10 Implement width. The horizontal distance perpendicular to the direction of travel between the outermost edges of the implement.

5.11 Incorporator, soil-additive. A machine used to mechanically incorporate or mix additives into soil.

5.12 Injector. An implement used to insert materials into the soil.

5.13 Jointer. A miniature plow attachment whose purpose is to turn over a small furrow slice directly ahead of the main plow bottom, to aid in covering trash.

5.14 Lateral tool spacing. The horizontal distance between corresponding reference points on adjacent tools when projected upon a vertical plane perpendicular to the direction of travel.

5.15 Line of travel. The line and direction along which the tillage implement travels.

5.16 Lister-planter. A combined tillage implement which is composed of a lister and planting attachment to permit a single listing-planting operation with the seed normally being placed in the furrow.

5.17 Longitudinal tool spacing. The horizontal distance between corresponding reference points of two tools when projected upon a vertical plane parallel to the direction of travel.

5.18 Moldboard plow clearances

5.18.1 Horizontal clearance. Distance measured between specified points on adjacent plows, e.g. diagonal (rake) (tip of share to tip of share), fore and aft, width of cut or furrow slice, throat width (min. distance from face of moldboard to projecting member of preceding bottom).

5.18.2 Vertical clearance. Distance measured from cutting edge of share to potentially obstructing member, e.g. main truss (backbone), frame, bean, release mechanism, etc.

5.19 Operating overlap. The distance perpendicular to the direction of travel that an implement reworks soil previously tilled.

5.20 Operating width. The horizontal distance perpendicular to the direction of travel an implement performs its intended function in one pass.

5.21 Orientation, tool. The position of the tool in a framework of cartesian coordinates which is usually oriented with the soil surface and the direction of travel. Orientation is specified in side, tilt, and lift angles as a minimum.

5.21.1 Lift (rake) angle. The angle, in a vertical plane parallel to the direction of travel, between a centralized tool or reference axis and the soil surface.

5.21.2 Side angle. The angle, in the soil surface plane, between a centralized tool or reference axis and a line which is perpendicular to the direction of travel.

5.21.3 Tilt angle. The angle, in a vertical plane perpendicular to the direction of travel, between a centralized tool or reference axis and the soil surface.

5.22 Protected zone. Soil and/or plant zone purposely protected by virtue of tool design, tool spacing or evasive tool movement.

5.23 Scouring (shedding). A soil-tool reaction in which soil slides over the surface of the tillage tool without significant adhesion.

5.24 Shank. A structural member primarily used for attaching a tillage tool to a beam or a standard.

5.25 Shovel. A spade-shaped, V-pointed soil working tool which is used for various plowstocks, cultivators, grain drills, and soil scarifiers.

5.26 Side force (side draft). The horizontal component of pull, perpendicular to the line of motion.

5.27 Soil opener. A tillage tool (e.g., disk, knife, runner) used to slice through soil and create an opening for the insertion of material (e.g., seeds, pesticides, fertilizers).

5.28 Soil roller. A rotating implement which pulverizes and/or firms or smooths soil by crushing and/or compacting.

5.29 Soil sliding path. The path along which one element of soil slides across a tillage tool.

5.30 Soil sliding path length. The length of the path along a tillage tool upon which soil slides.

5.30.1 Soil ascending angle. The angle between the sliding path and the horizontal at any point along the sliding path.

5.30.2 Soil sliding angle. The angle at any point on the surface of a tool between the soil sliding path and a horizontal contour line constructed through the surface of the tool.

5.31 Soil-tool geometry. The configuration of the soil-tool boundary. The overall shape is usually oriented with the direction of travel of the tool and the soil surface.

5.32 Soil-working surfaces. Portions of tillage tools which are designed to be in contact with soil in order to effect proper manipulation.

5.33 Specific draft (unit draft). The draft of an implement in terms of force per unit area of tilled cross section.

5.34 Standard (beam). An upright support which connects the shank to a tillage implement frame.

5.35 Sweep. A type of cultivator shovel which is wing-shaped. It is used principally for weed control.

5.36 Teeth. Projections on tillage tools which serve to penetrate, grip, cut, or tear soil.

5.37 Tillage implement (machine). Single, or groups of, soil-working tools together with power transmission structure, control, and protection systms present as an integral part of the machine.

5.38 Tillage tool. An individual soil-working element.

5.39 Tillage tools, complex. Tools which rotate or move so they present a varying boundary and contact area to the soil. Examples: Clod breakers, notched disks, rotary hoes.

5.40 Tillage tools, dynamic. Tillage tools which are powered so some of their movements are in directions other than along the line of travel.

5.41 Tillage tools, multipowered. Tillage tools powered by more than one form of power, such as draft and rotating power, or draft and electrical power.

5.42 Tillage tools, simple. Tools which present a reasonable constant boundary area to the soil.

5.43 Tool clearance. The minimum distance in a specified direction between a point on the tool and another potentially obstructing implement element.

5.44 Tool operating width. The maximum horizontal distance perpendicular to the line of motion over which a tool performs its intended function.

5.45 Tool overlap. The distance perpendicular to the direction of travel in which a tool operating width coincides with the operating width of another tool.

5.46 Tool skip area. The area of soil surface left undisturbed during passage of a tool.

5.47 Tool width. Maximum horizontal projection of a tool in the soil perpendicular to the line of motion.

5.48 Vertical tool spacing. The vertical distance between corresponding points on adjacent tools when projected upon a vertical plane parallel to the direction of travel.

5.49 Wings. Projections attached to the sides of tillage tools to increase the volume of soil which can be disturbed, or to control the nature and distance of soil movement. Wings usually have lift, tilt, and side angles which are different from those found in the orientation of the main tool and standard.

SECTION 6—SOIL REACTION NOMENCLATURE

6.1 Soil reaction. The nature of soil reaction caused by the application of mechanical forces to soil.

6.1.1 Soil abrasion. The scratching, cutting, or abrading of materials caused by the action of soil.

6.1.2 Soil adhesion. The sticking of soil to foreign materials such as tillage tools or wheels.

6.1.3 Soil compaction. A soil reaction in which a reduction in specific volume of soil occurs by mechanical manipulation.

6.1.4 Soil cutting. Soil separation of a soil mass by a slicing action.

6.1.5 Soil failure. The alteration or destruction of a soil structural condition by mechanical forces such as in shearing, compression, or tearing.

6.1.6 Soil heaving. The lifting or swelling of soil resulting from natural forces such as freezing.

6.1.7 Soil shatter (pulverization). The general fragmentation of a soil mass resulting from the action of tillage forces.

6.1.8 Soil sliding. The sliding of soil across a surface.

6.1.9 Throw. The movement of soil in any direction as a result of kinetic energy imparted to the soil by the tillage tool.

SECTION 7—SOIL NOMENCLATURE

7.1 Additive, soil. Foreign materials, other than seeds, which are added to and/or incorporated in soil for directly influencing the soil condition or environment. (These include pesticides, fertilizers, mulches, or conditioners, but not foreign bodies such as drain tiles which have an indirect influence).

7.2 Adhered soil bodies. Masses of soil which adhere on soil-working surfaces and act as a part of the tool. Soil bodies may be stationary or in a relatively slow motion (i.e., soil cone, an adhered soil body which resembles a cone; soil sheet, an adhered soil body which covers a large area of a tool like a sheet; soil wedge, an adhered soil body which resembles a narrow wedge).

7.3 Clods. Soil blocks or masses that are cut, sheared, or broken loose by tillage tools.

7.4 Compacted layer (plow pan, trafic pan, plow sole). A dense layer of soil created by mechanical pressure, as contrasted to a hardpan formed by cementation.

7.5 Concretions. Soil concretions are soil structural units which are irreversibly cemented together.

7.6 Covering depth. The thickness of soil with which materials are covered by an implement.

7.7 Foreign materials. All material added to or mixed into soil, including residues, soil additives, and foreign bodies that have not originated in the soil's development.

7.8 Mechanical impedance. The resistance to the movement of plant parts or tillage tools through soil that is caused by the mechanical strength of the soil.

7.9 Mechanical stability (strength). The degree of resistance of soil to deformation.

7.10 Residues. See foreign materials, paragraph 7.7, and additives, paragraph 7.1.

7.11 Shear blocks (or clods). The blocks of soil which are sheared loose from the main soil mass by tillage tool action.

7.12 Shear surfaces. Failure surfaces occurring where the soil has sheared.

7.12.1 Primary shear surfaces. The initial and distinct surfaces appearing during failure which are caused mainly by shear.

7.12.2 Secondary shear surface. Shear surfaces which result from the twisting, pushing, or tumbling of the soil after or during the initial displacement. Secondary shear surfaces are often perpendicular to the primary shear surfaces.

7.13 Soil aggregates (peds). Agglomerations of primary soil particles which are produced by natural processes.

7.14 Tillability. The degree of ease with which a soil may be manipulated for a specific purpose.

SECTION 8—SOIL AND SURFACE CHARACTERISTICS

8.1 With the increase of bedding equipment, nomenclature is suggested for all types of beds. The terminology should be general. (See Fig. 1)

8.1.1 Furrow depth (ditch, pit, trench). Depth of depression below a specified (initial or subsequent) soil surface.

8.1.2 Ridge height (bed, hill, windrow). Height of soil above a specified (initial or subsequent) soil surface.

8.1.3 Root bed. The soil profile modified by tillage or amendments for use by plant roots.

8.1.4 Root zone. That part of the soil profile exploited by the roots of plants.

8.1.5 Seed bed. The soil zone which affects the germination and emergence of seeds.

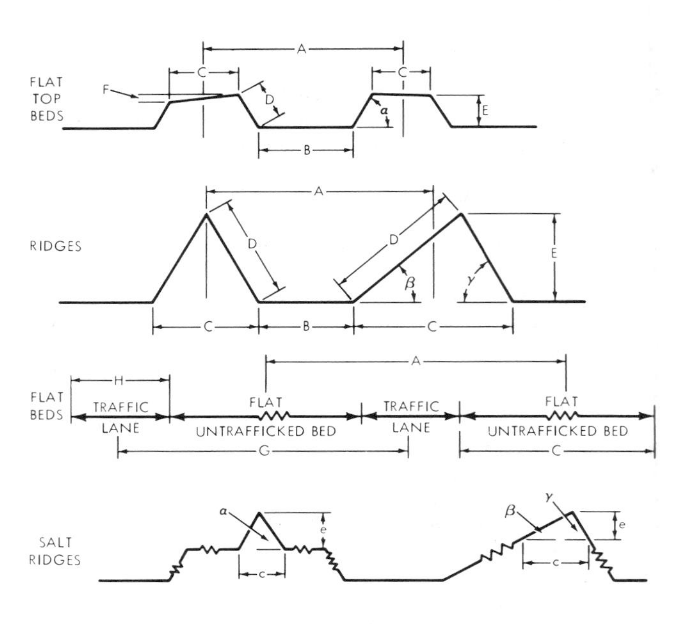

A	=	Bed spacing	G	=	Traffic lane spacing
B	=	Furrow width	H	=	Traffic lane width
C	=	Bed (ridge) width	c	=	Salt ridge width
D	=	Bed side	e	=	Salt ridge height
E	=	Bed (ridge) height	α	=	Bed (ridge) angle
F	=	Bed fall	β, γ	=	Bed (ridge) angles

FIG. 1—BED GEOMETRY

ASAE Standard: ASAE *S295.2 (SAE J709d)

AGRICULTURAL TRACTOR TIRE LOADINGS, TORQUE FACTORS, AND INFLATION PRESSURES

*Corresponds in substance to previous revision SAEJ709c

Developed by the Society of Automotive Engineers May 1959; revised June 1963; approved by ASAE Power and Machinery Division Technical Committee; adopted by ASAE June 1965; reconfirmed December 1970; revised March 1978; reconfirmed December 1982, December 1987.

SECTION 1—PURPOSE

1.1 This Standard establishes loadings, torque factors, and inflation pressure; relationships for tire sizes and ply ratings currently used on agricultural tractors. Primary purpose of this Standard is for use with ASAE Standard, ASAE S209, Agricultural Tractor Test Code.

1.2 The performance of the tractor is materially affected by tire loadings, torque transmitting ability, and inflation pressures; therefore, it becomes desirable to publish this information.

SECTION 2—GENERAL PRINCIPLES

2.1 All agricultural tractor tire loads shown in Tables 1, 2, and 3 are expressed in pounds.

2.2 All agricultural tractor tire inflation pressures shown in Tables 1, 2, and 3 are expressed in pounds per square inch gage, at approximately the prevailing atmospheric temperature at time of test and do not include any inflation pressure build-up due to vehicle operation.

2.3 Tire inflation pressures shall be recorded with the valve at the bottom position.

2.4 The maximum individual tire loading shall not exceed the tire load versus inflation pressure for its respective ply rating shown in Tables 1, 2, and 3. The minimum tire inflation pressure shall not be less than that shown in Tables 1, 2, and 3 for the respective tire size.

2.5 Ply Rating is used to identify a given tire with its maximum recommended load. It is an index of tire strength and does not necessarily represent the number of actual plies in the tire.

2.6 If the tire used is not recorded in Tables 1, 2, and 3, its load and inflation should be in accordance with the latest standards or experimental practices of The Tire and Rim Association, Inc.

Cited Standard:
ASAE S209, Agricultural Tractor Test Code

TABLE 1—AGRICULTURAL DRIVE WHEEL TRACTOR TIRES USED AS SINGLES (MAXIMUM SPEED—20 MPH)

Tire Size	Ply Rating	Tire Load Limits At Various Cold Inflation Pressures, lb									Torque Factor
		12 psi	14 psi	16 psi	18 psi	20 psi	22 psi	24 psi	26 psi		
8.3-24	4	970	1060	1150	1230	1310	1380(4)				0.94
9.5-16	4	970	1000	1080	1150	1230(4)					0.86
9.5-24	4	1210	1330	1430	1540	1630(4)					0.92
11.2-24	4	1470	1600	1740	1860(4)						0.90
11.2-28	4	1560	1710	1850	1980(4)						0.92
12.4-24	4, 6, 8	1760	1920	2080(4)	2230	2370	2510	2640(6)	2760	3120 @ 32(8)	0.89
12.4-28	4	1880	2050	2220(4)							0.91
13.6-24	4		2270(4)								0.87
13.6-28	4, 6		2420(4)	2620	2810	2980	3160(6)				0.89
13.6-38	4, 6		2810(4)	3040	3250	3460	3660(6)				0.94
13.9-36	6		2730	2960	3170	3370(6)					0.94
14.9-24	6		2700	2920	3130	3330(6)					0.86
14.9-26	6, 8		2790	3020	3230	3440(6)	3640	3830	4010(8)		0.87
14.9-28	6		2880	3120	3340	3550(6)					0.88
14.9-30	6		2980	3220	3450	3670(6)					0.89
15.5-38	6, 8		3160	3410	3660	3890(6)	4110	4330	4540(8)		0.94
16.9-24	6, 8			3540	3800(6)	4040	4270	4490(8)			0.84
16.9-26	6, 8			3660	3920(6)	4170	4410	4640(8)			0.86
16.9-28	6, 8			3780	4050(6)	4310	4560	4790(8)			0.87
16.9-30	6			3900	4180(6)						0.88
16.9-34	6			4140	4440(6)						0.90
16.9-38	6, 8			4380	4700(6)	5000	5280	5560(8)			0.91
18.4-16.1	6			2810(6)							0.79
18.4-26	6, 8, 10			4390(6)	4700	5000(8)	5290	5560	5830(10)		0.85
18.4-28	6			4530(6)							0.86
18.4-30	6, 8, 10			4680(6)	5010	5330(8)	5630	5930	6210(10)		0.87
18.4-34	6, 8, 10			4970(6)	5320	5660(8)	5980	6290	6600(10)		0.89
18.4-38	6, 8, 10, 12			5250(6)	5630	5990(8)	6330	6660	6980(10)	7880 @ 32(12)	0.90
20.8-34	8			6010	6440(8)						0.87
20.8-38	8, 10			6360	6820(8)	7250	7670(10)				0.89
23.1-26	8, 10			6280(8)	6730	7160(10)					0.82
23.1-30	8			6700							0.84
23.1-34	8			7110(8)							0.86
24.5-32	10				8180	8700(10)					0.84
17.5L-24	6			3390(6)							0.86
28L-26	10				7800(10)						0.82

NOTES:

1. Figures in parentheses denote ply rating for which loads and inflations are maximum.
2. When used on tractors with mounted implements or on self-propelled implements (except Combines) and operated at speeds not exceeding 10 mph, Drive Wheel tire loads may be increased up to 20 percent with no increase in inflation pressure. Tire loads should include full bins or tanks, all ballast, accessories, etc.
3. When used on tractors or self-propelled implements (except Combines) and operated at speeds not exceeding 5 mph in service which does not require sustained high torque, Drive Wheel tire loads may be increased up to 30 percent with inflation pressures increased 4 psi. Tire loads should include full bins or tanks, all ballast, accessories, etc. Wheel and rim strength must be adequate for maximum load and inflation of a given tire size and ply rating.
4. When used on tractors and operated at speeds not exceeding 15 mph in service which does not require sustained high torque, drive wheel tire loads may be increased up to 7 percent with no increase in inflation pressures.
5. For shipping purposes, tire inflation pressures may be increased to 30 psi. This higher inflation pressure must be reduced to operating inflation BEFORE the tractor is removed from the carrier.

TABLE 2—AGRICULTURAL AND INDUSTRIAL TRACTOR STEERING WHEEL TIRES (MAXIMUM SPEED—20 MPH)

Tire Size	Ply Rating	Tire Load Limits at Various Cold Inflation Pressures, lb											
		20 psi	24 psi	28 psi	32 psi	36 psi	40 psi	44 psi	48 psi	52 psi	56 psi	60 psi	64 psi
4.00-12SL	4	330	370	400	430	470	490	520	550(4)				
5.00-15SL	4	540	600	660	710	760	810(4)						
5.50-16SL	4	660	740	810	870	940(4)							
6.00-14SL	4, 6	680	760	830	900(4)	960	1030	1080	1140(6)				
6.00-16SL	4, 6	760	840	920	1000(4)	1070	1140	1200	1260(6)				
6.50-16SL	4, 6	850	950	1040	1130(4)	1210	1280	1360(6)					
7.50-16SL	4, 6, 8	1100	1220	1340(4)	1450	1550	1650(6)	1740	1830	1920(8)			
7.50-18SL	4, 6	1190	1330	1450(4)	1570	1680	1790(6)						
7.50-20SL	6	1280	1430	1560	1690	1810	1930(6)						
9.00-10SL	4	1100	1230(4)										
10.00-16SL	6, 8	1750	1950	2130(6)	2310	2470	2630(8)						
11.00-16SL	6, 8	2070	2300	2520(6)	2720	2920(8)							
		Low Section Height Tires											
7.5L-15SL	6, 8	1060	1180	1290	1390	1490	1590(6)	1680	1770	1850(8)			
9.5L-15SL	6	1290	1440	1580	1700(6)								
11L-15SL	6	1570	1740	1910(6)									
14L-16.1SL	6	2560	2850(6)										

NOTES:
1. Figures in parentheses denote ply rating for which loads and inflations are maximum.
2. For intermittent service only at maximum speeds of 5 mph, Steering Wheel maximum operating tire loads may be increased up to 50 percent with no increase in inflation pressures.
3. Steering Wheel tire loads may be increased up to 35 percent with no increase in inflation, when used on tractors with mounted implements or on self-propelled implements, and operated at speeds not exceeding 10 mph.
4. Steering Wheel tire loads may be increased up to 15 percent with no increase in inflation when used on tractors with mounted implements or on self-propelled implements, and operated at speeds not exceeding 15 mph.
5. Tire loads should include full bins or tanks, all ballast, accessories, etc.
6. Shipping inflation pressures shall not exceed the maximum pressures shown in Table.

TABLE 3—AGRICULTURAL IMPLEMENT TIRES (MAXIMUM SPEED—20 MPH)

Tire Size	Ply Rating	Tire Load Limits at Various Cold Inflation Pressures, lb									
		20 psi	24 psi	28 psi	32 psi	36 psi	40 psi	44 psi	48 psi	52 psi	56 psi
3.50-12SL	4	370	410	450	490	520	550	580(4)			
4.00-8SL	4	340	370	410	440	470	500(4)				
4.00-12SL	4	450	500	540	590	630	670(4)				
4.00-18SL	2, 4	580(2)	650	710	770	820	880(4)				
5.00-15SL	4	730	810	880	960(4)						
5.50-16SL	4	900	1000	1090	1180(4)						
5.90-15SL	4	850	950	1040	1120(4)						
6.00-16SL	4, 6	1020	1140	1240(4)	1340	1440	1530(6)				
6.40-15SL	4, 6	960	1060	1160(4)	1260	1350	1430	1520(6)			
6.50-16SL	4, 6, 8	1150	1280(4)	1400	1520	1630	1730(6)	1830	1930	2020(8)	
6.70-15SL	4, 6	1070	1190	1300(4)	1400	1500	1600(6)				
7.50-16SL	4, 6, 8, 10	1480	1650(4)	1810	1950(6)	2090	2230	2350(8)	2480	2590	2710(10)
7.50-18SL	4	1540	1720(4)								
7.50-20SL	4, 6	1590	1770(4)	1940	2100(6)						
7.60-15SL	4, 6, 8	1250	1390(4)	1530	1650	1770(6)	1880	1990	2090(8)		
9.00-16SL	8, 10	1960	2180	2380	2580	2760	2930(8)	3100	3260(10)		
9.00-24SL	6, 8	2530	2820(6)	3080	3330	3570(8)					
11.25-24SL	8	3300	3680	4020(8)							
11.25-28SL	10	3420	3810	4170	4510	4830(10)					
13.50-16.1SL	8	3500	3900	4270(8)							

NOTES:
1. For speeds not exceeding 5 mph, the above loads may be increased up to 18 percent with a 4 psi increase in inflation pressure.
2. For speeds not exceeding 10 mph, the above loads may be increased up to 12 percent with 4 psi increase in inflation pressure.
3. Figures in parentheses denote ply rating for which loads and inflations are maximum.
4. Tire loads should include full bins or tanks, all ballast, accessories, etc.
5. Maximum shipping pressures are the maximum inflation pressures for the tire sizes and ply ratings shown.

LOW SECTION HEIGHT IMPLEMENT TIRES

Tire Size	Ply Rating	Tire Load Limits at Various Cold Inflation Pressures, lb						
		20 psi	24 psi	28 psi	32 psi	36 psi	40 psi	44 psi
8.5L-14SL	6, 8	1400	1560	1710	1850(6)	1980	2100	2220(8)
9.5L-14SL	6, 8	1580	1760	1930(6)	2090	2230	2380(8)	
9.5L-15SL	8	1660	1840	2020	2180	2340	2480(8)	
11L-14SL	6	1850	2060(6)					
11L-15SL	6, 8, 10	1930	2150(6)	2350	2550(8)	2730	2900(10)	
11L-16SL	6, 8, 10	2010	2240(6)	2450	2650(8)	2840	3020(10)	
12.5L-15SL	6, 8	2290	2540(6)	2780	3010(8)			
12.5L-16SL	8	2380	2650	2900	3130(8)			
16.5L-16.1SL	6, 8, 10	3990(6)	4440(8)	4860	5250(10)			

NOTES:
1. Figures in parentheses denote ply rating for which loads and inflations are maximum.
2. For speeds not exceeding 5 mph in free rolling applications only, the above loads may be increased up to 18 percent with a 4 psi increase in inflation pressure.
3. For speeds not exceeding 10 mph in free rolling applications only, the above loads may be increased up to 12 percent with 4 psi increase in inflation pressure.
4. Tire loads should include full bins or tanks, all ballast, accessories, etc.
5. Maximum shipping pressures are the maximum inflation pressures for the tire sizes and ply ratings shown.

UNIFORM TERMINOLOGY FOR TRACTION OF AGRICULTURAL TRACTORS, SELF-PROPELLED IMPLEMENTS, AND OTHER TRACTION AND TRANSPORT DEVICES

Proposed by the ASAE Tractive and Transport Efficiency Committee; approved by the ASAE Power and Machinery Division Technical Committee; adopted by ASAE as a Recommendation June 1966; revised February 1970; reconfirmed December 1975; revised and reclassified as a Standard December 1976; reconfirmed December 1981; revised April 1987.

SECTION 1—PURPOSE AND SCOPE

1.1 The purpose of this terminology is primarily to assist in the standardized reporting of information on traction and transport devices. It is possible that data cannot always be reported using this terminology, but in such cases it is recommended that the terms used be clearly defined. Unless otherwise indicated, all definitions refer to single traction or transport devices (not the entire vehicle) operating on a level supporting surface.

SECTION 2—TERMINOLOGY FOR ALL TYPES OF TRACTION AND TRANSPORT DEVICES*

2.1 Ballast: Mass that can be added or removed for the purpose of changing total load or load distribution.†

2.2 Flotation: Ability to resist sinkage into the medium being traversed.

2.3 Load, dynamic [W]: Total force normal to the reference plane of the predisturbed supporting surface exerted by the traction or transport device under operating conditions. This force may result from ballast and/or applied mechanical forces (see Fig. 1).

2.4 Load, static: Total force normal to the reference plane of the predisturbed supporting surface exerted by the traction or transport device while stationary with zero net traction and zero input torque.

2.5 Load transfer: The change in normal forces on the traction and transport devices of the vehicle under operating conditions, as compared to those for the static vehicle.

2.6 Motion resistance ratio [ρ = MR/W]‡: Ratio between motion resistance and dynamic load.

2.7 Motion resistance of traction device [MR = GT − NT]: The difference between gross traction and net traction (see Fig. 2).

2.8 Motion resistance of transport device: Force required in the direction of travel to overcome the resistance from the supporting surface and the internal resistance of the device.

2.9 Power, drawbar [DP = P·V]: The product of drawbar pull and vehicle velocity in the direction of travel.

2.10 Power, input [T·Ω]: The product of input torque and angular velocity of the driving axle of a traction device.

2.11 Power, output [NT·V]: The product of net traction and forward velocity of a traction device.

2.12 Pull, drawbar [P]: Force in the direction of travel produced by the vehicle at the drawbar or hitch.§

2.13 Rolling circumference, zero conditions: Distance traveled per revolution of the traction device when operating at the specified zero condition. Rolling circumference may be determined on the test surface or another surface.

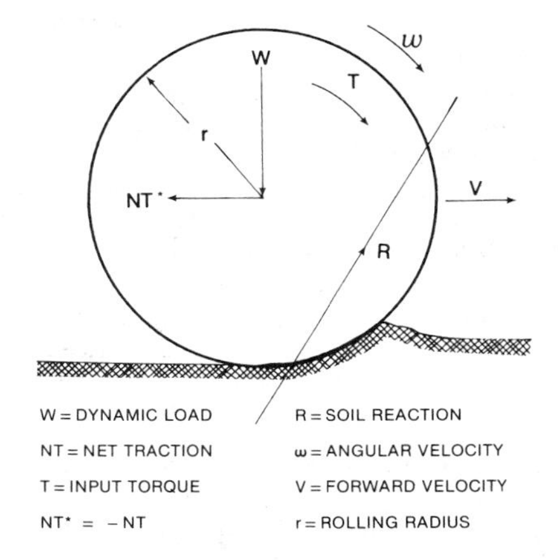

FIG. 1—BASIC VELOCITIES AND FORCES ON A SINGLE WHEEL WITH RESULTANT SOIL REACTION FORCE

2.14 Rolling radius [r]: Rolling circumference divided by 2π (r_o when zero conditions are specified.)

2.15 Slip [s]: A measure of relative movement at the mutual contact surface of the traction or transport device and the surface which supports it, expressed in percent and computed as one minus the travel ratio when the rolling circumference is defined at the self-propelled condition on a hard surface or test surface at the test load and inflation pressure.#

2.16 Soil reaction force [R]: The resultant of all forces acting on the traction device originating in the supporting surface (see Fig. 1).

2.17 Thrust: Total force in the direction of travel as determined from tangential stress measurements at the soil-traction interface.

2.18 Torque, input [T]: The driving moment in the axle of the traction device (see Fig. 1).

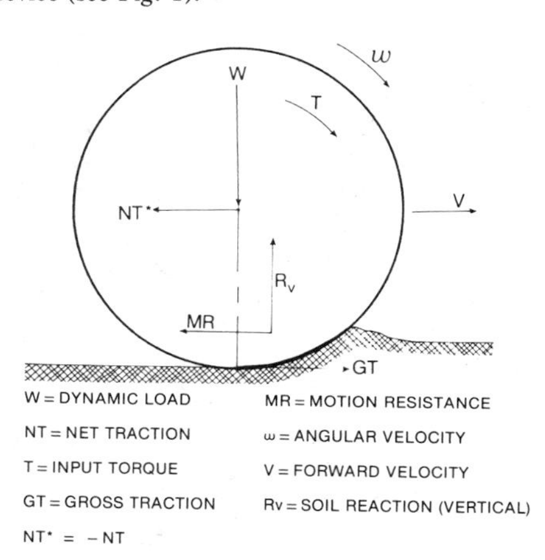

FIG. 2—BASIC VELOCITIES AND FORCES ON A SINGLE WHEEL WITH COMPONENT SOIL REACTION FORCES (MR and Rv act at soil-tire interface)

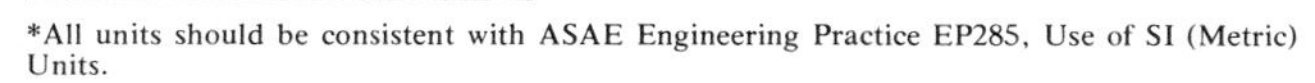

*All units should be consistent with ASAE Engineering Practice EP285, Use of SI (Metric) Units.

†ASAE Standard S209, Agricultural Tractor Test Code, includes a similar definition for ballast that does not recognize transport systems.

‡Motion resistance ratio is the preferred term and is sometimes called coefficient of rolling resistance.

§Draft and drawbar pull are used synonymously. See ASAE Data D230, Agricultural Machinery Management Data.

#Slip and travel reduction are sometimes used synonymously, and are expressed in percent. See ASAE Standard S209, Agricultural Tractor Test Code.

||Net traction ratio is the preferred term and is sometimes called coefficient of net traction.

2.19 Traction device: A device for propelling a vehicle using the reaction forces from the supporting surface.

2.20 Traction, gross [GT = T/r_o]: Total force in the direction of travel as defined by the input torque divided by the rolling radius, at a specified zero condtion.

2.21 Traction, net [NT = P + motion resistance of any unpowered traction or transport device]: Force in direction of travel developed by the traction device and transferred to the vehicle (see Fig. 1).

2.22 Traction ratio, dynamic: Ratio of drawbar pull of the vehicle to the dynamic load on the traction devices of the vehicle.

2.23 Traction ratio, gross [μ_{gt} = GT/W]: Ratio of gross traction to dynamic load.

2.24 Tractive ration, net [μ_{nt} = NT/W]||: Ratio of net traction to dynamic load.

2.25 Traction ratio, vehicle: Ratio of the drawbar pull of the vehicle to the gross vehicle load.

2.26 Tractive efficiency [TE = NT·V/(T·Ω) or DP/(T·Ω)]: Ratio of output power to input power.

2.27 Transport device: Element which supports a vehicle on a surface during travel over that surface and which has zero input torque (sometimes called towed device).

2.28 Travel ratio: Ratio of distance traveled per revolution of the traction device when producing output power to the rolling circumference under the specified zero conditions.

2.29 Travel reduction: One minus travel ratio.#

2.30 Zero conditions: Zero conditions may be those of zero net traction, or zero input torque for the traction device as well as zero drawbar pull for the vehicle. Other zero conditions might also be used. The specific zero conditions should always be stated.

SECTION 3—TERMINOLOGY FOR TRACK-TYPE TRACTION DEVICES

3.1 Angle of approach: The angle between the ground and that section of track between the front bogie wheel and the front idler or sprocket.

3.2 Angle of departure: The angle between the ground and that section of track between the rear bogie wheel and the rear idler or sprocket.

3.3 Grouser height: Vertical distance from the track shoe face to the tip of the grouser.

3.4 Grouser width: Overall width of grouser.

3.5 Grouser spacing or pitch: The distance between corresponding points on adjacent grousers when the shoe surfaces are parallel to each other.

3.6 Nominal ground contact length: The horizontal distance between centers of front and rear sprockets, bogies or idlers which carry a part of the vertical load of the vehicle.

3.7 Track pitch: Distance between corresponding points on adjacent track shoes. On band tracks, it is the distance between corresponding points on adjacent drive lugs.

3.8 Track width: Overall width of an individual track.

SECTION 4—TERMINOLOGY FOR TRACTOR TIRES AND RIMS

4.1 Bias-ply tire: A tire in which the cords of the body plies run diagonally from bead to bead.

4.2 Breaker plies: Plies of cord material, in bias-ply tires, that do not tie into the beads.

4.3 Cinch band or belt: Plies of cord material under the tread area of a tire having the cords nearly parallel to the centerline of the tire (see Fig. 3). These cords do not tie into the tire beads but furnish circumferential strength for the tire.

4.4 Deflection, tire: [δ = (tire diameter/2)—(radius at a static load)].

4.5 Deflection, percent tire: Tire deflection divided by that portion of the tire section height beyond the rim flange, expressed as a percentage.

4.6 End-of-lug clearance: Distance from trailing side of a lug to the end of the lug that follows (see Fig. 3).

4.7 Inflation pressure: For air-filled tires, it is the gauge pressure measured with the valve in any position. For tires containing liquid, it is the gauge pressure measured with an air-water gauge and with the valve in the bottom position.

4.8 Loaded radius (static): Distance from the center of the axle to the supporting surface for a tire when inflated to recommended pressure, mounted on an approved rim and carrying recommended load.

4.9 Lug angle: The angle between the centerline of the lug face and the centerline of the tire (see Fig. 3).

4.10 Lug base: The projected thickness of width of the lug at the points where the projected planes of the leading and trailing sides meet the projected undertread face (see Fig. 3, sec. A-A).

4.11 Lug face: The outermost surface of the lug (see Fig. 3, sec. A-A).

4.12 Lug fillet: The curved section which blends the lug sides into the undertread face (see Fig. 3, sec. A-A).

4.13 Lug height: Distance measured from the undertread face to the lug face (see Fig. 3, sec. A-A).

4.14 Lug length: Distance measured from end to end along the centerline of the lug face (see Fig. 3).

4.15 Lug pitch: Center-to-center spacing (circumferential) of lugs on one side at the centerline of the tire at the lug face (see Fig. 3).

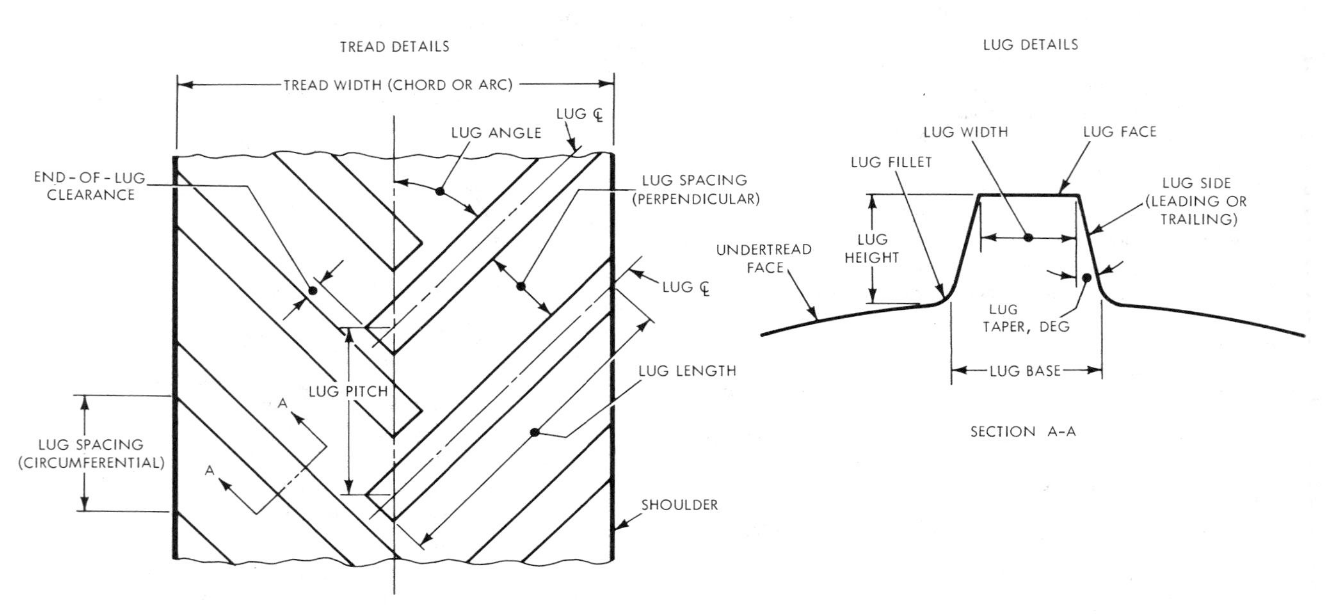

FIG. 3—TRACTOR TIRE LUG AND TREAD DIAGRAM

4.16 Lug side: The lug surface between the undertread face and the lug face (see Fig. 3, sec. A-A).

4.17 Lug spacing, circumferential: The distance from the leading side of a lug to the trailing side of the lug ahead of it, measured parallel to the centerline of the tire at the lug face (see Fig. 3).

4.18 Lug spacing, perpendicular: The distance, measured perpendicularly, from the leading side of a lug to the trailing side of the lug ahead of it at the lug face (see Fig. 3, sec. A-A).

4.19 Lug taper (for leading or trailing side of lug): The angle the lug side makes with a line parallel to the radius that extends from the centerline of the lug to the center of the wheel (see Fig. 3, sec. A-A).

4.20 Lug width: Width of lug face measured perpendicularly to the centerline of the lug face (see Fig. 3, sec. A-A).

4.21 Overall width: The width of a new tire, including normal growth caused by inflation, and including protective side ribs, and decorations (see Fig. 4).

4.22 Ply rating: Identification of a given tire with its maximum recommended load when used in a specific type service. It is an index of tire strength and does not necessarily represent the number of cord plies in the tire.

4.23 Radial-ply tire: A tire in which the cords of the body plies run radially from bead to bead.

4.24 Rim diameter: The nominal diameter at the intersection of the bead seat and the vertical portion of the rim flange (see Fig. 4).

4.25 Tangential pull value: Maximum horizontal pull that the tire can continuously withstand, excluding momentary and occasional peak loads.

4.26 Tire diameter: Tire circumference measured over the lugs in the center plane divided by π, with the tire mounted on its recommended rim and inflated to recommended operating pressure in an unloaded condition (see Fig. 4).

4.27 Tire section width: The width of a new tire, including normal growth caused by inflation and including normal side walls but not including protective side ribs, bars, or decorations (see Fig. 4).

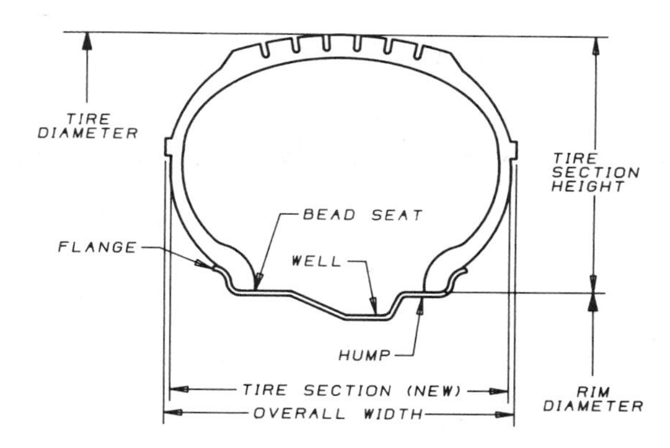

FIG. 4—NEW TIRE AND RIM DIMENSIONS

4.28 Tire section height: The height of a new tire, including normal growth caused by inflation, measured from the rim diameter to the point of maximum radius, on the lug face (see Fig. 4).

4.29 Tread radius: The radius of curvature of the lug faces measured at right angles to the center plane of the tire with the tire mounted on its recommended rim and inflated to recommended pressure (usually not constant).

4.30 Tread width: The distance from shoulder to shoulder (see Fig. 3).

4.31 Undertread face: The outermost surface of the rubber on the carcass where no lugs are located (see Fig. 3, sec. A-A).

Cited Standards:

ASAE S209, Agricultural Tractor Test Code
ASAE D230, Agricultural Machinery Management Data
ASAE EP285, Use of SI (Metric) Units

ASAE Standard: ASAE S300.1

TERMINOLOGY FOR MILKING MACHINE SYSTEMS

Proposed by the ASAE Milk Handling Equipment Committee; approved by the Electric Power and Processing Division Technical Committee; adopted by ASAE as a Recommendation June 1966; reconfirmed December 1970, December 1975; revised and reclassified as a Standard December 1977; reconfirmed December 1982, December 1987.

SECTION 1—PURPOSE AND SCOPE

1.1 This Standard is intended to establish common terms and definitions for milking machine systems.

1.2 The terminology has taken into account International Standard ISO 3918, Milking Machine Installations—Vocabulary, and International Dairy Federation Standard FIL-IDF 56A—1974, Milking Machine Installations, Definitions and Terminology, both of which in turn were based in part on an earlier version of this Standard.

SECTION 2—TERMS AND DEFINITIONS

2.1 **Air-admission hole:** (See air vent)

2.2 **Air pipeline:** (See vacuum pipeline)

2.3 **Air tube:** There are two types of air tubes.

2.3.1 **Short air tube:** A tube connecting pulsation chamber and claw air nipple or claw-mounted pulsator or pulsator mounted on the cover of a suspension unit.

2.3.2 **Long air tube:** A tube connecting the claw air manifold and pulsator or the stall cock and milking unit.

2.4 **Air vent:** A small opening in the milking unit or elsewhere admitting air at atmospheric pressure into the milk transport system.

2.5 **Airflow meter:** An instrument for measuring the quantity of air at a given pressure and temperature per unit time admitted into the milking machine system via this instrument.

2.6 **Automatic washer:** The assemblage of piping, automatic valves, detergent and sanitizer reservoirs and control devices needed for cleaning and/or sanitizing the internal surfaces of a milking system through a prescribed sanitary procedure without personal attention.

2.7 **Bulk milk tank:** A sanitary container located in the farm milk room to cool or store raw milk, or to do both.

2.8 **Claw:** The sanitary manifold that spaces and connects the four teat cup assemblies into a milking unit.

2.9 **Claw bowl:** A milk reservoir adjoining the claw between the milk tubes and the milk hose.

2.10 **Clean-in-place (CIP):** The capability to clean the milking system by circulating appropriate solutions through it without disassembly.

2.11 **Cluster:** (See milking unit)

2.12 **Controller response:** The variation in vacuum controller opening associated with corresponding change in vacuum level.

2.13 **Controller sensitivity:** The change in vacuum level from "just open" to "full open" position of the vacuum controller.

2.14 **Controller, vacuum:** An automatic valve mechanism designed to maintain a steady vacuum level.

2.15 **Cup, milk:** (See claw bowl)

2.16 **Delivery pipeline:** (See pipelines)

2.17 **Drain cock:** A valve device usually located at a low spot in the vacuum system to allow moisture to be drained.

2.18 **Expanded air:** Air at ambient atmospheric temperature at a specified vacuum level.

2.19 **Free air:** Air at 20 °C and standard atmospheric pressure.

2.20 **Gravity filter:** A filter device wherein the milk is admitted above the filter pad and it drains through by the force of gravity only.

2.21 **Inflation:** (See liner)

2.22 **In-line filter:** An enclosed filter device installed in the milk pipe or hose.

2.23 **Interceptor:** A vessel in the vacuum line, ahead of the pump, to intercept liquid and foreign matter.

2.24 **Liner (teat cup liner):** The rubber part of the milking machine in actual contact with the cow's teat. There are several types of liners.

2.24.1 **Liner, extruded:** A type of teat cup liner manufactured by an extrusion process.

2.24.2 **Liner, molded:** A type of teat cup liner manufactured by a molding process.

2.24.3 **Liner, multiple piece:** Teat cup and milk tube assemblies in which the teat cup liner and the milk tube are usually separate. Mouthpart is usually formed with a metal ring.

2.24.4 **Liner, narrow bore:** A liner with an inside diameter of 20 mm or less when under tension.

2.24.5 **Liner, single piece:** Mouthpiece and milk tube are integral with the liner.

2.24.6 **Liner, stretch:** A liner which, when properly installed, is under tension.

2.24.7 **Liner, wide bore:** A liner with an inside diameter of more than 20 mm when under tension.

2.25 **Manometer:** An instrument for measuring the relative pressure of two mediums by balancing them with a column of liquid.

2.26 **Master pulsator:** A pulsator which operates one or more milking units.

2.27 **Milk hose:** A hose connecting claw or milking unit to bucket or milk pipeline or weigh jar.

2.28 **Milk inlet:** A nipple on the milk line for attaching the milk hose. There are two special types of milk inlet.

2.28.1 **Tangential milk inlet:** A milk inlet (valve) which enters a pipe or vessel tangentially.

2.28.2 **Top-entry milk inlet:** A milk nipple that opens into the upper half of the milk line.

2.29 **Milk inlet valve:** An on-off valve incorporated in the milk inlet.

2.30 **Milk line washer:** A vessel used with some types of pipelines to effect washing by alternating vacuum and gravity flow.

2.31 **Milk meter:** A device between the milking unit and the milk pipeline for measuring a cow's milk production in either weight or volume.

2.32 **Milk pipe:** (See pipelines)

2.33 **Milk tube (milk hose):** A flexible-hose which conducts milk and air simultaneously. There are two special types of milk tube.

2.33.1 **Short milk tube:** A tube connecting the liner body to a claw nipple or a milk nipple of the cover of a suspension unit.

2.33.2 **Long milk tube:** A hose connecting claw or milking unit to bucket or pipeline or weigh jar.

2.34 **Milking machine:** A device, composed of several parts, which, when properly assembled and supplied with a source of energy will remove milk from an animal's udder. There are several types of milking machines.

2.34.1 **Bucket milking machine:** A milking machine in which milk flows from the milking unit into a portable milk receiving bucket connected to the vacuum system.

2.34.2 **Direct-to-can milking machine:** A milking machine in which milk flows from the milking unit into the transport can, which is connected to the vacuum system.

2.34.3 **Independent air and milk transport milking machine:** A milking machine in which air and milk are separated immediately below the teat cups and then transported in separate pipelines.

2.34.4 **Pipeline milking machine:** A milking machine in which milk flows from the milking unit into a pipeline that has the dual function of providing milking vacuum and conveying milk to a milk receiver.

2.34.5 **Recorder milking machine:** A milking machine in which milk flows from the milking unit into a weigh jar under vacuum from an air/vacuum pipeline. Milk is discharged when required from the weigh jar into a transfer pipeline to a milk receiver or collecting vessels under vacuum.

2.35 **Milking time:** Time used for milk removal from an animal's udder.

2.36 **Milking unit:** The portable portion of a milking machine for removing milk from individual udders.

2.37 **Mouthpiece:** The upper part or piece of the liner which forms a seal with the shell and udder.

2.38 **Nipple:** Usually a short pipe projection from the claw, pulsator, milking machine lid, or other part of the milking apparatus.

2.39 **Overmilking time:** The time the machine is left on the cow after the milk flow has ceased.

2.40 **Pipelines:** There are several types of pipelines.

2.40.1 **Delivery pipeline:** A pipeline in which milk flows under positive pressure from a releaser milk pump or recording vessel to a storage vessel.

2.40.2 **Milk pipeline:** A pipeline which carries milk and air during milking and has the dual function of providing milking vacuum and conveying milk to a milk receiver. In independent air and milk transport milking machines, only milk is conveyed. There are two main types of milk pipeline.

2.40.2.1 **High-level milk pipeline:** A system wherein the milk inlet is higher than the claw bowl so milk must be lifted to the milk pipeline.

2.40.2.2 **Low-level milk pipeline:** A system wherein the milk inlet is lower than the claw bowl so milk can flow by gravity, from the claw to the milk pipeline.

2.40.3 **Pipeline system:** The milk-conveying components of a milking machine system.

2.40.4 **Transfer pipeline:** A pipeline through which milk is transferred under vacuum from the weigh jar to the milk receiver or a collecting vessel under vacuum.

2.40.5 **Vacuum pipeline (air pipeline):** A pipeline which forms part of the fixed installation and which carries only air during milking; the pipeline may also act as part of the cleaning circuit. May supply milking vacuum, pulsation vacuum, or both.

2.41 **Plate cooler:** A totally enclosed plate-type heat exchanger used for rapid cooling, usually with water.

2.42 **Pressure filter:** A filter device wherein the milk is forced through the filter material by pressure difference created by a force other than gravity.

2.43 **Pulsation:** Cyclic opening and closing of a teat cup liner. There are two main types of pulsation in common use.

2.43.1 **Simultaneous pulsation:** When cyclic movement of all liners of a milking unit is synchronous.

2.43.2 **Alternate pulsation:** The cyclic movement of half the liners of a milking unit alternates with the movement of the other half (180 deg out of phase).

2.44 **Pulsation chamber:** The annual space between the liner and the shell.

2.45 **Pulsation chamber vacuum record:** Each cycle of the record of pulsation chamber vacuum is described as having four phases: (1) increasing vacuum phase, (2) maximum vacuum phase, (3) decreasing vacuum phase and (4) minimum vacuum phase. The duration of each phase as a percentage of the total cycle is measured between the points at which the record intersects abscissae drawn at 4 kPa below maximum vacuum and above minimum vacuum.

2.46 **Pulsation cycle:** One complete cycle of pulsation chamber vacuum.

2.47 **Pulsation rate:** The number of pulsation cycles per minute.

2.48 **Pulsator:** A device for producing cyclic pressure change.

2.49 **Pulsator controller:** A mechanism to operate pulsators, either integral with a single pulsator (self-contained pulsator) or a system controlling several pulsators. (See timer converter)

2.50 **Pulsator ratio:** (See ratio)

2.51 **Ratio:** A measure of the relative portion of time spent with the machine "milking" and "resting". There are several measures of ratio.

2.51.1 **Liner ratio:** The time ratio of the liner more than half open and less than half open.

2.51.2 **Milking ratio:** The time ratio of milk flow to no milk flow during each pulsation cycle.

2.51.3 **Pulsator ratio:** The time ratio of increasing and maximum vacuum phase to all four phases. (See pulsation chamber vacuum record)

2.52 **Receiver (milk):** The device that receives milk from the pipeline, and is the source of vacuum for the milk pipe.

2.53 **Recorder jar:** (See weigh jar)

2.54 **Regulator:** (See controller, vacuum)

2.55 **Regulator, differential:** A valve in the vacuum supply system intended to produce a controlled lower vacuum in that part of the system beyond the valve.

2.56 **Releaser:** A mechanism for removing milk from under vacuum and discharging it to atmospheric pressure.

2.57 **Releaser milk pump:** A device for pumping milk out of the vacuum system.

2.58 **Sanitary trap:** A flow vessel that separates the sanitary (milk) side of a milking machine system from the unsanitary (vacuum supply) side, to keep milk and fluids out of the vacuum system and to prevent back-flow of contaminated moisture.

2.59 **Shell:** The cylindrical rigid part of a teat cup.

2.60 **Short-tube milker:** A suspension-type unit.

2.61 **Sluggish action:** Comparatively gradual pulsated pressure changes or slow liner closing and opening.

2.62 **Snappy action:** Abrupt, pulsated pressure changes or fast liner closure and opening characterized by square shoulders on a vacuum recorder chart.

2.63 **Solution rack:** A rack-like part of the clean-in-place wash system which includes a teat-cup washer, supporting manifold and auxiliary devices, usually located directly above the washing sink.

2.64 **Stall cock:** The valve device on the vacuum line to which the vacuum hose or long air tube is attached.

2.65 **Strip cup or plate:** A cup with a broad flanged lip, or a plate, usually black, onto which fore-milk (the first one or two streams) is squirted to observe for abnormality.

2.66 **Surcingle:** The strap used to support a suspended-cup or suspended-pail milker.

2.67 **Suspension unit:** A type of milking unit suspended from a mechanical arm, surcingle, string or chain.

2.68 **Teat cup:** (See teat cup assembly)

2.69 **Teat cup assembly:** The liner, shell, air tube, or milk tube.

2.70 **Timer converter:** A control unit for operating electromagnetic pulsators.

2.71 **Tube cooler:** A tubular heat exchanger in which circulating water is used to absorb heat from milk.

2.72 **Vacuum:** Any pressure below atmosphere pressure, measured as the reduction below ambient atmospheric pressure. The point of measurement should be stated, e.g., liner vacuum, pulsation chamber vacuum, claw vacuum.

2.73 **Vacuum bulk tank:** An airtight, bulk milk tank which is subject to vacuum.

2.74 **Vacuum gage:** An instrument used to indicate the level of vacuum in a system.

2.75 **Vacuum hose:** A hose which conducts air only.

2.76 **Vacuum, differential:** The difference in vacuum between two parts of a system either by design or caused by pipeline friction.

2.77 **Vacuum-gravity washer:** A tank-like device of sanitary construction and with self-actuating valving, attached to the outer (upper) end of the single milk pipe for CIP cleaning by the non-circulating, back and forth process.

2.78 **Vacuum level:** (See vacuum)

2.79 Vacuum pipelines: (See pipelines)

2.80 Vacuum pump: An air pump which produces vacuum in the system.

2.81 Vacuum pump capacity: The air removal rate of the vacuum pump when it has attained working temperature, at a specified pump speed and vacuum level at the inlet. American Society of Mechanical Engineers Performance Test Code PTC 9, Displacement Compressors, Vacuum Pumps, and Blowers, specifies that capacity tests are referred to inlet conditions; common practice is to rate vacuum pumps in terms of air at either standard atmospheric pressure and temperature (American standard) or at one-half atmospheric pressure (New Zealand standard).

2.82 Vacuum pumping reserve: The additional air-moving capacity of the vacuum pump after the requirements of the milking units, bleeder-holes, operating accessories, and air leaks have been met. It is equal to the volume of air entering through the controller. This is measurable by two alternate techniques.

2.82.1 Effective reserve: Reserve vacuum pump capacity measured by admitting air near the regulator to lower the vacuum 2.0 kPa (approximately 15 mm Hg) below that existing when all units (with the liners stoppered) and accessories (including the vacuum regulator) are operating.

2.82.2 Manual reserve: Reserve vacuum pump capacity measured by admitting air near the regulator to lower the vacuum 2.0 kPa (approximately 15 mm Hg) below that existing when all units (with the liners stoppered) and accessories (excluding the vacuum regulator) are operating.

2.83 Vacuum recorder: An instrument which records vacuum level-time relationships.

2.84 Vacuum reserve tank: A tank, vessel, or chamber in the vacuum system between the pump and point of air admission intended to reduce and stabilize pressure differentials. it is used to store or maintain a state of vacuum.

2.85 Vacuum stability: The degree to which vacuum in any part of the milking system varies from the vacuum set point as a result of normal milking operations.

2.86 Washing manifold: A sanitary extension of the milk line or wash line into which are fabricated nipples or milk inlet valve parts and to which the milking units are attached for cleaning in place.

2.87 Weigh jar: A graduated or scale suspended glass receptacle located between the milking unit and pipeline for measuring a cow's milk production in volume or weight.

Cited Standards:

ASME PTC 9, Displacement Compressors, Vacuum Pumps, and Blowers

IDFS FIL-IDF 56A—1974, Milking Machine Installations, Definitions, and Terminology

ISO 3918, Milking Machine Installations—Vocabulary

FRONT-END AGRICULTURAL LOADER RATINGS

Developed by ASAE Farm Loader Subcommittee; approved by ASAE Power and Machinery Division Technical Committee; adopted by ASAE as a Standard December 1966; revised editorially December 1969; reconfirmed December 1974; revised December 1975; reconfirmed December 1980; revised March 1982; reconfirmed December 1986.

SECTION 1—PURPOSE AND SCOPE

1.1 The purpose of this Standard is to provide a uniform method of rating front-end agricultural loaders.

SECTION 2—DESCRIPTION OF EQUIPMENT RATED

2.1 Loader model: Manufacturer's model number.

2.1.1 Bucket: Specify size of material bucket.

2.1.1.1 Width: Overall width of bucket, in millimeters (inches).

2.1.1.2 Length: Horizontal distance, in millimeters (inches), from cutting edge to rearmost inner face.

2.1.1.3 Height: Overall vertical height, in millimeters (inches).

2.1.1.4 Capacity: Specify capacity per Society of Automotive Engineers Standard J742, Front End Loader Bucket Rating.

2.1.2 Bucket control: Specify whether mechanical or hydraulic.

2.1.3 Loader mass: Specify mass in kilograms (pounds) of all components necessary to provide an operational unit, including the bucket.

2.2 Tractor model: Specify manufacturer, model number, and other information necessary to define tractor.

2.2.1 Tires: Specify front and rear tire sizes.

2.2.2 Wheelbase: Specify wheelbase in millimeters (inches).

2.3 Hydraulic system: Specify whether tractor or independent system.

2.3.1 Rated flow: Delivery of hydraulic fluid to loader in liters per minute (gallons per minute) at rated pressure.

2.3.2 Rated pressure: Specify in pascals (pounds per square inch) at rated flow.

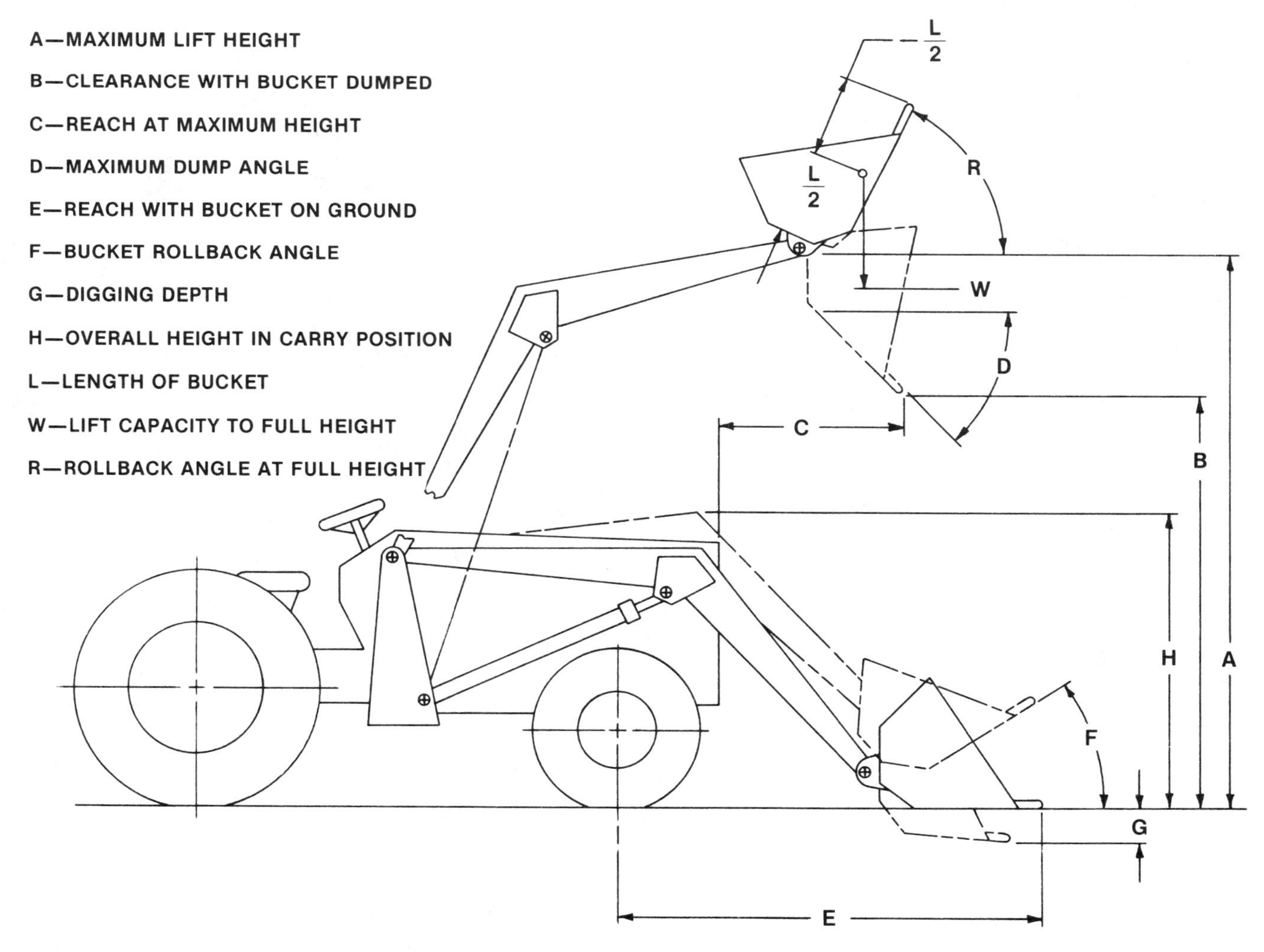

FIG. 1—FRONT-END AGRICULTURAL LOADER DIMENSIONAL SPECIFICATIONS

SECTION 3—SPECIFICATIONS, DIMENSIONAL

3.1 Dimensional specifications are to be determined with the bucket empty and without assembly changes or adjustments to loader or bucket. See Fig. 1.

3.1.1 Maximum lift height (A): The vertical distance, in millimeters (inches), from the ground to the lowest point on the bucket with loader fully raised.

3.1.2 Clearance with bucket dumped (B): Vertical distance, in millimeters (inches), from the ground to the lowest point on the bucket, when dumped 45 deg and loader fully raised. (If maximum dump angle is less than 45 deg, specify angle.)

3.1.3 Reach at maximum height (C): Horizontal distance, in millimeters (inches), from the tip of the cutting edge to the foremost part of the tractor or loader frame with loader fully raised and bucket dumped 45 deg. (If maximum dump angle is less than 45 deg, specify angle.)

3.1.4 Maximum dump angle (D): The angle, in degrees, that the bucket will rotate below horizontal with loader fully raised.

3.1.5 Reach with bucket on ground (E): The distance, in millimeters (inches), from center line of tractor front wheels to tip of the bucket cutting edge when bucket is level on ground.

3.1.6 Maximum rollback angle (F): The angle, in degrees, that the bucket will rotate above horizontal starting with the bucket at groundline.

3.1.7 Digging depth (G): Distance, in millimeters (inches), that tip of bucket cutting edge will move below ground level from horizontal position at ground line using lift cylinder(s) only.

3.1.8 Overall height in carry position (H): Vertical distance, in millimeters (inches), to the highest point on the loader with loader raised to provide 305 mm (12 in.) ground clearance.

SECTION 4—SPECIFICATIONS, OPERATIONAL

4.1 Operational specifications are to be determined at rated hydraulic flow and pressure.

4.1.1 Lift capacity to full height (W): The net load, in kilograms (pounds), located at bucket midpoint that loader will lift to full height, using lift cylinder(s) only, with bucket horizontal at ground level. If the rollback angle (R) at full height exceeds 75 deg, specify angle (mechanical bucket control) or adjust rollback angle to 75 deg (hydraulic bucket control). The bucket midpoint is located on the floor of the bucket at the midpoint of both width and length dimensions as described in Section 2—Description of Equipment Rated.

4.1.2 Breakout force: Vertical force, in newtons (pounds), loader will exert at tip of the bucket cutting edge, using lift cylinder(s) only, when bucket is horizontal at ground level.

4.1.3 Raising time: Time, in seconds, required to raise empty bucket from ground level to maximum lift height.

4.1.4 Lowering time: Time, in seconds, required to lower empty bucket from full lift height to ground level.

4.1.5 Bucket dumping time: Time, in seconds, required to rotate empty hydraulically controlled bucket from full rollback to full dump position.

4.1.6 Bucket rollback time: Time, in seconds, required to rotate empty hydraulically controlled bucket from full dump to full rollback position.

Cited Standard:

SAE J742, Front End Loader Bucket Rating

TEST PROCEDURE FOR SOLIDS-MIXING EQUIPMENT FOR ANIMAL FEEDS

Developed by American Feed Association committee that included representatives from ASAE Technical Committees PM-03 and EPP-03, ASAE Grain and Feed Processing and Storage Committee EPP-38; coordinated and approved by EPP-03 with joint review and approval by PM-03; adopted by ASAE as a Tentative Standard December 1966; revised by Animal Feed Processing Implements Subcommittee and approved by EPP-03 as a full Standard, December 1969; reconfirmed December 1973, December 1978, December 1983; revised December 1984.

SECTION 1—PURPOSE

1.1 This Standard is intended to:

1.1.1 Promote uniformity and consistency in the terms used to describe and evaluate animal feed mixers.

1.1.2 Improve the quality of animal feed mixtures.

SECTION 2—SCOPE

2.1 This Standard is applicable to equipment used to prepare animal feed mixtures and includes both batch type and continuous type. It covers mixers intended for the addition of liquid ingredients as well as dry ingredients.

2.2 Within the scope of this Standard, a mixer may include required auxiliary equipment which would normally be required to operate the mixer. Auxiliary equipment for dry and liquid ingredients may include feeders, surge bins, integral discharge augers, etc., but would not include normal conveying equipment used to convey material beyond the surge bins under quick discharge mixers.

SECTION 3—STANDARD PERFORMANCE CRITERIA

3.1 The following criteria are used to judge the performance of mixing equipment:

3.1.1 Uniformity of dispersion of the ingredients throughout the entire batch or run.

3.1.2 Time required for batch mixing.

3.1.3 Throughput (feed rate or discharge rate) of continuous mixers and residence time in mixer.

3.1.4 Starting and operating power or torque requirements when electric motors are applicable.

SECTION 4—STANDARD TEST CONDITIONS

4.1 Standard feed product formulas

4.1.1 Batch mixers. The standard formula for testing the performance of a batch mixer shall consist of a mixture of 98% ground shelled corn, U.S. Grade No. 2 of less than 14% moisture, and 2% sodium chloride salt (see paragraph 13.1.1). The corn shall be ground to a fineness of geometric mean diameter of 0.85 ± 0.15 mm and a geometric standard deviation of 2.0 ± 0.50 (see ASAE Standard ASAE S319, Method of Determining and Expressing Fineness of Feed Materials by Sieving). The salt shall have a geometric mean diameter of 0.45 ± 0.10 mm and a standard deviation of 1.5 ± 0.25. Formulation in addition to the standard formula may be tested. The particle size distribution and density of all ingredients comprising more than one percent of the formula, or the entire formula excluding any tracers, shall be measured for particle size distribution and density. The particle size distribution and density of each tracer material shall be reported.

4.1.2 Continuous mixers. Continuous mixers shall be tested using the standard formula listed under paragraph 4.1.1. In addition, mixers designed for the application of molasses shall be tested using a mixture of 80% wheat bran and 20% molasses mixture. The molasses mixture shall consist of 95% cane blackstrap molasses having a Brix of 78 and a viscosity of 300 to 100 mPa·s (300 to 1000 centipoises) at 43 °C (110 °F) (method of measurement to be specified), 2.5% potassium chloride and 2.5% ammonium chloride. Mixers designed for mixing heated molasses shall be tested with the molasses heated to 43 °C (110 °F) while those mixers designed for mixing unheated molasses shall be tested with molasses at 20 ± 5 °C (70 ± 10 °F). The dry feed ingredients shall enter the mixer at 20 ± 5 °C (70 ± 10 °F). Other formulations may be tested and the test report shall describe the materials used, particle size and density of dry ingredients, and the density and viscosity of the liquid at the temperature of addition.

4.2 Mixer characteristics. A description of the equipment used should include the make, model, and serial numbers. The following specifications shall be measured and reported:

4.2.1 Major vessel dimensions and total calculated volume. The maximum and minimum working volume of the mixer, as stated by the manufacturer, shall be reported.

4.2.2 Any special modifications to mixer made for testing.

4.3 Charge mass and ingredient proportions. The following information shall be reported:

4.3.1 Mass of each ingredient added.

4.3.2 Total volume occupied by charge in idle mixer.

4.4 Mixing conditions. The following information shall be reported.

4.4.1 Method, sequence, place and rate of adding each ingredient. Note at what point during the charging cycle the mixer is started. For continuous mixing, check feed rates prior to and following the test. Note conveyors between feeders and mixer which may contribute to mixing.

4.4.2 Mixing time in batch mixers or throughput and residence time of continuous mixers.

SECTION 5—STANDARD PROCEDURES

5.1 Mixer operations

5.1.1 Batch mixing. After the mixer is filled with all ingredients, other than the tracer material, the tracer shall be added to the mixer, the mixer started and mixing time shall start. Mixing time ends when discharge begins. In horizontal mixers the tracer shall be added to one end. In vertical mixers it shall be placed on the top of the charge or at the point of normal charging. If mixer is specifically designed for filling during operation, mixing tests shall be duplicated by adding the tracer to the filled mixer while it is run ning. The mixing time shall start at the time when the tracer is added. Final results shall be reported on the basis of at least 10 samples drawn at approximately uniform time intervals during the mixer discharge. Samples shall be drawn from the discharge stream, or the discharge from surge bins if such equipment is included as a part of the mixer. These samples shall be taken at the end of an uninterrupted mixing cycle. At least three trials should be made and results reported separately.

5.1.2 Continuous mixing. A minimum of 10 samples shall be taken from the discharge of the mixer during each of 3 test periods. A minimum test period should be at least 3 times and preferably more than 10 times the average residence time of solids in the mixing equipment. Continuous mixers are highly dependent upon the accuracy of associated feeders. If such feeders are an integral part of the mixing device and if the mixer is intended for continuous, automatic operation without the presence of an operator, it should be equipped with devices to stop the equipment when material flow from supply bins is interrupted or reduced. The operation of such devices should be checked separately over an extended period of time, and the sensitivity in terms of ability to respond to a reduced feed rate should be reported.

5.2 Sampling

5.2.1 Sampling batch mixers. Samples should be taken from the discharge of the mixer by cutting through the discharge stream of the mixer if possible. If this is not possible, as in a drop bottom mixer, then samples should be collected from the discharge of a suitable surge bin. Samples should be taken when discharged at the end of the recommended operating time of the mixer. If internal sampling is to be done, samples should be taken which will represent a wide cross section of the mixer (e.g. three samples from various depths of each quarter of the length of a horizontal mixer).

5.2.2 Sampling continuous mixers. Samples should be taken at equal time intervals during the mixer discharge. No samples should be taken until the mixer has been running for a period of at least twice the residence time plus time required to fill the mixer.

5.2.3 Sample size. At least 10 samples of about 0.5 kg (1 lb) should be taken and assayed separately. In general, the size of samples should be large enough to contain a minimum of 1000 tracer particles (the active ingredient which will be assayed). If this size of sample must be reduced for assay purposes, it should be ground before dividing to a fineness so that the assay sample will again be expected to contain at least 1000 tracer particles.

5.2.4 Sampling methods

5.2.4.1 Samples from the discharge should remove a cross section of the entire stream. Sampling the discharge will aid in locating segregation effects caused by emptying.

5.2.4.2 Internal probe sampling of batch mixers can be used to obtain data to plot a curve for mixing time or to locate points of non-uniformity. Sampling should disturb the mixture as little as possible.

5.2.4.3 A sampling thief, of 25 mm (1 in.) or larger diameter adapted to withdraw about the sample size desired, should be inserted into the batch with a minimum of disturbance, with the sample holes covered. The holes should be covered after the sample is taken and before the probe is withdrawn. Sample thieves may cause segregation and are not well adapted to the sampling of mixtures containing large particles or large amounts of liquids, such as molasses.

SECTION 6—METHODS OF MEASUREMENT

6.1 Analysis of samples

6.1.1 The standard formula shall be analyzed for level of salt, using a chemical procedure. The value of the analytical error of the assay should be determined and included in the test report. Other tracers may be used, but the chemical component of the tracer material should not be found in large amounts in other ingredients. An assay value showing the background level of the active component of the tracer in the other ingredient should be included for comparison. The mixing quality of a molasses feed shall be assayed by assaying for chloride ion concentration.

6.2 Power measurements

6.2.1 Total power input (watts) to the motor should be obtained with a wattmeter having a response time of 1/5 s, an accuracy of ± 5% of full scale reading. Motor power output may then be calculated by multiplying the wattmeter data by the motor efficiency as taken from the motor manufacturer's efficiency curve. Average and start ing power requirements should be shown. The starting power requirements are to be obtained when starting the mixer after it is loaded with the standard formula.

6.2.2 Suitable torquemeters may be used to measure the torque input to the mixer when starting and under normal load. Torque and speed data may be reported in lieu of electrical power measurements after converting to power requirements. The accuracy of the torque and speed measurements should be reported.

SECTION 7—TEST REPORT FOR FEED MIXER USING STANDARD FORMULA

Feed Characteristics, Solids

Material	Particle Size		Moisture %, w.b.
	Geometric mean dia, mm	Geometric std. dev.	

Feed Characteristics, Liquids

Material	Temp. °C (°F)	Density kg/L (g/cc)	Viscosity mPa·s (cps)	pH	Other

Mixer Characteristics

Make__________; Model__________; Serial No.__________

Volume: Total__________m^3 (ft^3)

Max recommended__________m^3 (ft^3)

Min recommended__________m^3 (ft^3)

Solids feed rate (continued)

Max recommended__________m^3/min (ft^3/min)

Min recommended__________m^3/min (ft^3/min)

Motor: Make______; kW (hp)______; Volts______; Type______

Drive__________

Special starter (if applicable):__________

Vessel or agitator speed during operation, rpm:__________

Feeders: Type__________; No.__________; Location______

Type of timing or safety devices to insure proper feeding:__________

Special modifications made for testing:__________

Charging Schedule for Full Batch
(listed in order of addition to mixers)

Ingredient	Batch mass, kg (lb)	Feed rate (continuous mixing) kg/min (lb/min)	Vol. m^3 (ft^3)	Method of addition	Point of addition	Time min.
Total						

Quality of Mixing

Characteristic assayed	Accuracy of assay	Background in ingredients other than tracer	Sample size assayed, g	Method of assay*

*Reference for the assay method

Coefficient of variation (CV) of level of tracer found in ten (if more-specify__________) discharge samples. Tracer assayed__________

Feeder no.	Material	Percent of normal

Mixing time in batch mixer__________minutes

Starting power__________kW; or torque__________N·m (lbf·ft)

Average power__________kW; or torque__________N·m (lbf·ft)

SECTION 8—AUGMENTED PERFORMANCE CRITERIA

8.1 In many cases when a mixer or mixing system is to be used to mix materials, whose physical characteristics are not similar enough to the standard product formula, more extensive testing may be required. Additional criteria which should be considered in evaluating the total performance of a mixer or system include:

8.1.1 Particle size reduction of friable ingredients.

8.1.2 Time required for filling and emptying the mixer.

8.1.3 Possibility of contamination of the product by lubricants, metals, or other materials.

8.1.4 Other facilities required such as ventilation.

8.1.5 Time required for cleanout.

8.1.6 Product loss during operation.

8.1.7 Effect of partial and overfilling on mixer performance.

8.1.8 Weight of residue left in the mixer after discharging.

SECTION 9—AUGMENTED TEST CONDITIONS

9.1 Feed characteristics. In addition to recording a complete identification of each feed components, such as source, grade, composition and previous history, the following physical properties shall be reported:

9.1.1 Particle size distribution.

9.1.2 Bulk density and particle density.

9.1.3 Observations of any unusual particle shape, surface characteristics or electrostatic properties.

9.1.4 Moisture content and temperature of solid materials.

9.1.5 Density, viscosity, pH, and temperature of any liquids added.

9.2 Mixer characteristics. A description of the equipment used should include the make, model, and serial numbers. The following specifications shall be measured and reported:

9.2.1 Major vessel dimensions and total calculated volume. The maximum and minimum working volume of the mixer, as stated by the manufacturer, shall be reported.

9.2.2 Dimensions, number, and type of agitator and/or baffle components, including end and side clearances between agitator, or other moving parts, and body of mixer. Adjustments available.

9.2.3 Size, location and type of access and discharge openings.

9.2.4 Motor power and type of drive.

9.2.5 Vessel or agitator speeds in revolutions per minute, and range of speeds, if any. Direction of rotation of moving elements.

9.2.6 Means provided for assuring completeness of discharge.

9.2.7 Type, location and number of ingredient feeders (including liquids). This does not include feeders to scale bins of batch mixers.

9.2.8 Details of equipment supports and vibration damping devices.

9.2.9 Specifications of any timing or other safety devices provided to insure the uniformity of the finished product.

9.2.10 Schematic diagram of mixer.

9.2.11 Any special modifications made to mixer for testing.

9.2.12 Grounding devices supplied to discharge static electricity.

9.3 Charge mass and ingredient proportions

9.3.1 Mass and calculated, bulk volume of each ingredient added.

9.3.2 Total volume occupied by charge in idle mixer.

9.4 Mixing conditions

9.4.1 Method, sequence, place and rate of adding each ingredient. Note at what point during the charging cycle the mixer is started. For continuous mixing, check feed rates prior to and following the test. Note conveyors between feeders and mixer which may contribute to mixing.

9.4.2 Measured operating speed, rpm or m/s (ft/s).

9.4.3 Mixing time in batch mixers or throughput and residence time of continuous mixers.

9.4.4 Visually note, if possible, any apparent changes in volume, points of accumulation or uneven motion of the mass.

9.4.5 Record method of collection, amount, physical characteristics and chemical characteristics of any material lost from the system.

9.4.6 Starting, or peak, and average operating power required. If torque measurements are made, the starting or peak and average operating torque may be reported.

9.4.7 Atmospheric ambient conditions, temperature and humidity.

9.4.8 Mixing time before addition of liquids.

SECTION 10—AUGMENTED PROCEDURES

10.1 In addition to the procedures described in paragraphs 5.11 and 5.1.2, the following tests may be made and reported.

10.1.1 Batch mixing operations. Trials should be repeated with the mixer loaded to 25, 75, and 110% of rated volumetric capacity and each loading condition repeated during three trials. For part-load trials only assay results need be reported. A batch mixer may be sampled internally by stopping it at intervals and removing probe samples. A plotted curve of the results obtained may be used to select the optimum, or an adequate mixing time. If such results are to be reported, 3 trials of at least 10 samples per trial should be made.

SECTION 11—AUGMENTED RESIDUE DATA

11.1 After a test is complete the equipment should be inspected and thoroughly cleaned. The weight of material removed by brushing and the amount of material removed by scraping should be recorded separately. The location and weight of all major accumulations should be recorded and appropriate tests made to identify the composition of the residue. Consistent patterns of residual accumulation of one or more ingredients should be reported.

SECTION 12—AUGMENTED TEST REPORT FORM FOR FEED MIXER USING STANDARD FORMULA

Feed Characteristics, Solids

Material	Particle Size		Density		Moisture %, w.b.	Temp. Deg. F
	Geometric mean dia., mm	Geometric std. dev.	Bulk kg/m^3 (g/cc)	True kg/m^3 (g/cc)		

Remarks: (Unusual particle shape, surface characteristics or electrostatic properties).___

____

Feed Characteristics, Liquids

Material	Temp. °C (°F)	Density kg/L (g/cc)	Viscosity mPa·s (cps)	pH	Other

Mixer Characteristics

Make__________; Model____________; Serial No.____________

Volume: Total_____________m^3 (ft^3)

Max recommended_________m^3 (ft^3)

Min recommended_________m^3 (ft^3)

Solids feed rate (continuous)

Max recommended_________m^3/min (ft^3/min)

Min recommended_________m^3/min (ft^3/min)

Motor: Make_______;kW (hp)______; Volts_______; Type_______

Drive____

Special starter (if applicable):_________________________

Vessel or agitator speed during operation, rpm:______________

Feeders: Type__________; No.___________; Location____________

Type of timing or safety devices to insure proper feeding:_________

____

Special modifications made for testing:_________________________

Charging Schedule for Full Batch
(listed in order of addition to mixers)

Ingredient	Batch mass, kg (lb)	Feed rate (continuous mixing) kg/min (lb/min)	Vol. m^3 (ft^3)	Method of addition	Point of addition	Time min.
Total						

Ambient conditions:________________ °C (°F)________________ % rh

Mixing time in batch mixer________________________________minutes

Starting power____________ kW; or torque____________N·m (lbf·ft)

Average power____________ kW; or torque____________N·m (lbf·ft)

Observation regarding volume increase during mixing:____________

Observation regarding residue left in mixer after discharging:________

Observations regarding material lost from system when mixing full batch or feed rate:

Point of loss	Amount, kg (lb)	Composition and how determined

Quality of Mixing

Characteristic assayed	Accuracy of assay	Background in ingredients other than tracer	Sample size assayed, g	Method of assay*

*Reference for the assay method

Coefficient of variation (CV) of level of tracer found in ten (if more-specify________________) discharge samples. Tracer assayed________

Test no.	25% Fill or rate, CV, %	75% Fill or rate, CV, %	110% Fill or rate, CV, %	Rated capacity or rate

Response of feeders to poor flow from bin.
(For each feeder measure rate of flow as a percent of normal rated setting at which safety device will stop operation or sound alarm.)

Feeder no.	Material	Percent of normal

Schematic of mixer: Show key dimensions, number of agitators (see paragraphs 4.2.1 to 4.2.9).

Note: The results reported here are for typical feed materials and may not necessarily apply to materials having unusual mixing properties.

SECTION 13—STATISTICAL CONCEPTS AND TESTS

13.1 Poisson distribution. The standard deviation due to a small number of tracer particles in a sample is given by:

$$S = \sqrt{N}$$

where
S = standard deviation
N = number of tracer particles in sample

Variation due to this distribution is minimized by using samples containing more tracer particles.

13.1.1 For design and development testing, it may be desirable to use lower levels of salt which will meet the normal animal feed requirements. The decreased number of salt particles may be compensated for by using a larger assay sample which will supply the same number of tracer particles.

13.2 Additive property of variances. Variances are additive and include mixing, sampling and assay variations. If the variances of each of a number of variables is known, then

$$S^2\ total = S_1{}^2 + S_2{}^2 \ldots Sn^2$$

13.3 Coefficient of variation. The coefficient of variation given by:

$$CV = \frac{S}{m}$$

where
CV = coefficient of variation
m = mean value of all samples

The coefficient of variation can be used to calculate the probability of a given percentage of samples falling within specified tolerance limits, if it is assumed that the distribution of assay values is normally distributed. Assume that the distribution of assay values is given by $f(y)$ with given m and S. It is desired to ascertain the probability that a given sample will fall within the range of (1-B) m to $(1 + B)\ m$. A transformation can be made to the standard normal distribution by:

$$Z = \frac{y - m}{S} \text{ with corresponding tolerance limits}$$

$$= \frac{B}{CV} < Z < \frac{B}{CV}$$

$$\text{or} = \frac{Bm}{S} < Z < \frac{Bm}{S}$$

where
B = tolerance

The areas under the distribution curve, $F(Z)$ for the indicated limits can be taken from any standard table, and this area represents the probability that any given sample will fall within the given tolerance.

13.4 F test for variances. It is frequently desirable to determine if there is a statistically significant difference between the results obtained under different operating conditions, e.g. mixing times. If the means of the two sets of data are the same, then the F test may be applied to the variances to determine if the degree of mixing is the same in both cases.

13.5 Sample calculations

Sample No.	Value of Sample X	Value of $(X\text{-}M)^2$
1	1.985	0.0204
2	1.625	0.0471
3	1.715	0.0161
4	1.625	0.0471
5	1.950	0.0012
6	1.800	0.0018
7	2.025	0.0337
8	1.950	0.0012
9	1.800	0.0018
10	1.800	0.0018
11	2.025	0.0337
12	1.625	0.0471
13	1.915	0.0053
	$23.840 = \Sigma X$	$0.2583 = \Sigma (X\text{-}M)^2$

n = number of samples taken in one test
x = measured value of each sample
M = mean value of all the samples taken
S = standard deviation of the samples, based on the normal curve. Sixty-eight percent of all samples should fall within this range.
CV = coefficient of variation. This is the percentage of the mean that is one standard deviation.
N = number of tracer particles in one sample

$$M = \frac{\Sigma X}{n} = \frac{23.840}{13} = 1.842$$

$$S^2 = \frac{\Sigma (X\text{-}M)^2}{n-1} = \frac{0.2583}{12} = 0.0214$$

$$S = 0.1464$$

CV based on one standard deviation due to test sample measurement:

$$CV = \frac{S}{M} \times 100 = \frac{0.1464}{1.8420} \times 100 = 7.95$$

CV based on one standard deviation due to the number of tracer particles in test sample.

$$N = 33{,}200$$

$$CV = \frac{1}{\sqrt{N}} \times 100 = \frac{1}{\sqrt{33{,}200}} \times 100$$

$$= \frac{1}{182.2} \times 100 = 0.55\%$$

References: Last printed in 1977 AGRICULTURAL ENGINEERS YEARBOOK; list available from ASAE Headquarters.

Cited Standard:

ASAE S319, Method of Determining and Expressing Fineness of Feed Materials by Sieving

ASAE Standard: ASAE S304.5 (ANSI/ASAE S304.5/SAE J389b)

SYMBOLS FOR OPERATOR CONTROLS ON AGRICULTURAL EQUIPMENT

Proposed by Engineering Committee of Farm and Industrial Equipment Institute; approved by ASAE Power and Machinery Division Technical Committee; adopted by ASAE June 1967; revised December 1967, December 1969, October 1970; approved by ANSI as an American National Standard July 1971; revised June 1972; reconfirmed December 1976; revised March 1978; reconfirmed December 1982; reaffirmed by ANSI June 1984; reconfirmed December 1987.

SECTION 1—PURPOSE AND SCOPE

1.1 These symbols provide a symbolic language for operator controls on agricultural equipment as defined in Section 2.

SECTION 2—DEFINITIONS

2.1 Agricultural equipment consists of agricultural tractors; self-propelled implements; implements; and combinations of tractors, implements, and self-propelled implements designed primarily for use in agricultural operations.

SECTION 3—GENERAL

3.1 Color combinations and sizes of all symbols should be adjusted to each particular application. It is desirable to use the universal language of color to indicate the urgency of action, such as red, amber, green.

3.2 Word captions as illustrated are for reference only and are not part of the recommendation. However, suitable descriptive words may be used to intially define the application of symbols. Twentieth Century Bold numerals should be used for clear readability.

3.3 For applications, where appropriate symbols are not shown, the principles illustrated herein should be used for guidance in developing symbols for the specific need.

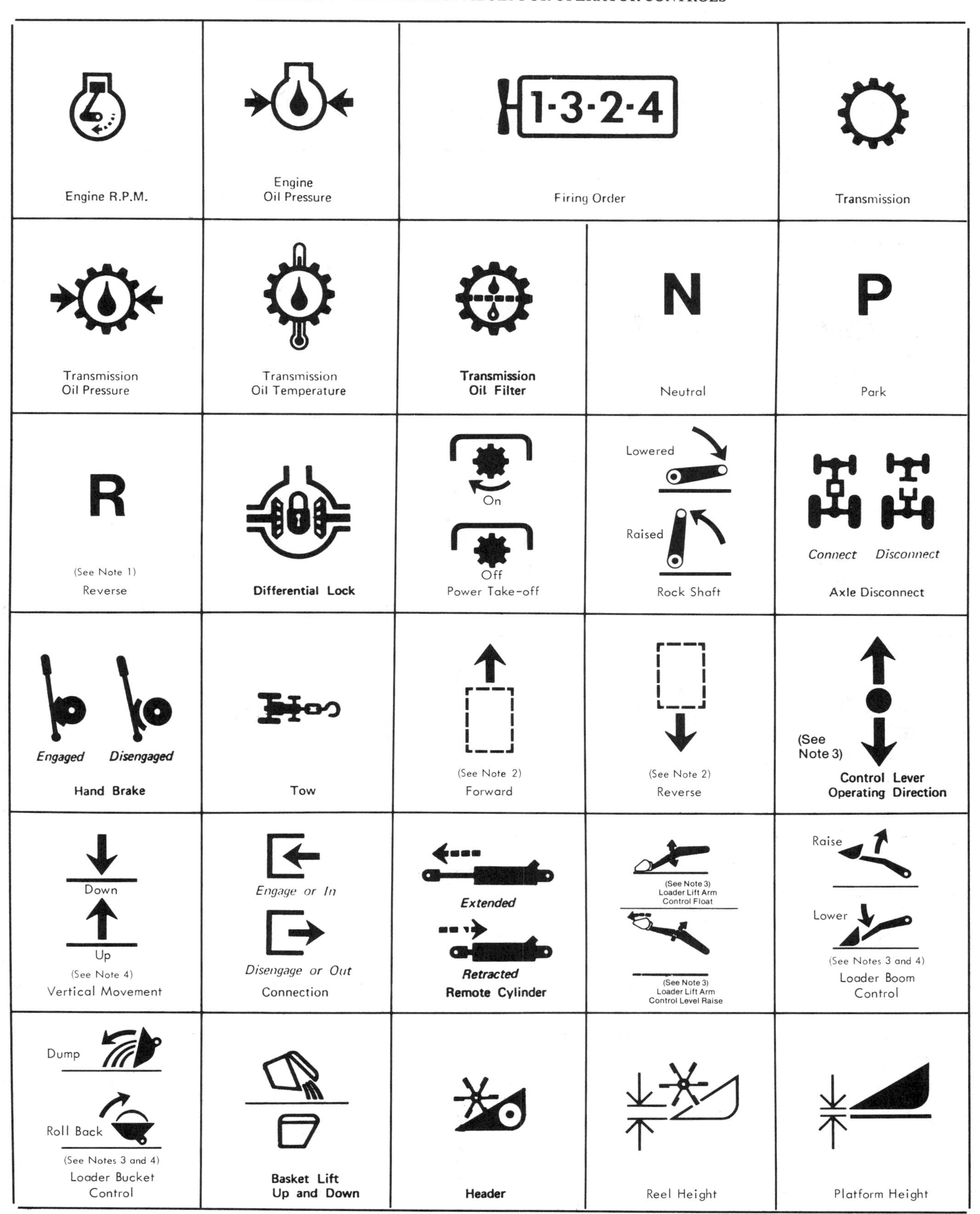

Note 1—Reverse symbol for transmission shift pattern.

Note 2—Replace block with appropriate machine symbol correctly oriented to arrow.

Note 3—Control Lever Operating Direction symbol may be used in conjunction with other symbols to designate lever motion.

Note 4—Captions within the symbols are for information, and should not be reproduced with the symbol.

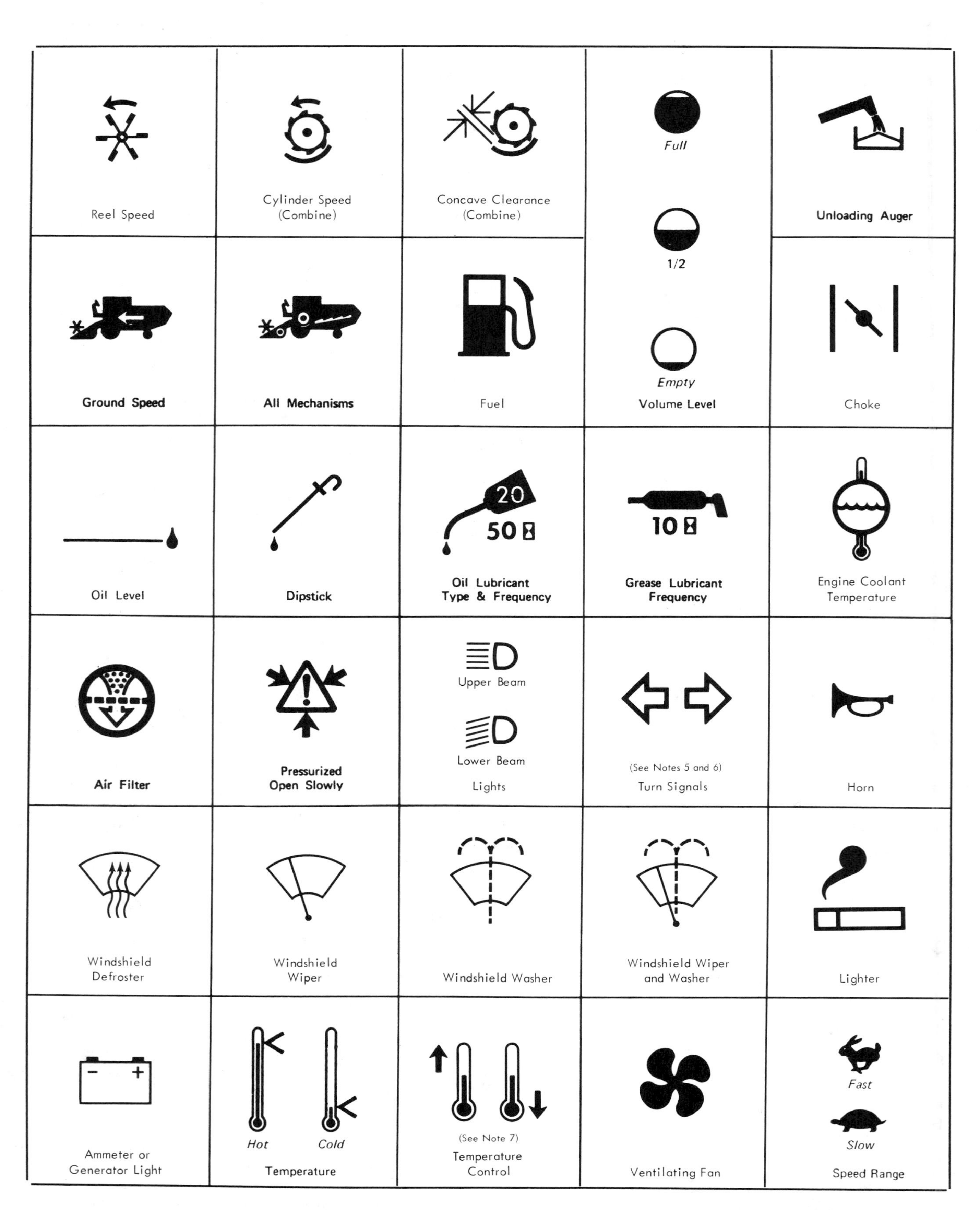

Note 5—Framed area of this symbol may be solid.

Note 6—It is permissible to separate the left and right arrows.

Note 7—Symbol for use at controls and not at temperature measurements.

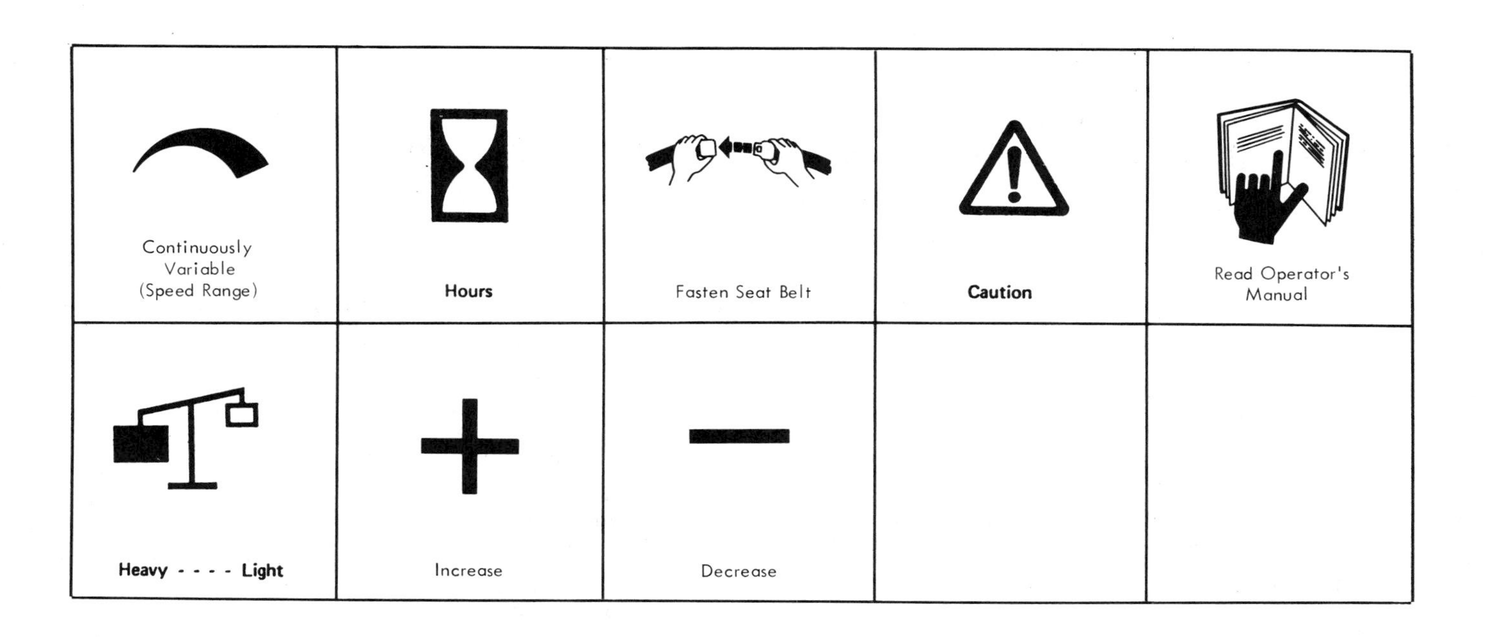

SECTION 5—TYPICAL ILLUSTRATIONS OF THE USE OF COLOR WITH UNIVERSAL SYMBOLS

Use of Colored Lights with Individual Symbols

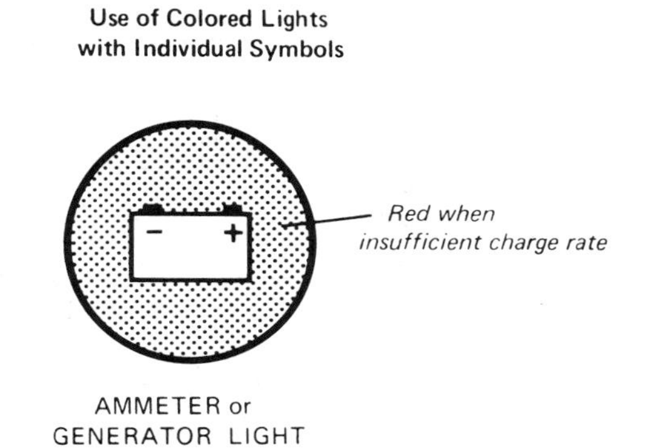

AMMETER or GENERATOR LIGHT

ENGINE OIL PRESSURE

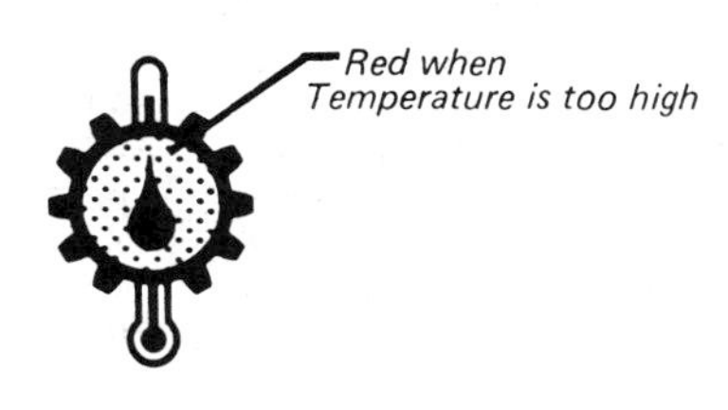

TRANSMISSION OIL TEMPERATURE

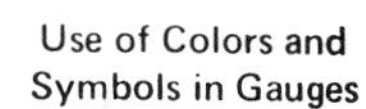

Use of Colors and Symbols in Gauges

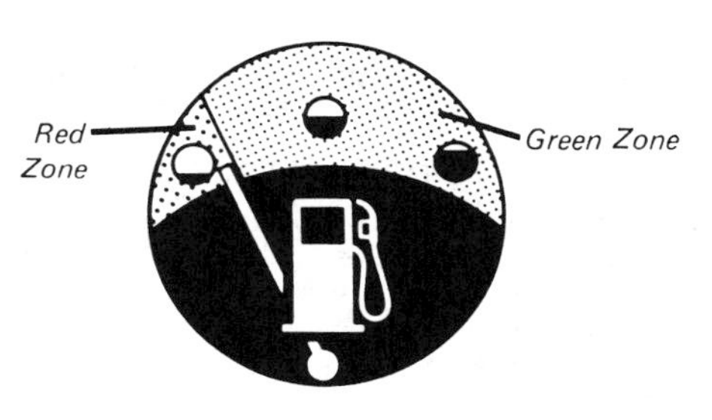

FUEL LEVEL

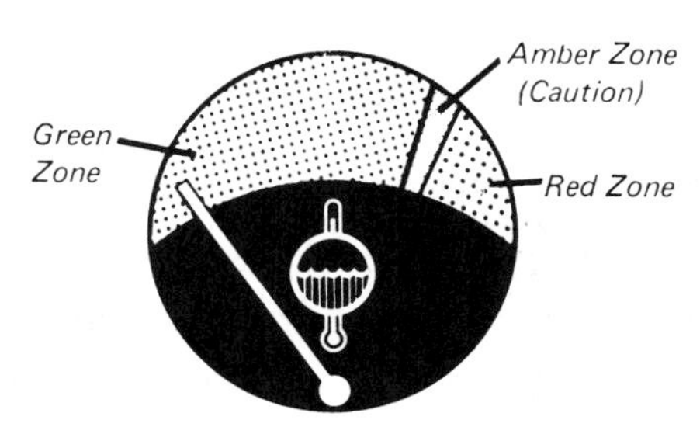

WATER TEMPERATURE

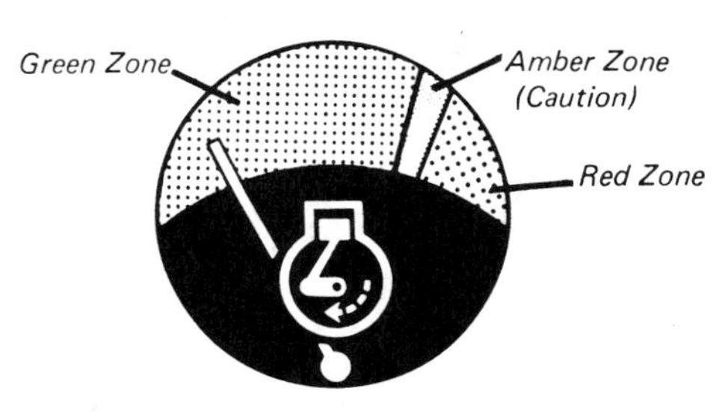

ENGINE R.P.M.

ASAE Standard: ASAE S310.3 (SAE J167 JUN81)

OVERHEAD PROTECTION FOR AGRICULTURAL TRACTORS—TEST PROCEDURES AND PERFORMANCE REQUIREMENTS

Proposed by the Engineering Policy Committee of the Farm and Industrial Equipment Institute; approved by the ASAE Power and Machinery Division Technical Committee; adopted by ASAE June 1968; revised February 1970, December 1972; revised editorially February 1974; revised editorially and reconfirmed December 1976; reconfirmed December 1977; revised December 1982; reconfirmed December 1987.

SECTION 1—PURPOSE

1.1 The purpose of this Standard is to establish the test and performance requirements of a protective frame or enclosure with overhead cover designed for wheel type agricultural tractors to minimize the frequency and severity of operator injury resulting from accidental upsets and overhead hazards during normal operation. General requirements for the protection of operators are specified in ASAE Standard ASAE S383, Roll-Over Protective Structures (ROPS) for Wheeled Agricultural Tractors.

SECTION 2—SCOPE

2.1 Fulfillment of the intended purpose requires conformance to Section 6—Test Procedures, and Section 7—Performance Requirements, of ASAE Standard ASAE S383, Roll-Over Protective Structures (ROPS) for Wheeled Agricultural Tractors, and the additional requirement of a drop test to verify the effectiveness of the overhead cover in protecting the operator from falling objects.

2.2 The test procedures and performance requirements outlined in this Standard are based on current available engineering data.

SECTION 3—DEFINITIONS

3.1 An agricultural tractor is defined in paragraph 3.1 of ASAE Standard ASAE S383, Roll-Over Protective Structures (ROPS) for Wheeled Agricultural Tractors.

3.2 Tractor mass is defined in paragraph 3.2 of ASAE Standard ASAE S383, Roll-Over Protective Structures (ROPS) for Wheeled Agricultural Tractors.

3.3 Seat Reference Point (SRP) is defined in paragraph 3.4 of ASAE Standard ASAE S383, Roll-Over Protective Structures (ROPS) for Wheeled Agricultural Tractors.

SECTION 4—DESCRIPTION

4.1 The protective frame or enclosure to which this Standard applies is a structure generally comprised of uprights mounted to the tractor and attached to an overhead cover extending above and over the operator's seat and conforming generally to Fig. 1.

SECTION 5—SPECIFICATIONS FOR OVERHEAD COVER

5.1 The overhead cover may be cons'ructed of a solid material. If grid or mesh is used, the largest permissible opening shall be such that a 38 mm (1.5 in.) diameter circle is the maximum circle that can be inscribed between the elements of the grid or mesh.

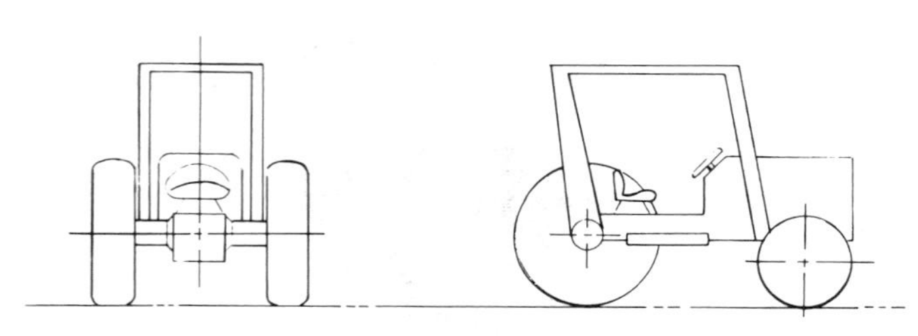

FIG. 1—TRACTOR WITH TYPICAL PROTECTIVE FRAME AND OVERHEAD COVER

SECTION 6—TEST PROCEDURES

6.1 General

6.1.1 Appropriate and applicable ROPS test procedures are specified in Section 6—Test Procedures, of ASAE Standard ASAE S383, Roll-Over Protective Structures (ROPS) for Wheeled Agricultural Tractors.

6.2 Drop test procedure

6.2.1 The overhead cover shall be subjected to a drop test using a solid steel sphere or material of equivalent spherical dimension with a mass of 45.4 kg (100 lb) dropped once from a height of 3 050 mm(10 ft) above the overhead cover.

6.2.2 The point of impact shall be on the overhead cover at a point within the zone of clearance as shown in Fig. 2 which is furthest removed from major structural members.

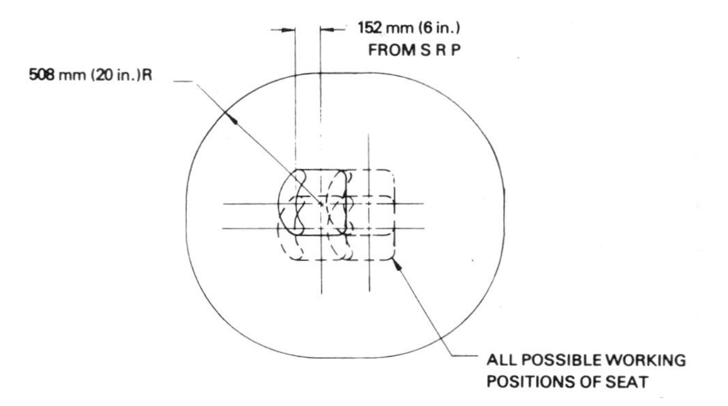

FIG. 2—ZONE OF CLEARANCE FOR DROP TEST

SECTION 7—PERFORMANCE REQUIREMENTS

7.1 General Requirements

7.1.1 Performance requirements specified in Section 7—Performance Requirements of ASAE Standard ASAE S383, Roll-Over Protective Structures (ROPS) for Wheeled Agricultural Tractors, shall apply.

7.1.2 Additional performance requirements, specified in paragraph 7.2 of this Standard, shall apply.

7.2 Drop test performance requirements

7.2.1 Instantaneous deflection due to impact of the sphere shall not enter the clearance zone as illustrated in Fig. 12 of ASAE Standard ASAE S383. Roll-Over Protective Structues (ROPS) for Wheeled Agricultural Tractors.

7.2.2 Minimum allowable dimensions are listed in paragraph 7.1.1 of ASAE Standard ASAE S383, Roll-Over Protective Structures (ROPS) for Wheeled Agricultural Tractors.

Cited Standard:

ASAE S383, Roll-Over Protective Structures (ROPS) for Wheeled Agricultural Tractors

ASAE Standard: ASAE S312.1

CAPACITY DESIGNATIONS FOR COMBINE GRAIN TANK SYSTEMS

Developed by ASAE Grain Harvesting Committee; approved by the ASAE Power and Machinery Division Technical Committee; adopted by ASAE December 1968; reconfirmed December 1973, December 1975, December 1980; revised March 1982; reconfirmed December 1986.

SECTION 1—PURPOSE AND SCOPE

1.1 This Standard is intended to provide a uniform method of determining and designating the capacity and unloading time of combine grain tank systems.

SECTION 2—GRAIN TANK CAPACITY

2.1 The grain tank capacity of a combine shall be the number of volumetric units of wheat that are unloaded by its own unloading system under the conditions described below. Grain tank capacity shall be expressed in liters and rounded to the nearest 100 L.

2.1.1 The wheat shall be U.S. Grade 2, in accordance with the Official Grain Standards of the United States, United States Department of Agriculture, Bulletin SRA-C & MS 177, or equivalent. The wheat shall have a moisture content not exceeding 13.5 percent.

2.1.2 During both filling and unloading of the grain tank, the combine shall be stationary, level and operating at rated speeds with both header and threshing mechanisms running.

2.1.3 To ensure that the grain tank and unloading system are effectively empty before starting the rating procedure, the unloading mechanism shall be operated for a minimum of 60 seconds after the main unloading stream of wheat has subsided.

2.1.4 The grain tank shall then be filled by means of its own loading system up to, but not beyond, the point of spillage.

2.1.5 The wheat shall then be unloaded by the combine's own unloading system operating at rated speed. The grain tank capacity is the volumetric quantity of wheat that is unloaded.

2.1.6 The number of volumetric units may be established by dividing the mass of wheat unloaded by its actual mass per liter.

SECTION 3—GRAIN TANK UNLOADING TIME

3.1 The unloading time is the minimum time in seconds required to unload the grain tank capacity as specified in paragraph 2.1.

ASAE Standard: ASAE S315.2 (ANSI/ASAE S315.2-1982)

TWINE FOR AUTOMATIC BALERS

Proposed by the Engineering Policy Committee of Farm and Industrial Equipment Institute; approved by ASAE Power and Machinery Division Technical Committee; adopted by ASAE December 1968; revised April 1974, December 1976; reconfirmed December 1981; approved by ANSI as an American National Standard December 1982; reconfirmed December 1986.

SECTION 1—PURPOSE AND SCOPE

1.1 The purpose of this Standard is to provide uniform twine specifications which will insure satisfactory performance in a properly adjusted baler knotter and have adequate durability in normal storage and handling of the baled material.

1.2 This Standard covers twines manufactured for use in automatic tie balers.

1.3 The intention of this Standard is to allow freedom in the use of materials and manufacturing processes.

SECTION 2—DEFINITIONS

2.1 For the purpose of this Standard the following definitions apply:

2.1.1 Ball: A cross wound cylindrical unit of twine.

2.1.2 Bale: A parcel of 2 balls.

2.1.3 Runnage: The normal measure of the size of twine, expressed in meters per kilogram (feet per pound).

2.1.4 Test sample: The specified number of tests to be carried out on one ball to determine the physical characteristics of the twine.

2.1.5 Test specimen: One test of the number comprising a sample.

2.1.6 Test batch: Quantity by number of balls from which a specified number of balls will be selected for testing.

SECTION 3—GENERAL REQUIREMENTS

3.1 The twine shall contain no substances which would prove harmful to the material being baled or to farm animals when normal diligence is maintained to prevent accidental consumption.

3.2 In order to meet the requirements for durability in handling and storage of the baled material, consideration shall be given to the effects of rodents, insects, rot and mildew. Appropriate treatments shall be applied to the twine if the materials used are susceptible to attack by these environmental factors.

3.3 The twine shall be uniform in cross-section throughout its length to insure satisfactory performance in a properly adjusted baler knotter.

3.4 Each ball shall be continuous and contain no more than one knot. If a knot is used, it shall be of proper size and strength to function satisfactorily in a baler.

3.5 The ball shall be wound so that the twine is drawn from the center.

3.6 If the twine is spun, a "Z" twist (right hand lay) as shown in Fig. 1 is preferred for improved knot strength.

SECTION 4—CLASSIFICATIONS

4.1 Twine shall be classified by minimum knot and tensile strength and designated as follows:

	Minimum Knot Strength		Minimum Tensile Strength*	
	lbf	N	lbf	N
Light	50	222	100	445
Medium	75	334	150	667
Heavy	100	445	200	890
Extra-Heavy	125	556	250	1112

*It should be recognized that the average tensile strength of the twine will normally be substantially greater than the minimum tensile strength for each classification.

SECTION 5—PACKAGING

5.1 Twine shall be furnished in balls conforming to the following dimensional requirements at the time of packaging:

	Category I	Category II
Height	254 ± 13 mm (10.0 ± 0.5 in.)	305 ± 13 mm (12.0 ± 0.5 in.)
Diameter	254 + 0 – 25 mm (10 + 0 – 1 in.)	286 + 0 – 25 mm (11.25 + 0 – 1 in.)

5.2 If balls are intended to be used in the baler in their shipping container, the outside dimensions of the container shall not be more than 19 mm (0.75 in.) greater than the maximum permissible ball dimensions.

5.3 Each ball shall be packaged in a suitable container to provide adequate support and prevent collapsing during normal use in the baler, and the trailing end of each ball shall be readily accessible.

5.4 Each ball container shall be labeled to indicate:

5.4.1 Compliance with this Standard by the clause, "Complies with ASAE Standard ASAE S315." The class designation, "light," "medium," "heavy" or "extra heavy" shall follow this clause.

5.4.2 An indication of the end from which the twine should be drawn.

5.4.3 Additional information at the manufacturer's option such as runnage, average tensile strength, bale weight, etc., and information necessary to meet legal requirements.

5.5 Each ball or each bale shall be packaged in a suitable container to insure protection of the ball during transit and handling. Protection from moisture should be provided if the nature of the twine is such that exposure to excessive moisture may adversely affect performance in the baler.

FIG. 1—"Z" TWIST ILLUSTRATION

5.6 Shipping containers shall be labeled to include the same information shown on each ball as stated in paragraph 5.4. In addition, each shipping container shall be labeled to show the name or brand name of the manufacturer or distributor and the nominal footage of twine in the container. The minimum permissible actual footage shall be 97 percent of the nominal footage stated on the package.

SECTION 6—INSPECTION AND TESTING

6.1 Where required, test samples shall be taken from 2 balls from each test batch of 100 balls.

6.2 Test samples for determining runnage, tensile strength and knot strength shall be obtained by drawing the specified number of test specimens from the center of the balls, after approximately 30.5 m (100 ft) has been withdrawn and discarded.

6.2.1 A test sample for the determination of minimum knot or tensile strength shall consist of 10 test specimens.

6.3 The minimum length of each test sample for determining runnage shall be 30.5 m (100 ft), and the runnage shall be determined by dividing the length of the test sample in feet (metres) by the weight of the test sample in pounds (newtons). The footage of each bale or ball in the test sample shall be determined by multiplying the runnage by the net weight of the bale or ball in pounds (newtons).

6.4 The strength of each test sample shall not be less than the applicable minimum tensile strength or minimum knot strength shown in Section 4. If any test specimen from any ball fails to meet the required minimum tensile strength or minimum knot strength, the balls sampled do not comply with this Standard, and a second sample of 2 balls from the 100 ball test batch shall be tested. If any one of the second test specimens fail, the test batch does not comply with this Standard.

6.5 Test specimens for minimum tensile strength and minimum knot strength shall be obtained by drawing the twine directly from the center of the ball to the clamps of the testing machine. Care should be taken to avoid loss of twist, other than that automatically lost in drawing out the twine.

6.6 Twine shall be tested on a constant rate of traverse tensile testing machine, preferably power driven, of appropriate capacity and type, and the use of such a machine for the purpose of this test shall be subject to the approval of the parties interested in the test. The speed of movement of the clamps shall be not less than 305 mm per min (12 in. per min) and not greater than 508 mm per min (20 in. per min).

6.7 Cord clamps specified by the Jute Association as 1000 lb, or clamps as described in Federal Specification Textile Test Methods, CCC-T-191b, Federal Standard Stock Catalog, Section IV, Part 6, shall be used to secure the ends of each test specimen.

6.8 The capacity of the machine shall be such that the specified value for the minimum tensile strength of the twine under test is not less than 10 percent of the capacity of the machine. The maximum permissible error of the testing machine shall not exceed 22 N (5 lb).

6.9 The free length of the test specimen at the start of the test shall not be less than 254 mm (10 in.) between the clamps of the machine.

6.10 The type of knot to be used to determine minimum knot strength shall be a hand-tied, doubled overhand knot as shown in Fig. 2 and not its mirror image.

6.11 If any test specimen slips in either clamp or breaks in, or at, either of the clamps at a load less than the appropriate minimum tensile strength specified, that test specimen shall be ignored and another specimen shall be tested.

6.12 The atmosphere for testing need not normally be controlled, but in case of dispute, repeat tests shall be carried out on test samples which have been conditioned for 48 hr in an atmosphere of 65 ± 2 percent relative humidity and at a temperature of 20 ± 1 °C (68 ± 2 °F).

6.13 Drop test for twine balls. This test shall be used under standard conditions to measure the axial elongation of the balls.

6.13.1 Apparatus. The equipment required for performing the drop test is a rule for measurement of the twine balls before and after being dropped. A carpenter's level or a vertical reference surface would aid in measuring the point of maximum height of the balls. The tests should be conducted on a smooth concrete floor or pavement.

6.13.2 Test specimens and conditions. For uniformity the tests shall be performed at the point of manufacture immediately after freshly manufactured twine has been packaged but prior to transfer to any conveyance for delivery to a shipping or storage area. No deviation from the normal method of handling from the ballers to the packaging station shall be permitted. If the drop tests cannot be conducted in the packaging area the specimens should be transported with care to the test location.

6.13.3 Preparation of test specimens. This test shall be conducted with the twine packaged in its normal shipping container. With one end of the package facing the observer, designate the top of the package as 1, the right side 2, the bottom 3, the left side 4, the near end 5, and the far end 6. A minimum of 10 packages shall constitute a test lot.

6.13.4 Test procedure. Each ball in the test shall be placed on a flat surface with core axis vertical, measured to determine the maximum length of the ball parallel to the core axis and then repacked in the original container. Each package shall be subjected to a flat drop from a height of 914 mm (36 in.) successively on sides 1 to 6 inclusive for a total of 6 drops. A man can hold the bales at this height for the flat drop by measurement of a convenient reference surface; for example, a work bench, a door or partition panel. Care should be taken in releasing the bale to avoid any rotation of the package while dropping. After the final drop the twine balls shall be measured in the same manner as at the start of the test.

6.13.5 The balls shall still be usable in an automatic tie baler after this drop test, and their height shall not exceed the maximum permissible height by more than 13 mm (0.5 in.).

Cited Standard:

FED. SPEC. No. 191b, Textile Test Methods

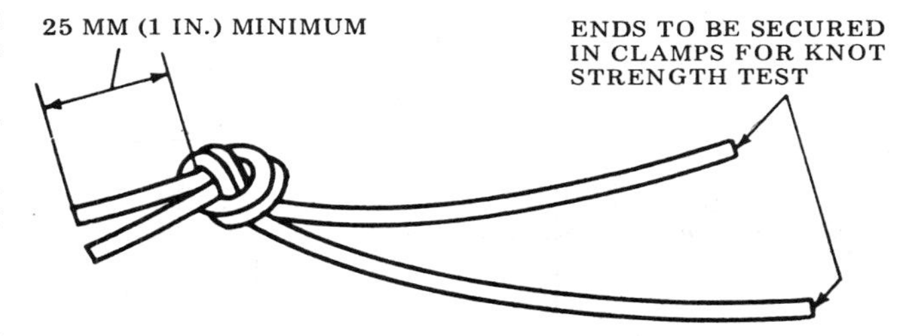

FIG. 2—DOUBLED OVERHAND KNOT

ASAE Standard: ASAE S316.1

APPLICATION OF REMOTE HYDRAULIC MOTORS TO AGRICULTURAL TRACTORS AND TRAILING-TYPE AGRICULTURAL IMPLEMENTS

Proposed by the Tractor and Implement Hydraulic Committee; approved by ASAE Power and Machinery Division Technical Committee; adopted by ASAE as a Tentative Standard December 1968; reconfirmed December 1969, December 1970, December 1971, December 1972, December 1973; revised and reclassified as a full Standard May 1975; reconfirmed December 1979, December 1984.

SECTION 1—PURPOSE AND SCOPE

1.1 The purpose of this standard is to establish common mounting and clearance dimensions for remote hydraulic motors and trailing-type agricultural implements, with such other specifications as are necessary to accomplish the following objectives:

1.1.1 To permit use of any make or model of trailing-type agricultural implement adapted for operation by a remote hydraulic motor, with any remote hydraulic motor furnished as part of any make or model of agricultural tractor in a drawbar horsepower size suitable for operating that implement.

1.1.2 To facilitate changing the hydraulic motor from one implement to another and to decrease the possibility of introducing dirt or other foreign material into the hydraulic system.

SECTION 2—DEFINITIONS

2.1 Mounting face. Finished surface on the shaft end of the motor, also called the front end of the hydraulic motor.

2.2 Hydraulic line connection. Motors with fluid lines attached on the side of the motor are designated as side connection, those with fluid lines attached to back of motor as end connection.

2.3 Starting torque. Torque available to start a load at zero rpm.

SECTION 3—CLASSIFICATION AND RATING

3.1 For the purpose of application of remote hydraulic motors, tractors are divided into categories as shown in Table 1.

3.2 A hydraulic motor with horsepower output as shown in Table 1 shall be considered as part of the tractor hydraulic system.

3.3 Operating speeds and drive shaft sizes shall be as shown in Table 1.

3.4 Horsepower output shall be based on the actual torque of the motor at 80 percent of the minimum specified relief valve setting.

SECTION 4—CLEARANCE DIMENSIONS

4.1 Implements shall provide clearance for the maximum motor size as shown in Fig. 1.

4.2 Hydraulic remote fluid motors with hoses shall be considered as part of the tractor.

SECTION 5—HOSE LENGTHS

5.1 The hose length shall be sufficient so that the remote fluid motor is operable when the drive shaft is located in accordance with dimensions in Table 1.

5.2 On tractors for which a standard power take-off shaft is available, the maximum spherical radius for locating the motor shall be measured from the agricultural tractor drawbar hitch point specified in ASAE Standard ASAE S203, Rear Power Take-Off for Agricultural Tractors.

5.3 On all track-type tractors without a standard power take-off shaft, the maximum spherical radius for locating the motor shall be measured from the drawbar hitch point specified in Society of Automotive Engineers Standard SAE J749, Industrial (Track Type) Tractor Drawbars. (See Fig. 3.)

0.76 (0.030) R MAX.
15 TOOTH, 16/32 PITCH FLAT ROOT SIDE FIT INVOLUTE SPLINE, PER ANSI B92.1
101.60 / 101.55 (4.000 / 3.998) DIA.
1.57 (0.062) X 45° CHAMFER
9.65 / 9.14 (0.380 / 0.360)
174.60 (6.874)
146.18 / 145.92 (5.755 / 5.745)
381 (15) MAX.
46.0 (1.81)
101.60 (4.000) MAX.
120.65 (4.750)
152.40 (6.000) MAX.
88.90 (3.500) MAX.
88.90 (3.500) MAX.
14.27 (0.562) DIA. HOLE 2 PLACES
DIMENSIONS ARE IN MILLIMETERS FOLLOWED BY INCH EQUIVALENT IN PARENTHESIS.

FIG. 1—CLEARANCE AND MOUNTING DIMENSIONS FOR REMOTE HYDRAULIC MOTORS

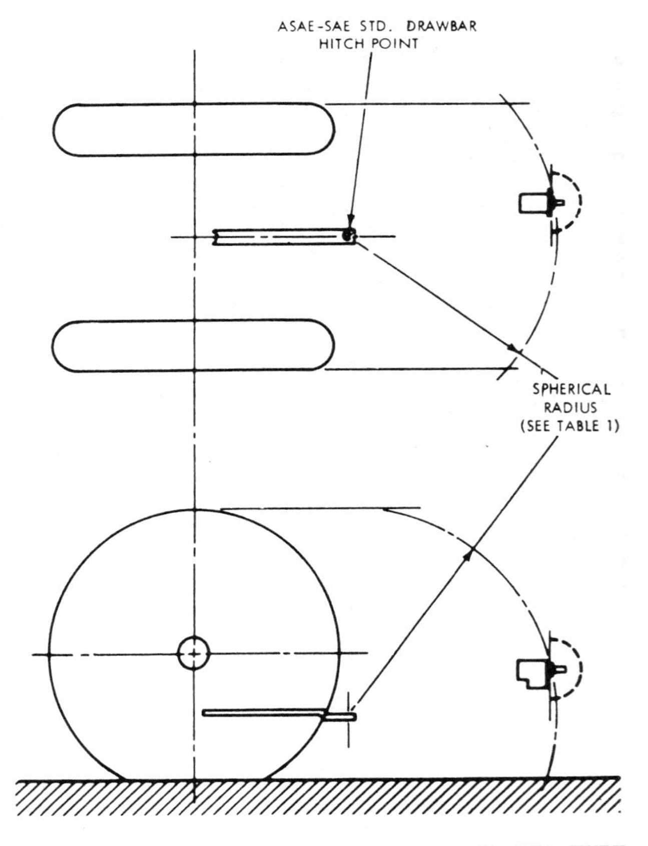

FIG. 2—HOSE LENGTH MEASURED FROM WHEEL TYPE TRACTOR HITCH POINT

5.4 On an implement attached to the tractor by means of a standard three-point free-link hitch, as specified in ASAE Standard ASAE S217, Three-Point Free-Link Attachment for Hitching Implements to Agricultural Wheel Tractors, and using a tractor hydraulic remote fluid motor, the maximum spherical radius for locating the motor drive shaft on the implement shall be measured from a point 178 mm (7 in.) horizontally ahead of the mid-point between the lower hitch points with lower links horizontal. (See Fig. 4.)

5.5 When a quick-attaching coupler as specified in ASAE Standard ASAE S278, Attachment of Implements to Agricultural Wheel Tractors Equipped with Quick-Attaching Coupler for Three-Point Free-Link Hitch, is provided with the tractor, remote hydraulic fluid motor hose length must provide for moving the implement 103.1 mm (4.06 in.) rearward.

SECTION 6—OTHER SPECIFICATIONS

6.1 Remote hydraulic motors should operate in either direction and be reversible from the operator's station.

6.2 Variable speed control, necessary in the application of hydraulic motors to some implements, shall be incorporated in the motor or hydraulic system and applied to reduce motor speed. Provisions shall be made on the implement to accommodate the full motor speed and torque without damage to the implement.

6.3 Smooth operation shall be possible down to 20 percent of rated motor speed.

6.4 No load speed shall not be greater than 140 percent of rated operating speed.

6.5 Hose support as required for hydraulic remote fluid motors shall be considered a part of the implement.

6.6 Hose connections shall enter the motor envelope at the back of remote hydraulic motors. Side connections shall be within the motor envelope.

6.7 If it is necessary to protect the hydraulic motor or hydraulic system from damage due to mismatch of implement and motor, this required device shall be considered a part of the motor or tractor hydraulic system.

6.8 Remote hydraulic motors shall be capable of withstanding a radial load applied at the end of the shaft as shown in Table 1.

6.9 The remote hydraulic motor shall be capable of operating in any position.

Cited Standards:

ASAE S203, Rear Power Take-Off for Agricultural Tractors
ASAE S217, Three-Point Free-Link Attachment for Hitching Implements to Agricultural Wheel Tractors
ASAE S278, Attachment of Implements to Agricultural Wheel Tractors Equipped with Quick-Attaching Coupler for Three-Point Free-Link Hitch
SAE J749, Industrial (Track-Type) Tractor Drawbars

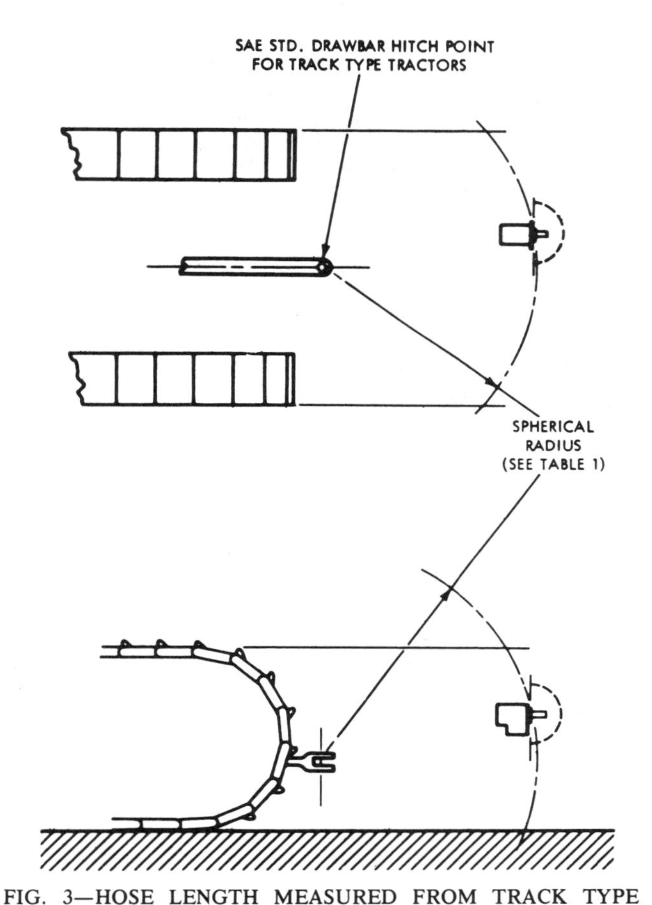

FIG. 3—HOSE LENGTH MEASURED FROM TRACK TYPE TRACTOR HITCH POINT

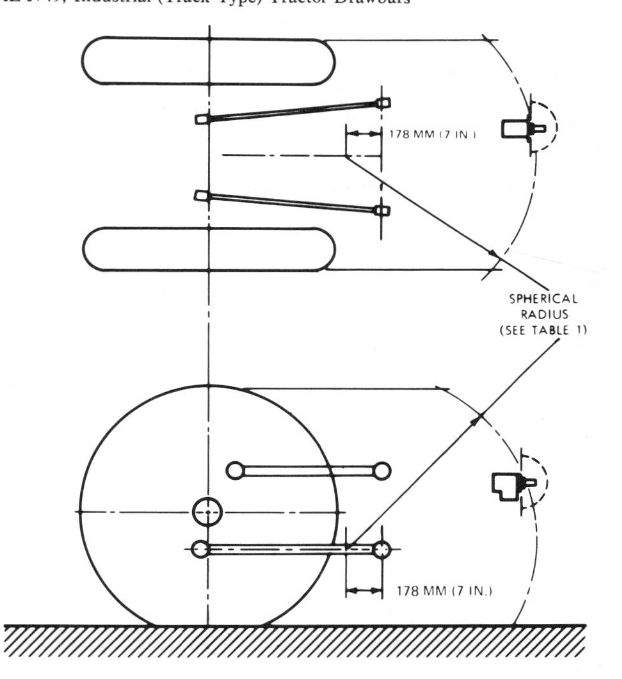

FIG. 4—HOSE LENGTH FOR THREE-POINT HITCH MOUNTED IMPLEMENTS

TABLE 1 — REMOTE HYDRAULIC MOTORS FOR TRAILING-TYPE AGRICULTURAL IMPLEMENTS

Category	Tractor* Drawbar Power		Minimum† Motor Power		Rated Operating speed,	Minimum Starting Torque		Nominal Motor Shaft‡ Outside Diameter		Side Load		Spherical Radius§ to Attaching Flange	
	kW	hp	kW	hp	rpm	lb-in.	Nm	in.	mm	lb	N	in.	mm
I	Up to 34	Up to 45	3.7	5	1000	263	29.7	1.000	25.40	132	587	60	1524
II	30-75	40 to 100	7.5	10	1000	525	59.3	1.000	25.40	262	1165	84	2134
III	Over 60	Over 80	11.2	15	1000	788	89.0	1.000	25.40	394	1752	96	2438

* Based on ASAE Standard ASAE S209, Agricultural Tractor Test Code.

† It is intended that implements be designed to require power below this figure.

‡ When an adapter is required to provide a 25.40 mm (1.000 in.) shaft, the adapter shall be part of the motor.

§ Refer to Fig. 2, 3, and 4.

ASAE Recommendation: ASAE R317

IMPROVING SAFETY ON ENCLOSED MOBILE TANKS FOR TRANSPORTING AND SPREADING AGRICULTURAL LIQUIDS AND SLURRY

Proposed by the ASAE Agricultural Safety Committee; approved by the Power and Machinery Division Technical Committee; adopted by ASAE December 1968; corrected editorially March 1973; reconfirmed December 1973, December 1975, December 1980, December 1985.

SECTION 1—PURPOSE AND SCOPE

1.1 This Recommendation has been developed to provide a guide for uniform practice and is intended to reduce the possibility of personal and public injury during normal servicing and operation of enclosed mobile tanks for transporting and spreading agricultural liquids and slurry.

1.2 This document applies to truck-mounted units, self-propelled units, towed units with integral axle, and towed units with single or multiple axle carriage.

SECTION 2—DEFINITIONS

2.1 For the purpose of this document the following definitions shall apply:

2.1.1 Enclosed Mobile Tanks: All tanks for transporting and spreading agricultural liquids or slurry in which the material is completely contained and covered, as opposed to open tanks which permit rapid escape of the material if the tank is tipped or suddenly halted.

2.1.2 Slurry: A mixture of solids and liquids which will flow as a liquid and create a surge force.

SECTION 3—TANK CONSTRUCTION

3.1 An enclosed mobile tank shall be so constructed as to prevent its load from leaking or spilling.

3.2 All openings shall be provided with covers and a means of securely fastening thereon.

3.3 Operating instructions shall be appropriately displayed on the tank and in the operator's manual directing the operator to securely fasten covers when transporting on a public road.

SECTION 4—WEIGHT AND WIDTH OF TANK UNITS

4.1 The gross weight and maximum width of an enclosed mobile tank and its appurtenances shall comply with applicable state laws when operated on public highways.

SECTION 5—SURGE CONTROL

5.1 Longitudinal Surge: All enclosed mobile tanks shall be so designed and constructed that when starting or stopping, the maximum resulting vertical force at the hitch point, including surge, shall not exceed the vertical dynamic load limitation defined by ASAE Standard S207, Operating Requirements for Tractors and Power Take-Off-Driven Implements.

5.2 Lateral Surge: Baffles or other appropriate means shall be provided to minimize the effects of lateral surge.

SECTION 6—GENERAL SAFETY

6.1 The following ASAE Standards shall apply:

ASAE S203, Rear Power Take-Off for Agricultural Tractors
ASAE S279, Lighting and Marking of Agricultural Field Equipment on Highways
ASAE S318, Safety for Agricultural Equipment

Cited Standards:

ASAE S203, Rear Power Take-Off for Agricultural Tractors
ASAE S207, Operating Requirements for Tractors and Power Take-Off-Driven Implements
ASAE S279, Lighting and Marking of Agricultural Field Equipment on Highways
ASAE S318, Safety for Agricultural Equipment

ASAE Standard: ASAE S318.10 (*ANSI/ASAE S318.6/SAE J208d)

SAFETY FOR AGRICULTURAL EQUIPMENT

*Corresponds in substance to previous document ASAE S318.6.

Supersedes ASAE R275, Improving Safety on Farm Implements, adopted June 1964; R280, Improving Safety on Farm Tractors, adopted December 1964; and S297T, Enclosure-Type Shielding of Forward Universal Joint and Coupling Means of Agricultural Implement Power Drive Lines, adopted June 1966. Proposed by the Engineering Policy Committee of the Farm and Industrial Equipment Institute; approved by the ASAE Power and Machinery Division Technical Committee; adopted by ASAE as a Recommendation December 1968; revised December 1969; revised and reclassified as a Standard February 1972; revised editorially June 1972; approved by ANSI as an American National Standard April 1973; revised December 1973, March 1977, March 1978; reconfirmed December 1982; revised March 1984, March 1985, March 1987, April 1988.

SECTION 1—PURPOSE AND SCOPE

1.1 This Standard is a guide to provide a reasonable degree of personal safety for operators and other persons during the normal operation and servicing of agricultural equipment.

1.2 This Standard does not apply to skid steer loaders, permanently installed grain dryers, and agricultural equipment covered by other safety standards, such as but not limited to permanently installed farmstead equipment, portable grain augers, and storage structures, except where specifically referenced by other standards.

SECTION 2—DEFINITIONS

T

2.1 Agricultural equipment and other terms: Refer to ASAE Standard S390, Classifications and Definitions of Agricultural Equipment.

2.2 Power take-off (PTO): Refer to ASAE Standard S203, Rear Power Take-Off for Agricultural Tractors.

2.3 Auxiliary power take-off (Aux. PTO): Refer to ASAE Standard S333, Agricultural Tractor Auxiliary Power Take-Off Drives.

E

2.4 Implement input drive line (IID): Two universal joints and their connecting member(s) and fastening means for transmitting rotational power from the tractor PTO to the implement input connection. A double Cardan, constant velocity joint is considered a single joint. The IID also includes integral shielding (guarding) where provided.

2.5 Implement input connection (IIC): The shaft or other connecting means to which the rear joint of the IID is attached on the implement.

SECTION 3—OPERATOR'S MANUAL

3.1 Operator's manuals shall be supplied with each piece of equipment.

3.2 Operator's manuals shall provide general safety instructions for normal operation and servicing of the equipment in accordance with ASAE Engineering Practice EP363, Technical Publications for Agricultural Equipment.

SECTION 4—OPERATOR CONTROLS

4.1 Location, movement, and identification of operator controls shall be in accordance with ASAE Standard S335, Operator Controls on Agricultural Equipment.

SECTION 5—SAFETY INTERLOCK PROVISIONS

5.1 For future design, agricultural tractors shall be equipped with devices which assure that the PTO and auxiliary PTO are disengaged when the engine is being started.

5.2 An interlock shall be provided to prevent starting the engine with recommended procedure unless the transmission is in a neutral position, the transmission clutch is disengaged, or the combination direction and speed control is in a neutral position.

SECTION 6—OPERATION AND SERVICING PROVISIONS

6.1 A suitable station shall be provided for each person required for the operation of the equipment.

6.2 All equipment shall have steps and handholds, or other means to facilitate entry and exit from the operating positions.

6.3 If the mounting step(s), ladder(s), or handhold(s) of the agricultural tractor or self-propelled machine are made inaccessible by the installation of equipment, alternate facilities shall be provided.

6.4 Handholds, hand rails, guard rails, or barrier type safeguards shall be provided, if necessary, to minimize falling during normal operation or servicing, unless means are provided by other parts of the equipment.

6.4.1 Guard rails when provided shall have a top rail 900 to 1050 mm (35.4 to 41.3 in.) above the working walkway or platform with a rail approximately midway between the platform and the top rail.

6.5 Step requirements for operating station

6.5.1 The height of the first step should not exceed 686 mm (27.0 in.), preferably 550 mm (21.6 in.).

6.5.2 The vertical interval between steps shall not exceed 406 mm (16.0 in.), preferably 300 mm (11.8 in.). Step intervals shall be reasonably uniform, except that the interval between the top step and platform may be less than between other steps.

6.5.3 The width of each step shall be not less than 250 mm (9.8 in.).

6.5.4 The combined step depth plus toe clearance shall be not less than 150 mm (5.9 in.).

6.5.5 Steps shall have slip-resistant surfaces and be designed to minimize the accumulation of debris.

6.6 Operator platforms shall have slip-resistant surfaces.

6.7 Shielding shall be provided on the back of steps or ladders wherever a protruding hand or foot may contact a moving machine part hazard.

6.8 Glazing material, such as glass or plastic, used in operator enclosures, shall meet the requirements of Society of Automotive Engineers Standard J674, Safety Glazing Materials—Motor Vehicles, except that:

6.8.1 Windshields may conform to tempered safety glass, Item 2 of American National Standard Z26.1, Safety Code for Safety, Glazing Materials for Glazing Motor Vehicles Operating on Land Highways.

6.8.2 Windshields may include zone toughened glass conforming to European Economic Community Regulation Number 43, Uniform Provisions Concerning the Approval of Safety Glazing and Glazing Materials for Installation on Power Driven Vehicles and Their Trailers.

6.8.3 Curved and flat glazing material larger than 1.4 m^2 (15.1 ft^2) for windshields and for side and rear windows of self-propelled machines may deviate from paragraph 6.8.1 requirements in that, when broken, what appears to be the 10 largest particles shall weigh no more than the equivalent weight of 64.5 cm^2 (10 $in.^2$) of the sample.

6.9 All sharp edges and corners at the operator's station shall be appropriately treated to minimize potential hazard to the operator.

6.10 Cabs for self-propelled agricultural field equipment shall include provision for emergency exit.

6.10.1 In addition to the primary door, at least one other exit shall be provided.

6.10.2 A second door, windshield(s), a roof panel, or window(s) not on the same wall as the primary door may be considered emergency exits if, from inside the cab, they can be opened or removed quickly without tools.

6.10.3 Emergency exits must have minimum dimensions of a rectangle 610 mm (24.0 in.) by 360 mm (14.2 in.).

SECTION 7—GUARDING REQUIREMENTS

7.1 Refer to ASAE Standard S493, Guarding for Agricultural Equipment.

SECTION 8—POWER TAKE-OFF (PTO), IMPLEMENT INPUT DRIVELINE (IID), IMPLEMENT INPUT CONNECTION (IIC), AND AUXILIARY POWER TAKE-OFF (AUX. PTO)

8.1 Tractors having PTO shafts shall be equipped with PTO shields in accordance with ASAE Standard S203, Rear Power Take-off for Agricultural Tractors. On future designs:

8.1.1 It is recommended that the PTO shield, or a portion of the shield, shall be movable without detachment from the tractor, if necessary, to facilitate attachment of the IID.

8.1.2 The movable portion of the shield shall be resistant to unintentional movement when in the operating position.

8.2 PTO driven implements that require removal of the tractor PTO shield shall include comparable shielding.

8.3 The tractor PTO shield, the IID guard(s), and the IIC shield constitute an interactive guarding system and shall meet the following provisions:

8.3.1 The IIC and adjacent end of the IID shall together provide guarding per Section 7—Guarding Requirements.

8.3.2 The IID shall be guarded such that driving members are free to rotate relative to the guard, and shall include telescoping provision where required.

8.3.3 The IID guard at the tractor PTO end shall extend to at least the rearmost point of the ear of the yoke which attaches to the PTO shaft, measured with no bend at the universal joint.

8.3.4 On future designs, for any device intended for installation between the tractor PTO shaft and the IID, guarding to maintain the interactive guarding system shall be provided.

8.4 The IID shall include a means to retain it on the tractor PTO shaft. The tractor end of the IID on future designs shall meet the following provisions:

8.4.1 Where yokes with lock pins are used, only one end of each lock pin shall protrude beyond the yoke body, and this end shall be smooth and shall trail the enclosed portion of the pin relative to the direction of rotation.

8.4.2 Where clamping is required, fastener end(s) in the direction of rotation shall be recessed in the yoke body.

8.4.3 All other mechanisms shall have no positively driven protrusions in the direction of rotation.

8.4.4 An alternative to the specifications of paragraphs 8.4.1, 8.4.2, and 8.4.3 is to provide IID shielding around the mechanisms described in those paragraphs.

8.5 Provisions per Section 7—Guarding Requirements, shall be made on the tractor to guard the auxiliary PTO when in use.

8.6 A safety sign(s) shall be provided at a prominent location on the implement specifying the normal PTO operating speed and that implement driveline shielding is to be kept in place.

8.7 PTO driven equipment designed to operate in a stationary position should be provided with a means to prevent separation of the IID.

SECTION 9—TRACTOR ROLL-OVER PROTECTION

9.1 Tractor roll-over protection meeting the requirements of ASAE Standard S383, Roll-Over Protective Structures (ROPS) for Wheeled Agricultural Tractors, shall be provided on wheeled agricultural tractors.

9.2 Batteries, fuel tanks, oil reservoirs, and coolant systems shall be constructed and located or sealed to reduce the possibility of spillage that might be injurious to the operator in the event of upset.

SECTION 10—LIFTED UNITS

10.1 Controls for lifted units (implements or components of implements and self-propelled machines) shall conform to ASAE Standard S335, Operator Controls on Agricultural Equipment.

10.2 Excluding 3-point mounted implements, a means shall be provided on implements to lock lifted components, such as:

10.2.1 Harvester headers and folding wings in the raised position for transport, servicing, and storage. The locking requirement on those components can be met by folding over-center, hydraulic cylinder locks, or other design solutions.

10.2.2 Tractor-mounted implements in a raised position where tractor maintenance is impractical when the implement is in the lowered position.

10.2.3 The mechanical locking device shall be difficult to remove if a load is on the locking device.

10.3 Instructions to securely support or block units including hitch-mounted implements, which must be in a raised position for servicing or adjustment, shall be included in operators manuals and on safety signs.

SECTION 11—TRAVEL ON HIGHWAYS

11.1 Lighting and marking of equipment shall conform to ASAE Standard S279, Lighting and Marking of Agricultural Field Equipment on Highways.

11.1.1 The operator's manual for the unit shall instruct the operator to turn on flashing warning lights whenever traveling on a highway except where such use in prohibited by law.

11.2 Agricultural tractors and self-propelled machines with operator enclosures (cabs) shall have at least one rear-view mirror to permit the operator to see the highway behind the machine.

11.3 Hitch pins and other hitching devices shall be provided with a retainer to prevent accidental unhitching.

11.4 Components that are retracted to decrease the width for transport shall have means for securement during transport.

11.5 Provisions shall be made for the use of auxiliary attaching systems per ASAE Standard S338, Safety Chain for Towed Equipment, on towing machines and on equipment where expected uses include towing on highways by single point attachment.

11.6 For towed equipment without brakes, the following information shall be provided:

Road Speed	Weight of fully equipped or loaded implement(s) relative to weight of towing machine
Up to 32 km/h (20 mph)	1 to 1, or less
Up to 16 km/h (10 mph)	2 to 1, or less
Do not tow	More than 2 to 1

SECTION 12—BRAKING AND PARKING REQUIREMENTS

12.1 For guidance in determining performance needs of agricultural equipment braking systems refer to ASAE Standard S365, Braking System Test Procedures and Braking Performance Criteria for Agricultural Field Equipment.

12.2 All towed equipment with a tongue imposing a vertical downward force at the hitch point of more than 245 N (55 lb) at a height of 406 mm (16 in.) when on level ground and any condition of loading shall be equipped with a means for attaching to the propelling machine without manual lifting.

SECTION 13—FIRE PROTECTION

13.1 Shields shall be provided for the engine exhaust manifolds, muffler and exhaust pipe when necessary to prevent contact with flammable crop materials.

13.2 Fuel sediment bowl assemblies used on gasoline engines shall be fire resistant.

SECTION 14—OVERHEAD POWER LINES

14.1 A safety sign(s) shall be provided to inform of possible overhead power line contact on agricultural equipment which exceeds 4.3 m (14 ft) in height during any mode of operation, transport, or preparation for transport.

SECTION 15—SAFETY SIGNS

15.1 Safety signs shall be appropriately displayed when necessary to alert the operator and others of the risk of personal injury during normal operations and servicing.

15.2 Safety signs shall conform to requirements of ASAE Standard S441, Safety Signs.

15.3 To distinguish from safety signs, instructional signs relating to equipment servicing and care should use signal words such as IMPORTANT or NOTICE, without the safety-alert symbols. The appearance of these signs should be different from safety signs.

Cited Standards:

ANSI Z26.1, Safety Code for Safety, Glazing Materials for Glazing Motor Vehicles Operating on Land Highways
ASAE S203, Rear Power Take-Off for Agricultural Tractors
ASAE S279, Lighting and Marking of Agricultural Field Equipment on Highways
ASAE S333, Agricultural Tractor Auxiliary Power Take-Off Drives
ASAE S335, Operator Controls on Agricultural Equipment
ASAE S338, Safety Chain for Towed Equipment
ASAE EP363, Technical Publications for Agricultural Equipment
ASAE S365, Braking System Test Procedures and Braking Performance Criteria for Agricultural Field Equipment
ASAE S383, Roll-Over Protective Structures (ROPS) for Wheeled Agricultural Tractors
ASAE S390, Classifications and Definitions of Agricultural Equipment
ASAE S441, Safety Signs
ASAE S493, Guarding for Agricultural Equipment
EEC Regulation Number 43, Uniform Provisions Concerning the Approval of Safety Glazing and Glazing Materials for Installation on Power Driven Vehicles and their Trailers
SAE J674, Safety Glazing Materials—Motor Vehicles

ASAE Standard: ASAE S322.1

UNIFORM TERMINOLOGY FOR AGRICULTURAL MACHINERY MANAGEMENT

Developed by the ASAE Farm Machinery Management Committee; approved by the Power and Machinery Division Standards Committee; adopted by ASAE June 1969; reconfirmed December 1974, December 1979; revised December 1983.

SECTION 1—PURPOSE AND SCOPE

1.1 This Standard is intended to establish uniform use of machinery management terms.

SECTION 2—TERMS USED PRIMARILY FOR SYSTEM ANALYSIS

2.1 Efficiency

2.1.1 Field: Ratio of effective field capacity to theoretical field capacity, expressed in percent.

2.1.2 Functional: Ratio of the actual effectiveness of a machine to its theoretical effectiveness, expressed in percent. Threshing efficiency of a combine is an example of a functional efficiency.

2.2 Field Capacity

2.2.1 Effective: Actual rate of performance of land or crop processed in a given time, based upon total field time.

2.2.2 Theoretical: Rate of performance obtained if a machine performs its function 100% of the time at a given operating speed using 100% of its theoretical width.

2.3 Field speed: Average rate of machine travel in the field during an uninterrupted period of functional activity. For example, functional activity would be interrupted when the plow is raised out of the soil.

2.4 Field time: The time a machine spends in the field measured from the start of functional activity to the time the functional activity for the field is completed.

2.5 Life of machine, economic: The useful service life of a machine before it becomes unprofitable for its original purpose due to obsolescence or wear.

2.6 Load factor, field: The ratio of engine power used in performing an operation to engine power available.

2.7 Management phases

2.7.1 Planning: Defining an objective for the system, selecting system components and predicting the expected performance of the system.

2.7.2 Scheduling: Determining the time when the various operations are to be performed. Availability of time, labor supply, job priorities, and crop requirements are some important factors.

2.7.3 Operating: Carrying out the operations with people and machines. The operator of an agricultural field machine tends to be self-supervised.

2.7.4 Controlling: Utilizing productivity measures and standards to control the system.

2.8 Motion-and-time study: Determining the time necessary to perform motions required for a particular job.

2.9 Operating width

2.9.1 Effective: The width over which the machine actually works. It may be more or less than the measured width of the machine.

2.9.2 Theoretical: The measured width of the working portion of a machine. For row crop machines, it is the average row width times the number of rows.

2.10 Operation

2.10.1 Individual: Operating one or more similar machines as one unit.

2.10.2 Parallel: Causing two or more similar machines to perform their differing functions simultaneously.

2.10.3 Series: Causing two or more machines to perform their respective functions in sequence; each machine operation, except the first, is dependent upon previous operations, and stopping one machine would halt all subsequent machines.

2.11 Subsystem, crop production: An ordered sequence of field machine operations performed in producing and harvesting a particular crop.

2.12 System, crop production: A combination of the various subsystems required for culture of all crops grown on a particular farm.

2.13 System, machines: An arrangement and use of two or more machines to achieve a desired output.

2.14 Timeliness: Ability to perform an activity at such a time that crop return is optimized considering quantity and quality of product (see ASAE Data D230, Agricultural Machinery Management Data.)

2.15 Timeliness coefficient: A factor used to estimate the reduction in crop return (quantity and quality) due to lack of timeliness in performing an activity (see ASAE Data D230, Agricultural Machinery Management Data.)

SECTION 3—TERMS ASSOCIATED WITH ECONOMICS

3.1 Cost accounting: A system of accounting in which records of all cash and non-cash costs chargeable to any enterprise and all cash and non-cash returns are kept for the purpose of preparing an account to show costs of production, returns, and net profit or loss on the enterprise. Examples are labor, power, machinery use, building use, fuel, and interest charges.

3.2 Costs, machine

3.2.1 Accumulated average cost: Total cost for the accumulated use of a machine divided by the number of accumulated time units. Usually the time units are years or hours.

3.2.2 Custom cost: The amount paid for hiring equipment and operator services to perform a certain task. Custom costs normally include a charge for the operation of the basic machine, and may or may not include supplemental labor and equipment for such tasks as handling into storage or transport of a harvested crop, transportation of seed or fertilizer to the field, etc. Charges may be determined on the basis of area, time, transport distance or quantity of crop processed.

3.2.3 Operating costs: Costs which depend directly on the amount of machine use. Examples are labor, fuel, lubrication, and repair and maintenance costs.

3.2.4 Ownership costs: The costs which do not depend on the amount of machine use. Examples are depreciation, interest on investment, taxes, insurance, and storage.

3.2.5 Total cost: The sum of ownership and operating costs.

3.3 Depreciation

3.3.1 Actual: Change in value of a machine.

3.3.2 Estimated: The change in value as determined by the difference between purchase price and estimated future value, both in constant dollars.

3.3.3 Straight line, declining balance, sum of years' digits: Methods to spread the change in machine value over the economic life of a machine. These methods may disagree with estimated depreciation.

3.4 Lease A lease is a contract for the use of machinery for an agreed period of time in return for periodic payments. Ownership remains with the lessor. The lessee acquires the right of temporary possession and use.

3.5 Obsolescence: The process of becoming obsolete.

3.6 Obsolete: The condition of a machine when it is either out of production and parts to repair or update it are not available from normal suppliers, or it can be replaced by another machine or method that will produce greater profit.

3.7 Price: Market value per unit. Examples are the price of grain, usually dollars per unit measure; of labor, dollars per hour; of machines, dollars per machine.

3.8 Rent: A rental agreement is a short-term contract that permits use of machinery in exchange for a fee.

3.9 Return for sale of a service or product

3.9.1 Gross: The value received for a service or product before expenses are deducted.

3.9.2 Net: The value received for a service or product, less all expenses except income taxes.

SECTION 4—MECHANICAL TERMS

4.1 Breakdown: An unexpected change in duty status from operational to non-operational, due to mechanical failure.

4.2 Continuous duty: A service requirement that demands operation for an indefinitely long period of time.

4.3 Conversion: Changing a machine from an arrangement suitable for the performance of one activity to one suitable for performance of another. An example is replacing a combine grain platform with a corn head.

4.4 Failure: The inability of a machine to perform its function under specified field and crop conditions.

4.5 Fuel consumption, specific: The fuel consumed by an engine to deliver a given amount of energy. (Kilograms of fuel per kilowatt-hour.)

4.6 Maintenance and service: Periodic activities to prevent premature failure and to maintain good functional performance. Examples are refueling, changing oil and filters, cleaning, lubricating and adjusting components.

4.7 Major overhaul: Extensive rebuilding which extends the useful life of a machine, increases its value or adapts the machine for a different use.

4.8 Repair: Restoring a machine to operative condition after breakdown, excessive wear, or accidental damage. Repairs are less extensive than major overhauls and normally do not alter the value of the machine.

Cited Standard:

ASAE D230, Agricultural Machinery Management Data

ASAE Standard: ASAE S324.1

VOLUMETRIC CAPACITY OF BOX TYPE MANURE SPREADERS—DUAL RATING METHOD

This Standard supersedes S237.1, Volumetric Capacities of Manure Spreaders, adopted January 1958. Developed by the Wagon Box, Forage Box, Manure Spreader and Farm Wagon Subcommittee; approved by Power and Machinery Division Standards Committee; adopted by ASAE as a Tentative Standard June 1969; reclassified as a full Standard February 1971; reconfirmed December 1975, December 1980; revised April 1986.

SECTION 1—PURPOSE AND SCOPE

1.1 This Standard is intended to provide a uniform method for calculating and expressing the volumetric capacity of a box type manure spreader.

SECTION 2—METHOD OF MEASURE

2.1 To conform to this Standard the volumetric capacity of box type manure spreaders shall be rated in cubic meters (cubic feet).

2.2 The volumetric capacity shall be calculated according to a struck level method and a heaped load method as described in Section 3.

2.3 Whenever spreader capacity is expressed according to this Standard, the two ratings shall be given, and the heaped load capacity shall be given before the struck level capacity.

SECTION 3—CALCULATING CAPACITY

3.1 Heaped load method. The volumetric capacity of a box type manure spreader calculated by the heaped load method shall include the heaped portion, Y, above the box in addition to the portion within the box, X, as shown in Fig. 1.

3.1.1 The portion within the box, X, shall be calculated as follows:

3.1.1.1 The height component shall be measured to the top of the sides or flareboards.

3.1.1.2 The length component shall be limited by vertical planes tangent to the restraining members of a spreading element or coinciding with a restraining surface.

3.1.1.2.1 The restraining member of the spreading or agitating element shall be understood as (a) the core tube or shaft in a paddle type beater; (b) a plane projecting vertically and contacting two adjacent finger mounting members in a finger type beater; (c) the flail mounting tube in a flail type beater.

3.1.1.3 The width component shall be limited by the side surfaces of the box.

3.1.2 The volume of the heaped portion above the top of the box, Y, shall be calculated as follows:

3.1.2.1 The height, H, shall be limited to a distance 381 mm (15 in.) above the outside diameter or tip of the upper-most spreading or agitating element of the spreader.

3.1.2.2 The length shall be limited at the front by a 60 deg angle of repose whose vertex is on the top edge of the front board or flareboard. The length shall be limited at the rear by a vertical plane tangent to the restraining member of the uppermost spreading element as defined in 3.1.1.2.1.

3.1.2.3 The width shall be limited by a 60 deg angle line drawn tangent to the outermost edge of the box sides or flareboards.

3.2 Struck level method. The volumetric capacity of a box type manure spreader calculated by the struck level method shall be the portion included within the box, X. It shall be calculated according to paragraphs 3.1.1.1, 3.1.1.2, and 3.1.1.3.

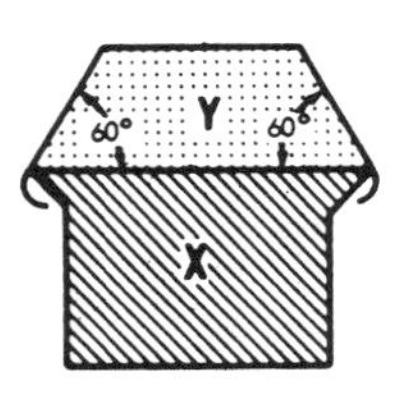

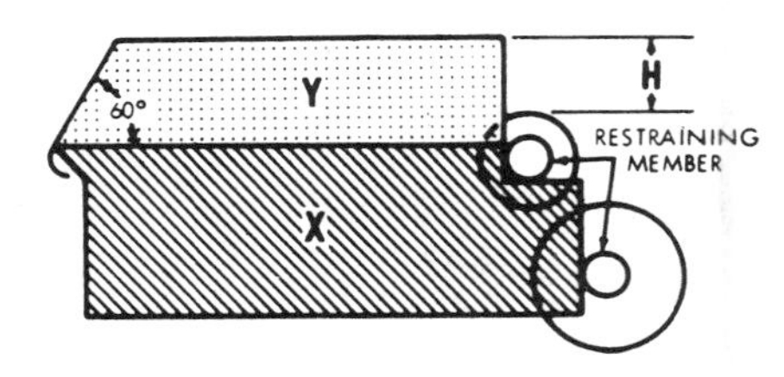

FIG. 1—CONFIGURATION FOR CALCULATING CAPACITY OF BOX TYPE MANURE SPREADERS

ASAE Standard: ASAE S325

VOLUMETRIC CAPACITY OF OPEN TANK TYPE MANURE SPREADERS

Developed by the Wagon Box, Forage Box, Manure Spreader and Farm Wagon Subcommittee; approved by Power and Machinery Division Standards Committee; adopted by ASAE June 1969; reconfirmed December 1973, December 1978, December 1983.

SECTION 1—PURPOSE

1.1 This standard is intended to provide a uniform method for calculating and expressing the volumetric capacity of open tank type manure spreaders.

SECTION 2—CAPACITY DESIGNATIONS

2.1 To conform to this standard the volumetric capacity of the spreader must be rated in cubic feet and calculated and expressed by the dual rating method shown below. Both ratings shall be used whenever expressing volumetric capacity.

2.1.1 Struck level capacity shall be calculated by using a struck level measure of volume A or where applicable volume A+B as shown in Fig. 1.

2.1.2 Heaped capacity shall be calculated by using a closed cylinder measure of volume 2A.

2.1.3 The volume to the nearest cubic foot of internal structural and functional members which lie within the areas expressed as capacity shall be subtracted from the total volume.

2.2 Equipment destined for export shall be rated in cubic meters.

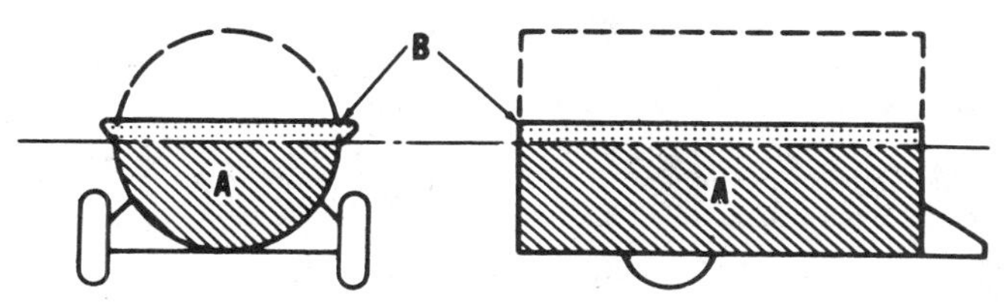

FIG. 1—CONFIGURATION FOR COMPUTING CAPACITY OF OPEN TANK TYPE SPREADERS

ASAE Standard: ASAE S326

VOLUMETRIC CAPACITY OF CLOSED TANK TYPE MANURE SPREADERS

Developed by the Wagon Box, Forage Box, Manure Spreader and Farm Wagon Subcommittee; approved by Power and Machinery Division Standards Committee; adopted by ASAE June 1969; reconfirmed December 1973, December 1978, December 1983.

SECTION 1—PURPOSE

1.1 This standard is intended to provide a uniform method for calculating and expressing the volumetric capacity of closed type manure (liquid or semi-liquid) spreaders.

SECTION 2—CAPACITY DESIGNATIONS

2.1 To conform to this standard the spreader shall be rated in U.S. gallons, and the volume shall be calculated using a closed tank measure.

2.2 The volume to the nearest gallon of internal structural and functional members which lie within the area expressed as capacity shall be subtracted from the total volume.

2.3 Equipment destined for export shall be rated in liters.

TERMINOLOGY AND DEFINITIONS FOR AGRICULTURAL CHEMICAL APPLICATION

Developed by the ASAE Agricultural Chemical Application Committee; approved by the Power and Machinery Division Standards Committee; adopted by ASAE December 1969; reconfirmed December 1974; revised March 1980; reconfirmed December 1984.

SECTION 1—PURPOSE AND SCOPE

1.1 The purpose of this Standard is to establish uniformity in terms used in the field of agricultural chemical application. Terms are adopted from related fields where applicable.

1.2 The terms are appropriate to any agricultural chemical application. Units in parentheses following a definition are meant as typical and are not exhaustive of all units available for the term.

SECTION 2—GENERAL CHEMICAL TERMS AND CHARACTERISTIC MEASUREMENTS

2.1 Air flow rate: The flow rate of air in a jet, expressed in volume per relevant unit (ft^3/min, m^3/s, m^3/tree).

2.2 Application rate: The amount of any material applied per unit treated.

2.2.1 Active chemical rate: The amount of active ingredient applied per unit treated, expressed in terms of mass per relevant unit treated. (For area treatment, kg/ha, lb/acre, or oz/1000 ft of row; for space application, mg/m^3, or oz/1000 ft^3; for individual units, kg/plant or animal)

2.2.2 Formulation rate: The amount of chemical formulation applied per unit treated, expressed in terms of mass or volume per relevant unit treated. (For area treatment, kg/ha, lb/acre, or oz/1000 ft of row; for space application, mg/m^3, or oz/1000 ft^3; for individual units, kg/plant or animal)

2.2.3 Spray rate: The amount of spray liquid applied to a unit receiving treatment, expressed in volume per unit treated. (For area treatment, L/ha, or gal/acre; for space treatment, mL/m^3, or oz/1000 ft^3; for individual units, L/plant, mL/animal, or gal/tree)

2.3 Carrier: A gas, liquid, or solid used to propel or transport a chemical.

2.4 Concentration: Amount of the active ingredient contained in the chemical formulation expressed as a percent or mass per relevant unit basis.

2.5 $\overline{D}_{pq}$, Mean droplet diameter: $\overline{D}_{pq}$ is represented by:

$$\overline{D}^{(p-q)} = \frac{\Sigma D_i^p}{\Sigma D_i^q}$$

where

D_i = the diameter of the ith particle, ΣD_i^o is the total number of drops in the sample.

Thus:

$\overline{D}_{10}$ = arithmetic mean diameter
$\overline{D}_{20}$ = surface mean diameter
$\overline{D}_{30}$ = volume mean diameter
$\overline{D}_{32}$ = Sauter mean diameter
$\overline{D}_{43}$ = mean diameter over volume

2.6 D_{xf}, Median droplet diameter: For cumulative distributions, D_{xf}, where x is V, A, L, or N, are diameters such that the fraction (f) of the total of volume, surface area, length of diameter, or number of drops respectively, is in drops of smaller diameter. Thus:

$D_{V.5}$ = volume median diameter
$D_{A.5}$ = area median diameter
$D_{L.5}$ = length median diameter
$D_{N.5}$ = number median diameter

Also, $D_{V.1}$ and $D_{V.9}$ = diameter of drop such that 10 percent and 90 percent respectively, of the liquid volume is in drops of smaller diameter.

2.7 Deposit rate: The amount of any material deposited per unit area.

2.7.1 Active chemical deposit rate: The amount of active ingredient deposited per unit area.

2.7.2 Formulation deposit rate: The amount of formulation deposited per unit area.

2.7.3 Spray deposit rate: The amount of spray liquid deposited per unit area.

2.7.4 Mean deposit rate: The average amount of deposit over the entire spray swath.

2.7.5 Effective spray deposit rate: The mean deposit from center to center of adjoining swaths.

2.8 Diluent: A gas, liquid, or solid used to reduce the concentration of active ingredient in a formulation or to reduce the concentration of a formulation for application.

2.9 Drift: The movement of chemicals outside the intended target by air mass transport or diffusion.

2.9.1 Particle drift deposits: The deposition of chemical particles outside the intended target.

2.9.2 Airborne drift: The dispersion of chemical particles to the atmosphere outside the intended target.

2.9.3 Vapor drift: The dispersion of vaporized chemical to the atmosphere and areas surrounding the target area during and following application.

2.10 Droplet size characterization: Classification of sprays.

2.10.1 Aerosols: Distribution of droplets with $D_{V.5} \leq 50\ \mu m$.

2.10.2 Mists: Distribution of droplets with $50\ \mu m < D_{V.5} \leq 100\ \mu m$.

2.10.3 Fine sprays: Distribution of droplets with $100\ \mu m < D_{V.5} \leq 400\ \mu m$.

2.10.4 Coarse sprays: Distribution of droplets with $D_{V.5} > 400\ \mu m$.

2.11 Formulation: The form in which a chemical is offered for sale to the user, and includes both the active and inert ingredients.

2.12 Sprayed width per nozzle: The effective width sprayed by a single nozzle. For broadcast spraying it is the nozzle spacing. For band spraying it is the band width. For row crop spraying it is the number of nozzles per row divided by the row width.

2.13 Swath, effective width: The center to center distance between overlapping broadcast applications.

SECTION 3—TYPES OF APPLICATION

3.1 Band application: Distribution of a chemical in parallel bands leaving the area between bands free of chemicals.

3.2 Basal application: Application of a chemical to the base of a tree in a circle around the trunk, or by injection into slashes or cuts.

3.3 Baseboard application: Application to a building on the lower portion of the inside walls.

3.4 Broadcast application: Application of a chemical over the entire area to be treated.

3.5 Crack and crevice application: Application in a building by a means which projects the material into cracks and crevices which are inaccessible to the normal inhabitants of the building.

3.6 Dip application: Application by direct immersion.

3.7 Directed application: Application of a chemical to a specific area such as a row, bed, or at the base of plants.

3.8 Foliar application: Application of a chemical to the stems, fruit, leaves, needles, or blades of a plant.

3.9 Pour on application: Application by pouring a chemical onto the target.

3.10 Saturation application: Application of a chemical with a sufficient volume of solution that it begins to flow from the plant or animal.

3.11 Space application: Dispersion of liquid or dry particles in an air space in such a manner that target pests are exposed to the chemical.

3.12 Spot treatment: Application of a chemical to a small restricted area, usually to control the spread of a pest.

3.13 Soil injection: Mechanical placement of a chemical beneath the soil surface with a minimum of mixing or stirring of the soil, as with an injection blade or knife.

SECTION 4—APPLICATION APPARATUS

4.1 Air carrier system: A sprayer apparatus consisting of a pressure source and controls for the spray liquid and a blower with suitable ducts to produce an air jet in which spray nozzles are located. Air from the blower carries the spray for a distance for deposition on the target being treated.

4.2 Boom sprayer: A sprayer apparatus consisting of a pressure source and controls, and specifically employing an over-the-crop (or vertical for tree and vine crops) boom with atomizers (hydraulic, rotary, or other) arranged to provide uniform coverage of the treated surfaces.

4.3 Bucket pump sprayer: A sprayer apparatus consisting of a manually operated pump which may be held or mounted in a bucket containing the spray solution. The pump is connected to an atomizing device which forms and distributes the spray.

4.4 Cold aerosol: Any device which produces a liquid dispersion by mechanical means having a volume diameter less than 50 μm.

4.5 Compressed air sprayer: A sprayer apparatus that uses air to pressurize the liquid or to siphon liquid from the container and an atomization device which forms and distributes the spray.

4.6 Granular applicator: An apparatus consisting of a hopper, a metering device and a device for spreading or placing the granules in the target area.

4.7 High clearance sprayer: An apparatus consisting of the components of a boom sprayer mounted on a self-propelled vehicle whose frame is constructed to permit the vehicle to pass over plants with minimal damage.

4.8 Hose end sprayer: An apparatus designed to be attached to standard garden hose, consisting of a hand-held container for spray mixture with an integral metering head through which water from the garden hose flows. The metering head utilizes water pressure, siphon effect, or some other water powered means to meter the spray mixture into the water stream which is then atomized.

4.9 Knapsack sprayer: A sprayer apparatus, carried on the operator's back, consisting of a spray solution tank, pressure source, and an atomizing device which forms and distributes the spray. Spray pressure is supplied by lever operated manual or engine powered pumps or a compressed air tank. Some knapsack sprayers have air carrier blowers to distribute the spray.

4.10 Slide pump sprayer: A sprayer consisting of a telescoping pump operated by both hands. On the outlet end of the pump is mounted a spray nozzle. On the inlet of the pump is attached a line leading to a container containing the spray solution.

4.11 Spray attachment: An apparatus consisting of the individual components of a sprayer (tank, pressure source, pressure controls, spray liquid lines, pressure nozzles, etc.) in an arrangement to permit its mounting on another implement to permit application of spray at the same time another field operation is performed.

4.12 Thermal aerosol: Any device using thermal energy which produces a liquid dispersion which has a volume median diameter less than 50 μm.

4.13 Thermal vaporizer: An apparatus consisting of a container for chemical and a heater to maintain the vessel at a temperature sufficiently high to accelerate evaporation or sublimination of the pesticide. The apparatus may contain a blower to disperse the pesticide vapor into a treated area, or may rely upon natural turbulent diffusion for dispersion.

4.14 Wheelbarrow sprayer: A sprayer consisting of a spray solution container mounted on a frame with wheelbarrow-type handles and one or two wheels, and a pressure source connected to hydraulic nozzles which forms and distributes the spray.

SECTION 5—ATOMIZERS

5.1 Blast atomizer: Devices that use kinetic energy of a gaseous stream for spray atomization. A jet of compressed gas or air impinges on a stream of spray liquid, or the liquid is aspirated into an air stream in such a way as to cause it to disintegrate into spray droplets.

5.2 Electrostatic atomizer: Devices that impart an electrical charge to a spray liquid for enhancement of atomization and/or deposition.

5.3 Pressure atomizers: Devices that use flow energy for spray atomization. Liquid may be transformed into jets, fan, conical, or other patterns.

5.3.1 Orifice atomizer: A pressure atomizer having a circular orifice, or converging flow nozzle that produces a simple liquid jet that breaks up at some distance from the nozzle.

5.3.2 Fan spray atomizer: A pressure atomizer having an orifice formed by the interception of a groove with a cone shaped chamber which produces a fan shaped liquid sheet that disintegrates into spray droplets at some distance from the nozzle.

5.3.3 Deflector atomizer: A pressure atomizer discharging a jet which impacts and spreads on a deflector plate. After leaving the edge of the plate it forms a fan-shaped liquid sheet which breaks into spray droplets.

5.3.4 Swirl spray atomizer: A pressure atomizer having a circular orifice, opening from a chamber with tangential inlets, a slotted distibutor, vanes, cones or other devices positioned to impart a swirl to the liquid. The spray is released in a hollow or solid conical pattern.

5.4 Rotary atomizers: Devices that use rotational kinetic energy for spray atomization.

5.4.1 Rotary disc atomizer: Devices that use rotational kinetic energy of a disc to produce fluid shear and surface tension forces that adjust the fluid sheet thickness leaving the disc for subsequent atomization. The disc may be flat or saucer shaped. Multiple discs may be used.

5.4.2 Rotary wheel atomizer: A rotary atomizer consisting of a rotating circular wheel constructed with vanes, bushings and perforated or porous openings. The liquid is fed to the inside of the wheel, moves through the constructed openings and is broken into droplets.

5.5 Vibrating atomizers: Devices that use kinetic energy of a vibrating member for spray atomization.

5.5.1 Acoustic atomizer: A vibrating atomizer that uses piezoelectric, magnetostrictive, other electrical, or gas driven devices at 6-20 kHz to control the atomization of jet streams.

5.5.2 Ultrasonic atomizer: A vibrating atomizer that uses electrically driven devices at 40-1000 kHz to control the atomization of jet streams.

5.5.3 Vibratory atomizer: A vibrating atomizer that uses vibrating blades, reeds, or rods to detach drops from liquid feeding through stationary capillary tubes.

ASAE Standard: ASAE S328.1

DIMENSIONS FOR COMPATIBLE OPERATION OF FORAGE HARVESTERS, FORAGE WAGONS AND FORAGE BLOWERS

Developed by the ASAE Farm Materials Handling Committee, PM-50; approved by the Power and Machinery Division Standards Committee; adopted by ASAE as a Tentative Standard December 1969; reconfirmed December 1970; revised February 1972; reconfirmed December 1972; reclassified as a full Standard December 1973; reconfirmed December 1978, December 1983.

SECTION 1—PURPOSE AND SCOPE

1.1 This Standard is intended as a guide to assist manufacturers of forage wagons, forage harvesters and forage blowers to make the operation of any one compatible with the others. The satisfactory performance of any of these machines is dependent upon dimensional compatability.

1.2 Dimensions specified in this Standard pertain to forage harvesters, forage wagons, and forage blowers when used in the following normal sequence:

1.2.1 A forage wagon is towed at the rear of a pull-type or self-propelled forage harvester. Harvested material is blown by the forage harvester into the forage wagon.

1.2.2 The forage wagon is then unhitched from the forage harvester and hitched to a tractor to transport the harvested material from the field to a storage area.

1.2.3 The forage wagon is then moved by the tractor to a position beside a forage blower. Harvested material is then transferred from the wagon into the forage blower and blown into an adjacent silo or storage area.

SECTION 2—FORAGE HARVESTER DIMENSIONS

2.1 The horizontal distance from the center of the rear drawbar hitch point to the tip of the spout shall be 72.0 in. (1829 mm) maximum as shown in Fig. 1.

2.2 The vertical distance from the ground to top of blower spout with cap level shall be 120.0 ± 6 in. (3048 ± 152 mm) when the forage harvester is in normal operating position as shown in Fig. 1. For delivery into trucks, auxiliary spout heights of 140.0 ± 6 in. (3556 ± 152 mm) and 160.0 ± 6 in. (4064 ± 152 mm) are recommended.

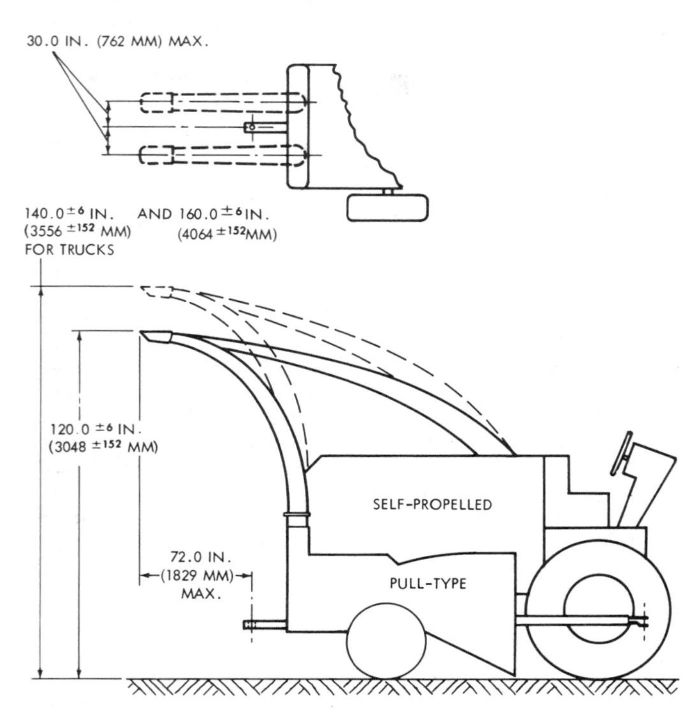

FIG. 1—FORAGE HARVESTER

2.3 The spout shall not be offset laterally more than 30.0 in. (762 mm) from the center line of the rear drawbar as shown in Fig. 1 for pull-type and self-propelled forage harvesters.

SECTION 3—FORAGE WAGON DIMENSIONS

3.1 The vertical distance from the ground to the top of the upper beater as shown in Fig. 2 shall be 108.0 in. (2743 mm) maximum.

3.2 To provide clearance under the spout when turning, the maximum height of that portion of the wagon sides within 80.0 in. (2032 mm) of the hitch pin shall be 112.0 in. (2845 mm) as indicated in Fig. 2.

3.3 The vertical distance from the ground to the bottom surface of the hood shall be 132.0 in. (3353 mm) minimum.

3.4 The minimum distance between the front edge of the hood and the hitch pin shall be 80.0 in. (2032 mm).

3.5 For wagons designed for rear unloading, the minimum distance between the ground and the bottom of any bed frame member shall be 25.0 in. (635 mm). This clearance shall exist from the rear of the wagon forward 6.0 in. (152 mm).

3.6 It must be possible to raise the wagon tongue hitch point above a 20 degree inclined plane extending forward from the intersection between the front wheel and ground line as shown in Fig. 2. The full swing to the right and left must be possible with the tongue raised to this height.

3.7 The minimum distance from the ground to the bottom of the cross conveyor as shown in Fig. 2 shall be 25.5 in. (648 mm).

3.8 The overall transport width shall not be more than 96.0 in. (2438 mm) as shown in Fig. 2. If the cross conveyor is the folding type, it should be folded for this measurement.

3.9 The maximum inside width of the cross conveyor discharge shall be 24.0 in. (610 mm) as shown in Fig. 3.

3.10 The wagon tongue length between the hitch point and vertical pivot bolt that permits horizontal pivot shall be a mimimum of 10.0 in. (254 mm) greater than the radius of an arc about the same pivot bolt that just clears the outermost point on the forage wagon as illustrated in Fig. 5.

3.11 The length of the wagon tongue and the length of the drive shaft are related. Also, the difference between minimum and maximum telescoping length is related to the wagon tongue length. The wagon tongue and drive shaft shall be of sufficient length to allow a tractor towing a forage wagon to turn right or left to the point where the portion of the drive shaft lying between the 2 universal joints has rotated 90

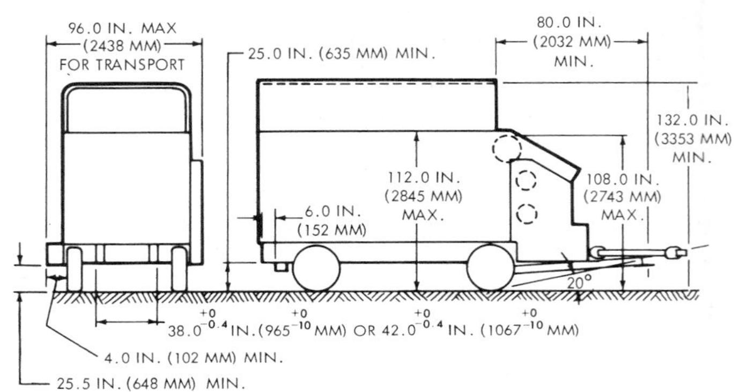

FIG. 2—FORAGE WAGON

degrees in a horizontal plane from a straight ahead position, and in this position the telescoping elements of the drive shaft shall have a minimum engaged length of 4.0 in. (102 mm) as shown in Fig. 3. This shall be possible on flat ground and with the power disengaged.

3.12 If the wagon box is made to mount on conventional running gear, the outside width of the support members that attach to the running gear must be either 38.0 in. +0 -0.4 in. (965 +0 -10 mm) or 42.0 in. + 0 -10 mm) or 42.0 in. +0 -0.4 in. (1067 +0 -10 mm) as shown in Fig. 2.

3.13 The minimum drive-through overlap of the cross conveyor over the blower hopper shall be 4.0 in. (102 mm). This is the distance from the outer end of the cross conveyor to the first obstruction on the wagon under the cross conveyor as shown in Fig. 2.

SECTION 4—FORAGE BLOWER DIMENSIONS

4.1 The maximum distance from the ground to the top of the conveyor sides shall be 23.0 in. (584 mm) as shown in Fig. 4.

4.2 The minimum drive-through conveyor length shall be 102.0 in. (2591 mm). This is the distance from outside of conveyor, when raised, to end of conveyor in operating position as shown in Fig. 4.

4.3 The minimum width of conveyor or blower hopper shall be 30.0 in. (762 mm) as shown in Figs. 4 and 5.

4.4 The minimum drive-through hopper overlap shall be 4.0 in. (102 mm). This is the distance from the side of the hopper to the first obstruction on the forage blower above the hopper as shown in Fig. 5.

4.5 No part of the blower below the hopper shall extend beyond the outside edge of the hopper.

4.6 The maximum distance from the ground to the top of the blower hopper sides shall be 24.5 in. (622 mm) as shown in Fig. 5.

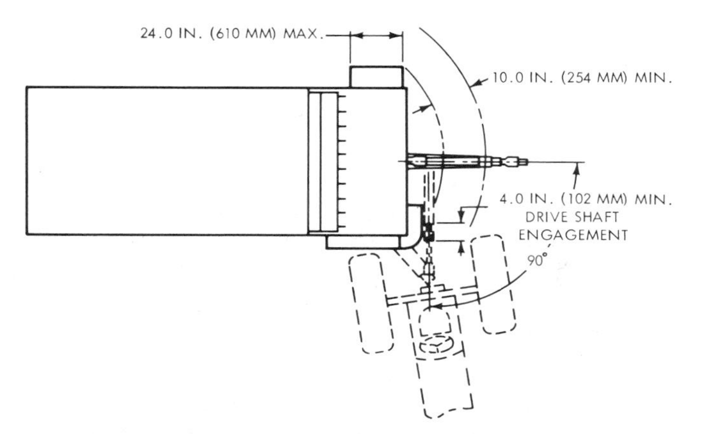

FIG. 3—FORAGE WAGON AND TRACTOR

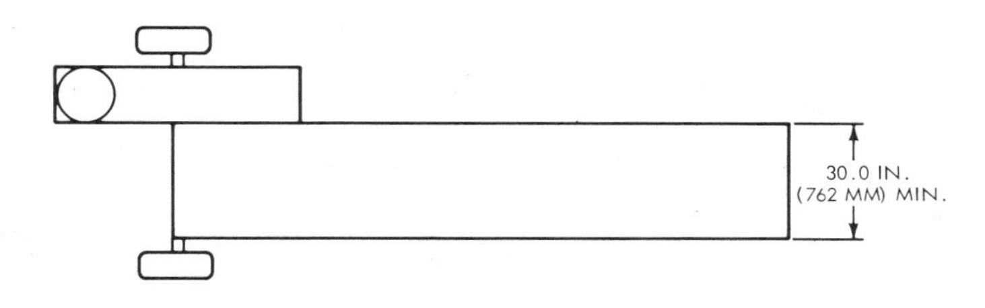

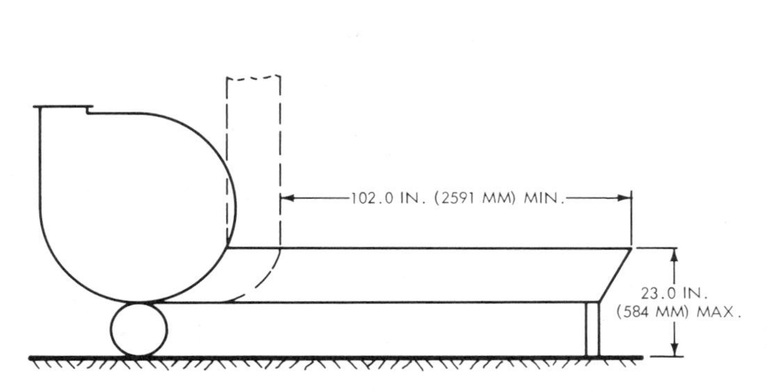

FIG. 4—CONVEYOR TYPE FORAGE BLOWER

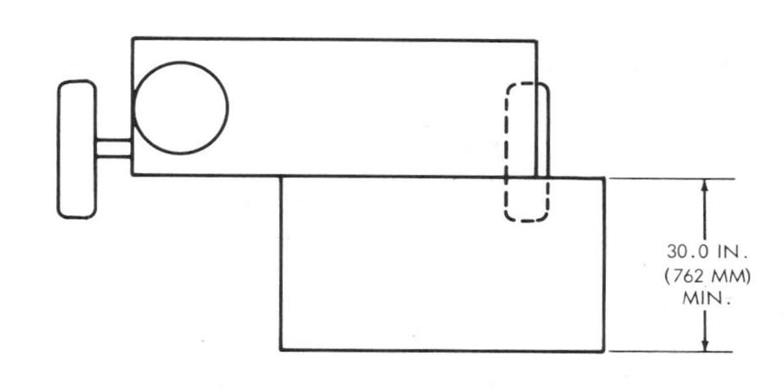

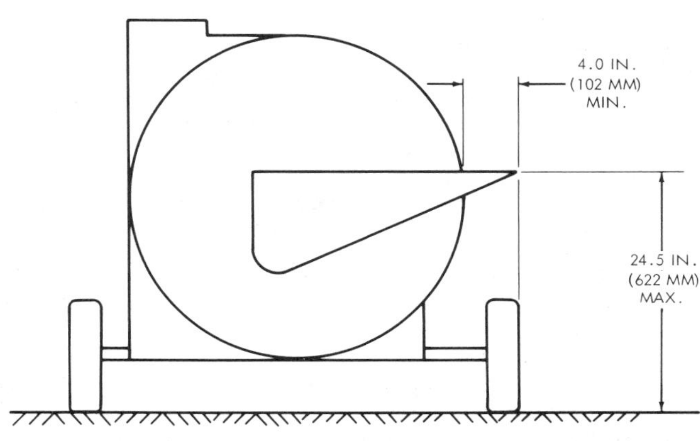

FIG. 5—HOPPER TYPE FORAGE BLOWER

IMPLEMENT POWER TAKE-OFF DRIVE LINE SPECIFICATIONS

Proposed by the Engineering Policy Committee of the Farm and Industrial Equipment Institute; approved by the ASAE Power and Machinery Standards Committee; adopted by ASAE as a Recommendation December 1969; reconfirmed December 1970; revised December 1971, December 1972; reconfirmed December 1976; reclassified as a Standard March 1978; revised March 1981; approved by ANSI as an American National Standard December 1982; reconfirmed December 1985.

SECTION 1—PURPOSE

1.1 The purpose of this Standard is to establish 6 categories of universal joint drive lines with two sub-sets of connecting members each, one heavy duty (HD) and one regular duty (RD).

1.2 The intended use of the drive lines is between tractor power take-off shafts and implement input shafts, or any universal joint application within the implement. The universal joint drive line from the tractor power take-off shaft to the implement shaft is considered a part of the implement.

1.3 This Standard does not provide for dimensional interchangeability from one implement to another.

SECTION 2—REQUIREMENTS

2.1 Seals used shall not be subject to damage from overlubrication.

2.2 Drive lines, when lubricated, shall be designed to permit lubrication of all sides of telescoping torque members.

2.3 Provisions shall be made in the telescoping torque members to assure correct phasing of universal joints.

SECTION 3—TEST SPECIFICATIONS

3.1 Drive lines shall meet the requirements of the static and dynamic tests. Separate assemblies shall be used for each test.

3.2 Static torsional tests

3.2.1 Torsional yield test for cross and bearing kit and yokes. With zero angularity, cross, bearings and yokes shall withstand the static torque shown in Table 1 without failure. Failure is defined as a break, crack or deformation greater than a 50 percent change in the slope of the torque deflection curve.

3.2.2 Torsional yield test for connecting members. Connecting members between universal joints shall withstand the static torque shown in Table 1 without failure. Failure is defined as any break, crack or permanent deformation exceeding 1 deg per lineal foot (1 deg per 305 mm). This does not apply to torque limiters used between joint centers.

3.3 Dynamic torsional tests

3.3.1 Constant angle durability test for needle bearing universal joints. Yokes, crosses, bearings, and connecting members shall operate with 10 deg angularity per joint, under a constant torque shown in Table 2 for 1,000,000 revolutions (31 hours) at a speed of 540 rpm without failure to the yokes and connecting members. A B10 life level, as defined in Anti-Friction Bearing Manufacturer's Association Standard, Method for Evaluating Load Ratings of Radial Roller Bearings and Thrust Roller Bearings, shall be used for crosses and bearings. Failure of a cross and bearing kit will be determined by the temperature of the needle cups. A temperature in excess of 350 °F (178 °C) is a failure. This test may be run without stopping, or in segments of 7 hours. Joints may be regressed during the test, but not more than three times.

3.3.2 Drive line fully-reversed torsional fatigue test. With zero angularity, yokes, crosses, bearings, and connecting members (excluding torque limiters) must withstand 375,000 fully reversed cycles under a torsional load shown in Table 2. A B10 life level shall be used. A failure is defined as a break or crack.

Cited Standard:

AFBMA, Method for Evaluating Load Ratings of Radial Roller Bearings and Thrust Roller Bearings

TABLE 1—STATIC TORSIONAL REQUIREMENTS

Category	Cross, Bearing and Yoke Assys.		Connecting Members			
			Regular Duty		Heavy Duty	
	lb - in.	N - m	lb - in.	N - m	lb - in.	N - m
1	10,000	1129.8	5,000	564.9	7,000	790.9
2	16,000	1807.8	10,000	1129.8	14,000	1581.8
3	22,000	2485.7	14,000	1581.8	19,000	2146.7
4	35,000	3954.5	17,000	1920.7	26,000	2937.6
5	50,000	5649.2	26,000	2937.6	37,000	4180.4
6	65,000	7344.0	33,000	3728.5	48,000	5423.3

TABLE 2—DYNAMIC TORSIONAL REQUIREMENTS

Category	Constant Angle Bearing Durability Test		Fully-Reversed Torsional Fatigue Test			
			Regular Duty		Heavy Duty	
	lb - in.	N - m	lb - in.	N - m	lb - in.	N - m
1	1,200	135.6	1,800	203.4	2,000	226.0
2	2,400	271.2	3,500	395.4	4,000	451.9
3	3,400	384.1	5,000	564.9	5,000	564.9
4	4,800	542.3	7,500	847.4	7,500	847.4
5	7,200	813.5	9,000	1016.9	11,000	1242.8
6	9,600	1084.7	13,500	1525.3	15,000	1694.8

ASAE Standard: ASAE S333.2 (*SAE J717 NOV84)

AGRICULTURAL TRACTOR AUXILIARY POWER TAKE-OFF DRIVES

*Corresponds in substance to previous revision S333.1.

Proposed by the Engineering Policy Committee of the Farm and Industrial Equipment Institute; approved by the ASAE Power and Machinery Division Standards Committee; adopted by ASAE as a Recommendation December 1969. Supersedes ASAE R206, Farm Tractor Auxiliary Power Take-Off Drives, and ASAE R311, Mid Power Take-Off for Farm and Light Industrial Tractors; title revised June 1972; reconfirmed December 1974; reclassified as a Standard March 1978; reconfirmed December 1979; reconfirmed and revised editorially December 1984; revised March 1987.

SECTION 1—PURPOSE AND SCOPE

1.1 This document establishes specifications for mid and side power take-off drives that will be helpful in designing implements which are front or side-mounted. Design of the implement attaching means must be tailored to each tractor, depending on attaching points available and the exact location of the auxiliary shaft.

SECTION 2—DEFINITION

2.1 Auxiliary power take-off (Aux. PTO): An external shaft on an agricultural tractor, other than the rear PTO, to provide rotational power to implements that are usually front or side mounted.

SECTION 3—SPECIFICATIONS

3.1 Auxiliary power take-off shafts shall conform to the dimensions of the 35 mm (1-3/8 in.) involute power take-off spline as specified in ASAE Standard S203, Rear Power Take-Off for Agricultural Tractors.

3.2 The normal speed of the auxiliary power take-off shaft shall conform to ASAE Standard S203, Rear Power Take-Off for Agricultural Tractors.

3.3 A minimum spherical clearance radius of 82.6 mm (3.25 in.) with sufficient openings to allow for equipment drives shall be provided as specified for the 35 mm (1-3/8 in.) involute power take-off spline in ASAE Standard S203, Rear Power Take-Off for Agricultural Tractors.

3.4 The ability of the auxiliary power take-off shaft to withstand side loading and torque shall conform to ASAE Standard S207, Operating Requirements for Tractors and Power Take-Off Driven Implements.

3.5 Provisions shall be made on the tractor to properly shield the power take-off shaft when it is connected to a drive assembly and also when it is not connected to an implement, similar to the power take-off master shield described in ASAE Standard S203, Rear Power Take-Off for Agricultural Tractors.

3.6 Machines driven from auxiliary power take-off drives shall be equipped with adequate shielding as specified in ASAE Standard S318, Safety for Agricultural Equipment.

3.7 Position, location, and direction of rotation of auxiliary power take-off shafts are shown in Table 1.

Cited Standards:

ASAE S203, Rear Power Take-Off for Agricultural Tractors
ASAE S207, Operating Requirements for Tractors and Power Take-Off Driven Implements
ASAE S318, Safety for Agricultural Equipment

TABLE 1—POSITION, LOCATION, AND ROTATION OF AUXILIARY POWER TAKE-OFF SHAFTS

	Mid	Right	Left
Position	Extends forward parallel to rear power take-off shaft.	Right side of tractor extending to right when viewed from driver's station.	Left side of tractor extending to left when viewed from driver's station.
Location	Within zone extending 254 mm (10 in.) to both the right and the left of the centerline of the tractor.	No recommendation.	No recommendation.
Direction of rotation	Clockwise when facing direction of forward travel.	Clockwise when viewed from outer end of shaft.	Counterclockwise when viewed from outer end of shaft.

ASAE Standard: ASAE S335.4

OPERATOR CONTROLS ON AGRICULTURAL EQUIPMENT

Supersedes ASAE R234, Operator Controls on Farm Tractors, and ASAE R235, Operator Controls on Self-Propelled Implements, adopted in 1962. Proposed by the Engineering Policy Committee of Farm and Industrial Equipment Institute; approved by the Power and Machinery Division Standards Committee; adopted by ASAE December 1969; revised June 1972; approved by ANSI as an American National Standard November 1976; reconfirmed December 1976; revised March 1980, December 1983; withdrawn as an ANSI Standard February 1985; revised April 1987.

SECTION 1—PURPOSE AND SCOPE

1.1 This Standard is intended to improve operator efficiency and convenience by providing guidelines for the uniformity of location and direction of motion of operator controls used on agricultural field and farmstead equipment. The controls covered are those located at the operator's normal position.

1.2 This Standard covers design and performance criteria for operator controls that have been shown to be technically and economically feasible, functionally practical and amenable to standardization.

1.3 Non-agricultural equipment used in agricultural applications is excluded from the scope of this Standard.

SECTION 2—DEFINITIONS

2.1 Refer to ASAE Standard S390, Classifications and Definitions of Agricultural Equipment, for definition of terms.

2.2 Right-hand, left-hand, and forward designations are those related to the operator when in the operating position or station.

2.3 Secondary motion, for purposes of this Standard, is met:

2.3.1 For mechanical controls, by an L-, Z-, or U-shaped slot, a slot latch, a thumb button, or other functional designs, but is not met by a simple detent alone.

2.3.2 For electrical control switches, by a partial shield, a spring-loaded shield over that portion of the switch which allows movement from the off position, or other designs meeting the intent.

SECTION 3—GENERAL

3.1 This Standard is based on the principle that a given direction of movement of any control should produce a consistent and expected effect.

3.2 Where confusion may result from the motion of the control, the effect from movement of the control shall be clearly and permanently identified. Refer to ASAE Standard S304, Symbols for Operator Controls on Agricultural Equipment.

3.3 All controls should be located convenient to the operator, with preference to those used more frequently or requiring higher forces to operate.

3.4 The inclusion of secondary motion requirement shall be considered in the design of all controls to reduce the likelihood of inadvertent actuation from the off or neutral position. Secondary motion shall be included:

3.4.1 Where inadvertent actuation of the control can create hazards associated with unexpected machine or component movement.

3.4.2 Only where technically feasible and functionally practical in operation.

3.5 Foot-operated controls shall have slip-resistant surfaces, be of adequate size and spacing, and be of appropriate configuration for proper operation.

3.6 Hand-operated controls shall be of appropriate configuration and size to permit adequate grasp and hand clearance throughout the operating range.

3.7 Controls not mentioned in this Standard, such as combinations, automatic, or new devices, are acceptable where applicable guidelines are followed.

3.8 Hand controls that are color coded shall conform to ASAE Engineering Practice EP443, Color Coding Hand Controls.

3.9 Operators manual information related to controls shall conform to ASAE Engineering Practice EP363, Technical Publications for Agricultural Equipment. In addition, operators manuals shall include:

3.9.1 Explanation of function and movement of each control.

3.9.2 Instructions for necessary actions to be taken before starting the machine's engine.

SECTION 4—BRAKE CONTROL

4.1 The service brake pedal(s) shall be actuated by the operator's right foot with the direction of motion forward and/or downward for engagement. Paragraph 5.2.1 describes an exception to right foot operation.

4.2 When separate brake pedals are provided for independent right-hand and left-hand brake control, it shall be possible to obtain combined and/or equalized control.

4.3 A control for a parking brake or parking device shall be provided.

4.3.1 Mechanical service brake(s) may be used as a parking brake where a locking device in the engaged position is provided.

4.3.2 Where a hand-operated parking brake is provided, pulling the control to apply brakes is preferred. A means to retain the control in the applied brake position shall be provided.

4.3.3 A mechanical lock on the drive train may be provided as a parking device.

SECTION 5—CLUTCH CONTROL

5.1 Traction control

5.1.1 When a foot pedal control is provided, it shall be actuated by the operator's left foot with the direction of motion forward and/or downward for disengagement.

5.1.2 When a hand-operated control is provided, it shall be moved toward the operator (generally rearward) for disengagement.

5.2 Combination traction clutch and brake control

5.2.1 When a foot-operated combination traction clutch and brake control is used, it shall be left foot operated with the direction of motion to be forward and/or downward to cause clutch disengagement and brake engagement.

5.3 Power take-off control

5.3.1 When a hand-operated clutch control is provided, its movement shall be generally rearward or downward for disengagement.

5.3.2 The combination traction clutch and constant running or transmission-driven power take-off clutch controls are governed by paragraph 5.1.1 and/or paragraph 5.1.2.

5.3.3 PTO engagement devices such as splined couplings and jaw clutches are excluded from the requirements of paragraph 3.4.

5.4 Field implement controls

5.4.1 When a hand-operated clutch control is provided, its movement shall be generally toward the operator or downward for disengagement, and away from the operator or upward for engagement.

5.4.2 When gathering mechanism rotation has been reversed on machines such as forage harvesters, the motion for hand control disengagement may be away from the operator.

5.4.3 When a foot-pedal clutch control is provided, it shall be governed by paragraph 5.1.1.

5.5 **Transportable or mobile farmstead equipment controls**

5.5.1 When a hand-operated clutch control is provided, its movement shall be generally away from the operator or downward for disengagement, and toward the operator or upward for enegagement.

5.5.2 A safety sign shall be provided near the control describing its activation; i.e., "Warning. Push to Stop".

SECTION 6—ENGINE SPEED CONTROL

6.1 The engine speed hand-operated control on agricultural field equipment shall be located to operate with the right hand.

6.1.1 On tractors with engine offset from the tractor centerline, left hand location and operation is permitted.

6.2 The direction of motion of the engine speed hand control shall be in a plane generally parallel to the machine's longitudinal axis. The direction of motion shall be generally forward or upward to increase engine speed, or

6.3 When the direction of motion of the engine speed hand control is in a plane parallel to the rim of the steering wheel, the direction of motion shall be rearward and/or downward to increase engine speed.

6.4 When a foot-operated engine accelerating control is provided, it shall be right foot operated with the direction of motion forward and/or downward to increase engine speed.

SECTION 7—GROUND SPEED AND DIRECTIONAL CONTROL

7.1 The shifting pattern(s) of the transmission speed selector lever(s) shall be clearly and permanently identified.

7.2 When a hand-operated, forward-reverse directional control lever (non-variable speed) is provided, it shall be moved forward for machine motion, and shall be moved generally rearward for rearward machine motion.

7.3 When a hand-operated, variable speed control is provided, it shall be moved generally forward and/or upward to increase speed.

7.4 Hand-operated combination direction and variable speed ratio control lever(s) shall be operated in one of the following patterns:

7.4.1 The lever shall be moved forward or away from the operator, from the neutral position, for forward travel and increasing forward speed. It shall be moved generally rearward or toward the operator, from the neutral position, for rearward travel and increasing rearward speed.

7.4.2 The lever shall be moved generally forward and away from the operator, from a neutral position, for forward travel and increasing speed. For rearward travel, the lever shall be moved laterally through a neutral position, and then forward and/or away from the operator for increased rearward speed.

7.5 Foot-operated combination directional and variable-speed control(s) shall be operated by the right foot and by one of the following configurations:

7.5.1 If two pedals are used, forward or downward motion on the outer pedal shall produce reverse motion, and forward or downward motion on the inner pedal shall produce forward motion. Forward or downward motion of either pedal shall increase speed, and both pedals shall return to neutral position upon release.

7.5.2 If a single pedal is used, it shall produce forward motion with a generally forward or downward toe motion on the pedal, and a rearward motion with a generally rearward or downward heel motion. Increased displacement of the pedal shall produce increased speed in either direction. The pedal shall return to neutral position upon its release.

SECTION 8—DIFFERENTIAL LOCK

8.1 When a differential lock control is provided it shall be moved forward or downward for engagement.

SECTION 9—STEERING CONTROL

9.1 When a steering wheel control is provided, a clockwise rotation shall effect a right turn and counterclockwise rotation shall effect a left turn.

9.2 When a single lever is used for steering, a lateral motion of the lever to the right shall effect a right turn and a lateral motion to the left shall effect a left turn.

9.2.1 If the single lever for steering also controls ground speed and direction of travel (forward/reverse), these functions shall be governed by paragraph 7.4.

9.3 When two levers are provided for steering by controlling the speed and direction of the driving elements the right-hand lever shall control the right-hand element and the left-hand lever shall control the left-hand element. The levers shall be moved forward or away from the neutral position, for forward travel and increasing forward speed. They shall be moved generally rearward or toward the operator, from the neutral position, for rearward travel and increasing rearward speed.

SECTION 10—ELECTRICAL CONTROLS

10.1 Rocker and toggle switch movements shall be consistent with the movement of a control lever used to control a similar function.

10.2 Rocker switches shall meet the following criteria, depending upon plane of installation:

10.2.1 For horizontal mounting, pushing on the forward or right portion shall turn on, start operation, move forward, increase speed, or cause movement downward.

10.2.2 For vertical mounting, pushing on the upper portion shall turn on, start operation, move forward, increase speed, or cause movement upward.

10.3 Switches mounted with side-to-side motion should only be used to control functions which move in corresponding planes.

10.4 All other types of switches shall follow above guidelines, where appropriate.

10.5 Because confusion may occur in up and down movements when the controlling switch is mounted on a sloping surface at or near 45 deg, such installation and use of switches only is not recommended.

SECTION 11—ENGINE STOP CONTROL

11.1 A key switch control is preferred. When used, the rotation to stop shall be counterclockwise.

11.2 When a mechanical push-pull control is used, the following shall apply:

11.2.1 The control shall be conspicuously located within 150 mm (6 in.) of the key switch.

11.2.2 Direction of movement shall be pull to stop.

11.2.3 The color of the control shall be red and contrast with the immediate background.

11.2.4 Labeling shall include the fuel stop symbol and an appropriate instruction, such as "pull to stop engine".

11.2.5 The control shall remain in the stop position without the application of sustained manual effort.

11.3 A mechanical stop control combined with engine speed control is provisionally acceptable, and the following shall apply:

11.3.1 The stop position shall be in the direction of and beyond low idle.

11.3.2 The stop position shall be identified with red and contrast with the immediate background. In addition, the fuel stop symbol or appropriate identification such as "engine stop" shall be applied at this position.

11.3.3 Labeling shall include an appropriate instruction for stopping the engine and be conspicuously located within 100 mm (4 in.) of the key switch.

SECTION 12—LIFT CONTROLS FOR IMPLEMENTS OR EQUIPMENT

12.1 Lift controls intended for movement from the operator's station:

12.1.1 Shall be clearly and permanently identified.

12.1.2 Should be located on the right-hand side of the operator.

12.1.3 When a hand control lever is provided the direction of motion shall be generally forward, downward or away from the operator to lower the implement or equipment; and rearward, upward, or toward the operator to raise the implement or equipment.

12.1.4 When a heel and toe foot control pedal is provided, the direction of motion of the forward part of the pedal shall be generally downward to lower and upward to raise the implement or equipment.

12.2 Operator controls for front-end agricultural loaders:

12.2.1 Shall be located on the right-hand side of the operator.

12.2.2 The lift arm control shall be located closest to the operator. The bucket control shall be located to the right of the lift arm control. If provided, the auxiliary function control shall be located to the right of the lift arm and bucket controls.

12.2.3 The direction of control motion shall be generally forward, downward, or away from the operator to lower the lift arm, and opposite directions to raise the lift arm. The direction of control motion shall be generally forward, downward, or away from the operator to dump the bucket, and opposite directions to roll back the bucket.

12.2.4 Should the bucket control be combined with the lift control for single lever operation, it shall move to the right to dump the bucket and to the left to roll back the bucket.

12.2.5 The direction of control motion for an auxiliary function shall be generally forward, downward, or away from the operator to lower or move forward the function; and rearward, upward, or toward the operator to raise or move rearward the function.

Cited Standards:

ASAE S304, Symbols for Operator Controls on Agricultural Equipment
ASAE EP363, Technical Publications for Agricultural Equipment
ASAE S390, Classifications and Definitions of Agricultural Equipment
ASAE EP443, Color Coding Hand Controls

ASAE Standard: ASAE S337.1

AGRICULTURAL PALLET BINS

Developed by the ASAE Farm Materials Handling Committee; approved by the ASAE Electric Power and Processing Division Standards Committee; adopted by ASAE December 1969; reconfirmed December 1974; reconfirmed with editorial revisions December 1975; reconfirmed December 1980; revised by the ASAE Transportation, Handling, and Warehousing of Agricultural Products Committee and approved by the ASAE Food and Process Engineering Institute Standards Committee February 1987.

SECTION 1—PURPOSE AND SCOPE

1.1 The purpose of this Standard is to provide uniform design and performance specifications for agricultural pallet bins (also known as, bulk bins, bin boxes, bulk crates, totes, field boxes and pallet boxes). This Standard is intended to help suppliers and users specify pallet bins that will facilitate handling, interchanging, storing, and transporting.

1.2 The scope of this Standard encompasses outside dimensions, capacity, materials, design features, performance tests, and specifications of pallet bins.

1.3 This Standard covers the general description of pallet bins for use in the agricultural industry. Additional recommendations may be required for specific applications. Detailed pallet bin design specifications are not included in order to permit alternate design concepts.

1.4 These guidelines are arranged in a manner that allows selection of those paragraphs applicable to specific cases. The purchaser or other user specifies applicable paragraphs and requirements. This Standard also functions as a checklist of characteristics that a user may wish to consider when purchasing pallet bins.

1.5 Terminology of construction is the same as defined in American National Standard MH 1.1.2, Pallet Definitions and Terminology. Sizes specified in ANSI Standard MH 1.2.2, Pallet Sizes, are used when appropriate.

SECTION 2—DIMENSIONS

2.1 Only overall outside dimensions are specified for length, width, and height. These dimensions shall not be exceeded by bulge or deflection from bin loading or by overfill.

2.2 One square cross section bin and one rectangular cross section bin are specified that give a projected area (footprint) with the dimensions shown. Two overall heights are specified. A separate stacking height may also be specified to give the distance measured from the base plane to the plane where the next pallet bin could rest.

2.2.1 The square bin shall be 120 ± 1 cm × 120 ± 1 cm (47.25 ± 0.39 in. × 47.25 ± 0.39 in.).

2.2.2 The rectangular bin shall be 120 ± 1 cm × 100 ± 1 cm (47.25 ± 0.39 in. × 39.37 ± 0.39 in.).

2.2.3 The overall height shall be 72 ± 1 cm (28.35 ± 0.39 in.) or 133 ± 1 cm (52.36 ± 0.39 in.).

2.3 Fork clearances shall be a minimum of 8.9 cm (3.5 in.) in opening height and a minimum of 22.90 cm (9.0 in.) in opening width, and shall be on 45.7 to 61.0 cm (18 to 24 in.) center spacings.

SECTION 3—CAPACITY

3.1 Since overall outside dimensions are specified, inside dimensions (and thus volume capacity) will vary with construction materials, bin design and inside depth.

3.2 Where volume capacity is required for measuring contents, quality control, piece work payments or other reasons, calibration marks shall be placed inside the bin.

3.3 To facilitate stacking and airflow for refrigerating and degreening, the bins should be loaded to provide a minimum of 5 cm (2 in.) of headspace. A mark (painted, stamped, etc.) shall be placed 5 cm (2 in.) below the top edge to provide a guide for loading.

SECTION 4—MATERIALS

4.1 Pallet bins may be constructed from any materials that have been shown to be suitable for the intended application. Common materials include wood, plastic, metal and fibreboard. The materials selected shall comply with the paragraphs in this section when they are specified.

4.2 Food Safety. Materials in contact with the product shall be approved by the U.S. Food and Drug Administration and/or the U.S. Department of Agriculture. Preservatives and other special treatment chemicals also must be approved.

4.3 Temperature range. The material shall withstand exposure to specified minimum and maximum temperatures without appreciable loss in strength or change in other characteristics.

4.4 Radiation. The material shall pass specified performance tests after specified exposure to sunlight (ultra-violet radiation).

4.5 Moisture. The material shall pass specified performance tests after specified exposures to high humidity or free moisture, e.g. hydrocooling.

4.6 Cleaning/sanitizing. The material shall be capable of being cleaned and/or sanitized by hot water, steam, chemical applications or other specified methods.

4.7 General requirements. Pallet bins shall be constructed from new materials that are free of harmful imperfections such as knots, bark, blow holes or other undesirable characteristics.

SECTION 5—DESIGN FEATURES

5.1 The following paragraphs describe bin characteristics which may be specified for specific applications.

5.1.1 Food safety. Design features shall be incorporated that prevent accumulations of mildew, bacteria and fungi. Surfaces in contact with food shall be free of rust, cracks, small radii, porosity and other conditions that prevent adequate cleaning and sanitation.

5.1.2 Handling safety. Pallet bin safety shall be enhanced by using designs that are without sharp edges, are resistant to splintering, are free of protruding parts and have other features that minimize the chance of injury.

5.2 General design features. The following design characteristics shall be specified when applicable.

5.2.1 Collapsible or knock down (KD). The design shall enable the pallet bin to be changed from a set-up configuration to a geometrically reduced configuration after use to promote efficient use of space while in storage or on back haul routes.

5.2.2 Expendable. The pallet bin shall have a life expectancy of one trip.

5.2.3 Reusable/returnable. The pallet bin shall have a life expectancy of more than one trip.

5.2.4 Lidded. The pallet bin shall have an uppermost surface capable of being closed by a lid.

5.2.5 Modular. The pallet bin shall be formed from smaller pallet bins which, when fit together, form a unit load equal in size to the base dimension of the parent pallet.

5.2.6 Nestable. The design shall enable one pallet bin to interpenetrate another of the same geometric form to reduce storage space when empty.

5.2.7 Strap guides. The design shall incorporate guides that prevent the inappropriate movement of the load during transport by forklift.

5.2.8 Top lift cleats. A provision for top cleats shall be provided to permit a loaded pallet bin to be lifted and/or dumped using its top rim.

5.3 Handling features. The following handling features shall be specified when applicable.

5.3.1 Forklift compatibility. Pallet bins shall be compatible with commonly used forklift, pallet trucks and other conveyances or lifting devices common to the agricultural industry. Handling by fork can be of the two-way or four-way entry style.

5.3.2 Stackability. Pallet bins shall be deisgned for safety in stacking during handling, transit, and storage.

5.3.3 Systems compatibility. Pallet bins shall be designed to be compatible with generally accepted conveyor systems, rack storage systems and other mechanical handling devices.

5.3.4 Runners. A bottom deck board shall be provided to enclose the fork entry opening and shall be designed into a given pallet bin base as a permanent structure. Runner design shall allow the load to be handled on powered or nonpowered conveyor systems.

5.4 Ventilation and hydrocooling. The pallet bin design shall have sufficient openings in its bottom, and sides if needed, to provide uniform cooling or warming of product in the bin and to permit timely drainage of water during hydrocooling. The location, size and quantity of openings shall be user specified; however, a minimum of 5% open area in the bin bottom is recommended.

5.5 Product protection. To minimize damage to the contents, the pallet bin shall have a smooth inside surface free of protrusions. Ventilation shots or holes shall have a 13 mm (0.5 in.) minimum radius or chamfer on their edges.

5.6 Labeling and identification. Provision shall be made on the pallet bin for identification of the product within the container, and where possible, the container itself shall be identified by using color codes, branding, bar codes, card slots or other suitable methods.

SECTION 6—PERFORMANCE TESTS

6.1 General. The purpose of performance and handling tests is to simulate (in an accelerated manner) the ordinary use that a pallet bin may be subjected to over its expected life. Performance criteria should be specified by the users and designed to meet their particular requirements. If only the pallet base is to be tested, the American National Standard ANSI/ASME MH1.4.1M, Procedures for Testing Pallets, may be used.

6.1.1 To assure confidence in test results, at least three pallet bins shall be subjected to each type of test. Also, if the test prescribes a loaded bin, the product for which the bin is intended shall be used as the test load.

6.1.2 One or more of the performance test procedures presented in this section may be specified for measuring expected performance against selected criteria.

6.2 Stacking test. The ability of the bins to endure expected stacking weights shall be determined by a stacking test. The test load may be determined by the following equation:

$$L = W \cdot n \cdot F$$

where

L = total load bottom pallet bins must withstand, N (lbf)
W = gross weight of loaded pallet bin, N (lbf)
n = expected number of overhead bins
F = safety compensating factor (1.5 is suggested)

The test bin shall be loaded with the same type bin or by a hydraulic compression tester. After the specified load has been applied, the container shall be examined for failure. The load requirement shall consider the effects of overload, time in storage and alignment.

6.3 Racking test. Measurement of the stiffness of a bin in the horizontal plane indicates resistance to racking. Holes shall be drilled in each top corner and a 0.64 cm (0.25 in.) diameter steel threaded rod with a dynamometer at the center shall be placed diagonally across the bin. Nuts shall be tightened on the rod from outside the bin until a force of 667 N (150 lbf) is recorded on the dynamometer. The distance from corner to corner is measured before any pull is exerted and again at the above specified force. The rods shall then be removed, placed across the other diagonal, and the same measurements repeated. The average change in the two diagonal measurements shall not exceed 7.0 cm (2.75 in.).

6.4 Rollover test. Racking resistance may also be tested by a rollover to simulate positioning of the empty bins in the field. Each test bin shall be subjected to a side-to-end rollover test on a strip of concrete pavement. The bin is tipped up on its side and pivoted up on one corner to just past the point of balance, and then shall be permitted to fall freely to the concrete. This procedure shall be repeated until 16 falls are completed or 4 revolutions of the bin. The difference in the racking measurements before and after the rollover test shall not exceed 10%.

6.5 Sag test. A sag test simulates the stress of placing full bins on uneven ground. For this test, a loaded bin shall be placed on a 10 × 10 cm (4 × 4 in.) timber so that the bin is oriented diagonally across the length of the timber. The sag of the unsupported corners in relation to the plane of the supporting timber shall be measured. The bin then is rotated 90 deg so that the other two corners are unsupported and the sag readings obtained. The sag shall not exceed 1.5 cm (0.6 in.).

6.6 Vibration test. The stress of vibration caused by truck or rail transport over rough roads shall be evaluated with a vibration test. For this test, a loaded bin shall be placed on a mechanically operated platform and vibrated at a frequency of 6 to 9 Hz for 30 min at peak acceleration of 4.90 m/s^2 (16.1 ft/s^2), ½ the acceleration of gravity. The vibration shall be stopped at 10 min intervals and bulge measurements taken on the sides and ends. Bulge shall not exceed 0.75 cm (0.30 in.). ASTM Standard D4169, Practice for Performance Testing of Shipping Containers and Systems, may be referred to for more detailed information on vibration testing.

6.7 Impact test. An impact test with a loaded bin shall be used to evaluate the resistance of the bin to severe lateral impacts. This test may be conducted using a pendulum or rolling incline device to horizontally impact the bin against solid steel or concrete bumpers at a velocity of 2 m/s (7 ft/s). The area of contact with the bumper shall be about 10 cm (4 in.) across the bottom of the bin. Four impacts are required, one on each side and end of the container. The side and ends will tend to rack out of square, and the resistance to this stress shall be determined by measuring the diagonals of the side or end impacted and calculating the change in dimensions due to impact. Change shall not exceed 0.5 cm (0.2 in.). ASTM Standard D4169, Practice for Performance Testing of Shipping Containers and Systems, may be referred to for more detailed information on impact testing.

6.8 Cornerwise-drop test. Resistance to dynamic racking stresses shall be evaluated by dropping a loaded bin on its corner. This is probably the most abusive test of the series and shall be conducted last. This test also is the most inclusive and shall be the single test to run if resources are not available to conduct the other tests in the series. This test shall be conducted by placing the loaded bin with one edge on a 15 cm (6 in.) high timber. One of the supported corners shall be blocked up with another 15 cm (6 in.) high block. The corner diagonally opposite the highest corner shall be lifted and set upon a block 15 cm (6 in.) high. This block then shall be jerked away allowing the unsupported side of the container to fall freely striking a steel plate embedded in a large mass of concrete. After the first fall, the bin shall be rotated 90 deg and the adjacent corner dropped in a similar manner. The other two corners shall be tested by the same procedure. The tests shall be repeated from heights of 30.5 cm (12 in.), 45.7 cm (18 in.), and 61 cm (24 in.). After a total of 16 cornerwise drops, 4 additional edgewise drops shall be made. These consist of 61 cm (24 in.) falls of one bottom edge while the other bottom edge is supported on just the 15 cm (6 in.) high timber. Evidence of deterioration or failure shall be recorded after each of the cornerwise and edgewise drops. Measurements shall also be taken on the amount of sag in the unsupported corners. Sag shall not exceed 9.0 cm (3.5 in.).

6.9 Top lift cleats. Pallet bins designed for handling with top lift cleats shall be tested by repeating a handling cycle that consists of filling, lifting and dumping. The bins tested shall be free of failure after 300 handling cycles.

6.10 Base deflection. Pallet bins loaded with 680 kg (1500 lb) of product shall not deflect more than 2.5 cm (1.0 in.) when lifted from a flat surface.

SECTION 7—SPECIFICATIONS

7.1 Bin specification. Purchasers and other users shall give specifications of pallet bins by identifying applicable paragraphs from Section 2—Dimensions, Section 3—Capacity, Section 4—Materials, Section 5—Design Features, through Section 6—Performance Tests. The specifications in paragraphs 7.1 through 7.6 shall be designated when applicable.

7.2 Application. State the intended application of the pallet bin including products, environments, storage durations, stacking heights, etc.

7.3 Dimensions. The first dimension stated shall be the length, the second shall be the width, and the third shall be the height. Any additional dimensions shall be clearly identified and shall be compatible with ANSI Standard MH 1.2.2, Pallet Sizes.

7.4 Capacity. State the internal volume required and the maximum mass to be contained. Provide details on any calibration or maximum level indicators.

7.5 Materials. State any requirements for food safety, temperature range, radiation levels, moisture cycles, cleaning and sanitizing, or other special needs.

7.6 Design. State design requirements such as strength, service life, smoothness, stackability, ventilation, pallet entry, design features for food safety or bin safety, etc. Construction details may also be specified.

7.7 Testing. State any requirements for evidence that the pallet bins provided are capable of performing satisfactorily when evaluated according to specified test procedures.

References:

1. Heebink, T.B. 1961. An evaluation of eleven bin-pallet designs. USDA Forest Products Lab., Madison, WI. Report No. 2216.
2. Maley, W.A. 1967. Systems for bulk handling of food crops. ASAE Paper No. 67-429. ASAE, St. Joseph, MI 49085.
3. National Safe Transit Association. 1973. Pre-shipment test procedures. Chicago, IL.
4. O'Brien, M. 1960. Designs for bulk fruit bins. California Agricultural Experiment Station, University of California, Davis. Circular 490.
5. Wardowski, W.F. and W. Grierson. 1978. Pallet boxes for Florida citrus. Florida Cooperative Extension Service, University of Florida, Gainesville. Circular 443.

Cited Standards:

ANSI MH1.1.2, Pallet Definitions and Terminology
ANSI MH1.2.2, Pallet Sizes
ANSI/ASME MH1.4.1M, Procedures for Testing Pallets
ASTM D4169, Practice for Performance Testing of Shipping Containers and Systems

ASAE Standard: ASAE S338.2 (*ANSI/ASAE S338.1)

T

SAFETY CHAIN FOR TOWED EQUIPMENT

*Corresponds in substance to previous document ASAE S338.1.

Proposed by the Engineering Policy Committee of Farm and Industrial Equipment Institute; approved by ASAE Power and Machinery Division Standards Committee; adopted by ASAE as a Recommendation December 1970; reconfirmed December 1975; reclassified as a Standard March 1978; reconfirmed December 1980; approved by ANSI as an American National Standard December 1982; revised April 1983; revision approved by ANSI June 1984; revised April 1988.

SECTION 1—PURPOSE AND SCOPE

1.1 This Standard covers the specifications for an auxiliary attaching system to retain a connection between towing and towed machines in the event of separation of the primary attaching system. It is intended as a guide for manufacturers and users of agricultural equipment towed on highways.

SECTION 2—DEFINITIONS

2.1 Highway: The entire width between the boundary lines of every way publicly maintained when any part thereof is open to use of the public for purposes of vehicular travel (Uniform Vehicle Code).

2.2 Towing machine: Any unit of equipment which furnishes the forces necessary to move and control a towed machine.

2.3 Towed machine: Any unit of equipment whose motive force and directional control is derived from one point of attachment to a towing machine. In a series of more than two machines, the same unit may serve as both a towed and a towing machine.

2.4 Primary attaching system: The means by which the motive and controlling forces are transferred from the towing to the towed machine under ordinary operating circumstances. The primary attaching point is the principal location of articulation in the system.

2.5 Auxiliary attaching system: A supplementary attaching means which will retain a connection between the towing and towed machines in the event of failure of the primary attaching system.

2.6 Safety chain: A chain, of designated size and strength, equipped with a latching hook on one end and a large pass through link on the opposite end.

SECTION 3—DESCRIPTION

3.1 An auxiliary attaching system includes a chain between the towed and towing machines. The system also includes attaching points at towed and towing machines. The auxiliary system shall be independent of, and in addition to, the primary means of connecting towed and towing machines.

3.2 Safety chain

3.2.1 The safety chain hook end shall pass through holes as small as shown in column entitled "Minimum fitting hole size" of Table 1.

3.2.2 The minimum ultimate tensile strength of the safety chain and its end fittings shall not be less than that shown in the system capacity requirements of Section 4—System Capacity Requirements.

3.2.3 The safety chain shall be fastened to the attaching points of the towed and towing machines by a positive means (hook latch) which cannot be separated inadvertently.

3.2.4 The safety chain shall have no more slack, when in use, than necessary to permit proper articulation of the machines.

3.2.5 The safety chain should have a permanent, non-corroding type identification tag attached in the link that fastens the large end link to the basic chain (see Fig. 3).

3.2.5.1 Tag size shall be 100 ± 6 mm (4.0 ± 0.25 in.) by 38 ± 6 mm (1.5 ± 0.25 in.) by 3 mm (0.12 in.) minimum thickness.

3.2.5.2 Tag information shall be stamped into the material surface and shall state the following: "ASAE S338. Agricultural safety chain for towed machine not exceeding ____ kg (____ lb) gross weight". The gross weight value shall match the chain capacity. Letters shall be 4 mm (0.16 in.) minimum height and easily readable.

3.2.6 The safety chain must be replaced if one or more links are broken, stretched, or otherwise damaged or deformed.

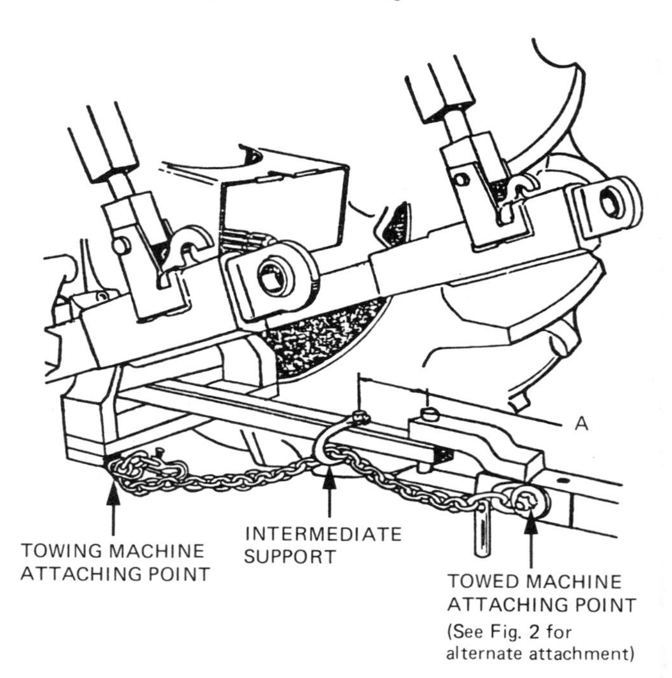

FIG. 1—TYPICAL AUXIILARY ATTACHING SYSTEM ON A TRACTOR

TABLE 1—CHAIN DATA

Minimum System Strength		Maximum Chain Link Diameter Fig. 3		Minimum Fitting Hole Size Fig. 3		Maximum A, B, C Dimension Figs. 1 & 2		Minimum Chain Length Fig. 3	
kN	lbf	mm	in.	mm	in.	mm	in.	mm	in.
29	6 400	32	1.3	32 x 67	1.3 x 2.6	230	9.0	1 500	59
45	10 100	32	1.3	32 x 67	1.3 x 2.6	230	9.0	1 500	59
90	20 200	38	1.5	45 x 102	1.8 x 4.0	230	9.0	1 500	59
135	30 400	45	1.8	45 x 102	1.8 x 4.0	230	9.0	1 625	64
180	40 500	50	2.0	45 x 102	1.8 x 4.0	280	11.0	1 750	69

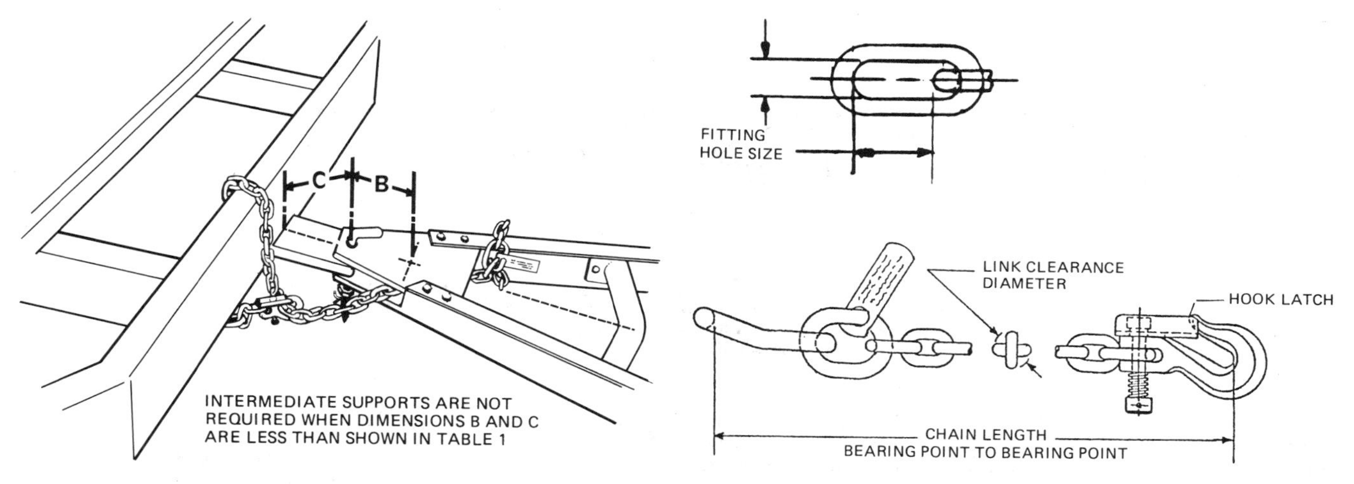

FIG. 2—TYPICAL AUXILIARY ATTACHING SYSTEM ON A TRUCK

FIG. 3—CHAIN LENGTH AND CLEARANCE

3.3 Attaching points (See Figs. 1 and 2)

3.3.1 Towing machine. The attaching point shall be of such form that it may be encompassed by the chain described in paragraph 3.2. The attaching point shall be located within 230 mm (9.0 in.) laterally and vertically from the projection along the line of travel of the primary attaching point and so located that no more than 900 mm (35.4 in.) of chain as measured from the primary attaching point will be required for fastening. Additional chain length, 125 mm (5 in.) and 250 mm (10 in.), is specified for the two largest chains (see Table 1) to accommodate the proportionately larger attaching structures. The attaching point must be located where it does not interfere with normal drawbar functions, such as a laterally swinging drawbar.

3.3.2 Towed machine. The attaching point (see Fig. 2) shall be of such form that it may be encompassed by the chain described in paragraph 3.2. The attaching point shall be located within 230 mm (9.0 in.) laterally and vertically from the projection along the line of travel of the primary attaching point. The attaching point shall be located so that no more than 600 mm (23.6 in.) of chain as measured from the primary attaching point will be required for fastening.

3.3.3 Intermediate support. When the B or C dimension on the towing or towed machines (see Fig. 2) exceeds the maximum intermediate support dimension shown in Table 1, an intermediate support shall be provided. The intermediate support shall be capable of providing vertical and lateral support as required in paragraph 4.3 and shall be located within the A, B, or C dimension from the primary attaching point.

SECTION 4—SYSTEM CAPACITY REQUIREMENTS

4.1 The safety chain and attaching parts shall have a minimum ultimate tensile strength equal to or greater than the gross weight of the towed machines and shall conform to dimensions and values shown in Table 1.

4.2 The attaching points on a towing machine shall have adequate strength to support the gross weight of the towing machine along the line of travel. The attaching point on a towed machine shall similarly support its individual gross weight. The location of the attaching points on the left side of the primary hitch is recommended for operational convenience.

4.3 The intermediate support points, if required, shall be capable of withstanding a force of one-half the required chain capacity, applied in any direction.

ASAE Standard: ASAE S341.2

T

PROCEDURE FOR MEASURING DISTRIBUTION UNIFORMITY AND CALIBRATING GRANULAR BROADCAST SPREADERS

Developed by Fertilizer Application Subcommittee of ASAE Agricultural Chemical Application Committee; approved by Power and Machinery Division Standards Committee; adopted by ASAE as a Tentative Standard December 1971; reclassified as a full Standard December 1972; revised March 1978; reconfirmed December 1982; revised March 1988.

SECTION 1—PURPOSE AND SCOPE

1.1 The purpose of this Standard is to establish a uniform method of determining and reporting performance data on broadcast spreaders designed to apply granular materials on top of the ground. Tests performed according to this Standard make it possible to predict field performance of the spreader and to compare spreader distribution patterns.

1.2 This Standard pertains to centrifugal, pendulum, and other types of broadcast spreaders designed for dry granular application while operating on the soil surface. Portions of the test procedures outlined herein are suitable for determining the delivery rate of gravity or drop spreaders; however, additional tests not covered in this Standard are needed to completely evaluate the performance of gravity spreaders. This standard does not cover dry pneumatic granular applicators.

SECTION 2—DEFINITIONS

2.1 Application rate: Application rates are as defined in ASAE Standard S327, Terminology and Definitions for Agricultural Chemical Application.

2.2 Single-pass application: An application method where the spreader applies one swath over the collection trays.

2.3 One-direction application: An application method where successive adjacent swaths are made in the same direction of travel (racetrack application). This method produces a right-on-left overlapping of adjacent patterns.

2.4 Progressive application: An application method where the spreader applies adjacent swaths in alternate directions (back and forth application). This method produces a right-on-right pattern overlap alternately with a left-on-left pattern overlap.

2.5 Swath spacing: The lateral distance between spreader centerlines for overlapping broadcast applications.

2.6 Effective swath width: The swath width that will produce acceptable field deposition uniformity for the intended application.

SECTION 3—TEST CONDITIONS

3.1 The spreader to be tested shall be in good mechanical condition and shall be properly adjusted.

3.2 Tests may be conducted on a spreader to evaluate an experimental model, to qualify a new production model, or to verify the performance of an existing production model. Tests may also be conducted to evaluate the performance of a spreader and to verify the adjustment of a spreader for a specific granular material being spread under conditions similar to actual field conditions.

3.3 The geometric specifications shall be checked with the machine standing on an impenetrable, horizontal surface in normal operating position. Dimensions of length and width shall be measured along horizontal lines and dimensions of height along vertical lines.

3.4 It is recommended that the manufacturer be notified well in advance of any test in which the manufacturer's current production model will be compared with those of other manufacturers. The manufacturer shall be entitled to have his representative present during the test.

3.5 The test may be conducted with a standardized test material such as uniform size spherical prills, a specific granular product or an inert product simulant. A description of the material including sieve size, moisture content, bulk density, and product name shall be specified in the test report (see ASAE Standard S281, Capacity Designation for Fertilizer and Pesticide Hoppers and Containers).

3.5.1 Results of a sieve analysis shall be reported. The percent of material retained on each screen shall be stated. The sieves used shall conform to American National Standard Z23.1, Specifications for Wire-Cloth Sieves for Testing Purposes.

3.5.2 A description of the test material's particle shape and surface texture shall be included in the test report.

SECTION 4—TEST PROCEDURE

4.1 Guidelines for test setup

4.1.1 The accuracy of the test can be influenced by wind, granule or particle size, rate of application, ground slope, travel speed, ground roughness and method of collecting samples. The following recommendations should be followed to maintain test accuracy:

4.1.1.1 All spreading shall be done when the wind velocity is less than 8 km/h (5 mile/h) at a height of 2.5 m (5 ft) above the ground. If a wind exists, the direction of travel shall be parallel (within ± 15 deg) to the direction of the wind.

4.1.1.2 Tests shall be conducted on a surface having a slope of less than 2%. If desired, the spreaders may also be tested on a sloping surface provided all spreaders in the comparison are tested on the same slope and the degree and direction of the slope reported.

4.1.1.3 The spreader should be operated for a period long enough for the flow or output to stabilize. This will vary with spreader design, however operating the spreader over a distance of 100 m (328 ft) is generally adequate.

4.1.1.4 If the test results are to be used for the specific purpose of adjusting the field performance of an individual spreader, it is recommended that the test be conducted under field conditions that represent normal use.

4.1.2 The spreader shall be filled the day of the test. A minimum of 10 min shall be allowed for settling before the test is conducted. If the test is not conducted within 4 h after filling, the spreader shall be emptied and refilled.

4.1.3 Tests shall be run with the spreader hopper or box filled and leveled to 40 to 50% of capacity as defined by ASAE Standard S281, Capacity Designation for Fertilizer and Pesticide Hoppers and Containers.

4.2 Collection devices

4.2.1 Width of each collecting tray (measured perpendicular to the direction of travel) shall not exceed 10% of the anticipated effective swath width. The length shall be equal to or greater than the width with a minimum length of 30 cm (1 ft). The maximum wall thickness of the tray sides shall be 2.3 mm (0.09 in.).

4.2.2 Trays should be of sufficient size to collect samples, from one pass of the spreader, that are large enough to accurately measure with available measuring equipment.

4.2.3 To decrease the possibility of particles ricocheting out of the trays, each tray should be divided into compartments. The maximum size of the compartments shall be 10 cm (4 in.) wide by 10 cm (4 in.) long. The minimum size of the compartments should be 5 cm (2.0 in.) by 5.0 cm (2.0 in.). The depth of the compartments shall be at least 50% of the maximum horizontal dimension. Precautions such as covering the tray floor with soft material may be taken if the granule being tested contains no small particles which may lodge in the covering material.

4.2.4 Sufficient trays shall be used to provide at least 10 trays within the effective swath width. Spacing of the trays shall be uniform, except that trays may be rearranged to allow passage of spreader and vehicle wheels. Additional trays shall be spaced out on either side of the effective swath width to a distance equal to at least 50% of the swath width.

4.2.5 During all tests, the tops of the trays shall not be over 10 cm (4 in.) above the ground level with the spreader in the normal operating position. If the height of the discharge point on the spreader is less than 0.5 m (20 in.), the tops of the trays shall be less than 5 cm (2 in.) above the ground level.

4.2.6 Power take-off driven units shall be operated at the speed specified by the spreader manufacturer. If the specified speed is 540 ± 10 rpm or 1 000 ± 25 rpm (see ASAE Standard S203, Rear Power Take-Off for Agricultural Tractors) or other, it shall be noted in the test report. For truck-mounted units the spinner shall be rotated at the speed recommended by the manufacturer. For electrically driven units, supply voltage shall conform to that recommended by the manufacturer; any deviation shall be stated in the report.

4.2.7 Relative travel speed between the spreader and the collection trays shall be in the range recommended in the manufacturer's literature, and kept constant during the conduct of the test. The actual speed of the spreader and/or collection trays at which the tests were conducted shall be reported.

4.3 Description of the test procedure. The test should consist of two parts: (1) determination of application rate and, (2) determination of the distribution pattern by measurement of applied materials from suitable collectors. Each part of the test should be replicated to account for random variation.

4.4 Determination of application rates

4.4.1 The preferred method of determining application rates is by measuring the amount of material exiting the spreader during operation over a known area.

4.4.2 Weight of material applied can be determined by collecting and weighing spreader output while traveling a measured distance or weighing the spreader and its contents before and after spreading material over a known distance.

4.4.3 Using the methods described in paragraph 4.4.2, the application rate should be calculated as follows:

$$R = QK/LW$$

where

R = application rate, kg/ha (lb/acre)
Q = weight applied, kg (lb)
L = distance spreader operated, m (ft)
W = swath spacing, m (ft)
K = constant, 10 000 (43 560)

4.4.4 An alternate method for determining application rates is by calculation from the amount of material collected in spread pattern tests (see paragraph 4.5.2). The accuracy of this method is influenced by collector design and type of surface around the collectors. If the collector surface is such that particles entering the collector bounce out, this method will yield rates that are lower than actual rates.

4.4.5 Application rates based on material collected in spread pattern tests should be calculated using the following equation:

$$R = KW/AE$$

where

R = application rate, kg/ha (lb/acre)
K = constant, 100 000 (13 829)
W = sample weight, g
A = area of collector opening, cm^2 ($in.^2$)
E = collector efficiency, 0%-100%

4.4.6 The volume (in cubic centimeters) of material collected in spread pattern tests can be used to determine application rates. Using the bulk density of the material being applied (see ASAE Standard S281, Capacity Designation for Fertilizer and Pesticide Hoppers and Containers), the weight of material collected can be calculated by the following equation:

$$W = DV/K$$

where

W = weight, g
D = bulk density, kg/m^3 (lb/ft^3)
V = volume, cm^3
K = constant, 1 000 (62.4)

4.5 Spread pattern test

4.5.1 Spread pattern tests indicate the degree of uniformity of distribution of material across the swath being spread.

4.5.2 The spread pattern test shall be accomplished by operating the spreader in a line perpendicular to a line of collection trays spaced equally on the ground. An odd number of trays should be used, and the spreader should be driven astride the center pan. Material collected in each tray should be weighed or measured volumetrically.

4.5.3 The actual delivery rate and the spreader settings used to achieve these rates shall be reported. All spreaders to be compared shall be tested at the same rate, if possible. If field performance of an individual spreader is desired, the application rate shall be selected based upon the agronomic requirements of the test. Application rates of approximately 25, 50, and 75% of the maximum application rate for the test material are suggested if multiple rates are to be used. Additional tests may be conducted and reported at other rates selected from within the manufacturer's recommended range of application rates.

4.5.4 Uniformity of distribution. The spreader tested shall be rated for uniformity of distribution. The coefficient of variation (CV) shall be used to determine and express the uniformity of distribution of applications. When overlapping of swaths occur, a simulated field application of multiple adjacent swaths shall be used to compute the CV. The simulated field distribution for each swath distribution pattern obtained in paragraph 4.5.2 is constructed by using the manufacturer's recommended swath width for the spreader being tested or the effective swath width (see paragraph 4.5.5) and accumulating the sample weights at each collection tray location. Individual replicates of the swath distribution pattern (not averages) shall be used. The method of spreading used shall be reported; i.e., either progressive (back and forth) or one direction (race track).

4.5.4.1 The mean value, standard deviation, and CV shall be determined as follows:

$$\text{Mean} = \overline{X} = \Sigma X_i/N$$

$$\text{Standard deviation} = \{\Sigma[(X_i-\overline{X})^2]/(N-1)\}^{1/2}$$

$$CV = (\text{standard deviation})\,(100)/\overline{X}$$

where

$\overline{X}$ = arithmetic mean
X_i = accumulated sample weight for each collector location for the overlapped swaths
N = number of collector locations used

4.5.4.2 Only the central portion of the simulated or measured overlapped distribution data shall be used to compute the CV. If the swath spacing is equal to or greater than one-half of the total spread pattern width, this shall include data from one spreader centerline to the next for the one direction application method or the data from the centerline of the first swath to the centerline of the third adjacent swath to the centerline of the third adjacent swath for the progressive pass method. If the swath spacing is less than one-half of the total spread pattern width, additional overlapped distribution data shall be added until the region for calculation as indicated above would be unaffected by the addition of distribution data resulting from additional overlapping swaths.

4.5.5 Effective swath width. The tests shall be run using the most effective swath width symmetrical about the centerline of travel.

4.5.5.1 Frequently, the effective swath width will be the distance between the points on either side of a single swath where the rate of deposit equals one-half of the effective application rate. The effective swath width shall be determined in a manner which will give the most uniform overall application rate. The manufacturer's recommendation shall be used as a guide in determining the most effective swath width.

4.5.5.2 Another method for determining the effective swath width of the spreader is by inspecting values for CV, computed from simulated overlapped distribution data, versus swath

TABLE 1—CV VALUES FOR DIFFERENT SWATH SPACINGS AND DRIVING METHODS

Swath spacing	One-direction application	Progressive pass application
(m)	(CV)	(CV)
4	0.5	1.3
6	2.1	2.4
8	7.7	7.8
10	1.4	1.3
12	17	17
14	24	24
16	22	22
18	13	13
20	2.6	2.9
22	21	21
24	42	42
26	59	59

width. Table 1 is a listing of CV's versus swath spacing and driving method. The largest swath width associated with the minimum acceptable CV shall be considered the effective swath width of the test.

4.5.6 The results of this test may also be presented graphically as shown in Fig. 1.

4.5.6.1 In plotting test data the vertical axis shall indicate the application rate in kg/ha (lb/acre or lb/1 000 ft^2) and the horizontal axis shall represent the spread width in m (ft).

4.5.6.2 The coefficient of variation and the setting at which it was determined shall be stated in each graph.

4.5.6.3 When indicating spreader overlap as shown in Fig. 1, it is necessary to report the method of spreading assumed; i.e., either progressive or one direction, and plot the graph accordingly.

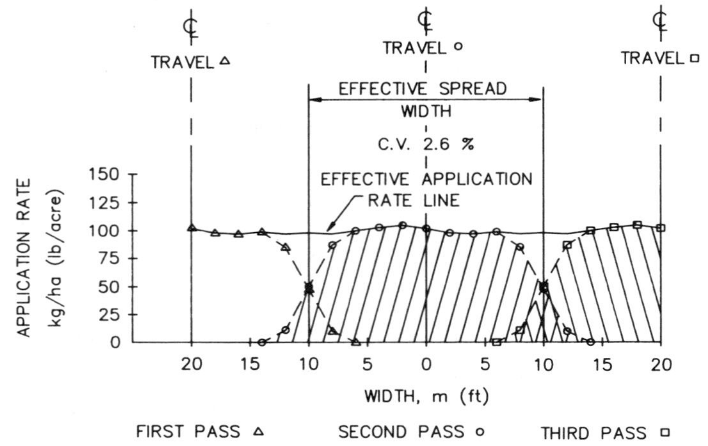

FIG. 1—GRAPHICAL PRESENTATION OF SPREAD PATTERN—ONE-DIRECTION APPLICATION METHOD

SECTION 5—METHOD OF REPORTING RESULTS

5.1 If the test has been conducted as described by this Standard, the test results should be identified as follows:

5.1.1 The following note shall be placed on each page on which test results appear: "These results have been obtained from a test made in accordance with ASAE Standard S341, Procedure for Measuring Distribution Uniformity and Calibrating Granular Broadcast Spreaders."

5.1.2 A descriptive statement shall be included in the report to explain the coefficient of variation. For example: This 18% variation means that at a setting of 100 kg/ha (89 lb/acre) the actual application rate would be expected to range between 82 and 118 kg/ha (73 to 105 lb/acre) on 68% of the area.

5.2 A brief description of the spreader shall precede the dimensions. If the spreader can be adjusted to deliver one side at a time (for headlands and row-crops), this should be mentioned. Adjustable spreader variables and their settings shall be reported. The following data should be included in the description:

Type (centrifugal, pendulum, etc.)
Manufacturer's name, model number, and year of manufacture: (Serial number if available)
Minimum and maximum application rates: kg/ha (lb/acre)
Minimum and maximum output rates: kg/min (lb/min)
Overall length: cm (in.)
Overall height: cm (in.)
Overall width: cm (in.)
Height of particle release above ground level (in operation): cm (in.)
Metering system used
Number of spinners or delivery points
Manufacturer's recommended spread width: m (ft)
Hopper capacity: m^3 and kg (ft^3 and lb)
Track width (c-to-c): cm (in.)
Number of wheels

5.3 All test results shall be stated as listed in paragraph 5.1 and include the following:

Application rate as indicated by the rate setting
Actual application rate and method used to measure actual rate
Standard deviation (for each application rate)
CV (for each application rate)
Material tested (analysis, bulk density, moisture content)
Forward travel speed
Relative travel speed
Effective swath width
Information concerning exceptions or additions which are peculiar to this test
Wind speed and direction relative to spreader line of travel (see paragraph 4.1.1.1)

Cited Standards:

ANSI Z23.1, Specifications for Wire-Cloth Sieves for Testing Purposes
ASAE S203, Rear Power Take-Off for Agricultural Tractors
ASAE S281, Capacity Designation for Fertilizer and Pesticide Hoppers and Containers
ASAE S327, Terminology and Definitions for Agricultural Chemical Application

ASAE Standard: ASAE S343.2

TERMINOLOGY FOR COMBINES AND GRAIN HARVESTING

T

Developed by the ASAE Grain Harvesting Committee; approved by the Power and Machinery Division Standards Committee; adopted by ASAE as a Tentative Standard February 1971; reclassified as a full Standard December 1971; reconfirmed December 1977; revised April 1981; reconfirmed December 1985; revised March 1988.

SECTION 1—PURPOSE AND SCOPE

1.1 The purpose of this Standard is to establish terminology pertinent to grain combine design and performance. It is intended to improve communication among engineers and researchers and to provide a basis for comparative listing of machine specifications.

SECTION 2—COMBINE COMPONENTS

2.1 Header: The portion of the combine comprising the mechanisms for gathering the crop.

2.1.1 Gathering width: The distance between the centerlines of the outermost divider points; expressed in meters to the nearest hundredth. Where adjustable dividers are used the maximum and minimum dimensions shall be stated.

2.1.2 Grain header width: The distance between the side sheets of the header measured immediately above the forward tips of the sickle sections; expressed in meters to the nearest hundredth.

2.1.3 Maize (ear corn) header width: The average distance between the centerlines of adjacent picking units multiplied by the number of units. Where the header width is adjustable, maximum and minimum distances between centerlines shall be stated, expressed in centimeters to the nearest whole centimeter. The maximum and minimum header widths shall then be expressed in meters to the nearest hundredth, and the number of picking units shall be stated.

2.2 Cutting mechanism: That device on the header for severing the plant stalks. May include reciprocating, rotary, continuous, scissor-type, or other mechanisms for severing.

2.2.1 Sickle: A cutting mechanism which uses a reciprocating cutter.

2.2.1.1 Sickle frequency: The number of cycles which the sickle makes in a given period of time. One cycle is the full movement of the sickle in one direction and its return to the starting point. Frequency shall be expressed in hertz.

2.2.1.2 Sickle stroke: The distance that a point on the sickle travels with respect to the centerline of a guard in one half cycle; expressed in millimeters.

2.3 Pickup: A device for gathering a crop from a windrow.

2.3.1 Pickup width: The minimum distance including the width of the outermost conveying elements but not including the gather of flared side sheets; expressed in meters to the nearest hundredth.

2.4 Cutting mechanism height: The height of the forward tip of any cutting blade or sickle section above the plane on which the machine is standing, measured under the following conditions and expressed in centimeters:

2.4.1 The maximum and minimum dimensions shall be in the highest point and the lowest point to which the cutterbar can be raised or lowered with the standard lift mechanism.

2.4.2 Tire and wheel equipment shall be stated, and tires shall be inflated to the field operating pressures recommended by the combine manufacturer.

2.4.3 The plane on which the combine is standing shall be substantially level.

2.4.4 The header installed at the time of measuring shall be stated.

2.4.5 The grain tank shall be effectively empty in accordance with ASAE Standard S312, Capacity Designations for Combine Grain Tank Systems.

2.5 Rotating threshing or separating elements

2.5.1 Threshing cylinder: A rotating element, which in conjunction with a stationary element adjacent to it, is fitted primarily to promote threshing. The crop being threshed is contained between rotating and stationary elements for less than 360 deg.

2.5.2 Threshing rotor: A rotating element similar to a threshing cylinder except that the crop is contained for 360 deg and may pass around the rotor axis one or more times.

2.5.3 Separating cylinder or rotor: Defined as for threshing cylinder or rotor, except that the terms "separating" and "separated" replace "threshing" and "threshed".

2.5.4 Rotary separator: An alternative term for a rotary device, similar to a cylinder, which is fitted to promote separation only.

2.5.5 Cylinder or rotor threshing or separating diameter: The diameter of the circle generated by the outermost point of the appropriate rotating element as it rotates about its own axis, dimension D, Figs. 1, 2, 3, and 4, expressed in millimeters.

2.5.6 Cylinder or rotor threshing or separating length: The length of the cylindrical volume generated by the outermost points of the cylinder or rotor elements, as the cylinder or rotor rotates about its own axis, and as appropriate to its threshing or separating section, dimension L, Fig. 2, expressed in millimeters.

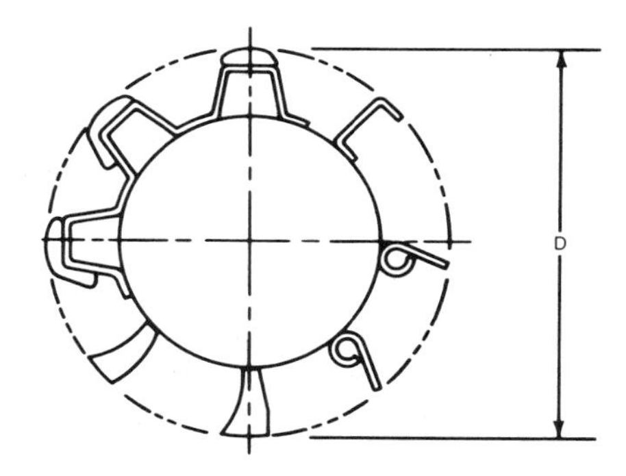

FIG. 1—CYLINDER OR ROTOR DIAMETER

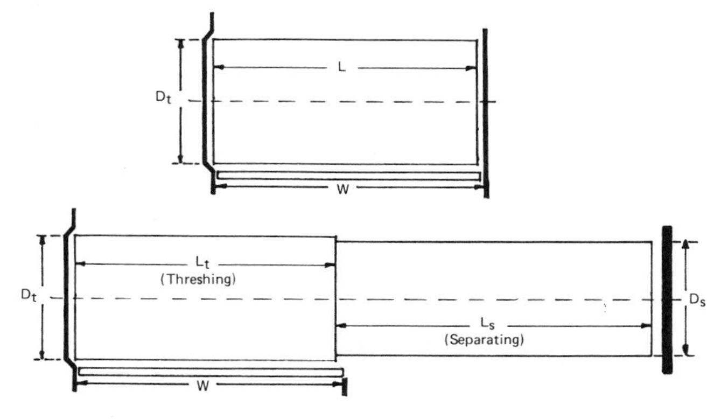

FIG. 2—CYLINDER, ROTOR, AND CONCAVE DIMENSIONS

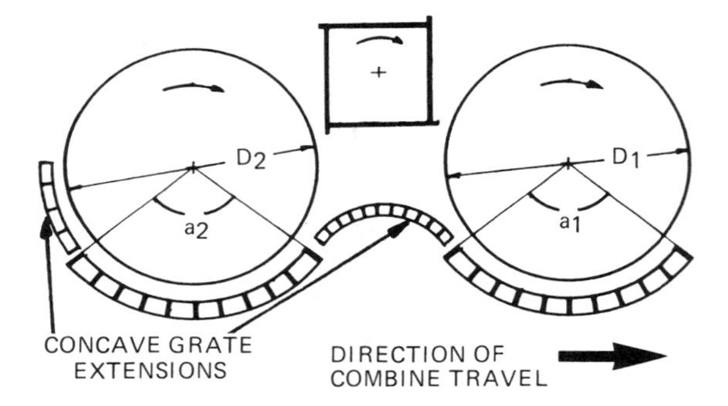

FIG. 3—LATERALLY-DISPOSED CYLINDERS OR ROTORS

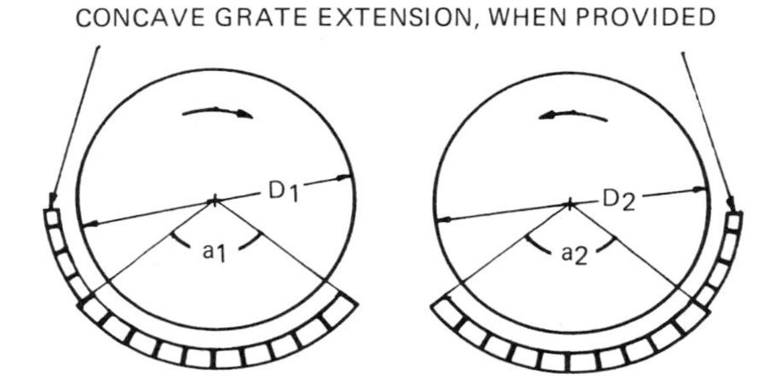

FIG. 4—LONGITUDINALLY-DISPOSED CYLINDERS OR ROTORS, VIEWED FROM REAR

2.5.7 Single or multiple cylinders or rotors may be disposed laterally (see Fig. 3) or longitudinally (see Fig. 4) within the combine. If multiple cylinders or rotors are used, the number shall be stated and the dimensions given as in Figs. 3 and 4.

2.6 Concave: A concave-shaped stationary element adjacent to the threshing cylinder or rotor fitted primarily to promote threshing. In the case of a concave that is permeable to grain flow, either in whole or in part, there is the secondary important function of primary separation.

2.6.1 Concave width or length: The outside dimension of the concave, measured parallel to the axis of its associated threshing cylinder or rotor, dimension W, Fig. 2, expressed in millimeters.

2.6.2 Concave arc length: The arc length dimension of the concave, including the first and last bars. This shall be measured in a plane perpendicular to the axis of its associated cylinder or rotor and around the contour formed by the inner surfaces of the concave bars, dimension A, Fig. 5. Concave arc length shall be expressed in millimeters.

2.6.3 Concave arc: A common alternative way of defining arc length in degrees. It shall be measured from the outside of the first bar to the outside of the last bar in a plane perpendicular to the axis of the associated cylinder or rotor, dimension a, Fig. 5.

NOTE: When using this means of defining the arc length of the concave, it is essential that the diameter of the associated cylinder or rotor be quoted also (see Fig. 1).

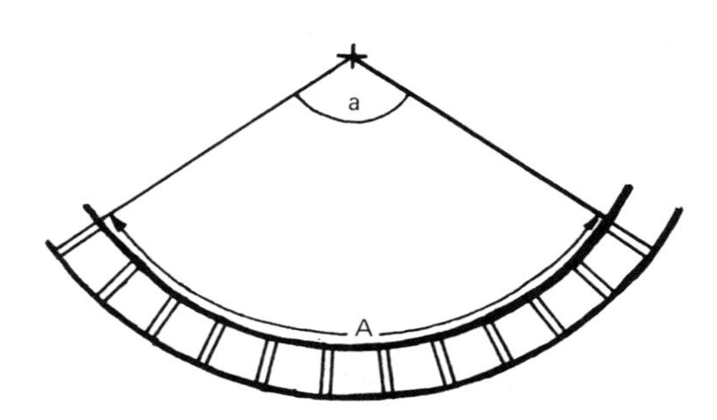

FIG. 5—CONCAVE ARC

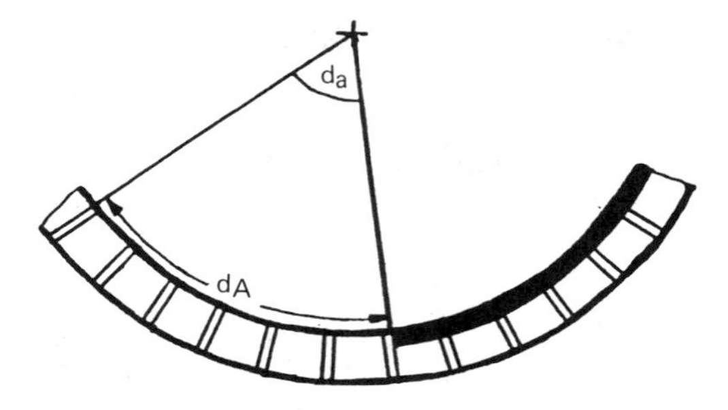

FIG. 6—CONCAVE GRATE ARC LENGTH

2.6.4 Concave area: The product of the concave width, or length, and arc length, expressed in square meters to the nearest hundredth.

2.6.5 If more than one concave is used, this shall be so stated, and the dimensions and areas shall be given separately.

2.6.6 Concave grate: That portion of the concave which is permeable for separation.

2.6.7 Concave grate width: As for concave width, W, Fig. 2, expressed in millimeters.

2.6.8 Concave grate length: Dimension dA, Fig. 6, expressed in millimeters.

2.6.9 Concave grate arc: That portion of the concave arc that corresponds to the concave grate length, dimension da, Fig. 6.

2.6.10 Concave grate area: The product of concave grate width, W, Fig. 2, and length, dA, Fig. 6, expressed in square meters to the nearest hundredth.

2.7 Concave grate extension: A permeable element, approximately concentric to the associated cylinder or rotor and generally forming an extension to concave, Fig. 4. This may exist merely as a gap between the concave and the transition grate shown as dimension G in Fig. 7.

2.7.1 Concave extension width: As for concave width, W, Fig. 2, measured in millimeters.

2.7.2 Concave extension length: Dimension G, Fig. 7, measured in millimeters.

2.7.3 Concave grate extension area: The product of the concave width, W, Fig. 2, and the concave grate extension length, G, Fig. 7, measured in square meters to the nearest hundredth.

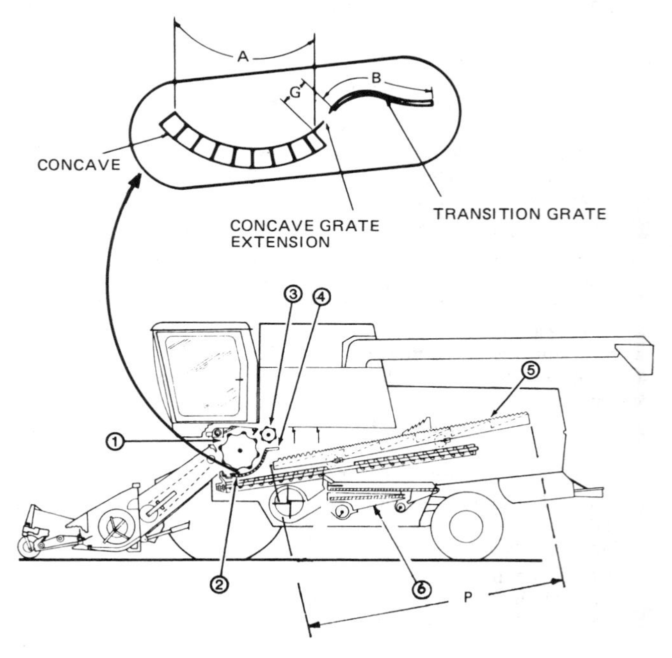

FIG. 7—TYPICAL WALKER TYPE COMBINE
(1) Cylinder (2) Concave (3) Beater (4) Transition Grate (5) Walkers (6) Shoe

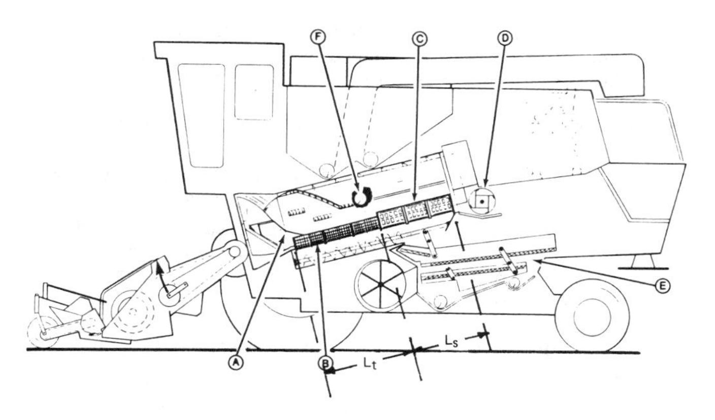

FIG. 8—COMBINE SIDE VIEW

(A) Rotor
(B) Threshing Concaves (L_t)
(C) Separating Concaves (L_s)
(D) Back Beater
(E) Shoe
(F) Tailings Return

2.8 Transition grate: A permeable element that provides transition from the concave grate extension to the next separating device.

2.8.1 Transition grate width: As for concave width, W, Fig. 2, measured in millimeters.

2.8.2 Transition grate length: The contour length of the upper surface of the transition grate, dimension B, Fig. 7, expressed in millimeters.

2.8.3 Transition grate area: Product of transition grate width, W, Fig. 2, and length, B, Fig. 7, expressed in square meters to the nearest hundredth.

2.9 Rotary or axial-rotor concave. These concaves are similar to the concaves described in paragraph 2.6 except that they may be longer, wider and/or oriented longitudinally within the combine (see Fig. 8). The dimensions are shown in Figs. 2 and 4. Areas are calculated in the same manner as described in paragraph 2.6.4; 2.6.10; 2.7.3; and 2.8.3.

2.10 Further separating devices

2.10.1 Straw walker: Multiple, permeable platforms mounted on rotating cranks which together fill the width of the separating body of the combine, dimension R, Fig. 9, expressed in meters to the nearest hundredth. These platforms shake and transport the straw rearward in the combine, separating the grain from the straw or plant.

2.10.1.1 Straw walker length: The distance from the front to the rear of the walker, dimension P, Fig. 7, expressed in meters to the nearest hundredth. If there are adjustable sections at the rear of the walkers, they should be in the fully extended position.

2.10.1.2 Straw walker area: The product of the width of the straw walker body, dimension R, Fig. 9, and the length of an individual walker, dimension P, Fig. 7, expressed in square meters to the nearest hundredth.

2.10.2 Axial or other rotary separating grates: The rear or final portion of these grates are designed to separate grain from the material-other-than-grain (mog) by centrifugal force (see Fig. 2).

2.10.2.1 Separating grate length: Dimension Ls, Fig. 2, expressed in millimeters. Where there is more than one rotor and associated grate, this shall be noted.

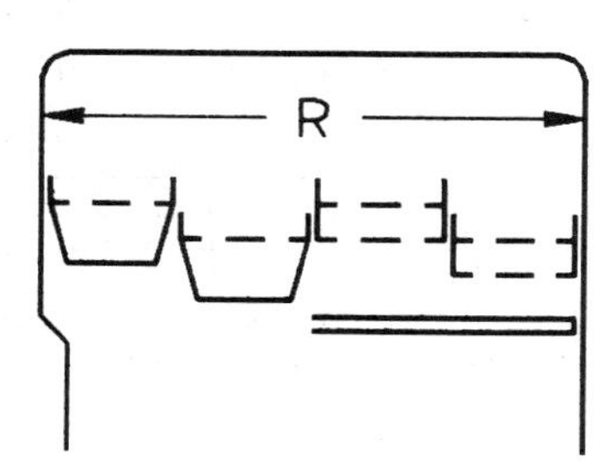

FIG. 9—WIDTH OF STRAW WALKERS

2.10.2.2 Separating grate arc length: Dimension A, Fig. 5. The method used to measure the concave arc length (see paragraph 2.6.2 and Fig. 5) shall be used to measure the separating grate arc length.

2.10.2.3 Separating grate area: The product of separating grate length and arc length expressed in square meters to the nearest hundredth. If there is more than one rotor with associated threshing and separating grates, this shall be stated. The threshing grate areas are additive as are the separating grate areas.

NOTE: The area that results from this calculation shall not be compared to the separating area that might be associated with a walker-type combine. The factors that cause separation to take place are radically different in the two systems, and presently there is no generally accepted factor, either empirical or theoretical, that can be used to relate the two.

2.10.3 Auxiliary separating devices: All devices such as rotors, forks, fingers, etc., that purport to augment the separating process may be mentioned but may not be used in calculating the walker or separating area of a combine.

2.10.4 Conveying devices: Devices that only convey material within the combine (e.g., grain pans, augers, raddles and other non-permeable conveyors) may contribute to good separation by stratifying the material, but they do no actual separating of grain from material-other-than-grain (mog). Their areas may not be classified as separating areas or added to the other separating areas described in paragraphs 2.6.4; 2.6.10; 2.7.3; and 2.8.3.

2.11 Cleaning devices: The main cleaning device is often referred to as the shoe. It is usually an oscillating mechanism, containing a number of adjustable and/or fixed sieves which, together with an air supply constitute the cleaning apparatus of a combine.

2.11.1 Pneumatic cleaning area: Where chaff and other mog are removed by aerodynamic means only, pneumatic cleaning area shall be calculated as the product of the width and depth of the air stream at the point of contact of air with the crop material. The area plane shall be normal to the mean direction of the air stream, and shall be expressed in square meters to the nearest hundredth. The location of the place where true pneumatic cleaning takes place is normally at the front of the shoe, prior to the point where the crop hits the top (or chaffer) sieve.

2.11.2 Cleaning sieve area: The area of each sieve shall be the product of its length and exposed width, measured in square meters to the nearest hundredth. These areas shall be additive to give the total cleaning sieve area of a combine.

2.11.3 Other cleaning areas: Dirt screens, recleaners, and other auxiliary cleaning devices provide cleaning areas. Permeable surfaces and pneumatic cleaning areas shall be calculated and expressed as specified in paragraphs 2.11.1 and 2.11.2.

2.11.4 For purposes of reporting combine specifications, the areas defined in paragraphs 2.6.4, 2.6.10, 2.7.3, 2.8.3, 2.10.1.2, 2.10.2.3, 2.11.1, 2.11.2, and 2.11.3 shall be listed separately and individually. Areas should not be used singly or in combination as a measure of a combine's capacity, performance or value.

2.12 Engine power: The corrected gross, rated, brake power in accordance with Society of Automotive Engineers Standard J816, Engine Test Code — Spark Ignition and Diesel, at the governed engine RPM which shall be stated. Where particular markets require the use of a different test code for determining engine power, the engine test code shall be stated.

2.13 Engine displacement: Engine displacement shall be expressed in liters to the nearest hundredth.

2.14 Turning radius: The distance from the turning center to the center of tire contact of the wheel describing the largest circle while the vehicle is executing its shortest turn without turning brakes in operation. The wheel base and guide wheel tread width shall be stated. Turning radius shall be expressed in meters to the nearest hundredth.

2.15 Clearance radius: The distance from the turning center to the outermost point of the combine executing its shortest turn without brakes in operation. If equipment options or attachments affect this dimension, such equipment shall be specified. The wheel base and guide wheel tread width shall be the same as in paragraph 2.14. Clearance radius shall be expressed in meters to the nearest hundredth.

2.16 Combine weight: The weight of a combine with the header removed, with the grain tank empty and with the fuel tank empty (or deduct weight of fuel). All accessories fitted, e.g., chopper, second cleaner, etc., shall be identified and their weight stated.

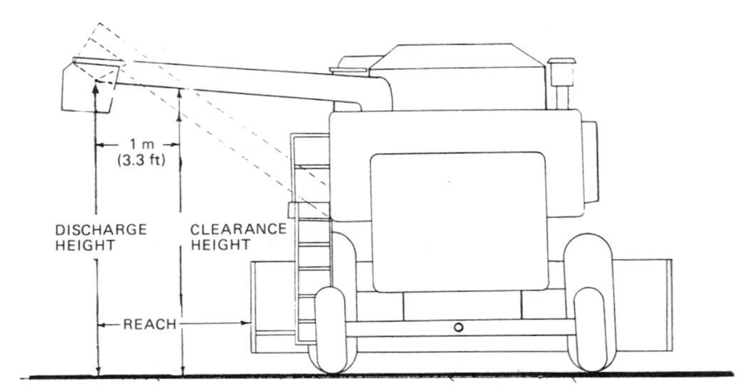

FIG. 10—GRAIN TANK UNLOADER SPECIFICATIONS

2.17 Combine length: The overall dimension from the foremost point to the rearmost point of the combine equipped for field operation measured parallel to the longitudinal centerline of the combine. The header installed shall be stated, and if other equipment options or attachments affect the length, such equipment shall be specified. Combine length shall be expressed in meters to the nearest hundredth.

2.18 Combine height: The vertical distance from the plane on which the combine is standing to the highest point on the combine. The height shall be measured under the conditions specified in paragraphs 2.4.2, 2.4.3, 2.4.4, and 2.4.5. The height with all components in position for transport and the height with all components in position for field operation shall be specified. Combine height shall be expressed in meters to the nearest hundredth.

2.19 Discharge height of unloader: The vertical distance from the plane on which the combine is standing to the lowest rigid point of the discharge opening with the unloader in operating position as shown in Fig. 10. The height shall be measured under conditions specified in paragraphs 2.4.2, 2.4.3, 2.4.4, and 2.4.5. Discharge height shall be expressed in meters to the nearest hundredth.

2.20 Clearance height of unloader: The vertical distance from the plane on which the combine is standing to a point on the underside of the unloader located at a horizontal distance of 1 m (3.3 ft) from the lowest point of the discharge opening as shown in Fig. 10. This height shall be measured under the conditions specified in paragraph 2.18, and expressed in meters to the nearest hundredth.

2.21 Reach of unloader: The horizontal distance measured perpendicular to the longitudinal centerline of the combine, from the lowest point of the unloader discharge opening to the outermost point of the header on the unloader side as shown in Fig. 10. The reach shall be measured under conditions specified in paragraph 2.18, and expressed in meters to the nearest hundredth.

SECTION 3—CROP TERMINOLOGY

3.1 For the purposes of this Standard, the term "grain" shall be taken as the general term of reference for the whole range of grains, seeds, legumes and fruits which are capable of being recovered from crops by a combine harvester.

3.2 Grain damage: For the purposes of this Standard, grain damage refers only to that attributable to the machine. It shall be expressed as the percentage by weight, to the nearest one-tenth, of damaged kernels in the sample.

3.2.1 Visible grain damage: Kernel damage where the seed coat appears broken to the naked eye.

3.2.2 Invisible grain damage: Kernel damage which requires instrumentation or special procedures for determination.

3.2.3 The applicable grain standard for the United States is Official United States Standards for Grain #810 published by the United States Department of Agriculture. Similar standards may apply in other countries.

3.3 Unthreshed heads: Any head, pod, cob, or part of same from which all or part of the seed has not been detached.

3.4 Returns: The material from the grain cleaning mechanism which is recirculated for reprocessing.

3.5 Material-other-than-grain (mog) to grain ratio: The total weight of material-other-than-grain (mog) divided by the total weight of grain in a sample.

3.6 Moisture content: Moisture content of the crop shall be expressed on the wet basis. The percentage moisture of the grain shall be determined from samples taken from the grain flow into the grain tank during the test runs. The mog samples shall be taken from the swath behind the combine applicable to a test run just made. Both samples shall be sealed in air-tight containers.

3.7 Plant length: The length of the plant from its base at ground level to its tip when the plant is straightened, expressed in millimeters.

3.8 Stubble length: The length of the straightened plant stalk still attached to the ground after the crop has been harvested expressed in millimeters. Where the stubble is lying flat, both its length and the height at which it was cut should be reported.

SECTION 4—COMBINE PERFORMANCE

4.1 Threshing: The detaching of seed from the head, cob or pod.

4.2 Separating: The isolating of detached seed, small debris, and unthreshed material from the bulk.

4.3 Cleaning: The isolating of desired seed from chaff, small debris and unthreshed material. The various stages in the cleaning process are listed in the order in which they occur.

4.3.1 Chaffing: The process of separating the grain from chaff and other mog on the top sieve by a combination of pneumatic and mechanical means.

4.3.2 Sieving: The isolation of desired seed by a mechanical device where the desired seed penetrates the device and the undesired material is carried over the device. Sieving is usually done on a lower sieve and may include the use of air.

4.3.3 Screening: The isolating of desired seed by a mechanical device, where the desired seed is carried over the device and the undesired material passes through it. This function is not employed in modern combines, except where rotary second cleaners and dirt/weed screens are used in elevators, grain pans, and auger troughs.

4.4 Feed rates

4.4.1 Grain feed rate: The weight of grain, including processing loss, passing through the combine per unit of time expressed in metric tons per hour (include leakage loss, if measured).

4.4.2 Material-other-than-grain (mog) feed rate: The weight of material-other-than-grain passing through the combine per unit of time expressed in metric tons per hour.

4.4.3 Total feed rate: The sum of grain feed rate and material-other-than-grain feed rate expressed in metric tons per hour.

4.5 Grain losses. Grain losses shall be classified according to their source, and shall include all losses attributable to the combine.

4.5.1 Gathering loss: The weight of grain and unthreshed grain that has been missed or dropped by the header or pick-up expressed as a percent of the sum of the grain feed rate and gathering loss feed rate. Care shall be exercised that natural losses caused by weather, birds, etc., prior to the harvesting of the crop, are not included in the gathering loss.

TABLE 1—CROPS AND CONDITIONS FOR CAPACITY TESTING

Crop	Acceptable range of mog to grain ratio	Range of moisture content, percent		Processing loss level, percent
		Grain	Mog	
Wheat	0.6 – 1.2	10-20	6-25	1
Barley	0.6 – 1.2	10-20	6-25	3
Rice	1.0 – 2.4	15-28	20-60	3
Sorghum	0.4 – 0.8	10-17	15-40	3
Corn/Maize	0.4 – 0.8	10-35	10-40	1
Rape/Canola	1.0 – 3.0	8-20	10-30	3
Soybeans	0.5 – 1.5	10-15	10-20	1

Note: If crop and/or climatic conditions do not permit test data to be obtained to the above standards, the circumstances shall be reported, and the actual results shall be recorded.

Because the functional characteristics of the conventional walker-type combines are so different from those of the axial, rotary, tangential-flow, or other non-conventional combines, no attempt shall be made to equate areas of these functional components from the one type of combine to another.

This is particularly dangerous when there is intent to use these areas to indicate, or imply, differences in capacity between combines of similar size.

4.5.2 Processing loss: The weight of threshed and unthreshed grain remaining in the material-other-than-grain, after the completion of the threshing, separation, and cleaning processes expressed as a percent of the grain feed rate.

4.5.3 Leakage loss: Any involuntary loss of grain from the combine, other than those described above, expressed as a percent of the grain feed rate.

4.6 Capacity

4.6.1 Combine capacity: The maximum sustained mog feed rate at which the processing loss level, with the combine infield operation on level ground, is as stated in Table 1. Capacity shall be expressed in metric tons per hour. Tests for determining combine capacity shall be conducted in accordance with ASAE Standard S396, Combine Capacity Test Procedure.

4.6.2 Grain output capacity: The maximum sustained rate of grain discharged from the combine's clean grain elevator at which the processing loss level, with the combine infield operation on level ground, is as stated in Table 1. Capacity shall be expressed in metric tons per hour.

Cited Standards:

ASAE S312, Capacity Designations for Combine Grain Tank Systems
ASAE S396, Combine Capacity Test Procedure
SAE J816, Engine Test Code—Spark Ignition and Diesel
USDA 810, Official United States Standards for Grain

ASAE Standard: ASAE S346.1 (SAE J884 NOV84)

LIQUID BALLAST TABLE FOR DRIVE TIRES OF AGRICULTURAL MACHINES

Developed by the Society of Automotive Engineers; administered through official liaison with SAE Tractor Tire Subcommittee; adopted by ASAE February 1972; revised March 1978; reconfirmed December 1982, December 1987.

SECTION 1—PURPOSE

1.1 This Standard is established for the purpose of providing average weight values for water or calcium chloride liquid filling solutions in drive tire sizes recommended in ASAE Standard ASAE S220, Tire Selection Tables for Agricultural Machines of Future Design. It is recommended that tires be filled to valve stem level while valve is located in the highest position.

Cited Standard:

ASAE S220, Tire Selection Tables for Agricultural Machines of Future Design

TABLE 1A—LIQUID BALLAST FOR DRIVE TIRES OF AGRICULTURAL MACHINES (CUSTOMARY UNITS)

Tire Size	Water		2 lb $CaCl_2$/gal H_2O			3-1/2 lb $CaCl_2$/gal H_2O			5 lb $CaCl_2$/gal H_2O		
	Slush Free, +32°F Solid, +32°F		Slush Free, +13°F Solid, −23°F			Slush Free, −12°F Solid, −52°F			Slush Free, −53°F Solid, −62°F		
	gal	Wt, lb	Water, gal	$CaCl_2$, lb	Total Wt, lb	Water, gal	$CaCl_2$, lb	Total Wt, lb	Water, gal	$CaCl_2$, lb	Total Wt, lb
9.5-16	12	100	11	22	114	10	35	118	10	50	133
12.4-16	21	175	19	38	196	18	63	213	17	85	227
18.4-16.1	49	409	44	88	455	42	147	497	39	195	520
8.3-24	13	108	11	22	114	11	39	131	10	50	133
9.5-24	17	142	16	32	165	15	53	178	14	70	187
11.2-24	24	200	22	44	227	20	70	237	19	95	253
12.4-24	30	250	28	56	290	26	91	308	25	125	333
13.6-24	38	317	34	68	352	32	112	379	30	150	400
14.9-24	47	392	43	86	445	40	140	474	38	190	507
16.9-24	61	509	55	110	569	52	182	616	49	245	654
17.5L-24	55	459	50	100	517	47	165	557	45	225	600
14.9-26	48	400	44	88	455	41	144	486	39	195	520
16.9-26	65	542	59	118	610	56	196	663	52	260	694
18.4-26	79	659	73	146	755	68	238	805	64	320	854
23.1-26	128	1068	117	234	1210	109	382	1291	103	515	1374
28L-26	157	1309	143	286	1479	134	469	1587	127	635	1694
11.2-28	27	225	25	50	258	24	84	284	22	110	293
12.4-28	35	292	32	64	331	30	105	355	28	140	374
13.6-28	43	359	39	78	403	37	130	439	35	175	467
14.9-28	53	442	49	98	507	46	161	545	43	215	574
16.9-28	69	575	63	126	651	59	207	699	56	280	747
18.4-28	84	701	77	154	796	72	252	852	68	340	907
14.9-30	57	475	52	104	538	48	168	568	46	230	614
16.9-30	73	609	67	134	693	63	221	746	59	292	787
18.4-30	89	742	82	164	848	77	270	912	72	360	960
23.1-30	143	1193	131	262	1355	123	431	1457	116	580	1547
24.5-32	170	1418	156	312	1613	146	511	1729	138	690	1841
16.9-34	82	684	75	150	775	70	245	829	66	330	880
18.4-34	100	834	91	182	941	85	298	1007	81	405	1081
20.8-34	128	1068	117	234	1210	109	382	1291	103	515	1374
23.1-34	159	1326	145	290	1499	136	476	1610	128	640	1708
13.9-36	51	425	47	94	486	44	154	521	42	210	560
13.6-38	57	475	52	104	538	49	172	581	46	230	614
15.5-38	66	550	60	120	620	56	196	663	53	265	707
16.9-38	90	751	82	164	848	77	270	912	73	365	974
18.4-38	110	917	100	200	1034	94	329	1113	89	445	1187
20.8-38	140	1168	128	256	1324	120	420	1421	114	570	1521

NOTE: This table is based on Type I Calcium Chloride (77 percent $CaCl_2$). For Type II (94 percent $CaCl_2$) reduce the "lb $CaCl_2$" weights by 25 percent.

TABLE 1B—LIQUID BALLAST FOR DRIVE TIRES OF AGRICULTURAL MACHINES (SI UNITS)

Tire Size	Water		239.7 g $CaCl_2$/1 H_2O			419.4 g $CaCl_2$/1 H_2O			599.3 g $CaCl_2$/1 H_2O		
	Slush Free, 0°C Solid, 0°C		Slush Free, −11°C Solid, −31°C			Slush Free, −24°C Solid, −47°C			Slush Free, −47°C Solid, −52°C		
	litre	Wt, kg	Water, l	$CaCl_2$, kg	Total Wt, kg	Water, l	$CaCl_2$, kg	Total Wt, kg	Water, l	$CaCl_2$, kg	Total Wt, kg
9.5–16	45	45	42	10	52	38	16	54	38	23	60
12.4–16	79	79	72	17	89	68	29	97	64	39	103
18.4–16.1	185	186	167	40	206	159	67	225	148	88	236
8.3–24	49	49	42	10	52	42	18	59	38	23	60
9.5–24	64	64	61	15	75	57	24	81	53	32	85
11.2–24	91	91	83	20	103	76	32	108	72	43	115
12.4–24	114	113	106	25	132	98	41	140	95	57	151
13.6–24	144	144	129	31	160	121	51	172	114	68	181
14.9–24	178	178	163	39	202	151	64	215	144	86	230
16.9–24	231	231	208	50	258	197	83	279	185	111	297
17.5L–24	208	208	189	45	235	178	75	253	170	102	272
14.9–26	182	181	167	40	206	155	65	220	148	88	236
16.9–26	246	246	223	54	277	212	89	301	197	118	315
18.4–26	299	299	276	66	343	257	108	365	242	145	387
23.1–26	485	485	443	106	549	413	173	586	390	234	623
28L–26	594	594	541	130	671	507	213	720	481	288	769
11.2–28	102	102	95	23	117	91	38	129	83	50	178
12.4–28	132	132	121	29	150	114	48	161	106	64	170
13.6–28	163	163	148	36	183	140	59	199	132	79	212
14.9–28	201	201	185	44	230	174	73	247	163	98	260
16.9–28	261	261	238	57	295	223	94	317	212	127	339
18.4–28	318	318	291	70	361	273	114	387	257	154	412
14.9–30	216	216	197	47	244	182	76	258	174	104	279
16.9–30	276	276	254	61	314	238	100	338	223	132	357
18.4–30	337	337	310	74	385	291	123	414	273	163	436
23.1–30	541	541	496	119	615	466	196	661	439	263	702
24.5–32	644	643	591	142	732	553	232	784	522	313	835
16.9–34	310	310	284	68	352	265	111	376	250	150	399
18.4–34	379	378	344	83	427	322	135	457	307	184	490
20.8–34	485	485	443	106	549	413	173	586	390	234	623
23.1–34	602	602	549	132	680	515	216	730	485	290	775
13.9–36	193	193	178	43	221	167	70	236	159	95	254
13.6–38	216	216	197	47	244	185	78	264	174	104	279
15.5–38	250	250	227	54	281	212	89	301	201	120	321
16.9–38	341	341	310	74	385	291	123	414	276	166	442
18.4–38	416	416	378	91	469	356	149	505	337	202	539
20.8–38	530	530	485	116	601	454	191	645	432	259	690

NOTE: This table is based on Type I Calcium Chloride (77 percent $CaCl_2$). For Type II (94 percent $CaCl_2$) reduce the "kg $CaCl_2$" weights by 25 percent.

ASAE Standard: ASAE S347.1

FLANGED STEEL BLOWER PIPE DIMENSIONS

Developed by Blower Pipe Subcommittee of ASAE Forage Harvesting Committee; approved by Power and Machinery Division Standards Committee; adopted by ASAE as a Tentative Standard February 1972; reconfirmed December 1972, December 1973, December 1974, December 1975, December 1976, December 1977, December 1978, December 1979; revised March 1981; reconfirmed December 1981, December 1982; reconfirmed and reclassified as a full Standard December 1983.

SECTION 1—PURPOSE AND SCOPE

1.1 This Standard establishes preferred dimensions for flanged steel blower pipe to facilitate interchange of equipment made by different manufacturers.

1.2 Pipe flanges for both bolted and clamped connections are included in this Standard. Configuration and construction of pipe, elbows, discharge deflectors and specialty pipe sections, other than inside pipe dimensions and connecting flange dimensions, are left to the discretion of the individual manufacturer.

SECTION 2—SPECIFICATIONS

2.1 Preferred dimensions for pipe and pipe flange connections are given in Table 1 and Fig. 1.

2.2 Bolted Connections

2.2.1 Four holes for the 230 mm (9 in.) pipe and 6 holes for the 280 mm (11 in.) pipe, or multiples thereof, (round, square or slotted at the discretion of the manufacturer) shall be used for each flange. The holes shall be of nominal diameter indicated in Table 1 and shall be equally spaced about the circumference defined by the bolt circle dimension in Table 1.

2.2.2 The minimum number and diameter of bolts for each pipe size connection are listed in Table 1. For stability and safety, the required number of bolts are to be used in holes equally spaced about the flange bolt circle.

2.3 Clamped Connections. Outside clamping devices shall be securely fastened so that contact with the silo, silo reinforcing bands, or other obstructions cannot result in release of the clamp.

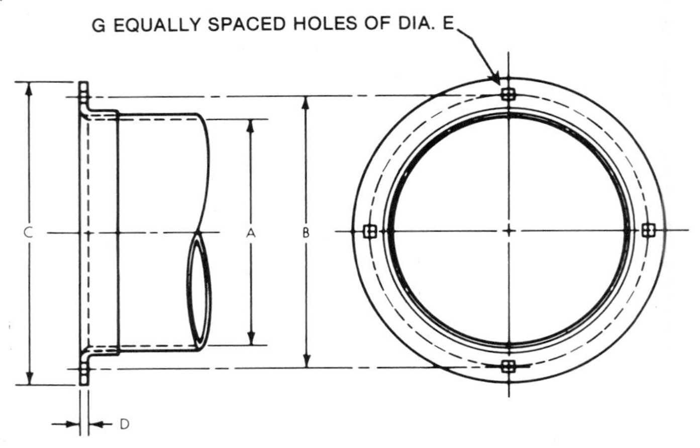

FIG. 1—TYPICAL PIPE AND PIPE FLANGE, See Table 1 for dimensions.

TABLE 1—DIMENSIONS FOR PIPE AND PIPE FLANGE CONNECTIONS

	Tolerance		230 mm (9 in.) Pipe		280 mm (11 in.) Pipe	
Dimension (See Fig. 1)	mm	(in.)	mm	(in.)	mm	(in.)
A — Pipe inside diameter	±1.5	(0.06)	230	(9.0)	280	(11.0)
B — Bolt circle diameter	±0.4	(0.015)	260	(10.2)	315	(12.4)
C — Flange outside diameter	±1.5	(0.06)	280	(11.0)	345	(13.6)
D — Flange thickness minimum			2.5	(0.10)	3.5	(0.14)
maximum			3.5	(0.14)	5.0	(0.19)
E — Bolt hole diameter	±0.4	(0.015)	9.0	(0.35)	9.0	(0.35)
F — Bolt diameter			8.0	(5/16)	8.0	(5/16)
G — Number of bolts			4		6	

TEST PROCEDURE FOR MEASURING HYDRAULIC LIFT FORCE CAPACITY ON AGRICULTURAL TRACTORS EQUIPPED WITH THREE-POINT HITCH

*Corresponds in substance to previous SAE J283.

Proposed by the Engineering Policy Committee of Farm and Industrial Equipment Institute; approved by ASAE Power and Machinery Division Standards Committee; adopted by ASAE February 1972; revised editorially May 1976; reconfirmed December 1976; revised April 1980; reconfirmed December 1984.

SECTION 1—PURPOSE AND SCOPE

1.1 The purpose of this Standard is to establish a common method of measuring and recording hydraulic lift force capacity referred to in ASAE Standard S217, Three-Point Free-Link Attachment for Hitching Implements to Agricultural Wheel Tractors (Reference: International Standard ISO 789/II, Test Code for Agricultural Tractors).

SECTION 2—DEFINITIONS

2.1 Terms used in this Standard are defined in ASAE Standard S217, Three-Point Free-Link Attachment for Hitching Implements to Agricultural Wheel Tractors.

SECTION 3—TEST PROCEDURE

3.1 With the lower links horizontal, set the mast height to 457 mm (18 in.) for Category I, to 483 mm (19 in.) for Category II, to 559 mm (22 in.) for Category III, and to 686 mm (27 in.) for Category IV. With the lower links horizontal, adjust the upper link such that the mast is vertical (See Fig. 1).

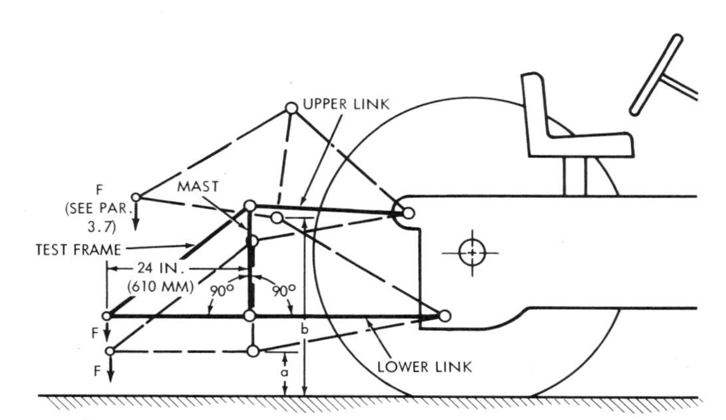

FIG. 1—TEST FRAME, See Table 1 for dimensions a, b.

3.2 A test frame shall be used which provides the mast height for each category in accordance with paragraph 3.1 and such that the lower hitch point spread is proper for the category being tested. [Category I, 681.0 to 684.3 mm (26.81 to 26.94 in.); Category II, 822.5 to 825.5 mm (32.38 to 32.50 in.); Category III, 963.7 to 966.7 mm (37.94 to 38.06 in.); Category IV, 1165 to 1168 mm (45.87 to 45.99 in.)]. The lift force application point shall be 610 mm (24 in.) to the rear of the lower hitch point with the lower links horizontal. The construction angle (the angle between the mast and the test frame load arm) shall be 90 deg (See Fig. 1). The center of gravity shall be at a point 610 mm (24 in.) to the rear of the hitch points, on a line at right angles to the mast and passing through the middle of the line joining the lower hitch points.

3.3 The tractor shall be rigidly supported at the height from ground level equivalent to the loaded radius of the largest R-1 rear tires and the corresponding largest front tires (See ASAE Standard S220, Tire Selection Tables for Agricultural Machines of Future Design). The tire sizes and loaded radii used shall be listed as part of the test data (Table 1). If the tractor tested cannot be conveniently supported at this radius, it is permissible to mathematically adjust the readings to the specified largest front and rear loaded radii.

3.4 The tractor should be operated at rated engine speed. The lift force capacity in this procedure shall be based upon the minimum specified system pressure. If the measured system pressure exceeds this value, mathematically adjust the lift force measured to the minimum specified system pressure level.

3.5 The tests shall be conducted so that lift force is measured throughout the total lift range as specified in Table 1 and in ASAE Standard S217, Three-Point Free-Link Attachment for Hitching Implements to Agricultural Wheel Tractors. Since the minimum power range is less than the total lift range, links may be adjusted during the tests to obtain lift force throughout the total lift range. If the links are readjusted to meet the specified lift range, this information shall be recorded in item 9 (See Fig. 2).

3.6 Tractors equipped with more than one upper link attaching point on the tractor should have the upper link attached to the category position recommended by the manufacturer for heavy draft implements such as mounted plows, etc.

3.7 Measure the vertical force, at static conditions, available from the 610 mm (24 in.) point on the test frame and include the weight of the test frame. This vertical lift force shall be measured at not less than 6 points equally spaced, including the maximum and the minimum points, throughout the total lift range.

3.8 The smallest value determined in paragraph 3.7 shall be the tractors minimum hydraulic lift force.

3.9 Data shall be recorded on a form similar to Fig. 2.

Cited Standards:

ASAE S209, Agricultural Tractor Test Code
ASAE S217, Three-Point Free-Link Attachment for Hitching Implements to Agricultural Wheel Tractors
ASAE S220, Tire Selection Tables for Agricultural Machines of Future Design
ISO 789/II, Test Code for Agricultural Tractors

TABLE 1—TRACTOR LIFT RANGE AND POWER RANGE

	Category I		Category II		Category III		Category IV	
	mm	in.	mm	in.	mm	in.	mm	in.
Maximum height for lowest position (Fig. 1 dimension a)	203	8	203	8	203	8	203	8
Minimum height for highest position (Fig. 1 dimension b)	813	32	914	36	1016	40	1120	44
Total lift range	610	24	711	28	813	32	914	36
Minimum power range	559	22	610	24	660	26	762	30

TRACTOR HYDRAULIC LIFT FORCE CAPACITY DATA

1. Tractor Make & Model: ____________________
2. Maximum Drawbar Power (Per ASAE S209): ____________________
3. Minimum Hydraulic Lift Force (See Par. 3.8): ____________________
4. Lift Force per Max. Drawbar Power (Line 3 ÷ Line 2): ____________________
5. Minimum Hydraulic Relief Valve Pressure Setting: ____________________
6. Tire Sizes: Front ____________________ Rear ____________________
7. Tire Loaded Radii (Use ASAE S220): Front ____________________ Rear ____________________
8. Category of Three-Point Hitch: Cat. I ☐ Cat. II ☐ Cat. III ☐ Cat. IV ☐
9. Force and Height Measurements of Par. 3.7:

Position of Hitch	Vertical Lift Force, N (lbf)	Measured Height of Balls Vertically Above Ground, mm (in.)
Lowest, 203 mm (8 in.)		
1/5 Point		
2/5 Point		
3/5 Point		
4/5 Point		
Highest		

FIG. 2—TYPICAL FORM FOR RECORDING LIFT FORCE DATA

ASAE Standard: ASAE S350 (ANSI/ASAE S350)

SAFETY-ALERT SYMBOL FOR AGRICULTURAL EQUIPMENT

Proposed by the Engineering Policy Committee of Farm and Industrial Equipment Institute; approved by ASAE Power and Machinery Division Standards Committee; adopted by ASAE as a Recommendation February 1972; reconfirmed and reclassified as a Standard March 1977; reconfirmed December 1981; approved by ANSI as an American National Standard December 1982; reconfirmed December 1986.

SECTION 1—PURPOSE

1.1 The purpose of this Standard is to establish a Safety-Alert Symbol for use on agricultural equipment, and to provide a symbol which means ATTENTION! BECOME ALERT! YOUR SAFETY IS INVOLVED!

SECTION 2—SCOPE

2.1 This Standard presents the general uses, limitations on use, and appearance of the Safety-Alert Symbol.

SECTION 3—DESCRIPTION

3.1 The Safety-Alert Symbol shall be an equilateral triangle with rounded corners and with an exclamation mark located in the center as shown in Fig. 1. The dimensions are optional.

3.2 The Safety-Alert Symbol should be of contrasting colors which will cause the exclamation mark to stand out.

SECTION 4—APPLICATION

4.1 The symbol should be used:

4.1.1 In conjunction with warning statements and signs.

4.1.2 In instruction manuals.

4.1.3 In connection with agricultural equipment safety standards.

4.1.4 On communications which concern agricultural equipment safety.

4.2 The symbol should *not* be used:

4.2.1 To indicate safety compliance or a safety characteristic.

4.2.2 Alone on equipment for safety purposes.

FIG. 1—SAFETY-ALERT SYMBOL

ASAE Standard: ASAE S354.2

SAFETY FOR PERMANENTLY INSTALLED FARMSTEAD EQUIPMENT

Proposed by the Farmstead Equipment Association of the Farm and Industrial Equipment Institute; approved by the ASAE Electric Power and Processing Division Standards Committee; adopted by ASAE as a Recommendation June 1972; revised and reclassified as a Standard April 1975; reconfirmed December 1979, December 1980, December 1981, December 1982; revised December 1983.

SECTION 1—PURPOSE AND SCOPE

1.1 This Standard is a guide to provide a reasonable degree of personal safety for operators and others involved during the normal operation and servicing of farmstead equipment. In addition to the design and configuration of equipment, hazard control and accident prevention are dependent upon the awareness, concern, and prudence of personnel involved in the operation and maintenance of equipment.

1.2 This Standard applies to powered farmstead equipment that is installed and used in one location. Excluded from this Standard are storage structures and equipment that can be classed as transportable or mobile or is periodically moved for use in another location. This Standard shall especially cover barn cleaners, cattle feeding equipment, farm feed conveyors, and silo unloaders which normally operate on top of the silage surface, however it is not necessarily limited to these items of farmstead equipment.

SECTION 2—DEFINITIONS

2.1 For the purpose of this Standard, the following definitions shall apply:

2.1.1 Farmstead equipment: Feed and manure handling equipment, including accessories, designed primarily for use in farmstead materials handling oeprations. This equipment normally is installed in a permanent manner and location.

2.1.2 Feed handling equipment: Equipment used to perform a variety of farm materials handling jobs associated with the placement of feed in storage, removal, conveying, mixing, grinding, processing and distributing feed for animal consumption.

2.1.3 Manure handling equipment: Equipment used to gather and convey animal wastes from the area where animals are confined, and equipment used to place, agitate and move animal wastes to or from a waste storage facility.

SECTION 3—SERVICING PROVISIONS

3.1 The following features shall be provided for the servicing of farmstead equipment:

3.1.1 Installation instructions shall recommend the inclusion of an electrical or mechanical means to positively prevent the application of electrical power to farmstead equipment.

3.1.2 An electrical or mechanical means to prevent and/or interrupt the application of power shall be provided at the location of handling equipment which is used on the top surface of materials within bulk storage structures.

3.1.3 Access shall be provided to areas requiring regular servicing, adjusting or inspection.

3.1.4 Installation instructions shall recommend units be installed to comply to American National Standard ANSI/NFPA 70, National Electrical Code.

SECTION 4—POWER DRIVEN EQUIPMENT

4.1 Power take-off and implement drive lines shall conform to applicable provisions of ASAE Standard S318, Safety for Agricultural Equipment, and ASAE Standard S207, Operating Requirements for Tractors and Power Take-Off-Driven Implements.

SECTION 5—SHIELDS OR GUARDS

5.1 Components shall be shielded in conformance with applicable provisions in ASAE Standard ASAE S318, Safety for Agricultural Equipment.

5.2 All revolving shafts, including projections such as bolts, keys, or set screws, shall be guarded with these exceptions:

5.2.1 Smooth shafts and shaft ends (without any projecting bolts, keys, or set screws) revolving at less than 10 rpm on feed handling equipment used on the top surface of materials in bulk storage facilities.

5.2.2 Smooth shaft ends protruding less than one-half the outside diameter of the shaft and its locking means.

5.3 Sweep arm material gathering mechanisms used on the top surface of materials within bulk material storage structures shall be provided with a bump guard or rail that is generally parallel to and extends the fullest practical length of the gathering mechanism. The bump guard or rail zone shall be within an area that is at least 152 mm (6 in.) but not greater than 762 mm (30 in.) in front of the gathering mechanism, and its height above the material surface shall be greater than one half but not higher than the horizontal distance in front of the gathering mechanism. Maximum bump guard or rail height shall not exceed 610 mm (24 in.) above the material surface. (See Fig. 1).

SECTION 6—SELECTION OF WINCHES AND WIRE ROPE FOR TOP SILO UNLOADER APPLICATIONS

6.1 Winch mechanisms shall include a self-braking feature to protect against inadvertent dropping of lifted units.

6.2 Winch drum diameter shall be no less than 7 times the wire rope diameter.

6.3 The root diameter of lifting pulleys shall be no less than 7 times the wire rope diameter.

6.4 Wire rope shall be corrosion resistance and selected for the design load, life and service intended. Wire rope used for lifting equipment shall be designed to provide a factor of safety of no less than 4 times when comparing lifted unit weight to breaking strength.

6.5 Testing: Wire rope testing shall be done by random selecting samples and loading each cable to one-fourth its minimum breaking

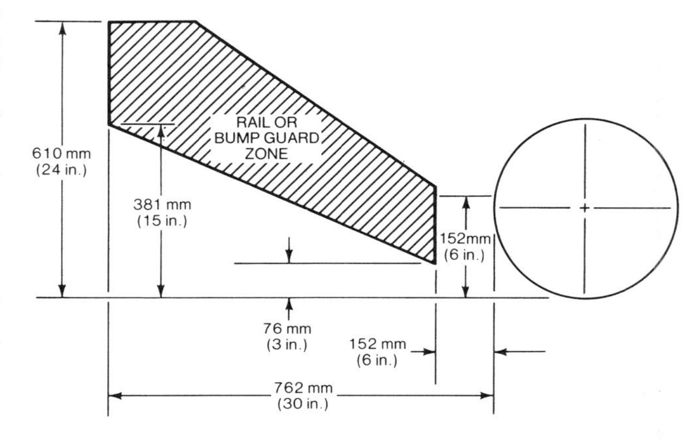

FIG. 1—BUMP GUARD OR RAIL ZONE

strength and by cycling it a minimum of 80 cycles at the approximate speed it will be used. This cable must pass over a pully (180 ° wrap) and winch (360 ° wrap) that is 7 times the cable diameter. After a minimum of 80 cycles, the tensil strength of the cable shall be equivalent to or greater than two times the working load of the cable.

SECTION 7—ELECTRICAL COMPONENTS

7.1 Electrical components including motors shall comply with ANSI/NFPA 70, National Electrical Code, and with the requirements of the authority having local jurisdiction.

7.2 Electric motors integrally equipped with thermal overload protection devices shall be of the manual reset type rather than the automatic time delay reset type, except where automatic reset is essential for functional requirements and no personal hazard is created.

SECTION 8—COMMUNICATIONS

8.1 Operator's manual shall be provided in accordance with ASAE Engineering Practice EP363, Technical Publications for Agricultural Equipment, and ASAE Standard S318, Safety for Agricultural Equipment, Section 3—Operator's Manual.

8.2 Safety signs shall be provided and displayed in accordance with ASAE Standard S318, Safety for Agricultural Equipment, Section 15—Safety Signs.

Cited Standards:

ASAE S207, Operating Requirements for Tractors and Power Take-Off-Driven Implements
ASAE S318, Safety for Agricultural Equipment
ASAE EP363, Technical Publications for Agricultural Equipment
ANSI/NFPA 70, National Electrical Code

ASAE Standard: ASAE S355.1

SAFETY FOR AGRICULTURAL LOADERS

Developed by ASAE Agricultural Loader Subcommittee; approved by the ASAE Power and Machinery Division Standards Committee; adopted by ASAE as a Recommendation June 1972; revised and reclassified as a Standard December 1976; reconfirmed December 1981, December 1986.

SECTION 1—PURPOSE AND SCOPE

1.1 The purpose of this Standard is to provide a uniform method of warning owners and operators of agricultural tractors equipped with loaders of the hazards encountered, and safety guides to be observed in the operation and servicing of such equipment.

1.2 In addition to the design and configuration of equipment, hazard control and accident prevention are dependent upon the awareness, concern and prudence of personnel involved in the operation, transport, and maintenance of equipment.

SECTION 2—INSTRUCTIONS FOR SAFE OPERATION OF LOADER EQUIPPED TRACTORS

2.1 A safety sign, Fig. 1, or a similar statement of guides that should be observed in the normal operation of loaders mounted on agricultural tractors, shall be affixed to a portion of the loader in such a manner that will be readily visible to the operator, either seated in the operating position or as he mounts the machine.

2.2 The safety sign shall be of the colors and material quality specified in ASAE Standard ASAE S318, Safety for Agricultural Equipment. The signal word CAUTION shall be used.

2.3 A more complete list of potential hazards and safety guides associated with loader operation and service shall be included in the instruction manual for the loader. Following is a typical list of hazards and safety guides. It is not intended to be an all-inclusive listing for all machines for all conditions. Individual statements may be used throughout the loader instruction manual.

2.3.1 Improper use of a loader can cause serious injury or death.

2.3.2 If the machine is equipped with a rollover-protective structure (ROPS), fasten seat belt prior to starting the engine.

2.3.3 Operate the loader from the operator's seat only.

2.3.4 Add recommended wheel ballast or rear weight for stability.

2.3.5 Move wheels to widest recommended settings to increase stability.

2.3.6 For better stability, use a tractor with wide front axle rather than narrow front wheels.

2.3.7 Do not lift or carry anyone on loader or in bucket or attachment.

2.3.8 Move and turn tractor at low speeds.

2.3.9 Carry loader arms at a low position during transport.

2.3.10 Avoid loose fill, rocks and holes; they can be dangerous for loader operation or movement.

2.3.11 Be extra careful when working on inclines.

2.3.12 Avoid overhead wires or obstacles when loader is raised.

2.3.13 Allow for the loader length when making turns.

2.3.14 Stop the loader arms gradually when lowering or lifting.

2.3.15 Use caution when handling loose or shiftable loads.

2.3.16 Do not walk or work under raised loader or bucket or attachment unless it is securely blocked or held in position.

2.3.17 Lower loader arms when parking or servicing.

2.3.18 Make sure all parked loaders on stands are on a hard, level surface and engage all safety devices.

2.3.19 Visually check for hydraulic leaks and broken, missing, or malfunctioning parts and make necessary repairs.

2.3.20 Escaping hydraulic oil under pressure can have sufficient force to penetrate the skin, causing serious personal injury. If injured by escaping fluid, obtain medical treatment immediately.

2.3.21 Before disconnecting hydraulic lines, relieve all hydraulic pressure.

2.3.22 Be certain anyone operating the loader is aware of safe operating practices and potential hazards.

Cited Standard:

ASAE S318, Safety for Agricultural Equipment

CAUTION

LOADER SAFETY GUIDES

1. MOVE AND TURN TRACTOR AT LOW SPEEDS.
2. CARRY LOADER ARMS AT A LOW POSITION DURING TRANSPORT.
3. LOWER LOADER ARMS, STOP ENGINE AND LOCK BRAKES BEFORE LEAVING OPERATOR SEAT.
4. DO NOT STAND OR WORK UNDER RAISED LOADER.
5. ADD RECOMMENDED WHEEL BALLAST OR REAR WEIGHT FOR STABILITY.
6. MOVE WHEELS TO WIDEST RECOMMENDED SETTINGS TO INCREASE STABILITY.
7. OBSERVE SAFETY RECOMMENDATIONS IN LOADER INSTRUCTION MANUAL.

FIG. 1—TYPICAL SAFETY SIGN FOR AGRICULTURAL LOADERS.

ASAE Standard: ASAE S356.1 (SAE J39 JUN83)

T-HOOK SLOTS FOR SECUREMENT IN SHIPMENT OF AGRICULTURAL EQUIPMENT

Proposed by Engineering Policy Committee of Farm and Industrial Equipment Institute; approved by ASAE Power and Machinery Division Standards Committee; adopted by ASAE June 1972; reconfirmed December 1976; revised March 1980; reconfirmed 1984.

SECTION 1—PURPOSE

1.1 To provide a means of securement in shipment of agricultural equipment on railroad flat cars and flat-bed trucks by utilizing a T-hook and slot method.

1.2 This proposal is intended to reduce the possibility of damage to material shipped and to provide increased economy in shipping.

SECTION 2—SCOPE

2.1 To provide standardized restraining slots, in accordance with product weight category and test requirements, to be incorporated in agricultural equipment that will be shipped, utilizing a T-hook and slot method of tiedown.

SECTION 3—SPECIFICATIONS

3.1 Slot dimensions are shown in Fig. 1.

3.2 Restraining slot shall be capable of withstanding impacts, in accordance with Association of American Railroads, Recommended Procedure for Conducting Field Impact Tests of Loaded Freight Cars.

3.3 T-hooks are designed to secure through product up to 19 mm (0.75 in.) thick. They shall be provided by the originating carrier.

3.3.1 Attached T-hook commonly used on rail equipped flat cars is shown in Fig. 2.

3.3.2 Grab T-hook commonly used on truck shipments is shown in Fig. 3.

References: Last printed in the 1981 AGRICULTURAL ENGINEERS YEARBOOK; list available from ASAE Headquarters.

Cited Standard:

AAR, Recommended Procedure for Conducting Field Impact Tests of Loaded Freight Cars

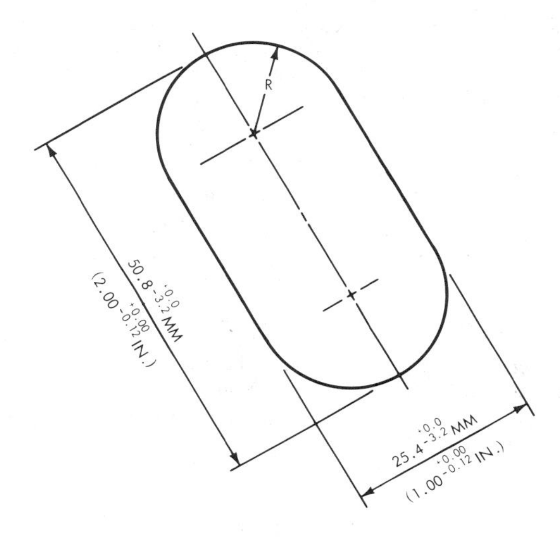

FIG. 1—RESTRAINING SLOT FOR T-HOOK

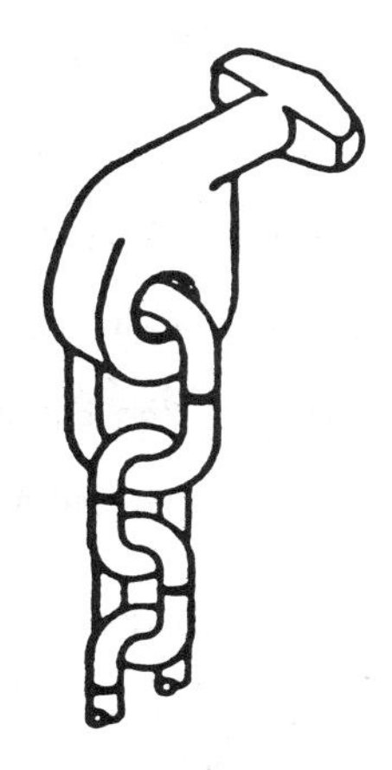

FIG. 2—ATTACHED T-HOOK

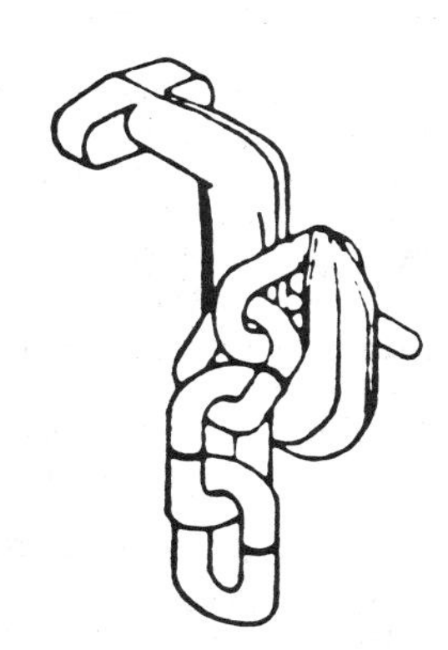

FIG. 3—GRAB T-HOOK

TEST PROCEDURE FOR DETERMINING THE LOAD CARRYING ABILITY OF FARM WAGON RUNNING GEAR

Developed by the ASAE Farm Wagon Subcommittee; approved by Power and Machinery Division Standards Committee; adopted by ASAE as a Tentative Standard February 1973; reconfirmed December 1973; reclassified as a full Standard December 1974; reconfirmed December 1979, December 1984.

SECTION 1—PURPOSE AND SCOPE

1.1 The Standard provides a uniform method for determining capacity ratings of farm wagon running gear. The Standard provides minimum acceptable performance criteria.

1.2 The Standard relates to farm wagon running gear having 2 or more axles. Steering means shall be auto-steer or fifth wheel type.

SECTION 2—GENERAL DESCRIPTION OF TEST

2.1 A wagon running gear is loaded with a load which is 110 percent of the load rating to be tested as outlined in Section 3. It is then run over a single ramp constructed as outlined in Section 7, in a manner described in Section 8, at a temperature not lower than 32 deg F (0 deg C). The wagon running gear is considered to pass the test for the rating indicated if it meets the criteria defined in Section 9.

SECTION 3—RATINGS AND LOADING METHODS

3.1 The ratings to be confirmed by testing shall be determined prior to testing to establish the test load. Published ratings shall be to the nearest 1000 lb or 500 kg.

3.2 The rating to be confirmed by testing shall include the weight of the wagon running gear, tires, and what they carry. For example, a wagon running gear rated for 10,000 lb (4536 kg) will weigh 10,000 lb (4536 kg) on the ground when loaded.

3.3 Wagon running gear to be tested shall be equipped with a framework for holding the load to be applied and shall be constructed:

3.3.1 To position the center of gravity of the applied load a minimum of 24 in. (610 mm) above the top of the bolster. (See Fig. 1).

3.3.2 To not reinforce the reach pole and to follow wagon running gear manufacturers' recommendations for securement to front and rear bolsters. The framework may be flexible enough to stay on the bolster throughout the test.

3.4 The applied load shall bring the weight upon the ground up to 110 percent of the rated load being tested.

3.5 Load distribution shall be equally distributed on all tires ± 5 percent.

3.6 Bow of the reach, toe-in and camber of all wheels when loaded as in paragraph 3.4 shall be measured and recorded.

SECTION 4—WAGON RUNNING GEAR AND EQUIPMENT

4.1 Wagon running gear shall be equipped as it is intended to be sold. Optional equipment that affects the load carrying ability of the wagon running gear shall be included if the wagon running gear is to be rated with this optional equipment.

4.2 Wheel base

4.2.1 Ratings 12,000 lb (5443 kg) and under shall be run at 96 ± 2 in. (2438 ± 51 mm).

4.2.2 Ratings over 12,000 lb (5443 kg) shall be run at 108 ± 2 in. (2743 ± 51 mm).

4.3 Wheel track width. The maximum setting recommended by the wagon running gear manufacturer shall be used.

4.4 Tires and tire pressure. Tires shall be selected in accordance with the recommended load limits and maximum inflation pressures published by the Tire and Rim Association for a maximum speed of 5 mph for implement tires and low section height implement tires used in field service. Tires shall be inflated to a cold inflation pressure adequate for carrying the loads determined in paragraph 3.5.

SECTION 5—TOWING VEHICLE

5.1 The towing vehicle shall be capable of pulling and backing the loaded farm running gear over the test ramp as outlined in Section 8.

5.2 Drawbar height shall be 15 to 17 in. (381 to 432 mm) above the ground.

SECTION 6—SPEED

6.1 The wagon running gear should be pulled and backed over the test ramp at a slow speed, approximately 0.5 mph (0.8 km per hr), to reduce the effect of impact. Faster speeds may be used, but impose high loads upon the wagon running gear.

SECTION 7—RAMP AND SURROUNDING APRON

7.1 A single ramp of durable material shall be constructed on relatively level ground to the dimensions shown in Fig. 2.

7.2 dimensions of the ramp relative to the surrounding apron shall be maintained.

SECTION 8—TEST PROCEDURE

8.1 The towing vehicle shall pull and back the loaded wagon running gear over the ramp in the following sequence as shown in Fig. 3.

8.1.1 Pull entire wagon running gear up 45 degree slope and down 30 degree slope at 90 ± 15 degree approach angle.

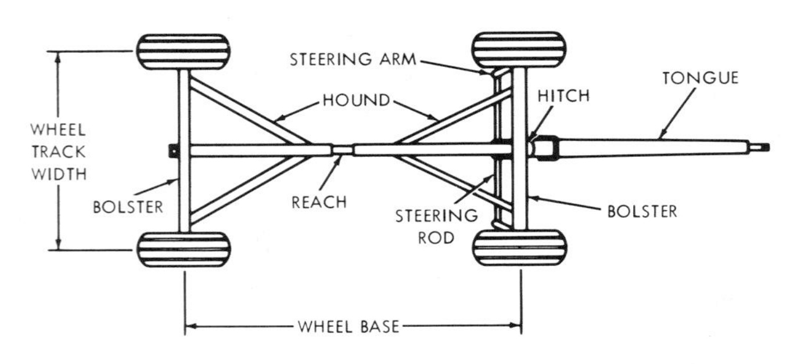

FIG. 1—NOMENCLATURE OF FARM WAGON RUNNING GEAR

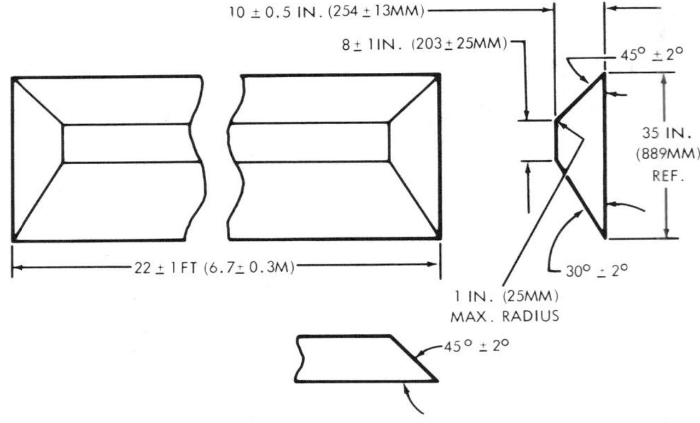

FIG. 2—TEST RAMP

8.1.2 Back entire wagon running gear up 30 degree slope and down 45 degree slope at 90 ± 15 degree approach angle.

8.1.3 Pull entire wagon running gear up 45 degree slope and down 30 degree slope at 45 ± 15 degree approach angle with right front wheel contacting ramp first.

8.1.4 Back entire wagon running gear up 30 degree slope and down 45 degree slope at 45 ± 15 degree approach angle with left rear wheel contacting ramp first.

8.1.5 Pull entire wagon running gear up 45 degree slope and down 30 degree slope at 45 ± 15 degree approach angle with left front wheel contacting ramp first.

8.1.6 Back entire wagon running gear up 30 degree slope and down 45 degree slope at 45 ± 15 degree approach angle with right rear wheel contacting ramp first.

8.1.7 Pull right wheels up end of ramp, go full length and down other end.

8.1.8 Pull left wheels up end of ramp, go full length and down other end.

8.2 The test sequence outlined in paragraph 8.1 shall be repeated 3 times for a total of 24 passes.

SECTION 9—CRITERIA FOR ACCEPTANCE

9.1 A farm wagon running gear is considered acceptable for the load rating being tested if it remains operational under the test load. The following limits shall not be exceeded after the test is completed. All measurements shall be made while the wagon running gear is still carrying the test load.

9.1.1 Bow of the reach shall be the original bow ± 1.0 in. (± 25 mm).

9.1.2 Tow-in shall not change more than 3 degrees from original toe-in.

9.1.3 Camber shall not change more than 3 degrees from original camber.

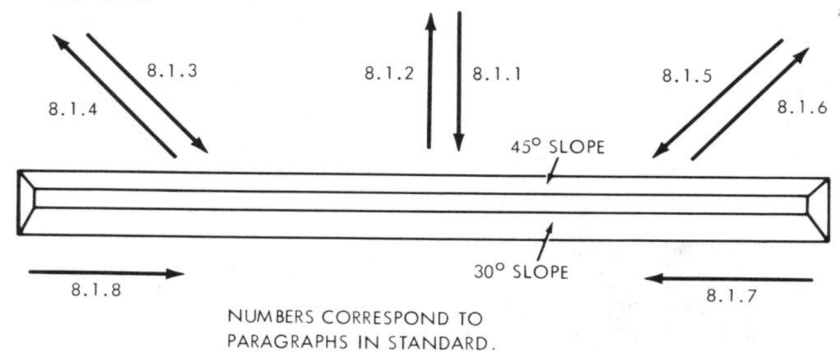

FIG. 3—TEST PROCEDURE SEQUENCE AND DIRECTIONS

ASAE Standard: ASAE S361.2

SAFETY FOR AGRICULTURAL AUGER CONVEYING EQUIPMENT

Proposed by the Auger Safety Committee of Farm and Industrial Equipment Institute; approved by ASAE Power and Machinery Division Standards Committee; adopted by ASAE as a Tentative Standard February 1973; reconfirmed December 1973; revised February 1975; reconfirmed December 1975, December 1976, December 1977, December 1978, December 1979, December 1981, December 1982; revised December 1983; reconfirmed December 1984; reconfirmed and reclassified as a full Standard December 1985.

SECTION 1—PURPOSE

1.1 The purpose of this Standard is to establish safety recommendations which will minimize the possibility of injury during normal operation of auger conveying equipment used to convey agricultural materials on farms.

SECTION 2—SCOPE

2.1 This Standard covers only portable farm augers and their related accessories designed primarily for conveying agricultural materials on farms. For definitions see ASAE Standard S374, Terminology and Specification Definitions for Agricultural Auger Conveying Equipment.

SECTION 3—ELECTRICAL SPECIFICATIONS

3.1 Conductors, connectors and grounding. The path to ground from motors, controls and electrical enclosures shall be continuous; shall have ample carrying capacity to conduct safely any currents liable to be imposed on it; and shall have impedance sufficiently low to limit the potential above ground and to facilitate operation of control and over-current devices in the circuit. Where separable connectors are used, the grounding conductor shall be the first to make contact and the last to break contact. Motor disconnect switches and conductor cables shall comply with American National Standard ANSI/NFPA 70, National Electrical Code, and any local codes which apply.

3.1.1 Flexible cords and cables may be used. Ratings of separable connectors and conductors shall be at least as great as motor-running current rating for motor leads, and at least as great as control current rating for control leads. Separable connectors shall be constructed and installed to guard against inadvertent contact with live parts.

3.2 Electrical controls and wiring

3.2.1 Motors operating at less than 30 volts. Controls shall be conveniently located and so arranged that the motor(s) will start only by button or lever actuation by the operator. Overload protection shall be of the "manual reset" type. Controls shall automatically go to the OFF mode in case of circuit interruption.

3.2.2 Motors operating at more than 100 volts. Controls shall be conveniently located and so arranged that the motor(s) will start only by button or lever actuation by the operator, except in automatic systems. Overload protection shall be the "manual reset" type. Motor starting devices shall be so arranged that controls will automatically go to the OFF mode in case of power interruption, conductor fault, low voltage, or circuit interruption. Overload reset and motor-starting control station should be so located that the operator can see that all personnel are clear of the equipment.

3.3 Electric motors

3.3.1 Motor rating. Motors should be selected for each auger unit with no less than the minimum mechanical power rating as recommended by the equipment manufacturer.

3.3.2 Motor enclosure. Motors should be enclosed to exclude entry of moisture, dirt and foreign objects but allow air fan cooling. Enclosures must suit the environment where the motor will be used. Special enclosures should be used in hazardous or explosive atmospheres, such as enclosed bins with grain dust suspended in the air.

SECTION 4—GENERAL SPECIFICATIONS

4.1 All shields and guards for belts, pulleys and sprockets shall conform to ASAE Standard S318, Safety for Agricultural Equipment.

4.2 Implement input driveline guards shall conform to ASAE Standard S318, Safety for Agricultural Equipment.

4.3 Drive shaft shielding. All power driven rotating drive shafts shall be guarded in accordance with ASAE Standard S318, Safety for Agricultural Equipment.

4.4 Hopper flighting guard. The exposed flighting shall be covered. Openings in grating type covers shall have their largest dimension no greater than 121 mm (4.75 in.). The area that includes such openings shall be no larger than 6450 mm^2 (10 in.^{2}). Slotted openings in solid baffle style covers shall be no wider than 38 mm (1.5 in.) or closer than 89 mm (3.5 in.) to the exposed flighting.

4.5 Intake guard. The intake shall be guarded or otherwise designed to provide a deterrent from accidental contact with the rotating flighting. The guard shall cover the top 180 deg of the inlet area and extend a minimum of 64 mm (2.5 in.) above and below the exposed flighting. Openings in the guard, for the free flow of material, shall have their largest dimension no larger than 121 mm (4.75 in.). The area of each opening shall be no larger than 6450 mm^2 (10 in.^{2}). The guard shall be no closer to the rotating flighting than 64 mm (2.5 in.) and be of sufficient strength to support a 136 kg (300 lb) person without permanent deformation. (See Fig. 1)

4.6 Drag auger shielding. Drag augers should have the handle end and the top half shielded by a guard. Openings in such guards shall have their largest dimension no greater than 121 mm (4.75 in.). The guard openings shall be located no closer to the rotating flighting than 64 mm (2.5 in.). The area of each opening shall be no larger than 8450 mm^2 (10 in.^{2}).

4.7 Lateral stability. The wheel tread width shall be of sufficient span that, with the auger in the lower transport position, static side tipping will not occur on slopes below the angle of 20 deg.

4.8 Tube restraint. To avoid inadvertent separation, a positive restraint shall be provided between the auger tube and the undercarriage lifting arm. Stops that restrict the maximum raised angle and the minimum lowered angle shall be provided.

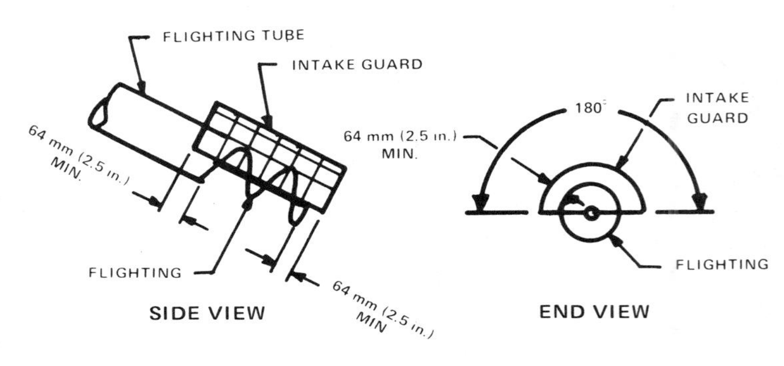

FIG. 1—INTAKE GUARD

4.9 Winch

4.9.1 Winch drum. The diameter of the winch drum center shall be no less than 10 times the wire rope diameter.

4.9.2 Hand winch. The hand winch shall be provided with a control which will hold the auger at any angle of inclination and respond only to handle actuation. It shall not be necessary to disengage such control to lower the auger. The force required on the handle to raise or lower the auger manually shall not exceed 222 N (50 lb).

4.9.3 Electric winch. Controls shall be conveniently located and so arranged that the winch will operate only during switch or lever actuation by the operator. The actuating switch or lever shall automatically return to the OFF mode when the operator releases the actuating switch or lever. Overload protection shall be provided.

4.10 Wire rope (cable). Wire rope shall be rust resistant and selected for the design load and service intended. Wire ropes and their anchors used for lifting the auger tube into the raised operating position shall be designed with a safety factor, working load compared to breaking strength, of no less than 5. Wire ropes and their respective anchors used as structural supports for the auger tube shall be designed with a safety factor, working load compared to breaking strength, of no less than 3. All wire rope fastening devices shall be in accordance with wire rope manufacturer's recommendations.

4.11 Pulleys. The wire rope lifting pulleys shall be grooved to fit the wire rope with which they are used. Their pitch diameter shall be no less than 10 times the wire rope diameter. A factor of safety of no less than 5 shall be used when comparing working load to breaking strength of the pulleys and pulley anchors.

SECTION 5—SAFETY SIGNS

5.1 Safety signs shall be appropriately displayed on the machinery to alert the operator and others of the risk of personal injury during normal operation and servicing, and should comply with ASAE Standard S318, Safety for Agricultural Equipment. A similar statement of warning should also be included in the operator's manual.

5.2 Typical caution sign information for portable augers is as follows:

CAUTION

1. Read and understand the operator's manual before operating.
2. Keep all safety shields and devices in place.
3. Make certain everyone is clear before operating or moving the machine.
4. Keep hands, feet and clothing away from moving parts.
5. Shut off power to adjust, service or clean.
6. Support discharge end or anchor intake end to prevent upending.
7. Disconnect power before resetting motor overload.
8. Empty auger before moving to prevent upending.
9. Lower auger to transport position for transporting.
10. Make certain electric motors are grounded.

5.3 Danger signs should be located at or near a serious hazard. Typical danger sign information for auger intakes is as follows.

DANGER

Keep hands and feet away from intake.

Cited Standards:

ASAE S318, Safety for Agricultural Equipment
ASAE S374, Terminology and Specification Definitions for Agricultural Auger Conveying Equipment
ANSI/NFPA 70, National Electrical Code

ASAE Engineering Practice: ASAE EP363.1

TECHNICAL PUBLICATIONS FOR AGRICULTURAL EQUIPMENT

Proposed by the Farm and Industrial Equipment Institute Engineering Policy Committee; approved by the ASAE Power and Machinery Division Standards Committee; adopted by ASAE as a Recommendation December 1973; reconfirmed and reclassified as an Engineering Practice December 1978; approved by ANSI as an American National Standard December 1982; revised December 1982; withdrawn as an ANSI Standard February 1985; reconfirmed December 1987.

SECTION 1—PURPOSE

1.1 The purpose of this Engineering Practice is to assist in the development of technical publications which will stress safety and uniformity of format and content.

1.1.1 This Engineering Practice should be accepted as a desirable guideline for production of techical publications based upon current knowledge, the state-of-the-art, and prevailing requirements.

SECTION 2—SCOPE

2.1 This Engineering Practice concerns technical publications which are directed to individuals responsible for the proper unloading, set-up, installation, pre-delivery inspection, operation and servicing of agricultural equipment (see ASAE Standard S390, Classifications and Definitions of Agricultural Equipment). Advertising and pre-purchase publications are not included.

SECTION 3—DEFINITIONS

3.1 The instructions defined below may be published under one cover, separately, or in any combination desired. The selection and grouping of instructions may be dictated by the type of equipment involved.

3.1.1 Pre-delivery set-up and/or installation instructions: Outline in detail the procedures for properly preparing and/or installing new equipment for delivery to the customer.

3.1.2 Operation instructions: Explain the procedures for proper operation of the equipment.

3.1.3 Maintenance Instructions: Explain the maintenance required and the methods of performing matintenance on the equipment, including recommendations for lubricants, fuels, and other fluids as appropriate.

3.1.4 Repair/overhaul instructions: Provide detailed procedures for properly diagnosing, repairing, or overhauling the equipment.

3.1.5 Installation and/or modification instructions: Outline in detail the procedures for properly installing components and/or performing inspections or modifications on existing equipment.

3.1.6 Parts information: Lists and identifies all replaceable parts needed to service the product, including safety signs and any optional equipment.

3.1.7 Other instructions: Publications required to supplement instructions listed in paragraphs 3.1.1 through 3.1.6.

SECTION 4—PUBLICATION SPECIFICATIONS

4.1 Publication identification. Each publication should identify the manufacturer's or distributor's name, the model designation of the machine, the name or type of publication, the part number or publication number by which it may be ordered, and the printing or publication date.

4.2 Elements. Where applicable, each publication should contain a foreword or introduction, a description and illustration of the machine, a table of contents listing major areas of the machine, a safety section, subject material, and an index for identifying individual components of the machine. (For details, refer to Section 6—Content of Publications.)

4.2.1 Each publication should contain a statement advising where to get assistance if items covered in the publication are not understood.

4.3 Page Layout. Illustrations and/or appropriate safety messages should be on the same page as related text or on adjacent pages and may be at other appropriate locations.

4.4 Media. Any publication may be issued in printed form and/or on microfiche.

4.4.1 The preferred size of printed publications is 210 X 297 mm or 8.5 X 11 in. Other sizes may be considered if more suitable.

4.4.2 The preferred microfiche size is 105 X 148 mm (4.1 in. X 5.8 in.).

4.4.3 The preferred microfiche reduction factor is 42X.

4.4.4 Each microfiche should have a header (title stripe). (Refer to paragraph 4.1.)

4.4.5 Microfiche should be suitable for magnification on commercially available equipment.

4.4.6 Microfiche images should read from left to right and from top down. The "A" row should be at the top. Images should be "right reading" with the header (title stripe).

4.5 Legibility. All information shall be presented in such a way that it can be easily read and understood, and where applicable contain illustrations to supplement the written information.

4.6 Durability. The medium shall have reasonable durability when exposed to conditions of rain, dampness, or grease during its intended use.

4.7 Shipment. Packing containers shall be provided which will reasonably protect the publication during shipment.

4.8 Emerging technology. Nothing in this Engineering Practice shall limit or restrict the use of developing technology for the publication, display, and/or communication of technical information.

SECTION 5—SAFETY CONTENT

5.1 All publications shall provide safety information, if applicable.

5.1.1 Safety-Alert Symbol. The Safety-Alert Symbol shall be displayed and defined as in ASAE Standard ASAE S350, Safety-Alert Symbol for Agricultural Equipment, and its use in the publication explained to the reader. This information shall precede the safety messages in the safety section as defined under paragraph 5.1.2.

5.1.2 Safety section. A safety section shall precede the primary text or other functional information in all applicable publications. This section shall contain safety-related information, and safety messages of a general nature (see ASAE Standard ASAE S318, Safety for Agricultural Equipment).

5.1.3 Specific safety messages. Safety messages that are specific to particular hazards shall appear in appropriate sections of the text, and may also be displayed in the safety section (see ASAE Standard S318, Safety for Agricultural Equipment). Specific safety messages shall precede functional instructions to which they apply.

5.1.4 Signal words. Each safety message shall be accompanied by the Safety-Alert Symbol and one of the signal words "Caution", "Warning", or "Danger"; however, when repeating a safety message from a "Caution", "Warning", or "Danger" safety sign on the machine, the same signal word used on the safety sign shall appear with the safety message in the applicable publication (see ASAE Standard ASAE S318, Safety for Agricultural Equipment).

5.1.5 Safety signs. Safety signs that appear on the equipment shall be reproduced in legible size in all applicable publications, either in the appropriate section of the text relative to the point of use, in the safety section, and/or in a separate safety sign section. Safety signs may be reproduced in black and white even if they appear in color on the machine (see ASAE Standard ASAE S318, Safety for Agricultural Equipment).

All applicable publications shall also include:

5.1.5.1 Instructions on the need for keeping safety signs legible.

5.1.5.2 Instructions that safety signs must be replaced if they are missing or illegible.

5.1.5.3 Instructions that new equipment components installed during repair shall include the current safety signs specified by the manufacturer, to be affixed to the replaced component.

5.1.5.4 Information on how to obtain replacement safety signs.

5.1.5.5 Instructions on how to affix safety signs.

5.1.5.6 Information on the location of each safety sign on the equipment.

SECTION 6—CONTENT OF PUBLICATIONS

6.1 Pre-delivery, set-up and/or installation instructions. These shall meet all applicable specifications as defined in Section 4—Publication Specification and Section 5—Safety Content. In addition, they may include any or all of the following specific information:

6.1.1 General description of the equipment/machinery.

6.1.2 Shipping information.

6.1.3 Unloading instructions.

6.1.4 Assembly and/or installation instructions.

6.1.5 Pre-delivery and delivery service check list.

6.1.6 Start-up and operation instructions.

6.2 Operation instructions. These shall meet all applicable specifications as defined in Section 4—Publication Specifications and Section 5—Safety Content. In addition, they may include any or all of the following specific information:

6.2.1 General specifications and description of the equipment/machinery.

6.2.2 Operator responsibility (may be defined, as appropriate) (see ASAE Standard ASAE S318, Safety for Agricultural Equipment).

6.2.3 Identification of indicators and controls (see ASAE Standard ASAE S304, Symbols for Operator Controls on Agricultural Equipment).

6.2.3.1 Illustrations identifying indicators and controls relative to the operator's positions.

6.2.3.2 Separate detailed illustrations and explanation if purpose, function, and/or usage is not readily evident.

6.2.3.3 Illustrations and explanations of all symbols used on the equipment/machinery.

6.2.4 Instructions for proper operation of the equipment/machinery, including:

6.2.4.1 Pre-start instructions.

6.2.4.2 Starting, stopping, shutdown instructions, and emergency procedures, if applicable.

6.2.4.3 Operation instructions, specifications, and adjustments.

6.2.5 Operations troubleshooting information.

6.2.6 Description of accessories and/or optional equipment.

6.2.7 Transporting instructions.

6.2.8 Storage instructions.

6.3 Maintenance instructions (for operators). These shall meet all applicable specifications as defined in Section 4—Publication Specifications and Section 5—Safety Content. In addition, they may include any or all of the following specific information:

6.3.1 Recommended fuels, lubricants, fluids, and other maintenance materials.

6.3.2 Refill capacities.

6.3.3 Maintenance intervals.

6.3.4 Maintenance charts.

6.3.5 Detailed maintenance procedures.

6.4 Repair/overhaul instructions. These shall meet all applicable specifications as defined in Section 4—Publication Specifications and Section 5—Safety Content. In addition, they may include the following specific information:

6.4.1 General description and specifications of the equipment/machinery.

6.4.2 Sectionalized, illustrated diagnostic and repair instructions by major assemblies and systems, including:

6.4.2.1 Detailed table of contents.

6.4.2.2 Theory of operation—charts and diagrams.

6.4.2.3 Diagnostic procedures.

6.4.2.4 Detailed segmental disassembly and assembly procedures including weights of major components (or specify handling equipment where appropriate) to assure safe handling.

6.4.2.5 Repair specifications such as allowable wear limits of parts, tolerances, torque values, adjustments, etc.

6.4.2.6 Information on special tools and fixtures.

6.4.3 Start-up and break-in procedures.

6.5 Installation and/or modification instructions. These shall meet all applicable specifications as defined in Section 4—Publication Specifications and Section 5—Safety Content. In addition, they may include any or all of the following specific information:

6.5.1 General description of the attachments and/or components to be installed, or the inspection and/or modification to be performed.

6.5.2 Detailed list of parts including parts illustrations, as required.

6.5.3 Inspection, modification, or installation procedures.

6.5.4 Start-up, check-out and/or test procedures.

6.6 Parts information. This publication shall meet all applicable specifications as defined in Section 4—Publication Specifications. In addition, it may include any or all of the following:

6.6.1 Introduction, table of contents, and/or alphabetical index.

6.6.2 Detailed illustrations showing all replaceable parts, including safety signs and any optional equipment.

6.6.3 Part number, name, and quantities used.

6.6.4 Numerical index listing all part numbers in the catalog keyed to the location where they are found.

6.7 Other publications. These shall meet applicable specifications as defined in Section 4—Publication Specifications and Section 5—Safety Content. In addition, they shall include information and/or instructions specific to the subjects covered.

Cited Standards:

ASAE S304, Symbols for Operator Controls on Agricultural Equipment
ASAE S318, Safety for Agricultural Equipment
ASAE S350, Safety-Alert Symbol for Agricultural Equipment
ASAE S390, Classifications and Definitions of Agricultural Equipment

ASAE Standard: ASAE S365.2 (*SAE J1041 MAR83)

T

BRAKING SYSTEM TEST PROCEDURES AND BRAKING PERFORMANCE CRITERIA FOR AGRICULTURAL FIELD EQUIPMENT

*Corresponds in substance to previous document S365.1T.

Proposed by the Engineering Committee of the Farm and Industrial Equipment Institute; approved by the ASAE Power and Machinery Division Standards Committee; adopted by ASAE as a Tentative Standard December 1973; reconfirmed December 1974, December 1975, December 1976, December 1977, December 1978, December 1979, December 1980, December 1981, December 1982; revised April 1983; reconfirmed December 1983, December 1984, December 1985, December 1986; revised and reclassified as a full Standard March 1988.

SECTION 1—PURPOSE

1.1 The purpose of this Standard is to establish test procedures for measurement of braking system performance and minimum performance criteria for braking systems on agricultural field equipment.

SECTION 2—SCOPE

2.1 The test procedures and performance criteria are directed to operation and parking of agricultural field equipment equipped with braking system(s) and having a maximum design speed exceeding 6 km/h (3.7 mph). Combinations of agricultural towing machines equipped with braking systems and towed agricultural machines without braking systems are included in this Standard.

SECTION 3—DEFINITIONS

3.1 Agricultural field equipment: Agricultural tractors, self-propelled machines, implements and combinations thereof designed primarily for agricultural field operations (per ASAE Standard S390, Classifications and Definitions of Agricultural Equipment).

3.2 Agricultural trailer: A transport machine used in agriculture which, according to its design, is suitable and intended for coupling to an agricultural tractor or self-propelled machine.

3.3 Agricultural tractor: A traction machine designed and advertised primarily to supply power to agricultural implements and farmstead equipment (per ASAE Standard S390, Classifications and Definitions of Agricultural Equipment).

3.4 Average deceleration: The retardation rate of a machine defined by the formula:

$$a = \frac{V^2}{2S}$$

where

a = average deceleration, m/s^2
V = initial speed, m/s
S = stopping distance, m

3.5 Braking control input force: The sum of all forces applied by the operator to the braking system control(s), as measured at the point of force application, in a line from the point of application through the operator's hip joint for foot pedal controls, or through the arm to shoulder joint for hand-operated controls.

3.6 Cold brakes: A brake is deemed to be cold if one of the following conditions is met:

3.6.1 The temperature measured at the periphery of the disc or on the outside of the drum is below 100 °C (212 °F).

3.6.2 In the case of totally enclosed brakes including oil-immersed brakes, the temperature measured on the outside of the housing is below 50 °C (122 °F) or within the manufacturer's specifications.

3.6.3 The brake has not been actuated in the previous 1 h.

3.7 Combination: Trailed equipment with or without braking systems coupled to an agricultural tractor or self-propelled machine.

3.8 Maximum gross mass: The maximum permissible mass (weight) of the test machine in accordance with manufacturer's recommendations regardless of travel speed restrictions, and including maximum ballast, equipment and material load recommended or permitted.

3.9 Maximum gross mass for stopping tests: The maximum permissible mass (weight) of the test machine in accordance with the manufacturer's recommendations for maximum transport speed or 25 km/h (15.5 mph) whichever speed is less.

3.10 Parking brake system: A means for holding a machine continuously in a parked position.

3.11 Secondary braking system: A braking system used for stopping a machine in the event of a malfunction in the operation or control of the service braking system.

3.12 Self-propelled agricultural machine: An implement designed with integral power unit to provide both mobility and power for performing agricultural operations (per ASAE Standard S390, Classifications and Definitions of Agricultural Equipment).

3.13 Service braking system: The primary system(s) used for retarding and stopping a machine.

3.14 Single-unit machine: A self-propelled machine not coupled to trailed equipment.

3.15 Special self-propelled agricultural machine: An implement designed with integral power unit with front drive axle and major mass on front axle. (Examples: combine, cotton picker, windrower, etc.)

3.16 Stopping distance: The distance traveled between the point at which the braking control is first moved and the point at which the machine comes to a stop.

3.17 Stopping time: The time elapsed between the first movement of the braking control and the instant at which the machine comes to a stop.

3.18 Test machine: The term used in this Standard to identify the agricultural machinery on which braking performance is measured by test.

3.19 Towed agricultural machine: An implement that is designed to perform agricultural operations and is pulled by an agricultural tractor or self-propelled agricultural machine. It is usually equipped with wheels for transport.

3.20 Towing or towed force: The force required to move a machine in a specified manner by another machine which has the motive power.

3.21 Unladen machine: A machine completely serviced with fuel, coolant, and lubricants, carrying a driver (if required) having a minimum weight of 75 kg (165 lb) but no optional accessories, weights, ballast or material load.

SECTION 4—CLASSIFICATIONS

4.1 For the purpose of this Standard, agricultural field equipment is classified in the following categories:

Category I - Agricultural Tractors
Category IIa - Self-Propelled Agricultural Machines
IIb - Special Self-Propelled Agricultural Machines
Category III - Agricultural Trailers
Category IV - Towed Agricultural Machines

SECTION 5—FACILITIES AND INSTRUMENTATION

5.1 The following facilities and instrumentation capabilities are required:

5.1.1 Ambient temperature. A means of measuring ambient temperature within ± 3 °C (± 5 °F).

5.1.2 **Braking system input force.** An instrument to measure the applied force to the braking control, with an accuracy of ± 5%.

5.1.3 **Stopping distance.** A means of measuring the stopping distance with an accuracy of ± 1%.

5.1.4 **Test course.** The test course shall be straight and consist of a clean swept, level dry concrete or other hard clean surface with equivalent friction characteristics and of adequate length to conduct the test. The approach shall be of sufficient length, smoothness and uniformity of grade to assure stabilized travel speed of the machine. The braking surface shall not have more than 1% grade in the direction of travel or more than 3% grade at right angles to the direction of travel.

5.1.5 **Test speed.** A means of measuring the test speed with an accuracy of ± 2%.

5.1.6 **Test mass (weight).** A means for determining wheel loads with an accuracy of ± 3%.

5.1.7 **Tire pressure.** A means of measuring tire inflation pressure with an accuracy of ± 5%.

5.1.8 **Towing or towed force.** An instrument to measure towing force (tension and compression) with an accuracy of ± 3%, and with a towing force indicator visible to the test machine operator.

5.1.9 **Towing machine.** A towing machine with sufficient power and mass (weight) to pull the test machine. The device connecting the towing machine to the test machine shall be horizontal within ± 4 deg when the machine is on a level surface.

5.1.10 **Wind velocity.** A means of measuring wind velocity with an accuracy of ± 3 km/h (± 2 mph).

5.2 **Instrumentation to measure the following is optional:**

5.2.1 **Brake temperature.** A temperature measuring system shall have ± 2% accuracy.

5.2.2 **Braking system fluid pressure** (for Category III and IV only). A means of measuring braking system fluid pressure with an accuracy of ± 5%.

5.2.3 **Deceleration.** An instrument to measure and record average deceleration with an accuracy of ± 3%.

5.2.4 **Time to stop.** A means of measuring the stopping time with an accuracy of ± 1%.

5.3 **Braking system test report.** A typical braking system test report form is shown in Fig. 1.

SECTION 6—TEST PROCEDURE FOR AGRICULTURAL MACHINES AND COMBINATIONS—CATEGORIES I AND II

6.1 General

6.1.1 Examine the test machine to assure conformance with the manufacturer's specifications and operational instructions. This examination includes all control systems, braking systems' components, lubricant type(s), reservoir levels, tires, and distribution of load. Set all tire pressures to the maximum of manufacturer's specifications for the test mass (weight).

6.1.2 Specify and describe all braking and retarding devices included in the service braking system, secondary braking system, and/or parking brake system, as applicable for the machine being tested. These descriptions should correspond with the manufacturer's designations of braking systems.

6.1.3 Auxiliary retarder, variable ratio drive, engine exhaust brake, or other auxiliary braking devices may be used in the braking performance tests if one of the following conditions is met. Description and use of such devices shall be noted in the test report.

6.1.3.1 The device is simultaneously actuated by the braking control.

Manufacturer's Name and Address ____________

Vehicle Specification

Make ____________ Model ____________

Type ____________ Serial No. ____________

Unladen Mass ____________ Maximum Gross Mass ____________ kg

Maximum Design Speed ____________ Maximum Gross Mass (Stopping Tests) ______ kg

Location and Description of Axle or Tractive Device	Tire Size	Maximum Mass Technically Permissible on Each Axle or Wheel	Mass on Each Axle or Wheel as Tested
		kg	kg
		kg	kg
		kg	kg
		kg	kg

Service Braking Device(s) Specification — Brief Description of Parking Brake

Make ____________ ____________

Type ____________ ____________

Secondary Braking Device(s) Specification ____________

TEST RESULTS

	Test Speed (km/h)	Brake Control Input Force (N)	Required Stopping Distance (m)	Measured or Calculated Stopping Distance (m)	Slope Percent (%) OR	Towing Force (N)	Brake Control Link Fluid Pressure Cat III or IV
Service Braking System							
Stopping Perform. Tests							
Maximum Gross Mass							
Unladen							
Fade Test							
Recovery Test							
Wetted Brake Test							
Recovery Test							
Holding Test (Forward)							
(Reverse)							
Parking Brake Holding (Forward)							
(Reverse)							
Secondary Braking System (Forward)							
(Reverse)							

Ambient Temperature ____________ Test Surface ____________

Wind Speed/Direction ____________ Date of Test ____________

Location of Tests ____________ Wind direction relative to test course ______

FIG. 1—BRAKING SYSTEM TEST REPORT

6.1.3.2 Retardation is automatic with normal braking procedure.

6.1.3.3 The manufacturer designates such device as a braking system (see paragraph 6.1.2).

6.1.4 Burnish or "break in" the test machine brakes as specified by the manufacturer or as required to establish consistent braking performance.

6.1.5 Record test machine gross mass (weight), (as defined in paragraphs 3.8, 3.9 or 3.21), individual axle or wheel load distribution, and vertical hitch point load where applicable.

6.1.6 Record ambient temperature, wind velocity and wind direction with respect to the test course. The test shall not be conducted when the wind velocity exceeds 32 km/h (20 mph) or when the ambient temperature is below −10 °C (14 °F) or above 35 °C (95 °F) unless it can be shown that the test results will not be significantly influenced.

6.1.7 Record a description of any unusual or erratic brake performance or noise characteristics.

6.1.8 All test stops conducted shall be accomplished without pull or swerve which would take the test machine out of a test lane 1.2 m (4 ft) wider than the maximum width of the test machine.

6.1.9 Safe operating procedures and practices should be observed during tests to control machines within their stability and control limitations.

6.2 Braking performance test procedure

6.2.1 The test machine brakes shall be cold, as defined by paragraph 3.6, prior to each stop. If the drum or the disc temperature cannot be conveniently measured, the machine manufacturer may specify an alternate point of measurement and a correction factor for that point.

6.2.2 Ballasted braking performance test shall be performed with the test machine at the maximum gross mass (weight) for stopping tests as defined in paragraph 3.9.

6.2.3 Unladen braking performance test shall be performed with the test machine as defined in paragraph 3.21.

6.2.4 Braking performance tests shall be conducted on the specified test course from an initial test machine speed of 25 km/h (15.5 mph) +5 −0% or maximum speed for the test machine recommended by the manufacturer, whichever is less, to measure minimum (shortest) stopping distance and corresponding braking control input force, without locking of the wheels.

6.2.5 Stopping distance (optionally, average deceleration rate) shall be measured from a stabilized travel speed with the engine not affecting the braking effort, except as defined in paragraph 6.1.3.

6.2.6 For service braking system performance tests, the foot-control input force shall not exceed 600 N (135 lbf). The hand-control input force shall not exceed 400 N (90 lbf).

6.2.7 Report the minimum stopping distance and corresponding braking control input force.

6.2.8 Stopping distances for initial speeds greater than 25 km/h (15.5 mph) may be corrected and reported with corresponding braking control input force by the formula:

$$S_{25} = \frac{(V_{25})^2\, S_1}{(V_1)^2} = \frac{(6.94)^2\, S_1}{(V_1)^2} = \frac{48.2\, S_1}{(V_1)^2}$$

where

S_{25} = stopping distance corrected to 25 km/h initial speed, m
S_1 = measured stopping distance, m
V_1 = measured initial speed, m/s
V_{25} = desired speed of 6.94 m/s or 25 km/h

6.3 Service braking system fade and recovery test procedure. This test is intended to evaluate performance of friction devices which are used for the service braking system. The procedure may be waived if the service braking system as designated in paragraph 6.1.2 does not include a friction retarding device, and if this procedure is not appropriate for measuring the ability of the designated service braking system to withstand prolonged input of retardation energy. The reasons for waiving this procedure shall be reported.

6.3.1 Start with cold brakes and the test machine at the maximum gross mass (weight) for stopping tests as defined in paragraph 3.9. The service brakes shall be heated by towing the braked test machine at a speed of 20 km/h (12.4 mph) +5 −0% or 80% of maximum travel speed, whichever is less, with engine clutch disengaged or transmission in neutral and with input to the service braking system sufficient to maintain a constant towed force (drawbar pull) equivalent to at least 10% of the machine gross mass (weight) over the specified distance (refer to paragraph 8.1.3) without locking of the wheels. Formula for minimum heating toward towing force is:

$$F_t = (0.1)\, m_1 g = 0.981\, m_1$$

where

F_t = towing force, N
m_1 = mass of test machine, kg
g = acceleration of gravity, 9.81 m/s^2

6.3.2 Immediately disconnect the towing machine and perform a one-stop performance test with the machine at test mass (weight), using the braking control input force for minimum stopping distance reported in paragraph 6.2.7. This test shall be completed within 3 min of the completion of the heating cycle specified in paragraph 6.3.1.

6.3.3 Cool the brakes to a cold brake condition and repeat the braking performance test procedure in paragraph 6.2 at the machine test mass (weight).

6.3.4 Report the stopping distances and corresponding braking control input forces for the fade test and the recovery test.

6.4 Wetted brake performance and recovery test procedure. This test is not applicable to fully sealed braking systems.

6.4.1 Immediately after flooding the test machine brakes with water for a minimum of 2 min, perform a one-stop wetted braking performance test using procedure in paragraph 6.2 for ballasted braking performance. Use the braking control input force for minimum stopping distance reported in paragraph 6.2.7.

6.4.2 Dry the wetted brakes. Perform recovery test using procedure in paragraph 6.2 for ballasted braking performance.

6.4.3 Report stopping distances and corresponding braking control input forces for wetted brake recovery test in paragraph 6.4.2.

6.5 Secondary braking system test procedure

6.5.1 Use the braking performance test procedure in paragraph 6.2 to measure the performance of the secondary braking system. The machine shall be laden to the maximum gross mass (weight) for stopping tests as defined in paragraph 3.9.

6.5.2 For secondary braking system performance tests, the foot-control input force shall not exceed 900 N (200 lbf). The hand-control input force shall not exceed 400 N (90 lbf).

6.6 Parking brake and service braking system holding test procedure

6.6.1 Holding tests shall be conducted on a level test course with the transmission of the test machine in neutral. The test machine shall be laden to the maximum gross mass (weight) as defined in paragraph 3.8. Apply a forward horizontal pull force to the test machine equivalent to the downgrade gravitational force of the test machine on a specified grade. Determine the braking control input force required to prevent the test machine from moving.

6.6.2 For braking system holding tests, the foot-control input force shall not exceed 600 N (135 lbf). The hand-control input force shall not exceed 400 N (90 lbf).

6.6.3 Report braking control input force, towing force, and equivalent percent of grade.

6.6.4 Repeat procedures in paragraphs 6.6.1, 6.6.2, 6.6.3 in reverse direction.

6.6.5 Holding tests may be optionally conducted on a specified grade. Report braking control input force and percent grade.

6.6.6 If the test machine is equipped with a positive transmission locking device (park position) rather than a parking brake, engage test machine parking device and perform applicable procedures in paragraphs 6.6.1 through 6.6.5. Report towing force and equivalent percent grade.

SECTION 7—TEST PROCEDURE FOR AGRICULTURAL MACHINES WITH BRAKING SYSTEMS—CATEGORIES III AND IV

7.1 General

7.1.1 Connect the towed test machine to a towing machine in a manner representing the manufacturer's recommendation for the towed test machine.

7.1.2 Perform all applicable steps in paragraph 6.1.

7.2 Braking performance test procedure

7.2.1 The towed test machine brakes shall be cold, as defined in paragraph 3.6, prior to each stop.

7.2.2 The braking performance test shall be performed with the towed test machine ballasted to maximum gross mass (weight) for stopping tests as defined in paragraph 3.9. All unbraked axle(s) shall be loaded to the permissible maximum mass (weight) specified by the manufacturer.

7.2.3 Braking performance tests shall be conducted on the specified test course from an initial towing machine and towed test machine speed of 25 km/h (15.5 mph) +5 −0% or maximum recommended speed for the towed test machine, whichever is less. Stopping distance (optionally average deceleration rate) shall be measured from a stabilized travel speed without locking the wheels. Determine minimum stopping distance by measuring stopping distances and corresponding braking control input forces.

7.2.4 For service braking system performance tests, the foot-control input force shall not exceed 600 N (135 lbf). The hand-control input force shall not exceed 400 N (90 lbf).

7.2.5 The towed test machine braking force and stopping distances are calculated from the following formulas:

where

a_1 = average deceleration of towing machine alone, m/s^2
a_3 = average deceleration of combination, m/s^2
m_2 = mass of towed test machine, kg
m_1 = mass of towing machine, kg
F_2 = braking force of towed test machine, N
S_2 = stopping distance (calculated) of towed test machine, m
S_3 = minimum stopping distance of combination, m
S_{25} = stopping distance from 25 km/h initial speed, m
V_3 = initial speed of combination, m/s
V_{25} = desired speed of 6.94 m/s or 25 km/h
g = acceleration of gravity, 9.81 m/s^2

7.2.5.1 The average deceleration of the test combination is:

$$a_3 = \frac{(V_3)^2}{2S_3}$$

7.2.5.2 If the brakes of the towed test machine only are applied, the towed test machine braking force is:

$$F_2 = (m_1 + m_2)\, a_3$$

7.2.5.3 If the towing machine and the towed test machine brakes are applied in combination, the towed test machine braking force is:

$$F_2 = (m_1 + m_2)\, a_3 - m_1\, a_1$$

where

a_1 = the average deceleration measured for the towing machine alone, using the same value of input force to braking control of towing machine as used to establish minimum stopping distance of the combination, m/s^2

7.2.5.4 The calculated stopping distance of the towed test machine is:

$$S_2 = \frac{M_2\, (V_3)^2}{2\, F_2}$$

7.2.5.5 The stopping distance for initial speeds greater than 25 km/h may be corrected by the formula:

$$S_{25} = \frac{(V_{25})^2\, S_2}{(V_3)^2} = \frac{(6.94)^2\, S_2}{(V_3)^2} = \frac{48.2\, S_2}{(V_3)^2}$$

7.2.6 Report the minimum calculated stopping distance and corresponding braking control input force for the towed test machine. If the braking device transmission link is not mechanical, report the relationship between stopping distance and an appropriate transmission parameter such as fluid pressure or voltage.

7.3 Service braking system fade and recovery test procedure

7.3.1 Modification of the braking device may be necessary to enable this test to be carried out. For example, overrun brakes may require an independent means of application for the heating phase.

7.3.2 Start with cold brakes and the towed test machine ballasted as specified in paragraph 7.2.2. The service brakes shall be heated by towing the test machine at a speed of 20 km/h (12.4 mph) +5 −0% or at 80% of maximum towed machine speed, whichever is less, with an input to towed test machine service braking system sufficient to maintain a constant towed force (drawbar pull) equivalent to at least 10% of the towed test machine mass (weight) for the specified distance (refer to paragraph 8.1.3) without locking of the wheels.

7.3.3 Immediately after heating the brakes by the procedure in paragraph 7.3.2, perform a one-stop braking performance test per paragraph 7.2 using the braking control input force for minimum calculated stopping distance reported in paragraph 7.2.6. This test shall be completed within 3 min of the completion of the heating cycle specified in paragraph 7.3.2.

7.3.4 Cool the brakes to a cold brake condition and repeat the braking performance test procedure in paragraph 7.2.

7.3.5 Report the minimum calculated stopping distances and braking control input forces for fade test in paragraph 7.3.3 and recovery test in paragraph 7.3.4.

7.4 Wetted brake performance and recovery test procedure. This test is not applicable to fully sealed braking systems.

7.4.1 Immediately after flooding the test machine brakes with water for a minimum of 2 min, perform a one-stop wetted braking performance test per paragraph 7.2 using the braking control input force for minimum calculated stopping distance reported in paragraph 7.2.6.

7.4.2 Dry the wetted brakes. Perform recovery test using braking performance procedure in paragraph 7.2.

7.4.3 Report minimum calculated stopping distances and corresponding braking control input forces for wetted brake recovery test in paragraph 7.4.2.

7.5 Secondary braking system test procedure

7.5.1 Tow the machine, at maximum permissible mass (weight) for stopping tests as defined in paragraph 3.9, at 25 km/h (15.5 mph) +5 −0% or maximum towed test machine speed, whichever is less, under braking performance test procedure conditions, and intentionally apply secondary brakes. Measure the secondary brake performance using paragraphs 7.2.1, 7.2.2, 7.2.3 and 7.2.5.

7.5.2 For secondary braking system performance tests, the foot-control input force shall not exceed 900 N (200 lbf). The hand-control input force shall not exceed 400 N (90 lbf).

7.5.3 Test the ability of braking systems using stored energy to hold the towed machine in the event of accidental uncoupling from the towing machine. After the towed test machine, at maximum permissible mass (weight) as defined in paragraph 3.8, has remained at rest for the specified time period (refer to paragraph 8.4) on the specified grade or with equivalent constant towing forces applied, apply additional forward towing forces until the machine moves to determine the holding performance of the secondary braking system.

7.5.4 Repeat procedures in paragraph 7.5.3 in reverse direction.

7.5.5 Report calculated stopping distance for secondary braking system test in paragraph 7.5.1 and holding performance tests in paragraphs 7.5.3 and 7.5.4 with corresponding secondary braking control input force, or level and type of energy source.

7.6 Parking brake and service braking system holding test procedure. Use parking brake and service braking system holding test procedure in paragraph 6.6.

7.7 Optional braking performance test procedure. The performance test for towed agricultural machines equipped with braking systems, Category III and IV, may be performed as a series of brake applications as follows in lieu of braking performance test procedure in paragraph 7.2.

7.7.1 Perform applicable steps as directed in paragraph 7.1.

7.7.2 A towing force indicator shall be inserted between the towed test machine and the towing machine in such a manner that only the towing forces are indicated and no vertical force transference is shown.

7.7.3 The towed test machine brakes shall be cold, as defined by paragraph 3.6, prior to start of the test.

7.7.4 The towed test machine shall be loaded or ballasted as specified in paragraph 7.2.2.

7.7.5 With the towing machine traveling at a constant test speed of 25 km/h (15.5 mph) +5 −0% or maximum recommended speed, whichever is less, and powered at a governed speed or constant throttle setting, apply the towed test machine braking system to cause an added towing force equivalent to at least 25% of mass (weight) of towed test machine for a period of 3 s. The braking control input force shall not exceed those values specified in paragraph 7.2.4. Repeat 6 times the successive 3 s braking system applications after towing machine has accelerated to the constant test speed. The maximum elapsed time between braking system applications in this series of applications shall not exceed 2 min. Formula for the minimum additional towing force is:

$$F_t = 0.25\ m_2 g = 2.45\ m_2$$

where

F_t = added towing force, N
m_2 = mass of towed test machine, kg
g = acceleration of gravity, 9.81 m/s^2

7.7.6 Record test speed, speed at end of each 3 s braking system application, added towing force and braking control input force.

7.7.7 Average the data of the 6 applications and report initial test speed, average braking control input force and average equivalent stopping distance by formula in paragraph 7.2.5.4.

7.8 Optional service braking system fade and recovery test procedure. Perform the following to test fade and recovery performance for test machines using the optional braking performance test procedure (see paragraph 7.7).

7.8.1 Perform brake heating procedure as specified in paragraph 7.3.2.

7.8.2 Immediately after heating the towed test machine brakes, accelerate the combination to a constant speed of 25 km/h (15.5 mph) +5 −0% or maximum recommended speed, whichever is less. Powered at a governed speed or constant throttle setting, apply the average braking control force reported in paragraph 7.7.7 one time for 3 s and measure braking force (added towing force), F_t.

7.8.3 Report initial test speed, braking control input force, and equivalent average stopping distance by formula in paragraph 7.2.5.4.

7.8.4 Cool the brakes to a cold brake condition and perform recovery test in accordance with the optional braking performance test procedure in paragraph 7.7.

SECTION 8—MINIMUM BRAKING PERFORMANCE CRITERIA

8.1 Service braking system

8.1.1 Holding performance. Agricultural field machines equipped with service braking systems shall have capability equivalent to holding the machine on dry swept concrete or a surface with equivalent friction characteristics under conditions specified:

Category	Grade	Condition
I	30%	Machine loaded to manufacturer's maximum mass (weight) rating and
II, III or IV	25%	distribution, as defined in paragraph 3.8.

8.1.2 Stopping performance. Agricultural field machines equipped with service braking systems shall have stopping performance as defined by the applicable formula:

(V = initial velocity, m/s)

Category	Maximum stopping distance, m
I, IIa, III, IV	$S = \frac{V^2}{6.27} + \frac{V}{3}$
IIb and combinations of braked and unbraked machines I and III, I and IV, IIa and III, IIa and IV	$S = \frac{V^2}{3.52} + \frac{V}{3}$

8.1.3 Fade and recovery performance

8.1.3.1 Agricultural field machines equipped with service braking systems shall have capability to absorb braking energy through the designated service braking system equivalent to an energy input rate defined in paragraph 6.3.1 over a distance of 1 km (0.62 mile), maintaining fade stopping performance as defined by the applicable formula:

(V = initial velocity, m/s)

Category	Maximum stopping distance, m
I, IIa, III, IV	$S = \frac{V^2}{4.38} + \frac{V}{3}$
IIb and combinations of braked and unbraked machines I and III, I and IV, IIa and III, IIa and IV	$S = \frac{V^2}{2.54} + \frac{V}{3}$

The machines shall have recovery stopping performance as defined by the applicable formulas in paragraph 8.1.2.

8.1.3.2 For machines with service braking systems which require both infinitely variable drive as primary retarding system and friction device to bring the vehicle to a complete stop, the friction device alone shall absorb braking energy at a rate defined in paragraph 6.3.1 over a distance of 0.1 km (0.06 mile), with the service braking system maintaining fade and recovery stopping performance as defined in paragraph 8.1.3.1.

8.1.4 Wetted brake recovery performance. Agricultural field machines equipped with service braking systems that have friction surfaces subject to wetting shall have wetted brake recovery stopping performance as defined by the applicable formula in paragraph 8.1.2.

8.1.5 Agricultural field machines equipped with service braking systems and having capability of carrying increasingly heavy loads with corresponding restriction to decreased maximum speeds shall have braking capability at speeds below 25 km/h (15.5 mph) and corresponding maximum gross mass as specified by stopping distance formulas in paragraph 8.1.2.

8.1.6 Any machine with a service braking system which an operator might disconnect during normal operation shall have other braking capability, tested independently, as specified by secondary braking systems in paragraph 8.3.

8.1.7 Agricultural field machines with service braking systems other than conventional friction type service brakes shall meet the performance specifications given in paragraphs 8.1.1, 8.1.2, 8.1.3, 8.1.4, and 8.1.5. Service braking systems which require more than one braking device in combination to meet the definition of service braking system shall meet the performance specifications given in paragraphs 8.1.1, 8.1.2, 8.1.3, 8.1.4 and 8.1.5.

8.2 Parking brake system

8.2.1 Agricultural field machines equipped with parking brake systems shall have the capability of holding the agricultural machine, at manufacturer's maximum mass (weight) rating as defined in paragraph 3.8, stationary in both forward and reverse direction on a 20% grade or equivalent condition of loading.

8.2.2 The parking brake system, when applied, shall remain in the applied position in compliance with paragraph 8.2.1 despite any contraction of the brake parts, exhaustion of the source of energy, or leakage of any kind.

8.3 Secondary braking system

8.3.1 Agricultural field machines equipped with secondary braking systems shall have stopping performance as defined by the applicable formula:

(V = initial velocity, m/s)

Category	Maximum stopping distance, m
I, IIa, IIb, III, IV	$S = \frac{V^2}{3.52} + \frac{V}{3}$
Combinations of braked and unbraked machines I and III, I and IV, IIa and III, IIa and IV	$S = \frac{V^2}{2.07} + \frac{V}{3}$

8.3.2 The secondary braking system for Categories I and II shall be capable of being applied by a person seated in the operator's seat. The system shall be arranged so that it cannot be released from the operator's seat after any application unless immediate reapplication can be made from the seated position.

8.4 Braking systems for Categories III and IV, which actuate automatically in the event of accidental uncoupling from the towing machine, shall have stopping capabilities specified in paragraph 8.3.1 and holding capabilities specified in paragraph 8.2.1 for 15 min after the stop. These brakes shall be released only by a specific action of the operator.

8.5 Service and secondary braking system maximum permissible stopping distances are summarized in Table 1, for a test speed of 25 km/h (15.5 mph).

TABLE 1—MAXIMUM PERMISSIBLE STOPPING DISTANCES FROM 25 km/h (15.5 mph) TEST SPEED FOR SERVICE AND SECONDARY BRAKING SYSTEMS

Categories and Combinations	Service Braking System Maximum Stopping Distance, m (ft)*		Secondary Braking System Maximum Stopping Distance, m (ft)
	Stopping Performance and Recovery	Fade	
I, IIa, III, IV	10 (33)	13 (43)	16 (53)
IIb	16 (53)	21 (69)	16 (53)
Combinations of braked and unbraked I and III, I and IV, IIa and III, IIa and IV	16 (53)	21 (69)	26 (85)

*Stopping distances are distances traveled from the first movement of the brake controls to the point at which the machine comes to a stop.

ASAE Standard: ASAE S366.1 (*SAE J1036 JAN85)

DIMENSIONS FOR CYLINDRICAL HYDRAULIC COUPLERS FOR AGRICULTURAL TRACTORS

*Coupler dimensions correspond.

Proposed by the Engineering Policy Committee of the Farm and Industrial Equipment Institute; approved by the ASAE Power and Machinery Division Standards Committee; adopted by ASAE as a Recommendation December 1973; reclassified as a Standard March 1978; reconfirmed December 1978; revised March 1982; reconfirmed and revised editorially December 1986.

SECTION 1—PURPOSE

1.1 The purpose of this Standard is to establish interface dimensions of cylindrical hydraulic couplers frequently used by the agricultural equipment industry to connect hydraulic remote cylinders and other hydraulic devices to agricultural tractors.

1.2 Also the purpose of this Standard is to permit interchangeable use of different makes of remote cylinders and other hydraulic devices on other makes of tractors when designed for this use.

1.3 This Standard is equivalent to International Standard ISO 5675-1981, Agricultural Tractors and Machinery—Hydraulic Couplers for General Purposes—Specifications.

SECTION 2—SCOPE

2.1 This Standard covers only cylindrical couplers when interchange between makes is recommended.

2.2 This Standard does not preclude other types of hydraulic couplers used for similar purposes.

SECTION 3—COUPLER DIMENSIONS

3.1 The male half of the coupler is considered a part of the cylinder or hydraulic device system and shall conform to dimensions given in Table 1.

3.2 The female half of the coupler is considered a part of the tractor hydraulic system and shall accept the male half which conforms to dimensions given in Table 1.

3.3 When a quick release, breakaway or other special feature is required, it shall be provided as part of the female coupler without interfering with interchangeability of the standard male coupler.

3.4 The travel of the valve in the male coupler will be limited as shown in Fig. 1, and the force required to fully open the valve when no pressure is in the coupler shall not exceed 44 N (10 lb). This allows double travel of the valve in the female coupler to be provided when necessary to facilitate coupling with pressure in the hydraulic lines.

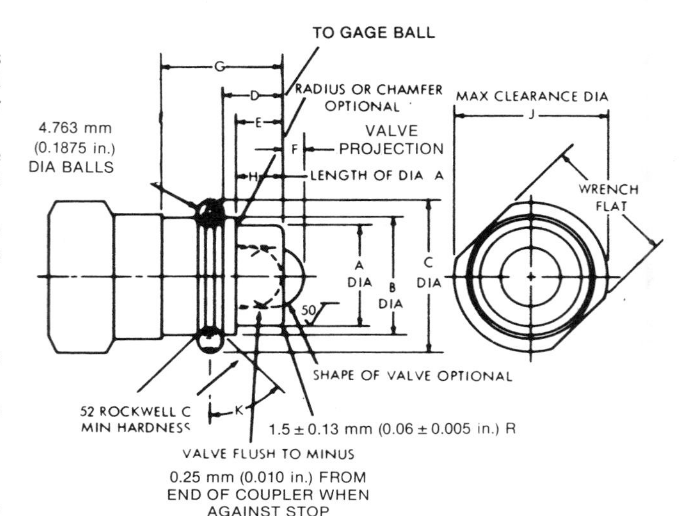

FIG. 1—DIMENSIONS OF CYLINDRICAL HYDRAULIC COUPLERS. (See Table 1)

TABLE 1—SPECIFICATIONS AND DIMENSIONS FOR CYLINDRICAL HYDRAULIC COUPLERS (SEE FIG. 1)

Rated Operating Pressure		Maximum Relief Valve Pressure		Rated Flow		Diameter A		Diameter B		Diameter C-Gage	
						◎ ϕ 0.025 mm (0.001 in.) B					
MPa	psi	MPa	psi	liter per min	gal per min	mm	in.	mm	in.	mm	in.
20.0	2900	25.0	3625	60	16	20.52 ±0.05	0.808 ±0.002	23.70 ±0.05	0.933 ±0.002	30.30	1.193

Dimension D		Dimension E		Maximum Valve Projecion F		Minimum Dimension G		Minimum Dimension H		Maximum Diameter J		Angle K
mm	in.	mm	in.	mm	in.	mm	in.	mm	in.	mm	in.	deg
11.79 +0.00 -0.13	0.464 +0.000 -0.005	9.3 ±0.1	0.366 ±0.004	4.32	0.170	24.0	0.95	8.4	0.330	31	1.22	45 ±1

SECTION 4—COUPLER LOCATIONS

4.1 For standard remote cylinder or other hydraulic device (Reference: ASAE Standards S201, Application of Hydraulic Remote Control Cylinders to Agricultural Tractors and Trailing-Type Agricultural implements, and S316, Application of Remote Hydraulic Motors to Agricultural Tractors and Trailing-Type Agricultural Implements) operation, the female coupler shall be located to the rear of the tractor within limits specified in Table 2 and Fig. 2.

4.2 For special cylinder or other hydraulic device operation, female couplers shall be located to the rear or on the right or left side within limits specified in Table 2 and Fig. 2.

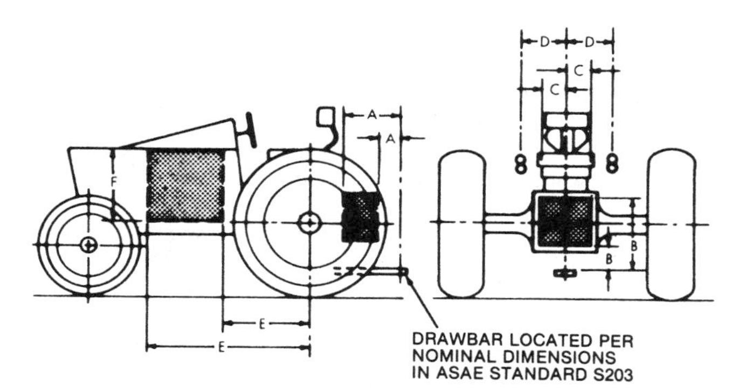

FIG. 2—LOCATION OF CYLINDRICAL HYDRAULIC COUPLERS. (See Table 2)

TABLE 2—LOCATION OF CYLINDRICAL HYDRAULIC COUPLERS

Dimensions (See Fig. 2)	Range mm	Range in.
For rear locations:		
A Horizontal distance from pin hole of drawbar	229 to 635	9 to 25
B Vertical distance from top of drawbar	305 to 914	12 to 36
C Horizontal distance from centerline of tractor—either side	254 max	10 max
For side locations:		
D Horizontal distance from centerline of tractor	381 max	15 max
E Horizontal distance from centerline of rear axle	762 to 1397	30 to 55
F Vertical distance from centerline of of rear axle	635 max	25 max

SECTION 5—PERFORMANCE REQUIREMENTS

5.1 The maximum pressure drop through the male half of the coupler shall not exceed 172.4 kPa (25 psi) at rated flow with mineral oil at a viscosity of 55 Saybolt Universal seconds minimum. When measuring restriction, use two male couplers in test fixture shown in Fig. 3. Use National Fluid Power Association Standard NFPA/T3.20.2R1, Test Procedure for Hydraulic Fluid Power Quick Disconnect Couplers, as a guide for the test setup and general procedure. Measure the pressure drop across the two male couplers with rated flow in each direction. Add these two measurements and divide by four to get the effective pressure drop across one male half.

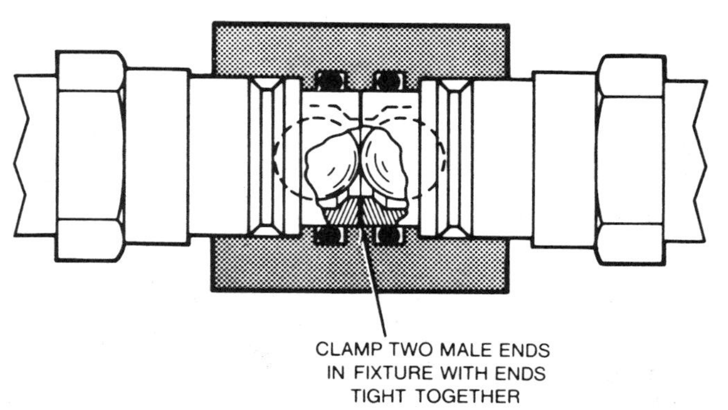

FIG. 3—PRESSURE DROP TEST SETUP

5.2 When the two halves of the coupler are locked together in the operating position, they shall remain locked together, without failure, when subjected to a proof pressure of 200% of the rated pressure.

5.3 The male half of the uncoupled coupler shall withstand a minimum burst pressure of 70 MPa (10,150 psi).

SECTION 6—COMPATIBILITY WITH FLUIDS

6.1 For the purpose of this Standard, it should be assumed that SAE 20 mineral oil will be used.

6.2 Couplers and cylinders, or other hydraulic devices, should be made with seals and other materials which are compatible with SAE 20 mineral oils and additives normally used in tractor hydraulic systems.

Cited Standards:

ASAE S201, Application of Hydraulic Remote Control Cylinders to Agricultural Tractors and Trailing-Type Agricultural Implements

ASAE S203, Rear Power Take-Off for Agricultural Tractors

ASAE S316, Application of Remote Hydraulic Motors to Agricultural Tractors and Trailing-Type Agricultural Implements

ISO 5675-1981, Agricultural Tractors and Machinery—Hydraulic Couplers for General Purposes—Specifications

NFPA/T3.20.2R1, Test Procedure for Hydraulic Fluid Power Quick Disconnect Couplers

GUIDE FOR PREPARING FIELD SPRAYER CALIBRATION PROCEDURES

Developed by the ASAE Agricultural Chemical Application Committee; approved by the Power and Machinery Division Standards Committee; adopted by ASAE December 1973; revised December 1975; reconfirmed December 1980, December 1985.

SECTION 1—PURPOSE

1.1 This Engineering Practice sets forth guidelines for those who prepare field sprayer calibration procedures. The purpose is to encourage practices that will improve uniformity, accuracy and safety of pesticide application with field sprayers.

SECTION 2—SCOPE

2.1 This Engineering Practice provides information on the calibration of boom type field sprayers used for broadcast, band or row applications. It is easily adapted to boomless and air-carrier type sprayers for broadcast applications and to rotary atomizers such as spinning disks. Because of the nature of the variables involved in pesticide application, a single calibration procedure applicable in all situations is not feasible.

SECTION 3—EQUIPMENT

3.1 Equipment needed will depend on the method of calibration. Common items include:

3.1.1 Measuring tapes of 25, 50, or 100 m length (50, 100, or 300 ft), or a land measuring wheel.

3.1.2 Stakes or flags for marking a measured course.

3.1.3 Liter and five liter containers graduated in 10 ml increments (quart and gallon containers graduated in liquid ounces).

3.1.4 Accurate pressure gage of proper range.

3.1.5 Watch with sweep second hand or a stop watch.

SECTION 4—SAFETY PRECAUTIONS

4.1 A caution statement, Fig. 1, for the safe handling, storage, and disposal of agricultural chemicals shall be included with applicator calibration procedures. (Reference: ASAE Standard ASAE S318, Safety for Agricultural Equipment.)

4.2 Use water alone to calibrate the sprayer unless the flow rate of the actual spray mixture varies more than 5 percent from the flow rate of water.

4.3 Calibration with actual spray mixture.

4.3.1 Wear suitable, approved safety equipment and protective clothing. Avoid contact with the spray.

4.3.2 Avoid contamination of area. Calibrate only when wind speed is below 8 km/h (5 mph).

SECTION 5—GENERAL

5.1 The volume of spray material applied to a given area depends on nozzle flow rate, ground speed of the sprayer and the sprayed width per nozzle. Each variable must be determined when developing a specific calibration procedure.

CAUTION

AGRICULTURAL CHEMICALS CAN BE DANGEROUS. IMPROPER SELECTION OR USE CAN SERIOUSLY INJURE PERSONS, ANIMALS, PLANTS, SOIL, OR OTHER PROPERTY. BE SAFE: SELECT THE RIGHT CHEMICAL FOR THE JOB. HANDLE IT WITH CARE. FOLLOW THE INSTRUCTIONS ON THE CONTAINER LABEL AND INSTRUCTIONS FROM THE EQUIPMENT MANUFACTURER.

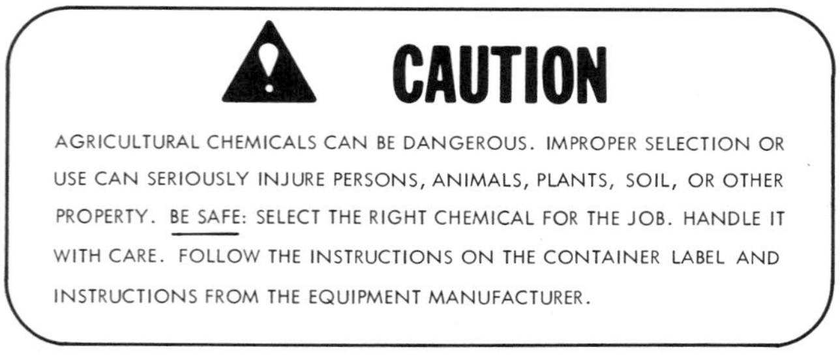

FIG. 1—SUGGESTED SAFETY SIGN

5.1.1 Nozzle flow rate. Nozzle flow rate varies with nozzle capacity, nature of the fluid and fluid pressure.

5.1.1.1 Nozzle capacity. Select the nozzle that will best fit the requirements of application volume, pressure and ground speed.

5.1.1.2 Nature of the fluid. If the spray mixture will be altered considerably by the addition of adjuvants, compare the flow rate of the spray mixture to that of water. If the rate difference is 5 percent or more, use the actual spray mixture in the calibration.

5.1.1.3 Fluid pressure. A constant pressure must be maintained to achieve uniform application. Flow rate is generally proportional to the square root of the pressure drop across the nozzle.

5.1.2 Ground speed of sprayer. Spray volume has an inverse relationship to the ground speed. Ground speed is the easiest factor to change for minor corrections in application rate. Ground speed must be constant for uniform application.

NOTE: This paragraph does not apply to ground driven pumps.

5.1.3 Sprayed width per nozzle. Calibration procedures should be used to determine the volume (liters or gallons) of liquid applied per unit area (hectare or acre) actually treated with the agricultural chemical. For band application, the area treated is the area in the band and not the area of crop land. Some pesticide recommendations are based on area of crop land rather than the area actually sprayed. In these cases, the spray width used in the calibration should be based on the row spacing and not on the band width.

SECTION 6—CALCULATIONS

6.1 Spray volume and nozzle output may be determined from the following formulas:

$$V = \frac{KQ}{SW} \quad [1]$$

$$Q = \frac{VWS}{K} \quad [2]$$

$$V = \frac{\text{quantity used}}{\text{area treated}} \quad [3]$$

V = spray volume, liters per ha (gal per acre)*
K = constant 60 000 (5940)
Q = output per nozzle(s) spraying on width W
S = speed, km/h (mph)
W = spray width, cm (in.)†
Quantity used = volume of liquid in liters (gallons) used by the nozzle or boom on a measured course or in the equivalent time
Area treated = area actually treated in hectares (acres)

* 1 gal per acre equals 9.354 liters per ha.
† (1) Nozzle spacing for boom spraying.
(2) Spray swath width for boomless spraying.
(3) Band width for band spraying.
(4) For some row crop plant application, the spray width per nozzle is equal to the row spacing (or band width) divided by the number of nozzles per row (or band).

6.2 There are several methods for determining spray volume and nozzle output other than the formulas in paragraph 6.1; such as nomographs, charts and special slide rules. Calculations are aided by selecting a measured course length so that an even decimal or fractional part of a hectare (acre) is covered.

SECTION 7—CALIBRATION PROCEDURES

7.1 Nozzle tip selection

7.1.1 Considering label recommendations and field conditions, select a spray volume and an operating speed.

7.1.2 From the spray width, speed and spray volume, determine the nozzle output, Q, required by paragraph 6.1.

7.1.3 Select a nozzle tip that will give the required output when operating within the recommended pressure range.

7.2 Pre-Calibration check. Be sure that all sprayer parts are free of foreign material and are functioning properly. Inspect nozzle tips and internal parts for obvious wear, defects, proper size and type. Check the flow rate of each nozzle using water at the planned operating pressure for uniform output, equal fan angle, and uniform appearance of spray pattern. Replace any nozzle tips having flow 5 percent more or less than the average of the other nozzles checked and/or having obviously different fan angles or patterns. Check the flow rate of new nozzles.

7.3 Adjustments to obtain desired spray volume

7.3.1 Normal operating speed may be changed to adjust volume if the volume change is under 25 percent.

7.3.2 Operating pressure may also be changed to adjust volume while operating within the recommended pressure range if the volume change is under 25 percent. A greater range affects drop size and pattern excessively.

7.3.3 Change nozzle tips to obtain volume changes greater then 25 percent.

7.4 Lay out a measured course to the nearest 0.1 m (nearest 0.5 ft) in the field where spraying will take place or in similar soil and terrain conditions. The course length depends on the travel speed, the spray width and spray volume. The course should be long enough that an accurate measure can be made of time (at least 15 seconds) or of spray volume (at least 10 percent of tank volume), whichever method is used.

7.5 Calibration methods

7.5.1 Method A. The liquid sprayed from one or more nozzles is accurately determined while operating (1) over the measured course; (2) over the period of time equivalent to the travel time over the course; or (3) for a given time such as one minute. Select representative nozzles and maintain the desired operating pressure. Determine spray volume by formula (3) in paragraph 6.1 if technique (1) or (2) above is used. Determine spray volume by paragraph 6.1 if technique (3) above is employed.

7.5.2 Method B. The amount required to refill the tank is measured after operation as in (1) or (2) of Method A above, or (3) for a suitable given time such as 5 minutes. Use an accurate liquid level mark, or fill the tank to overflowing before a calibration run then measure the amount required to refill to overflowing after the run. The boom and lines should be full before and after operation. For accurate measurement the sprayer must be in exactly the same position (preferably level) before and after operation. Determine spray volume by formula (3) in paragraph 6.1 if technique (1) or (2) above is used. Determine spray volume by formula (1) in paragraph 6.1 if technique (3) above is employed.

7.6 Recalibration is required periodically becasue of nozzle wear and to compensate for any changes in the variables listed in Section 5.

Cited Standard:

ASAE S318, Safety for Agricultural Equipment

ASAE Engineering Practice: ASAE EP371

PREPARING GRANULAR APPLICATOR CALIBRATION PROCEDURES

Developed by the ASAE Agricultural Chemical Application Committee; approved by the Power and Machinery Division Standards Committee; adopted by ASAE December 1974; reconfirmed December 1979, December 1984.

SECTION 1—PURPOSE AND SCOPE

1.1 This Engineering Practice is intended as a guide for those who prepare granular applicator calibration procedures. The purpose is to encourage practices that will improve uniformity and accuracy of application with granular applicators. Attainment of accurate and more uniform application can reduce the quantity of active ingredient required for a given degree of control.

1.2 This Engineering Practice pertains to the calibration of row, band, and broadcast applicators. Because of the nature of the variables involved in chemical applications, a single calibration procedure applicable in all situations is not feasible.

SECTION 2—GENERAL

2.1 Safety

2.1.1 An appropriate warning for safe handling of chemicals shall be included with calibration instructions. (See ASAE Standard ASAE S318, Safety for Agricultural Equipment, and sample safety sign in Fig. 1)

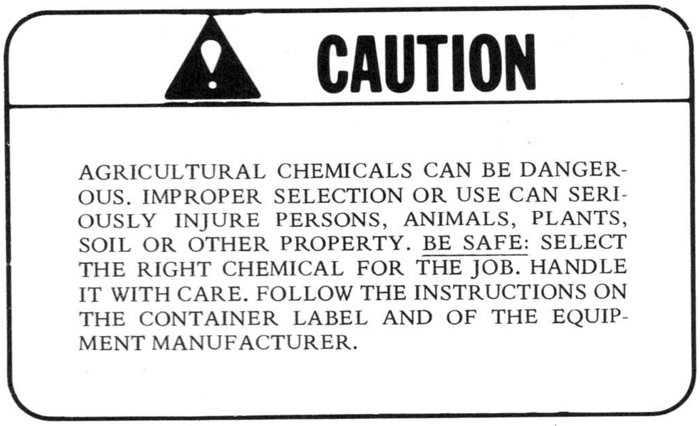

CAUTION

AGRICULTURAL CHEMICALS CAN BE DANGEROUS. IMPROPER SELECTION OR USE CAN SERIOUSLY INJURE PERSONS, ANIMALS, PLANTS, SOIL OR OTHER PROPERTY. BE SAFE: SELECT THE RIGHT CHEMICAL FOR THE JOB. HANDLE IT WITH CARE. FOLLOW THE INSTRUCTIONS ON THE CONTAINER LABEL AND OF THE EQUIPMENT MANUFACTURER.

FIG. 1—SAMPLE SAFETY SIGN

2.2 The amount of granular material applied to a given area depends on the following:

2.2.1 Orifice area. An adjustable orifice is used on most applicators to regulate the flow rate. The orifice area is determined by the orifice setting. Adjustment in the orifice setting should always be made in one direction only, such as from closed to open, to reduce variability in output for a particular setting.

2.2.2 Ground speed of the applicator. A constant speed must be maintained for uniform application. Even on applicators that use a rotor whose speed varies with ground speed, the flow of the granules through the orifice is not necessarily proportional to ground speed.

2.2.3 Product characteristics. The wide variation in size, density, and shape of the particles and the nature of the inert materials require that calibration be made for each chemical applied. (See ASAE Engineering Practice ASAE EP372, Granular Pesticide Guidelines)

2.2.4 Width of treatment. Most calibration procedures are used to determine the amount (kg or lb) applied per unit area of surface (hectare or acre) actually covered with chemical. For band application the area actually treated is the area in the band and not the area of cropland covered. Some row crop recommendations are based on treating in-the-row rather than the area actually covered. In these cases recommendations are generally based on amount of granules per row length; generally kilograms per 100 m of row (ounces per 1000 ft).

NOTE: If a chemical manufacturer lists the application rate for in-the-row application as pounds per 40-in. row acre, it should be converted to ounces per 1000 ft of row by multiplying times 1.22.

2.2.5 Rotor speed. Manufacturer's literature should be consulted for recommended rotor speeds that will give maximum efficiency.

2.2.6 Climatic conditions. Temperature and humidity can change the flow rate of the granules through the orifice.

2.3 The granular application rate may be determined from the following:

2.3.1
$$\text{Granular application rate} = \frac{\text{Quantity used (kg or lb)}}{\text{Area covered (hectare or acres)}}$$

Granular application rate, as calculated, is the amount of granules per unit area. The quantity used is the total quantity of granules (kg or lb) used by the applicator on a measured course. The area covered is the area (hectares or acres) actually treated on the measured course.

2.3.2 In-the-row applications are determined directly by measuring quantity in kilograms per 100 m (ounces per 1000 ft).

2.3.3 There are other good methods for calculating the granule application rate. Tables, charts, and special slide rules are examples. Special factors can be used if the length of the measured course is selected so that a known decimal or fractional part of an acre or hectare is covered with the application width used.

SECTION 3—EQUIPMENT

3.1 The type and amount of equipment needed will depend on the method calibration. Common items of equipment include:

3.1.1 Tape, 25, 50, or 100 m (50, 100, or 300 ft) or land measuring wheel.

3.1.2 Stakes or flags for marking a measured course.

3.1.3 Containers to collect granules (paper, cloth, or plastic bags).

3.1.4 Scales for weighing granules (kg, lb, ounces).

SECTION 4—CALIBRATION PROCEDURES

4.1 Several different methods are satisfactory for calibration of granular applicators. In each method there are several basic steps that must be performed.

4.1.1 Lay out a measured course in the field where application will take place or in an area having similar soil and terrain conditions. To minimize the error in collecting granules, the acceleration and deceleration distances should be as short as possible, and the total distance driven should be as long as is practical. Sufficient quantity of material must be collected during the test to allow for accurate weighing on the scales that are available.

4.1.2 Determine the ground speed that will be used. Except for the orifice setting, ground speed is the most significant factor affecting the application rate. Therefore, it is important that a uniform speed be maintained during calibration and application.

4.1.3 Determine and set the initial orifice setting. Both chemical and equipment manufacturers furnish application rate charts which are intended to be a starting point for determining the correct orifice setting.

4.1.4 Fill the hoppers at least half full and operate the applicator until the granules are feeding from all units.

NOTE: for broadcast applicators fill the hopper to an easily determined level and omit paragraph 4.1.5.

4.1.5 Attach a collection container under the openings.

4.1.6 Operate the applicator over the measured course at the speed determined in paragraph 4.1.2.

4.1.7 Weigh the granules collected from each unit of the applicator.

NOTE: For broadcast applicators determine the quantity of material (kg or lb) required to refill the hopper to the original level.

4.1.8 If desired application rate is amount per unit area, use paragraph 2.3.1 to calculate the rate. For in-the-row applications calculate kilograms per 100 m of row (ounces per 1000 ft).

4.1.9 Compare the amount collected or calculated with the recommended rate. If necessary, reset the applicators and recalibrate until the desired application rate is obtained.

4.1.10 Recalibrate frequently to compensate for any changes due to the variables listed in paragraph 2.2.

Cited Standards:

ASAE S318, Safety for Agricultural Equipment
ASAE S372, Granular Pesticide Guidelines

ASAE Engineering Practice: ASAE EP372

GRANULAR PESTICIDE GUIDELINES

Developed by the ASAE Agricultural Chemical Application Committee; approved by the Power and Machinery Division Standards Committee; adopted by ASAE December 1974; reconfirmed December 1979, December 1984.

SECTION 1—PURPOSE AND SCOPE

1.1 The purpose of this Engineering Practice is to encourage practices that will improve uniformity of application with surface distributors. A further purpose is to discourage very high application rates for economics. Another objective is to reduce the range of and the use of widely varying density of carrier materials.

1.2 Special purpose applications may require granule sizes or application rates outside the range recommended.

SECTION 2—GENERAL

2.1 Indicated mesh sizes are from American Society for Testing and Materials Standard E-11, Specifications for Wire-Cloth Sieves for Testing Purposes.

2.2 All indicated rates in this Engineering Practice are total product applied on a broadcast basis.

SECTION 3—GRANULAR CARRIER GUIDELINES

3.1 To improve uniformity of application

3.1.1 Use granules from 14 mesh thru 50 mesh. Material finer than 50 mesh increases drift.

3.1.2 Reduce the size range of granules in a formulation to a maximum spread of three sizes. (Examples: 14 thru 18, 16 thru 20, 18 thru 25, 20 thru 30, etc.)

3.1.3 Avoid application rates of less than 3.4 kg per hectare (3 lb per acre) by decreasing the concentration of active ingredient in the formulation.

3.1.4 Use materials with a static angle of repose less than 55 degrees, to avoid bridging in applicator hoppers.

3.2 Maximum application rate

3.2.1 Increase the concentration of active ingredient for pesticides now requiring higher application rates to conform, and to reduce the carrier, freight and handling costs of these formulations. Unless special circumstances such as safety or registration clearance warrant, limit application rates to a maximum of 34 kg per hectare (30 lb per acre).

3.3 Density

3.3.1 Use of carrier materials having bulk densities between 416 and 1522 kg/m^3 (26 and 95 lb/ft^3).

Cited Standard:

ASTM E11, Specifications for Wire-Cloth Sieves for Testing Purposes.

SAFETY FOR SELF-UNLOADING FORAGE BOXES

Developed by the ASAE Wagon Box, Forage Box, Manure Spreader and Farm Wagon Subcommittee; approved by the Power and Machinery Division Standards Committee; adopted by ASAE March 1975; reconfirmed December 1979, December 1984.

SECTION 1—PURPOSE

1.1 This Standard is intended to improve the degree of personal safety for operators and others involved during normal operation and servicing of self-unloading forage boxes used to move agricultural material on farms.

SECTION 2—SCOPE

2.1 This Standard is for wagon and truck-mounted, self-unloading forage boxes designed primarily for use in transporting and unloading forage.

2.2 In addition to the design and configuration of equipment, hazard control and accident prevention are dependent upon the awareness, concern, and prudence of personnel involved in the operation, transport, maintenance, and storage of equipment.

SECTION 3—POWER DISCONNECT DEVICE

3.1 A device shall be provided to disengage power from all moving mechanisms except the main power source.

3.2 The device shall be readily accessible from areas in or near the unloading chamber where the operator is exposed to functional mechanisms that are not otherwise shielded because of necessary function.

3.3 The force required to activate the device shall not exceed 178 N (40 lb). After activation, movement of the device in any direction shall not reengage power until it has been manually reset at a location remote from the disconnect device.

SECTION 4—CONTROLS

4.1 Controls for the forage box shall be arranged so they can be readily reached from the position where the unloading operation is normally observed.

SECTION 5—SHIELDING

5.1 All moving parts shall be shielded in accordance with ASAE Standard ASAE S318, Safety for Agricultural Equipment.

5.2 Slatted bed chains extending beyond the rear tailgate shall be shielded the entire width of the box.

SECTION 6—SAFETY SIGNS

6.1 A safety sign shall alert the operator and others of the risk of personal injury in normal operation and servicing. The sign shall be prominently affixed near the controls.

6.2 A safety sign shall be prominently affixed near the side unloading opening indicating potential hazard of moving parts in this area.

6.3 All safety signs shall conform to ASAE Standard ASAE S318, Safety for Agricultural Equipment.

6.4 A safety sign shall be provided near the power disconnect device describing its activation, i.e. "Warning! Push for Emergency Stop".

SECTION 7—HIGHWAY TRAVEL

7.1 Items related to safety for travel on highways shall conform to the following ASAE Standards:

- S318, Safety for Agricultural Equipment
- S276, Slow-Moving Vehicle Identification Emblem
- S277, Mounting Brackets and Socket for Warning Lamp and Slow-Moving Vehicle (SMV) Identification Emblem
- S279, Lighting and Marking of Agricultural Field Equipment on Highways
- S338, Safety Chain for Towed Equipment.

E

SECTION 8—GENERAL

8.1 Forage boxes shall meet all other applicable portions of ASAE Standard ASAE S318, Safety for Agricultural Equipment, not specifically mentioned in this Standard.

Cited Standards:

ASAE S276, Slow-Moving Vehicle Identification Emblem
ASAE S277, Mounting Brackets and Socket for Warning Lamp and Slow-Moving Vehicle (SMV) Identification Emblem
ASAE S279, Lighting and Marking of Agricultural Field Equipment on Highways
ASAE S318, Safety for Agricultural Equipment
ASAE S338, Safety Chain for Towed Equipment

TERMINOLOGY AND SPECIFICATION DEFINITIONS FOR AGRICULTURAL AUGER CONVEYING EQUIPMENT

Proposed for the Auger and Elevator Manufacturers Council of the Farm and Industrial Equipment Institute; approved by the Power and Machinery Division Standards Committee; adopted by ASAE March 1975; reconfirmed December 1979, December 1984.

SECTION 1—PURPOSE AND SCOPE

1.1 The purpose of this Standard is to provide uniform terminology, performance specifications and dimensional specifications for portable farm augers and their related accessories designed primarily for conveying agricultural materials on farms.

SECTION 2—PERFORMANCE SPECIFICATIONS

2.1 Capacity. The capacity of units shall be expressed in bushels per hour (one bushel = 1.244 cubic feet) or liters per hour.

2.2 Power. The power requirements shall be expressed in horsepower or watts.

SECTION 3—DIMENSIONAL SPECIFICATIONS (See Fig. 1)

3.1 Auger length: The length of the tube assembly including any intake but not including any intake hopper or head drive components (dimension A).

3.2 Intake length: The length of the visible flighting with the control gate (if unit is so equipped) in the full open position (dimension B).

3.3 Transport angle: The angle included between the auger tube and the ground when the unit is in the lowest recommended transport position and with hitch on the ground (dimension C).

3.4 Maximum operating angle: The angle included between the auger tube and the ground when the unit is in the highest recommended operating position, and with the hitch on the ground (dimension D).

3.5 Auger size: The outside diameter of the auger tube (dimension E).

3.6 Reach at maximum height: The horizontal distance from the foremost part of the undercarriage to the center of the discharge end when the unit is at the maximum recommended operating angle with hitch on ground (dimension F).

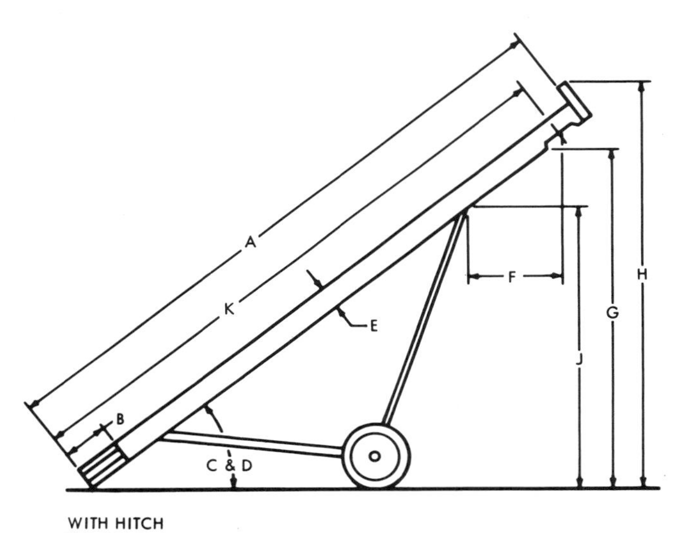

FIG. 1—TYPICAL PORTABLE AUGER

3.7 Maximum lift height: The vertical distance from the ground to the lowest point of the discharge (excluding down spout attachments) when the unit is raised to the maximum recommended operating angle and with the hitch on the ground (dimension G).

3.8 Transport height: The vertical distance from the ground to the uppermost portion with the unit in the lowest transport position and with the hitch on the ground (dimension H).

3.9 Eave clearance: The vertical distance from the ground to the foremost component of the undercarriage when the unit is at the maximum raised height (dimension J).

3.10 Discharge length: The total length of conveying from the outer end of the exposed flighting assembly at the intake to the centerline of the discharge (dimension K).

SECTION 4—TERMINOLOGY

4.1 Types of machines

4.1.1 Auger: A conveyor with screw type flighting in a tubular shaped enclosure with auxiliary accessories, to be usable in conveying recommended materials by rotating the flighting in relation to the enclosure.

4.1.2 Distributing auger: An auger capable of discharging material to one or more locations.

4.1.3 Feeding auger: An auger which releases conveyed material essentially uniformly along a substantial portion of its length.

4.1.4 Portable auger: An auger whose accessories include a suitable support system which provides mobility.

4.1.5 Stationary auger (fixed): An auger essentially permanently installed on a particular site without mobility capability.

4.1.6 Tube auger: An auger in which the enclosure is essentially a cylinder.

4.1.7 Utility auger: A mobile auger which is not equipped with an auxiliary transport support.

4.2 Types of final drives

4.2.1 Center drive: When final drive location is between the intake and discharge.

4.2.2 Final drive: Where torque is applied to the auger flighting assembly.

4.2.3 Intake or bottom drive: When final drive location is at the intake end.

4.2.4 Top or head drive: When final drive location is at the discharge end.

4.3 General terminology

4.3.1 Automatically controlled: Operated by the action of its mechanism being initiated by some impersonal influence, such as being controlled by low-level and high-level indicators.

4.3.2 Axle: Portion of undercarriage support to which wheels are attached.

4.3.3 Connecting stub shaft: Connector between two flighting assemblies.

4.3.4 Discharge: The area where conveyed material is discharged from the machine.

4.3.5 Drive shaft: Shaft that transmits power between the power source and the final drive.

4.3.6 Drive shaft bearing support: Drive shaft bearing holder.

4.3.7 Enclosed: Moving parts are so guarded that physical contact is precluded as long as the guard remains in place. This does not prohibit use of access areas for inspection or lubrication as long as the access is otherwise inaccesible by use of substantial covers designed for removal and replacement.

4.3.8 Flighting: Helicoid screw.

4.3.9 Flighting assembly: Flighting shaft with flighting attached.

4.3.10 Flighting shaft: Shaft on which flighting is mounted.

4.3.11 Guarded: Shielded fenced, or otherwise protected by means of suitable deterrent, or by nature of location so as to remove foreseeable risk of personal injury from accidental contact or approach.

4.3.12 Guarded by location: Moving parts are so protected by their location with reference to frame, foundation, or structure as to remove the foreseeable risk of accidential contact by persons or objects.

4.3.13 Guarding not possible: Wherever conditions prevail which if guarded would render the auger unusable.

4.3.14 Head stub shaft: Connector between flighting assembly and head drive or head bearing.

4.3.15 Hitch: Device for connecting to a towing vehicle.

4.3.16 Installer: Management in effective control of putting equipment in place and in operating condition.

4.3.17 Intake: The area where material to be conveyed enters the machine.

4.3.18 Intake guard: Safety device for exposed intake flighting assembly.

4.3.19 Intake stub shaft: Connector between intake flight assembly and intake drive or bearing.

4.3.20 Lift arm: Undercarriage support member located nearest discharge end.

4.3.21 Lower arm: Undercarriage support member located nearest to the intake end.

4.3.22 Manager: The management in effective control of the operation after installation.

4.3.23 Moving parts: Parts which have motion during operation of the machine.

4.3.24 Nip point (pinch point): That point at which a machine element moving in line or rotating meets another element in such a manner that it is possible to nip, pinch, squeeze, or entrap objects coming into contact with one or both of the members.

4.3.25 Operator: Designated agent(s) of the owner or manager.

4.3.26 Safety device: Mechanism or an arrangement for the specific purpose of improving the degree of personal safety for the operator and others involved during the normal operation and servicing of a portable farm auger.

4.3.27 Shield: Device used to enclose or guard to reduce the possibility of accidental contact with portions of a machine which, if contacted, could cause personal injury.

4.3.28 Track: Guide for lift arm for undercarriage support.

4.3.29 Trough: Flighting assembly housing which is open at the top and essentially "U" shaped in cross section.

4.3.30 Truss: Structural supporting framework.

4.3.31 Trrss anchors: End attaching point for truss.

4.3.32 Truss rod or cable: Tie between truss anchors and truss support.

4.3.33 Truss support: Stand off brace for truss.

4.3.34 Tube: Flighting assembly housing which is essentially round in cross section.

4.3.35 Undercarriage: Assembly that supports auger and provides mobility.

4.3.36 Winch: Drum type lifting device to achieve desired angle of elevation.

4.3.37 Winch cable: Wire rope used for raising and lowering the auger.

4.4 Accessories or auxiliary equipment:

4.4.1 Control gate: Device to adjust the intake areas for capacity control.

4.4.2 Discharge spout: Means for guiding released material from the discharge in a desired direction to a desired receptacle.

4.4.3 Drag auger: A device with a screw type flighting in contact with material, pivotally attached to the auger intake to draw material toward the intake when rotated. Rotating power is through the pivotal attachment.

4.4.4 Drag auger handle: Portion of drag auger assembly used to control position of drag auger.

4.4.5 Gravity hopper: Device for receiving and directing material into the rotating flighting without power driven components.

4.4.6 Hopper: Device for receiving and directing material into the rotating flighting.

4.4.7 Powered hopper: A device for receiving material to by conveyed at a point near the auger rotating flighting and for conveying the material to the rotating auger flighting. Method of conveying is usually by rotating flighting in the hopper, but is not confined to this conveying method.

4.4.8 Swing away hopper: Powered hopper which swings to one or both sides to clear a driving lane through the normal operating position.

4.4.9 Tilting hopper: Powered hopper which raises to approximately vertical to clear a driving lane through the normal operating position.

T CAPACITY RATINGS AND UNLOADING DIMENSIONS FOR COTTON HARVESTER BASKETS

Developed by the Farm Materials Handling Committee; approved by the Power and Machinery Division Standards Committee; adopted by ASAE as a Tentative Standard February 1975; reconfirmed December 1975, December 1976, December 1977, December 1978, December 1979, December 1980, December 1981, December 1982, December 1983, December 1984; revised editorially and reclassified as a full Standard June 1986; revised March 1988.

SECTION 1—PURPOSE AND SCOPE

1.1 This Standard is intended to provide uniformity in the method of expressing the following information relative to cotton strippers and cotton pickers.

1.1.1 Capacity of basket.

1.1.2 Unloading height of basket.

1.1.3 Lip height of raised basket.

1.1.4 Unloading angle of basket.

1.1.5 Maximum basket height.

SECTION 2—CAPACITY

2.1 Capacity of cotton stripper and cotton picker baskets shall be expressed as the total volume enclosed by the basket and lid. This volume shall be expressed in cubic meters (cubic feet) and may be determined from dimensional layout drawings.

2.2 In addition to the volumetric capacity rating, it shall be permissible to express a nominal capacity in kilograms (pounds). This rating shall be determined based on a density of 80.1 kg/m^3(5.0 lb/ft^3) for baskets without a compaction device.

2.3 If a mechanical compaction device is used, the capacity shall be the total weight obtained by actual tests conducted by the manufacturer. It shall be expressed in kilograms (pounds) and be based on cotton having an uncompacted density of 80.1 kg/m^3 (5.0 lb/ft^3).

SECTION 3—UNLOADING HEIGHT

3.1 Unloading height H is defined as the maximum height of a 51 mm (2 in.) thick vertical side of a container which has clearance to all aspects of the machine when the basket is being unloaded into the container, under the following conditions and as shown in Fig. 1.

3.1.1 Machine is stationary.

3.1.2 Basket is in fully raised position.

3.1.3 The lip location is in any vertical plane falling within the container.

3.2 Unloading height may be determined from dimensional layout drawings, and shall be expressed in meters to the nearest hundredth (feet to the nearest tenth).

SECTION 4—LIP HEIGHT

4.1 Lip height H_L is defined as the vertical dimension to the lip of the basket with the basket in fully raised position. This dimension may be determined from dimensional layout drawings, and shall be expressed in meters to the nearest hundredth (feet to the nearest tenth).

SECTION 5—UNLOADING ANGLE

5.1 The unloading angle of cotton picker or cotton stripper baskets is the angle α between the floor of the basket (unloading surface) and the horizontal when the basket is in its fully raised position. This angle is shown in Fig. 1 and shall be specified in degrees.

SECTION 6—MAXIMUM HEIGHT

6.1 Maximum basket height H_M is defined as the maximum vertical dimension which can be attained by any component on the basket during the unloading cycle with the assumption that the machine is stationary and on level ground as shown in Fig. 1.

6.2 Maximum basket height may be determined from dimensional layout drawings and shall be expressed in meters to the nearest hundredth (feet to the nearest tenth).

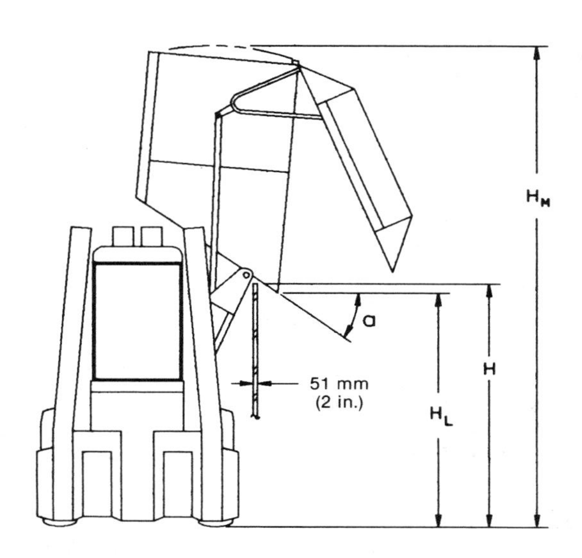

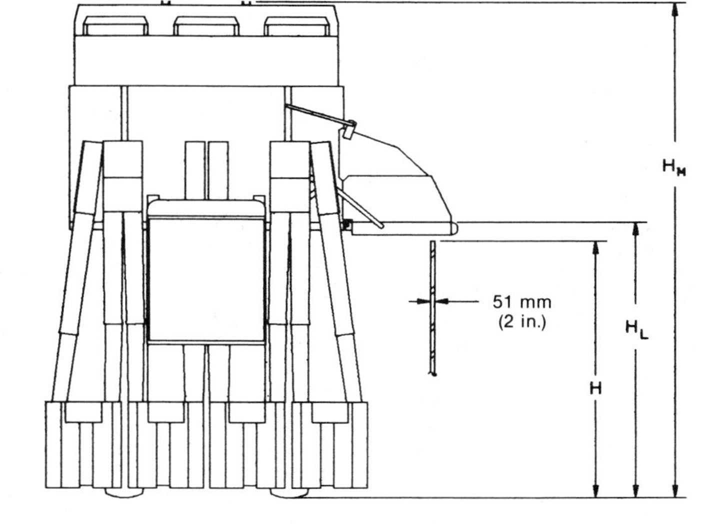

FIG. 1—TYPICAL COTTON HARVESTERS

ASAE Standard: ASAE S380

TEST PROCEDURE TO MEASURE MIXING ABILITY OF PORTABLE FARM BATCH MIXERS

Developed by the ASAE Animal Feed Processing Implements Subcommittee of the Power and Machinery Division Standards Committee; approved by the Power and Machinery Division Standards Committee; adopted by ASAE as a Tentative Standard December 1975; reconfirmed December 1976, December 1977, December 1978, December 1979; reclassified as a full Standard March 1981; reconfirmed December 1985.

SECTION 1—PURPOSE AND SCOPE

1.1 The purpose of this Standard is to define a uniform test procedure and measurement to evaluate the mixing ability of on the farm, portable batch mixers.

1.2 This Standard is applicable to batch mixers mixing dry granular ingredients of different particle size and density. The test shall consist of measuring the mix of an ingredient of smaller particle size and/or one of larger particle size with one base material.

SECTION 2—PERFORMANCE CRITERIA

2.1 The performance criteria is the uniformity of dispersion of the ingredients throughout the entire batch.

SECTION 3—TEST PROCEDURE

3.1 Test formula. The formula for testing the performance of a batch mixer shall consist of a mixture of 0.5 percent salt and/or 5.0 percent whole kernel shelled corn by weight mixed into a base material of ground shelled corn, U.S. Grade No. 2 of less than 14 percent moisture. The base material, corn, shall be ground to a fineness of a log mean diameter within 0.90 to 1.30 mm and a logarithmic standard deviation of 2.50 or less. (See ASAE Standard ASAE S319, Method of Determining and Expressing Fineness of Feed Materials by Sieving) Most hammer mills should obtain this grind with a 9.7 mm (0.38 in.) screen. The salt tracer shall have a log mean diameter of within 0.35 to 0.55 mm and a logarithmic standard deviation of 1.75 or less. The whole kernel corn tracer shall have no more hulls and cracked kernels than allow 2.5 percent to pass through a 4.70 mm (0.185 in.) screen.

3.2 Mixing procedure. The mixing procedure shall be the same as that recommended to the customer by the manufacturer such as optimum mixing time and batch size. The exact mixing procedure used to obtain the results for this standard test procedure shall be reported. This shall include:

3.2.1 Sequence when ingredients were added.

3.2.2 Machine location where ingredients were added.

3.2.3 Mixer speed such as full rated speed or portion thereof.

3.2.4 Tracer, if only one is used.

3.2.5 Total time tracer element(s) was circulated in mixer including grinding time and circulating time after mixer is filled. Both times shall be reported if two tracers are used.

3.2.6 Mixer speed while unloading; such as full rated speed or portion thereof.

3.2.7 Batch size by weight.

3.2.8 Percentage of full tank.

3.3 Method of sampling. Samples shall be taken from the discharge of the mixer. Fifteen samples shall be taken, one 15 seconds after unloading begins, one when a 0.17 m^3 (5 bu) or less is still in the tank and the remaining equally spaced between.

3.4 Sample size. The sample size shall be at least 150 g (5.3 oz) for salt and at least 4000 g (8.8 lb) for shelled corn.

3.5 Analysis of samples

3.5.1 Salt tracer analysis. The salt may be measured by a laboratory using a chemical procedure. The salt may be measured by any other documented reliable method. Documentation of measurement reliability must be included in the report.

3.5.2 Shelled corn tracer analysis. The corn shall be measured by sifting through a sieve. The sieve must be a size that allows all of the ground corn to pass through and retains all but 0.25 percent of the whole kernel corn.

3.6 Method of reporting. The following must be reported in the results of the test performed.

3.6.1 Machine model, size and serial number.

3.6.2 Date test performed.

3.6.3 Measure of mixing ability as expressed by a coefficient of variation for each of the tracers.

3.7 Sample calculations for coefficient of variation

Sample No.	Value of Sample, X	Value of $(X-M)^2$
1	0.590	0.000 784
2	0.560	0.000 004
3	0.625	0.003 969
4	0.560	0.000 004
5	0.560	0.000 004
6	0.560	0.000 004
7	0.530	0.001 024
8	0.590	0.000 784
9	0.560	0.000 004
10	0.530	0.001 024
11	0.560	0.000 004
12	0.520	0.001 764
13	0.570	0.000 064
14	0.560	0.000 004
15	0.560	0.000 004
	$\Sigma X = 8.435$	$\Sigma(X-M)^2 = 0.009\ 445$

$$M = \frac{\Sigma X}{n} = \frac{8.435}{15} = 0.562$$

$$S^2 = \frac{\Sigma(X-M)^2}{n-1} = \frac{0.009\ 445}{14} = 0.000\ 674\ 6$$

$$S = 0.026$$

$$CV = \frac{S}{M} \times 100 = \frac{0.026}{0.562} \times 100 = 4.63 \text{ percent}$$

where

n = number of samples
X = percent of tracer in sample
M = mean value of samples, X
S = one standard deviation
CV = coefficient of variation

Cited Standard:

ASAE S319, Method of Determining and Expressing Fineness of Feed Materials by Sieving

ROLL-OVER PROTECTIVE STRUCTURES (ROPS) FOR WHEELED AGRICULTURAL TRACTORS

*Corresponds in substance to previous revision S383.

Supersedes S305, Operator Protection for Wheel Type Agricultural Tractors, adopted August 1967; S306, Protective Frame for Agricultural Tractors—Test Procedures and Performance Requirements, adopted August 1967; and S336, Protective Enclosures for Agricultural Tractors—Test Procedures and Performance Requirements, adopted February 1970. Proposed by the Engineering Policy Committee of Farm and Industrial Equipment Institute; approved by the Power and Machinery Division Standards Committee; adopted by ASAE March 1977; reconfirmed December 1981; revised December 1983.

SECTION 1—PURPOSE

1.1 The purpose of this Standard is to establish the test and performance requirements of a roll-over protective structure (ROPS) designed for wheel-type agricultural tractors to minimize the frequency and severity of operator injury resulting from accidental upsets. All self-propelled implements are excluded.

SECTION 2—SCOPE

2.1 Fulfillment of the intended purpose requires testing as follows:

2.1.1 A laboratory test, under repeatable and controlled loading, to permit analysis of the ROPS for compliance with the performance requirements of this Standard. Either the static test as described in paragraph 6.1 or the dynamic test as described in paragraph 6.2 shall be conducted.

2.1.2 A crush test to verify the effectiveness of the deformed ROPS in supporting the tractor in an upset attitude.

2.1.3 A field upset test under reasonably controlled conditions, both to the rear and side, to verify the effectiveness of the protective system under actual dynamic conditions. (See paragraph 6.4.1 for requirements for the omission of this test.)

2.1.4 In addition to the laboratory and field loading requirements, there is a temperature-material requirement (see paragraph 7.1.2).

2.2 The test procedures and performance requirements outlined in this Standard are based on currently available engineering data.

SECTION 3—DEFINITIONS

3.1 An agricultural tractor, for purposes of this Standard, is defined as a 2 or 4-wheel drive type machine of more than 15 kW (20 hp) net engine power as defined by Society of Automotive Engineers Standard J816, Engine Test Code—Spark Ignition and Diesel, designed primarily to provide the tractive power to pull, push, carry, propel, and/or provide power to implements designed primarily for agricultural usage.

3.2 The tractor mass is defined as the maximum gross machine mass determined by the manufacturer or a minimum ratio of mass to maximum power take-off power at rated engine speed of 67 kg/kW (110 lb/hp), whichever is greater. The mass includes the ROPS, all fuels, and other components required for normal use. (When power take-off power is not available, use 95% of net engine flywheel power.)

3.3 A roll-over protective structure (ROPS) is a cab or frame for the

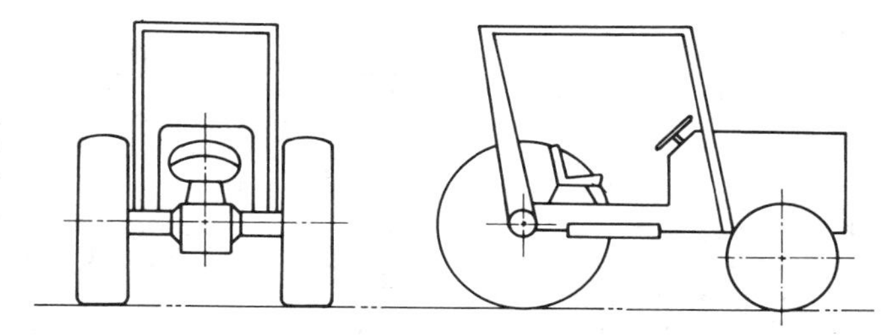

FIG. 2—TRACTOR WITH TYPICAL 4 POST ROPS

protection of operators of wheeled agricultural tractors to minimize the possibility of serious operator injury resulting from accidental upsets. The protective structure is characterized by providing space for the clearance zone inside the envelope of the structure or within a space bounded by a series of straight lines from the outer edge of the structure to any part of the tractor that might come in contact with flat ground and is capable of supporting the tractor in that position if the tractor overturns (see Figs. 1, 2 and 3 for typical configurations).

3.4 Seat reference point (SRP in Figs. 4, 5 and 6) is that point where a vertical line tangent to the most forward point at the longitudinal seat centerline of the seat back and a horizontal line tangent to the highest point of the seat cushion intersect in the longitudinal seat centerline section. The SRP is determined with the seat unloaded and adjusted to the highest and most rearward position provided for seated operation of the tractor.

SECTION 4—GENERAL REQUIREMENTS

4.1 Refer to ASAE Standard S318, Safety for Agricultural Equipment, for safety requirements.

4.2 Rear input energy tests (static, dynamic, or field upset) need not be performed on ROPS applied to tractors having four driven wheels and where the static vertical force reaction at the front wheels is greater than the static vertical force reaction at the rear wheels since this type of tractor is not prone to rearward upset.

4.3 The tractor mass used shall be that of the heaviest tractor model on which the ROPS is to be used.

4.4 New ROPS and mounting connections of the same design shall be used for conducting the tests as described in the static (paragraph 6.1), the dynamic (paragraph 6.2), or the field upset procedures (paragraph 6.4).

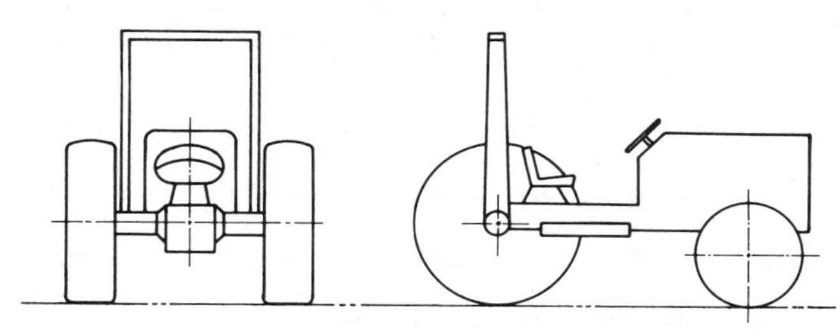

FIG. 1—TRACTOR WITH TYPICAL 2 POST ROPS

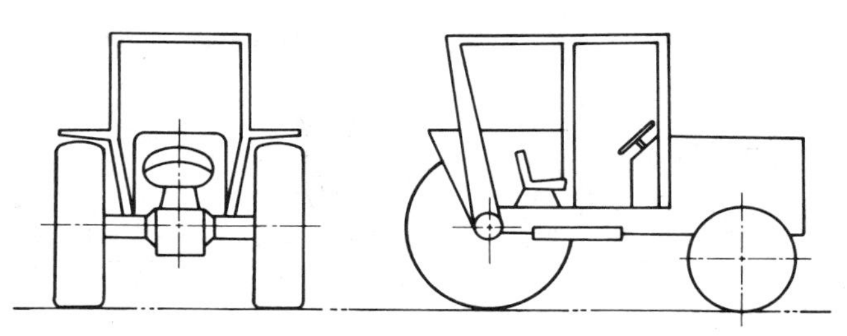

FIG. 3—TRACTOR WITH TYPICAL ROPS ENCLOSURES

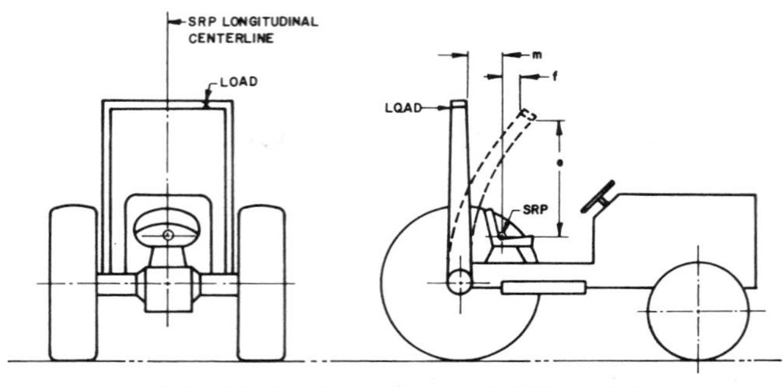

FIG. 4—TYPICAL REAR LOAD APPLICATION

4.5 In case of an offset seat, the ROPS loading shall be on the side with the least space between the centerline of the seat and the protective structure.

4.6 Accuracy of measurement:

MEASUREMENT	ACCURACY
Deflections of enclosure	± 5% of deflection measured
Tractor mass	± 5% of mass measured
Force applied to frame	± 5% of force measured
Dimensions of critical zone	± 12.7 mm (0.5 in.)

4.7 Where movable or normally removable portions of the ROPS add to structural strength, they shall be placed in configurations that contribute least to the structural strength during the test.

4.8 If an overhead weather shield is available as an optional attachment to the protective structure, it may be in place during tests, provided it does not contribute to the strength of the protective structure.

4.9 If an overhead falling object protective cover is available as an optional attachment to the protective structure, it may be in place during tests, provided it does not contribute to the strength of the protective structure.

4.10 No repairs or adjustments shall be made during the tests.

4.11 The protective structure shall meet the performance requirements established in Section 7-Performance Requirements.

SECTION 5—SEAT AND SEAT BELT REQUIREMENTS

5.1 ROPS equipped tractors shall be fitted with seat belt assemblies (Type 1) conforming to SAE Standard J4c, Motor Vehicle Seat Belt Assemblies, except as noted hereafter.

5.2 Where a suspended seat is used, the seat belt shall be fastened to the movable portion of the seat to accommodate the ride motion of the operator.

5.3 The seat belt anchorage shall be capable of withstanding a static tensile force of 4 448 N (1000 lbf) at 45 deg to the horizontal equally divided between the anchorages. The seat mounting shall be capable of withstanding this force plus a force equal to four times the force of gravity on the mass of all applicable seat components applied 45 deg to the horizontal in a forward and upward direction. In addition, the seat mounting shall be capable of withstanding 2 224 N (500 lbf) belt force plus two times the force of gravity on the mass of all applicable seat components both applied at 45 deg to the horizontal in an upward and rearward direction. Floor and seat deformation is acceptable provided there is no structural failure or release of the seat adjuster mechanism or other locking device. The seat adjuster or locking device need not be operable after application of the test load.

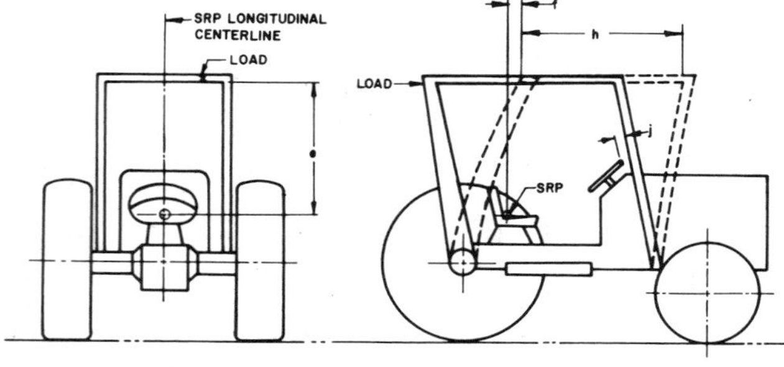

FIG. 5—TYPICAL REAR LOAD APPLICATION

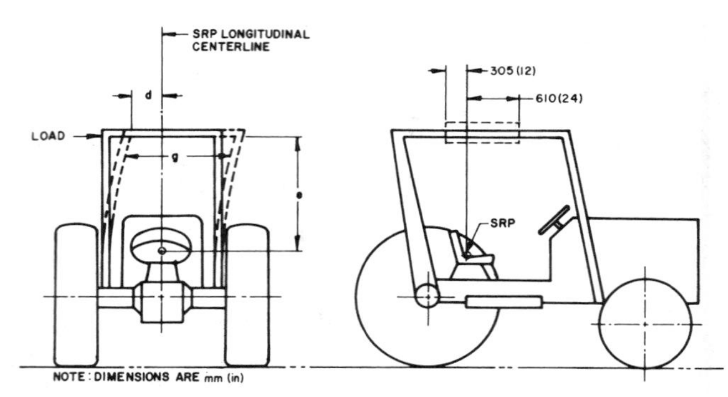

FIG. 6—TYPICAL SIDE LOAD APPLICATION

SECTION 6—TEST PROCEDURES

6.1 Static test (optional to paragraph 6.2)

6.1.1 Test conditions

6.1.1.1 The ROPS mounting base shall be the tractor chassis or the equivalent for which the ROPS is designed to assure the integrity of the entire system.

6.1.1.2 The ROPS shall be instrumented with the necessary equipment to obtain the required load deflection data at the location and direction specified in Figs. 4, 5 and 6. The measuring devices shall be located to record the force and deflection at the point of and along the line of loading. Load and deflection points shall be plotted in increments of deflection no greater than 13 mm (0.5 in.). The rate of application of deflection (load) shall be such that it can be considered static.

6.1.2 Definition of terms

M = tractor mass as defined in paragraphs 3.2 and 4.3
Units: M in kg
M′ in lb

E_{is} = energy input to be absorbed during side loading
E_{is} = 980 + 1.2 M units: J and kg
E_{is} = 8676 + 4.8 M′ units: in.-lbf and lb

E_{ir} = energy input to be absorbed during rear loading
E_{ir} = 1.4 M units: J and kg
E_{ir} = 5.64 M′ units: in.-lbf and lb

F = static load
Units: F in N
F′ in lbf

D = deflection under F
Units: D in mm
D′ in inches

F-D = static force deflection curve

E_u = strain energy absorbed by the structure
E_u = area under F-D curve
Units: E_u in J
E_u' in in.-lb

6.1.3 Static test procedures

6.1.3.1 Apply the rear load per Figs. 4 or 5 and record F and D simultaneously. Rear load application shall be uniformly distributed along a projected dimension no greater than 686 mm (27 in.) and an area no greater than 0.1032 m² (160 in.²) normal to the direction of load application. The load shall be applied to the upper extremity of the ROPS at the point which is midway between the center of the ROPS and the inside of the ROPS upright. If no structural cross member exists at the rear of the ROPS, a substitute test beam which does not add strength to the ROPS may be utilized to complete this test procedure (see paragraph 6.4.1.1 if field upset is omitted). Stop the test when (a) the strain energy absorbed by the structure is equal to or greater than the required input energy E_{ir} (paragraph 6.1.2); or (b) deflection of the structure exceeds the allowable deflection (paragraph 7.1.1).

6.1.3.2 Using data obtained in paragraph 6.1.3.1, construct the F-D curve as shown typically in Fig. 8 and calculate E_u.

6.1.3.3 Apply the side load as shown in Fig. 6 and record F and D simultaneously. Static side load application shall be uniformly distributed along a projected dimension no greater than 686 mm (27 in.) and an area no greater than 0.1032 m²

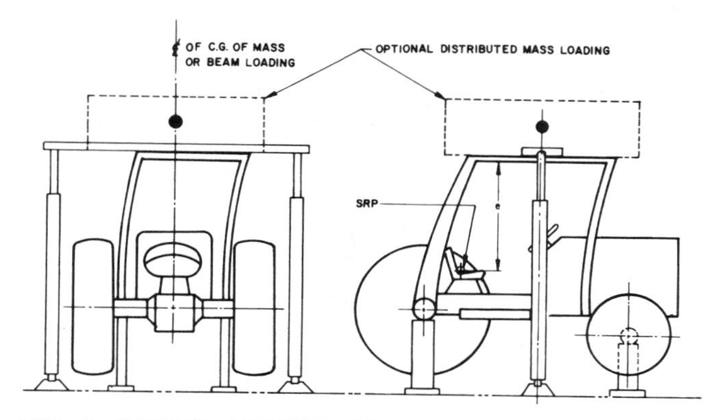

FIG. 7—TYPICAL METHOD OF LOAD APPLICATION FOR CRUSH TEST

(160 in.²) normal to the direction of load application. Side load application shall be at a 90 deg angle to the centerline of the vehicle. The center of side load application shall be located between a distance 610 mm (24 in.) forward, and a distance 305 mm (12 in.) rearward of the seat reference point to best utilize the structural strength (see Fig. 6). If the ROPS is a one or two post design, the side load shall be applied in line with the upper cross member. This side load shall be applied to the longitudinal side farthest from the point of rear load application. (See paragraph 6.4.1.1 if field upset is omitted). Stop the test when (a) the strain energy absorbed by the structure is equal to or greater than the required input energy E_{is} (paragraph 6.1.2); or (b) deflection of the structure exceeds the allowable deflection (paragraph 7.1.1).

6.1.3.4 Using data obtained in paragraph 6.1.3.3 construct the F-D curve as shown typically in Fig. 8 and calculate E_u.

6.2 Dynamic test (optional to paragraph 6.1)

6.2.1 Test conditions

6.2.1.1 The tractor shall be ballasted to achieve the mass as specified in paragraph 3.2 so that the static vertical force reaction at the front wheels shall be at least 33% of the static vertical force reaction at the rear wheels. The wheel tread setting, where adjustable, shall be at the position nearest to half-way between the minimum and maximum settings obtainable on the tractor. Where only two settings are obtainable, the minimum setting shall be used provided the tires do not interfere with structure deflection. The tires shall have no liquid ballast and shall be inflated to the maximum operating pressure recommended by the manufacturer.

6.2.1.2 The dynamic loading shall be produced by use of a 2 000 kg (4410 lb) mass acting as a pendulum. The impact face of the mass shall be 686 ± 25 x 686 ± 25 mm (27 ± 1 x 27 ± 1 in.) and shall be constructed so that its center of gravity is within 25 mm (1 in.) of its geometric center. The mass shall be suspended from a pivot point 5.5 to 6.7 m (18 to 22 ft) above the point of impact on the ROPS and shall be conveniently and safely adjustable for height (see Fig. 9).

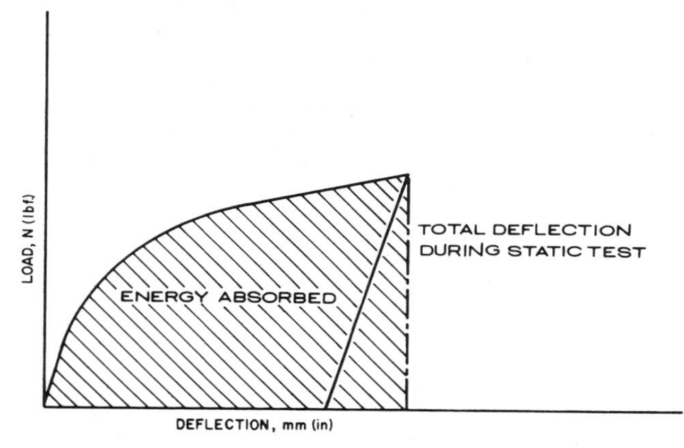

FIG. 8—FORCE DEFLECTION (F-D) CURVE

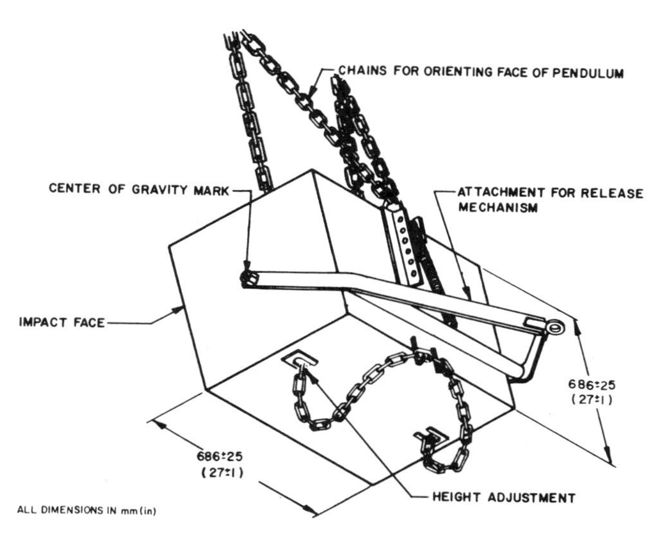

FIG. 9—PENDULUM

6.2.1.3 For each phase of testing, the tractor shall be restrained from moving when the dynamic load is applied. The restraining members shall have strength no less than and elasticity no greater than that of a 12.7 mm (0.50 in.) diameter steel cable. Points of attaching restraining members shall be located an appropriate distance behind the rear axle and in front of the front axle to provide 15 to 30 deg angle between a restraining cable and the horizontal. For the impact from the rear, the restraining cables shall be located in the plane in which the center of gravity of the pendulum will swing or, alternatively, two sets of symmetrically located cables may be used at convenient lateral locations on the tractor. For the impact from the side, restraining cables shall be used as shown in Figs. 10 and 11.

6.2.1.4 The restraining cable(s) shall be tightened to provide tire deflection of 6 to 8% of nominal tire section width. After the tractor is properly restrained, a beam no smaller than 150 x 150 mm (6 x 6 in.) in cross section shall be driven tightly against the appropriate wheels and clamped. For the test to the side, an additional beam of sufficient strength to prevent rim displacement shall be placed as a prop against the wheel rim nearest the operator's station on the side opposite the pendulum impact and shall be secured to the base so it is held tightly against the wheel rim during impact. The length of this beam shall be chosen so that it is at an angle of 25 to 40 deg to the horizontal when it is positioned against the wheel rim (see Figs. 10 and 11).

6.2.1.5 Means shall be provided for indicating the maximum instantaneous deflection relative to the SRP and parallel to the

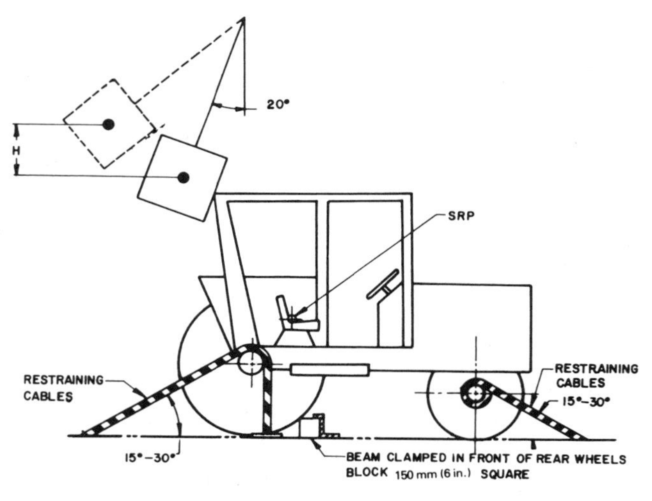

FIG. 10—TYPICAL REAR IMPACT APPLICATION

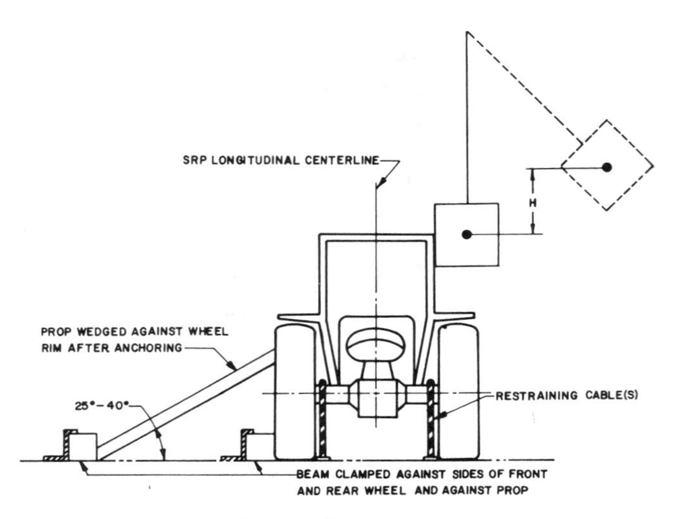

FIG. 11—TYPICAL SIDE IMPACT APPLICATION

vertical plane of the pendulum swing. A simple friction device is illustrated in Fig. 13.

6.2.1.6 If any cables, props, or blocking shift or break during the test, the test shall be repeated.

6.2.2 Definition of terms

M = tractor mass as defined in paragraphs 3.2 and 4.3
Units: M in kg
M′ in lb

H = vertical height of the pendulum mass
Units: H in mm
H′ in inches

The pendulum mass shall be 2 000 kg (4410 lb).

The pendulum mass shall be pulled back so that the height of its center of gravity above the point of impact is as follows:

$H = 125 + 0.107\ M$ Units: mm and kg

$H' = 4.92 + 0.0019\ M'$ Units: in. and lb

6.2.3 Dynamic test procedures

6.2.3.1 The ROPS shall be evaluated by imposing dynamic loading from the rear, followed by a load to the side on the same ROPS. The pendulum swinging from the height determined by paragraph 6.2.2 imposes the dynamic load. The position of the pendulum shall be selected so that the initial point of impact on the ROPS shall be in line with the arc of travel of the center of gravity of the pendulum. A quick-release mechanism should be used but shall not influence the attitude of the pendulum.

6.2.3.2 Impact at rear. The tractor shall be properly restrained per paragraphs 6.2.1.3 and 6.2.1.4. The tractor shall be positioned so that the supporting chains of the pendulum are at an angle of 20 deg to the vertical when striking the structure as shown in Fig. 10. If the angle of the cab or frame member at the point of contact is greater than 20 deg forward of the vertical, the angle of the face of the pendulum shall be further adjusted by any convenient means so that the striking face of the pendulum and the cab or frame member are parallel. The

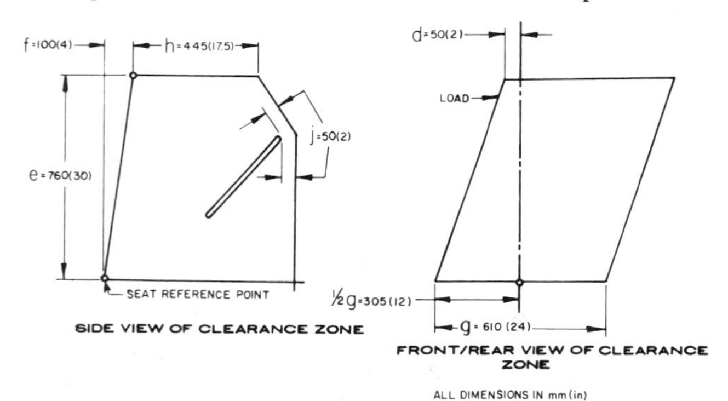

FIG. 12—PICTORIAL REPRESENTATION OF CLEARANCE ZONE

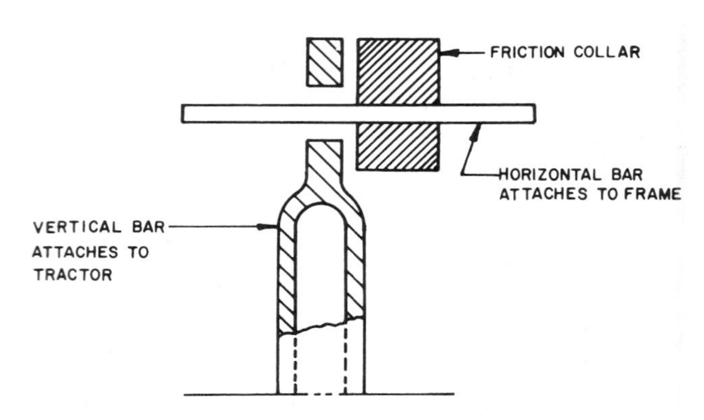

FIG. 13—TYPICAL METHOD OF MEASURING DEFLECTION

impact shall be applied to the upper extremity of the ROPS at the point which is midway between the centerline of the ROPS and the inside of the ROPS upright. If no structural cross member exists at the rear of the ROPS, a substitute test beam which does not add to the strength of the structure may be utilized to complete the test procedure.

6.2.3.3 Impact at side. The tractor shall be properly restrained as per paragraphs 6.2.1.3 and 6.2.1.4. The tractor shall be positioned so that the supporting chains of the pendulum are vertical when striking the structure as shown in Fig. 11. If the structural member at the point of impact is not vertical, the angle of the face of the pendulum shall be adjusted by any convenient means so that the striking face of the pendulum and the cab or frame member are parallel. The point of impact shall be at the upper extremity of the ROPS at a 90 deg angle to the centerline of tractor and located between a distance 610 mm (24 in.) forward, and a distance 305 mm (12 in.) rearward of the seat reference point to best utilize the structural strength (see Figs. 6 and 11). If the ROPS is a one or two post design, the side impact shall be applied in line with the upper cross member. The side impact shall be applied to the longitudinal side farthest from the point of rear impact.

6.3 Crush test procedure

6.3.1 After the ROPS has been subjected to either static or dynamic loads, the same ROPS shall be subjected to a static crush test.

6.3.2 The test load shall be 1.5 times the gravity force of the tractor mass as defined and explained in paragraph 3.2. The manner of distributing this load on the ROPS shall be such to best utilize those structural members in the fore and aft plane which will support the tractor in an upset position. The resultant of the initial crushing force or load shall be in the vertical directon and shall be in a vertical plane passing through the SRP and parallel to the longitudinal axis of the tractor (see Fig. 7). The tractor chassis shall be rigidly supported during this test.

6.4 Field upset test

6.4.1 Test conditions

6.4.1.1 The field upset test may be omitted if laboratory test (static or dynamic) results indicate compliance at an energy application of 115% or more of the requirements defined in paragraph 6.1.2 or 6.2.2 and the ROPS meets the temperature-material requirements (see paragraph 7.1.2).

6.4.1.2 The tractor shall be ballasted to achieve the mass as specified in paragraph 3.2 so that the static vertical force reaction at the front wheels shall be at least 33% of the static vertical force reaction at the rear wheels. The wheel tread setting, where adjustable, shall be at the position nearest to halfway between the minimum and maximum settings of the shortest axles obtainable on the tractor. Where only two settings are obtainable, the minimum setting shall be used provided the tires do not interfere with structure deflection. The tires may have liquid ballast and shall be inflated to the maximum operating pressure recommended by the manufacturer.

6.4.1.3 The tests shall be conducted on a firm soil bank as shown in Figs. 14 and 15. The soil in the impact area shall have an average cone index in the 0 to 152 mm (0 to 6 in.) layer not less than 150. (Cone index is defined in ASAE Standard S313,

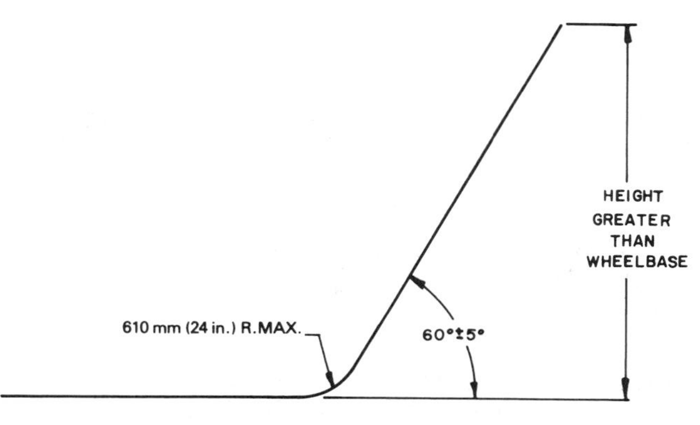

FIG. 14—TYPICAL REAR OVERTURN BANK

Soil Cone Penetrometer, or SAE J106, Soil Type and Strength Classification.) The path of tractor travel shall be 12 ± 2 deg to the top edge of the bank. A 457 mm (18 in.) high ramp as described and located in Fig. 15 shall be used to assist in upsetting the tractor to its side.

6.4.1.4 A means shall be provided for indicating the maximum instantaneous deflection of the structure during upset. A simple friction device is illustrated in Fig. 13.

6.4.2 Field upset test procedures

6.4.2.1 Rear upset. Rear upset shall be induced by engine power with the tractor operating in a gear to obtain 4.8 to 8.0 km/h (3 to 5 mph) at maximum governed engine rpm preferably by driving forward directly up a minimum slope of 60 ± 5 deg as shown in Fig. 14. The engine clutch may be used to aid in inducing the upset.

6.4.2.2 Side upset. Side upset shall be induced by driving the tractor under its own power along the specified path of travel at a minimum speed of 16 km/h (10 mph) or at maximum tractor speed if under 16 km/h (10 mph) and over the ramp as described in paragraph 6.4.1.3 (see Fig. 15.)

SECTION 7—PERFORMANCE REQUIREMENTS

7.1 General requirements

7.1.1 The ROPS, overhead weather shield, overhead protection, fenders, cab sheet metal, or related ROPS parts, outside of, but near the operator area may be deformed in tests as described in paragraph 6.1, 6.2, 6.3 or 6.4 but shall not leave sharp edges exposed to the operator or intrude on the clearance zone described by the dimensions shown in Figs. 4, 5 and 6 as follows:

d = 50 mm (2 in.) inside of frame upright to vertical centerline of seat

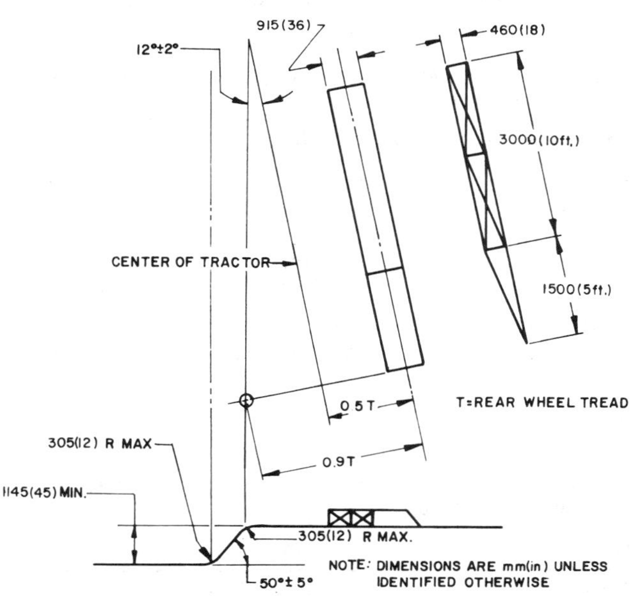

FIG. 15—SIDE OVERTURN BANK AND RAMP

TABLE 1—MINIMUM CHARPY V-NOTCH IMPACT STRENGTHS†

Specimen Size	Impact Strength	
mm	J	ft-lb
10 x 10*	11.0	8.0
10 x 9	10.0	7.5
10 x 8	9.5	7.0
10 x 7.5*	9.5	7.0
10 x 7	9.0	6.5
10 x 6.7	8.5	6.5
10 x 6	8.0	6.0
10 x 5*	7.5	5.5
10 x 4	7.0	5.0
10 x 3.3	6.0	4.5
10 x 3	6.0	4.5
10 x 2.5*	5.5	4.0

*Indicates preferred size. Specimen size shall be no less than the largest preferred size that the material will permit.
†Reference: ASTM A 370-76, Standard Methods and Definitions for Mechanical Testing of Steel Products.

e = 760 mm (30 in.) at the longitudinal centerline
f = no greater than 100 mm (4 in.) to rear edge of crossbar measured forward of the SRP
g = 610 mm (24 in.) min (see Fig. 6)
h = 445 mm (17.5 in.) min (see Fig. 5 cab or 4 post ROPS.)
j = 50 mm (2.0 in.) measured from outer periphery of steering wheel (see Fig. 5)
m = not greater than 305 mm (12 in.) measured from SRP to forward edge of crossbar (see Fig. 4, 1 or 2 post ROPS only)

Fig. 12 is a pictorial representation of the clearance zone.

7.1.2 The temperature material requirements will be met if the ROPS passes either of the dynamic tests (paragraph 6.2 or 6.4) at a metal temperature of -18 °C (0 °F) or below. The temperature-material requirements will also be met if the ROPS passes either the dynamic test (paragraph 6.2 or 6.4), or the static test (paragraph 6.1) at ambient temperature provided structural members are from material which exhibit Charpy V-notch impact strengths at -30 °C (-20 °F) shown in Table 1. Specimens are to be "longitudinal" and taken from flat stock, tubular sections, or structural sections before forming or welding for use in the ROPS. Specimens from tubular or structural sections are to be taken from the middle of the side of greatest dimension, not to include welds. There is no Charpy requirement for steel 2.6 mm (0.10 in.) or less in thickness and a maximum carbon content of 0.20%.

7.1.3 Fasteners used to attach the ROPS to the tractor frame and to connect structural parts of the ROPS shall be SAE grade 5 through 8 or equivalent (see SAE Standard J429, Mechanical and Material Requirements for Externally Threaded Fasteners, and SAE Standard J995, Mechanical and Material Requirements for Steel Nuts).

7.2 Static test performance requirements. The structural requirements will be met if the required rear and side energy input levels are reached or exceeded, and the dimensions of the zone of clearance in paragraph 7.1.1 are adhered to in rear and side loading.

7.3 Dynamic test performance requirements. The structural requirements will be met if the dimensions of the zone of clearance in paragraph 7.1.1 are adhered to in both rear and side loading.

7.4 Crush test performance requirements. The structural requirements will be met if the dimensions of the zone of clearance in paragraph 7.1.1 are adhered to.

7.5 Field upset performance requirements. The structural requirements will be met if the dimensions of the zone of clearance in paragraph 7.1.1 are adhered to in both rear and side upsets.

Cited Standards:

ASAE S313, Soil Cone Penetrometer
ASAE S318, Safety for Agricultural Equipment
ASTM A370, Standard Methods and Definitions for Mechanical Testing of Steel Products
SAE J4c, Motor Vehicle Seat Belt Assemblies
SAE J106, Soil Type and Strength Classification
SAE J429, Mechanical and Material Requirements for Externally Threaded Fasteners
SAE J816, Engine Test Code—Spark Ignition and Diesel
SAE J995, Mechanical and Material Requirements for Steel Nuts

COMBINE HARVESTER TIRE LOADING AND INFLATION PRESSURES

Developed by the ASAE Implement Tire Subcommittee; approved by the Power and Machinery Division Standards Committee; adopted by ASAE as a Tentative Standard April 1977; reconfirmed December 1977; revised April 1979; reconfirmed December 1979; reclassified as a full Standard January 1981; revised April 1983; reconfirmed December 1987.

SECTION 1—PURPOSE AND SCOPE

1.1 This Standard establishes loading and inflation pressure relationships for the tire sizes and ply ratings listed in ASAE Standard ASAE S295, Agricultural Tractor Tire Loadings, Torque Factors, and Inflation Pressures, when those tires are fitted to self-propelled combine harvesters, except hill-side combines.

1.2 This Standard is based on the tire testing data used to establish ASAE Standard ASAE S295, Agricultural Tractor Tire Loadings, Torque Factors, and Inflation Pressures, but adjusted to reflect the different operating duty cycle of the combine harvester compared to the agricultural tractor. That is, the harvester has a minimal requirement for torque transmission, and its total weight fluctuates appreciably as the grain tank is repeatedly filled and emptied. Harvesting normally takes place at approximately 8 km/h (5 mph).

1.3 This Standard also includes a tire usage statement to be fixed to each combine harvester to ensure that operators are notified of correct loadings and inflation pressures.

1.4 In addition to the design and configuration of equipment, hazard control and accident prevention are dependent upon the awareness, concern, and prudence of personnel involved in the operation, transport, maintenance, and storage of equipment or in the use and maintenance of facilities.

SECTION 2—GENERAL PRINCIPLES

2.1 Maximum individual tire loading shall not exceed the tire load for its respective ply rating and inflation pressure as given in Tables 1 and 2 for drive wheel tires and Tables 3 and 4 for steering wheel tires.

2.2 If the drive wheel tire used is not recorded in Tables 1 and 2, its load and inflation should not exceed 170 percent of the latest standards of the Tire and Rim Association, Inc. for agricultural drive wheel tractor tires used as singles (maximum speed—20 mph) and with 4 psi increased inflation pressure. If the steering wheel tire used is not recorded in Tables 3 and 4, its load and inflation should not exceed the latest standards of the Tire and Rim Association, Inc. for steering wheel tires used on tractors or harvesting equipment with cyclic loading service and maximum speed of 5 mph.

2.3 Maximum tire loading shall be based on combine weight equipped for field operation per paragraph 2.17 of ASAE Standard ASAE S343, Terminology for Combines and Grain Harvesting, but with full grain tank per ASAE Standard ASAE S312, Capacity Designations for Combine Grain Tank Systems, including capacity extensions as specified by the combine manufacturer.

2.4 The combine manufacturer may fit various tire sizes and/or ply ratings to the same basic model combine to suit various headers and/or accessories, provided that the limitations are indicated on the tire usage statement (see Section 3).

2.5 Wheel and rim strength shall be adequate for the maximum load and inflation of the greatest ply rating offered in that size by the combine manufacturer.

SECTION 3—TIRE USAGE STATEMENT

3.1 To account for the wide variety of headers and accessories which may be used with a basic combine, each combine shall display a tire usage statement showing the following:

3.1.1 Make and model of combine.

3.1.2 Maximum grain tank capacity with approved extensions.

3.1.3 Limitations on header-accessory-tire combinations. The manufacturer may list as many combinations on the tire usage statement as considered necessary, or the manufacturer may reference the operator's manual.

3.1.4 Inflation pressure information.

3.1.5 Road travel limitations.

Cited Standards:

ASAE S295, Agricultural Tractor Tire Loadings, Torque Factors, and Inflation Pressures
ASAE S312, Capacity Designations for Combine Grain Tank Systems
ASAE S343, Terminology for Combines and Grain Harvesting

TABLE 1—COMBINE HARVESTER DRIVE WHEEL TIRE LOADING AND INFLATION PRESSURES—(Customary Units) (Excluding Hillside Combines)—Maximum Speed—5 mph

	Maximum Tire Loading in Pounds for Indicated Inflation Pressures in PSI										
Tire Size	18 PSI	20	22	24	25	26	28	30	32	34	36
13.6-26	3980 lb (4)*	4300	4605	4915		5185(6)					
14.9-26	4746	5135	5510	5850(6)							
16.9-26		6220	6665(6)	7090		7495	7890(8)				
18.4-26		7445(6)	7900	8500(8)		8975	9450	9910(10)			
18.4-30		7940(6)	8515	9060(8)		9570	10065	10555(10)			
18.4-34		8430(6)	9045	9605(8)		10165	10695	11205(10)			
18.4-38		8925(6)	9570	10165(8)		10760	11320	11865(10)	12375	12885	13400(12)
20.8-38		10810	11595(8)	12325		13040					
23.1-26		10675(8)	11440	12170(10)		12870	13530(12)				
23.1-30		11375(8)	12190	12970(10)							
23.1-34		12085(8)	12935	13750(10)							
24.5-32		12990	13905	14790(10)		15640	16455(12)	17240	18005	18750(16)	
28L-26		12375	13260(10)	14095(12)		14910	15675(14)				
30.5L-32			15505(10)	17665(12)		18685	19650	20585(16)			
67x34.00-30 NHS†		14110(8)			16090(10)			17900(12)			

*Figures in parentheses denote tire ply rating for which underlined load is maximum.
†NHS—Not for highway service.

TABLE 2—COMBINE HARVESTER DRIVE WHEEL TIRE LOADING AND INFLATION PRESSURES—(SI Units)
(Excluding Hillside Combines)—Maximum Speed 8 km/h

Tire Size	124 kPa	138	152	165	172	179	193	207	221	234	248
	Maximum Tire Loading in Kilograms for indicated inflation Pressures in Kilopascals										
13.6-26	1805 kg (4)*	1950	2089	2229		2352(6)					
14.9-26	2152	2329	2499	2654(6)							
16.9-26		2821	3023(6)	3216		3400	3579(8)				
18.4-26		3377(6)	3624	3856(8)		4071	4287	4495(10)			
18.4-30		3602(6)	3862	4110(8)		4341	4565	4788(10)			
18.4-34		3824(6)	4103	4357(8)		4611	4851	5083(10)			
18.4-38		4048(6)	4341	4611(8)		4881	5135	5382(10)	5613	5845	6078(12)
20.8-38		4903	5259(8)	5591		5915(10)					
23.1-26		4842(8)	5189	5520(10)		5838	6137(12)				
23.1-30		5160(8)	5529	5883(10)							
23.1-34		5482(8)	5867	6237(10)							
24.5-32		5892	6307	6709(10)		7094	7464(12)	7820	8167	8505(16)	
28L-26		5613	6015(10)	6393(12)		6763	7110(14)				
30.5L-32			7033(10)	8013(12)		8475	8913	9337(16)			
67x34.00-30 NHS†		6400(8)			7298(10)			8119(12)			

*Figures in parentheses denote tire ply rating for which underlined load is maximum.
†NHS—Not for highway service.

TABLE 3—COMBINE, FREE ROLLING, STEERING WHEEL TIRE LOADING AND INFLATION PRESSURE—(Customary Units)
(Excluding Hillside Combines)—Maximum Speed—5 mph

Tire Size	24 PSI	28	32	36	40	44	48	52	56	60	64	68
	Maximum Tire Loading in Pounds for Indicated Inflation Pressures in PSI											
7.50-16SL†	1650 lb	1830	2010(4)*	2170	2320	2470(6)	2610	2750	2880(8)	3010	3130	3250(10)
7.50-18SL	1790	1990	2180(4)	2350	2520	2680(6)						
7.50-20SL	1930	2140	2340(4)	2530	2720	2890(6)						
7.50-24SL	2230	2480(4)										
9.00-24SL	3360(4)	3740	4090	4420(6)	4740	5040(8)						
9.5-16	1840	2050(4)	2240	2430	2600(6)							
9.5-24	2450	2730(4)	2980	3230	3460(6)							
9.50-20SL	2770	3080	3380	3650(6)	3910	4160(8)						
9.50-24SL	3150(4)	3500	3830	4140(6)								
10.00-16SL	2630	2920	3200(6)	3460	3710	3940(8)						
11.00-16SL	3100	3450	3780(6)	4080	4370(8)	4650	4920	5180	5420	5660(12)		
11L-16SL	2440	2720	2980(6)	3220	3450	3670(8)	3880	4080(10)				
11.2-24	2970(4)	3300	3610	3910(6)								
11.25-24SL	4380	4870(6)	5330	5760	6170(8)							
12.4-16	2730	3030	3320(6)	3590	3840	4090(8)	4320	4550(10)	4770	4980	5180(12)	
12.4-24	3550(4)	3950	4330(6)	4680	5010	5330(8)						
12.5L-16SL	2670	2970	3250(6)	3520	3770(8)	4010	4240(10)	4460	4670(12)			
13.6-16.1	3240	3610	3950(6)	4270	4580(8)							
14L-16.1SL	3830	4270(6)	4670	5050(8)	5410	5750(10)						
14.9-24	5000	5560(6)	6080	6580(8)								
16.5L-16.1SL	4860(6)	5410	5920(8)	6400	6860(10)	7290	7710(12)					
18.4-16.1	5760(6)	6410(8)										

*Figures in parentheses denote tire ply rating for which underlined load is maximum.
†Service limited—maximum speed 20 mph.

TABLE 4—COMBINE, FREE ROLLING, STEERING WHEEL TIRE LOADING AND INFLATION PRESSURE—(SI Units)
(Excluding Hillside Combines)—Maximum Speed 8 km/h

Tire Size	165 kPa	193	221	248	276	303	331	359	386	414	441	469
	Maximum Tire Loading in Kilograms for Indicated Inflation Pressures in Kilopascals											
7.50-16SL†	748 kg	830	912(4)*	984	1052	1120(6)	1184	1247	1306(8)	1365	1420	1474(10)
7.50-18SL	812	903	989(4)	1066	1143	1216(6)						
7.50-20SL	875	971	1061(4)	1148	1234	1311(6)						
7.50-24SL	1012	1125(4)										
9.00-24SL	1524(4)	1696	1855	2005(6)	2150	2286(8)						
9.5-16	835	930(4)	1016	1102	1179(6)							
9.5-24	1111	1238(4)	1352	1465	1569(6)							
9.50-20SL	1256	1397	1533	1656(6)	1774	1887(8)						
9.50-24SL	1429(4)	1588	1737	1878(6)								
10.00-16SL	1193	1325	1452(6)	1569	1683	1787(8)						
11.00-16SL	1406	1565	1715(6)	1851	1982(8)	2109	2232	2350	2459	2567(12)		
11L-16SL	1107	1234	1352(6)	1461	1565	1665(8)	1760	1851(10)				
11.2-24	1347(4)	1497	1637	1774(6)								
11.25-24SL	1987	2209(6)	2418	2613	2799(8)							
12.4-16	1238	1374	1506(6)	1628	1742	1855(8)	1960	2064(10)	2164	2259	2350(12)	
12.4-24	1610(4)	1792	1964(6)	2123	2273	2418(8)						
12.5L-16SL	1211	1347	1474(6)	1597	1710(8)	1819	1923(10)	2023	2118(12)			
3.6-16.1SL	1470	1637	1792(6)	1937	2200(8)							
14L-16.1SL	1737	1937(6)	2118	2291(8)	2454	2608(10)						
14.9-24	2268	2522(6)	2758	2985(8)								
16.5L-16.1SL	2204(6)	2454	2685(8)	2903	3112(10)	3307	3497(12)					
18.4-16.1	2613(6)	2908(8)										

*Figures in parentheses denote tire ply rating for which underlined load is maximum.
†Service limited—maximum speed 30 km/h.

ASAE Standard: ASAE S386.2

T

CALIBRATION AND DISTRIBUTION PATTERN TESTING OF AGRICULTURAL AERIAL APPLICATION EQUIPMENT

Developed by the ASAE Agricultural Chemicals Application Committee; approved by the ASAE Power and Machinery Division Standards Committee; adopted by ASAE as a Tentative Standard June 1977; reconfirmed December 1978, December 1979, December 1980, December 1981; revised and reclassified as a full Standard January 1983; revised February 1988.

SECTION 1—PURPOSE AND SCOPE

1.1 This Standard establishes uniform procedures for measuring and reporting application rates and distribution patterns from agricultural aerial application equipment.

1.2 The procedures covered deal with both fixed and rotary wing aircraft equipped with either liquid or dry material distribution systems.

1.3 These procedures and the statistics reported do not imply optimum conditions for satisfying biological requirements.

SECTION 2—DEFINITIONS

2.1 For the purpose of this Standard the following definitions shall apply:

2.1.1 Application rate: Application rates are as defined in ASAE Standard S327, Terminology and Definitions for Agricultural Chemical Application.

2.1.2 Deposit rate: Deposit rates are as defined in ASAE Standard S327, Terminology and Definitions for Agricultural Chemical Application.

2.1.3 Single-pass application: An application method where the aircraft applies one swath over the sample line.

2.1.4 One-direction application: An application method where successive adjacent swaths are made in the same direction of travel (racetrack application). This method produces a right-on-left wing overlap pattern.

2.1.5 Progressive application: An application method where the aircraft applies adjacent swaths but travels in alternate directions for each swath (back and forth application). This method produces a right-on-right wing overlap alternately with a left-on-left wing overlap pattern.

2.1.6 Swath spacing: The lateral distance between the aircraft centerlines for overlapping broadcast applications.

2.1.7 Effective swath width: The swath spacing that will produce acceptable field deposition uniformity for intended application.

2.1.8 Nozzle orientation: The angle of spray discharge from the nozzles measured relative to the local airflow in flight. A nozzle orientation of 90 deg denotes spray discharge perpendicular to the direction of the local airflow while a nozzle orientation of 0 deg denotes spray discharge that is parallel and to the rear.

2.1.9 Indicated airspeed: The speed as indicated by the airspeed indicator of the aircraft in flight.

SECTION 3—TEST CONDITIONS

3.1 The physical characteristics of the liquid or dry material have an effect on the application rate and the distribution patterns. If inert test solutions or materials are substituted for the material to be applied, they must have physical characteristics similar to those of the material to be applied.

3.2 The distribution equipment to be tested should be in good mechanical condition and properly equipped and adjusted for the type of application to be simulated.

3.3 The tests should be conducted when wind speeds are less than 16 km/h (10 mph) measured at 2.5 m (8.2 ft) above the land surface or crop canopy. The distribution pattern test flights should be made parallel to or within 15 deg of the direction of the wind to minimize errors due to crosswinds. Output rate test flights should be made considering both headwind and tailwind components to minimize the effects of wind velocity on the ground speed of the aircraft.

3.4 The following provisions will help assure that the tests will be carried out in a safe and efficient manner.

3.4.1 Test site. The test site should be selected where the aircraft can have a minimum unobstructed approach (from power lines, buildings, trees, fences, etc.) and departure distance to and from the sample line of 300 m (1 000 ft). The site should allow orientation of a 30 m (100 ft) sample line at a right angle to the prevailing wind. The site should be located in an area where there is a minimum of other flying aircraft. Local airport and/or FAA authorities should be informed of scheduled activity so that proper notification can be made to other aircraft operating in the immediate area.

3.4.2 Toxic materials. When toxic materials are used, all safety precautions prescribed by the manufacturer and regulating authorities for handling, loading, application and disposal shall be observed.

3.4.3 Residues. Distribution equipment previously used in field applications should be cleaned and flushed of any residue prior to starting the test procedure. Special cleaning agents may be necessary to neutralize previously used pesticides or additives.

3.4.4 Safety precautions. Prior to initiating any tests, the pilot and all test site personnel shall be thoroughly briefed on test procedures. At the flight test area, all personnel shall stand clear of the aircraft flight path. Special safety precautions shall be observed when stationary aircraft tests are conducted with the engine running to prevent serious injury by a moving propeller or rotor. If toxic materials are used, personnel should remain clear of application and drift areas, and appropriate precautions shall be taken to prevent contamination of test personnel and test site.

SECTION 4—TEST DESCRIPTION AND PROCEDURE

4.1 A test shall consist of four parts: (1) determination of the output rate from the aircraft, (2) determination of the swath distribution pattern by measurement of the applied materials from suitable collectors, (3) determination of the maximum effective swath width and the corresponding uniformity of distribution for overlapped swaths and, (4) determination of application rate. Each part of the test shall be replicated to account for random variation.

4.2 Output rate test

4.2.1 Liquid materials. The output rate should be determined by measuring the amount of liquid discharged from the tank for a measured time interval while the aircraft is operated under normal conditions. The time interval should be sufficient to permit accurate measurement and minimize errors due to turning the system on and off (at least 30 s) and should be measured to the nearest 0.1 s. The amount of liquid used shall be measured by either refilling the tank to the initial level or by measuring the amount remaining in the tank and subtracting from the initial amount. Care must be taken to position the aircraft in exactly the same position on a level surface for the measurement and refilling operations. Measurement precision should be $\pm$ 1% of the amount output. These data may also be used to calibrate flow meters that may already be a part of the system. If the liquid dispersal system can be operated normally with the aircraft stationary, the test can be accomplished without actually flying the aircraft. Output rate shall be expressed in L/min (gpm).

4.2.2 Dry materials. The output rate should be determined by measuring the amount of material discharged from the hopper over a given time interval while in normal flight. The time interval length and measurement precision specified in paragraph 4.2.1 shall also apply to the determination of the output rate for dry materials. The test shall be run with the aircraft hopper filled to at least 25% of capacity. Output rate shall be expressed in kg/min (lb/min).

4.3 Swath distribution pattern test. This test shall be accomplished by flying the aircraft over the center of a target sample line placed at a right angle to the line of flight. The center of the sample line shall be marked, and any deviation of the aircraft line of flight from the sample line center shall be noted. The sample line may be placed on the land surface, at crop height or at any other height consistent with the purpose of the test. The aircraft shall be flown at a height suited to the type of material applied and the purpose of the application. Actual aircraft height shall be measured and recorded. The airspeed shall be that recommended for the particular type of application, and the aircraft should be flown straight and level through the entire test course. The sample line should extent beyond the ends of the pattern being tested. Ordinarily, the sample line will be oriented so that the aircraft will be flying directly into the wind to minimize the effects of crosswind on the distribution pattern. However, once an acceptable distribution pattern has been obtained, a crosswind series may be run to establish the distribution pattern under this operating condition. Ambient temperature, humidity, horizontal wind speed and wind direction (with respect to the direction of flight) shall be measured at a height of 2.5 m (8.2 ft) above the sample line. The dispersing equipment in the aircraft shall be turned on at least 200 m (660 ft) prior to crossing the sample line and shall continue operating the same distance beyond. For tests utilizing granular fertilizer this distance may be reduced by one-half. Care must be taken to turn off the dispensing system before the pull up at the end of the test course. Evaluation shall be based on at least three replications of the test. Where possible, each replication shall be made with a single pass of the aircraft in the same direction of travel.

4.3.1 Spray test procedure and sample collectors. An inert or dye tracer material may be added to the contents of the spray tank, or the active chemical may be used as a tracer for the spray pattern tests. Blank formulations or suitable amounts of emulsifier, spreader-stickers and other solvents and carriers shall be included to closely simulate the physical properties of the material to be applied.

4.3.1.1 Quantitative distribution pattern measurement. Distribution pattern measurement techniques may employ discrete sampling targets or a narrow continuous sampling surface placed across the aircraft line of flight. Quantitative analysis of these samples may involve washing techniques or electronic scanning of the sample surface. Collectors shall be selected on the basis of collection efficiency, size and ratio of collection area to accuracy. Collector detail should be reported as outlined in paragraph 6.1.10.

4.3.1.1.1 The pattern may be determined from the amount of tracer material on the targets. Target surfaces that are analyzed by washing techniques should permit all or a constant percentage of the tracer to be removed by a suitable solvent. Washing techniques should insure that only the interior part of containers with raised edges are washed. If the tracer degrades because of exposure to sunlight, passage of time or other factors, the test procedure shall correct for the degradation. Degradation shall be based on tests of the recovery of tracer from targets to which known amounts of the spray liquid have been applied. The exposed surface of individual flat targets shall have an area of at least 50 cm^2 (7.8 in.^{2}). Spacing of the targets across the swath shall not exceed 1 m (3.3 ft). Total length of the sample line resulting from the use of either discrete targets or a continuous surface shall be a minimum of 30 m (100 ft).

4.3.1.1.2 In the event the sample targets are positioned at any angle other than horizontal, all the targets or the entire sample line should feature the same angle of inclination, and this angle should be reported as outlined in paragraph 6.1.10. Care shall be exercised when using sample targets with raised edges to minimize the shadowing effect (the spray droplets approach the target at less than 90 deg) and to make a sample target area correction when converting data to a field area basis.

4.3.1.1.3 For samples that are electronically scanned to measure deposition on the sample surface based on droplet size and numbers, an appropriate area must be scanned to obtain a true representation of the droplet-size distribution in the sample. Also, the spread factor versus droplet size function should be reported for the sampling surface material and the test liquid under test conditions (temperature and relative humidity).

4.3.1.2 Qualitative distribution pattern measurement. A qualitative measure of the distribution pattern may be used to diagnose and correct distribution system deficiencies (plugged or worn nozzles, improper size nozzles, system leaks, improperly placed nozzles, etc.). Qualitative distribution pattern measurement techniques may employ discrete sample targets or a continuous collector placed across the flight line of the aircraft. The measurement technique used should provide a relative or absolute measure of the deposition on the sample surfaces across the flight line.

4.3.2 Dry material test procedure and collectors. Care must be taken to prevent granular materials such as pellets and seeds from bouncing out of or into the collectors. This can be accomplished by collector design or by lining the collectors with material which prevents bouncing and elevating the collectors to prevent granules from bouncing into them. Dust or other small particles may be collected on greased boards or other sticky surfaces or in shallow pans. The area of the top opening of the collectors shall be 0.1 m^2 (1 ft^2) or larger as required to provide a representative sample of the deposit. Spacing of collectors along the swath shall not exceed 1 m (3.3 ft). Particles of material caught may be counted, weighed, or dissolved in a solute for analysis as appropriate.

4.4 Sample analysis and conversion of swath distribution pattern data

4.4.1 Spray pattern test

4.4.1.1 Sample analysis of any type that is compatible with the spray tracer may be used. Examples are validated methods using photoelectric colorimetry, absorption or emission spectroscopy and liquid or gas chromatography. The sensitivity of the analysis shall be at least 1 part per million (ppm). The concentration of tracer in the solvent after a collector is washed in accordance with paragraph 4.3.1.1 may be determined by use of a standard calibration curve developed for the tracer and analytical method employed. The rate of spray deposit on the target collectors in L/ha (gal/acre) may then be determined for each location across the line of target collectors as follows:

$$\text{Target deposit rate} = \frac{K_1 V_t C_t}{C_s A}$$

where

Target deposit rate, L/ha (gal/acre)
K_1 = constant 10^5 (1 657)
A = collector area, cm^2 (in.^{2})
V_t = volume of solvent used to wash tracer from collector, mL
C_t = concentration of tracer washed from collector, mg/L
C_s = concentration of tracer in original spray solution, mg/L

4.4.1.2 The electronic technique used in the image scanning of discrete or continuous sample surfaces shall result in a droplet-size distribution having a minimum of 20 droplet-size classes. A droplet-size versus spread-factor function over the size range encountered under test conditions for the sample surface material and test liquid shall be developed and used in calculating the deposit volume per unit of area.

4.4.2 Dry material pattern test

4.4.2.1 If the dry material deposited in the collector at each location across the line of collectors is weighed, the deposit rate may be determined in kg/ha (lb/acre) as follows:

$$\text{Deposit rate} = \frac{K_2 W}{AE}$$

where

Deposit rate, kg/ha (lb/acre)
K_2 = constant, 10^5 (13 829)
W = weight collected, g
A = area of collector opening, cm^2 (in.^{2})
E = collector efficiency, 0-100%

The collectors used should be described as discussed in paragraph 6.1.10.

4.4.2.2 If the physical characteristics of the material collected make counting of individual particles desirable, the results should be expressed as the number of particles per unit area. Areas should be reported in metric (customary) units. Material collected and weighed should be expressed as weight per unit area. If the material collected is a dust, it may be advantageous to use greased boards or other sticky surfaces or shallow pans filled with a solute for the collectors. Procedures similar to those outlined in paragraph 4.4.1 may be used for analysis of dust deposits collected in solute provided the dust itself can serve as the tracer material or a suitable tracer material is mixed with the dust. The deposit rate should be determined as kg/ha (lb/acre) at each location across the line of collectors.

SECTION 5—TEST RESULTS

5.1 General. Data from the test shall be subjected to a statistical analysis to characterize the distribution pattern uniformity and shall also be presented graphically. Once the maximum effective swath width has been determined, the application rate can be determined.

5.2 Distribution pattern graphing. The data for individual distribution patterns obtained in Section 4—Test Description and Procedure, shall be first graphed as single swath patterns to enable relating the pattern centerline with the aircraft centerline and to show the distribution pattern characteristics. The single swath patterns shall then be graphed as multiple adjacent swaths with additive deposits in the overlapped regions to obtain a composite graph showing simulated field distribution. Since the distribution patterns frequently are not perfectly symmetrical, graphics shall be prepared for both the progressive pass (back and forth) and one-direction pass (racetrack) application methods. If the single swath patterns are skewed due to crosswind, multiple swath graphics may indicate artificial irregularities in the simulated field distribution. Separate graphs shall be prepared for each replication as averaging may mask significant pattern variations. As an alternative to simulated field deposits, actual replicated flight tests may be conducted to present field deposition data from multiple pass applications. Field tests shall be conducted in a suitable manner to construct the multiple pass distribution patterns as described in paragraphs 5.2.2 and 5.2.3.

5.2.1 Single-pass distribution pattern. The measured deposit or relative deposition shall be plotted on the ordinate scale and the target or sample positions on the abscissa scale to the right and left of the aircraft centerline. The graph should show the pattern as the pilot would see it, i.e., with the left-most deposits to the left of the aircraft centerline.

5.2.2 One-direction application distribution pattern. The combined patterns of one-direction passes (racetrack) shall be plotted with the ordinate scale showing the measured deposit rate or relative deposition and the abscissa scale showing the position of the aircraft centerlines and deposit locations. Single-pass distribution patterns shall be plotted around the aircraft centerlines with the pattern centerline being moved a distance equal to the effective swath width. The single-pass distributions and the composite pattern should be shown with each line clearly labeled. Enough patterns should be overlapped to ensure a representative simulated field deposition at least two swath widths in length that would be unaffected by additional swaths (a minimum of four swaths would be needed if the distribution pattern tails extended beyond the centerline of adjacent swaths). The graph should also show the test identification as well as the swath width and coefficient of variation (CV) for the swath width graphed.

5.2.3 Progressive application distribution pattern. The progressive pass distribution shall be plotted with the ordinate scale showing the measured deposit rate of relative deposition and the abscissa scale showing the position of the aircraft centerlines and deposit locations. The graph shall be prepared by the accumulation of deposition at each deposit location for multiple adjacent swaths with the pattern centerline being moved a distance equal to the effective swath width. Every other swath will be the reverse of the single-pass distribution pattern to reflect the field deposits of the back and forth application procedure. Sufficient single swath patterns should be overlapped following this procedure to ensure that the simulated field distribution would be unaffected by additional swaths. A minimum of five swaths would be needed if the distribution pattern tails extend beyond the centerline of adjacent swaths. The single-pass distribution patterns and composite patterns shall be shown with each line clearly labeled. The graph should show the test identification, the swath width and the coefficient of variation (CV) for the swath width graphed. The words "right/right" and "left/left" should also appear to indicate the pattern overlap orientation between the successive pass centerlines.

5.3 Uniformity of distribution. The coefficient of variation (CV) shall be used to determine and express the uniformity of distribution of applications resulting from multiple adjacent swaths. A simulated field application of multiple adjacent swaths using the single-pass distribution patterns obtained in Section 4—Test Description and Procedure, or samples of the deposition from multiple adjacent swaths obtained from actual flight tests will be used to compute the CV.

5.3.1 The mean value, standard deviation, and coefficient of variation (CV) shall be determined as follows:

$$\text{Mean} = \overline{X} = \frac{\Sigma X_i}{n}$$

$$\text{Standard deviation} = \left[\frac{n(\Sigma X_i^2) - (\Sigma X_i)^2}{n(n-1)}\right]^{1/2}$$

$$\text{Coefficient of variation} = \frac{\text{standard deviation x 100}}{\overline{X}}$$

where

$\overline{X}$ = arithmetic mean
X_i = quantified deposit for one collector location for the combined swaths
n = number of collector locations used

5.3.2 Only the central portion of the simulated or measured overlapped field distribution data shall be used to compute the coefficient variation (CV). If the swath spacing is equal to or greater than one-half of the total spread pattern width, this shall include data from one swath centerline to the next for the one direction method of application or the data from the centerline of the first swath to the centerline of the third adjacent swath for the progressive pass method. If the swath spacing is less than one-half of the total spread pattern width, additional overlapped distribution data shall be added until the region for calculation as indicated above would be unaffected by the addition of distribution data resulting from additional overlapping swaths.

5.3.3 Coefficient of variation (CV) calculations. Prior to preparing graphs for paragraphs 5.2.2 and 5.2.3, the coefficient of variation (CV) shall be calculated for both application methods (one-direction pass and progressive pass) for swath centerline spacings ranging from one sampling interval width to the total width of the single swath pattern. Swath increments for this calculation shall not be greater than the sampling interval (or one meter for continuous sampling) across the swath.

5.3.4 Effective swath width. The effective swath width for each method of application can be determined from an inspection of a table as described in paragraph 5.3.3. The largest swath width associated with the minimum acceptable coefficient of variation (CV) shall be considered the effective swath width for the test. An alternative method of determining effective swath width is to determine the distance between the points on either side of the pattern where the rate of deposit equals one-half peak height of that single-pass distribution pattern. Swath width determined by this method should be so stated and the CV computed and reported for both one-direction and progressive pass as outlined in paragraphs 5.2.2 and 5.2.3.

5.3.5 Measured field distributions. If actual flight tests are used to determine field distributions, the line of target collectors must extend at least 3 widths of the effective swath determined in paragraph 5.3.4 and at least 5 adjacent swaths must be flown over the collectors.

5.4 Application rate. The application rate shall be calculated as follows using average values of output rate, ground speed and effective swath width:

$$R = \frac{QK_3}{VS}$$

where

R = application rate, L/ha or kg/ha (gal/acre or lb/acre)
Q = output rate, L/min or kg/min (gal/min or lb/min)
K_3 = constant, 600 (495)
V = ground speed, km/h (mile/h)
S = effective swath width, m (ft)

SECTION 6—REPORTING RESULTS

6.1 Aircraft and application data. The following information should be reported:

6.1.1 Aircraft model, type, manufacturer, and year manufactured.

6.1.2 Wing span and type of wing tip spoilers, if used, or any modification of wings.

6.1.3 Engine manufacturer, power rating (normal sea level flight), type of propeller, and engine speed.

6.1.4 Hopper capacity.

6.1.5 Amount of material (dry or liquid) in hopper.

6.1.6 Gross aircraft mass (total mass of the aircraft including pilot, fuel, oil, material, etc., at the time of flight over the line of collectors).

6.1.7 Height of flight above the land surface or crop canopy.

6.1.8 Aircraft indicated airspeed and measured ground speed.

6.1.9 Weather data from paragraphs 3.3 and 4.3.

6.1.10 Size, shape, orientation, material, spacing, number, collection efficiency and height of collectors above or below the land surface or crop canopy.

6.1.11 Type, size, and general description of ground cover or crop where tests are conducted.

6.2 Liquid dispersal. The following information shall be reported:

6.2.1 Spray mixture ingredients and proportion for each including physical properties.

6.2.2 Type and size of pump.

6.2.3 Method of driving pump.

6.2.4 Size, type and location of control valves and liquid filters.

6.2.5 Size and type (i.e., round, airfoil, etc.) of boom.

6.2.6 Length of boom and position relative to the trailing edge of the aircraft wing or rotor.

6.2.7 Number, size, type, orientation, arrangement, and condition of atomizing devices. If the atomizer arrangement is not symmetrical, a diagram showing the position of each atomizer relative to the centerline of the aircraft shall be included.

6.2.8 Position of atomizing devices relative to the boom.

6.2.9 Spray pressure and location measured at the boom.

6.3 Dry material dispersal. The following information shall be reported:

6.3.1 Manufacturer, type and model of distribution equipment plus a description of any modifications.

6.3.2 Rotor size, configuration and speed of positive metering system.

6.3.3 Dimensions of venturi spreaders such as frontal opening, throat width and depth, width and depth of discharge end, and overall length.

6.3.4 Vane adjustments for venturi spreaders and any modifications from original design.

6.3.5 Rotor size, speed, configuration, and location for centrifugal spreaders.

6.3.6 Manufacturer, type and size of metering gate and gate opening for gravity feed devices and/or agitators.

6.3.7 Name of product, source of supply, particle size, particle shape, bulk density and moisture content of material applied.

6.4 Calibration and distribution pattern data. The following information shall be reported:

6.4.1 Method of pattern measurement as described in Section 4—Test Description and Procedure.

6.4.2 Output rate (mean) as determined in paragraphs 4.2.1 or 4.2.2.

6.4.2.1 Length of time interval during material discharge for each test.

6.4.2.2 Amount of liquid or dry material discharged during test run.

6.4.2.3 Amount of liquid or dry material discharged during total test interval.

6.4.3 Graphical presentation of individual single and multiple swath patterns indicated in paragraph 5.2.

6.4.4 Effective swath width as determined in paragraph 5.3.4.

6.4.5 Application rate as determined in paragraph 5.4.

6.4.6 Field application uniformity data, i.e., arithmetic mean, standard deviation and coefficient of variation as determined in paragraph 5.3 (mean values from replicated tests).

Cited Standard:

ASAE S327, Terminology and Definitions for Agricultural Chemical Application

TEST PROCEDURE FOR MEASURING DEPOSITS AND AIRBORNE SPRAY FROM GROUND SWATH SPRAYERS

Developed by the ASAE Agricultural Chemical Application Committee; approved by the Power and Machinery Division Standards Committee; adopted by ASAE as a Tentative Standard June 1977; reconfirmed December 1978, December 1979, December 1980, December 1981; revised and reclassified as a full Standard January 1983; reconfirmed for one year December 1987.

SECTION 1—PURPOSE AND SCOPE

1.1 This Standard establishes a test procedure for use in measuring and reporting in-swath and out-of-swath deposits and airborne spray from sprayers.

1.2 This Standard pertains to moving, ground sprayers applying agricultural chemicals in a swath. It does not pertain to orchard hand gun, air carrier systems, foggers, or other sprayers where the spray is not intended to be deposited in a swath on the land surface or crop canopy. It does not pertain to aerial applicators.

SECTION 2—ESSENTIAL ELEMENTS OF A TEST

2.1 A standard reference spray application, as described in paragraph 2.1.1, shall be included with each test, because it is not desirable to specify all details of this sprayer test procedure precisely. For example, it is not desirable to limit all users of this procedure to one size, composition, and placement of collectors. The standard reference application provides a basins of comparison between any two spray tests.

2.1.1 Standard reference application for ground sprayers. Eight flat fan nozzles rated at 0.76 L/min (0.2 gpm) at 275 kPa (40 psi) gage pressure with an 80 deg spray angle shall be used. The nozzles shall be pointed vertically downward with the fans in a plane 85 deg to the direction of travel to prevent mixing in the overlap region. The nozzle spacing shall be 50 cm (20 in.). The nozzle gage pressure shall be 300 kPa (43.5 psi). The nozzle orifice shall be approximately 50 cm (20 in.) above the land or crop surface. The speed of travel shall be 5 km/h (3.1 mile/h). The same spray liquid shall be used in the standard reference application as was used in the sprayer test except when spray carriers are being compared. The carrier for the standard reference application shall be water.

2.1.2 Through use of a dual sprayer (consisting of two independent sprayers carried on the same chassis), the standard reference application may be made simultaneously with the test being reported to insure equality of meteorological conditions. If the standard reference application is not made simultaneously with the test being reported, it shall be made when site and meteorological conditions are similar to those existing at the time of the test being reported. Site conditions include those described in paragraph 3.1. Meteorological conditions during the standard reference application shall not differ from those existing during the test being reported by more than the amounts given in Table 1.

TABLE 1—TOLERANCES ON METEOROLOGICAL FACTORS

Factor	Tolerance
Horizontal wind speed	± 10% of base* wind speed
Dry bulb temperature	± 3 °C (± 5.4 °F)
Relative humidity	± 4% of base relative humidity
	± 25% of base stability ratio

*Base refers to meteorological conditions existing at the time of the test being reported.

SECTION 3—TEST SITE AND METEOROLOGICAL CONDITIONS

3.1 The test site (see Fig. 1) shall include a sample line along which collectors shall be placed to sample spray deposits, and a spray line along which the centerline of the sprayer is operated. The sample line shall be approximately parallel to the wind direction (within ± 30 deg), and the spray line shall be perpendicular to the sample line. A sketch and a description of the test site shall be included with the test report and the sketch should include natural barriers (e.g. roads, trees, etc.). The description shall include the kind and height of crop cover on the test surface, the direction of crop rows, if any, relative to the sample line, the approximate degree of land slope both parallel to and perpendicular to the sample line, and the size and location of any wind obstructions within 0.8 km (0.5 mi) upwind of the test area. The spray line shall be at least 0.6 times as long as the sample line. The description of the test site shall also include the location, relative to the spray and sample lines, of the weather station and, if used, a tower for sampling airborne spray.

3.2 At least one weather station shall be provided for measurement of meteorological conditions and for determining stability ratio, as described in paragraph 3.3. The following meteorological conditions shall be measured and time averaged for the duration of each spray pass or 2 minutes, whichever is greater (all elevations are relative to the ground).

3.2.1 The temperature difference between 2.5 m (8.2 ft) and 10 m (32.8 ft) elevation, measured within 0.1 °C (0.18 °F). (When temperature is greater at 2.5 m than at 10 m, this difference is referred to as a temperature lapse, and shall be reported as a negative quantity.)

3.2.2 The dry bulb air temperature at 2.5 m (8.2 ft) elevation, measured within 0.1 °C (0.18 °F).

3.2.3 Either the wet bulb temperature measured within 0.2 °C (0.36 °F) or the percent relative humidity measured within 2% at 1.6 m (4.9 ft) elevation. If the wet bulb temperature is measured, the results shall be converted to relative humidity and the percent relative humidity shall be reported.

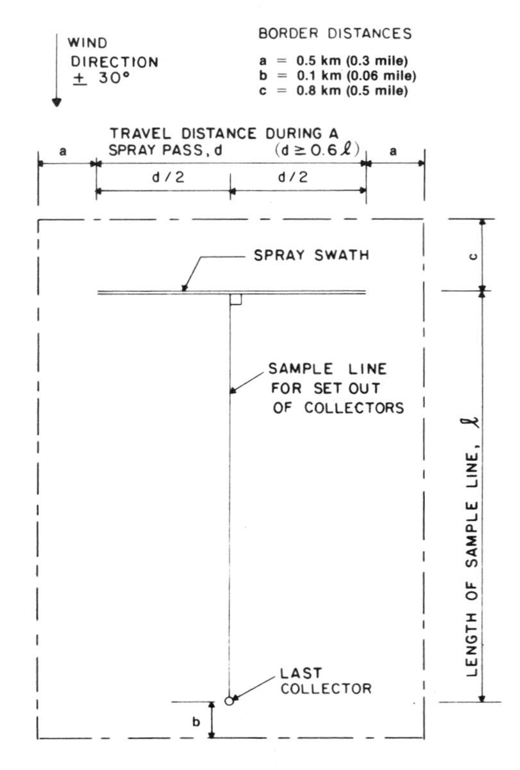

FIG. 1—TYPICAL TEST SITE CONFIGURATION.

3.2.4 Horizontal wind speed at 5 m (16.4 ft) elevation, measured within 0.2 m/s (0.45 mile/h).

3.2.5 The vertical turbulence intensity, defined as the standard deviation of the vertical air speed divided by the average horizontal wind speed with both speeds to be measured at 2.5 m (8.2 ft) elevation. The standard deviation of the vertical air speed, σ_w, may be calculated from the following equation:

$$\sigma_W = \sqrt{\frac{1}{T}\int_o^T (W_i - W)^2 \, dt}$$

where

T = averaging time as described in paragraph 3.2

W_i = instantaneous values of vertical air speed, measured within 0.1 m/s (0.22 mile/h)

W = time-averaged value of W_i

3.2.6 The mean vertical flow W shall be reported as a positive quantity if the mean flow is directed downward and as a negative quantity if the mean flow is directed upward. If the quantity σ_w is computed using numerical integration, the time increments between ordinates shall not exceed 0.5 s.

3.3 The stability ratio shall be calculated from the following equation and shall be reported as an indication of atmospheric stability:

$$SR = \frac{10(T_{10} - T_{2.5})}{U_5^2} \quad \ldots\ldots [1]$$

where

SR = stability ratio (°C·s²/m²)

$T_{10} - T_{2.5}$ = temperature difference between 10 m and 2.5 m elevation (°C)

U_5 = horizontal wind speed difference between 5 m elevation (m/s)

SECTION 4—SPRAY DEPOSITS

4.1 Collectors shall be provided for sampling the quantity of spray material depositing at various locations along the sample line. The size, shape, orientation, and material of the collectors, and their vertical elevvation relative to the land surface of crop canopy shall be reported. If flat collectors are used, the exposed surface shall be level. Spacing of the collectors along the sample line shall be in accordance with paragraphs 4.1.1 and 4.1.2.

4.1.1 Within the intended swath area, identical collectors shall be equally spaced along the sample line beginning at the upwind edge of the intended swath and continuing to the downwind edge of the intended swath. The intended swath width and the distance from the spray line to the downwind edge of the intended swath shall be reported. The theoretical application shall be calculated as the quantity of spray per unit area in the swath if all of the spray has deposited uniformly across the intended swath width. The amount deposited on each collector in the swath shall be reported as a percentage of the theoretical application and may also be reported in quantity of spray per unit area. The theoretical application shall be reported.

4.1.2 In the area downwind from the intended swath, logarithmic spacing of the collectors along the sample line is recommended. The number of collectors and their respective distances from the spray line shall be reported. The quantity of spray deposited on each collector shall be reported as a percentage of the theoretical application and may also be reported in quantity of spray per unit area.

4.1.3 When the angle between the mean wind direction and the sample line is not zero, deposits shall be corrected to account for the greater distance the spray cloud must travel in reaching the sample line. The following procedure should be used each pass for correcting wind deviation. Using the method of least squares, the following equation should be fitted to the data:

$$\log D_k = b_{1k} + b_{2k} U + b_{3k} U^2 \quad \ldots\ldots [2]$$

where

D = deposit at the kth collector downwind of the spray line, expressed as a percent of the theoretical application

$$U = \log\left(\frac{X}{\cos\phi}\right) \quad \ldots\ldots [3]$$

X = distance from the spray line to the kth collector, measured along the sample line

ϕ = mean angle of deviation between the sample line and the wind direction during the test; the deviation should not exceed 30 deg

b_{ik} = coefficients used to obtain the best fit between the equation and the measured data.

After the equation for log D_k is established, U is replaced by X and the equation with X is used to calculate wind-corrected deposits along the sample line.

4.1.4 There are several tracers and measuring devices that can be used to measure the deposit on the collectors. Tests should be run to determine the background level of the tracer in the test area; as well as the degradation rate, if any, and recovery from the collector of the tracer being used. Test results shall be corrected to the basis of zero background, zero degradation and 100% tracer recovery.

4.2 Measurement of droplet sizes is optional. If droplet sizes are to be measured, collectors suitable for measuring the droplet size spectrum shall be provided at various locations along the sample line. The test report shall include a description of the method used for obtaining the droplet sizes, the locations of the collectors, and the following data for each collector.

4.2.1 The total number of droplets (N_t) which deposit on the collector during the spray run.

4.2.2 The percent of N_t that fall within each of a range of droplet size classes.

4.2.3 For each class, the size class limits expressed as droplet diameters in micrometers.

SECTION 5—AIRBORNE SPRAY

5.1 Portions of the spray may follow a trajectory that precludes deposition within the test area. Therefore, at least one vertical line of aspirated air samplers should be provided to detect quantities of airborne spray. The air samplers should be isokinetic for accurate measurement of spray concentration in the air. If non-isokinetic samplers are used, the air velocity through the sampler inlet shall be reported, as well as the horizontal wind speed at each sampler elevation. The vertical line of air samplers should be near the sample line, and may be coincident with the weather station described in paragraph 3.2. The uppermost air sampler should be above the top edge of the spray cloud. Samplers between the uppermost sampler and the ground should be equally spaced. The test procedure should include the capture and reporting of the amount of airborne spray reaching each sampler during the passage of the entire spray cloud. The number and type of air samplers used, and their respective heights above the ground, should also be reported.

5.2 Measurement of airborne droplet sizes is optional. If droplet sizes are to be measured, isokinetic air samplers suitable for measuring the droplet size distribution in the spray shall be interspersed with the air samplers described in paragraph 5.1. The test report shall include a description of the methods used for obtaining the droplet sizes and data for each sampler, as indicated in paragraphs 4.2.1 through 4.2.3.

SECTION 6—REPORTING OF RESULTS

6.1 A complete test report shall include the following:

6.1.1 The model, size, number, spacing and orientation of the atomizers used; the combined and individual flow rate of all atomizers, measured within 1%; all other relevant atomizer variables, such as pressure at the nozzle, spinning disc speed and diameter, etc.; the height of the atomizers above the land or crop surface; and the speed of the sprayer.

6.1.2 Type and proportion by volume of the carrier medium, the pesticide type and concentration of active ingredient and any additives contained in the spray liquid. For non-liquid pesticides or additives, the mass shall be given per unit volume of carrier medium.

6.1.3 Test side and meteorological conditions as described in Section 3—Test Site and Meteorological Conditions.

6.1.4 Spray deposits, as described in Section 4—Spray Deposits.

6.1.5 Airborne spray, as described in Section 5—Airborne Spray.

AUGER FLIGHTING DESIGN CONSIDERATIONS

Developed by the Auger Flight Subcommittee of the ASAE Power and Machinery Division Standards Committee; approved by the Power and Machinery Division Standards Committee; adopted by ASAE December 1977; revised editorially December 1980, January 1982; reconfirmed December 1982; revised March 1988.

SECTION 1—PURPOSE AND SCOPE

1.1 This Engineering Practice is a guide for designing conveyor augers using steel helicoid flighting and for specifying helicoid flighting as generally used in agricultural equipment.

SECTION 2—FLIGHTING MATERIAL

2.1 Physical properties generally limit the number of materials that can be formed into helicoid flighting.

2.2 Present day helicoid manufacturing machines are designed for ferrous metals that are ductile and are therefore generally not used in manufacture of auger flighting from non-ferrous and plastic materials. Other means, such as molding, extruding, forming, and casting, must be developed for these materials. The heavy deformation of the material as it goes through the forming rolls of the flighting machine demands a material that is very ductile and which will stand large amounts of elongation and bending without fracture and wrinkling.

2.3 Hot rolled and cold rolled carbon steels with specifications in the area of AISI (American Iron and Steel Institute) 1006, 1008, 1010, and 1012 and hardnesses of not more than 115 to 140 BHN (Brinell Hardness Number) may be used satisfactorily. The more difficult the flighting is to make, the lower the carbon content must be in the strip. In many cases, it is necessary to use fully killed, fine grained steel. Stainless steel requires special consideration. Consult manufacturer before designing.

SECTION 3—FLIGHTING DIMENSIONS

3.1 Illustrations and standard tolerances for appropriate dimensions are shown in Fig. 1 and Table 1.

3.2 Inside diameter (I.D.). The selection of the nominal inside diameter will depend on the type of fit desired. Inside diameters less than 1/5 of the outside diameter should be avoided. Tolerance of shaft diameter used must be considered in determining the inside diameter.

TABLE 1—FLIGHTING TOLERANCES mm(in.)

Inside Diameter		
0 to 40 (1.6)	+ 3.0 (0.12)	- 0.0
40 (1.6) to 70 (2.8)	+ 5.0 (0.20)	- 0.0
70 (2.8) and over	+ 7.0 (0.28)	- 0.0
Strip Width		
20 (0.8) to 50 (2.0)	+ 0.8 (0.03)*	- 0.8 (0.03)*
50 (2.0) to 120 (5.0)	+ 1.2 (0.05)*	- 1.2 (0.05)*
120 (5.0) to 250 (10.0)	+ 1.5 (0.06)*	- 1.5 (0.06)*
250 (10.0) to 300 (12.0)	+ 2.4 (0.09)*	- 2.4 (0.09)*
Pitch		
0 to 149 (6.0)	+ 15.0 (0.59)	- 15.0 (0.59)
150 (6.0) to 249 (10.0)	+ 20.0 (0.79)	- 20.0 (0.79)
250 (10.0) to 349 (14.0)	+ 25.0 (1.00)	- 25.0 (1.00)
350 (14.0) and over	+ 40.0 (1.57)	- 40.0 (1.57)
Outside Diameter (welded assembly)		
0 to 200 (8.0)	+ 3.0 (0.12)	- 3.0 (0.12)
200 (8.0) to 350 (14.0)	+ 5.0 (0.20)	- 5.0 (0.20)
350 (14.0) and over	+ 7.0 (0.28)	- 7.0 (0.28)
Length		
0 to 1500 (60.0)	+ 0.0	- 13.0 (0.51)
1500 (60.0) and over	+ 0.0	- 20.0 (0.79)
Strip Thickness	Standard mill thickness for thickness specified.	

*Mill edge tolerance. If tighter tolerances are required, they should be specified on the print and have the concurrence of the supplier.

T

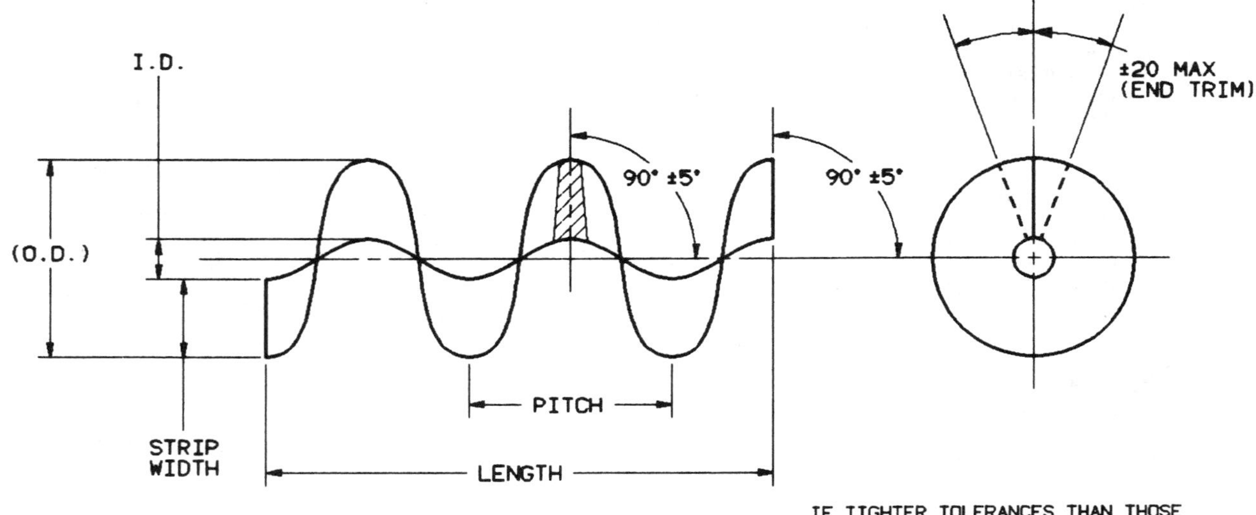

T FIG. 1—STANDARD AUGER FLIGHTING FEATURES

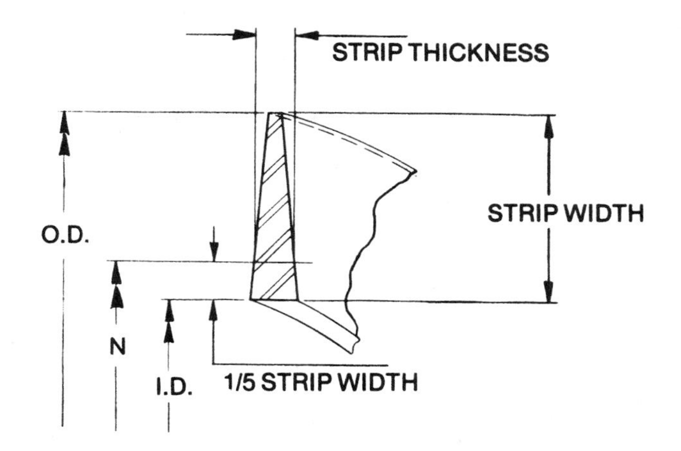

FIG. 2—FLIGHTING DIMENSION DEFINITIONS

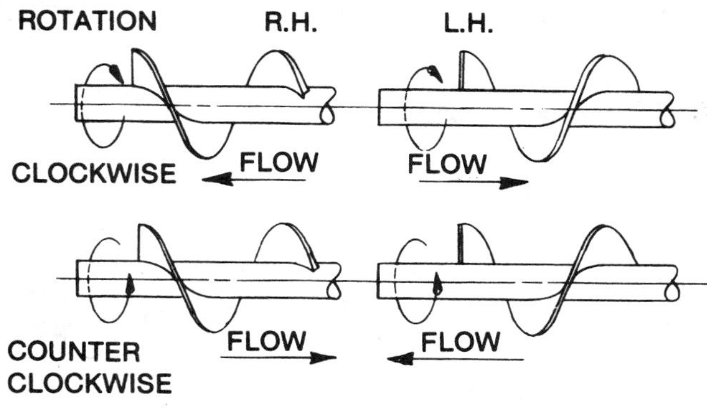

FIG. 3—FLOW

3.3 Outside diameter (O.D.) The outside diameter is determined by adding 2 times the strip width to the I.D. The O.D. is not specified except as a reference dimension for the loose helicoid flighting. O.D. tolerances on the completed welded assembly are given in Table 1.

3.4 Pitch of flighting. The most economical pitch is equal to the flighting outside diameter. Pitches less than 0.9 O.D. and more than 1.5 O.D. are generally not recommended. Pitches shorter than 0.9 O.D. or longer than 1.5 O.D. should only be used after verifying that the supplier has the capability of making the part.

3.5 Strip thickness. The strip thickness may be determined from the outer edge thickness desired. The outer edge thickness is approximately one half the strip thickness, but if greater accuracy is required, it may be calculated. In rolling flighting the material is stretched and thinned along the outer edge and compressed and thickened slightly on the inner edge. The neutral axis is out about 1/5 the strip width from the I.D. (see Fig. 2). Calculate the edge and strip thicknesses as follows:

D = O.D.
d = I.D.
P = pitch
W = strip width $= (D - d)/2$
C = O.D. circumference $= \pi D$
N = neutral axis diameter $= d + 2W/5$
A = neutral axis circumference $= \pi N$
H = length of helix at N in one pitch $= \sqrt{A^2 + p^2}$
L = length of helix at O.D. in one pitch $= \sqrt{C^2 + p^2}$
T = strip thickness $= (L\ E)/H$
E = outer edge thickness $= (H\ T)/L$

Strip thickness tolerance should be standard mill tolerance for strip thickness specified.

3.6 Cup or lean of flighting. In most cases the flighting should be perpendicular to the auger axis within the tolerances shown in Fig. 1. There are situations where deliberate cupping or leaning of the flighting toward the material flow will improve performance. The maximum degree or amount of lean is dependent on the capabilities of the supplier but generally should be less than 15 deg. Angular tolerances with the auger axis should be the same as with perpendicular flighting.

3.7 End shearing. Flighting should be sheared on a radial line passing through the centerline of the auger flighting. Allowable cut variations from the radial are given in Fig. 1.

SECTION 4—HAND OF FLIGHTING

4.1 Flighting may be rolled with either a right hand or left hand helix as shown in Fig. 3.

SECTION 5—FLIGHTING SIZES

5.1 Flighting is specified by the I.D., strip width, strip thickness, pitch, and length. O.D. and outer edge thickness are reference only.

5.2 Strip width should be specified in multiples of 10 mm; for example: 40 mm, 50 mm, 60 mm, etc.

5.3 Strip thickness should be specified in accordance with American National Standard B32.3, Preferred Metric Sizes for Flat Metal Products. Sizes of whole 1 mm increments are preferred; for example: 4 mm, 5 mm, 6 mm, etc.

5.4 The strip width should not exceed 25 times the strip thickness; for example: 4 mm thick—100 mm maximum width.

Cited Standard:

ANSI B32.3, Preferred Metric Sizes for Flat Metal Products

ASAE Standard: ASAE S390.1 (*SAE J1150 MAY83)

CLASSIFICATIONS AND DEFINITIONS OF AGRICULTURAL EQUIPMENT

*Corresponds to earlier draft of S390.

Proposed by the Engineering Policy Committee of the Farm and Industrial Equipment Institute; approved by the ASAE Power and Machinery Division Standards Committee; adopted by ASAE March 1978; reconfirmed December 1982; revised February 1984.

SECTION 1—PURPOSE AND SCOPE

1.1 This Standard provides classifications and definitions of agricultural equipment† designed primarily for use in agricultural operations for the production of food and fiber.

1.2 This Standard is intended to establish uniformity in terms used for agricultural equipment in standards, technical papers, specifications and in general usage.

SECTION 2—CLASSIFICATIONS AND DEFINITIONS

2.1 Agricultural field equipment: Agricultural tractors, self-propelled machines, implements, and combinations thereof designed primarily for agricultural field operations.

2.1.1 Agricultural tractor: A traction machine designed and advertised primarily to supply power to agricultural implements and farmstead equipment. An agricultural tractor propels itself and provides a force in the direction of travel to enable attached soil engaging and other agricultural implements to perform their intended functions.

2.1.2 Agricultural implement: An implement that is designed to perform agricultural operations.

2.1.2.1 Towed implement: An implement that is pulled by a tractor and is usually equipped with wheels for transport.

2.1.2.2 Mounted implement: An implement which is mounted directly on the tractor and is carried by the tractor during transport.

2.1.2.3 Semi-mounted implement: An implement which is partially mounted on the tractor and partially carried on wheels during operation and/or transport.

†Other terms commonly used are farm machinery, farm implements, implements of husbandry, and agricultural machinery.

2.1.3 Self-propelled machine: An implement designed with integral power unit to provide both mobility and power for performing agricultural operations. Definitions for some self-propelled machines follow.

2.1.3.1 Self-propelled beet harvester: A self-propelled machine which digs and conveys sugar beets to an attached bin or into an accompanying truck or wagon.

2.1.3.2 Self-propelled combine: A self-propelled machine for harvesting a wide variety of grain crops. Normally the machine gathers the crop.

2.1.3.3 Self-propelled cotton picker: A self-propelled machine for collecting cotton from open bolls on the stalk usually consisting of picking heads equipped with revolving spindles or other picking means, a conveying means and a bin for carrying the picked cotton.

2.1.3.4 Self-propelled forage harvester: A self-propelled machine which gathers and chops forage crops. The machine usually has a blower type discharge which loads the chopped material into accompanying wagons and trucks.

2.1.3.5 Self-propelled hay baler: A self-propelled machine which picks up and compresses loose hay into compact bales and secures them with wire or twine. Bales are discharged to the ground or to an accompanying wagon or truck.

2.1.3.6 Self-propelled high clearance sprayer: A self-propelled machine which carries a storage tank, pump and spray heads for spraying crops which require high clearance. Crop clearance of the machine is usually over 1220 mm (48 in.).

2.1.3.7 Self-propelled windrower: A self-propelled machine which cuts and gathers standing hay or grain into uniform rows for drying and pickup. In haying operations, the machine normally includes a conditioning attachment designed to crimp the hay and thus decrease the drying time required.

2.2 Farmstead equipment: Equipment, other than agricultural field equipment, used in agricultural operations for the production of food and fiber (examples include livestock feeding systems, livestock watering and waste handling systems, crop dryers, milling systems, material handling equipment, etc.).

ASAE Engineering Practice: ASAE EP391.1

AGRICULTURAL MACHINERY MANAGEMENT

Proposed by the ASAE Farm Machinery Management Committee; approved by the ASAE Power and Machinery Division Standards Committee; adopted by ASAE December 1977; revised editorially September 1979; reconfirmed December 1982; revised December 1983.

SECTION 1—PURPOSE

1.1 This Engineering Practice is intended to provide those who manage agricultural machinery operations with assistance in using available data to determine optimum practices.

1.2 It is intended that corresponding sections in ASAE Data D230, Agricultural Machinery Management Data, be used in Sections 3 through 8 of this Engineering Practice. Terms used in this Engineer ing Practice are defined in ASAE Standard S322, Uniform Terminology for Agricultural Machinery Management.

SECTION 2—SCOPE

2.1 This Engineering Practice includes information helpful in making management decisions involving machine power requirements, capacities, cost, selection and replacement.

SECTION 3—TRACTOR PERFORMANCE

3.1 Tractors use internal combustion engines to power farm machines. Power losses are experienced in exerting power through the drive wheels, the PTO shaft, and the hydraulic system. Fig. 1 illustrates the maximum mechanical power performance expected from a two-wheel, rear axle drive tractor with rubber tires on a level concrete sur face.

3.2 Slippage of drive wheels (see ASAE Standard S296, Uniform Terminology for Traction of Agricultural Tractors, Self-Propelled Implements, and Other Traction and Transport Devices) on soil surfaces is a power loss. This travel reduction, or slip (sl) is measured as follows:

$$\text{Slip (sl)} = \frac{\text{Advance under no load per wheel or track revolution} - \text{Actual Advance per revolution}}{\text{Advance under no load per wheel or track revolution}}$$

Expected slip values for a single drive wheel can be calculated from ASAE Data D230, Agricultural Machinery Management Data, Section 3—Tractor Performance, when the net pull and the dynamic wheel load are known.

3.3 A drawbar power to axle power ratio, called tractive efficiency (TE), can be estimated for a complete tractor from Fig. 1 in ASAE Data D230, Agricultural Machinery Management Data, when the slip is known. Maximum TE is obtained within optimum slip ranges of:

4—8% for concrete
8—10% for firm soil
11—13% for tilled soil
14—16% for soft soils and sands

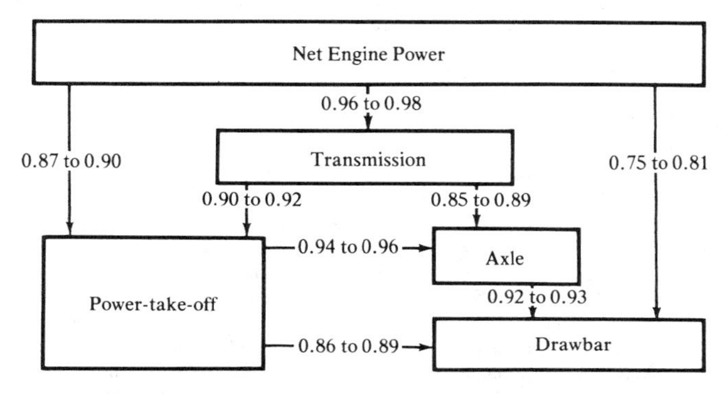

FIG. 1—MAXIMUM MECHANICAL POWER PERFORMANCE EXPECTED

TE (and tire efficiency) of a single drive wheel can be predicted from ASAE Data D230, Agricultural Machinery Management Data, Section 3—Tractor Performance. The performance of a tractor is made up of the sum of the in dividual wheel performances. For a two-wheel rear drive tractor, the rolling resistances of the front wheels subtract from the net pulls ob tained from the two drive wheels to produce the net tractor drawbar pull.

SECTION 4—POWER REQUIREMENT

4.1 Implement (machine) power components

4.1.1 Drawbar power is that developed through the drive wheels or tracks to move the tractor and/or implement through or over the crop or soil. Draft is the total force parallel to the direction of travel required to propel the implement. It is the sum of the soil and crop resistance and the implement rolling resistance.

4.1.1.1 Soil and crop (S & C) resistance is the force parallel to the direction of travel resulting from the contact between the soil or crop and the working components of the implement. Typical values of unit soil and crop resistance forces are given in ASAE Data D230, Agricultural Machinery Management Data, Section 4—Draft and Power Requirements.

S & C resistance, N = (unit draft, N/m)(machine width, m)
[S & C resistance, lb = (unit draft, lb/ft)(machine width ft)]

4.1.1.2 Rolling resistance (RR) (see ASAE S296, Uniform Terminology for Traction of Agricultural Tractors, Self-Propelled Implements, and Other Traction and Transport Devices) becomes appreciable when heavy implements are used in soft or loose soils. Values for the coefficient of rolling resistance are predicted by ASAE Data D230, Agricultural Machinery Management Data, Section 3—Tractor Performance, Tire parameters and wheel loadings must be known or assumed.

Machine RR = the sum of the individual RR's of the wheels supporting the machine's mass
RR, N = (coefficient of RR)(mass on wheel, kg)(Acceleration of gravity, 9.8 m/s^2)
[RR, lb = (coefficient of RR)(weight on wheel, lb)]

4.1.1.3 Drawbar power for tractor-powered implements (and propulsion power for self-propelled implements) is computed as follows:

Drawbar kW = (draft, kN)(speed, km/h)/3.6
[Drawbar hp = (draft, lb)(speed, mi/h)/375]

4.1.2 Power-takeoff (PTO) power is that required by the implement from the PTO shaft of the tractor or engine. Typical values of unit rotary power requirements are given in ASAE Data D230, Agricultural Machinery Management Data, Section 4—Draft and Power Requirements.

PTO kW = (Unit power requirement, kW/m)(machine width, m)
[PTO hp = (Unit power requirement, hp/ft)(machine width, ft)]

4.1.3 Hydraulic power is the fluid power required by the implement from the hydraulic system of the tractor or engine.

Hydraulic kW = (flow, L/s) (pressure, kPa)/1000
[Hydraulic hp = (flow, gal/min) (pressure, lb/in^2)/1714]

4.1.4 Electric power is required to operate components of some implements.

Electric kW = (amperage, I) (voltage, E)/1000
[Electric hp = kW/0.746]

4.2 Total power requirement for operating implements (drawn or self-propelled) is the sum of implement power components converted to equivalent PTO power.

PTO equivalent kW = DB kW/(0.96 x TE)
[PTO equivalent hp = DB hp/(0.96 x TE)]
Total implement kW = PTO kW + PTO equiv. kW + hydraulic kW + electric kW
[Total implement hp = PTO hp + PTO equiv. hp + hydraulic hp + electric hp]

4.3 Total engine power must be greater than the total implement power required. Additional power is required to accelerate and overcome changes in topography, soil, and crop conditions. Additional power is also required for operator-related equipment such as hydraulic control systems, air conditioning, etc.

SECTION 5—FIELD MACHINE PERFORMANCE

5.1 Field efficiency (see ASAE Standard S322, Uniform Terminology for Agricultural Machinery Management) is the ratio between the productivity of a machine under field conditions and the theoretical maximum productivity. Field efficiency accounts for failure to utilize the theoretical operating width of the machine; time lost because of operator capability and habits and operating policy; and field characteristics. Travel to and from a field, major repairs, preventive maintenance, and daily service activities are not included in field time or field efficiency. Field efficiency is not a constant for a particular machine, but varies with the size and shape of the field, pattern of field operation, crop yield, moisture, and crop conditions. The following activities account for the majority of time lost in the field:

- Turning and idle travel
- Materials handling
  - Seed
  - Fertilizer
  - Chemicals
  - Water
  - Harvested material
- Cleaning clogged equipment
- Machine adjustment
- Lubrication and refueling (besides daily service)
- Waiting for other machines

5.2 Effective field capacity (C) is a function of field speed (S), machine working width (w), field efficiency (ef), and unit yield of the field (y). Material capacity is expressed as metric tons (t) per hour [tons (T) per hour]. Area capacity is expressed as:

C, ha/h = (S, Km/h) (w, m) (ef, decimal)/10
[C, acre/h = (S, mi/h) (w, ft) (ef, decimal)/8.25]

Material capacity is expressed as:

C, t/h = (S, km/h) (w, m) (ef, decimal) (y, metric tons/ha)/10
[C, T/h = (S, mi/h) (w, ft) (ef, decimal) (y, tons/acre)/8.25]

Typical ranges of field efficiency and field speed can be found in ASAE Data D230, Agricultural Machinery Management Data, Section 5—Machine Performance. Theoretical field capacity can be determined by using a field efficiency of 1.0.

5.3 The ability of a manager to make good use of his own working hours and those of his employees is evaluated as scheduling efficiency. For example, if a usual workday is 10 h long and 8 h are used effectively, the scheduling efficiency is 80 percent. Ineffective scheduling requires larger capacity machines than are really necessary and increases capital investment.

5.4 Performance of machines operated by ground wheel drives depends on the amount of slippage experienced between the tire and the ground surface. A correction for slippage may be needed to predict performance of planters, grain drills, and other metering and rate application machines.

SECTION 6—COST OF USE

6.1 **Cost factors.** The total cost of using a field machine includes charges for ownership and operation. Ownership costs are seemingly independent of use and are often called fixed costs or overhead costs. Costs for operation vary directly with the amount of use and are often called variable costs.

6.2 **Ownership costs**

6.2.1 **Depreciation.** This cost reflects the reduction in value of an asset with use and time. The actual total depreciation can never be known until the equipment has been sold; however, an estimated depreciation can be predicted from any of several methods. Different computational methods are used depending upon the objective. (For more detailed information consult an engineering economics textbook.)

6.2.1.1 To predict costs for crop production accounting, depreciation may be spread evenly over the accumulated use of the equipment in hectares (acres) or hours. Simple annual depreciation is determined by subtracting the salvage value from the purchase price and dividing by the anticipated length of time owned.

6.2.1.2 A current market value helps estimate depreciation. Two publications report such values, Official Tractor and Farm Equipment Guide, published by the National Farm and Power Equipment Dealer's Association, and National Farm Tractor and Implement Blue Book, published by National Market Reports, Inc. These on-farm, remaining values are approximated as percentages of the list price for the end of each year (see ASAE Data D230, Agricultural Machinery Management Data, Section 6—Costs of Use). Inflation and equipment shortages and surpluses in the market place cause wide variation in these predicted remaining values. The price of used equipment after being reconditioned by a dealer may be 1.3 times the on-farm value.

6.2.2 **Interest.** An interest charge for the use of the money in a machine investment is an ownership cost. Simple interest on the average investment over the life of the machine can be added to the annual depreciation to estimate the yearly capital costs of ownership (see paragraph 6.2.4). A method for determining the capital costs of ownership which includes the time value of money makes use of a Capital Recovery Factor (CRF). The investment in the machine is multiplied by the proper CRF to give a series of equal payments over the life of the machine which includes both the cost of depreciation and interest.

$$R - (P - S) \left[\frac{i/q(1 + i/q)^{nq}}{(1 + i/q)^{nq} - 1} \right] + S\, i/q$$

where

R = one of a series of equal payments due at the end of each compounding period, q times per year
P = principal amount
i = interest rate as compounded q times per year
n = life of the investment in years
S = salvage value

6.2.3 **Other ownership costs.** Taxes, housing and insurance can be estimated as percentages of the purchase price. If the actual data are not known, the following percentages can be used.

Taxes	1.00
Housing	0.75
Insurance	0.25
Total	2.00% of purchase price

6.2.4 **Total ownership costs.** A simple estimate of total ownership costs is given by multiplying the purchase price of the machine by the ownership cost percentage. For a 10-yr machine life, a 10% salvage value and an interest rate of 8%, ownership cost is:

$$100 \left[\frac{P - 0.1P}{10} + \frac{P + 0.1P}{2} (0.08) + 0.02P \right] \quad \text{or}$$

approximately 16% P

6.3 Operating costs

6.3.1 **Repair and maintenance.** Expenditures are necessary to keep a machine operable due to wear, part failures, accidents, and natural deterioration. The costs for restoring a machine are highly variable. Good management may keep costs low. Indices of repair and maintenance costs for Midwest conditions are shown in ASAE Data D230, Agricultural Machinery Management Data, Section 6—Costs of Use. The size of the machine, as reflected by its price, and the amount of use are factors affecting the costs. Both the use and costs are expressed in an accumulated mode to reduce variability. In times of rapid inflation, the purchase price used must be multiplied by $(1+i)^n$ where i is the average inflation rate and n is the age of the machine in question. Accumulated repair and maintenance costs at a typical field speed can be determined with the following relationship using the repair and maintenance factors RF1 and RF2 (see ASAE Data D230, Agricultural Machinery Management Data, Section 6→Costs of Use) and the accumulated use of the machine X [X = accumulated h/1000]. The purchase price must be in current dollars. In time of rapid inflation, the original purchase price must be multiplied by $(1 + i)^n$ where i is the average inflation rate and n is the age of the machine in question.

$$\text{Accumulated R \& M} = P\,[RF1\,(X)^{RF2}]$$

As an example of the use of these indices, the accumulated repair and maintenance costs for a $4000 moldboard plow used 1200 h in 5 yr would be:

$$4000\,[0.43\,(1.2)^{1.8}] = \$2388 \text{ for an average-to-date of \$1.99/h.}$$

6.3.2 Fuel

6.3.2.1 Average fuel consumption for tractors. Annual average fuel requirements for tractors may be used in calculating overall machinery costs for a particular enterprise. However, in determining the cost for a particular operation such as plowing, the fuel requirement should be based on the actual power required.

6.3.2.1.1 Average annual fuel consumption for a specific make and model tractor can be approximated from the Nebraska Tractor Test Data, which is available from the Department of Agricultural Engineering, University of Nebraska, Lincoln, NE 68583. Average gasoline consumption over a whole year can be estimated by the following formula:

Avg. gasoline consumption, L/h = 0.305 x max PTO kW

[Avg. gasoline consumption, gal/h = 0.06 x max PTO hp]

6.3.2.1.2 A diesel tractor will use approximately 73% as much fuel in volume as a gasoline tractor, and liquified petroleum (LP) gas tractors will use approximately 120% as much.

6.3.2.1.3 Fuel consumption for engines not tested by the Nebraska Tractor Test Laboratory may be estimated by the above formula by substituting the advertised PTO kW [hp] for the maximum PTO kW [hp], or by comparing them with a tractor engine of similar displacement.

6.3.2.2 Fuel consumption for a specific operation. Predicting fuel consumption for a particular operation requires determination of the total tractor kW [hp] for that operation (see Section 4—Power Equipment). The equivalent PTO kW [hp] is then divided by the rated maximum to get a percent load for the engine. The fuel consumption at that load is obtained from ASAE Data D230, Agricultural Machinery Management Data, Section 3-Tractor Performance.

L/h = L/kW·h x equivalent PTO kW required

[gal/h = gal/hp·h x equivalent PTO hp required]

A fuel consumption of 15% above that for Nebraska Tractor Tests is included for loss of efficiency under-field conditions.

6.3.3 Engine oil consumption is based on 100-h oil change intervals. The consumption rate of oil ranges from 0.0378 to 0.0946 L/h (0.01 to 0.025 gal/h) depending upon the volume of the engine's crankcase capacity. If oil filters are changed every second oil change, total engine lubrication cost approaches 15% of total fuel cost. Usually the cost of filters and the cost of oil other than crankcase oil is included as maintenance cost. For oil consumption as related to engine size see ASAE Data D230, Agricultural Machinery Management Data, Section 3—Tractor Performance.

6.3.4 Labor cost. The cost of labor varies with geographic location. For owner-operators, labor cost should be determined from alternative opportunities for use of time. For hired operators, a constant hourly rate is appropriate. In no instance should the charge be less than a typical, community labor rate.

6.3.5 A portion of the tractor fixed costs must be included in the cost of use of implements. Tractor fixed costs are recovered by assessing those operations that use the tractor. Assessing may be done on an energy use basis, but is more commonly done on a time basis. For example: If a tractor has $1000 of fixed costs per year and is used 500 h annually, a $2 charge is made against the implement operation for each hour the tractor powers the implement. Implements not using a tractor (self-propelled) do not have such a cost.

SECTION 7—REALIABILITY

7.1 Operational reliability is defined as the statistical probability that a machine will function satisfactorily under specified conditions at any given time. The operational reliability is computed as one minus the probability for downtime when both probabilities are in decimal form. The reliability probability for the next minute of machine operation is essentially one, but decreases when the time span under consideration lengthens. The probability of having a complex machine continually operational for several seasons on a large farm is essentially zero.

7.2 The reliability of a combination of components or machines is the product of the individual probabilities. Complex machines with many components must have very high individual component reliabilities to achieve satisfactory operational reliability.

7.3 Surveys of field breakdowns as reported in ASAE Data D230, Agricultural Machinery Management Data, Section 7—Realiability, indicate expected reliability for several field operations.

SECTION 8—SELECTION OF FIELD MACHINE CAPACITY

8.1 Simple capacity selection (C), ha/h (acre/h), is made by estimating the number of days (D), in the time span within which the operation should be accomplished, and by determining the probability of a working day in this time span (see ASAE Data D230, Agricultural Machinery Management Data). The required capacity for an area (A) is:

$$C = \frac{A}{D\,H\,pwd}$$

where

H = the expected hours available for field work each day

pwd = the probability of a working day.

8.2 Economic selection finds that capacity (C) which produces the lowest net cost. The increased fixed costs of high capacity machines are balanced against the increased operation costs and timeliness costs of low capacity machines.

8.2.1 A price function (p) must be determined which reflects the increased price of one unit of increased capacity. On many machines, the price of a unit increase in effective width is linear and directly related to price per capacity. Price per capacity is obtained by dividing the price per meter of increased width by forward speed in km/h and by ef/10.

8.2.2 If fuel, oil, and repair and maintenance costs can be assumed to be functions of field area covered, they are not pertinent to the selection problem and can be ignored. Only a labor cost (L), $ per h, and a tractor fixed cost (T), $ per h, are important to the operations costs. T = zero for self-propelled machine.

8.2.3 The timeliness cost (Tc), $ per yr, is estimated from a time liness coefficient (K) obtained from ASAE Data D230, Agricultural Machinery Management Data, Section 8—Working Days, Timeliness. Annual timeliness costs for an operation of (A) area having a yield of (y) per area and a value of (V) per yield is:

$$Tc = \frac{K A^2 y\,V}{X\,H\,pwd\,C}$$

where X has a value of 4 if the operation can be balanced evenly about the optimum time and a value of 2 if the operation either commences or terminates at the optimum time.

8.2.4 The optimum capacity (C*) is found from the first differential of the annual cost with respect to (C):

$$C^* = \sqrt{\frac{100A}{FC\,p}\left(L + T + \frac{KAyV}{X\,H\,pwd}\right)}$$

where

Fc = fixed cost percentage

As an example, consider the required capacity for a field cultivator having a unit price function of $400 per ha per h, a 16% fixed cost percentage, used on 200 ha twice each year. Labor and tractor fixed costs per h are $3 each. The crop has a value (yV) of $150 per ha, there are 10 field working hours per day at a probability of 0.8. The operations will be balanced about the optimum time (X = 4) and the value of K is selected as 0.0002.

$$C^* = \sqrt{2\left[\frac{100 \times 200}{16 \times 400}\left(3 + 3 + \frac{0.0002 \times 200 \times 150}{4 \times 10 \times 0.8}\right)\right]}$$

$$= 6.2 \text{ ha/h}$$

If field speed is expected to be 8 km/h and the field efficiency 0.80, the width of the field cultivator would be

$$6.2 = 8 \text{ w } 0.80/10 \quad \therefore \quad w = 9.7 \text{ m}$$

8.3 The selection of a machine system involves the selection of tractors and their associated machines which give least cost when both timeliness and scheduling factors are considered. Except for the simplest of systems, such selections are accomplished with digital computer programs.

SECTION 9—REPLACEMENT

9.1 Machines employed in production may need to be replaced for one or more reasons.

9.1.1 A machine suffers accidental damage such that the cost of renewal is so great that a new machine is more economic.

9.1.2 The capacity of the existing machine is inadequate because of increased scale of production.

9.1.3 The machine is obsolete (see ASAE Standard S322, Uniform Terminology for Agricultural Machinery Management).

9.1.4 The machine is not expected to operate reliably. (Suffers considerable unanticipated downtime from random part failures.)

9.1.5 The cost of making an anticipated repair would increase the average unit accumulated cost (see ASAE Standard S322, Uniform Terminology for Agricultural Machinery Management) above the expected minimum. Only capital costs and actual repair and maintenance costs need be accumulated. For example, a $3000 machine is used 100 ha annually. It experiences the following end-of-year depreciation, interest (8% simple interest on average investment), and actual repair and maintenance costs listed in Table 1. Year 9 has the lowest unit cost and indicates the machine should be replaced with a similar machine at the end of year 9 if not before for other reasons. Inflation effects must be considered in making replacement decisions. Annual depreciation charges may be quite low or even negative in times of rapid inflation producing a premature minimum unit accumulated cost. In such instances replacement is better indicated by comparing the unit accumulated cost of the present machine with the projected costs for a potential successor machine. Optimum replacement time may be delayed beyond that time determined under more stable economic conditions.

Cited Standards:

ASAE D230, Agricultural Machinery Management Data
ASAE S296, Uniform Terminology for Traction of Agricultural Tractors, Self-Propelled Implements, and Other Traction and Transport Devices
ASAE S322, Uniform Terminology for Agricultural Machinery Management

TABLE 1—AVERAGE UNIT ACCUMULATED COSTS

End of Year	Remaining Value	R&M Costs	Depr.	Int.	Acc. Depr.	Acc. Int.	Acc. R&M	Tot. Acc. Costs, $	Acc. Use, ha	Unit Acc. Costs, $/ha
1	2000	10	1000	200	1000	200	10	1210	100	12.10
2	1400	50	600	136	1600	336	60	1996	200	9.98
3	1000	70	400	96	2000	432	130	2562	300	8.54
4	700	100	300	68	2300	500	230	3030	400	7.58
5	500	200	200	48	2500	548	430	3478	500	6.96
6	350	300	150	34	2650	582	730	3962	600	6.60
7	225	350	125	23	2775	605	1080	4460	700	6.37
8	125	450	100	14	2875	619	1530	5024	800	6.28
9	100	550	25	9	2900	628	2080	5608	900	6.23
10	75	600	25	7	2925	635	2680	6240	1000	6.24

COTTON MODULE BUILDER STANDARD

Developed by the Module Building Standards Subcommittee of the ASAE Farm Materials Handling Committee; approved by the ASAE Power and Machinery Division Standards Committee; adopted by ASAE March 1978; reconfirmed December 1982, December 1987.

SECTION 1—PURPOSE AND SCOPE

1.1 The purpose of this Standard is to provide a uniform size cotton module builder and to establish recommendations for module builders and transporters. Uniformity will allow interchangeability of module builders and transporters between manufacturers. There is a need for cotton module covers that are adequate for cotton protection and a long wearing life. Safety is a factor in transporting modules on highways.

SECTION 2—COTTON MODULE BUILDER DIMENSIONS

2.1 All dimensions are to be determined with the module builder in the field position. The heights of the sides are measured from the ground level to the highest horizontal projection which it will be necessary to clear in order to dump cotton into the module builder. This height (A) in Fig. 1 shall be either 2.74 m (9 ft) or 3.35 m (11 ft). The 2.74 m (9 ft) model is designed for use with cotton pickers, the 3.35 m (11 ft) model can be used with cotton strippers that have a higher dumping height.

2.2 The width of a module builder is the inside dimension measured at the base of the module builder. The width (B) in Fig. 1 shall be 2.13 to 2.21 m (7.0 to 7.25 ft).

2.3 Length (C) in Fig. 1 of the module shall be 9.75 m (32 ft) or 7.32 m (24 ft) measured at the base of the module.

2.4 Taper (D) in Fig. 1 on the two sides and two ends of the module builder should be 25.4 mm (1.0 in.) inward for each 304.9 mm (12 in.) of vertical rise.

2.5 The maximum tread width (E) shall be such that tires do not project beyond the upper sides of the module builder. This will allow cotton harvesting machines to get close enough to minimize spilled cotton while dumping.

2.6 The height (K) for the module is measured from the ground to the point at which the sides flare outward. This height (K) in Fig. 1 shall be either 2.29 m (7.5 ft) for the 2.74 m (9 ft) module builder used with cotton pickers or 2.59 m (8.5 ft) for the 3.35 m (11 ft) module used with cotton strippers.

SECTION 3—TRANSPORTER DIMENSIONS

3.1 Maximum width (F) in Figs. 2 and 3 should be 2.44 m (8 ft) or less.

3.2 The length (G) in Figs. 2 and 3 of transporter and wheel base shall be adequate to meet Department of Transportation requirements and all state and federal requirements.

3.3 The minimum transporter height (H) recommended is 762 mm (30 in.) measured from the ground to the bottom of the pallet.

3.4 Loading angle (J) shall be 12 deg (or less) measured from horizontal to the loading surface.

3.5 Van bodies if used shall be 2.44 m (8 ft) or less in width. Adequate marking and lighting shall be provided. Vertical clearance shall comply with all applicable state and federal laws.

SECTION 4—CANVAS MODULE COVER SPECIFICATIONS

4.1 Final finish size shall be 3.1 × 8.2 m (10 × 27 ft) for the 24 ft module or 3.1 × 10.7 m (10 × 35 ft) for the 32 ft module. Seams are to run the short direction with width of material to be 914 mm (36 in.) to 1220 mm (48 in.) wide.

4.2 Covers shall meet as a minimum standard of quality the requirements of Canvas Products Association International Standard CPAI-63, Canvas Tarpaulin Specification, as amended or revised.

4.3 Canvas covers are to be made of 310 gm/m^2 (10 oz/yd^2) single-filled cotton duck before treatment.

4.4 Treatment of covers shall be full penetration (not surface covered) for water and mildew resistance so as to weigh approximately 496 gm/m^2 (16 oz/yd^2). The treated weight should be 465 gm/m^2 (15 oz/yd^2) minimum and 574 gm/m^2 (18.5 oz/yd^2) maximum.

4.5 Fabrication of covers

4.5.1 Flat lapped seams shall run the short direction. Three rows of parallel chain stitching with synthetic thread shall be used, and the cross seams are to be locked into edge (hem) seams.

4.5.2 Hems (edge seams) shall be triple thickness on all four sides; two rows of lock stitch with synthetic thread.

4.6 Reinforcements on canvas covers

4.6.1 The four corners shall be reinforced with a 142 mm (6 in.) doubled triangular patch of self material securely sewn into each corner.

4.6.2 All areas where the grommets are to be secured shall have five thicknesses of material. In areas other than the corners and at width seams, a 6 in. doubled triangular patch is to be securely sewn into the cover and into the hem for reinforcement of the grommet.

4.6.3 Grommets shall be brass, rust resistant steel or aluminum rolled rim spur grommets #2 minimum size to #4 maximum size. Grommets shall be located on 3 ft minimum centers and 4 ft maximum centers.

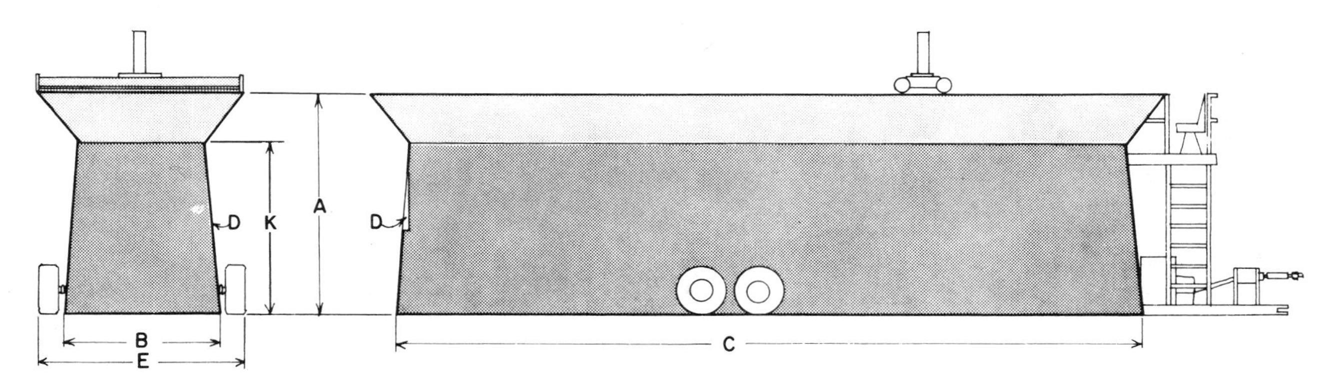

FIG. 1—MODULE BUILDER

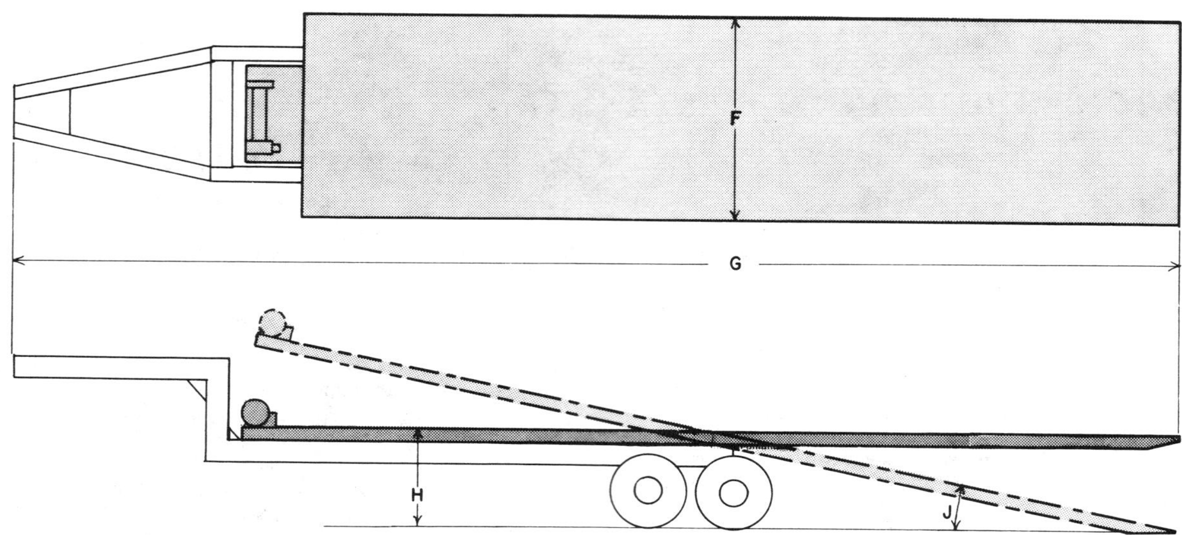

FIG. 2—MODULE TRANSPORTER

SECTION 5—SAFETY

5.1 Items related to safety shall conform to the following ASAE Standards.

5.1.1 ASAE Standard S207, Operating Requirements for Tractors and Power Take-Off Driven Implements

5.1.2 ASAE Standard S276, Slow-Moving Vehicle Identification Emblem

5.1.3 ASAE Standard S277, Mounting Brackets and Socket for Warning Lamp and Slow-Moving Vehicle (SMV) Identification Emblem

5.1.4 ASAE Standard S279, Lighting and Marking of Agricultural Field Equipment on Highways

5.1.5 ASAE Standard S335, Operator Controls on Agricultural Equipment

5.1.6 ASAE Standard S338, Safety Chain for Towed Equipment

5.1.7 Cotton module builders and module transporters shall meet all applicable portions of ASAE Standard S318, Safety for Agricultural Equipment, not specifically mentioned in this Standard.

Cited Standards:

ASAE S207, Operating Requirements for Tractors and Power Take-Off Driven Implements
ASAE S276, Slow-Moving Vehicle Identification Emblem
ASAE S277, Mounting Brackets and Socket for Warning Lamp and Slow-Moving Vehicle (SMV) Identification Emblem
ASAE S279, Lighting and Marking of Agricultural Field Equipment on Highways
ASAE S318, Safety for Agricultural Equipment
ASAE S335, Operator Controls on Agricultural Equipment
ASAE S338, Safety Chain for Towed Equipment

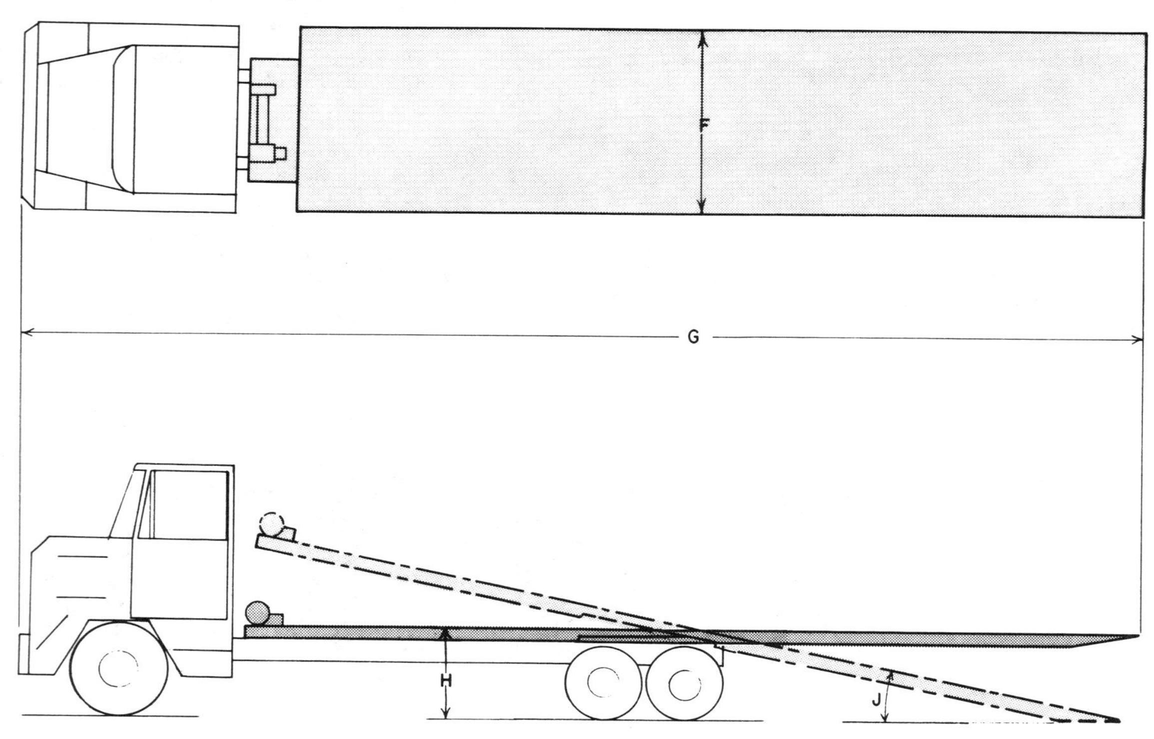

FIG. 3—MODULE MOVER TRUCK

ASAE Standard: ASAE S396.1

T

COMBINE CAPACITY TEST PROCEDURE

Developed by the ASAE Grain Harvesting Committee; approved by the Power and Machinery Division Standards Committee; adopted by ASAE as a Tentative Standard March 1979; reconfirmed December 1979; reclassified as a full Standard February 1981; reconfirmed December 1985; revised March 1988.

SECTION 1—PURPOSE AND SCOPE

1.1 This Standard is intended to provide the basic requirements for a uniform procedure for measuring and reporting combine capacity, as defined in ASAE Standard S343, Terminology for Combines and Grain Harvesting. Because crop conditions are variable and uncontrollable, the procedure provides only for the comparative testing of one combine, or one combine configuration, relative to another, in a particular crop condition.

SECTION 2—TERMINOLOGY

2.1 Test combine: The combine or combine configuration to be tested.

2.2 Comparison combine: The combine or combine configuration with which the test combine is being compared.

2.3 Test run: The events necessary to record a single set of measurements.

2.4 Test: All the events and data of the test combine and the comparison combine that define their performance during the test runs on each combine.

2.5 Catch: Any material collected during a test run.

SECTION 3—SELECTION OF CROP

3.1 Capacity tests should preferably be conducted in crops and conditions listed in ASAE Standard S343, Terminology for Combines and Grain Harvesting. Where test data indicate that such requirements have not been adhered to, the reasons for departures shall be stated in the test report.

3.2 Crops used for tests shall be reasonably uniform, free of disease, weeds and other crops. They should in general be standing well. If local climatic conditions or local practices lead to different conditions typical of the locality (for instance widespread lodging or the windrowing of crops), the circumstances should be stated in the test report.

SECTION 4—MACHINES

4.1 The test combine shall be fully identified as to make, model, year and other pertinent information. The terminology and methods of measurement defined by ASAE Standard S343, Terminology for Combines and Grain Harvesting, shall be used where relevant.

4.2 The comparison combine shall be similarly identified, functionally sound and one which has been available on the open market for a continuous period of at least one year prior to the date of the test.

4.3 At the time of the test, both test and comparison combines shall be in good working order with all working surfaces free of thick or sticky paint, rust, grease, or other impediments to smooth operation. A run-in period of 25 h is required to ensure this condition on a new combine, or one that has not been used recently.

4.4 Immediately prior to testing, both test and comparison combines should be adjusted for optimum performance in the same piece of crop that will be used for the tests. Pretest adjustments should aim for optimum performance with harvesting conditions typical of normal practice in the relevant locality. Care should be taken to ensure that the dockage level in the clean grain sample of the test combine is similar to that of the comparison combine. Persons responsible for adjusting the machines shall be given adequate time and opportunity to do so, with regard to the time requirements of Section 6—Test Procedure for conduct of the test proper. They shall also be responsible for deciding when the optimum adjustments have been attained.

4.5 No operator adjustment of the threshing, separating, or cleaning devices and header height shall be permitted during a test on either combine.

SECTION 5—COLLECTION OF SAMPLES

5.1 Apparatus for catching crop material discharged from the combine shall be built and operated so that:

5.1.1 The whole of the effluent from the combine is caught during the catch period.

5.1.2 The grain catch may be delayed after initiation of material-other-than-grain (mog) catch to compensate for the time the grain spends in the conveyor. This circumstance must be reported.

5.1.3 Catches are started and stopped without interrupting combine mechanisms or forward travel.

5.1.4 The apparatus does not significantly interfere with the combine's normal operation; for instance, with the flow of air from the cleaning mechanism.

5.1.5 Catches are taken from the points of normal discharge from the combine's separating and primary cleaning mechanisms at the normal rate of discharge. Conveyor-type apparatus for improving the accessibility of the catch points is permitted, but must be such as to cause no change in the condition of the crop material as discharged from the combine.

5.1.6 Samples of grain for analysis can be taken by passing a container through the stream of grain at the catching point during or immediately after taking the main catch. Storage containers shall be completely filled, and air-tight.

SECTION 6—TEST PROCEDURE

6.1 Prior to each catch period, the combine shall operate for such distance as the test supervisor may deem necessary to ensure that conditions have become stabilized throughout the relevant mechanisms.

6.2 A test shall consist of at least 5 test runs, and preferably not less than 7.

6.3 During this period and during the catching periods, the full gathering width of the header shall be utilized. If the crop is windrowed, the windrow(s) should preferably be symmetrical and shall be picked up wholly and smoothly to ensure flow of crop across substantially the full width of the thresher.

6.4 Field speed and compensatory header adjustments shall be made only between test runs. The only adjustments permitted are reel speed, height, or fore-and-aft settings, or pick-up speed to accommodate differences in ground speed. No adjustment of header height is permitted either during or between test runs within a specific test.

6.5 Field speeds for test runs should be selected to start below the speed used during the pre-test adjustment period and should be carried high enough to perceptibly indicate that maximum feasible feed rate has been reached. The limiting factor on feed rate such as engine overload, feeding difficulty, etc., should be noted and recorded in the test report.

6.6 The time of day selected for the test shall be when crop conditions are most stable, usually the several hours after noon. Comparative tests shall be conducted as closely as possible to the same time and location in the field. Differing circumstances should be reported.

6.7 The catch may be timed for purposes of control, but must be at least 9 m (30 ft) in length. Exceptions shall be noted, and reasons shall be given.

6.8 The test supervisor may discard attempted test runs at the time of test if in his judgment there is obvious reason to do so; e.g., functional failure, detrimental foreign objects entering combine, overfilling or spillage from catch receptacle, etc. Otherwise the results of all test runs made shall be entered on the test report, and comments shall be included on any unusual circumstance.

6.9 At least 3 samples of not less than 1 L shall be taken through the test period for grain analysis.

6.10 Straw samples for moisture testing shall be at least three in number and taken throughout the test period. They should be taken from the straw discharged immediately after the end of the catch period and stored in completely-filled, air-tight containers until analyzed. Requirements are similar with regard to straw moisture measurements made by portable meter.

SECTION 7—PROCESSING OF CATCHES

7.1 Separating and cleaning should be as fully mechanized as practicable to ensure consistency. Feeding of crop should aim for relatively low feed rates, to retain at least 99% of the free grain contained in the samples before processing. Manual sorting of cleaned grain should be minimized.

7.2 In the event that USDA grade No. 2 grain or seed standards are not met, the sample should be evaluated by USDA standard techniques. These techniques may be found in USDA publication, Official United States Standards for Grain #810. In countries other than the United States, a similar government procedure should be followed.

SECTION 8—TEST REPORTING

8.1 The test file should include all original data and measurements recorded for both test and comparison combines. These data shall include the following:

8.1.1 Location of test site.

8.1.2 Date and time of starting and stopping tests.

8.1.3 Combine and header identifications.

8.1.4 Crop, variety, crop conditions, and a reasonable estimate of average yield.

8.1.5 Combine settings and adjustments, particularly those relevant to the crop-handling mechanisms.

8.1.6 Details from the test runs recorded as follows:

8.1.6.1 Duration in seconds to nearest 0.1 s.

8.1.6.2 Field speed in kilometers per hour (miles per hour) to the nearest 0.1 km/h (0.1 mph).

8.1.6.3 Grain catch in kilograms (pounds) to the nearest 0.2 kg (0.5 lb).

8.1.6.4 Separator mechanism catch in kilograms (pounds) to the nearest 0.05 kg (0.1 lb).

8.1.6.5 Cleaning mechanism catch in kilograms (pounds) to the nearest 0.05 kg (0.1 lb).

8.1.6.6 Free grain from separator in kilograms (pounds) to the nearest 0.005 kg (0.01 lb) up to 2 kg (5 lb) and to nearest 0.05 kg (0.1 lb) thereafter.

8.1.6.7 Free grain weight from cleaning mechanism in kilograms (pounds) to the nearest 0.005 kg (0.01 lb) up to 2 kg (5 lb) and to the nearest 0.05 kg (0.1 lb) thereafter.

8.1.6.8 Grain weight from rethreshed mog catches in kilograms (pounds) to the nearest 0.005 kg (0.01 lb) up to 2 kg (5 lb) and to the nearest 0.05 kg (0.1 lb) thereafter.

8.1.7 Moisture content of grain and straw samples expressed on the wet basis to the nearest whole percentage, and method of measurement.

8.1.8 Sample constituents if required by paragraph 7.2.

8.2 The report shall include sections in which the test supervisor shall record details specified above together with notes of unusual changes in weather or other conditions during the tests, and general comments on the performance of the combines and the conduct of the tests.

SECTION 9—CALCULATIONS

9.1 Calculations for each test run on each combine shall include these items as defined in ASAE Standard S343, Terminology for Combines and Grain Harvesting:

9.1.1 The total feed rate, mog feed rate, and grain feed rate in metric tons per hour (pounds per minute).

9.1.2 The processing losses recorded to the nearest 0.1%.

9.1.3 The mog/grain ratio for the crop, and the mean value from all runs for each combine.

9.1.4 The mean moisture contents of the grain and mog samples.

9.2 The resulting calculations shall be tabulated in the test report.

SECTION 10—PRESENTATION OF RESULTS

10.1 Graphs with linear scales are preferred for presentation of processing loss results, with total feed rate, mog feed rate or grain feed rate specified as the horizontal coordinate and percentage loss as the vertical coordinate. The data points for each test run shall be plotted on the graphs. The mog feed rate is preferred in small cereal grains. However, if there is good reason for expressing results otherwise, e.g., the clean grain elevator was plugged with grain thus affecting shoe performance, a full explanation is required, and the graphs supplied to illustrate the effects of this phenomenon shall be in addition to those plotted against mog feed rate.

10.2 The capacity of each combine shall be the feed rate level at which the loss curve intersects the specified level of loss, as defined in ASAE Standard S343, Terminology for Combines and Grain Harvesting.

Cited Standards:

ASAE S343, Terminology for Combines and Grain Harvesting
USDA 810, Official United States Standards for Grains

ASAE Standard: ASAE S399.1

PREFERRED METRIC DIMENSIONS FOR AGRICULTURAL IMPLEMENT DISK BLADES

Proposed by the Engineering Committee of the Farm and Industrial Equipment Institute; approved by the ASAE Power and Machinery Division Standards Committee; adopted by ASAE as an Engineering Practice February 1980; revised March 1981; reconfirmed December 1985; reclassified as a Standard October 1986.

SECTION 1—PURPOSE AND SCOPE

1.1 This Standard establishes preferred nominal metric dimensions for the design of agricultural implement disk blades. The range of preferred metric blade thicknesses and diameters is intended to aid in selection of disk blades for future designs as metric sized components become available and to allow practical interchangeability with prior usage.

1.2 The preferred nominal metric dimensions are intended for all metric designed agricultural implement disk blades regardless of shape; i.e., flat, conical, concave, cut-out, etc.

SECTION 2—PREFERRED NOMINAL METRIC DIMENSIONS

2.1 Preferred nominal metric disk blade thicknesses include:

2.0 mm	5.0
2.5	6.5
3.0	8.0
3.5	10.0
4.0	12.0
4.5	16.0

2.2 Preferred nominal metric disk blade diameters include:

200 mm	450	700
250	500	800
300	550	900
350	600	1000
400	650	1100

2.3 Disk blades larger than 1100 mm shall have diameters in increasing increments of 200 mm.

ASAE Tentative Standard: ASAE S404T

METRIC ROW SPACINGS

Developed by the Metric Row Spacings Subcommittee of the ASAE Cultural Practices Equipment Committee; approved by the ASAE Power and Machinery Division Standards Committee; adopted by ASAE as a Tentative Standard March 1980; reconfirmed December 1981, December 1982, December 1983, December 1984, December 1985, December 1986; reconfirmed for one year December 1987.

SECTION 1—PURPOSE AND SCOPE

1.1 This Tentative Standard establishes preferred nominal metric (SI) row spacings of crops for the worldwide mutual benefit of farmers, agricultural equipment manufacturers, and others.

1.2 This Tentative Standard includes planted crops which may also be cultivated and/or harvested as rows along with drilled crops which are not cultivated. It does not include number of rows per machine or the related value, working width.

SECTION 2—COMPATIBILITY

2.1 This Tentative Standard considers agronomic and economic needs by:

2.1.1 Providing sufficient choice of row spacings to permit optimum yields for various crops, climates, soils, etc.

2.1.2 Providing row spacings which are practical for planting, cultivating, and harvesting as row crops.

2.1.3 Providing row spacings compatible with various makes and models of agricultural equipment.

2.1.4 Providing nominal metric row spacings on future equipment which is compatible with row spacings on most currently-owned equipment.

2.1.5 Minimizing the number of row spacings to permit greater use of common planting, cultivating, or harvesting equipment for various crops.

2.1.6 Minimizing the number of row spacings to reduce farmer costs through lower production and inventory costs for the manufacturer.

2.2 Preferred row spacings are given in Table 1.

TABLE 1—PREFERRED ROW SPACINGS

First Preference		Second Preference	
cm	(in.)	cm	(in.)
	—	100	(39.4)
95	(37.4)		—
90	(35.4)		—
75	(29.5)		—
	—	70	(27.6)
	—	55	(21.7)
50	(19.7)		—
	—	45	(17.7)
40	(15.7)		—
	—	38	(15.0)
30	(11.8)		—
20	(7.9)		—
18	(7.1)		—
15	(5.9)		—
	—	12	(4.7)

References: Last printed in 1985 STANDARDS; list available from ASAE Headquarters.

ASAE Standard: ASAE S414

TERMINOLOGY AND DEFINITIONS FOR AGRICULTURAL TILLAGE IMPLEMENTS

Developed by the ASAE Cultural Practices Equipment Committee in cooperation with the Tillage Equipment Council of the Farm and Industrial Equipment Institute; approved by the ASAE Power and Machinery Division Standards Committee; adopted by ASAE March 1982; reconfirmed December 1986.

SECTION 1—PURPOSE AND SCOPE

1.1 The purpose of this Standard is to provide uniform terminology and definitions for tillage implements designed primarily for use in the production of food and fiber. It does not include implements designed for earth movement and transport.

1.2 Dimensions, spacings, depths of operations, widths or velocities may be used as a part of the implement specifications. These in no way should be considered as performance specifications for any type of design or publication.

SECTION 2—TILLAGE IMPLEMENT CATEGORIES

2.1 Primary tillage: Tillage which displaces and shatters soil to reduce soil strength, and to bury or mix plant materials and fertilizers in the tilled layer. Primary tillage is more aggressive, deeper, and leaves a rougher surface relative to secondary tillage.

2.1.1 Examples of primary tillage implements:

- Plows
  - Moldboard
  - Chisel
  - Combination chisel with cutting blades
  - Wide-sweep
  - Disk
  - Bedder
- Moldboard listers
- Disk bedders
- Subsoilers
- Disk harrows
  - Offset disk
  - Heavy tandem disk
- Powered rotary tillers

2.2 Secondary tillage: Tillage that works the soil to a shallower depth than primary tillage, provides additional pulverization, levels and firms the soil, closes air pockets, and kills weeds. Seedbed preparations are the final secondary tillage operations.

2.2.1 Examples of secondary tillage implements:

- Harrows
  - Disk
  - Spring, spike, coil, or tine tooth
  - Knife
  - Roller
  - Powered oscillatory spike tooth
  - Packer
  - Ridger
  - Leveler
  - Rotary ground-driven
- Cultivators
  - Field or field conditioner
- Rod weeders
- Roller harrows
- Powered rotary tillers
- Bed shapers
- Rotary hoes

2.3 Cultivating tillage: Shallow post-planting tillage whose principle purpose is to aid the crop by either loosening the soil and/or by mechanical eradication of undesired vegetation.

2.3.1 Examples of cultivating implements:

- Row crop cultivators
  - Rotary ground-driven
  - Spring tooth
  - Shank tooth
- Rotary hoes
- Rotary tillers—strip type, power driven

SECTION 3—IMPLEMENT HITCH CLASSIFICATIONS

3.1 Pull

3.1.1 Wheel mounted

3.1.2 Drag

3.1.3 Squadron

3.2 Semi-mounted (semi-integral)

3.3 Rear mounted (three-point integral)

3.4 Front mounted

3.5 Center mounted

3.6 All tillage tools are not produced in all classifications.

SECTION 4—IMPLEMENT FRAME CONFIGURATIONS

4.1 Rigid

4.2 Rigid with rigid wings

4.3 Single folding wing

4.4 Dual folding wings

4.5 Multiple folding wings

4.6 Hinged

4.7 Sectional

4.8 Endways transported

4.9 Wing styles may have mechanical, hydraulic, or no folding assistance.

SECTION 5—DEFINITIONS AND ILLUSTRATIONS

5.1 Disk harrow: A primary or secondary tillage implement consisting of two or four gangs of concave disks. Adjustment of gang angle controls cutting aggressiveness. Disk harrow hitches are either rear mounted or pull type. Types of disk harrows are shown in Figs. 1-7.

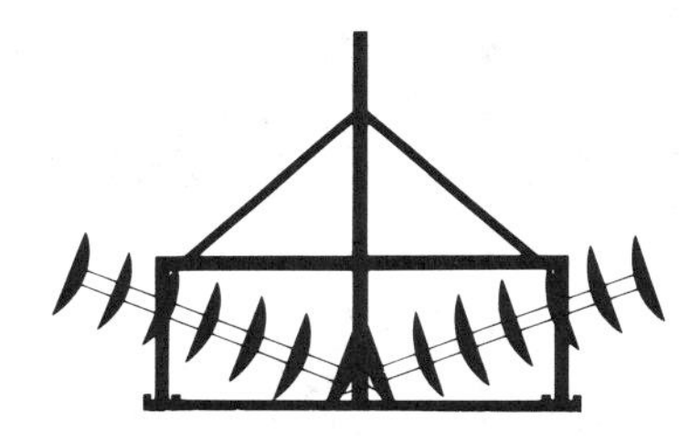

FIG. 1—SINGLE DISK HARROW—TWO GANGS OF DISKS SET TO OPPOSE EACH OTHER. DRAG TYPE PULL HITCH

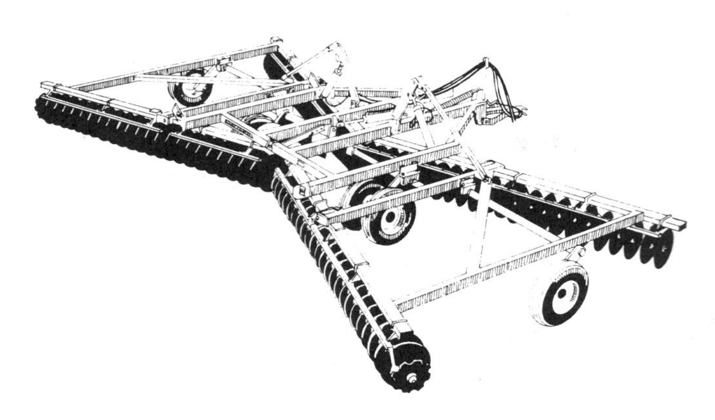

FIG. 2—TANDEM DISK HARROW—IN LINE—FOUR GANGS OF DISKS WITH THE TWO FRONT GANGS SET AS A SINGLE DISK HARROW AND THE TWO REAR GANGS IN TANDEM TO THOSE IN FRONT. THE REAR GANGS THROW SOIL IN THE OPPOSITE DIRECTION TO THAT FROM THE FRONT GANGS. WHEEL MOUNTED PULL HITCH

FIG. 5—TANDEM DISK HARROW—DUAL FOLDING WINGS. WHEEL MOUNTED PULL HITCH

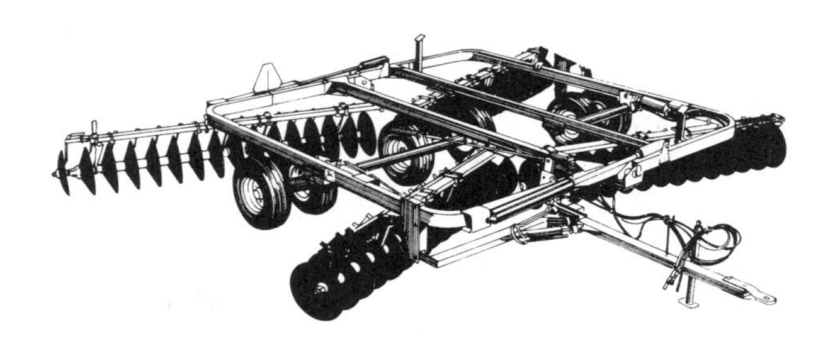

FIG. 3—TANDEM DISK HARROW—FRONT GANGS OFFSET—THE INNER END OF ONE FRONT GANG TRAVELS BEHIND THE INNER END OF THE OTHER FRONT GANG. WHEEL MOUNTED PULL HITCH

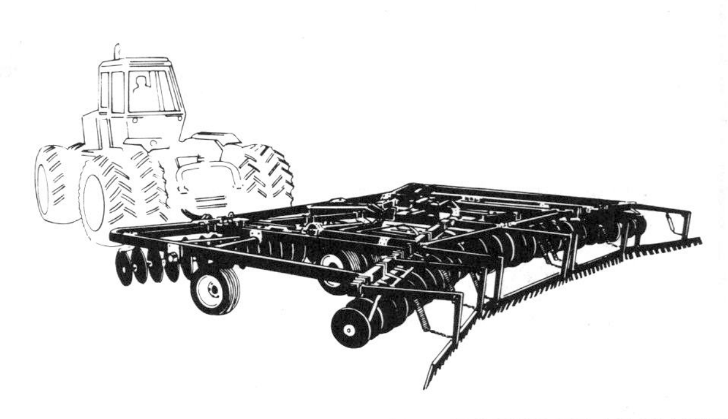

FIG. 6—TANDEM DISK HARROW—TOOTH DRAG HARROW ATTACHMENT. WHEEL MOUNTED PULL HITCH

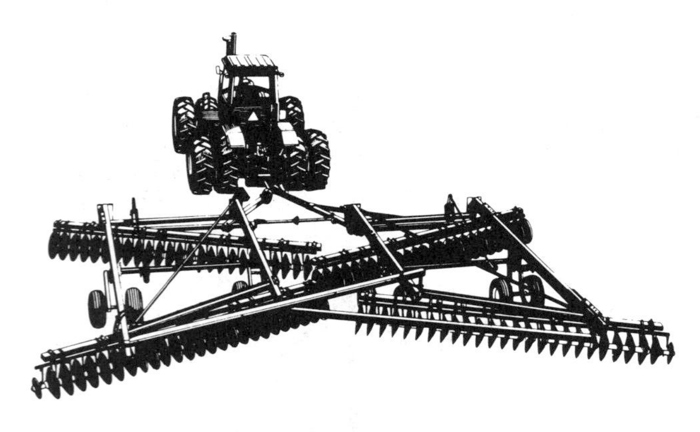

FIG. 4—TANDEM DISK HARROW—DOUBLE OFFSET—THE INNER END OF ONE GANG TRAVELS BEHIND THE INNER END OF THE OTHER GANG IN THE SAME RANK. WHEEL MOUNTED PULL HITCH

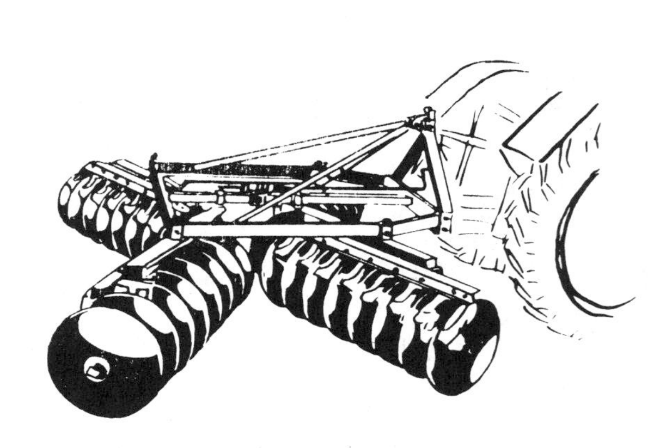

FIG. 7—TANDEM DISK HARROW—REAR MOUNTED HITCH

5.2 Offset disk harrow: A primary or secondary tillage implement consisting of two gangs of concave disks in tandem. The gangs cut and throw soil in opposite directions. Types of offset disk harrows are shown in Figs. 8-10.

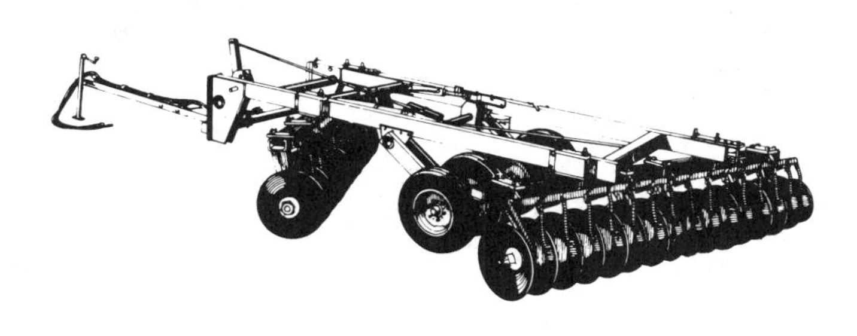

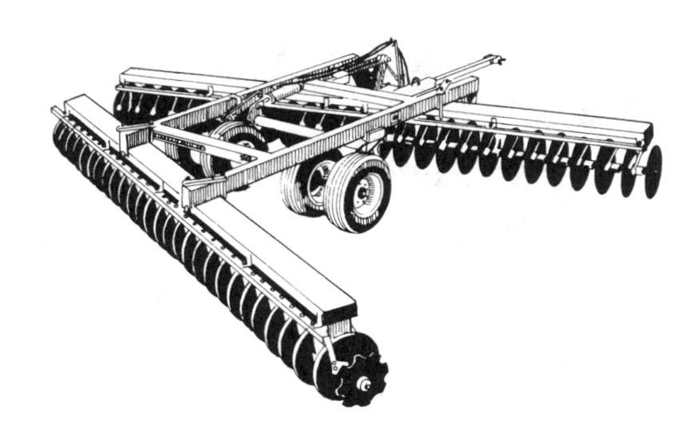

FIG. 8—OFFSET DISK HARROW—WHEEL MOUNTED PULL HITCH

FIG. 9—OFFSET DISK HARROW—DRAG TYPE PULL HITCH

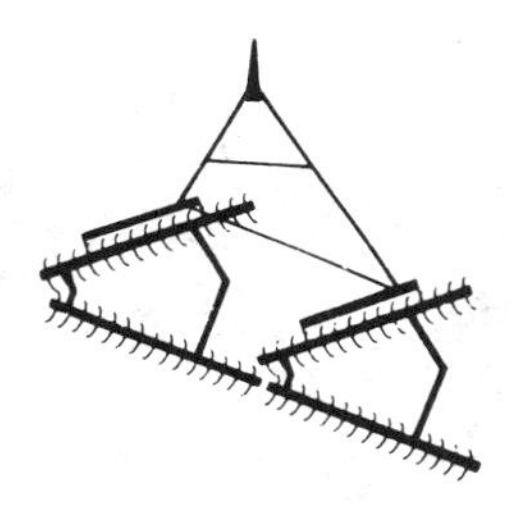

FIG. 10—OFFSET DISK HARROW—SQUADRON, DRAG TYPE PULL HITCH

5.3 One-way disk harrow: A tillage implement equipped with one gang of concave disks. When mounted in short flexible gang units, the harrow conforms to uneven soil surfaces. Types of one-way disk harrows are shown in Figs. 11-12.

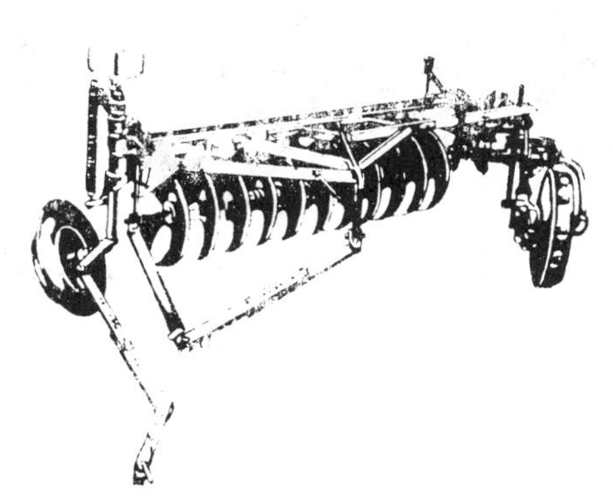

FIG. 11—ONE-WAY DISK HARROW—RIGID FRAME. WHEEL MOUNTED PULL HITCH

FIG. 12—ONE-WAY DISK HARROW—EQUIPPED WITH SEEDER ATTACHMENT. SQUADRON, WHEEL MOUNTED PULL HITCH

5.4 Moldboard plow: A primary tillage implement which cuts, partially or completely inverts a layer of soil to bury surface materials, and pulverizes the soil. The part of the plow that cuts the soil is called the bottom or base. The moldboard is the curved plate above the bottom which receives the slice of soil and inverts it. Moldboard plows are equipped with one or more bottoms of various cutting widths. Bottoms are commonly right-hand that turn all slices to the right. Two-way moldboard plows are equipped with right-hand and left-hand bottoms that are alternately used to turn all slices in the same direction as the plow is operated back and forth across the field. Types of moldboard plows are shown in Figs. 13-16.

FIG. 13—MOLDBOARD PLOW—SIX BOTTOM. WHEEL MOUNTED PULL HITCH

FIG. 14—MOLDBOARD PLOW—FIVE BOTTOM. SEMI-MOUNTED HITCH

FIG. 15—MOLDBOARD PLOW—FIVE BOTTOM. REAR MOUNTED

FIG. 16—MOLDBOARD PLOW—TWO-WAY. REAR MOUNTED

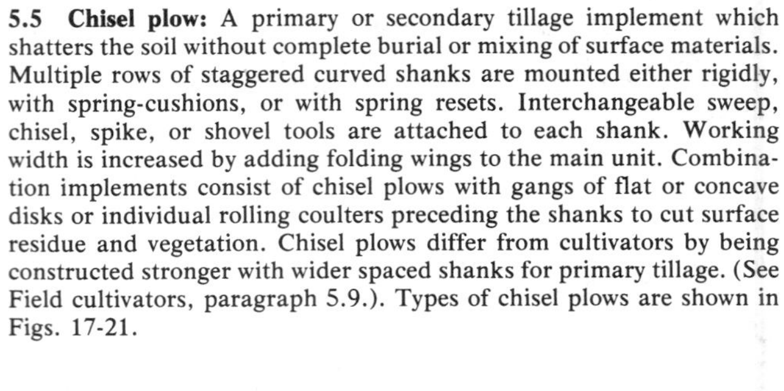

5.5 Chisel plow: A primary or secondary tillage implement which shatters the soil without complete burial or mixing of surface materials. Multiple rows of staggered curved shanks are mounted either rigidly, with spring-cushions, or with spring resets. Interchangeable sweep, chisel, spike, or shovel tools are attached to each shank. Working width is increased by adding folding wings to the main unit. Combination implements consist of chisel plows with gangs of flat or concave disks or individual rolling coulters preceding the shanks to cut surface residue and vegetation. Chisel plows differ from cultivators by being constructed stronger with wider spaced shanks for primary tillage. (See Field cultivators, paragraph 5.9.). Types of chisel plows are shown in Figs. 17-21.

FIG. 17—CHISEL PLOW—RIGIDLY MOUNTED SHANKS. WHEEL MOUNTED PULL HITCH

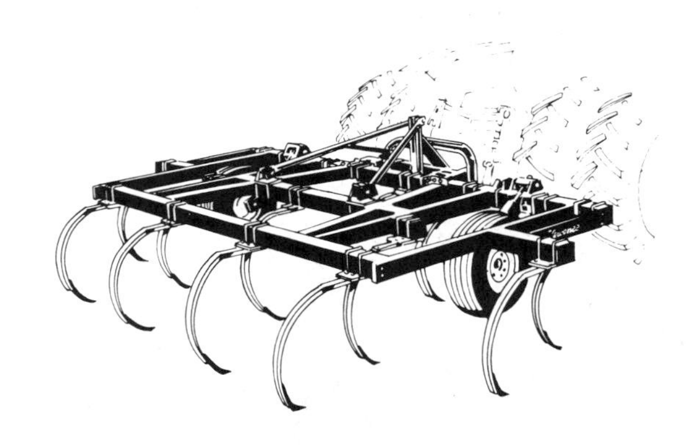

FIG. 18—CHISEL PLOW—WHEELS ARE FOR DEPTH CONTROL. REAR MOUNTED

FIG. 19—CHISEL PLOW—DUAL FOLDING WINGS. WHEEL MOUNTED PULL HITCH

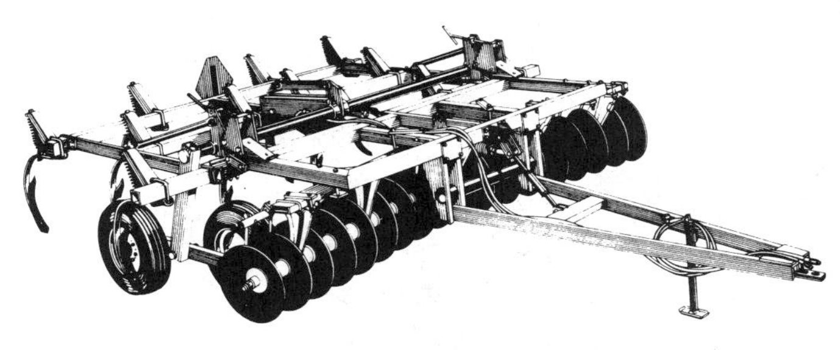

FIG. 20—COMBINATION CHISEL PLOW—TWO OPPOSING GANGS OF CONCAVE DISKS, SPRING-CUSHIONED SHANK MOUNTS. WHEEL MOUNTED PULL HITCH

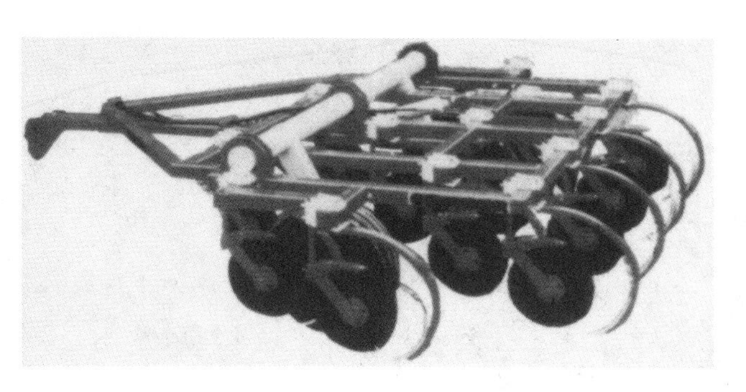

FIG. 21—COMBINATION CHISEL PLOW—INDIVIDUAL ROLLING COULTERS PRECEDING EACH SHANK. WHEEL MOUNTED PULL HITCH

5.6 Disk plow: A primary tillage implement with individually mounted concave disk blades which cut, partially or completely invert a layer of soil to bury surface material, and pulverize the soil. Blades are attached to the frame in a tilted position relative to the frame and to the direction of travel for proper penetration and soil displacement. Penetration is increased by the addition of ballast weight. Disk plows are equipped with one or more blades of diameter corresponding to intended working depth. Disk plows are commonly right-hand, but two-way plows are equipped with right-hand and left-hand blades. Types of disk plows are shown in Figs. 22-24.

FIG. 22—DISK PLOW—TWO-WAY. WHEEL MOUNTED PULL HITCH

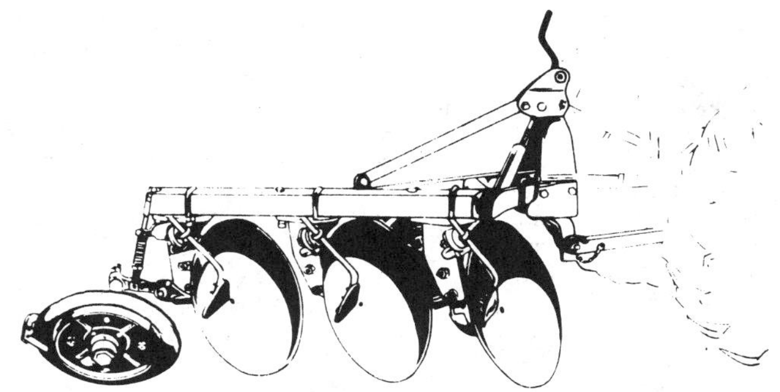

FIG. 23—DISK PLOW—THREE BLADE. REAR MOUNTED

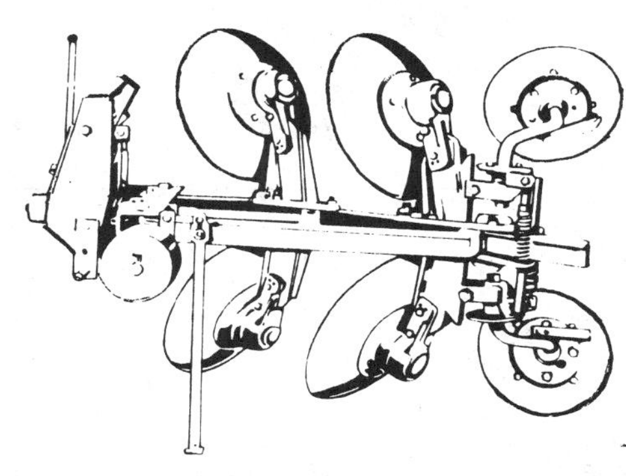

FIG. 24—DISK PLOW—TWO-WAY, TWO BLADE. REAR MOUNTED

5.7 Subsoiler: A primary tillage implement for intermittent tillage at depths sufficient to shatter compacted subsurface layers. Subsoilers are equipped with widely spaced shanks either in-line or staggered on a V-shaped frame. Subsoiling is commonly conducted with the shank paths corresponding to subsequent crop rows. Strong frame and shanks are required for deep operation. Types of subsoilers are shown in Figs. 25-26.

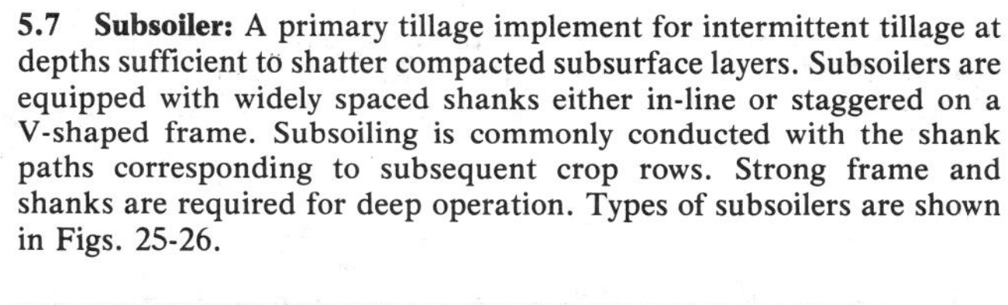

FIG. 25—SUBSOILER—V-FRAME. WHEEL MOUNTED PULL HITCH

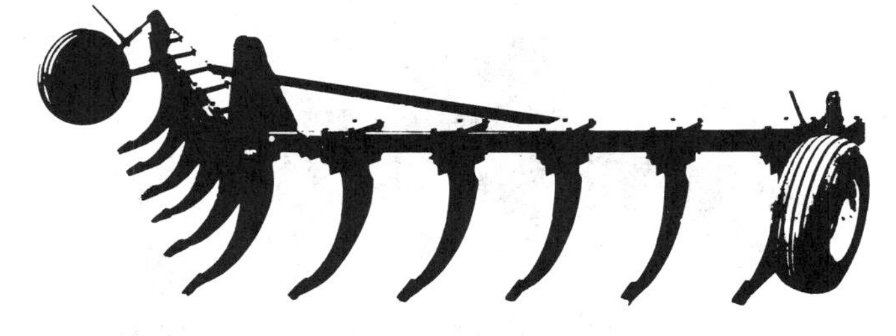

FIG. 26—SUBSOILER—V-FRAME. REAR MOUNTED

5.8 Bedder-ridger: A primary tillage implement or a secondary tillage implement for seedbed forming. Bedder tools are either moldboard lister bottoms which simultaneously throw soil in both right-hand and left-hand directions or short disk gangs with two or more disks of equal or varying diameters. Each disk gang throws soil in one direction and is followed by another disk gang throwing soil in the opposite direction to form a furrow. Planting attachments are sometimes added behind a bedder for planting either on top of the beds or in the furrows. Types of bedder-ridgers are shown in Figs. 27-31.

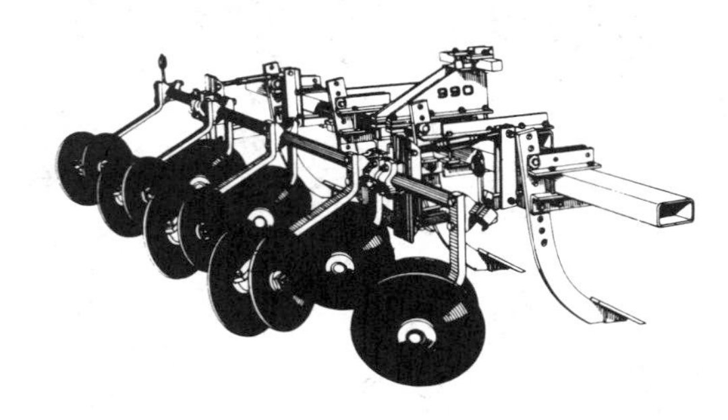

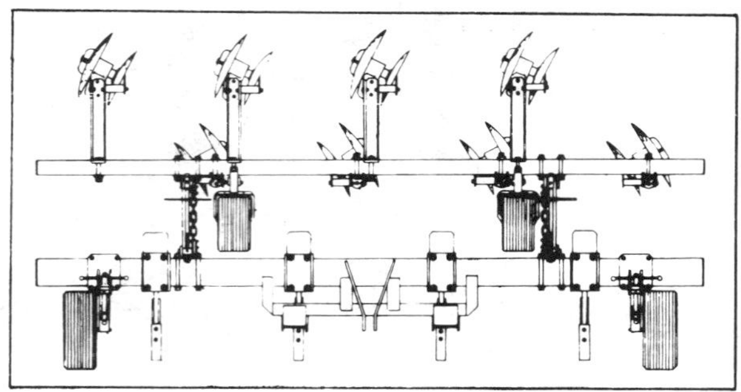

FIG. 27—SUBSOIL BEDDER—IN-LINE SHANKS, DISK BEDDER WITH TWO DISKS PER INDIVIDUALLY MOUNTED GANG UNIT. DISK GANGS ARE STAGGERED. REAR MOUNTED

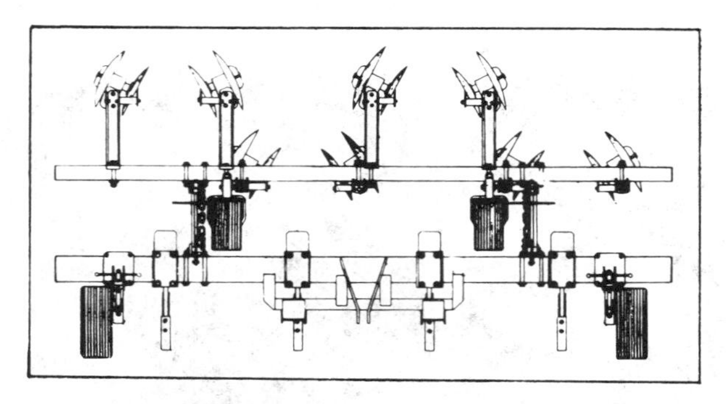

FIG. 28—SUBSOIL BEDDER—SAME AS FIG. 27 EXCEPT THAT DISK GANGS ARE OPPOSED

FIG. 29—ROW BEDDER—DISK BEDDER, GAUGE WHEELS FOR DEPTH CONTROL. REAR MOUNTED

FIG. 30—ROW BEDDER—MOLDBOARD LISTER BOTTOMS, GAUGE WHEELS FOR DEPTH CONTROL. REAR MOUNTED

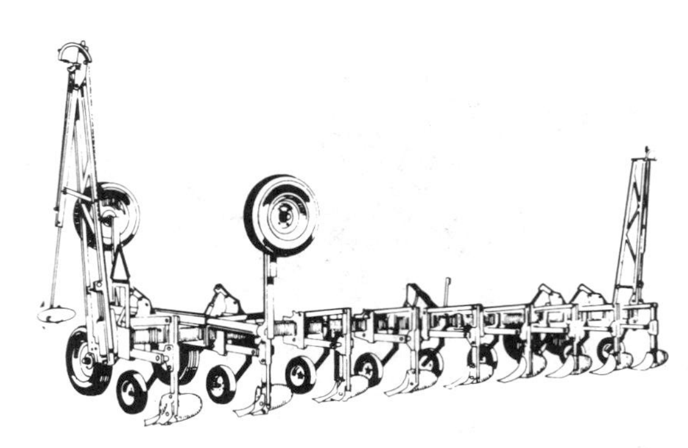

FIG. 31—ROW BEDDER—LISTER BOTTOMS MOUNTED ON PARALLEL LINKAGE WITH INDIVIDUAL DEPTH CONTROL. REAR MOUNTED, ENDWAYS TRANSPORTED HITCH

5.9 Field cultivator: A secondary tillage implement for seedbed preparation, weed eradication, or fallow cultivation subsequent to some form of primary tillage. Field cultivators are equipped with spring steel shanks or teeth which have an integral forged point or mounting holes for replaceable shovel or sweep tools. Teeth are generally spaced 15-23 cm (6-9 in.) in a staggered pattern. Frame sections are folded upwards or backwards for transport. Types of field cultivators are shown in Figs. 32-33.

FIG. 32—FIELD CULTIVATOR—DUAL FOLDING WINGS. WHEEL MOUNTED PULL HITCH

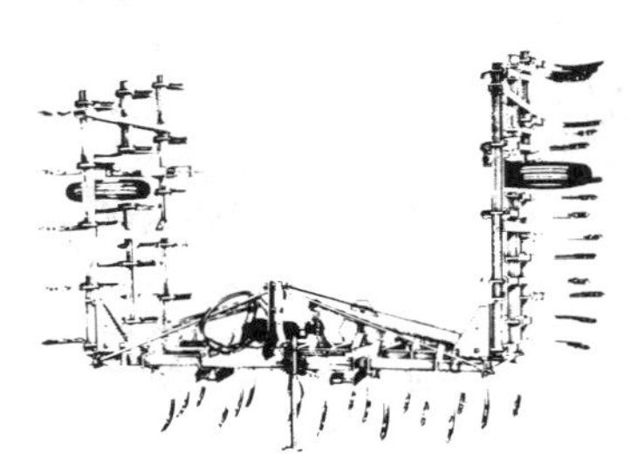

FIG. 33—FIELD CULTIVATOR—DUAL WINGS FOLDED FOR TRANSPORT. REAR MOUNTED

5.10 Row crop cultivator: A secondary tillage implement for tilling between crop rows. The frame and cultivating tools are designed to adequately pass through standing crop rows without crop damage. Gangs of shanks are often independently suspended on parallel linkages with depth-controlling wheels to provide flotation with the soil surface. Tool options are shanks with shovels or sweeps, spring teeth, and ground-driven rotary finger wheels. Types of row crop cultivators are shown in Figs. 34-40.

FIG. 34—ROW CROP CULTIVATOR—FOUR-ROW, SHANKS ON PARALLEL LINKAGES WITH GAUGE WHEELS, ROTARY CROP SHIELDS. FRONT MOUNTED

FIG. 35—ROW CROP CULTIVATOR—FOUR-ROW. REAR MOUNTED

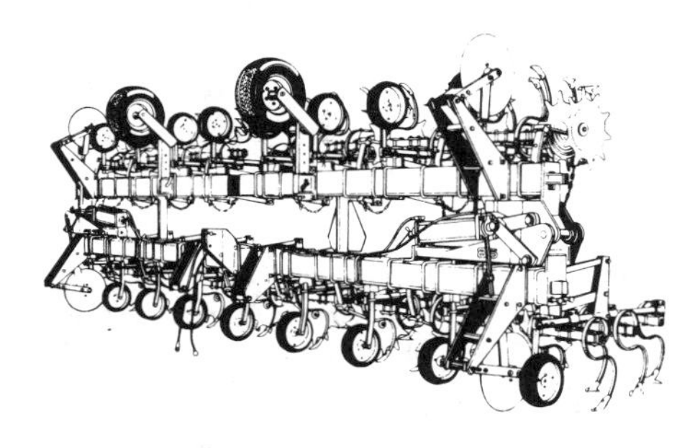

FIG. 36—ROW CROP CULTIVATOR—SPRING TEETH, DUAL WINGS FOLDED FOR TRANSPORT. REAR MOUNTED

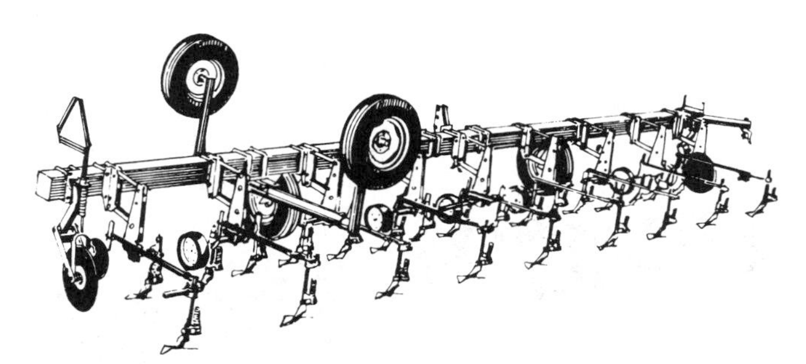

FIG. 37—ROW CROP CULTIVATOR—SWEEP TOOLS ON SHANKS GANGED ON PARALLEL LINKAGES. REAR MOUNTED, ENDWAYS TRANSPORTED HITCH

FIG. 38—ROW CROP CULTIVATOR—ONE-ROW, SPRING TEETH, CROP SHIELDS. REAR MOUNTED

FIG. 39—ROW CROP CULTIVATOR—SAME AS FIG. 38 EXCEPT TWO-ROW, ROLLING COULTER FOR LATERAL STABILITY

FIG. 40—ROW CROP CULTIVATOR—ROTARY GROUND-DRIVEN GANGS OF FINGER WHEELS. REAR MOUNTED

5.11 Harrows: Tillage implements used for seedbed preparation and in some cases light surface cultivation after the seed is planted and before or after the crop emerges. Harrows level the soil surface, enhance moisture retention, pulverize surface clods, and disturb the germination of small weeds. Harrows have staggered teeth of either rigid spikes, coil-spring round wires, flat-spring bars, or S-shaped spring bars. Types of harrows are shown in Figs. 41-46.

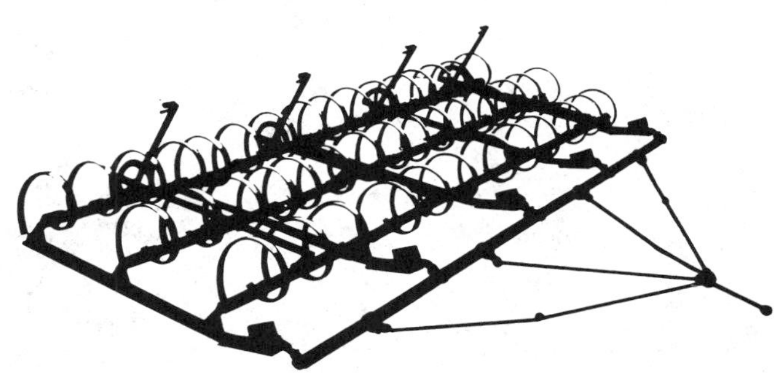

FIG. 41—HARROW—SPRING TEETH, HINGED FRAME SECTIONS. DRAG TYPE PULL HITCH

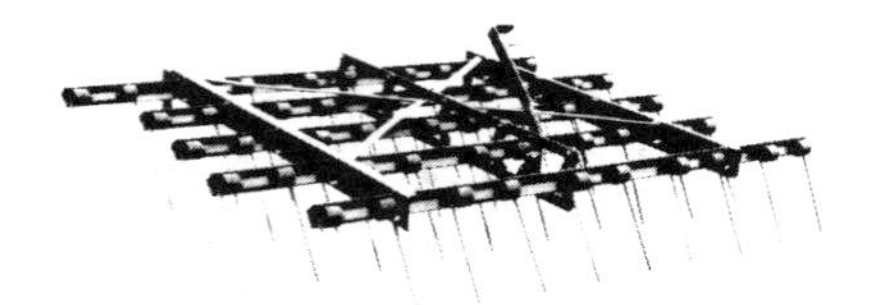

FIG. 42—HARROW—SECTION, ROUND-WIRE TEETH

FIG. 45—HARROW—WHEELS CONTROL DEPTH; S-SHAPED SPRING TEETH FOLLOWED BY COIL-SPRING WIRE TEETH. REAR MOUNTED

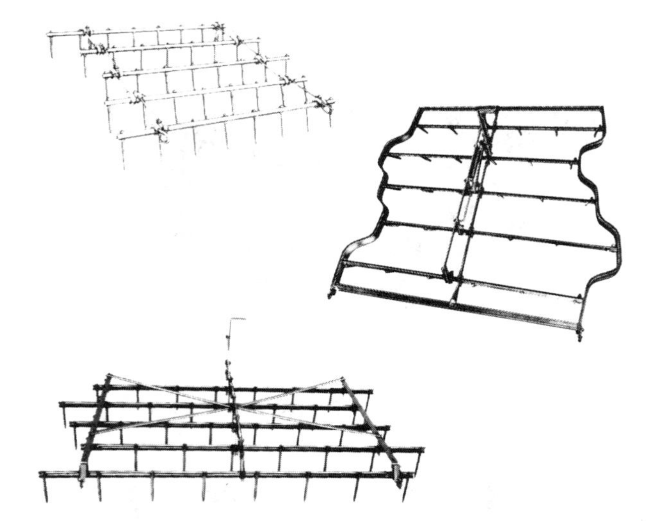

FIG. 43—HARROW—SECTIONS, SPIKE TEETH

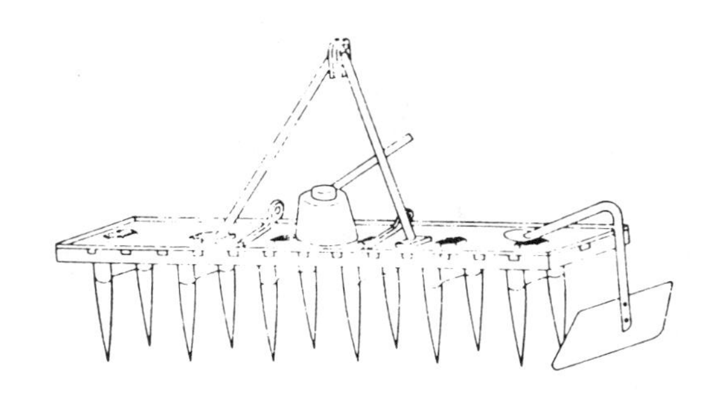

FIG. 46—HARROW—POWER-OSCILLATED SPIKE TEETH. REAR MOUNTED

FIG. 44—HARROW—FOUR SECTIONS, SPIKE TEETH. SQUADRON HITCH

5.12 Rotary hoe: A secondary tillage implement for dislodging small weeds and grasses and for breaking soil crust. Rotary hoes are used for fast, shallow cultivation before or soon after crop plants emerge. Rigid curved teeth mounted on wheels roll over the soil, penetrating almost straight down and lifting soil as they rotate. Hoe wheels may be mounted in multiple gangs or as short gangs on spring loaded arms suspended from the main frame. Types of rotary hoes are shown in Figs. 47-49.

FIG. 47—ROTARY HOE—IN-LINE SECTIONS ON SPRING LOADED ARMS. REAR MOUNTED

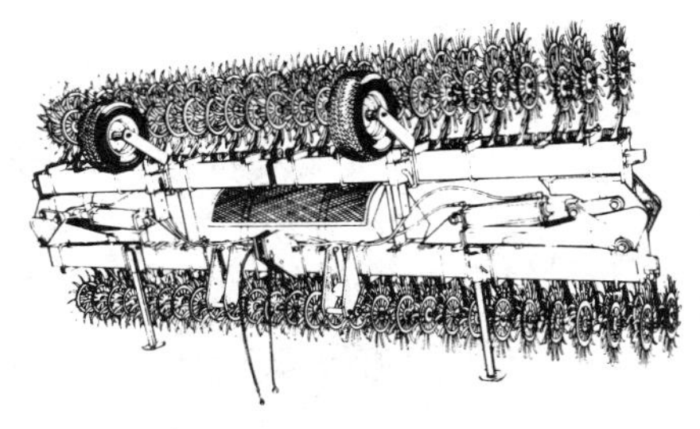

FIG. 48—ROTARY HOE—DUAL FOLDING WINGS FOR TRANSPORT. REAR MOUNTED

FIG. 49—ROTARY HOE—REAR MOUNTED, ENDWAYS TRANSPORT HITCH

5.13 Seedbed conditioner: A combination secondary tillage implement for final seedbed preparation. Typical purpose is to smooth and firm the soil surface for flat-planting. A seedbed conditioner is shown in Fig. 50.

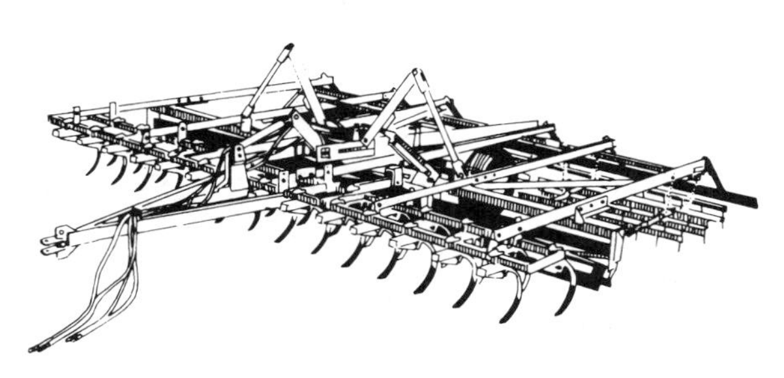

FIG. 50—SEEDBED CONDITIONER—COMPRISED OF FIELD CULTIVATOR TEETH, ROLLING CUTTER-MIXER BLADES, SPIKE TOOTH HARROW SECTIONS, AND A DRAG BAR. WHEEL MOUNTED PULL HITCH

5.14 Roller harrow: A secondary tillage implement for seedbed preparation which crushes soil clods and smooths and firms the soil surface. It consists of an in-line gang of ridged rollers, followed by one or more rows of staggered spring cultivator teeth, followed by a second in-line gang of ridged rollers. Types of roller harrows are shown in Figs. 51-52.

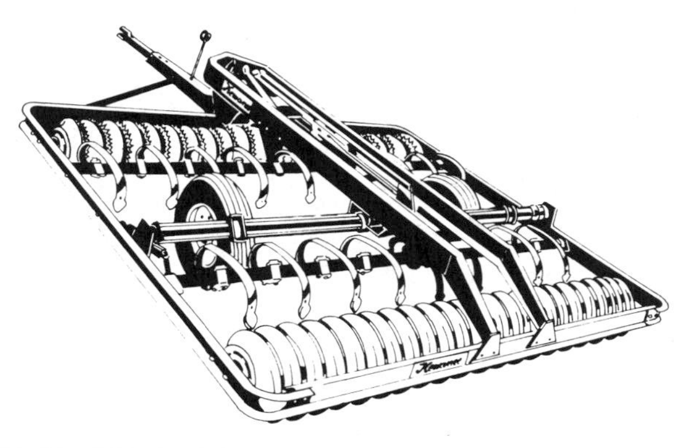

FIG. 51—ROLLER HARROW—WHEEL MOUNTED PULL HITCH

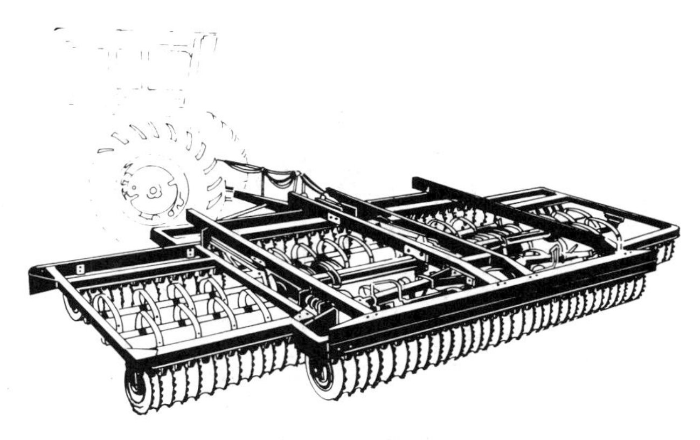

FIG. 52—ROLLER HARROW—WITH DUAL FOLDING WINGS. WHEEL MOUNTED PULL HITCH

5.15 Packer: A secondary tillage implement for crushing soil clods and compacting the soil. Packers consist of one or two in-line gangs of rollers. Roller sections may be lugged wheels or any one of various shaped ridged wheels. Types of packers are shown in Figs. 53-56.

FIG. 53—SINGLE PACKER ROLLER—IN-LINE ROLLER GANG. DRAG TYPE PULL HITCH

FIG. 54—SINGLE PACKER ROLLER—OFFSET ROLLER GANGS. DRAG TYPE PULL HITCH

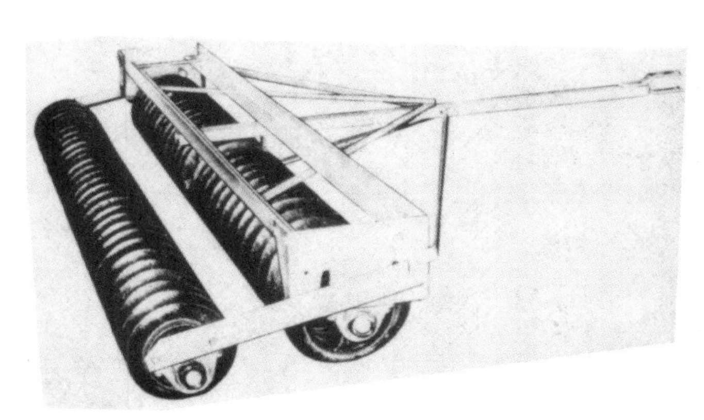

FIG. 55—TANDEM PACKER ROLLER—LEADING ROLLER WHEELS ARE LARGER DIAMETER THAN TRAILING ROLLER WHEELS. DRAG TYPE PULL HITCH

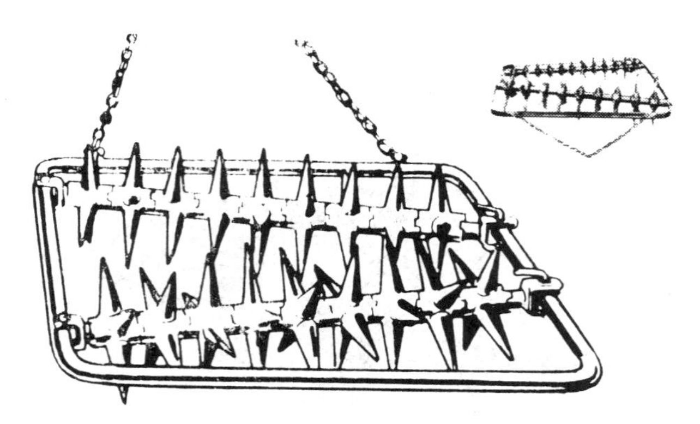

FIG. 56—CLOD BUSTER ROLLER—TWO ANGLED GANGS OF SPIKED ROLLER WHEELS. TYPICALLY PULLED BEHIND A PLOW. DRAG TYPE PULL HITCH

5.16 Rotary tiller: A primary or secondary tillage implement used for broadcast or strip tillage. Rotary tillers are also used as chemical incorporators prior to planting and as row crop cultivators. They consist of a power-driven shaft, transverse to the direction of travel, equipped with curved knives that slice through the soil, chop surface residue, and mix all materials in the disturbed layer. Types of rotary tillers are shown in Figs. 57-58.

FIG. 57—ROTARY TILLER—BROADCAST FULL WIDTH TILLING. REAR MOUNTED

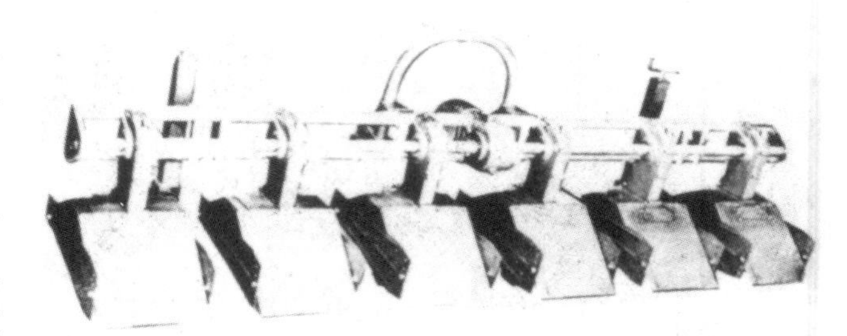

FIG. 58—ROTARY TILLER—STRIP TILLER WITH CROP SHIELDS FOR ROW CROP CULTIVATION. REAR MOUNTED

ASAE Standard: ASAE S441

SAFETY SIGNS

Proposed by the Farm and Industrial Equipment Institute; approved by the ASAE Power and Machinery Division Standards Committee; adopted by ASAE December 1983.

SECTION 1—PURPOSE AND SCOPE

1.1 This Standard establishes uniformity of safety signs to promote safety of persons associated with agricultural and powered lawn and garden equipment.

1.2 This Standard establishes signal words, color combinations, letter sizes and durability requirements for permanently and temporarily affixed safety signs for agricultural equipment as defined in ASAE Standard S390, Classifications and Definitions of Agricultural Equipment; and powered lawn and garden equipment as defined in ASAE Standard S323, Definitions of Powered Lawn and Garden Equipment. This Standard is suitable for application to tools, machines and machinery used in the specified categories.

SECTION 2—DEFINITIONS

2.1 Hazard: A source of danger to a person on or near a machine.

2.2 Safety signs: Information affixed to a machine to alert persons to hazards which can cause personal injury.

2.2.1 Permanent safety sign: Information affixed to the machine to warn against a hazard inherent in the machine or a hazard that might be created during reasonably anticipated machine usage. The sign is to be permanently attached to the machine so that it cannot be readily removed.

2.2.2 Temporary safety sign: A sign affixed to the machine to warn against a temporary hazard created by situations such as shipment or repair. The sign is temporarily attached to the machine and can be readily removed when the hazard is eliminated.

2.3 Panel: Area of a safety sign having a distinctive background color different from the other areas of the sign, or an area which is clearly delineated by a line or border.

2.3.1 Signal word panel: Distinctive area of a safety sign that alerts the viewer to a hazard.

2.3.2 Message word panel: Area of a safety sign which explains the hazard and/or provides instructions to avoid the hazard.

2.3.2.1 Specific safety message: A message that is specific to a particular hazard.

2.3.3 Pictorial panel: Area of a safety sign that explains pictorially the hazard and/or provides instructions to avoid the hazard.

2.4 Signal word: Distinctive word on a safety sign that alerts the viewer to the existence and relative degree of a hazard.

2.4.1 DANGER: Denotes an extreme intrinsic hazard exists which would result in high probability of death or irreparable injury if proper precautions are not taken. A sign with this signal word is to be limited to a specific safety message and to the most extreme situation.

2.4.2 WARNING: Denotes a hazard exists which can result in injury or death if proper precautions are not taken. A sign that has a specific safety message normally is to use this signal word.

2.4.3 CAUTION: Denotes a reminder of safety practices, or directs attention to unsafe practices which could result in personal injury if proper precautions are not taken. A sign with this signal word may be used with multiple instructions condensed from the user's manual and placed for convenient viewing by the operator, but should not be used for a specific safety message.

2.5 Safety sign colors: Specified colors used on safety signs which are approximately the same color as follows:

Color	Reference Munsell Number
Black	N2.0/or lower
Red	7.5R4.0/14
White	N9.0/or higher
Yellow	5.0Y8.0/12

2.6 Safety alert symbol: A pictorial symbol which indicates that the sign involves personal safety. The symbol is to conform with ASAE Standard S350, Safety-Alert Symbol for Agricultural Equipment.

SECTION 3—FORMAT

3.1 A safety sign shall consist of a signal word panel plus a message panel and/or a pictorial panel. These panels are to be arranged to permit the reader to logically proceed from the general to the specific information (from a signal word panel to a pictorial panel to a message panel).

3.2 The safety-alert symbol shall precede the signal word in the signal word panel. The minimum height of the safety alert-symbol shall equal or exceed the signal word letter height.

3.3 The safety sign shall be distinctive on the machine. A contrasting margin or border may be used on the sign to achieve the distinctiveness.

3.4 The message panel, if used, should include a message that is concise and readily understood.

SECTION 4—SAFETY SIGN COLORS

4.1 Signs with DANGER signal word

4.1.1 The signal word panel shall have a red background and white lettering. The background may be any suitable shape as long as it is predominant.

4.1.2 The message panel shall have a white background and red lettering.

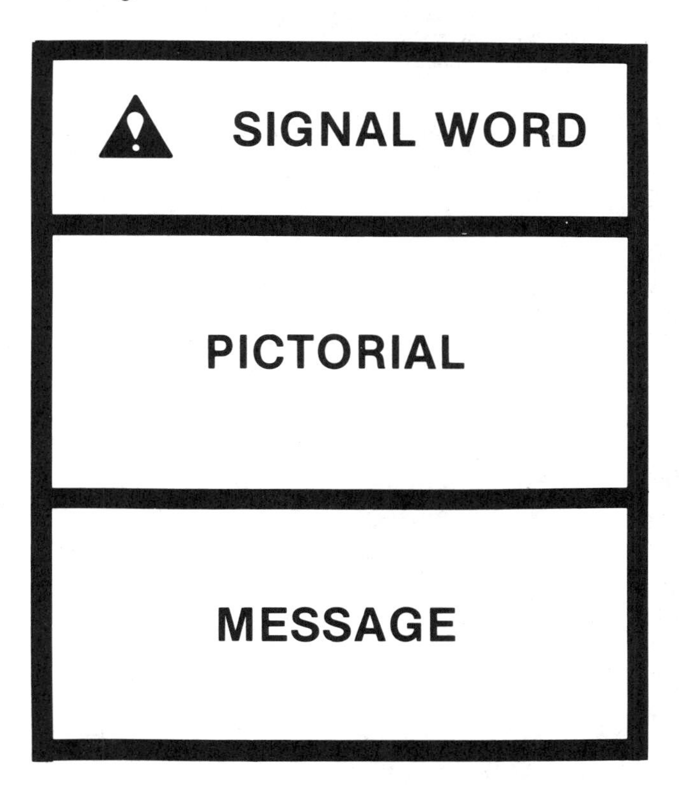

FIG. 1—EXAMPLE OF SAFETY SIGN FORMAT (Panel arrangement and size not specified)

4.2 Signs with WARNING signal word

4.2.1 The signal word panel shall have a black background and yellow lettering.

4.2.2 The message panel shall have a yellow background and black lettering.

4.3 Signs with CAUTION signal word

4.3.1 The signal word panel whall have a black background and yellow lettering.

4.3.2 The message panel shall have a yellow background and black lettering.

4.4 Safety-alert symbol. The triangular portion shall be the same color as the signal word panel lettering, and the exclamation mark portion shall be the same color as the signal word panel background.

4.5 Pictorial panel. Any combination of the colors defined in paragraph 2.5 may be used.

4.6 Color options

4.6.1 If the effectiveness of the safety sign is significantly improved due to background contrast, the background and letter colors specified in paragraphs 4.1, 4.2 and 4.3 may be reversed.

4.6.2 The message panel background and letter colors may be any combination of the colors defined in paragraph 2.5.

SECTION 5—LETTER SIZE AND STYLE

5.1 Sans-Serif Gothic upper case lettering is recommended, such as Standard Medium, Folio Medium, News Gothic Bold, or equivalent.

5.2 The minimum letter height should be one unit in height for every 500 units of safe viewing distance from the safety sign.

5.2.1 The minimum letter height shall not be less than 6 mm for a signal panel word nor less than 3 mm for a message panel word.

5.2.2 The safe viewing distance for the signal word panel should include a reasonable hazard avoidance reaction time. The message safe viewing distance may be different than the distance for a signal word panel.

SECTION 6—DURABILITY

6.1 Permanent signs

6.1.1 Permanent safety signs shall have reasonable life for the anticipated machine operating environment. Users should be instructed to ensure safety signs are installed, properly maintained, and replaced as necessary.

6.1.2 Permanent safety signs are considered to have a reasonable life if, when viewed at distances described in paragraph 5.2, the sign has good color and legibility for a period of at least five years. Exterior durability is based on vertical exposure tests for weatherability in the extreme climates of the United States, or a 2000 h carbon-arc test per American Society for Testing and Materials Standard G 23, Recommended Practice for Operating Light- and Water-Exposure Apparatus (Carbon-Arc Type) for Exposure of Nonmetallic Materials, or a 2000 h xenon-arc test per ASTM Standard G 26, Recommended Practice for Operating Light-Exposure Apparatus (Xenon-Arc Type) With and Without Water for Exposure of Nonmetallic Materials. Other test methods may be used providing the method ensures equal or superior sign durability.

6.1.3 Permanent sign installation on the machine should be in accordance with the sign manufacturer's recommended procedure considering the selection and preparation of the substratum and the environment in which the machine operates.

6.2 Temporary signs

6.2.1 Temporary signs shall have durability limits sufficient to cover the time during machine storage, repair or shipment.

6.2.2 Instructions should be included on the temporary sign to remove it when the hazard no longer exists.

SECTION 7—PLACEMENT

7.1 Safety signs shall be in the immediate vicinity of the hazard and readily visible so the viewer can avoid the hazard or take appropriate action, except that multiple item safety signs may be placed for convenient viewing by the operator. Where possible, sign placement should protect the sign from abrasion, damage, or obstruction from mud, dirty oil, etc.

Cited Standards:

ASAE S323, Definitions of Powered Lawn and Garden Equipment
ASAE S350, Safety-Alert Symbol for Agricultural Equipment
ASAE S390, Classifications and Definitions of Agricultural Equipment
ASTM G 23, Recommended Practice for Operating Light- and Water-Exposure Apparatus (Carbon-Arc Type) for Exposure of Nonmetallic Materials
ASTM G 26, Recommended Practice for Operating Light-Exposure Apparatus (Xenon-Arc Type) With and Without Water for Exposure of Nonmetallic Materials.

ASAE Engineering Practice: ASAE EP443

COLOR CODING HAND CONTROLS

Proposed by the Farm and Industrial Equipment Institute; approved by the ASAE Power and Machinery Division Standards Committee; adopted by ASAE February 1984.

SECTION 1—PURPOSE AND SCOPE

1.1 The purpose of this Engineering Practice is to define a system of color coding hand controls which will aid in identification by operators. Controls are described in ASAE Standard S335, Operators Controls for Agricultural Equipment.

1.2 It is intended that manufacturers have the option of whether or not to color code hand controls. Where controls are color coded, the provisions of this uniform system of color coding should apply.

1.3 This Engineering Practice concerns color coding hand controls on agricultural equipment as defined in ASAE Standard S390, Classifications and Definitions of Agricultural Equipment.

SECTION 2—GENERAL

2.1 For the purpose of this Engineering Practice, hand controls include, but are not limited to, levers, switches, knobs, handles, and buttons, which the operator manipulates to activate or control machine functions.

2.2 When new types of hand controls are adopted or combination controls are used, the color should be selected in accordance with the primary function.

2.3 If it is not practical to color code the control, it is sufficient to color code either the area surrounding the control or the identification of that control rather than the control.

2.4 Color coding does not replace the need for symbols identification. Controls for functions that are not obvious should be identified in accordance with ASAE Standard S304, Symbols for Operator Controls on Agricultural Equipment.

SECTION 3—COLOR CODE

3.1 Red (General Services Administration Federal Standard #595, color #11105) shall be used only for single-function engine stop controls. Where key switches, ignition switches or hand throttles are used to stop the engines, the "off" or "stop" positions shall be indicated with red lettering and/or symbols.

3.2 Orange (GSA FED. STD. # 595, color #12246) shall be used only for machine ground motion controls, such as engine speed controls, transmission controls, parking brakes or park-locks, and independent emergency brakes.

3.2.1 Where the engine speed and engine stop controls are combined, the controls may be red.

3.2.2 Steering wheels or other steering controls may be black or any color other than red or yellow.

3.3 Yellow (GSA FED. STD. #595, color #13538) shall be used only for function controls which involve the engagement of mechanisms, such as power take-offs, separators, cutterheads, feed rolls, picking units, elevators, spray pumps, winches and unloading augers.

3.4 Black (GSA FED. STD. #595, color #17038) or some other dark color to harmonize with decor selected by the manufacturer shall be used for all other controls, such as these positioning and adjusting functions.

3.4.1 Component lift or position; such as remote hydraulic control, implement hitch, boom lift and swing, bucket dipper, header height, blade lift, blade shift, boom-bucket, stabilizer, wheel lean, frame steer and reel lift.

3.4.2 Control of unloading components; such as spout cap, unloading auger swing, bin dump, ejector gate and elevator lift.

3.4.3 Mechanism setting and adjustment; such as chokes, cylinder speed, concave space, seat adjustment, steering column, transmission disconnect, concave lock, lift stops, rockshaft stops, reel speed and flow dividers.

3.4.4 Machine lights; such as headlights, work or floodlights, tail lights, flashers and turn signals.

3.4.5 Cab comfort; such as pressurizer, cooling, heating and windshield wipers.

Cited Standards:

ASAE S304, Symbols for Operator Controls on Agricultural Equipment
ASAE S335, Operators Controls on Agricultural Equipment
ASAE S390, Classifications and Definitions of Agricultural Equipment
GSA FED. STD. 595

ASAE Engineering Practice: ASAE EP445

TEST EQUIPMENT AND ITS APPLICATION FOR MEASURING MILKING MACHINE OPERATING CHARACTERISTICS

Proposed by the Milking Machine Manufacturers Council of the Farm and Industrial Equipment Institute in conjunction with the ASAE Milk Handling Equipment Committee; approved by the ASAE Electric Power and Processing Division Standards Committee; adopted by ASAE March 1985.

SECTION 1—PURPOSE AND SCOPE

1.1 The Engineering Practice provides a procedure for measuring milking system operating characteristics including test equipment specifications and designated locations for its use.

1.2 This Engineering Practice is limited to static and dynamic measurements of the vacuum and airflow characteristics of a milking system.

SECTION 2—DEFINITIONS

2.1 For definitions, see ASAE S300 Terminology for Milking Machine Systems.

SECTION 3—TEST EQUIPMENT AND SPECIFICATIONS

3.1 Vacuum recorders. Vacuum recorders are essential in evaluating the performance of milking machines for proper system operation. The recommended operating ranges and performance parameters for vacuum recorders and for other instruments which provide graphic records for field use are as follows:

3.1.1 Operating range. The minimum operating vacuum (pressure) range is 0 to 51 kPa (0 to 15 in. Hg). Equipment must withstand up to 68 kPa (20 in. Hg) without damage to the recorder.

3.1.2 Response time. In response to an abrupt change in input vacuum (pressure), from one level within the operating range to any other level within the operating range, the recorder shall indicate 90% of total change within 0.2 s.

3.1.3 Accuracy (vacuum level). The accuracy should be ± 3% of the recorder's operating range.

3.1.4 Chart speed (high). A chart speed sufficient to achieve a representative pulsation graph; a minimum of 20 mm/s (0.79 in./s).

3.1.5 Chart speed (low). A chart speed sufficient to achieve a representative vacuum level recording of system or component that is being evaluated.

3.1.6 Chart speed accuracy. The chart speed should remain accurate to ± 2% of recorder specification.

3.1.7 Chart paper. Lined or graduated paper as recommended by the recorder manufacturer, and a minimum of 40 mm (1.6 in.) in width, should be used.

3.1.8 Trace presentation. Two or more are recommended and should be easily readable.

3.1.9 Temperature (operating). The recorder should be operable and maintain its rated accuracy in an ambient temperature range of 0 °C to 50 °C (32 °F to 122 °F).

3.1.10 Temperature (storage). The allowable storage temperature should cover the range of −40 °C to 66 °C (−40 °F to 151 °F).

3.1.11 Humidity. The recorder should operate and maintain accuracy up to a maximum of 95% relative humidity.

3.1.12 Calibration. The recorder should have a zero adjustment and recommended method of vacuum (pressure) level verification.

3.1.13 Portability. The recorder should be rugged and readily portable.

3.1.14 Electric power. The recorder may operate on battery or 115 VAC power. The 115 VAC recorder must be properly grounded for use in wet and humid conditions.

3.2 Electronic pulse testers

3.2.1 Operating range. The tester should have an operating range of 40 to 80 pulsation cycles/min and a minimum operating vacuum range of 0 to 51 kPa (0 to 15 in. Hg). Equipment must withstand up to 68 kPa (20 in. Hg) without damage to the recorder.

3.2.2 Accuracy. The accuracy should be ± 2% of a pulsation cycle.

3.2.3 Temperature (operating). The tester should be operable and maintain its rated accuracy in an ambient temperature range of 0 °C to 50 °C (32 °F to 122 °F).

3.2.4 Temperature (storage). Allowable storage temperature should cover the range of −40 °C to 66 °C (−40 °F to 151 °F).

3.2.5 Humidity. The tester should operate and maintain accuracy up to a maximum of 95% relative humidity.

3.2.6 Portability. The tester should be rugged and readily portable.

3.3 Vacuum (pressure) gauges. For the purpose of the Engineering Practice, gauges can be divided into two categories according to common usage.

3.3.1 Checking type. Checking-type gauges are to be of rugged construction to serve as a portable means of checking on-farm gauges, or for milking equipment analysis. A protective carrying case is required.

3.3.2 Calibration type. A mercury manometer should be used for calibrating other gauges or instruments.

3.3.3 Range. 0 to 100 kPa (0 to 30 in. Hg).

3.3.4 Accuracy. Checking type: ± 1% error maximum full scale. Calibration type: ± 0.5% error maximum full scale.

3.3.5 Calibration adjustment. Dial types shall be adjustable. Manometer types shall have provision for scale adjustment or set point for mercury level, or both.

3.3.6 Environment. Checking type must be of rugged construction for field use. Gauges should have an operating temperature range of 0 °C to 50 °C (32 °F to 122 °F).

3.4 Orifice airflow meters. These are designed and calibrated to operate at a fixed differential pressure of 51 kPa (15 in. Hg). They are constructed with a vacuum gauge and the airflow rate reading is taken when the gauge registers 51 kPa (15 in. Hg) vacuum.

3.4.1 Capacity. Capacity and range of the meter will vary with manufacturer; adapters or additional meters may be used to increase ranges.

3.4.2 Accuracy. ± 5% of measured airflow at 51 kPa (15 in. Hg).

3.4.3 Meter vacuum gauge. This gauge must comply with paragraph 3.3.1 and must be calibrated at 51 kPa (15 in. Hg).

3.5 Combination instruments. Any instrument which combines two or more of the above functions shall meet each of the above applicable specifications.

3.6 Instruction manuals. Manuals should be provided with the measuring equipment and contain complete specifications as well as specific information on the proper use, calibration, care, cleaning, and storage of the equipment.

SECTION 4—MILKING SYSTEM TEST PREPARATION

4.1 Warmup and stabilization period. The milking system must be allowed to run for a minimum of 5 min or until operating characteristics stabilize before taking measurements.

4.2 Visual inspection. A visual check should be made to determine that the equipment is in good running order, and that all controls are set per manufacturer's recommendations.

SECTION 5—MEASUREMENT POINTS AND PRACTICES

5.1 The recording equipment must be properly calibrated and checked prior to any testing, per manufacturer's specifications.

5.1.1 Two types of tests are described in this section:

5.1.1.1 Static tests. These are tests with the milking unit in a simulated milking position, and the teat cups sealed per manufacturer's recommendations.

5.1.1.2 Dynamic tests. These are tests with the milking unit attached to a cow in its prescribed manner during milking and noting the number of other units operating.

5.1.2 Recorder attachments. Equipment for attaching the recorder to the measuring points shall contain proper size tees or adapters that will not reduce the internal diameter of the tube under test. Manufacturer's directions for proper hose size and length from adapters to recorders must be followed.

5.1.3 Short air tube tests

5.1.3.1 Short air tube, static. Adapters shall be attached between the claw air manifold and teat cup shell. Vacuum (pressure) variations shall be recorded for a minimum of 10 complete pulsation cycles.

NOTE: The length of a short air tube should not be increased by the addition of the adapters.

5.1.3.2 Short air tube, dynamic. The procedure is the same as in paragraph 5.1.3.1, but with the milking machine used as described in paragraph 5.1.1.2.

5.1.4 Pulsator tests

5.1.4.1 Pulsator, static (rate and ratio). Adapters shall be attached between the pulsator and claw air manifold within 5 cm (2 in.) from the end of the pulsator outlet nipple. Vacuum variations shall be recorded for 10 complete pulsation cycles.

5.1.4.2 Pulsator, dynamic. The procedure is the same as in paragraph 5.1.4.1, but with the milking machine used as described in paragraph 5.1.1.2.

5.1.5 Claw tests

5.1.5.1 Static. The adapter shall be attached directly to the claw bowl. This may require special or modified equipment. Vacuum (pressure) variations shall be recorded for a minimum of 10 complete pulsation cycles or for 15 s.

5.1.5.2 Dynamic. The procedure is the same as in paragraph 5.1.5.1, but with the milking machine used as described in paragraph 5.1.1.2. Vacuum shall be recorded for a minimum of 1 min at various milk flow rates.

NOTE: Regarding teat end vacuum tests: As of the adoption date of this Engineering Practice, with field equipment as described herein, there is not an accurate way to measure true teat end vacuum and/or vacuum fluctuation at the teat end.

5.1.6 Claw outlet (static). The adapter shall be attached within 10 cm (4 in.) of the claw outlet and vacuum (pressure) variations shall be recorded for a minimum of 10 complete pulsation cycles or for 15 s.

5.1.7 Weighing or metering devices (dynamic test only)

5.1.7.1 Milk meters and weigh jars. The adapter shall be attached within 10 cm (4 in.) of the vacuum supply inlet. Vacuum variations shall be recorded for a minimum of 1 min.

5.1.8 Milk pipeline (dynamic test only). Adapters shall be attached at the milk inlet farthest from the vacuum (pressure) source with all other milking units in operation under what would be considered peak operating conditions for that system. Vacuum variations shall be recorded for a minimum of 1 min.

5.1.9 Vacuum pulsator line (supplying pulsation vacuum), static and dynamic

5.1.9.1 The adapter shall be attached at the pulsator stall cock farthest from the vacuum (pressure) source with all other milking units operating. Vacuum (pressure) variations shall be recorded for a minimum of 1 min.

5.1.9.2 Adapters shall be attached downstream from the pulsator nearest the vacuum (pressure) source, and the vacuum (pressure) level shall be recorded for a minimum of 1 min.

5.1.10 Vacuum distribution tank, static and dynamic. Adapters shall be attached directly to the vacuum distribution tank. Variations shall be recorded for a minimum of 1 min.

5.2 Airflow measurement

5.2.1 Vacuum pump capacity. The vacuum pump shall be disconnected from the system, leaving the exhaust system intact. Airflow shall be measured at the vacuum pump inlet.

5.2.2 Reserve air capacity. Airflow measurement can be made at one or more of the following points: milk receiver, milk pipeline, or vacuum pulsator line. Units in simulated milking position, pulsation on, teat cups sealed per manufacturer's recommendation with either the regulator sealed or with it removed and the openings plugged. Airflow shall then be measured.

5.2.3 Effective reserve capacity. The procedure is the same as in paragraph 5.2.2, but with regulators installed and operating normally. The system vacuum level shall be measured with all orifice(s) on the airflow meter closed. Orifice(s) shall be opened until the vacuum level falls 2.0 kPa (0.6 in. Hg) or other value recommended by the manufacturer.

SECTION 6—EVALUATION OF RESULTS (FOR MILKING MACHINE INSTALLATION)

6.1 The measurements should be compared to those published by the manufacturer in their manuals to evaluate the milking machine's performance.

6.2 The operating characteristics for milking machines do vary due to different manufacturers' designs. Therefore these characteristics must be taken into consideration when evaluating the results.

6.3 Records of tests performed should indicate values for each measurement, and the instrument used for each test performed.

Cited Standard:

ASAE S300, Terminology for Milking Machine Systems

ASAE Engineering Practice: ASAE EP456

TEST AND RELIABILITY GUIDELINES

Developed by the ASAE Testing and Reliability Committee; approved by the ASAE Power and Machinery Division Standards Committee; adopted by ASAE December 1986; editorially corrected June 1987.

SECTION 1—PURPOSE AND SCOPE

1.1 This Engineering Practice shows how product life can be specified in probabilistic terms, how life data should be analyzed, and presents the statistical realities of life testing.

1.2 Life tests are required to be conducted on a representative sample to estimate the expected life of the population. Analysis of the results is essentially a statistical problem. Life test planning should include these four specifications: 1) test environment, 2) number of test units, 3) life goals, 4) definition of failure, that is, what failure modes constitute the end of life. Calendar test time and cost may affect the priority of these specifications.

1.3 This Engineering Practice is not intended to be a comprehensive study on the large variety of subjects encompassed in the analysis of life testing, but rather a guide to demonstrate the power, usefulness and methodology of reliability analysis.

SECTION 2—BACKGROUND

2.1 A key characteristic of a product is its life. In the agricultural equipment industry a great deal of testing is conducted on materials, components and complete machines to determine expected life. Life tests are conducted under controlled environments in the laboratories and in the more variable environments of the field. When tests on several nearly identical parts are replicated under even the most carefully controlled conditions, variability or scatter in the cycles or time to failure is observed.

2.2 For example, extreme care is taken to insure dimensional and metallurgical control during manufacture of precision ball or roller bearings, so virtually no discernible difference can be found between bearings. However, when several "identical" bearings are tested to failure under the same carefully controlled test conditions, variations in life of 10 to 1 are not uncommon. As a result of the inevitable variations, product life needs to be specified in probabilistic terms. This mandates that test programs require enough replications to minimize the risk of erroneous test conclusions. The designer and test engineer must establish a test program to balance test time and cost against risk of a product failure. Fig. 1 shows a histogram for 100 ball bearings tested to failure under identical laboratory conditions. Note that the individual bearing lives varied from 1000 to 5000 h. This variability is a fundamental reason for planning life tests and interpreting the results.

2.3 The results of repetitive life tests conducted on the bearings at three different environmental or stress levels are shown in Fig. 2. At each of the three levels, bearing life would be distributed as shown by the dotted lines. Connecting the point on each dotted distribution line at which 5% of the bearings failed, establishes a 5% failure line. Similarly the 50 to 95% failure lines can be drawn. The diagram of Fig. 2 defines the life characteristics of the bearings for these test conditions. Due to variability, statistical tests must be performed to estimate the products probability of success or reliability.

SECTION 3—DEFINITIONS

3.1 Confidence: Relative frequency that the statistically derived interval contains the true value being estimated.

3.2 Distribution: A mathematical function giving the cumulative probability that a random quantity, such as a component's life, will be less than or equal to any given value.

3.3 Failure mode: An accurate description of the type of failure, i.e., seal ball bearing typical failure modes are 1) spalled inner race, 2) spalled outer race, 3) spalled ball, 4) seal failure, 5) etc.

3.4 Histogram: A graphical representation of a frequency distribution by a series of rectangles where the width of the rectangle represents the range of the variable and the height represents the frequency of occurrence.

3.5 Life goals: Desired cycles or hours to (×)% failure in a specific operating environment.

3.6 Population: The total quantity of parts to be produced and put into service.

3.7 Probability: Likelihood of occurrence based on significant tests.

3.8 Reliability: The probability that a part, assembly or system will perform satisfactorily for a specified period of time under specified operating conditions.

3.9 Sample: A small number of parts which will be considered as representative of the total population.

3.10 Suspension: A test or operational unit that has not failed by the mode under consideration at the time of the life data analysis.

SECTION 4—RELIABILITY FIGURES OF MERIT

4.1 Two easily determined reliability figures of merit are failure rate, λ, and mean time between failure (MTBF). The failure rate is the

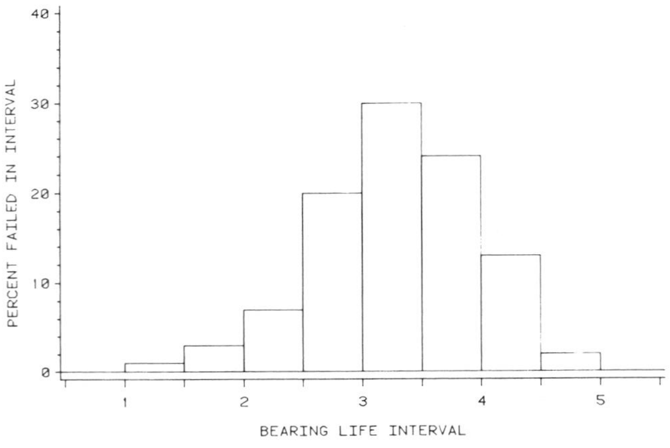

FIG. 1—HISTOGRAM OF 100 BEARINGS TESTED IN A CONSTANT ENVIRONMENT

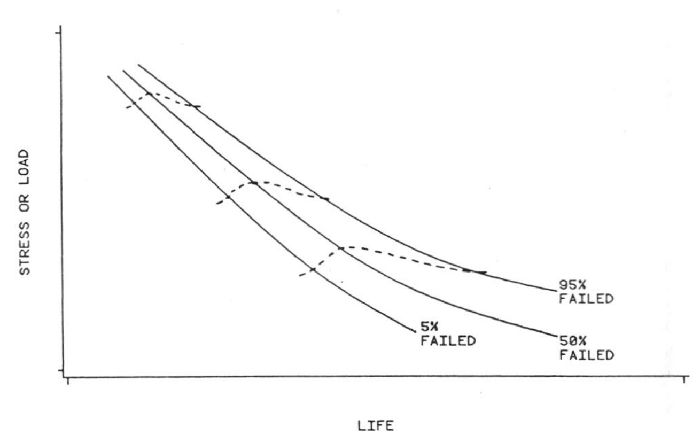

FIG. 2—EFFECT OF VARYING STRESS ON BEARING LIFE

number of failures of an item per unit time. For example, the failure rate for a tractor experiencing 5 failures after 500 h of operation would be determined as follows:

Failure rate = number failures/units of cumulative time
= 5 failures/500 h
= 0.01 failures per hour

4.2 The other reliability figure of merit is mean time between failure (MTBF). MTBF is the total operating time of a population of a product divided by the total number of failures. Using the same values as the tractor in the previous example we have:

MTBF = units of cumulative time/number failures
= 500 h/5 failures
= 100 h

4.3 The MTBF is the inverse of the failure rate. Which figure of merit is utilized depends on the point of view that is taken.

SECTION 5—SERIES AND PARALLEL RELIABILITY

5.1 Any system can be modeled by combining its elements. Reliability modeling allows the designer to determine the effect of a component on the system.

5.2 The series model is more typical of mechanical agricultural machinery, in which a failure will disable the function of the unit. In this case the failure rate of the system comprising of n components is the sum of the component failure rates.

$$\lambda_{system} = \lambda_1 + \lambda_2 + \cdots + \lambda_n$$

5.3 As a simple example of the impact of component failure rates on a system, consider a hydraulic cylinder. The cylinder is composed of four components with corresponding failure rates as illustrated in Fig. 3.

$$\lambda_{cylinder} = \lambda_{barrel} + \lambda_{piston} + \lambda_{rod} + \lambda_{seal}$$

$$= 0.000002 + 0.000001 + 0.000001 + 0.000004$$

$$= 0.000008$$

5.4 On agricultural equipment a parallel reliability system model is likely to be utilized where redundancy is required. A parallel system assures that a system failure does not occur as a result of a single component failure. The system failure rate for a two component parallel system is determined from the following equation:

$$\lambda_{system} = \frac{1}{\frac{1}{\lambda_1} + \frac{1}{\lambda_2} - \frac{1}{(\lambda_1 + \lambda_2)}}$$

5.5 As an example of the parallel system failure rate as impacted by the component rates, consider an engine air cleaner system consisting of a primary filter and a secondary filter. If the failure rate for the primary and secondary filters is 0.000008, we have a system as illustrated in Fig. 4.

$$\lambda_{system} = \frac{1}{\frac{1}{\lambda_{primary}} + \frac{1}{\lambda_{secondary}} - \frac{1}{(\lambda_{primary} + \lambda_{secondary})}}$$

$$= \frac{1}{\frac{1}{0.000008} + \frac{1}{0.000008} - \frac{1}{(0.000008 + 0.000008)}}$$

$$= \frac{1}{125{,}000 + 125{,}000 - 62{,}500}$$

$$= \frac{1}{187{,}500} = 0.00000533$$

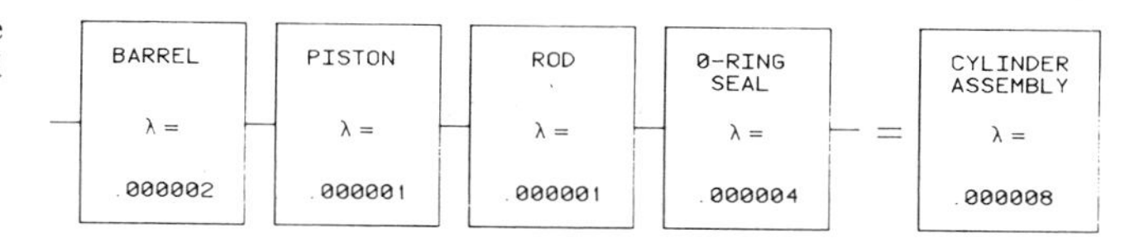

FIG. 3—SERIES FAILURE MODEL

5.6 Failure rate data are available from several sources:

Government and industrial data exchange program, GIDEP

Failure rate data handbook, SP-63-470, from the Bureau of Navy Weapons

Reliability prediction of electronic equipment, MIL-HDBK-217D

5.7 A more accurate prediction of system failure rate might be determined by making use of warranty data, service records, or service parts movement history of previous similar products. A formal procedure for analyzing systems not only by component failure rate, but by component failure modes is known as a failure mode and effects analysis (FMEA). This procedure is widely used in private industry and is a U.S. Department of Defense requirement before prototype hardware is produced.

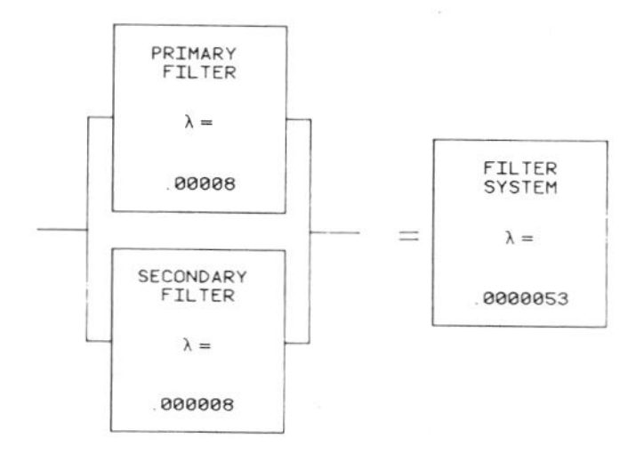

FIG. 4—PARALLEL FAILURE MODEL

SECTION 6—PRODUCT LIFE

6.1 During the life of a product, the failure rate typically varies during its period of use. If the failure rate were determined at each occurrence throughout its life and plotted against the accumulated hours of product use, a curve similar to Fig. 5 is obtained. This is often referred to as the "bathtub" curve. A high, but decreasing failure rate characterizes the initial hours of the product's life. This is the period of time the product experiences "break in". Attempts are made by quality assurance and product distribution organizations to screen the problems with inspection and pre-delivery as well as other techniques. The adjoining area is one of constant failure rate, which afforts the customer the maximum usefulness of the product. During the "useful life" period the design engineer is challenged to select materials and components that provide adequate strength for the stress the customer applies to the product. The region of increasing failure rate to the right is the result of wearout, in which the designed strength deteriorates due to failure modes such as fatigue or wear. The engineer must identify short-lived parts and revise them to assure their lives extend into the wearout region.

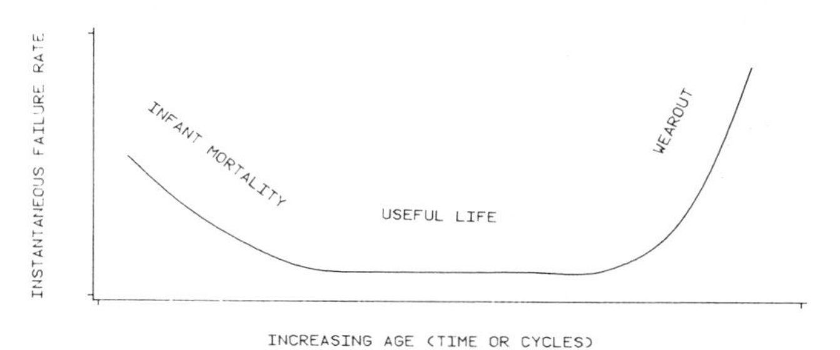

FIG. 5—LIFE CHARACTERISTIC CURVE

SECTION 7—EXPONENTIAL DISTRIBUTION

7.1 The reliability of a product during the useful life period is described by single parameter exponential distribution:

$$R(t) = e^{-\lambda t}$$

where

$R(t)$ = reliability
e = base of the natural logarithm (2.7182 . . . .)
λ = failure rate

7.2 To illustrate how the exponential distribution characterizes a product of constant failure rate consider the times to failure for the five units in Fig. 6. Rearrangement of all data in order of failure time length is presented in Fig. 7. By connecting the tails of the lengths, the observed exponential reliability function is apparent. In Fig. 8, the theoretical reliability function is shown to fit the observed data reasonably well. It should also be apparent that with less data available, the less likely the exponential distribution can be determined. In many cases, the exponential is assumed.

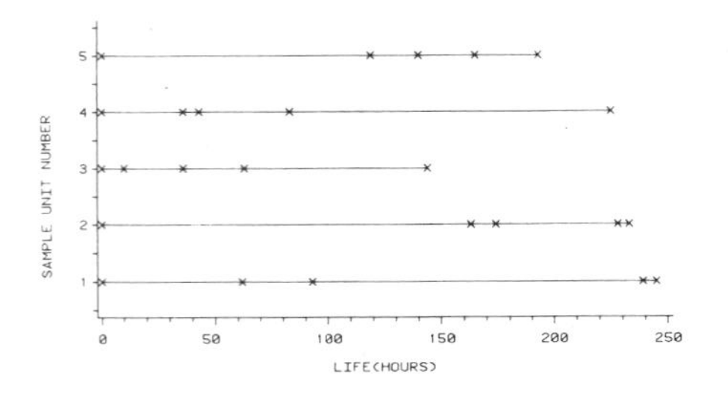

FIG. 6—TIME TO FAILURE FOR 5 UNITS

7.3 The reliability can be computed for any time once the failure rate is known and the population is determined or assumed to follow the exponential. This relationship is illustrated in the following example:

An engineer has five prototype grain augers on a reliability demonstration test. After accumulating 1000 h on each prototype, 5 failures have occurred.

Determine the failure rate:

$$\lambda = \text{Total Failures/Total Time} = 5/(5)(1000) = 0.001 \text{ failure/hour}$$

Determine the reliability at 1000 h:

$$R = e^{-\lambda t} = e^{-(0.001)(1000)} = 0.368$$

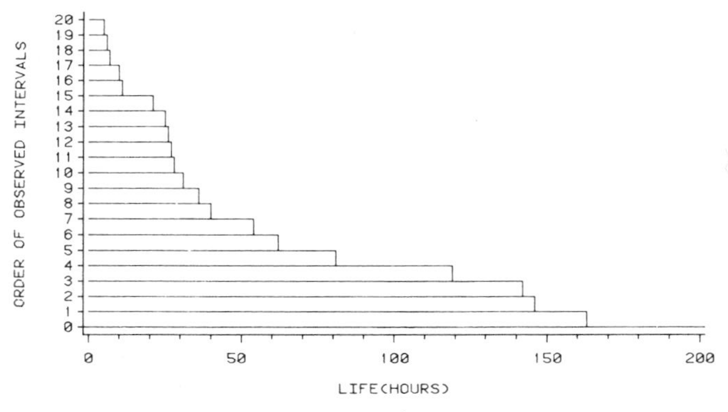

FIG. 7—TIME TO FAILURE FOR 5 UNIT SAMPLE IN ORDER OF LIFE

7.4 The product manager is interested in determining the number of augers with a problem during the warranty period of 300 h.

Determine the reliability at 300 h:

$$R = e^{-\lambda t} = e^{-(0.001)(300)} = 0.741$$

That is, 74.1% of the population will survive 300 h. Conversely, 25.9% of the population would be expected to fail during the one-year warranty period.

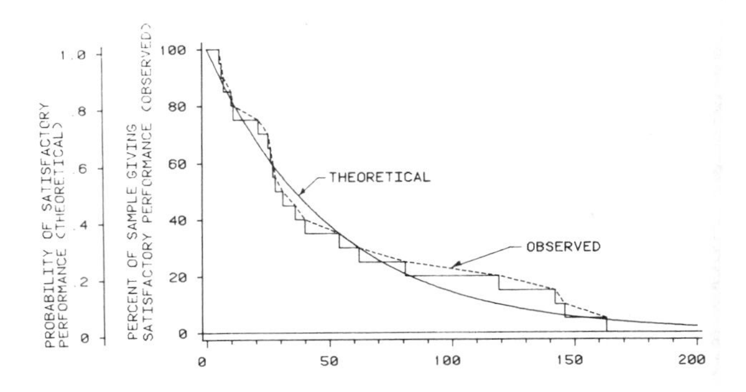

FIG. 8—OBSERVED AND THEORETICAL RELIABILITY FUNCTIONS WITH SAME MEAN VALUE FOR A SAMPLE OF FIVE UNITS

SECTION 8—WEIBULL DISTRIBUTION

8.1 A more general failure model than the exponential distribution is provided by the Weibull distribution. Weibull analysis is a popular methodology due to its multi-distribution capability and the ability to plot much failure-time data as a straight line on Weibull graph paper. The relationship of the Weibull distribution to the "bath tub" curve is illustrated in Fig. 9. The reliability of a product which has a Weibull distribution is described by this function.

$$R(t) = e^{-(\lambda t)^b} \quad \text{(2 parameter form)}$$

where

b = Weibull slope

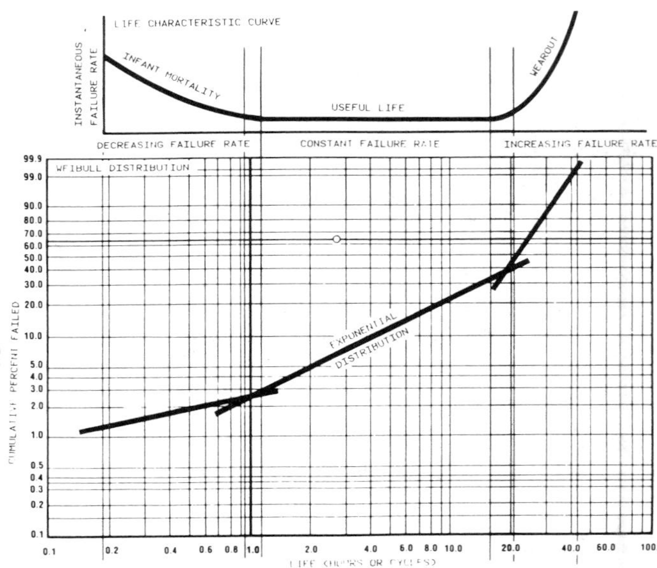

FIG. 9—WEIBULL DISTRIBUTION AND "BATHTUB" CURVE RELATION

8.2 Complete data. Following is an example of utilization of the Weibull analysis on complete data. Complete data are data in which all the samples are tested to failure.

8.2.1 Example 1. Five V-belts were tested to failure as shown in Table 1. To plot the data in Table 1:

TABLE 1—BELT TEST COMPLETE DATA

Test numbers	Hours to failure
1	145
2	52
3	310
4	95
5	230

8.2.1.1 Order the data in ascending hour level (Column C, Table 2).

8.2.1.2 Assign percent median ranks to each data point (Column D, Table 2). Median ranks are taken from the column headed 5 (5 = sample size) of Table 3.

8.2.1.3 Plot the median ranks on Weibull graph paper as shown in Fig. 10.

8.2.1.4 Fit a best fit straight line to the data.

TABLE 2—BELT TEST COMPLETE DATA

(A) Order number	(B) Test number	(C) Life, hours	(D) Percent median rank	*(E) 5% Confidence rank	*(F) 95% Confidence rank
1	2	52	12.9	1.0	45.0
2	4	95	31.3	7.8	65.7
3	1	145	50.0	18.9	81.0
4	5	230	68.6	34.2	92.3
5	3	310	87.0	54.9	98.9

*See Section 9—Confidence

The resultant line represents an estimate of the distribution of life expected for the population from which the sample of five was taken. The percent failure rate at any operating time can be estimated from Fig. 10. For example, about 20% of the population will fail by 67 h when operated under the test environment.

8.3 Suspended data. Frequently, life tests do not produce all failures as in Example 1. Items may be removed from test for a variety of reasons. For example, a test may be stopped if the test stand breaks down, or total allowable test time runs out or more items may have been placed on test than needed to fail in order to decrease testing time. These test survivors are called suspended data.

8.3.1 Example 2. Ten V-belts were tested. The results are shown in Table 4. Analysis of suspended data is handled analogously to a complete data analysis except median rank table values cannot be used. The basic procedure is to calculate order numbers and percent median ranks accounting for the suspended items. The new order numbers reflect the probability that a suspended item would fail either before the next ordered failure or sometime after, if the test had been continued. To plot the data in Table 4, proceed as follows:

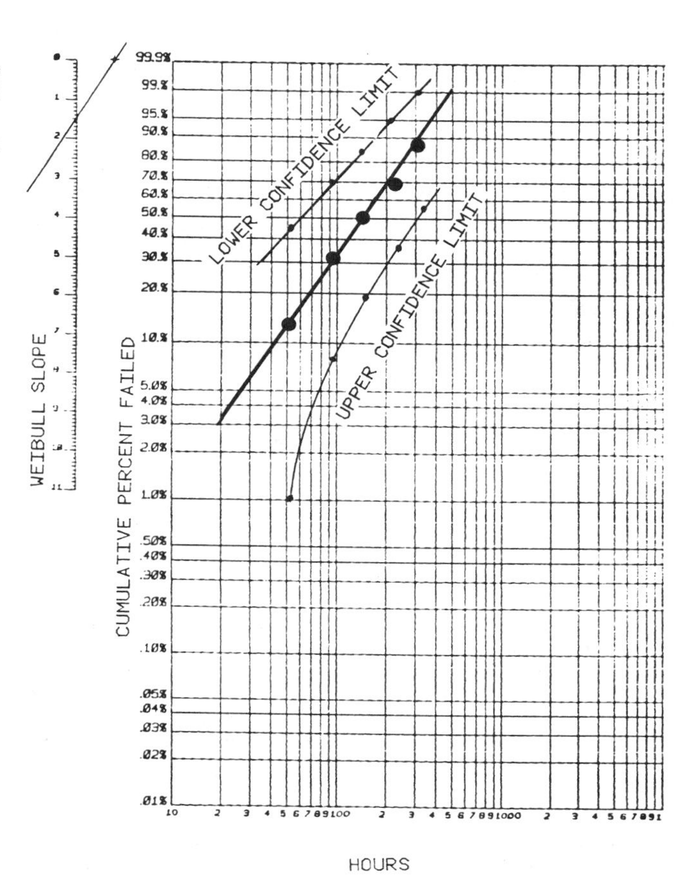

FIG. 10—WEIBULL PLOT OF 5 V-BELTS

8.3.1.1 Rank all the data and assign order numbers to each failure. Order numbers before suspensions are the rankings.

TABLE 4—BELT TEST SUSPENDED DATA

Test hours	Failure/Suspension
554	F
1099	F
663	F
1084	F
897	F
1202	S
802	S
939	S
827	S
914	F

TABLE 3—MEDIAN RANKS

Ranked order number	Sample size 1	2	3	4	5	6	7	8	9	10	11
1	50.0	29.2	20.6	15.9	12.9	10.9	9.4	8.3	7.4	6.6	6.1
2		70.7	50.0	38.5	31.3	26.4	22.8	20.1	17.9	16.2	14.7
3			79.3	61.4	50.0	42.1	36.4	32.0	28.6	25.8	23.5
4				84.0	68.6	57.8	50.0	44.0	39.3	35.5	32.3
5					87.0	73.5	63.5	55.9	50.0	45.1	41.1
6						89.0	77.1	67.9	60.6	54.8	50.0
7							90.5	79.8	71.3	64.4	58.8
8								91.7	82.0	74.1	67.6
9									92.5	83.7	76.4
10										93.3	85.2
11											93.8

Order numbers after suspensions are calculated by: Order number = last order number + i

where

$$i = \frac{(n+1) - (\text{previous order number})}{1 + (\text{number of items following suspended set})}$$

n = sample size

8.3.1.2 Calculate percent median ranks by:

$$\text{Percent median rank} = \frac{j - 0.3}{n + 0.4} \times 100$$

where

j = order number
n = sample size

8.3.1.3 Plot the results on Weibull paper. The results for steps 1 and 2 are shown in Table 5.

TABLE 5—BELT TEST SUSPENDED DATA

(A) Rank	(B) Life, hours	(C) Order number	(D) Percent median rank	*(E) 5% Confidence rank	*(F) 95% Confidence rank
1	544	1	6.6	0.5	25.8
2	663	2	16.2	3.6	39.4
3	802				
4	827				
5	897	3.29	28.8	10.6	53.5
6	914	4.58	41.1	19.2	65.9
7	939				
8	1084	6.18	56.5	32.0	79.0
9	1099	7.78	71.9	47.1	89.9
10	1202				

*See Section 9—Confidence

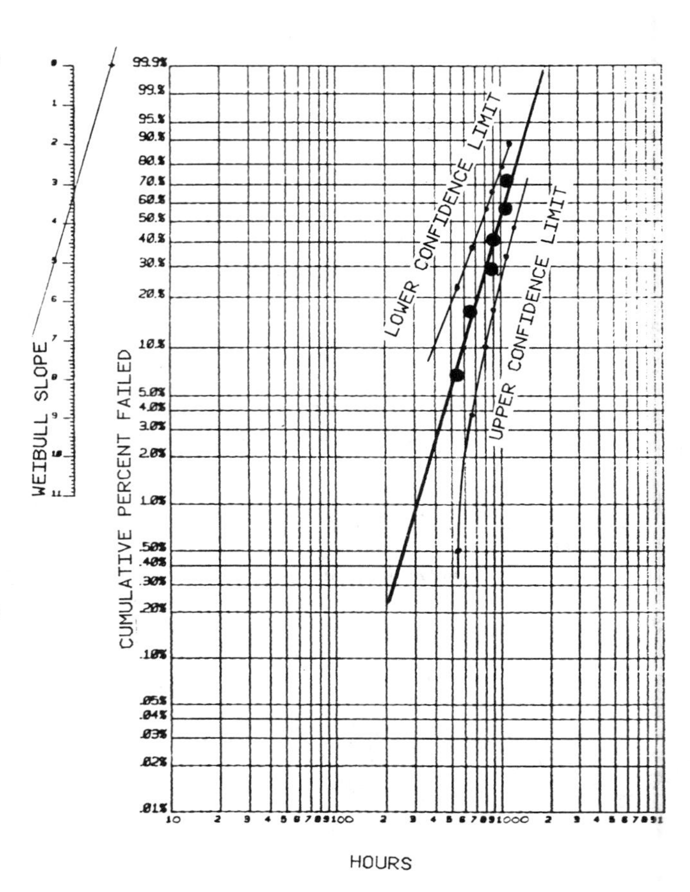

FIG. 11—WEIBULL PLOT OF 10 V-BELTS

8.4 Interpreting data. Figs. 10 and 11 give two significant parameters relating to the life of the product. One is the slope which is a measure of the scatter or variability of the data and the other is the characteristic life, which is a parameter of the scale of the data. The characteristic life is defined as the point at which 63.2% have failed on

TABLE 6—5 AND 95 PERCENT CONFIDENCE RANKS

Ranked order number	Sample size 1	2	3	4	5	6	7	8	9	10	11
						5% Rank					
1	5.0	2.5	1.6	1.2	1.0	0.8	0.7	0.6	0.5	0.5	0.4
2		22.3	13.5	9.7	7.8	6.2	5.3	4.6	4.1	3.6	3.3
3			36.8	24.8	18.9	15.3	12.8	11.1	9.7	8.7	7.8
4				47.2	34.2	27.1	22.5	19.2	16.8	15.0	13.5
5					54.9	41.8	34.1	28.9	25.1	22.2	19.9
6						60.6	47.9	40.0	34.4	30.3	27.1
7							65.1	52.9	45.0	39.3	34.9
8								68.7	57.0	49.3	43.5
9									71.6	60.5	52.9
10										74.1	63.5
11											76.1
						95% Rank					
1	95.0	77.6	63.1	52.7	45.0	39.3	34.8	31.2	28.3	25.8	23.8
2		97.4	86.4	75.1	65.7	58.1	52.0	47.0	42.9	39.4	36.4
3			98.3	90.2	81.0	72.8	65.8	59.9	54.9	50.6	47.0
4				98.7	92.3	84.6	77.4	71.0	65.5	60.6	56.4
5					98.9	93.7	87.1	80.7	74.8	69.6	65.0
6						99.1	94.6	88.8	83.1	77.7	72.8
7							99.2	95.3	90.2	84.9	80.0
8								99.3	95.8	91.2	86.4
9									99.4	96.3	92.1
10										99.4	96.6
11											99.5

the Weibull graph and in Fig. 10 this is 190 h. The slope is taken from the nomograph and is 1.4. These two parameters define the probabilistic life as a function of time. As the number tested approaches the number in the population, there is more confidence that the sample statistics are characteristic of the population. Therefore, extremely small sample sizes should be avoided, or the results should be treated with caution.

SECTION 9—CONFIDENCE

9.1 Consideration of the degree of confidence in a reliability test plan is important in establishing the risk of making a wrong decision in assessing the reliability of a product. A simple methodology for evaluating the effect of sample size on possible error in estimating the Weibull population line is to assign percent confidence rank to the point data of the Weibull distribution population line. The percent confidence ranks for the point data given in Table 2 were taken from the 5 and 95% ranks in Table 6. The 5% ranks are assigned in Column E and the 95% ranks in Column F. The 5 and 95% confidence ranks define the 90% confidence band. A 90% confidence band is interpreted as a 90% chance that the population falls within the confidence band shown. If fewer samples were tested to failure, the 90% confidence band would have been wider. If more samples had been tested the 90% confidence band would have been narrower.

9.2 Previously from Fig. 10, it had been determined that 20% of the V-belts would be expected to survive to 67 h. Utilizing the 5 and 95% confidence boundaries, it can now be said with a greater accuracy and 90% confidence that the percentage of survivors at 67 h is between 53% and 3.4%.

9.3 The procedure for determining percent confidence ranks for suspended data is similar to complete data. However, it is necessary to interpolate between the order numbers provided in Table 6. These values have been included in Columns E and F of Table 5 and have been plotted in Fig. 11.

References:

1. ASM. 1974. Source book in failure analysis. American Society for Metals, Metals Park, OH.
2. ASTM. 1975. Manual on statistical planning and analysis for fatigue experiments. ASTM Special Publication 588. R.E. Little, Editor. American Society for Testing and Materials, Philadelphia, PA.
3. Ang and Tang. 1975. Probability concepts in engineering planning and design. John Wiley & Sons, New York, NY.
4. Kapur, K.C. and Lamberson, L.R. 1977. Reliability in engineering design. John Wiley & Sons, New York, NY.
5. Nelson, W. 1985. Weibull analysis of reliability data with few or no failures. Journal of Quality Technology (17)3:140-146.
6. Pratt and Whitney Aircraft. 1983. Weibull Analysis Handbook. Pratt and Whitney Aircraft, Government Products Division, United Technologies Incorporated.
7. Wetzel, R.M. 1977. Fatigue under complex loading: Analysis and experiments. Advances in Engineering (6). Society of Automotive Engineers, Warrendale, PA.

ASAE Standard: ASAE S472 (ANSI/ASAE S472-1988)

TERMINOLOGY FOR FORAGE HARVESTERS AND FORAGE HARVESTING

Developed by the ASAE Forage Harvesting and Utilization Committee; approved by the ASAE Power and Machinery Division Standards Committee; adopted by ASAE December 1986; revised editorially February 1988; approved by ANSI as an American National Standard March 1988.

SECTION 1—PURPOSE AND SCOPE

1.1 The purpose of this Standard is to establish terminology and specifications pertinent to forage harvester design and performance. It is intended to improve communication among engineers and researchers and to provide a basis for comparative listing of machine specifications.

SECTION 2—FORAGE HARVESTER BASIC DESIGNS

2.1 Forage harvesters are used to harvest and cut crops into short particle lengths. The chopped product may be preserved in storage by ensiling or dehydrating, or it may be fed directly to livestock. Forage harvesters may be tractor mounted, pull-type (towed), or self-propelled.

2.2 There are two basic types of forage harvesters; precision cut and non-precision cut.

2.2.1 Precision cut forage harvesters: A forage harvester that uses a feeding mechanism to meter the crop into the cutting or shearing mechanism at a uniform velocity; thus, the crop is cut off at regular, "uniform" particle lengths generally ranging from 3 to 50 mm.

2.2.2 Non-precision cut forage harvesters: A forage harvester that generally uses a rotary impact cutting device to cut standing crop or windrows directly into shorter pieces. The chopped particle length distribution is not uniform due to the random cutting process of the crop stems. A secondary shearing device may be incorporated into the crop blower to recut the crop into more uniform lengths. Typical particle lengths generally exceed those of precision cut forage harvesters.

SECTION 3—FORAGE HARVESTER COMPONENT TERMINOLOGY AND SPECIFICATIONS

3.1 Crop gathering headers: Devices used to gather the crop into the forage harvester. They are usually detachable from the forage harvester.

3.1.1 Row crop header: A device used to cut off and gather row crops. Cutting of the plant usually takes place near ground level.

3.1.2 Maize (ear corn) header: A device used to harvest and gather only the ears of corn (maize).

3.1.3 Pickup header: A device for picking up a previously cut crop. The crop may be in a swath or a windrow.

3.1.4 Direct cut header: A device capable of cutting a standing crop across its full width and conveying the cut crop directly into the forage harvester.

3.2 Header harvesting widths

3.2.1 Row crop and ear corn header harvesting widths: The average distance between the centerlines of adjacent row units multiplied by the number of row units. For single row headers, the harvesting width equals the row centerline spacing. Preferred row spacings are specified in ASAE Standard S404T, Metric Row Spacings. Where row crop unit width is adjustable, maximum and minimum distances between row centerlines shall be stated, expressed in centimeters to the nearest whole centimeter. The maximum and minimum harvesting widths shall be expressed in meters to the nearest hundredth, and the number of row units shall be stated.

3.2.2 Pickup header harvesting width: The minimum distance between the outermost conveying elements, up to and including the header side sheets but not the side sheet flared portions, expressed in meters to the nearest hundredth.

3.2.3 Direct cut header harvesting width: The minimum distance between the side sheets of the harvesting unit measured directly above the forward tips of the sickle sections. For rotary impact cutters, the cutting width is equal to the distance between the innermost and outermost disk/drum centerlines plus one disk/drum diameter. The width shall be expressed in meters to the nearest hundredth.

3.3 Header cutting mechanism: A device on the header used to cut off the standing crop from its root system. The cutting device may be a sickle, rotary impact knives, rotary disk(s), oscillating scissor, or other devices for cutting.

3.3.1 Sickle: A cutting device which uses a reciprocating cutter to cut the standing crop.

3.3.1.1 Sickle frequency: The number of cycles which the sickle makes in a given time period. One cycle is the full movement of the sickle in one direction and its return to the starting point. Frequency shall be expressed in hertz.

3.3.1.2 Sickle stroke: The distance that a point on the sickle travels with respect to the centerline of a guard in one half cycle expressed in millimeters.

3.3.2 Rotary impact knives: A rotary cutting device using high velocity knives driven about a vertical or horizontal axis to impact cut the standing crop (no stationary knife used).

3.3.2.1 Disk cutter: A multiple disk device, using two or more blades per disk, driven about vertical axes from beneath at sufficiently high rotational speeds to achieve impact cutting.

3.3.2.2 Drum cutter: A multiple drum device, using two or more blades per drum, driven about vertical axes from above at sufficiently high rotational speeds to achieve impact cutting. The blades are located at or near the drum bottom.

3.3.2.3 Flail cutter: A device using multiple, radially mounted blades that are pivotally mounted on a horizontal rotor to impact cut. The rotor is positioned transverse to the direction of travel.

3.3.3 Rotary disk(s): One or two disks per row crop header row unit are used to shear off the crop. The one disk rotary knife system requires a stationary knife to shear the crop against. The two disk rotary knife system requires either a stationary knife or disks that are overlapped and rotated in the opposite directions such that the crop is sheared off at the forward intersection of the two disk peripheries.

3.3.4 Oscillating scissor: A device consisting of one pivoting knife with two cutting surfaces per row crop header row unit. The knife reciprocates in a semi-circular arc and cuts the crop off against one of two stationary knives. Knife frequency and knife stroke at the forward most cutting point shall be determined as specified in paragraphs 3.3.1.1 and 3.3.1.2.

3.4 Header cutting height

3.4.1 Sickle, oscillating scissor, and rotary impact disk or drum cutter: The cutting height shall be measured at the forward tip of the cutting element above the plane on which the machine is standing.

3.4.2 Rotary flail cutter: The cutting height shall be established at the minimum distance between the cutting blade rotating periphery and the plane on which the machine is standing.

3.4.3 Rotary disk(s): The cutting height shall be determined at the point that the stalks are sheared, either at the forward tip of the stationary knife or at the forward intersection of the two rotary knife disks.

3.4.4 The header cutting heights shall be established under the following conditions and expressed to the nearest 5 millimeters.

3.4.4.1 The maximum and minimum heights shall be measured in the highest and the lowest positions to which the cutting device can be raised or lowered with the standard lift mechanism.

3.4.4.2 Tire and wheel rim sizes and axle mounting positions shall be stated, and tires shall be inflated to the field operating pressures recommended by the forage harvester manufacturer.

3.4.4.3 For pull-type (towed) forage harvesters, the tractor drawbar height shall be stated in millimeters.

3.4.4.4 The plane upon which the forage harvester is standing shall be hard and level.

3.4.4.5 The type of header installed at the time of measuring shall be stated.

3.4.4.6 The crop container of a self-propelled forage harvester, if so equipped, shall be empty.

3.5 **Header mass:** The mass of the complete header equipped for field operation. If other equipment options affect the mass, such equipment shall be specified. The header mass shall be expressed to the nearest 10 kilograms.

3.6 Feeding mechanism to cutterhead (precision cut)

3.6.1 **Feedrolls:** One or more cylindrical rolls (generally with protrusions or flutes) used to gather, compress and meter the crop into the cutterhead.

3.6.2 **Feedroll throat area:** The throat area shall be calculated by multiplying the feedroll housing inside width at the plane formed by centerlines of the upper and lower rear feedrolls nearest the cutterhead times the maximum height between the outside diameters of the rear feedrolls, expressed in square centimeters.

3.6.3 **Theoretical length of cut:** A mathematically derived expression which calculates the theoretical stem or particle lengths produced by a forage harvester. In the equation, it is assumed that there is no slippage of crop between the feedrolls and that the stems are cut perpendicular to the longitudinal axis of the stem. Theroretical length of cut, TLOC, shall be expressed in mm/knife. The crop mat velocity represented by the terms in the numerator of the TLOC equation is the average peripheral velocity of the upper and lower rear feedrolls.

$$TLOC = \frac{\pi \times (D1 \times N1 + D2 \times N2)/Z}{N \times K}$$

where

D1 = upper rear feedroll effective feeding diameter*, mm
D2 = lower rear feedroll effective feeding diameter*, mm (if used)
N1 = upper rear feedroll speed, r/min
N2 = lower rear feedroll speed, r/min (if used)
N = cutterhead speed, r/min
K = number of cutterhead knives per revolution passing by a fixed point
Z = number of feedrolls used in the TLOC numerator. (Z=1 for machines with only one feedroll, otherwise Z=2.)

*Feedroll Effective Feeding Diameter:

A. **Smooth feedroll (no protrusions):** The effective feeding diameter is equal to the roll OD (outside diameter).

B. **Feedrolls with unnotched protrusions:** The effective feeding diameter is equal to the roll protrusion OD.

C. **Feedrolls with notched protrusions:** The effective feeding diameter is equal to the roll protrusion OD minus the depth of one protrusion notch.

3.7 **Precision cut cutterhead:** A device intended to shear the crop into "uniform" short lengths.

3.7.1 **Cylinder type cutterhead:** Knives mounted on cylindrical mountings such that the knives are essentially parallel to the axis of rotation. The number of knives per revolution passing by a fixed point and rotational speed (r/min) shall be specified. The cutterhead diameter and width shall be expressed in millimeters.

3.7.2 **Flywheel type cutterhead:** Knives mounted essentially radially with the cutting edges describing a plane perpendicular to the axis of rotation. The number of knives, the number of forage impeller blower paddles, if so equipped, and the rotational speed (r/min) shall be specified. The inner and outer effective knife cutting diameters about the axis of rotation shall be expressed in millimeters. The diameter formed by the blower paddles, if so equipped, and the blower housing inside width shall be expressed in millimeters.

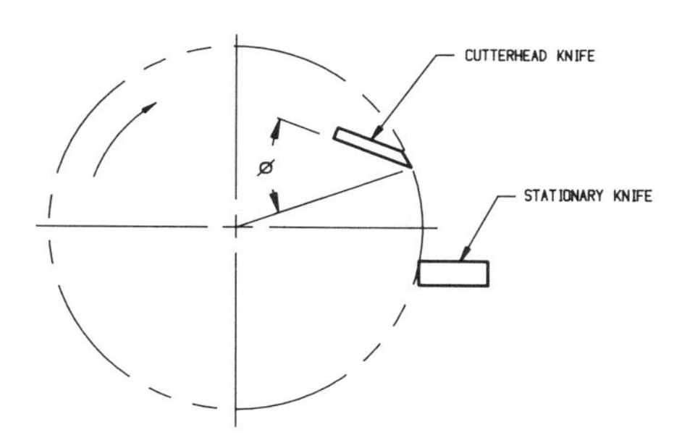

FIG. 1—CYLINDER CUTTERHEAD KNIFE RAKE ANGLE, Φ

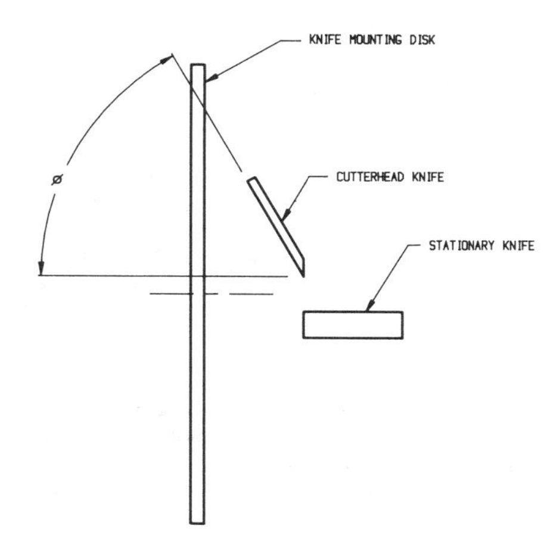

FIG. 2—FLYWHEEL CUTTERHEAD KNIFE RAKE ANGLE, Φ

3.7.3 **Stationary knife:** A knife that provides a stationary edge for the cutterhead knives to shear the crop against.

3.7.4 Cutterhead knife rake angle

3.7.4.1 **Cylinder cutterheads:** The knife rake angle is the included angle between the knife leading surface at the cutting tip and a radial line passing through the knife cutting tip (see Fig. 1).

3.7.4.2 **Flywheel cutterheads:** The knife rake angle is the included angle between the knife leading surface at the cutting tip and a line parallel to the cutterhead axis of rotation (see Fig. 2).

E

3.7.5 **Cylinder cutterhead knife helix angle:** The knife helix angle is the arc tangent of the arc length described by the knife cutting edge divided by the cylinder width (see Fig. 3).

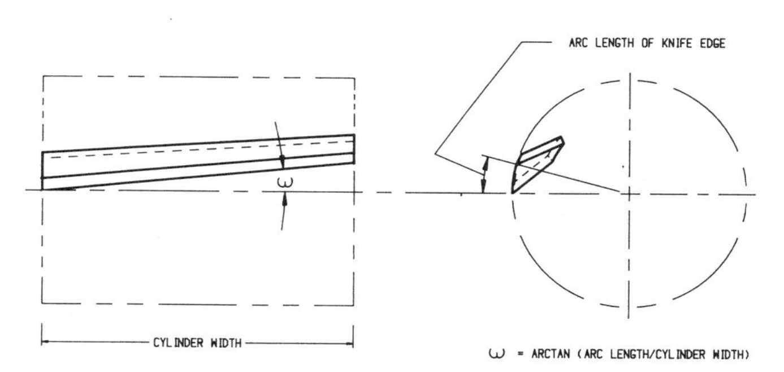

FIG. 3—CYLINDER CUTTERHEAD KNIFE HELIX ANGLE, ω

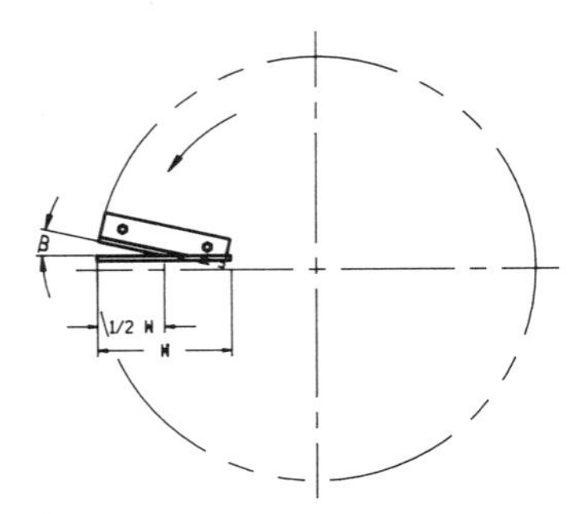

FIG. 4—FLYWHEEL CUTTERHEAD KNIFE SHEAR ANGLE, β

3.7.6 Flywheel cutterhead knife shear angle: The included angle between the cutterhead knife and the stationary knife in a transverse direction (see Fig. 4). The average shear angle shall be established with the cutterhead knife cutting edge positioned at the midspan of the stationary knife.

3.7.7 Recutter screen: A semi-cylindrical band (with holes) mounted concentric with a cylinder cutterhead. The screen starts beyond the stationary knife and continues around the discharge opening. The chopped crop is recut by the cutterhead knives as the crop passes through the screen holes. The recutter screen is used primarily to reduce particle lengths that are substantially beyond the theoretical length of cut. The screen hole dimensions shall be expressed in millimeters.

3.8 Non-precision cut flail chopping rotor (cutterhead): Multiple, radially mounted flail blades pivotally mounted on a rotor positioned transverse to the direction of travel and parallel to the ground. The swath or windrow or standing crop is cut directly by the flail blades into shorter, "random" lengths by impact cutting (no stationary knife used). The number of rotor flail banks per revolution, total number of flails, and rotational speed (r/min) shall be specified. The rotor diameter shall be expressed in millimeters. The rotor width shall be expressed in meters to the nearest hundredth.

3.9 Crop delivery devices: A mechanism used to propel the chopped crop from the forage harvester through a converging section to the transport container. Typical propelling devices are:

3.9.1 Cylinder or flywheel cutterhead

3.9.2 Flail chopping rotor

3.9.3 Cylinder impeller blower: A device consisting of multiple rows of radial, fixed or free swinging paddles mounted on a transverse rotor. The crop is fed essentially tangentially to the rotor. The number of paddle banks, total number of paddles, and rotational speed (r/min) shall be specified. The rotor diameter and width shall be expressed in millimeters.

3.9.4 Flywheel impeller blower: A device generally using one row of paddles mounted essentially radially to the axis of rotation. The crop is fed into the blower essentially parallel to the axis of rotation. The number of paddles and rotational speed (r/min) shall be specified, and the blower diameter and blower housing inside width shall be expressed in millimeters.

3.10 Self-propelled forage harvester specifications

3.10.1 Engine power: The corrected gross, rated, brake power, kW, measured in accordance with Society of Automotive Engineers Standard SAE J1349, Engine Power Test Code—Spark Ignition and Diesel, at the governed engine rpm which shall be stated. Where particular markets require the use of different test code for determining engine power, the engine test code shall be stated; however, the engine power determined by SAE Standard J1349 shall also be included.

3.10.2 Engine displacement: Engine volumetric displacement shall be expressed in liters to the nearest hundredth.

3.10.3 Turning radius: The distance from the turning center to the center of tire contact of the wheel describing the largest circle while the vehicle is executing its shortest turn without turning brakes in operation. The measurement shall be made on a hard, level surface. The wheel base and guide wheel tread width shall be stated. Turning radius shall be expressed in meters to the nearest hundredth.

3.10.4 Clearance radius: The distance from the turning center to the outermost point of the forage harvester executing its shortest turn without turning brakes in operation. If equipment options or crop headers affect this dimension, such equipment shall be specified. The crop header shall be fully raised. The wheel base and guide wheel tread width shall be the same as in paragraph 3.10.3. Clearance radius shall be expressed in meters to the nearest hundredth.

3.10.5 Self-propelled forage harvester mass: The mass of the complete machine equipped for field operation, but without the crop header mounted unless it is an integral part of the machine. The mass shall be determined under the conditions specified in paragraphs 3.4.4.2, 3.4.4.4, 3.4.4.6, and the fuel tank shall contain a maximum of 20 liters fuel. If equipment options or such items as static weights, tire ballast, additional fuel, etc., are included in the mass, such items shall be specified. Self-propelled forage harvester mass shall be expressed to the nearest 10 kilograms.

3.10.6 Self-propelled forage harvester length: The overall dimension from the foremost point to the rearmost point of the machine with and without crop header(s) measured parallel to the longitudinal centerline of the forage harvester. If other equipment options or attachments affect the length, such equipment and related dimensions shall be specified. The crop header shall be fully raised. The length shall be expressed in meters to the nearest hundredth.

3.10.7 Self-propelled forage harvester width: The overall side to side dimension of the machine with and without crop header(s). If other equipment options or tire sizes and axle positions affect the width, such equipment and related dimensions shall be specified. The width shall be expressed in meters to the nearest hundredth.

3.10.8 Self-propelled forage harvester height: The vertical distance from the plane on which the machine is standing to the highest point on the machine. The height shall be measured under the conditions specified in paragraphs 3.4.4.2, 3.4.4.4, 3.4.4.5, 3.4.4.6, and with a maximum of 20 liters fuel in the fuel tank. The height with all components in position for transport and the height with all components in position for field operation shall be specified. If optional equipment affects height, such equipment and related dimensions shall be specified. Machine height shall be expressed in meters to the nearest hundredth.

3.10.9 Self-propelled forage harvester spout discharge height: The vertical distance from the plane on which the machine is standing to the top of the spout cap when horizontal. Where spout height is adjustable, maximum and minimum height shall be specified. The height shall be measured under the conditions outlined in paragraph 3.10.8, expressed in meters to nearest hundredth. Spout discharge heights shall conform to ASAE Standard S328, Dimensions for Compatible Operation of Forage Harvesters, Forage Wagons, and Forage Blowers.

3.10.10 Fuel tank capacity: The usable capacity of the fuel tank shall be expressed in liters to the nearest whole liter.

3.11 Side mounted and pull-type (towed) forage harvester specifications

3.11.1 Forage harvester power rating: The maximum and minimum tractor PTO (power take-off) power levels at which the forage harvester was designed to be operated. The power shall be expressed in kilowatts and the PTO speed shall be stated as 540 or 1000 r/min.

3.11.2 Forage harvester PTO category size: The forage harvester PTO hookup size (category 1, 2, 3, 4, 5, or 6) shall be specified as defined by American National Standard ANSI/ASAE Standard S331, Implement Power Take-Off Drive Line Specifications. The tractor PTO shaft diameter and number of spline teeth required shall be stated.

3.11.3 Forage harvester mass: The mass of the complete machine equipped for field operation, but without the crop header mounted unless it is an integral part of the machine. The mass shall be determined under the conditions specified in paragraphs 3.4.4.2 and 3.4.4.4. If other equipment options affect the mass, such equipment shall be specified. The mass shall be expressed to the nearest 10 kilograms.

3.11.4 Forage harvester length: Length shall be determined as outlined in paragraph 3.10.6

3.11.5 Forage harvester width: The overall side to side dimension of the machine with and without crop headers set for transport and for field operation. If other equipment options or tire sizes and axle positions affect the width, such equipment and related dimensions shall be specified. The width shall be expressed in meters to the nearest hundredth.

3.11.6 Forage harvester spout discharge height: Height shall be determined as outlined in paragraphs 3.10.9, 3.4.4.2, 3.4.4.3, 3.4.4.4, and 3.4.4.5.

3.11.7 Side mounted forage harvester hitch category size: The tractor three-point hitch category required to lift and operate the machine shall be specified per American National Standard ANSI/SAE J715/ASAE S217, Three-Point Free-Link Attachment for Hitching Implements to Agricultural Wheel Tractors.

SECTION 4—CROP TERMINOLOGY

4.1 Crops that are typically harvested with forage harvesters are grasses, alfalfa, clover, or mixtures of grass, alfalfa, and clover, forage cereals, corn (maize), and sorghum.

4.2 Moisture content: Moisture content of the crop shall be expressed on the wet basis. The moisture content shall be determined per ASAE Standard S358, Moisture Measurement—Forages.

4.3 Plant length: The length of the unharvested plant from ground level to its tip when the plant is straightened, expressed in centimeters.

4.4 Stubble height: The length of the plant stalk attached to the ground immediately after harvesting, expressed to the nearest whole centimeter.

SECTION 5—FORAGE HARVESTER PERFORMANCE SPECIFICATIONS

5.1 Items defining forage harvester performance specifications are capacity, machine specific energy, crop throw distance, particle length distribution, and corn kernel fracture. When these items are measured, the crop and its moisture content, theoretical length of cut, and recutter screen (if used) shall also be stated.

5.1.1 Capacity: The forage harvester capacity shall be expressed in metric tonnes (wet mass) per hour.

5.1.2 Machine specific energy: The forage harvester specific energy shall be expressed as power divided by capacity of kW·h/t (wet mass).

5.1.3 Crop throw distance: The horizontal distance which the forage harvester propels the crop to the left or right, 90 deg to the direction of travel, in no wind conditions. The distance shall be measured from the spout's vertical centerline at its plane of rotation to the center of crop material mass on the ground, expressed to the nearest meter. The measurement shall be done on a level plane. Spout discharge height shall be stated as specified in paragraph 3.10.9 or paragraph 3.11.6.

5.1.4 Particle length distribution: The chopped crop particle length distribution shall be determined per ASAE Standard S424, Method of Determining and Expressing Particle Size of Chopped Forage Materials by Screening, or by hand sorting.

5.1.4.1 Hand sort method: The procedures outlined in ASAE Standard S424, Method of Determining and Expressing Particle Size of Chopped Forage Materials by Screening, shall be used to determine particle length distribution except that the sample shall be sorted by hand into the following lengths, L:

ASAE "screen sizes*	Particle length†, mm
1	$L>27$
2	$18<L\leqslant27$
3	$9<L\leqslant18$
4	$5.6<L\leqslant9$
5	$2<L\leqslant5.6$
6	$0\leqslant L\leqslant2$

*Corresponds to screen number in ASAE Standard S424, Method of Determining and Expressing Particle Size of Chopped Forage Materials by Screening.

†Particle length sizes correspond to diagonal length of holes in ASAE screens.

5.1.5 Corn (maize) kernel fracture: This is a measure of a forage harvester's ability to abrade, crack, or break the kernels. It shall be expressed as the percentage to the nearest one-half percent of undamaged kernels to the total field kernel population. Dye penetrants may be used to determine invisible kernel fracture.

Cited Standards:

ANSI/ASAE S331, Implement Power Take-Off Drive Line Specifications
ANSI/SAE J715/ASAE S217, Three-Point Free-Link Attachment for Hitching Implements to Agricultural Wheel Tractors
ASAE S328, Dimensions for Compatible Operation of Forage Harvesters, Forage Wagons, and Forage Blowers
ASAE S343, Terminology for Combines and Grain Harvesting
ASAE S358, Moisture Measurement—Forages
ASAE S404T, Metric Row Spacings
ASAE S424, Method of Determining and Expressing Particle Size of Chopped Forage Materials by Screening
SAE J1349, Engine Power Test Code—Spark Ignition and Diesel

AGRICULTURAL ROTARY MOWER SAFETY

Proposed by the Industrial/Agricultural Mower Manufacturers' Council of the Farm and Industrial Equipment Institute; approved by the ASAE Agricultural Safety Committee and by the ASAE Power and Machinery Division Standards Committee; adopted by ASAE April 1987.

SECTION 1—PURPOSE

1.1 The purpose of this Standard is to establish the safety requirements for agricultural type rotary mowers whose intended use falls within the scope of this Standard.

SECTION 2—SCOPE

2.1 The safety specifications in this Standard apply to towed, semi-mounted or mounted mowers, with one or more blade assemblies of 78 cm (30.5 in.) blade tip circle diameter or over, intended for marketing as agricultural mowing equipment and designed for cutting and/or shredding grass and other growth and crop residue and for spreading the cut material to meet varying requirements while powered by an agricultural tractor as defined in ASAE Standard S390, Classifications and Definitions of Agricultural Equipment, or similar machine of at least 15 kW (20 hp).

2.2 These specifications do not apply to:

2.2.1 Turf care equipment primarily designed for personal use, consumption or enjoyment by a consumer in or around a permanent or temporary household or residence.

2.2.2 Equipment designed primarily for commercial purposes such as along highways or other populated areas, but which may be used for agricultural use.

2.2.3 Self-powered or self-propelled mowers or mowing machines.

2.3 Where other standards are referenced, such references apply only to the document identified and not to revisions thereof.

SECTION 3—DEFINITIONS

(See also ASAE Standard S390, Classifications and Definitions of Agricultural Equipment)

3.1 Arm type mower: Mowers which are intended to be used frequently with the cutter portion not adjacent or parallel to the ground.

3.2 Agricultural mowing: Mowing agricultural areas such as pasture clipping, crop residue shredding and disposal, heavy brush cutting for land clearing or waterways, or right-of-way maintenance along power or gas lines, etc., but not along highways or roadways.

3.3 Functional component: A working mechanism of an attachment or implement designed to perform a specific task such as the cutting blade of a rotary mower.

3.4 Guarded by location: A hazard is guarded when it is covered by other parts or components of the machine or when, because of its remote location, inadvertent contact is minimized during normal operation or servicing.

3.5 Hit: Rupture of the first layer of the target material by a test projectile. By definition, punctures are also hits.

3.6 Inadvertent contact: Contact between a person and a moving machinery part hazard, or other type of hazard, resulting from the person's unplanned actions during normal operation or servicing.

3.7 Machinery hazard: A source of potential injury created by machinery parts which can cause serious injury upon contact or by entanglement of personal apparel. This includes, but is not limited to, the pinch points of power driven gears, run-on points of belts and chains, and projections on rotating parts.

3.8 Normal operator position: The space within the operator zone occupied by the operator while operating a mower. The operator is sitting on the seat with hands on the steering controls and with feet on controls or areas provided for foot placement. For operator zone for thrown object test, see Figs. 1, 3 and 4 and paragraph 7.5.1.

3.9 Power take-off (PTO): An external shaft on the rear of a tractor to provide rotational power to implements. (Reference ASAE Standard S203.10, Rear Power Take-Off for Agricultural Tractors)

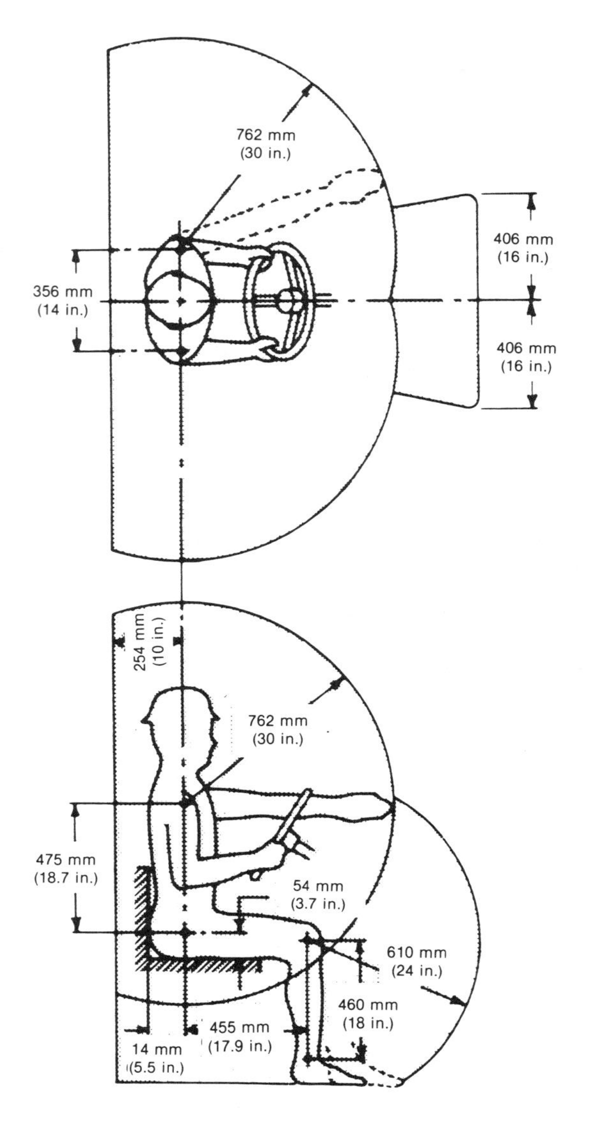

OPERATOR ZONE

The operator zone is the area into which the extremities of a 95th percentile male can reach from the normal operator position.

The zone is established with the seat in the rearward adjustment position. A 10 cm (4 in.) forward adjustment will approximate the position of a 5th percentile male or a 50th percentile female.

All barriers within the operator zone will reduce the zone by the space occupied and protected by the barrier.

The operator zone includes the maximum range of all operator control movement but is not intended to represent preferred operator control positions. SAE Standard J898a, Control Locations for Off-Road Work Machines, is recommended for determining desirable hand and foot control locations and space allocations. SAE Standard J833 DEC83, USA Human Physical Dimensions, is a source of male and female physical dimensions.

FIG. 1—OPERATOR ZONE

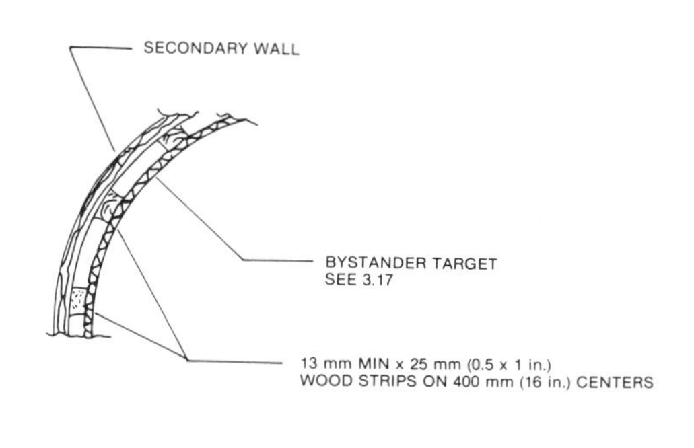

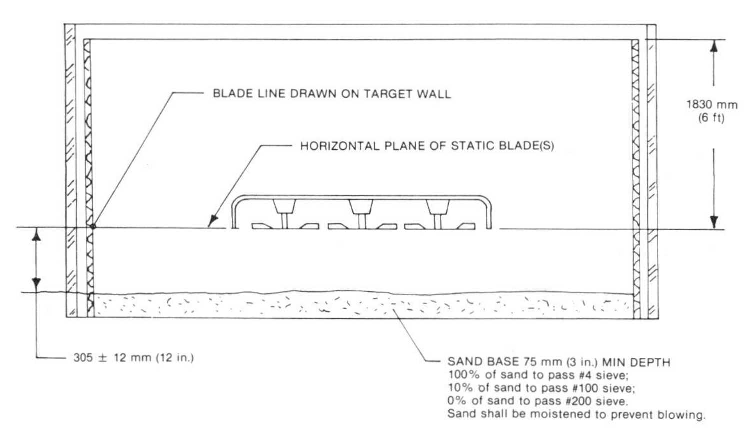

FIG. 2—TARGET WALL CONSTRUCTION AND MOWER VERTICAL POSITION

3.10 Implement input driveline (IID): Two universal joints and their connecting member(s) and fastening means for transmitting rotational power from the tractor PTO to the implement input connection. A double Cardan, constant velocity joint is considered a single joint. The IID also includes integral shielding where provided. (Reference ASAE Standard S207.11, Operating Requirements for Tractors and Power Take-Off Driven Implements)

3.11 Label: A durable label used as a safety sign or for instruction or identification. Labels shall meet or exceed the requirements of paragraph 4.2.3.

3.12 Propelling machine: A tractor or self-propelled machine used to operate towed, semi-mounted or mounted rotary mowers.

3.13 Puncture: The rupture of all layers of the target material by a test projectile.

3.14 Rotary mower: A power mower in which one or more functional components cut or shear by impact and rotate about an axis perpendicular to the cutting plane.

3.15 Standard test operator: A person weighing 120±5 kg (265±11 lb) and standing 188±5 cm (74±2 in.) tall.

3.16 Shield (or guard): A barrier which minimizes inadvertent personal contact with hazards.

3.17 Target material: 350# uncoated corrugated board, specification paper weight 90# - 26# - 90# B flute as shown below.

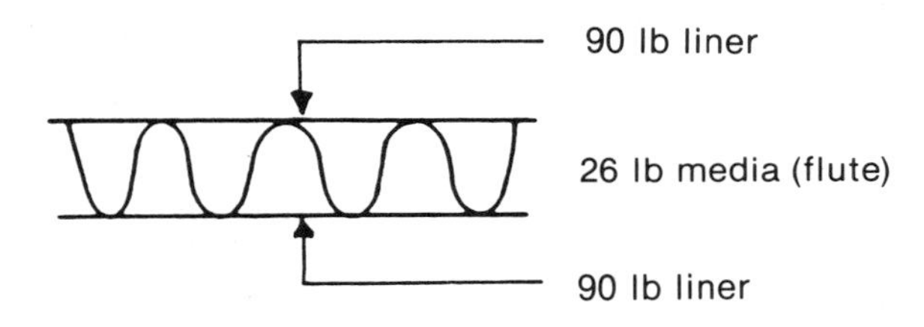

SINGLE WALL CORRUGATED BOARD

3.18 Test projectile: An uncoated six (6) penny steel box nail with the following dimensional limits:

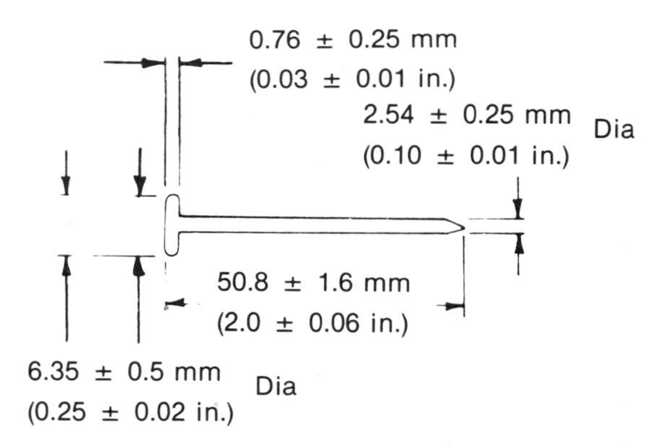

SECTION 4—GENERAL REQUIREMENTS

4.1 Guarding and shielding

4.1.1 Inadvertent contant with hazards shall be minimized by guarding and shielding, to the maximum extent permitted by the intended function of the component, during normal mounting, starting, operating or dismounting of the equipment.

4.1.2 The following are some of the potential hazard areas:

4.1.2.1 Pinch points of gears and the run-on point where a belt or chain contacts a sheave, sprocket or idler

4.1.2.2 Outside faces of pulleys, sheaves, sprockets and gears on rotating drives

4.1.2.3 Rotating parts with projections such as exposed bolts, keys or set screws

4.1.2.4 Revolving shafts, except smooth (without keyways, splines, etc.) shaft ends protruding less than one-half the diameter of the rotating element

4.1.2.5 Implement in-put driveline assembly

4.1.2.6 Functional components

4.1.3 Shields and their supports shall remain functional and withstand the forces that a 120 kg (265 lb) individual, leaning on or falling against the shield would exert on that shield in normal operation or servicing. Those shields designed to be used as steps, or where that use can be anticipated, shall remain functional if used as a step by a 120 kg (265 lb) operator.

4.1.4 Equipment on which hands, feet or legs could come in contact with rotating components or equipment with access doors and shields which can be opened or removed while components continue to rotate more than 7 s after the power is disengaged shall have: (1) visible or audible indication of rotation and (2) a suitable safety sign near the opening per paragraph 4.2.3.

4.1.5 Access doors, guards and shields which must be opened for normal servicing shall be easily opened and closed.

4.2 Labels and instructions

4.2.1 Control identification. The controls furnished with the mower and their direction of motion for stopping, starting, speed control and operation, whose functions are not obvious, shall be identified by a label per paragraph 4.2.3. Symbols as provided in ANSI/ASAE Standard S304, Symbols for Operator Controls on Agricultural Equipment, may be used for control identification.

4.2.2 Machine identification. The mower shall be provided with identification per paragraphs 4.2.3.3 or 4.2.3.4 giving model number, serial number and the name and address of the source of replacement parts and service.

4.2.3 Labels. Labels, and name plates provided on units shall meet the following minimum requirements:

4.2.3.1 Labels shall form a durable bond with the base material surface and shall show no appreciable loss of adhesion during weathering exposure. Labels shall not curl at the edges. Labels shall not lose legibility or suffer appreciable loss of adhesion when exposed to occasional contact with gasoline or oil.

4.2.3.2 Labels shall meet the weathering requirements of ASAE Standard S441, Safety Signs, for permanent signs.

4.2.3.3 Embossed, indented, cast or molded labels shall be considered sufficient to meet the requirements of this paragraph.

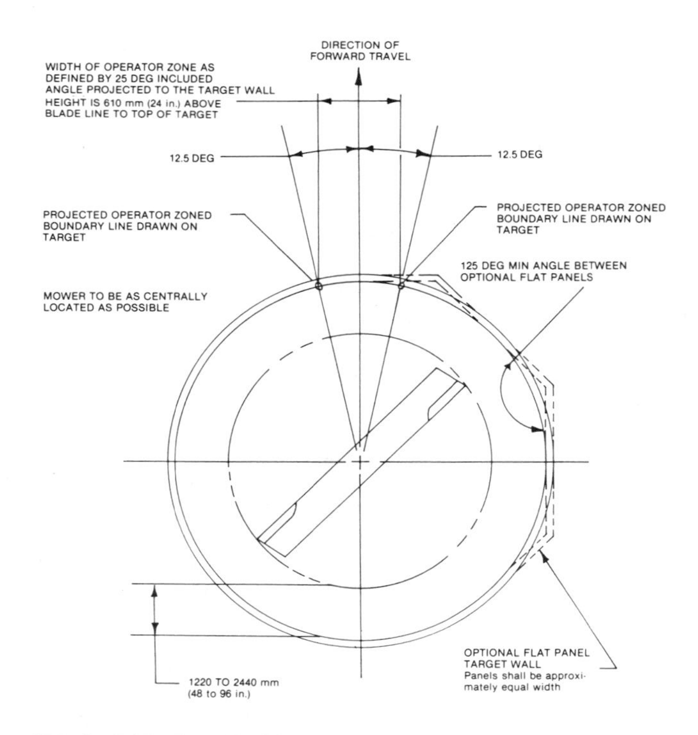

FIG. 3—TARGET CONFIGURATION AND OPERATOR ZONE FOR SINGLE BLADE MOWERS

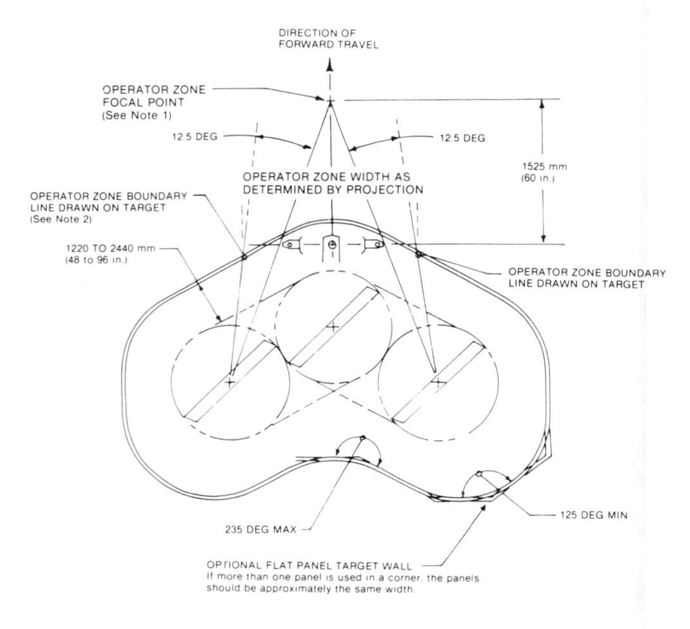

NOTES

(1) The operator zone focal point shall be on a line in the forward direction of travel midway between the hitch pins or on the centerline of the tongue hole.
(2) Operator zone height is 610 mm (24 in.) above the blade line to the top of the target wall.
(3) The mower position shall be as symmetrical as possible within the target enclosure.

FIG. 4—TARGET CONFIGURATION AND OPERATOR ZONE FOR MULTI-BLADE MOWERS

4.2.3.4 Metal plates over 0.48 mm (0.019 in.) thick with embossed or etched lettering and fastened with rivets or equivalent fastening means shall be considered sufficient to meet the requirements of this paragraph.

4.2.4 Operation, service and maintenance instructions

4.2.4.1 General requirements. Written instructions shall be provided with the equipment describing proper operation of the equipment plus proper operational and service procedures and necessary maintenance procedures to avoid potential hazards. (See also ASAE Engineering Practice EP363.1, Technical Publications for Agricultural Equipment) The manual shall also advise that, "In addition to the design and configuration of equipment, hazard control and accident prevention are dependent upon the awareness, concern, prudence and proper training of personnel involved in the operation, transport, maintenance and storage of equipment."

4.2.4.1.1 Written safety instructions shall include identification of the need for personal protective equipment; such as but not limited to, protection for the eyes, ears, feet, hands and head.

4.2.4.1.2 The operator's manual shall contain instructions so that a person unfamiliar with the mower will have required information to prepare the mower for operation and to adjust, start, operate, transport, stop, park and unhitch the mower.

4.2.4.2 Stored energy devices. Any stored energy device such as, but not limited to, spring-loaded mechanisms, elevated components and pressurized fluid systems, such as hydraulic accumulators, which can be disconnected, disassembled or freed in such a way to release energy or material in a hazardous manner shall have an appropriate safety sign on or near the device. The safety sign shall include instructions for de-energizing and proper disassembly or shall include a reference to instructions to be provided in the operator's manual.

4.2.4.3 **Hydraulic hazards.** When applicable, the operator's service and maintenance manuals shall contain:

4.2.4.3.1 A warning that hydraulic fluid escaping under pressure may have sufficient force to penetrate skin and cause serious injury, and that if foreign fluid is injected into the skin, it may result in gangrene if the fluid is not surgically removed within a few hours by a doctor familiar with this form of injury.

4.2.4.3.2 Information cautioning to keep body and hands away from pin holes or nozzles which eject hydraulic fluid under high pressure and to use paper or cardboard and not hands or other body parts to search for leaks.

4.2.4.3.3 Information cautioning the operator to make sure all hydraulic fluid connections are tight and all hydraulic hoses and lines are in good condition before applying pressure to the system.

4.2.4.3.4 Information explaining how to minimize the hazard during the relieving of all pressure or force in each system before disconnecting the lines or performing work on the system.

4.3 Operator zone - pressurized components

4.3.1 Hydraulic hoses furnished with the mower shall meet the requirements of the applicable section of Society of Automotive Engineers Standard J517 JUN81, Hydraulic Hose, based on the working pressure of each system.

4.3.2 Pressurized hoses, lines and components furnished with the mower shall be located or shielded so that in the event of rupture, a stream of fluid is not discharged directly onto the operator from within the operator zone when the operator is in the operator zone (See Figs. 1, 3 and 4).

SECTION 5—MOWER REQUIREMENTS

5.1 **Tongue.** Any trailed unit should avoid a hitch connection characteristic which could move uncontrollably upward when disconnected. If this condition exists, the trailed unit shall have a conspicuous label per paragraph 4.2.3 in the vicinity of the hitch point identifying this condition.

5.2 **Attachment means.** Three-point hitch mounted and semi-mounted mowers shall be attached to the propelling machine by one or a combination of the ASAE standardized attachment means described in ASAE Standard S217.10, Three-Point Free-Link Attachment for Hitching Implements to Agricultural Wheel Tractors, and ASAE Standard S278.6, Attachment of Implements to Agricultural Wheel Tractors Equipped with Quick-Attaching Coupler.

SECTION 6—FUNCTIONAL COMPONENT REQUIREMENTS

6.1 **Power disconnect.** Multi-section mowers shall have the means to disconnect power to any section which can be carried in a raised or transport position while mowing continues with another section.

6.2 **Guarding and shielding.** Functional components which must be exposed for proper function shall be shielded sufficiently to meet the thrown object test provisions of this Standard, except arm-type mowers. Any movable or removable guard(s) shall conform to the following:

6.2.1 A safety sign shall be affixed to the mower in a prominent location stating that the mower shall not be operated without guard(s) in place.

6.2.2 The operator's manual shall state that the mower shall not be operated without guard(s) in place.

6.2.3 If any guard or shield which is offered as an option (because of special requirement such as alternate types available, replaceability requirements or agricultural uses where it may interfere with function or create problems) is required for the mower to comply with the tests in Section 7—Test for Mower Components, this fact shall be prominently noted at each of the following locations: in the price list and sales literature, in the operating instructions, and by a safety sign located on the mower.

6.2.4 If a guard is constructed of woven fabric or other such material which may be subject to rapid wear or deterioration, the mower shall bear a label per paragraph 4.2.3 stating that it may require frequent inspection and possible replacement.

6.3 **Cutting elements.** The components which are used to attach the cutting elements shall not become worn or fail in a hazardous manner before the elements themselves are worn beyond practical use.

SECTION 7—TEST FOR MOWER COMPONENTS

7.1 Test conditions, general

7.1.1 **Assembly.** The mower shall be completely assembled and mounted on or attached to its propelling machine except for tests where mounting on a suitable test fixture is designated, or, where necessary, the mower unit may be tested while separated from the power unit and power provided by some other means. However, speeds must be the same as when on or attached to the propelling machine, and parts which extend into the trajectory area should be duplicated as nearly as practicable. Adjustable guards shall be set in the most open position for the test.

7.1.2 **Mower position.** The mower shall rest on a horizontal surface and in a horizontal position that is flat within 2 deg.

7.1.3 **Test speed.** Except for static tests, the mower shall be operated at the manufacturer's maximum recommended operating speed.

7.1.4 **Number of tests.** All tests shall be run once for each blade assembly of the mower except where otherwise herein designated. A new mower may be used for each test, except for the tests of paragraphs 7.4 and 7.5.

7.1.5 **Restraints.** Resilient restraints may be used to keep the mower in position during the test.

7.2 **Blade impact test.** (May be conducted before or after any other test.)

7.2.1 **Test equipment.** The mower shall be completely encircled at the time of test by a wall of target material, per paragraph 3.16, resting on the floor. The wall shall be approximately 1220-2440 mm (4-8 ft) from the blade tip circle with a minimum height of 1830 mm (6 ft) above the horizontal plane of the blade tip circle (see Figs. 2, 3 and 4). A protective barrier shall be provided to protect the test operator.

7.2.2 **Test conditions.** The mower shall be adjusted for approximately 75 mm (3 in.) height of cut or the cutting height setting closest to 75 mm (3 in.)

7.2.3 **Test procedure.** The mower shall be positioned over the fixture described in Fig. 5 and dropped onto the test rod such that the blade makes positive contact with the rod. The mower shall be dropped fast enough so that mower speed is not materially reduced by glancing contact before solid contact is made. The mower shall be dropped onto the rod and allowed to continue for a minimum of 2 s before disengaging the power or lifting the mower. The test shall be conducted once in each of the two following manners: (1) The mower positioned so that the contact between the blade and the rod is at a point as close to the blade holder connection as possible; (2) The mower positioned so that the contact between the blade and centerline of the rod is approximately 25 mm (1 in.) from the outer tip of the blade.

7.2.4 **Acceptance criteria.** The test shall be completed without loss of any part of the mower or failure of any mower component in a manner that could be hazardous to the operator or bystanders. Any target puncture by any part of the mower or blades shall constitute failure.

7.3 Blade unbalance test

7.3.1 **Test equipment.** (See paragraph 7.2.1.)

7.3.2 **Test conditions.** For multi-blade construction; remove all blades or cutting elements and fasteners from one blade attaching point. For one-piece blade construction, remove the beveled or sharpened length of the blade on one end only.

7.3.3 **Test procedure.** The mower shall be run for 2 min before shutoff. For multi-spindle mowers, the test may be conducted on all spindles concurrently.

7.3.4 **Test acceptance.** The test shall be completed without loss of any part of the unit or failure of any component in a manner that could be hazardous to the operator or bystanders. Any target puncture by any part of the mower shall constitute failure.

7.4 Structural integrity test

7.4.1 **Test equipment.** Use a test fixture per Fig. 2 and Figs. 3 or 4 as applicable. Use low carbon, hot finish sawed or sheared end test rods as follows: For mowers with blades up to 1220 mm (48 in.) tip circle diameter, use test rods 10 mm (0.375 in.) diameter x 50 mm (2 in.) long. For mowers with one or more blades over 1220 mm (48 in.) tip circle diameter, use test rods 13 mm (0.5 in.) diameter x 50 mm (2 in.) long. Rod lengths shall be ± 3 mm (0.125 in.).

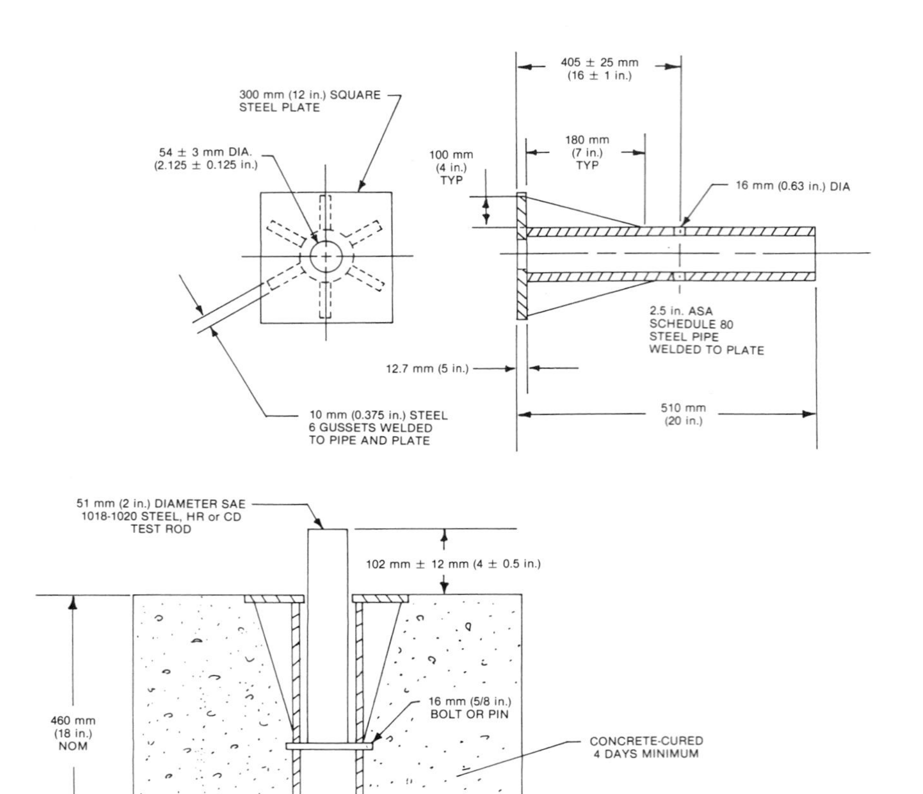

FIG. 5—IMPACT TEST FIXTURE

7.4.2 Test conditions. The mower should be positioned so the cutting edge of a stationary blade is 305 ± 13 mm (12 ± 0.5 in.) above sand base. When supports are necessary to position the mower such that the cutting edge of the blade is 305 mm (12 in.) above the sand base, the supports shall be of round steel bars or tubing no larger than 40 mm (1.6 in.) in diameter, and no more than 6 shall be used per frame unit. The supports shall be placed as necessary under wheels, side skids or other structural components which normally rest on the ground if the mower were at the minimum cutting height. If additional supports are needed, they should be located at least 150 mm (5.9 in.) outside the blade enclosure. The mower may also be supported from above.

7.4.3 Test procedure. The test shall consist of vertical downward introduction of test rods inserted into each of 8 equally spaced holes for each blade assembly in accordance with Fig. 7. The test rods shall be introduced through the tube and funnel arrangement as specified by Fig. 6 or through a similar arrangement with air or mechanical assist. A sufficient number of test rods shall be dropped into each of the 8 positions so that a blade contacts at least 12 test rods per position.

7.4.4 Test acceptance. The mower shall remain in compliance with all applicable requirements of this Standard. The test rods shall not break through the blade housing or blade enclosure but may escape through deflector-type shields such as chain shielding as long as no shielding failure is caused.

7.5 Thrown object test. (To be conducted after the structural integrity test). (Does not apply to arm-type mowers) (See Fig. 9 reporting form)

7.5.1 Test equipment. Use test projectiles per paragraph 3.18 and test fixture per Fig. 2 and Figs. 3 or 4 as applicable. For undermounted units, a 915 mm (36 in.) diameter vertical cylinder of target material shall be placed in the operator zone such that the back of the cylinder shall be 76 mm (3 in.) behind the back of the operator's seat or 76 mm (3 in.) behind the rear position of an actual operator in the event there is no back support on the seat. The target cylinder shall extend from the operator's normal foot position to a height of 1 m (39 in.) above the operator's seat.

NOTE: Provisions must be made to protect the operator during test.

7.5.2 Test conditions. (See paragraph 7.1)

7.5.3 Test procedure. The test shall consist of vertical downward introduction of 75 test projectiles head first and 75 test projectiles point first inserted into each of 8 equally spaced holes for each blade assembly in accordance with Fig 7. The test projectiles shall be introduced through the tube and funnel arrangement as specified by Fig. 6 or through a similar arrangement with air or mechanical assist. The introduction shall be repeated 3 times for each hole (450 per hole) for a total of 3600 per blade assembly. The drop velocity should remain relatively constant and be adjusted to

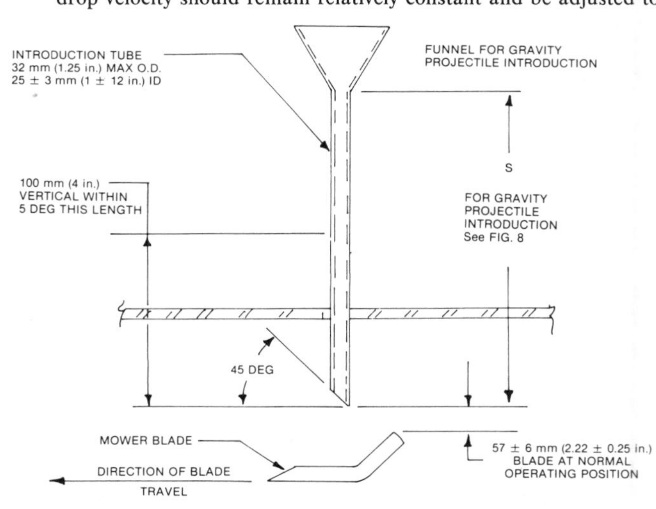

FIG. 6—INTRODUCTION TUBE CONFIGURATION

ensure that between 5 and 15% of the test objects drop through the blade without making blade contact (to ensure that the entire length of the test object is exposed to the blade). After each 150 test projectiles are introduced, the projectiles in a 610 mm (24 in.) diameter circle under the introduction hole shall be counted to verify that between 5 and 15% of the test projectiles passed through the blade without making blade contact. See Fig. 8 for height, s, calculation. On some mowers, it may not be possible to prevent more than 15% of the test projectiles from passing through the blade path without contact. In this case, the 150 quantity must be increased to assure that at least 127 projectiles do make blade contact each test. This can be determined either by sound or by counting the pass-throughs.

7.5.4 Scoring. After every 150 nails has been introduced, record the number of nails contacted by the blade, and record the marks on the wall above the blade line in the following groups: (1) hits in the operator zone; (2) punctures in the operator zone; (3) hits outside the operator zone; (4) punctures outside the operator zone. Total the number of marks in each of these 4 categories individually to obtain individual totals for the particular blade spindle. Divide each sum by the total number of blade-nail contacts for that spindle.

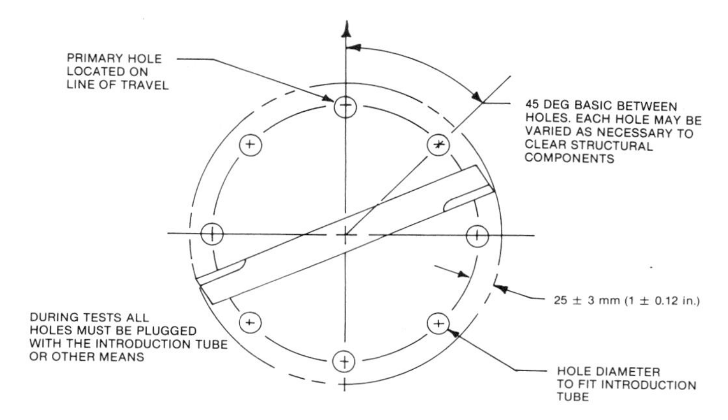

FIG. 7—TYPICAL INTRODUCTION TUBE LOCATION

7.5.5 Test acceptance. For each blade spindle, none of the composite individual spindle scores shall exceed the following acceptance criteria: (1) 2% hits in the operator zone; (2) 0.5% punctures in the operator zone; (3) 20% hits outside the operator zone; (4) 10% punctures outside the operator zone. Failure of any of the 4 acceptance criteria shall constitute failure of the machine. In the event the machine fails the test, it may be retested. The scores are then computed on the sum of the two tests. If the score still exceeds the acceptance criteria, the machine has failed the test.

7.5.5.1 Where it is not practical to meet the thrown object acceptance criteria (3) and (4) and still perform the job intended for the mower, these two criteria need not be met. However, a prominent "DANGER" label per paragraph 4.2.3 shall be provided on the mower advising that the mower may throw objects several hundred feet and that the operator must cease operation whenever anyone other than the operator comes within the area.

$S = 5.66 \times 10^{-8} n^2 L^2$

N = BLADE REVOLUTIONS PER MINUTE
L = LENGTH OF PROJECTILE AND BLADE HEIGHT IN MILLIMETERS. SEE BELOW
S = DROP HEIGHT IN MILLIMETERS. SEE FIG. 6

$S = 1.438 \times 10^{-6} n^2 L^2$

N = BLADE REVOLUTIONS PER MINUTE
L = LENGTH OF PROJECTILE AND BLADE HEIGHT IN INCHES. SEE BELOW
S = DROP HEIGHT IN INCHES. SEE FIG. 6

PROJECTILE

L

FLAT MOWER BLADE

MOWER BLADE WITH FIN

THE DROP HEIGHT CALCULATED FROM THE ABOVE FORMULA MAY HAVE TO BE ADJUSTED TO OBTAIN THE PASS-THROUGH CRITERIA DUE TO ADVERSE AIRFLOW UP THE INTRODUCTION TUBE GENERATED BY THE MOWER BLADE AND DRAG OF THE PROJECTILE AGAINST THE INSIDE OF THE INTRODUCTION TUBE. AIR OR MECHANICAL ASSIST MAY BE REQUIRED FOR MOWER BLADE DIAMETERS NEAR 775 mm (30.5 in.)

FIG. 8—DROP HEIGHT, S, CALCULATION

7.5.5.2 On some specialty mowers it may not be practical to meet criteria (1) or (2) and still perform the intended function of the mower. In this case a prominent "DANGER" label per paragraph 4.2.3 shall be provided on the mower advising of this fact and advising that protective shielding must be provided for the operator.

7.5.5.3 This paragraph pertains only to agricultural use of machines which do not meet the thrown object and foot probe requirements for industrial use per SAE Standard J232 DEC84, Industrial Rotary Mowers. A note shall be provided in the operator's manual and a prominent warning label per paragraph 4.2.3 shall be provided on these machines advising that this machine was designed to meet agricultural uses and safety requirements and must not be used in areas where bystanders may be present and could be injured, or where property could be damaged by thrown objects.

Cited Standards:

ANSI/ASAE S304, Symbols for Operator Controls on Agricultural Equipment
ASAE S203.10, Rear Power Take-Off for Agricultural Tractors
ASAE S207.11, Operating Requirements for Tractors and Power Take-Off Driven Implements
ASAE S217.10, Three-Point, Free-Link Attachment for Hitching Implements to Agricultural Wheel Tractors
ASAE S278.6, Attachment of Implements to Agricultural Wheel Tractors Equipped with Quick-Attaching Coupler
ASAE EP363.1, Technical Publications for Agricultural Equipment
ASAE S390, Classifications and Definitions of Agricultural Equipment
ASAE S441, Safety Signs
SAE J232 DEC84, Industrial Rotary Mowers
SAE J517 JUN81, Hydraulic Hose
SAE J833 DEC83, USA Human Physical Dimensions
SAE J898a, Control Locations for Off-Road Work Machines

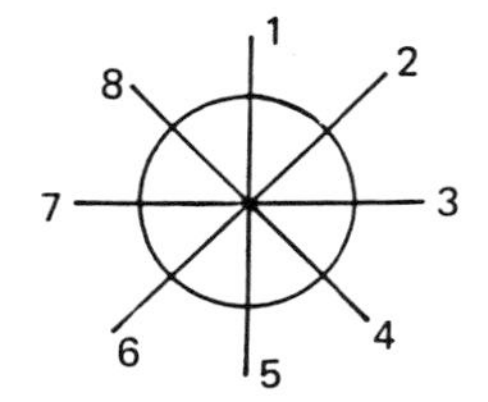

AGRICULTURAL ROTARY MOWER THROWN OBJECT TESTING

HIT: RUPTURE OF THE FIRST LAYER OF THE THROWN OBJECT TARGET MATERIAL BY A TEST PROJECTILE.

PUNCTURE: RUPTURE OF ALL LAYERS OF THE THROWN OBJECT TARGET MATERIAL BY A TEST PROJECTILE.

MACHINE MODEL: ______________________

SERIAL NO.: ______________________

BLADE SPINDLE (If Multiple): ______________________

BLADE PT. NO.: ____________ **TYPE:** ____________

SHIELDING TYPE (If used): ______________________

SHIELDING PT. NO.: ______________________

CONDUCTED BY: ______________________

DATE: ______________________

%Based on test object hits or punctures divided by number of objects hit by blade

Hole Position	Run No.	No. of Test Objects Dropped (150 Min)	No. of Test Objects Hit by Blade (127 Min)	Outside Operator Zone				In Operator Zone			
				Object Target Hits		Object Target Punctures		Object Target Hits		Object Target Punctures	
				No.	% of Hits	No.	% of Punct.	No.	% of Hits	No.	% of Punct.
1.	1.										
	2.										
	3.										
Total =1											
2.	1.										
	2.										
	3.										
Total =2											
8.	1.										
	2.										
	3.										
Total =8											
Total of all 8 holes											
					(20% Max)		(10% Max)		(2% Max)		(0.5% Max)

E

FIG. 9—THROWN OBJECT TEST REPORTING FORM

ASAE Tentative Standard: ASAE S483T

ROTARY MOWER BLADE DUCTILITY TEST

Proposed by the Industrial/Agricultural Mower Manufacturer's Council of the Farm and Industrial Equipment Institute; approved by the ASAE Agricultural Safety Committee and the ASAE Power and Machinery Division Standards Committee; adopted by ASAE as a Tentative Standard April 1987; reconfirmed for one year December 1987.

SECTION 1—PURPOSE AND SCOPE

1.1 The purpose of this Tentative Standard is to identify production lots of blades, from which samples were subjected to destructive testing, that will bend beyond a usable shape without breaking.

1.2 This is a Tentative Standard for industrial and agricultural rotary mower blades which includes a blade bend performance test. The blade bend test is a destructive test to be used in conjunction with other normal quality control and consistency testing procedures and shall be performed on all blade lots following the minimum sampling requirements herein.

SECTION 2—DEFINITIONS

2.1 Rotary mower: A power mower in which one or more functional components cut or shear by impact and rotate about an axis perpendicular to the cutting plane. (Ref. Society of Automotive Engineers Standard J990, Nomenclature-Industrial Mowers)

2.2 Lot: A group of blades from one heat treat run and one mill heat all of which come from only one production run.

2.3 Permanent set angle: The angle formed by the back of the blade in the area of the bend after removal from the die (see Fig. 1).

2.4 Total deflection angle: The angle which is the sum of the permanent set angle and the estimated springback. This angle is for ease of fixture design and not acceptance criteria.

SECTION 3—GENERAL REQUIREMENTS

3.1 Identification. Every blade shall have a vendor identification and a date, lot number or production run number stamped or otherwise permanently affixed in a non-critical stress area that will be readable on a used blade.

3.2 Sampling procedure. Pull random samples from each lot at the minimum rate of one for each 200 blades, but not less than two blades from any one lot. A change in mill heat run shall start a new lot.

3.3 Hardness requirements. A specific hardness is not a requirement of this Tentative Standard as long as the bend test can be passed satisfactorily. However, if the hardness range in the high-stress area of the blade (except the cutting area when it has been selectively hardened) varies more than five points on the Rockwell "C" scale total range, either within the same blade or between sample blades, then double sampling is required.

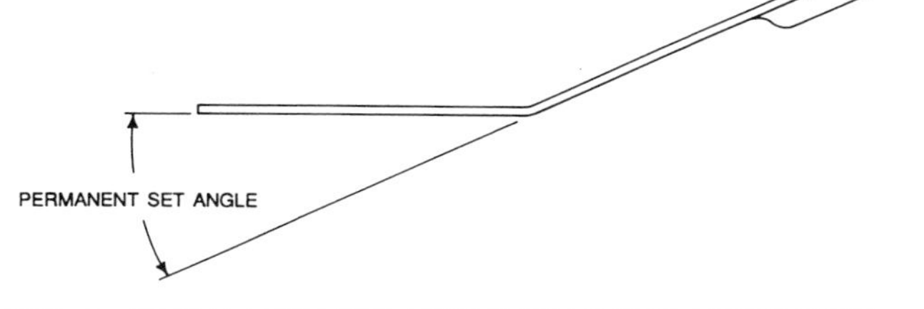

FIG. 1—PERMANENT SET ANGLE OF BLADE AFTER TEST

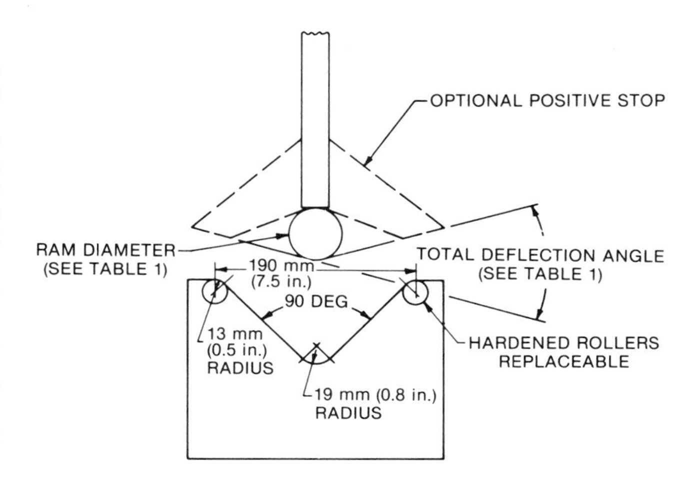

FIG. 2—BLADE BEND TEST FIXTURE

SECTION 4—BEND TEST REQUIREMENTS

4.1 Test fixture

4.1.1 All parts selected for bend testing shall be tested on the fixture specified in Fig. 2 (except as noted in paragraph 4.3) using the ram diameter selected from Table 1 based on the thickness of the blade in the area to be bent.

4.1.2 Stops may be used on the ram which will stop against the roller area of the bottom die, but they shall not make contact in the area of the bend in the blade. The stops shall be positioned to provide the minimum permanent set angle per Table 1. Suggested stop angles (total deflection angles) are specified in Table 1.

TABLE 1—BLADE BEND ANGLES

Blade thickness, mm	Blade thickness, in.	Ram diameter, mm	Ram diameter, in.	Minimum permanent set angle, deg	*Approximate total deflection angle, deg
Under 5.73	Under 0.225	38	1.5	25	36
5.74 8.52	0.226 0.335	50	2.0	25	36
8.53 10.68	0.336 0.420	70	2.75	25	36
10.69 14.49	0.421 0.570	89	3.5	25	36
over 14.50	over 0.571	108	4.25	15	23

*For reference only.

4.2 Blade bend test procedure. All parts shall be placed flat on the bend test die and bent at least enough to give the permanent set, total included angle over a time period of not to exceed 5 min and at a maximum temperature of 49 °C (120 °F). The blade area placed between the support points in the die should not be, if possible, in a fin area, near a mounting hole or in an area of other bends or distortions. When the beginning of a crack or break is indicated, stop the test and remove the blade.

4.2.1 If cracked, bend the blade back enough to close the crack on the tension side of the blade to measure the outside bend angle at the time of failure. (Permanent set, total included angle)

4.2.2 If the blade breaks completely, hold broken pieces together in a position such that material on the tension side of the blade comes together or closes the crack on that side at least 75% of the distance across the back of the blade. Then measure the outside angle of bend. It may be necessary to tap parts back together because of distortion of the mating surfaces.

4.3 Alternate procedure. The object of this test fixture and procedure is to produce 14.5% permanent elongation in the surface of the metal all the way across the blade. This is based on the assumption that the neutral axis of the blade is located in from the inside bend surface a distance equal to 40% of the thickness. Any other form of blade bend test fixture and procedure may be substituted as long as this objective is achieved.

4.4 Acceptance criteria. If any blade from a lot breaks or incurs a crack visible to the naked eye before reaching the permanent set angle specified in Table 1, that entire lot shall be totally rejected, and no blades from that lot may be used without corrective measures.

4.5 Corrective measures. If a corrective measure such as annealing or reheat treating is performed on the entire lot, the lot shall then be retested, but the sample size must be doubled. If all sample blades pass the retest, then the lot may be accepted.

Cited Standard:

SAE J990, Nomenclature-Industrial Mowers

ASAE Standard: ASAE S489

HYDRAULIC PRESSURE AVAILABLE ON AGRICULTURAL TRACTORS FOR REMOTE USE WITH IMPLEMENTS

Developed by the ASAE Tractor and Implement Hydraulics Committee; approved by the Power and Machinery Division Standards Committee; adopted by ASAE March 1988.

SECTION 1—PURPOSE AND SCOPE

1.1 This Standard has the following two purposes:

1.1.1 To establish the pressure available through the pair of hydraulic couplers frequently used to connect hydraulic remote cylinders and other hydraulic devices on implements to agricultural tractors.

1.1.2 To permit interchangeable use of various makes of implements using remote cylinders and other hydraulic devices on various makes of agricultural tractors when both are designed for this use.

1.2 This Standard is applicable only to implements and agricultural tractors intended for interchangeable use.

1.3 This Standard specifies only available hydraulic pressure. Coupler dimensions are specified in ASAE Standard S366, Dimensions for Cylindrical Hydraulic Couplers for Agricultural Tractors.

SECTION 2—DEFINITIONS

2.1 Coupler pair: A pair of female hydraulic couplers compatible with male couplers as specified in ASAE Standard S366, Dimensions for Cylindrical Hydraulic Couplers for Agricultural Tractors. Coupler pairs are mounted on agricultural tractors and are connected to the hydraulic system to allow flow from one coupler and to simultaneously allow flow into the other coupler.

2.2 Available differential pressure: The steady state difference in hydraulic pressure between two male couplers connected to a coupler pair.

2.3 Rated operating pressure: The maximum steady state hydraulic pressure at either male coupler connected to a coupler pair.

2.4 Relief valve pressure: The set point of a relief valve when one is used in a tractor hydraulic system to limit the pressure at the coupler pair.

2.5 Peak pressure: The peak hydraulic pressure at either male coupler connected to a coupler pair.

SECTION 3—SPECIFICATIONS

3.1 The available differential pressure at minimum available flow shall be at least 16 000 kPa (2 320 psi).

3.2 The rated operating pressure shall not be greater than 20 000 kPa (2 900 psi).

3.3 The maximum relief valve pressure shall not be greater than 25 000 kPa (3 625 psi).

3.4 The peak pressure shall not be greater than 29 000 kPa (4 205 psi).

SECTION 4—TEST CONDITIONS

4.1 The tractor shall be connected to a double acting cylinder without a cushion through a pair of 2 500 ± 100 mm (98.4 ± 3.9 in.) long sections of hydraulic hose having a nominal inside diameter of 10 mm (0.375 in.) and being of a suitable construction such as hose SAE 100R2 as specified in SAE Standard J517, Hydraulic Hose. The cylinder dimensions being 80 ± 5 mm (3.15 ± 0.20 in.) bore, 30 ± 5 mm (1.18 ± 0.20 in.) rod and a 200 ± 10 mm (7.87 ± 0.39 in.) stroke.

4.2 Fig. 1 shows the hydraulic test setup which includes the following:

4.2.1 A pressure transducer with a high frequency response sensing within 100 mm (3.94 in.) of the male coupler connecting the cap end of the hydraulic cylinder to the tractor.

4.2.2 A differential pressure transducer or other suitable pressure transducer sensing within 100 mm (3.94 in.) of each male coupler.

4.2.3 A flow metering valve, having less than 100 kPa (14.5 psi) drop at a flow of one L/s (15.85 gpm) when fully open, mounted in the flow path between the cap end of the cylinder and its associated male coupler.

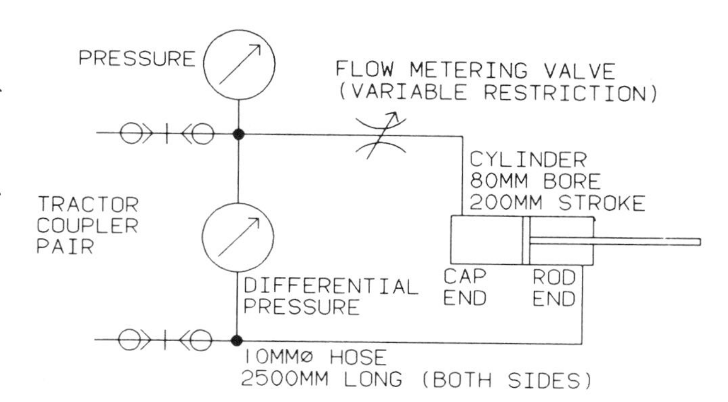

FIG. 1—HYDRAULIC TEST SETUP

4.3 The hydraulic fluid shall be that recommended by the tractor manufacturer.

4.4 The throttle or governor control lever shall be adjusted to maintain the rated engine speed.

4.5 The temperature in the hydraulic reservoir shall be 65 ± 10 °C (149 ± 18 °F).

4.6 All tractor-mounted flow controls shall be fully open.

4.7 After connecting the cylinder and attaining the other test conditions, the cylinder shall be fully extended and retracted at least 10 times to remove air and to stabilize conditions.

SECTION 5—TEST PROCEDURE

5.1 A test cycle shall consist of extending the cylinder its total stroke. The operator should use the valve detent when practical. When the tractor has no detent or the detent releases, the operator should hold the tractor valve fully open. The travel shall be stopped by the cylinder coming to the end of its stroke. The flow metering valve shall be adjusted to attain the desired cycle time. A test run shall be made at each of the following cycle times:

5.1.1 Full extension in 30 ± 5 s.

5.1.2 Full extension in the minimum possible time.

5.2 The results shall be recorded as follows:

5.2.1 The available differential pressure shall be the steady state average differential pressure observed during the test run specified in paragraph 5.1.1.

5.2.2 The rated operating pressure shall be the average steady state pressure observed during the test run specified in paragraph 5.1.1 near the coupler attached to the cap end of the cylinder.

5.2.3 The peak pressure shall be the maximum instantaneous pressure observed during test runs specified in paragraphs 5.1.1 or 5.1.2.

5.3 The following additional information shall be recorded:

5.3.1 Tractor make and model

5.3.2 Relief valve set point (if applicable)

5.3.3 Actual stroke times

5.3.4 Cylinder extend displacement

5.3.5 Reservoir temperature

Cited Standards:

ASAE S366, Dimensions for Cylindrical Hydraulic Couplers for Agricultural Tractors
SAE J517, Hydraulic Hose

ASAE Standard: ASAE S493

GUARDING FOR AGRICULTURAL EQUIPMENT

Proposed to ASAE by the Farm and Industrial Equipment Institute; developed and approved by the ASAE Agricultural Safety Committee and by the ASAE Power and Machinery Division Standards Committee; adopted by ASAE April 1988.

SECTION 1—PURPOSE AND SCOPE

1.1 This Standard provides guarding guidelines to minimize the potential for personal injury from hazards associated with agricultural equipment.

1.2 This Standard applies to agricultural equipment as identified in ASAE Standard S390, Classifications and Definitions of Agricultural Equipment.

SECTION 2—DEFINITIONS

2.1 Guard: A protective device designed and fitted to minimize the possibility of inadvertent contact with machinery hazards, as well as to restrict access to other hazardous areas. There are three types of guards, each consistent with the requirements of safety distance as defined in paragraph 2.4 below:

2.1.1 Shield: A guard that, alone or with other parts of the machine, provides protection from the side(s) covered.

2.1.2 Enclosure: A guard that, alone or with other parts of the machine, provides protection on all sides.

2.1.3 Barrier: A guard such as a rail, fence, frame, or the like.

2.2 Inadvertent contact: Unplanned contact between a person and a hazard, resulting from the person's actions during normal operation or servicing of equipment.

2.3 Hot surface: A surface which reaches operating temperatures in excess of 130 °C (266 °F) and which could involve injury by inadvertent contact.

2.4 Safety distance guarding: A means of providing guarding where the possibility of inadvertent contact with the hazard is minimized by the combination of the guard configuration (including openings) and the distance between the guard and the hazard. An additional aspect includes separation dimensions of pinch points in relation to body parts.

2.5 Machinery hazard: Machinery parts which can cause injury upon direct contact or by entanglement of personal apparel. This includes, but is not limited to, pinch points, nip points, and projections on rotating parts.

2.6 Guarding by location: A hazard is guarded by location when it is guarded by other parts or components of the machine that are not themselves guards, or when the hazard is beyond the safety distance.

2.7 Nip-point: A type of pinch point characteristic of components such as meshing gears and the run-on point where a belt, chain or cable contacts a sheave, sprocket or idler.

2.8 Ground-driven components: Components which are powered by the forward or rearward motion of equipment traveling over the ground.

SECTION 3—GUARDING REQUIREMENTS

3.1 Components which must be exposed for proper function, drainage or cleaning shall be guarded to the maximum extent that is practical and reasonable as permitted by the intended operation or use.

3.2 Where paragraph 3.1 does not apply and where hazard elimination through design is not both technically feasible and functionally practicable, machinery hazards shall be guarded by location, or with guard(s), or by safety distance guarding as described in Section 5—Safety Distance Guarding. Examples of such hazards are:

3.2.1 Moving traction elements in relation to the operator's station.

3.2.2 Revolving engine components.

3.2.3 Nip-points.

3.2.4 Outside faces of pulleys, sheaves, sprockets and gears.

3.2.5 Revolving shafts, universal joints, and other revolving parts with projections such as exposed bolts, keys, pins or set screws. Revolving shafts excluded are:

3.2.5.1 Smooth shafts revolving at less than 10 rpm.

3.2.5.2 Smooth shaft ends protruding less than one half the outside diameter of the shaft.

3.2.6 Surfaces which create shearing or pinching hazards.

3.2.7 Ground-driven components, if operating personnel are required to be in the area while the drives are in motion.

3.2.8 Hot surfaces.

3.3 Guarding, where required, shall minimize inadvertent contact with machinery hazards during normal mounting, starting, operating, dismounting, and servicing of the equipment.

3.4 Machines with access doors or guards which can be opened or removed to expose machine elements which continue to rotate or move after the power is disengaged shall have, in the immediate area, a readily visible evidence of rotation, or an audible indication of rotation, or a suitable safety sign.

3.5 A safety sign(s) per ASAE Standard S441, Safety Signs, and/or operating instructions shall be provided stating that guards must be kept in place, and/or that the machine should not be operated with guards removed.

SECTION 4—GUARD CONSTRUCTION

4.1 Guards shall have no sharp edges, shall be weather resistant, and shall retain required strength under expected climatic and operational conditions for their intended use.

4.2 Guards shall normally be permanently attached, which includes the use of threaded fasteners, split pins or other means that can be dismantled with common hand tools.

4.3 Guards and access doors which must be opened for routine or daily service, inspection or cleaning shall:

4.3.1 Be easy to open and close.

4.3.2 Remain attached; for example, by means of a hinge, slide, linkage, tether or other suitable means.

4.3.3 Include a convenient and effective means to keep them closed.

4.4 Guards shall remain functional under the forces that could be applied by a 123 kg (270 lb) person leaning on or falling against them in normal operation or servicing of equipment.

4.5 Those guards designed to be used as steps, or where that use can be anticipated, shall remain functional if used as steps by a 123 kg (270 lb) person in normal operation or servicing of equipment.

SECTION 5—SAFETY DISTANCE GUARDING

5.1 Where guarding is required (see Section 3—Guarding Requirements), openings in welded or rigid mesh or grille shall not exceed and distances between guards and hazards shall not be less than the dimensions shown in the following paragraphs of this section unless adherence to these dimensions would interfere with the intended operation or use.

5.2 It is possible to circumvent the protection provided by a safety distance as specified in paragraph 5.3 by the misuse of steps, ladders, boxes, chairs, etc., but the general principle of a safety distance is acceptable provided the following criteria are met so that the hazards are out of reach.

TABLE 1—DOWNWARD AND SIDEWARD SAFETY DISTANCE

a, in mm	b*, in mm							
	2 400	2 200	2 000	1 800	1 600	1 400	1 200	1 000
	c, in mm (min.)							
2 400	—	100	100	100	100	100	100	100
2 200	—	250	350	400	500	500	600	600
2 000	—	—	350	500	600	700	900	1 100
1 800	—	—	—	600	900	900	1 000	1 100
1 600	—	—	—	500	900	900	1 000	1 300
1 400	—	—	—	100	800	900	1 000	1 300
1 200	—	—	—	—	500	900	1 000	1 400
1 000	—	—	—	—	300	900	1 000	1 400
800	—	—	—	—	—	600	900	1 300
600	—	—	—	—	—	—	500	1 200
400	—	—	—	—	—	—	300	1 200
200	—	—	—	—	—	—	200	1 100

*Values of b < 1 000 mm do not increase the reach. Moreover, the danger arises of falling towards the hazard.

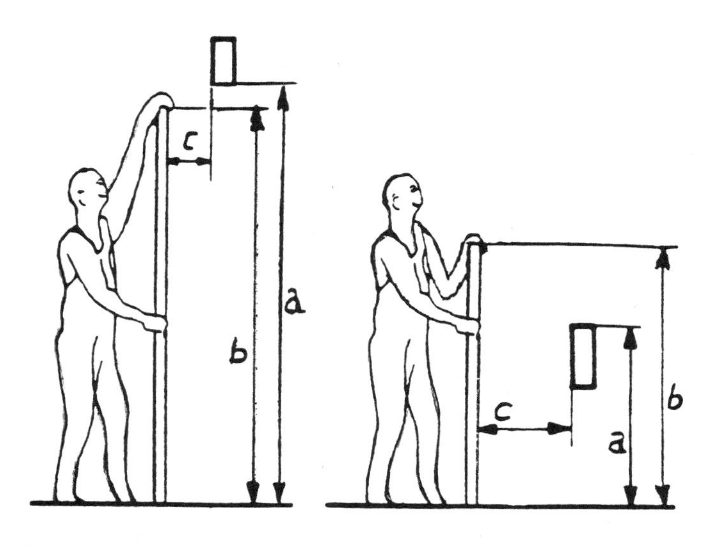

FIG. 1—PRINCIPLES FOR DETERMINING THE DISTANCE REQUIRED FROM A BARRIER TO THE HAZARD

TABLE 2—EXTENT OF REACH

Limb		Illustration	Safety distance, r, in mm
From	To		
Finger base	Finger tip		r > 120
Wrist	Finger tip		r > 230
Elbow	Finger tip		r > 550
Shoulder	Finger tip		r > 850

TABLE 3—REACH DIMENSIONS THROUGH RECTANGLE OR SLOT

Limb	Illustration	Width of aperture (rectangle or slot), a, in mm	Safety distance to hazard, b, in mm
Finger tip		$4 < a \le 8$	$b \ge 15$
Finger	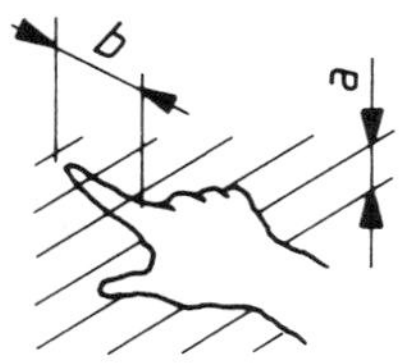	$8 < a \le 20$	$b \ge 120$
Hand	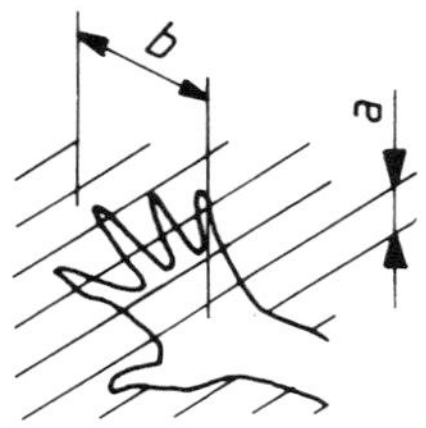	$20 < a \le 30$	$b \ge 200$
Arm	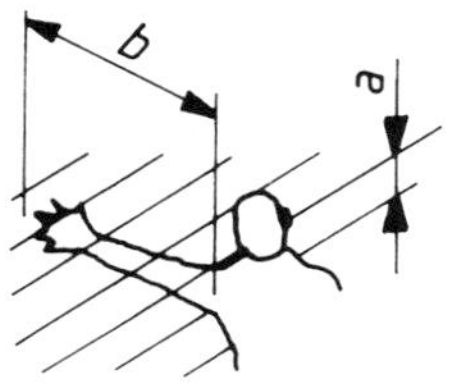	$30 < a \le 135$*	$b \ge 850$

*When the width is greater than 135 mm, part of the body can also pass through the aperture. In this case, safety distances as specified in 5.4 shall be observed.

TABLE 4—REACH DIMENSIONS THROUGH MESH OR GRILLE

Limb	Illustration	Width of aperture (diameter or lateral length), a, in mm	Safety distance to hazard, b, in mm
Finger tip		$4 < a \le 8$	$b \ge 15$
Finger	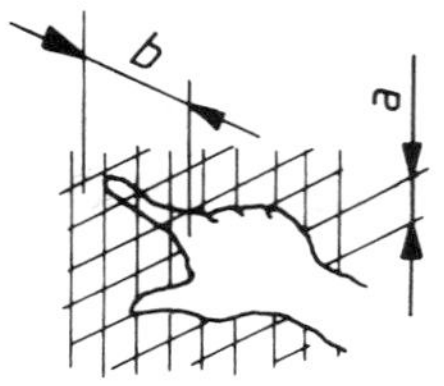	$8 < a \le 25$	$b \ge 120$
Hand	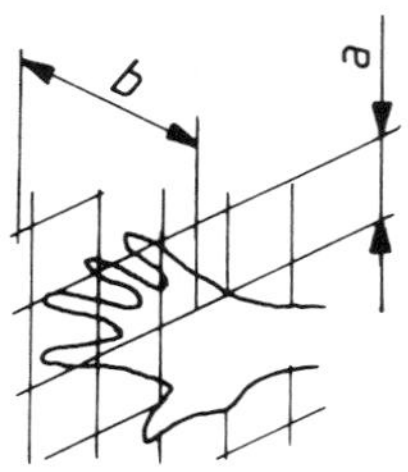	$25 < a \le 40$	$b \ge 200$
Arm	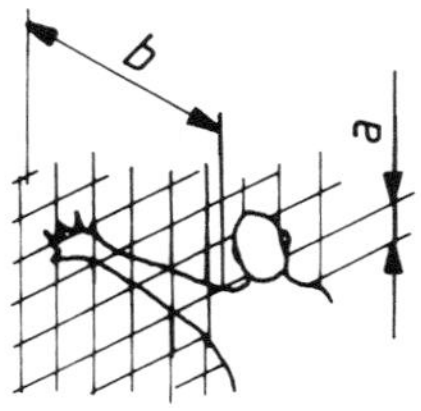	$40 < a \le 250$	$b \ge 850$

TABLE 5—MINIMUM SEPARATION DISTANCES FOR PINCHING POINTS

Limb	Illustration	Minimum separation distance required, mm
Finger		25
Hand Wrist Fist		100
Arm		120
Foot		120
Leg		180
Body		500

5.3 Safety distance from a hazard to a guard is based on measurements from the location which a person can occupy with reference to the hazard.

5.3.1 Safety distance for upward reach is 2 500 mm for persons standing upright.

5.3.2 No safety distance is specified where it is possible to reach below a safety barrier, unless the aperture is small enough to be considered only in relation to finger, hand or arm access in which case the requirements of paragraph 5.3.5 apply.

5.3.3 Reach-over barriers

5.3.3.1 Barriers shall be 1 000 mm in height, minimum.

5.3.3.2 The safety distance for sideward or downward reach-over barriers of 1 000 mm or greater height depends on:

- the distance from the ground level to the hazard
- the height of the barrier
- the horizontal distance between the hazard and the barrier.

The dimensions in Table 1 shall be met (see Fig. 1). The dimension c is a minimum.

5.3.4 **Round reach.** Table 2 shows the extent of reach around guards which can be attained, taking into account the aperture and the distance from other obstructions. Potential hazards shall be beyond these limits if they are not independently guarded.

5.3.5 **Inside reach through guards.** The safety distances depend on the shape of the openings. The openings shall not exceed the size appropriate to the distance of the guard from the potential hazard (see Tables 3 and 4).

5.3.5.1 **Polygonal openings.** Polygonal openings, where the diameter of the largest circle that can be inscribed is not less than half the distance between the two apexes that are the furthest apart, shall meet the same requirements as for round openings. The diameter of the inscribed circle shall be regarded as the size of the opening. All other polygonal openings shall be regarded as slots.

5.4 **Pinching points.** A pinching point is considered a potential hazard or the parts of the body illustrated in Table 5 if the appropriate minimum separation distance is not maintained. The design of the machine shall ensure that the next bigger part of the body cannot pass through.

Cited Standards:

ASAE S390, Classifications and Definitions of Agricultural Equipment
ASAE S441, Safety Signs

POWERED LAWN AND GARDEN EQUIPMENT

ASAE Notation:

The letter S preceding numerical designation indicates ASAE Standard; EP indicates Engineering Practice; D indicates Data. A decimal and numeral following the file number indicate the number of times a document has been revised. Thus ASAE S201.4 indicates Standard number 201, four times revised. The letter T after the designation indicates tentative status. Always refer to ASAE documents by complete designation to avoid confusion with standards of other organizations. For example: ASAE S201.4.

The symbol **T** preceding or in the margin adjacent to section headings, paragraph numbers, figure captions, or table headings indicates a technical change was incorporated in that area when this document was last revised. The symbol **T** preceding the title of a document indicates essentially the entire document has been revised. The symbol **E** used similarly indicates editorial changes or corrections have been made with no intended change in the technical meaning of the document.

DRAWBAR FOR LAWN AND GARDEN RIDE-ON TRACTORS

Developed by the ASAE Small Tractor and Power Equipment Committee; approved by ASAE Power and Machinery Division Technical Committee; adopted by ASAE as a Tentative Standard June 1966; reclassified as a full Standard December 1968; reconfirmed December 1973; revised editorially December 1975; reconfirmed December 1978; revised February 1984.

SECTION 1—PURPOSE AND SCOPE

1.1 The purpose of this Standard is to establish dimensions which permit all makes of towed equipment designed for operation in conjunction with lawn and garden tractors to be operated with all makes of lawn and garden tractors which are designed to pull such equipment.

Cited Standard:

ASAE S370, 2000-RPM Power Take-Off for Lawn and Garden Ride-On Tractors

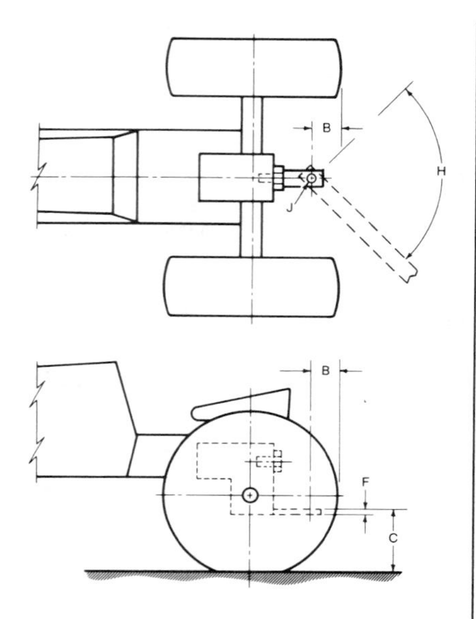

FIG. 1—CONFIGURATION FOR NON—PTO DRIVEN DRAFT LOADS

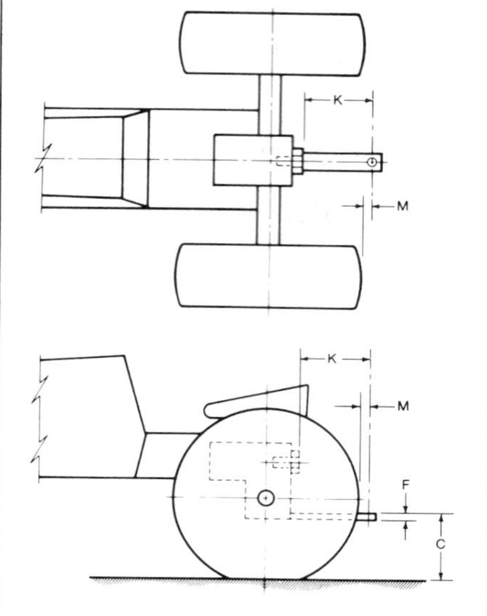

FIG. 2—EXTENDED DRAWBAR CONFIGURATION FOR PTO DRIVEN TRAILING ATTACHMENTS

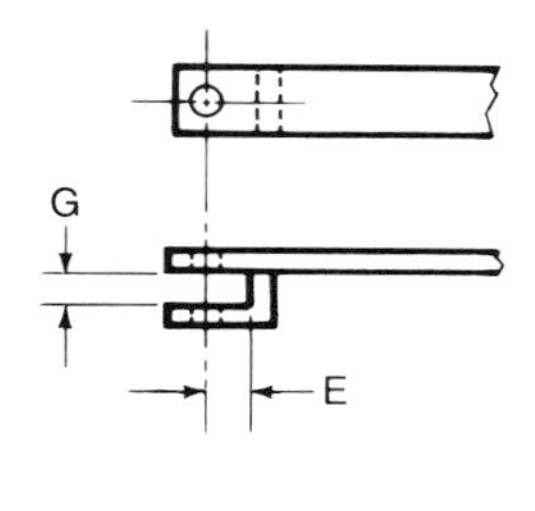

FIG. 3—IMPLEMENT DRAWBAR DETAIL

TABLE 1—DIMENSIONS ASSOCIATED WITH TRACTOR AND IMPLEMENTS

	Dimensions designated in detail	Millimeters		Inches	
		Min	Max	Min	Max
(B)	Drawbar hitch hole location along tractor centerline (distance ahead of largest outside dia rear tire)	—	102	—	4
(C)	Drawbar height (distance from ground to top of drawbar with largest rear tire)	178	229	7	9
(E)	Clearance, relative to implement drawbar	38	—	1.5	—
(F)	Drawbar thickness	6.4	19.1	0.25	0.75
(G)	Implement drawbar clevis opening	38.1	—	1.50	—
(H)	Drawbar swinging clearance	Tongue end of the tractor drawbar shall clear an implement clevis (50.8 mm) 2 in. wide, through a max 90 deg swing, either right or left of the tractor drawbar centerline or implement tongue interference with the tractor tire, whichever occurs first.			
(J)	Drawbar hitch hole dia	13.5	22.2	0.53	0.88
(K)	PTO location, horizontal distance	(See ASAE Standard S370, 2000-RPM Power Take-Off for Lawn and Garden Ride-On Tractors.)			
(M)	Hitch hole location along PTO centerline (PTO) applications)	(See ASAE Standard S370, 2000-RPM Power Take-Off for Lawn and Garden Ride-On Tractors.)			

ASAE Standard: ASAE S320.1 (*ANSI/ASAE S320)

CATEGORY "O" THREE-POINT FREE-LINK ATTACHMENT FOR HITCHING IMPLEMENTS TO LAWN AND GARDEN RIDE-ON TRACTORS

*Corresponds in substance to previous revision S320.

Developed by the ASAE Small Tractor and Power Equipment Committee; approved by the Power and Machinery Division Technical Committee; adopted by ASAE as a Tentative Standard December 1968; reclassified as a full Standard February 1971; reconfirmed December 1975; approved by ANSI as an American National Standard October 1976; reconfirmed December 1980; revised February 1984; previous revision S320 reaffirmed and redesignated by ANSI June 1984.

SECTION 1—SCOPE

1.1 This specification sets forth requirements for the attachment of three-point hitch implements or equipment to the rear of lawn and garden ride-on tractors by means of a three-point free-link in association with a power lift.

1.2 Lawn and garden ride-on tractors to which this Standard applies are defined in ASAE Standard S323, Definitions of Powered Lawn and Garden Equipment.

1.3 In order to assure proper performance of certain implements, standard dimensions for mast height, mast pitch adjustment and implement leveling adjustment are included. Location of link-attachment points is not restricted and is, therefore, left to the discretion of the tractor designer.

1.4 If draft links are used for trailing power take-off implements, a means shall be included for locking the draft links in a fixed position, and a drawbar hitch point shall be positioned in conformance with power take-off standards.

SECTION 2—DEFINITION OF TERMS (see Figs. 1-4)

2.1 Linkage: The combination of 1 upper link and 2 lower links, each articulated to the tractor and the implement at opposite ends in order to connect the implement to the tractor.

2.2 Upper link, lower link: Elements in the linkage.

2.3 Hitch point: The articulated connection between a link and the implement. For geometrical analysis, the hitch point is established as the center of the articulated connection between a link and the implement.

2.4 Link point: The articulated connection between a link and the tractor. For geometrical analysis, the link point is established as the center of the articulated connection between a link and the tractor.

2.5 Upper hitch point: The articulated connection between the upper link and the implement.

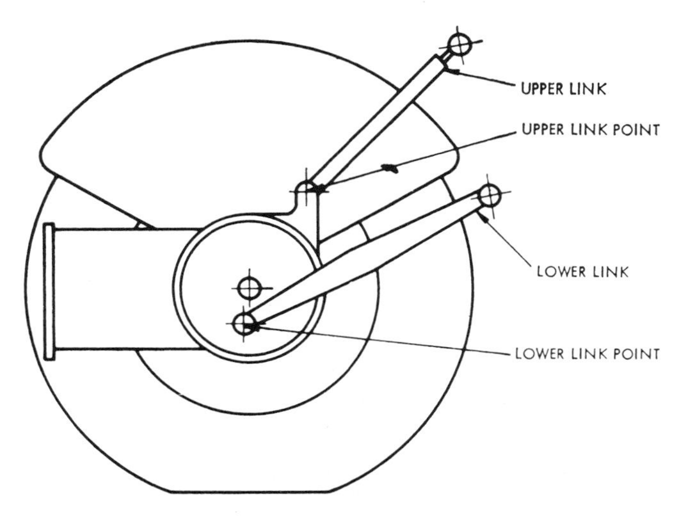

FIG. 1—TRACTOR LINKAGE

2.6 Upper link point: The articulated connection between the upper link and the tractor.

2.7 Lower hitch point: The articulated connection between a lower link and the implement.

2.8 Lower link point: The articulated connection between a lower link and the tractor.

2.9 Upper hitch pin: The pin that connects the upper link to the implement.

2.10 Upper link pin: The pin that connects the upper link to the tractor.

2.11 Lower hitch stud or pin: The stud or pin, attached to the implement, on which a lower link is secured.

2.12 Linchpin: The retaining pin used in the hitch pins or studs.

2.13 Mast: The member that provides attachment of the upper link to the implement.

2.14 Mast height: The perpendicular distance between the upper hitch point and common axis of the lower hitch points.

2.15 Mast adjustment: The usable range of movement of the mast in a vertical plane. It is measured as the maximum and minimum heights of the lower hitch points above the ground between which a mast of standard height can be adjusted to any inclination between vertical and 5 deg from vertical towards the rear. Adjustment of the mast controls the pitch of the implement. Specifying the mast adjustment to be provided enables the tractor designer to determine the minimum acceptable adjustment of the length of the top link in relation to the point of attachment of the linkage. It also permits the implement designer to determine the range of operating depths of the implement over which pitch adjustment can be obtained.

2.16 Leveling adjustment: The adjustment of the lower links so that the one lower hitch point may be moved vertically with respect to the other lower hitch point to provide an inclination of the implement.

2.17 Lower hitch-point spread: The distance between lower hitch points measured at the base of the lower hitch stud, or the distance between the innermost restraining means provided on the implement.

2.18 Linchpin hole distance: The distance between the linchpin hole centerline and the lower link stud base.

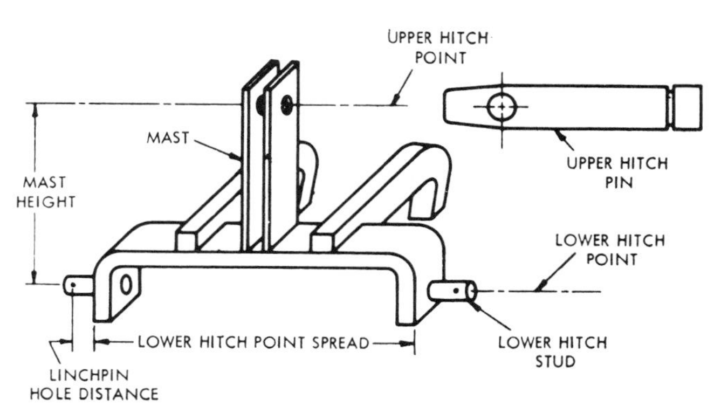

FIG. 2—DIMENSIONS ASSOCIATED WITH IMPLEMENT

2.19 Lift linkage: The connecting linkage that transmits force to the lower links for raising and lowering.

2.20 Lift range: The range of movement of the lower hitch points utilizing the extent of manual adjustment provided in the lift linkage in conjunction with the power range, expressed as the maximum and minimum possible heights of the lower hitch points above ground level, the lower hitch point axis being maintained horizontal to the ground.

2.21 Power range: The total vertical movement of the lower hitch point excluding any adjustment in the linkage or lift linkage.

2.22 Lower hitch-point tire clearance: Clearance expressed as a radial dimension from the lower hitch point to the outside diameter of the tire with the implement in raised position and all side sway removed from the links.

2.23 Lower hitch-point tractor clearance: The horizontal dimension between the rearmost parts of the tractor in the area between the two draft links and the horizontal line through the two lower hitch points throughout the range of vertical movement of the hitch points (see Fig. 4).

Cited Standard:

ASAE S323, Definitions of Powered Lawn and Garden Equipment

ASAE S370, 2000-RPM Power Take-Off for Lawn and Garden Ride-On Tractors

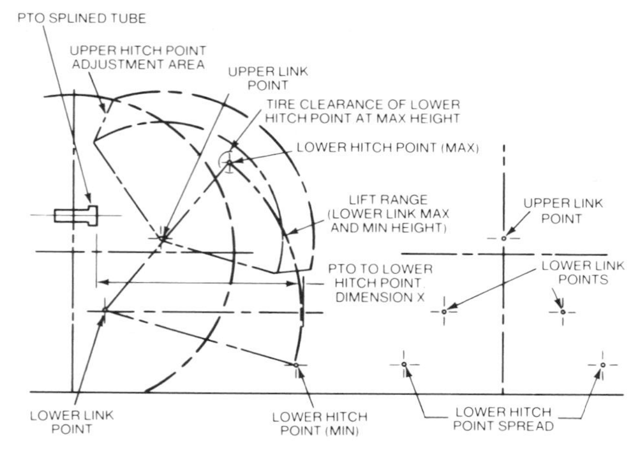

FIG. 3—DIMENSIONS ASSOCIATED WITH TRACTOR

TABLE 1—DIMENSIONS ASSOCIATED WITH IMPLEMENT

	Millimeters		Inches	
	Min	Max	Min	Max
Upper Hitch Point				
Width inside	28.45	—	1.12	—
Width outside	—	49.28	—	1.94
Clearance radius for upper link	38.10	—	1.50	—
Hitch pin hole diameter	16.26	16.51	0.64	0.65
Lower Hitch Point				
Stud diameter	15.75	16.00	0.62	0.63
Linchpin hole distance	36.58	—	1.44	—
Linchpin hole diameter	7.11	7.37	0.28	0.29
Lower hitch point spread	501.65	508.00	19.75	20.00
Clearance radius for lower link	50.80	—	2.00	—
Implement encroachment in front of lower hitch point if implement extends laterally behind tire	—	12.70	—	0.50
Implement Mast Height*	—	304.80	—	12

*The mast height is not necessarily a mechanical dimension on the implement itself. It is a figure used in design and if properly used for design of both implement and tractor, a well-performing interchangeable implement and tractor combination will be achieved. This standard makes it possible to produce tractors and implements that will give good performance in any combination; therefore, consideration to hitch geometry is essential. This makes it desirable to establish a standard mast height and a standard mast adjustment within a working range, because these items influence the position of hitch points that are common to both the implement and the tractor.

Mast height is one of the essential factors in establishing the virtual hitch point of the free-link system, draft signal for the draft-responsive system, loads on the linkage and hitch points, changes in implement pitch corresponding to changes in working depth, implement pitch when the implement is in transport position, clearance of the implement with the tractor, especially in transport position, and clearance of the hitch links with the implement or with the tractor, especially in the transport position.

When an implement mast height is made different than standard to accomplish some specific performance feature, care should be exercised to insure that the desired performance is secured with tractors likely to operate the implement.

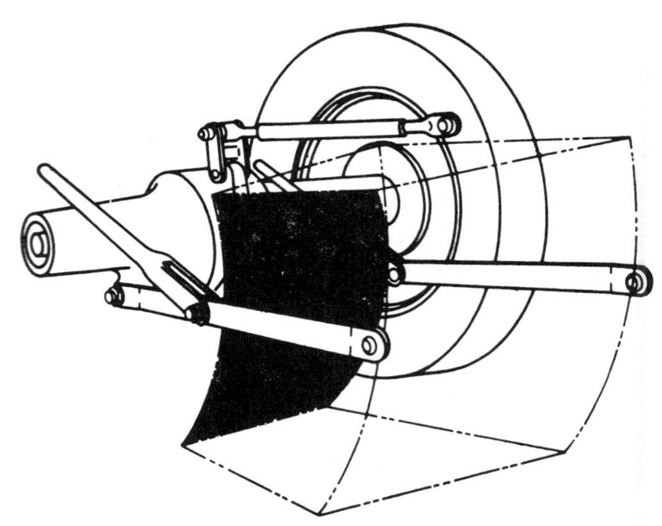

FIG. 4—LOWER HITCH-POINT TRACTOR CLEARANCE

TABLE 2—DIMENSIONS ASSOCIATED WITH TRACTOR

	Millimeters		Inches	
	Min	Max	Min	Max
Upper Link				
Width at hitch point	—	26.92	—	1.06
Radius at hitch point	—	25.40	—	1.00
Hitch pin hole diameter	16.26	16.51	0.64	0.65
Side sway at hitch point*				
Upper Hitch Pin				
Diameter	15.75	16.00	0.62	0.63
Distance from head to centerline of linchpin hole	53.85	—	2.12	—
Linchpin hole diameter	7.11	7.37	0.28	0.29
Lower Link				
Width at hitch point	—	26.92	—	1.06
Radius at hitch point	—	38.10	—	1.50
Stud hole diameter	16.26	16.51	0.64	0.65
Lower hitch point tire clearance with largest rear tire offered	76.20	—	3	—
Lower hitch point tractor clearance	203.20	—	8	—
Side sway at hitch point each side of center position w/draft links horizontal†	50.80	—	2	—
Horizontal distance from PTO splined tube to lower hitch point with lower link horizontal, dimension X, Fig. 3 (See ASAE S370)				
Lift Range, Power Range, Adjustments and Tractor Lift Force Capacity				
Lift Range				
Max height for lowest position	—	177.80	—	7
Min height for highest position	508.00	—	20	—
Power range	254.00	—	10	—
Leveling adjustment				
Higher	50.80	—	2	—
Lower	50.80	—	2	—
Mast adjustment				
Min height for highest position	381.00	—	15	—
Max height for lowest position	—	177.80	—	7
Tractor lift force capacity: A minimum lift force of 2.00 kN (450 lbf) shall be available at a distance of 305 mm (12 in.) to the rear of the lower hitch point and throughout the power range using 80% of the minimum hydraulic relief valve pressure setting.				

*Side sway of the upper link must be compatible with that provided at the lower links plus necessary additional allowance for lateral leveling adjustment.

†Means should be provided to lock the draft links in a rigid lateral position for operations where side sway cannot be tolerated and when the hitch is raised to the transport position. No maximum dimensions for side sway are specified; this must be limited in each individual application so that hitch or implement components will not come in contact with the tractor tires.

ASAE Standard: ASAE S323.2

DEFINITIONS OF POWERED LAWN AND GARDEN EQUIPMENT

Developed by the Definitions Subcommittee of the ASAE Small Tractor and Power Equipment Committee; approved by the Power and Machinery Division Standards Committee; adopted by ASAE as a Tentative Standard June 1969; reclassified as a full Standard February 1971; revised March 1974; reconfirmed December 1978; revised June 1983.

1.1 The purpose of this Standard is to classify and define various types of machines and terms so that these definitions may be used in future ASAE Standards and to aid in clear-cut communication.

SECTION 2—DEFINITION OF TERMS

2.1 Lawn and garden ride-on tractor: A self-propelled machine, designed and advertised for general purpose lawn and garden work, having the following characteristics:

2.1.1 Designed to supply power for home lawn, home garden and yard maintenance implements.

2.1.2 Generally designed for mowing lawns.

2.1.3 Must have all implements separate from the tractor.

2.1.4 Provides means to lift an implement such as a moldboard plow, tiller, cultivator, snow thrower, sweeper, dozer blade, etc.

2.2 Lawn ride-on tractor: A self-propelled machine, designed and advertised for general purpose lawn work, having the following characteristics:

2.2.1 Designed to supply power for home lawn and yard maintenance implements.

2.2.2 Generally designed for mowing lawns.

2.2.3 May have implements separate from the tractor.

2.2.4 May provide for means to lift an implement such as a sweeper or snow thrower.

2.3 Ride-on lawn mower: A self-propelled machine, designed and advertised for mowing lawns, having the following characteristics:

2.3.1 Has a seat upon which the operator rides and guides the mower.

2.3.2 Has cutting assemblies which are an integral part of the unit.

2.4 Lawn and garden walk-behind tractor: Generally a self-propelled, single-axle, wheel-driven machine designed and advertised for general purpose lawn and garden work having the following characteristics:

2.4.1 Controlled from an operator position located at the rear of the machine by an operator who is on foot or on a trailing sulky.

2.4.2 Must have all implements separate from the tractor.

2.5 Mowing height: The minimum static-condition distance between the cutting edge of the blade and a plane-smooth floor while the mowing unit is resting upon that floor. Tire pressure shall be adjusted to the manufacturer's recommendation. For ride-on machines a 91 kg (200 lb) operator (or equivalent) shall be seated upon the seat. The cutting height shall be measured to the nearest 5 mm (0.25 in.).

2.6 Working width: The actual measured width of the working element of the machine, measured at a right angle to the direction of travel. If the machine housing determines the working width, then the measured width of that housing shall be the working element.

2.7 Garden ride-on tractor: A self-propelled machine, designed and advertised for general purpose garden work, having the following characteristics:

2.7.1 Designed to supply power for home garden implements.

2.7.2 Generally designed for gardening work.

2.7.3 Must have all implements separate from the tractor.

2.7.4 Provides means to lift an implement such as a moldboard plow, tiller, cultivator, snow thrower, sweeper, dozer blade, etc.

2.7.5 May provide for a mower attachment for mowing lawns.

2.8 Grounds maintenance tractor: A self-propelled machine, designed and advertised for general purpose grounds maintenance work, having the following characteristics:

2.8.1 Designed to supply power for grounds maintenance implements.

2.8.2 Generally designed for mowing lawns.

2.8.3 Must have all implements separate from the tractor.

2.8.4 Provides means to lift an implement such as a moldboard plow, tiller, snow thrower, sweeper, dozer blade, etc.

2.9 Home garden: An area prepared, planted, cultivated and harvested for the personal consumption and enjoyment of a family unit (a non-commercial enterprise).

2.10 Grounds maintenance: The mowing, grooming and general care of non-home lawns and grounds, such as industrial parks, schools, cemeteries and golf courses.

2.11 Powered unit: A machine having an electric motor or internal combustion engine as one of its components, for the purpose of providing power to the machine and, when required, to attachments for such machine.

2.12 Power take-off (PTO): A power outlet such as a shaft, tube, pulley, etc., to provide rotational power to an implement or attachment.

2.13 Attachment: Any machine or implement which is connected to a powered unit for the purpose of accomplishing work such as mowing, plowing, etc.

2.14 Accessory: A part or mechanism that may be added to a functional powered unit, not utilizing its mechanical power, for the purpose of enhancing the performance of that unit.

2.15 Attachment drive: The pulleys, belts, chains, shafts, universal joints, connectors and fasteners provided with the attachment to transmit rotational power from the PTO of the powered unit to the first driven component on the attachment, such as a gear set, pulley, sprocket, or flywheel.

2.16 Ground driven components: Components powered by the turning motion of a wheel as the equipment travels over the ground.

2.17 Control: A device which will control the operation of a powered unit, attachment, accessory, or any specific operating function thereof.

2.18 Rotary mower: A power mower in which one or more elements rotate about a vertical axis and cut by impact.

2.19 Snow thrower: A powered machine or attachment designed for throwing snow.

2.20 Service brake: The primary braking system used for decelerating and stopping wheel motion of the machine.

2.21 Parking brake: The system used to hold one or more brakes or braking means continuously in the applied position.

2.22 Functional component: A working mechanism of an attachment or implement designed to perform a specific task, such as the rotary blades of rotary mowers or the reel and shear bar or reel mowers.

ASAE Standard: ASAE S348.1

ONE-POINT TUBULAR SLEEVE ATTACHMENT FOR HITCHING IMPLEMENTS TO LAWN AND GARDEN RIDE-ON TRACTORS

Developed by Implement Attachment Subcommittee of ASAE Small Tractor and Power Equipment Committee; approved by Power and Machinery Division Standards Committee; adopted by ASAE as a Tentative Standard February 1972; reconfirmed December 1972; reclassified as a full Standard December 1973; reconfirmed December 1978; revised February 1984.

SECTION 1—SCOPE

1.1 This Standard sets forth requirements for the attachment of one-point hitch implements or equipment to the rear of lawn and garden ride-on tractors by a one-point (single pin connection) hitch in association with a manual or power lift system. (See Fig. 1).

1.2 Lawn and garden ride-on tractors to which this Standard applies are defined in ASAE Standard S323, Definitions of Powered Lawn and Garden Equipment.

1.3 To assure performance and attachment of certain implements, standard dimensions for hitch point location, hitch tube and implement yoke are included.

SECTION 2—DEFINITION OF TERMS (See Figs. 1 and 2)

2.1 Hitch point: The pivotal point of connection of hitch to tractor. Also, actual hitch point of implement to tractor when implement is attached.

2.2 Implement attachment point: The connection between the hitch and implement.

2.3 Hitch pin: The pin that connects implement to hitch.

2.4 Hitch tube: The tube portion on the hitch which receives the implement yoke.

2.5 Hitch bail: A portion of the hitch assembly containing hitch tube, stabilizer bolts, and holes for attachment to tractor frame at hitch point.

2.6 Implement yoke: The clevis shaped part attached to implement, constructed to fit loosely to hitch tube and secured by hitch pin.

2.7 Stabilizer bolts: The bolts that are used for adjusting clearance between hitch bail and implement yoke.

2.8 Lift range: The range of vertical adjustment plus the power range. Adjustment may be provided in the lift linkage, hitch point or both.

2.9 Power range: The total vertical movement of the hitch measured at the hitch tube and excluding any vertical adjustment in the hitch lift linkage.

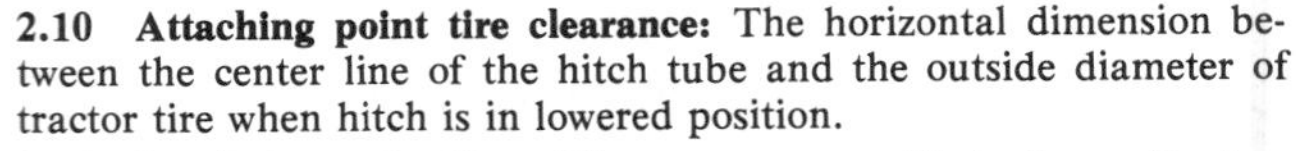

2.10 Attaching point tire clearance: The horizontal dimension between the center line of the hitch tube and the outside diameter of tractor tire when hitch is in lowered position.

2.11 Implement leveling: The means provided for adjusting implement for level operation in a working position, generally accomplished by adjustments built into implement where required. Implement leveling may also be provided within the adjustments of the lift linkage and hitch point.

Cited Standard:

ASAE S323, Definitions of Powered Lawn and Garden Equipment

TABLE 1—DIMENSIONS ASSOCIATED WITH IMPLEMENT

Implement Yoke Dimensions (See Fig. 2)	Millimeters		Inches	
	Min	Max	Min	Max
Q Center of hole to inside edge	31.8	35.1	1.25	1.38
R Hole dia. - 2 holes in line	16.5	17.3	0.65	0.68
S Vertical depth inside	88.9	90.4	3.50	3.56
T Width	101.6*	—	4.00*	—
U Thickness	9.52†	12.7	0.375†	0.50

*Min. dimension applies only to implements requiring side stabilization.
†Recommended thickness.

NOTE: Implement leveling—Means should be provided in implement design to allow leveling of implement when at its normal working depth. Implement weight—When implement weight adversely affects lengthwise stability of tractor, front end ballast should be made available and use recommended.

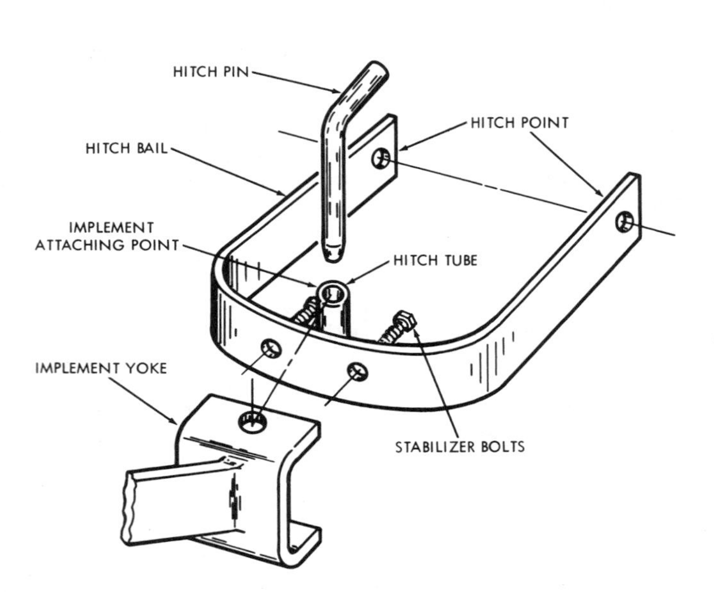

FIG. 1—TYPICAL ONE-POINT TUBULAR SLEEVE HITCH

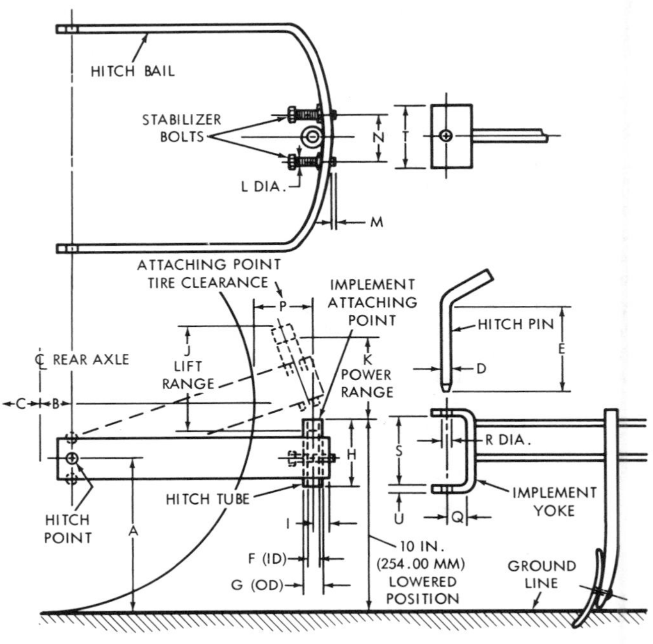

FIG. 2—HITCH AND YOKE DIMENSIONS, See Tables 1 and 2

TABLE 2—DIMENSIONS ASSOCIATED WITH TRACTOR HITCH

	Dimensions (See Fig. 2)	Millimeters		Inches	
		Min	Max	Min	Max
HITCH POINT:					
Vertical height from ground, 203 mm (8 in.) recommended	A	177.8	228.6	7.00	9.00
Horizontal distance from centerline rear axle — Rearward	B	—	254.0	—	10.00
Forward	C	—	76.2	—	3.00
HITCH PIN:					
Diameter	D	15.62	15.88	0.615	0.625
Straight portion length	E	127.0	—	5.00	—
HITCH TUBE:					
Inside diameter	F	16.5	17.3	0.65	0.68
Outside diameter	G	25.4	—	1.00	—
Length	H	84.12	85.72	3.312	3.375
CENTER OF HITCH TUBE TO REAR OF BAIL	I	—	25.4	—	1.00
LIFT RANGE:					
Recommended — Only including power range	J	139.7	—	5.50	—
POWER RANGE:					
Vertical travel from top of hitch tube when top of hitch tube is 254.0 mm (10.00 in.) from ground in lowered position	K	88.9	—	3.50	—
STABILIZER BOLTS:*					
Diameter	L	15.8	—	0.62	—
Adjustment	M	12.7	—	0.50	—
Spacing	N	63.5	69.9	2.50	2.75
ATTACHING POINT TIRE CLEARANCE:	P	76.2	—	3.00	—
TRACTOR LIFT FORCE CAPACITY:	A minimum lift force of 1.78 kN (400 lb) shall be available at the implement attaching point.				

*Locking means should be provided such as jam nut, etc.

2000-RPM POWER TAKE-OFF FOR LAWN AND GARDEN RIDE-ON TRACTORS

Developed by the ASAE Small Tractor and Power Equipment Committee; approved by the Power and Machinery Division Standards Committee; adopted by ASAE December 1974; reconfirmed December 1975, December 1976, December 1977; revised December 1978; reconfirmed December 1979, December 1980, December 1981, December 1982; revised and reclassified as a full Standard February 1984.

SECTION 1—PURPOSE AND SCOPE

1.1 This Standard establishes the specifications that are essential in order that a 2000-rpm power take-off-driven machine may be operated with any make of lawn and garden ride-on tractor equipped with an equivalent size 2000-rpm power take-off drive.

1.2 This Standard does not in inself insure adequate telescoping of the power line or safety shielding.

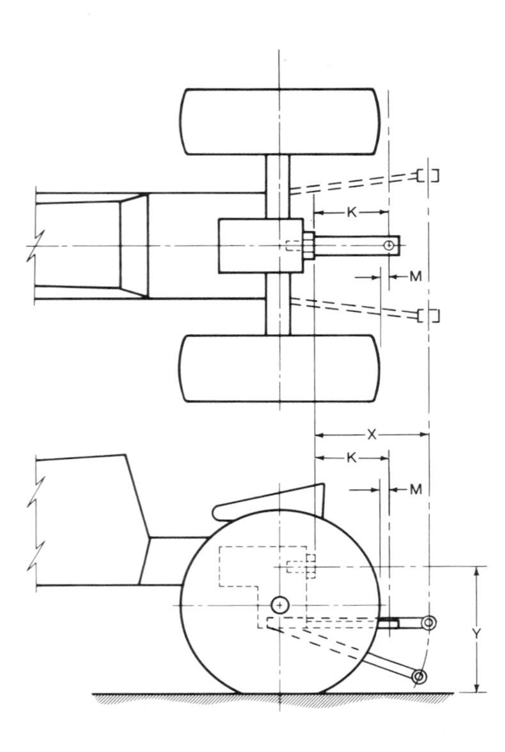

FIG. 1—RELATIONSHIP OF TRACTOR DRAWBAR, PTO AND CATEGORY "O" LIFT LINKS

TABLE 1—DIMENSIONAL RELATIONSHIPS OF TRACTOR DRAWBAR, PTO, AND CATEGORY "O" LIFT LINKS

Dimensions shown in Fig. 1		Millimeters		Inches	
		Min	Max	Min	Max
(K)	PTO location, horizontal distance	267	292	10.5	11.5
(M)	Hitch hole location along PTO centerline, PTO applications (with largest outer dia. tire specified by manufacturer)	102	—	4.0	—
(X)	Horizontal distance from PTO splined tube to lower hitch point (with lower link horizontal)	254	356	10.0	14.0
(Y)	Vertical position of PTO splined tube	356	483	14.0	19.0

SECTION 2—SPECIFICATIONS

2.1 The hitch hole shall be directly in line with the top view centerline of the power take-off tube (see Fig. 1), and provisions shall be made on the tractor for locking the drawbar in this position.

2.2 The dimensional relationship between the power take-off splined tube and a category "O" three point hitch, lower hitch point shall be as given in Fig. 1 and Table 1.

2.3 The location of the tractor power take-off tube shall be within the limits of 25 mm (1.0 in.) to the right or left of the centerline of the tractor, tractor centerline being the recommended location.

TABLE 2—POWER TAKE-OFF TUBE AND SHAFT SPLINE DIMENSIONS

Flat root side fit*	Tractor tube (Internal)	Attachment shaft (External)
Number of teeth	15	15
Pitch	16/32	16/32
Pressure angle	30 deg	30 deg
Base dia.	20.622230 mm (0.8118988 in.)	20.622230 mm (0.8118968 in.)
Pitch dia.	23.812 mm (0.9375 in.)	23.812 mm (0.9375 in.)
Major dia.	26.06 mm max. (1.026 in.) max.	25.40/24.84 mm (1.00/0.978 in.)
Form dia.	25.50 mm (1.004 in.)	22.15 mm (0.872 in.)
Minor dia.	22.40/22.28 mm (0.882/0.877 in.)	21.34 mm min. (0.840 in.) min.
Circular space width (Tol. class 5):		
Max. actual	2.563 mm (0.1009 in.)	—
Min. effective	2.494 mm (0.0982 in.)	—
Circular tooth thickness:		
Max. effective	—	2.456 mm (0.0967 in.)
Min. actual	— —	2.388 mm (0.0940 in.)
Measurement over pins	—	28.19 mm ref. (1.110 in.) ref.
Pin dia.	—	3.05 mm (0.120 in.)
Measurement between pins	19.807 mm ref. (0.7798 in.) ref.	—
Pin dia.	2.743 mm (0.1080 in.)	—

*Same as specifications in American National Standard B92.1, Involute Splines and Inspection, except with modified tooth thickness.

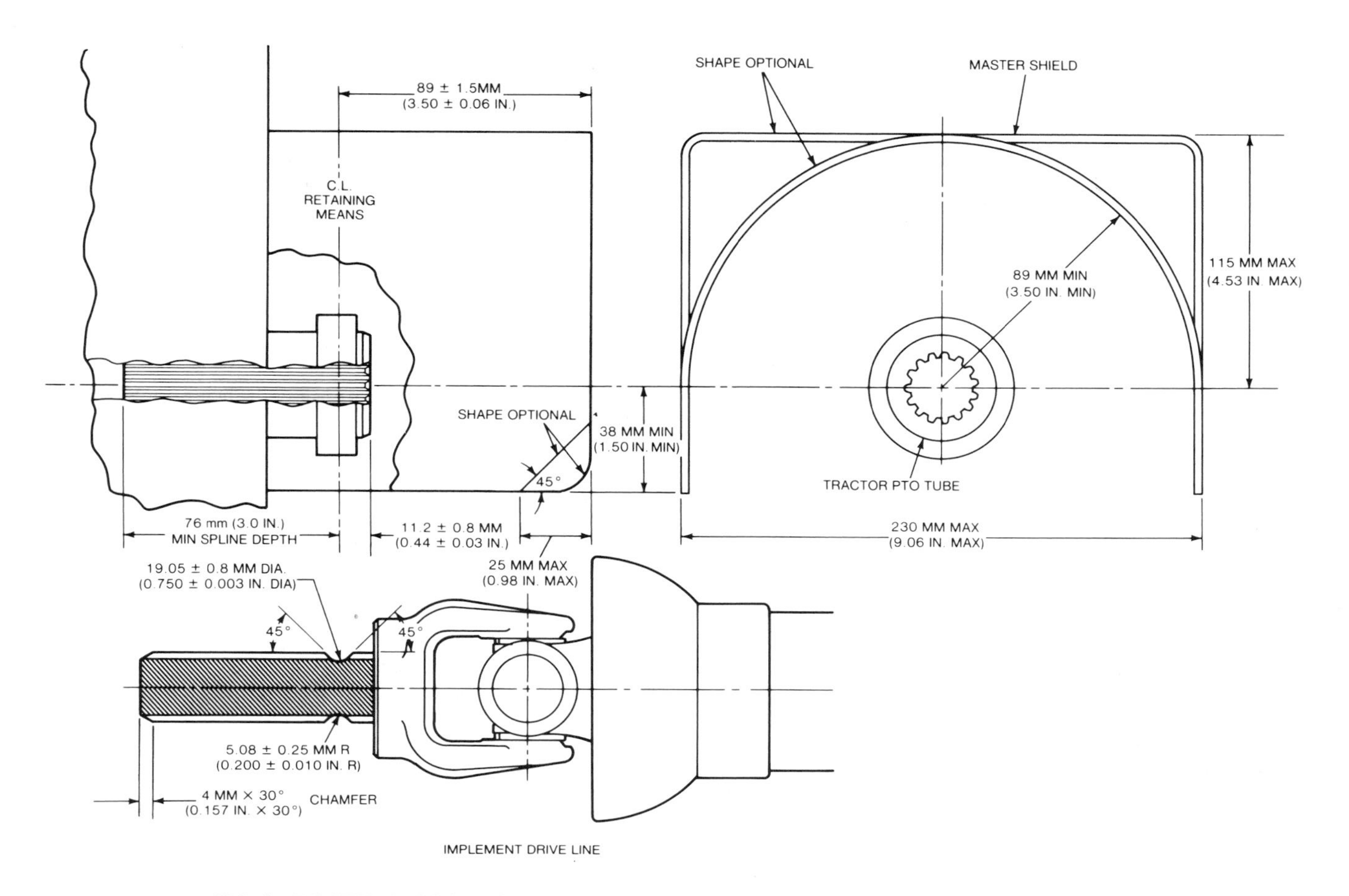

FIG. 2—DIMENSIONS FOR POWER TAKE-OFF RETAINING MEANS AND MASTER SHIELD

2.4 The power take-off tube and shaft for power take-off drives extending from the tractor to the rear shall have the spline dimensions shown in Table 2. Dimensions required for tube and shaft retaining means are shown in Fig. 2.

2.5 The normal speed of the rear power take-off tube, when operating at rated engine speed, shall be 2000 rpm ± 50 rpm. The direction of the rotation shall be clockwise when facing in the direction of forward travel.

2.6 All attachments must be capable of operating at 10% over normal speed (2000 rpm). In addition tractors as described above shall be equipped with means to prevent the operator from inadvertently operating the power take-off tube in excess of 2300 rpm under no load conditions.

2.7 The attachment clutch on the tractor shall be independent of the traction clutch.

2.8 The tractor shall be equipped with a power take-off master shield with dimensions which conform to Fig. 2.

2.9 Although not intended as a step, the master shield shall not permanently deform if used as a step by a 120 kg (265 lb) operator.

2.10 Power take-off driven implements shall be equipped with shielding for the implement drive line and its connectors or fasteners to minimize inadvertent personal contact with hazards created by positively driven parts.

2.10.1 If removal of the master shield is required for integral power take-off driven implements, shielding shall be provided with the implement to provide protection as specified in paragraphs 2.8, 2.9 and 2.10.

2.11 The master shield shall not be removable without the use of tools.

Cited Standard:

ANSI B92.1, Involute Splines and Inspection

APPLICATION OF REMOTE LINEAR CONTROL DEVICES TO LAWN AND GARDEN RIDE-ON TRACTOR ATTACHMENTS AND IMPLEMENTS

Developed by the ASAE Small Tractor and Power Equipment Committee; approved by the Power and Machinery Division Standards Committee; adopted by ASAE December 1974; revised editorially December 1975; reconfirmed December 1979; reconfirmed and revised editorially December 1984.

SECTION 1—PURPOSE AND SCOPE

1.1 The purpose of this Standard is to establish common mounting and clearance dimensions for remote linear control devices as applied to lawn and garden ride-on tractor attachments and implements with such other specifications as are necessary to accomplish the following objectives:

1.1.1 To permit use of any make or model of attachment or implement adapted for control by a remote linear control device.

1.1.2 To facilitate changing the remote linear control device from one attachment or implement to another.

SECTION 2—DEFINITIONS

2.1 Body end: The end containing the means of powering the device.

2.2 Reciprocating end: The moveable end of the device opposite the body end.

2.3 Motor housing: The housing which is mounted to the body end of the electric linear actuator.

2.4 Attaching pins: Fixed trunion pins located on the body end and a loose pin on the reciprocating end of the device for attachment of device to tractor attachments or implements.

2.5 Remote installation: Neither end of the linear control device is attached to the basic tractor.

SECTION 3—CLASSIFICATION AND RATING

3.1 For purposes of application, remote linear control devices are divided into two categories, hydraulic cylinders and electric linear actuators as shown in Figs. 1 and 2.

3.2 Specifications for mounting and capacities are shown in Table 1.

SECTION 4—CLEARANCE DIMENSIONS

4.1 Implements and attachments shall provide clearance for maximum hydraulic cylinder size as shown in Fig. 1.

4.2 Implements and attachments shall provide clearance for maximum electric linear actuator size as shown in Fig. 2.

4.3 Hose connections to cylinders shall provide clearance for trunion attachment means or rod end attachment means.

SECTION 5—STANDARD HOSE AND WIRE LENGTHS FOR REMOTE LINEAR CONTROL DEVICES

5.1 The hose or wire length shall be sufficiently long to insure proper operation of the remote linear control device in its application.

TABLE 1—DATA FOR REMOTE LINEAR CONTROL DEVICES FOR LAWN & GARDEN RIDE-ON TRACTOR IMPLEMENTS & ATTACHMENTS (See Figs. 1 and 2)

SPECIFICATION	HYDRAULIC CYLINDER* mm	HYDRAULIC CYLINDER* in.	ELECTRIC ACTUATOR* mm	ELECTRIC ACTUATOR* in.
Stroke	101.6	4.00 Nominal	101.6	4.00 Nominal
Minimum Thrust	2.22 kN	500 lb	2.22 kN	500 lb
Maximum Thrust	6.67 kN	1500 lb	6.67 kN	1500 lb
Time for Stroke Max.		8.0 s		8.0 s
Time for Stroke Min.		2.0 s		2.0 s
A Between Pin Extended	304.8+6.4-0.0	12.00+0.25-0.00	304.8+6.4-0.0	12.00+0.25-0.00
B Between Pin Retracted	203.2+0.0-6.4	8.00+0.00-0.25	203.2+0.0-6.4	8.00+0.00-0.25
C Reciprocating End, Pin Hole to End of Rod	19.1	0.75	19.1	0.75
D Centerline Trunion to Front of Body End	177.8	7.00	177.8	7.00
E Centerline Trunion to Rear of Body End	19.1	0.75	38.1 min.	1.50 min.
F Reciprocating End Dia.	25.4±0.5	1.00±0.02	25.4±0.5	1.00±0.02
G Reciprocating End, Pin Hole Dia.	13.21±0.15	0.520±0.006	13.21±0.15	0.520±0.006
H Trunion Dia.	12.57±0.13	0.495±0.005	12.57±0.13	0.495±0.005
J Centerline to End of Trunion	50.8±0.8	2.00±0.03	50.8±0.8	2.00±0.03
K Width Across Trunion	101.6±1.5	4.00±0.06	101.6±1.5	4.00±0.06
L Outside Radius of Cylinder	38.6	1.52		
M Centerline to Body Side			38.6	1.52
N Centerline to Body Top and Body Bottom			40.6	1.60
P Centerline Trunion to Motor Housing Front			76.2	3.00
R Centerline to Motor housing Bottom			44.5 min.	1.75 min.
S Centerline to Motor Housing Top			209.6	8.25
T Centerline to Motor Housing Side			63.5	2.50
U Centerline Trunion to Rear of Motor Housing			139.7	5.50

*All dimensions maximum unless otherwise specified.

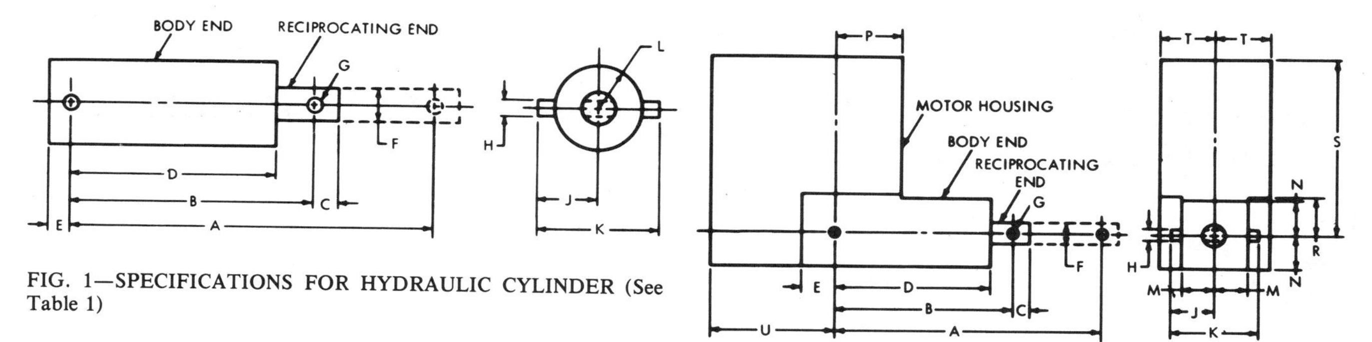

FIG. 1—SPECIFICATIONS FOR HYDRAULIC CYLINDER (See Table 1)

FIG. 2—SPECIFICATIONS FOR ELECTRIC ACTUATOR (See Table 1)

DIMENSIONS FOR CYLINDRICAL HYDRAULIC COUPLERS FOR LAWN AND GARDEN TRACTORS

Developed by the ASAE Small Tractor and Power Equipment Committee; approved by the ASAE Power and Machinery Division Standards Committee; adopted by ASAE April 1983; corrected editorially June 1984; reconfirmed December 1987.

SECTION 1—PURPOSE

1.1 The purpose of this Standard is to establish interface dimensions of cylindrical hydraulic couplers frequently used by the equipment industry to connect hydraulic remote cylinders and other hydraulic devices to lawn and garden tractors.

1.2 The purpose of this Standard is also to permit interchangeable use of remote cylinders and other hydraulic devices on different makes of tractors when designed for this use.

SECTION 2—SCOPE

2.1 This Standard covers only cylindrical couplers where interchange between makes is recommended.

2.2 This Standard does not preclude other types of hydraulic couplers used for similar purposes.

SECTION 3—COUPLER DIMENSIONS

3.1 The male half of the coupler is considered part of the cylinder or hydraulic device and shall conform to dimensions given in Table 1.

3.2 The female half of the coupler is considered part of the tractor hydraulic system and shall accept the male half which conforms to dimensions given in Table 1.

3.3 When a quick release, breakaway, or other special feature is required, it shall be provided as part of the female coupler without interfering with interchangeability of the standard male coupler.

3.4 The travel of the valve in the male coupler will be limited as shown in Fig. 1, and the force required to open the valve when no pressure is in the coupler shall not exceed 44 N (10 lb).

SECTION 4—COUPLER LOCATIONS

4.1 For standard remote cylinder or other hydraulic device operation (see ASAE Standards S201, Application of Hydraulic Remote Control Cylinders to Agricultural Tractors and Trailing-Type Agricultural Implements, and ASAE S316, Application of Remote Hydraulic Motors to Agricultural Tractors and Trailing-Type Agricultural Implements), the female coupler shall be located to the rear of the tractor within limits specified in Table 2 and Fig. 2.

4.2 For special cylinder or other hydraulic device operation, female couplers shall be located to the rear, on the right or left side, or in front, within limits specified in Table 2 and Fig. 2.

SECTION 5—PERFORMANCE REQUIREMENTS

5.1 The maximum pressure drop through the male half of the coupler shall not exceed 172.4 kPa (25 psi) at rated flow with mineral oil at a viscosity of 55 Saybolt Universal Seconds minimum. When measuring restriction, the sealing valve shall be depressed flush with the end of the valve body.

5.2 When the two halves of the coupler are locked together in the operating position, they shall remain locked together, without failure, when subjected to a proof pressure of 200% of the rated pressure.

5.3 The male half of the uncoupled coupler shall withstand a minimum burst pressure of 400% of the rated pressure.

SECTION 6—COMPATIBILITY WITH FLUIDS

6.1 For the purposes of this Standard, it should be assumed that SAE 20 mineral oil will be used.

6.2 Couplers and cylinders, or other hydraulic devices, should be made with seals and other materials which are compatible with SAE 20 mineral oils and additives normally used in tractor hydraulic systems.

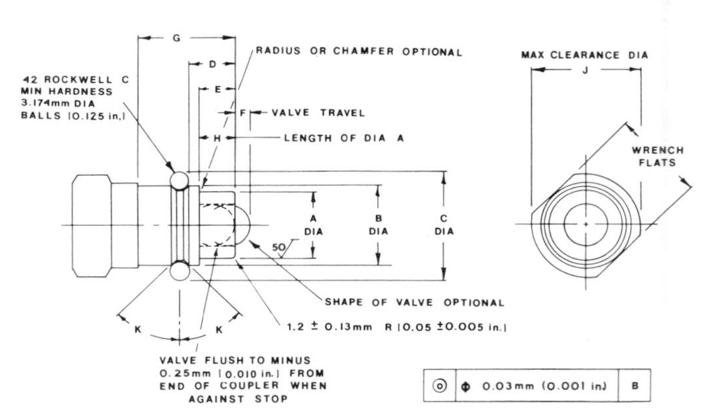

FIG. 1—DIMENSIONS OF CYLINDRICAL HYDRAULIC COUPLERS (SEE TABLE 1)

TABLE 1—SPECIFICATIONS AND DIMENSIONS FOR CYLINDRICAL HYDRAULIC COUPLERS (SEE FIG. 1)

Rated operating pressure		Maximum relief valve pressure		Rated flow		Diameter A		Diameter B		Diameter C-gage	
MPa	psi	MPa	psi	L/min	gal/min	mm	in.	mm	in.	mm	in.
17.2	2500	20.7	3000	11.25	3	14.17 ±0.05	0.558 ±0.002	16.87 ±0.05	0.664 ±0.002	21.33	0.840

Dimension D		Dimension E		Maximum valve travel F		Minimum dimension G		Minimum dimension H		Maximum diameter J		Angle K
mm	in.	mm	in.	mm	in.	mm	in.	mm	in.	mm	in.	deg
9.70 +0.00 −0.13	0.382 +0.005 −0.005	7.93 ±0.13	0.312 ±0.005	3.17	0.125	18.41	0.725	6.60	0.260	24.13	0.95	45 ±1

TABLE 2—LOCATION OF CYLINDRICAL HYDRAULIC COUPLERS (SEE FIG. 2)

	Dimensions from Fig. 2	Range mm	Range in.
For rear locations:			
A	Horizontal distance from pin hole of drawbar	50 to 635	2 to 25
B	Vertical distance from top of drawbar	100 to 914	4 to 36
C	Horizontal distance from centerline of tractor - either side	254 max	10 max
For side locations:			
D	Horizontal distance from centerline of tractor	254 max	10 max
E	Horizontal distance from centerline of front wheel spindle	152 to 406	6 to 16
F	Vertical distance from centerline of front wheel spindle	100 to 355	4 to 14
For front locations:			
G	Horizontal distance either side of tractor centerline	254 max	10 max
H	Horizontal distance ahead of front wheel spindle centerline	100 to 355	4 to 16
J	Vertical distance above front wheel spindle centerline	0 to 355	0 to 14

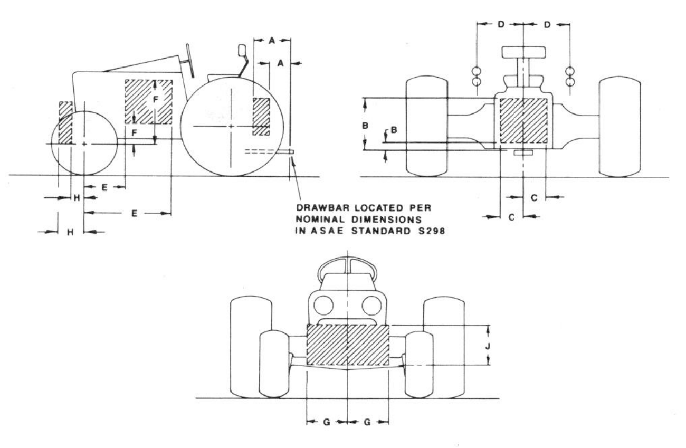

FIG. 2—LOCATION OF CYLINDRICAL HYDRAULIC COUPLERS (SEE TABLE 2)

Cited Standards:

ASAE S201, Application of Hydraulic Remote Control Cylinders to Agricultural Tractors and Trailing-Type Agricultural Implements
ASAE S298, Drawbar for Lawn and Garden Ride-On Tractors
ASAE S316, Application of Remote Hydraulic Motors to Agricultural Tractors and Trailing-Type Agricultural Implements

ASAE Standard: S440.1

SAFETY FOR POWERED LAWN AND GARDEN EQUIPMENT

Developed by the ASAE Small Tractor and Power Equipment Committee; approved by the Power and Machinery Division Standards Committee; adopted by ASAE February 1984; revised April 1987.

SECTION 1—PURPOSE AND SCOPE

1.1 This Standard is intended to provide a reasonable degree of personal safety for operators and other persons during the normal operation and servicing of powered lawn and garden equipment; such as garden tractors, ride-on and walk-behind mowers, tillers, snow removal machines, edgers, trimmers, plows, cultivators, and other related machines or combinations thereof designed primarily for consumer yard and garden operations.

1.2 In addition to the design and configuration of equipment, hazard control and accident prevention are dependent upon the awareness, concern, prudence, and proper training of personnel involved in the operation, transport, maintenance, and storage of equipment.

SECTION 2—DEFINITIONS

2.1 See ASAE Standard S323, Definitions of Powered Lawn and Garden Equipment, for additional definitions.

2.2 Guarded by location: A hazard is guarded by location when it is covered by other parts or components of the machine, or because of its remote location, inadvertent contact (see ASAE Standard S318, Safety for Agricultural Equipment) is minimized.

2.3 Nip point: The pinch point of gears and run-on point where a belt or chain contacts a sheave, sprocket or idler.

2.4 Safety sign: Information affixed to a machine to alert persons to hazards, their severity, consequences, and how to avoid them.

2.5 Shield (or guard): A barrier which minimizes inadvertent personal contact with hazards.

SECTION 3—OPERATOR'S MANUAL

3.1 Specific written instructions shall be provided with the equipment describing proper operation of the machine, safety instructions, and the necessary maintenance procedures. (see ASAE Engineering Practice EP363, Technical Publications for Agricultural Equipment)

SECTION 4—CONTROLS

4.1 Operator controls shall be designed and located to allow safe and easy operation. Human factors principles, such as direction of motion, length of reach, normal body strength, and clearances for appropriate clothing such as artic mittens and boots, shall be considered.

4.2 Location, movement, and identification of operator controls shall conform with the requirements of American National Standard B71.1, Safety Specifications for Power Lawn Mowers, Lawn and Garden Tractors, and Lawn Tractors; ANSI Standard B71.3, Safety Specifications for Snow Throwers, or ANSI Standard B71.4, Safety Specifications for Commercial Turf Care Equipment, as applicable.

4.3 Controls that are color coded shall conform to ASAE Engineering Practice EP443, Color Coding Hand Controls.

SECTION 5—OPERATOR'S STATION

5.1 A suitable station shall be provided for the operator of each powered unit.

5.2 All powered units shall be designed to permit safe and easy ingress to or egress from the operator's position.

5.3 Foot supports and steps shall have a slip-resistant means.

5.4 Glazing material, such as glass or plastic, used in operator enclosures shall be in accordance with Society of Automotive Engineers Recommended Practice J674, Safety Glazing Materials—Motor Vehicles.

SECTION 6—SERVICING

6.1 Components, attachments or accessories, which must be raised for normal servicing or adjusting and which might be hazardous if dropped, should be provided with a means to hold them in the raised position. Such means may consist of either the mechanism to hold the load and/or instructions (such as "see operator's manual") for an operator supplied support using provisions on the machine necessary for securely installing such support. The operator's manual shall contain instructions on how to securely support or block raised components, attachments, or accessories before servicing or adjusting.

6.2 Spring-loaded mechanisms, pressurized fluid systems like cooling and hydraulics, and other sources of stored energy which can be released in a hazardous manner, shall be indicated by a safety sign which includes instructions for de-energizing and proper disassembly or refers to such instructions in the operator's manual.

SECTION 7—STABILITY

7.1 Ride-on powered units shall conform to stability requirements of the applicable ANSI Standard B71.1, Safety Specifications for Power Lawn Mowers , Lawn and Garden Tractors, and Lawn Tractors; ANSI Standard B71.3, Safety Specifications for Snow Throwers; or ANSI Standard B71.4, Safety Specifications for Commercial Turf Care Equipment.

SECTION 8—SHIELDS AND WARNINGS

8.1 Shielding shall be provided to minimize inadvertent contact with the hazards created by the following unless the hazard is guarded by location or an operator presence control stops the motion that could cause a hazard to the operator:

8.1.1 Moving traction elements.

8.1.2 Exposed nip points.

8.1.3 Outside faces of pulleys, sheaves, sprockets, and gears on drives that rotate when the engine is running with all clutches disengaged.

8.1.4 Rotating parts with projections such as exposed universal joint yokes, bolts, keys or set screws.

8.1.5 Rotating shafts, including their locking means, except smooth shaft ends protruding less than one-half the diameter of the shaft.

8.1.6 Ground-driven components, if the operator is exposed to them while the drives are in motion.

8.1.7 Surfaces with temperatures above 129 °C (264 °F) at 21 °C (70 °F) ambient, except surfaces of equipment intended primarily for winter use, which shall be 129 °C (264 °F) at 5 °C (41 °F) ambient. All surfaces which exceed 65.5 °C (150 °F) at 21 °C (70 °F) ambient and which might be contacted by the operator during normal starting, mounting, operating, or refueling shall be indicated by a safety sign located on or adjacent to the surface.

8.1.7.1 Surface temperature tests. Equipment accuracy shall be ±3°C and ±5°F. The engine shall be operated at its maximum idle no load speed for one (1) hour or until surface temperatures stabilize. Test shall be conducted in the shade. Temperatures are to be determined by correcting the observed temperature by the difference between the specified ambient and test ambient temperature.

8.1.8 A power take-off (PTO) when it is not connected to an attachment drive.

8.2 Functional components of powered units and attachments which must be exposed for proper function shall be shielded to the maximum extent permitted by the intended function of the component. Functional components of mowers shall also be protected in accordance with the following applicable product standards: Consumer Product Safety Commission Standard 16 CFR Part 1205, Safety Standard for Walk-Behind Power Lawn Mowers; ANSI Standard B71.1, Safety specifications for Power Lawn Mowers, Lawn and Garden Tractors, and Lawn Tractors; or ANSI Standard B71.4, Safety Specifications for Commercial Turf Care Equipment. Snow throwers shall also be protected in accordance with ANSI Standard B71.3, Safety Specifications for Snow Throwers.

8.3 All shields and guards required to be opened for normal service or maintenance shall be easily opened and closed. Shields and guards not required to be opened for normal service or maintenance shall be attached or secured to prevent removal without the use of tools.

8.4 Equipment with access doors and shields which can be opened or removed while components continue to move or rotate after the power is disengaged shall have (1) visible or audible indication of rotation and (2) a suitable safety sign.

SECTION 9—BRAKING REQUIREMENTS

9.1 For walk-behind units, ANSI Standard B71.4, Safety Specifications for Commercial Turf Care Equipment, shall be used for service and parking brake performance.

9.2 For ride-on units. ANSI Standard B71.1, Safety Specifications for Power Lawn Mowers, Lawn and Garden Tractors, and Lawn Tractors shall be used for service and parking brake performance.

9.2.1 Brake performance criteria shall be met with the heaviest normal use combination of attachment and accessories not including towed equipment. Load carrying attachments shall be loaded to the manufacturer's rated load, except front-end loader bucket, forklift, and backhoe attachments shall be tested empty. Towed equipment is exempt from this requirement.

SECTION 10—FIRE PROTECTION

10.1 Shielding shall be provided to prevent the collection of flammable materials such as grass and chaff on the exhaust manifold(s), muffler(s), and exhaust pipe(s).

10.2 Electrical requirements

10.2.1 Electrical wiring shall conform to SAE Recommended Practice J1292, Automobile, Truck, Truck-Tractor, Trailer, and Motor Coach Wiring, where applicable.

10.2.2 Circuits, except starting motor and ignition circuits, shall have appropriate overload protective devices on the battery feed side of switches. Overload protective devices shall comply with requirements of SAE Standard J156, Fusible Links; SAE Recommended Practice J553, Circuit Breakers; SAE Standard J554, Electric Fuses (Cartridge Type); or SAE Standard J1284, Blade Type Electric Fuses.

10.2.3 Terminals and uninsulated electrical parts shall be protected against shorting during normal maintenance and operation.

SECTION 11—SAFETY SIGNS

11.1 Safety signs shall conform to ASAE Standard S441, Safety Signs. (See CPSC Standard 16CFR Part 1205, Safety Standard for Walk-Behind Power Lawn Mowers, for required safety sign.)

Cited Standards:

ANSI B71.1, Safety Specifications for Power Lawn Mowers, Lawn and Garden Tractors, and Lawn Tractors
ANSI B71.3, Safety Specifications for Snow Throwers
ANSI B71.4, Safety Specifications for Commercial Turf Care Equipment
ASAE S318, Safety for Agricultural Equipment
ASAE S323, Definitions of Powered Lawn and Garden Equipment
ASAE EP363, Technical Publications for Agricultural Equipment
ASAE S441, Safety Signs
ASAE EP443, Color Coding Hand Controls
CPSC 16 CFR Part 1205, Safety Standard for Walk-Behind Power Lawn Mowers
SAE J156, Fusible Links
SAE J553, Circuit Breakers
SAE J554, Electric Fuses (Cartridge Type)
SAE J674, Safety Glazing Materials—Motor Vehicles
SAE J1284, Blade Type Electric Fuses
SAE J1292, Automobile, Truck, Truck-Tractor, Trailer, and Motor Coach Wiring

ELECTRICAL AND ELECTRONIC SYSTEMS

ASAE Notation:

The letter S preceding numerical designation indicates ASAE Standard; EP indicates Engineering Practice; D indicates Data. A decimal and numeral following the file number indicate the number of times a document has been revised. Thus ASAE S201.4 indicates Standard number 201, four times revised. The letter T after the designation indicates tentative status. Always refer to ASAE documents by complete designation to avoid confusion with standards of other organizations. For example: ASAE S201.4.

The symbol **T** preceding or in the margin adjacent to section headings, paragraph numbers, figure captions, or table headings indicates a technical change was incorporated in that area when this document was last revised. The symbol **T** preceding the title of a document indicates essentially the entire document has been revised. The symbol **E** used similarly indicates editorial changes or corrections have been made with no intended change in the technical meaning of the document.

ASAE Engineering Practice: ASAE EP258.2

INSTALLATION OF ELECTRIC INFRARED BROODING EQUIPMENT

Adopted by ASAE 1955; reconfirmed December 1963; revised December 1964, February 1971; reconfirmed December 1975, December 1976; revised editorially, reclassified as an Engineering Practice, and reconfirmed December 1977; reconfirmed December 1982, December 1987.

SECTION 1—PURPOSE AND SCOPE

1.1 This Engineering Practice is intended to promote safety through the proper design and installation of livestock and poultry brooding systems which utilize infrared energy; and to provide guidance for the design and installation of electrical wiring systems and equipment used primarily for brooding purposes.

1.2 This Engineering Practice covers only that equipment and those installations which use 120 volt, 250 watt or smaller reflector heat lamps or the R-40 or G-30 type as specified in Illuminating Engineering Society, IES Lighting Handbook.

SECTION 2—ELECTRICAL SERVICE

2.1 All permanent buildings in which electric infrared units are used for brooding should be equipped with service equipment as required by the National Fire Protection Association Standard No. 70, National Electrical Code (American National Standard C1, National Electrical Code), local requirements and requirements of power suppliers. The service should include adequate capacity to handle the anticipated load. Wire size from meter to the building distribution panel should be selected to limit voltage drop to a maximum of 2 percent under load.

2.2 A portable building requiring only one circuit, or one sharing a circuit with a similar building, may be served from a service-equipment center. The capacity of the center and the size of the wire from the meter to the center should be in accordance with paragraph 2.1.

SECTION 3—CIRCUITS

3.1 Any branch circuit serving electrical brooding units and equipment should be permanently installed. The receptacles should be grounded and should be located within 6 ft. (1.8m) of the brooder locations and above the height that animals can reach. The installation should be in accordance with NFPA Standard No. 70, National Electrical Code, local regulations, and requirements of power suppliers.

3.2 Type NMC non-metallic sheathed or Type UF cable as specified in NFPA Standard No. 70, National Electrical Code, cable should be used unless local regulation require other wiring methods.

3.3 The conductors should not be smaller than No. 12 AWG (American Wire Gage) and should be large enough to hold circuit voltage drop within 2 percent.

3.4 Each circuit should be protected by fuses or circuit breakers in accordance with NFPA Standard No. 70, National Electrical Code, and local regulations or requirements.

3.5 The maximum load should not exceed 80 percent of the rated capacity of the circuit.

3.6 Three-wire circuits (two "hot" wires with a common neutral) should be used when practical with an equal number of outlets connected to each "hot" wire to balance the load. Such three-wire circuits should be connected through double-pole breakers or double-pole, pull-out fuse blocks.

SECTION 4—BROODING UNITS, GENERAL

4.1 The heating unit should be protected adequately from moisture and mechanical injury by means of a metal reflector, shield, or guard. The reflector or guard should be large enough to prevent animals from contacting the heating unit. The guard should be of such a size and shape that it will prevent the unit from igniting litter or other flammable material if it should fall from its supporting means.

4.2 A screw-shell lamp receptacle or lampholder of the switchless or keyless type and made of porcelain or other material of equivalent temperature characteristics should be used for all units employing screw-base heat lamps. Screw-shell lamp receptacles should not be used with infrared lamps or equivalent heat units with a rating greater than 250 w unless they are especially approved for the purpose.

4.3 The unit should be suspended securely by chain or bracket. The cord leading to the receptacle should not be used to provide support for the unit.

4.4 The cord on each unit should not exceed 8 ft (2.4m) in length and, where practical, should be of such length that it will become disconnected if the unit falls from its support. The cord should be a heater type designed for use in damp locations. Type HSJ cord as specified in NFPA Standard No. 70, National Electrical Code, is an example of this type cord.

4.5 The protection of a jacket is required on the supply cord wherever the supply cord is outside of the unit. Therefore, the jacket should be in place from the attachment plug cap to a point inside the device or unit where the individual conductors are split. The purpose of the jacket on the cord is to prevent moisture from coming in contact with the wiring and to provide resistance to mechanical abuse. This cord is not intended to be immersed in water.

4.6 Wire size for brooder units should not be less than:
No. 18 AWG for single 250-w lamp unit, or equivalent
No. 16 AWG for two or three 250-w lamp unit, or equivalent
No. 14 AWG for four or a five 250-w lamp unit, or equivalent
No. 12 AWG for a seven 250-w lamp unit, or equivalent six or seven lamp unit.

4.7 Asbestos covered wire should be used for the interior wiring of metal frame brooders. Where the temperature of the frame exceeds 140 F, the interior wiring should be installed with individual conductors having suitable supplementary insulation to hold the asbestor in place.

4.8 Animals and poultry should be given adequate room for moving out of the heated area.

SECTION 5—PIG BROODERS

5.3 If suspended over the sow at farrowing time, the heating unit should not be less than 30 in. (762 mm) above the bedding, nor less than 6 in. (152 mm) above the standing animal.

5.2 When suspended over the litter of pigs separated from the sow, the heating unit may be lowered to a minimum of 18 in. (457 mm) above the standing animals being brooded, provided a barricade of sufficient strength to restrain the sow, constructed to pen height, is installed and securely anchored.

SECTION 6—POULTRY BROODERS

6.1 Brooders should be suspended so that the surface of the heating unit is not less than 15 in. (381 mm) above the litter. Generally, when lamps are used, the brooder should be placed so that the bottom surface of the lamp is about 18 in. (457 mm) from the litter.

SECTION 7—BROODING CALVES, LAMBS, AND OTHER

7.1 The heating unit should be suspended a minimum of 6 in. (152 mm) higher than the animal can reach.

7.2 For lambs it may be necessary to suspend the heating unit considerably higher than 6 in. (152 mm) above the animal's reach to prevent damaging the skin since wool prevents loss of heat.

7.3 Ewes should be fenced away from the heating unit, but not separated from the lambs. A partition of sufficient strength to restrain the ewe and designed to permit the lamb to pass under is necessary.

Cited Standard:

NFPA No. 70, National Electrical Code

COMPUTING ELECTRICAL DEMANDS FOR FARMS

Corresponds to Article 220-D of the National Electrical Code, NFPA No. 70-1984, published by National Fire Protection Association, and published by the American National Standards Institute as C1-1984. This document was developed by ASAE's Farm Wiring Committee; approved by the Electric Power and Processing Division Standards Committee; adopted by ASAE in December 1965; reconfirmed with editorial changes December 1970; revised editorially January 1972, December 1975; reconfirmed December 1980; reconfirmed with editorial changes December 1985.

SECTION 1—PURPOSE

1.1 This Engineering Practice provides formulas for computing the minimum capacity of feeders supplying farm buildings and the minimum capacity of service conductors and service equipment at the in point of delivery to farms.

1.2 The Engineering Practice was developed by the Farm Wiring Committee for inclusion in National Electrical Code, the recognized authority on electrical wiring safety. It was approved by the National Fire Protection Association on May 20, 1965, and appears in Article 220-D of NFPA Standard No. 70-1984, National Electrical Code.

SECTION 2—SCOPE

2.1 Farm wiring systems meeting the requirements of this Engineering Practice should be safe but not necessarily adequate or convenient. Detailed suggestions for farmstead wiring design are contained in the ASAE endorsed Agricultural Wiring Handbook available from the National Food and Energy Council, 409 Vandiver West, Suite 202, Columbia, Missouri 65202.

SECTION 3—FARM LOADS—BUILDINGS AND OTHER LOADS

3.1 Dwellings. The feeder or service load of a farm dwelling shall be computed in accordance with the provisions for dwellings in Part B or C of Article 220-D, NFPA Standard No. 70-1984, National Electrical Code.

3.2 Other than dwellings. For each farm building or load supplied by two or more branch circuits the load for feeders, service-entrance conductors, and service equipment shall be computed in accordance with demand factors not less than those indicated in Table 1.

3.3 Overhead conductors. See Section 230-21 of NFPA Standard 70-1984, National Electrical Code, for overhead conductors from a pole to a building or other structure.

SECTION 4—FARM LOADS—TOTAL

4.1 Total load. The total load of the farm for service-entrance conductors and service equipment shall be computed in accordance with the farm dwelling load and demand factors specified in Table 2. Where there is equipment in two or more farm buildings or for loads having the same function, such loads shall be computed in accordance with Table 1 and may be combined as a single load in Table 2 for computing the total load.

4.2 Overhead conductors. See Section 230-21 of NFPA Standard 70-1984, National Electrical Code, for overhead conductors from a pole to a building or other structure.

Cited Standard:

NFPA No. 70-1984, National Electrical Code

TABLE 1—METHOD FOR COMPUTING FARM LOADS FOR OTHER THAN DWELLINGS

Ampere Load at 230 Volts	Demand Factor Percent
Loads expected to operate without diversity, but not less than 125 percent full-load current of the largest motor and not less than the first 60 amperes of load	100
Next 60 amperes of all other loads	50
Remainder of other load	25

TABLE 2—METHOD FOR COMPUTING TOTAL FARM LOAD

Individual Loads Computed in Accordance With Table 1	Demand Factor Percent
Largest load	100
Second largest load	75
Third largest load	65
Remaining loads	50

To this total load, add the load of the farm dwelling computed in accordance with Part B or C of Article 220-D, NFPA No. 70-1984 National Electrical Code.

SINGLE-PHASE RURAL DISTRIBUTION SERVICE FOR MOTORS AND PHASE CONVERTERS

Developed by the ASAE Farm Wiring Committee; approved by the Electric Power and Processing Division Standards Committee; adopted by ASAE December 1969; revised editorially January 1972; reclassified as an Engineering Practice December 1974; reconfirmed December 1979, December 1984.

SECTION 1—PURPOSE

1.1 This Engineering Practice is intended to serve the following purposes:

1.1.1 To provide minimum utility service regulations for motors on single-phase rural distribution systems in order that rural consumers, equipment manufacturers, and others concerned will know the single-phase motor sizes and types that will be served in areas where three-phase power is not available.

1.1.2 To eliminate confusion in motor applications where motor load is well above nameplate rating because of service regulations based on rated horsepower of the motor.

1.1.3 To call to the attention of power suppliers the need by farmers for larger motor-operated equipment and to encourage the use of such equipment.

1.1.4 To encourage uniform criteria of maximum inrush current and minimum voltage to enable equipment and motor manufacturers to design and produce satisfactory electric motor-operated equipment for present-day farms.

1.1.5 To encourage the use of magnetic motor controllers for 3.7 kW (5 hp) and larger motors.

SECTION 2—SCOPE

2.1 This Engineering Practice concerns 3.7 kW (5 hp) and larger single-phase motors, and three-phase motors with phase converters, that are started infrequently (generally with less than 6 starts in a 24-hour period and not more than 1 start between 6 p.m. and midnight) and that serve loads with non-pulsating power requirements. Some examples are motors for fans on crop driers, feed grinders, silo unloaders, irrigation pumps, and auger feeders.

2.2 Excluded from this Engineering Practice are those motors used on pulsating loads and loads started frequently. Typical examples are motors on some deep-well piston pumps, reciprocating compressors, rock crushers, stamping machines, automatically controlled water pumps, and refrigeration systems.

SECTION 3—RURAL DISTRIBUTION SYSTEM CAPABILITY

3.1 Single-phase motors meeting the requirements of paragraph 2.1 shall be permitted anywhere on a distribution system, if the design locked-rotor current at 230 volts is no more than 260 amperes and if no more than 260 amperes is required at any time during the starting cycle. This allows the use of conventional Design L, 7.5 kW (10 hp) motors and, with reduced starting current devices, 11.2 kW (15 hp) or larger motors.

3.2 Single-phase motors with design locked-rotor currents greater than 260 amperes should be permitted on systems and parts of systems where the primary voltage drop can be held to acceptable limits at the time of the motor start.

3.3 Phase converters supplying three-phase motors shall be permitted anywhere on a system if the design inrush current to the converter does not exceed 260 amperes at 230 volts.

3.4 Phase converters with design inrush currents greater than 260 amperes should be permitted on systems and parts of systems where the primary voltage drop can be held to acceptable limits at the time of the motor start.

3.5 Generally, no limit should be placed by the power supplier on the number of motors that may be operated at one time, provided that the motors are started individually and comply with paragraphs 3.1 through 3.4. If motors are started in combination, the maximum total inrush current shall be no greater than allowed for a single motor in paragraphs 3.1 through 3.4.

3.6 Under starting conditions, voltages at terminals of motors and phase converters for the conditions specified in paragraph 3.1 shall not be less than 184 volts. At this voltage, single-phase motor starting torque is approximately 60 percent of normal and starting loads should be sized accordingly. A maximum of 10 percent of nominal voltage shall be allowed for feeder and branch circuit voltage drop under starting conditions. For a nominal voltage of 230 volts, a minimum of 207 volts would be required at the meter terminals during the starting cycle.

3.7 Under running conditions, voltages at terminals of 230-volt motors on rural distribution systems shall not exceed 253 volts, and shall not be less than 207 volts at rated frequency, to meet National Electrical Manufacturers Association Standard MG1-12.43, Variation from Rated Voltage. Voltage variation at other than rated frequency shall be in accordance with NEMA Standard MG1-12.45, Combined Variation of Voltage and Frequency. Under the provisions of paragraph 3.6, the maximum voltage drop for the feeder and branch circuit under running conditions should be within 5 percent as recommended in Section 210-6(d) of National Fire Protection Association, NFPA Standard, No. 70-1971, National Electrical Code (Also American National Standard ANSI C1-1971). For a nominal voltage of 230 volts and a 5-percent voltage drop, a minimum of 218 volts would be required at the meter terminals.

3.8 Each rural consumer with a motor 3.7 kW (5 hp) or larger should be served by a separate distribution transformer. The consumer will be expected to tolerate the light flicker caused by the starting of his own motors.

SECTION 4—MOTOR REQUIREMENTS

4.1 Single-phase motors without reduced voltage starting shall be designed to start on 184 volts (80 percent of 230-volt rating). The reduced torque available at this voltage (see paragraph 3.6) must be considered by equipment manufacturers or installers in matching the mechanical load to the motor. Reduced-voltage-start motors with equal inrush currents will have more than 184 volts available because of the larger transformer and wiring required for running conditions.

4.2 The motors described in paragraph 4.1 shall operate satisfactorily at rated mechanical load on 207 volts (90 percent of 230-volt rating) at the motor terminals.

SECTION 5—EQUIPMENT REQUIREMENTS

5.1 Loading of single-phase motors and three-phase motors operated with phase converters shall be such that the starting, pullup, and running torques under the starting and operating voltage specified in paragraph 3.6 and 3.7 will be adequate to operate the equipment.

5.2 Single-phase motors and three-phase motors operated with phase converters shall be so loaded that they accelerate to rated speed (motor current reduced to rated value at rated voltage) without producing excessive motor heating or flicker problems.

5.3 Motors 3.7 kW (5 hp) and larger shall operate with a magnetic motor controller equipped with an undervoltage release.

Cited Standards:

NEMA MG1-12.43, Variation from Rated Voltage
NEMA MG1-12.45, Combined Variation of Voltage and Frequency
NFPA No. 70-1971, National Electrical Code

SAFETY FOR ELECTRICALLY HEATED LIVESTOCK WATERERS

Developed by the ASAE Farm Wiring and Electic Utilization Committee; approved by the Electric Power and Processing Division Standards Committee; adopted by ASAE December 1971; reconfirmed December 1976; revised editorially, reclassified as an Engineering Practice, and reconfirmed December 1977; reconfirmed December 1982, December 1987.

SECTION 1—PURPOSE AND SCOPE

1.1 This Engineering Practice is intended to promote safety and to conform to National Fire Protection Association Standard No. 70, National Electrical Code, as it may apply to the construction and installation of electrically heated livestock waterers (hereafter called waterers).

1.2 This Engineering Practice is intended to serve as guidelines for the manufacture and the installation of waterers.

SECTION 2—GENERAL RECOMMENDATIONS

2.1 Waterers shall be manufactured to facilitate installation in accordance with NFPA Standard No. 70, National Electrical Code.

2.2 Materials and construction of waterers shall be approved for wet locations.

2.3 Waterers shall be installed in accordance with NFPA Standard No. 70, National Electrical Code, and state or local regulations.

2.4 Waterers shall be installed on a firm foundation.

2.5 Every waterer shall have a disconnecting switch with over-current protection located within sight of the waterer (less than 50 ft) or shall be equipped with a mechanical lockout and shall be protected not over 125 percent of the ampere rating of the heating element.

2.6 Switches and conductors shall be protected from physical damage.

2.8 Metal fences near the waterer should be bonded to the grounding conductor.

SECTION 3—CONSTRUCTION-ELECTRICAL COMPONENTS

3.1 Waterers shall be equipped with a readily identifiable grounding terminal to which the grounding conductor of the supply circuit shall be connected.

3.2 All metal parts of the waterer shall be bonded together in positive electrical contact with the grounding terminal. Where the bowl or trough is of non-conducting material, either provision shall be made to ground the water in the bowl or trough, or the waterer shall be so constructed that the water in the bowl or trough is isolated from the ground.

3.3 The heating element shall be constructed and attached as follows:

3.3.1 It shall have no exposed part other than the terminals to be energized.

3.3.2 It shall be constructed to prevent the entrance of moisture into the area between the heating element and its protective cover or sheath.

3.3.3 It shall be attached to the waterer so that all exposed metal parts except the terminals will be in electrical contact with the other metal parts of the waterer to provide a grounding circuit under a fault condition.

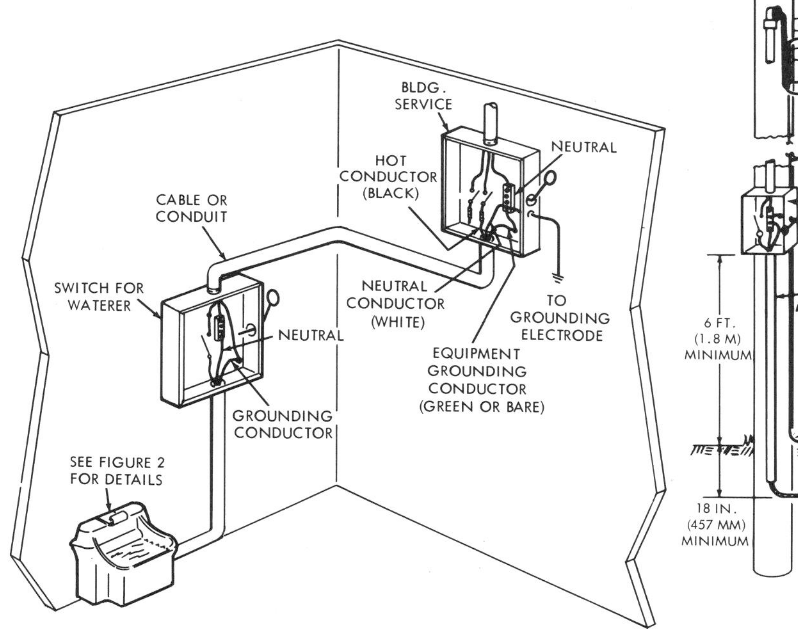

FIG. 1—INSTALLATION IN OR NEAR A BUILDING

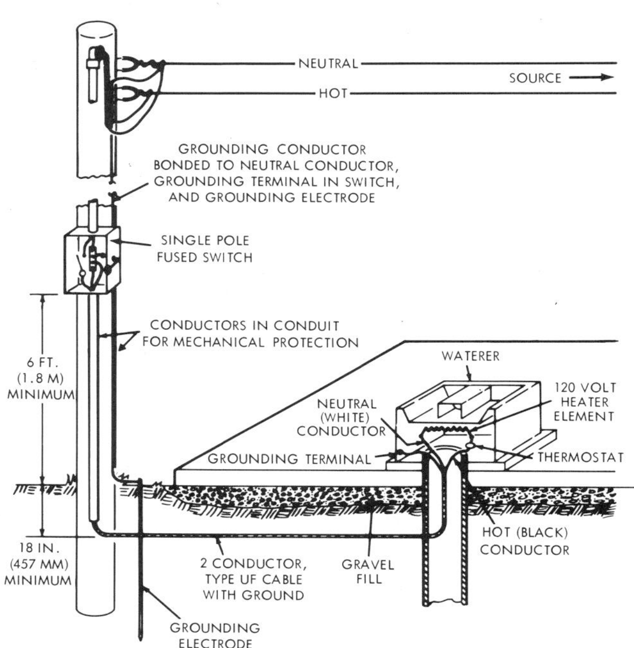

FIG. 2—INSTALLATION SERVED BY INDIVIDUAL ELECTRICAL SERVICE

SECTION 4—ELECTRICAL SERVICE

4.1 Service conductors shall conform to the following:

4.1.1 Conductors shall have sufficient ampacity for the load to be served.

4.1.2 Overhead conductors shall have mechanical strength for the distance spanned. For spans of over 50 ft (15 m), No. 8 American Wire Gage conductors or larger shall be used and for spans under 50 ft (15 m), No. 10 AWG conductors may be used.

4.1.3 Buried conductors shall be of Type USE (or UF when fused as a feeder or branch circuit) and shall be buried at least 18 in. (457 mm) (24 in. (610 mm) depth recommended in unpaved feedlots) below the ground surface.

4.2 Waterers installed near or in a building may be served from the electrical service of the building.

SECTION 5—SUPPLY CIRCUITS

5.1 Waterers installed within or near a building and served from the electrical service of the building shall be connected as in Fig. 1.

5.1.1 The grounding conductor shall originate at the electrical service equipment of the building.

5.1.2 The grounding conductor shall be connected to the switch box serving the waterer and to the grounding terminal of the waterer. It shall be insulated from the grounded (neutral) conductor at every point beyond the service equipment, including the disconnecting switch serving the waterer.

5.2 An individual waterer installed in a lot and served by an individual service shall conform to the following: (See Fig. 2)

5.2.1 Overhead conductors shall be firmly attached to support points and conform to paragraphs 4.1.1 and 4.1.2. Buried conductors shall conform to paragraph 4.1.3.

5.2.2 The switch enclosure, the grounding conductor and the grounded circuit conductor serving the waterer shall be connected to an imbedded 1/2 in. × 8 ft (12.7 mm × 2.4 m) nonferrous or other approved electrode.

5.2.3 The grounding conductor shall be installed with the circuit conductors from the switch to the waterer.

5.2.4 The grounding conductor shall be connected to the grounding terminal of the waterer.

SECTION 6—MULTIPLE INSTALLATIONS

6.1 When more than one waterer is to be served from a central service, the service equipment should be located near the load center.

6.2 Each waterer shall be connected as specified in paragraph 5.2.

Cited Standard:

NFPA No. 70, National Electrical Code

WIRING AND EQUIPMENT FOR ELECTRICALLY DRIVEN OR CONTROLLED IRRIGATION MACHINES

Developed by a standards subcommittee of the ASAE Agricultural Wiring and Utilization Committee; approved by the Electric Power and Processing Division Standards Committee; adopted by ASAE June 1973; revised April 1978; revised January 1983; reconfirmed December 1987; approved by ANSI as an American National Standard April 1988.

SECTION 1—PURPOSE AND SCOPE

1.1 This Standard provides detailed information for the application of electrical apparatus to electrically driven or controlled irrigation machines. The purpose of this Standard is to improve the degree of personal safety in operation and application of products and materials under a reasonable range of conditions.

1.2 This Standard covers all electrical equipment, apparatus, components, and wiring necessary for electrically driven or controlled irrigation machines, from the point of connection of electric power to the machine.

1.3 Provisions of this Standard apply to electrical equipment for use on circuits operating at voltages between 30 and 600 V.

SECTION 2—DEFINITIONS AND GLOSSARY OF TERMS

2.1 For purposes of this Standard the following definitions shall apply:

2.1.1 Irrigation machine: (Hereinafter called "machine.") An electrically driven or controlled machine, not portable by hand, used primarily to transport and distribute water for agricultural purposes.

2.1.2 Power conductors: Those conductors that carry current to provide electric power from the machine disconnect to the drive motors.

2.1.3 Control conductors: Those conductors that carry current to control devices necessary for movement of the machine.

2.1.4 Auxiliary conductors: Those conductors used to carry current to devices that are not required for the movement of the machine.

2.1.5 Main control panel: An enclosure containing the main controller and other control devices necessary for starting and stopping the machine.

2.1.6 Auxiliary panels: Enclosed assemblies of auxiliary control devices for the machine, such as motor controllers, relays, switches, and transformers, but not including the main controller or main start and stop function switches. A junction box is not an auxiliary panel.

2.2 The following glossary of terms applies to this Standard:

2.2.1 Bonded: A reliable connection to assure the required electrical conductivity between metal parts required to be electrically connected.

2.2.2 Collector ring: An assembly of slip rings for transferring electrical energy from a stationary to a rotating member.

2.2.3 Front-mounted: Mounting of replaceable devices so that they may be individually replaced from the front of an enclosure without removing subpanels or numerous other devices, gaining access to hidden nuts, or removing a rear enclosure cover.

2.2.4 Front-wired: Mounting of replaceable devices so that they may be individually wired from the front of an enclosure without removing the device, subpanels, or rear enclosure covers to gain access to electrical connections.

2.2.5 Grounded: Connected to earth or to some conducting body which serves in place of earth.

2.2.6 Grounded conductor: A circuit conductor which is intentionally grounded.

2.2.7 Grounding conductor: The conductor used to connect noncurrent-carrying metal parts of a machine to a service-grounded conductor and/or a grounding electrode conductor.

2.2.8 Machine disconnect: A disconnecting means at the point of connection of electrical power to the machine.

2.2.9 Metal-to-metal: Attachment of metal parts to the machine with bolts and screws to provide an adequate contact for bonding purposes where all paint and dirt have been removed from under the bearing surface area of bolt or screw heads.

2.2.10 Readily accessible: Capable of being opened quickly for maintenance, repair, or inspection. May be fastened by latches or held by mechanical interlocks or similar apparatus. A cover firmly attached by two or more screws is not considered readily accessible.

2.2.11 Weatherproof: Constructed or protected so that exposure to the environment will not interfere with successful operation.

SECTION 3—GENERAL REQUIREMENTS

3.1 Disconnecting means

3.1.1 A machine disconnect with overcurrent protection capable of being locked in the OFF position shall be provided at the point of connection of electric power to the machine.

3.2 Enclosures

3.2.1 All enclosures shall be constructed to meet at least the minimum standards for a National Electrical Manufacturers Association Type 3R enclosure as specified in NEMA Standard IS1.1, Controls and Systems.

3.2.2 Entrances into, exits from, and penetrations of the enclosure shall be made in such a manner as to reduce the possibility of the collection of water or contaminants at the point of connection and to preserve the minimum NEMA rating of the enclosure as specified in paragraph 3.2.1.

3.2.3 Dimensions of enclosures shall not be less than the maximum dimensions of the enclosed equipment plus the required electrical clearances as specified in National Fire Protection Association Standard No. 70, National Electrical Code (American National Standard, C1).

3.2.4 Where used, gaskets shall be securely attached to one of the mating parts.

3.2.5 Enclosures should be mounted in such a manner as to minimize the possibility of subsequent physical damage to the enclosure.

3.3 Interlocking

3.3.1 Where personal hazard or property damage may be caused by the failure of any one device to function properly, protective interlocks shall be provided. Where practicable, these interlocks shall interrupt all operations, providing such interruption will not create a hazardous condition.

3.4 Automatic or remote starting

3.4.1 Automatic restarting shall be connected in such a manner that restarting will occur only on resumption of full single-phase or three-phase voltage, as needed, following a power outage.

3.4.1.1 Automatic restarting shall not be construed to encompass the functions of a device or circuit designed for reversing the direction of travel of the machine at a preset point or points.

3.4.2 Machines equipped with automatic restarting or remote starting shall have affixed to the machine a safety sign indicating that the machine can start automatically.

3.4.2.1 This safety sign shall be affixed to the machine disconnect. It shall also be affixed to the main control panel if the machine disconnect is not part of or adjacent to the main control panel.

3.4.2.2 Wording and size of this sign are specified in Section 10—Safety Signs.

3.5 All conductors within an enclosure shall be clearly marked or color coded for identification as specified in paragraph 8.6.

3.6 Transformers shall be of the isolated type with proper overcurrent protection for the transformer, conductors, or control devices it serves.

3.7 Terminal blocks or strips shall be sized to accommodate the number and size of the conductors terminated or connected thereto and shall be rated for the voltage and current transmitted on the connected or terminated conductors.

3.8 Control devices such as relays, limit switches, and similar equipment shall be suitable for the application and shall be capable of handling the voltage and current imposed on or through the devices.

3.9 Panel-mounted devices shall be front-mounted and front-wired for convenient servicing.

3.10 Current ratings

3.10.1 **Machines.** Where a machine is multimotored and individual motors are controlled by alignment switches or similar devices, and where intermittent duty is inherent, the equivalent continuous-current rating and the peak-current rating shall be determined as follows:

3.10.1.1 The equivalent continuous-current rating shall be calculated as 125% of the full-load current rating of the largest motor plus a quantity equal to the sum of the full-load current ratings of all remaining motors multiplied by the maximum percent duty cycle at which the remaining motors can continuously operate.

3.10.1.2 The peak-current demand shall be equal to the sum of the locked-rotor current of the two largest motors plus 100% of the full-load current ratings of all other motors on the circuit.

3.10.2 **Center pivot machines.** Where a machine is multimotored and individual motors are controlled by alignment switches or similar devices, where the machine operates in a circle, and where intermittent duty is inherent, the continuous-current rating and the peak-current rating shall be determined as follows:

3.10.2.1 The equivalent continuous-current rating shall be calculated as 125% of the full-load current rating of the largest motor plus 60% of the sum of the full-load current ratings of the remaining motors.

3.10.2.2 The peak-current rating shall be calculated as two times the locked-rotor current of the largest motor plus 80% of the sum of the full-load current ratings of the remaining motors.

SECTION 4—GROUNDING

4.1 A grounding conductor used for no other purposes than machine grounding shall be provided.

4.1.1 The grounding conductor shall be bonded to the machine within the main control panel enclosure and each auxiliary panel enclosure.

4.1.2 The ground conductor shall be within the same sheath, jacket, or conduit as the power, control, or auxiliary conductors.

4.1.3 The grounding conductor may be bare or insulated as specified in paragraph 8.6.2.1.

4.1.4 Metallic sheath of any cable or conduit shall not be used as the primary grounding conductor.

4.2 Metal-to-metal contact with a part which is bonded to the grounding conductor and the noncurrent-carrying parts of the machine shall be considered as an adequate grounding path.

4.3 Metallic sheath of cable or conduit, where used, shall be grounded.

4.4 External metal parts which could become inadvertently energized shall be grounded.

4.5 Motor frames shall be bonded to the grounding conductor.

4.6 Metal frames of devices such as switches, solenoids, and junction boxes shall be bonded to the grounding conductor or to the noncurrent carrying metal parts of the machine.

4.7 The grounding conductor shall not have less ampacity than its associated power conductors.

4.7.1 When there is a reduction of conductor size because of the use of an interposing relay, controller, or similar device, the grounding conductor may also be reduced to the same size as the power, control, or auxiliary conductors originating at the interposing device.

4.8 Where a machine has a stationary point, such machine shall include provision for the connection of the machine-grounding conductor to the grounding-electrode conductor. Proper installation of the grounding-electrode conductor shall complete an electrical path from the machine-grounding conductor to the grounding electrode.

4.8.1 A prominently displayed, durable sign shall be permanently affixed to the main control panel which indicates the need for proper grounding.

4.8.2 The installation and operating instructions for the machine shall include specific recommendations for connecting or installing the grounding electrode in accordance with the applicable sections of the NFPA Standard No. 70, National Electrical Code.

4.9 A common connection point for all grounding conductors shall be provided.

SECTION 5—MAIN CONTROL PANEL

5.1 The enclosure for the main panel shall conform to paragraph 3.2.

5.2 Disconnecting means

5.2.1 If the machine disconnect is not in or adjacent to the main control panel, a disconnecting means capable of being locked in OFF position shall also be provided in or adjacent to the main control panel.

5.2.2 If the main control panel is remote from the machine, a disconnecting means capable of being locked in the OFF position shall be provided at the machine for removing power from all circuits of 30 V or more.

5.2.3 If the main control panel contains a disconnecting means, it shall be interlocked with the enclosure opening.

NOTE: Means may be provided for qualified persons to gain access to enclosures without removing power if the interlocking is reactivated automatically when the enclosure is closed.

5.2.4 If a disconnecting means is adjacent to the main control panel and live parts of components are not readily accessible, the enclosure opening is not required to contain an interlock. It is recommended, however, that interlocking be provided.

5.2.5 Equipment within an enclosure receiving electrical energy from more than one source shall not be required to have a disconnecting means for the additional source provided its voltage is 30 V or less or meets the requirements of Section 725-31 of the NFPA Standard No. 70, National Electrical Code.

5.3 Hinged doors for enclosure openings shall have a mechanical stop to prevent strain on conductors and termination points contained in or on the door.

5.3.1 Conductors to components or terminations contained in or on the hinged door shall be fastened so that any conductor flexing occurs at the fastener and not at conductor terminals.

5.4 The main control panel shall not be used as a raceway.

5.5 Attachment plugs and receptacles shall be of a locking type of NEMA configuration to prevent "accidental disconnections" and shall be of proper design for the voltage, current, and environment for which they are used.

5.5.1 Where two or more mating receptacles are installed on a machine, they shall not be interchangeable unless both receptacles serve the same purpose, have the same voltage and current rating, and interchangeability will not affect the operation or safety of the machine. Mating receptacles shall be wired so that no energized terminals are exposed when the receptacles are disconnected.

5.6 Operator control devices

5.6.1 All operator-controlled devices such as selector switches and start and stop controls shall be clearly marked as to their function.

5.6.2 All push buttons and selector switches, indicating lights, etc., which are exterior-mounted, shall be of the weatherproof type and shall be installed to conform to paragraph 3.2.2.

5.6.3 "Stop" push buttons shall be red in color. Red color shall not be used for push buttons having functions other than "stop".

5.6.4 All controls shall be protected from possibility of accidental operation by normal servicing or by normal movement of the machine.

5.7 Overcurrent protective devices shall be provided and sized in accordance with the requirements of the device or devices served, but shall not exceed the current allowed by NFPA Standard No. 70, National Electrical Code, for the conductors employed.

5.7.1 The preferred main or master overcurrent device should consist of properly sized fuses (single or dual element) complete with the required holders or fuse blocks.

5.8 If the machine operates where intermittent duty is inherent, a controller in the main control panel, which is used to manually or automatically start and stop the complete machine, shall have an interrupting capability not less than the peak current as determined in paragraphs 3.10.1.2 or 3.10.2.2 and a continuous current rating not less than determined in paragraphs 3.10.1.1 or 3.10.2.1.

5.9 A basic wiring diagram of the machine shall be affixed inside the main control panel with components properly identified.

5.10 The main control panel shall contain a nameplate stating the manufacturer's name, the design voltage, phase and frequency of the incoming power supply, and the current rating of the recommended overcurrent protection for the main power circuit.

SECTION 6—AUXILIARY PANELS

6.1 Enclosures for auxiliary panels shall conform to paragraph 3.2.

6.2 A disconnecting means (one or more) shall be provided where an enclosure contains relays, controllers, switches or other similar devices that may need maintenance or repair and where such enclosures are farther than 9 m (30 ft) from the machine disconnect.

6.3 Enclosure covers shall be secured in a manner that will prevent accidental opening or removal during or resulting from normal operation of the machine.

6.4 Each auxiliary panel shall have affixed to the panel a safety sign indicating the possible presence of a hazardous voltage.

6.4.1 Although paragraph 6.2 provides for disconnection of power in an auxiliary panel, the safety sign is specified to dissuade casual opening of the auxiliary panel and to indicate possible danger should the panel be opened for servicing.

6.4.2 Wording and size of this safety sign are specified in Section 10—Safety Signs.

SECTION 7—MOTORS AND MOTOR CONTROLLERS

7.1 Motors

7.1.1 Motors shall be so constructed or protected so that exposure to the operating environment will not interfere with successful operation. It is recommended that the motors be labeled by the motor manufacturer to indicate that they are specifically designed for use on an irrigation machine in addition to the nameplate marking specified in NFPA Standard No. 70, National Electrical Code.

7.1.2 Motors shall be rodent proof.

7.1.3 If a junction box is supplied, the usable volume of the housing should be a minimum of 200 cm^3 (12 $in.^3$) with 50 mm (2 in.) minimum dimension of the opening. The junction box shall be equipped with an easily accessible, internal frame grounding terminal.

7.2 Motor controllers

7.2.1 Controllers shall be marked with the manufacturer's name or identification, voltage, horsepower rating, and such other data as may be needed to properly indicate the motors for which they are suitable.

7.2.2 Running overcurrent protection

7.2.2.1 Motor running overcurrent protection shall be provided to protect each motor, motor controller, and motor feeder conductor against excessive heating due to motor overload or failure to start.

7.2.2.2 To provide the maximum possible protection, motor running overcurrent devices should be sized smaller than the maximum allowable wherever possible.

7.2.2.3 The minimum number and location of running overcurrent protection devices shall be determined from Table 1.

TABLE 1—RUNNING OVERCURRENT DEVICES

Kind of motor	Supply system	Number and location of overcurrent units (such as trip coils, relays, etc.)
1-phase ac or dc	2-wire, 1-phase ac or dc, one conductor grounded	1 in ungrounded conductor
1-phase ac or dc	3-wire, 1 phase ac or dc, grounded neutral	1 in either ungrounded conductor
3-phase ac	any 3-phase	3, one in each phase

7.2.2.4 Thermal devices or systems mounted in the motor and sensitive to the temperature of the motor, or to both motor temperature and current, may be used in lieu of externally mounted overload units. The thermal devices or systems shall be capable of protecting the motor against stalled conditions and repeated starting under locked-rotor conditions.

7.2.2.5 Automatic resetting of an overload protective device shall not restart the motor where restarting could damage the machine or result in unsafe operation.

SECTION 8—CONDUCTORS

8.1 General

8.1.1 All conductors shall be within either an enclosure, a raceway, or a jacketed cable.

8.1.2 Conductors within 2.6 m (8 ft) of the ground shall be protected from physical damage by enclosure in rigid metal conduit, liquid-tight flexible metal conduit, jacketed metallic-sheathed cable, or other suitable means.

8.1.3 Mechanical protection for conductors may be provided by utilization of the machine structure.

NOTE: It is the intent of paragraphs 8.1.2 and 8.1.3 to protect conductors from physical damage by livestock or hazards encountered in normal usage of the machine.

8.2 Size

8.2.1 The size of power conductors shall be based on the assumption that the maximum voltage at any motor will be the design motor voltage plus 5% with allowance for a 10% voltage drop using this assumed maximum and the average continuous current calculated from paragraphs 3.10.1.1 or 3.10.2.1.

8.2.1.1 In addition to voltage drop considerations, the ampacity of power conductors shall equal or exceed that permitted by NFPA Standard No. 70, National Electrical Code.

8.2.2 Control and auxiliary conductor size shall provide sufficient ampacity to carry the total current drawn by the devices served and shall not be smaller than shown in Table 2.

NOTE: Conductors smaller than No. 18-gage are included to anticipate use of solid-state control devices which may operate at currents of 1 ma or smaller. Conductors smaller than No. 18-gage shall not be used for other than circuits serving solid-state control devices.

8.2.3 Conductors supplying a motor shall have an ampacity not less than 125% of the motor full-load current.

TABLE 2—COPPER CONDUCTOR AMPACITY SINGLE CONDUCTOR CONSTRUCTION (CONTROL OR AUXILIARY CIRCUITS)

Conductor size AWG	Ampacity in		Minimum standing
	cable or raceway	control enclosure	
24	2	2	7
22	3	3	7
20	5	5	7
18	7	7	7
16	10	10	7
14	15	20	7
12	20	25	7
10	30	40	7
8	40	55	7

8.3 Stranding

8.3.1 Conductors used to connect or interconnect control devices that are physically displaced in normal operation of the machine shall be of stranded construction.

8.3.2 All power, control, and auxiliary conductors should be annealed stranded copper with minimum stranding as shown in Table 2.

8.3.3 Crimp-type connectors shall not be used if solid conductor wires are employed.

8.4 Insulation

8.4.1 Conductor insulation should be flame, moisture, and corrosion resistant and should be suitable for operation within temperatures of −10 ° to 60 °C. Conductor insulation shall be rated at not less than 75 °C wet location.

8.4.2 Conductors used to connect or interconnect components within an enclosure shall be rated at not less than 600 V when any circuits utilize voltage over 300 Vac and less than 600 Vac.

8.4.3 All conductors within an enclosure shall be insulated with the exception of the grounding conductor which may be bare.

8.4.4 Conductor insulation in a cable, conduit, or tubing, where different voltages are present, shall be rated at or above the highest voltage carried in the cable, conduit, or tubing.

EXCEPTION: Control or auxiliary conductors may have a lesser voltage insulation providing that the control or auxiliary computors are within an inner cable with an outer protective covering rated at or above the highest voltage of the power conductors in the cable, conduit, or tubing, and also providing that the conductor insulation within this protective covering is rated at or above the highest voltage utilized within the inner multiconductor cable.

8.5 Jacketing. Cable jacketing where the cable is attached to a metal frame and operated under a sprinkler system need not be flameresistant. However, it should be resistant to sunlight, moisture, and corrosion. The jacketing should also provide some mechanical protection, flexibility, and be suitable for operation within temperatures of −10 ° to 60 °C, and should not be attractive to livestock. Nonwicking filler material should be used. The jacketing construction should allow proper termination to provide a mechanically strong, weatherproof connection that can be field-installed with standard electrical tools.

8.6 Identification

8.6.1 All conductors within an enclosure shall be identified in some manner at each termination, and this identifying means shall be consistent throughout the machine.

8.6.2 Color coding for identification

8.6.2.1 Grounding conductors shall be either bare, green, or green with a yellow stripe.

8.6.2.2 Grounded conductors shall be white or natural gray.

8.6.2.2.1 Where grounded conductors of different systems are in the same cable, conduit, or tubing, additional grounded conductors may be identified by white with a colored stripe other than green.

8.6.2.3 Where color coding is used for power conductors, the colors of black, red, and blue should be used for ungrounded conductors.

8.6.2.4 All ungrounded conductors of the same identification shall be connected to the same circuit.

8.6.2.5 All conductors for systems of different voltages shall be of different colors.

SECTION 9—COLLECTOR RINGS

9.1 Enclosure

9.1.1 Enclosures for a collector ring shall conform to paragraph 3.2.

9.1.2 Enclosures for collector rings are not required to have an interlock.

9.2 Ampacity

9.2.1 Slip rings transmitting current for power functions shall have a continuous-duty rating not less than that determined in paragraph 3.10.

9.2.2 The remaining slip rings for control and auxiliary purposes shall have a continuous-duty rating not less than 125% of the full-load current of the largest device served plus the sum of the full-load currents of all other devices served.

9.2.3 The current-carrying capacity of the slip ring used for machine grounding shall be of the same ampacity as the slip ring with the greatest ampacity in the assembly.

9.2.4 When slip rings or slip-ring terminals are of different ampacity, each ring or terminal shall be identified with a durable marking which indicates the ampacity.

SECTION 10—SAFETY SIGNS

10.1 Automatic starting signs

10.1.1 The safety sign for automatic starting specified in paragraph 3.4.2 shall have the following wording:

CAUTION

This machine may start automatically. Do not service until machine disconnect is in "OFF" position and locked.

10.1.2 The upper portion of this sign shall have the word "CAUTION" in all capital letters in black on a yellow background. The lower portion for the remaining wording may have uppercase or lowercase letters in yellow on a black background.

10.2 Auxiliary panel signs

10.2.1 The auxiliary panel safety sign specified in paragraph 6.5 shall have the following wording:

DANGER

_____VOLTS. Do not open until machine disconnect is in "OFF" position and locked.

10.2.2 The upper portion of the auxiliary panel safety sign shall have the word "DANGER" in all capital letters in white on a red background. Lower portion for the remaining sign wording shall have uppercase letters for the words "_VOLTS." The other words may be in uppercase or lowercase letters. Words in the lower portion shall have letters in red on a white background.

10.2.3 The underscore in paragraph 10.2.2 shall be replaced by the design voltage of the machine as specified on the nameplate described in paragraph 5.10.

10.3 Design and letter size

10.3.1 The ratio of width to height of the upper portion of the sign shall fall within the range of 2:1 and 5:1.

10.3.2 The lower portion of the sign shall be the same width as the upper portion.

10.3.3 The height of the lower portion of the sign shall not be less than the height of the upper portion.

10.3.4 Minimum height of letters shall be 13 mm (0.5 in.) in the upper portion and 6.4 mm (0.25 in.) in the lower portion.

10.4 Additional safety signs should comply with ASAE Standard ASAE S318, Safety for Agricultural Equipment, or American National Standard Z35.1, Specifications for Accident Prevention Signs.

10.5 The safety-alert symbol specified in ASAE Standard ASAE S350, Safety-Alert Symbol for Agricultural Equipment, should be used with each safety sign to indicate that safety is involved.

Cited Standards:

ASAE S318, Safety for Agricultural Equipment
ASAE S350, Safety-Alert Symbol for Agricultural Equipment
ANSI Z35.1, Specifications for Accident Prevention Signs
NEMA IS1.1, Controls and Systems
NFPA No. 70, National Electrical Code

INSTALLATION AND MAINTENANCE OF FARM STANDBY ELECTRIC POWER

Corresponds to IMFS—1—1974 published by Electrical Generating Systems Marketing Association; developed by ASAE Standby Electric Power Committee; approved by Electric Power and Processing Division Standards Committee; adopted by ASAE as a Recommendation December 1973; reclassified as an Engineering Practice December 1977; reconfirmed December 1978, December 1983.

SECTION 1—PURPOSE AND SCOPE

1.1 The purpose of this Engineering Practice is to provide information to assist installers, maintenance personnel, operators and others in the proper installation and maintenance of farm standby electrical systems.

1.2 The scope of this Engineering Practice covers both engine-driven and tractor-driven generators for farm standby electrical power service as defined in Electrical Generating Systems Marketing Association Standards EGS 1-1970, Standard Specifications for Standby Engine Driven Generator Sets, and TDGS 1-1972, Standard Specifications for Farm Standby Tractor Driven PTO Generators. The terms generator and alternator may be used interchangeably in this Engineering Practice.

SECTION 2—DEFINITIONS

2.1 Alternators: A device for converting mechanical energy into electrical energy in the form of alternating current. Also called an AC generator.

2.2 Transfer switch: An automatic or manual device for transferring one or more conductor connections from one power source to another.

2.3 Electric utilities: Enterprises engaged in the production and/or distribution of electricity for use by the public.

2.4 Engine-driven generators: An electrical generator so constructed that its rotor is driven by an engine, and generates a voltage.

2.5 Generator: A general name for a device that converts mechanical energy into electrical energy. The electrical energy may be direct current (DC) or alternating current (AC). An AC generator may be called an alternator.

2.6 PTO: Abbreviation used for power take-off.

2.7 Standby power: The power to be delivered by a generator during periods when there is an outage of utility or prime power sources or when these sources are outside the acceptable limits of quality or capacity.

2.8 Tractor-driven generator: An electric generator so constructed that its rotor is driven by the PTO mounted on a farm tractor through a PTO speed changer and drive shaft.

2.9 Additional terminology and definitions are included in EGSMA Standard GTD 3-1975, Glossary of Standard Industry Terminology and Definitions.

SECTION 3—INSTALLATION

3.1 The generator plant should be carefully inspected on delivery for evidence of possible damage. If the damage appears to be of a major nature, the generator should not be operated until the fault has been corrected.

3.2 The manufacturer's installation and operator's manual should be thoroughly reviewed before attempting to install the standby generating system, whether it is an engine-driven or tractor-driven type.

3.3 Location

3.3.1 The plant should be located in an atmosphere that is free from excessive dust, wind-blown particles, high and/or low temperatures and corrosive fumes.

3.3.2 Allowance should be made for a minimum clearance of 3 ft (0.9 m) around the set for service accessibility.

3.4 Mounting

3.4.1 Permanently installed, engine-driven electric plants should be mounted on a concrete base in accordance with the manufacturer's instructions

3.4.2 Vibration dampening pads should be placed between the skid, or mounting base, and the floor or concrete base to minimize the transfer of vibration to other equipment.

3.4.3 Tractor-driven generators are normlly either stationary mounted or mobile mounted on a trailer or a three point hitch.

3.4.3.1 Stationary mounting should follow the manufacturer's recommendations of mounting on concrete pads.

3.4.3.2 In the event of mobile mounting, such as a trailer or a three point hitch, the mounting should be of sufficient size and stability to withstand pulling over rough terrain and to withstand torque stresses experienced when full loads are applied to the generator through the power take-off drive.

3.4.3.3 Alternators should be mounted on the trailer or three point hitch such that power drive lines are in as good alignment as possible.

3.5 The power take-off and power take-off drive lines for PTO tractor-driven generators should conform to safety provisions of ASAE Standard S318, Safety for Agricultural Equipment, and ASAE Standard S207, Operating Requirements for Tractors and Power Take-Off-Driven Implements.

3.6 Ventilation for engine-driven generators. It is imperative that the engine and generator have an adequate supply of air for combustion and cooling.

3.6.1 Duct work should be used when necessary from the radiator to the outlet to prevent hot air recirculation.

3.6.2 There should be a flexible section at any connection to the plant.

3.6.3 The air inlet must be larger in size than the air outlet. The air inlet should be 2 times the size of the radiator frontal area or as recommended by the manufacturer.

3.7 Exhaust. When an engine-driven generator is installed inside a building or other enclosure, means shall be provided for exhausting the gases from the engine exhaust system out of the building or enclosure.

3.7.1 Where the exhaust pipe passes through a wall of combustible material, the wall shall be shielded by a metal thimble at least 12 in. (305 mm) larger in diameter than the exhaust pipe.

3.7.2 The end of the exhaust pipe should be equipped with a rain cap, and should be located a suitable distance away from th air inlet of the engine so that exhaust gases cannot be drawn back through the air inlet.

3.7.3 Flexible steel tubing shall be used for connecting the engine exhaust fittings to a rigid pipe or muffler to eliminate the possibility of breakage.

3.7.4 Exhaust piping should be as short as possible with minimum bends and restrictions and large enough to prevent back pressures higher than those recommended by the engine manufacturer.

3.7.5 A drain cock should be installed at the lowest point in the exhaust piping system for draining any moisture accumulation as a result of condensation.

3.8 Fuel. Engine-driven generator sets shall be provided with fuel lines and fuel equipment to comply with the following standards of the National Fire Protection Association:

NFPA No. 31, Standard for Installation of Oil Burning Equipment.

NFPA No. 54, Standard for Installation of Gas Appliances and Gas Piping.

NFPA No. 58, Standard for Storage and Handling of Liquified Petroleum Gases.

3.8.1 The fuel line from the natural gas meter to the fuel inlet of the engine shall be sized and installed to meet American National Standard Z83.1, Installation of Gas Piping and Gas Equipment on Industrial Premises and Certain Other Premises.

3.9 Electrical. The following safety rules should be observed at all times:

3.9.1 All circuits not known to be "dead" must be considered "live" and dangerous.

3.9.2 All exposed conductors, terminals, and components should be regarded as energized and treated accordingly.

3.9.3 Metal tools, flashlights, metallic pencils, and other exposed conducting objects should not be used while working near energized electrical equipment.

3.9.4 Precautions should be taken to avoid grounding your body while using electrical measuring apparatus or making adjustments to energized electrical equipment.

3.9.5 All equipment should be de-energized before connecting or disconnecting leads.

3.9.6 Electrical wiring shall be in accordance with the recommendations of the manufacturers and comply with local regulations and NFPA No. 70, National Electrical Code (ANSI C1).

3.9.7 Electric generators on engine or tractor-driven units shall be grounded in accordance with Article 250, NFPA No. 70, National Electrical Code (ANSI C1).

3.10 Automatic transfer switches or manual transfer switches shall be installed so that the generator and the electric utility's power cannot be energized at the same time. Transfer switches shall be installed in accordance with Article 750, NFPA No. 70, National Electrical Code (ANSI C1).

3.11 Shields and guards

3.11.1 Provisions shall be made to minimize the possibility of personal contact with rotating components by installing suitable shields and guards.

3.11.2 A guard or shield shall be provided for heat protection to minimize the possibility of inadvertent contact with any exposed element which may cause burns during normal operation or servicing.

3.11.3 Loose clothing should not be worn while working near rotating equipment.

SECTION 4—PTO GENERATORS—INITIAL STARTUP

4.1 The PTO generator owner's manual should be thoroughly read and followed.

4.2 Prior to operation the PTO speed changer shall be filled to the specified level with oil recommended by the manufacturer.

4.3 Electrical wiring should be checked for loose connections and for loose or missing cap screws, bolts and nuts.

4.4 Starting tractor-driven PTO generators

4.4.1 When so equipped, set the load circuit breaker on the generator to "off."

4.4.2 Connect the PTO drive member, or PTO shaft, securely to the generator and the tractor. Install all safety shields. Position PTO shaft as near to a straight line as possible, not to exceed a 10 degree angle.

4.4.3 Adjust power take-off speed to 540 to 1000 rpm (depending on type) by advancing the tractor throttle, so that PTO generator is producing 60 hertz, as indicated on the frequency meter. If no frequency meter is available adjust to proper voltage as indicated on the voltmeter.

4.4.4 Place power cable coming from the manual transfer switch in the correct power plugs or receptable on the generator control panel.

4.4.5 Place transfer switch in the emergency position.

4.4.6 When so equipped, place the load circuit breaker in the "on" position.

4.4.7 Maintain frequency meter at 60 hertz by adjusting tractor speed. If frequency meter is not available, keep voltmeter at correct voltage by adjusting tractor PTO speed.

4.4.8 When normal power has been restored and after a sufficient time has elapsed to assure that power restoration is not temporary, return the manual transfer switch to normal or utility power.

4.4.9 Place the generator circuit breaker in "off" position. Do not remove the connecting plug before opening the breaker. A shock hazard may result if the plug is removed before opening the breaker.

4.4.10 Slowly reduce power take-off speed to a minimum and disengage power take-off lever.

SECTION 5—MANUAL START, ENGINE-DRIVEN GENERATOR—INITIAL STARTUP

5.1 The engine generator set owner's manual should be thoroughly read and followed.

5.2 Prior to operation the engine crankcase shall be filled to the specified level with oil recommended by the manufacturer.

5.3 Liquid cooled, engine-driven generator sets are normally shipped without liquid in the cooling system. If freezing temperatures are encountered use antifreeze with liquid cooling systems. Fill radiator with proper mixture of water and antifreeze to specified level in accordance with manufacturer's recommendations.

5.4 Check all electrical wiring for loose connections and for loose or missing cap screws, bolts and nuts.

5.5 Starting batteries may be either the standard or dry charge type. If dry charged, fill each battery cell with battery electrolyte. Do not allow battery acid to contact engine or generator and control parts.

5.6 Starting manual-start, engine-driven generators

5.6.1 Place the generator circuit breaker in the "off" position, or open the line switch to disconnect the load.

5.6.2 When so equipped, turn the ammeter and voltmeter selector switches to the desired position.

5.6.3 When equipped with a voltage regulator, rotate the voltage regulator adjusting rheostat counterclockwise (decrease) to minimum voltage.

5.6.4 Turn the voltage regulator "on-off" switch to the "off" position (regulator will be damaged otherwise).

5.6.5 If the engine is equipped with a speed control (throttle), set it for idle.

5.6.6 Place the engine stop switch in the "on" position.

5.6.7 Place the start switch button and hold in until the engine starts.

IMPORTANT: Do not energize the starting motor for a period exceeding 30 seconds without pausing for two minutes to allow it to cool. To do so may cause overheating and damage to the motor windings.

5.6.8 Observe the oil pressure gauge, water temperature gauge, and DC (battery charging) ammeter if so equipped. As soon as the engine attains normal operating temperature, adjust the speed control, if required, to produce 60 hertz operation by viewing frequency meter or tachometer if supplied. Set for 1800 rpm for 4-pole generators or 3600 rpm for 2-pole generators.

5.6.9 Turn the voltage regulator "on-off" switch to the "on" position. Rotate the voltage regulator adjusting rheostat, clockwise (increase) to obtain rated AC voltage indicated on the AC voltmeter.

5.6.10 Apply the load to the generator by placing the generator circuit breaker in the "on" position, or by closing the line switch.

5.6.11 Observe the monitoring meters and adjust the generator voltage to the desired output.

5.6.12 The generator may now be stopped and started without making further adjustments.

SECTION 6—AUTOMATIC START, ENGINE-DRIVEN GENERATOR EQUIPPED WITH AUTOMATIC SWITCH—INITIAL STARTUP

6.1 Perform the functions in paragraphs 5.1 through 5.5.

6.2 Place the generator circuit breaker in the "off" position.

6.3 When so equipped, turn the ammeter and voltmeter selector switches to the desired position.

6.4 When equipped with a voltage regulator, rotate the voltage regulator adjusting rheostat counterclockwise (decrease).

6.5 Turn the voltage regulator "on-off" switch to the "off" position (regulator will be damaged otherwise).

6.6 Place the engine start switch in the "on" or start position to energize the cranking motor (starter).

6.7 Observe the oil pressure gauge, water temperature gauge, and DC (battery charging) ammeter if so equipped. As soon as the engine attains normal operating temperature, adjust the speed control if required to produce 60 hertz operation by viewing frequency meter or tachometer if supplied, and set for 1800 rpm for 4-pole generators or 3600 rpm for 2-pole generators.

6.8 Turn the voltage regulator "on-off" switch to the "on" position. Rotate the voltage regulator adjusting rheostat, clockwise (increase), to obtain rated AC voltage indicated on the AC voltmeter.

6.9 Place the generator circuit breaker in the "on" position. Turn the 4-position switch on the automatic line transfer switch to "auto" position. Turn the main utility service breaker or switch to "off" (thus simulating a power outage). The automatic line transfer switch should operate, start the engine, and transfer the load to the generator set.

6.10 Observe AC voltmeter and adjust voltage to desired output if required.

6.11 Turn the main utility service breaker to "on" position. The automatic transfer switch should operate, transferring the load to the normal utility power, and the engine-driven generator should shut down. The automatic genrator set should start and accept load without further adjustment upon transfer of the automatic transfer switch whenever a power outage occurs.

SECTION 7—EXERCISING OF STANDBY UNITS

7.1 Tractor-driven PTO generators should be exercised at least every 90 days to blow out dust, to dry out moisture which may have entered the generator, to remove oxidation from sliprings, and to familiarize the operator with the necessary steps to take during an actual emergency. If possible, at least 50 percent of its normal load should be applied to the generator during the exercise period.

7.2 Some automatics engine-driven generator sets are equipped with an automatic exerciser which starts the unit automatically once per week, runs the unit for a pre-determined time and then shuts the unit down automatically. If the engine-driven generator is not equipped with an automatic exerciser, manual exercising should be required at least every 7 days. Exercising periods should be long enough to enable the engine to obtain normal operating temperature while carrying, if possible, at least 50 percent of its normal load. To exercise engine-driven generator sets, proceed as follows:

7.2.1 Before starting engine, check lubricating oil and coolant levels. Make complete visual inspection of unit to be sure it is in operating condition.

7.2.2 Start and run engine 5 minutes with no load.

7.2.3 Run engine at rated speed with whatever load is available up to full load for the period of time required to reach normal operating temperature. Continue to operate engine for 30 minutes or more dependent upon manufacturer's recommendations. Check and correct any coolant or oil leaks.

7.2.4 Run engine with no load for 5 minutes to allow combustion chamber temperature to decrease gradually to a minimum before stopping.

SECTION 8—MAINTENANCE

8.1 Maintenance includes those functions and activities that will keep the standby generating system in peak operating condition and prevent unneccessary trouble from developing.

8.2 Batteries used for starting engine-driven generator sets should never be allowed to stand for a long period of time in a discharged condition. It is recommended that standby engine-driven generator sets be equipped with a rectifier type battery charger. The level of the battery electrolyte should be checked weekly under constant use and monthly with intermittent use. Refill to proper level with distilled water. Keep battery clean. Do not allow battery acid to contact engine or generator and control parts.

8.3 Generators

8.3.1 Brushless generators require little maintenance. Dust and dirt should be blown from interior of generator by means of low pressure compressed air (30 psi or under). Most generator bearings are of the sealed type and require no re-lubrication.

8.3.2 Generators equipped with brushes should be checked in accordance with the manufacturer's recommendations, but not to exceed 500 hours of initial running and every 100 hours thereafter. If brush length is less than 0.5 in. (13 mm), the brush should be replaced.

8.3.3 On units with a commutator, an even brown film on the commutator denotes a desirable condition, and it should not be removed. If the commutator becomes rough or dirty, it can be cleaned by using a fine grade of sandpaper. Emery cloth should not be used since it contains metallic parts which could short out the commutator segments.

8.4 Engine maintenance should be in accordance with the engine manufacturer's recommendations and instructions regarding the types of lubricants used and the frequency of application.

8.5 An owner's service record should be maintained showing hours of operation for scheduled tune up and changing of lubricating oil, oil filters, air cleaners, fuel filters, etc.

8.6 Preventive maintenance

8.6.1 Troubles that occur in standby generator operation are avoided when those responsible for maintenance adhere to an adequate program of lubrication, inspection and maintenance.

8.6.2 The time and expense involved in a good maintenance program are only a fraction of that incurred when poor maintenance practice results in a major malfunction or breakdown at the very time that the standby generator is required to provide power during an emergency.

Cited Standards:

ANSI Z83.1, Installation of Gas Piping and Gas Equipment on Industrial Premises and Certain Other Premises
ASAE S207, Operating Requirements for Tractors and Power Take-Off Driven Implements
ASAE S318, Safety for Agricultural Equipment
EGSMA EGS1-1970, Standard Specifications for Standby Engine Driven Generator Sets
EGSMA GTD2-1971, Glossary of Standard Industry Terminology and Definitions
EGSMA TDGS1-1972, Standard Specifications for Farm Standby Tractor Driven PTO Generators
NFPA No. 31, Standard for Installation of Oil Burning Equipment
NFPA No. 54, Standard for Installation of Gas Appliances and Gas Piping
NFPA No. 58, Standard for Storage and Handling of Liquified Petroleum Gases
NFPA No. 70, National Electrical Code

ELECTRICAL SERVICE AND EQUIPMENT FOR IRRIGATION

Proposed by the Nebraska Inter-Industry Electrical Council and The Irrigation Association; reviewed by the ASAE Soil and Water Division Standards Committee; approved by the Electric Power and Processing Division Standards Committee; adopted by ASAE as a Tentative Standard December 1978; reconfirmed December 1979, December 1980, December 1981, December 1982, December 1983, December 1984; reclassified as a full Standard and revised December 1985.

SECTION 1—PURPOSE

1.1 The purpose of this Standard is to provide a common document for use by all those involved in electrical irrigation systems; such as electricians, power suppliers, well drillers, irrigation dealers and manufacturers, extension specialists and irrigators.

SECTION 2—SCOPE

2.1 This Standard applies to three-phase, 240 V, or 480 V service, the most commonly used irrigation service voltages for irrigation pump motors, irrigation machines, and auxiliary equipment. The Standard is in accordance with the National Fire Protection Association Standard No. 70-1984, National Electrical Code (American National Standard, C1), the 14th Edition of Canadian Electrical Code, Part I—1982 where applicable (Canadian Standards Association Standard, C22.1-1982). All materials shall conform to Article 100 of NFPA Standard No. 70, National Electrical Code, and in Canada shall conform to Section 2-024 of Canadian Electrical Code.

SECTION 3—GENERAL EQUIPMENT REQUIREMENTS

3.1 The minimum irrigation electrical installation consists of a circuit disconnecting means (safety switch), a motor controller (starter), a raceway or conduit for conductors, and a pump motor.

3.2 In many installations, equipment such as irrigation machines, injector pumps, compressors, lights, magnetic oilers, time switches, time-delay restart relays, and timers are used which require additional control and protective equipment.

3.3 Clearance. Sufficient access and working space shall be provided and maintained around all electrical equipment. A minimum of 91 cm (3 ft) is required by Article 110-16, NFPA Standard No. 70, National Electrical Code, and Section 2-308, Canadian Electrical Code, for voltages greater than 150 V. Location of enclosures, motors, and irrigation piping is important in maintaining proper access.

3.4 Support. Electrical installations shall be designed and constructed to provide adequate support and protection for equipment and services.

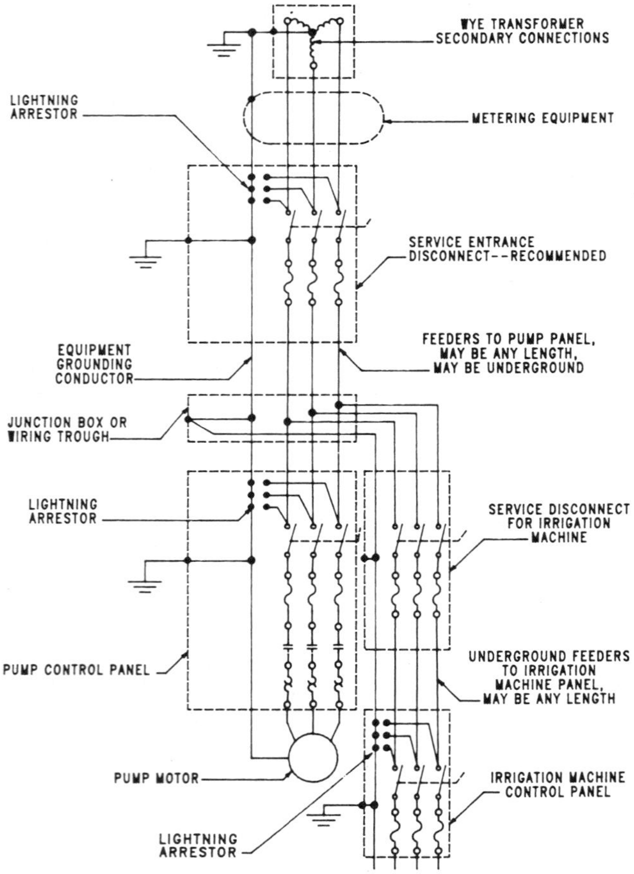

FIG. 1—RECOMMENDED EQUIPMENT AND GROUNDING FOR WYE TRANSFORMER SECONDARY CONNECTIONS

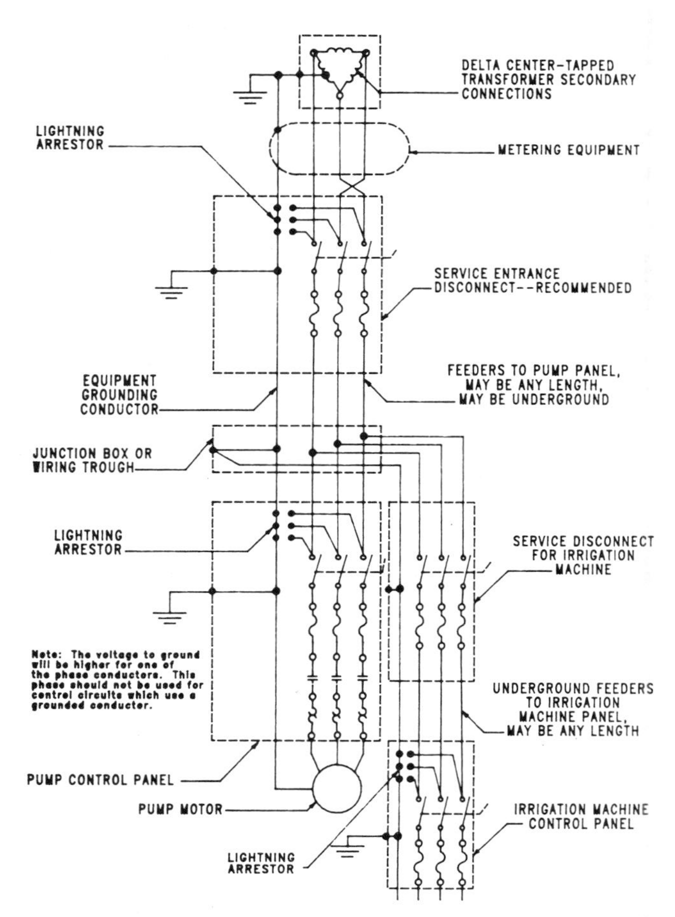

FIG. 2—RECOMMENDED EQUIPMENT AND GROUNDING FOR DELTA CENTER-TAPPED TRANSFORMER SECONDARY CONNECTIONS

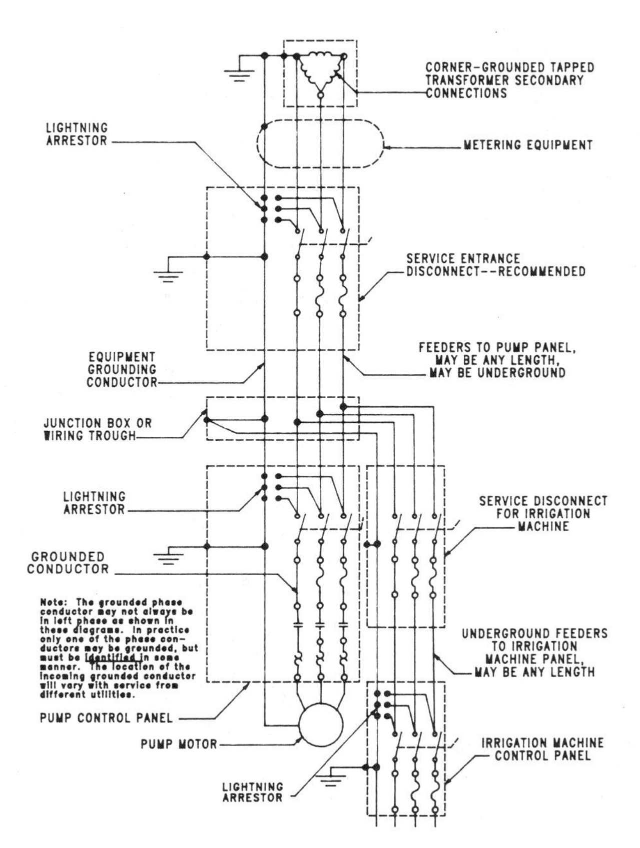

FIG. 3—RECOMMENDED EQUIPMENT AND GROUNDING FOR A CORNER-GROUNDED DELTA TRANSFORMER SECONDARY CONNECTIONS

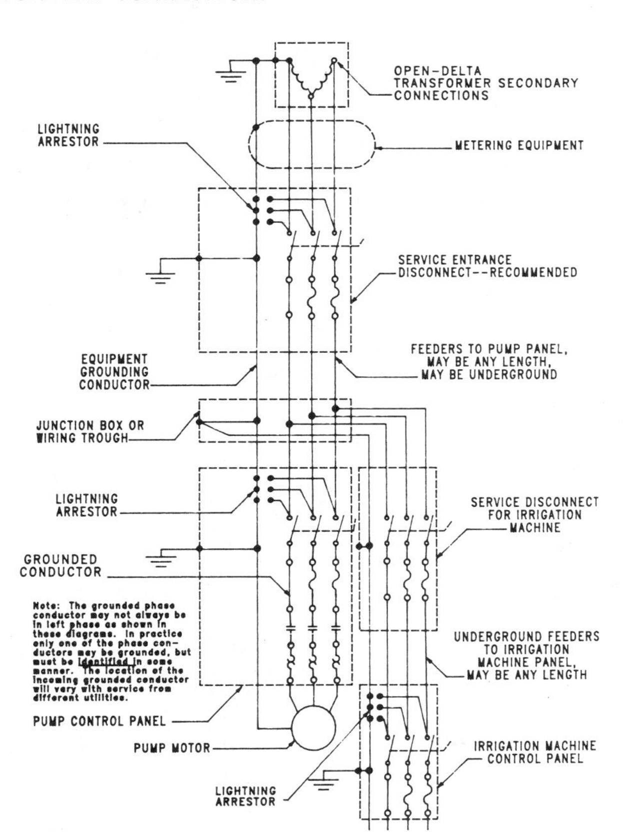

FIG. 4—RECOMMENDED EQUIPMENT AND GROUNDING FOR A GROUNDED PHASE OPEN-DELTA TRANSFORMER SECONDARY CONNECTIONS

3.5 Before purchasing and installing electrical equipment, determine the type of electrical service which will be supplied. Figs. 1-7 show the proper equipment to go with each specified service. Canadian Electrical Code does not permit service as shown in Figs. 3, 4, and 7.

SECTION 4—CIRCUIT DISCONNECTING MEANS

4.1 Circuit disconnecting means shall be a fused safety switch, not a circuit breaker.

4.1.1 NFPA Standard No. 70, National Electrical Code, and Canadian Electrical Code permit circuit breakers. Because of infrequent operation, breakers may be adversely affected by dust and moisture. Therefore a fused switch is recommended for more positive fault protection.

4.2 Disconnecting means shall be four-wire, three-pole, solid neutral, horsepower rated and rated as service equipment.

4.2.1 Where two fuses are used, three fuses shall be used when service has all ungrounded phase conductors (see Figs. 1, 2, 4 and 6). Two fuses shall be used when service has a grounded phase (see Figs. 3, 4 and 7).

4.3 Enclosure. Enclosures shall meet National Electrical Manufacturers Association 3R rating for outdoor installation and NEMA 1 rating for indoor installation (Ref. NEMA Standard ICS 6-1978, Enclosures for Industrial Controls and Systems).

4.4 All service disconnects shall be permanently labeled as to their function in accordance with Article 230-70, NFPA Standard No. 70, National Electrical Code, and Canadian Electrical Code Rule 6-200.

SECTION 5—MOTOR CONTROLLERS (STARTERS)

5.1 Enclosure. Enclosures shall meet NEMA 3R rating for outdoor installation and NEMA 1 rating for indoor installation (Ref. NEMA Standard ICS 6-1978, Enclosures for Industrial Controls and Systems).

5.2 Type. Magnetic, manual, or solid state.

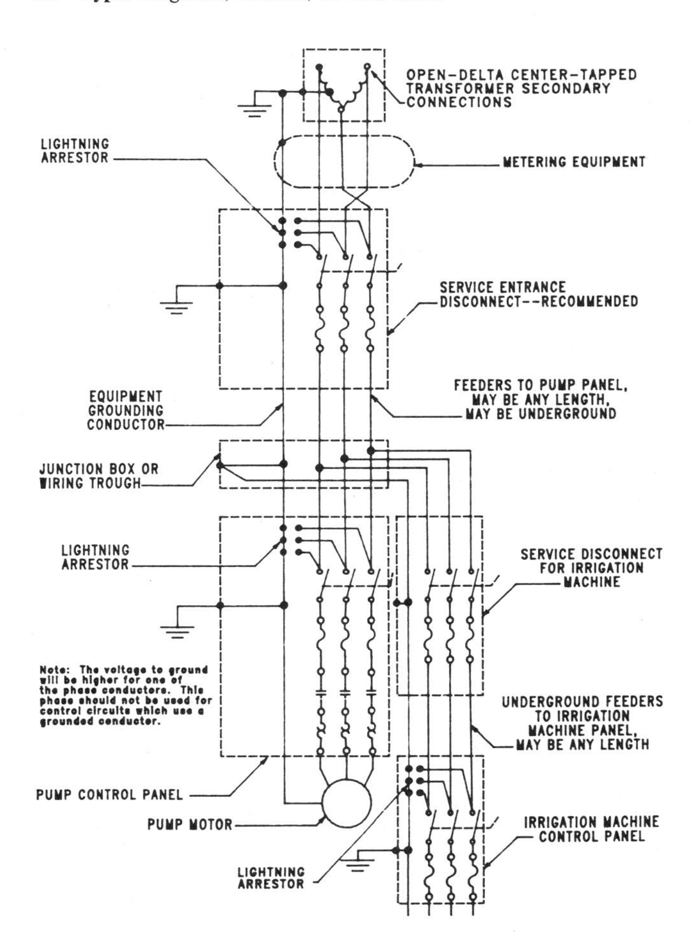

FIG. 5—RECOMMENDED EQUIPMENT AND GROUNDING FOR OPEN-DELTA, CENTER-TAPPED TRANSFORMER SECONDARY CONNECTIONS

5.3 Overload relays. Ambient-compensated with three overload elements (heaters). In submersible pump applications, "fast-trip" heater elements may be required.

5.4 Heater element selection. Ambient-compensated overload heaters should be selected from the full-load current rating of the motor, and the controller manufacturer literature.

5.4.1 Where capacitors for power factor correction are added on the load side of a controller, the overload heater rating should be reduced according to the running current measured after capacitors are installed (see paragraph 11.2).

5.5 Internal motor protective devices may be used.

SECTION 6—PUMP PANELS

6.1 A circuit disconnecting means and motor controller may be mounted in a single enclosure which has been approved as an irrigation pump controller (Ref. NEMA Standard ICS 2-449-1978, AC Automatic Combination Irrigation Pump Controllers).

SECTION 7—PUMP MOTORS

7.1 General specifications. Most pump motors are three phase, 60 hertz, squirrel cage induction, normal starting torque, 40 °C rise, with 1.15 service factor, or 50 °C rise with 1.0 service factor if non-submersible.

7.1.1 Loading. Motors shall be selected for the load so that the full-load current does not exceed the service factor.

7.2 Deep-well turbine pump vertical-hollow-shaft motors. These motors shall be provided with bearings of adequate thrust capacity to equal or exceed the total thrust imposed by the pump and shall be equipped with a nonreverse ratchet to prevent operation in reverse rotation. These motors shall meet NEMA weather-protected type 1 specifications, MG 1-1.25, Motors and Generators.

7.3 Other motors. Motors shall be equipped with bearings suitable for the application and shall be selected from frame sizes, facing, and shaft dimensions recommended by NEMA Standard MG 1-Part 11, Motors and Generators.

7.3.1 Horizontal motors must meet NEMA specifications for drip-proof motors, MG 1-1.25, Motors and Generators.

7.3.2 When motors are used in the vertical position, they shall meet NEMA specifications for weather-protected type 1 motors, MG 1-1.25, Motors and Generators, and be suitable for such operation.

7.4 Rodent screens. Motors shall be protected from rodents by factory-installed screens (Ref. NEMA Standard MG 1-14.09, Motors and Generators). All unused knock-outs on motor and control enclosures shall be closed.

7.5 Guarding. Guards shall be installed to adequately protect persons from accidental contact with belts, pulleys, or other rotating equipment in accordance with ASAE Standard S318, Safety for Agricultural Equipment.

SECTION 8—OTHER MOTORS

8.1 Other motors, such as those used on injector pumps, hydraulic pumps, and compressors shall be suitable for use in the intended environment.

8.2 Guarding for other motors shall be in accordance with paragraph 7.5.

SECTION 9—EQUIPMENT PROTECTION AND SIZING

9.1 Recommended sizing for fuses, switches, starters, conductors, and conduit are listed in Table 1 for 230 V motors and in Table 2 for 460 V motors. The circuit location of switches and protective devices is shown in all the figures.

9.1.1 Electrical service is nominally 240 or 480 V at the transformer power supply. This voltage provides the proper range for 230 and 460 V motors.

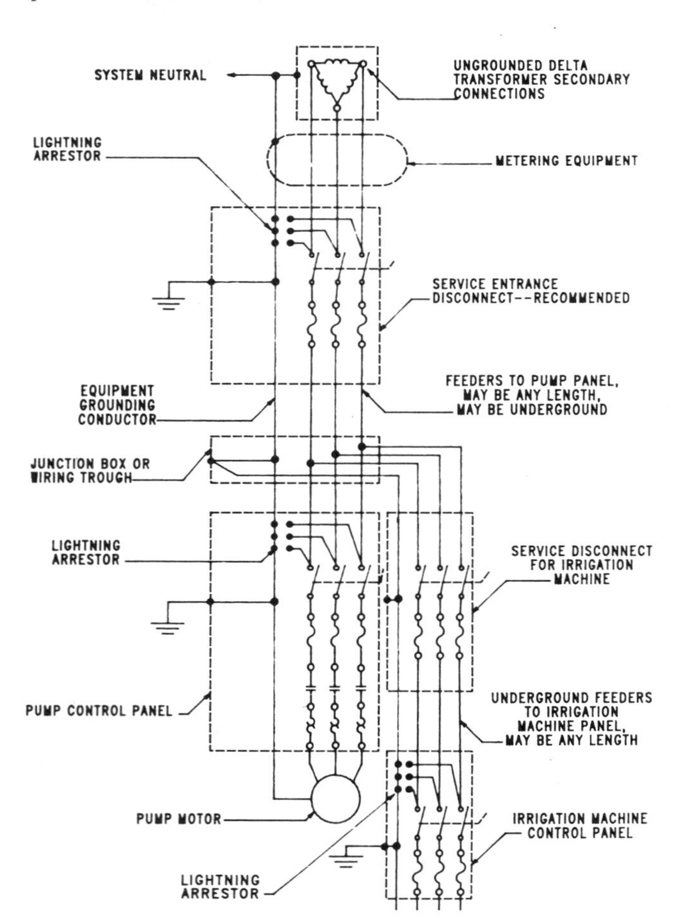

FIG. 6—RECOMMENDED EQUIPMENT AND GROUNDING OF THREE-PHASE, THREE-WIRE SERVICE WHEN THE TRANSFORMER SYSTEM IS UNGROUNDED

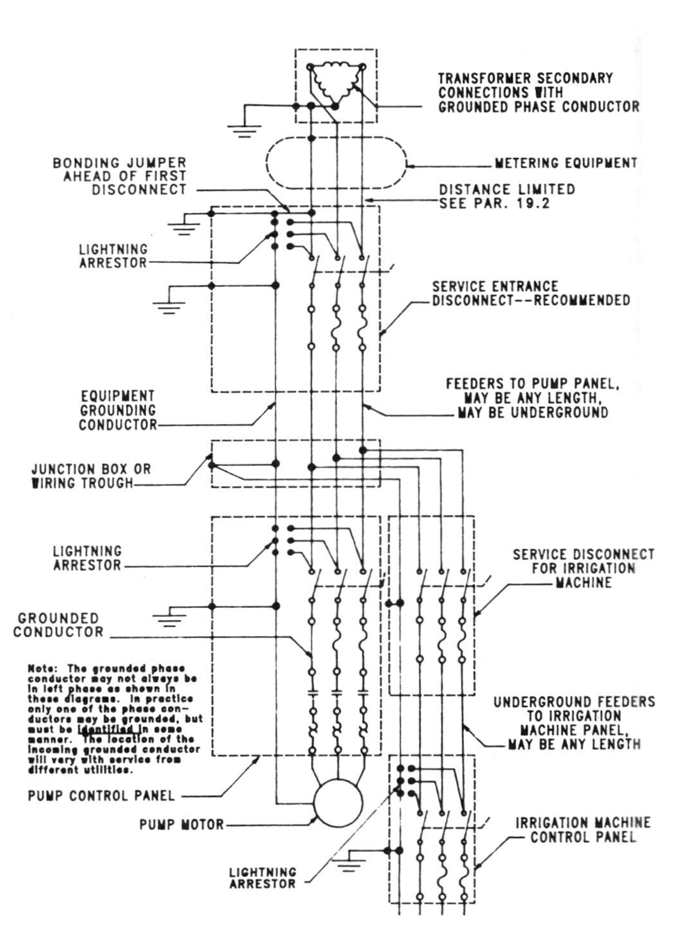

FIG. 7—RECOMMENDED EQUIPMENT AND GROUNDING OF THREE-PHASE, THREE-WIRE SERVICE WHEN ONE OF THE PHASE CONDUCTORS IS GROUNDED

9.2 Auxiliary devices are recommended which protect motors from either phase failure or low voltage.

SECTION 10—GROUNDING

10.1 A grounding means shall be installed at the meter and shall serve as a service ground. If the service disconnecting means is not adjacent to the meter, a separate service ground rod shall be installed for connecting a grounding conductor. See Figs. 1-7 for proper grounding connections.

10.2 A grounding conductor shall be provided to serve as the interconnection between equipment grounds, the service grounds, and the transformer ground.

10.3 The necessity for maintaining the integrity of the grounding connection in irrigation equipment dictates that grounding conductors be required when motors, auxiliary enclosures or equipment are involved (Figs. 1-7). This is similar to requirements for marine applications in Article 555-7, NFPA Standard No. 70, National Electrical Code, except that grounding conductors are not required to be insulated.

10.4 Grounding-electrode conductors should be routed in the most direct manner, and without sharp bends, to the ground rod.

10.5 The equipment grounding conductor shall not be used as a current carrying conductor on the load side of the first disconnecting means.

SECTION 11—POWER FACTOR CORRECTION

11.1 Capacitors for power factor correction are recommended for motors 7.5 kW (10 hp) and larger. The recommended size of the capacitor that should be installed is shown in Table 3. Capacitors shall be installed on the motor side of the running-overcurrent device (Figs. 8 and 9) or shall be protected by a disconnecting means and overcurrent protection in accordance with Article 460-8, NFPA Standard No. 70, National Electrical Code.

11.2 Size of overload heaters may need to be reduced when power factor correction is installed (see paragraph 5.4.1).

SECTION 12—LIGHTNING (SURGE) ARRESTORS

12.1 Secondary lightning (surge) arrestors should be used. When used, the first surge arrestor shall be installed on the supply side, main service entrance disconnect. Additional arrestors may be installed on load side or line side of equipment (see Figs. 1-7).

12.2 Install arrestors on the exterior of enclosures.

SECTION 13—IRRIGATION MACHINES

13.1 Irrigation machines should comply with Article 675, NFPA Standard No. 70, National Electrical Code, and ASAE Standard S362, Wiring and Equipment for Electrically Driven or Controlled Irrigation Machines.

13.2 A disconnecting means shall be provided for the main control panel of an irrigation machine.

13.3 Disconnecting means shall be provided at each supply point when a single irrigation machine is moved from one point to another.

13.4 Figs. 10 and 11 indicate the recommended equipment, connections, protection and grounding for service to phase converters from 240 or 480 V supply.

TABLE 1—RECOMMENDED PROTECTION AND EQUIPMENT SIZING FOR THREE-PHASE 230 V MOTORS AND CIRCUITS

Size of Motor			Dual-Element Fuse for Motor Overload Protection (These fuses also provide branch circuit protection.)			Branch Circuit Protection (Short-Circuit Protection Only) (These fuses do not give motor overload protection)					Minimum Size of Starter	Minimum Size of Copper Wire*		Minimum Size of Trade Conduit†
Kilowatts	Horsepower	Ampere Rating	Motor Rated Not Over 40 °C or Not Less Than 1.15 S.F. (Max. Fuse 125 Percent)	All Other Motors (Max. Fuse 115 Percent)	Switch or Fuseholder Size	Class for Motor Starting Inrush and Code Letter	Dual-Element Fuse (Time Delay)	Switch or Fuseholder Size	Non-Time-Delay Fuse	Switch or Fuseholder Size	NEMA Size	AWG or MCM	aTHW (75 °C) bTHWN (75 °C) cTHHN (90 °C) dXHBH (90 °C)	Inch
0.4	1/2	2	2 1/2	2 1/4	30	Any	4	30	15	30	00	14	a,b,c,d	1/2
0.6	3/4	2.8	3 1/2	3 2/10	30	Any	4	30	15	30	00	14	a,b,c,d	1/2
0.7	1	3.6	4 1/2	4	30	Any	6 1/4	30	15	30	00	14	a,b,c,d	1/2
1.1	1 1/2	5.2	6 1/4	5 6/10	30	Any	8	30	15	30	00	14	a,b,c,d	1/2
1.5	2	6.8	8	7	30	1	10	30	25	30	0	14	a,b,c,d	1/2
						2	10	30	20	30				
						3-4	10	30	15	30				
2.2	3	9.6	12	10	30	1	15	30	30	30	0	14	a,b,c,d	1/2
						2	15	30	25	30				
						3	15	30	20	30				
						4	15	30	15	30				
3.7	5	15.2	17 1/2	17 1/2	30	1	25	30	50	60	1	12	a,b,c,d	1/2
						2	25	30	40	60				
						3	25	30	35	60				
						4	25	30	25	30				
5.6	7 1/2	22	25	25	30	1	35	60	70	100	1	10	a,b,c,d	1/2
						2	35	60	60	60				
						3	35	60	45	60				
						4	35	60	35	60				
7.5	10	28	35	30	60	1	40	60	90	100	2			
						2	40	60	70	100		8	a,d	3/4
						3	40	60	60	60		8	b,c	1/2
						4	40	60	45	60				
11.2	15	42	50	45	60	1	60	60	125	200	2			
						2	60	60	110	200		6	a	1
						3	60	60	90	100		6	b,c,d	3/4
						4	60	60	70	100				
14.9	20	54	60	60	100	1	80	100	175	200	3			
						2	80	100	150	200		4	a,b	1
						3	80	100	110	200		6	c,d	3/4
						4	80	100	90	100				
18.7	25	68	80	70	100	1	100	100	225	400	3	4	a,b,c,d	1
						2	100	100	175	200				
						3	100	100	150	200				
						4	100	100	110	200				
22.4	30	80	100	90	100	1	125	200	250	400	3			
						2	125	200	200	200		3	a	1 1/4
						3	125	200	175	200		3	b,c,d	1
						4	125	200	125	200				

*Equipment in general use is made primarily for connection to copper wire. Wire size shown does not compensate for voltage drop (From Table 310-16 NFPA Standard No. 70, National Electrical Code).

†Rigid metal, intermediate grade, and electrical metallic conduit are not approved for direct burial unless corrosive protection is provided. Galvanized rigid metal conduit and rigid nonmetallic conduit are recommended for direct burial.

13.5 Fig. 12 indicates the recommended equipment connections, protection, and grounding for a three-phase generator serving an irrigation machine.

SECTION 14—INTERLOCKING

14.1 When personal hazard or property damage may be caused by the failure of any one device (such as a fertilizer injector or an irrigation machine) to function properly, protective interlocks shall be provided. When practical these interlocks shall interrupt all operations provided that such interruption will not create a hazardous condition.

TABLE 2—RECOMMENDED PROTECTION AND EQUIPMENT SIZING FOR THREE-PHASE 460 V MOTORS AND CIRCUITS

Size of Motor			Dual-Element Fuse for Motor Overload Protection (These fuses also provide branch circuit protection.)			Branch Circuit Protection (Short-Circuit Protection Only) (These fuses do not give motor overload protection)					Minimum Size of Starter	Minimum Size of Copper Wire*		Minimum Size of Trade Conduit†
Kilowatts	Horsepower	Ampere Rating	Motor Rated Not Over 40 °C or Not Less Than 1.15 S.F. (Max. Fuse 125 Percent)	All Other Motors (Max. Fuse 115 Percent)	Switch or Fuseholder Size	Class for Motor Starting Inrush and Code Letter	Dual-Element Fuse (Time Delay)	Switch or Fuseholder Size	Non-Time-Delay Fuse	Switch or Fuseholder Size	NEMA Size	AWG or MCM	aTHW (75 °C) bTHWN (75 °C) cTHHN (90 °C) dXHBH (90 °C)	Inch
0.4	1/2	1	1 1/4	1 1/8	30	Any	2	30	15	30	00	14	a,b,c,d	1/2
0.6	3/4	1.4	1 6/10	1 6/10	30	Any	2 1/2	30	15	30	00	14	a,b,c,d	1/2
0.7	1	1.8	2 1/4	2	30	Any	3 2/10	30	15	30	00	14	a,b,c,d	1/2
1.1	1 1/2	2.6	3 2/10	2 8/10	30	Any	4	30	15	30	00	14	a,b,c,d	1/2
1.5	2	3.4	4	3 1/2	30	Any	5	30	15	30	00	14	a,b,c,d	1/2
2.2	3	4.8	5 6/10	5	30	Any	8	30	15	30	0	14	a,b,c,d	1/2
3.7	5	7.6	9	8	30	1	15	30	25	30	0	14	a,b,c,d	1/2
						2	15	30	20	30				
						3-4	15	30	15	30				
5.6	7 1/2	11	12	12	30	1	20	30	35	60	1	14	a,b,c,d	1/2
						2	20	30	30	30				
						3	20	30	25	30				
						4	20	30	20	30				
7.5	10	14	17 1/2	15	30	1	20	30	45	60	1	12	a,b,c,d	1/2
						2	20	30	35	60				
						3	20	30	30	30				
						4	20	30	25	30				
11.2	15	21	25	20	30	1	30	30	70	100	2	10	a,b,c,d	1/2
						2	30	30	60	60				
						3	30	30	45	60				
						4	30	30	35	60				
14.9	20	27	30	30	60	1	40	60	90	100	2	8	a,d	3/4
						2	40	60	70	100		8	b,c	1/2
						3	40	60	60	60				
						4	40	60	45	60				
18.7	25	34	40	35	60	1	50	60	110	200	2	8	a,d	3/4
						2	50	60	90	100		8	b,c	1/2
						3	50	60	70	100				
						4	50	60	60	60				
22.4	30	40	50	45	60	1	60	60	125	200	3	6	a	1
						2	60	60	100	100		6	b	3/4
						3	60	60	80	100		8	c	1/2
						4	60	60	60	60		8	d	3/4
29.8	40	52	60	60	60	1	80	100	175	200	3	6	a	1
						2	80	100	150	200		6	b,c,d	3/4
						3	80	100	110	200				
						4	80	100	80	100				
37.3	50	65	80	70	100	1	100	100	200	200	3	4	a,b,c,d	1
						2	100	100	175	200				
						3	100	100	150	200				
						4	100	100	100	100				
44.8	60	77	90	80	100	1	125	200	250	400	4	3	a	1 1/4
						2	125	200	200	200		3	b,c,d	1
						3	125	200	175	200				
						4	125	200	125	200				
56.0	75	96	110	110	200	1	150	200	300	400	4	1	a,b	1 1/4
						2	150	200	250	400		2	c,d	1
						3	150	200	200	200				
						4	150	200	150	200				
74.6	100	124	150	125	200	1	200	200	400	400	4	2/0	a,b	1 1/2
						2	200	200	350	400		1/0	c,d	1 1/4
						3	200	200	250	400				
						4	200	200	200	200				
93.3	125	156	175	175	200	1	250	400	500	600	5	3/0	a,b	2
						2	250	400	400	400		3/0	c,d	1 1/2
						3	250	400	350	400				
						4	250	400	250	400				
114.9	150	180	225	200	400	1	300	400	600	600	5	4/0	a,b,c,d	2
						2	300	400	450	600				
						3	300	400	400	400				
						4	300	400	300	400				
149.2	200	240	300	250	400	1	400	400			5	350	a,b	2 1/2
						2	400	400	600	600		300	c,d	2
						3	400	400	500	600				
						4	400	400	400	400				

*Equipment in general use is made primarily for connection to copper wire. Wire size shown does not compensate for voltage drop (From Table 310-16 NFPA Standard No. 70, National Electrical Code).

†Rigid metal, intermediate grade, and electrical metallic conduit are not approved for direct burial unless corrosive protection is provided. Galvanized rigid metal conduit and rigid nonmetallic conduit are recommended for direct burial.

SECTION 15—MISCELLANEOUS REQUIREMENTS, PUMPING PLANTS

15.1 Control circuit for magnetic starters. The starter control circuit shall be wired with three-wire control. Two-wire control shall not be used unless "on-delay" relay protection is provided. Conductors of motor control circuits shall be protected against overcurrent in accordance with their ampacities as specified in Article 420-72(a), NFPA Standard No. 70, National Electrical Code. If the start-stop station is not located in the cover of the starter or nippled to the starter, the extended control circuit wires shall be enclosed in a raceway, and overcurrent protection shall be provided when the rating of the protective device is more than 300 percent of the ampacity of control circuit conductors as specified in Article 430-72(a), NFPA Standard No. 70, National Electrical Code, or Rule 28.500, Canadian Electrical Code.

TABLE 3—RECOMMENDED CAPACITOR SIZES (kV Ar) FOR POWER FACTOR CORRECTION FOR 230 AND 460 V MOTORS*

		T-Frame Motors			U-Frame Motors		
		Motor Speed, rpm			Motor Speed, rpm		
kW	hp	1800	1200	900	1800	1200	900
7.5	10	4	5	5	2	4	5
11.2	15	5	7.5	10	4	4	5
14.9	20	5	7.5	10	4	5	7.5
18.7	25	7.5	7.5	10	7.5	7.5	10
22.4	30	7.5	10	15	7.5	7.5	10
29.8	40	10	15	20	10	10	10
37.3	50	20	15	25	10	10	15
44.8	60	20	30	30	10	15	20
56.0	75	25	30	40	15	15	20
76.6	100	30	30	50	20	25	30
93.3	125	35	40	50	20	30	30
111.9	150	35	45	50	35	35	45
149.2	200	50	55	70	40	50	70
186.5	250	55	70	85	50	50	80

*For other voltages or motor speeds consult capacitor manufacturers literature. Check size of overload heaters when capacitors are installed (paragraph 5.4.1).

15.2 Magnetic oilers. When used, magnetic oilers shall be wired from the motor terminals (see Figs. 8 and 9). (Rule 14-100, Canadian Electrical Code). The conductors shall be protected by an approved raceway.

15.3 Pilot lights. A pilot light to indicate whether the motor is on or off may be used.

15.3.1 A two-pole, fused disconnect switch shall be installed in the circuit to permit de-energizing the circuit while the pump motor is running. This circuit shall be fused properly. The lampholder shall be of the type approved for outdoor installation. The conductors shall be enclosed in an approved raceway.

15.3.2 On 240 V installations the pilot lamp should be connected as shown in Fig. 8.

15.3.3 If a 240 or 120 V power source is not available, a 480/240 V to 120 V transformer shall be used with a fused primary and secondary to supply a standard 120 V lamp [Ref. Article 450-3(b), NFPA Standard No. 70, National Electrical Code and Rule 10-106, Canadian Electrical Code].

15.4 Lights and other 120 V equipment. When lighting or convenience outlets are required, a transformer to reduce the supply voltage from either 240 or 480 V to 120 V shall be installed. An autotransformer shall not be used. This transformer shall be fed from the line side of the service. It shall have a properly fused disconnecting means. Wire size shall not be smaller than No. 14. The switch rating, fuse size, conductor size on the load side of the transformer, and transformer capacity will be determined by the size of the load to be served (Figs. 8 and 9).

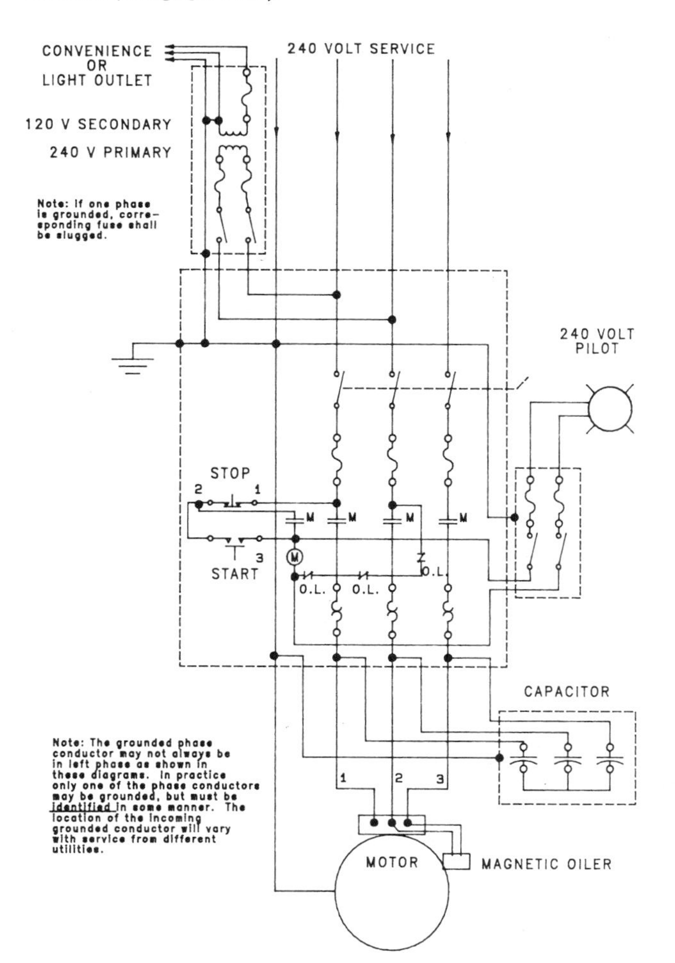

FIG. 8—WIRING DIAGRAM FOR CAPACITOR INSTALLATION AND AUXILIARY EQUIPMENT FOR 240 V SERVICE

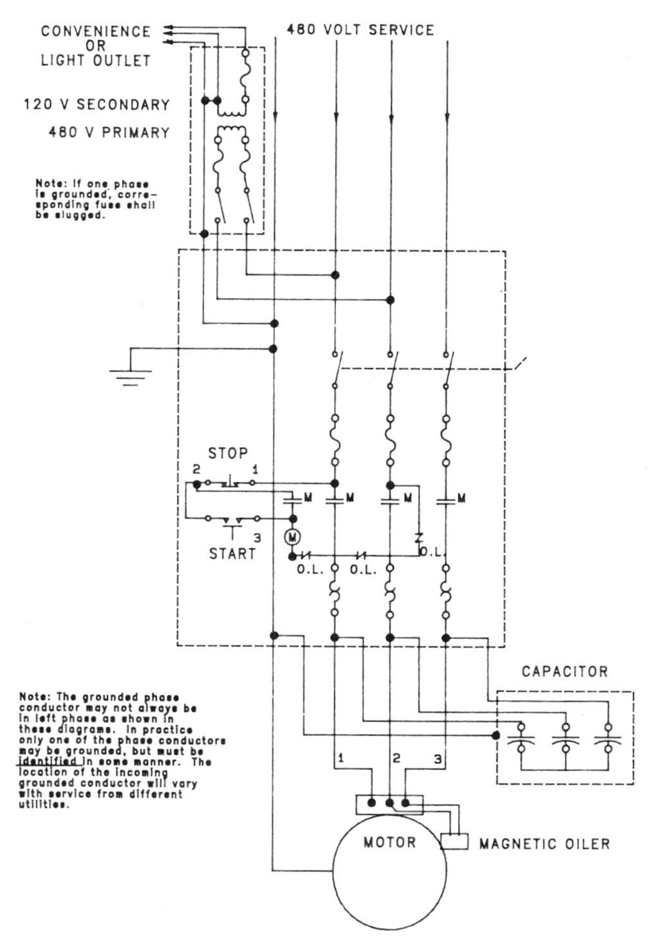

FIG. 9—WIRING DIAGRAM FOR CAPACITOR INSTALLATION AND AUXILIARY EQUIPMENT FOR 480 V SERVICE

15.5 Time switch, time delay, and other automatic control. Automatic starting is permitted provided that devices for starting are installed in a manner that conforms with the requirements for interlocking in accordance with paragraph 14.1. The setting of "on-delay" relays required in all "two-wire" control circuits may be determined by the electric utility. Proper setting will provide "on-delay" for staggered starting of groups of motors when a power interruption occurs as well as protection for the irrigation equipment. For siphon-tube irrigation in which "on-delay" restart is used, an "off-delay" timer is also recommended for ditch protection when power is off long enough to require repriming of siphon tubes.

15.5.1 For safety when automatic controls are used, a sign stating "caution automatic starting" shall be posted near the motor.

15.6 Weather protection for motors and controls. Motors and controls may be installed outdoors or in an enclosure. All electrical equipment for outdoor installation shall be in accordance with NEMA Standard ICS 6-1978, Enclosures for Industrial Controls and Systems.

15.6.1 Protective structures. Structures may be closed or have open sides and function only as a shade. Structures should be removable to facilitate service and repair of motor and well.

15.6.1.1 Size. Structures shall be large enough to allow servicing of the motor and equipment from all sides. Clearances are specified in paragraph 3.3.

15.6.1.2 Base. For a closed structure the base should be the same size as the structure, constructed of concrete, and drained to the outside. For a shade only, the base does not need to be the size of the shade.

15.6.1.3 Ventilation. Cross ventilation of closed structures shall be provided by doors or louvers. All doors shall have positive catches to hold the door open.

15.6.1.4 Control mounting. Controls should be protected from the heat of the sun and from inclement weather. When possible, controls should be mounted on a north wall with at least 25 mm (1 in.) of air space between the control and the outside covering of the structure.

SECTION 16—MAIN DISCONNECTING MEANS

16.1 Installation of a disconnecting means ahead of the main pump panel is recommended for safety and convenience. Many accessories such as irrigation machines and fertilizer injectors are often added to today's irrigation systems. These accessories shall be connected to the service side of the pump motor disconnect and shall be provided with a disconnecting means and overcurrent protection. These accessory disconnects can be installed and serviced simply if a main disconnecting means is provided ahead of the pump disconnect. A wiring trough may be added to accommodate additional connections required for accessory disconnects.

16.2 The pump motor disconnect is approved as a single motor disconnect and controller, not as a distribution panel or load center; therefore, equipment such as fertilizer injectors or irrigation machines shall not be supplied from the load side of the disconnect. Each disconnecting means becomes the service disconnect for the equipment it serves.

SECTION 17—TWO FUSES IN THREE-WIRE, THREE-PHASE EQUIPMENT

17.1 The schematic drawings in Figs. 3, 4, and 7 do not show a fuse in series with the grounded phase conductor. Article 240-22 of NFPA Standard No. 70, National Electrical Code, specifies that no overcurrent device shall be connected in series with any conductor that is intentionally grounded unless the overcurrent device opens all conductors of the circuit or unless, as specified in Article 430-36, the overcurrent device is used for motor overload protection. However, Article 430-36 (Part C of Article 430) does not apply when the fuses are sized for

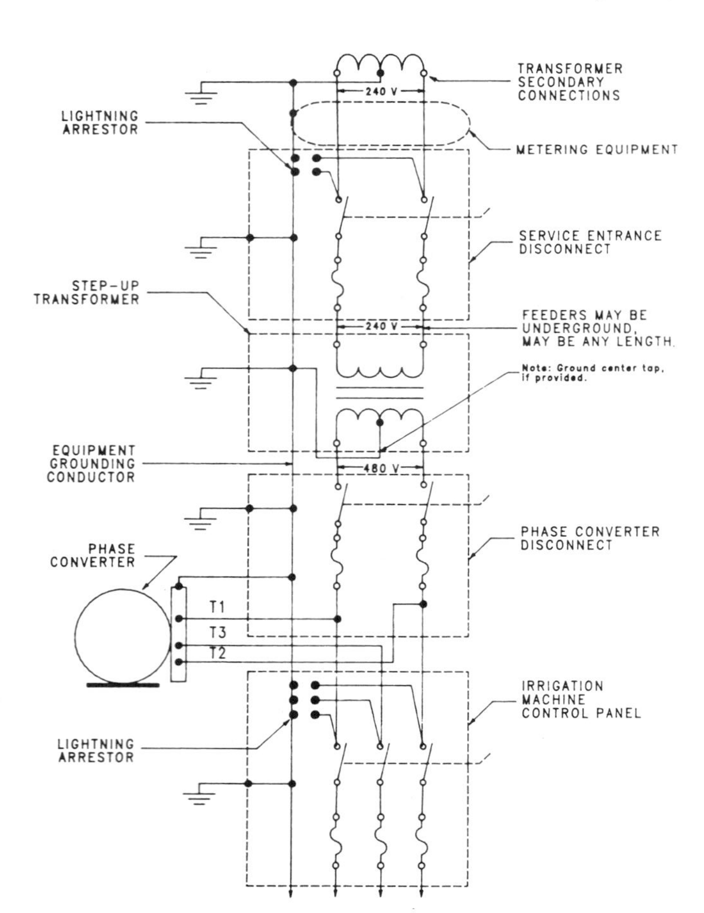

FIG. 10—RECOMMENDED EQUIPMENT AND GROUNDING FOR SERVICE TO A PHASE CONVERTER SUPPLYING AN IRRIGATION MACHINE FROM 240 V, SINGLE-PHASE TRANSFORMER SECONDARY CONNECTIONS

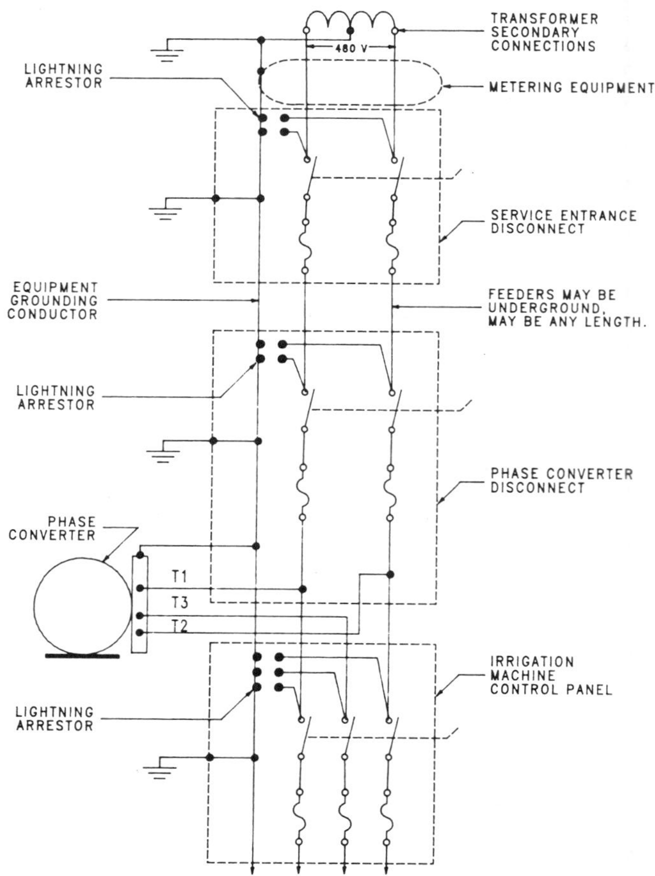

FIG. 11—RECOMMENDED EQUIPMENT AND GROUNDING FOR SERVICE TO A PHASE CONVERTER SUPPLYING AN IRRIGATION MACHINE FROM 480 V, SINGLE-PHASE TRANSFORMER SECONDARY CONNECTIONS

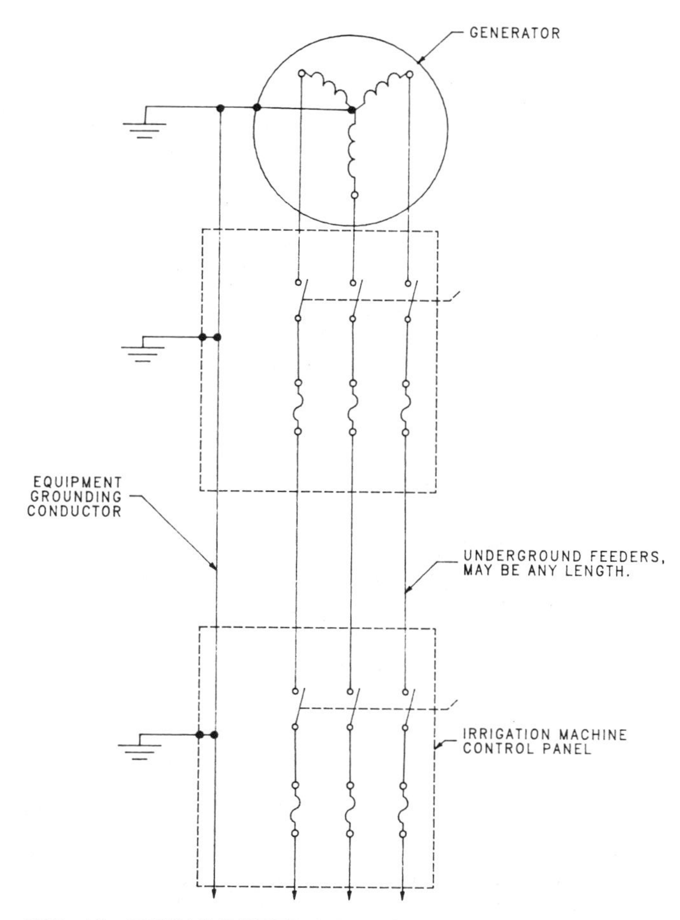

FIG. 12—RECOMMENDED EQUIPMENT AND GROUNDING FOR A GENERATOR SERVING AN IRRIGATION MACHINE

branch circuit protection (Part D of Article 430). In Figs. 1-7 the schematic drawings indicate that motor overload protection is provided by the motor controller. Part D of Article 430, 430-56, NFPA Standard No. 70, National Electrical Code, which deals with branch circuit protection, refers to Article 240-20 which requires overcurrent protection in series with each ungrounded conductor.

SECTION 18—CONDUCTORS

18.1 Irrigation equipment in general use is made primarily for connection to copper wire. Copper wire should be used between the starter and the pump motor.

18.2 Aluminum wire, when used, requires special care in terminating. Manufacturers' recommendations should be followed.

SECTION 19—THREE-WIRE SERVICE

19.1 When systems are provided with three-phase, three-wire service with a grounded phase, a bonding jumper shall be installed at the first disconnecting means as shown in Fig. 7. The bonding jumper shall connect the grounding conductor to the grounded (identified) conductor ahead of the disconnecting means.

19.2 When the pump motor is more than 15 to 30 m (50 to 100 ft) from the transformer bank, a three-phase, three-wire service with a grounded phase should not be used.

SECTION 20—WORKMANSHIP

20.1 All materials and equipment shall be installed in a neat and workmanlike manner.

Cited Standards:

ASAE S318, Safety for Agricultural Equipment
ASAE S362, Wiring and Equipment for Electrically Driven or Controlled Irrigation Machines
CSA Standard C22.1-1982, Canadian Electrical Code
NEMA ICS 2-449-1978, AC Automatic Combination Irrigation Pump Controllers
NEMA ICS 6-1978, Enclosures for Industrial Controls and Systems
NEMA MG 1, Motors and Generators
NFPA No. 70, National Electrical Code

SPECIFICATIONS FOR ALARM SYSTEMS UTILIZED IN AGRICULTURAL STRUCTURES

Developed by the Agricultural Alarm Subcommittee of the ASAE Electrical Controls for Farmstead Equipment Committee; approved by the ASAE Electric Power and Processing Division Standards Committee; adopted by ASAE February 1983; revised March 1985.

SECTION 1—PURPOSE AND SCOPE

1.1 The purpose of this Standard is to establish specifications for fixed installation alarm systems utilized in agricultural structures.

1.2 This Standard defines characteristics and requirements for components, wiring and service of fixed installation alarm systems in agricultural structures.

SECTION 2—DEFINITIONS

2.1 Agricultural alarm system: A fixed installation alarm system that is utilized in an agricultural structure. This system shall include the following components (see Fig. 1):

2.1.1 Control console: The electronic/electrical assembly which monitors all sensors and activates auxiliary alarm devices upon sensing an alarm condition.

2.1.1.1 Zones: Individual sensor input terminals on a control console are defined as zones provided the console is capable of differentiating (through either audible or visual means) the specific input line that triggered the alarm. If there is **no** differentiation capability, the console is defined as a single-zone device. For example, a console with one input line for fire sensors, a second line for door sensors, but no means to indicate which line caused the alarm, would be considered a single-zone device.

2.1.2 Sensors: Any sensing devices which are monitored by the system.

2.1.2.1 Normally open sensor (NO): A sensing device that is electrically equivalent to an open switch which closes or completes a circuit in an alarm condition.

2.1.2.2 Normally closed sensor (NC): A sensing device that is electrically equivalent to a closed switch which opens or breaks a circuit in an alarm condition.

2.1.2.3 Special sensor: A sensing device that does not fall in either of the two categories defined in paragraphs 2.1.2.1 and 2.1.2.2.

2.1.3 Power source: The means for providing the energy for the system to operate.

2.1.4 Auxiliary alarm device: A component usually, but not necessarily, remotely located from the console which provides an audible and/or visual indication of an alarm state.

SECTION 3—SYSTEM CLASSIFICATION CODE

3.1 Agricultural alarm systems shall be classified according to a three-category classification code. Category 1 shall relate to the input capabilities of the console. Category 2 shall relate to the power source utilized by console. Category 3 shall relate to the type of auxiliary alarm devices utilized.

3.1.1 Category 1 (input capabilities) shall be defined as follows:

3.1.1.1 Type A(N): Utilizes NC and/or NO sensors wired directly to the control console on one or more zones. This system category shall be capable of accepting **both** NC and NO sensors on one wire pair/console input. (N) indicates the number of individual inputs that the console is capable of monitoring and differentiating. For example, A(1) would denote a one-zone system, while A(4) would denote a four-zone system.

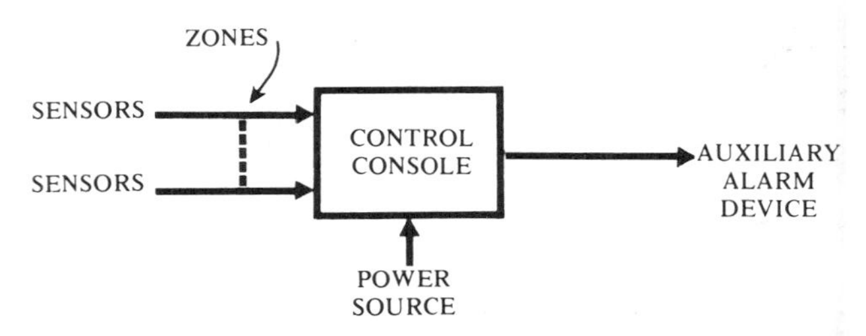

FIG.1—SYSTEM BLOCK DIAGRAM

3.1.1.2 Type B(N): Utilizes NC and/or NO sensors wired directly to the central console on one or more zones. This system category shall be capable of accepting **either, but not both,** NC or NO sensors on one wire pair/console input. (N) indicates the number of individual inputs that the console is capable of monitoring and differentiating.

3.1.1.3 Type C(N): Utilizes NC and/or NO sensors which return an alarm condition to the console by some means other than a dedicated wire pair. (N) indicates the number of individual input channels that the console is capable of monitoring and differentiating.

3.1.1.4 Type G(N): Utilizes a type of sensor/zone configuration not covered in any of the categories defined in paragraphs 3.1.1.1, 3.1.1.2, or 3.1.1.3.

3.1.2 Category 2 (power source) shall be defined as follows:

3.1.2.1 Type I: Utilizes primary AC power with a standby non-rechargeable battery.

3.1.2.2 Type II: Utilizes primary AC power with a standby rechargeable battery.

3.1.2.3 Type III: Utilizes self-contained non-rechargeable battery only.

3.1.2.4 Type IV: Utilizes self-contained rechargeable battery only.

3.1.2.5 Type V: Utilizes primary AC power source only.

3.1.2.6 Type X: Utilizes none of the power sources given in paragraphs 3.1.2.1 to 3.1.2.5.

3.1.3 Category 3 (auxiliary alarm devices) shall be defined as follows:

3.1.3.1 Type R: Internal—activates an alarm on the console.

3.1.3.2 Type S: Local—activates an alarm on the building premises where the console is located.

3.1.3.3 Type T: Remote/direct—activates an alarm via a direct-wire link at some location other than where the console is located.

3.1.3.4 Type U: Remote/telephone—activates an alarm via telephone link at some location other than where the console is located.

3.1.3.5 Type V: Remote/RF—activates an alarm via RF (radio frequency) link at some location other than where the console is located (see Section 10—RF-Transmitting Devices).

3.1.3.6 Type W: Central—reports alarm condition for several different alarm systems to a central point (usually a receiving computer).

3.1.3.7 **Type Y:** Visual—activates a visual alarm indicator.

3.1.3.8 **Type Z:** Utilizes a type of auxiliary alarm device not covered in paragraphs 3.1.3.1 to 3.1.3.7.

3.2 **Code delineation:** The complete classification code shall be recorded in the following manner: multiple types within an individual category will be separated by a slash (/) while major categories will be separated by a dash (—). For example, A(1)/G—I—R/T is a system with one Type A zone and one Type G zone, powered by primary AC power with a standby non-rechargeable battery, with both an internal alarm and a direct-wired remote alarm.

3.3 **Customer/installer notification:** A descriptive sheet or label shall be included with each agricultural alarm system indicating the classification code of the system along with an explanation of that particular code as defined in this document.

SECTION 4—ENVIRONMENTAL CONDITIONS

4.1 **Component Utilization Groups:** The appropriate environmental operating conditions for individual agricultural alarm system components as given in Section 2—Definitions (sensors, power sources, consoles auxiliary alarm devices) shall be defined by a specific component utilization group for **each component.** These groups shall be defined as follows:

4.1.1 **Group I:** Designated for use inside a living area or work area where a dust-free, noncorrosive environment exists, such that the building **cannot** be classified as meeting the dust/moisture/corrosion conditions of American National Standard ANSI/NFPA Standard No. 70, National Electrical Code article 547, -1(a), -1(b) [1984 edition]. Components rated for this environment shall be capable of operation within a temperature range of +10 °C to +40 °C.

4.1.2 **Group II - (A or B):** Designed for use inside a storage area or outbuilding where a **relatively** dust-free, moisture-free, noncorrosive environment exists, such that the building **cannot** be classified as meeting the criteria of ANSI/NFPA Standard No. 70, National Electrical Code, article 547, -1(a), -1(b) [1984 edition]. Components rated for this group shall meet the requirements of paragraph 4.2.2 and shall be futher classified as either (A) or (B) according to the operating temperatures of the environment as specified below:

4.1.2.1 **(A):** Components shall be operational over the temperature range of 0 °C to +50 °C, provided that the temperature of the environment surrounding the component will not exceed these limits. (This may be controlled by a suitable insulating enclosure.)

4.1.2.2 **(B):** For all other temperature extremes not meeting the conditions of paragraph 4.1.2.1, the components shall be capable of operating over the temperature range of −30 °C to +50 °C.

4.1.3 **Group III - (A or B):** Designed for use inside a building which **can** be classified as meeting the dust/moisture/corrosion criteria of ANSI/NFPA Standard No. 70, National Electrical Code, article 547, -1(a), -1(b) [1984 edition]. Components rated for this group shall meet the requirements of paragraph 4.2.2 and shall be further classified as either (A) or (B) according to the operating temperatures of the environment as specified in paragraphs 4.1.2.1 and 4.1.2.2.

4.1.4 **Group IV:** Components designated for outdoor use. Components utilized in an outdoor environment shall be assembled in a housing that minimizes the entrance of moisture and dust and shall be certified weatherproof. These components should be capable of operation over the temperature range of −40 °C to +60 °C.

4.1.5 **Substitution:** A component designed for a higher-number group may be utilized in a lower-number group, provided that it meets all the requirements of the lower-number group.

4.1.6 **Identification:** The Component Utilization Group will be clearly indicated on each component where possible; otherwise, this must be provided on a separate label/tag attached to the item. This identification label/tag will include both the group designation and an explanation of that group designation as defined in this document.

4.2 **Component protection—additional requirements**

4.2.1 **Printed circuit board requirements:** Printed circuit boards utilized in any agricultural alarm system component (console, sensor, power source, or auxiliary alarm device) shall be copper-clad, glass-epoxy, material G10-FR4 or equivalent, the manufacturer of which shall meet or exceed MIL-P-13949 (Revision F), General Specification For Plastic Sheet, Laminated, Metal-Clad (For Printed Wiring). The fabricated printed wiring board shall meet or exceed all the requirements of Institute for Interconnecting and Packaging Electronic Circuits Acceptance Specification IPC-D-320 (Revision A), Specification for Printed Board, Rigid, Single- and Double-Sided, End Product. The fabricated board shall also be tinned as per IPC-D-320 (Revision A).

4.2.2 **All electronic circuitry** (in any Component Utilization Group other than Group I) shall be protected against environmental conditions by at least one of the following:

4.2.2.1 **Outer housing** as per ANSI/NFPA Standard No. 70, National Electrical Code, article 547, -3(a) [1984 edition]. If the basic equipment housing does not meet this criteria, it may be placed in a supplemental housing which does meet the criteria, or coated as per paragraph 4.2.2.2 below.

4.2.2.2 **Conformal coating** of printed circuit boards and components with a product specifically recommended by the manufacturer for this application. The coating shall be capable of withstanding temperature extremes at least as great as those of the Component Utilization Group temperature range for which the component is qualified.

4.2.2.3 **Internally mounted mechanical/electro-mechanical components** such as tape transports, relays, etc., shall be protected in a manner which meets the requirements of paragraph 4.2.2.1.

SECTION 5—CONSOLE REQUIREMENTS

5.1 **Environmental:** All consoles shall meet the appropriate environmental requirements of Section 4—Environmental Conditions.

5.2 **Transient protection:** The console electronics shall be protected against over-voltage/over-current conditions through the use of surge arrestors and transient suppressors.

5.2.1 All console connections to primary AC power lines shall be equipped with transient protection circuitry rated at 20 J minimum.

5.2.2 All consoles designated for 115 VAC operation shall have a fuse, circuit breaker or other similar current-limiting device in the 115 VAC input line.

5.2.3 All other console input/output lines which are not automatically disconnected during periods of non-use shall have transient protection circuitry rated for at least the following minimum transient power handling level: 2 kW peak power @ 10μs duration (rectangular pulse).

5.3 **Power loss/battery requirements:** Consoles utilized to monitor power failure or operate during periods of power failure shall have the following capabilities:

5.3.1 115 VAC primary systems must have a standby power source.

5.3.2 In the event of the loss of AC power, systems shall provide an audible alarm which can be easily heard.

5.3.3 For systems of Category 2—Type I or III (see paragraph 3.1.2), non-rechargeable battery: upon detection of a low voltage condition of the standby power source, systems shall activate a periodic audible alarm which can be easily heard. This periodic audible alarm shall be distinctive from the power failure audible alarm. This low voltage alarm shall be capable of sounding at least once every 5 min, with a minimum time duration of 0.5 s, for at least a 30-day period.

5.3.4 For systems of Category 2—Type II or IV (see paragraph 3.1.2), rechargeable battery: since monitoring terminal voltage is not a reliable method of testing these batteries, the following must be specified to the owner at the time of purchase:

5.3.4.1 A weekly system test procedure under full load for a specified period of time with primary power removed from the system.

5.3.4.2 Replacement of the battery at a specified time interval.

5.4 Console sensor inputs: Any console inputs designated to receive either normally open (NO) or normally closed (NC) sensors shall be "fully supervised" (i.e., capable of detecting **both** shorted and open conditions in the sensor wire pair and sounding an audible alarm if either of these conditions exist).

5.5 Equipment reliability: All consoles shall be designed using current engineering practices. All systems shall be provided to the owner with written warranty and service program information.

SECTION 6—SENSOR AND AUXILIARY ALARM DEVICE REQUIREMENTS

6.1 Environmental requirements: All sensors and auxiliary alarm devices shall meet the requirements of their respective Component Utilization Group as delineated in paragraph 4.1.

6.2 Additional requirements: All sensors and auxiliary alarm devices with a utilization designation other than Group I shall be designed to withstand temperature/moisture/corrosion extremes by at least one of the following:

6.2.1 A sealed housing as per ANSI/NFPA Standard No. 70, National Electrical Code, article 547, -4 [1984 edition].

6.2.2 Conformal coating as per paragraph 4.2.2.2.

6.2.3 Hermetically sealed assembly/components.

6.2.4 Exception to the above subparagraphs: when a particular portion of a sensor **must** be exposed directly to the elements for proper operation, then only the remaining portions of the sensor need to be protected according to the above criteria.

SECTION 7—WIRING REQUIREMENTS

7.1 This section applies only to wiring where the maximum nominal voltage does not exceed 30 VDC or 30 VAC RMS, and the maximum nominal shortcircuit current does not exceed 2 A. Alarm system wiring which does not meet both of these criteria should be referred to applicable sections of the ANSI/NFPA Standard No. 70, National Electrical Code, articles 720, 725, 760, and 800 (1984 edition).

7.2 Inside wiring:

7.2.1 All inside wiring for connection to sensors/auxiliary alarm devices shall be copper wire at least No. 22 AWG.

7.2.2 Any inside wiring lines meeting the requirements of paragraph 7.1 may be installed without conduit.

7.3 Wiring connections:

7.3.1 All wire-to-wire connections shall be connected by either crimped-type connectors or twisted connections using wire nuts. These connections shall be sufficiently wrapped with protective tape.

7.3.2 When connecting wires to a terminal on a device, they shall be protected by one of the following methods:

7.3.2.1 Connection **inside** the device itself: protected with an effective moisture/dust barrier at the point of entry, which meets the intent of ANSI/NFPA Standard No. 70, National Electrical Code, article 547, -3(a) [1984 edition].

7.3.2.2 Connections made on the **outside** of the device: (a) soldered directly to the terminal (use rosin-core solder only), or (b) protected with a moisture boot or with an electrically neutral lubricant. This lubricant shall be required on all screw-type terminals, including battery lugs.

SECTION 8—LINE-CARRIER DEVICES

8.1 A line-carrier device is an assembly which transmits an alarm signal to a remote receiver via existing 115/230 VAC building wiring.

8.2 Any line-carrier device which is a primary alarm device must continually transmit a signal from the transmitter to the receiver when in the "normal" state, and sound an alarm upon the interruption of this signal.

SECTION 9—TELEPHONE DIALING DEVICES

9.1 A telephone dialer is an assembly that will generate one or more outgoing calls on an existing telephone line in the event of an alarm condition. Dialers can be classified as one of the following:

9.1.1 Tape-type dialer: Mechanical assembly; multiple call capability.

9.1.2 Digital dialer: Solid-state assembly; single call capability.

9.1.3 Digital communicator: Solid-state assembly; makes call directly to computer.

9.2 Dialing devices which are primary alarm devices shall only be utilized on single-party telephone lines.

9.3 A system utilizing a telephone dialer must also have an auxiliary local alarm device on the premises.

SECTION 10—RF-TRANSMITTING DEVICES

10.1 An RF-transmitting device is an assembly that will transmit alarm condition to a remote receiver by means of a radio frequency transmission. All RF-transmitting devices must be Federal Communication Commission (FCC) approved.

10.2 A system utilizing an RF-transmitting device must also have an auxiliary local alarm device on the premises.

SECTION 11—INSTALLATION, TESTING, AND SERVICE

11.1 Installation: Agricultural alarm systems shall be installed in a workmanlike manner and in accordance with applicable standards and the manufacturer's recommendations. In the design of a particular system layout, applicable codes for specific sensors shall be utilized (for example, for fire protection systems, American National Standard ANSI/NFPA Standard No. 72E, Automatic Fire Detectors).

11.1.1 To facilitate installation, an adequate installation/instruction manual shall be furnished. If system is designed to be installed by the user, it should include a troubleshooting/checkout procedure.

11.2 Testing program: An agricultural alarm system must have a preventative maintenance testing program designed to be performed on a regular basis by one of the following:

11.2.1 Testing on premises: This should be conducted at least once each month by activating one of the sensors or a test switch and verifying that all auxiliary alarm devices are functional. For this type of testing to be acceptable, a system test record must be provided to record and verify the results of the testing.

11.2.2 Testing via telephone lines from a remote point: This testing must be conducted at least once a month and must test the console, sensors, and required auxiliary alarm devices. Written verification of the results of these tests should be provided to the owner.

11.3 Warranty and service information: An agricultural alarm system shall have accompanying documents specifying:

11.3.1 Warranty program: The specific warranty program covering the system.

11.3.2 Service program: Details on how, where, and by whom required servicing of the system is accomplished.

References: Last printed in 1983 AGRICULTURAL ENGINEERS YEARBOOK; list available from ASAE Headquarters.

Cited Standards:

ANSI/NFPA No. 70, National Electrical Code
ANSI/NFPA No. 72E, Automatic Fire Detectors
IPC-D-320 (Revision A), Specification for Printed Board, Rigid, Single-And Double-Sided, End Product
MIL-P-13949 (Revision F), General Specification For Plastic Sheet, Laminated, Metal-Clad (For Printed Wiring)

FOOD AND PROCESS ENGINEERING

ASAE Notation:

The letter S preceding numerical designation indicates ASAE Standard; EP indicates Engineering Practice; D indicates Data. A decimal and numeral following the file number indicate the number of times a document has been revised. Thus ASAE S201.4 indicates Standard number 201, four times revised. The letter T after the designation indicates tentative status. Always refer to ASAE documents by complete designation to avoid confusion with standards of other organizations. For example: ASAE S201.4.

The symbol **T** preceding or in the margin adjacent to section headings, paragraph numbers, figure captions, or table headings indicates a technical change was incorporated in that area when this document was last revised. The symbol **T** preceding the title of a document indicates essentially the entire document has been revised. The symbol **E** used similarly indicates editorial changes or corrections have been made with no intended change in the technical meaning of the document.

ASAE Data: ASAE D241.3

DENSITY, SPECIFIC GRAVITY, AND WEIGHT-MOISTURE RELATIONSHIPS OF GRAIN FOR STORAGE

Approved by the ASAE Committee on Technical Data; adopted by ASAE 1948; revised 1954, 1962; revised by Electric Power and Processing Division Technical Committee, December 1967; reconfirmed December 1972; revised December 1973; revised editorially March 1975; reconfirmed December 1978, December 1983; revised April 1987.

TABLE 1—BULK DENSITIES OF GRAIN AND SEEDS BASED ON WEIGHTS AND MEASURES USED IN THE U.S. DEPARTMENT OF AGRICULTURE

Grain or Seed	kg/m³	lb/bu*
Alfalfa	772	60
Barley	618	48
Beans:		
lima, dry	721	56
lima, unshelled	360-412	28-32
snap	360-412	28-32
other, dry	772	60
Bluegrass	180-386	14-30
Broomcorn seed	566-644	44-50
Buckwheat	618	48
Castor beans	528	41
Clover seed	772	60
Corn:		
ear, husked	901	70†
shelled	721	56
Cottonseed	412	32
Cowpeas	772	60
Flaxseed	721	56
Hempseed	566	44
Hickory nuts	644	50
Kapok seed	451-515	35-40
Lentils	772	60
Millet	618-644	48-50
Mustard seed	747-772	58-60
Oats	412	32
Orchardgrass seed	180	14
Peanuts, unshelled:		
Virginia type	219	17
runners, southeastern	270	21
Spanish:		
southeastern	322	25
southwestern	322	25
Perillas seed	476-515	37-40
Popcorn:		
ear, husked	901	70†
shelled	721	56
Poppy seed	592	46
Rapeseed	644 and 772	50 and 60
Redtop seed	644 and 772	50 and 60
Rice, rough	579	45
Rye	721	56
Sesame	592	46
Sorgo seed	644	50
Sorghum grain	721	56
Soybeans	772	60
Spelt (p.wheat)	515	40
Sudangrass seed	515	40
Sunflower seed	309 and 412	24 and 32
Timothy seed	579	45
Velvet beans, hulled	722	60
Vetch	772	60
Walnuts, black	644	50
Wheat	772	60

*Source of lb/bu weights: USDA, 1985. Table of Weights and Measures, pg. V in: Agricultural Statistics, 1985. U.S. Government Printing Office, Washington, DC. A standard U.S. bushel contains 1.244456 ft³ (2,150.42 in.³).

†The standard weight of 70 lb is usually recognized as being about 2 measured bushels of corn, husked, on the ear, because 70 lb would normally yield one bushel, or 56 lb, of shelled corn.

TABLE 2—SPECIFIC GRAVITY AND PERCENTAGE OF VOIDS IN BULK GRAIN

Grain	Variety	Moisture Content, Percent (wet basis)	Air Space or Voids in Bulk, Percent	Kernel Specific Gravity
Barley	Coast (6 rows)	10.3	57.6	1.13
Barley	Hannchen	9.7	44.5	1.26
Barley	Synasota	9.8	45.4	1.21
Barley	Trebi (6 rows)	10.7	47.9	1.24
Barley	White hulless	10.4	39.5	1.33
Buckwheat	Japanese	10.1	41.0	1.10
Corn, mixed	Yellow and white	9.0	40.0	1.19
Corn, shelled	Yellow dent	25.0	44.0	1.27
Corn, shelled	Yellow dent	15.0	40.0	1.30
Flaxseed		5.8	34.6	1.10
Grain sorghum	Blackhull kafir	9.9	36.8	1.26
Grain sorghum	Yellow milo	9.5	37.0	1.22
Millet	Siberian	9.4	36.8	1.11
Oats	Iowar	9.7	51.4	0.95
Oats	Kanota	9.4	50.9	1.06
Oats	Red Texas	10.3	55.5	0.99
Oats	Victory	9.8	47.6	1.05
Rice	Honduras	11.9	50.4	1.11
Rice	Wataribune	12.4	46.5	1.12
Rye	Common	9.7	41.2	1.23
Soybeans	Manchu	6.9	36.1	1.18
Soybeans	Wilson	7.0	33.8	1.13
Wheat, hard	Turkey, winter	9.8	42.6	1.30
Wheat, hard	Turkey, winter (yellow)	9.8	40.1	1.29
Wheat, soft	Harvest queen	9.8	39.6	1.32

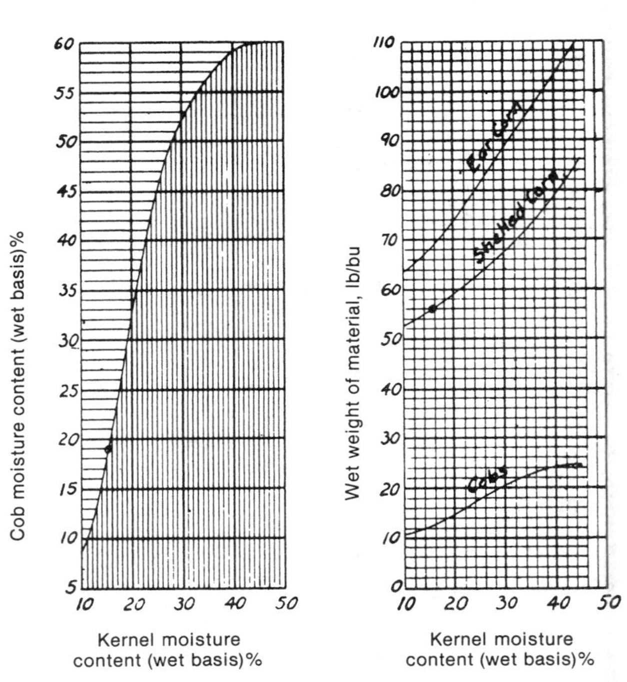

FIG. 1—POUNDS OF EAR CORN REQUIRED TO YIELD 56 LB (1 BU) SHELLED CORN CONTAINING 15.5 PERCENT MOISTURE
(Based on dry matter weight of 47.32 lb for grain and 9.94 lb for cobs, and cob-grain moisture shown)

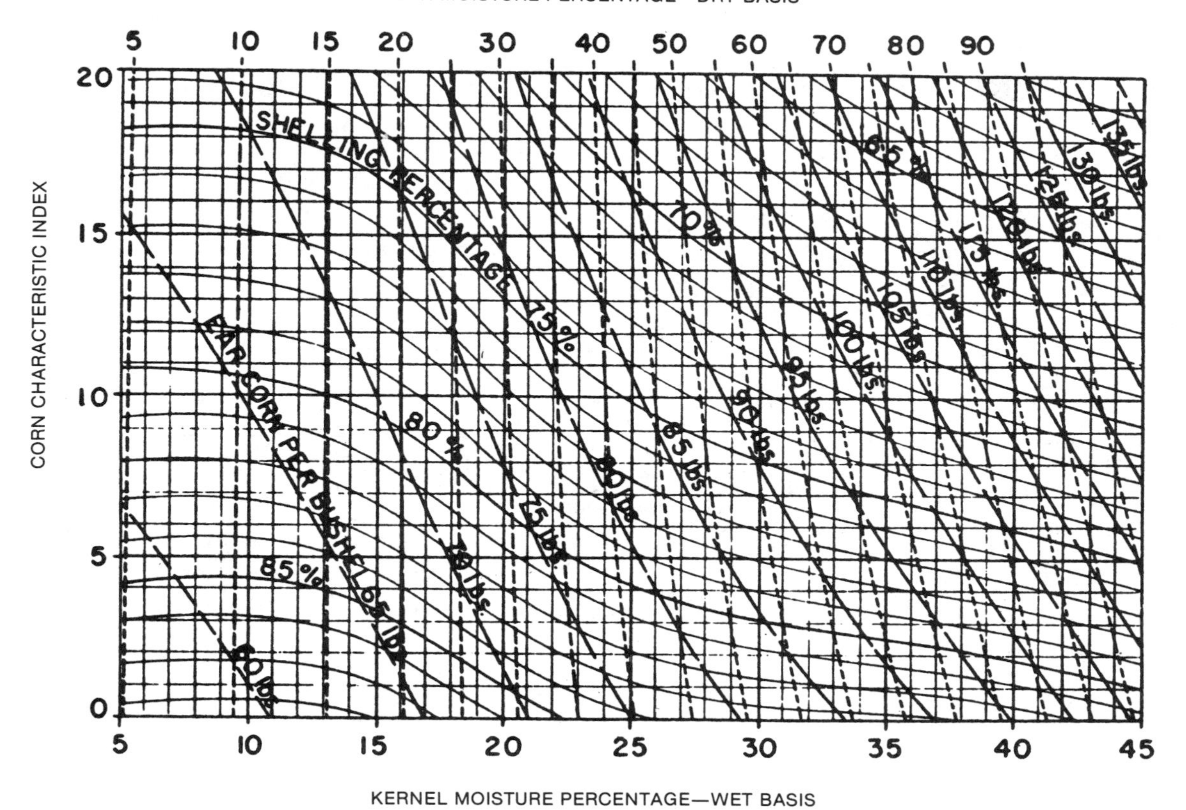

FIG. 2—RELATION OF KERNEL MOISTURE CONTENT AND SHELLING PERCENTAGE TO TOTAL MOISTURE CONTENT OF EAR CORN AND TO POUNDS OF EAR CORN REQUIRED TO YIELD ONE BUSHEL (56 LB) OF SHELLED CORN WITH 15.5 PERCENT MOISTURE CONTENT

NOTE: For any lot of corn the characteristic index number may be determined by measuring kernel moisture and shelling percentage. The corn characteristic index is useful when it is desirable to get successive samples from a field as the corn matures and dries. For a given lot of corn standing in the field the index remains approximately constant as the moisture content drops and other factors change. Once established for a field of corn the index permits making estimates from measurement of kernel moisture only. Corn with a characteristic index of 4 or 5 has well-filled ears. Corn with ears only partly filled or having extensive insect damage will have a higher index number. Fig. 1 of ASAE D241 will apply to corn having a characteristic index of 5 to 7.

THERMAL PROPERTIES OF GRAIN AND GRAIN PRODUCTS

Approved by the ASAE Committee on Technical Data; adopted by ASAE 1948; revised 1954, 1962; revised by Electric Power and Processing Division Technical Committee, December 1967; revised editorially March 1972; revised December 1973; reconfirmed December 1978; revised editorially April 1982; reconfirmed December 1983; revised December 1984.

Grain or grain product	Moisture content, percent, wet basis	Temperature range, °C	Mean temperature, °C	Specific heat*, kJ/(kg·K)	Conductivity*, W/(m·K)	Diffusivity*, m^2/h	Reference
Corn, yellow dent	0.9	12.2 to 28.8	20.5	1.532	0.1405	0.000367	(5)
Corn, yellow dent	5.1	for sp. heat	for sp. heat	1.691	0.1466	0.000354	(5)
Corn, yellow dent	9.8	8.7 to 23.3	13.8	1.834	0.1520	0.000335	(5)
		for diffusivity	for diffusivity				
Corn, yellow dent	13.2	26.7 to 31.1			0.1765	—	(8)
Corn, yellow dent	14.7	12.2 to 28.8	20.5	2.026	0.1591	0.000326	(5)
Corn, yellow dent	20.1	for sp. heat	for sp. heat	2.223	0.1636	0.000312	(5)
Corn, yellow dent	24.7	8.7 to 23.3	13.8	2.374	0.1700	0.000320	(5)
Corn, yellow dent	30.2	for diffusivity	for diffusivity	2.462	0.1724	0.000333	(5)
Corn, yellow dent	—	—	20.5	1.465 +0.0356M	—	—	(5)
Corn, yellow dent	—	—	35.2	—	0.1409 +0.00112M	—	(8)
Flour, wheat	—	—	—	1.662	—	—	(12)
Oats	9.1	—	—	—	0.0640	—	(2)
Oats	12.7	26.7 to 31.1	—	—	0.1298	—	(8)
Oats	27.7	—	—	—	0.0929	—	(2)
Oats	11.7 to 17.8	—	—	1.277 +0.0327M	—	—	(4)
Rice, rough	10.2 to 17.0	—	—	1.110 +0.0448M	—	—	(4)
Rice, shelled	9.8 to 17.6	—	—	1.202 +0.0381M	—	—	(4)
Rice, finished	10.8 to 17.4	—	—	1.181 +0.0377M	—	—	(4)
Rice, rough, medium	10 to 20	—	—	0.921 +0.0545M	0.0866 +0.00133M	0.000486 +0.00000897M	(13)
Rice, rough, short	10 to 20	—	—	—	—	0.000451 -0.00000585M	(6)
Rice, rough, short	11 to 24	—	—	1.269 +0.0349M	0.10000 +0.00111M	—	(6)
Sorghum	2 to 30	—	—	1.3971 +0.0322M	—	—	(11)
Sorghum	1 to 23	24 to 29	—		0.0976 +0.00148M	—	(11)
Soybeans	19.7	24 to 54	—	1.97†	—	—	(9)
Soybeans	24.5	23 to 88	—	2.05†	—	—	(5)
Starch, wheat	8.6	22 to 50	—	1.34‡	—	—	(10)
Starch, wheat	22.6	22 to 50	—	1.59‡	—	—	(10)
Wheat, hard	9.2	—	—	1.549	0.1402	0.000414	(1)
Wheat, hard red	9.6	22 to 50	—	1.63	—	—	(7)
Wheat, hard red	12.5	—	30.6	—	0.1281	—	(7)
Wheat, hard red	12.5	—	36.2	—	0.1367	—	(7)
Wheat, hard red	14.0	—	25.4	—	0.1367	—	(7)
Wheat, hard red	14.0	—	32.9	—	0.1419	—	(7)
Wheat, hard red	21.3	22 to 51	—	2.14	—	—	(7)
Wheat, hard red	23.0	—	26.3	—	0.1501	—	(7)
Wheat, hard red	23.0	—	32.0	—	0.1542	—	(7)
Wheat, hard red	23.0	—	37.6	—	0.1601	—	(7)
Wheat, soft white	0.1 to 33.6			1.240 +0.0362M‡			(3)
Wheat, soft white	0.7	10.7 to 32.2	21.6	1.398	0.1170	0.000334	(5)
	to	for sp. heat	for sp. heat	+0.0409M‡	+0.00113M‡	-0.00000245M‡	
	20.3	9.1 to 23.2	13.9				
		for diffusivity	for diffusivity				

*Where regression equations are presented, M represents moisture content in percent, wet basis.
†Apparent specific heats based on respiration data.
‡Computed for indicated temperature range from published constants and equations.

References: Last printed in 1985 STANDARDS; list available from ASAE Headquarters.

MOISTURE RELATIONSHIPS OF GRAINS

Approved by the ASAE Committee on Technical Data; adopted by ASAE 1948; revised 1954, 1962, 1964; revised by Electric Power and Processing Division Technical Committee December 1968, April 1974, December 1978; revised March 1980; reconfirmed December 1984, December 1985.

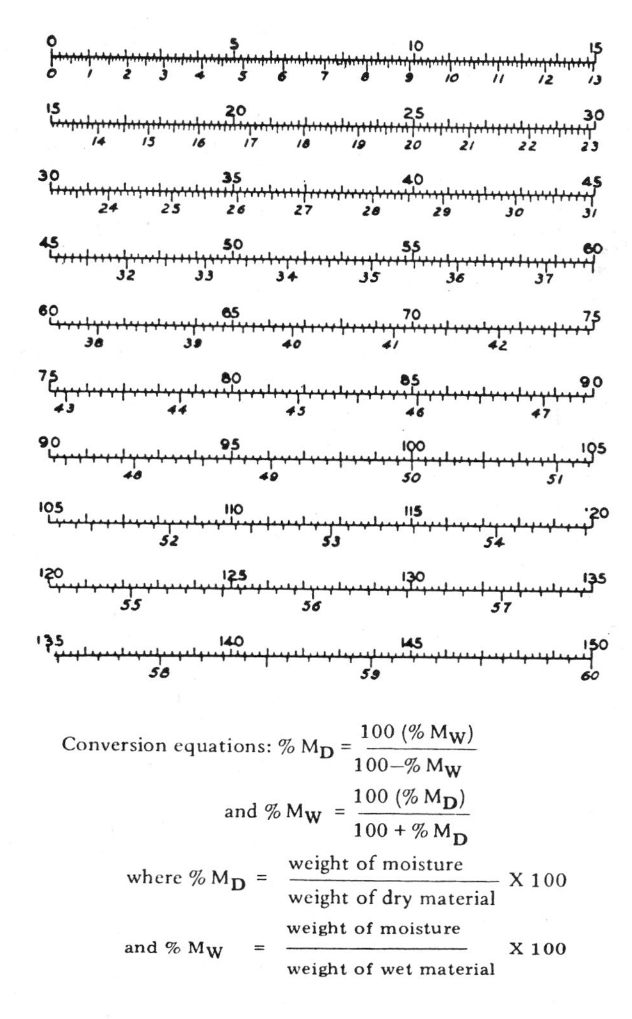

FIG. 1—EQUIVALENT MOISTURE PERCENTAGES Upper scale: $\% M_D$—moisture content (dry basis) %. Lower scale: $\% M_w$—moisture content (wet basis) %.

TABLE 1—EQUILIBRIUM MOISTURE CONTENT OF GRAINS AND SEEDS (PERCENT WET BASIS)

	Temp		Relative Humidity (Percent)									
	°C	°F	10	20	30	40	50	60	70	80	90	100
Barley	25	77	4.7	6.9	8.4	9.6	10.6	11.9	13.4	15.7	19.2	26.5
Buckwheat	25	77	5.6	7.7	9.2	10.2	11.2	12.4	13.9	15.9	19.1	24.1
Cottonseed	25	77				6.9	7.8	9.1	10.1	12.9	19.6	
Dry beans, Michelite	4	40						12.8	14.4	17.0		
	10	50						13.6	15.3	18.1		
	38	100						12.0	14.2	17.1		
	54	130						12.5	14.3	18.6		
	25	77	5.5	7.4	8.5	9.5	11.0	12.6	14.9	18.2*		
Dry beans, red Mexican	25	77	5.8	7.5	8.7	9.7	10.1	12.7	15.2	18.6*		
Dry beans, great Northern	25	77	5.9	7.8	8.4	9.5	10.9	12.6	14.8	17.9*		
Dry beans, light red kidney	25	77	5.9	7.7	9.0	9.8	11.0	12.6	15.0	18.6*		
Dry beans, dark red kidney	25	77	5.3	7.1	8.2	9.5	11.0	12.7	15.0	17.9*		
Dry beans, flat, small white	25	77	5.8	7.4	8.3	9.5	10.8	12.6	14.9	18.3*		
Dry beans, pinto	25	77	5.7	7.5	8.8	9.7	11.1	12.8	15.1	18.4*		
Flaxseed	25	77	3.8	5.0	5.5	6.1	6.7	7.7	9.2	11.2	14.9	21.1
	25	77				6.1	6.8	7.9	9.3	11.4	15.7	
Bromegrass, rescue	- 1	30			6.7	8.3	9.8	11.0	12.2	13.6		
	49	120			5.8	7.8	9.5	11.1	12.9	15.3		
Legume, crimson clover	- 1	30	5.9	7.4	8.6	9.3	10.1	10.8	11.3	12.0		
	16	60	5.1	6.7	7.8	8.8	9.5	10.2	11.0	11.8		
	32	90	4.3	5.8	7.0	7.9	8.6	9.4	10.3	11.4		
	49	120	3.1	4.9	6.1	7.1	7.7	8.6	9.5	10.6		
Legume, blue lupine	- 1	30	5.0	7.7	9.3	10.8	12.1	13.3	14.5	15.6		
	16	60		6.0	7.9	9.4	10.8	12.0	13.3	14.9		
	32	90			6.2	7.7	9.3	10.5	11.9	13.5		
	49	120				6.0	7.4	8.8	10.2	11.7		
Oats	25	77	4.5	6.6	8.2	9.4	10.3	11.4	12.8	15.0	18.2	23.9
Peanuts, whole pods	10	50		4.2	5.4	6.6	7.7	8.9	10.3	12.0	14.6	
	21	70		4.0	5.1	6.2	7.2	8.4	9.6	11.2	13.4	
	32	90		3.6	4.6	5.6	6.6	7.7	8.9	10.5	12.7	
Peanuts, kernels	10	50		4.0	4.9	5.7	6.4	7.2	8.1	9.1	10.5	
	21	70		3.4	4.3	5.1	5.9	6.7	7.7	8.9	10.6	
	32	90		3.0	3.9	4.7	5.6	6.5	7.5	8.8	10.6	
Peanuts, hulls	10	50		7.3	9.1	10.7	12.3	13.9	15.8	18.0	21.2	
	21	70		6.2	7.9	9.4	11.0	12.6	14.5	16.8	20.1	
	32	90		5.6	7.1	8.6	10.2	11.8	13.6	15.9	19.3	
Rice, whole grain	25	77	5.9	8.0	9.5	10.9	12.2	13.3	14.1	15.2	19.1	
	38	100	4.9	7.0	8.4	9.8	11.1	12.3	13.3	14.8	19.1	
Rice, milled	25	77	4.9	7.7	9.5	10.3	11.0	12.0	13.4	15.3	18.3	23.3
Rice, rough	27	80					10.2	11.7	13.2	14.9	17.2	
	32	90						11.2	12.9	14.6	16.8	
	44	111						10.3	12.3	14.3	16.5	
	0	32		8.2	9.9	11.1	12.3	13.3	14.5	16.6	19.2	
	20	68		7.5	9.1	10.4	11.4	12.5	13.7	15.2	17.6	
	30	86		7.1	8.5	10.0	10.9	11.9	13.1	14.7	17.1	
	25	77	4.6	6.5	7.9	9.4	10.8	12.2	13.4	14.8	16.7	
	23	73	4.9	7.3	8.7	9.8	10.9	12.4	13.5	15.9	19.0	
Rye	25	77	5.3	7.4	8.8	9.8	10.8	12.2	13.9	16.3	19.6	25.7
Shelled corn, YD	10	50	6.6	8.0	9.3	10.8	12.2	13.8	15.2	17.5	21.8	
	32	90	4.9	6.6	7.7	9.3	10.8	12.4	14.0	16.2	19.3	
	49	120				8.6	10.0	11.2	13.1	14.9		
	68	155				7.4	8.4	10.0	11.5	12.2		
	- 1	30	3.8	7.0	9.1	11.2	13.0	14.5	17.8			
	16	60		4.9	7.2	9.0	10.7	12.3	14.1			
	32	90			5.3	7.1	8.7	10.1	11.6	13.3		
	49	120				5.2	6.7	7.9	9.3	10.7		
	0	32				11.0	12.4	14.0	15.7	18.0	21.8	
	30	86					10.8	12.2	13.8	15.8	20.0	
	4	40	6.4	8.6	9.9	11.2	12.6	13.9	15.6	17.7	21.4	
	16	60	5.6	7.8	9.3	10.5	11.6	12.6	14.2	16.2	19.8	
	27	80	4.2	6.4	7.9	9.2	10.3	11.5	12.9	14.8	17.5	
	38	100	4.2	6.2	7.5	8.5	9.8	11.3	12.5	14.4	16.9	
	50	122	3.6	5.7	7.0	8.1	9.3	10.5	11.9	13.8	16.3	
	60	140	3.0	5.0	6.0	7.0	7.9	8.8	10.3	12.1	14.6	
	- 7	20					11.8	13.3	15.0	16.6		
	0	32				10.0	11.3	12.6	14.1	15.7		
	10	50				9.2	10.7	12.1	13.7	15.4		
	21	70				8.4	9.8	11.5	13.2			
	25	77	5.0	7.1	8.8	10.0	11.0	12.4	14.0	16.1	19.0	23.9
Shelled corn, WD	25	77	5.2	7.4	8.9	10.1	11.0	12.2	13.7	15.9	19.1	24.5
Shelled popcorn	25	77	5.8	7.5	8.4	9.2	10.2	11.4	13.1	15.1	18.2	22.7
Sorghum	- 1	30	6.1	8.3	10.0	11.3	12.4	13.4	14.6	15.8		
	16	60	5.4	7.7	9.5	10.7	11.9	13.0	14.1	15.2		
	32	90	4.7	7.1	8.8	10.1	11.3	12.4	13.5	14.7		
	49	120		6.5	8.2	9.5	10.7	11.7	12.9	14.1		
Sorghum, kafir	4	40	6.8	8.5	9.7	11.0	12.3	13.7	15.3	17.3		
	21	70	6.0	7.7	9.1	10.4	11.5	12.8	14.2	16.0		
	32	90	5.0	7.0	8.4	9.6	10.8	12.0	13.2	14.7		

*Unreliable because of mold growth.

TABLE 1—EQUILIBRIUM MOISTURE CONTENT OF GRAINS AND SEEDS (PERCENT WET BASIS) (cont'd)

	Temp		Relative Humidity (Percent)									
	°C	°F	10	20	30	40	50	60	70	80	90	100
Soybeans	5	41	5.2	6.3	6.9	7.7	8.6	10.4	12.9	16.9	22.4	
	15	59	4.3	5.7	6.5	7.2	8.1	10.1	12.4	16.1	21.9	
	25	77	3.8	5.3	6.1	6.9	7.8	9.7	12.1	15.8	21.3	
	35	95	3.5	4.8	5.7	6.4	7.6	9.3	11.7	15.4	20.6	
	45	113	2.9	4.0	5.0	6.0	7.1	8.7	11.1	14.9		
	55	131	2.7	3.6	4.2	5.4	6.5	8.0	10.6			
	25	77				7.0	8.0	10.1	12.2	16.0	20.7	
	25	77		5.5	6.5	7.1	8.0	9.3	11.5	14.8	18.8	
Sugar beet seeds	4	40			10.0	11.5	12.7	13.9	15.3	17.6	22.6	
	16	60			9.0	10.0	11.5	12.5	14.1	16.2	19.9	
	27	80			8.0	9.1	10.4	11.6	12.9	14.7	18.0	
	38	100			7.0	8.3	9.2	10.4	11.5	13.2	15.8	
Wheat, soft red winter	- 7	20					12.8	14.1	15.6	17.0		
	0	32				11.0	12.2	13.5	14.7	16.2		
	10	50				10.2	11.7	13.1	14.4	16.0		
	21	70				9.7	11.0	12.4	14.0			
	25	77	4.8	7.0	8.6	9.8	10.8	12.1	13.6	15.8	19.3	25.7
Wheat, hard red winter	25	77	5.0	7.2	8.2	9.9	10.9	12.1	13.8	16.0	19.4	25.4
Wheat, hard red spring	25	77	5.3	7.2	8.4	9.5	10.7	12.2	14.0	16.3	19.6	25.2
Wheat, white	25	77	5.4	7.1	8.2	9.3	10.4	11.9	13.7	16.2	19.6	25.9
Wheat, durum	25	77	5.3	7.3	8.5	9.4	10.4	11.7	13.4	15.8	19.5	26.7
Wheat	25	77	5.7	7.6	9.0	10.5	11.9	13.1	14.7	16.7	17.1	
	50	122	3.9	5.6	6.9	8.3	9.6	10.9	12.7	15.1	19.0	
	- 1	30		7.1	9.1	10.6	12.1	13.5	14.7	16.5		
	16	60		6.0	8.2	9.7	11.3	12.6	13.9	15.6		
	32	90		5.1	7.1	8.8	10.4	11.7	13.0	14.7		
	49	120			6.2	7.9	9.5	10.8	12.1	13.8		
	20	68	5.5	7.0	8.2	9.6	10.9	12.0	13.4	14.8	17.1	
	40	104	5.3	6.0	7.4	8.6	9.7	11.0	12.3	14.0	16.3	
	80	176	2.4	3.6	4.5	5.5	6.7	7.8	9.6	11.4	13.9	
	0	32		8.6	10.0	11.5	13.0	14.6	16.2	18.0	21.5	
	20	68			9.1	10.5	12.0	13.5	15.2	16.7	21.0	
	40	104			8.2	9.4	10.9	12.3	14.0	16.0	20.0	
	10	50		8.7	9.9	10.9	12.0	13.3	14.8	16.9		

HYGROSCOPICITY OF SEEDS

NOTES:

These curves represent average or typical data on hygroscopicity of seeds. Specific seed moisture—relative humidity relationships depend on several factors. For a given relative humidity, the equilibrium moisture content will be higher during drying (desorption) and lower during rewetting (absorption). The moisture content difference between desorption and absorption decreases with repeated wetting-drying cycles. High speed drying may increase the relative humidity that grain of a given moisture content will support.

Other variables in these relationships include grain variety, maturity, and history; the relative-humidity and moisture measuring techniques used; the degree of seed deterioration, the oil and protein content of the seed and other uncontrollable biological factors.

Figures 2 through 12 show the relationship between moisture, relative humidity and temperature; these were plotted using the Chung equation and constants of Table 2.

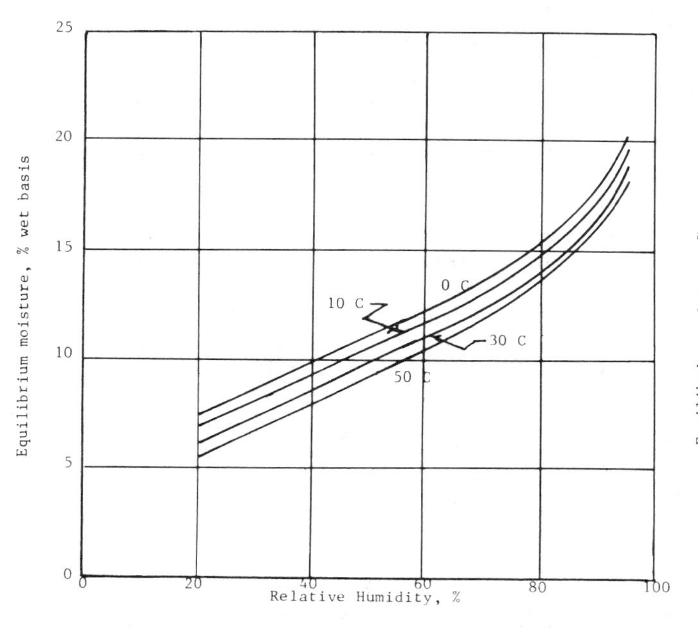

FIG. 2—EQUILIBRIUM MOISTURE CONTENT, BARLEY

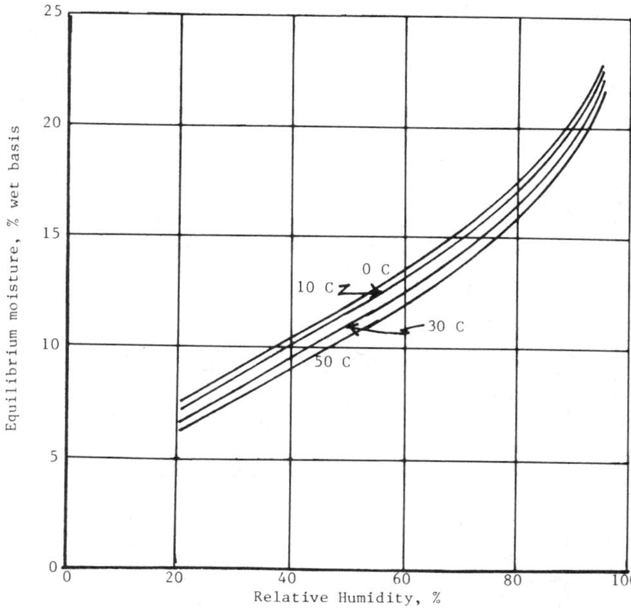

FIG. 3—EQUILIBRIUM MOISTURE CONTENT, EDIBLE BEANS

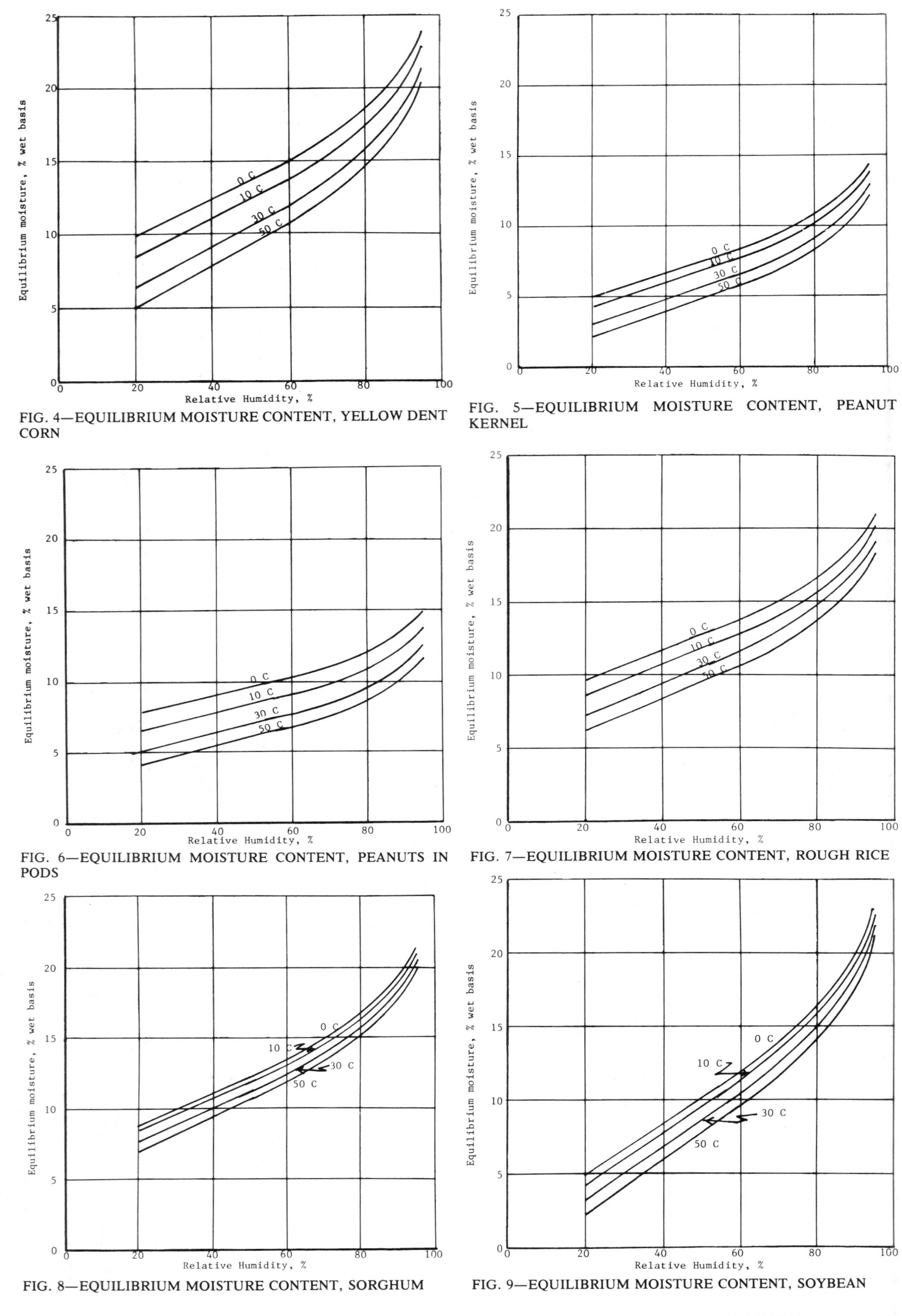

FIG. 4—EQUILIBRIUM MOISTURE CONTENT, YELLOW DENT CORN

FIG. 5—EQUILIBRIUM MOISTURE CONTENT, PEANUT KERNEL

FIG. 6—EQUILIBRIUM MOISTURE CONTENT, PEANUTS IN PODS

FIG. 7—EQUILIBRIUM MOISTURE CONTENT, ROUGH RICE

FIG. 8—EQUILIBRIUM MOISTURE CONTENT, SORGHUM

FIG. 9—EQUILIBRIUM MOISTURE CONTENT, SOYBEAN

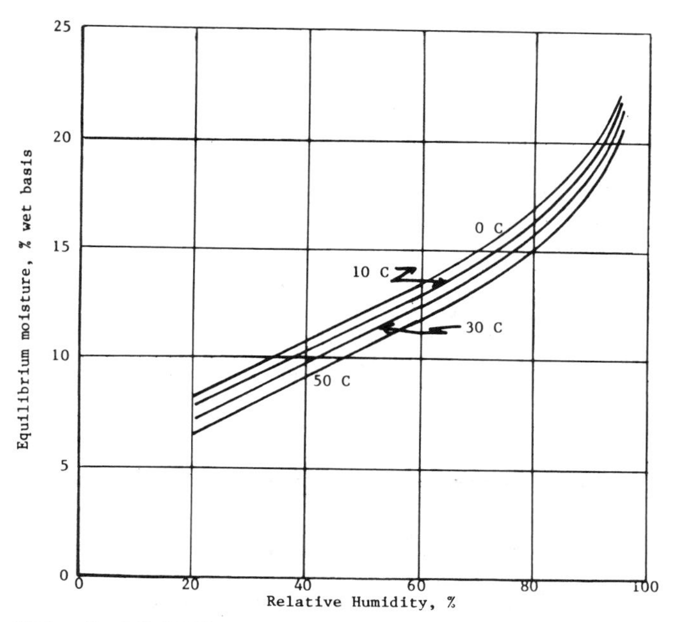

FIG. 10—EQUILIBRIUM MOISTURE CONTENT, DURUM WHEAT

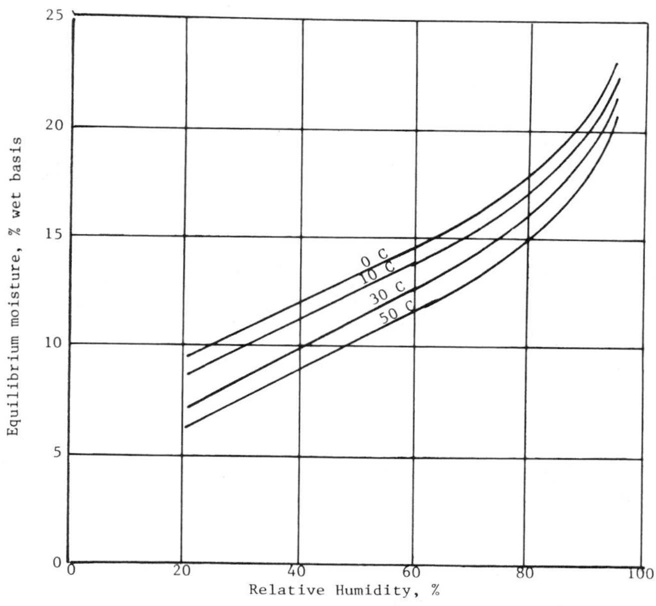

FIG. 11—EQUILIBRIUM MOISTURE CONTENT, HARD WHEAT

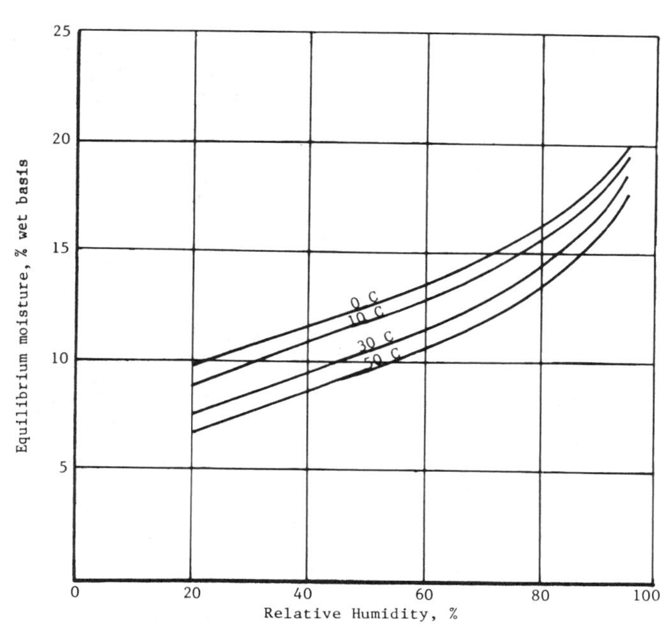

FIG. 12—EQUILIBRIUM MOISTURE CONTENT, SOFT WHEAT

References: Last printed in 1979 AGRICULTURAL ENGINEERS YEARBOOK; list available from ASAE Headquarters.

TABLE 2—EQUILIBRIUM MOISTURE CONTENT EQUATIONS AND CONSTANTS

Modified Henderson Equation

$$M = \frac{\left[\frac{\ln(1 - RH)}{-K\cdot(T + C)}\right]^{\frac{1}{N}}}{100}$$

$$RH = 1 - \exp[-K\cdot(T + C)\cdot(100\cdot M)^N]$$

Grain	K	N	C	Standard Error Moisture
Barley	2.2919×10^{-5}	2.0123	195.267	0.0080
Beans, edible	2.0899	1.8812	254.23	0.0138
Corn, yellow dent	8.6541	1.8634	49.810	0.0127
Peanut, kernel	65.0413	1.4984	50.561	0.0126
Peanut, pod	6.6587	2.5362	23.318	0.0303
Rice, rough	1.9187	2.4451	51.161	0.0097
Sorghum	0.8532	2.4757	113.725	0.0087
Soybean	30.5327	1.2164	134.136	0.0173
Wheat, durum	2.5738	2.2110	70.318	0.0068
Wheat, hard	2.3007	2.2857	55.815	0.0071
Wheat, soft	1.2299	2.5558	64.346	0.0122

Chung Equation

$$M = E - F\cdot\ln[-(T + C)\cdot\ln(RH)]$$

$$RH = \exp\left[\frac{-A}{(T + C)}\exp(-B\cdot M)\right]$$

Grain	A	B	C	E	F	Standard Error Moisture
Barley	761.66	19.889	91.323	0.33363	0.050279	0.0055
Beans, edible	962.58	15.975	160.629	0.43001	0.062596	0.0136
Corn, yellow dent	312.30	16.958	30.205	0.33872	0.058970	0.0121
Peanut, kernel	254.90	29.243	33.892	0.18948	0.034196	0.0133
Peanut, pod	522.01	37.903	12.354	0.16510	0.026383	0.0322
Rice, rough	594.61	21.732	35.703	0.29394	0.046015	0.0096
Sorghum	1099.67	19.644	102.849	0.35649	0.050907	0.0086
Soybean	328.30	13.917	100.288	0.41631	0.071853	0.0191
Wheat, durum	921.65	18.077	112.350	0.37761	0.055318	0.0057
Wheat, hard	529.43	17.609	50.998	0.35616	0.056788	0.0061
Wheat, soft	726.49	23.607	35.662	0.27908	0.042360	0.0147

M = Grain Moisture, decimal dry basis
RH = Relative Humidity, decimal
T = Temperature, °C

ASAE Standard: ASAE S248.3

CONSTRUCTION AND RATING OF EQUIPMENT FOR DRYING FARM CROPS

Developed and approved by Crop Dryer Manufacturers Council of Farm and Industrial Equipment Institute July 18, 1962; subsequently approved by ASAE Electric Power and Processing Steering Committee August 1962; adopted by ASAE as a Tentative Standard December 1962; editorial revisions April 1964; revised and reclassified as a full Standard December 1964; revised February 1971, March 1976; reconfirmed December 1980, December 1981, December 1982; reconfirmed for one year December 1987.

SECTION 1—PURPOSE

1.1 This Standard is intended to promote uniformity and consistency in the terms used to describe, rate and evaluate crop dryers, dryer components, and crop dryer operation.

1.2 This Standard is intended to promote efficiency and safety in the design, construction, and use of heated-air crop dryers.

1.3 This Standard gives minimum requirements. Where local codes are more stringent, they should be followed.

SECTION 2—SCOPE

2.1 This Standard refers to equipment in which heated air or artificially generated radiation is used to reduce the moisture content of grain, hay, and other farm crops. These units, hereinafter referred to as dryers, include the batch type units, continuous-flow type units, heated-air units and supplemental-heat units and fans.

2.2 Within the scope of this Standard a dryer shall be understood to include the grain-holding compartment in which the product is confined while being dried, together with its source of heat or radiation, fuel supply, fuel transmission system, burner, vaporizer, heat exchanger, air moving device (AMD), control system, power source, materials handling system, and AMD tube or other discharge ducts which are integral parts of the basic dryer.

SECTION 3—COMPLIANCE DEFINITIONS

3.1 Compliance definitions. The accepted definitions of "shall," "should," and "approved," as included in this Standard, are:

3.1.1 "Shall" is intended to indicate requirements.

3.1.2 "Should" is intended to indicate recommendations, or that which is advised but not required.

3.1.3 "Approved" or "approval" refers to listing by a nationally recognized testing laboratory or agency.

SECTION 4—GENERAL CONSTRUCTION

4.1 General. Dryers and related equipment shall be built with regard to the hazard inherent in the equipment operating at elevated temperatures, the hazard occasioned by overheating the product, the hazard of open flames, the hazard in incomplete combustion of direct-fired devices, the hazard to the operator from mechanical equipment and high temperatures, and with regard to the promotion of safety and reliable operation over the expected life of the equipment.

4.1.1 Combustible materials shall not be used in the construction of dryers where that material will be subjected to sustained temperatures in excess of 74 °C (165 °F).

4.1.2 Dryers designed for outdoor use shall be so constructed or anchored that they will safely withstand wind pressures or snow loads to which they may be normally subjected.

4.1.3 Dryers shall be equipped with appropriate safety shields, guards, and screens. Intake openings for drying air shall be equipped with not greater than 13 mm (0.5 in.) screen openings which shall be readily accessible for cleaning.

4.1.4 Dryers of the direct-fired type shall have combustion-air intake openings so located that foreign material is not likely to enter the combustion chamber. Where that is not practical, such openings shall be provided with a screen. Where foreign material is likely to enter the primary air to the burner, this intake should be screened. Such screens shall be readily accessible for cleaning.

4.1.5 All dryers designed to recirculate a portion of the exhaust air shall employ an approved means of removing combustible material from the air in the recirculation duct. In a direct-fired dryer using fuel oil or solid fuels, a target plate or other effective means shall be used to minimize the possibility of burning materials entering the drying chamber.

4.1.6 Interior surfaces of all dryers shall be designed to facilitate cleaning.

4.1.7 Access doors or openings shall be provided to permit inspection, cleaning, maintenance, and the effective use of extinguishers or hose streams in all parts of the dryer and connecting spouts or conveyors. All access doors which permit personnel entry shall be provided with hardware which will permit manual opening without tools from either side of access doors, except for access openings in spray dryers and cyclone-type dryers where the use of inside hardware will interfere with the proper operation of the dryer.

4.1.8 All wiring shall comply with National Fire Protection Association Standard NFPA No. 70, National Electrical Code (American National Standard, C-1, National Electrical Code). Electric motors shall be installed so that adequate ventilation is provided, in accordance with NFPA No. 70, National Electrical Code.

4.1.9 Clearance requirements and means of venting furnaces shall be in accordance with standard practice.

NOTE: For provisions on clearances in venting, refer to: National Board of Fire Underwriters, National Building Code; National Fire Protection Association, NFPA No. 90-A, Installation of Air Conditioning and Ventilating Systems of Other Than Residence Type; NFPA No. 211 Chimneys, Flues and Vents; Underwriters Laboratories, Inc. UL-731 Oil Fired Heaters, and UL-795 Commercial Industrial Gas Heating Equipment.

4.1.10 Provisions for expansion or contraction shall be provided if necessary to prevent damage.

4.1.11 Outside surface temperatures of dryer housing should be maintained at less than 93 °C (200 °F) where adjacent to combustible material. Where insulating material is used, it shall be so installed that it will not absorb moisture.

4.2 Noise level. Consideration should be given to design factors influencing noise emissions and every effort should be made to keep noise levels at a minimum.

4.2.1 Each manufacturer shall have available the sound power level rating in decibels and shall make available octave sound power levels on request.

4.2.2 The sound power level rating shall be obtained from octave sound power levels measured in accordance with Air Moving and Conditioning Association Standard AMCA 300-67, Test Code for Sound Rating Air Moving Devices.

4.3 Pollutant emission level. Consideration should be given to design factors influencing particulate emissions and every effort should be made to keep particulate emission levels at a minimum. Each manufacturer shall have available the particulate emission level rating as soon as reliable test standards are developed.

SECTION 5—HEAT-PRODUCING DEVICES

5.1 General clearance requirements and means of venting furnaces shall be in accordance with standard practice.

NOTE: For provisions on clearances in venting, refer to: The National Board of Fire Underwriters, National Building Code; National Fire Protection Association NFPA No. 90-A, Installation of Air Conditioning and Ventilating Systems of Other Than Residence Type; NFPA No. 211 Chimneys, Flues and Vents; Underwriters Laboratories, Inc., UL-731 Oil Fired Heaters, and UL-795 Commerical Industrial Gas Heating Equipment.

5.2 Gas-fired devices

5.2.1 Gas burners shall be designed so that the normal function of the automatic gas-control valves cannot be manually bypassed.

5.2.2 Gas burners and associated mixing equipment shall be suitable for the service intended as follows:

5.2.2.1 For heat content of the gas used.

5.2.2.2 For the operating pressures listed on nameplate.

5.2.2.3 Capable of maintaining flame stability throughout the turn-down range.

5.2.2.4 Designed with the required safety interlocks.

5.2.3 Approved methods of ignition shall be employed.

5.2.4 Electrical operation of the valves shall be of common characteristics of cycle, frequency and voltage.

5.2.5 Gas-fired infrared generators may be used. Adequate clearances shall be provided to assure safe surface temperatures for the product being dried.

5.2.6 Methods of ignition of gas-fired devices

5.2.6.1 Pilots

5.2.6.1.1 Continuous and interruptible gas pilots

5.2.6.1.1.1 If a continuous or interruptible pilot is employed, it shall be suitably interlocked to insure a pre-purge of at least 15 seconds or 4 complete air changes of the plenum chamber and duct prior to an ignition attempt of the pilot.

5.2.6.1.1.2 Trial for ignition shall not exceed 90 seconds.

5.2.6.1.1.3 The construction of all safety pilots shall be such that in the event of breakage or failure of the flame sensing device they shall fail safe.

5.2.6.1.1.4 The input to the pilot(s) shall not exceed 21.10 MJ/h (20 000 BTU/h) or 3 percent of the maximum input to the main burner as fired, whichever is greater.

5.2.6.1.1.5 Pilot burners not automatically lighted shall be fixed in place so that they can be safety lighted manually.

5.2.6.1.1.6 Main flame supervision is mandatory where any part of the main burner is more than 0.91 m (3 ft) from the point of supervision of the pilot along the path of flame travel or run. Main flame supervision shall also be required if the main flame cannot be observed from the point of manual start up. The main flame supervision is necessary with any burner that can supply gas through a channel or part that is not immediately adjacent to a proved pilot.

5.2.6.1.1.7 The pilot shall be proved before gas is admitted to the main burner.

5.2.6.1.1.8 Flame response timing shall not exceed 90 seconds.

5.2.6.1.2 Standing pilot. Pilot supervision is mandatory.

5.2.6.2 Direct spark ignition

5.2.6.2.1 If direct spark ignition is employed, it shall be suitably interlocked to insure a pre-purge of at least 15 seconds or 4 complete air changes of the plenum chamber and duct, prior to an ignition attempt.

5.2.6.2.2 The high voltage ignition spark shall ignite the main flame within 4 seconds after the gas reaches the main burner(s) parts.

5.2.6.2.3 The trial for ignition of the main flame shall not exceed 90 seconds.

5.2.6.2.4 Flame failure response timing shall not exceed 90 seconds.

5.2.6.2.5 The ignition transformer shall not be energized until combustion air supply and circulating air flow is proved. Air flow sensing devices such as pressure switches, sail switches, and differential switches shall be of the approved type.

5.2.6.2.6 Main flame supervision shall be located so as to prove ignition across the entire face of the burner.

5.2.7 The gas or vapor handling components shall have a working pressure rating in accordance with the following:

5.2.7.1 For natural gas and liquid propane gas, twice the maximum available safety monitored supply pressure.

5.2.7.2 For liquid propane systems, at least 1.72 MPa (250 psi) from point of supply connection to and including the pressure regulator downstream of the vaporizer.

5.2.7.3 All components downstream from the vaporizer shall be designed to withstand the maximum vapor temperature that is encountered with the system.

5.3 Electrically heated devices. All types of heating devices that use electrical energy for heat shall comply with the following requirements:

5.3.1 Induction or dielectric heating systems which employ high-frequency electrical energy may be used for dryers.

5.3.2 Infrared heating systems may be used for dryers.

5.3.2.1 Infrared lamps shall not be used where combustible dust may be present or where accumulation of combustible dust may form on surfaces of the lamps.

5.3.2.2 Infrared lamps of suitable focal length shall be used to assure safe surface temperatures for the product being dried. Instruction concerning replacement of lamps should include necessary warning and be a part of the permanent label.

5.3.3 Resistance-type heating systems may be used for dryers.

5.3.4 All internal wiring of electrically heated devices shall conform to the requirements contained in NFPA No. 70, National Electrical Code.

5.4 LP-gas vaporizers and vaporizer-burners for direct-fired application. The vaporizer section of vaporizer-burners used for dryers shall be constructed as follows:

5.4.1 Vaporizer-burners shall have a minimum design pressure of 1.72 MPa (250 psi) with a factor of safety of 5.

5.4.2 The vaporizer section shall be protected by a hydrostatic relief valve, located where it shall not be subjected to temperatures in excess of 60 °C (140 °F), and with a pressure setting such as to protect the components involved but not lower than 1.72 MPa (250 psi). The relief valve discharge shall be directed upward and away from the component parts of the vaporizer burners. Fusible plugs shall not be used.

5.4.3 A means shall be provided for manually turning off the gas to main burner and pilot.

5.4.4 Vaporizing-burners shall be provided with an automatic safety device to shut off the flow of gas to the main burner and pilot in the event pilot is extinguished. See Prevention of Fire and Dust Explosions in Grain Elevators and Bulk Grain Handling Facilities, NFPA No. 61B, for ignition and combustion controls applicable to vaporizing-burners associated with grain dryers.

5.4.5 Pressure regulating and control equipment shall be so protected as not to be subject to temperatures above 60 °C (140 °F), unless it is designed and recommended for use by the manufacturer for a higher temperature.

5.4.6 Pressure regulating and control equipment located downstream of the vaporizing section shall be designed to withstand the maximum discharge temperature of the hot vapor.

5.4.7 The vaporizer section of vaporizer-burners shall not be provided with fusible plugs.

5.4.8 Dryers utilizing vaporizing burners shall be equipped with automatic devices both upstream and downstream of the vaporizing section. These devices shall be installed and connected to shut off in the event of excessive temperature, flame failure and if applicable, insufficient air flow.

5.4.9 Vaporizing-burners shall not raise the fuel pressure above the design pressure of the vaporizing section.

5.4.10 The vaporizing-burner or the appliance in which it is installed shall be permanently and legibly marked with the maximum burner input J/h (BTU/h) and the name or symbol of the manufacturer.

SECTION 6—LP GAS SYSTEMS

6.1 Piping, tubing and fittings

6.1.1 Pipe. Pipe shall be wrought iron or steel (black or galvanized), brass or copper and shall comply with 6.1.1.1 through 6.1.1.4.

6.1.1.1 Wrought iron pipe. ANSI B36.10, Wrought-Steel and Wrought-Iron Pipe.

6.1.1.2 Steel pipe. ASTM A53, Specification for Welded and Seamless Steel Pipe.

6.1.1.3 Brass pipe. ASTM B43, Specification for Seamless Red Brass Pipe, Standard Sizes.

6.1.1.4 Copper pipe. ASTM B42, Specification for Seamless Copper Pipe, Standard Sizes.

6.1.2 Tubing. Tubing shall be steel, brass or copper and shall comply with 6.1.2.1.1 through 6.1.2.1.3.

6.1.2.1 Steel tubing. ASTM A539, Specification for Electric-Resistance-Welded Coiled Steel Tubing for Gas and Fuel Oil Lines.

6.1.2.2 Brass tubing. ASTM B135, Specification for Seamless Brass Tube.

6.1.2.3 Copper tubing

6.1.2.3.1 Type K or L: ASTM B88, Specification for Seamless Copper Water Tube

6.1.2.3.2 ASTM B280, Specification for Seamless Copper Tube for Air Conditioning and Refrigerator Field Service.

6.1.3 Pipe and tubing fittings. Fittings shall be steel, brass, copper, malleable iron or ductible (nodular) iron, and shall comply with 6.1.3.1 and 6.1.3.2.

6.1.3.1 Pipe joints in wrought iron, steel, brass or copper pipe may be screwed, welded or brazed.

6.1.3.1.1 Fitting used with liquid LP-Gas, or with vapor LP-Gas at operating pressures over 0.86 MPa (125 psi) shall be suitable for working pressure of 1.72 MPa (250 psi).

6.1.3.1.2 Fittings used with vapor LP-Gas at pressures not exceeding 0.86 MPa (125 psi) shall be suitable for a working pressure of 0.86 MPa (125 psi).

6.1.3.1.3 Soldering or brazing filler material shall have a melting point exceeding 538 °C (1000 °F).

6.1.3.2 Tubing joints in steel, brass or copper tubing shall be made with approved gas tubing fittings and may be flared, soldered or brazed.

6.1.3.2.1 Fittings used with liquid LP-Gas, or with vapor LP-Gas at operating pressures over 0.86 MPa (125 psi) shall be suitable for a working pressure of 1.72 MPa (250 psi).

6.1.3.2.2 Fittings for use with vapor LP-Gas at pressures not exceeding 0.86 MPa (125 psi) shall be suitable for a working pressure of 0.86 MPa (125 psi).

6.1.3.2.3 Soldering or brazing filler material shall have a melting point exceeding 538 °C (1000 °F).

6.1.4 Installation of piping, tubing, and fittings

6.1.4.1 LP-Gas normally is transferred into containers as a liquid, but may also be conveyed as a liquid or vapor under container or lower regulated pressure. Piping shall comply with the following:

6.1.4.1.1 Vapor LP-Gas piping with operating pressures in excess of 0.86 MPa (125 psi) and liquid LP-Gas piping shall be suitable for a working pressure of at least 1.72 MPa (250 psi).

6.1.4.1.2 Vapor LP-Gas piping subject to pressures of not more than 0.86 MPa (125 psi) shall be suitable for a working pressure of at least 0.86 MPa (125 psi).

6.1.4.2 Pipe joints may be threaded, flanged, welded or brazed using pipe and fittings complying with 6.1.1 and 6.1.3. Threaded joints, including joints threaded and back welded, shall comply with the following:

6.1.4.2.1 For LP-Gas vapor at pressures in excess of 0.86 MPa (125 psi) or LP-Gas liquid, Schedule 80 or heavier pipe or nipples shall be used.

6.1.4.2.2 For LP-Gas vapor at pressures of 0.86 MPa (125 psi) or less, Schedule 40 or heavier pipe shall be used.

6.1.4.2.3 Welded joints shall be made with suitable types of welding fittings or flanges for the service in which they are to be used and shall be at least Schedule 40 pipe.

6.1.4.3 Tubing joints may be flared, soldered or brazed using tubing and fittings, and solder or brazing material complying with 6.1.2 and 6.1.3.

6.1.4.4 Piping in systems shall be run as directly as is practicable from one point to another, and with as few restrictions, such as ells and bends, as conditions will permit, giving consideration to provisions of 6.2.

6.1.4.5 Provision shall be made in piping including interconnecting of permanently installed containers, to compensate for expansion, contraction, jarring and vibration, and settling. Where necessary, flexible connectors complying with 6.2 should be used.

6.2 Hose, hose connections and flexible connectors

6.2.1 Hose, hose connections and flexible connectors shall be fabricated of materials resistant to the action of LP-Gas both as liquid and vapor. If wire braid is used for reinforcement it shall be of corrosion resistant material such as stainless steel.

6.2.2 The correctness of design, construction and performance of hose shall be determined by Underwriters Laboratories, Inc., their nationally recognized testing laboratory or approval of authority having jurisdiction.

6.2.3 Hose, hose connections, and flexible connectors used for conveying LP-Gas liquid or vapor at pressures of 0.0345 MPa (5 psi) shall comply with 6.2.3.1 and 6.2.3.2.

6.2.3.1 Hose shall be designed for a minimum bursting pressure of 120.66 MPa (1750 psi)[2.41 MPa (350 psi) working pressure] and shall be identified "LP-Gas" or "LPG" at not greater than 3 m (10 ft) intervals.

6.2.3.2 Hose assemblies after the application of connections shall be capable of withstanding a test pressure of not less than 4.83 MPa (700 psi).

6.2.4 Flexible connections used in piping systems shall comply with 6.2.1, 6.2.2, and 6.2.3 for the service in which they are used and shall be installed in accordance with manufacturers instructions. Flexible connectors in lengths up to 914 mm (36 in.) may be used for liquid or vapor piping, or portable or stationary tanks to compensate for expansion, contraction, jarring and vibration, and settling. This is not to be construed to mean that flexible connectors shall be used if provisions were incorporated in the design to compensate for these effects.

6.2.5 Hose and hose connections subjected to regulated pressures not greater than 0.345 MPa (5 psi) shall be designed for a bursting pressure of not less than 0.86 MPa (125 psi).

6.2.6 A shut-off valve shall be provided in the piping immediately upstream of the inlet connection of the hose. When more than one such appliance shut off is located near another precautions shall be taken to prevent operation of the wrong valve.

6.3 Relief valves

6.3.1 A hydrostatic relief valve shall be installed in each section of piping (including hose) in which liquid LP-Gas can be isolated betwen shut-off valves so as to relieve to a safe atmosphere the pressure which could develop from the trapped liquid. The start-to-discharge pressure setting of such relief valves shall not be in excess of 3.45 MPa (500 psi).

6.3.2 Shut-off valves shall not be located between a safety relief device and the container, unless the arrangement is such that the relief device capacity flow will be achieved through additional relief devices which remain operative.

6.3.3 Relief shall be away from component parts of the equipment and operating personnel, and shall not terminate in or beneath any part of the equipment if below 1.8 m (6 ft).

6.3.4 Safety relief valves shall be so designed that the possibility of tampering will be minimized. Externally set or adjusted valves shall be provided with an approved means of sealing the adjustment.

6.3.5 Fail safe automatic safety shut-off valve

6.3.5.1 Two automatic safety shut-off valves shall be installed in series and wired in parallel if one fail-safe type is not used.

6.3.5.2 On vaporizer-burners the fail-safe valve shall be installed up-stream of the vaporizer section.

6.4 Testing. All piping, tubing and hose shall be tested after assembly and proved free from leaks at not less than normal operating pressure.

SECTION 7—CONTROL EQUIPMENT

7.1 General. Control equipment shall be of such construction and design and so arranged that required conditions of safety for the operation of the heat-producing device, the dryer, and the ventilating equipment used will be maintained.

7.2 Minimum fire-safety controls for dryers

7.2.1 Automatic recycling-type control systems shall not be used. Control systems shall be of the type that requires manual resetting after shut-down for an unsafe condition.

7.2.2 Dryers in which the dried product moves automatically from the dryer to the storage building shall have an approved maximum temperature thermostat located in the exhaust air stream, or other approved device, which will prevent the transfer of fire from the dryer to the storage building.

7.2.3 Dryers in which the dried product moves manually from the dryer into a storage building should have one or more approved maximum temperature thermostats located in the exhaust air stream, or other approved devices, which will prevent the transfer of fire from the dryer to the storage building. The operation of these controls shall shut off all heat being supplied to the dryer and should stop the movement of air through the dryer, interrupt the flow of the product into and away from the dryer; and sound an audible alarm.

7.2.4 Dryers of combustible construction or containing combustible trays shall have a thermostat set to prevent the combustible material from reaching sustained temperatures in excess of 74 °C (165 °F). Such controls shall shut off all heat being supplied to the dryer; shall permit the continued movement of unheated air through the dryer and shall sound an audible alarm.

7.2.5 A control device of suitable design shall be provided which will cut off all heat being supplied to the dryer should the movement of air through the dryer be stopped. For AMD driven directly by electric motors, this may be an electrical device at the motor, and in all other instances a device which proves air flow. Drum or rotary dryers which do not employ air flow are exempt from this requirement.

7.2.6 A maximum temperature thermostat shall be located between the heat producing device and the dryer or shall be located in the heat producing device.

7.3 Minimum fire safety controls for heat producing devices

7.3.1 The circuitry shall be so designed that there will not be an automatic restart after an unsafe shutdown.

7.3.2 All safety devices and safeguarding sequences shall be automatic in operation following initial ignition of the pilot and main flame for devices burning liquid or gaseous fuel. Trial for ignition of main flame or flame failure shall not exceed 90 seconds.

7.3.3 The fuel supply to gas pilots shall be automatically shut off in the event the pilot is extinguished.

7.3.4 Circuitry for safety shut-off valves, combustion safeguards, pressure switches, maximum-temperature thermostats, electrical interlocks, and other controls, shall be arranged to provide fail-safe operation.

7.3.5 All electric power for the operation of the dryer and its controls shall be supplied by the same source, or shall be interlocked to accomplish the same as a single circuit or feeder.

7.3.6 Pre-ventilation of the combustion chamber and dryer shall be accomplished by a control installed on the heating device, for direct-fired dryers employing gaseous fuels, which will prevent ignition and flow of fuel before the fan has provided 4 air changes of the plenum chamber.

7.4 Manual controls for heat producing devices. In addition to the automatic controls required, each installation using either gaseous or liquid fuels shall have a quick-acting, manually operable valve in the fuel supply line ahead of all dryer fuel equipment and outside any enclosures.

7.5 Electrical control enclosures

7.5.1 **Exposed control.** All controls directly exposed to the elements shall be of weather-resistant construction.

7.5.2 **Housing of electrical controls**

7.5.2.1 All electrical controls not inherently of the weather-proof type shall be enclosed in weather-resistant construction.

7.5.2.2 The control box shall provide for separation of the electrical controls from all gas controls. Gas valves, strainers, reliefs, etc., shall not be located in the electrical control box.

SECTION 8—HOLDING CAPACITY OF BATCH DRYERS

8.1 The holding capacity of a batch dryer shall be defined as the volumetric capacity of the structure containing the product to be dried when the structure is filled to the normal operating level. It may be expressed in m^3 (bushels), based on 28.4 bushels/m^3.

SECTION 9—BASIS FOR STATING DRYING CAPACITY OF BATCH AND CONTINUOUS-FLOW GRAIN DRYERS

9.1 Drying capacity shall be stated as the number of wet bushels of grain dried per hour, wet basis, using clean, yellow corn as the standard grain and shall include the following:

9.1.1 Moisture reduction from 25.5 percent to 15.5 percent, or 20.5 percent to 15.5 percent.

9.1.2 The drying air temperature used in rating the dryer shall be stated. Based on a 10 °C (50 °F) ambient, the rating temperature shall not exceed that at which the dryer will operate continuously without damage to grain or machine.

9.1.3 Inclusion of time to cool to 15.6 °C (60 °F)(i.e. within 5.6 °C (10 °F) of ambient temperature).

9.1.4 Loading and unloading time shall not be included in the rating for batch dryers. This shall be so stated in all published data.

9.1.5 Ratings of other grains shall be on a similar basis.

SECTION 10—BURNER RATINGS

10.1 Direct fired LP and natural gas

10.1.1 Maximum inlet gas pressure shall be stated on the nameplate.

10.1.2 The burner rating shall be indicated on the nameplate and in descriptive material in joules (BTU) per hr input at rated manifold gas pressure with a specified orifice.

10.2 Convertible units (LP to natural)

10.2.1 **Natural gas—LP vapor burner.** On burners designed to burn either natural gas or LP vapor, the changing of the orifice shall constitute the only conversion necessary to switch from one gas supply to the other.

10.2.2 **Natural gas—liquid LP burners.** Dryers utilizing liquid burners shall be designed so that the burner and fuel train assembly can be dismounted and a natural gas burner and fuel manifold be mounted without major alterations to the remainder of the dryer.

SECTION 11—TESTING AND RATING OF AIR MOVING DEVICES (AMD)

11.1 All air moving devices (AMD) used for crop drying or aeration shall be tested and rated in accordance with Air Moving and Conditioning Association, Inc., Bulletin 210, Standard Test Code for Air Moving Devices.

11.2 Testing AMD's only. Only AMD's as equipped for production shall be used in tests. AMD's intended for use as individual or separate units shall be tested and rated with all screens, guards, mounting structure, etc., in place as equipped for operation.

11.3 Testing AMD's for heated air units. AMD's intended for use in heated air units shall be tested and rated with all screens, guards, fan tube extension, burner, plumbing, controls, and all other equipment which comes in contact with the air stream, in place as used in operation. All optional equipment shall also be in normal operating arrangement.

11.4 Rating curves and charts on AMD's and heated-air units

11.4.1 All rating curves shall show cfm air flow, static pressure, brake horsepower, total and basic efficiency per AMCA Standard Test Bulletin No. 210-67, and shall specify which regular and optional equipment was installed.

11.4.2 All published performance charts shall show cfm air flow and brake horsepower of each static pressure point shown.

11.4.3 Ratings expressed on compliance certificates shall be as noted in Section 13.

11.5 Log sheets. Log sheets are to be used in such form as to suit the testing procedure, as long as they are in keeping with the code requirements.

11.6 Test supervision. All AMD's and heated-air unit tests should be supervised and certified by a registered engineer or recognized test authority.

SECTION 12—ELECTRIC MOTOR RATINGS AND APPLICATIONS

12.1 Motor types

12.1.1 Single-phase motors shall be capacitor-start induction-run, capacitor-start capacitor-run, or repulsion-start induction-run on belted applications only.

12.1.2 Three-phase motors shall be squirrel cage induction motors.

12.2 Motor enclosures. The following motor enclosures, as approved by National Electrical Manufacturers Association Standard, NEMA MG 1, Motors and Generators, may be used:

12.2.1 Drip-proof, equipped with rodent screens.

12.2.2 Totally enclosed non-ventilated.

12.2.3 Totally enclosed, fan-cooled.

12.2.4 Totally enclosed, air-over.

12.2.5 Dustproof or explosion-proof.

12.3 Motor loading

12.3.1 The maximum load imposed on motors shall be such that the temperature in the windings shall not exceed that permitted in NEMA standards. The motor horsepower loading shall not exceed the motor horsepower rating on nameplate multiplied by service factor.

12.3.2 For air-over applications, the motor nameplate shall clearly state that air-over motor must be installed in air streams.

12.4 Motor controls. Motor controls shall meet the requirements of NFPA No. 70, National Electrical Code.

12.4.1 Each motor shall be protected against running over-current (overload) by one of the following means:

12.4.1.1 A separate over-current (overload) device, such as dual-element fuses, adjustable or non-adjustable circuit breakers, thermal relays or thermal-trip motor switches which are responsive to motor current shall be used. This device shall be rated or selected to trip at not more than 115 percent of the maximum-load current of the motor.

12.4.1.2 A thermal protector, integral with the motor, approved for use with the motor which it protects on the basis of service factor and prevention of dangerous over-heating of the motor due to overload or failure to start shall be used. If the motor-current-interrupting device is separate from the motor and its control circuit is operated by a protective device integral with the motor, it shall be so arranged that the opening of the control circuit will result in interruption of current to the motor.

12.4.2 Where devices other than dual element fuses are used for motor-running overload protection, the minimum allowable number and location of over-current units, such as trip coils, relays, or thermal cutouts, shall be as follows:

12.4.2.1 One in either ungrounded conductor to all single-phase motors.

12.4.2.2 One in each of the ungrounded conductors to all three-phase motors.

12.4.3 Magnetic starter size shall be selected on the basis of maximum horsepower or ampere rating of the motor.

12.4.4 Enclosures

12.4.4.1 Controls located outdoors shall be in Naitonal Electrical Manufacturers Association, Standard "III" or "IV" enclosures or NEMA "I" enclosures if mounted in a weather-proof housing or shelter.

12.4.4.2 Controls attached to units and mounted in weather-proof housing may be NEMA "O" open starters.

12.5 Electric wiring

12.5.1 **Internal wiring.** All conducting non-current-carrying material and enclosures attached to or part of the unit shall be electrically bonded together and connected to the proper grounding conductor of the electric supply system.

12.5.2 Electric supply wiring and protective devices should comply with NFPA Standard No. 70, National Electrical Code.

12.5.2.1 Wire, service entrance, and fusing shall be sized according to maximum operating current of unit.

12.5.2.2 Voltage drop between motor and power meter should be limited to 3 percent.

12.5.2.3 All units shall be grounded directly to service neutral through power cable: Single-phase, 120-v and 240-v, with 3-wire connections; and three-phase with 4-wire connections.

12.6 Electric generators on engine or tractor-driven units

12.6.1 Generators shall be equipped with copper grounding wire or strip solidly bonded to generator frame and attached to copper ground rod by approved clamp or other fastener, with a suitable label or tag stating that ground rod must be driven into ground when unit is operating. Grounding shall be in accordance with NFPA No. 70, National Electrical Code.

12.6.2 When pre-wired electric outlets are provided on the generator, they shall be of the 3-wire, polarized type with ground connections.

12.7 Motor nameplates

12.7.1 A nameplate shall be provided for each electric motor showing the following minimum data for the maximum loading to which the motor is subjected in the particular application:

MAX RATED HP . . . . @ RPM . . . .

PHASE . . . . CYCLES . . . . . VOLTS . . . . .

AMPS . . . . @ FULL LOAD S. F. . . . .

MOTOR RATINGS FOR AIRSTREAM

APPLICATION ONLY* . . . .

. . . .

*Where applicable fill in type of application (airstream only—intermittent duty.)

SECTION 13—COMPLIANCE CERTIFICATES

13.1 Compliance certificates shall be applied only by the original equipment manufacturer. When a certificate of compliance is displayed on a crop dryer or AMD, information specified in paragraphs 13.2 thru 13.5 shall be included. Data may be in the form of decals, metal plates, or other suitable materials. Size, color, etc., shall be established by the manufacturer. Arrangement of data shall be as shown in paragraphs 13.2, 13.3, 13.4.

13.2 Data for batch and continuous-flow dryers. The following minimum data shall be included with compliance certificates for batch or continuous-flow dryers as applicable:

MODEL . . . . .TYPE . . . .

GRAIN HOLDING CAPACITY . . .CU FT . . . . . . .BUSHELS

TYPE OF FUEL . . . .

BURNER RATING . . . . BTU/HR MAX,

OR . . . . GAS FUEL/HR MAX

OR: . . . . .GAS FUEL/HR AT 200° F TEMP RISE

FAN RATING . . . . CFM AT 2 IN. WC PRESSURE IN PLENUM

RATED FAN SPEED . . . . RPM

BRAKE HORSEPOWER REQUIREMENTS AT RATED SPEED:

MAX AMD ONLY . . . . HP

MAX CONVEYOR SYSTEM . . . . HP

TOTAL MAX . . . . HP

TOTAL ELECTRICAL POWER

REQUIREMENTS . . . . .KW MAX

. . . . . . .AMP MAX AT . . .VOLTS

13.3 Data for heated-air units and supplemental-heat units. The following minimum data shall be included with compliance certificates for heated-air units as applicable. Where burners and AMD units are furnished separately, individual certificates shall be applied to each and shall include only the appropriate data.

MODEL . . . .

TYPE OF FUEL . . . .

BURNER RATING . . . . BTU/HR MAX FUEL INPUT

OR: . . . . GAS FUEL/HR MAX

AMD RATING . . . . CFM AT 1 IN. OR 2 IN. WC PRESSURE AT DISCHARGE

RATED FAN SPEED . . . RPM

AMD BRAKE HP MAX: HP FROM 0 IN. WC TO . . . . IN. WC

TOTAL ELECTRICAL POWER REQUIREMENT . .KW MAX

. . . . .AMP MAX AT . . . . .VOLTS MAX

13.4 Data for drying and aeration AMD's. The following data shall be included on compliance certificates for all fans used with unheated air for drying or aeration:

MODEL . . . .

AMD RATING . . . . .CFM @ 1 IN. OR 2 IN. WC PRESSURE AT DISCHARGE

RATED SPEED . . . . RPM

AMD BRAKE HP MAX: HP FROM 0 IN. TO . . . . . . IN. WC

TOTAL ELECTRICAL POWER REQUIREMENT . .KW MAX

. . . . .AMP MAX AT . . . . .VOLTS MAX

13.5 Motor nameplates. Nameplate data as specified in paragraph 12.7 shall appear for each motor used on drying and aeration equipment. This data may be included with the compliance certification data, or may be attached separately to each motor in a location clearly visible to the operator.

13.6 Instructions. Manufacturers whose equipment bears a compliance certificate shall provide printed instructions for installation and operation of units.

SECTION 14—INSTALLATION RECOMMENDATIONS

14.1 Fuel supplies and electrical connections, up to the point of connection with the dryer where applicable shall comply with the following standards of the National Fire Protection Association:

NFPA NO. 31, Standard for the Installation of Oil Burning Equipment.

NFPA No. 54, Standard for the Installation of Gas Appliances and Gas Piping.

NFPA No. 58, Standard for the Storage and Handling of Liquified Petroleum Gases.

NFPA No. 70, National Electrical Code.

14.2 The fuel line from the natural gas meter to the fuel train inlet of the dryer shall be sized and installed to meet American National Standard, Z83.1, Installation of Gas Piping and Gas Equipment on Industrial Premises and Certain Other Premises.

APPENDIX A

A.1 Definitions are intended to relate only to this Standard and its application:

Air-moving device (AMD): A revolving, wheel-type, mechanical device used to move air for drying or aeration. For the purposes of this Standard, an AMD shall include the wheel or blade assembly, mounting structure and casing, but may or may not include a power source.

Aeration: The purposeful movement of air at a low rate through a product to maintain or improve product quality. Air flow rates usually do not exceed 0.08 cubic meters per minute per cubic meter of product (0.1 cubic feet per minute per bushel of product) through a dry product and 0.80 cubic meters per minute per cubic meter of product (1 cubic foot per minute per bushel of product) through a wet product.

Ambient temperature: The temperature of the surrounding air.

Approved: The use of "approved" or "approved-type" refers to listing by a recognized testing agency or laboratory.

Available heat: In drying a harvested crop, the quantity of heat in air that can be utilized in evaporating water from the product.

Batch: A quantity of a harvested crop put into a bin or container on a repetitive basis specifically for treatment, such as drying.

Blending: The process of mixing two or more different products together, such as grains and supplements, to obtain desired food ratios, or the process of mixing different quantities of the same product with different moisture contents to obtain a final mass with a uniform moisture content.

Bushel:

One bushel by volume equals 1.25 cubic feet.

One dry bushel is that weight of grain defined by government grain standards as a standard unit for trading at a specific moisture (e.g. 56 pounds at 15.5 percent w.b. for No. 2 corn).

A wet bushel is the weight of wet grain which when dried will give one dried bushel as specified in paragraph above (e.g. 63.5 lb of 25.5 percent w.b. corn).

Casing: The outer enclosure surrounding the entire heat exchanger and confining the air being heated.

Combustible material: Combustible refers to a material or structure which can burn. Combustible is a relative term. Many materials which will burn under one set of conditions will not burn under other. For example, structural steel is not combustible, but fine steel wool is combustible. The term combustible is not related to any specific ignition temperature.

Conduction: Transmission through or by means of a conductor, distinguished in the case of heat from convection and radiation.

Control: Any component of a dryer, or dryer heat source, so designed to affect or limit any normal or abnormal condition of the drying operation.

Convection: Transference of heat or electricity by moving masses of matter, as by currents in gases and liquids caused by differences in density, or by electrically charged particles across a spark gap.

Cooling stage: The time required to move a cooling zone entirely through a product mass.

Cooling zone: That portion of the product mass in storage where the product temperature of the crop is falling during aeration.

Cubic meters per minute (CFM): Volumetric measure of quantity of flow. (CFM is the most common customary measure.)

Curing: A form of conditioning as opposed to simple drying in which a chemical change occurs, such as in tobacco, sweet potatoes, etc., to prepare the crop for storage or use.

Cycling burner: Type of operation wherein application of maximum heat is periodic such as cycling between high fire and low fire, cycling between high fire and a constant pilot; and cycling from high fire to "off", then restarted to high fire by constant or intermittent ignition.

Dehydration: The rapid removal of moisture, usually to a very low level.

Depth factor (Df): When drying with air, a depth which would contain enough product that, if all the theoretical heat available for drying could be used, it would all dry to equilibrium in a period of time equal to the time required for the fully exposed product to dry half-way to equilibrium.

Dryer: A unit which provides the conditions for removing moisture from a product.

Dryer, batch: Any dryer wherein the product to be dried is placed in the dryer, the complete drying or drying and cooling operation performed, and then removed for storage or futher processing. Usually it is self-contained comprising a drying compartment, either horizontal or vertical, to contain the product being dried while heated air is forced or drawn through it from a central portion usually called a plenum chamber. Cooling is usually carried out in the same manner. Integral conveying equipment provides for loading and unloading the dryer. Also included are the AMD, burner, and control system. Batch dryers are usually portable, but may be stationary.

Dryer, circulating: A batch dryer equipped to circulate or mix the product during the drying and cooling period.

Dryer, concurrent flow: A type of continuous flow dryer wherein the product being dried moves in the same direction as drying air. Sometimes referred to as parallel flow.

Dryer, continuous-flow: Any dryer wherein the product to be dried is in continuous movement through the dryer and air movement is continuous, in contrast to batch operation.

Dryer, counter-flow: A type of continuous-flow dryer wherein the product being dried moves in one direction and the drying air moves in the opposite direction.

Dryer, cross-flow: A type of dryer wherein the flow of air is transverse to the direction of flow of the product being dried.

Dryer, direct-fired: Type of dryer in which the products of combustion come into direct contact with the product being dried.

Dryer, fluidized or spouted bed: A dryer where the product is in suspension, or is moved through the dryer by the drying and/or the cooling air.

Dryer, indirect-fired: Type of dryer in which the products of combustion do not come in contact with the products being dried.

Dryer, in-storage: A dryer in which the drying bin or compartment is also used to store the product after it is dried.

Dryer, self-contained: Any dryer manufactured as a package unit consisting of the drying and cooling chamber, necessary heat or radiation source, all AMD's and duct work, along with the necessary controls and product handling equipment. These dryers may be either fixed or portable.

Dryer, tunnel: A type of dryer wherein the product being dried is conveyed through a tunnel-like chamber. It may be continuous or batch-type.

Drying: The removal of moisture from a product, usually to some predetermined moisture content.

Drying air: The air being passed through the product which is being dried.

Drying air temperature: The temperature of the air entering the product being dried.

Drying front: The divisional layer between the dried and undried products in drying systems.

Drying time maximum permissible: The maximum elapsed time that may be used to complete the drying of any portion of the product without undesirable change in quality.

Drying zone: The band or layer of product in which most of the drying is occurring at any instant.

Electrical induction: Act or process by which an electric conductor becomes electrified when near a charged body.

Equilibrium moisture content: The moisture content of a product when it is in equilibrium with the surrounding atmosphere.

Equilibrium relative humidity: The relative humidity of air surrounding a product in equilibrium with a given moisture content. The air and product are at the same temperature.

Fail-safe control: A control designed so that a malfunction of any of its components will stop the operation of the device or equipment controlled by it.

Fuel train: The fuel train, mounted on the grain dryer structure and connected to the gas burner, includes all piping components of fuel flow control and safety shut-off valves.

Heat exchanger: A device used to transfer heat from one fluid stream to another without intermixing.

Heating-air drying: Use of forced ventilation with the addition of heat for removing moisture.

Heated-air unit: Basic heated-air-producing unit including AMD, burner system, and electrical system. It is usually coupled to drying structure by means of a flexible duct. It may have transport chassis and wheels for portability.

Insulating fitting: A type of fitting designed to prevent galvanic current flow when used between two dissimilar metals.

Joule (BTU): A measure of quantity of heat. One joule will raise the temperature of one gram of water one degree Celsius. (One BTU will raise the temperature of one pound of water one degree Fahrenheit.)

Line pressure: The pressure of the fuel in the supply line to the dryer.

LP-gas: A mixture of gaseous petroleum products normally stored and transported as a liquid under pressure. The principal constituents are propane and butane.

Modulate (as applied to crop drying in reference to regulation of fuel in continuous flow): Automatically governing the rate of fuel flow by a control which is temperature-sensitive in order to maintain a constant temperature at the location of the sensing device.

Moisture content (dry basis): For products, expressed as percentage, by weight, of water in the product divided by dry matter.

$$\text{Moisture content, percent} = (100) \times \frac{\text{weight of water in product}}{\text{weight of dry matter}}$$

Moisture content (wet basis): For agricultural producers, usually expressed as percentage, by weight, of water in the product, wet basis.

$$\text{Moisture content, percent} = (100) \times \frac{\text{weight of water in product}}{\text{weight of dry matter + water}}$$

Natural gas: A gaseous hydrocarbon, odorless and flammable, found in its natural state in particular geologic formation as a product of decomposition of organic matter. The composition is chiefly of the methane series with varying amounts of other components such as carbon dioxide, hydrogen, and helium often being present.

Orifice: The opening through which gas is admitted to the burner.

Pascal (psi—pounds per square inch or inches of water column): A measure of pressure or stress. Assumed to be gage pressure unless followed by the term "absolute".

Pilot-continuous: The pilot remains lit during the time power is supplied to the burner.

Pilot-interruptible: The pilot is lit during the time combustion is being established. When combustion is established the pilot is de-energized.

Pilot-standing: The pilot is manually lit and remains lit until manually shut off.

Plenum: An air chamber maintained under pressure (positive or negative) usually connected to one or more distributing ducts in a drying or aeration system. The term is also used to designate the air chamber under the perforated floor in a grain bin and the pressure chamber between grain columns in some types of batch or continuous dryers.

Pressure regulator: A mechanical device which reduces the fluid (liquid or gas) pressure to a relatively constant delivery pressure while the inlet pressure may vary and while the volume of gas may also vary.

Pressure system: Method of air movement in which air is forced through the product with the air duct or ducts at a pressure above atmospheric pressure. It is called a pushing or forcing system of air movement.

Pre-ventilation: As applied to crop dryers, the term refers to clearing or purging the plenum chamber or duct of any volatile gases prior to ignition of the burner. It is usually accomplished by a device which insures that the fan must operate for a certain period of time before ignition will be permitted.

Radiation: The process by which energy is emitted from molecules and atoms owing to the internal changes. Also the combined processes of emission, transmission, and absorption of radiant energy.

Recirculation: As applied to crop dryers, the term refers to the return of a portion of the exhaust air to the air intake of the dryer, or to the return of underdried grain to the dryer or container from which it was removed.

Relative humidity: A measure of the moisture content of air expressed as a percentage. It is the ratio of the weight of water vapor in a given volume of air at a given temperature to the maximum quantity of water vapor which the same volume of air could hold at the same temperature.

Static pressure: A measure of air pressure usually expressed in pascals [inches of water column (WC)].

Steady state: Condition when the operation or process reaches equilibrium.

Suction system: Method of air movement in which the air is moved through the product with the air duct or ducts at a pressure lower than atmospheric. It is also called an exhaust system of air movement.

Supervision: Continuous monitoring to react automatically to flame failure so as to shut off gas flow to the unit.

Supplemental heat: Any heat added to that already present in the atmosphere to obtain a limited temperature rise, usually less than 11 °C (20 °F), to accomplish drying within the maximum permissible drying time to prevent spoilage.

Temperature rise: As applied to crop drying, the term refers to the difference between ambient temperature and the temperature of the drying air resulting from the addition of heat by the dryer burner.

Tempering: Equilization of moisture or temperature throughout the product. Bringing a product to a desired moisture content or temperature for processing.

Time of drying: The elapsed time from the start of the drying process to the instant the drying front arrives at any point or place in the product.

Time of one-half response: Time required to dry fully exposed products halfway to equilibrium.

Turning: The process of moving a product through the air within a bin or storage structure, or from one bin or storage structure to another.

Valve, excess-flow: A check valve which permits flow of fluid in either direction but which limits excessive flow in one direction. If the designated flow rate is exceeded, the valve automatically closes.

Valve, pressure-relief: A valve designated as a safety device to open, and remain open, to discharge a fluid whenever the fluid pressure reaches the start-to-discharge setting of the valve. When the fluid pressure drops below this setting, the relief valve automatically closes.

Valve, quick-acting: A manually operated valve specially designed to accomplish rapid shut-off to fuel flow to dryer.

Valve, solenoid: A valve which is opened or closed by a solenoid (electromagnet). In the normally closed type, for example, the valve is opened by the solenoid but closed by a return spring and held closed by the fluid pressure upstream from the valve.

Vapor pressure (fuel): Commonly taken to mean saturated vapor pressure which is the vapor pressure of a vapor in contact with its liquid form. An example is the pressure in an LP-gas storage tank. The term is also used for the pressure of the vaporized fuel being fed to the burner orifice.

Vaporizer: In an LP-gas system, the vaporizer is a heat exchanger wherein heat is supplied to change the liquid fuel to vapor, ready for combustion. The vaporizer may be integral with the burner so that part of the heat of combustion is used for vaporization.

Vaporizer-burner: An integral vaporizer-burner unit, dependent upon the heat generated by the burner as the source of heat to vaporize the liquid fuel.

Velocity of air flow for conditioning:

Apparent velocity in meters per minute (feet per minute): The rate of air flow determined by dividing the quantity of air flow in cfm by the cross-sectional area.

Average velocity: The rate of air travel through product void space. It is determined by dividing the apparent velocity by the product void space expressed as a decimal. The average velocity is always greater than the apparent velocity.

Traverse time: The in-product travel time of air from entrance to any point, usually expressed in seconds.

Ventilation: Air movement through space.

Ventilation front: The locus of all points of equal traverse time in the product being conditioned.

Void space: The space between particles in a bulk of stored crop, usually expressed as percent of total volume.

Weatherproof: So constructed or protected that exposure to the weather will not interfere with safe operation.

A.2 A partial list of testing laboratories of agencies follows:

A.2.1 All testing laboratories approved by Air Moving and Conditioning Association, Inc., 205 West Touhy Avenue, Park Ridge, Illinois 60068.

A.2.2 Canadian Standards Association, 235 Montreal Road, Ottawa 7, Ontario, Canada.

A.2.3 Factory Mutual Engineering Corporation, 1151 Boston-Providence Turnpike, Norwood, Massachusetts 02062.

A.2.4 Underwriters Laboratories, Inc., 207 E. Ohio St., Chicago, Illinois 60611.

Cited Standards:

AMCA 210-67, Standard Test Code for Air Moving Devices
AMCA 300-67, Test Code for Sound Rating Air Moving Devices
ANSI B36.10, Wrought-Steel and Wrought-Iron Pipe
ANSI Z83.1, Installation of Gas Piping and Gas Equipment on Industrial Premises and Certain Other Premises
ASTM A53, Specification for Welded and Seamless Steel Pipe
ASTM A539, Specification for Electric-Resistance-Welded Coiled Steel Tubing for Gas and Fuel Oil Lines
ASTM B42, Specification for Seamless Copper Pipe, Standard Size
ASTM B43, Specification for Seamless Red Brass Pipe, Standard Sizes
ASTM B88, Specification for Seamless Copper Water Tube
ASTM B135, Specification for Seamless Brass Tube
ASTM B280, Specifications for Seamless Copper Tube for Air Conditioning and Refrigerator Field Service
NBFU, National Building Code
NEMA MG1, Motors and Generators
NFPA No. 31, Standard for Installation of Oil Burning Equipment
NFPA No. 54, Standard for Installation of Gas Appliances and Gas Piping
NFPA No. 58, Storage and Handling of Liquified Petroleum Gases
NFPA No. 61B, Prevention of Fire and Dust Explosions in Grain Elevators and Bulk Grain Handling Facilities
NFPA No. 70, National Electrical Code
NFPA No. 90A, Installation of Air Conditioning and Ventilating Systems of Other than Residemce Type
NFPA No. 211, Chimneys, Flues and Vents
UL 731, Oil Fired Heaters
UL 795, Commercial Industrial Gas Heating Equipment

FRICTION COEFFICIENTS OF CHOPPED FORAGES

Approved 1956 by the ASAE Committee on Technical Data; reconfirmed December 1963, December 1968, December 1973; reconfirmed and revised editorially December 1974; revised March 1980; reconfirmed December 1984, December 1985.

SECTION 1—SCOPE

1.1 These data cover the friction coefficients between chopped forages (grass, corn, hay, straw) and metal surfaces. Since several variables must be considered in selecting an appropriate friction coefficient, the data are presented in six sections. Each section shows the effect of one variable.

SECTION 2—EFFECT OF CONDITION OF METAL SURFACE

2.1 Static coefficient. Considering the effect of surface condition of a galvanized steel sheet on the static-friction coefficient of chopped dry hay and straw, Fig. 1 shows that the friction tests themselves reduced the coefficient by polishing the surface. Also, intermediate polishing was done with abrasives and a reciprocating machine using chopped hay as the polishing medium. The intermediate polishing plus the polishing caused by the tests themselves finally reduced the coefficient to 0.17, beyond which it could not be reduced. The original condition was a slightly weathered zinc surface, whereas the final condition was a highly polished zinc surface. Chopped grass and corn silage showed little or no decrease in static-friction coefficient due to such surface polishing.

2.2 Sliding coefficient. For sliding coefficients of chopped grass (71 to 73 percent moisture-w.b.) at low sliding velocities, the following data were reported:

Surface	Friction coefficient
Polished galvanized steel	0.68
Stainless steel No. 4 polish	0.65

SECTION 3—EFFECT OF NORMAL PRESSURE

3.1 Static coefficient. Figs. 2 and 3 show the effect of normal pressure intensity on static-friction coefficients on highly polished galvanized steel.

3.2 Sliding coefficient. The following tabulation shows the effect of normal pressure intensity on sliding-friction coefficients on polished galvanized steel, with sliding velocities 0.025 to 1.62 m/s (5 to 320 ft/min).

Material	Normal pressure, kPa (lb/ft^2) 0.27 (5.6)	0.43 (8.9)	0.68 (14.2)	1.36 (28.4)	2.73 (57)	Average
Chopped Straw			0.30	0.30	0.30	0.30
Corn Silage (73% moisture)	0.70	0.70	0.68	0.66		0.68
Grass Silage (73% moisture)	0.63	0.66	0.70	0.72		0.68

(Moisture content figured on wet basis)

SECTION 4—EFFECT OF VELOCITY

4.1 Velocities below 1.62 m/s (320 ft/min). In the range of 0.025 to 1.62 m/s (5 to 320 ft/min) there was no significant effect caused by changes in sliding velocity. Average coefficients on galvanized steel were:

Chopped straw	0.30
Corn silage (73% moisture w.b.)	0.68
Grass silage (73% moisture w.b.)	0.68

4.2 Velocities from 5.1 to 30.5 m/s (1000 to 6000 ft/min). In this range friction coefficients for fresh chopped alfalfa and corn were lower than at slow velocities, but did not change significantly within the range. Average coefficients on stainless steel, No. 4 polish, were:

Chopped alfalfa	0.49
Chopped corn	0.49

Below 5.1 m/s (1000 ft/min), results were variable but gave generally higher coefficients, tending to confirm the figures under paragraph 4.1.

SECTION 5—EFFECT OF MOISTURE CONTENT

5.1 Sliding coefficient. Fig. 4 shows the effect of moisture content on sliding-friction coefficient of chopped alfalfa on stainless steel, No. 4 polish. Velocity and pressure not reported. Length of cut, 25 to 102 mm (1 to 4 in.).

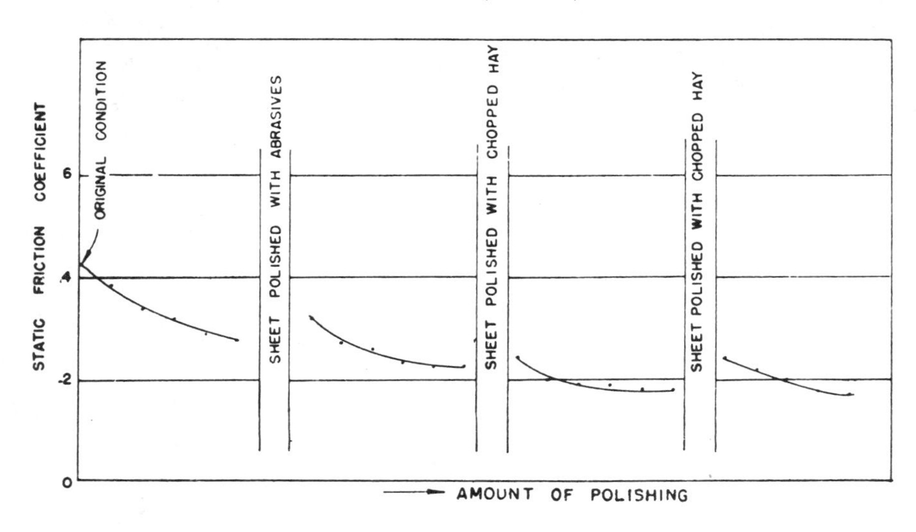

FIG. 1—EFFECT OF SURFACE CONDITION OF GALVANIZED STEEL SHEET ON THE STATIC-FRICTION COEFFICIENT OF CHOPPED DRY HAY AND STRAW

SECTION 6—EFFECT OF LENGTH OF CUT

6.1 Sliding coefficient. Sliding-friction coefficients of chopped alfalfa (moisture, 51.8 to 73.4 percent-w.b.):

Length of Cut, mm (in.)	Average Friction Coefficient
25 (1)	0.652
51 (2)	0.647
76 (3)	0.677
102 (4)	0.622

SECTION 7—INTERNAL FRICTION COEFFICIENTS

7.1 Static coefficient. Figs. 5 and 6 show results of shear tests of corn and grass silage. The curves are indicative of internal static friction in a plane parallel with the major fiber axes.

References: Last printed in 1963 AGRICULTURAL ENGINEERS YEARBOOK; list available from ASAE Headquarters.

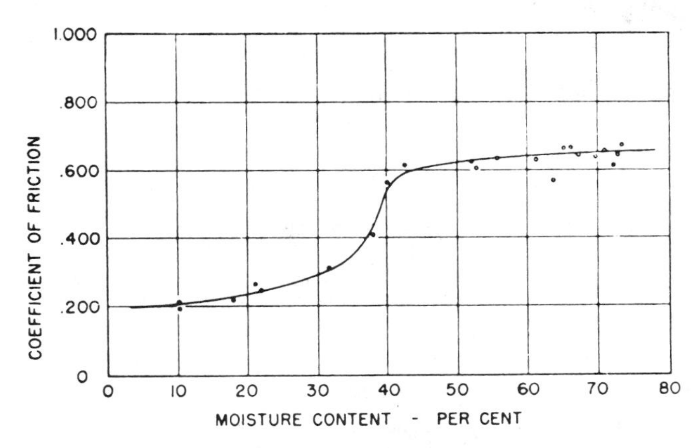

FIG. 4—EFFECT OF MOISTURE CONTENT (W.B.) ON SLIDING-FRICTION COEFFICIENT OF CHOPPED ALFALFA ON STAINLESS STEEL

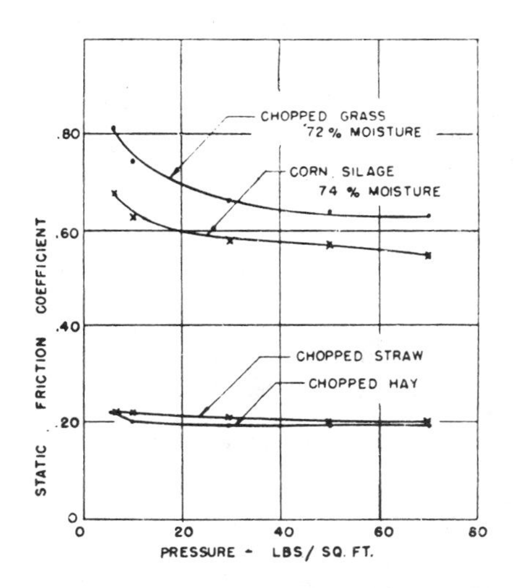

FIG. 2—EFFECT OF NORMAL PRESSURE INTENSITY ON STATIC-FRICTION COEFFICIENTS ON HIGHLY POLISHED GALVANIZED STEEL (MOISTURE CONTENT FIGURED ON WET BASIS); ENGLISH UNITS

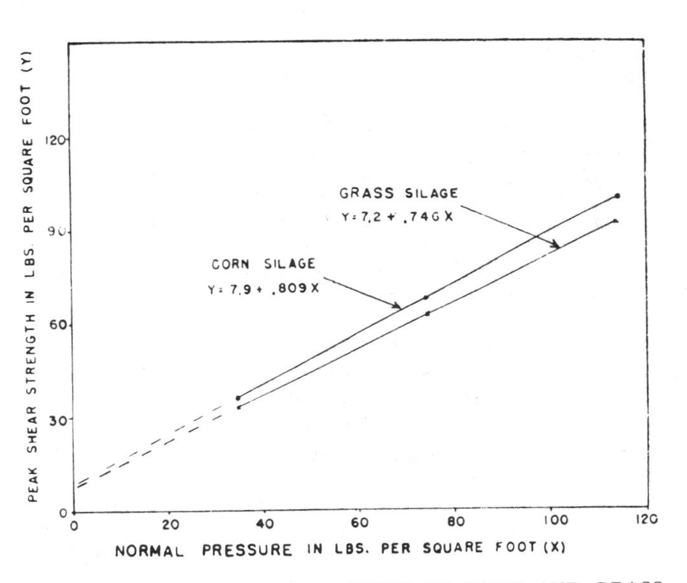

FIG. 5—RESULTS OF SHEAR TESTS OF CORN AND GRASS SILAGE; ENGLISH UNITS

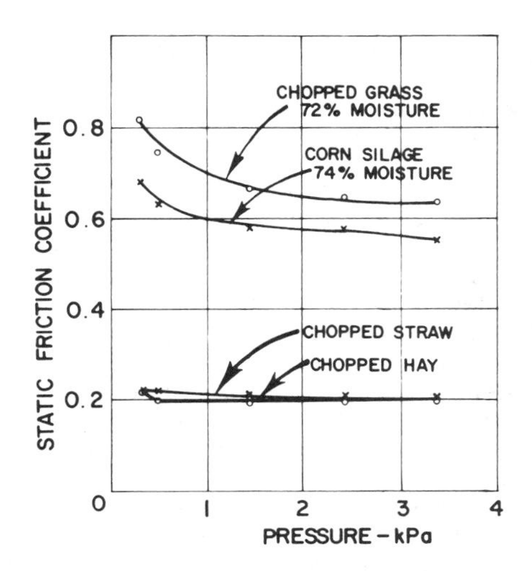

FIG. 3—EFFECT OF NORMAL PRESSURE INTENSITY ON STATIC-FRICTION COEFFICIENTS ON HIGHLY POLISHED GALVANIZED STEEL (MOISTURE CONTENT FIGURED ON WET BASIS); SI UNITS

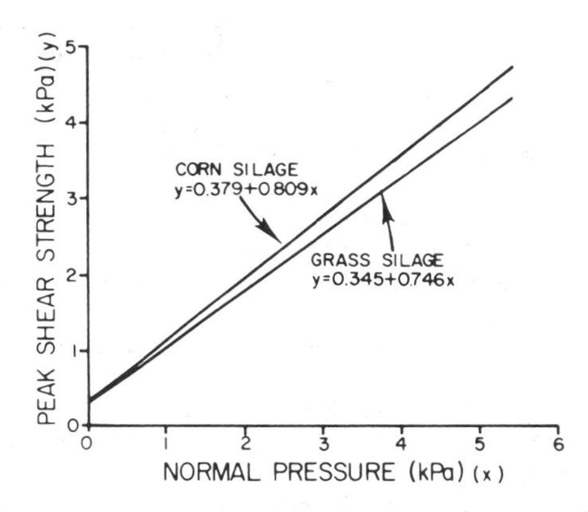

FIG. 6—RESULTS OF SHEAR TESTS OF CORN AND GRASS SILAGE; SI UNITS

ASAE Standard: ASAE S254.3

UNIFORM TERMINOLOGY FOR BULK MILK HANDLING

Developed by the ASAE Bulk Milk Cooling Systems Committee; approved by the ASAE Electric Power and Processing Division Technical Committee; adopted by ASAE as a Recommendation January 1959; reconfirmed December 1963; reconfirmed with minor editorial change December 1966; revised December 1968; reconfirmed December 1973; reclassified as a Standard December 1977; reconfirmed December 1978; revised June 1979; reconfirmed December 1984; revised March 1985.

SECTION 1—PURPOSE AND SCOPE

1.1 This terminology is intended to establish uniformity in terms used in the field of bulk milk handling. Terms and definitions were adopted from related fields where applicable. This Standard is intended as a reference for those responsible for, or concerned with, the specifications of use of bulk milk handling equipment.

SECTION 2—TERMS AND DEFINITIONS

2.1 Agitator: A mechanical or air means, provided with the milk tank, for stirring the milk to facilitate cooling and to provide a uniform product for sampling.

2.2 Blend temperature: Mass average temperature of milk in the milk tank as warm milk is added to cold milk.

2.3 Breast: That portion of the metal which joins the inside lining and extends to the outside shell of the milk tank. The breast is considered to be a milk contact surface and therefore a part of the inside lining.

2.4 Bridge: The part of the milk tank, usually 610 mm (24 in.) or less in width, which extends from one side of the tank to the other across the top. It is usually placed in the center of the tank, and also is used to provide support for the agitator, measuring device, etc.

2.5 Bulk milk tank: (See milk tank)

2.6 Bulk milk tanker: (See milk tanker)

2.7 Bulk milk trailer: (See milk trailer)

2.8 Calibration: Affixing a volumetric equivalent to the graduations on the gauge rod or surface gauge.

2.9 Calibration chart: A chart prepared for each tank from information derived by calibration. Its purpose is to correlate a volumetric equivalent for the graduations on the gauge rod or surface gauge.

2.10 Calibration rod: (See gauge rod)

2.11 Center-reading tank: A tank designed so that the gauge rod or surface gauge, when properly positioned for use, will be approximately in the vertical axis of the tank and centrally positioned with respect to the tank walls.

2.12 Compressor: That part of a refrigeration unit in which the vapor from the evaporator is compressed and delivered to the condenser.

2.13 Concentric tube cooler: (See tube cooler)

2.14 Condenser: That part of a refrigeration unit in which the refrigerant changes from a vapor to a liquid. The condenser may be air cooled or water cooled.

2.15 Condensing unit: A unit containing the compressor, condenser and associated electrical and refrigeration equipment.

2.16 Cooling capacity: The rate of heat removal, W (Btu/h).

2.17 Direct expansion cooling: A method of cooling the milk by a direct transfer of heat from the milk to the refrigerant contained in the evaporator.

2.18 Evaporator: That part of the refrigeration system in which the refrigerant absorbs heat from the milk or water and changes from a liquid to a vapor.

2.19 Every-day and every-other-day pickup: (Abbreviated ED and EOD) Terms denoting the frequency of collection of milk from the farm milk tank.

2.20 Finish: The smoothness of metal surfaces.

2.21 Gallonage chart: (See calibration chart)

2.22 Gauge brackets: A substantial, rigid element, permanently affixed to the milk tank, designed for supporting and positioning the gauge rod or surface gauge.

2.23 Gauge rod: A graduated measuring device designed to be immersed in the product (dip stick).

2.24 Gauge tube: A vertical glass or plastic tube, externally mounted on the milk tank, for the purpose of reading the level of milk within the tank.

2.25 Gauge tube scale: A metal plate, with graduations, mounted adjacent to the gauge tube.

2.26 Hose port: An opening in the exterior milkhouse wall with a self-closing door, for the transfer hose and electric motor cord during milk transfer.

2.27 Ice bank cooling: A system of cooling whereby ice is formed around evaporator coils submerged in a water bath thus forming an ice bank. Water is then circulated around the ice and to the cooling device.

2.28 In-level: That known and reproducible position of the milk tank which is in the normal, specified operating position.

2.29 In-line-cooler: A cooling device placed in the milk transfer system between the milk receiver and milk tank. Approved well water, spring or iced water is used as the coolant. Product may be partially or totally cooled.

2.30 Inside lining: That part of the milk tank with which the milk is in contact (inner tank) (inner liner) (product contact lining).

2.31 Intermittent agitation: The agitator operates for short periods as controlled by a timer between milkings, irrespective of milk temperature.

2.32 Level-indicating means: Reference points, marks or appurtenances on the milk tank used to reestablish the calibration axes. Examples are two-way or circular levels, plumb bobs, scribe or punch marks.

2.33 Meniscus: The curved upper surface of a column of liquid.

2.34 Milk sample: (See universal sample)

2.35 Milk tank and/or cooling tank: A sanitary container or vat, usually located in the farm milkroom, used to cool and/or store milk in bulk.

2.36 Milk tanker: A vehicle, consisting of an engine, chassis and milk tank used to collect and transport milk. It may or may not be equipped with a meter to measure picked-up quantities of milk.

2.37 Milk trailer: A tank mounted on wheels and moved by truck or tractor for collecting and transporting milk.

2.38 Multiple-tube cooler: (See tube cooler)

2.39 Outer shell: The exterior or outside portion of the milk tank. The outer shell is separated from the lining by insulation and integral cooling surfaces.

2.40 Outlet valve: A device for controlling flow when removing milk from a tank.

2.41 Package unit: A unit in which the condensing unit is mounted by the manufacturer as an integral part of the milk cooling tank.

2.42 Plate cooler: An in-line cooler which uses plates to separate milk and coolant, which flow through alternate spaces between the series of plates.

2.43 Pounce: Dust or powder applied to the gauge rod to obtain a water-level indication on the gauge.

2.44 Remote condensing unit: A mechanical unit not affixed to the frame of the milk cooling tank. The unit is connected to the tank proper by refrigerant lines, electric cable, etc.

2.45 Sampling dipper: A device for retrieving a sample of milk.

2.46 Sampling container: A vial or bag.

2.47 Sanitizer: A container for carrying a dipper.

2.48 Sample case: An insulated carrying case for samples.

2.49 Scribe marks: Permanent marks placed on the inside lining or outer shell, usually at or near the four corners. The marks, if placed on the inside lining, are used to indicate a level plane of the milk tank when the surface of a volume of water touches all four marks simultaneously. The marks placed on the outer shell of the tank are used as a guide in placing a standard spirit level.

2.50 Surface gauge: A depth measuring device consisting of a stationary element, and a movable, graduated element designed to be moved into contact with the liquid surface from above.

2.51 Three A (3A) standards: A list of detailed requirements and specifications for most milk handling equipment, formulated by the International Association of Milk, Food and Environmental Sanitarians; the United States Public Health Service/Food and Drug Administration; and the Dairy Industry Committee (which includes The Dairy and Food Industries Supply Association).

2.52 Transfer hose: A flexible plastic hose, used to convey the milk from the milk tank to the pump of the milk tanker.

2.53 Transport tanks: (See milk tanker)

2.54 Tube cooler: (See also in-line cooler). A type of in-line cooler which utilizes a series of small tubes arranged within a larger tube or jacket. Milk flows inside the smaller tubes, and coolant flows between the small tubes and larger outside tube. A concentric tube cooler utilizes three different size tubes all having the same center axis. Milk flows between first and second tubes and coolant flows in both center and outer most tubes.

2.55 Universal sample: A representative sample of milk from each farm milk tank prior to transfer to truck or other transport container.

2.56 Vacuum milk tank: A milk tank which is designed to withstand the vacuum of the milking system.

2.57 Volumetric standard: A container which is used to calibrate the milk cooling tank and other volumetric measuring devices. The containers are usually checked and approved by "Weights and Measures" agencies to deliver an accurate volume of liquid (measuring can) (standard measure) (standard can).

ASAE Standard: ASAE S269.3

WAFERS, PELLETS, AND CRUMBLES—DEFINITIONS AND METHODS FOR DETERMINING DENSITY, DURABILITY AND MOISTURE CONTENT

Proposed by subcommittee on Standards of Forage Harvesting Committee of ASAE Power and Machinery Division; approved by Power and Machinery Division Technical Committee; adopted by ASAE December 1963; revised December 1965; reconfirmed December 1970; Section 7 revised December 1972; reconfirmed December 1975, December 1980; revised March 1982; reconfirmed December 1986.

SECTION 1—PURPOSE

1.1 Several methods are currently being used to measure the unit density, bulk density, durability, and moisture content of wafers, pellets, and crumbles commonly used for animal feeds. Because of the wide variation in methods, it is very difficult to compare these characteristics on a uniform basis. It is, therefore, desirable to define wafers, pellets and crumbles and to establish universal methods and procedures for testing the physical qualities. Standardization of methods and procedures will permit assimilation of data from all research workers using this Standard. Furthermore, standard procedures, precisely set forth, will eliminate the necessity for new researchers to spend valuable time working out satisfactory procedures.

SECTION 2—SCOPE

2.1 This Standard defines wafers, pellets, and crumbles and establishes methods and procedures for measuring unit specific density, bulk density, durability, and moisture content.

SECTION 3—DEFINITIONS

3.1 Wafer: An agglomeration of unground ingredients in which some of the fibers are equal to or greater than the length of the minimum cross-sectional dimension of the agglomeration.

3.2 Pellet: An agglomeration of individual ground ingredients, or mixture of such ingredients, commonly used for animal feeds.

3.3 Crumbles: Pelleted feed reduced to granular form.

SECTION 4—UNIT DENSITY OF WAFERS

The unit density of wafers shall be determined by the following procedure:

4.1 Select a representative sample of at least five wafers. This provides two spare wafers in case a leaky bag (paragraph 4.4) permits damage to one or two samples.

4.2 Weigh each wafer in air and record the mass, W_1.

4.3 Place the empty water container on the scale, fill two-thirds with water, and tare the scale to zero. The container must be transparent, approximately 300 mm (12 in.) high, and its diameter large enough to accommodate a wafer in a plastic bag.

4.4 Place each wafer in a 0.03 mm (1.25 mil) plastic bag. Use a 2mm (0.06 in.) diameter metal rod with a ring formed at one end to control the wafer from outside the plastic bag. The neck of the bag is to be placed through the ring of the rod. Bags are to be no larger in diameter than necessary to contain a wafer without stretching the bag.

4.5 With the top of the bag extending above the water surface, use the rod to control the wafer while submerging it to the bottom of the container. If the wafer density is greater than 998 kg/m³ (62.3 lb/ft³), the force required to submerge it will be negative; if water density is less than 998 kg/m³ (62.3 lb/ft³), the force will be positive. Gently agitate to expel air from the bag so it will fit snugly around the wafer, under water.

4.6 Raise the wafer in the water column to within no less than 50 mm (2 in.) of the water surface, and read the scale immediately. Repeat the procedure of paragraph 4.5 to verify scale reading, and record the mass, W_2. Also repeat procedure of paragraph 4.2, verifying W_1, as a check against the wafer absorbing water through a leaky bag.

4.7 Calculate the unit density:

$$SW_i = \left(\frac{W_1}{W_3}\right) 998\ \text{kg/m}^3 \qquad \left[SW_i = \left(\frac{W_1}{W_3}\right) 62.3\ \text{lb/ft}^3\right]$$

where

SW_i = density of test wafer
W_1 = wafer mass in air
W_2 = mass of displaced water
W_3 = W_2 minus the mass of water displaced by bag and rod when they are immersed to the depth at which W_2 reading was taken
998 kg/m³ (62.3 lb/ft³) = density of water at 21 °C (70 °F)

NOTE: It may be that W_3 should be further corrected by subtracting a small value, for example 7 g (0.25 oz), to compensate for error in the fit of the bag to the wafer. This can be approximated by constructing an artificial, waterproof block similar in size, shape, and density to the wafers being tested and determining the weight of water displaced by the bare block. The bag-fit error is approximately the difference between the weight of water displaced by the bare block, as determined above, and W_3 for the block.

4.8 This unit density will then be corrected to zero percent moisture content by:

$$SW_c = SW_i \frac{(\%\ DM)}{100}$$

where

SW_c = density at zero percent moisture content
SW_i = density indicated by test wafer
$\%\ DM$ = percent dry matter in test wafer

4.9 Because of a tendency for wafers to expand for some time after forming, both the time interval between forming and this measurement, and the moisture content at the time of this measurement, should be specified.

SECTION 5—BULK DENSITY

5.1 Wafers. A cylindrical container, 380 mm (15 in.) in diameter and 495 mm (19.5 in.) high (inside dimensions), shall be used. Multiply the net mass of the material by 17.81 (0.50) to obtain bulk density in kg/m³ (lb/ft³).

5.2 Pellets and crumbles. A cylindrical container 300 mm (12 in.) in diameter and 310 mm (12.25 in.) high (inside dimensions) shall be used. Multiply the net mass of the material by 45.64 (1.25) to obtain the bulk density in kg/m³ (lb/ft³).

5.3 Filling sample container. The container shall be filled by pouring from a height of 610 mm (2 ft) above the top edge of the container. The container shall then be dropped five times from a height of 150 mm (6 in.) onto a non-resilient surface to allow settling. In the case of small pellets and crumbles, the material shall be struck off level with the top surface. In the case of wafers and large pellets, remove the wafers or large pellets which have more than one-half their volume above the top edge of the container, leaving in the container those wafers or large pellets with more than one-half their volume below the top edge of the container.

5.4 Density corrected. The density determined by this method shall be corrected to zero percent moisture content by the use of the equation in paragraph 4.8 under *Unit Density.*

5.5 Because of the tendency for wafers to expand for some time after forming, both the time interval between forming and this measurement, and the moisture content at the time of this measurement, should be specified when dealing with wafers.

SECTION 6—DURABILITY

6.1 Wafers. The durability of wafers shall be determined by the following procedure:

6.1.1 Device. Durability of wafers shall be determined by tumbling the test sample for 3 minutes at 13 r/min. The outside dimensions of the angle iron frame of the tumbler are shown in Fig. 1. The covering shall be 12.5 mm (0.5 in.) mesh hardware cloth applied taut to the outside of the frame. Interior projections, such as screw heads, should be kept to a minimum and should be well rounded. The box shall be mounted on a diagonal axis (two planes) with two stub shafts terminating at the exterior of the angle iron frame. These may be hollow shafts for ease of fabrication. There will be a hinged triangular door 300 × 300 × 430 mm (12 × 12 × 17 in.) on each end. The axis of rotation shall be horizontal (Fig. 1).

6.1.2 Determining durability of wafers. Wafers shall be tested by tumbling a representative 10-wafer sample, whose individual mass does not vary over ± 10 percent of the average original mass, in the manner described in paragraph 6.1.1.

6.1.3 After this tumbling test, the total mass of all particles each weighing more than 20 percent of the average initial wafer mass shall be recorded and designated as wafer size material (*WSM*). The durability rating for wafers is expressed as the percentage of *WSM*.

6.1.4 Using the original average wafer mass, compute five mass classes, each expressing 20-percent increments of the original average wafer mass. Separate the wafer pieces remaining after tumbling into piles prescribed by the five mass classes. Pieces weighing more than the average original wafer mass will be included in the highest mass class. The percentage of material in each class shall then be determined by dividing the total mass in each class by the total mass before testing.

TABLE 1—EXAMPLE AND SUGGESTED WORK SHEET

Sample Number	Mass Classes of Wafers* % original mass in each class					Size-Distribution Index	Durability Rating
	50-40g	40-30g	30-20g	20-10g	10-0g		
1	92	—	—	—	8	368	92
2	40	48	—	—	12	304	88
3	8	47	15	10	20	213	80
4	—	40	20	18	22	178	78

*Example assumes original average weight per wafer is 50 grams.
NOTE: Wafers formed by extrusion may hang together in clusters of two or three. These wafers should be separated at their cleavage planes into reasonable lengths. Each resulting unit should be treated as an individual wafer.

TABLE 2—SCREEN SIZES FOR PELLET AND CRUMBLES DURABILITY TESTS

Diameter of pellets or crumbles		Required screen size		
mm	in.	Size*	mm	in.
	All Crumbles	No. 12	1.7	0.066
	Pellets			
2.4	0.094	No. 10	2.0	0.079
3.2	0.125	No. 7	2.8	0.111
3.6	0.141	No. 6	3.4	0.132
4.0	0.156	No. 6	3.4	0.132
4.8	0.188	No. 5	4.0	0.157
5.2	0.203	No. 4	4.8	0.187
6.4	0.250	No. 3½	5.7	0.223
7.9	0.313	0.265	6.7	0.265
9.5	0.375	5/16	7.9	0.313
12.7	0.500	7/16	11.1	0.438
15.9	0.625	0.530	13.5	0.530
19.0	0.750	5/8	15.9	0.625
22.2	0.875	3/4	19.0	0.750
25.4	1.00	7/8	22.2	0.875

*American Society for Testing and Material Standard E11-61, Specifications for Wire-Cloth Sieves for Testing Purposes.

6.1.5 A size-distribution index of the durability is then obtained by multiplying the percentage of material in each of the five classes from highest to lowest, respectively, by four (4), three (3), two (2), one (1), zero (0), and then calculating the summation of products obtained by this method. A perfect index is 400. See Table 1.

6.1.6 The time interval between forming and durability testing should be specified. The moisture content of all the pieces, as a group sample taken immediately after the durability test, should also be specified.

6.2 Pellets and crumbles. The durability of pellets and crumbles shall be determined by the following procedure:

6.2.1 Device. Durability of pellets and crumbles shall be determined by tumbling the test sample for 10 minutes at 50 r/min., in a dust-tight enclosure. The construction of this device is illustrated in Fig. 2. The device is rotated about an axis which is perpendicular to and centered in the 300 mm (12 in.) sides. A 230 mm (9 in.) long baffle is affixed symmetrically to a diagonal of one 300 × 300 mm (12 × 12 in.) side of the box. One leg of this formed angle baffle extends 50 mm (2 in.) into the box and the other leg is securely fastened to the back of the box. A door may be placed in any side and

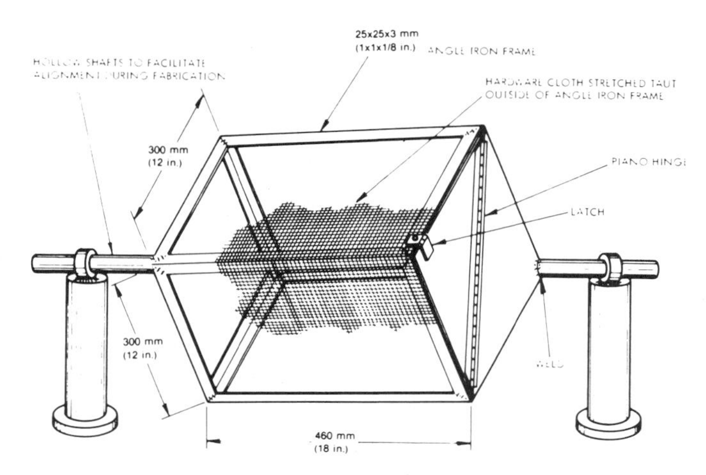

FIG. 1—DURABILITY TESTER FOR WAFERS

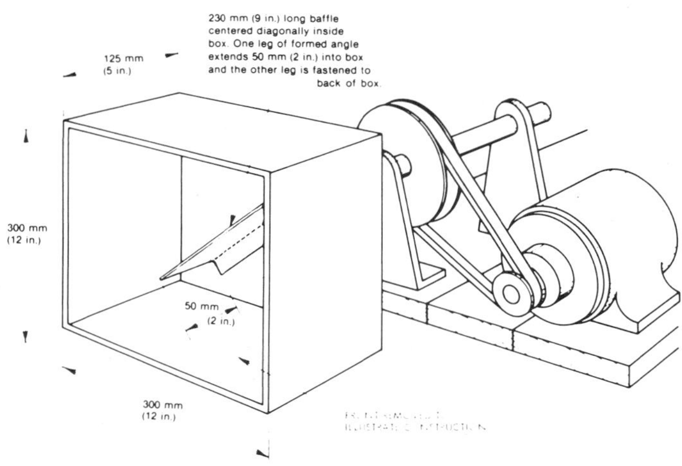

FIG. 2—DURABILITY TESTER FOR PELLETS AND CRUMBLES

should be dustproof. Projections, such as rivets and screws, shall be kept to a minimum and well rounded.

6.2.2 Screens. Fines shall be determined by screening a sample on a wire sieve having openings just smaller than the nominal pellet diameter. Table 2 shows the recommended sieves for crumbles and pellets of various diameters.

6.2.3 Test procedure. A sample of pellets or crumbles to be tested will be sieved on the appropriate sieve to remove fines. If pellets of 12.5 mm (0.5 in.) diameter, or larger, are being tested, select pellets which are between 30 mm (1.25 in.) and 40 mm (1.5 in.) in length. Place a 500 g (1.1 lb) sample of sieved pellets or crumbles in the tumbling box device. After tumbling for 10 min., the sample will be removed, sieved, and the percent of whole pellets or crumbles calculated. Pellet and crumbles durability will be defined as:

$$\text{Durability} = \frac{\text{mass of pellets or crumbles after tumbling}}{\text{mass of pellets or crumbles before tumbling}} (100)$$

Normally pellets will be tested immediately after cooling. When the temperature of the pellets falls within ± 5 °C (±10 °F) of ambient, they are considered cool. If tested at a later time, the time, in hours *after* cooling, will be indicated as a subscript of the durability. For example, if the pellet durability tested 95 after a 4 h delay from the time of cooling, then the results will be expressed as $(95)_4$. If pellets are tested *before* cooling, there will be a significant weight loss caused by water vaporization, and the apparent durability will be decreased by this loss of water vapor. The loss of water vapor must be determined by making moisture content determinations before and after tumbling and compensating the final mass accordingly. When this procedure is followed, the durability would be expressed as $(95)_{-1}$.

SECTION 7—MOISTURE CONTENT

7.1 Samples. Moisture content shall be determined in accordance with procedures of ASAE Standard ASAE S358, Moisture Measurement—Forages. The time of sampling relative to formation of wafers, pellets, or crumbles should be specified.

7.2 To speed up the drying time, wafers and pellets may be broken into small chunks making certain no fines are lost in this operation.

7.3 Moisture content will be expressed on a wet basis.

Cited Standards:

ASAE S358, Moisture Measurement—Forages
ASTM E11-61, Specifications for Wire-Cloth Sieves for Testing Purposes

RESISTANCE TO AIRFLOW OF GRAINS, SEEDS, OTHER AGRICULTURAL PRODUCTS, AND PERFORATED METAL SHEETS

Approved by the ASAE Committee on Technical Data; adopted by ASAE 1948; revised 1954, 1962; reconfirmed without change by Electric Power and Processing Division Technical Committee December 1968, December 1973, December 1978, December 1979; revised December 1980; reconfirmed December 1985; revised March 1987.

SECTION 1—PURPOSE AND SCOPE

1.1 These data can be used to estimate the resistance to airflow of beds of grain, seeds, and other agricultural products, and of perforated metal sheets. An estimate of this airflow resistance is the basis for the design of systems to dry or aerate agricultural products.

1.2 Data are included for common grains, seeds, other agricultural products, and for perforated metal sheets, over the airflow range common for aeration and drying systems.

SECTION 2—EMPIRICAL CURVES

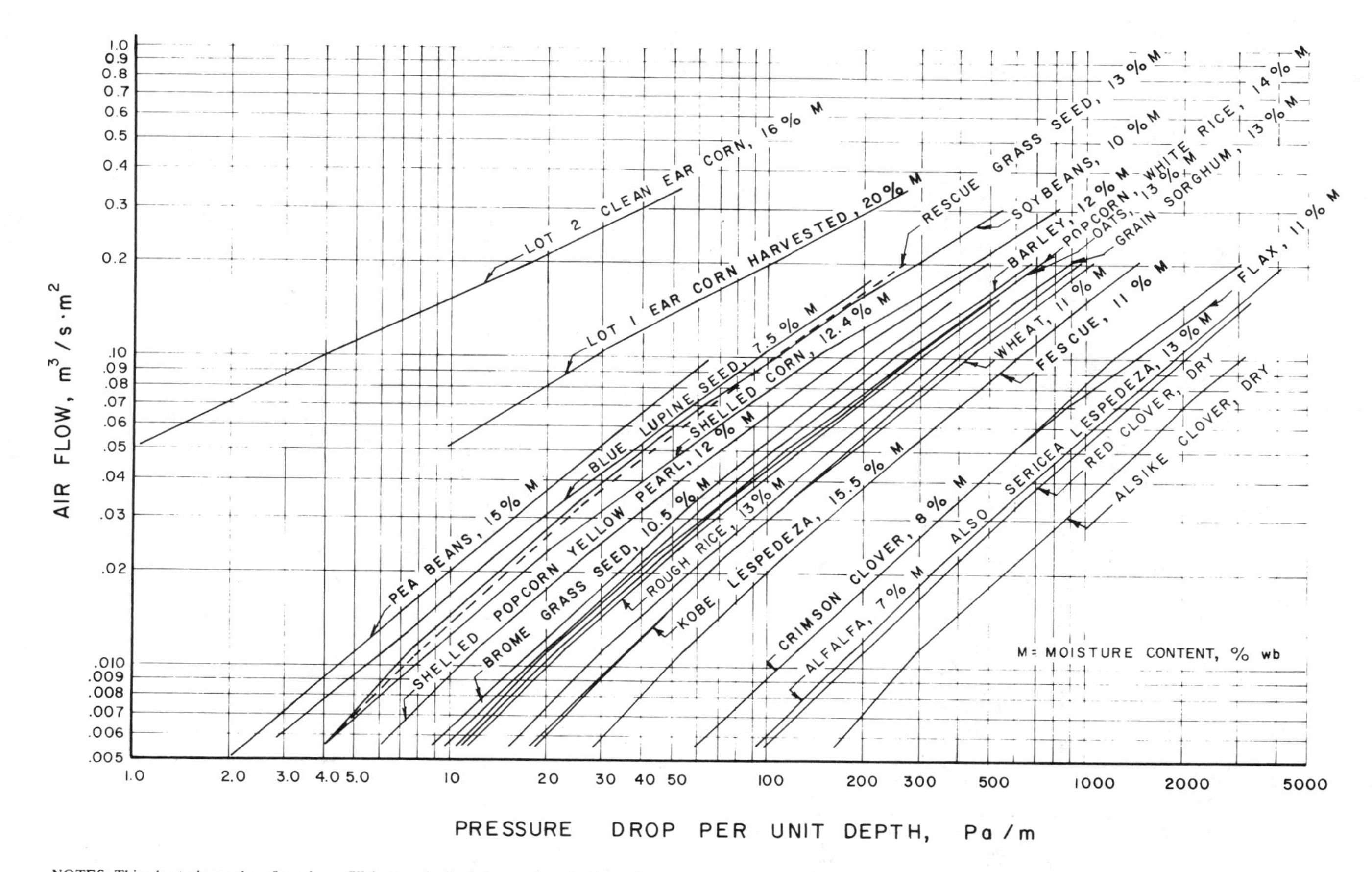

NOTES: This chart gives values for a loose fill (not packed) of clean, relatively dry grain.
For a loose fill of clean grain having high moisture content (in equilibrium with relative humidities exceeding 85 percent), use only 80 percent of the indicated pressure drop for a given rate of air flow.
Packing of the grain in a bin may cause 50 percent higher resistance to air flow than the values shown.
White rice is a variety of popcorn.

FIG. 1—RESISTANCE TO AIRFLOW OF GRAINS AND SEEDS (SI Units) (Shedd's data)

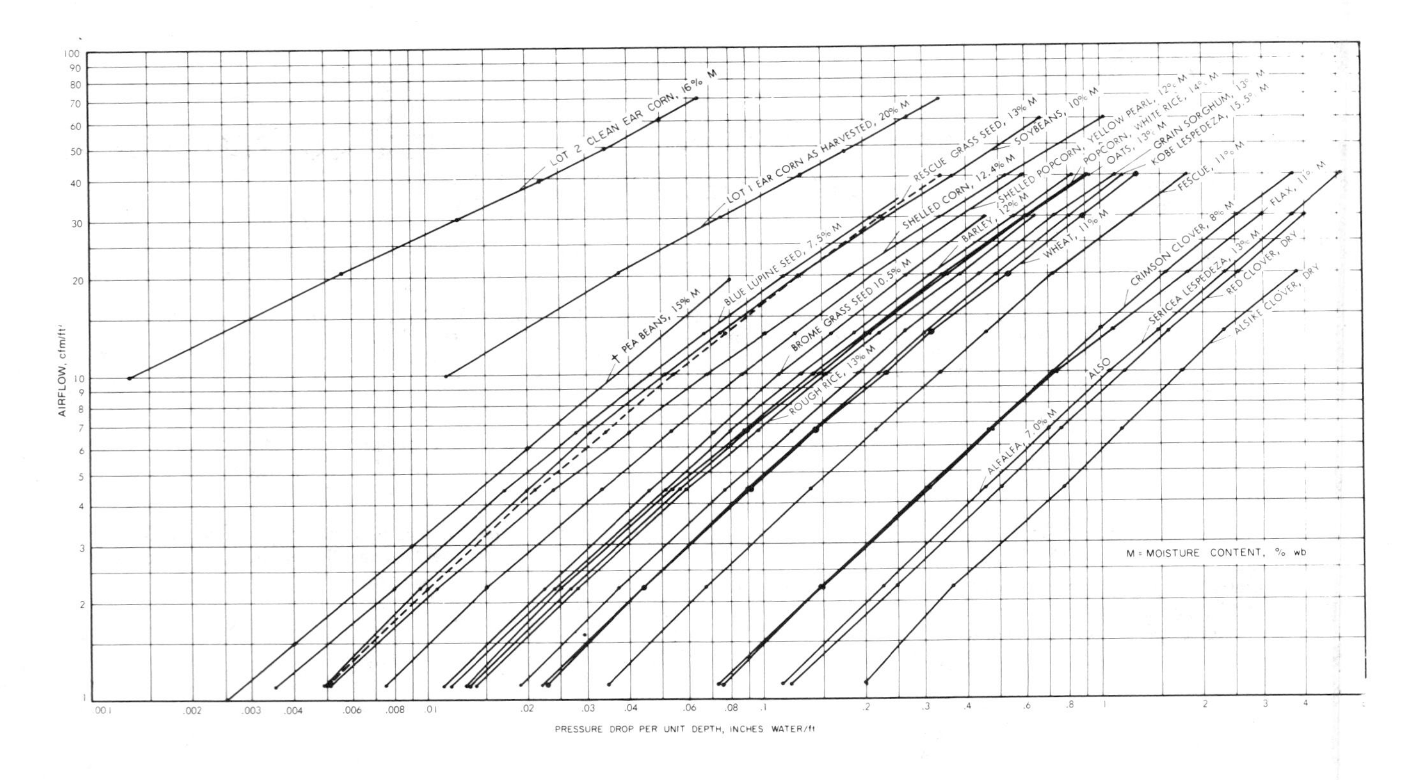

NOTES: This chart gives values for a loose fill (not packed) of clean, relatively dry grain.
For a loose fill of clean grain having high moisture content (in equilibrium with relative humidities exceeding 85 percent), use only 80 percent of the indicated pressure drop for a given rate of air flow.
Packing of the grain in a bin may cause 50 percent higher resistance to air flow than the values shown.
When foreign material is mixed with grain no specific correction can be recommended. However, it should be noted that resistance to air flow is increased if the foreign material is finer than the grain, and resistance to air flow is decreased if the foreign material is coarser than the grain.
White rice is a variety of popcorn.

FIG. 2—RESISTANCE TO AIRFLOW OF GRAINS AND SEEDS (Inch-pound units) (Shedd's data)

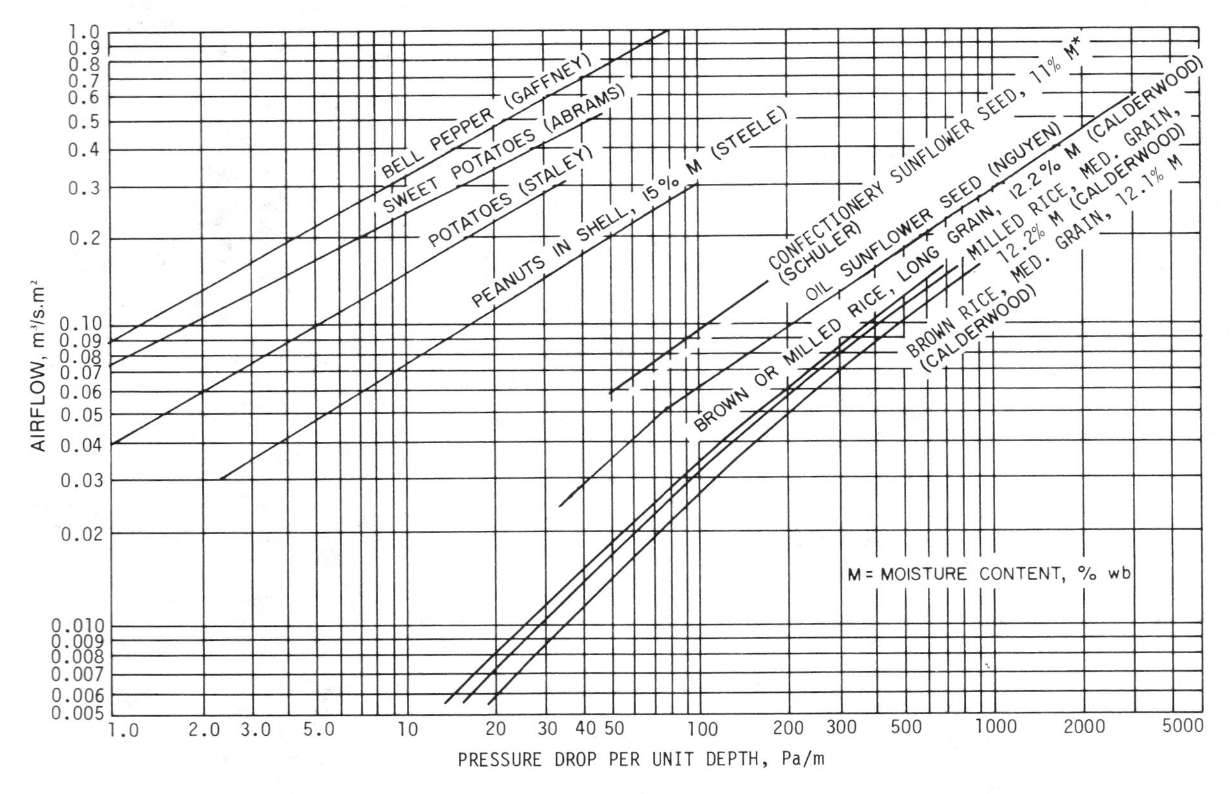

NOTES: Rice: Clean, loose-fill. A packing operation which raised the bulk density by 14-17 percent resulted in pressures 2.3 to 3.4 times those for loose fill.

FIG. 3—RESISTANCE TO AIRFLOW FOR OTHER AGRICULTURAL PRODUCTS (SI units)

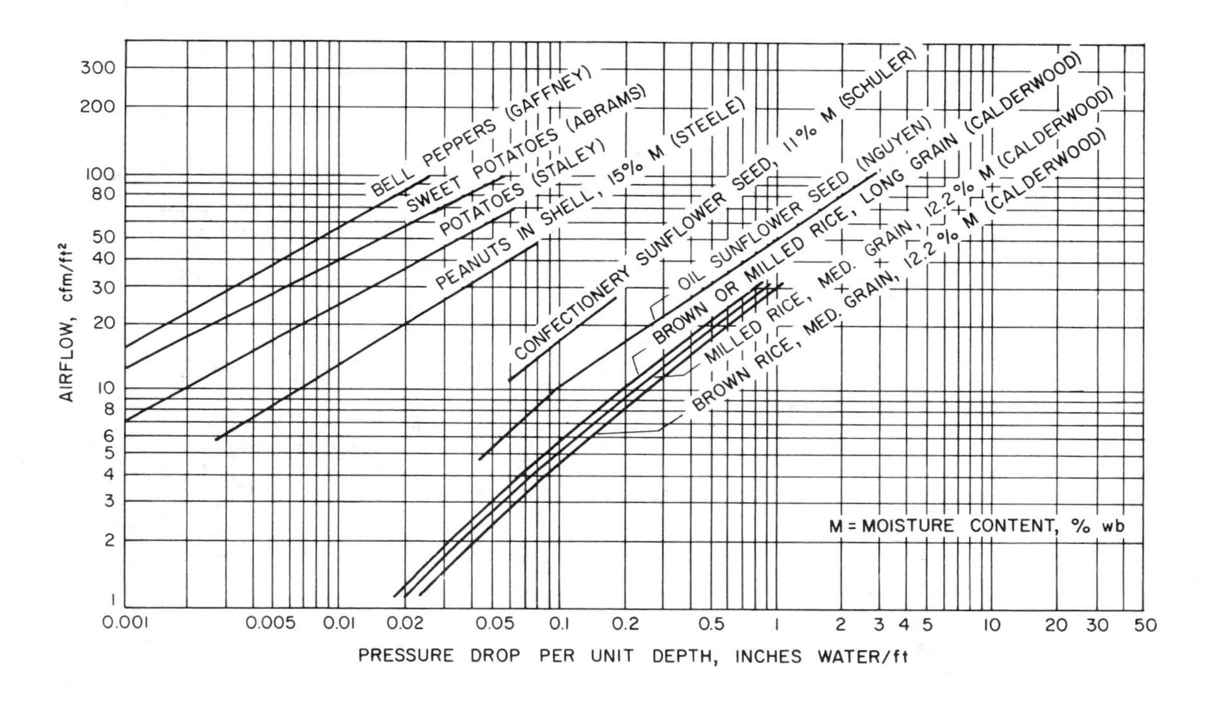

NOTES: Rice: Clean, loose-fill. A packing operation which raised the bulk density by 14 to 17 percent resulted in pressures 2.3 to 3.4 times those for loose fill.

FIG. 4—RESISTANCE TO AIRFLOW OF OTHER AGRICULTURAL PRODUCTS (Inch-pound units)

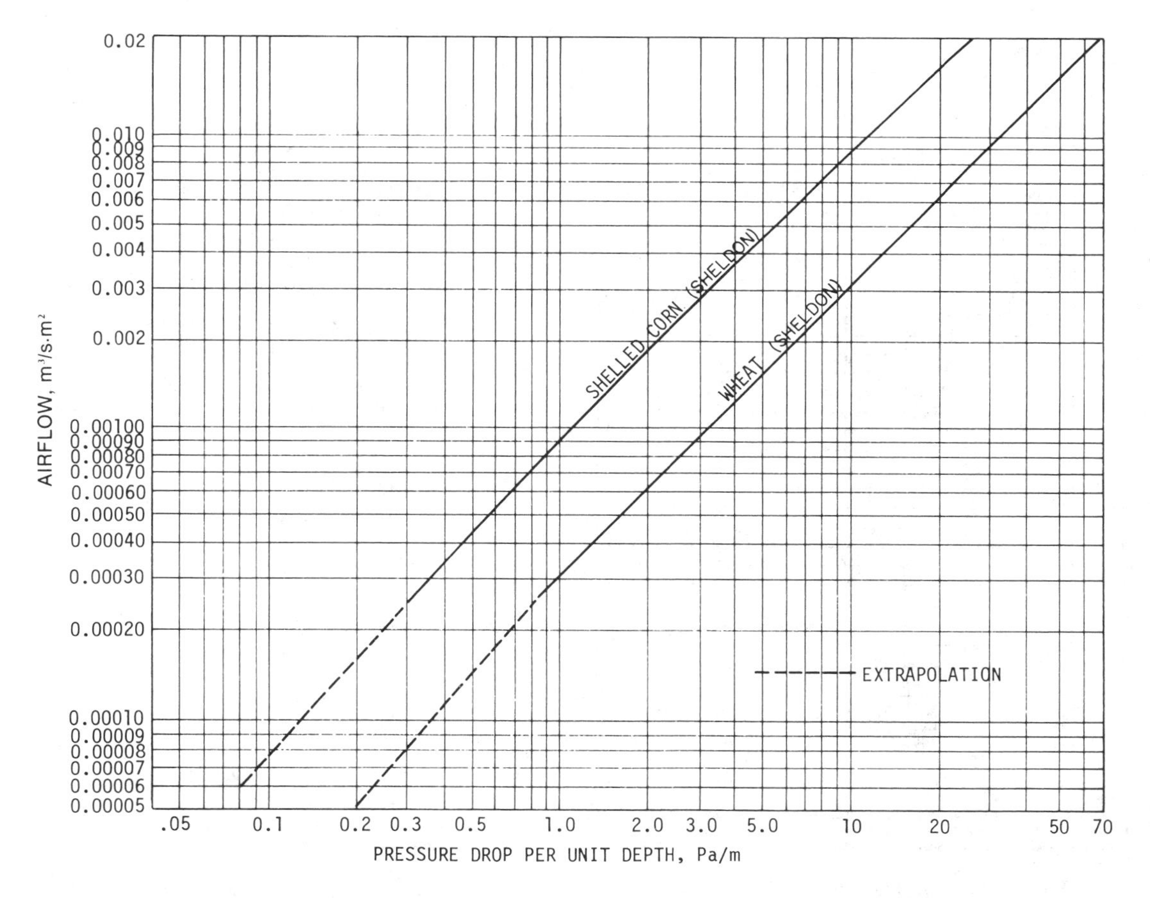

FIG. 5—RESISTANCE TO AIRFLOW OF SHELLED CORN AND WHEAT AT LOW AIRFLOWS (SI units)

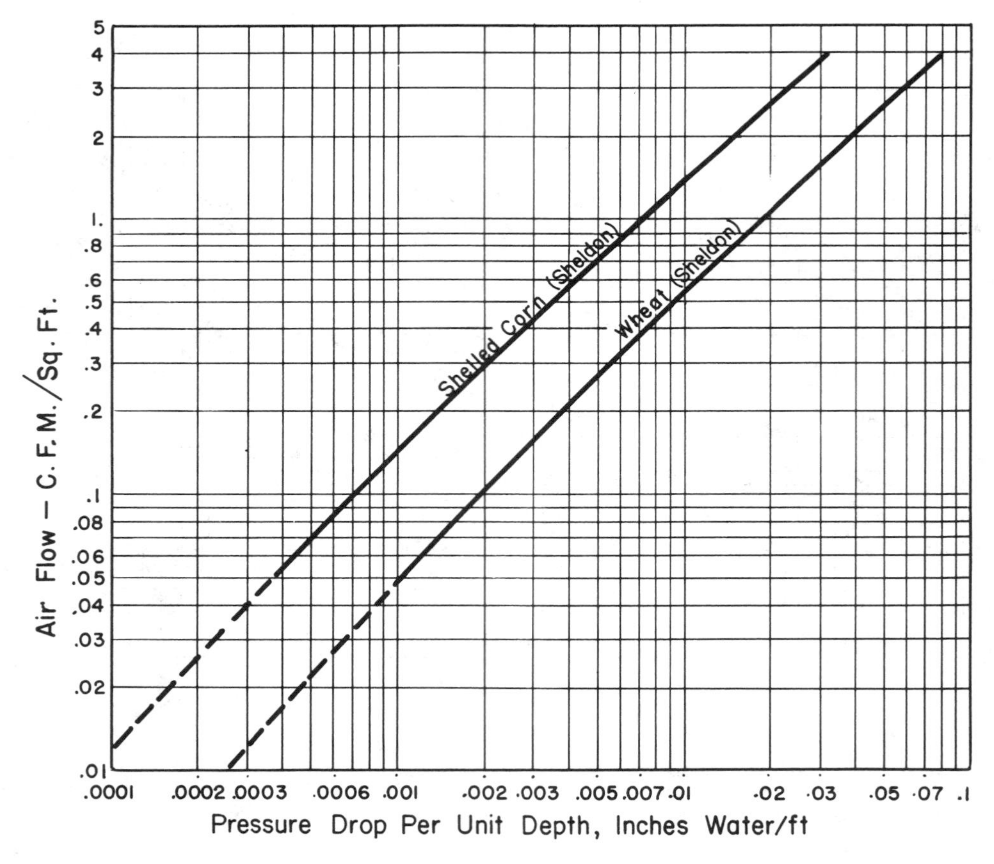

FIG. 6—RESISTANCE TO AIRFLOW OF SHELLED CORN AND WHEAT AT LOW AIRFLOWS (Inch-pound units)

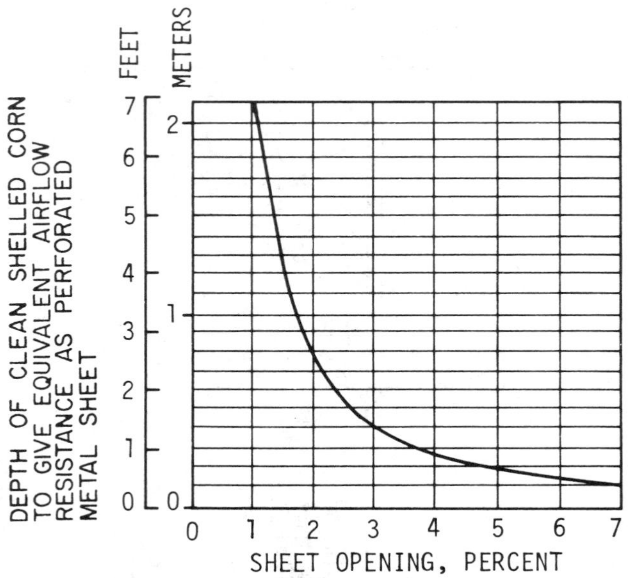

DEPTH OF CLEAN SHELLED CORN TO GIVE EQUIVALENT AIRFLOW RESISTANCE AS PERFORATED METAL SHEET

FEET

METERS

SHEET OPENING, PERCENT

NOTES: When sheet openings amount to 20 percent, no additional resistance to airflow is produced.
A large number of small perforations is preferred to a smaller number of large perforations for the same amount of opening.
The curve shown is based on tests of sheets having width of perforations from 1 to 3.3 mm (0.04 to 0.13 in.).

FIG. 7—RESISTANCE TO AIRFLOW OF PERFORATED METAL SHEETS WHEN SUPPORTING GRAIN (Henderson)

SECTION 3—AIRFLOW RESISTANCE EQUATION

$$\frac{\Delta P}{L} = \frac{a\,Q^2}{\log_e (1+bQ)}$$

where
- ΔP = pressure drop, Pa or inches of water
- L = bed depth, m or ft
- a = constant for particular grain (see Table 1)
- Q = airflow, $m^3/s{\cdot}m^2$ or cfm/ft^2
- b = constant for particular grain (see Table 1)

SECTION 4—EFFECT OF FINES ON RESISTANCE TO AIRFLOW OF SHELLED CORN

4.1 An effect of adding fines to shelled corn is an increase in the airflow resistance of the corn. The pressure drop per unit bed depth can be corrected to account for fines using this equation:

SI units: $$\left(\frac{\Delta P}{L}\right)_{corrected} = \left(\frac{\Delta P}{L}\right)_{clean} (1 + (14.5566 - 26.418Q)(fm))$$

Customary units: $$\left(\frac{\Delta P}{L}\right)_{corrected} = \left(\frac{\Delta P}{L}\right)_{clean} (1 + (14.5566 - 0.1342Q)(fm))$$

where
- ΔP = pressure drop, Pa or inches of water
- L = bed depth, m or ft
- Q = airflow, $m^3/s{\cdot}m^2$ or cfm/ft^2
- fm = decimal fraction of fines, by weight

NOTES: Range of applicability: 0.076 to 0.20 $m^3/s{\cdot}m^2$ (15 to 40 CFM/FT^2) and $0 \leqslant fm \leqslant 0.2$. Broken grain and other matter which passed through a 4.76-mm (12/64-in.) round-hole sieve are defined as fines. (Hague)

SECTION 5—EFFECT OF BULK DENSITY ON RESISTANCE TO AIRFLOW OF SHELLED CORN

5.1 An increase in bulk density causes an increase in the airflow resistance per unit bed depth of the corn. The pressure drop per unit bed depth can be predicted as a function of airflow rate and corn bulk density by use of this empirical equation:

$$\frac{\Delta P}{L} = X_1 + X_2 \frac{\left(\frac{\rho_b}{\rho_k}\right)^2 Q}{\left(1 - \frac{\rho_b}{\rho_k}\right)^3} + X_3 \frac{\left(\frac{\rho_b}{\rho_k}\right) Q^2}{\left(1 - \frac{\rho_b}{\rho_k}\right)^3}$$

where

ΔP = pressure drop, Pa or inches of water
L = bed depth, m or ft
ρ_b = corn bulk density, kg/m^3 or lb/ft^3
ρ_k = corn kernel density, kg/m^3 or lb/ft^3
Q = airflow, $m^3/s \cdot m^2$ or cfm/ft^2
X_1, X_2, X_3 = constants (see Table 2 or Table 3.)

TABLE 2—VALUE FOR CONSTANTS (SI UNITS) FOR EQUATION IN PARAGRAPH 5.1

Airflow Range, $m^3/s \cdot m^2$	X_1	X_2	X_3
$0.027 \leqslant Q \leqslant 0.13$	-0.998	88.8	511
$0.13 \leqslant Q \leqslant 0.27$	-10.9	111	439
$0.27 \leqslant Q \leqslant 0.60$	-76.5	163	389

TABLE 3—VALUES FOR CONSTANTS (INCH-POUND UNITS) FOR EQUATION IN PARAGRAPH 5.1

Airflow range, cfm/ft^3	X_1	X_2	X_3
$5.3 \leqslant Q \leqslant 26.3$	-0.0012	5.53×10^{-4}	1.62×10^{-5}
$26.3 < Q \leqslant 52.5$	-0.013	6.94×10^{-4}	1.39×10^{-5}
$52.5 < Q \leqslant 117$	-0.094	10.2×10^{-4}	1.23×10^{-5}

NOTES: Range of applicability: 732 to 799 kg/m^3 (45.7 to 49.9 lb/ft^3) (corn bulk density) 0.027 to 0.60 $m^3/s \cdot m^2$ (5.3 to 117 cfm/ft^2). (Bern)

TABLE 1—VALUES FOR CONSTANTS IN AIRFLOW RESISTANCE EQUATION

Material	Value of a ($Pa \cdot s^2/m^3$)	Value of b ($m^2 \cdot s/m^3$)	Range of Q (m^3/m^2s)	Reference
Alfalfa	6.40×10^4	3.99	0.0056 — 0.152	Shedd
Barley	2.14×10^4	13.2	0.0056 — 0.203	Shedd
Brome grass	1.35×10^4	8.88	0.0056 — 0.152	Shedd
Clover, alsike	6.11×10^4	2.24	0.0056 — 0.101	Shedd
Clover, crimson	5.32×10^4	5.12	0.0056 — 0.203	Shedd
Clover, red	6.24×10^4	3.55	0.0056 — 0.152	Shedd
Corn, ear (lot 1)	1.04×10^4	325.	0.051 — 0.353	Shedd
Corn, shelled	2.07×10^4	30.4	0.0056 — 0.304	Shedd
Corn, shelled (low airflow)	9.77×10^3	8.55	0.00025 — 0.0203	Sheldon
Fescue	3.15×10^4	6.70	0.0056 — 0.203	Shedd
Flax	8.63×10^4	8.29	0.0056 — 0.152	Shedd
Lespedeza, Kobe	1.95×10^4	6.30	0.0056 — 0.203	Shedd
Lespedeza, Sericea	6.40×10^4	3.99	0.0056 — 0.152	Shedd
Lupine, blue	1.07×10^4	21.1	0.0056 — 0.152	Shedd
Oats	2.41×10^4	13.9	0.0056 — 0.203	Shedd
Peanuts	3.80×10^3	111.	0.030 — 0.304	Steele
Peppers, bell	5.44×10^2	868.	0.030 — 1.00	Gaffney
Popcorn, white	2.19×10^4	11.8	0.0056 — 0.203	Shedd
Popcorn, yellow	1.78×10^4	17.6	0.0056 — 0.203	Shedd
Potatoes	2.18×10^3	824.	0.030 — 0.300	Staley
Rescue	8.11×10^3	11.7	0.0056 — 0.203	Shedd
Rice, rough	2.57×10^4	13.2	0.0056 — 0.152	Shedd
Rice, long brown	2.05×10^4	7.74	0.0055 — 0.164	Calderwood
Rice, long milled	2.18×10^4	8.34	0.0055 — 0.164	Calderwood
Rice, medium brown	3.49×10^4	10.9	0.0055 — 0.164	Calderwood
Rice, medium milled	2.90×10^4	10.6	0.0055 — 0.164	Calderwood
Sorghum	2.12×10^4	8.06	0.0056 — 0.203	Shedd
Soybeans	1.02×10^4	16.0	0.0056 — 0.304	Shedd
Sunflower, confectionery	1.10×10^4	18.1	0.055 — 0.178	Shuler
Sunflower, oil	2.49×10^4	23.7	0.025 — 0.570	Nguyen
Sweet Potatoes	3.40×10^3	6.10×10^8	0.050 — 0.499	Abrams
Wheat	2.70×10^4	8.77	0.0056 — 0.203	Shedd
Wheat (low airflow)	8.41×10^3	2.72	0.00025 — 0.0203	Sheldon

NOTES: The parameters given were determined by a least square fit of the data in Fig. 1-6. To obtain the corresponding values of (a) in inch-pound units ($in.H_2O\ min^2/ft^3$) divide the above a-values by 31635726. To obtain corresponding values of (b) in inch-pound units (ft^2/cfm) divide the above b-values by 196.85. Parameters for the Lot 2 Ear Corn data are not given since the above equation will not fit the data.

Although the parameters listed in this table were developed from data at moderate airflows, extrapolations of the curves for shelled corn, wheat, and sorghum agree well with available data (Stark) at airflows up to 1.0 $m^3/s \cdot m^2$.

References:

1. Abrams, C. F. and J. D. Fish, Jr. 1982. Air flow resistance characteristics of bulk piled sweet potatoes. TRANSACTIONS of the ASAE 25(4):1103-1106.
2. Bern, C. J. and L. F. Charity. 1975. Airflow resistance characteristics of corn as influenced by bulk density. ASAE Paper No. 75-3510. ASAE, St. Joseph, MI 49085.
3. Calderwood, D. L. 1973. Resistance to airflow of rough, brown and milled rice. TRANSACTIONS of ASAE 16(3):525-527, 532.
4. Gaffney, J. J. and C. D. Baird. 1975. Forced air cooling of bell peppers in bulk. ASAE Paper No. 75-6525. ASAE, St. Joseph, MI 49085.
5. Hague, E., G. H. Foster, D. S. Chung, and F. S. Lai. 1978. Static pressure across a corn bed mixed with fines. TRANSACTIONS of the ASAE 21(5):997-1000.
6. Henderson, S. M. 1943. Resistance of shelled corn and bin walls to airflow. AGRICULTURAL ENGINEERING 24(11):367-369.
7. Hukil, W. V. and N. C. Ives. 1955. Radial air flow resistance of grain. AGRICULTURAL ENGINEERING 36(5):332-335.
8. Nguyen, V. T. 1981. Airflow resistance of sunflower seed. Unpublished term project for AE568 at Iowa State University, Ames, IA. Under the direction of Carl J. Bern.
9. Schuler, R. T. 1974. Drying-related properties of sunflower seeds. ASAE Paper No. 74-3534. ASAE, St. Joseph, MI 49085.
10. Shedd, C. K. 1953. Resistance of grains and seeds to air flow. AGRICULTURAL ENGINEERING 34(9):616-619.
11. Sheldon, W. H., C. W. Hall, and J. K. Wang. 1960. Resistance of shelled corn and wheat to low airflows. TRANSACTIONS of the ASAE 3(2):92-94.
12. Staley, L. M. and E. L. Watson. 1961. Some design aspects of refrigerated potato storages. Canadian Agricultural Engineering 3(1):20-22.
13. Stark, B. and K. James. 1982. Airflow characteristics of grains and seeds. National Conference Publication N:82/8. 242-243. The Institution of Engineers, Australia.
14. Steele, J. L. 1974. Resistance of peanuts to airflow. TRANSACTIONS of the ASAE 17(3):573-577.

AIRFLOW RELATIONSHIPS FOR CONVEYING GRAIN AND OTHER MATERIALS

Approved by the ASAE Committee on Technical Data; adopted by ASAE 1948; revised 1954, 1962; reconfirmed by the Electric Power and Processing Division Technical Committee, December 1967, December 1972; reconfirmed December 1977; December 1982.

FIG. 1 AVERAGE VELOCITIES AND AIR VOLUMES FOR LOW PRESSURE PNEUMATIC CONVEYING OF MATERIAL

(Copyright 1946 by the Westinghouse Electric Corp., Sturtevant Division. Reproduced by permission.)

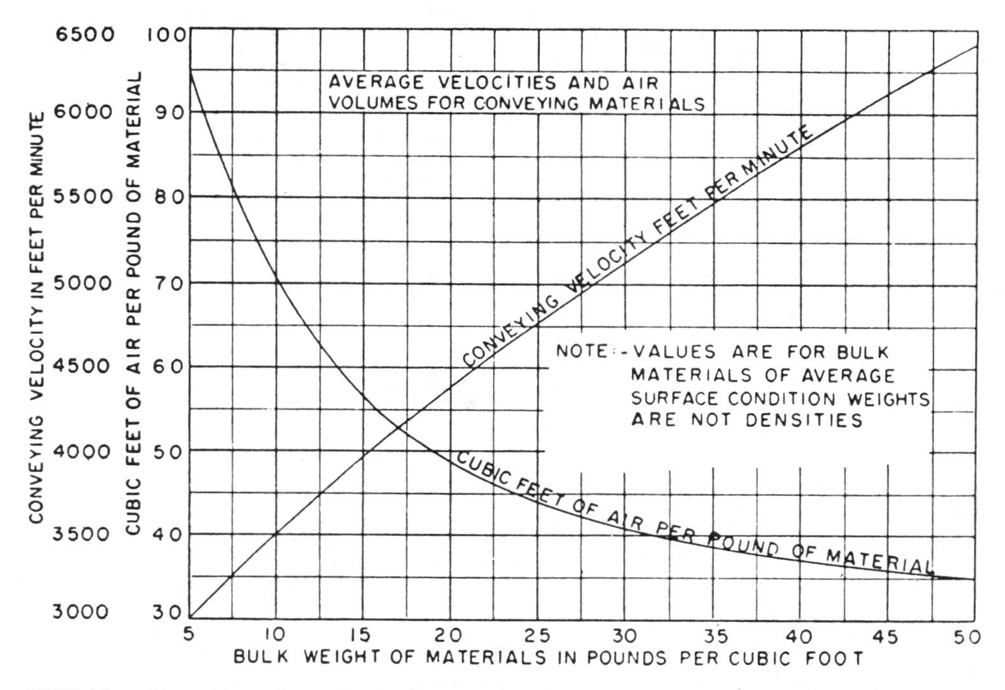

NOTE: The weight used is not the true density of material but is the average bulk weight of a cubic foot of material in the conditions in which it is to be conveyed. If material is moist, the bulk weight of the wet material must always be used. The curves apply for a uniform feed of material to the system. If the feed is uneven, sufficient air volumes must be provided to convey the material at the maximum rate of feed.

TABLE 1—AIR VELOCITIES REQUIRED TO CAUSE LOOSE GRAIN TO MOVE (SLIDE OR ROLL) ALONG THE BOTTOM OF A DUCT*

		Air velocities (fpm) to move clean whole seeds when placed along bottom of a glass tube†		
Grain	Bulk weight, lbs/cu ft	Several single seeds	Several small piles	Continuous thin layer
Oats	22.1	1100-1350	1800-2200	2100-2300
Wheat	43.5	1600-1800	2200-2700	2000-2400
Soybeans	41.4	1200-1600	2300-2600	2700-3000
Corn	44.6	1900-2050	2600-2700	2800-3100

*Data from tests made in USDA Grain Storage Laboratory, Ames, Iowa, to determine air velocities that might cause plugging of unscreened ventilating ducts.

†Lower velocities produced very slow and intermittent movement of seeds, while higher velocities produced rapid and more uniform movement of seeds.

ASAE Data: ASAE D274

FLOW OF WHEAT THROUGH ORIFICES

Approved by the ASAE Committee on Technical Data; adopted by ASAE 1948; revised 1954, 1962; reconfirmed by the Electric Power and Processing Division Technical Committee December 1968, December 1973, December 1978, December 1983.

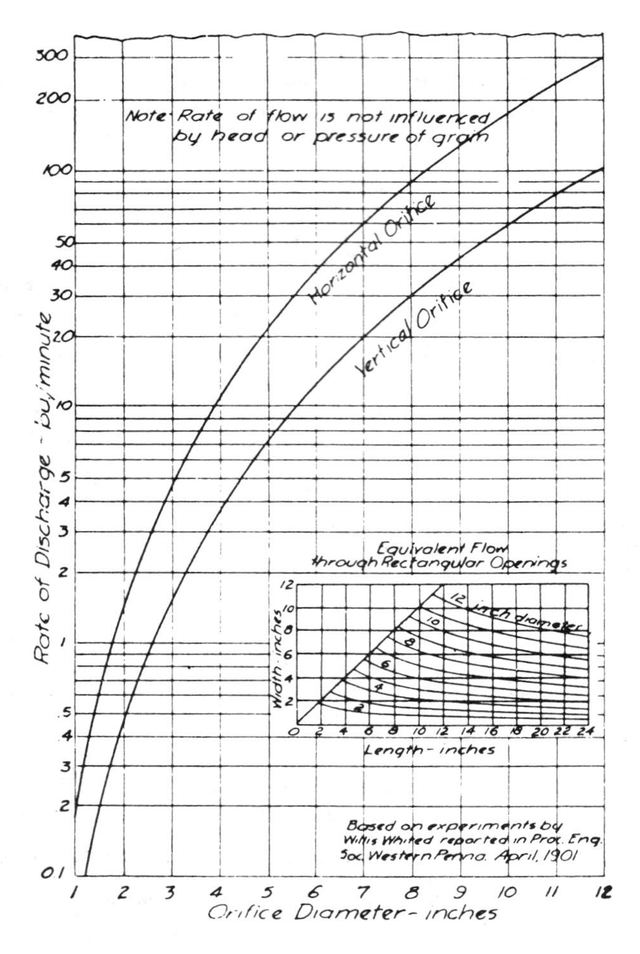

FIG. 1 FLOW OF WHEAT THROUGH OPENINGS

ASAE Data: ASAE D293.1

DIELECTRIC PROPERTIES OF GRAIN AND SEED

Based on research by S. O. Nelson and others; proposed by ASAE's Electric Power and Processing Division Technical Committee (EPP-03); reviewed by ASAE's Physical Properties of Agricultural Products Committee; approved by EPP-03; adopted by ASAE, December 1965; reconfirmed December 1970, January 1972, December 1976, December 1977, December 1978; revised February 1984; revised editorially March 1985.

SECTION 1—PURPOSE AND SCOPE

1.1 These data are provided for use in the design of equipment for radiofrequency (RF) or microwave dielectric heating applications, for applications in grain and seed moisture measurement by electrical methods, and for use in the development of other capacitive or RF-energy-absorption sensing devices. The data are also useful in predicting the dielectric heating behavior of grain and seed subjected to high-frequency and microwave electromagnetic fields.

1.2 Frequency ranges for which detailed dielectric-properties data are available on several kinds of grain and seed include those from 1 to 50 MHz and 250 Hz to 20 kHz, and such data are provided over a range of moisture contents for some materials. Wide-frequency-range data (250 Hz to 12 GHz) are available on only a few commodities. Data for this wide frequency range are presented graphically when dependence on moisture content is also known. Tabular data are provided for single moisture contents on some commodities. Dependence of the dielectric properties of grain and seed on temperature is not well documented, and the only information included relates to shelled corn. An equation is included that takes into account changes in the dielectric constant of a given grain or seed lot that result from a change in bulk density. The dielectric properties presented are given with particular wet-basis moisture contents and test weights, when available. Although most values are for specific varieties and origins, these data are believed to be representative, and crop origin is of minor significance. With materials as variable as grain and seed, some deviation from listed values can be expected in dielectric properties of particular lots, even though moisture content, test weight, and bulk density may be similar.

SECTION 2—DIELECTRIC-PROPERTIES DATA

2.1 Definition of terms and symbols. Values presented are those of the dielectric constant, ε_r', and the dielectric loss factor, ε_r'', respectively, the real and imaginary parts of the complex relative permittivity, $\varepsilon_r^* = \varepsilon_r' - j\varepsilon_r''$ (Nelson, 1973). Values for the loss tangent, $\tan \delta = \varepsilon_r''/\varepsilon_r'$ (where δ is the loss angle of the dielectric) and the a-c conductivity, σ, are not presented, because they can be calculated from the ε_r' and ε_r'' values presented. The dielectric constant, loss factor, and loss tangent (sometimes called the dissipation factor) are dimensionless quantities. The conductivity, $\sigma = \omega\varepsilon_o\varepsilon_r''$, where ω is the angular frequency, $2\pi f$, and ε_o is the permittivity of free space (8.854×10^{-12} farad/m), can be calculated as $\sigma = 55.63\ f\ \varepsilon_r'' \times 10^{-12}$ siemens/m (100 S/m = 100 mho/m = 1 mho/cm). Power factor can be calculated as $\tan \delta/(1 + \tan^2\delta)^{1/2}$.

2.2 Graphical data. Wide-frequency-range data on the dielectric properties of hard red winter wheat and shelled, yellow-dent field corn are presented as contour plots of ε_r' and ε_r'' as functions of frequency and wet-basis moisture content in Figs. 1 and 2. Dielectric properties of several kinds of grain and seed are shown at several moisture contents over the frequency range from 1 to 50 MHz in Figs. 3 to 11 (Nelson, 1965).

2.3 Tabular data. Wide-frequency-range (1 kHz to 10 GHz) data on the dielectric properties of a few kinds of grain and seed at single moisture contents are presented in Table 1. Data in the frequency range between 1 and 50 MHz are presented for a much wider range of commodities, and for different moisture contents in many instances, in Table 2. Data on several kinds of grain and seed are presented for frequencies from 250 Hz to 20 kHz in the audiofrequency range in Table 3.

2.4 Equations. Equations 1, 2, and 3 provide estimated values for the dielectric constants of shelled, yellow-dent field corn as functions of moisture content, bulk density and temperature at frequencies of 20, 300 and 2450 MHz, respectively. Equation 4 predicts the dielectric constant of granular material, such as grain and seed, at any bulk density, when the dielectric constant of the material is known at any given bulk density. If the density of the individual kernels or seeds is known, the dielectric constant of the kernels or individual seeds can be estimated with equation 4 from the dielectric constant of the bulk material at a known density.

Equations for the Dielectric Constant of Grain and Seed

The dielectric constant of shelled, hybrid, yellow-dent field corn may be estimated at frequencies of 20, 300 and 2450 MHz by the following equations (Nelson, 1979):

$$\varepsilon'_{r20} = 3.51 + 0.132(M - 10) + (\rho_b - \bar{\rho}_b)(0.839 - 0.086M + 0.027M^2) + 0.012(T - 24) \quad [1]$$

$$\varepsilon'_{r300} = 2.89 + 0.098(M - 10) + (\rho_b - \bar{\rho}_b)(0.460M - 2.16) + 0.015(T - 24) \quad [2]$$

$$\varepsilon'_{r2450} = 2.48 + 0.099(M - 10) + (\rho_b - \bar{\rho}_b)(0.387M - 3.22) + 0.013(T - 24) \quad [3]$$

where

M = Moisture content, percent, wet basis (range: 10 to 35%)
ρ_b = specific gravity of the bulk material, or density in g/cm^3, (range: 0.64 to 0.74)
T = temperature of the grain or seed, °C (range: 25 to 60 °C)
$\bar{\rho}_b$ = $0.6829 + 0.01422M - 0.000979M^2 + 0.0000153M^3$

The dielectric constant of a granular material at bulk density ρ_2 is predicted by the following equation (Nelson, 1983):

$$\varepsilon'_{r2} = [((\varepsilon'_{r1})^{1/3} - 1)\rho_2/\rho_1 + 1]^3 \quad [4]$$

where

ε'_{r1} = dielectric constant of the bulk material at bulk density ρ_1.

References: Last printed in 1985 STANDARDS; list available from ASAE Headquarters.

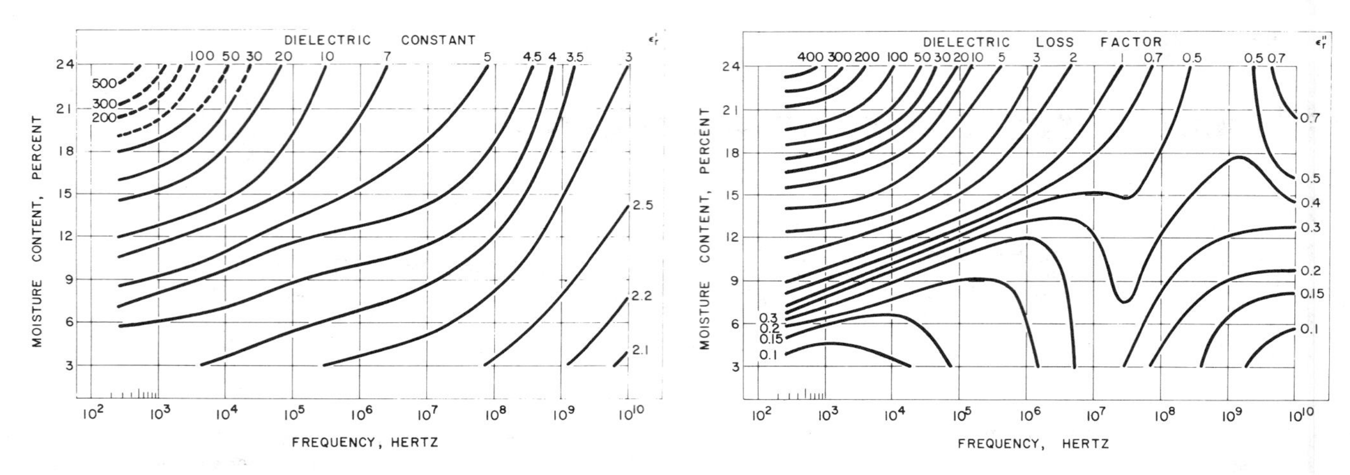

FIG. 1—DIELECTRIC PROPERTIES OF HARD RED WINTER WHEAT (*TRITICUM AESTIVUM* L.) AT 24 °C (Averages for seven lots of sound clean wheat, Nelson (1982)).

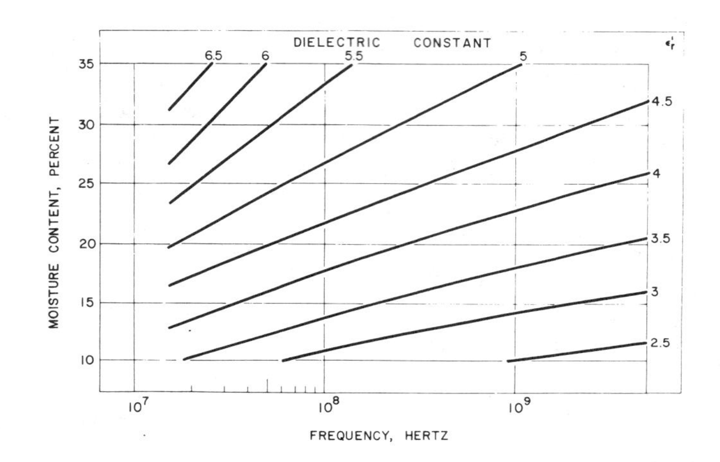

FIG. 2—DIELECTRIC CONSTANT OF SHELLED, HYBRID, YELLOW-DENT FIELD CORN (*ZEA MAYS* L.) AT 24 °C (Averages for 21 lots of sound, clean corn—condensed from Nelson (1979). Corrected and redrawn from Nelson (1982).

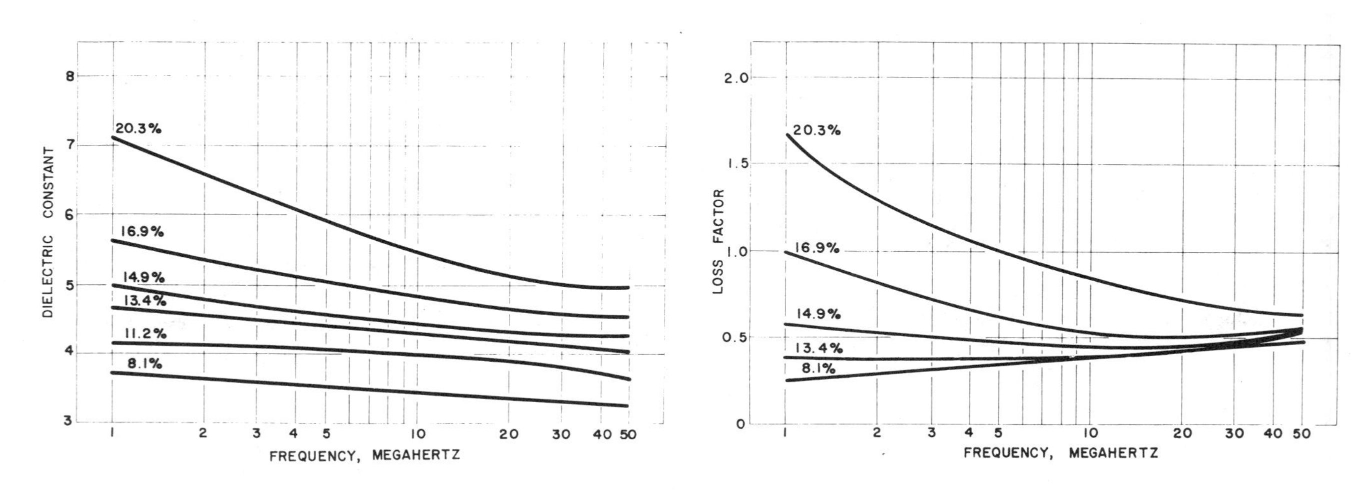

FIG. 3—DIELECTRIC PROPERTIES OF 1959 NEBRASKA-ORIGIN 501 SHELLED, YELLOW-DENT FIELD CORN (*ZEA MAYS* L. - UNGRADED KERNELS) AT 24 °C AND INDICATED MOISTURE CONTENTS. Test weight: 761 kg/m^3 (59.1 lb/bu) at 11.2% moisture. Natural drying from high moisture at harvest.

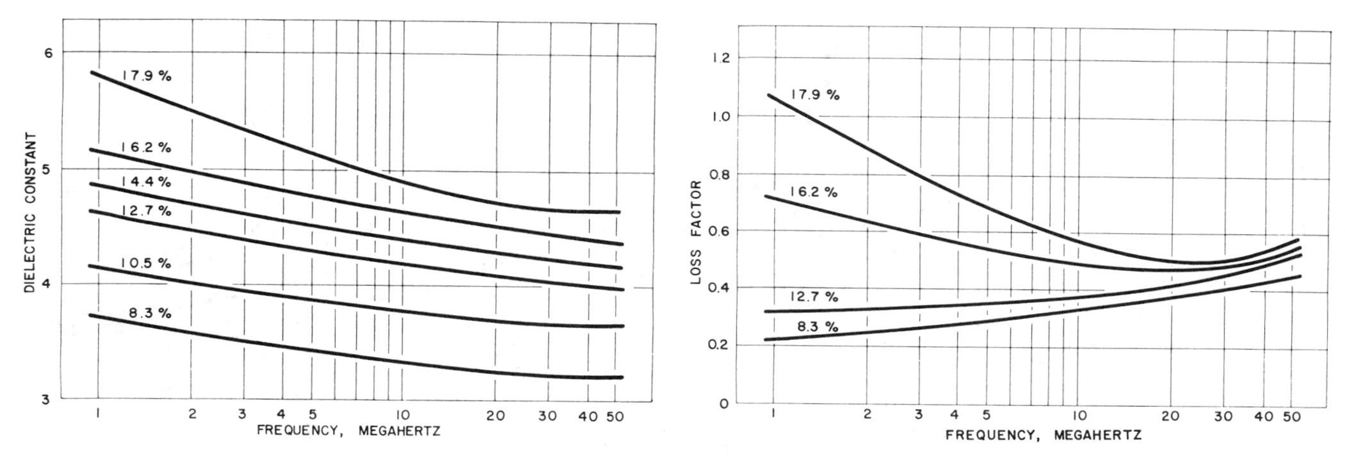

FIG. 4—DIELECTRIC PROPERTIES OF 1958 NEBRASKA-ORIGIN NEBRED HARD RED WINTER WHEAT (*TRITICUM AESTIVUM* L.) AT 24 °C AND INDICATED MOISTURE CONTENTS. Test weight: 768 kg/m³ (59.7 lb/bu) at 13.0% moisture. Water added for moisture contents above 12.7%.

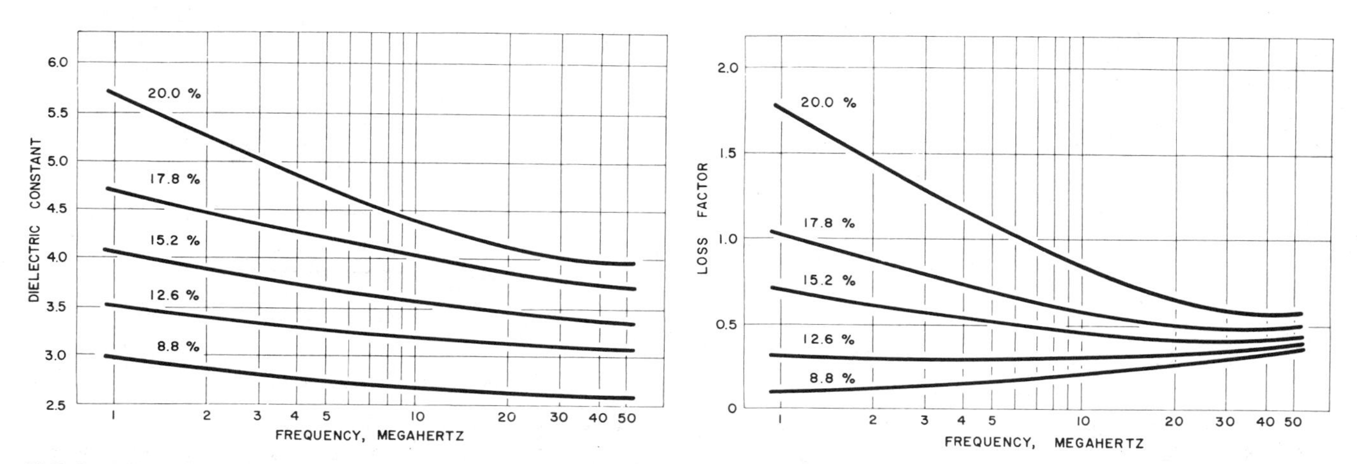

FIG. 5—DIELECTRIC PROPERTIES OF 1961 NEBRASKA-ORIGIN CHASE WINTER BARLEY (*HORDEUM VULGARE* L.) AT 24 °C AND INDICATED MOISTURE CONTENTS. Test weight: 597 kg/m³ (46.4 lb/bu) at 12.6% moisture. Water added for moisture contents above 12.6%.

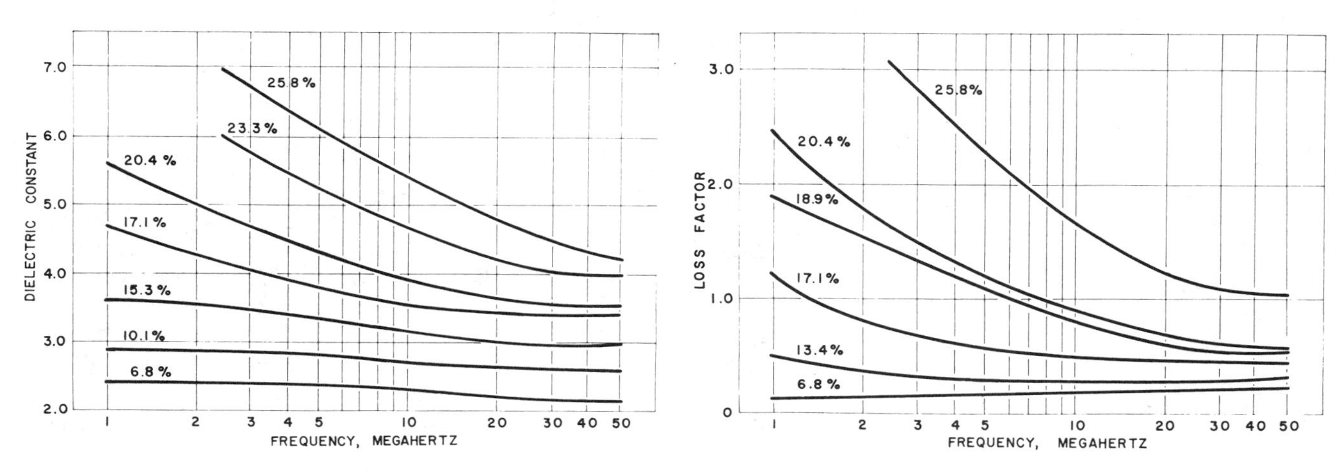

FIG. 6—DIELECTRIC PROPERTIES OF 1959 NEBRASKA-ORIGIN CHEROKEE SPRING OATS (*AVENA SATIVA* L.) AT 24 °C AND INDICATED MOISTURE CONTENTS. Test weight: 519 kg/m³ (40.3 lb/bu) at 10.1% moisture. Water added for moisture contents above 10.6%.

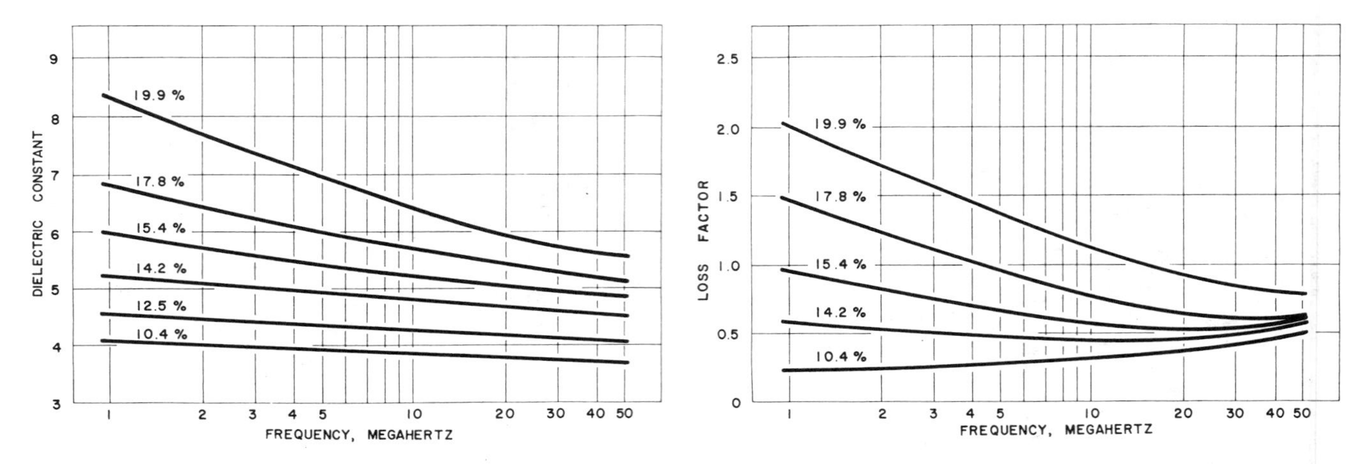

FIG. 7—DIELECTRIC PROPERTIES OF 1955 NEBRASKA-ORIGIN MARTIN MILO GRAIN SORGHUM (*SORGHUM BICOLOR* (L.) MOENCH) AT 24 °C AND INDICATED MOISTURE CONTENTS. Water added for moisture contents above 10.4%.

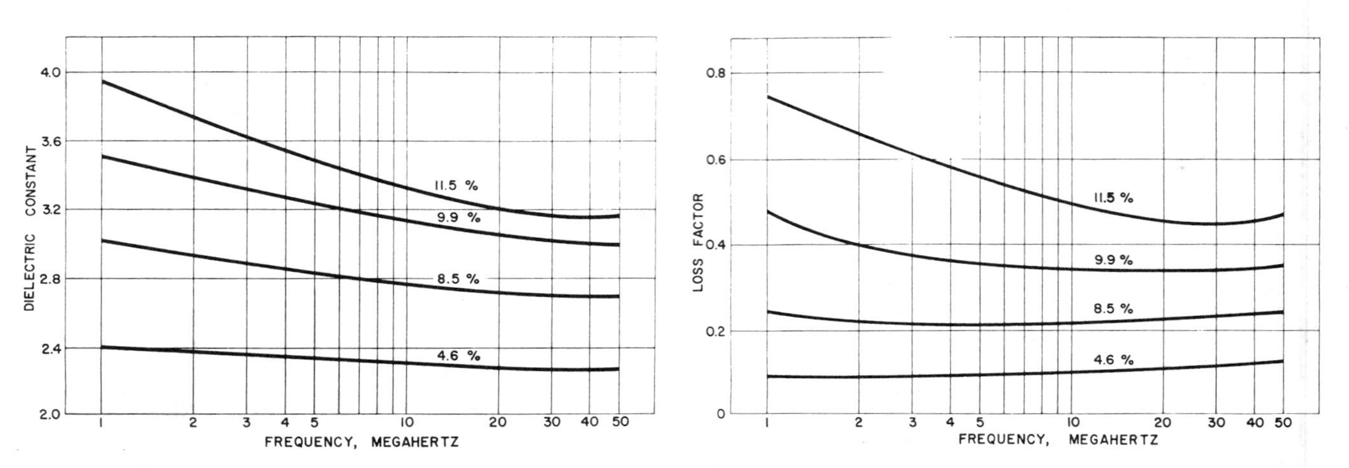

FIG. 8—DIELECTRIC PROPERTIES OF 1958 NEBRASKA-ORIGIN HAWKEYE SOYBEANS (*GLYCINE MAX* (L.) MERRILL) AT 24 °C AND INDICATED MOISTURE CONTENTS. Test weight: 738 kg/m³ (57.3 lb/bu) at 7.5% moisture. Water added for moisture contents above 9.9%.

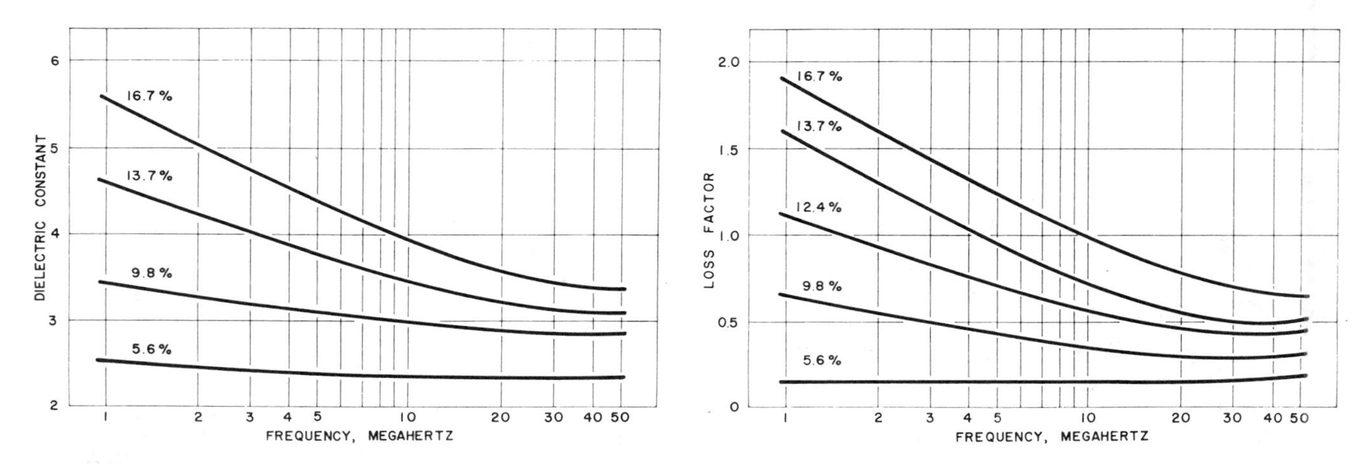

FIG. 9—DIELECTRIC PROPERTIES OF 1959 NEW MEXICO-ORIGIN ACALA 1517D ACID-DELINTED COTTONSEED (*GOSSYPIUM HIRSUTUM* L.) AT 24 °C AND INDICATED MOISTURE CONTENTS. Test weight: 618 kg/m³ (48.0 lb.bu) at 6.8% moisture. Water added for moisture contents above 6.8%.

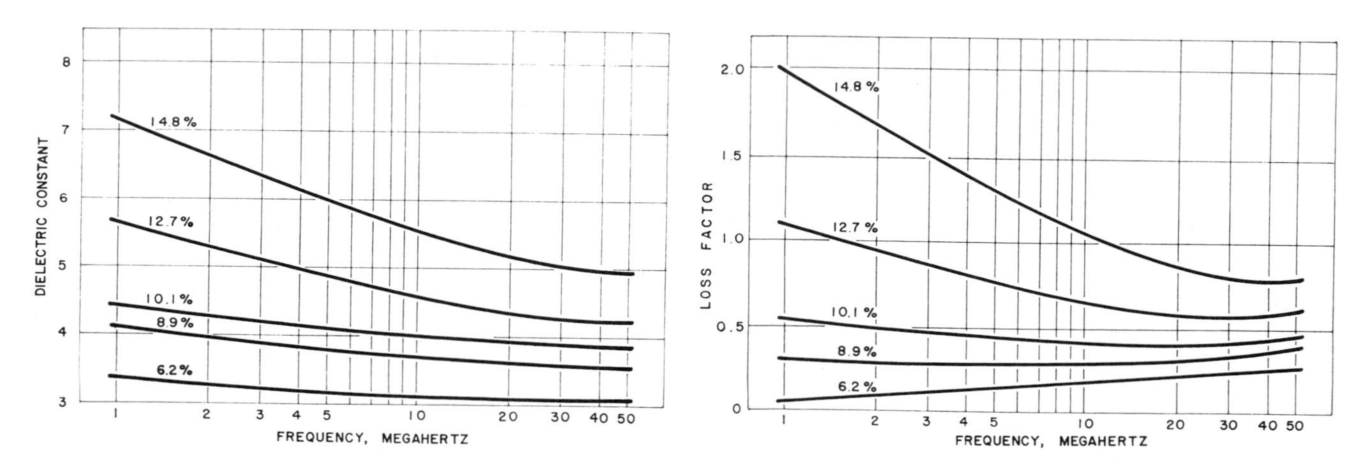

FIG. 10—DIELECTRIC PROPERTIES OF 1959 WASHINGTON-ORIGIN RANGER ALFALFA (*MEDICAGO SATIVA* L.) SEED AT 24 °C AND INDICATED MOISTURE CONTENTS. Test weight: 804 kg/m^3 (62.5 lb/bu) at 7.7% moisture. Water added for moisture contents above 7.7%.

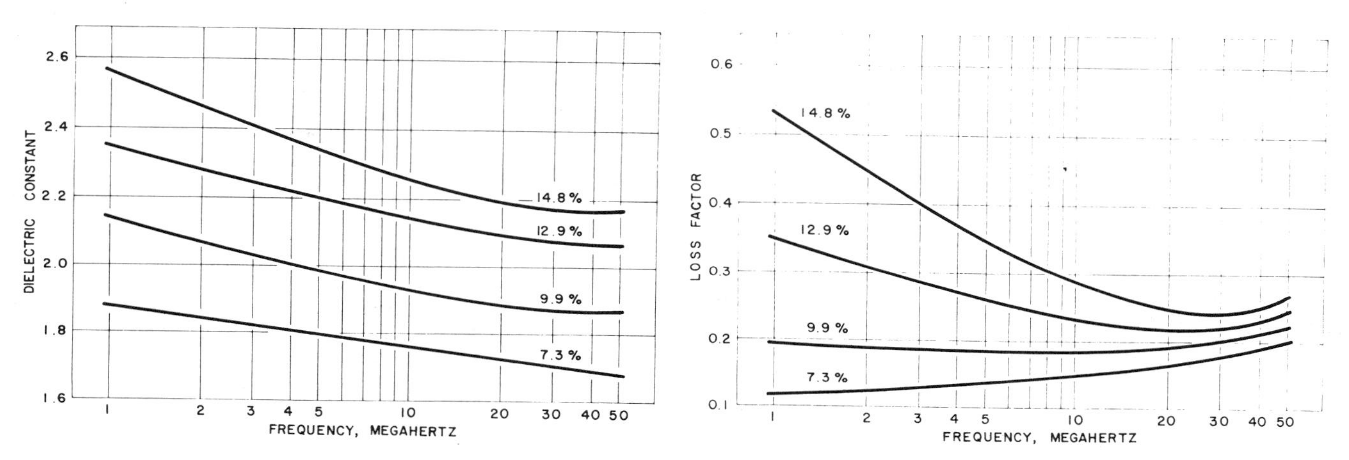

FIG. 11—DIELECTRIC PROPERTIES OF 1959 SOUTH DAKOTA-ORIGIN KENTUCKY BLUEGRASS (*POA PRATENSIS* L.) SEED AT 24 °C AND INDICATED MOISTURE CONTENTS. Test weight: 323 kg/m^3 (25.1 lb/bu) at 9.9% moisture. Water added for moisture contents above 9.9%.

TABLE 1—WIDE-FREQUENCY-RANGE DIELECTRIC PROPERTIES OF GRAIN AND SEED AT 24° C*

Grain or seed	Moisture, content, %	Bulk density kg/m^3	Bulk density lb/bu		Dielectric constant (ϵ_r') and dielectric loss factor (ϵ_r'') — Frequency, Hz: 10^3	10^4	10^5	10^6	10^7	10^8	10^9	10^{10}
Alfalfa, 'Ranger' (*Medicago sativa* L.)	7.5	779	60.5	ϵ_r'	4.9	4.0	3.6	3.4	3.2	3.0	2.6	2.5
				ϵ_r''	1.8	0.49	0.21	0.17	0.24	0.27	0.21	0.17
Grain sorghum (*Sorghum bicolor* (L.) Moench)	11.4	768	59.7		6.7	5.1	4.6	4.5	4.2	3.6	2.9	2.6
					1.9	0.71	0.27	0.21	0.38	0.42	0.29	0.23
Oats, spring, 'Neal' (*Avena sativa* L.)	10.7	532	41.3		5.0	3.6	3.1	3.0	2.8	2.4	2.2	1.9
					2.2	0.72	0.22	0.12	0.20	0.23	0.18	0.14
Soybean, 'Wayne' (*Glycine max* (L.) Merrill)	8.5	732	56.9		5.0	3.6	3.6	3.4	3.2	3.0	2.6	2.4
					2.4	0.82	0.35	0.29	0.24	0.20	0.15	0.13

*Nelson and Stetson (1975)

TABLE 2—DIELECTRIC PROPERTIES OF GRAIN AND SEED AT 24°C

Grain or seed	Moisture content, %	Bulk density			Dielectric constant (ϵ_r') and loss factor (ϵ_r'')					Reference*
					Frequency, MHz					
		kg/m³	lb/bu		1	10	27	40	50	
Barley, spring (*Hordeum vulgaris* L.)	12.9	602	46.8	ϵ_r'	3.5	3.2	3.0	3.0	3.0	1
				ϵ_r''	0.34	0.25	0.33	0.38	0.41	
	15.2				3.9	3.4	3.3	3.3	3.3	
					0.56	0.35	0.35	0.40	0.47	
	20.0				5.1	4.2	4.0	3.9	3.8	
					1.3	0.70	0.50	0.50	0.50	
Bean, dry edible, Great Northern (*Phaseolus vulgaris* L.)	8.7	720	55.9			3.0		2.9		1
						0.25		0.30		
climbing green, 'Kentucky Wonder 191'	7.1	783	60.8			3.0		2.8		1,6
						0.26		0.29		
Beet, sugar (*Beta vulgaris* L.)	9.4				2.4	2.1	2.1	2.1	2.1	1,6
					0.34	0.24	0.24	0.23	0.23	
	12.1				3.4	2.5	2.4	2.3	2.3	
					1.1	0.47	0.35	0.32	0.32	
Brome, 'Lyon' (*Bromus arvensis* L.)	7.7				1.4	1.3	1.3	1.3	1.3	1,6
					0.21	0.17	0.17	0.18	0.18	
	10.6	158	12.3		1.6	1.5	1.4	1.4	1.4	
					0.15	0.16	0.18	0.19	0.21	
	15.9				2.0	1.7	1.7	1.7	1.7	
					0.50	0.29	0.24	0.24	0.25	
Cabbage, 'Globe YR' (*Brassica oleracea* var. *capitata* L.)	6.1	718	55.8			2.7		2.7		6
						0.16		0.20		
Cantaloupe (*Cucumis melo* L.)	6.5	419	32.6			2.1		2.1		1,6
						0.15		0.17		
Castor bean, 'Lynn' (*Ricinus communis* L.)	4.9	551	42.8			2.1		2.1		6
						0.12		0.17		
Clover, red (*Trifolium pratense* L.)	2.5							2.7		1
								0.14		
	6.0							2.9		
								0.22		
	7.4							3.1		
								0.27		
Cotton, gin-run (*Gossypium hirsutum* L.)	6.5	391	30.4					1.7		6
								0.12		
mechanically-delinted	7.4							1.9		1
								0.18		
Cucumber, 'Model' (*Cucumis sativus* L.)	6.5	602	46.8			2.5		2.5		1,6
						0.14		0.18		
Dichondra (*Dichondra repens* Forst.)	8.0					3.7		3.5		1
						0.29		0.38		
Eggplant, 'Florida Market' (*Solanum melongena* L. var. *esculentum* Nees)	7.1	547	42.5			2.6		2.6		6
						0.20		0.26		
Lettuce, 'Great Lakes' (*Lactuca sativa* var. *capitata* L.)	6.1	467	36.3			2.1		2.1		6
						0.12		0.14		
Millet, German (*Setaria italica* (L.) Beauv.)	9.8	720	55.9			3.2		3.1		1,6
						0.28		0.35		
Okra (*Hibiscus esculentus* L.)	7.4	687	53.4			2.9		2.8		1,6
						0.29		0.27		
Onion (*Allium cepa* L.)	8.4	486	37.8			2.3		2.3		6
						0.19		0.23		
Pea, garden, 'Dwarf Gray Sugar' (*Pisum sativum* L.)	5.4							2.9		1
								0.30		
'Wando'	7.6							2.6		1
								0.29		

*Numbers refer to References listed for ASAE Data D293, Dielectric Properties of Grain and Seed.

TABLE 2—DIELECTRIC PROPERTIES OF GRAIN AND SEED AT 24°C (cont'd)

Grain or seed	Moisture content, %	Bulk density		Dielectric constant (ϵ_r') and loss factor (ϵ_r'')					Reference*
				Frequency, MHz					
		kg/m^3	lb/bu	1	10	27	40	50	
Peanut, shelled, Spanish 'Starr' (*Arachis hypogaea* L.)	4.7	626	48.6		2.3 0.12		2.2 0.14		1,6
	11.0			4.6 0.98	4.2 0.95	3.3 0.85	2.7 0.74	2.2 0.64	10
	30.0			7.2 1.9	6.7 1.8	5.9 1.8	5.3 1.7	4.8 1.6	10
Pepper, 'Yolo Wonder A' (*Capsicum frutescens* L.)	6.4	514	39.9		2.4 0.19		2.3 0.22		6
'Pine, eastern white (*Pinus strobus* L.)	5.8	509	39.5				2.2 0.19		6,9
loblolly (*P. taeda* L.)	7.4	522	40.6		2.6 0.16		2.6 0.20	2.6 0.22	6,9
ponderosa (*P. ponderosa* Laws. var. *ponderosa*)	6.9	441	34.3		2.1 0.15		2.1 0.18		1,6
sugar (*P. lambertiana* Dougl.)	6.8	492	38.2		2.1 0.16		2.1 0.19		1,6
Rice, rough, 'Zenith' (*Oryza sativa* L.)	10.0	594	46.2			3.2 0.30			11
	15.4					3.7 0.40			
	21.1					4.5 0.87			
Rye, winter, 'Balbo' (*Secale cereale* L.)	12.7	731	56.8				4.0 0.52		1
Ryegrass (*Lolium multiflorum* Lam.)	9.4				2.1 0.21		1.9 0.25		1
Safflower (*Carthamus tinctorius* L.)	5.7	515	40.0	2.3 0.12	2.2 0.14	2.2 0.15	2.2 0.16	2.2 0.17	1
Spinach, 'Hybrid 424' (*Spinacia oleracea* L.)	8.6				2.4 0.26		2.3 0.28		1
'Early Hybrid 7'	9.4	532	41.3				2.5 0.28		6
Sweetclover, common yellow (*Melilotus officinalis* (L.) Lam.	9.7			4.6 0.43	4.1 0.36	3.9 0.39	3.9 0.40	3.9 0.43	1,6
Switchgrass (*Panicum virgatum* L.)	9.8	701	54.5	3.7 0.17	3.5 0.23	3.4 0.31	3.3 0.34	3.3 0.36	1
	13.4			4.5 0.45	4.1 0.34	4.0 0.36	4.0 0.39	4.0 0.42	1
Tobacco, burley, 'Kentucky 16' (*Nicotiana tobacum* L.)	5.9				2.0 0.14		2.0 0.08		1,6
Maryland, 'Catterton'	5.3	440	34.2	2.0 0.12	1.9 0.12	1.9 0.14	1.9 0.14	1.9 0.15	1,6
shade, 'Dixie Shade'	6.4	420	32.6				2.0 0.17		6
Tomato, 'Homestead 24' (*Lycopersicon esculentum* Mill.)	6.9	381	29.6		1.8 0.12		1.9 0.16		1,6
Trefoil, birdsfoot (*Loftus corniculatus* L.)	2.4						2.7 0.16		1
	6.4						3.0 0.26		
	9.7						3.7 0.40		
Wheatgrass, crested (*Agropyron desertorum* (Fisch.) Schult.)	6.5			1.5 0.10	1.5 0.13	1.5 0.15	1.5 0.16	1.5 0.17	1
	10.5	242	18.8	1.7 0.12	1.6 0.15	1.6 0.17	1.6 0.18	1.6 0.19	
	16.1			2.2 0.34	2.0 0.22	1.9 0.21	1.9 0.22	1.9 0.23	
western (*Agropyron smithii* Rydb.)	10.2						1.5 0.22		

*Numbers refer to References listed for ASAE Data D293, Dielectric Properties of Grain and Seed.

TABLE 3—AUDIOFREQUENCY DIELECTRIC PROPERTIES OF GRAIN AND SEED AT 24°C*

Grain or seed	Moisture content, %	Bulk density			Dielectric constant (ϵ_r') and loss factor (ϵ_r'')				
					Frequency, kHz				
		kg/m^3	lb/bu.		0.25	1	5	10	20
Alfalfa, 'Ranger'	6.8	804	62.5	ϵ_r'	5.5	4.3	4.0	3.8	3.7
(*Medicago sativa* L.)				ϵ_r''	3.33	1.48	0.53	0.39	0.26
	7.8	802	62.3		10.4	6.0	4.4	4.2	4.0
					5.9	3.6	1.4	0.92	0.60
Bluegrass, Kentucky	8.8	295	22.9		4.3	3.0	2.4	2.3	2.2
(*Poa pratensis* L.)					3.0	1.6	0.72	0.52	0.38
	10.4	298	23.2		9.5	5.5	3.7	3.0	2.8
					5.6	4.0	2.0	1.4	0.95
Corn, field, yellow-dent	12.0	699	54.3		12.0	8.5	6.3	5.6	5.3
(*Zea mays* L.)					4.4	3.6	2.0	1.5	1.1
	14.2	687	53.4		17.8	13.6	9.6	8.3	7.2
					6.1	5.1	3.6	3.0	2.6
Cotton, acid-delinted	7.9	567	44.0		10.5	8.1	4.8	3.9	3.4
(*Gossypium hirsutum* L.)					2.2	3.5	2.8	2.0	1.5
	9.9	553	43.0		11.9	10.6	7.8	6.2	5.0
					2.6	2.4	3.2	3.1	2.6
Grain sorghum	12.0	783	60.8		11.2	8.6	6.2	5.8	5.4
(*Sorghum bicolor* (L.) Moench)					2.5	3.0	1.8	1.3	0.96
	15.1	785	61.0		14.2	13.9	12.4	11.1	9.4
					0.80	1.1	2.6	3.0	3.1
Oats, spring 'Neal'	12.6	603	46.8		15.9	13.5	9.1	7.1	5.6
(*Avena sativa* L.)					2.6	3.9	4.3	3.8	3.9
	14.0	558	43.3		18.7	16.9	13.1	11.1	8.8
					3.0	3.4	4.5	4.6	4.3
Soybean, 'Wayne'	7.8	678	52.7		4.9	3.8	3.3	3.2	3.1
(*Glycine max* L.)					2.4	1.3	0.62	0.46	0.34
	9.5	671	52.1		11.0	8.2	5.5	4.8	4.4
					2.8	3.2	2.2	1.7	1.3
Wheatgrass, western	8.5	214	16.6		2.4	2.0	1.9	1.8	1.8
(*Agropyron smithii* Rydb.)					1.2	0.63	0.31	0.22	0.15
	10.0	212	16.5		4.8	3.1	2.4	2.2	2.1
					3.4	2.1	1.1	0.74	0.48

*Stetson and Nelson (1972)

ASAE Standard: ASAE S319.1

METHOD OF DETERMINING AND EXPRESSING FINENESS OF FEED MATERIALS BY SIEVING

Proposed by a subcommittee of the American Feed Manufacturers Association; approved by the ASAE Electric Power and Processing Division Technical Committee; adopted by ASAE December 1968; reconfirmed December 1973, December 1978, December 1983; revised March 1985.

SECTION 1—PURPOSE AND SCOPE

1.1 The purpose of this Standard is to define a test procedure to determine the fineness of feed ingredients and to define a method of expressing the particle size of the material. The particle size determined can be used to calculate surface area and number of particles per unit weight.

1.2 This Standard shall be used to determine the fineness of feed ingredients where the reduction process yields particles which are essentially spherical or cubical. It is not adequate to define the particle size of materials such as steamed and rolled grains which are a flaked product, or products, such as chopped hay, where substantial fraction consists of elongated particles.

SECTION 2—TEST EQUIPMENT

2.1 A set of woven-wire cloth sieves having a diameter of 203 mm (8 in.) shall be used. With the most common shaking equipment, sieves having a height of 25 mm (1 in.) or half-height sieves are most suitable to avoid the necessity of resieving the finer fraction. A set of sieves as specified in American Society for Testing and Materials Standard E11, Specifications for Wire-Cloth Sieves for Testing Purposes, shall consist of the following sizes:

U.S. Standard Sieve No.	Nominal Sieve Opening mm	in.*
4	4.76	0.187
6	3.36	0.132
8	2.38	0.0937
12	1.68	0.0661
16	1.19	0.0469
20	0.841	0.0331
30	0.595	0.0234
40	0.420	0.0165
50	0.297	0.0117
70	0.210	0.0083
100	0.149	0.0059
140	0.105	0.0041
200	0.074	0.0029
270	0.053	0.0021
Pan		

2.2 A suitable sieve shaker, such as a Ro-tap†, is required.

2.3 A balance having an accuracy of at least ±0.1 grams shall be used.

2.4 Sieve agitators such as plastic or leather rings, or small rubber balls may be required to break up agglomerates on finer sieves, usually those smaller than U.S. No. 50.

2.5 A dispersion agent‡ should be available to facilitate sieving of high fat or similar materials.

2.6 Sieve openings must be kept free of feed particles so that normal sieving can be accomplished. A stiff bristle sieve cleaning brush, or compressed air, is useful for cleaning sieves which have become clogged with feed particles. Sieves must be cleaned periodically to remove oil. Oil can be removed by washing with water containing a detergent. Sieves must be dried before use.

SECTION 3—METHOD OF SIEVING

3.1 A sample size of 100 g should be used although smaller samples may be used if extra care is taken to recover all material from the sieves.

3.2 Place the sample on the top sieve of the set of sieves and shake until the weight of material, on the smallest sieve which contains any material, reaches equilibrium. Equilibrium shall be determined by inspecting and weighing at 5 min intervals after an initial sieving time of 10 min. If the weight on the smallest sieve containing any material changes by 0.2% or less of the total sample weight during a 5 min period the sieving shall be considered complete at the onset of the previous period.

3.3 Material on all sieves shall be weighed and recorded.

3.4 If a dispersing agent is required, it should be added at a level of 0.5%, and its effect on particle size need not be recorded.

3.5 If 20% or more of the material by weight passes the smallest sieve, the fine material shall be subjected to a nonsieving particle size analysis, such as microscopic measurement or sedimentation testing, and such analysis shall be reported separately.

SECTION 4—DATA ANALYSIS

4.1 Analysis of weight distribution data of all ground feeds and feed ingredients are based on the assumption that these distributions are logarithmic normally distributed.

4.2 Calculation of particle size

4.2.1 The size of particles shall be reported in terms of geometric mean diameter and geometric standard deviation by weight.

4.2.2 Calculated values are obtained as follows:

$$d_{gw} = \log^{-1} \left[\frac{\Sigma(W_i \log \bar{d}_i)}{\Sigma W_i} \right]$$

$$S_{gw} = \log^{-1} \left[\frac{\Sigma W_i (\log \bar{d}_i - \log d_{gw})^2}{\Sigma W} \right]^{1/2}$$

where

d_i = diameter of sieve openings of the i'th sieve

d_{i+1} = diameter of openings in next larger than i'th sieve (just above in a set)

d_{gw} = geometric mean diameter

$\bar{d}_i$ = geometric mean diameter of particles on i'th sieve

= $(d_i \times d_{i+1})^{1/2}$

S_{gw} = geometric standard deviation

W_i = weight fraction on i'th sieve

4.2.3 Material passing U.S. Sieve No. 270 shall be considered to have a mean diameter of 44 μm.

---

*Only approximately equivalent to the values given in millimeters.

†Registered trade name.

‡Dispersion agents include Cab-O-Sil MS available from the Cabot Corp., Boston; Ziolex 23A and Zeofree 80 available from the J. M. Huber Corp., New York; and Flo-Gard available from the Pittsburgh Plate Glass Co., St. Louis.

4.2.4 Graphical solutions for geometric mean diameter and log-normal geometric standard deviation may be obtained by plotting results on logarithmic probability graph paper. Fig. 1 shows an example where:

$d_{gw} = d_{50} =$ particle diameter at 50% probability

$S_{gw} = \frac{d_{84}}{d_{50}} = \frac{d_{50}}{d_{16}} =$

particle size at 84% probability/$d_{gw} = d_{gw}$/ particle diameter at 16% probability

or

$$d_{gw} = 350\ \mu m$$

and

$$S_{gw} = \frac{640}{350} = \frac{350}{191} = 1.83$$

Cited Standard:

ASTM E11, Specifications for Wire-Cloth Sieves for Testing Purposes

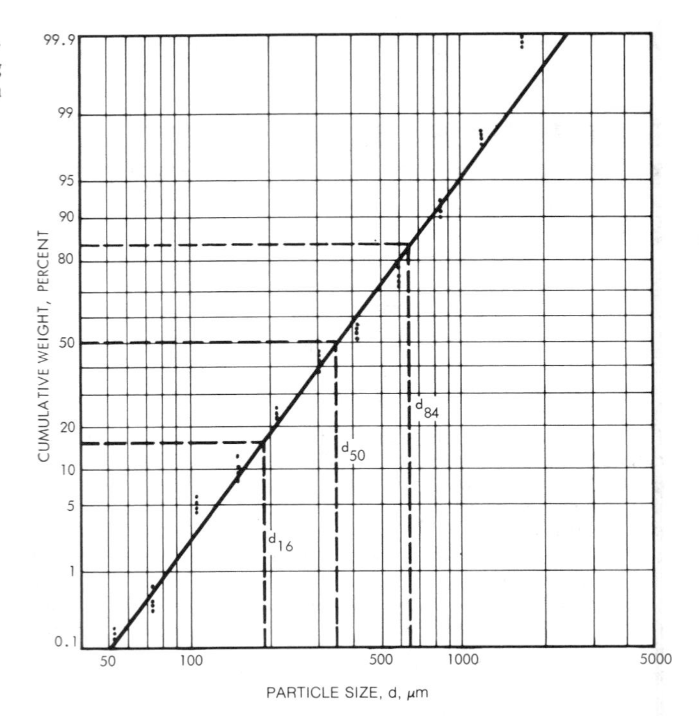

FIG. 1—LOGNORMAL DISTRIBUTION FOR SORGHUM GRAIN GROUND THROUGH 3.18 mm (1/8 in.) SCREEN

T

MOISTURE MEASUREMENT—UNGROUND GRAIN AND SEEDS

Developed by the ASAE Physical Properties of Agricultural Products Committee; approved by Electric Power and Processing Division Standards Committee; adopted by ASAE February 1972; reconfirmed December 1976, December 1977; revised December 1982, April 1988.

SECTION 1—PURPOSE AND SCOPE

1.1 This Standard is to provide a uniform method for determining the moisture content of unground samples of agricultural seeds. Only those seeds are included for which documented comparisons with the Karl Fischer method are available. These techniques should become common practice and their use referenced in all technical presentations where moisture content determinations have been a factor. Deviations from these methods should be reported.

1.2 Specifying sampling procedures is not within the scope of this recommendation. It will be assumed that the sample used for analysis is representative of the quantity from which it was taken.

SECTION 2—APPARATUS

2.1 Moisture dishes. Moisture dishes should be made of heavy gauge aluminum that does not dent readily. The dishes should be provided with tightly fitting covers. Both the dish and its cover should be identified by the same number. Before using, dry the moisture dish for one hour at the drying temperature to be used and obtain the tare weight. Other similar containers may be used provided they are tared before and after drying and their use is mentioned in referring to the moisture measurement techniques used.

2.2 Desiccator. Desiccator should be airtight and should contain activated alumina or other equally suitable desiccant. Silica gel and anhydrous calcium chloride are not suitable desiccants.

2.3 Oven. Oven may be of gravity-convection or mechanical-convection (forced-draft) type. It should be well insulated, maintain a reasonably uniform temperature throughout the chamber and maintain specified temperature at shelf level. A properly ventilated oven, equipped with removable perforated or wire shelves and a suitable thermometer accurate to within 0.5 °C, is required. To insure uniformity of heating, ovens should be in operation for several hours prior to the drying operation.

2.4 Balance. An analytical balance should be used for all weighings in making moisture content determinations. Weighings should be made to the nearest 1 mg.

SECTION 3—PROCEDURE

3.1 Using the sample size specified in Table 1, place a representative sample of the unground grain or seed in each of two or more tared moisture dishes. Weigh the covered dishes and contents. Subtract the weight of each dish from the total weight and record the weight of the portion. Uncover the dishes and place them with their covers in the oven. The oven temperature and heating period depend on the seed as shown in Table 1.

3.2 The dishes should be placed with the bulb of the oven thermometer as close to them as possible. At the end of the heating period, cover the dishes as soon as possible and place them in a desiccator. Weigh the dishes when they reach room temperature. Calculate the percentage of moisture by dividing the loss in weight due to heating by the weight of the original sample and multiply by 100. Replicate determinations should check within 0.2% moisture.

TABLE 1—OVEN TEMPERATURE AND HEATING PERIOD FOR MOISTURE CONTENT DETERMINATIONS

Seed	Oven temperature, ± 1 °C	Heating time		Sample size, g	Reference
		hr	min		
Alfalfa	130	2	30	10	2
Barley	130	20	0	10	2
Beans, edible	103	72	0	15	1
Bentgrass	130	1	0	10	2
Bluegrass	130	1	0	5	2
Bluestem, yellow	100	1	0	1	2
Bromegrass, smooth	130	0	50	4	2
Cabbage	130	4	0	10	2
Carrott	100	1	40	10	2
Clover	130	2	30	10	2
Collard	130	4	0	10	2
Corn	103	72	0	15 or 100*	1,3
Fescue	130	3	0	5	2
Flax	103	4	0	5-7	1,3
Kale	130	4	0	10	2
Mustard	130	4	0	10	2
Oats	130	22	0	10	2
Onion	130	0	50	10	2
Orchardgrass	130	1	0	5	2
Parsley	100	2	0	10	2
Parsnip	100	1	0	10	2
Radish	130	1	10	10	2
Rape (Canola)	130	4	0	10	2
Rye	130	16	0	10	2
Ryegrass	130	3	0	5	2
Safflower	130	1	0	10	3
Sorghum	130	18	0	10	2
Soybeans	103	72	0	15	†
Sunflower	130	3	0	10	3
Timothy	130	1	40	10	2
Turnip	130	4	0	10	2
Wheat	130	19	0	10	2

*Use 100 g if moisture exceeds 25%.
†No official method exists for unground soybeans.

References:

1. AACC 44-15A. 1981. Moisture-air oven methods. American Association of Cereal Chemists, St. Paul, MN.

2. Hart, J. R., L. Feinstein and C. Golumbic. 1959. Oven methods for precise measurement of moisture in seeds. Marketing Research Report No. 304. U.S. Government Printing Office, Washington, D.C.

3. USDA-FGIS. 1986. Air-oven methods. Chapter 4, Moisture Handbook. U.S. Department of Agriculture-Federal Grain Inspection Service, Washington, D.C.

ASAE Standard: ASAE S353

MOISTURE MEASUREMENT—MEAT AND MEAT PRODUCTS

Corresponds to Moisture Measurement Standards 24.001, 24.002, and 24.003 of Association of Official Analytical Chemists; proposed by ASAE Physical Properties of Agricultural Products Committee; approved by Electric Power and Processing Division Standards Committee; adopted by ASAE March 1972; reconfirmed December 1976; reconfirmed and revised editorially December 1977; reconfirmed December 1982, December 1987.

SECTION 1—PURPOSE AND SCOPE

1.1 This Standard establishes a common procedure for determining the moisture content of meat and meat products to which all other moisture content determination procedures should be correlated, and provides a basis upon which all reported moisture content values can be correlated.

SECTION 2—PREPARATION OF SAMPLE

2.1 To reduce moisture loss during preparation and subsequent handling, the use of small samples is not recommended. Keep ground material in glass or similar containers with air and water tight covers. Prepare samples for analysis in the following manner:

2.1.1 Fresh meats, dried meats, cured meats, smoked meats, etc. Separate as completely as possible from any bone; pass rapidly three times through food chopper with plate openings less than or equal to 3 mm (1/8 in.), mixing thoroughly after each grinding; and begin all determinations promptly. If any delay occurs, chill sample to inhibit decomposition.

2.1.2 Canned meats. Pass entire contents of can through food chopper, continue as described in paragraph 2.1.1.

2.1.3 Sausages. Remove from casings and pass through food chopper, continue as described in paragraph 2.1.1.

SECTION 3—MOISTURE CONTENT DETERMINATION

3.1 Drying in vacuum. Dry quantity of sample representing approximately 2g dry material to constant weight at 95 to 100 deg C under absolute pressure not greater than 13.3 kPa (100 mm Hg) for about 5 hours. Use covered aluminum dish at least 50 mm in diameter and not greater than 40 mm deep. Report loss in weight as moisture. This procedure is not suitable for high fat products such as pork sausage. For these products, use one of the air drying methods.

3.2 Air drying

3.2.1 With lid removed, dry sample representing approximately 2g dry material for 16 to 18 hours at 100 to 102 deg C in air oven (mechanical convection preferred). Use covered aluminum dish at least 50 mm in diameter and not greater than 40 mm deep. Cover, cool in desiccator and weigh. Report loss in weight as moisture.

3.2.2 With lid removed, dry sample representing approximately 2g dry material to constant weight (2 to 4 hours depending on product) in mechanical convection oven at approximately 125 deg C. Use covered aluminum dish at least 50 mm in diameter and not greater than 40 mm deep. Avoid excessive drying. Cover, cool in desiccator, and weigh. Report loss in weight as moisture.

References: Last printed in 1983 AGRICULTURAL ENGINEERS YEARBOOK; list available from ASAE Headquarters.

ASAE Standard: ASAE S358.1

MOISTURE MEASUREMENT—FORAGES

Developed by the ASAE Physical Properties of Agricultural Products Committee; approved by Electric Power and Processing Division Standards Committee; adopted December 1972; reconfirmed December 1977; revised April 1979; reconfirmed December 1983.

SECTION 1—PURPOSE AND SCOPE

1.1 This Standard establishes uniform methodology for determining the moisture content of forage products in their various forms.

1.2 Specification of sampling procedures is not within the scope of this Standard. It is assumed that the portion of a collected sample used for moisture content determination is representative of the entire sample. A sample dried at the higher temperature will be of limited value for subsequent chemical analysis.

SECTION 2—APPARATUS

2.1 Moisture cans. Moisture cans are required for containing the part of a collected sample that is to be dried. They should be constructed of nonhygroscopic material (preferably aluminum) and have a depth/diameter ratio less than 0.80. If they are fitted with covers, the covers shall also be nonhygroscopic and vaportight.

2.2 Desiccator. The desiccator should be airtight and should contain a sufficient exposure of suitable desiccant to quickly lower the relative humidity of its atmosphere after loading. Activated alumina is a recommended disiccant.

2.3 Drying oven. The drying oven must be properly ventilated (mechanical convection preferred) and must have temperature sensitivity and uniformity specifications better than ±2 °C, in the range of 50 to 150 °C.

2.4 Balance. A balance with a sensitivity of 0.1 g or better is essential for determining moisture contents of materials at a 5 percent moisture level with less than 5 percent error using 25 g samples. If more accurate moisture determinations are necessary, a more sensitive balance will be required. Analytical balances with sensitivities near 0.0001 g in the 50 to 100 g load range are preferred.

SECTION 3-PROCEDURE

3.1 Weigh the moisture can.

3.2 Select a representative sample of at least 25 g. It is not necessary to chop or grind the selected samples except in the case of compacted forage products (e.g., cubes, wafers, et.) which should be reduced in size without moisture loss so that at least one dimension of a particle is less than 15 mm.

3.3 Place the sample in the moisture can. Ensure that no moisture is lost or gained by the sample between the time that it was collected and when it is weighed in a moisture can. This can be done by using moisture cans with covers to collect samples or by rapid transfer techniques from other vaportight containers.

3.4 Weigh the moisture can plus sample.

3.5 Place the moisture can plus sample in the drying oven. Covers must be removed while the sample is being dried in the oven.

3.6 Dry the sample in the oven.

3.6.1 If the sample is for moisture determination only, dry at 103 °C for 24 h.

3.6.2 If the sample will be used for additional analysis, dry at 65 °C for 72 h.

3.7 On removing samples from the oven, container covers must be replaced as quickly as possible, or the containers must be placed in a desiccator as quickly as possible and the sample allowed to cool to the ambient temperature of the balance.

3.8 Weigh the mositure can plus dried samples.

3.9 Record loss in weight as moisture.

3.10 Calculate moisture content as follows:

$$MC(\text{wb percent}) = \frac{(\text{Wt of Wet Sample - Wt of Dry Sample})\text{x}100}{\text{Weight of Wet Sample}}$$

or

$$MC(\text{db percent}) = \frac{(\text{Wt of Wet Sample - Wt of Dry Sample})\text{x}100}{\text{Weight of Dry Sample}}$$

where

MC(wb percent) = Moisture Content Wet Basis, percent
MC(db percent) = Moisture Content Dry Basis, percent

3.11 Indicate whether the moisture content figure reported is wet basis or dry basis.

ASAE Standard: ASAE S368.1

COMPRESSION TEST OF FOOD MATERIALS OF CONVEX SHAPE

Reviewed by the ASAE Physical Properties of Agricultural Products Committee; approved by the Food Engineering Division Standards Committee; adopted by ASAE as a Recommendation December 1973; revised and reclassified as a Standard December 1979; reconfirmed and equation corrected in Paragraph 3.9.1 December 1984; reconfirmed for one year December 1987.

SECTION 1—PURPOSE

1.1 This Standard is intended for use in determining mechanical attributes of food texture, resistance to mechanical injury, and force-deformation behavior of food materials of convex shape, such as fruits and vegetables, seeds and grains, and manufactured food materials.

SECTION 2—SCOPE

2.1 Compression tests of intact biological materials provide an objective method for determining mechanical properties significant in quality evaluation and control, maximum allowable load for minimizing mechanical damage, and minimum energy requirements for size reduction.

2.2 Realizing the shortcomings associated with subjective methods, the use of fully automatic testing machines has become popular in recent years. With the use of testing machines, the need for a recommendation for testing, interpretation of data, and reporting of results has become evident.

2.3 Determination of compressive properties requires the production of a complete force-deformation curve. From the force-deformation curve, stiffness; modulus of elasticity; modulus of deformability; toughness; force and deformation to points of inflection, to bioyield, and to rupture; work to point of inflection, to bioyield, and to rupture, and maximum normal contact stress or a stress index at low levels of deformation can be obtained. Any number of these mechanical properties can, by agreement, be chosen for the purpose of evaluation and control of quality.

SECTION 3—DEFINITIONS

3.1 Bioyield point: A point such as shown in Fig. 1 where an increase in deformation results in a decrease or no change in force.

3.2 Force-deformation curve: A diagram plotted with values of deformation as abscissae and values of force as ordinates.

3.3 Modulus of elasticity: A modulus given by the equations shown in Fig. 2. The equations are based on the Hertz problem of contact stresses in solid mechanics and assume very small and elastic deformations. To separate elastic and plastic deformations, an unloading force-deformation curve is required.

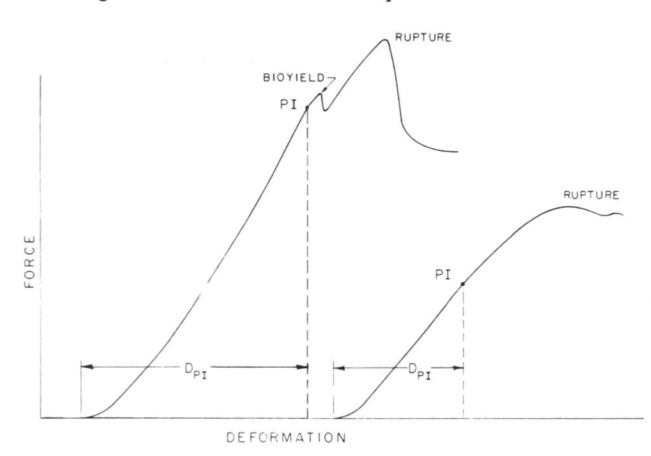

FIG. 1—FORCE-DEFORMATION CURVES FOR MATERIALS WITH AND WITHOUT BIOYIELD POINT. PI = point of inflection, D_{PI} = deformation at point of inflection.

3.4 Modulus of deformability: A modulus defined by the same equations as those shown in Fig. 2 for modulus of elasticity except that D in these expressions is the sum of both elastic and plastic deformations. No unloading force-deformation curve is required in this case.

3.5 Point of inflection: A typical force-deformation curve is first concave up and then concave down (Fig. 1). The point at which the rate of change of slope (second derivative) of the curve becomes zero is called the point of inflection. This point, designated as PI, can be found by using a straight edge to follow the change of slope of the curve and to determine the point at which the slope begins to decrease.

3.6 Radius of the circle of contact: The radius of the contact area between the compression tool and the convex-shape specimen at any given load and deformation. For a rigid spherical compression tool on convex specimens and very small deformations (Fig. 2c), the radius of the circle of contact can be approximated by the following formula:

$$a^2 = \frac{D}{\frac{1}{R} + \frac{2}{d}}$$

where

a = radius of the circle of contact
D = elastic deformation
d = diameter of spherical indenter
R = average radius of the convex body

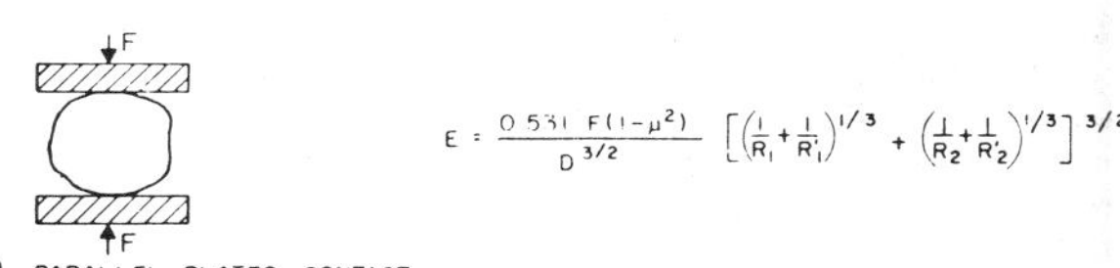

(a) PARALLEL PLATES CONTACT

$$E = \frac{0.531\ F(1-\mu^2)}{D^{3/2}} \left[\frac{1}{R_1}+\frac{1}{R_1'} \right]^{1/2}$$

(b) SINGLE PLATE CONTACT

$$E = \frac{0.531\ F(1-\mu^2)}{D^{3/2}} \left[\frac{1}{R_1}+\frac{1}{R_1'}+\frac{4}{d} \right]^{1/2}$$

(c) SPHERICAL INDENTER ON CURVED SURFACE

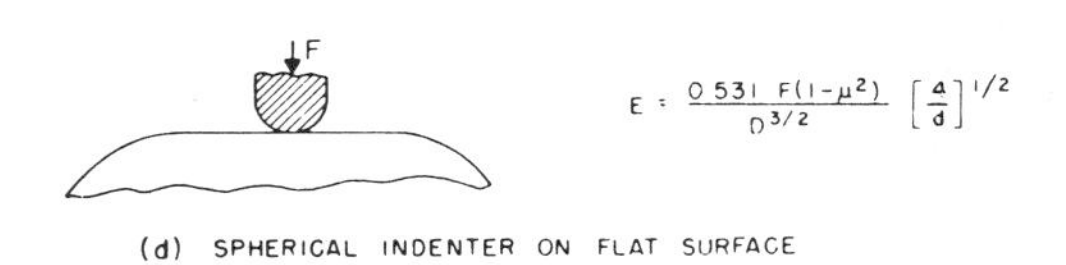

(d) SPHERICAL INDENTER ON FLAT SURFACE

FIG. 2—MODULUS OF ELASTICITY CALCULATED FROM FORCE AND DEFORMATION DATA. E = modulus of elasticity, Pa (psi); F = force in N (lbf); D = elastic deformation at both loading and supporting points of contact, m(in.); μ = Poisson's ratio; R_1, R_1', R_2, R_2' = radii of curvature of the convex body at the points of contact, m(in.); d = diameter of the spherical indenter, m(in.); 0.531 = constant valid for the case where the angle between the normal planes containing the principal curvatures of the convex body is 90 deg and the difference between the curvatures in each plane is small.

For a rigid spherical tool on the flat surface of a specimen (Fig. 2d), the radius of the circle of contact can be approximated by the following formula:

$$a^2 = \frac{dD}{2}$$

3.7 Rupture point: The point on the force-deformation curve at which the loaded specimen shows a visible or invisible failure in the form of breaks or cracks. This point is detected by a continuous decrease of the load in the force-deformation diagram (Fig. 1).

3.8 Stiffness: The ratio of force to deformation at the point of inflection.

3.9 Stress index: The maximum stress which the material is capable of sustaining before a given deformation can occur. This deformation shall be one-half the deformation corresponding to the point of inflection.

3.9.1 For a convex body loaded with a steel flat plate (Figs. 2a, 2b), stress index is computed from:

$$S_i = 0.365 \left[\frac{F}{A^2} (1/R_1 + 1/R_1')^2 \right]^{1/3}$$

For a convex body loaded with a smooth, steel spherical indenter (Fig. 2c), stress index is computed from:

$$S_i = 0.365 \left[\frac{F}{A^2} (1/R_1 + 1/R_1' + 4/d_2)^2 \right]^{1/3}$$

where

S_i = stress index; Pa (psi)
F = force, N (lbf)
R_1, R_1' = radii of curvature of convex body at the point of contact, m (in.)
d_2 = diameter of spherical indenter, m (in.)
A = an elastic constant, 1/Pa (1/psi) given by

$$A = \frac{(1-\mu^2)}{E}$$

μ = Poisson's ratio
E = modulus of elasticity of material, Pa (psi)*

---

*For the purpose of this Standard, modulus of deformability shall be used for E.

SECTION 4—APPARATUS

4.1 Testing machine. Any suitable testing machine capable of constant-rate-of-crosshead movement and comprising essentially the following:

4.1.1 A drive mechanism for imparting to the crosshead a constant velocity with respect to the base. The crosshead velocity shall also be repeatable with an accuracy of ± one percent.

4.1.2 A load indicating mechanism capable of showing the total compressive load carried by the test specimen. The mechanism shall be essentially free from inertia-lag at the specified site of testing and shall indicate the load with an accuracy of ± one percent of the maximum anticipated value of the load on the specimen. The accuracy of the testing machine shall be verified at least once a year in accordance with American Society for Testing and Materials Standard E-4, Load Verification of Testing Machines.

4.1.3 A suitable instrument capable of determining the change in distance between the point of loading and a fixed member of the base of the testing machine at any time during the test. It is desirable that this instrument automatically record the change of distance as a function of the load on the test specimen. The instrument shall be essentially free of inertia-lag at the specified rate of loading and shall conform to the requirements for a class B-2 extensometer as defined in American Society for Testing and Materials Standard E-83, Verification and Classification of Extensometers. Realizing that in the case of test specimens of food materials of convex shape the use of extensometers on the specimen is difficult and in most cases impractical, the second alternative shall be the use of an extensometer on the movable crosshead of the testing machine, sensing the true displacement of the crosshead. The third alternative shall be the use of crosshead velocity as an indirect method of measuring deformation.

4.2 Compression tool. A compression tool with smooth and polished surface is used for applying the load to the specimen. The type of the compression tool used for each test shall depend on the size of the specimen, the nature of the test, and the extent of information expected from that test. Examples of suitable compression tools are shown in Fig. 3. Flat plate and spherical tools may be used for both soft and hard materials. A plunger with end machined to a known radius without rounding the edge (Fig. 3b) is suitable for soft materials such as fruits and vegetables, particularly those which exhibit a bioyield point.

4.2.1 The rounded tip of the plunger in Fig. 3b is the segment of a sphere with a diameter d calculated from equations in paragraph 3.6. The radius of the circle of contact, a, in this case is one-half of the diameter of the cylindrical portion of the compression tool when full contact is established. Knowing the magnitude of the desired deformation, the appropriate radius of the spherical tool can be computed from the equations. To obtain a distinct bioyield point on the recording chart in case of fruits and vegetables, it is desirable to bring the full diameter of the cylinder in contact with the specimen with as little deformation as possible. For example, if the skin of the fruit is removed and the tool in Fig. 3b is used, the full diameters of the cylindrical portion of the plunger will come in contact with the specimen at about 0.64 mm (0.025 in.) deformation. Such value of deformation for fruits such as apples is below the usual range of deformation reported for bioyield point.

4.2.2 If the modulus of deformability or stress index is to be computed, the force and deformation to be used in the equation shall be those corresponding to the values at one-half the deformation at the point of inflection on the force-deformation curve (Fig. 1). Also, the ratio of the radius of the circle of contact to the radius of the spherical indenter shall not be greater than 1:10.

4.2.3 The 8 mm (5/16 in.) diameter cylindrical plunger is recommended for use on fruits because this size corresponds to one of the two plungers of the Magness-Taylor fruit pressure tester commonly used for quality evaluation of such fruits as apples, pears, and peaches. If one of these plungers is used as the loading tool in conducting a compression test on fruits, it is recommended to continue the compression beyond and through the point of

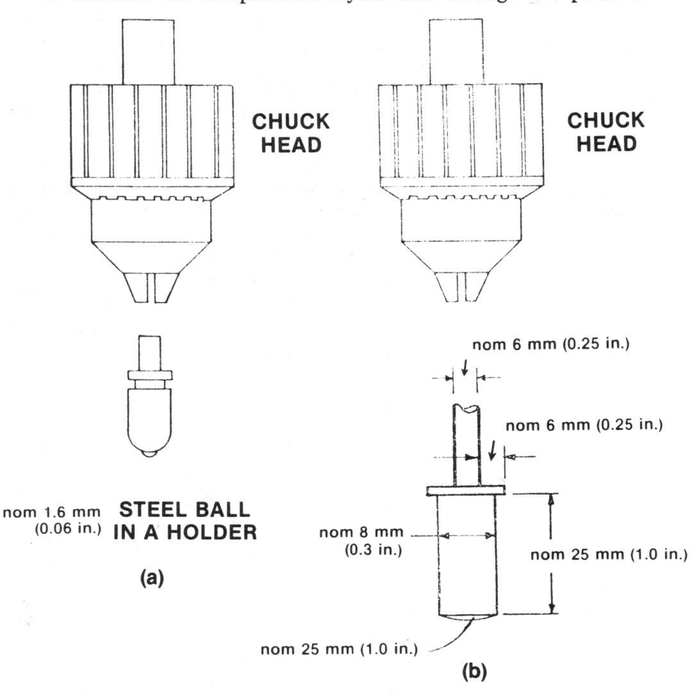

FIG. 3—EXAMPLES OF COMPRESSION TOOLS WITH SPHERICAL ENDS

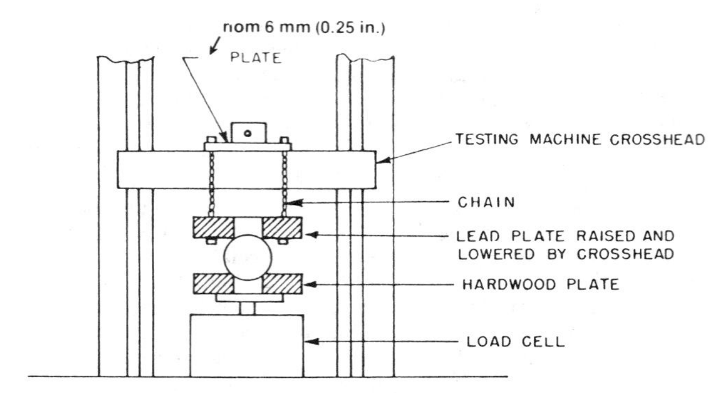

FIG. 4—FIXTURE FOR PRE-SETTING OF THE SPECIMEN

rupture. In this manner the force reading at the point of rupture can serve also as the Magness-Taylor pressure reading of the specimen.

4.3 Supporting jig. For small hard specimens such as seeds and grains, a hardened metal plate with smooth surface finish can be used as supporting jig. The test specimens shall be mounted on the plate by using a thin layer of quick hardening cement or glue. The flatter side of the grain is first lightly sanded before being cemented. If the grain specimens have been equilibriated for moisture contents higher than normal storage conditions, it is necessary to use a special glue which will harden at high-humidity atmosphere.

4.3.1 For larger and softer specimens such as fruits and vegetables, it is essential to provide a support to ensure that the deformation of the whole specimen at the support is negligible. This can be accomplished by providing some pre-setting arrangement such as that shown in Fig. 4. It consists of a lower wooden plate and an upper heavy plate, each provided with a rounded circular hole. The specimen shall be placed between the two plates and pre-set by the weight of the upper plate on the top of the specimen. The weight of the upper plate required for pre-setting shall be determined by the maximum load which the specimen can carry without leaving an objectionable mechanical damage where the rounded circular hole of the heavy plate contacts the specimen.

SECTION 5—TEST SPECIMENS

5.1 Specimens shall be tested in their natural form and size after being placed in their appropriate supporting jigs. For larger convex-bodies such as fruits and vegetables, testing the specimen in its natural form permits taking several readings on the same specimen non-destructively. This would be in contrast with the case where the specimen is cut in the form of a segment of a sphere. If skin is sliced off at the point of compression, the details of test specimen with or without skin shall be given.

5.2 If only relative measurements of force and deformation are to be made, it is desirable to make measurements on shape, size, weight, color and even such constituents as sugar and starch before testing the specimen. For more exact tests where equations given in Fig. 2 are expected to be used, in addition to the above data, the radii of curvature of the convex-shape specimen shall also be determined by either using a radius of curvature meter such as shown in Fig. 5 or by calculation as in Fig. 6. If the loading tool is rounded at the end, the radius of curvature of the tip shall also be specified.

SECTION 6—CONDITIONING

6.1 The test specimen shall be conditioned for the desired temperature and relative humidity before testing. Fruits and vegetables normally require a few hours to adjust their temperatures from field conditions to room temperature. Hard seeds and grains may have to be conditioned in an atmosphere of given relative humidity for several days if they are to be brought up to equilibrium moisture contents other than the ordinary storage conditions. In that case the conditioning chamber shall be brought into the testing room before testing so that moisture gains or losses of the specimen can be minimized.

SECTION 7—NUMBER OF TEST SPECIMENS

7.1 Because of the large variance inherent in biological materials, each experiment shall be statistically designed with sufficient number of replications to result in an acceptable level of confidence insofar as significant differences are concerned. The variation due to shape, size, age and cellular structure are normally such that at least a minimum of twenty specimens is required to be tested for each sample.

SECTION 8—SPEED OF TESTING

8.1 Speed of testing shall be relative rate of motion of the compression tool during the test. Rate of motion of the driven compression tool when the machine is running idle may be used if it can be shown that the resulting speed of testing is within the limits of variations allowed. The recorder response shall be such that the rate of rise of force on the chart shall be at most 1/3 of the maximum theoretical rate of rise of the recorder pen.

8.2 The speed of testing shall be chosen on the basis of the sensitiveness of the specimens to loading rate. For most hard fruits and vegetables the speed of 25 mm per minute ± 20 percent shall be specified as the standard speed. For seeds and grain, the speed of 1.25 mm per minute ± 50 percent shall be specified.

SECTION 9—TESTING PROCEDURE

9.1 Conduct the test under laboratory atmosphere of constant relative humidity and temperature. If possible, tests should be conducted under laboratory conditions of 20 °C ± 5 °C and 50 percent relative humidity ± 5 percent.

9.2 Measure the major, minor and intermediate diameters of the specimen to the nearest 10 percent of the dimensions.

9.3 Weigh the specimen and record the weight together with some observation of color and appearance.

9.4 If modulus of deformability is to be calculated, measure radius of curvature at the loading point to the nearest 10 percent.

9.5 Record information on variety, age, maturity and the history of the material prior to testing.

9.6 Select the compression tool according to the requirements given in paragraph 4.3, and install the tool in the testing machine.

9.7 If the compression tool is one with a spherical end, record the radius of the rounded end. Set the speed control at the desired rate and calibrate the recording chart for load and displacement.

9.8 Place the specimen in the testing machine under the compression tool, taking care to align the center of the tool with the peak of the curvature of the test specimen.

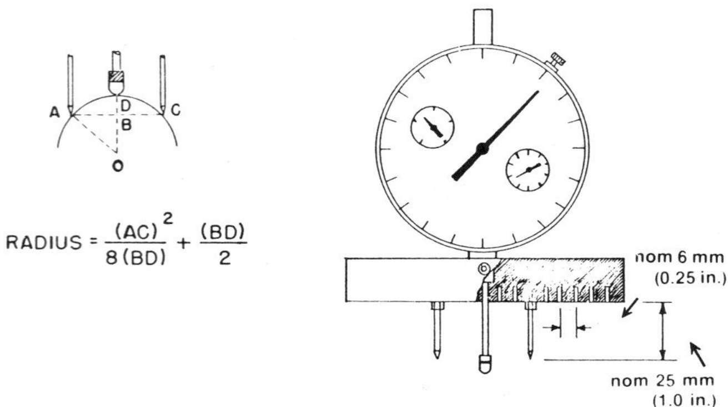

FIG. 5—RADIUS OF CURVATURE METER

H

R_1

$$R_1 \simeq H/2$$

$$R_1' \simeq \frac{H^2 + L^2/4}{2H}$$

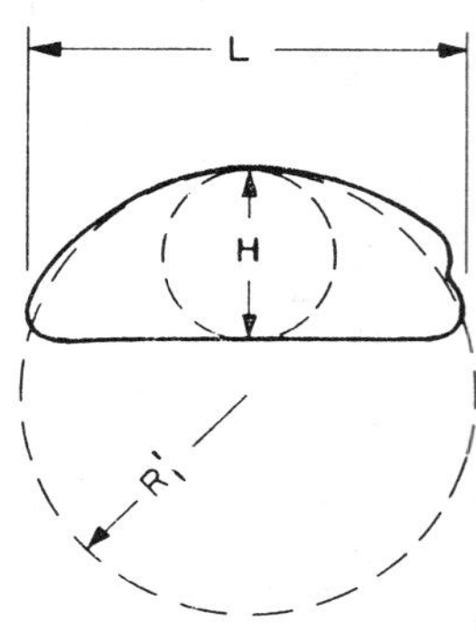

FIG. 6—APPROXIMATION OF R_1 AND R_1' FOR CONVEX BODIES

9.9 Place the specimen over the support and lower the pre-setting weight as shown in Fig. 4. For specimens of seeds and grains which have been softened in an atmosphere of high relative humidity, a similar pre-setting arrangement can be devised to insure that the deformation of the whole specimen in comparison with the deformation at the point is negligible.

9.10 Start the machine and record the complete force-deformation curve through the point of rupture.

9.10.1 When testing small, hard specimens such as grains, the deflections of most load cells cannot be considered negligible. For this reason either proof shall be given that the load cell deflection is negligible or the deflection is determined and deducted from the recorded deformation.

SECTION 10—CALCULATIONS

10.1 Force and deformation to bioyield and to rupture. These values are read directly from the chart and recorded along with the type of compression tool used.

10.2 Point of inflection. Determine the point of inflection by using a straight edge on the force-deformation curve to locate the point where the slope of the curve begins to decrease.

10.3 Modulus of deformability. Calculate the modulus of deformability by using the appropriate equation in Fig. 2 given for modulus of elasticity, except that for the values of F and D use the force and deformation corresponding to the point of the force-deformation curve at one-half of the total deformation up to the point of inflection, as shown in Fig. 1. Express the results in Pa (psi).

10.4 Stress index. Calculate stress index by using equations in paragraph 3.9.1, depending on the type of compression tool used. For the force value F use the force at one-half of total deformation for the point of inflection. Calculate the elastic constant using the modulus of deformability in place of E. Express the results in Pa (psi).

10.5 For each series of tests, calculate the mean and standard deviation.

SECTION 11—REPORT

11.1 The report shall include include the following:

11.1.1 Complete identification of the material tested including source, previous history, time of the day samples were harvested, weight, color and appearance, shape and size (as defined by major, minor, intermediate diameters), and if calculations of mechanical properties are reported, the radii of curvature at the point of testing.

11.1.2 Complete identification of the compression tool used including the diameter of the cylindrical part of the tool and the radius of the spherical end.

11.1.3 Method of preparing and pre-setting the specimen for test.

11.1.4 Conditioning procedure used.

11.1.5 Atmospheric conditions in testing room.

11.1.6 Number of specimens used.

11.1.7 Speed of testing.

11.1.8 Date of test.

11.1.9 Type of testing machine used.

11.1.10 Mean and standard deviation for force and deformation to bioyield and to rupture.

11.1.11 Mean and standard deviation for modulus of deformability and stress index, if of interest.

References: Last printed in 1977 AGRICULTURAL ENGINEERS YEARBOOK; list available from ASAE Headquarters.

Cited Standards:

ASTM E4, Load Verification of Testing Machines
ASTM E83, Verification and Classification of Extensometers

ASAE Standard: ASAE S410.1

MOISTURE MEASUREMENT—PEANUTS

Developed by the ASAE Special Crops Processing Committee; approved by the Electric Power and Processing Division Standards Committee; adopted by ASAE as a Tentative Standard December 1981; revised and reclassified as a full Standard December 1982; reconfirmed December 1987.

SECTION 1—PURPOSE AND SCOPE

1.1 This Standard provides uniform methods for determining the moisture content of unground samples of peanuts. The primary method is for determining the moisture content of hulls and kernels separately and then calculating the pod moisture content from the component moisture contents. An alternate method for determining pod moisture content directly is also specified. These techniques should become common practice and their use referenced in all technical presentations where moisture determinations have been a factor. Deviations from these methods should be reported.

SECTION 2—APPARATUS

2.1 Sample containers. Sample containers shall be metal pans with sufficient size to allow the sample to be dried in a layer no more than three kernels deep. Before using, dry the sample containers for one hour at the drying temperature and obtain the tare weight.

2.2 Oven. The drying oven shall be the forced-draft type and have temperature sensitivity and uniformity specifications throughout the drying zone better than ± 3 °C, in the range of 100 to 130 °C.

2.3 Balance. An external balance shall have a sensitivity of 0.1 g or better.

2.4 Sample divider. A suitable divider shall be used for obtaining a random sample of the appropriate size. A riffle divider having eight 4.4 cm slots is recommended.

SECTION 3—PREPARATION OF SAMPLE

3.1 Remove all foreign matter from the sample by spreading the sample on a clean, dry surface and hand picking the foreign material.

3.2 Reduce the sample through the divider to a size sufficient for the number of tests that will be required.

SECTION 4—PROCEDURE

4.1 Whole pods

4.1.1 Shell a known mass of at least 200 g.* Keep shells and kernels separate and save all portions.

4.1.2 Weigh each portion separately and calculate the percentage of each as directed in Section 5—Calculations.

4.1.3 Oven dry kernels and hulls in separate sample containers using one of the following time and temperature schemes:

4.1.3.1 130 ± 3 °C for 6 h

4.1.3.2 100 ± 3 °C for 72 h

4.1.4 Remove from oven, weigh immediately, and calculate moisture as directed in Section 5—Calculations.

4.2 Shelled stock—kernels

4.2.1 Dry and weigh the kernels as directed in paragraphs 4.1.3 and 4.1.4 above.

SECTION 5—CALCULATIONS

5.1 Percent shells (A)

$$A = \frac{100\ (\text{Initial Mass of Shells})}{(\text{Initial Mass of Pods})}$$

5.2 Percent kernels (B)

$$B = \frac{100\ (\text{Initial Mass of Kernels})}{(\text{Initial Mass of Pods})}$$

5.3 Moisture content of shells, percent wet basis (C)

$$C = \frac{100\ (\text{Loss in Mass of Shells})}{(\text{Initial Mass of Shells})}$$

5.4 Moisture content of kernels, percent wet basis (D)

$$D = \frac{100\ (\text{Loss in Mass of Kernels})}{(\text{Initial Mass of Kernels})}$$

5.5 Moisture content of whole pods, percent wet basis (E)

$$E = \frac{D(B) + C(A)}{100}$$

SECTION 6—ALTERNATE WHOLE POD METHOD

6.1 Purpose of alternate method. This alternate method may be used in situations where moisture contents of individual components are not required and shelling of peanuts too time consuming.

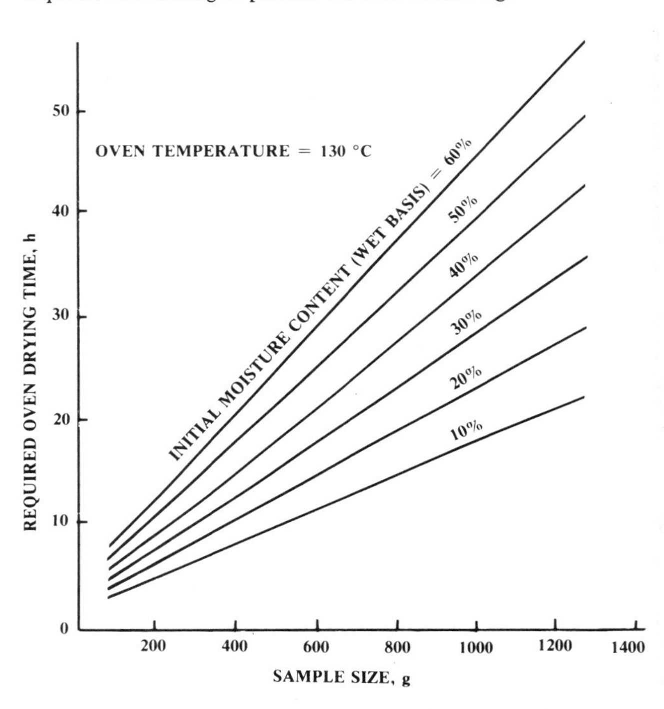

FIG. 1—OVEN DRYING TIMES REQUIRED FOR WHOLE PEANUT POD MOISTURE DETERMINATION WITH VARIOUS SAMPLE SIZES AND ESTIMATED INITIAL MOISTURE CONTENTS

*A sample size of 200 g is estimated to yield precisions (error/true mean) at the 95% confidence level of ± 5%, ± 8%, and ± 6% for kernels, hulls and pods, respectively, in the 20% moisture range. Greater precision may be obtained by using multiple samples and averaging results.

6.2 Procedure. Whole pods are oven dried at 130 ± 3 °C, for a time to be determined from Fig. 1. The time of drying depends both on sample size and initial moisture content. A reasonably good estimate of initial moisture content is necessary to determine the oven drying time to be used.

6.3 Calculations. Whole pod moisture content, percent wet basis (E):

$$E = \frac{100 \text{ (Loss in Mass of Pods)}}{\text{(Initial Mass of Pods)}}$$

References: Last printed in 1982 AGRICULTURAL ENGINEERS YEARBOOK; list available from ASAE Headquarters.

ASAE Standard: ASAE S424

METHOD OF DETERMINING AND EXPRESSING PARTICLE SIZE OF CHOPPED FORAGE MATERIALS BY SCREENING

Developed by the ASAE Forage Harvesting and Utilization Committee; approved by the ASAE Power and Machinery Division Standards Committee; adopted by ASAE April 1986; revised editorially March 1988.

SECTION 1—PURPOSE AND SCOPE

E **1.1** The purpose of this Standard is to define a test procedure to determine the particle size distribution of chopped forage materials and to define a method of expressing the particle length of the material. The determined particle size distribution can be used to evaluate forage harvesting machine and handling equipment variables and to define forage physical length in animal feeding trials.

1.2 This Standard shall be used to determine the particle size of chopped forage materials where the reduction process yields particles such as that material produced by shear-bar type forage harvesters. It is not intended for use on material produced by flail-type harvesters where substantial fractions of the material may be extremely long.

1.3 This Standard is intended for use in the field as well as in the laboratory. It is intended to separate chopped forage samples without drying them first.

SECTION 2—TEST EQUIPMENT

E **2.1** For particle measuring purposes, a set of square-hole screens having widths of 406 mm (16.0 in.), lengths of 565 mm (22.25 in.) and specifications shown in Table 1 shall be used. The screens shall be supported in frames with depths of 63.5 mm (2.50 in.) and arranged horizontally in a stack such that the screen with the largest opening size is at the top. Those with smaller openings shall be arranged with progressively smaller hole sizes below each other. If screens with different size openings from those listed in Table 1 are used, the actual dimensions for the openings shall be used in the data analysis and shall be reported.

E **2.2** A suitable screen shaker is required. The shaker shall oscillate the screen stack in a horizontal plane. The center of one end of the screen stack shall oscillate in a straight horizontal line on a slider block. The opposite end of the screen stack shall be supported on horizontal crank arms, the crank end centers of which are located 765.2 mm (30.12 in.) from the center of the slider block pivot located on the other end of the screen stack. The centers of the arms shall travel in a horizontal circle with a diameter of 117 mm (4.62 in.) (see Figs. 1 & 2).

NOTE: Information on plans for constructing such a screen shaker may be obtained from the American Society of Agricultural Engineers.

E **2.3** The screen shaker shall drive the screen stack at a frequency of 2.4 ± 0.08 H_z (144 ± 5 cycles/min).

2.4 The shaker should be operated with the screens level.

2.5 A weighing balance having an accuracy of at least ± 0.5 g shall be used for weighing the fractions.

SECTION 3—METHOD OF SCREENING

E **3.1** Uncompressed samples of 9 to 10 L of forage should be used. Samples of 2 to 3 L of material may be used if extra care is taken to recover the material from each screen. For field work, the larger samples are usually preferred. The sample volume size should be reported with the data.

TABLE 1—DIMENSIONS OF SQUARE-HOLE SCREENS FOR TESTING PURPOSES

Screen No.	Nominal size opening mm	(in.)	Square hole diagonal mm	(in.)	Screen thickness mm	(in.)	Open area %
1	19.0	0.75	26.9	1.06	12.7	0.50	45
2	12.7	0.50	18.0	0.71	9.6	0.38	33
3	6.3	0.25	8.98	0.35	4.8	0.19	33
4	3.96	0.156	5.61	0.22	3.1	0.12	39
5*	1.17	0.046	1.65	0.065	0.64	0.025	41.5
Pan	—	—	—	—	—	—	—

E *14 mesh woven wire cloth with 0.64 mm (0.025 in.) diameter wires. All others are aluminum sheets or plates.

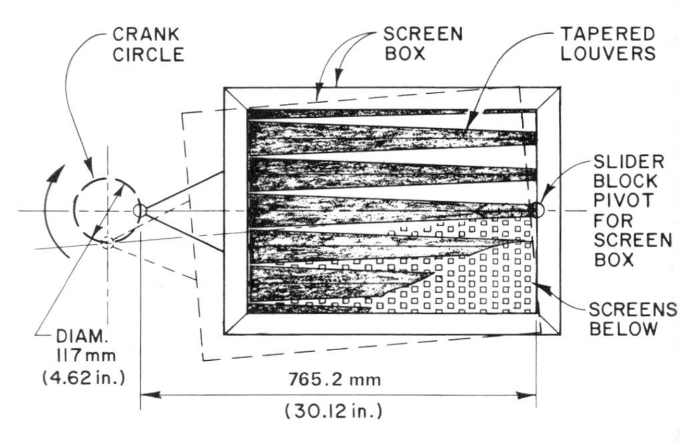

FIG. 1—TOP VIEW OF FORAGE PARTICLE SEPARATOR SHOWING SCREEN MOTION AND FEEDER POSITION E

3.2 Place the sample on the top of the sample feeder (near the closed end) above the top screen of the screen set and operate the shaker for 120 s.

3.3 The tapered louvers on the feeder should be preset to feed the sample to the top screen in 20 to 30 s. The louvers may be inclined to change the feeding time. This adjustment is made using successive trials with practice samples similar to the crop material to be evaluated for particle size distribution. E

3.4 Material on each screen and bottom pan shall be weighed and recorded.

3.5 If the amount retained on the top screen exceeds 1% of the total sample mass, representative subsamples should be obtained from this screen and measured manually. The average length may then be used in the data analysis as geometric mean length $\overline{X}_1$. E

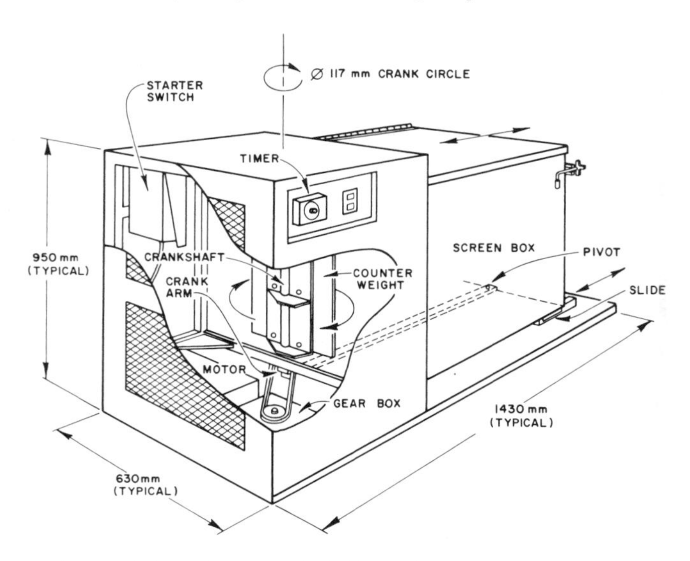

FIG. 2—SCHEMATIC DIAGRAM OF FORAGE PARTICLE SEPARATOR E

3.6 The screening process should be repeated to produce 3 sets of data. These data sets may be averaged or analyzed separately with the procedure specified in the report.

3.7 A representative sample of the unscreened material shall be used for moisture content determination. Moisture content (wet basis) shall be reported along with particle size data.

E 3.8 Screen openings must be kept free of forage particles so that effective screening can be accomplished. A stiff bristle cleaning brush, or compressed air, is useful for cleaning screens which have become clogged with forage particles. Screens may need to be cleaned periodically to remove plant residue materials. Plant residues may be removed by washing with water containing a detergent. Screens and the particle separator must be air dried before use.

E 3.9 If static electricity becomes a problem when separating dry forage materials, liquid laundry static control may lightly be sprayed onto the sample before separating.

SECTION 4—DATA ANALYSIS

4.1 Analysis of mass distribution of all chopped forage materials is based on the assumption that these distributions are logarithmic normally distributed.

4.2 Calculation of particle size

4.2.1 The size of particles shall be reported in terms of geometric mean length (X_{gm}) and standard deviation (S_{gm}) by mass.

4.2.2 Calculated values are obtained as follows:

$$X_{gm} = \log^{-1} \frac{\Sigma(M_i \log \overline{X}_i)}{\Sigma M_i} \quad \text{[1]}$$

$$S_{gm} = \log^{-1} \left[\frac{\Sigma M_i(\log \overline{X}_i - \log X_{gm})^2}{\Sigma M_i}\right]^{1/2} \quad \text{[2]}$$

where

X_i = diagonal of screen openings of the ith screen

$X_{(i-1)}$ = diagonal of screen openings in next larger than the ith screen (just above in a set)

X_{gm} = geometric mean length

$\overline{X}_i$ = geometric mean length of particles on i^{th} screen = $[X_i \times X_{(i-1)}]^{1/2}$ ..... [3]

M_i = mass on i^{th} screen (actual mass or percent of total, decimal or percent form)

S_{gm} = standard deviation

E NOTE: $\overline{X}_1$ is measured manually as described in paragraph 3.5. If it is less than 1% of the total, it is treated as zero.

4.2.3 Material passing through screen No. 5 and collected in the pan shall be considered to have a geometric mean length of 0.81 mm (0.032 in.). This becomes $\overline{X}_6$ in equations [1] and [2].

4.2.4 An example of how the equations may be used to find geometric mean particle length and standard deviation for a sample data set follows:

PERCENT MASS DISTRIBUTION OF A CHOPPED ALFALFA SAMPLE

Screen No.	Screen diagonal, mm	Percent total mass on screens, %	Cumulative undersize %
1	26.9	3.8	96.2
2	18.0	8.1	88.1
3	8.98	25.1	63.0
4	5.61	26.9	36.1
5	1.65	34.2	1.9
Pan	—	1.9	—
		100.0	

The average measured length of the particles on the top screen (No. 1) was 48 mm. This becomes $\overline{X}_1$ in equations [1] and [2]. Equation [3] and the above information are used to obtain the following mean lengths for particles in each fraction

$\overline{X}_1 = 48$

$\overline{X}_2 = (18.0 \times 26.9)^{1/2} = 22.0$

$\overline{X}_3 = (8.98 \times 18.0)^{1/2} = 12.7$

$\overline{X}_4 = (5.61 \times 8.98)^{1/2} = 7.10$

$\overline{X}_5 = (1.65 \times 5.61)^{1/2} = 3.04$

$\overline{X}_6 = 0.82$

Equations [1] and [2] are used to obtain

$$X_{gm} = \log^{-1}\left[\frac{0.038 \log(48) + 0.081 \log(22) + 0.251 \log(12.7) + 0.269 \log(7.1) + 0.342 \log(3.04) + 0.019 \log(0.82)}{0.038 + 0.081 + 0.251 + 0.269 + 0.342 + 0.019}\right] = 6.95 \text{ mm}$$

$$S_{gm} = \log^{-1}\left\{\frac{0.038\left[\log\left(\frac{48}{6.95}\right)\right]^2 + 0.081\left[\log\left(\frac{22}{6.95}\right)\right]^2 + 0.251\left[\log\left(\frac{12.7}{6.95}\right)\right]^2 + 0.269\left[\log\left(\frac{7.1}{6.95}\right)\right]^2 + 0.342\left[\log\left(\frac{3.04}{6.95}\right)\right]^2 + 0.019\left[\log\left(\frac{0.82}{6.95}\right)\right]^2}{0.038 + 0.081 + 0.251 + 0.269 + 0.342 + 0.019}\right\}^{1/2} = 2.26$$

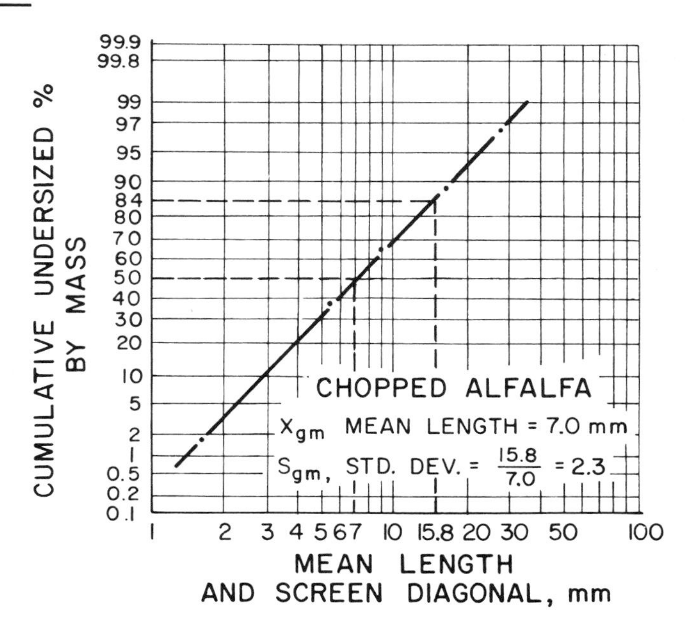

FIG. 3—CUMULATIVE PERCENT UNDERSIZED PARTICLES VERSUS SCREEN DIAGONAL OPENING SIZE FOR ALFALFA FOR GRAPHIC DETERMINATION OF MEAN LENGTH AND STANDARD DEVIATION E

E

4.2.5 Graphical solutions for geometric mean length and standard deviation may be obtained by plotting the results on logarithmic normal probability graph paper. Fig. 3 shows an example of a plot of the data from paragraph 4.2.4.

where

$X_{gm} = X_{50} =$ particle length at 50% probability

$S_{gm} = \frac{X_{84}}{X_{50}} =$ standard deviation

$X_{84} =$ particle length at 84% cumulative probability

References:

1. Finner, M. F., J. E. Hardzinski and L. L. Pagel. 1978. Evaluating particle length of chopped forages. ASAE Paper No. 78-1047. American Society of Agricultural Engineers, St. Joseph, Michigan.

2. Stockham, J. D. and E. G. Fochtman. 1977. Particle size analysis. Ann Arbor Science Pub., Inc.

E

ASAE Standard: ASAE S487

MOISTURE MEASUREMENT—TOBACCO

Developed by the ASAE Special Crops Processing Committee; approved by Food and Process Engineering Institute Standards Committee; adopted by ASAE December 1987.

SECTION 1—PURPOSE AND SCOPE

1.1 This Standard establishes a uniform method for determining the moisture content of tobacco lamina, midrib or stalk materials at various stages of growth, harvest or curing. These techniques should be used and referenced in all technical presentations where moisture content determination is a factor. Deviation from these methods should be reported.

1.2 Specifying sampling procedures is not within the scope of this Standard. It is assumed that representative samples are collected for analysis.

SECTION 2—APPARATUS

2.1 Sample containers. Sample containers shall be made of screen wire with a cross section of 5 x 20 cm and a height of about 15 cm. The bottoms of the containers shall be made of heavy-gauge sheet metal firmly attached to the screen wire to provide a rigid base. A suitable sheet metal or screen wire cover shall be used to prevent loss of light pieces of lamina during drying.

2.2 Oven. The drying oven shall be of the forced draft type, properly ventilated and having temperature and uniformity specifications equal or better than ± 2 °C in the range of 100 to 110 °C.

2.3 Balance. A balance with a sensitivity of 0.01 g or better shall be used. Calibration or zeroing of the balance should be repeated frequently during use to assure accuracy.

SECTION 3—PREPARATION OF SAMPLE

3.1 Separate tobacco into lamina, midrib and stalk portions, being careful to remove any foreign material.

3.2 Chop or slice lamina, midrib and stalk portions to produce small pieces having the largest dimension no greater than 5 cm and the smallest, other than leaf or midrib thickness, no less than 1 cm. Fines shall not exceed 10% of sample weight.

SECTION 4—PROCEDURE

4.1 Preheat oven to 101 ± 2 °C.

4.2 Weigh the sample container (tare weight).

4.3 Select a sample of about 50 to 100 g of chopped or sliced material (lamina, midrib or stalk).

4.4 Place sample material into wire container without forcible packing. Lamina, stem and stalk portions must be dried in separate containers.

4.5 Weigh container plus sample (weight of wet sample plus tare weight).

4.6 Place the filled container into the drying oven. No more than three containers shall be placed in a line along the path of air movement.

4.7 Dry the sample in the oven at 101 ± 2 °C for 24 ± 2 h or until the change in mass between successive weighings at intervals of at least 1 h is less than 0.1% of the final specimen mass.

4.8 Remove sample from oven and immediately weigh container plus dried sample (weight of dry sample plus tare weight). If several samples are to be weighed in succession, precautions shall be taken to insulate the scale pan from the hot sample containers. In-oven weighing procedures may be substituted if equipment is available.

4.9 Calculate moisture content as follows:

$$\text{MC (wb percent)} = \frac{\text{weight of wet sample - weight of dry sample}}{\text{weight of wet sample}} \times 100$$

or

$$\text{MC (db percent)} = \frac{\text{weight of wet sample - weight of dry sample}}{\text{weight of dry sample}} \times 100$$

where

MC (wb percent) = moisture content wet basis, percent
MC (db percent) = moisture content dry basis, percent

Reference:

1. Bunn, J. M., W. H. Henson, Jr. and L. R. Walton. 1980. Moisture measurement in biologically active material-green burley tobacco. TRANSACTIONS of the ASAE 23(4):1012-1015.

STRUCTURES, LIVESTOCK, AND ENVIRONMENT

ASAE Notation:

The letter S preceding numerical designation indicates ASAE Standard; EP indicates Engineering Practice; D indicates Data. A decimal and numeral following the file number indicate the number of times a document has been revised. Thus ASAE S201.4 indicates Standard number 201, four times revised. The letter T after the designation indicates tentative status. Always refer to ASAE documents by complete designation to avoid confusion with standards of other organizations. For example: ASAE S201.4.

The symbol **T** preceding or in the margin adjacent to section headings, paragraph numbers, figure captions, or table headings indicates a technical change was incorporated in that area when this document was last revised. The symbol **T** preceding the title of a document indicates essentially the entire document has been revised. The symbol **E** used similarly indicates editorial changes or corrections have been made with no intended change in the technical meaning of the document.

ASAE Engineering Practice: ASAE EP250.2

SPECIFICATIONS FOR FARM FENCE CONSTRUCTION

Developed by the ASAE Farm Construction Standards Committee; adopted by ASAE as a Recommendation December 1960; revised June 1966, December 1968; reconfirmed December 1972; reclassified as an Engineering Practice and reconfirmed December 1977; reconfirmed December 1982, December 1983, December 1984.

SECTION 1—PURPOSE AND SCOPE

1.1 This Engineering Practice is intended primarily to guide those responsible for, or concerned with, the specification of non-electric farm fencing.

1.2 The specifications for materials are not intended to preclude new and improved materials, alloys, etc., of proven equal or better performance.

1.3 Common assemblies are detailed to show current practice under normal conditions.

SECTION 2—MATERIALS

2.1 Woven wire fabric shall conform to the requirements of the current American Society for Testing and Materials. ASTM A116, Specifications for Zinc-Coated (Galvanized) Iron or Steel Farm-Field and Railroad Right-of-Way Wire Fencing. The wire shall be coated with Class-1 zinc coating, Class-3 zinc coating for heavy-duty installations, or aluminum coating of not less than 0.25 oz per sq ft (0.0076 g per sq cm) and as per Table IV of ASTM A474, Aluminum-Coated Steel Wire Standard.

2.2 Barbed wire shall be composed of two main strands of number 12½-gage wire with 14-gage round barbs. If four-point barbed wire is specified, barbs shall be spaced on approximately 5-in. (12.7-cm) centers. If two-point barbed wire is specified, barbs shall be spaced on approximately 4-in. (10.2 cm) centers. Barbed wire shall conform to the requirements of the current ASTM A121, Specifications for Zinc-Coated (Galvanized) Steel Barbed Wire. The wire shall be coated with Class-1 zinc coating, or Class-3 zinc coating for heavy-duty installations, or aluminum coating of not less than 0.25 oz per sq ft (0.0076 g per sq cm) as per Table IV of ASTM A474.

2.3 Strand (cable) shall be 0.38 in. (0.95 cm) 1 x 7 steel wire strand (ASTM A475* and A474) common grade, minimum breaking strength 4250 lb (1930 kg). Coating shall be Type-1 zinc, or aluminum of not less than 0.32 oz per sq ft (0.0098 g per sq cm).

2.4 Smooth wire for braces shall be galvanized 0.40 oz per sq ft (0.012 g per sq cm) or aluminum-coated 0.34 oz per sq ft (0.01 g per sq cm) No. 9 gage steel wire, minimum tensile strength 45,000 psi (3164 kg per sq cm).

2.5 Wire ties, clamps and staples; nails, bolts, and other fence hardware

2.5.1 Wire ties, clamps and staples shall be coated equivalent to fence or barbed wire specified. Staples shall be 9-gage, and 1 in. (2.5 cm) long for use in dense hardwoods and 1.5 in. (3.8 cm) long for use in preservative-treated softwoods.

2.5.2 Nails, bolts, and other fence hardware shall be hot-dipped galvanized as per ASTM A153, Specifications for Zinc Coating (Hot Dip) on Iron and Steel Hardware.

2.6 Steel corner and end posts, line posts, and braces shall be hot-dipped galvanized with not less than 2 oz per sq ft (0.06 g per sq cm) of zinc coating in accordance with ASTM A123 Specifications for Zinc (Hot Galvanized) Coatings on Products Fabricated from Rolled, Pressed, and Forged Steel Shapes, Plates, Bars, and Strip. The plans may, as an alternate, call for painted steel posts and/or braces. Painted posts shall be cleaned of all loose scale before finishing, and one or more coats of high-grade, weather-resistant, special steel paint or enamel shall be applied and baked. Weight specifications for steel posts and braces are ±5 percent for galvanized steel and ±3 percent for painted steel.

2.6.1 Steel corner and ends posts shall be 2.5 x 2.5 x 0.25-in. angle, 4.10 lb per ft (6.35 x 6.35 x 0.63 cm, 0.061 kg per cm), or 2.5-in. (6.35 cm) OD pipe or tubular steel, 3.65 lb per ft (0.054 kg per cm), or equivalent. Length shall be 7 ft (213 cm) minimum, or as specified on the plans.

2.6.2 Steel line posts shall be studded or punched *T, U,* or *Y*-shaped posts, with anchor plates or acceptable alternate, 1.33 lb per ft (0.020 kg per cm). Length shall be 6.5 ft (198 cm) minimum, and spacing shall be as specified on the plans.

2.6.3 Steel braces shall be 2 x 2 x 0.25-in. angle, 3.19 lb per ft (5.08 x 5.08 x 0.63 cm, 0.047 kg per cm), or 1.62-in. (4.1 cm) OD pipe or tubular steel, 2.27 lb per ft (0.034 kg per cm), or equivalent. Lengths shall be as specified in Section 3—Assemblies, or in the plans.

2.7 Concrete, non-reinforced, reinforced, or prestressed shall have a minimum 28-day test strength of 4000 psi (281 kg per sq cm) and shall have a minimum of 7 days of curing before loading. All reinforcing or prestressing steel shall have a 0.75-in. (1.9-cm) minimum concrete coverage.

2.7.1 Concrete for post and brace embedment shall be placed as specified in the plans. The top of embedment piers shall extend 2 in. (5 cm) above grade at the post or brace and shall slope to grade at the edge of the concrete. If failure occurs when fencing is stretched, pier or piers shall be replaced.

2.7.2 Concrete corner and end posts shall be a minimum of 48 sq in. (310 sq cm) cross section, with a minimum of 0.44 sq in. (2.8 sq cm) reinforcing steel.

*ASTM A475, Specifications for Zinc-Coated Steel Wire Strand.

TABLE 1—MINIMUM WEIGHT OF COATING ON ZINC-COATED WIRE

Size, Steel Wire Gage	Nominal Diameter, Coated Wire		Minimum Weight of Coating Per Unit Area of Uncoated Wire Surface					
			Class 1		Class 2		Class 3	
	in.	cm	oz/sq ft	g/sq cm	oz/sq ft	g/sq cm	oz/sq ft	g/sq cm
			Woven Wire Fabric, ASTM A116					
7	0.177	0.450	0.40	0.0122	0.60	0.0183	0.80	0.0244
9	0.148	0.376	0.40	0.0122	0.60	0.0183	0.80	0.0244
10	0.135	0.343	0.30	0.0092	0.50	0.0152	0.80	0.0244
11	0.120	0.305	0.30	0.0092	0.50	0.0152	0.80	0.0244
12	0.105	0.267	0.30	0.0092	0.50	0.0152	0.80	0.0244
12½	0.099	0.251	0.30	0.0092	0.50	0.0152	0.80	0.0244
14½	0.076	0.193	0.20	0.0061	0.40	0.0122	0.60	0.0183
			Barbed Wire, ASTM A121					
12½	0.099	0.251	0.30	0.0092	0.50	0.0152	0.80	0.0244
14	0.080	0.203	0.25	0.0076	0.45	0.0137	0.65	0.0198
16	0.0625	0.159	0.15	0.0046	0.35	0.0107	0.50	0.0152

2.7.3 Concrete line posts shall be a minimum of 16 sq in. (103 sq cm) in cross section with a minimum of 0.11 sq in. (0.7 sq cm) reinforcing steel.

2.7.4 Prestressed concrete posts of similar strength design are an acceptable alternate to conventionally reinforced concrete posts.

2.8 Wood posts and braces shall be pressure-preservative treated according to Federal Specification TT-W-571, Wood Preservation: Treating Practices, latest revision and may be round or square. Decay-resistant species may be used untreated if specified. Minimum nominal sizes and lengths shall be as follows, or as specified in the plans:

2.8.1 Wood corner and end posts shall be a minimum 5-in. (12.7-cm) top diameter or square and 8-ft (244-cm) length.

2.8.2 Wood brace posts shall be a minimum 4-in. (10.2 cm) or 5-in. (12.7-cm) (see Section 3—Assemblies) top diameter or square and 8-ft (244-cm) length.

2.8.3 Wood braces shall be 3.5-in. (8.9 cm) minimum end diameter or square, and 10-ft (305-cm) maximum length.

2.8.4 Wood line posts shall be a minimum 3-in. (7.6-cm) top diameter or square, a minimum 6.5-ft (198 cm) length, and shall be set a minimum of 2 ft (61 cm) deep. Post spacing shall be as specified on the plans. Posts set in muck, peat, or soils on which water stands, should be 8 ft (244 cm) long and set a minimum of 3.5 ft (107 cm) deep. Posts pointed for driving shall be shaped before preservative treatment.

2.9 Wood board and plank shall be decay-resistant species, or preservative-treated according to Federal Specification TT-W-571, latest revision, for lumber not in contact with the ground, or USDA Farmers' Bulletin 2049, Preservative Treatment of Fence Posts and Farm Timbers.

SECTION 3—ASSEMBLIES

3.1 Ends and corners. Corner assemblies are constructed as two end assemblies with a single end post. End and corner posts shall be set within a maximum tolerance of ±1 in. (2.5 cm) of vertical.

3.1.1 Braced steel end or corner posts (Fig. 1) shall be embedded at least 3 ft (91 cm) deep in a concrete pier either 12 in. (30 cm) circular or square. The brace shall be fastened to the corner post with a positive connector such as a clamp or bolt, and it shall extend 12 in. (30 cm) into a brace block which is at least 16 in. (41 cm) square and 12 in. (30 cm) deep.

3.1.2 Horizontal brace assemblies (Fig. 2) shall have the end or corner, and brace, posts set a minimum of 3.5 ft (107 cm) deep. Brace posts shall be spaced to accommodate 8-ft (244-cm)-long braces. Horizontal braces shall be mounted 12 in. (30 cm) below the top of the end post.

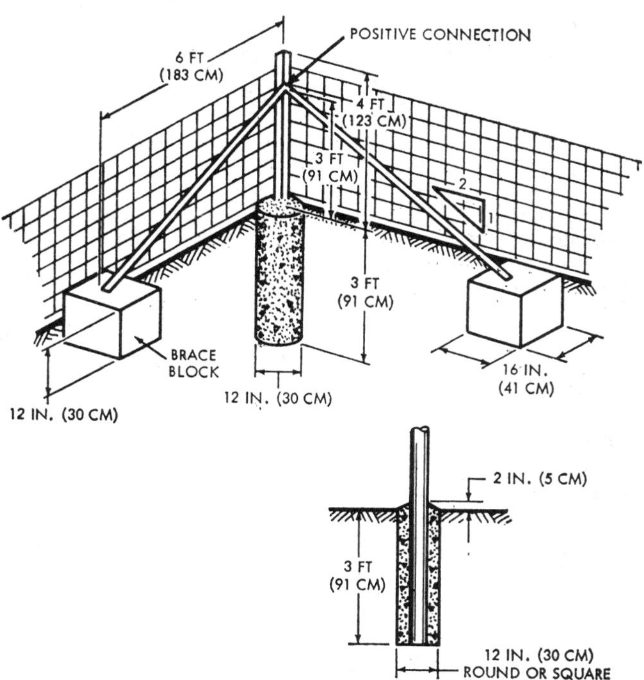

FIG. 1—BRACED STEEL END OR CORNER POST

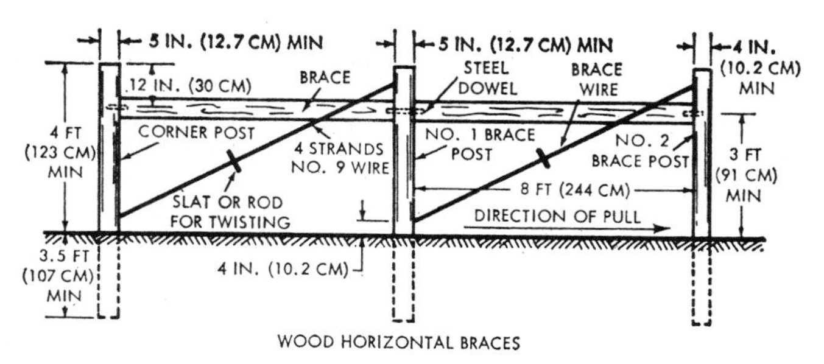

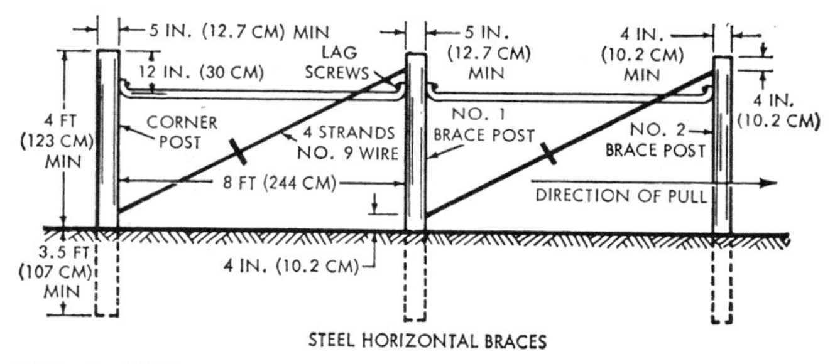

FIG. 2—HORIZONTAL BRACE ASSEMBLIES

3.1.2.1 Wood braces shall be connected with a 0.38 x 4-in. (0.95 x 10.2 cm) steel dowel at each end extending 2 in. (5 cm) into the brace and 2 in. (5 cm) into the post.

3.1.2.2 Steel braces shall be connected with a 0.38 x 3-in. (0.95 x 7.6 cm) lag screw at each end, or shall be inserted into 1-in. (2.5 cm)-deep, prepared sockets at each end.

3.1.2.3 Wire braces shall be four strands of 9-gage steel wire positively fastened 4 in. (10.2 cm) below the top of the post and 4 in. (10.2 cm) above grade. They shall be tightened (twisted) with a 0.75 x 1-in. (1.9 x 2.5 cm)-wood slat or 0.38-in. (0.95 cm)-diameter steel rod until the entire assembly is rigid. Slats or rods shall be left in position.

3.1.3 Diagonal brace assemblies (Fig. 3), recommended in soft soils, shall have the end or corner, and brace, posts set a minimum of 3.5 ft (107 cm) deep. Brace posts shall be spaced to accommodate 8-ft (244 cm)-long braces. Wood or steel braces shall be attached

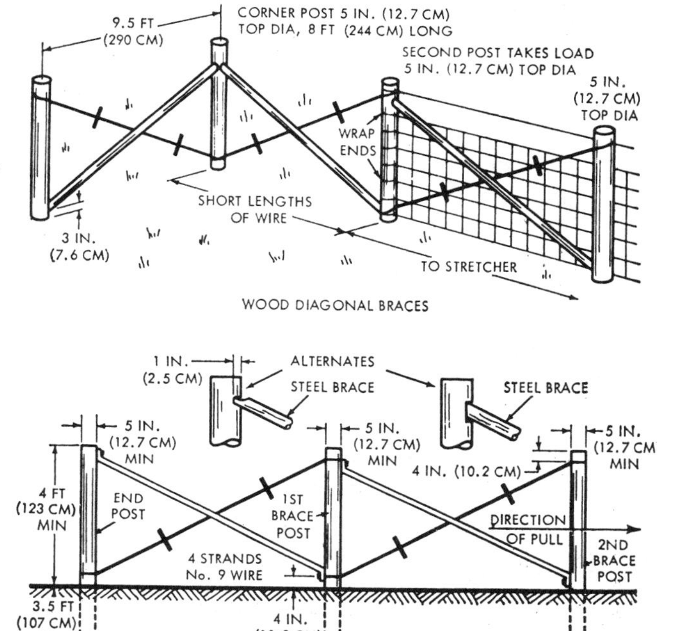

FIG. 3—DIAGONAL BRACE ASSEMBLIES

according to paragraphs 3.1.2.1, 3.1.2.2, and 3.1.2.3 above. Fencing is stretched from the first brace post. Ends are filled in after wire is attached.

3.2 Pull-post assemblies (Fig. 4) shall be placed a maximum of 40 rods (200 m) apart in straight runs and at the top and bottom (ridge and valley) of appreciable slope changes. Construction will follow the specifications of Section 3.1. The center post is specified as an *end or corner post* (2.6.1, 2.7.2, 2.8.1), and the two side posts as *No. 2 brace posts*. Smooth wire braces shall be placed as in paragraph 3.1.2.3, above, to form an *X* in both spans.

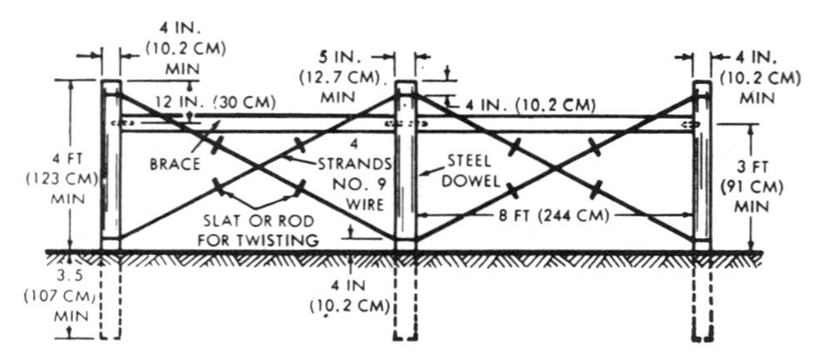

FIG. 4—PULL-POST ASSEMBLY

3.3 Assembly at change in vertical alignment shall anchor fencing with two steel fence posts, or equivalent, of at least 4-ft (122 cm) length, as shown in Fig. 5 where change in vertical-alignment exceeds 10 in. per 14 ft (1:8 slope) between line posts.

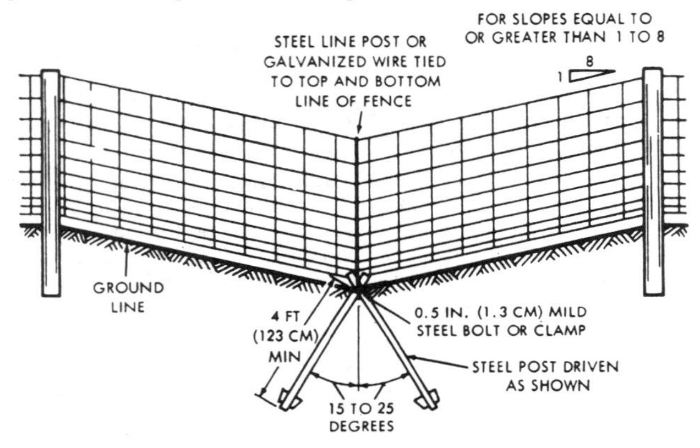

FIG. 5—ASSEMBLY AT CHANGE OF VERTICAL ALIGNMENT

3.4 Curve or contour fence shall have maximum post spacing as follows:

Fence Curvature		Post Spacing	
in.	cm	ft	cm
4 or less	10	14	427
5-6	13-15	12	366
7-8	18-20	10	305
9-14	23-36	8	244
15-20	38-51	7	213

3.4.1 Curvature shall be measured in inches from a stake along the proposed fence line, to a string connecting two other stakes along the fence line each 14 ft (427 cm) away from the first post (Fig. 6).

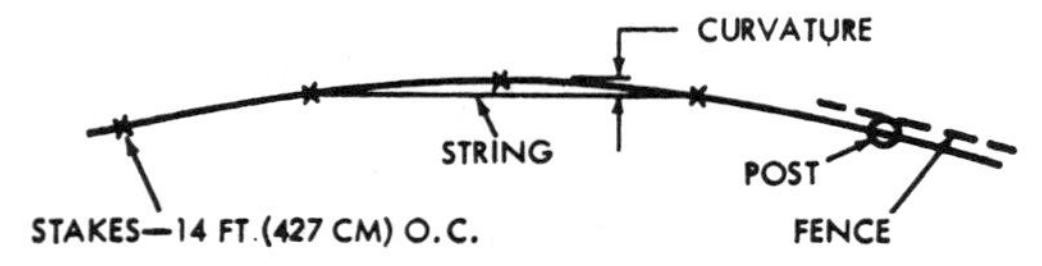

FIG. 6—FENCE CURVATURE

3.4.2 Post spacing shall be measured horizontally.

3.4.3 Posts shall lean outward from the curve approximately 2 in. (5 cm) at the top, and fencing shall be applied to the side of the posts outside the curve and stretched to straighten the posts approximately 1 in. (2.5 cm).

SECTION 4—CONSTRUCTION

4.1 One barbed wire shall be first stretched at the bottom to determine alignment of line posts and shall be temporarily fastened to end posts. This barbed wire may be left at the bottom when needed.

4.2 End (or corner) post assemblies in Fig. 2. The fence shall be attached to one end (or corner) post and the fence stretchers attached to the opposite end (or corner) post (or pull-post assembly). The fence at the stretcher end is then attached directly to the pull-post corner or end.

4.3 End (or corner) post assemblies in Fig. 3. The fence or stretchers shall be attached to the first brace post in the assembly. Its design provides for maximum strain taken at this point. A slack span of fence fabric is used between the end (or corner) post and the first brace post after stretching is completed.

4.4 Pull-post assembly (Fig. 4). The fence fabric shall be extended past the first post and attached to the middle post. The wires shall be cut and wrapped around the post.

4.5 The tension for stretching the woven-wire fence shall be applied at two points on the clamp bar for all fences over 32 in. (81 cm) high by using stretchers designed and manufactured for that purpose. Stretchers shall be so designed that tension can be applied to both ends of the bar at the same time. All splices in the fabric shall be securely made, with a Western Union splice or commercial splicing device approved by the engineer.

4.6 The tension for stretching the barbed wire shall be applied by use of single-wire stretchers designed and manufactured for that purpose, and in accordance with the wire manufacturer's recommendations.

4.7 Fence fabric shall be fastened to steel line posts with five or more wire ties or clamps as provided by the post manufacturer. Fence fabric shall be fastened to wood posts with staples at alternate line wires. The staples shall be driven at an angle to the grain and with points slightly downward.

4.8 The fence fabric shall be on the side of the posts where livestock will be kept except for:

4.8.1 Fences constructed on curves

4.8.2 Right-of-way fence. The fabric may be put on the road side provided the top line wire of the woven wire is tied to the wood posts with No. 9, galvanized, smooth wire.

4.9 If barbed wire is used, it shall be fastened to all posts in the same manner as the fabric.

SECTION 5—GROUNDING WIRE FENCING

5.1 To provide livestock protection against lightning, all wires of the fence shall be securely fastened, with galvanized wire ties, to fence grounding electrodes at intervals of not more than 150 ft (46 m) for normally dry, rocky soil; and not more than 300 ft (91 m) for normally, moist or damp soils. (United States of America Standards Institute, C5-1, Lightning Protection Code, also National Fire Protection Association Bulletin No. 78, indicate fences built with metal posts set in earth are as safe from lightning as it is practical to make them, especially if the electrical continuity is broken.)

5.2 Electrodes for grounding fences shall be driven into firm earth to a minimum depth of 3 ft (0.91 m) and shall be either a standard galvanized steel post or a ¾-inch (trade size) galvanized steel pipe with spacing as per paragraph 5.1, above.

5.3 The continuity of wire fences shall be broken at maximum intervals of 1000 ft (305 m) by means of a wooden gate, wooden panel section, insulating material, or by wood strips measuring 2 x 2 x 24 in. (51 x 51 x 610 mm) or their equivalents in insulation and mechanical strength properties.

Cited Standards:

ASTM A116, Specifications for Zinc-Coated (Galvanized) Iron or Steel Farm-Field and Railroad Right-of-Way Wire Fencing
ASTM A121, Specifications for Zinc-Coated (Galvanized) Steel Barbed Wire
ASTM A123, Specifications for Zinc (Hot Galvanized) Coatings on Products Fabricated from Rolled, Pressed, and Forged Steel Shapes, Plates, Bars, and Strip
ASTM A153, Specifications for Zinc Coating (Hot Dip) on Iron and Steel Hardware
ASTM A474, Specifications for Aluminum-Coated Steel Wire Strand
ASTM A475, Specifications for Zinc-Coated Steel Wire Strand
FED. SPEC. TT-W-571, Wood Preservation, Testing Practices
NFPA No. 78, Lightning Protection Code

TOWER SILOS: UNIT WEIGHT OF SILAGE AND SILO CAPACITIES

Approved by Steering Committee of ASAE Structures and Environment Division; adopted by ASAE February 1962; reconfirmed by the ASAE Structures and Environment Division Technical Committee December 1967, December 1972, December 1973, December 1974, December 1979, December 1980; revised March 1982; reconfirmed December 1986; reconfirmed for one year December 1987.

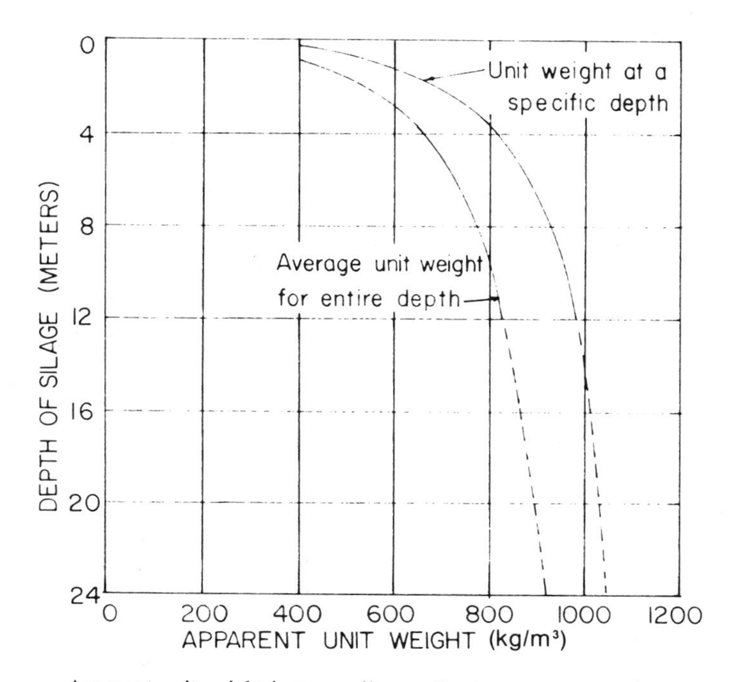

Apparent unit weight in tower silos applies to corn or grass silage, from 68 to 72 percent moisture (wet basis). Data were obtained from silos up to 6.1 m in diameter and to 12.2 m high. The curves were extrapolated beyond 12.2 m. Curves assume normal treatment in harvesting and in filling silos. Kilograms of dry matter per cubic meter = apparent unit weight × (100-percent moisture)/100.

FIG. 1—APPARENT UNIT WEIGHT OF SILAGE IN TOWER SILOS, SI UNITS

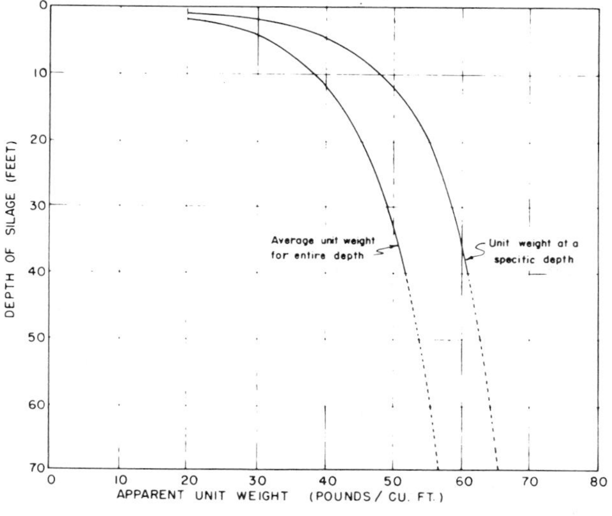

Apparent unit weight in tower silos applies to corn or grass silage, from 68 to 72 percent moisture (wet basis). Data were obtained from silos up to 20 ft in diameter and to 40 ft high. The curves were extrapolated beyond 40 ft. Curves assume normal treatment in harvesting and in filling silos. Pounds of dry matter per cubic foot = apparent unit weight × (100-percent moisture)/100.

FIG. 2—APPARENT UNIT WEIGHT OF SILAGE IN TOWER SILOS, CUSTOMARY UNITS

TABLE 1—APPROXIMATE DRY MATTER CAPACITY OF SILOS IN TONNES*

Silo height, m	Silo diameter, m										
	3.0	3.7	4.3	4.9	5.5	6.1	6.7	7.3	7.9	8.5	9.1
6.1	7	11	14	19	24	30	36	43	51	59	67
7.3	10	14	19	24	31	39	47	55	66	76	88
8.5	12	17	24	32	40	48	58	69	82	94	108
9.8	14	21	29	37	47	59	71	84	99	115	132
11.0	17	25	34	44	56	69	84	99	117	136	156
12.2	20	29	40	52	65	81	97	115	136	157	180
13.4		34	45	59	74	92	112	133	156	181	208
14.6		38	51	67	84	104	127	151	177	205	236
15.8			58	75	95	117	142	169	199	230	264
17.1			64	84	106	131	158	188	220	256	294
18.3			71	92	117	144	174	207	248	280	324
19.5					129	158	190	227	270	308	355
20.7					141	172	207	247	294	336	386
21.9								266	318	363	416
23.2								285	341	387	444
24.4								303	356	413	472

*Capacities allow 0.3 m unused depth for settling in silos up to 10 m high, and 0.3 additional m for each 3 m beyond 10 m height.
Tonnes of moist silage = tonnes of dry matter × dry matter factor
Dry matter factor = 100/(100-percent moisture)

TABLE 2—APPROXIMATE DRY MATTER CAPACITY OF SILOS IN TONS*

Silo height, ft	Silo diameter, ft										
	10	12	14	16	18	20	22	24	26	28	30
20	8	12	16	21	27	33	40	47	56	65	74
24	11	15	21	27	34	43	52	61	72	83	96
28	13	19	26	35	44	53	64	76	90	104	119
32	16	23	32	41	52	65	78	93	109	127	145
36	19	28	37	48	62	76	92	109	129	150	172
40	22	32	44	57	72	89	107	127	150	173	199
44		37	50	65	82	102	123	147	172	200	229
48		42	56	74	93	115	140	166	195	226	260
52			64	83	105	129	157	186	219	254	291
56			71	93	117	144	174	207	243	282	324
60			78	102	129	159	192	228	273	309	357
64					142	174	210	250	298	340	391
68					155	190	228	272	324	370	425
72								293	350	400	458
76								314	376	427	489
80								334	392	455	520

*Capacities allow one foot unused depth for settling in silos up to 30 ft high, and one additional foot for each 10 ft beyond 30 ft height.
Tons of moist silage = tons of dry matter × dry matter factor
Dry matter factor = 100/(100-percent moisture)

TABLE 3—CAPACITY OF SILOS FOR 30 PERCENT MOISTURE SHELLED CORN, IN TONNES*

Silo height, m	Silo diameter, m						
	3.0	3.7	4.3	4.9	5.2	5.5	6.1
6.1	33	47	64	84	95	107	132
6.7	36	53	72	93	105	118	145
7.3	40	57	78	102	114	129	159
7.9	43	62	84	111	125	140	172
8.5	46	67	91	119	134	151	186
9.1	50	72	98	128	144	162	200
9.8	53	76	104	136	153	172	212
10.4		82	111	146	165	183	226
11.0		86	118	153	173	194	240
11.6		92	124	162	183	205	253
12.2		96	131	171	192	216	267
12.8			138	180	202	227	280
13.4			144	188	212	239	294
14.0			151	197	222	250	308
14.6			158	206	232	260	321
15.2				214	241	271	335
15.8				223	252	283	349
16.5				232	262	294	364
17.1				241	272	306	377
17.7				253	283	318	392
18.3				260	294	329	406

*Tonnage will increase approximately 5 percent with moisture content from 25 to 35 percent.

TABLE 4—CAPACITY OF SILOS FOR 30 PERCENT MOISTURE SHELLED CORN, IN TONS*

Silo height, ft	Silo diameter, ft						
	10	12	14	16	17	18	20
20	36	52	71	93	105	118	145
22	40	58	79	103	116	130	160
24	44	63	86	112	126	142	175
26	47	68	93	122	138	154	190
28	51	74	100	131	148	166	205
30	55	79	108	141	159	178	220
32	58	84	115	150	169	190	234
34		90	122	161	182	202	249
36		95	130	169	191	214	264
38		101	137	179	202	226	279
40		106	144	188	212	238	294
42			152	198	223	250	309
44			159	207	234	263	324
46			166	217	245	275	339
48			174	227	256	287	354
50				236	266	299	369
52				246	278	312	385
54				256	289	324	401
56				266	300	337	416
58				276	312	350	432
60				287	324	363	448

*Tonnage will increase approximately 5 percent with moisture content from 25 to 35 percent.

References: Last printed in 1983 AGRICULTURAL ENGINEERS YEARBOOK; list available from ASAE Headquarters.

DESIGN OF VENTILATION SYSTEMS FOR POULTRY AND LIVESTOCK SHELTERS

Developed by the Joint ASAE Structures and Environment—Electric Power and Processing Division Committee on Animal Shelter Ventilation; approved by the Electric Power and Processing and Structures and Environment Division Steering Committees; adopted by ASAE as a Data February 1963; revised June 1966, December 1968, March 1970; revised April 1975 to incorporate and supersede D249, Effect of Thermal Environment on Production, Heat and Moisture Loss and Feed and Water Requirements of Farm Livestock which was originally adopted by ASAE 1955; reconfirmed December 1979, December 1980, December 1981, December 1982, December 1984, December 1985; revised and reclassified as an Engineering Practice December 1986.

SECTION 1—PURPOSE AND SCOPE

1.1 This information is provided for the convenience of agricultural engineers and other professional persons serving the agricultural industry. It includes basic information and technical data, supported by research, for use in designing and/or evaluating ventilation systems for livestock or poultry shelters. Much of the information presented in Section 3—General Design Information will apply to the design of ventilation systems for any structure housing livestock or poultry.

1.2 In addition to the design and configuration of equipment, hazard control and accident prevention are dependent upon the awareness, concern, and prudence of personnel involved in the operation, transport, maintenance, and storage of equipment or in the use and maintenance of facilities.

SECTION 2—DEFINITION

2.1 Ventilation as used herein is defined as a system of air exchange which accomplishes one or more of the following:

2.1.1 Provides a desired amount of fresh air, without drafts, to all parts of the shelter. A draft is an air speed in excess of 0.15-0.3 m/s (29.5-59 ft/min) in cool weather (speed depends on animal size). In summer these limits may be exceeded except for small animals. High air speeds, besides causing wind chill, may also create excessive dust levels.

2.1.2 Maintains temperatures in the shelter within desired limits.

2.1.3 Maintains relative humidity in the shelter within desired limits for the animal species (see Section 4—Ventilation for Dairy Cows and Calves, Section 5—Ventilation for Beef Cattle, Section 6—Ventilation for Swine, Section 7—Ventilation for Broiler Chickens and Young Turkeys, Section 8—Ventilation for Laying Hens, through Section 9—Effect of Environment on Sheep for relative humidity requirements by species) unless ammonia levels are beyond desired limits for operating personnel.

2.1.4 Maintain ammonia levels in the shelter at less than 25 mg/m^3 for operating personnel. This limit was set by the American Conference of Government Industrial Hygienists for personnel working no more than four intervals of 15 min each per day in the shelter. The 8 h threshold limit is 18 mg/m^3.

SECTION 3—GENERAL DESIGN INFORMATION

3.1 Weather data

3.1.1 Design temperature. The outdoor design temperature should be selected for the particular area where the animal shelter will be located. The cold weather temperature is normally used for determining heat loss from a building, insulation requirements, minimum continuous air exchange rate and supplemental heating requirements. The summer design temperature is used to determine the maximum required ventilation capacity.

3.1.2 Cold weather conditions. Cold outdoor air contains relatively small amounts of moisture. As the cold ventilating air enters the building and warms up, it has the capacity to absorb relatively large quantities of vaporized moisture from within the building. Outside design temperature suitable for calculating winter minimum air exchange rates and heat loss through the building components are shown in Fig. 1. Tabular data on outside winter design temperatures can be found in several books and periodicals. Most are derived from the American Society of Heating, Refrigerating and Air-Conditioning Engineers Handbook of Fundamentals or other ASHRAE publications. The 97.5% values are recommended for animal shelter use.

3.1.3 Hot weather conditions. In buildings which depend upon mechanical ventilation during hot weather, air exchange must be sufficient to keep the inside temperature only slightly warmer than the outside temperature. Summer design temperatures suitable for calculating maximum air exchange rates are shown in Fig. 2. The design maximum ventilation capacity can be computed using equation 3.3.1-6 and an assumed temperature difference of 1-2°C (2-4°F) or equation 3.3.1-4, whichever gives the greater exchange rate.

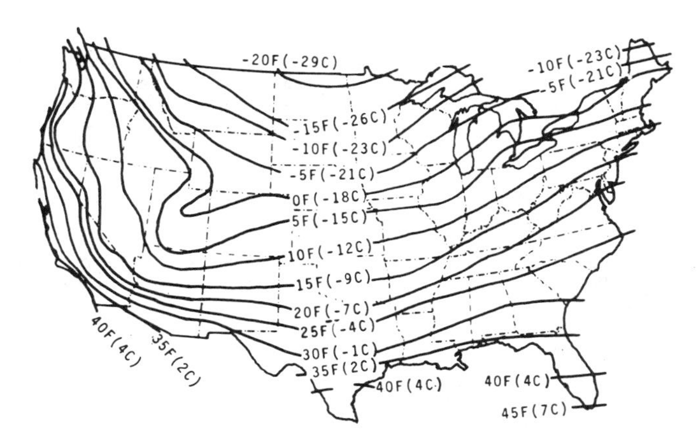

FIG. 1—WINTER (DEC.-FEB.) TEMPERATURE THAT IS EXCEEDED MORE THAN 97.5% OF THE TIME

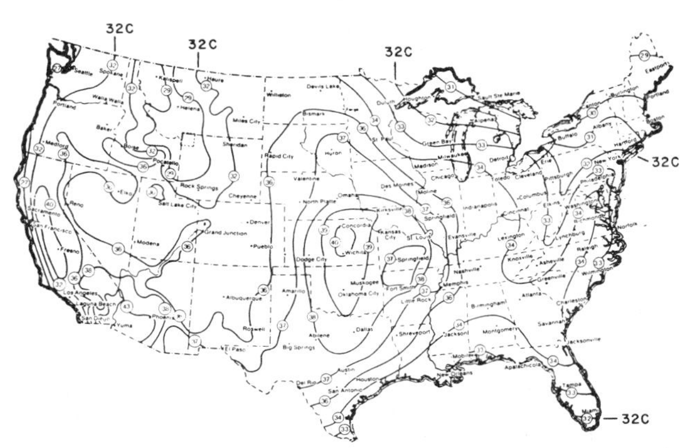

FIG. 2—SUMMER DRY-BULB TEMPERATURE DATA. The dry-bulb temperatures shown will be exceeded not more than 5% of the 12 h during the middle of the day in June to September inclusive. (Ausburger, et al.)

3.2 Building requirements

3.2.1 Heat transmission through building materials. The rate of heat transmission through building materials depends upon the characteristics of the material. A comprehensive list of heat transmission coefficients of building materials is given in the ASHRAE Handbook of Fundamentals.

3.2.2 Need for vapor barriers. Animal shelters of the totally enclosed and ventilated type may be subjected to high moisture conditions. For example, a shelter with inside air temperature of 10°C (50°F) and relative humidity of 75% and with outside air temperature of -12°C (10°F) and relative humidity of 85% will be subjected to a winter vapor pressure difference of approximately 0.746 kPa (3 in. H_2O). A vapor pressure difference of this magnitude forces water vapor out through the components of the enclosure. Water vapor transfer is not a problem so long as the saturated vapor pressure gradient within the enclosure components remains above the actual vapor pressure gradient. Condensation will occur at or near the point in the building component where these gradients intersect. The condensate may migrate to other parts of the enclosure due to gravitational force or capillary attraction. Condensed water vapor promotes decomposition of building components and reduces the insulating capacity of many insulations. This situation increases structural heat loss which in turn hinders moisture control by ventilation. To keep insulation dry and free from condensed vapor, a vapor barrier should be applied on or near the side of the insulation with the highest vapor pressure, which is usually the warm side. A vapor barrier is a material with a high resistance to water vapor flow. Vapor barriers should be as continuous as possible. Thus, joints and holes in the vapor barrier should be minimized.

3.3 Ventilation requirements

3.3.1 Livestock and poultry produce best when their environment is within an optimum zone. The housing and environmental control system should be arranged to achieve these conditions. Design of the environmental control begins with analysis of the heat flows in the building using equation 3.3.1-1.

$$Q_S + Q_E + Q_{supp} = Q_M + Q_B + Q_V + Q_{stored} \quad \text{[3.3.1-1]}$$

where

Q_S = sensible heat loss rate from animals, J/s
Q_E = heat production rate by equipment, i.e., motors, lights, etc., J/s
Q_{supp} = supplemental heat production rate, J/s
Q_M = heat required to evaporate moisture, J/s
Q_B = rate of heat loss by conduction through walls, floor and ceiling, J/s
Q_V = rate of sensible heat exchanged in ventilation air, J/s
Q_{stored} = rate heat is stored or released by building material, especially by concrete floors, J/s (This term is zero in steady state analysis.)

When a building is ventilated, heat is added or removed. This heat flow rate can be expressed as:

$$Q_V = M\, c_p \Delta T \quad \text{[3.3.1-2]}$$

where

M = mass flow rate, kg/s
c_p = specific heat of air, 1005 J/kg·K (approx)
ΔT = temperature difference between the inside and outside, K

Fans are rated by volumetric flow rate, and equation 3.3.1-2 is based on mass flow rate. Ventilation rate should be obtained by multiplying the mass flow rate by the specific volume obtained from a psychrometric chart with conditions chosen to represent those on the intake side of the fan.

Each animal produces heat and moisture. Tables 1 and 2 show the production of these constituents for beef cattle, dairy cattle, swine, sheep and poultry. Moisture in the form of latent heat (vapor) and liquid in the waste is produced. Moisture in the liquid form must be vaporized for removal by ventilation, and the heat required for this process is given by:

$$Q_M = m\, N\, h_{fg} \quad \text{[3.3.1-3]}$$

where

m = rate of moisture production in ventilated space, kg/s·animal
N = number of animals
h_{fg} = latent heat of water vaporization, J/kg

The mass flow rate of dry air required to remove moisture is found using:

$$M = \frac{m\, N}{W_i - W_o} \quad \text{[3.3.1-4]}$$

where

M = ventilation rate, kg/s
W_i = humidity ratio of inside air, kg water vapor/kg dry air
W_o = humidity ratio of outside air, kg water vapor/kg dry air

The rate of sensible heat produced by the animals, Q_s, is found using data in Tables 1 and 2. Multiply the sensible heat loss rate per animal unit by the number of animal units. Electrically generated heat, Q_E, is generally small and can often be neglected. The rate of supplemental heating, Q_{supp}, is either known from the capacity rating of any heaters within the airspace, or it is the unknown to be solved for in equation 3.3.1-1 (when the ventilation rate is known).

The rate heat is conducted through the walls, floor and ceiling of a building can be calculated for steady state conditions using the following expression:

$$Q_B = U\, A\, T \quad \text{[3.3.1-5]}$$

where

U = overall heat transmission coefficient, J/(m²·k·s)
A = area of surface to which U and T apply, m²
T = temperature difference between inside and outside, K

Performing a heat balance of all sources and sinks of sensible and latent heat is a necessary first step in design of any environmental control system. The mass flow rate of dry air required to remove heat is found using:

$$M = \frac{Q_S + Q_E + Q_{supp} - Q_M - Q_B - Q_{stored}}{c_p\, T} \quad \text{[3.3.1-6]}$$

3.3.1.1 Relative humidities. Moisture ventilation rates are proportional to the humidity ratio difference between the inside and outside air ($W_i - W_o$, equation 3.3.1-4). The magnitude of W_i depends both upon temperature and relative humidity. Design relative humidites for most confinement housing systems should be in the desired range for the animal species, but higher humidities can be allowed for short periods (6 to 12 h) when maintaining lower humidities would result in large supplemental heating requirements.

3.3.2 Types of ventilation systems. Ventilation occurs in a building because of a difference in static pressure between the inside and outside of the building. If the ventilation fan forces air into the building through the air inlets, the static pressure in the building is greater than outside and is commonly referred to as a positive pressure system. If the fan removes air from the structure, the static pressure in the building will be less than that outside, and air will flow in through the inlets. This system is referred to as a negative pressure system. Each system has characteristics that should be considered when deciding which system to use.

TABLE 1—MOISTURE PRODUCTION (MP), SENSIBLE HEAT LOSS (SHL) AND TOTAL HEAT LOSS (THL) OF LIVESTOCK BASED ON CALORIMETRIC AND HOUSING SYSTEMS STUDIES

Livestock	Building temperature °C	MP g H_2O / kg·h	SHL W / kg	THL W / kg	Reference
Cattle					
Dairy cow					(Yeck and Stewart, 1959)
500 kg	-1	0.77	1.9	2.4	
500 kg	10	1.0	1.5	2.2	
500 kg	15	1.3	1.2	2.1	
500 kg	21	1.3	1.1	2.0	
500 kg	27	1.8	0.6	1.9	
Beef cattle					(Hellickson et al., 1974)
500 kg	4	2.5	1.5	2.8	
Calves					
Ayrshire male*					(Gonzalez-Jimenez and Blaxter, 1962)
39 kg (8 days)	3	0.7	2.5	2.9	
40 kg (14 days)	3	0.7	2.4	2.8	
45 kg (25 days)	3	0.7	2.6	3.0	
39 kg (8 days)	23	0.7	2.0	2.4	
40 kg (14 days)	23	0.7	1.9	2.3	
44 kg (24 days)	23	0.7	2.0	2.4	
British Friesian male*					(Holmes and Davey, 1976)
47 kg	20	––	––	2.3	
Jersey*					(Holmes and Davey, 1976)
28 kg	20	––	––	2.6	
British Friesian male*					(Webster and Gordon, 1977)
2 days to 8 wk	5	––	––	(7.9)†	
2 days to 8 wk	10	––	––	(7.3)†	
2 days to 8 wk	15	––	––	(7.3)†	
Cattle					
Hereford x Friesian male*					(Webster and Gordon, 1977)
2 days to 8 wk	5	––	––	(7.3)‡	
2 days to 8 wk	10	––	––	(6.5)‡	
2 days to 8 wk	15	––	––	(6.5)‡	
British Friesian male*					(Webster and Gordon, 1976)
90-180 kg	5	(2.2)†	(6.8)‡	(8.2)‡	
100 kg	5	0.75	2.1	2.6	
90-180 kg	10	(4.4)†	(6.3)‡	(9.3)‡	
100 kg	10	1.4	2.0	2.9	
90-180 kg	15	(6.2)†	(4.6)‡	(8.7)‡	
100 kg	15	2.0	1.4	2.8	
90-180 kg	20	(7.1)†	(4.4)‡	(9.1)‡	
100 kg	20	2.3	1.4	2.9	
Brown Swiss Holstein					(Yeck and Stewart, 1960)
16 wk	10	2.0	2.3	3.7	
32 wk	10	1.2	1.5	2.4	
48 wk	10	1.0	1.5	2.2	
16 wk	27	3.0	1.5	3.5	
32 wk	27	2.2	1.1	2.6	
48 wk	27	1.9	1.0	2.2	

TABLE 1—MOISTURE PRODUCTION (MP), SENSIBLE HEAT LOSS (SHL) AND TOTAL HEAT LOSS (THL) OF LIVESTOCK BASED ON CALORIMETRIC AND HOUSING SYSTEMS STUDIES *(con't)*

Livestock	Building temperature °C	MP g H_2O / kg·h	SHL W / kg	THL W / kg	Reference
Jersey					(Yeck and Stewart, 1960)
16 wk	10	2.4	2.5	4.1	
32 wk	10	1.5	1.8	2.8	
48 wk	10	1.3	1.6	2.5	
16 wk	27	3.8	1.4	3.9	
32 wk	27	2.5	1.0	2.7	
48 wk	27	2.3	0.8	2.3	
Shorthorn, Brahman Santa Gertrudis					(Yeck and Stewart, 1959)
25 wk	10	1.6	1.4	2.5	
40 wk	10	1.1	1.0	1.8	
55 wk	10	1.0	1.0	1.6	
25 wk	27	2.8	0.7	2.6	
40 wk	27	2.1	0.5	1.9	
55 wk	27	1.8	0.5	1.7	
Swine					
Sow and litter (solid floor)					(Bond et al., 1959)
177 kg (0 wk)	16-27	1.8	1.3	2.6	
181 kg (2 wk)	16-27	2.4	1.7	3.3	
186 kg (4 wk)	16-27	2.6	1.7	3.5	
200 kg (6 wk)	16-27	2.7	1.7	3.5	
227 kg (8 wk)	16-27	2.6	2.1	3.9	
Nursery pigs*					(Ota et al., 1975)
4-6 kg	29	1.7	2.2	3.3	
6-11 kg	24	2.2	3.1	4.5	
11-17 kg	18	2.2	3.5	5.0	
4 kg	15	—	—	6.4	(Cairnie and Pullar, 1957)
(single)	20	—	—	5.2	
	25	—	—	4.5	
	30	—	—	4.0	
6.8 kg	15	—	—	6.0	(Cairnie and Pullar, 1957)
(single)	20	—	—	5.6	
	25	—	—	5.2	
	30	—	—	5.1	
10-12 kg	15	—	—	5.8	
(single)	20	—	—	5.6	
	25	—	—	5.6	
	30	—	—	5.6	
Growing-finishing pigs (solid floor)					(Bond et al., 1959)
20 kg	5	2.5	4.2	5.9	
	10	2.7	4.0	5.4	
	15	3.1	3.0	5.0	
	20	3.7	2.3	4.8	
	25	4.7	1.6	4.8	
	30	6.3	0.6	4.8	
40 kg	5	1.5	3.0	4.0	
	10	1.6	2.5	3.6	
	15	1.9	2.0	3.3	
	20	2.2	1.6	3.1	
	25	2.8	1.2	3.0	
	30	3.6	0.6	3.0	
60 kg	5	1.2	2.5	3.3	
	10	1.3	2.0	2.9	
	15	1.4	1.7	2.6	
	20	1.7	1.3	2.4	
	25	2.0	1.0	2.3	
	30	2.7	0.5	2.3	

TABLE 1—MOISTURE PRODUCTION (MP), SENSIBLE HEAT LOSS (SHL) AND TOTAL HEAT LOSS (THL) OF LIVESTOCK BASED ON CALORIMETRIC AND HOUSING SYSTEMS STUDIES *(con't)*

Livestock	Building temperature °C	MP $\frac{g\ H_2O}{kg \cdot h}$	SHL $\frac{W}{kg}$	THL $\frac{W}{kg}$	Reference
80 kg	5	1.1	2.2	2.9	
	10	1.1	1.8	2.5	
	15	1.2	1.5	2.3	
	20	1.4	1.2	2.1	
	25	1.7	0.85	2.0	
	30	2.2	0.49	1.9	
100 kg	5	0.94	2.0	2.6	
	10	1.0	1.6	2.3	
	15	1.1	1.3	2.0	
	20	1.2	1.1	1.9	
	25	1.4	0.80	1.8	
	30	1.8	0.49	1.7	
Gilts, sows, and boars (solid floor)					
140 kg	5	0.79	1.8	2.3	
	10	0.79	1.5	2.0	
	15	0.84	1.2	1.8	
	20	0.93	1.0	1.6	
	25	1.1	0.77	1.5	
	30	1.3	0.54	1.4	
180 kg	5	0.64	1.7	2.1	
	10	0.63	1.4	1.8	
	15	0.65	1.2	1.6	
	20	0.70	0.97	1.4	
	25	0.80	0.80	1.3	
	30	0.96	0.63	1.3	
Sheep					
Mature* 60 kg (maintenance diet) Fleece length					(Armstrong et al., 1960)
Shorn	8	0.33	2.4	2.6	
	20	0.40	1.5	1.7	
	32	0.79	0.81	1.3	
3 cm	8	0.34	1.2	1.4	
	20	0.52	0.94	1.3	
	32	1.3	0.45	1.3	
6 cm	8	0.40	1.0	1.3	
	20	0.76	0.68	1.2	
	32	1.4	0.29	1.2	
(1.3 x maintenance diet) Fleece length					(Alexander, 1974)
12 cm	13	1.0	0.85	1.5	
	23	1.5	0.54	1.5	
Lambs					
1-14 days	10	—	—	6.7	(Alexander, 1961)
(dry - no wind)	23-27	—	—	5.2	

*Figures are obtained from calorimetric studies. Therefore the moisture production in livestock housing systems will be increased with a corresponding decrease in sensible heat loss.

†Figures in ( ) are $\frac{gH_2O}{h \cdot kg^{0.75}}$

‡Figures in ( ) are $\frac{W}{kg^{0.75}}$

TABLE 2—MOISTURE PRODUCTION (MP), SENSIBLE HEAT LOSS (SHL) AND TOTAL HEAT LOSS (THL) OF POULTRY BASED ON CALORIMETRIC AND HOUSING SYSTEMS STUDIES

Poultry	Building temperature °C	MP $\frac{g\ H_2O}{kg \cdot h}$	SHL $\frac{W}{kg}$	THL $\frac{W}{kg}$	Reference
Laying hen					
Leghorn*	8	2.1	4.5	5.8	(Ota and McNally, 1961)
	12	2.8	4.0	5.9	
	18	2.9	3.9	5.8	
	28	3.8	3.2	5.8	
Broilers*					
0.1 kg	29	4	12	14	(Longhouse, et al., 1968)
0.7 kg	25	3	7	9	
1.1 kg	19	2	7	9	
1.6 kg	19	2	6	7	
2.0 kg	19	1.5	5	6	
Broilers					
0.1 kg	29	22.0	4.5	19.5	(Reece and Lott, 1982)
0.4 kg	24	12.5	6.5	15.0	
0.7 kg	16	10.5	6.0	13.0	
	27	10.5	3.0	10.0	
1.0 kg	16	8.0	5.0	10.5	
	27	9.5	3.0	9.5	
1.5 kg	16	7.5	4.5	9.5	
	27	9.0	3.0	9.0	
2.0 kg	16	6.5	4.0	8.5	
Turkeys					
Large white*					
toms					
0.1 kg	35	15.6	6.1	16.8	(DeShazer et al., 1974)
0.2 kg	32	9.8	6.3	12.9	
0.4 kg	29	7.2	5.4	10.2	
0.6 kg	27	4.1	5.9	8.7	
1.0 kg	24	2.4	6.3	7.9	
15 kg	25	1.7	1.1	2.2	(Shanklin et al., 1977)
hens					
8.2 kg	25	1.4	1.4	2.4	(Shanklin et al., 1977)
Worlstad white*					(Buffington et al., 1974)
toms					
2.2 kg	21 light	—	—	9.1	
	21 dark	—	—	6.6	
3.0 kg	21 light	—	—	8.1	
	21 dark	—	—	6.1	
3.4 kg	21 light	—	—	7.7	
	21 dark	—	—	5.8	
3.9 kg	21 light	—	—	7.1	
	21 dark	—	—	5.5	
hens					
1.6 kg	21 light	—	—	9.4	
	21 dark	—	—	6.8	
2.4 kg	21 light	—	—	8.1	
	21 dark	—	—	6.1	
2.8 kg	21 light	—	—	7.5	
	21 dark	—	—	5.7	
3.2 kg	21 light	—	—	6.9	
	21 dark	—	—	5.4	

TABLE 2—MOISTURE PRODUCTION (MP), SENSIBLE HEAT LOSS (SHL) AND TOTAL HEAT LOSS (THL) OF POULTRY BASED ON CALORIMETRIC AND HOUSING SYSTEMS STUDIES *(con't)*

Poultry	Building temperature °C	MP g H_2O / kg·h	SHL W / kg	THL W / kg	Reference
Beltsville white*					(Ota and McNally, 1961)
toms					
8.9 kg	18 light	1.8	2.6	3.8	
	18 dark	1.4	2.2	3.2	
	25 light	2.2	1.8	3.3	
	25 dark	1.7	1.6	2.8	
hens					
4.4 kg	18 light	1.7	2.7	3.9	
	18 dark	1.0	2.0	2.6	
	25 light	4.4	1.2	4.2	
	25 dark	3.0	0.7	2.8	
Broad-breasted Bronze*					(Shanklin et al., 1977)
toms					
17 kg	10	0.35	1.9	2.1	
	15	0.53	1.7	2.0	
	20	0.66	1.6	2.1	
16 kg	25	0.88	1.5	2.1	
17 kg	30	1.1	0.81	1.5	
16 kg	35	1.5	0.44	1.4	
hens					
9.8 kg	10	0.31	2.1	2.2	
9.5 kg	15	0.75	1.9	2.4	
9.5 kg	20	0.70	1.6	2.0	
9.3 kg	25	0.83	1.5	2.1	
9.1 kg	30	1.0	1.0	1.6	
8.7 kg	35	1.3	0.64	1.4	

*Data are obtained from calorimetric studies. Therefore, the moisture production in poultry housing systems will be increased with a corresponding decrease in sensible heat loss.

3.3.2.1 Positive pressure systems. The pressure in the building forces the humid air out through planned outlets, if any, and through leaks in the walls and ceiling. If the air is humid, moisture may condense within the walls and ceiling in winter. This may cause deterioration of building materials and reduce effectiveness of insulation. Since the air is forced from the building, the positive pressure system is well suited to applications where ventilation air must be filtered to prevent contaminants and pathogens from entering the building. The 20 to 30% of the energy used by fan motors that is rejected as heat is added to the building; an advantage in winter but a disadvantage in summer.

3.3.2.2 Negative pressure systems. Because the ventilation fans are located in the air exhaust, they are exposed to dust, ammonia, other corrosive gases and high humidity. Allowances should be made for reduced fan efficiency as some dust accumulation on fan blades cannot be avoided. Fan motors should be totally enclosed and should be cleaned periodically to prevent overheating. The air distribution system is less complex and costly since simple openings and slots in walls function to control and distribute air in the building. However, at low airflow rates, negative pressure systems may not provide good, uniform air distribution due to air leaks and wind pressure effects.

3.3.3 Air distribution and intake-infiltration data. Ventilation is accomplished in an exhaust system by reducing the pressure within the structure below outside pressure, causing fresh air to enter wherever openings exist. Pressure differences across walls in ventilated shelters should range between 5 and 30 Pa (0.02 to 0.12 in. H_2O). The distribution of the fresh air is affected by the location, number and cross-sectional area of the openings. Ideally the amount of fresh air entering a section of the shelter should be a function of the amount of heat and moisture produced by the animals in that particular section. Satisfactory air distribution can be obtained by location of the inlets and by varying intake cross-sectional area to fit the animal distribution. Automated motorized inlet controllers have been developed to adjust the slot intake opening depending on the pressure difference between inside and outside the building. These units adjust the inlet baffle opening in relation to the static pressure difference between inside and outside, thus better controlling the fresh air distribution within the shelter. Another method of intake-distribution uses a fan connected to a perforated polyethylene air distribution tube or a rigid duct which allows a combination of heating, circulation, and ventilation to be designed into one system. Equipment manufacturers should be consulted for proper application of this type of equipment. These air distribution ducts may be used in either negative or positive pressure ventilation systems.

3.3.4 Inlet design. The rate of air exchange in a building depends upon the ventilation capacity. The uniformity of air distribution throughout the building, however, depends primarily upon the location and size of the air inlet.

3.3.4.1 Location of inlet. For many exhaust ventilation systems, slot inlets are used around the perimeter of the ceiling (except at end walls and near the fan). The slots may be continuous or intermittent. A slot wide enough for high air flow requirements during the summer can be partially closed with an adjustable baffle during the winter periods. Air coming from the attic space may be desirable for winter ventilation because of the wind protection and warming effect of the attic but could be undesirable during the summer if flow rate is low and the air experiences appreciable heat gain.

3.3.4.2 Size of inlet. The system characteristic technique can be used to design slotted inlet systems and provide understanding of the interaction between slotted inlet systems and fan system. This technique determines the operating points for the ventilation rate and pressure difference across the inlets. Data to graph fan characteristics (airflow rate as a function of pressure difference across the fan) should be available from the manufacturer, and should apply to the fan tested as installed.

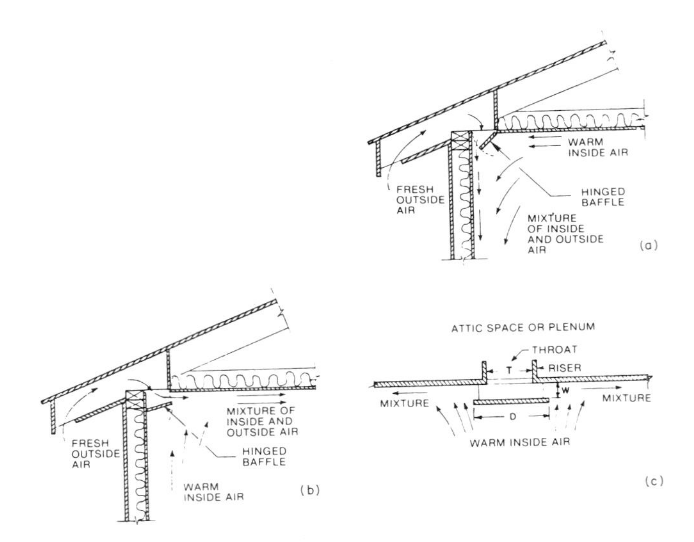

FIG. 3—EXAMPLES OF SLOTTED INLET CONFIGURATIONS

Data to graph inlet characteristics are available for hinged baffle, and center-ceiling flat baffle slotted inlets (see Fig. 3). For case (a) in Fig. 3, airflow rates are calculated by (Ref. Albright, 1976, 1978, 1979):

$$Q = 0.0011\ W^{0.98}\ P^{0.49} \qquad [3.3.4\text{-}1]$$

For case (b):

$$Q = 0.00071\ W^{0.98}\ P^{0.49} \qquad [3.3.4\text{-}2]$$

For case (c):

$$Q = 0.0065\ W^{0.98}\ P^{0.49}\ (D/T)^{0.08} \exp[-0.867\ W/T] \qquad [3.3.4\text{-}3]$$

NOTE: Air enters through two sides of the center-ceiling slot inlet system (see Fig. 3c). Airflow per unit of slot inlet length is obtained by multiplying Q (airflow rate per meter length of slot opening) by two.

where
- Q = airflow rate, m^3/s per meter length of slot opening
- W = slot width, mm
- P = pressure difference across the inlet, Pa
- D = baffle width, mm (see Fig. 3c)
- T = width of ceiling opening, mm (see Fig. 3c)

Another inlet system uses intermittently placed openings, usually rectangular in shape. The airflow rate as a function of pressure difference across the opening should be obtained from the manufacturer since pressure loss due to entrance and discharge coefficients varies with each configuration. Intermittent openings are useful where long building lengths or posts/trusses make continuous slots less practical to construct. Compared to a slot inlet, a rectangular shape will achieve further penetration of the air into the ventilated space since its entraining edge, which is directly related to drag on the airstream, is much less than a slot.

In addition to airflow through the inlet, infiltration airflow must be included to provide a true description of the characteristics of the system. Fig. 4 illustrates infiltration rates for two types of dairy barn construction. The infiltration rates can be described as:

Tight construction:

$$I = 0.017\ P^{0.67} \qquad [3.3.4\text{-}4]$$

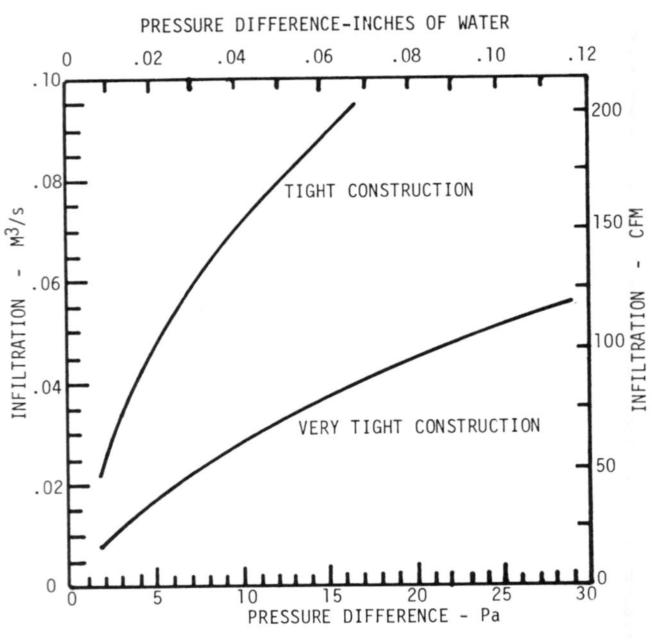

FIG. 4—INFILTRATION RATES AS A FUNCTION OF PRESSURE DIFFERENTIAL FOR TWO EXAMPLE PENNSYLVANIA DAIRY BARNS

Very tight construction:

$$I = 0.006\ P^{0.67} \qquad [3.3.4\text{-}5]$$

where
- I = the infiltration rate, m^3/s per 500 kg animal unit

Example: A dairy barn, housing 110 animal units (500 kg units) is to be ventilated with hinged baffle, wall flow, slotted inlets (see Fig. 3a). A total inlet length of 140 m is available. The barn is considered to be very tight construction. Combining equations 3.3.4-1 and 3.3.4-5 the total fresh air ventilation rate is described by:

$$Q = (0.154W^{0.98}\ P^{0.49}) + (0.66\ P^{0.67}) \qquad [3.3.4\text{-}6]$$

This characteristic equation is graphed in Fig. 5 for the range of pressures characterizing most dairy barn ventilation applications, and for seven inlet widths.

Next, assume a tentative system of fans with their controls staged based on temperature. Assume previous calculations (based on field recommendations) have shown a required maximum ventilation rate of 16.4 m^3/s in the summer and a minimum continuous ventilation rate of 2 m^3/s in the winter. Based on these needs, a fan system is chosen with one small fan to operate continuously to provide the minimum ventilation rate, and 4 additional larger fans to be staged thermostatically until all 5 fans operate to provide the maximum ventilation rate. This results in 5 possible modes of fan system operation, which are also graphed on Fig. 5. The increments between the fan stages are not meant to imply an optimum system. In practice it is usually better to make the increment for the lower temperature stages smaller with the final one or two increments quite large.

The intersection points of the fan characteristic and inlet (planned and unplanned) characteristic curves indicate potential operating points of the complete system. For the situation shown in the graph, the minimum winter ventilation rate is attainable with the inlets closed completely. Even then the pressure difference will be only 5 Pa making wind effects significant. A minimum pressure difference of 10 to 15 Pa is needed to limit wind effects. The maximum summer ventilation

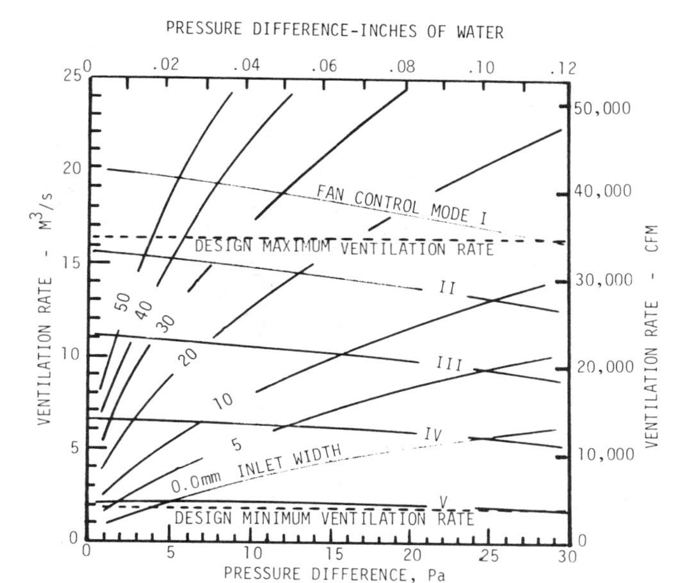

FIG. 5—SLOTTED INLET DESIGN EXAMPLE (SEE PARAGRAPH 3.3.4.2) SYSTEM CHARACTERISTICS CURVES

rate can be attained with an inlet opening of 20 mm and a pressure difference of 19 Pa. If inlet width is not automatically controlled by an inside to outside pressure differential sensor, the system characteristic graph provides insight into the strategy to choose proper inlet width.

3.3.5 Modulation of ventilation rate. A means of regulating ventilation rate must be provided to permit control of temperature and humidity in the building under various climatic conditions and to accommodate variations in size and numbers of livestock. The following basic methods of varying ventilation rate are used, either singly or in combination:

3.3.5.1 Variable-speed fan. Special permanent split-capacitor motors designed for high slip operation permit modulating fan volume smoothly from maximum down to as low as 10% of maximum by regulating the fan speed. Manufacturers should be consulted for actual speed ranges available on their fans. Speed is regulated by varying the RMS voltage to the motor by a variable transformer or solid-state control, either manually or thermostatically. The motors must be direct-coupled to the fan since they do not develop enough torque to start a belt-driven fan at low speed.

3.3.5.2 Multi-speed fan. Large fans, 373 W (½ hp) and larger, can be equipped with two-speed motors that provide two ventilation rates, the lower one about 60% of the maximum rate. Small direct-driven fans can be equipped with permanent split-capacitor motors with up to 5 speeds.

3.3.5.3 Intermittent fan operation. A minimum ventilation rate can be provided for intermittent operation of fans sized for the maximum ventilation rate required for a system. The intermittent operation can be controlled by a percentage timer that typically operates for a set period of time on a ten-minute interval or controlled by a thermostat, or by a combination of the two. Temperature stratification and odor and bacterial buildup can be a greater problem when intermittent fan operation is used, especially during cold weather. Fans should be sized to operate at least 50% of the time with frequent cycles.

3.3.6 Exhaust fan and thermostat locations. Fans should be located so they will not exhaust against prevailing winds. If structural or other factors make it necessary to install fans on the windward side, it is importatnt to select fans rated to deliver the required capacity against at least 30 Pa (0.12 in. H_2O) static pressure and having a relatively flat power curve. Without wind protection such as a weatherhood, the fan must be equipped with a motor of sufficient size to withstand a wind velocity of 14 m/s (30 mph), equivalent to a static pressure of 100 Pa (0.4 in. H_2O) without overloading beyond the service factor of the motor.

Since vane-axial or propeller fans normally used as exhaust fans do not operate efficiently or reliably at this static pressure, a centrifugal fan would be necessary. Due to cost and lack of noncorrosive materials on these fans, the wind-breaking weatherhood is a more viable alternative. The fan performance should be examined to determine if its output is maintained fairly consistently through static pressures as high as 31 Pa (0.125 in. H_2O).

Locate each temperature-control thermostat at a point which represents the temperature sensed by the animals. Locate thermostats away from potential physical damage. They should not be placed near an animal, a water pipe, a light, heater exhaust, outside wall or any other object which will affect their action. Because the temperature near the ceiling will be somewhat higher than near the floor, the proper setting of the thermostat should be made with reference to temperature indicated by a thermometer at the level occupied by the animals. A recommended location for the thermostat may often be near the fan exhaust, especially when thermometers are used in the animal microenvironment to determine the final setting of the thermostat.

3.3.7 Wiring for electric ventilating fans. All wiring must conform to Article 547 and other appropriate articles in American National Standard ANSI/NFPA No. 70, National Electrical Code.

3.3.8 Emergency warning system. The confinement of animals in high density, windowless shelters in a mechanically controlled environment involves considerable financial risk in the event of power or equipment failure. The failure of ventilating equipment can result in serious impairment of animal health or mortality from heat prostration. An adequate alarm system to indicate failure of the ventilation equipment is highly recommended for mechanically ventilated structures. An automatic standby electric generator should also be considered. There are many types of alarm systems for detecting failure of the ventilation system. These range from inexpensive "power off" alarms to more expensive systems for sensing interruptions of airflow, temperature extremes and certain gases. Automatic telephone dialing systems are effective as alarms and are relatively inexpensive for the protection provided. Refer to ASAE Standard S417, Specifications for Alarm Systems Used in Agricultural Structures. Alarm system components must meet the requirements of ASAE Standard S417, Specifications for Alarm Systems Used in Agricultural Structures, and Article 547 of ANSI/NFPA No. 70, National Electrical Code. Each building should have a minimum of two ventilation fans with each one on a separate circuit. If both fans were on the same circuit, the blowing of a fuse or the tripping of a circuit breaker could stop the flow of electricity to both fans. Power failure alarm systems should be capable of monitoring both "hot legs" of an electrical service. To insure that the selection and installation of an alarm system will be dependable, it is recommended that the assistance of the manufacturer or other qualified person be sought.

3.3.9 Ventilation for disease control. The airborne route of disease transmission can be blocked by filtering the incoming air and maintaining a positive pressure inside the house. More data are needed, but reported evidence suggests the following:

3.3.9.1 Air filters having an efficiency of 95% (based on ASHRAE Standard 52-76, Method of Testing Air-Cleaning Devices Used in General Ventilation for Removing Particulate Matter) are probably adequate.

3.3.9.2 Positive pressure of 60 to 70 Pa (0.25 to 0.30 in. H_2O) inside the building (relative to outside static pressure) is probably adequate where the outside wind speed seldom exceeds 11.2 m/s (25 mph).

3.3.9.3 To block other routes of disease transmission, it is necessary to decontaminate the building before the livestock or poultry are put in and to prevent organisms from being brought in with the animals, feed and other supplies, or the caretaker.

3.4 Supplemental heating and cooling

3.4.1 Supplemental heat. Brooders or heat lamps are usually required for young chicks, ducklings, poults, quail, and pigs, and may be beneficial for calves, lambs, and other young animals that are housed apart from their dams. Various types of heating equipment can be incorporated into ventilation systems if desired. Supplemental and/or primary heating systems must be installed in accordance with applicable codes and the manufacturers installation instructions. Check with other ASAE standards that might apply; i.e., ASAE Engineering Practice EP258 Installation of Electric Infrared Brooding Equipment. Be sure to check with local building code authorities and the insurance carrier to see if other restrictions apply.

3.4.2 Supplemental cooling. Cooling, although not as commonly used as supplemental heating, can be of economic benefit in hot climates for dairy cows, swine and poultry. Under intensive housing conditions, supplemental cooling may be necessary during heat waves to prevent heat prostration and mortality or serious losses in production and reproduction. The cooling equipment, like heating equipment, may be incorporated into the ventilation system if desired, and can consist of supplemental air movement, evaporative cooling or earth-tempered cooling.

3.4.2.1 Air movement. Increased air movement during heat stress conditions can increase growth and decrease water consumption of chicks, improve heat tolerance of chickens, improve feed efficiency of cattle, and increase growth of swine.

3.4.2.2 Evaporative cooling. Cooling by evaporation can be applied directly to the animals as with foggers for poultry or sprinklers for swine; or can be used to cool the air, as with pad-and-fan or packaged evaporative coolers for dairy cattle shelters or poultry houses.

3.4.2.3 Mechanical refrigeration. Mechanical refrigeration systems can be designed for effective animal cooling but are considered economically impractical for most production systems.

SECTION 4—VENTILATION FOR DAIRY COWS AND CALVES

4.1 Environmental requirements for mature dairy cattle. Mature dairy animals can adapt to wide variations in environmental conditions without exhibiting significant production decreases. Small losses in milk production occur within a temperature range of 2 to 24°C (35 to 75°F) with coincident relative humidities from 40 to 80%. Increases in feed consumption occur as temperatures drop below 10°C (50°F) to compensate for added body heat loss. Fig. 6 illustrates air temperature effects on milk yield for Holstein and Jersey cows. The Temperature-Humidity Index, THI, provides a reasonable measure of the combined effects of humidity with air temperature above 21°C (70°F) (Ref. Berry, et al., 1964) such that:

$$MPD = 1.08 - 1.736NL + 0.02474(NL)(THI) \quad [4.1\text{-}1]$$

where

MPD = absolute decline in milk production, kg/day·cow
NL = normal level of production, kg/day·cow
THI = daily mean value of temperature humidity index, obtained from the dry-bulb temperature (t_{db}, °C) and dewpoint temperature (t_{dp}, °C) according to the relation:

$$THI = t_{db} + 0.36\, t_{dp} + 41.2 \quad [4.1\text{-}2]$$

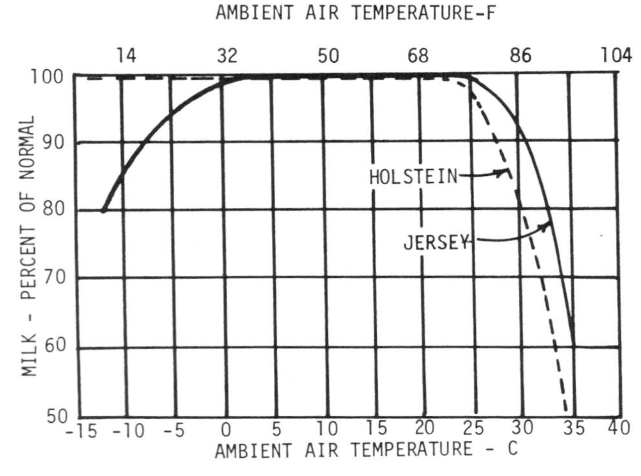

FIG. 6—EFFECT OF AMBIENT AIR TEMPERATURE ON MILK PRODUCTION IN HOLSTEIN AND JERSEY CATTLE. Relative humidity ranged from 55 to 70%. (Yeck and Stewart, 1959)

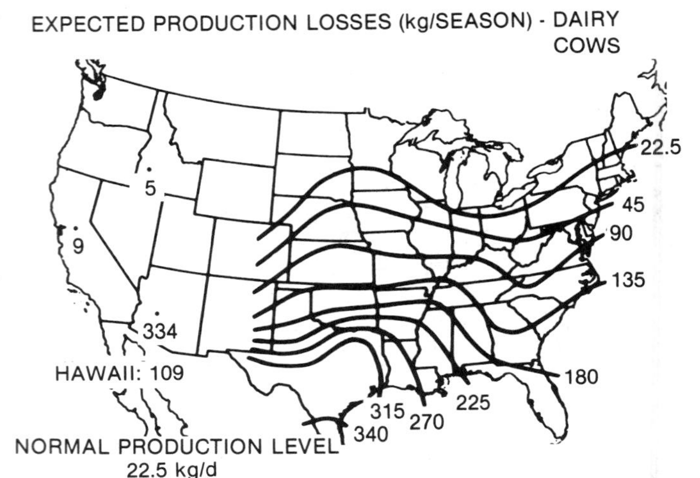

FIG. 7—EXPECTED SEASONAL MILK PRODUCTION LOSSES (kg/cow) FOR 122 DAY SUMMER PERIOD FOR COWS OF 22.5 kg/d (50 lb/day) PRODUCTION LEVEL. (Hahn and Osburn, 1969)

Below 2°C (35°F), production efficiency declines and management problems increase.

4.2 Heat and moisture produced by dairy cattle. Table 1 provides an estimate of heat and moisture produced by dairy cattle under conditions with varying ambient temperatures.

4.3 Winter ventilation requirements. Ventilation rates for maintaining heat balances or for removal of excess moisture can be determined using procedure described in paragraph 3.3.1.

4.4 Summer ventilation requirements. Ventilation rates for enclosed buildings should be adequate to hold inside air temperatures within 1 to 2°C (2 to 4°F) of outside temperatures during hot weather. The required ventilation rate is a function of animal size and density, building construction and waste management practices. Refer to paragraph 3.3 for ventilation rate calculation procedures.

4.5 Aids for summer cooling. At temperatures above 21°C (70°F), cooling can increase productivity, conception rates and feed efficiency (see paragraph 4.1). Predicted milk production losses for June - September, inclusive, with only shades provided are shown in Fig. 7 for cows of 23 kg/d (50 lb/day) normal production level.

4.5.1 Evaporative cooling. Adequately designed and maintained systems using wetted pads and fans have potential for economic application in large areas of the U.S. Fig. 8 shows predicted

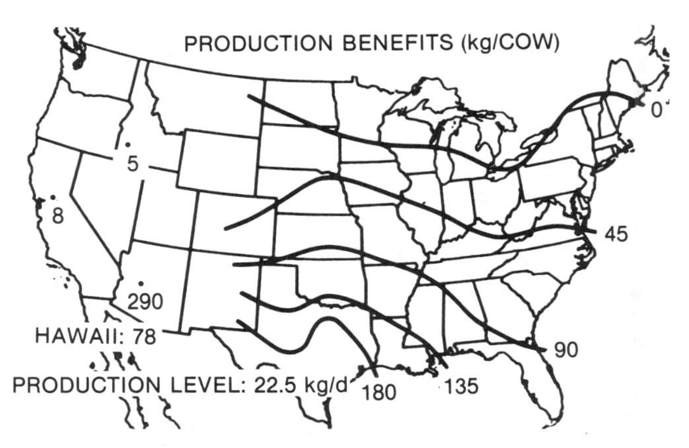

FIG. 8—EXPECTED SEASONAL PRODUCTION BENEFITS FROM EVAPORATIVE COOLING DURING 122 DAY SUMMER PERIOD FOR COWS OF 22.5 kg/d (50 lb/day) PRODUCTION LEVEL. (Hahn and Osburn, 1970)

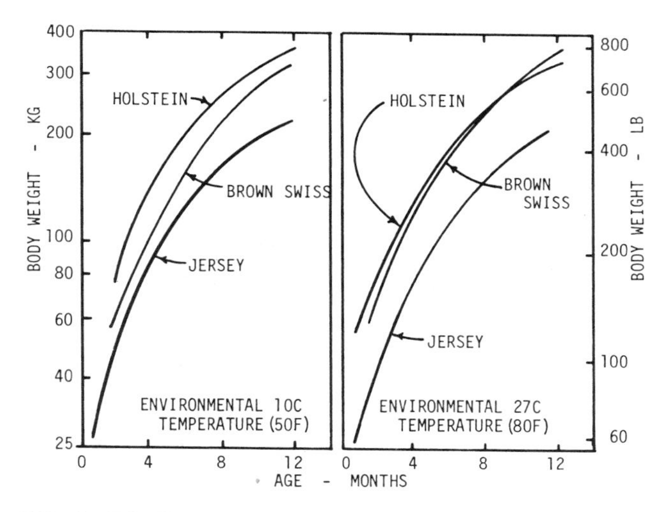

FIG. 9—BODY WEIGHT VS. AGE OF HOLSTEIN, BROWN SWISS, AND JERSEY CALVES AT 10°C (50°F) AND 27°C (80°F)

production benefits. System design information is available in University of Arizona Report P-25 (Ref. Stott, et al., 1972).

4.5.2 Other water cooling methods. Specific heat and latent heat of vaporization of water can be utilized directly for increased animal comfort in hot weather by such methods as spraying with sprinklers or foggers and supplying cooled drinking water.

4.5.3 Cooling inhalation air. Outside air cooled by refrigeration to 15°C (60°F), supplied to head enclosures at the rate of 0.7-0.85 m^3/min (25-30 ft^3/min) per cow, has been shown to benefit milk production for cows in hot environments (Ref. Hahn, et al., 1965).

4.5.4 Air conditioning of total space. Air conditioning of dairy cow housing can require 2500 or more J/s of refrigeration per cow, depending on local design conditions and the individual situation. Close attention must be given to air filtration, adequate ventilating air and to maintenance. Feasibility analyses indicate limited application only to high-producing cows in hot humid areas of the U.S.

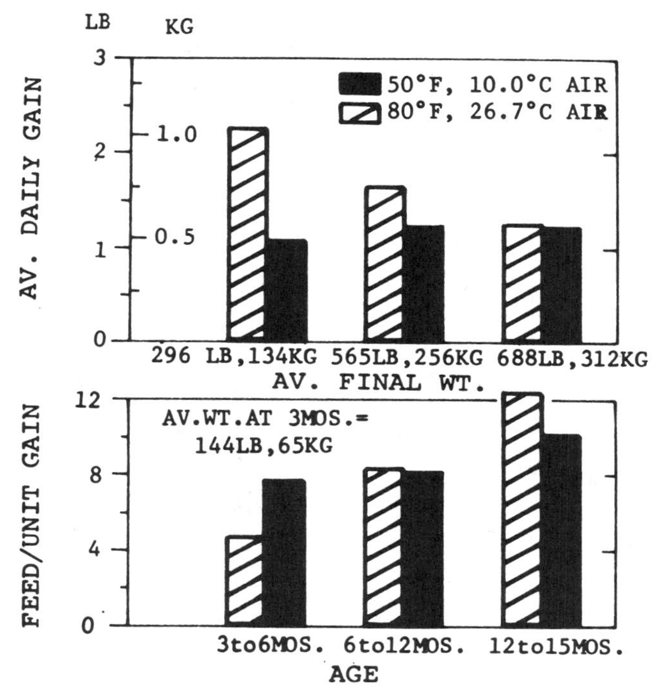

FIG. 10—RATE OF GROWTH AND FEED UTILIZATION (UNIT FEED PER UNIT GAIN) OF SHORTHORNS. (Ragsdale, Cheng and Johnson, 1985)

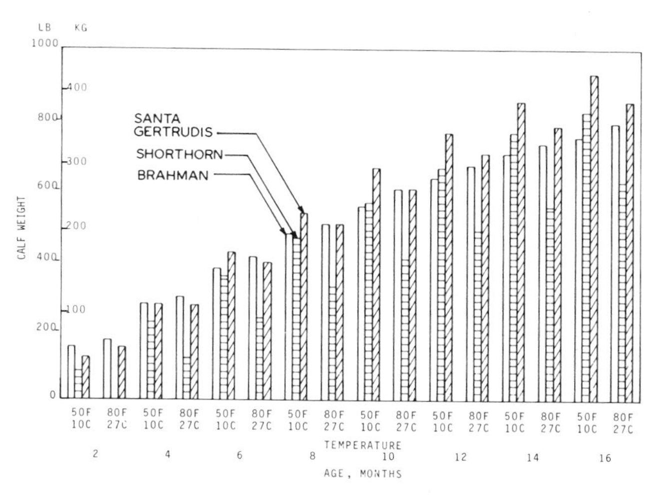

FIG. 11—EFFECTS OF AGE AND TEMPERATURE ON WEIGHT OF BRAHMAN, SHORTHORN AND SANTA GERTRUDIS CALVES. (Ragsdale, Cheng and Johnson, 1958)

4.6 Dairy calf ventilation. Moisture and heat production rates differ between calves and mature animals necessitating the use of different system design parameters. Both research and field experience have shown that calves have a high tolerance for temperature variations so long as the variations are not abrupt. Fig. 9 shows the effects of selected environmental temperatures on the growth rate, and Table 1 shows the moisture production rates for several breeds of dairy calves. Calves are more susceptible to the effects of manure gases and airborne organisms than are mature cows. This must be taken into account when designing minimum ventilation rates.

4.7 Free-stall barn ventilation. Free-stall barns used to house dairy cattle may be of the cold-enclosed, cold three-sided or warm-enclosed design. Cold barns, barns where no conscious effort is made to maintain a minimum temperature in the barn during winter months, are usually ventilated non-mechanically whereas warm barns, barns that are enclosed and insulated in an attempt to keep interior winter temperatures above freezing, are mechanically ventilated. Research data are not available for the design of mechanical ventilation systems for free-stall barns. However, experience suggests that the broad alleys and associated greater area of exposed wet surface increase moisture evaporation by 10-15% over a tie-stall type barn with narrow gutters, thereby necessitating appropriate increases in ventilation rates to maintain relative humidity levels below 65-70%. Wide free-stall barns and vaulted ceilings make mechanical ventilation difficult.

4.8 Non-mechanical ventilation. Although data herein are primarily directed at providing guidance in the design of powered or mechanically ventilated structures, it should be understood that in many cases dairy cattle structures can be satisfactorily ventilated by non-mechanical ventilation systems (also called non-powered, gravity or natural ventilation) if appropriate features are built into the structure. This approach has found widespread use in free-stall barns. While some work has been done in this area, most information is based on models and computer simulations. The influence of animal heat on ventilation by non-mechanical means is not yet clearly understood.

SECTION 5—VENTILATION FOR BEEF CATTLE

5.1 Definition of the confinement growing of beef cattle: The housing of beef within a structure which is enclosed on all sides (with or without windows or window openings), ventilated, insulated and possibly heated to allow some control of interior temperatures.

5.2 Environmental requirement for beef cattle

5.2.1 Temperature. Temperature influences the feed conversion and weight gain of beef cattle. Figs. 10 and 11 show the type of response to temperature found under research conditions where temperature is controlled. Fig. 12 illustrates the effect of temperature on feed requirements.

5.2.2 Relative humidity. Relative humidity has not been shown to influence animal performance except when accompanied by thermal stress. Relative humidities consistently below 40% may contribute to excessive dustiness, and above 80% may increase building and equipment deterioration.

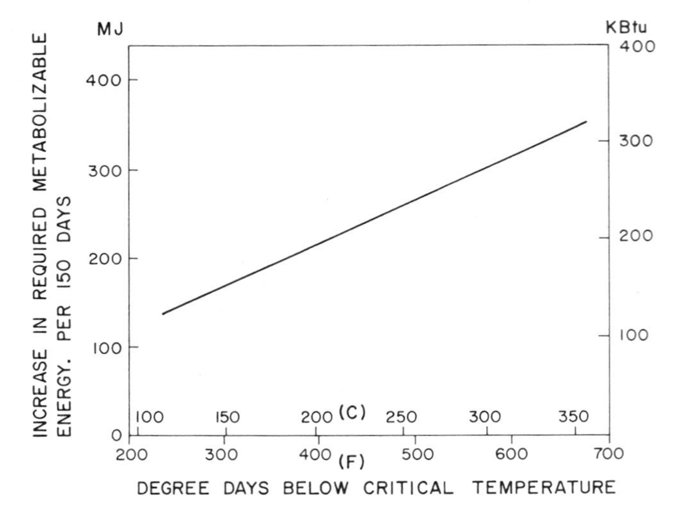

FIG. 12—EFFECT OF DEGREE-DAYS BELOW CRITICAL TEMPERATURE, -14°C (6°F) ON METABOLIZABLE ENERGY REQUIREMENTS OF A 400 kg (900 lb) BEEF STEER OVER 150 DAY PERIOD. (Webster, 1971)

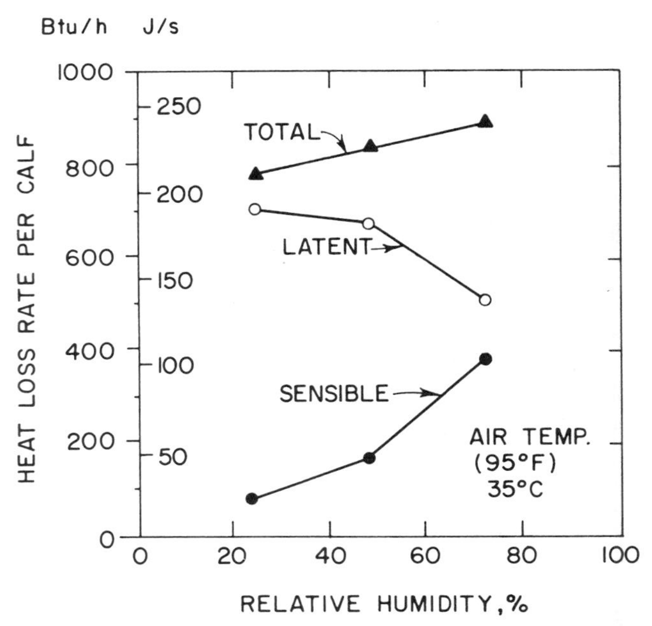

FIG. 14—EFFECT OF HUMIDITY ON HEAT LOSSES OF AYRSHIRE BULL CALVES 6 TO 10 MONTHS OF AGE. (McLean, 1963)

5.3 Heat and moisture produced by beef cattle. Figs. 13 and 14 give total, latent and sensible heat production values for Ayrshire bull calves ranging in age from 6 to 10 months as affected by temperature and relative humidity. Table 1 provides total, latent and sensible heat production from Shorthorn, Brahman and Santa Gertrudis calves at temperatures of 10 to 27°C (50 to 80°F). These data represent values for the animals on full feed and include heat and moisture transfer effects between the animal environment and a bedded concrete floor. For design purposes the sum of sensible heat from the room and latent heat from the room is the total heat loss from the beef cattle. However, the ratio of sensible heat to latent heat from the beef cattle may be quite different from the ratio of sensible heat from the room to latent heat from the room because of the utilization of some sensible animal heat to vaporize moisture from the floor. The data can be used for normal management practices in solid floor buildings. Daily average temperature should be used in estimating heat and moisture production. Fig. 15 illustrates the effect of ration on hourly heat production at environmental temperatures of 20, 30 and 40°C (68, 86 and 104°F).

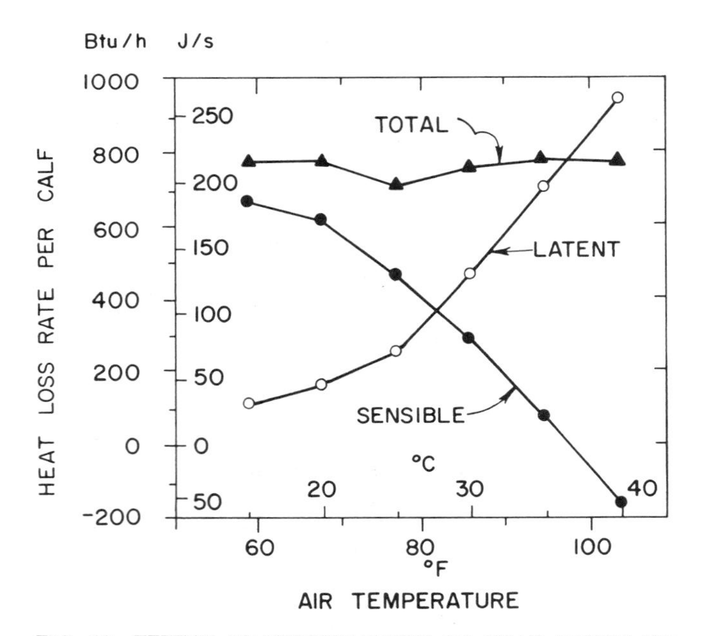

FIG. 13—EFFECT OF TEMPERATURE ON HEAT LOSSES OF THREE AYRSHIRE BULL CALVES 6 TO 12 MONTHS OF AGE. Vapor pressure 1066 Pa, dewpoint temperature 8°C (46°F). (McLean, 1963)

5.4 Winter ventilation requirements

5.4.1 Design values. Inside air temperature, 7 to 18°C (45 to 65°F). Inside relative humidity, 40 to 80%.

5.4.2 Procedure. For determining ventilation rates for heat balance and moisture removal, see paragraph 3.3.1.

5.5 Summer ventilation requirements

5.5.1 Air change rates. Air change rates are adjusted to remove heat produced by the cattle, plus other heat gains, and to minimize inside temperature rise above the outside air temperature. For specific application, the ventilation rate should be calculated based on procedures in paragraph 3.3. Consideration must also be given to air distribution and velocity to aid in animal comfort and feed efficiency. Baffles should be employed to direct airflow at animal level. This may be accomplished by baffled center ceiling inlets or wall baffles.

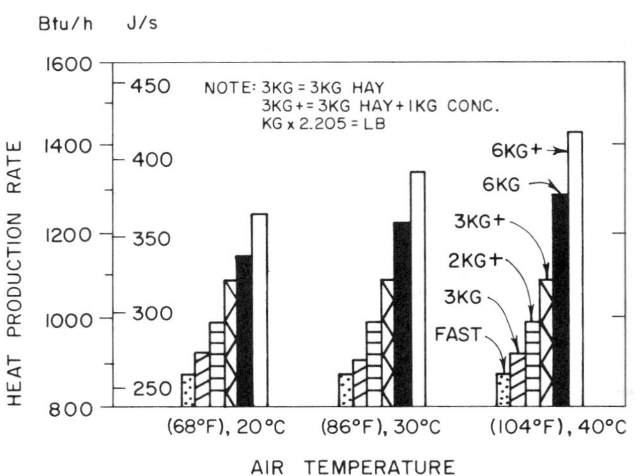

FIG. 15—EFFECT OF RATION ON HEAT PRODUCTION OF A GRADE STEER 20 TO 24 MONTHS OF AGE. Hay was poor quality. Concentrate was cotton cake and barley meal. Fasting was for 72 h. The "+" sign indicates addition of 1 kg concentrate to ration. (Rogerson, 1960)

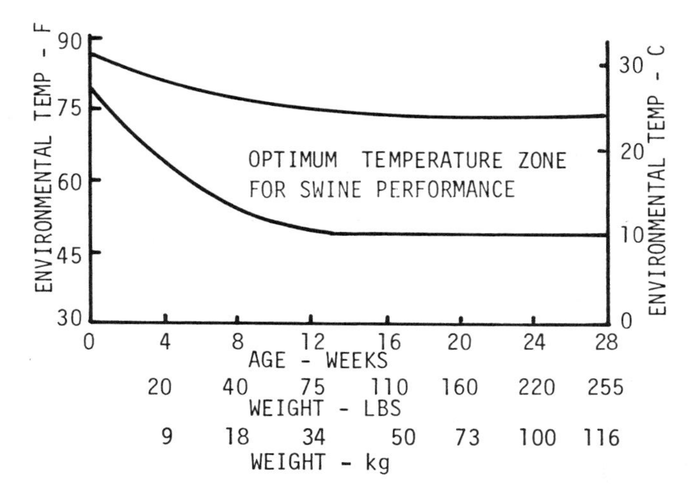

FIG. 16—APPARENT OPTIMUM TEMPERATURE ZONE FOR SWINE

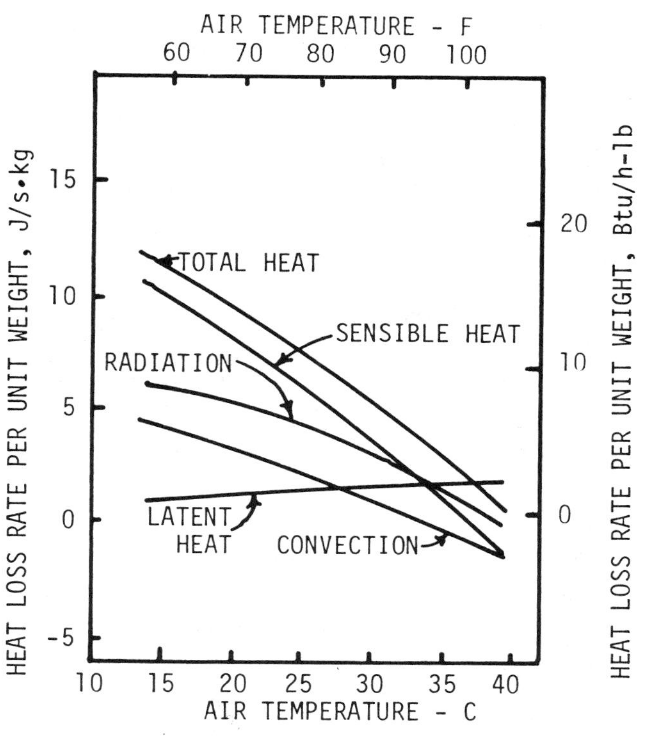

FIG. 17—HEAT LOSS RATE PER UNIT WEIGHT OF NEW-BORN PIGS VS. AIR TEMPERATURE. (Butchbaker and Shanklin, 1964)

SECTION 6—VENTILATION FOR SWINE

6.1 Environmental requirements for swine. The following physical environment factors should be considered when designing a ventilation system for swine:

6.1.1 Temperature. Different size swine have different air temperature ranges for maximum performance (see Fig. 16). Small pigs, up to 18 kg (40 lb), require a much higher temperature than larger pigs. Pigs larger than 18 kg (40 lb) have an optimum feed efficiency when temperatures are between 20-24°C (50-70°F). Temperatures in excess of 32°C (90°F) will reduce the fertility of sows, gilts and boars.

6.1.2 Air movement. Pigs will generally benefit from air movement if the air temperature is between 32-39°C (90-103°F). Above 39°C (103°F), large rates of air movement will increase heat load on the pig. Pigs prefer still air or, at most, a very low rate of air movement, 0.15 m/s (30 ft/m) when the air temperature is at or below the optimal level for the pig.

6.1.3 Relative humidity. Relative humidity has not been shown to influence animal performance except when accompanied by thermal stress. Relative humidities consistently above 80% can cause condensation on building and equipment surfaces during cold weather. This condensation enhances pathogenic organism survival and promotes building and equipment deterioration. Relative humidity consistently below 40% may contribute to excessive dustiness.

6.1.4 Air quality. Contaminants in the air that hogs breathe can influence swine health. These include:

6.1.4.1 Dust. Dust can act as a respiratory irritant, but by itself has not been shown to cause disease in swine. Dust acts as a conveyance mechanism for pathogens to the respiratory tract of animals.

6.1.4.2 Pathogens. Pathogenic bacteria and viruses can be transmitted through the air.

6.1.4.3 Manure gases. Manure gases, especially ammonia, act to irritate the respiratory tract and defeat the immune response system of animals. Waste agitation below a slotted floor can bring hydrogen sulfide to toxic concentrations for both humans and pigs. These hydrogen sulfide concentrations have exceeded 950 mg/m^3. Studies have shown that animals exposed continuously to levels of about 24 mg/m^3 of hydrogen sulfide develop fear of light, loss of appetite and nervousness. Symptoms at levels between 59 and 238 mg/m^3 have included vomiting, nausea and diarrhea; however, winter ventilation rates designed to remove moisture from the building have been found to maintain the hydrogen sulfide level well below 24 mg/m^3 and closer to 1.2 mg/m^3. A tight structure is especially dependent on the reliability of the ventilating system to remove toxic gases.

6.2 Heat and moisture produced by swine. Table 1 provides an estimate of moisture production, sensible heat loss and total heat loss of pigs. Fig. 17 shows the heat loss at various air temperatures for newborn pigs. Fig. 18 gives sensible and latent heat production in hog buildings at various air temperatures for 18 and 100 kg (40 and 220 lb) hogs. These data were collected in a chamber having unbedded, solid concrete floor scraped twice daily. For design purposes, the sum of sensible heat and latent heat from the room is the total heat loss from the hogs. However, the ratio of sensible heat to latent heat from the hogs may be quite different from the ratio of sensible heat from the room to latent heat from the room because some sensible animal heat is expended to vaporize moisture from the floor. The data in Table 1 can be used for normal management practices in solid floor buildings. The

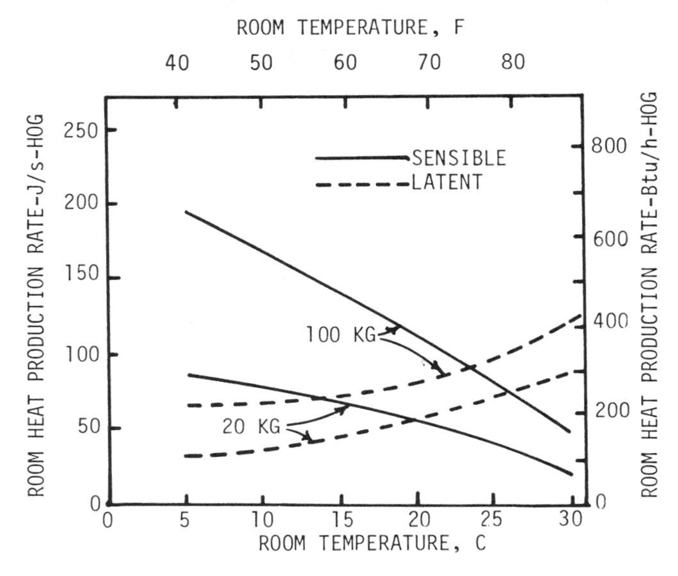

FIG. 18—ROOM SENSIBLE AND LATENT HEAT PRODUCTION RATE IN A HOG HOUSE (SOLID FLOOR)

latent heat from buildings may increase as much as one-third, (with a corresponding decrease in sensible heat) because of floor flushings, water wastage, temperature, ventilation rate, inside relative humidity, air velocity over the floor surface, and less frequent scraping of floors. Under similar conditions, less moisture is removed from slatted-floor structures. Slatted floors reduce the amount of wet surface exposed to the air compared to solid floors. About 40% as much water is evaporated from a totally slatted floor as from a solid concrete floor, while partial slats (approx. 35% of floor area slatted) produce 75% as much evaporation as solid floors. Daily average temperatures should be used in estimating heat and moisture production from Table 1.

6.3 Winter ventilation requirements

6.3.1 Minimum ventilation. For determining ventilation rates for heat balance and moisture removal, see paragraph 3.3.1. Table 3 shows ventilation rate guidelines for determining fan capacity. Use both calculated rates (see paragraph 3.3.1 and Table 3) in determining the actual minimum ventilation rate. If the calculated rate is lower than the recommended value in Table 3, the rate in Table 3 should be used to assure the removal of stale, odorous air.

6.3.2 Winter maximum ventilation. The winter maximum ventilation rate is normally adequate for temperature control as the outside temperature approaches 10°C (50°F).

6.4 Summer ventilation requirements. Air exchange rates are adjusted to remove the heat produced by the hogs plus other building heat gains to limit inside air temperature rise to 1-2°C (2-4°F) above the outside air temperature. Consideration must be given to air distribution and velocity to help animal comfort, sanitary training of animals, and feed efficiency.

SECTION 7—VENTILATION FOR BROILER CHICKENS AND YOUNG TURKEYS

7.1 Definition for broiler/poultry housing: This section applies to structures used for brooding and/or growing of broiler chickens or young turkeys. These structures are generally well-insulated, fully enclosed structures. Some may have curtains over part of the sidewall. The structures permit some modification of the interior environment through use of supplemental heating and ventilating equipment. Supplemental evaporative cooling is often employed. Ventilation can be accomplished by a mechanical ventilation system or a natural ventilation system supplemented by stirring and mixing fans. A common method is to use controlled mechanical ventilation during brooding and/or cold weather and to use natural, curtain ventilation during warm weather when high air exchange rates are necessary. This type of system is currently referred to as flex housing (Ref. Timmons and Baughman, 1983). Lighting programs can be imposed in enclosed structures or in opaque curtain-walled houses. The growth rate of broilers today is considerably faster than just a few years ago. Broilers are currently marketed at 7 weeks of age (1.8 kg). A decade ago broilers required about 8 weeks to reach market weight. A comparison of typical growth rates for broilers for 1967 and 1978 data is presented in Fig. 19. Some of the results presented in the following sections are based on poultry research conducted when growth rates were slower. However, one is able to evaluate the relative effects of temperature, temperature cycles, humidity, light, air movement, etc., on the overall performance of the birds based on the materials presented.

TABLE 3—VENTILATION-RATE GUIDELINES FOR DETERMINING FAN CAPACITY

Species	Winter minimum ventilation rate, $(m^3/s \cdot sow)10^{-2}$	Winter maximum ventilation rate, $(m^3/s \cdot sow)10^{-2}$	Summer ventilation rate, $(m^3/s \cdot sow)10^{-2}$
Swine			
Sow and litter	0.95	3.7	23.6
Growing pigs:			
9-18 kg	0.095	0.71	1.7
18-45 kg	0.24	0.95	2.3
45-68 kg	0.33	1.18	3.4
68-95 kg	0.47	1.65	4.7
Gilt, sow, or boar			
91-114 kg	0.47	1.65	5.7
114-136 kg	0.57	1.89	8.5
136-227 kg	0.71	2.12	11.8

Reference: (Midwest Plan Service, 1982.)

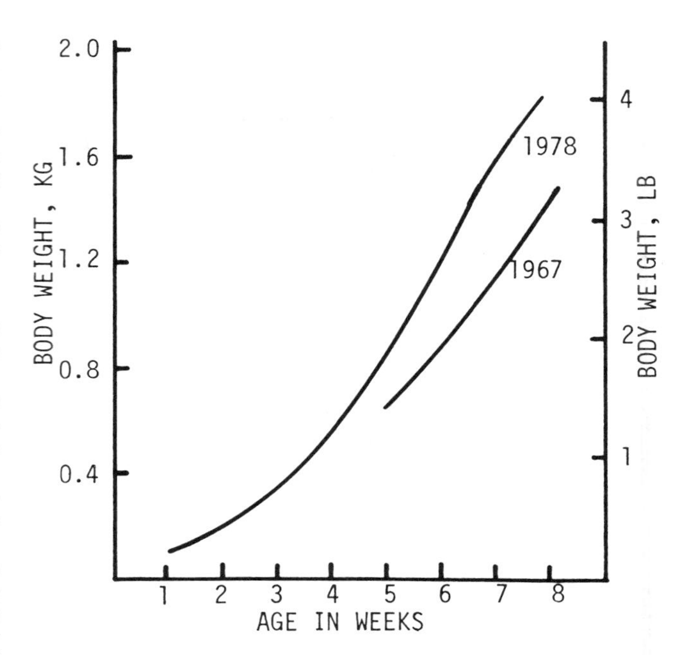

FIG. 19—COMPARISON OF TYPICAL GROWTH RATES OF BROILERS IN 1967 AND 1978

7.2 Environmental requirements for broilers. Requirements presented herein cover the growing/finishing period (over 4 weeks of age). They reflect the current knowledge of how the immediate environment affects the economic efficiency of broiler meat production, hereafter referred to as performance. Most of the data in this section apply to birds reared on litter floors and may need to be adjusted for birds in cages or coops. Where data are given for only one sex, it should be noted that by 7 weeks of age the average weight of male broiler chickens is about 1.3 times that of females and that environmental conditions tend to exert somewhat greater influence on growth rates of males than of females.

7.2.1 Temperature. The influence of various constant temperatures on feed efficiency, rate of gain and sensible and latent heat production for mixed sex broilers is given in Figs. 20-26. Feed efficiencies (gain/feed) during growout increase almost linearly with increasing ambient temperatures from 7 to 15.6°C (45 to 60°F). The percentage increase is different depending upon whether the birds are floor reared where huddling behavior decreases low temperature effects (0.42% per °C)(Ref. Deaton, et al., 1977) or if the birds are cage reared (1.0% per °C) (Ref. Prince, et al., 1961). Growth rates decrease and mortality increases for daily temperatures exceeding 35°C (95°F) (Ref. Griffin and Vardaman, 1970) and growth rates may begin to decrease below 10°C (50°F).

7.2.2 Humidity. During the first 2 to 3 weeks of brooding, chicks, especially those from smaller eggs, may require relative humidities of 60% or higher, but during growout the performance of commercial broilers is only slightly affected by relative humidities between 30 and 80% when the dry-bulb temperature is below 29°C (85°F), see Fig. 21. At 7 weeks of age broilers are severely heat stressed by wet-bulb temperatures above 27.5°C (81.5°F) when dry-bulb temperatures are above 29°C (85°F). At stocking densities of 0.07 to 0.093 m^2 (0.75 to 1.0 ft^2) per broiler on litter floors, the litter becomes too wet (slick, caked) when the average relative humidity remains consistently above 80% and too dry (dusty) when the relative humidity remains below 40%. At 0.05 m^2 (0.5 ft^2) per broiler, it is difficult to maintain satisfactory litter condition regardless of the humidity.

7.2.3 Air movement. Air speed greater than 0.3 m/s (60 ft/min) at broiler height is not recommended for broilers under 2 weeks of age. Increasing air speeds from 0.2 to 2.5 m/s (40 to 500 ft/min) around broilers has been shown to improve the weight gains and water-use efficiencies of 1.4 kg (3 lb) broilers when the temperature was cycled diurnally from 21 to 36°C (70 to 96°F). Similar increases in air speed alleviated heat stress in 8-week old broilers that were exposed to environment temperatures less than the chicken's body

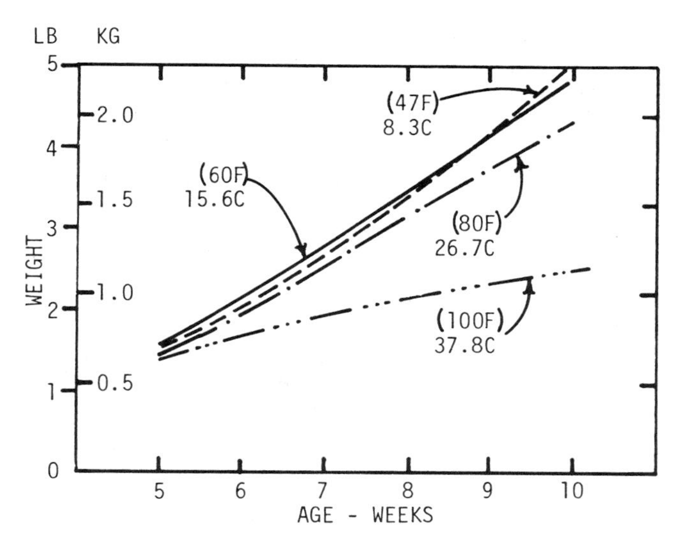

FIG. 20—EFFECT OF AIR TEMPERATURE ON BODY WEIGHT OF MALE BROILERS. Relative humidity 60% at all temperatures exeept 80% at 8.3°C. (Winn and Godfrey, 1967)

temperature of 40.6°C (105°F). At environmental temperatures above body temperatures, an increase in wind speed will exacerbate the heat stress.

7.2.4 Radiant heat. As stated in paragraph 7.2.1, supplemental radiant heat is beneficial to broilers in cool surroundings, but in hot, humid weather the radiant heat from an uninsulated roof can increase mortality in larger broilers. Radiant heat exchange between broilers and the inside surfaces of well-insulated houses is probably of little consequence in most situations.

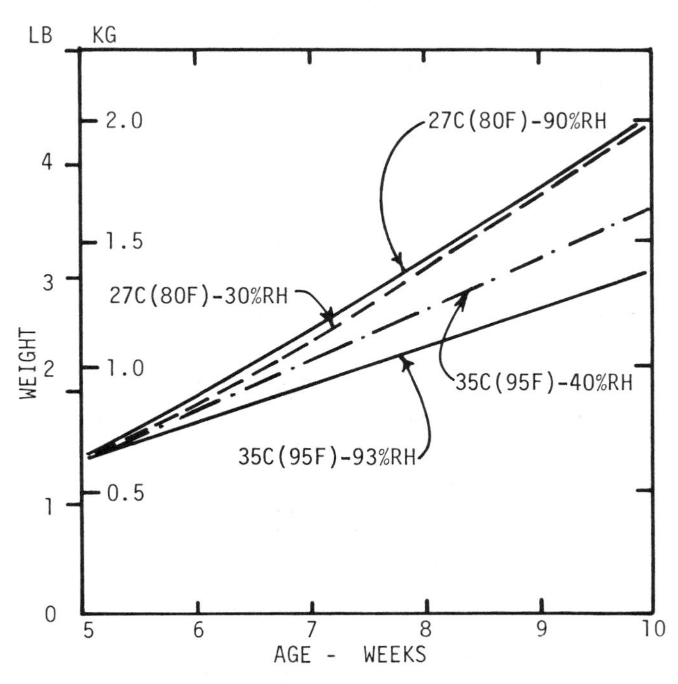

FIG. 21—EFFECT OF TEMPERATURE AND HUMIDITY ON MALE BROILER WEIGHT. (Winn and Godfrey, 1967). Constant low (30-40%) or high (80-90%) relative humidity has little effect on performance if air temperature is below 29°C (84°F).

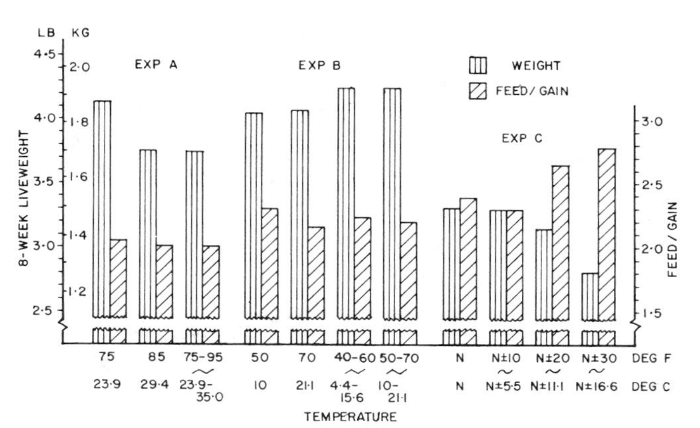

FIG. 22—CONSTANT AND CYCLIC TEMPERATURE EFFECTS ON GROWTH AND FEED CONVERSION OF MALE BROILER CHICKENS. In Experiments A and B commercial chicks were reared on litter floors at normal brooding temperatures for 3 weeks, then held at the indicated constant temperature or 24 h linear cyclic temperature for 5 weeks. In Experiment C, non-commercial chicks were reared on grid floors for 8 weeks at a normal (N) temperature schedule which started at 32°C (90°F) and decreased 2.8°C (5°F) per week to final temperature of 21°C (70°F) or at one of three 24 h sinusoidally cyclic temperature schedules; e.g., in the N ± 10 schedule the temperature was varied daily from 5.6°C (10°F) above to 5.6°C (10°F) below the normal (N) temperature for that week.

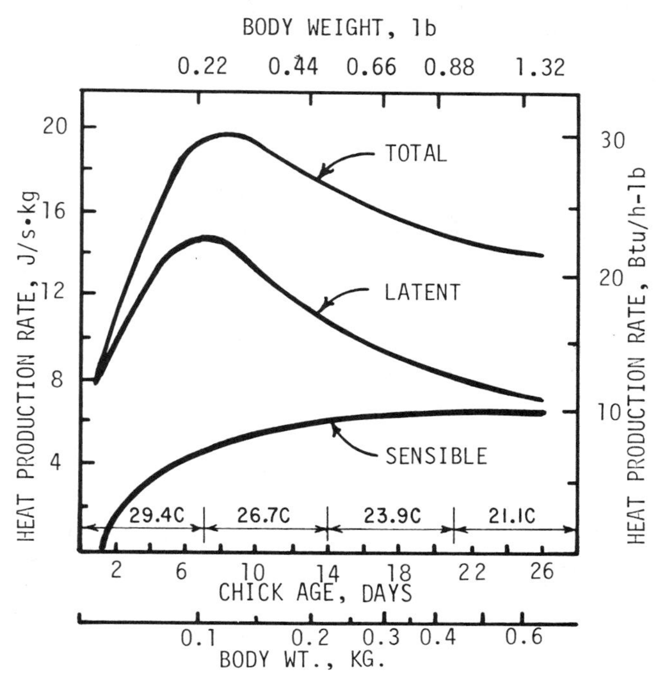

FIG. 23—SENSIBLE, LATENT AND TOTAL HEAT PRODUCTION RATES FOR BROILER CHICKENS DURING BROODING ON LITTER. (Reece and Lott, 1982a)

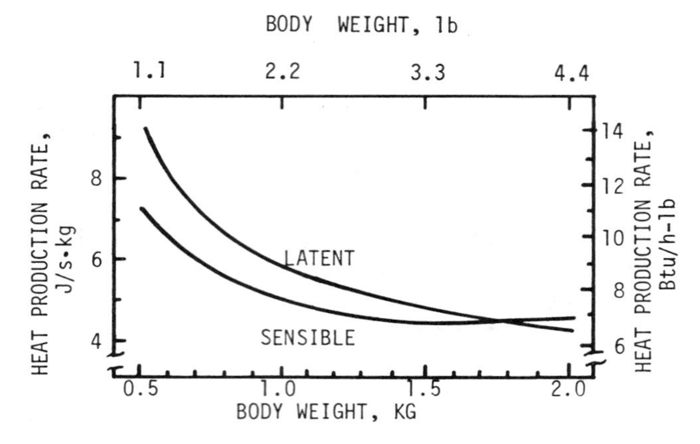

FIG. 24—SENSIBLE AND LATENT HEAT PRODUCTION RATES FOR BROILER CHICKENS GROWN AT 15.6°C (60°F) ON LITTER. (Reece and Lott, 1972b)

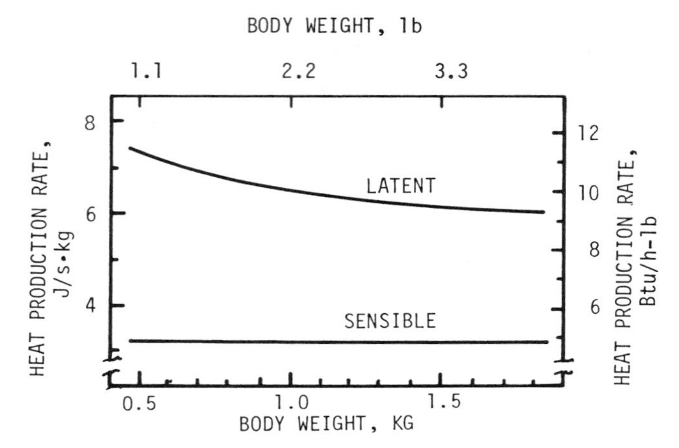

FIG. 25—SENSIBLE AND LATENT HEAT PRODUCTION RATE FOR BROILER CHICKENS GROWN AT 26.7°C (80°F) ON LITTER. (Reece and Lott, 1972b)

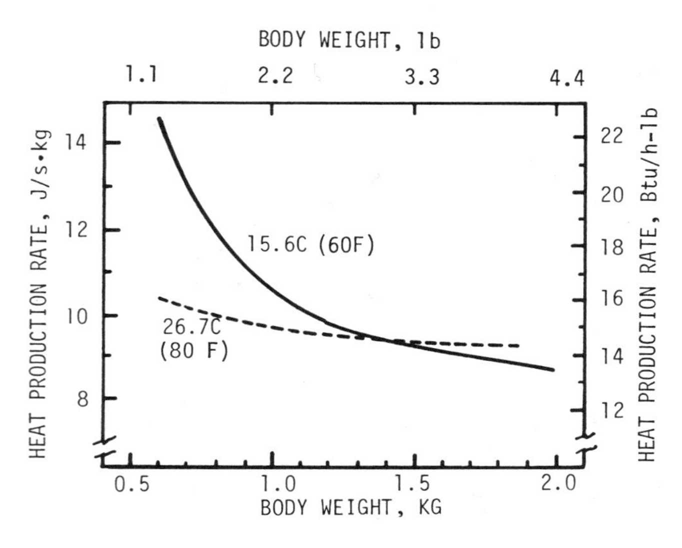

FIG. 26—COMPARISON OF TOTAL HEAT PRODUCTION RATE OF BROILER CHICKENS GROWN AT 15.6°C (60°F) and 26.7°C (80°F) ON LITTER. (Reece and Lott, 1972b)

7.3 Heat and moisture production

7.3.1 Broiler chickens. Net sensible, latent and total heat production of broilers grown on litter at stocking rates typical of partial-house brooding are shown in Figs. 23, 24, 25, and 26. The effects of heat and moisture absorption or release by the litter are included in the results presented in the figures and those studies cited in Table 2 that are based on poultry housing systems.

7.3.2 Turkeys. Total and latent heat production by turkeys are presented in Table 2. The influence of various constant temperatures upon growth and feed conversion of young turkeys is shown in Fig. 27.

7.4 Cold weather ventilation requirements

7.4.1 Brooding period. A ventilation rate of 0.18 m^3/h (0.1 ft^3/min) per chick from 1 to 10 days of age is adequate to remove moisture and supply the necessary fresh air when outside daily mean dewpoint temperatures do not exceed 3°C (38°F). For higher dewpoints and for ages past 10 days, use the sensible and latent heat balances as presented in paragraph 3.3.1 to determine necessary ventilation rates.

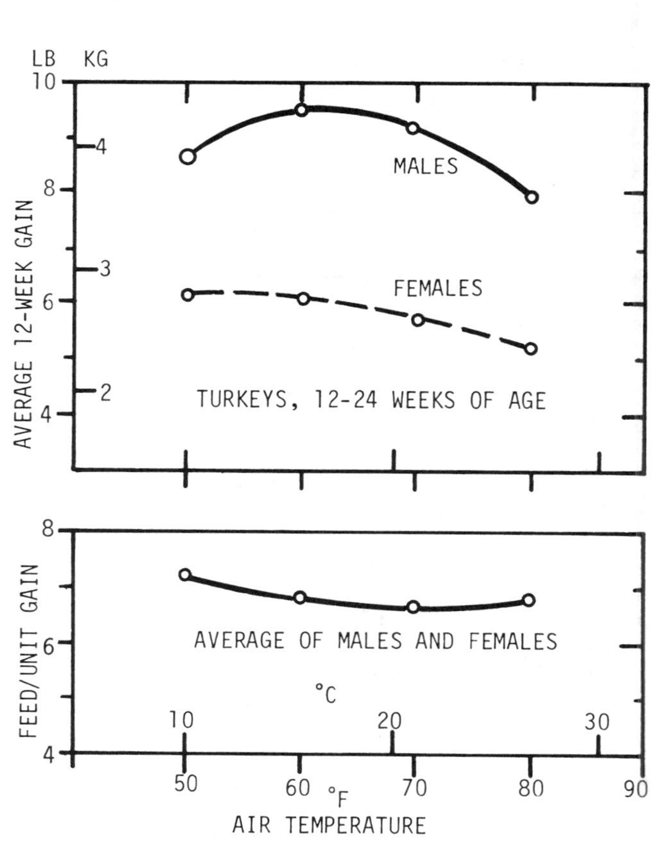

FIG. 27—EFFECT OF CONSTANT AIR TEMPERATURE ON WEIGHT GAIN AND FEED EFFICIENCY CONVERSION OF BROAD-BREASTED WHITE AND BROAD-BREASTED BRONZE TURKEYS BETWEEN 12 AND 24 WEEKS OF AGE. Relative humidity about 50% and 16 h daylength. (Hellickson, et al., 1967)

7.4.2 Growing period. Equations in paragraphs 3.3.1 are satisfactory for determining the rate of ventilation for heat balance and moisture balance. Refer to Fig. 24, Table 2, and other sources that might be available, for values of sensible and latent heat. An example to illustrate the use of the latent heat balance equation is presented below:

Example: Determine the necessary minimum ventilation rate for 7-week old broilers so that the inside dewpoint temperature will not exceed 12°C when the outside temperature is 2°C.

From Fig. 19, the body weight of 1.3 kg is read for 7-week old broilers. The latent heat production of a 1.3 kg broiler is 5.3 J/s·kg at 15.6°C (from Fig. 24) and 6.3 J/s·kg at 26.7°C (from Fig. 25) or an average of 5.8 J/s·kg. For a 1.3 kg broiler, the latent heat production is then 7.5 J/s. The latent heat of vaporization of water can be considered to have a constant value of 2.4408 MJ/kg water (1050 Btu/lb). Using the substitution of one joule is one watt·second, one can calculate the moisture production from equation 3.3.1-3 as:

$$m = 7.5 \text{ J/s}/(2.4408 \times 10^6 \text{ J/kg water})$$

$$= 0.0031 \text{ g water/s}$$

Obtaining values of 0.008857 and 0.002286 kg water per kg dry air for inside and outside humidity ratios, respectively, from a psychrometric chart, the required ventilation rate can then be calculated from equation 3.3.1-4 as:

$$M = \frac{0.0031 \text{ g } H_2O/\text{s}}{(0.008857 - 0.002286 \text{ g } H_2O/\text{g dry air}}$$

$$= 0.47 \text{ g dry air/s}$$

Considering a standard specific volume 0.83 m³/kg dry air, the volumetric ventilation rate for latent heat balance for each 1.3 kg broiler is:

$$\frac{0.47 \text{ g dry-air}}{\text{s}} \times \frac{\text{kg dry air}}{0.83 \text{ m}^3} \times \frac{1 \text{ kg}}{1000 \text{ g}}$$

$$= 3.9 \times 10^{-4} \text{ m}^3/\text{s}$$

7.5 Summer ventilation requirements. Housing for broilers and turkeys may be either open, curtain-sided housing (dominant style in the south) or totally enclosed housing (dominant style in northern climate). Ventilation requirements from a practical standpoint must be addressed dependent upon the housing style. Ventilation requirements for open-sided housing should be based upon providing air movement on the birds and relying upon natural ventilation for air exchange which will maintain inside house temperatures usually within 1°C (1.8°F) of air temperature. The use of stirring fans distributed throughout a house is the most efficient means to provide air stirring, approximately 1.8 m³/h·kg (0.5 ft³/min·lb). Design against extreme outside air temperatures by providing some form of evaporative cooling (see paragraph 7.6). For enclosed housing, ventilation rates can be based upon paragraph 3.3.1 and specifying a maximum inside to outside temperature difference of 2-4°C (4-7°F). In either housing style, ventilation should always be adequate to remove latent heat as fast as it is produced within the house. Heat and moisture production by poultry are included in Table 2.

7.6 Aids for summer cooling. During heat waves, when outside dewpoint temperatures exceed 20°C (70°F) accompanied by dry bulb temperatures exceeding 35°C (95°F), a high ventilation rate alone may not prevent heat prostration of 7-week old broilers, and some type of supplemental cooling will be needed. Fog nozzles periodically spraying water directly on the chickens provide an economical means of reducing mortality due to heat. The routine use of supplemental cooling (see paragraph 3.4.2) to relieve the milder heat stress that occurs on many summer days should also be considered. Evaporative coolers, if used, should be designed with sufficient ventilation so the temperature rise through the building does not exceed 1 to 2°C (1.8 to 3.6°F). Typical air face velocity through the cooling pads should be less than 1.0 m/s (200 ft/min) for conventional home coolers or 0.8 m/s (150 ft/min) for aspen pad-and-fan coolers. For the other new pad materials on the market today, one needs to obtain the operational characteristics from the manufacturer to design the system.

SECTION 8—VENTILATION FOR LAYING HENS

8.1 Definition of the confinement of laying hens: The housing of laying hens within a structure which is enclosed on all sides (with or without windows) and is ventilated and/or insulated.

8.2 Environmental requirements of laying hens. Maximum egg output for white birds (Leghorns) is 48.10 g/d and occurs at a temperature of 24°C (75°F) (Ref. Charles, 1984). Temperature influences feed and water consumption as shown in Table 4.

8.3 Heat and moisture produced by laying hens. The total heat produced by laying hens at various liveweights and temperatures is shown in Fig. 28. Moisture production by White Leghorn laying hens is presented in Table 5.

TABLE 4—CONSTANTS FOR DETERMINING WATER CONSUMPTION, FECAL AND WATER ELIMINATION OF LAYING HENS IN RELATION TO FEED CONSUMPTION. (Longhouse, Ota and Ashby, 1960)

	Ambient Temperature			
	-6.7-4.4 °C (20-40 °F)	10.0-15.6 °C (50-60 °F)	15.6-26.7 °C (60-80 °F)	26.7-37.8 °C (80-100 °F)
Water to feed ratio	1.5-1.7	1.7-2.0	2.0-2.5	2.5-5.0
Water-plus-feed to feces ratio	1.7	1.7*	1.8*	1.9*
Percent water content of feces	75	75	77	80
Percent water content of eggs	65	65	65	65
Egg size, g per doz (ounces per dozen)	680 (24)	680 (24)	680 (24)	680 (24)
Percent free, hygroscopic and metabolizable water in feed	54	54	54	54
Approx. heat in respired moisture, cal per g (Btu per lb)	611 (1100)	611 (1100)	611 (1100)	611 (1100)
Ratio of respired water to water input:				
S.C. White Leghorn hens	0.30-0.33	0.33-0.40	0.40-0.45	0.45-0.55
Rhode Island Red hens	0.22-0.35	0.35	0.35-0.42	0.42-0.55
New Hampshire and Cornish hens	0.25	0.25-0.35	----	----

*For S.C. White Leghorns add 0.03 to these values.

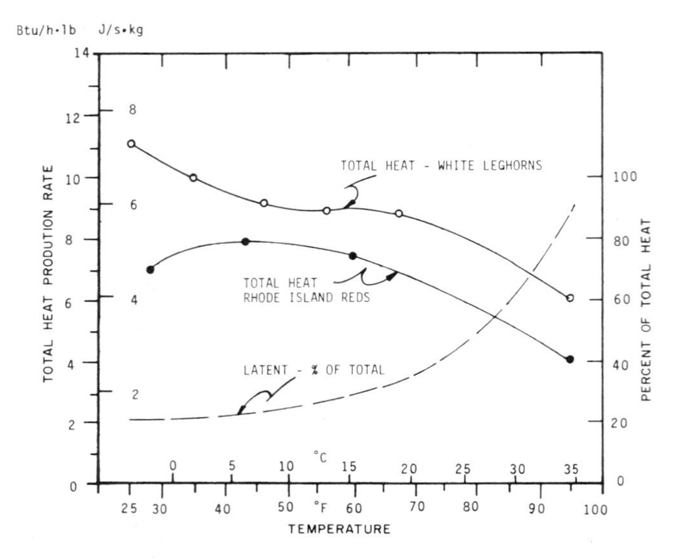

FIG. 28—EFFECT OF AIR TEMPERATURE ON TOTAL AND LATENT HEAT LOSSES OF RHODE ISLAND RED AND WHITE LEGHORN LAYING HENS. (Ota and McNally, 1961)

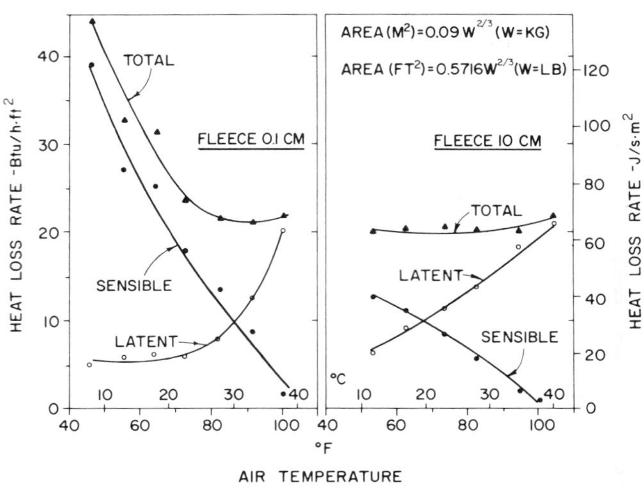

FIG. 29—EFFECT OF ENVIRONMENT AND FLEECE LENGTH ON HEAT LOSSES FROM SHEEP (HALF BRED × DOVER CROSS ADULT WETHERS). Relative humidity 45 to 54%. Airflow rate on 0.002 m^3/s (4 ft^3/min). (Blaxter, McC. Graham and Wainman, 1959)

8.4 Summer ventilation rates. Summer ventilation rates can be calculated using equations in paragraph 3.3.1 with values for sensible and latent heat loads from Fig. 28.

SECTION 9—EFFECT OF ENVIRONMENT ON SHEEP

9.1 Sheep housing. Sheep are not customarily housed in structures enclosed on all sides so no data are available on ventilation requirements.

9.2 Environment influences. Effect of environment and fleece length on heat losses from Half-Bred X Dover Cross adult wethers is shown in Fig. 29.

References:

1 Albright, L. D. 1976. Air flow through hinged baffle slotted inlets. TRANSACTIONS of the ASAE 19(4):728-732, 735.

2 Albright, L. D. 1978. Air flow through baffled, center-ceiling slotted inlets. TRANSACTIONS of the ASAE 21(5):944-947, 952.

3 Albright, L. D. 1979. Designing slotted inlet ventilation by the system characteristic technique. TRANSACTIONS of the ASAE 22(1):158-161.

4 Alexander, G. 1961. Temperature regulation in the new-born lamb. III. Effect of environment temperature on metabolic rate, body temperature and respiratory quotient. Australian Journal of Agricultural Research 12(6):1151-1174.

5 Alexander, G. 1974. Heat loss from sheep. In: heat loss in animals and man. Proceedings 20th Easter School in Agricultural Science, Univ. Nottingham. (Eds.) J. L. Montheith and L. E. Mount. 173-203. Butterworth Group, London.

6 Armstrong, D. G., K. L. Blaxter, J. L. Clapperton, N. McC. Graham and F. W. Wainman. 1960. Heat production and heat emission of two breeds of sheep. Journal of Agricultural Science 55:395-401.

TABLE 5—AVERAGE HOURLY MOISTURE PRODUCTION OF 1000 1.8 kg (4-lb) WHITE LEGHORN LAYING CHICKENS AT VARIOUS TEMPERATURES. (Ota, 1966)

Temperature		Respired		Defecated		Total*	
°C	°F	kg	lb	kg	lb	kg	lb
-3.9	25	2.9	6.3	6.6	14.5	10.4	22.8
1.7	35	3.8	8.3	6.6	14.5	11.3	24.8
7.2	45	3.8	8.4	5.9	12.9	10.8	23.7
15.6	60	5.2	11.4	5.8	12.7	12.0	26.4
26.7	80	6.5	14.3	6.4	14.4	14.3	31.6
35.0	95	9.1	20.0	4.7	10.3	15.3	33.7

*Includes drinking water wasted by hens estimated at 10% of water consumed from -3.9 to 36.7 °C (25 to 80 °F) and 15% at 35.0 °C (95 °F).

7 ASHRAE Handbook of Fundamentals. 1985. American Society of Heating, Refrigerating and Air-Conditioning Engineers, Atlanta, GA.

8 Augsberger, N. D., H. R. Bohganon and J. H. Hensley. Environmental control for livestock confinement. ACME Handbook Form C15M-482.

9 Berry, I. L., M. D. Shanklin, and H. D. Johnson. 1964. Dairy shelter design based on milk production decline as affected by temperature and humidity. TRANSACTIONS of the ASAE 7(3):329-331.

10 Blaxter, K. L., M. McC. Graham, and F. W. Wainman. 1959. Environmental temperature, energy metabolism and heat regulation in sheep. III. The metabolism and thermal exchanges of sheep with fleeces. Journal of Agricultural Science 52:41-49.

11 Bond, T. E., C. F. Kelly and H. Heitman, Jr. 1959. Hog house air conditioning and ventilation data. TRANSACTIONS of the ASAE 2(1):1-4.

12 Buffington, D. E., K. A. Jordan, W. A. Junnila and L. L. Boyd. 1974. Heat production of active, growing turkeys. TRANSACTIONS of the ASAE 17(3):542-545.

13 Butchbaker, A. F. and M. D. Shanklin. 1964. Partitional heat loss of newborn pigs as affected by air temperature, absolute humidity, age and body weight. TRANSACTIONS of the ASAE 8(1):118-121.

14 Carr, L. E. and T. A. Carter. 1978. Fuel savings by reducing broiler brooding temperatures. TRANSACTIONS of the ASAE 21(6):1189-1192.

15 Cairnie, A. B. and J. D. Pullar. 1957. The metabolism of the young pig. Journal of Physiology 139:15P.

16 Charles, D. R. 1984. A model of egg production. British Poultry Science 25:309-321.

17 Deaton, J. W., F. N. Reece and C. W. Bouchillin. 1969. Heat and moisture production of broilers. Part 2- Winder conditions. Poultry Science 48:1579.

18 Deaton, J. W., F. N. Reece, B. D. Lott, L. F. Kubena, and J. D. May. 1972. The efficiency of cooling broilers in summer as measured by growth and feed utilization. Poultry Science 51:69-71.

19 Deaton, J. W., F. N. Reece, L. F. Kubena, and J. D. May. 1973. The effect of low versus moderate rearing temperature on broiler performance. Poultry Science 52:1175-1178.

20 Deaton, J. W., F. N. Reece and J. L. McNaughton. 1977. The effect of temperature during the growing period as broiler performance. Poultry Science 57:785-788.

21 DeShazer, J. A., L. L. Olson and F. B. Mather. 1974. Heat losses of large white turkeys — 6 to 36 years of age. Poultry Science 53(6):2047-2054.

22 Gonzales-Jimenez, E. and K. L. Blaxter. 1962. The metabolism and thermal regulation of calves in the first month of life. British Journal of Nutrition 16:199-212.

23 Griffin, J. G. and T. H. Vardaman. 1970. Diurnal cycle versus daily constant temperatures for broiler performance. Poultry Science 49:387-391.

24 Hahn, LeRoy, H. D. Johnson, M. D. Shanklin and H. H. Kibler. 1965. Inspired-air cooling for lactating dairy cows in a hot environment. TRANSACTIONS of the ASAE 8(3):332-334, 337.

25 Hahn, LeRoy, and D. D. Osburn. 1969. Feasibility of summer environmental control of dairy cattle based on expected production losses. TRANSACTIONS of the ASAE 12(4):448-451.

26 Hahn, LeRoy and D. D. Osburn. 1970. Feasibility of evaporative cooling for dairy cattle based on expected production losses. TRANSACTIONS of the ASAE 13(3):289-291, 294.

27 Harwood, F. W. and F. N. Reece. 1974. Summer ventilation design for windowless broiler houses in the southern regions of the U.S. Poultry Science 53:2148-2152.

28 Hellickson, M. A., A. F. Butchbaker, R. L. Witz and R. L. Bryant. 1967. Performance of growing turkeys as affected by environmental temperature. TRANSACTIONS of the ASAE 10(6):793-795.

29 Hellickson, M. A., H. G. Young and W. B. Witmer. 1974. Ventilation design for closed beef buildings. In: Livestock Environment. Proceedings of the International Livestock Environment Symposium. SP-0174. ASAE, St. Joseph, MI, pp. 123-129.

30 Holmes, C. W. and A. W. F. Davey. 1976. The energy metabolism of young Jersey and Friesian calves fed fresh milk. Animal Production 23:43-53.

31 Johnson, H. D. 1965. Environmental temperatures and lactation, Interntional Journal of Biometeorology 9:103-106.

32 Johnson, H. D. and A. C. Ragsdale. 1959. Effects of constant environmental temperatures of 50° and 80°F on the growth response of Holstein, Brown Swiss, and Jersey calves. Agricultural Experiment Station Research Bulletin No. 705. University of Missouri.

33 Longhouse, A. D., H. Ota and W. Ashby. 1960. Heat and moisture design data for poultry housing. AGRICULTURAL ENGINEERING 41(9):567-576.

34 Longhouse, A. D., H. Ota, R. E. Emerson and J. O. Heishman. 1968. Heat and moisture design data for broiler houses. TRANSACTIONS of the ASAE 11(5):694-700.

35 McLean, J. A. 1963. The partition of insensible losses of body weight and heat from cattle under various climatic conditions. Journal of Physiology 167:427-477.

36 Meyer, V. M. and L. VanFossen. 1971. Effects of environment on pork production. Iowa State University Research Review AE 1063.

37 Ota, H. and E. H. McNally. 1961. Heat and moisture production of Beltsville white turkeys. Poultry Science 40(5):1440.

38 Ota, H. and E. H. McNally. 1961. Poultry respiration calorimetric studies of laying hens. U.S. Dept. of Agriculture. Bulletin ARS 42-43.

39 Ota, H. 1966. The physical control of environment for growing and laying birds. Symposium on Environmental Control in Poultry Production. British Egg Marketing Board, London.

40 Ota, H., J. A. Whitehead and R. J. Davey. 1975. Heat production of male and female piglets. Journal of Animal Science 41(1):436.

41 Prince, R. P., L. M. Potter and W. W. Irish (1961). Response of chickens to temperature and ventilation environments. Poultry Science 40:102-108.

42 Ragsdale, A. C., C. S. Cheng and H. D. Johnson. 1958. Effects of environmental temperatures of 50 and 80°F on the growth responses of Santa Gertrudis and Shorthorn calves. Missouri Agricultural Experiment Station Research Bulletin 642.

43 Reece, F. N. and B. D. Lott. 1982a. Heat and moisture production of broiler chickens during brooding. Poultry Science 61:4; 661-666.

44 Reece, F. N. and B. D. Lott. 1982b. The effect of environmental temperature on sensible and latent heat production in broilers. Poultry Science 61:8 1590-1594.

45 Rogerson, A. 1960. The effect of environmental temperature on the energy metabolism of cattle. Journal of Agricultural Science 55:359-363.

46 Shanklin, M. D., R. K. Malhotra, G. L. Hahn and H. V. Biellier. 1977. Predicted equations for heat loss of mature Broad Breasted Bronze turkeys. TRANSACTIONS of the ASAE 20(1):148, 149, 154.

47 Siegel, H. S. and L. N. Drury. 1970. Broiler growth in diurnally cycling temperature environments. Poultry Science 49:238-244.

48 Stott, G. H., F. Wiersma and O. Lough. 1972. Consider cooling possibilities: The practical aspects of cooling dairy cattle. University of Arizona Agricultural Experiment Station Report P-25.

49 Timmons, M. B. and G. R. Baughman. 1983. The flex house: A new concept in poultry housing. TRANSACTIONS of the ASAE 26(2):529-532.

50 Webster, A. J. F. 1971. Prediction of heat losses from cattle exposed to cold outdoor environments. Journal of Applied Physiology 30(5) 684-690.

51 Webster, A. J. F., J. G. Gordon and J. S. Smith. 1976. Energy exchanges of veal calves in relation to body weight, food intake and air temperature. Animal Production 23:35-42.

52 Webster, A. J. F. and J. G. Gordon. 1977. Air temperature and heat losses from calves in the first weeks of life. Animal Production 24:142.

53 Winn, P. H. and E. F. Godfrey. 1967. The effect of temperature and moisture on broiler performance. University of Maryland Experiment Station Bulletin No. A-153.

54 Yeck, R. G. and R. E. Stewart. 1959. A ten-year summary of the phychoenergetic laboratory dairy cattle research at the University of Missouri. TRANSACTIONS of the ASAE 2(1):71-77.

55 Yeck, R. G. and R. E. Stewart. 1960. Stable heat and moisture dissipation with dairy calves at temperatures of 50 and 80°F. Missouri Agricultural Experiment Station Research Bulletin No. 759.

ASAE Engineering Practice: ASAE EP282.1

DESIGN VALUES FOR LIVESTOCK FALLOUT SHELTERS

Developed by the ASAE Emergency Preparedness Committee; approved by the Structures and Environment Division Standards Committee; adopted by ASAE as a Recommendation December 1964; withdrawn December 1970; revised and reinstated February 1973; reclassified as an Engineering Practice December 1977; reconfirmed December 1982; reconfirmed for one year December 1987.

SECTION 1—PURPOSE

1.1 This Engineering Practice provides a guide for the functional design of livestock fallout shelters, and for establishing feed and water supply storage requirements for an emergency and during recovery from an emergency.

SECTION 2—SCOPE

2.1 This Engineering Practice applies to livestock fallout shelters, and refers to human fallout shelter used only in conjunction with livestock fallout shelter for livestock management.

2.2 The concepts and design values apply to protection of livestock from nuclear fallout, and to maintenance of environmentally adequate conditions for animal survival.

SECTION 3—GENERAL CONSIDERATIONS

3.1 Human shelter should be incorporated with a shelter for housing animals requiring frequent attention, such as dairy animals. (For human fallout shelters see Office of Emergency Preparedness TR-39 Codes and Standards for Fallout Shelters.)

3.2 High level protection for a few valued breeding animals may be more desirable than low level protection for many production animals.

3.3 Farm buildings intended for fallout protection should function efficiently in normal use.

3.4 Feed and water supplies must be protected from contamination by radioactive particles.

SECTION 4—WATER AND FEED REQUIREMENTS

4.1 Water supply and requirements

4.1.1 Water supplies from covered wells or springs, free of surface contamination, or from closed storage should be provided.

4.1.2 If water must be stored, provide two week's capacity. Animals can survive two days without water. Water requirements are shown in Table 1.

TABLE 1—WATER REQUIREMENTS PER ANIMAL PER DAY*

Animal	Ample Supply		Limited Supply†	
	Liters	Gallons	Liters	Gallons
Cattle	64.0	17.0	26.5	7.0
Hogs	9.5	2.5	4.8	1.2
Sheep	5.8	1.5	3.8	1.0
Poultry				
Layers and Broilers	0.24	0.06	0.20	0.05
Turkeys	1.26	0.30	0.50	0.12

*Average requirements at a temperature of 27 deg C (80 deg F)
†Water rationing facilities required

4.1.3 Emergency electric power for pumping is recommended unless artesian or gravity flow is available.

4.2 Feed requirements. At least two week's feed storage capacity should be provided. Limited feed requirements are shown in Table 2.

4.2.1 Healthy animals can survive on limited rations for several months.

4.2.2 Supplemental proteins, vitamins, and minerals will be needed for extended limited feeding.

4.2.3 Feedstuffs susceptible to damage from high humidity should be isolated from humid air in the livestock shelter.

TABLE 2—LIMITED FEED REQUIREMENTS FOR LIVESTOCK*

Animal	Feed	Amount of Feed Per Day Percent of body wt
Cattle		
Cow, lactating	hay	2
Cow, dry	hay	1
Calf, less than 9 mo of age	hay	1
	plus 40% protein supplement	0.2
Sheep		
Ewe	alfalfa hay	1
Lamb, 27 kg (60 lb)	alfalfa hay	1.5
Swine		
Sow, pregnant	corn	0.4
	plus 35% protein supplement	0.2
Sow, lactating	corn	1
	plus 35% protein supplement	0.2
Hog, 45 kg (100 lb)	corn	1.5
91 kg (200 lb)	corn	1
Poultry		
Laying hen	mash	2
Turkey, 5 kg (10 lb)	mash	1.7
11 kg (25 lb)	mash	1.3

*Equivalent feeds may be substituted. Hay should be at least one-half legume or equivalent in protein content.

SECTION 5—VENTILATION REQUIREMENTS

5.1 Air functions are to:

5.1.1 Supply oxygen for food metabolism. Oxygen levels should not drop below 16 percent by volume.

5.1.2 Dilute carbon dioxide concentrations. Maximum concentration should not exceed 2 percent by volume.

5.1.3 Act as a vehicle for removal of water vapor.

5.1.4 Act as a vehicle for removal of heat.

5.2 Air intakes should be hooded, screened, and designed to keep air intake velocities to less than 45.7 m per min (150 fpm) to minimize ingress of fallout particles. Filters are required if face velocity at the intake exceeds 45.7 m per min (150 fpm).

5.3 Heat and moisture design outputs should be calculated according to ASAE Engineering Practice EP270, Design of Ventilation Systems for Poultry and Livestock Shelters.

SECTION 6—SPACE REQUIREMENTS

6.1 Basic requirements for floor area for livestock in fallout shelters are given in Table 3. Additional space should be allowed for movement of men caring for animals.

SECTION 7—RADIATION ATTENUATION REQUIREMENTS

7.1 The level of fallout protection provided should be based on the future reproductive capability of the livestock. Suggested levels of protection are shown in Table 4.

7.2 Protection factor should be determined by U.S. Department of Defense design procedures.

7.3 Poultry withstand more radiation than other livestock but equal protection is recommended because of their importance as emergency food.

7.4 Geographic locations near potential target areas should be assigned relatively high protection factors.

SECTION 8—EMERGENCY ELECTRICAL POWER

8.1 Power should be available for ventilation fans, water supply, minimum lighting, and other equipment required for survival.

8.1.1 Other equipment may include milking machines, refrigeration equipment, and mechanical feeding equipment.

8.1.2 Minimum general lighting is 5.4 to 10.8 lumens per sq m (1/2 to 1 foot-candle). Higher local intensities may be required.

8.2 Emergency power equipment should be located where it can be serviced without unduly exposing the operator. Heat dissipation from the engine and removal of exhaust fumes must be considered.

8.3 Electrical connections adequate to serve emergency equipment must be provided before an emergency.

References: Last printed in 1983 AGRICULTURAL ENGINEERS YEARBOOK; list available from ASAE Headquarters.

TABLE 3—LIMITED SPACE FOR ANIMALS IN FALLOUT SHELTERS

Animal	Space per animal	
	Sq M	Sq Ft
Cattle		
Cow	1.9	20
Calf	1.1	12
Sheep		
Ewe	0.93	10
Lamb, 27 kg (60 lb)	0.37	4
Swine		
Sow, lactating	3.0	32
Hog, 45 kg (100 lb)	0.37	4
91 kg (200 lb)	0.56	6
Poultry		
Chicken	0.06	0.7
Turkey, 5 kg (10 lb)	0.14	1.5
11 kg (25 lb)	0.19	2

TABLE 4—PROTECTION FACTORS

Animal Use	Protection Factors*
Highly developed breeding stock	40 - 25
Normal breeding stock	25 - 15
Production stock	15 - 5
Stock for consumption	5 - 1

*Protection factor indicates the relative reduction in the amount of radiation that would be received in a protected location compared to the amount that would be received in an unprotected location. A protection factor of 40 means that 1/40 as much radiation would be received as compared to that received in an unprotected location.

Cited Standard:

ASAE EP270, Design of Ventilation Systems for Poultry and Livestock Shelters

AGRICULTURAL BUILDING SNOW AND WIND LOADS

Supersedes ASAE S288 documents published during the period June 1965-February 1983. Developed by the ASAE Snow and Wind Loads Subcommittee; approved by the Structures and Environment Division Standards Committee; adopted by ASAE December 1986; revised editorially July 1987.

SECTION 1—PURPOSE AND SCOPE

1.1 This Engineering Practice presents minimum snow and wind design loads and methods of applying them for agricultural building design.

1.2 This Engineering Practice does not include values for dead loads and live loads, nor the combinations used for building design. It does not include valves for tornado wind loads.

SECTION 2—TERMINOLOGY

2.1 Terms used in this Engineering Practice are defined as follows:

2.1.1 Agricultural building: A shelter for farm animals or crops; or when incidental to agricultural production, a shelter for processing or storing products of farm animals or crops, or for storing or repairing agricultural implements. An agricultural building is not for use by the public, for human habitation, or for commercial retail trade or warehousing.

2.1.2 Snow loads: Vertical loads from the weight of snow applied to the horizontal projection of a roof.

2.1.3 Wind loads: Loads from wind coming from any horizontal direction.

2.1.4 Dead loads: Vertical loads due to the permanent weight of building construction materials such as floor, roof, framing, and covering.

2.1.5 Live loads: Both static and dynamic loads resulting from the use or occupancy of the building. Static loads result from the weight and/or pressure from equipment, livestock, and stored products and the materials used for construction and maintenance activities. Dynamic loads result from the dynamic effect of cranes, hoists and materials handling equipment.

SECTION 3—GENERAL DESIGN CONSIDERATIONS

3.1 The values on the ground snow load and wind speed maps in Section 4—Snow Loads, and Section 5—Wind Loads, have an annual probability of being exceeded equal to 0.02 (50-yr mean recurrence interval). Importance factors, paragraphs 4.2 and 5.4, are used to adjust the annual probability of the snow and wind load values to 0.04 (25-yr mean recurrence interval).

3.2 Minimum roof design load shall be 0.57 kPa (12 psf) in addition to the dead load. (Exception: The minimum roof load need not apply to livestock sunshades, plastic-covered greenhouses, and similar structures.)

3.3 If a code that specifies design loads applies to agricultural buildings, that code shall be followed. An alternative to the design load criteria of this Engineering Practice is to apply design loads required by local building code.

SECTION 4—SNOW LOADS

4.1 Ground snow loads, P_g

4.1.1 Design ground snow loads, P_g, for the contiguous United States are shown in Figs. 1, 2, and 3.

4.1.2 Snow load values for shaded areas on the map are not for areas such as high country, where snow loads vary considerably. Consult local sources for data.

4.1.3 Black areas on the maps have local variation in snow loads so extreme as to preclude meaningful mapping. Consult Table 8 or local sources for data.

4.1.4 Snow load values for Alaska are in Table 1. In Alaska extreme local variations preclude state-wide mapping.

4.1.5 Snow loads for Hawaii are zero.

4.2 Balanced roof snow loads, P_s

4.2.1 Calculate the snow load, P_s, on the horizontal projection of an unobstructed roof as follows:

$$P_s = R \cdot C_e \cdot I \cdot C_s \cdot P_g$$

where

R = Roof snow factor (relates roof snow to ground snowpack)
= 1.0 for ground snowpack of 0.72 kPa (15 psf) or less
= 0.7 for ground snowpack of 0.96 kPa (20 psf) or more in contiguous U.S.
= 0.6 in Alaska
= or the site specific coefficient of Table 7

C_e = exposure factor
= 0.8 in windy areas with roof exposed on all sides with no shelter afforded by terrain, higher structures or trees
= 1.0 where wind cannot be relied on to reduce roof loads because of terrain, higher structures, or several trees nearby
= 1.1 where terrain, higher structures, or trees will shelter the building increasing the potential for drifting

I = importance factor
= 1.0 (0.02 probability) for agricultural buildings that require a greater reliability of design to protect property or people
= 0.8 (0.04 probability) for agricultural buildings that present a low risk to property or people

C_s = slope factor (angles are from horizontal)
= 1.0 for 0 to 15 deg slope

$$= 1.0 - \frac{(\text{slope} - 15)}{55} \text{ for 15 to 70 deg slope}$$

= 0 for slopes > 70 deg

P_g = ground snow load from Figs. 1, 2, 3.

Example: Snow load on a low-risk agricultural building located near Des Moines, IA with a 4:12 (18.4 deg) sloped roof in a highly sheltered area:

P_s = 0.7 x 1.1 x 0.8 x 0.94 x 0.048 x 25
= 0.69 kPa
P_s = 0.7 x 1.1 x 0.8 x 0.94 x 25
= 14.5 psf

TABLE 1— GROUND SNOW LOADS, P_g, FOR ALASKAN LOCATIONS

kPa	psf		kPa	psf		kPa	psf	
0.96	20	Adak	3.12	65	Galena	6.24	130	Petersburg
2.16	45	Anchorage	2.88	60	Gulkana	2.16	45	St. Paul Isl.
3.60	75	Angoon	2.16	45	Homer	3.64	55	Seward
1.44	30	Barow	3.36	70	Juneau	0.96	20	Shemya
2.88	60	Barter Isl.	2.64	55	Kenai	2.16	45	Sitka
1.68	35	Bethel	1.44	30	Kodiak	8.40	175	Talkeetna
2.88	60	Big Delta	3.36	70	Kotzebu	3.64	55	Unalakleet
0.96	20	Cold Bay	3.36	70	McGrath	8.16	170	Valdez
4.80	100	Cordova	2.64	55	Nenana	19.2	400	Whittier
2.64	55	Fairbanks	3.84	80	Nome	3.36	70	Wrangell
3.36	70	Ft. Yukon	2.40	50	Palmer	8.40	175	Yakutat

(ANSI A58.1-1982)

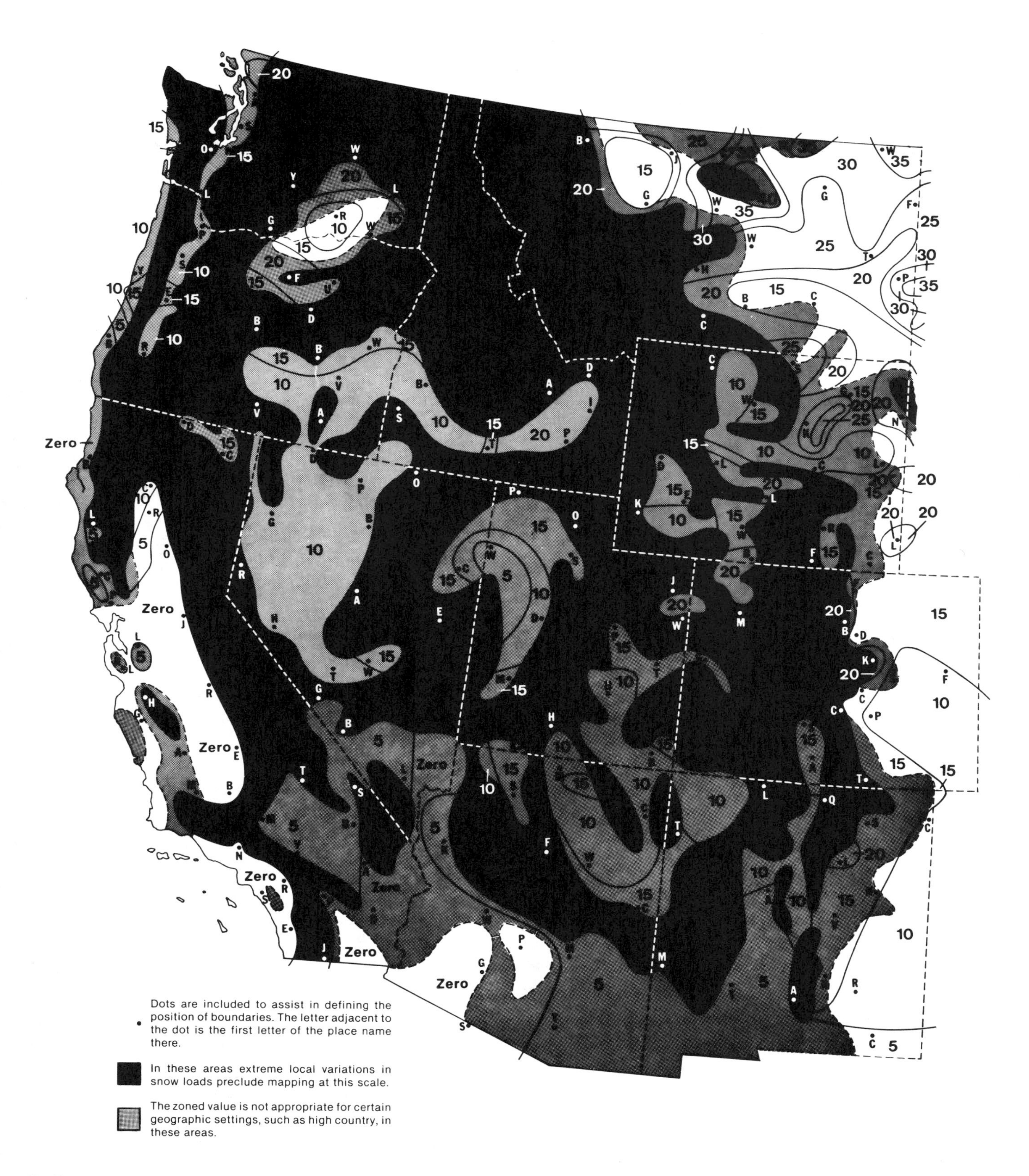

E FIG. 1—GROUND SNOW LOADS, P_g, IN psf FOR THE WESTERN UNITED STATES. Multiply by 0.048 for ground snow load in kPa. Values are loads associated with an annual probability of 0.02 (50-yr recurrence interval). (ANSI A58.1-1982, modified at the Montana-North Dakota border.)

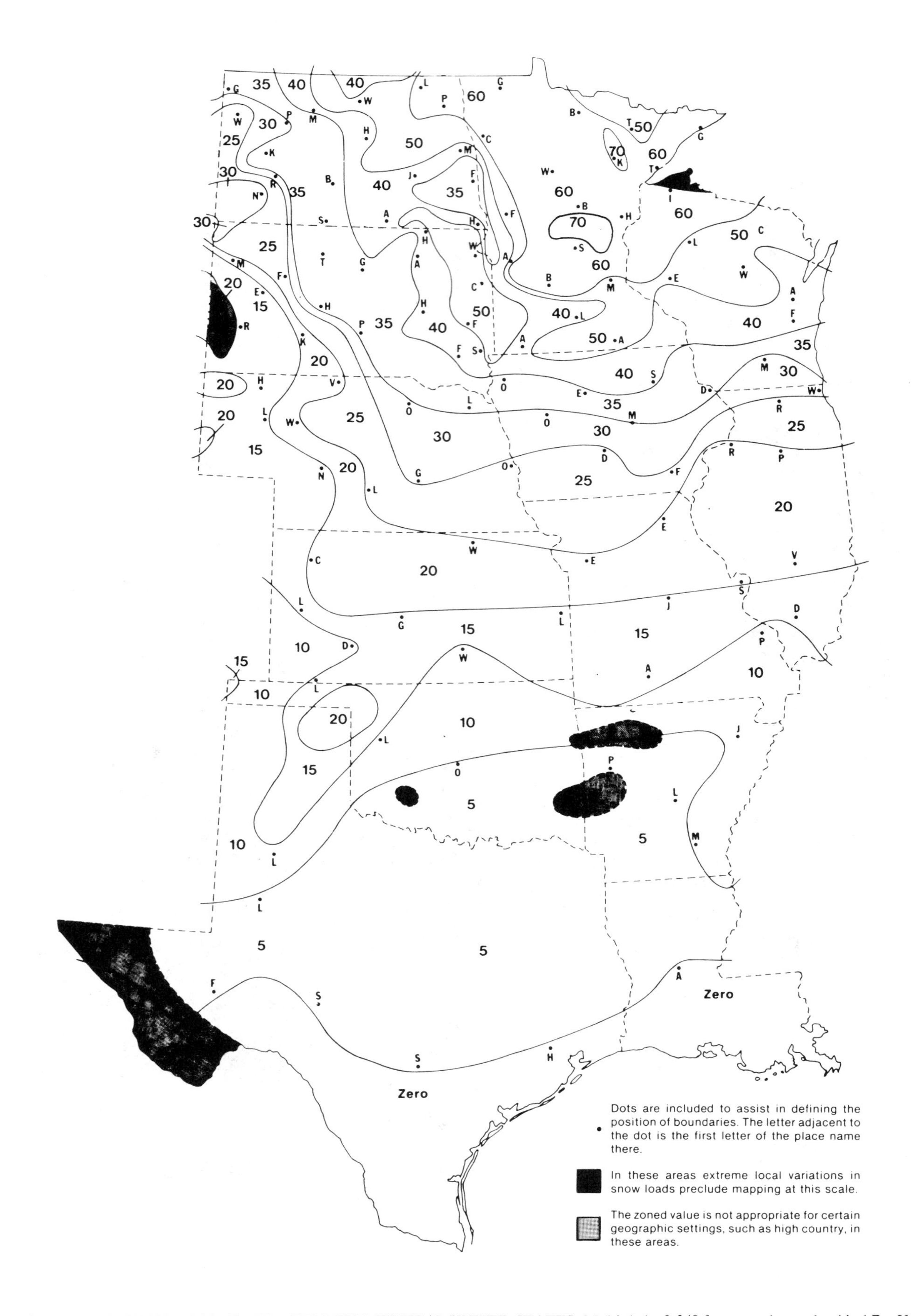

E FIG. 2—GROUND SNOW LOAD, P_g, IN psf FOR THE CENTRAL UNITED STATES. Multiply by 0.048 for ground snow load in kPa. Values are loads associated with an annual probability of 0.02 (50-yr recurrence interval). (ANSI A58.1-1982, modified in North and South Dakota, Minnesota and Wisconsin.)

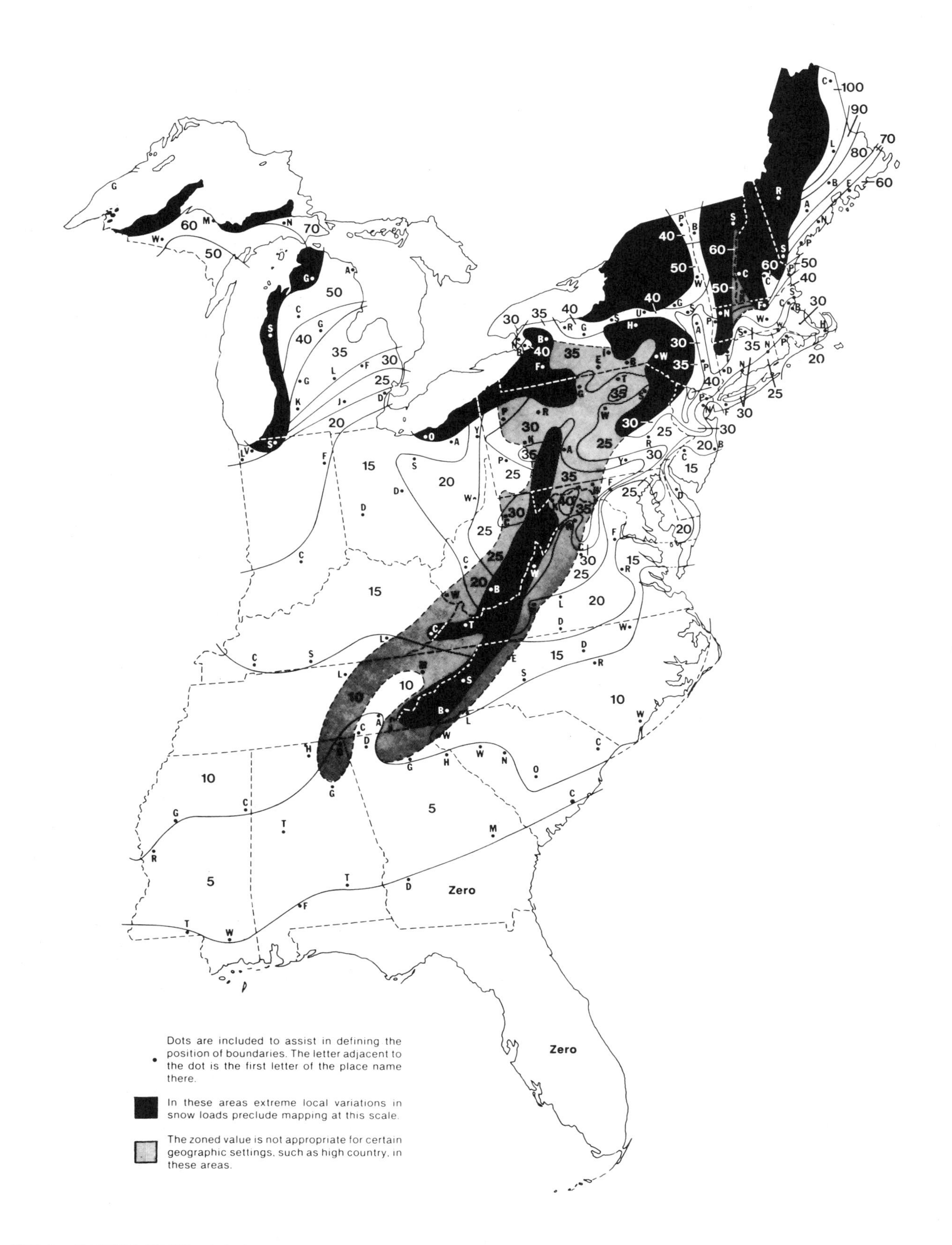

FIG. 3—GROUND SNOW LOADS, P_g, IN psf FOR THE EASTERN UNITED STATES. Multiply by 0.048 for ground snow load in kPa. Values are loads associated with an annual probability of 0.02 (50-yr recurrence interval). (ANSI A58.1-1982)

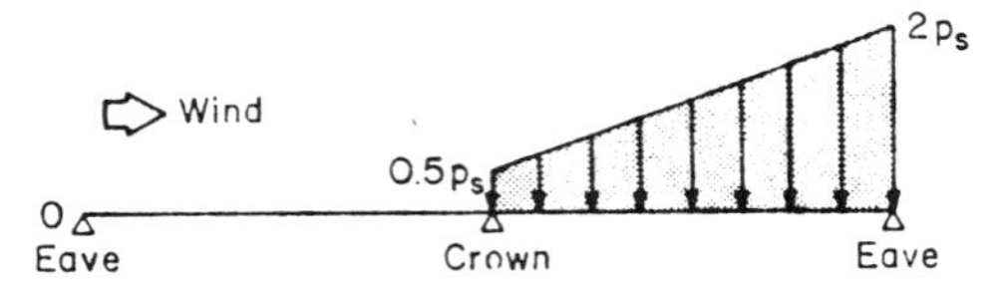

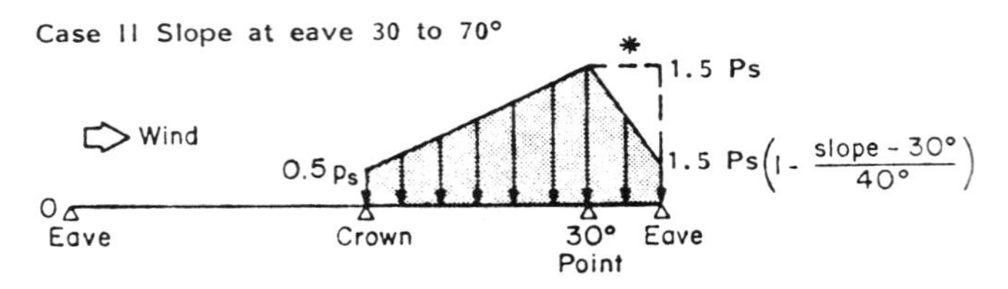

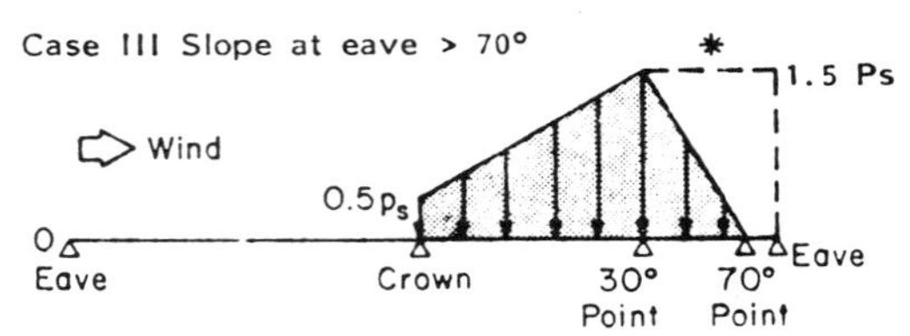

FIG. 4—UNBALANCED LOADING CONDITIONS FOR CURVED ROOFS. (ANSI A58.1-1982)

4.2.2 Curved Roofs. Consider slopes exceeding 70 deg free from snow load.

4.2.3 Multiple gable, sawtooth, and barrel vault roofs. Do not apply a reduction in snow load because of slope.

4.3 Unbalanced roof snow loads

4.3.1 Hip and gable roofs. For roofs with a slope less than 15 deg or more than 70 deg, unbalanced snow loads need not be considered. For slopes between 15 and 70 deg, design the structure to sustain an unbalanced uniform snow load on the lee side of 1.5 P_s while the windward side is free of snow.

4.3.2 Curved roofs. Consider the windward side and the portions of curved roofs having a slope exceeding 70 deg to be free of snow load. Determine unbalanced loads according to the loading diagrams in Fig. 4. If the ground or another roof abuts a Case II or a Case III arched roof structure within 1 m (3 ft) of its eave, do not decrease the snow load below 1.5 P_s for the area between the 30 deg point and the eave, shown by dashed lines in Fig. 4.

4.3.3 Multiple gable, sawtooth, and parallel vault roofs. Consider the valleys as filled with snow of density d, as defined in Table 2.

4.4 Drifts on lower roofs (aerodynamic shade)

4.4.1 Design roofs to support snow that drifts in the wind shelter of adjacent structures, terrain features, and higher portions of the same structure.

4.4.2 Lower roof of a structure. The geometry and magnitude of the drift load shall be determined as follows, except that drifting shall not be considered, when $(H_r - H_b) / H_b < 0.2$.

$$H_d = \frac{W_b \cdot P_s}{8d}^{0.5}$$

where

H_d = drift height, m (ft)
W_b = horizontal dimension of upper roof normal to the line of the change in the roof level, m (ft)
P_s = roof load of upper roof, kg/m^2 (lb/ft^2)
d = density of snow, Table 2
H_r = height differences between the upper and lower roofs, m (ft)
H_b = lower roof balanced snow load height, m (ft) = P_s/d
$H_d + H_b$ cannot exceed H_r

The width of the drift, W, is equal to 4 H_d or 4 $(H_r - H_b)$, whichever is less. Maximum intensity of the load, P_d, at the height change is as follows and shall not exceed $d \cdot H_r$:

$$P_d = d\,(H_d + H_b)$$

4.4.3 Adjacent structures and terrain features. Use paragraphs 4.4.1 and 4.4.2 to establish surcharge loads on a roof within 6 m (20 ft) of a higher structure or terrain feature that could cause drifting on it. However, the separation distance, s, between the two structures reduces drift loads on the lower roof. Apply the factor 1-s/6, m (1-s/20, ft) to the maximum drift load intensity, P_d, to account for spacing.

4.5 Roof projections. Consider as triangular the loads caused by a drift around a projection. Compute a surcharge load for all sides of any obstruction longer than 4.6 m (15 ft). Determine the drift surcharge load and the width of the drift as for lower roofs in paragraph 4.4.

4.6 Sliding snow. Snow may slide from a sloped roof to a lower roof creating extra loads on the lower roof. Assume that snow accumulated on the upper roof under the balanced loading condition slides onto the lower roof. Determine the total extra load available from the upper roof by the equation in paragraph 4.2.1. Where a portion of the upper roof is expected to slide clear of the lower roof, reduce the extra load on the lower roof accordingly.

4.7 Ponding loads. Consider roof deflections caused by snow loads when determining possible ponding of rain-on-snow or snow meltwater.

SECTION 5—WIND LOADS

5.1 Estimated extreme winds shown in Fig. 6 are for open terrain with scattered obstructions less than 16 m (53 ft) high. Adjust map values for special winds such as the Santa Ana; for unusual exposures causing channeling or uplift such as ocean promontories, mountains, or gorges; and for exposures or elevations where wind records or experience indicate illustrated wind speeds are inadequate.

5.2 Recommendations in this section apply to agricultural buildings up to about 20 m (66 ft) tall and tall, slender structures (silos and grain legs) up to about 30 m (100 ft) in height.

5.3 Overturning moment due to wind is resisted by stabilizing moments of 2/3 of the dead load unless the structure is anchored to resist excess moment. When friction is insufficient to prevent sliding, anchor the building to resist sliding forces.

5.4 Effective velocity pressure, g, is calculated as follows:

$$q = 0.00061 \cdot V^2 \cdot K_zG \cdot I^2 \text{ (kPa)}$$
$$q = 0.00256 \cdot V^2 \cdot K_zG \cdot I^2 \text{ (psf)}$$

where

V = wind velocity for extreme winds at 10 m (33 ft) elevation for an 0.02 annual probability (50-yr recurrence interval), m/s (mph) from Fig. 6
K_zG = combined building height, location exposure and gust factor from Table 3

I^2 = importance factor
= 1.0 (0.02 probability) for agricultural buildings that require greater reliability of design to protect property or people
= 0.9 (0.04 probability) for agricultural buildings that present a low risk to property or people

TABLE 2—DENSITIES FOR ESTABLISHING DRIFT LOADS

Ground snow load, P_g*		Snow density, d	
kPa	psf	kg/m^3	lb/ft^3
0.5-1.4	11-30	240	15
1.5-2.9	31-60	320	20
3.0+	>60	400	25

*Drifting is not considered where ground snow load, P_g, is less than 0.5 kPa (10 psf).

TABLE 3—COMBINED EXPOSURE AND GUST FACTORS, K_zG

Height m	ft	K_zG, exposed	K_zG, sheltered
4.5	15	1.07	0.62
6	20	1.14	0.68
9	30	1.24	0.77
12	40	1.32	0.84
15	50	1.39	0.91
18	60	1.45	0.96
21	70	1.50	1.01
24	80	1.54	1.06
27	90	1.58	1.10
30	100	1.62	1.14

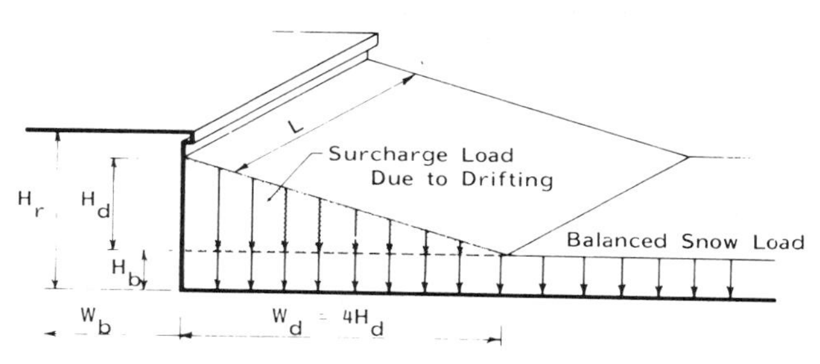

FIG. 5—SHAPE OF DRIFT ON LOWER ROOFS (Metal Building Systems Manual, 1986).

TABLE 4—PRESSURE COEFFICIENTS, C_p, FOR COMPONENTS AND CLADDING

Location	Enclosed building	Open building or overhang
Wall	+0.9; -1.0	+0.9; -1.5
Roof	-1.0	-1.5
Roof edges and ridge	-2.2	-2.2

(Where two values are shown, check both cases.)

Silo or tank Ht/diam ≤ 5	Angle B, deg	Local C_p	Angle B, deg	Local C_p
	0	+1.0	105	-1.5
	15	+0.8	120	-0.8
	30	+0.1	135	-0.6
	45	-0.8	150	-0.5
	60	-1.5	165	-0.5
	75	-1.9	180	-0.5
	90	-1.9		

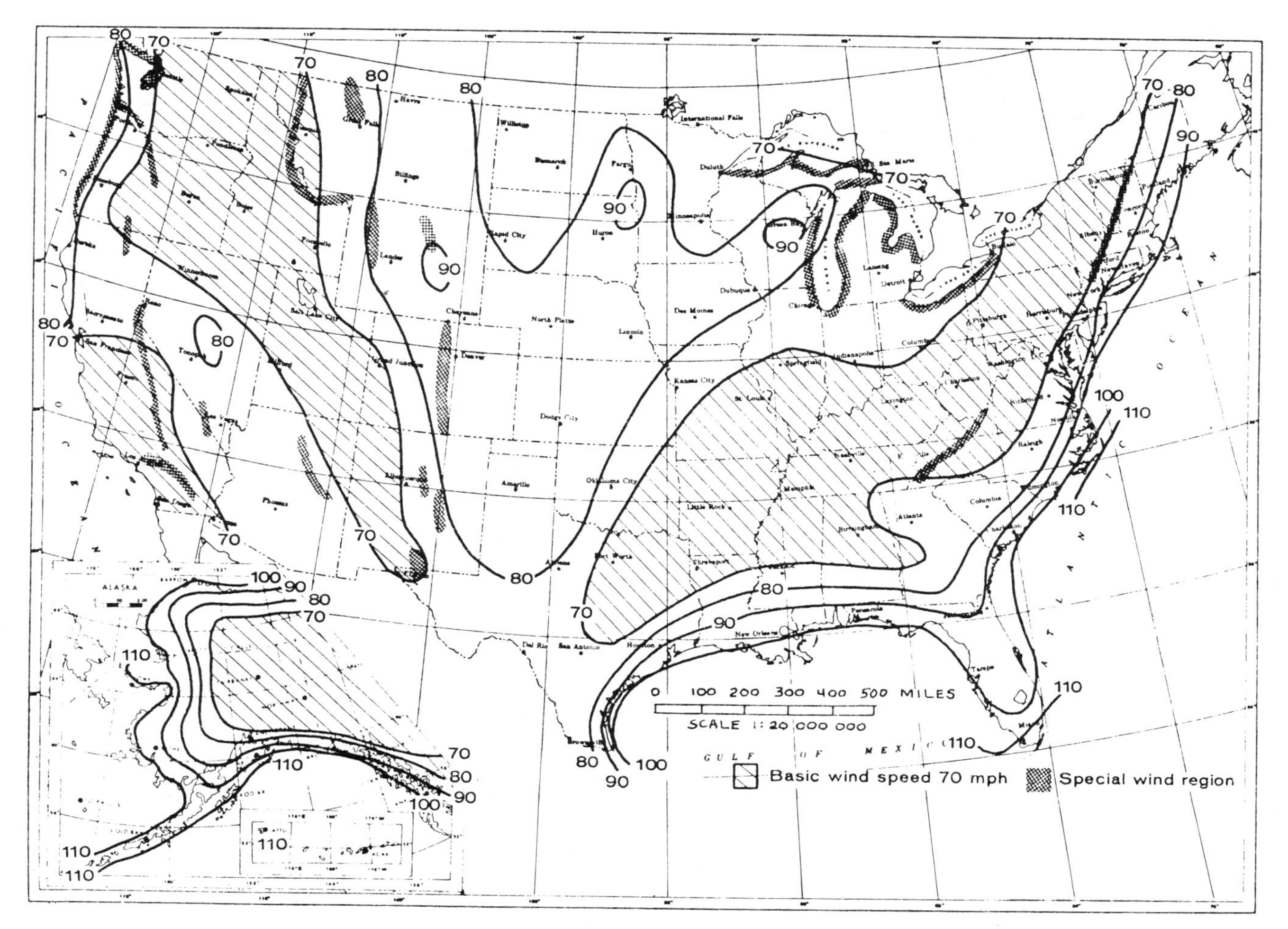

FIG. 6—WIND SPEED, V, in mph. Multiply by 0.45 for wind speed in m/s. Values are fastest-mile speeds at 10 m (33 ft) elevation for rural exposure with an annual probability of 0.02 (50-yr recurrence interval). (ANSI A58.1-1982)

Example: Effective velocity pressure for a 14.6 m (48 ft) wide, 4:12 sloped roof, 4.9 m (16 ft) high side wall, low-risk building located near Des Moines, IA, in an exposed location:

$$q = 0.00061 \times (0.45 \times 80)^2 \times 1.14 \times 0.9$$
$$= 0.81 \text{ kPa}$$
$$q = 0.00256 \times 80^2 \times 1.14 \times 0.9$$
$$= 16.8 \text{ psf}$$

5.4.1 Effective velocity pressure, q, reflects dry air mass density at 15 °C (59 °F) and 101 kPa (29.92 in·Hg).

5.4.2 The building height and location exposure factor, K_z, applies both for sheltered (wooded areas or terrain with closely spaced obstructions) and exposed (flat, open country with scattered obstructions) locations.

$$K_z = 2.6\,(z/z_g)^{2a}$$

where

z = design height of building in meters (feet) equal to the mid-elevation of the roof. All heights are measured from the average grade line. Minimum design height is 4.6 m (15 ft)

z_g = the gradient height at which the influence of surface friction is negligible. Gradient height is assumed at 274 m (900 ft) for exposed locations and 366 m (1200 ft) for sheltered locations

a = power law exponent. The recommended values for sheltered and exposed locations are 1/4.5 and 1/7, respectively

5.4.3 The gust factor, G, is calculated by:

$$G = 0.65 + 3.65\,(T_z)$$

where

T_z = an exposure factor which accounts for the degree of turbulence at height z and is expressed as:

$$T_z = 2.35\,\frac{D_o^{0.5}}{(z/h)^a}$$

TABLE 5—PRESSURE COEFFICIENTS, C_p, USED TO CALCULATE HORIZONTAL WIND FORCES ON MISCELLANEOUS STRUCTURES

Structure or part thereof	Description	C_p factor, any direction
Chimneys, tanks and silos	Square	1.4
	Round	0.6
Signs, flagpoles, lightpoles		1.4
Fences, latticed framework*	Flat sided members	1.7
	Round members	0.9

*Ratio of solid area to gross area is 0.7 or less; for greater than 0.7, use C_p factor of 1.4.

TABLE 6—PRESSURE COEFFICIENTS, C_p, FOR AGRICULTURAL STRUCTURES*

Wind direction →	Wall	Windward Lower roof	Upper roof	Upper roof	Leeward Lower roof	Wall	End walls
Gambrel roof	+0.7	+0.5	-0.8	-0.6	-0.6	-0.4	-0.7

Multispan gable roof (h/w ≈ 0.25, ≈ 0.40)	Wind ward wall	Span: First a	First b	Second c	Second d	Other m	Other n	Other wall
h/w ≈ 0.25	+0.6	+0.3	-1.0	-1.0	-0.6	-0.4	-0.5	-0.4
h/w ≈ 0.40	+0.6	-0.7	-1.0	-1.0	-0.6	-0.4	-0.5	-0.4

Arched roof	Wind ward wall	Windward quarter	Roof Center half	Leeward quarter	Lee wall	End wall
On elevated structure	+0.8	2 h/w-0.4	-0.7	-0.6	-0.5	-0.7
Springing from ground	—	1.2 h/w	-0.7	-0.6	—	-0.7

Gable roof: Roof slopes from 10° to 30° (2:12 to 7:12) (Two cases illustrate wide variation in data for load on windward roof.)

Wind direction →	Windward Wall	Windward Roof	Leeward Roof	Leeward Wall	End or side walls
Gable roof					
Case I	+0.7	+0.2	-0.7	-0.5	-0.7
Case II	+0.7	-0.7	-0.7	-0.5	-0.7
Gable Roof	+0.7	-0.7	-0.7	-0.5	-0.5
Open gable roof					
	—	-0.9	-1.2	-1.1	-1.4
	+1.3	+0.2	-0.2	—	-0.3

Wind direction →	Windward Wall	Windward Roof	Leeward Roof	Leeward Wall	End or side walls
Monoslope roof					
	+0.7	-0.7		-0.7	-0.8
	+0.6	0		-0.6	-0.9
Monoslope Roof	+0.7	-0.7		-0.4	-0.6
Open monoslope roof					
	—	-1.3		-1.3	-1.3
	+1.1	+0.5		—	-0.4
Umbrella roof					
	—	+0.6	-0.6	—	—
Stored Product (0.8 h)	—	-1.0	-1.1	—	—

*The listed coefficients are the vector sum of both internal and external pressure. Positive coefficients are directed inward; negative coefficients are directed outward. 0.2 internal bursting coefficient is included.

where

D_o = surface drag coefficient. Use 0.01 for sheltered locations and 0.005 for exposed locations

h = 10 m (33 ft)

a = power law exponent (see paragraph 5.4.2)

z = design height (see paragraph 5.4.2)

5.5 Design pressure calculations

5.5.1 Calculate design pressure normal to the surface by multiplying the effective velocity pressure, q, by a dimensionless pressure coefficient, C_p, from Tables 4, 5, and 6. Positive pressure coefficients indicate inward pressure. Negative signs indicate outward pressure. Consider wind from all directions.

5.5.2 Pressure coefficients in Table 6 are for main structural systems such as rigid and braced frames, braced trusses, posts, poles, and girders.

5.5.3 Pressure coefficients for components and cladding, such as purlins, studs, girts, curtain walls, sheathing, roofing, and siding, are in Table 4.

5.5.3.1 An enclosed building is one that has a perimeter of solid walls. These "solid" walls can have openings, but openings must be protected by doors or windows.

5.5.3.2 An open building is one that has more than 20% of the windward surface open. A building or shed open on one side while the rest of the surfaces are fairly airtight, is an example.

5.5.3.3 Design pressure coefficients assume a uniform straight line pressure distribution except for ridge edges, perimeter eaves and canopy corners.

5.5.3.4 Consider higher wind loading at roof edges, ridge and corners due to turbulence. The area affected is 0.1 times the minimum width or 0.4 times the design height of the building, whichever is smaller, but not less than 1 m (3 ft) wide.

5.5.4 Coefficients for miscellaneous structures are in Table 5.

References

1. ANSI. 1982. Minimum design loads for building and other structures. American National Standard A58.1-1982. American National Standards Institute, New York, NY. (Figures and tables as noted were reproduced from A58.1-1982 with permission from ANSI. Copies of A58.1-1982 are available from ANSI.)

2. MBMA. 1986. Metal Building Systems Manual. Metal Building Manufacturers Association, Cleveland, OH.

3. O'Rourke, M.J. and Steifel, U. 1983. Roof snow loads for structural design. ASCE Journal of Structural Engineering 109(7):1527-1537.

TABLE 7—ROOF SNOW FACTORS

Location	Importance factor		Location	Importance factor		Location	Importance factor		Location	Importance factor	
	0.8	1.0		0.8	1.0		0.8	1.0		0.8	1.0
Flagstaff AZ	0.576	0.597	Kalispell MT	0.568	0.587	Topeka KS	0.563	0.580	Youngstown OH	0.615	0.645
Blue Canyon CA	0.592	0.616	Missoula MT	0.590	0.613	Wichita KS	0.567	0.586	Burns OR	0.587	0.609
Mt Shasta CA	0.567	0.586	Grand Island NE	0.555	0.571	Covington KY	0.581	0.603	Allentown PA	0.584	0.606
Alamosa CO	0.567	0.586	Norfolk NE	0.563	0.581	Lexington KY	0.577	0.597	Erie PA	0.608	0.636
Colorado Spr CO	0.584	0.606	North Platte NE	0.579	0.600	Caribou ME	0.586	0.609	Harrisburg PA	0.584	0.606
Denver CO	0.606	0.633	Omaha NE	0.574	0.594	Portland ME	0.585	0.608	Philadelphia PA	0.566	0.584
Grand Junction CO	0.556	0.573	Scottsbluff NE	0.587	0.610	Boston MA	0.595	0.620	Pittsburg PA	0.575	0.595
Pueblo CO	0.617	0.647	Valentine NE	0.563	0.581	Nantucket MA	0.583	0.606	Scranton PA	0.597	0.623
Bridgeport CT	0.572	0.591	Asheville NC	0.583	0.605	Worcester MA	0.615	0.645	Williamsport PA	0.594	0.618
Hartford CT	0.606	0.634	Concord NH	0.574	0.594	Alpena MI	0.593	0.617	Providence RI	0.587	0.610
New Haven CT	0.586	0.608	Atlantic City NJ	0.583	0.605	Detroit MI	0.586	0.609	Aberdeen SD	0.565	0.583
Wilmington DE	0.587	0.610	Neward NJ	0.583	0.605	Detroit AP, MI	0.587	0.610	Huron SD	0.552	0.568
Boise AP, ID	0.614	0.644	Albany NY	0.619	0.649	Detroit WR, MI	0.557	0.573	Rapid City SD	0.586	0.609
Pocatella AP, ID	0.632	0.655	Binghamton NY	0.593	0.618	Flint MI	0.568	0.587	Sioux Falls SD	0.560	0.577
Chicago/O-H, IL	0.599	0.625	Buffalo NY	0.574	0.594	Grand Rapids MI	0.569	0.588	Amarillo TX	0.570	0.589
Chicago IL	0.574	0.594	NYC-LA Guard NY	0.560	0.578	Houghton Lk MI	0.575	0.596	Milford UT	0.589	0.613
Moline IL	0.596	0.621	Rochester NY	0.594	0.618	Lansing MI	0.567	0.586	Salt Lake C UT	0.663	0.703
Peoria IL	0.576	0.597	Syracuse NY	0.601	0.627	Marquette MI	0.680	0.725	Washington DC	0.608	0.636
Rockford IL	0.568	0.587	Bismarck ND	0.569	0.588	Muskegon MI	0.581	0.603	Burlington VT	0.584	0.607
Springfield IL	0.566	0.585	Fargo ND	0.563	0.581	Sault S Marie MI	0.611	0.640	Spokane WA	0.562	0.580
Ft Wayne IN	0.571	0.591	Williston ND	0.580	0.601	Duluth MN	0.617	0.647	Stampede Pass WA	0.663	0.704
Indianapolis IN	0.568	0.587	Clayton NM	0.586	0.609	Int Falls MN	0.625	0.657	Yakima WA	0.555	0.571
South Bend IN	0.569	0.588	Elko NV	0.562	0.579	Minn-St Paul MN	0.558	0.575	Green Bay WI	0.566	0.585
Burlington IA	0.577	0.598	Ely NV	0.638	0.673	Rochester MN	0.554	0.570	La Crosse WI	0.552	0.567
Des Moines IA	0.569	0.588	Winnemucca NV	0.603	0.629	St Cloud MN	0.562	0.580	Madison WI	0.555	0.571
Dubuque IA	0.565	0.583	Akron-Canton OH	0.602	0.629	Columbia Reg MO	0.556	0.572	Milwaukee WI	0.568	0.587
Sioux City IA	0.557	0.574	Cleveland OH	0.596	0.621	Kansas City MO	0.582	0.604	Charlestone WV	0.549	0.564
Waterloo IA	0.559	0.576	Columbus OH	0.592	0.616	Springfield MO	0.574	0.594	Elkins WV	0.584	0.606
Concordia KS	0.556	0.572	Dayton OH	0.587	0.610	Billings MT	0.613	0.642	Huntington WV	0.571	0.591
Dodge City KS	0.575	0.595	Mansfield OH	0.572	0.592	Glasgow MT	0.579	0.600	Casper WY	0.615	0.645
Goodland KS	0.585	0.607	Toledo, Expr. OH	0.609	0.637	Great Falls MT	0.616	0.646	Cheyenne WY	0.590	0.613
						Havre MT	0.572	0.591	Lander WY	0.620	0.650
						Helena MT	0.610	0.639	Sheridan WY	0.563	0.581

NOTE: The roof snow factors are probability sensitive and apply only with the importance factors listed. (Abstracted from O'Rourke, M.J. and Stiefel, U., Roof Snow Loads for Structural Design.)

TABLE 8—GROUND SNOW LOAD AT NATIONAL WEATHER SERVICE LOCATIONS*

Location	Years of Record	Ground Snow Load (lbf/ft²) Maximum Observed	2% Annual Probability
ALABAMA			
Huntsville	18	7	7
ARIZONA			
Flagstaff	28	88	48
Prescott	5	2	3
Winslow	25	12	7
ARKANSAS			
Fort Smith	22	4	5
Little Rock	22	6	6
CALIFORNIA			
Blue Canyon	18	213	255
Mt. Shasta	28	62	69
COLORADO			
Alamosa	28	14	15
Colorado Springs	27	16	14
Denver	28	14	15
Grand Junction	28	18	16
Pueblo	26	7	7
CONNECTICUT			
Bridgeport	27	19	23
Hartford	28	23	29
New Haven	17	11	15
DELAWARE			
Wilmington	27	12	13
GEORGIA			
Athens	24	5	5
Macon	28	8	8
IDAHO			
Boise	26	6	6
Lewiston	24	6	9
Pocatello	28	9	7
ILLINOIS			
Chicago–O'Hare	20	25	18
Chicago	26	37	22
Moline	28	21	17
Peoria	28	27	16
Rockford	14	31	25
Springfield	28	20	23
INDIANA			
Evansville	27	11	12
Fort Wayne	28	22	17
Indianapolis	28	19	21
South Bend	28	58	44
IOWA			
Burlington	11	15	17
Des Moines	28	22	22
Dubuque	28	34	38
Sioux City	26	28	33
Waterloo	21	25	36
KANSAS			
Concordia	17	12	23
Dodge City	28	10	12
Goodland	27	12	14
Topeka	27	18	19
Wichita	26	8	11
KENTUCKY			
Covington	28	22	12
Lexington	28	11	12
Louisville	26	11	11
MAINE			
Caribou	27	68	100
Portland	28	51	62
MARYLAND			
Baltimore	28	20	17
MASSACHUSETTS			
Boston	27	25	30
Nantucket	16	14	18
Worcester	21	29	39
MICHIGAN			
Alpena	19	34	53
Detroit City	14	6	9
Detroit Airport	22	14	17
Detroit–Willow Run	12	11	21
Flint	25	20	28
Grand Rapids	28	32	37
Houghton Lake	16	33	56
Lansing	23	34	42
Marquette	16	44	53
Muskegon	28	40	43
Sault Ste. Marie	28	68	80
MINNESOTA			
Duluth	28	55	64
International Falls	28	43	43
Minneapolis–St. Paul	28	34	50
Rochester	28	30	50
St. Cloud	28	40	53
MISSISSIPPI			
Jackson	27	3	3
MISSOURI			
Columbia	27	18	21
Kansas City	27	18	18
St. Louis	25	26	16
Springfield	27	9	14
MONTANA			
Billings	28	21	17
Glasgow	28	18	17
Great Falls	28	22	16
Havre	26	22	24
Helena	28	15	18
Kalispell	17	27	53
Missoula	28	24	23
NEBRASKA			
Grand Island	27	24	30
Lincoln	8	15	20
Norfolk	28	28	29
North Platte	26	16	15
Omaha	25	23	20
Scottsbluff	28	8	11
Valentine	14	15	22
NEVADA			
Elko	12	12	20
Ely	28	9	9
Reno	25	9	11
Winnemucca	24	5	6
NEW HAMPSHIRE			
Concord	28	36	66
NEW JERSEY			
Atlantic City	24	7	11
Newark	27	17	15
NEW MEXICO			
Albuquerque	25	6	4
Clayton	25	8	10
Roswell	22	6	8
NEW YORK			
Albany	28	26	25
Binghamton	28	30	35
Buffalo	28	41	42
NYC–Kennedy	7	7	18
NYC–LaGuardia	28	23	18
Rochester	28	33	38
Syracuse	28	32	35
NORTH CAROLINA			
Asheville	16	7	12
Cape Hatteras	22	5	5
Charlotte	28	8	10
Greensboro	26	14	11
Raleigh-Durham	22	13	10
Wilmington	24	7	9
Winston-Salem	12	14	17
NORTH DAKOTA			
Bismarck	28	27	25
Fargo	27	24	34
Williston	28	25	25
OHIO			
Akron–Canton	28	16	15
Cleveland	28	27	16
Columbus	27	9	10
Dayton	28	18	11
Mansfield	18	31	17
Toledo Express	24	8	8
Youngstown	28	14	12
OKLAHOMA			
Oklahoma City	24	5	5
Tulsa	21	5	8
OREGON			
Burns City	28	19	24
Eugene	22	22	17
Medford	25	6	8
Pendleton	28	9	11
Portland	25	10	10
Salem	27	5	7
PENNSYLVANIA			
Allentown	28	16	23
Erie	20	20	19
Harrisburg	19	21	23
Philadelphia	27	13	16
Pittsburgh	28	27	22
Scranton	25	13	16
Williamsport	28	18	20
RHODE ISLAND			
Providence	27	22	21
SOUTH CAROLINA			
Columbia	24	9	12
Greenville–Spartanburg	12	4	4
SOUTH DAKOTA			
Aberdeen	16	23	42
Huron	28	41	43
Rapid City	28	14	14
Sioux Falls	28	40	38
TENNESSEE			
Bristol	27	7	8
Chattanooga	27	5	6
Knoxville	25	10	8
Memphis	27	7	5
Nashville	23	5	8
TEXAS			
Abilene	23	6	6
Amarillo	26	15	10
Dallas	22	3	3
El Paso	24	5	5
Fort Worth	24	5	6
Lubbock	27	9	10
Midland	25	2	2
San Angelo	22	3	3
Wichita Falls	23	4	5
UTAH			
Milford	14	23	16
Salt Lake City	28	9	8
Wendover	13	2	3
VERMONT			
Burlington	28	43	37
VIRGINIA			
Dulles Airport	17	15	19
Lynchburg	27	13	16
National Airport	27	16	18
Norfolk	25	9	9
Richmond	28	10	12
Roanoke	27	14	17
WASHINGTON			
Olympia	24	23	24
Quillayute	13	21	24
Seattle-Tacoma	28	15	14
Spokane	28	36	41
Stampede Pass	27	483	511
Yakima	27	19	25
WEST VIRGINIA			
Beckley	8	20	51
Charleston	26	21	20
Elkins	20	22	21
Huntington	18	13	15
WISCONSIN			
Green Bay	28	37	36
La Crosse	16	23	32
Madison	28	32	32
Milwaukee	28	34	32
WYOMING			
Casper	28	9	10
Cheyenne	28	18	15
Lander	27	26	20
Sheridan	28	20	25

*Multiply table load by 0.048 for load in kPa
(ANSI A58.1-1982)

ASAE Standard: ASAE S292.4

UNIFORM TERMINOLOGY FOR RURAL WASTE MANAGEMENT

Proposed by the ASAE Rural Waste Disposal Committee; reviewed by the ASAE Structures and Environment Division Technical Committee, ASAE Water Treatment and Use Committee, and the USDA North Central Region Committee on Farm Waste Disposal; approved by the Structures and Environment Division Technical Committee; adopted by ASAE as a Recommendation December 1965; reconfirmed December 1970, December 1971; revised March 1973; revised and reclassified as a Standard December 1977; reconfirmed December 1982; revised June 1983; revised March 1987.

SECTION 1—PURPOSE AND SCOPE

1.1 The terminology reported herein is intended to establish uniformity in terms used in the field of rural waste management and to serve as a focal point for the development of useful new terms and definitions. Terms and definitions were adopted from related fields where applicable.

1.2 Standard procedures for the determination of values for many of the terms defined herein may be found in Standard Methods for the Examination of Water and Wastewater, American Public Health Association, Washington, DC. A source of additional wastewater terms can be found in the Glossary—Water and Waste Water Control Engineering, Water Pollution Control Federation, Washington, DC.

SECTION 2—DEFINITIONS

2.1 Activated sludge process: A biological wastewater treatment process in which a mixture of wastewater and biological solids or activated sludge is agitated and aerated. The activated sludge is subsequently separated from the treated wastewater (mixed liquor) by sedimentation and wasted or returned to the process as needed.

2.2 Adsorption: (1) The adherence of dissolved, colloidal, or finely divided solids on the surfaces of solid bodies with which they are brought into contact. (2) Action causing a change in concentration of gas or solute at the interface of a two-phase system.

2.3 Aerobic bacteria: Bacteria that require free elemental oxygen for their growth. Oxygen in chemical combination will not support aerobic organisms.

2.4 Aerobic decomposition: Reduction of the net energy level of organic matter by aerobic microorganisms.

2.5 Aerobic lagoon: (See lagoon.)

2.6 Aeration: A process causing intimate contact between air and a liquid by one or more of the following methods: (a) spraying the liquid in the air, (b) bubbling air through the liquid, and (c) agitating the liquid to promote absorption of oxygen through the air liquid interface.

2.7 Aeration unit: A tank or lagoon in which sludge, wastewater, or other liquid is aerated.

2.8 Aerosol: A system of colloidal particles dispersed into air or gas, e.g., smoke or fog.

2.9 Agitation: The turbulent mixing of liquid and solids.

2.10 Agricultural wastes: Wastes normally associated with the production and processing of food and fiber on farms, feedlots, ranches, ranges, and forests which may include animal manure, crop residues, and dead animals; also agricultural chemicals, fertilizers and pesticides which may find their way into surface and subsurface water.

2.11 Agricultural residue: A term normally associated with the production and processing of food and fiber on farms, feedlots, ranches, ranges and forests which may include animal manure and crop residues.

2.12 Algae: Primitive plants, one- or many-celled, usually aquatic, and capable of synthesizing their foodstuffs by photosynthesis.

2.13 Alkalinity: The capacity of water to neutralize acids, a property imparted by the water's content of carbonates, bicarbonates, hydroxides, and occasionally borates, silicates, and phosphates. It is expressed in milligrams per liter of equivalent calcium carbonate.

2.14 Ammonification: The biochemical process whereby ammoniacal nitrogen is released from nitrogen-containing organic compounds.

2.15 Anaerobic bacteria: Bacteria not requiring the presence of free or dissolved oxygen. Facultative anaerobes can be active in the presence of dissolved oxygen, but do not require it.

2.16 Anaerobic decomposition: Reduction of the net energy level of organic matter by anaerobic microorganisms in the absence of oxygen.

2.17 Anaerobic digestion: Conversion of organic matter in the absence of oxygen under controlled conditions to gases such as methane and carbon dioxide.

2.18 Bacteria: A group of universally distributed, rigid, essentially unicellular procaryotic microorganisms. Bacteria usually appear as spheroid, rod-like or curved entities, but occasionally appear as sheets, chains, or branched filaments.

2.19 Biochemical oxygen demand (BOD): The quantity of oxygen used in the biochemical oxidation of organic matter in a specified time, at a specified temperature, and under specified conditions. Normally 5 days at 20 °C unless otherwise stated. A standard test used in assessing the biodegradable organic matter in municipal wastewater.

2.20 Biogas: Gaseous product of anaerobic digestion that consists primarily of methane and carbon dioxide.

2.21 Biological oxidation: The process whereby living organisms convert organic matter into a less complex or a mineral form.

2.22 Biological wastewater treatment: Forms of wastewater treatment in which bacterial or biochemical action is intensified to stabilize or oxidize the unstable organic matter present. Oxidation ditches, aerated lagoons, anaerobic lagoons and anaerobic digesters are examples.

2.23 Biomass: A term used to describe organic matter which has been grown by photosynthetic conversion of solar energy.

2.24 Carbon-nitrogen ratio (C/N): The weight ratio of carbon to nitrogen in organic matter.

2.25 Cesspool: A partially lined or unlined underground pit into which raw animal and/or household wastewater is discharged and from which liquid seeps into the surrounding soil.

2.26 Chemical oxidation: Oxidation of organic substances without benefit of living organisms. Examples are by thermal combustion or by oxidizing agents such as chlorine.

2.27 Chemical oxygen demand (COD): A measure of the oxygen-consuming capacity of inorganic and organic matter present in water or wastewater. It is expressed as the amount of oxygen consumed from a chemical oxidant in a specified test. It does not differentiate between stable and unstable organic matter and thus does not necessarily correlate with biochemical oxygen demand.

2.28 Chlorination: The application of chlorine to water, sewage, or industrial wastes, generally for the purpose of disinfection.

2.29 Coagulation: In water and wastewater treatment, the aggregation of colloidal, finely divided suspended matter and/or bacterial cells by the addition of a floc-forming chemical or by biological processes.

2.30 Coliform-group bacteria: A group of bacteria predominantly inhabiting the intestines of man or animal, but also found in soil. It includes all aerobic and facultative anaerobic, gram-negative, nonspore-forming bacilli that ferment lactose with production of gas. This group of "total" coliforms includes Escherichia coli which is considered the typical coliform of fecal origin.

2.31 Colloidal matter: Finely divided solids which will not settle but may be removed by coagulation or biochemical action or membrane filtration.

2.32 Composting: Biological degradation of organic matter under aerobic conditions to a relatively stable humus-like material called compost.

2.33 Contamination: Any introduction into the environment (water, air or soil) of microorganisms, chemicals, wastes, or wastewater in a concentration that makes the environment unfit for its intended use.

2.34 Dehydration: The chemical or physical process whereby water in chemical or physical combination with other matter is removed.

2.35 Denitrification: The reduction of oxidized nitrogen compounds (such as nitrates) to nitrogen gas or nitrous oxide gas.

2.36 Detention pond: An earthen structure constructed to store runoff water and other wastewater until such time as the liquid may be recycled onto land. Sometimes called holding ponds or waste storage ponds.

2.37 Detention time: The time wastes are subjected to a stabilization process or held in storage.

2.38 Deoxygenation: The depletion of the dissolved oxygen in a liquid, through the biochemical oxidation of organic matter present or by chemical addition.

2.39 Digestion: Usually refers to the breakdown of organic matter in water solution or suspension into simpler or more biologically stable compounds, or both. In anaerobic digestion organic matter may be decomposed to soluble organic acids or alcohols and subsequently converted to such gases as methane and carbon dioxide. Complete decomposition of organic solid materials to gases and water by bacterial action alone is never accomplished.

2.40 Disinfection: Killing the larger portion of microorganisms in or on a substance with the probability that all pathogenic bacteria are killed by the agent used.

2.41 Dissolved oxygen (DO): The molecular oxygen dissolved in water, wastewater, or other liquid, usually expressed in milligrams per liter, parts per million, or percent of saturation.

2.42 Earthen storage basin: An earthen structure usually with sloping sides and a flat floor, constructed to store semi-solid, slurry or liquid manure. Also called a waste storage pond.

2.43 Effluent: The discharge of wastewater or other liquid, treated or untreated.

2.44 Electrical conductivity: A measure of a solution's ability to carry an electrical current; varies both with the number and type of ions contained by the solution.

2.45 Escherichia coli (E. Coli): One of the species of bacteria in the intestinal tract of warm-blooded animals. Its presence is considered indicative of fresh fecal contamination.

2.46 Evaporation rate: The quantity of water evaporated from a given water surface per unit of time. It is usually expressed in millimeters (inches) depth per day, month per year.

2.47 Facultative bacteria: Bacteria which can use either free oxygen or reduced carbon compounds as electron acceptors (as in organic substrates like sugars, starches, etc.) in their metabolism.

2.48 Fertilizer value: An estimate of the value of commercial fertilizer elements (N, P, K) that can be replaced by manure or organic waste material. Usually expressed as dollars per ton of manure or quantity of nutrients per ton of manure.

2.49 Fixed solids: The portion of the total solids remaining as an ash or residue when heated at a specific temperature and time (usually 600 °C for at least one hour).

2.50 Food to microorganisms ratio (F/M): The weight ratio of biodegradable organic matter (BOD) to microorganisms.

2.51 Filtration: The process of passing a liquid through a filtering medium, such as activated carbon, sand, magnetite, diatomaceous earth, finely woven cloth, unglazed porcelain, or specially prepared paper for the removal of suspended or colloidal matter.

2.52 Flocculation: In water and wastewater treatment, an operation which promotes the coalescence of suspended particles by increasing contact with each other for the purpose of removal by sedimentation, filtration or flotation. This operation may involve mechanical, physical or biological mechanisms.

2.53 Flushing system: A system that collects and transports or moves waste material with the use of water such as in washing of pens and flushing confinement livestock systems.

2.54 Grassed infiltration area: An area with vegetative cover where runoff water infiltrates into the soil.

2.55 Gasification: The transformation of organic materials into gas through biological or physical processes.

2.56 Holding pond: (See detention pond.)

2.57 Humus: The dark or black carboniferous relatively stable residue resulting from the decomposition of organic matter.

2.58 Hydraulic settling: Removal of water-transported particles by gravity.

2.59 Hydraulic settling basin: (See sedimentation tank.)

2.60 Incineration: The rapid oxidation of solids within a specially designed combustion chamber.

2.61 Incubation: Maintenance of viable organisms in or on a nutrient substrate at constant temperature for a growth period.

2.62 Infiltration rate: The rate at which water enters the soil or other porous material under a given condition, expressed as depth of water per unit time, usually in millimeters per hour.

2.63 Influent: Water, wastewater, or other liquid flowing into a reservoir, basin, or treatment plant, or any unit thereof.

2.64 Inoculum: Living organisms, or an amount of material containing living organisms (such as bacteria or other microorganisms) which are added to initiate or accelerate a biological process, e.g., biological seeding.

2.65 Lagoon: An earthen structure for the storage and biological treatment of wastewater. Lagoons can be aerobic, anaerobic, or facultative depending on their loading and design.

2.66 Land application: Application of manure, sewage sludge, municipal wastewater and industrial wastes to land either for ultimate disposal or for reuse of the nutrients and organic matter for their fertilizer value.

2.67 Leaching: (1) The removal of soluble constituents such as nitrates or chlorides from soils or other material by water. (2) The removal of salts and alkali from soils by irrigation combined with drainage. (3) The disposal of a liquid through a nonwatertight artificial structure, conduit, or porous material by downward or lateral drainage, or both, into the surrounding permeable soil.

2.68 Liquefaction: Act or process of hydrolysis, rendering or becoming liquid; reduction to a liquid state.

2.69 Liquid manure: (See manure.)

2.70 Litter: The bedding material used for poultry and livestock.

2.71 Livestock residue: All livestock waste.

2.72 Livestock waste: (See manure.)

2.73 Loading rate: The quantity of material added per unit volume or unit area per unit time.

2.74 Manure: The fecal and urinary excretion of livestock and poultry. Sometimes referred to as livestock waste. This material may also contain bedding, spilled feed, water or soil. It may also include wastes not associated with livestock excreta, such as milking center wastewater, contaminated milk, hair, feathers, or other debris. Manure may be described in different categories as related to solids and moisture content. These categories are related to handling equipment and storage types. Fig. 1 gives relationships between

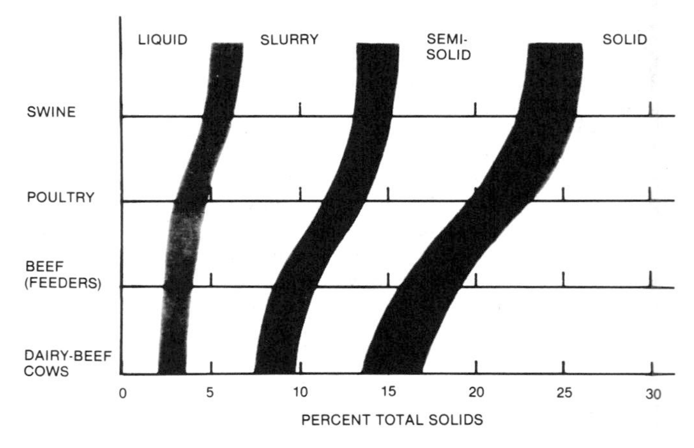

FIG. 1—RELATIVE HANDLING CHARACTERISTICS OF DIFFERENT TYPES OF MANURE AND PERCENT TOTAL SOLIDS

percent solids and categories of manure for major livestock species and poultry. The shaded lines are broad because the transition from one category to another, e.g., SLURRY to SEMI-SOLID, is not sharply defined. The transition does not depend on percent solids alone but is affected by type livestock manure and their feed ration, type and amount of bedding, feed spillage, and other residues in the manure.

2.74.1 Liquid manure (thin slurry): Manure which has had sufficient water added so that it can be pumped easily. Normally fibrous material such as chopped straw or waste hay is not present.

2.74.2 Slurry manure: Manure in which the precent total solids approximates that of excreted manure. The total solids content could vary by a few percent depending on whether water is added or a slight drying occurs. Slurry manure can be handled with conventional, centrifugal manure pumps and equipment.

2.74.3 Semi-solid manure: Manure which has had some bedding added or has received sufficient air drying to raise the solids content such that it will stack but has a lower profile than solid manure and seepage may collect around the outside. It may be pumped with positive displacement pumps or be handled with a front-end loader.

2.74.4 Solid manure: Manure which has had sufficient bedding or soil added, or has received sufficient air drying to raise the solids content to where it will stack with little or no seepage. It is best handled with a front-end loader.

2.75 Manure flume: Any restricted passageway, open along its full length to the atmosphere, through which liquid moves by gravity.

2.76 Manure storage: A storage unit to keep manure contained for some period of time prior to its ultimate utilization or disposal. Manure storages are usually classified by type and form of manure stored and/or construction of the storage, e.g., above or below ground liquid manure tank, earthen storage basin, solid manure storage.

2.77 Manure tank: A storage structure with vertical side walls and an inpervious floor, constructed to store semi-solid, slurry and liquid manure. The tank may be located either in-ground or aboveground. The in-ground tank usually is constructed of concrete, and may have a sloping access ramp. The aboveground tank is usually constructed of concrete or steel, and contains provisions for agitation and pumping.

2.78 Mechanical solids separation: The process of separating suspended solids from a liquid-carrying medium by trapping the particles on a mechanical screen or sieve, or by centrifugation.

2.79 Mesophilic bacteria: Bacteria which are found in a temperature range of 25-40 °C. They are most active around 35 °C.

2.80 Milking center wastes: The wastewater containing milk residues, detergents, and manure which is generated in a milking center.

2.81 Nitrification: The biochemical oxidation of ammoniacal nitrogen to nitrate.

2.82 Odor threshold: The lowest concentration of an odorant (odorous substance) in air which can be detected by the human olfactory sense.

2.83 Organic matter: Chemical substances of animal or vegetable origin, consisting of hydrocarbons and their derivatives.

2.84 Oxidation ditch: A modified form of the activated sludge process. An aeration device supplies oxygen and circulates the liquid in a circular open channel.

2.85 Oxidation pond: An earthen pond or lagoon for the biological oxidation of organic material by natural or mechanical aeration.

2.86 pH: The logarithm (base 10) of the reciprocal of the hydrogen-ion concentration expressed in moles per liter. For example, a pH value of 7 indicates a hydrogen-ion concentration of 10^{-7} moles/L.

2.87 Percolation rate: The rate of movement of water under hydrostatic pressure down through the interstices of rock, soil, or filtering media except movement through large openings such as caves.

2.88 Permeability: The property of a material which permits movement of water through it when saturated and actuated by hydrostatic pressure of the magnitude normally encountered in natural subsurface water.

2.89 Pollution: The presence in a body of water (or soil or air) of a substance (pollutant) in such quantities that it impairs the body's usefulness or renders it offensive to the senses of sight, taste, or smell. In general, a public-health hazard may be created, but in some instances only economic or aesthetics are involved as when foul odors pollute the air.

2.90 Population equivalent (P.E.): A means of expressing the strength of a pollutant in wastewater relative to the strength of human sewage on such basis as BOD, total solids, suspended solids, or nitrogen. It has no real value in terms of handling animal wastes.

2.91 Porous dam: A runoff control structure which reduces the velocity of the runoff so the solids will settle out in the settling basin or terrace. It may consist of rock, expanded metal, narrow wood slots, etc. The liquid passing through the porous dam is normally collected and stored in a detention pond or earthern storage basin. The settled solids are generally handled as a semi-solid or solid manure.

2.92 Putrefaction: Biological decomposition of nitrogenous organic matter with the production of foul-smelling products associated with anaerobic degradation of proteins.

2.93 Rural wastes: Wastes produced in rural areas. These wastes normally include animal manure, crop residues and dead animals. Residual fertilizers, pesticides, inorganic salts and eroded soil may also be classified as rural wastes when they are in nonurban areas. Domestic solid refuse, human sewage and industrial wastes generated and handled in the rural environment are also considered rural wastes.

2.94 Sediment: Any material carried in suspension or bedload in water which will ultimately settle after the water loses velocity.

2.95 Sedimentation tank: A unit in which water or wastewater containing settleable solids is retained to remove by gravity a part of the suspended matter. Also called sedimentation basin, settling basin, settling tank or settling terrace.

2.96 Seepage: (1) Percolation of water through the soil. (2) The slow movement of water through small cracks, pores, interstices, of a material. (3) The loss of liquid by infiltration from a canal, reservoir, manure tank or manure stack. It is generally expressed as flow volume per unit time.

2.97 Semi-solid manures: (See manure.)

2.98 Septage: Septic tank pumpings; the mixed liquor and solid contents pumped from septic tanks and dry wells used for receiving domestic type sewage.

2.99 Septic tank: A settling tank in which settled solid matter is removed from the wastewater flowing through the tank and the organic solids are decomposed by anaerobic bacterial action.

2.100 Settling basin: (See sedimentation tank.)

2.101 Settling terrace: (See sedimentation tank.)

2.102 Settleable solids: (1) That matter in wastewater which will not stay in suspension during a preselected settling period, such as one hour. (2) In the Imhoff cone test, the volume of matter that settles to the bottom of the cone.

2.103 Settling tank: (See sedimentation tank.)

2.104 Sewage: The spent water of a community. Term now being replaced in technical usage by wastewater.

2.105 Silt: (1) Soil particles which constitute the physical fraction of a soil between 0.005 mm and 0.05 mm in diameter. (2) Fine particles of soil carried in suspension by flowing water. (3) Deposits of waterborne material in a reservoir, on a delta or on overflowed lands.

2.106 Slotted floor: The floor surface of a building which has open spaces, cracks or slots to allow manure and other waste material to pass through the floor.

2.107 Sludge: (1) The precipitate or settled solids from treatment, coagulation, or sedimentation of water or wastewater. (2) Deposits on bottoms of streams or other bodies of water.

2.108 Slurry manure: (See manure.)

2.109 Solid manure: (See manure.)

2.110 Solid manure storage: A storage unit in which accumulation of solid manure are stored before subsequent handling and field spreading. The manure is generally stacked on a concrete slab ("stacking slab") but may also be simply stacked on the soil for short term storage. Liquids, including urine and precipitation, may or may not be drained from the unit.

2.111 Solids content: (1) The sum of the dissolved and suspended constituents in water or wastewater. (2) The residue remaining when the water is evaporated away from a sample of sewage, other liquids, or semi-solid masses of material and the residue is then dried at a specified temperature (usually 103 °C for 24 h); usually stated in milligrams per liter or percent solids.

2.112 Specific conductance: (See electrical conductivity.)

2.113 Stabilization pond: (See oxidation pond.)

2.114 Sterilization: The killing of all living microorganisms, ordinarily through the use of heat or some chemical.

2.115 Supernatant: The liquid standing above a sediment or precipitate after settling or centrifuging.

2.116 Suspended solids: (1) Solids that are in water, wastewater, or other liquids, and which are largely removable by filtering or centrifuging. (2) The quantity of material filtered from wastewater in a laboratory test, as prescribed in APHA Standard Methods for the Examination of Water and Wastewater.

2.117 Thermophilic bacteria: Bacteria which are found in a temperature range of 40-75 °C. They are most active in a temperature range of 49-60 °C.

2.118 Thermophilic digestion: Anaerobic digestion in the temperature range of 45-60 °C.

2.119 Total solids: (See solids content.)

2.120 Toxic waste: A waste containing a material that either directly poisons living things or alters their environment so that they die.

2.121 Trickling filter: A biological treatment unit consisting of an artificial bed of coarse material, such as broken stone, clinkers, slate, slats brush, or plastic materials, over which wastewater is distributed or applied and through which it trickles to the underdrains, giving opportunity for the formation of biological slimes which oxidize organic matter in the wastewater.

2.122 Volatile acids: Fatty acids containing six or less carbon atoms, which are soluble in water and which can be steam-distilled at atmospheric pressure. Volatile acids are commonly reported as equivalent to acetic acid.

2.123 Volatile solids: That portion of the total solids driven off as volatile (combustible) gases at a specified temperature and time (usually 600 °C for at least 1 h).

2.124 Volatile suspended solids (VSS): That portion of the suspended solids driven off as volatile (combustible) gases at a specified temperature and time (usually 600 °C for at least 20 min).

Cited Standard:

APHA, Standard Methods for the Examination of Water and Wastewater

ASAE Data: ASAE D321.2

DIMENSIONS OF LIVESTOCK AND POULTRY

Proposed by the ASAE Livestock Dimension and Configuration Factors Subcommittee; approved by the ASAE Structures and Environment Division Technical Committee; adopted by ASAE December 1968; reconfirmed December 1973; revised December 1978; reconfirmed December 1983, December 1984; revised March 1985.

SECTION 1—PURPOSE AND SCOPE

1.1 The purpose of these Data is to provide information for use in the design of buildings and equipment for livestock and poultry. Summaries of available information are presented in the form of tables and curves of weight versus age and physical dimensions versus weight for various classes of livestock and poultry.

SECTION 2—BEEF

2.1 Average weight versus age for 68 steers of mixed European beef breeds is shown in Fig. 1. The steers were used to provide the dimensional data in paragraph 2.2. The regression equation for Fig. 1 is

$$WT = 15.33 + 0.8233\,(AGE) - 0.0001805\,(AGE)^2$$

where

WT = weight in kg
AGE = animal age in days
Percent error = 18.8

Cross breeding, creep feeding, and other management practices produce greater weight gains than shown in Fig. 1. Table 1 shows the average weight at specific ages for calves from matings of Hereford and Angus dams to Hereford, Angus, Jersey, South Devon, Limousin, Charolais and Simmental sires. From 62 to 399 animals were observed for each average. After weaning and preconditioning, the steers received diethylstilbestrol implants and were fed a ration of 60% corn silage, 33% grain concentrate and 7% protein supplement. The heifers were fed 50% corn silage and 50% grass haylage, with supplemental protein and minerals. The heifers were bred at 430 days of age or later. Growth curves for the steers during the feeding period, beginning at 225 days of age, are shown in Fig. 2.

TABLE 1—WEIGHT OF BEEF CALVES AT SPECIFIC AGES, (kg)

Breed Group*	All Calves		Steers		Heifers	
	Birth Weight	Weaning Weight	180-day feeding period			
	0 days	215 days	225 days	405 days	400 days	550 days
HH	34.7	182	214	419	262	306
AA	31.0	190	225	420	275	306
HAx	33.7	194	230	430	286	325
Jx	29.4	183	221	408	255	298
SDx	35.8	194	227	443	296	337
Lx	36.2	197	233	428	291	324
Cx	38.6	207	246	470	305	344
Sx	38.0	204	239	463	305	349

*H = Hereford, A = Angus, J = Jersey, SD = South Devon, L = Limousin, C = Charolais, S = Simmental, HA = Hereford sires by Angus dams; HAx = HA + AH, Jx = JH + JA, etc.

2.2 Physical dimensions versus weight for 68 steers of mixed European beef breeds are shown in Fig. 3 and Table 2. The regression equations for the curves in Fig. 3 are given by

$$DIM = 10[A + B\,(WT) - C\,(WT)^2]$$

where

DIM = dimension in mm
WT = weight in kg
A, B, and C = constants listed in Table 2.

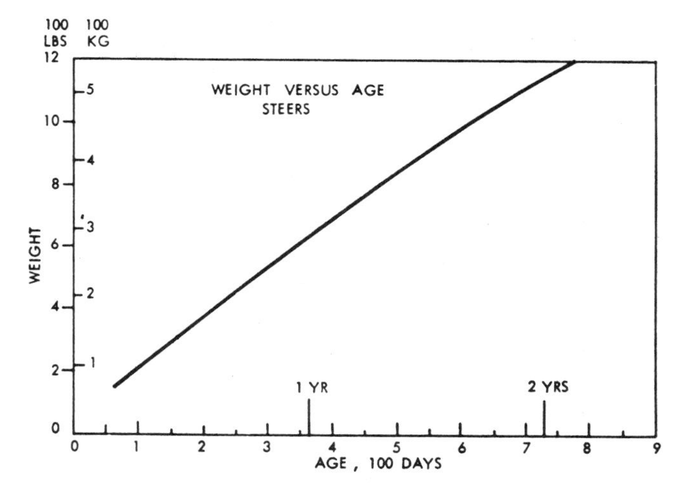

FIG. 1—WEIGHT VERSUS AGE FOR STEERS OF EUROPEAN BREEDS

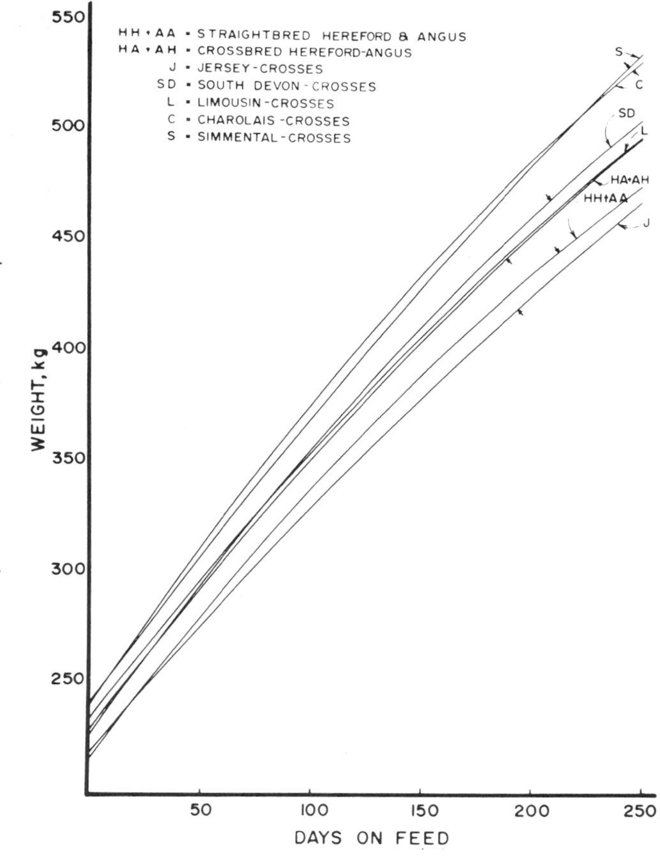

FIG. 2—WEIGHT OF CROSS BRED STEERS ON FEED BEGINNING AT 225 DAYS OF AGE

The reference points for Fig. 3 are defined by

Point	Description
25	Length coordinate, X = 0 Rear tangent of the animal
1	Height coordinate, Y = 0 Floor
12	Thickness coordinate Z = 0 Center of back over shoulder. X coordinate not definite
2	Thickest point of hindquarter
6	Center of back above hindquarter. X coordinate not meaningful
8	Bulge of the belly
10	Bottom of chest highest above ground
11	Thickest point of shoulder
14	Pronounced break in brisket. Not accurately located
20	Center of foreleg at knee
21	Center of foreleg at dew-claw
22	Center of rear leg at dew-claw
23	Center of rear leg at hock.

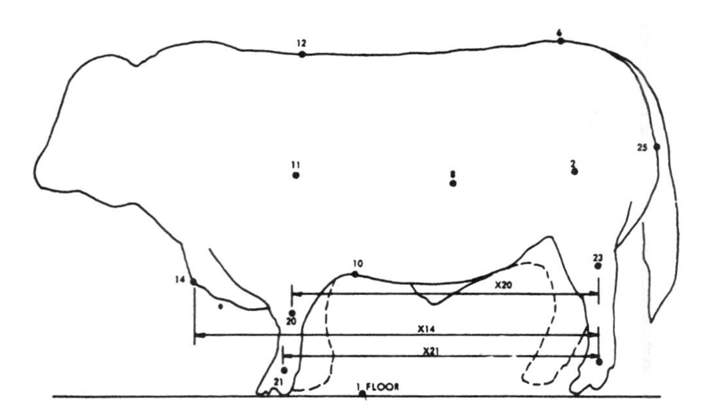

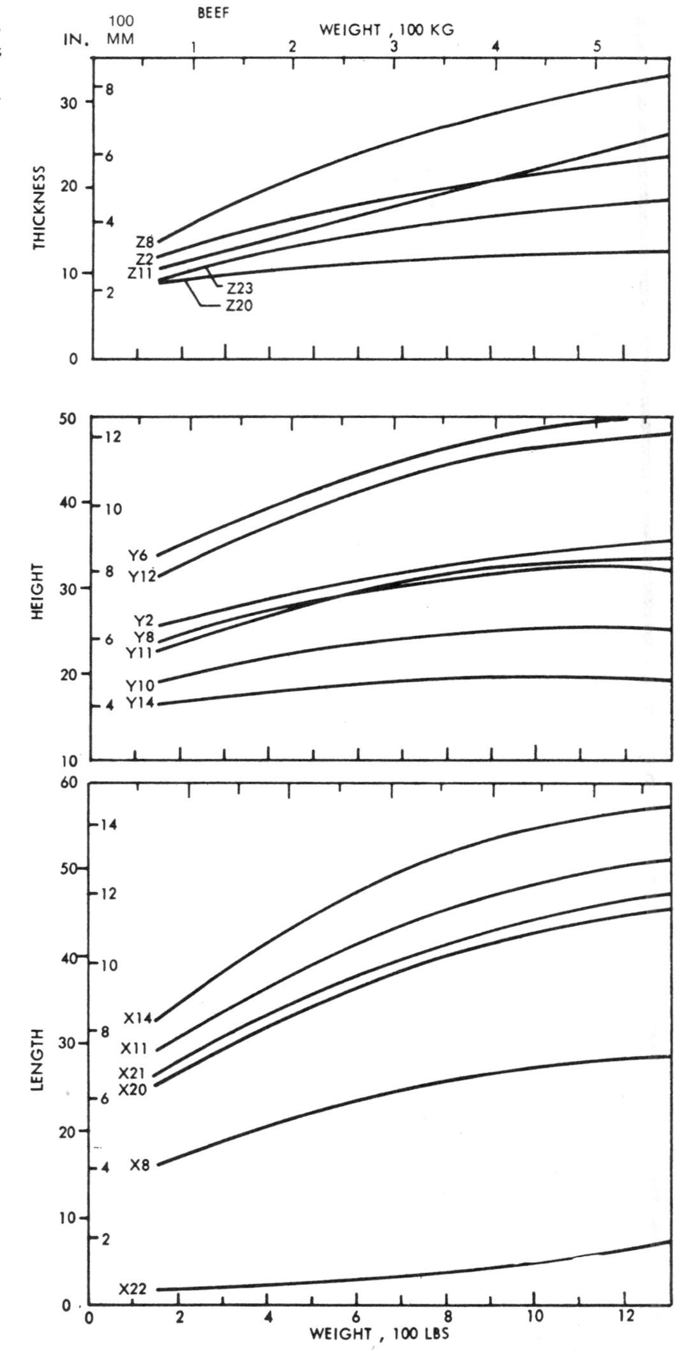

FIG. 3—PHYSICAL DIMENSIONS VERSUS WEIGHT OF STEERS

TABLE 2—CONSTANTS FOR REGRESSION EQUATIONS

Dimensions	A	B	C x 10^6	Percent Error
X 8	32.67	0.1241	96.66	5.2
11	60.38	0.2098	156.78	3.7
12	57.30	0.1836	132.83	4.2
14	67.44	0.2466	193.57	13.9
20	52.11	0.1850	132.09	5.8
21	54.97	0.1864	129.25	21.0
22	4.62	0.0007	37.78	33.1
Y 2	59.58	0.0842	56.66	6.7
6	75.01	0.1629	128.39	3.7
8	52.66	0.1079	90.86	6.1
10	43.19	0.0796	76.66	6.9
11	50.27	0.1211	116.04	6.6
12	68.51	0.1678	131.47	4.7
14	35.85	0.0620	69.01	10.9
Z 2	24.17	0.0955	61.48	23.2
8	24.98	0.1552	95.06	11.3
11	21.92	0.0751	0.12	10.9
20	20.42	0.0367	30.02	20.9
23	17.85	0.0884	68.27	28.6

SECTION 3—DAIRY

3.1 Weight versus age for selected breeds of dairy cattle is shown in Fig. 4.

3.2 Physical dimensions versus weight for Holstein, Ayrshire, Jersey and Guernsey females are shown in Figs. 6, 7, 8 and 9, respectively. Reference points for the dimensions are shown in Fig. 5. The dimensions are defined as

Height above floor (Y)

Y1 Height to pinbones (Y1), hips, (Y2) and withers (Y3) are nearly the same for all breeds and ages. The pinbones are generally lower by up to 5%, especially in mature animals.

Lengths (X)

X1-6 Pinbone to front of shoulder (body)
X1-2 Pinbone to hip (rump)
X1-3 Pinbone to wither (back)
X3-4 Wither to pole (neck)
X4-5 Pole to muzzle (face)

Depths (Y)

Y7 Chest, at heart girth
Y8 Barrel

Depth at chest and barrel are nearly the same at birth. By six months of age the following approximations apply:

Y7 = Y8-30 mm for Holstein and Guernsey
Y7 = Y8-25 mm for Jerseys
Y7 = Y8-45 mm for Ayrshire

Circumferences (girths) (C)

C5 Muzzle
C7 Chest, at heart girth
C8 Barrel

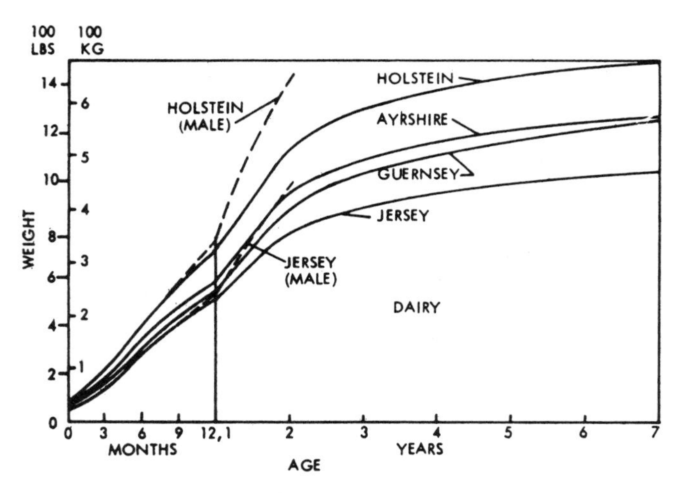

FIG. 4—WEIGHT VERSUS AGE OF DAIRY CATTLE

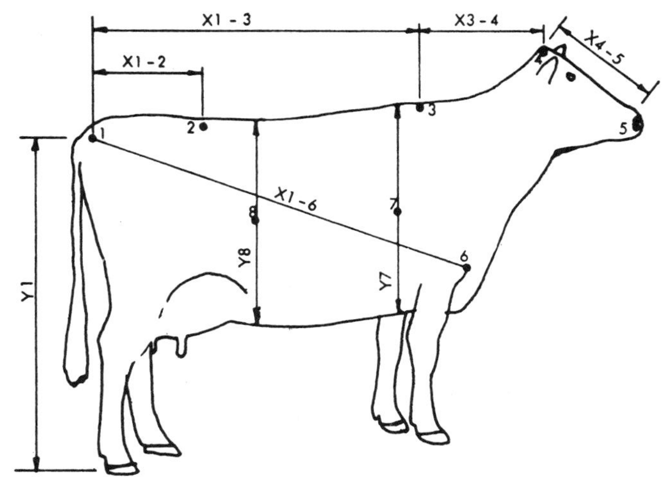

FIG. 5—REFERENCE POINTS FOR DIMENSIONS OF DAIRY CATTLE

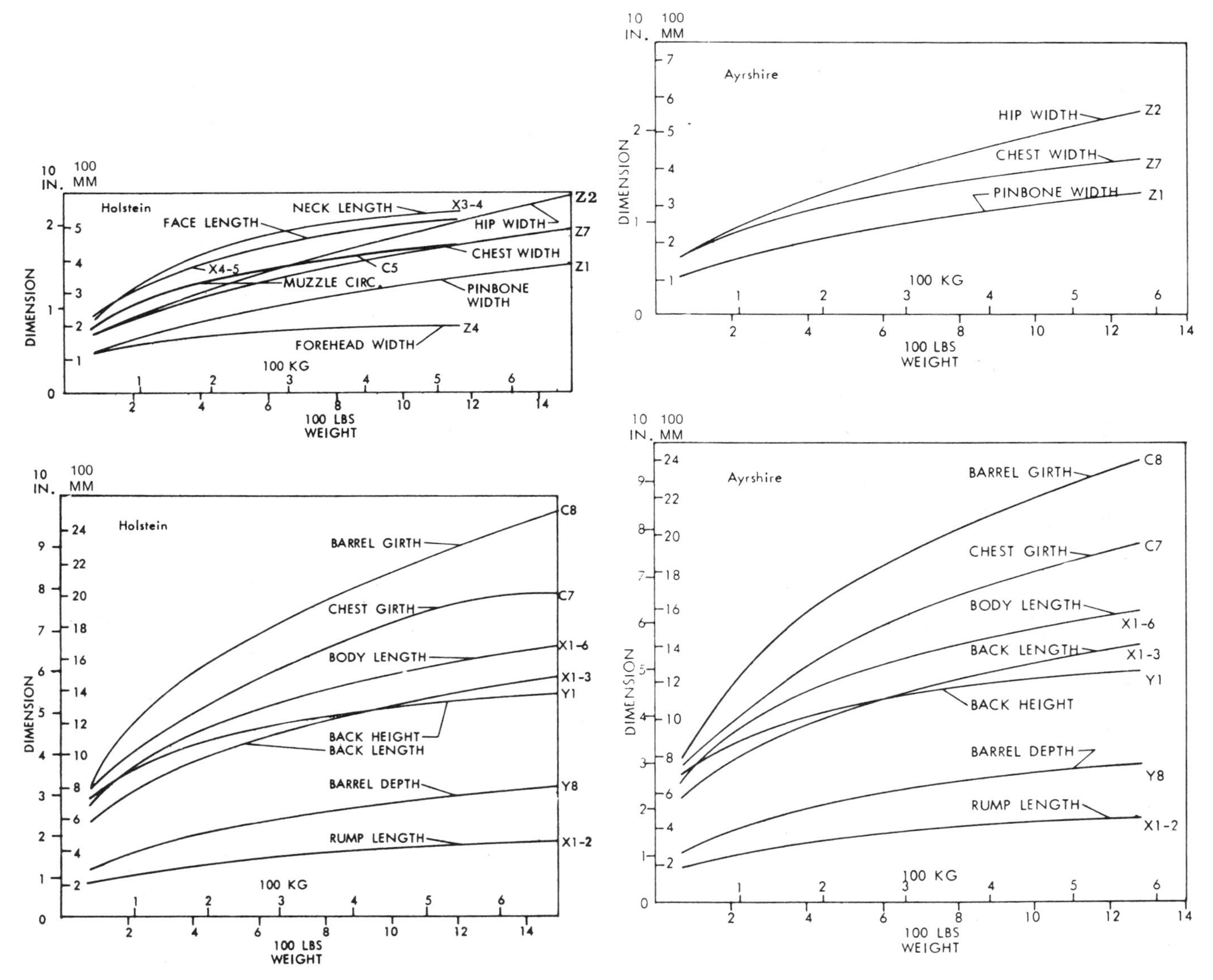

FIG. 6—PHYSICAL DIMENSIONS VERSUS WEIGHT OF HOLSTEIN FEMALES

FIG. 7—PHYSICAL DIMENSIONS VERSUS WEIGHT OF AYRSHIRE FEMALES

Widths (Z)

Z1 Pinbones

Z2 Hips

Z4 Forehead

Z7 Chest, at heart girth

Height and heart girth of Holstein and Jersey males are shown in Fig. 10. The reach of mouth versus animal length and feed table height are shown in Table 3.

TABLE 3—REACH OF MOUTH VERSUS ANIMAL LENGTH AND FEED TABLE HEIGHT (mm)

Body length, mm (X1-6 in Fig. 5)	Horizontal reach of mouth from front legs (mm) at indicated height (mm) of table from level of feet				
	0	100	200	300	400
1100	750	800	840	910	940
1200	780	820	880	940	970
1300	800	840	910	960	990
1400	830	860	950	990	1020
1500	850	880	980	1010	1040
1600	880	900	1020	1030	1070
1700	900	920	1060	1060	1090
1800	930	940	1090	1080	1120

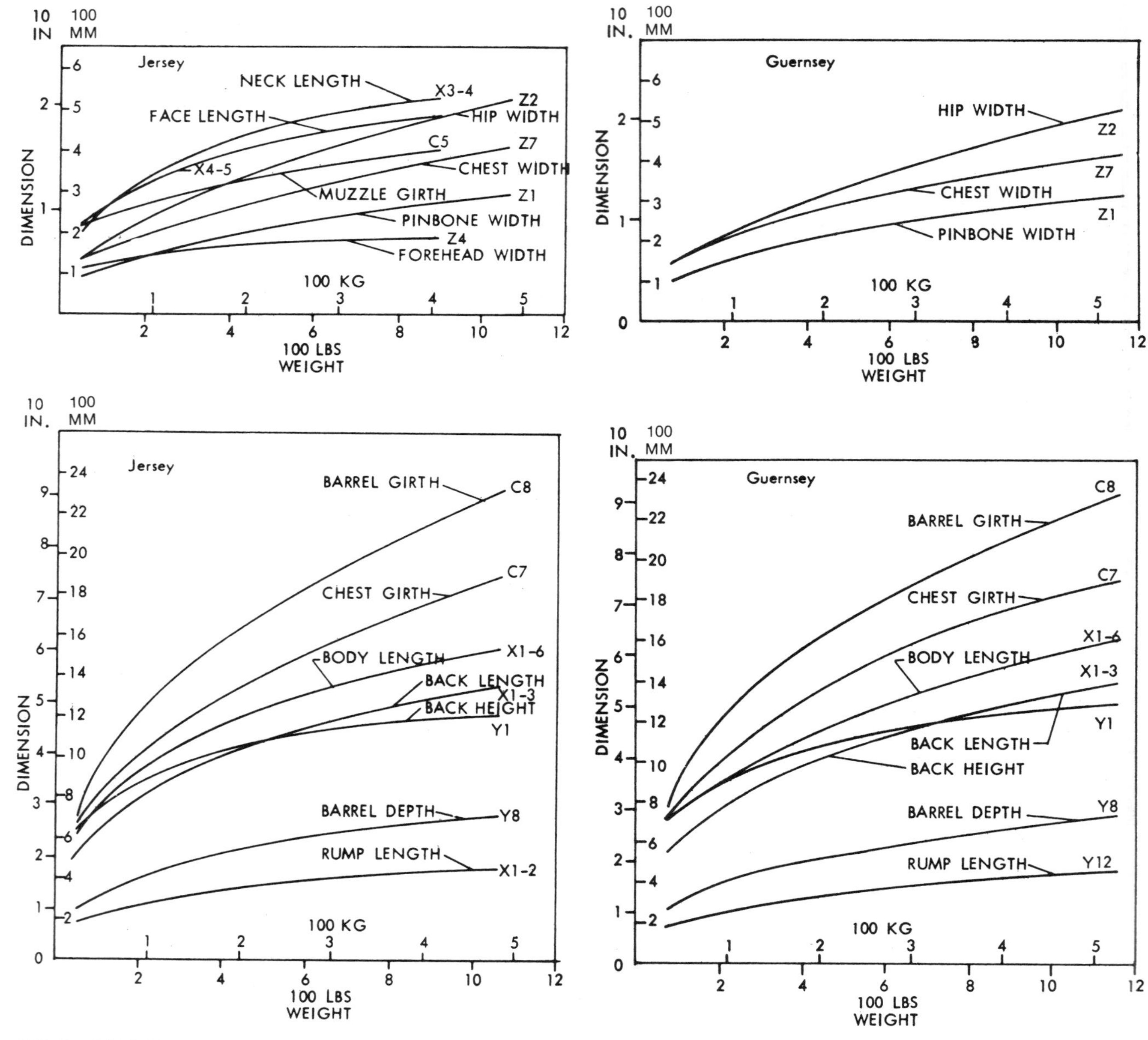

FIG. 8—PHYSICAL DIMENSIONS VERSUS WEIGHT OF JERSEY FEMALES

FIG. 9—PHYSICAL DIMENSIONS VERSUS WEIGHT OF GUERNSEY FEMALES

SECTION 4—SWINE

4.1 The weight of pigs is related to age in Fig. 11. This curve, and those in Fig. 12, are representative of typical crossbred pigs. Some deviation will exist due to breed, sex, environment, and individual pigs.

4.2 Physical dimensions of pigs are shown in Fig. 12. These dimensions are defined as follows:

L = length, total overall
H = height from floor to top of back
D = depth from lowest point on belly to top of back
W = width at shoulders. Width across middle of belly and across hams approximately equals this dimension.
FL = front leg length from floor to point between front legs
HH = head weight measured from point in front of front legs to top of head between ears
J = jowl length from nose to front legs
F = face length from nose to ear
TL = taper length from nose to widest point on shoulders

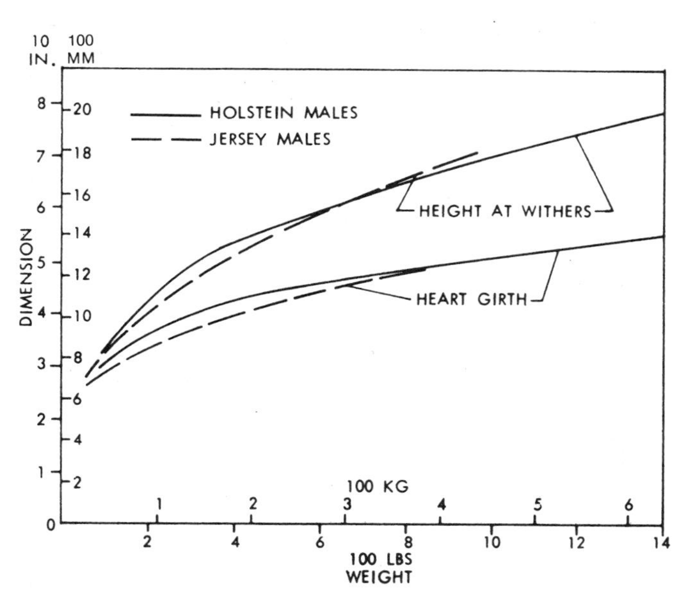

FIG. 10—PHYSICAL DIMENSIONS VERSUS WEIGHT OF DAIRY MALES, BIRTH TO 2 YEARS OLD

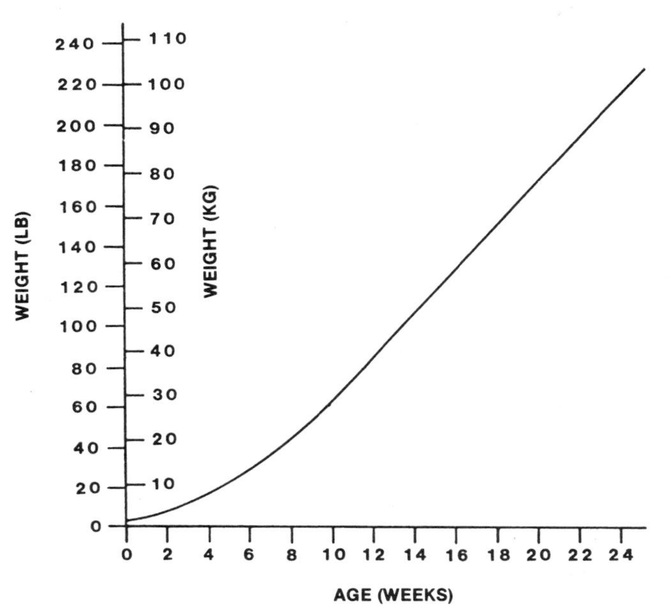

FIG. 11—WEIGHT VERSUS AGE OF SWINE

SECTION 5—SHEEP

5.1 Weight versus age of 860 purebred ram lambs from 7 breeds are shown in Fig. 13. The spring lambs and their dams were placed on cool season pastures and a creep ration on April 10 or later, depending on the date of lambing. The lambs were weaned at ages ranging from 56 to 98 days and after a 2-week adjustment period, were self-fed a feedlot ration of about 12.9% crude protein and about 73.5% total digestible nutrients.

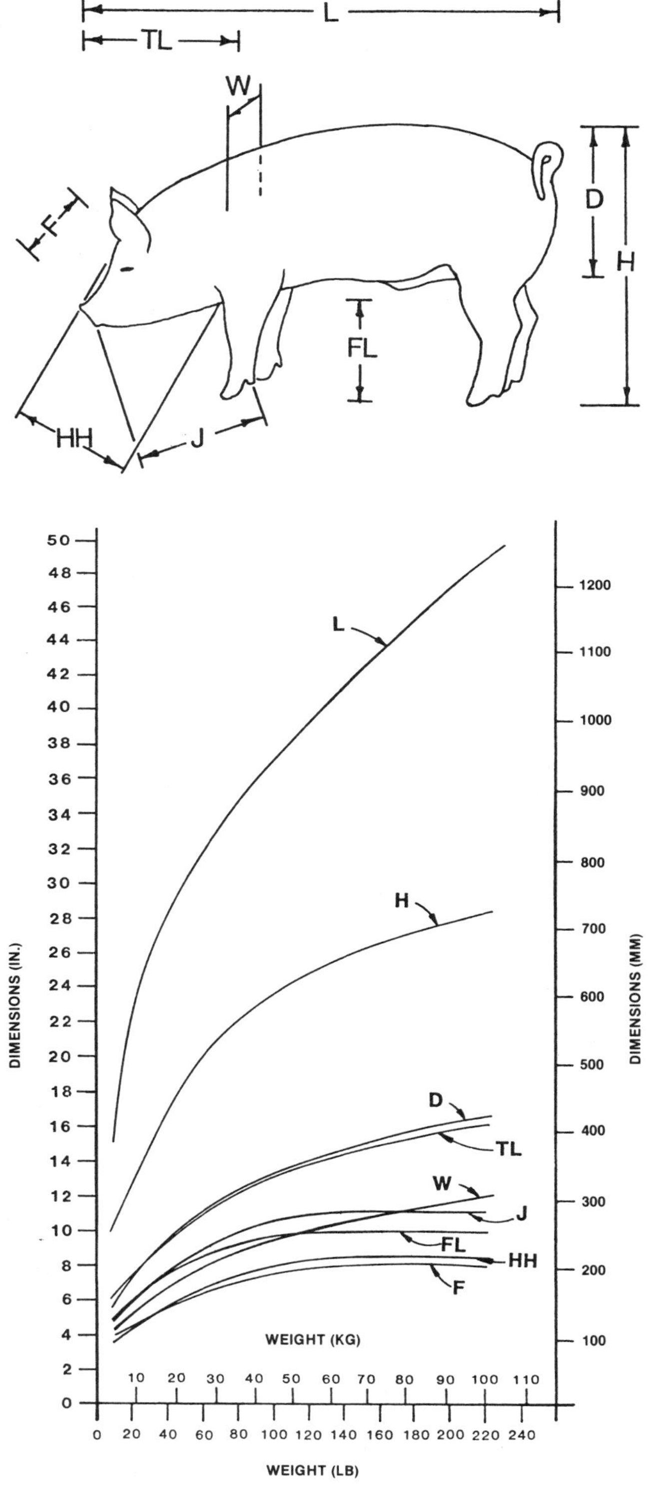

FIG. 12—PHYSICAL DIMENSIONS VERSUS WEIGHT OF SWINE

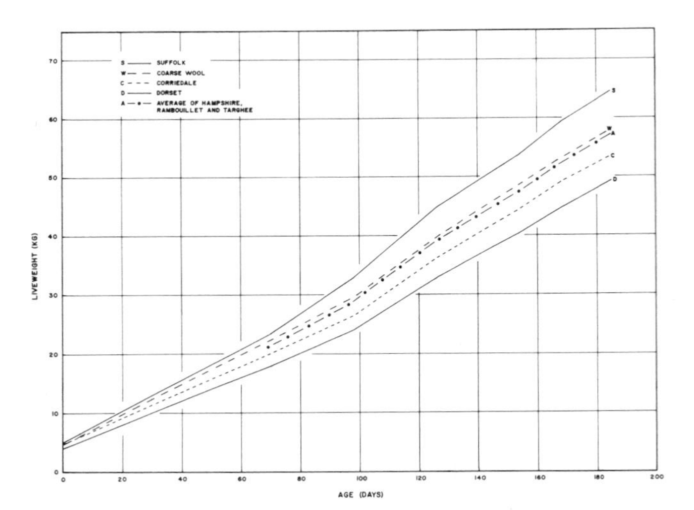

FIG. 13—WEIGHT VERSUS AGE OF RAM LAMBS IN FEED-LOTS

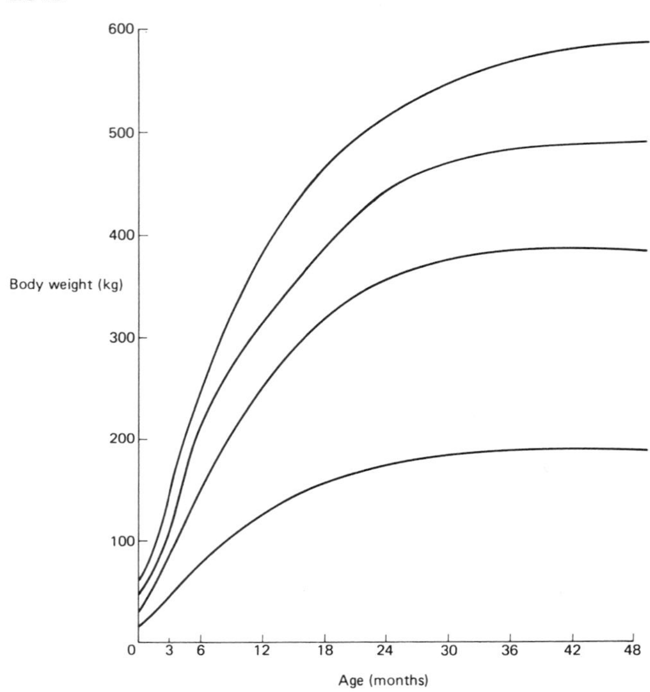

FIG. 14—WEIGHT VERSUS AGE FOR HORSES OF VARIOUS MATURE WEIGHTS

SECTION 6—HORSES

6.1 Body weight versus age for horses of various mature weights is shown in Fig. 14. The curves are based on limited data and represent examples rather than models of growth for horses.

SECTION 7—POULTRY

7.1 Weight

7.1.1 The weight of White Plymouth Rock, White Leghorn and New Hampshire pullets versus age is shown in Fig. 15. The weight of several strains of broilers versus age is shown in Fig. 16. A growth curve for an improved commercial broiler strain is shown in Fig. 17.

7.1.2 The weight of White Pekin ducks versus age is shown in Fig. 18. The curves are typical of several breeds of ducklings.

7.1.3 The weight of turkeys versus age for large type and small white toms and hens is shown in Fig. 19. Mean weight and 95% confidence limits versus age for over 900 Wrolstad White turkeys are shown in Figs. 20 and 21.

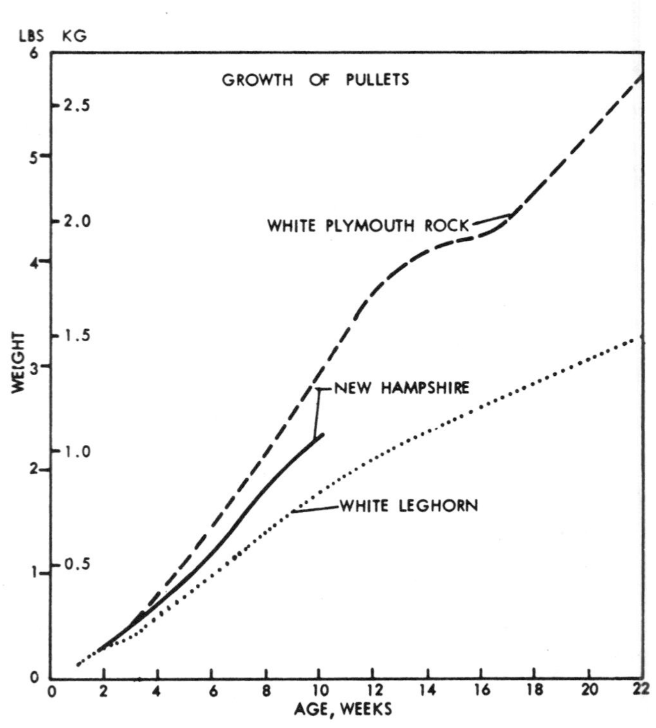

FIG. 15—WEIGHT VERSUS AGE OF PULLETS

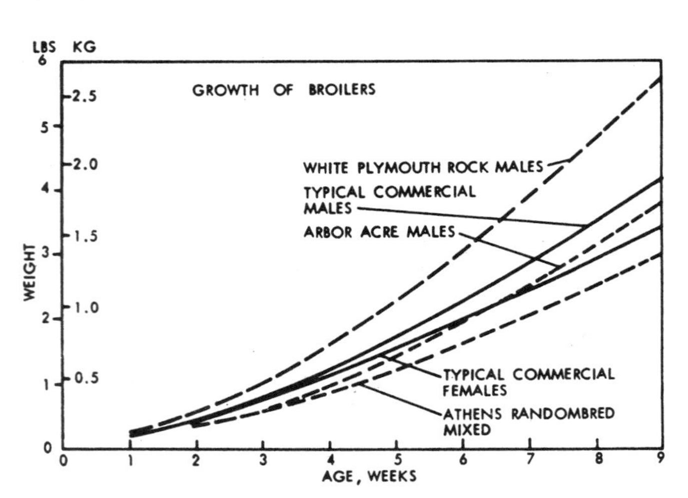

FIG. 16—WEIGHT VERSUS AGE OF BROILERS

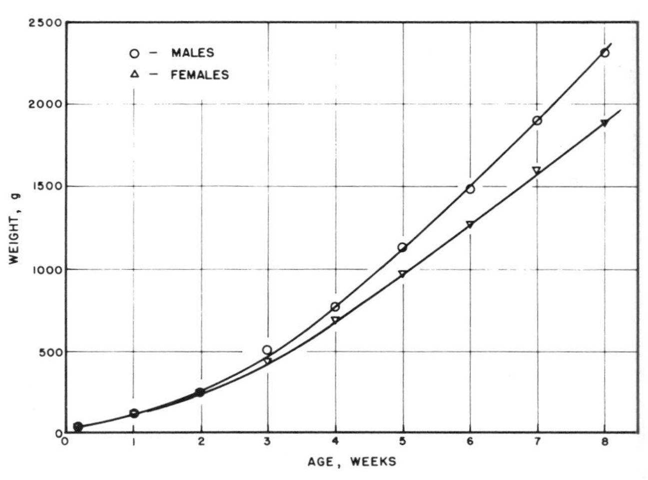

FIG. 17—WEIGHT VERSUS AGE OF BROILERS

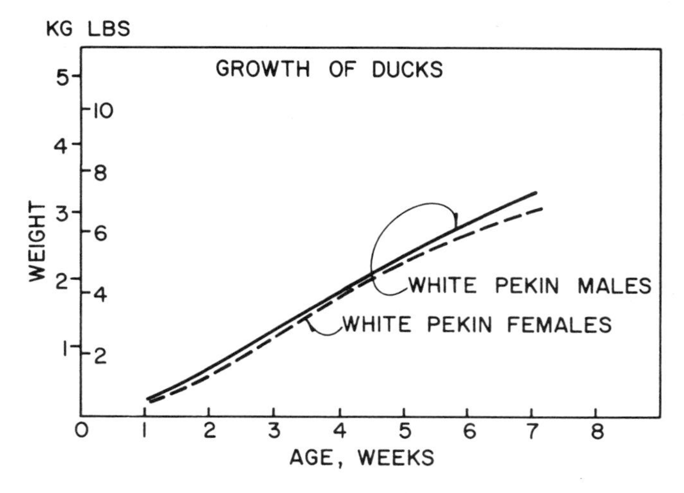

FIG. 18—WEIGHT VERSUS AGE OF DUCKLINGS

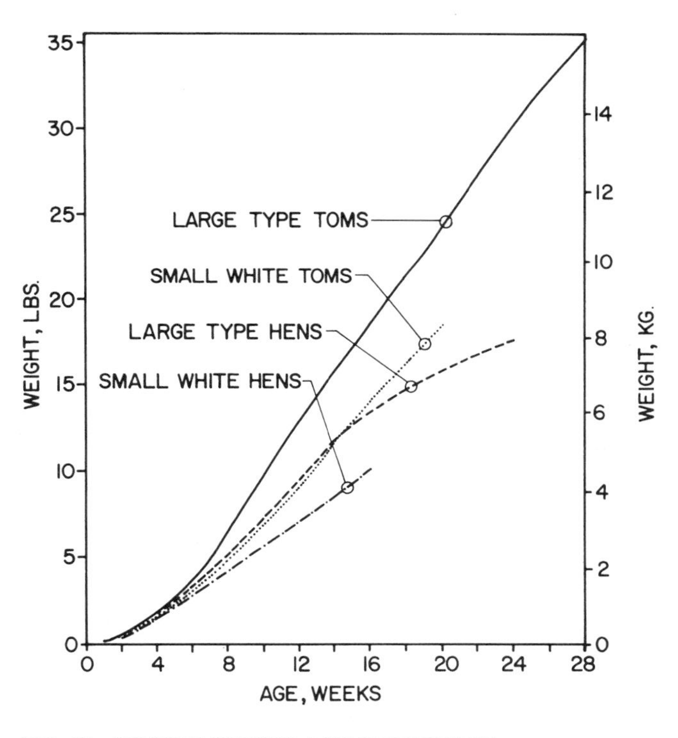

FIG. 19—WEIGHT VERSUS AGE OF TURKEYS

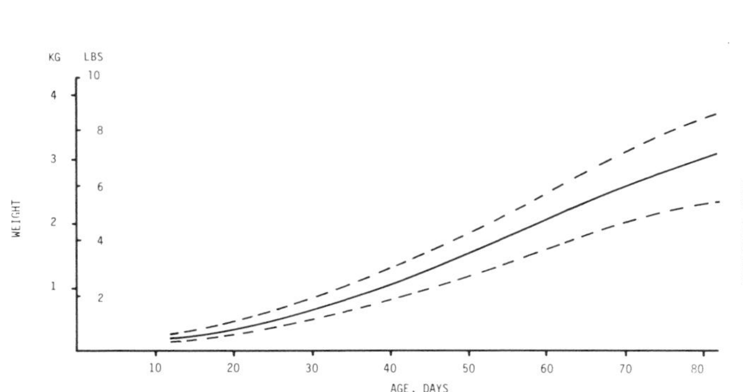

FIG. 21—MEAN WEIGHTS AND 95 PERCENT ENVELOPING WEIGHT LIMITS OF FEMALE WROLSTAD WHITE TURKEYS

7.2 Dimensions. The standing height and the sitting height versus weight of commercial broilers are shown in Fig. 22. The heights were measured vertically from the floor to the highest point along the pro-dorsal region, with the chickens standing and sitting with their heels resting on the floor. For standing chickens,

$$H_1 = 22.58 \; WT^{0.3236}$$

and for sitting chickens,

$$H_2 = 10.26 \; WT^{0.3152}$$

where

WT = weight in g
H_1 = standing height in mm
H_2 = sitting height in mm

References: Last printed in 1981 AGRICULTURAL ENGINEERS YEARBOOK; list available from ASAE Headquarters.

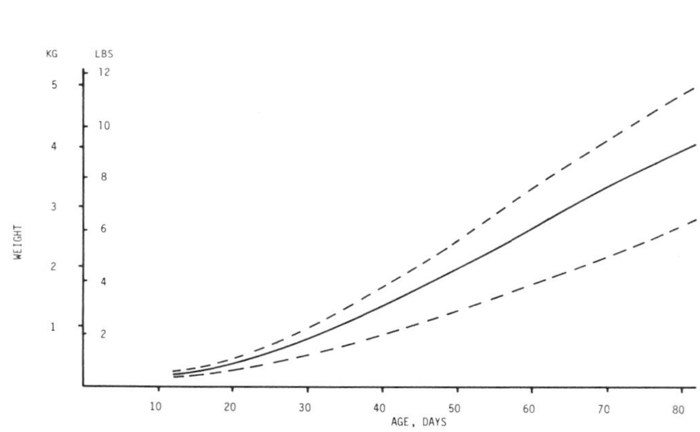

FIG. 20—MEAN WEIGHTS AND 95 PERCENT ENVELOPING WEIGHT LIMITS OF MALE WROLSTAD WHITE TURKEYS

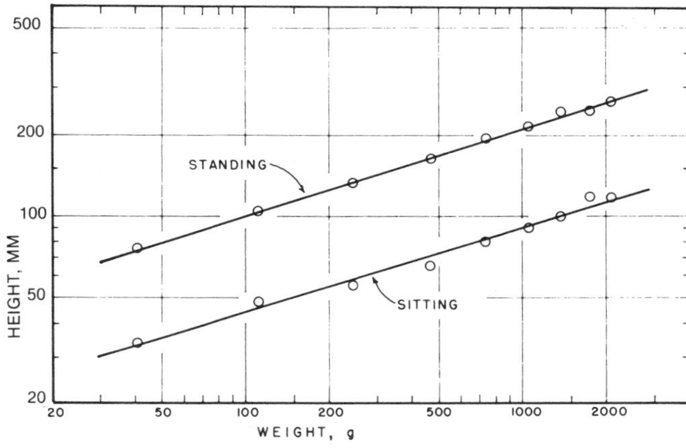

FIG. 22—STANDING HEIGHT AND SITTING HEIGHT VERSUS WEIGHT OF BROILERS

ASAE Engineering Practice: ASAE EP344.1

LIGHTING FOR DAIRY FARMS AND THE POULTRY INDUSTRY

This Engineering Practice combines and therefore supersedes ASAE R286, Lighting for Dairy Farms, and ASAE R332, Poultry Industry Lighting, developed by the joint Illuminating Engineering Society—ASAE Farm Lighting Committee, EPP-46. R286 was adopted by ASAE June 1965; R332 was adopted by ASAE December 1969. This document was approved by the ASAE Electric Power and Processing Division Standards Committee and adopted by ASAE February 1971 as ASAE Recommendation R344; revised editorially and reclassified as an Engineering Practice December 1975; reconfirmed December 1980; revised March 1982; reconfirmed July 1986, December 1987.

SECTION 1—PURPOSE AND SCOPE

1.1 This Engineering Practice is intended to guide those responsible for or concerned with, the design of lighting installations on dairy farms, on poultry farms, and within the poultry industry.

1.2 This Engineering practice applies only to the safety and effective performance of workers as they accomplish specific tasks requiring various levels of illumination as defined in paragraph 3.1. It does not apply to lighting installations used for changing the physiological or biological properties of poultry or livestock to alter their production capabilities.

1.3 The lighting recommendations are based on information obtained from search of current literature, from people and organizations active in this field, and from field measurements of lighting requirements for difficult seeing tasks. The portable Visual Task Evaluator from the Illuminating Engineering Society (IES) headquarters was used to establish the lighting requirements for many of the visual tasks. This document is in accordance with the latest knowledge and practice of the lighting field, and conforms to all official IES reports. However, future progress in agriculture and lighting will undoubtedly make revisions desirable.

SECTION 2—GENERAL RECOMMENDATIONS

2.1 Quantity of illumination

2.1.1 Recommended illumination levels

2.1.1.1 Recommended illumination levels for the poultry industry are in Table 1. Recommended illumination levels for dairy farms are in Table 2. Recommended illumination levels for general areas associated with these facilities are in Table 3. These recommended levels are commensurate with the difficulty of the seeing tasks of each area and with the general cost of lighting.

2.1.1.2 The illumination values shown in Tables 1, 2, and 3 are intended to be minimum on the task regardless of the plane in which they are located. Supplemental luminaires may be used in combination with general lighting to achieve those illumination levels on the work plane.

2.1.1.3 Recommended illumination levels should be maintained at all times when luminaires are in use. Initial values must be greater by a percentage sufficient to compensate for the normal depreciation of illumination expected in luminaire service.

2.1.2 Daylighting

2.1.2.1 The source of illumination is not limited to electric lighting. Illumination from daylight should also be considered. Many farm buildings where seeing tasks are not difficult can be adequately illuminated during daytime by the judicious placement of windows. In areas where there are difficult seeing tasks, windows should be located so glare from the direct sunlight will not likely cause a problem. Many poultry buildings must eliminate all daylighting where photo period is controlled for stimulation of poultry.

2.1.2.2 Shielding should be provided on windows directly facing the sun and should be located in areas of difficult seeing tasks. Electric illumination is often preferred over daylight illumination in these areas because of difficulty in controlling daylight illumination and the variability of daylighting.

TABLE 1—RECOMMENDED ILLUMINATION FOR POULTRY FARM INDUSTRY TASKS

Areas and visual tasks	Minimum light on task at any time lx	fc	Explanation
Brooding, production, and laying houses			
Feeding, inspection, and cleaning	200	20	Provided by a lighting circuit separate from the circuit used to stimulate production and growth.
Charts and records	300	30	Localized lighting is needed where charts and records are kept.
Thermometers, thermostats, and time clocks	500	50	Localized lighting is needed to accurately determine readings or setting.
Hatcheries			
General area and loading platform	200	20	Needed for operators to move about readily and safely. Needed for cleanliness of the general area.
Inside incubators	300	30	Portable or localized lighting is needed for inspection and cleaning inside incubators.
Dubbing station	1500	150	Needed to prevent excessive cuts and injury. Supplemental light in addition to general lighting should be used.
Sexing	10 000	1000	Needed for sex sorting of baby chicks. Supplemental light should be used in a closed area to prevent excessive brightness ratio between the task area and the immediate surrounding areas.
Egg handling, packing, and shipping			
General cleanliness	500	50	General illumination is needed to keep area clean and to detect any unsanitary conditions.
Egg quality inspection	500	50	Needed to examine and grade eggs. Candling and other special grading equipment are used as separate devices for examining and grading eggs.
Loading platform, egg storage area, etc.	200	20	Needed for operator to move about readily and safely, and for safe operation and mechanical and loading equipment.
Egg processing			
General lighting	700	70	Must meet the requirements of cleanliness for food preparation area. Includes liquid processing, pasteurizing, and freezing of raw eggs.

TABLE 1—RECOMMENDED ILLUMINATION FOR POULTRY FARM INDUSTRY TASKS

Areas and visual tasks	Minimum light on task at any time lx	fc	Explanation
Fowl processing plant			
General (excluding killing and unloading area)	700	70	General lighting for cleanliness, inspection, and sanitation. Must meet requirements of food preparation areas.
Government inspection station and grading stations	1000	100	Needed to detect diseases and blemishes. Vertical illumination is needed if birds are hanging.
Unloading and killing area	200	20	Needed to move about readily and safely.
Feed storage			
Grain, feed rations	100	10	Needed to read labels, scales and detect impurities and spoilage in feed.
Processing	100	10	Needed for operator to move about readily and safely, read labels, scales, and equipment dials. Supplemental light would be needed if machine repairs are necessary.
Charts and records	300	30	If detailed records or charts are kept in the feed room, localized lighting in this area would be needed.

TABLE 2—RECOMMENDED ILLUMINATION FOR DAIRY FARMS

Areas and visual tasks	Minimum light any time lx	fc	Explanation
Milking operation area (milking parlor & stall barn)			
General	200	20	Required to determine cleanliness of cow, detect undesirable milk, handle milking equipment readily, and to detect dirt and foreign objects on the floor. Should be available at cow-edge of gutter, on floor.
Cow's udder	500	50	Supplemental, to determine cleanliness of udder, to clean udder, to examine udder.
Milk handling equipment and storage area (milk house or milk room)			
General	200	20	Required for operator to move about readily and safely, and to determine floor cleanliness.
Washing area	1000	100	Necessary to detect dirt and other impurities on the milk handling equipment. Supplementary, portable, ultraviolet fixture should be available in this area to aid in detecting milkstone on the equipment.
Bulk tank interior	1000	100	Necessary to adequately inspect tank for cleanliness. Additional spots may be required to illuminate dip-stick or scale.
Loading platform	200	20	Required for operator to move about readily and safely.
Feeding area (stall barn feed alley, pens, and loose housing feed area)	200	20	Required for detecting foreign objects in grain, hay, or silage.
Feed storage area, forage			
Haymow	30	3	Required for safety of the operator in moving about.
Hay inspection area	200	20	Required for detecting foreign objects in grain, hay, or silage.
Ladders and stairs	200	20	
Silo	30	3	Luminaires should be mounted at the top of the silo, near the ladder chute, for ease in cleaning and lamp replacement.
Silo room	200	20	Required for detecting foreign objects in grain, hay, or silage.
Feed storage area, grain and concentrate			
Grain bin	30	3	Required to inspect amount and condition of grain. When grain is suspected of being moldy, containing foreign objects or otherwise contaminated, samples should be inspected under higher illumination levels.
Concentrate storage area	100	10	Required to read labels. Higher illumination levels are required for critical inspection for impurities and spoilage.
Feed processing area	100	10	Required for operator to move about readily and read labels, scales, and equipment dials. Additional light must be supplied by portable luminaires or daylighting if machine repairs are necessary.
Livestock housing area (community, maternity, individual calf pens, and loose-housing holding and resting areas)	70	7	Required to observe the condition of the animals and to detect hazards to the livestock and operator. Portable, supplementary lighting units can be used to examine or treat individual animals when required.

2.1.2.3 Efficient operations are often performed when only electric illumination is available. Therefore, these recommendations provide desirable amounts of electric illumination without dependence on daylight. For information on techniques for utilizing and controlling daylight, refer to IES Recommended Practice RP-5, Recommended Practice of Daylighting.

2.2 Quality of illumination. Specifying the quantity of light in no way establishes quality of a lighting installation. Important factors affecting lighting installation are uniformity, glare, color, and environment.

2.2.1 Uniformity

2.2.1.1 Uniformity of illumination is expressed as a ratio of a maximum lux (footcandle) value to the minimum lux (footcandle) value over an area. Satisfactory uniformity ratios vary from 1.5:1 for difficult seeing tasks to 5:1 for less difficult tasks. In general, better uniformity of illumination is achieved with greater mounting heights and closer spacing of luminaires.

TABLE 3—RECOMMENDED ILLUMINATION FOR GENERAL AREAS ASSOCIATED WITH DAIRY AND POULTRY FACILITIES

Areas and visual tasks	Minimum light on task at any time lx	fc	Explanation
Machine storage			
Garage and machine shed	50	5	Needed to move machinery safely. Supplemental lighting is needed for minor equipment repair.
Farm shop			
Active storage area	100	10	Needed for operator to move about readily and safely.
General shop	300	30	Machinery repair, rough sawing.
Rough bench machine work	500	50	Painting, small parts, storage, ordinary sheet metal work, welding, medium bench work. May use localized lighting.
Miscellaneous			
Farm office	700	70	
Restrooms	300	30	
Pumphouse	200	20	
Exterior			
General inactive areas	2	0.2	Recommended to discourage prowlers and predatory animals.
General active areas (paths, rough storage, barn lots)	10	1	Needed for operator to move about safely.
Service areas (fuel storage, shop, feed lots, building entrances)	30	3	Needed for servicing machinery.

2.2.1.2 Light sources should be located to minimize shadows cast on the work by workers and obstructions. Objects should receive illumination from more than one direction to minimize the density of shadows and to provide uniform illumination.

2.2.2 Glare. Glare is any brightness within the field of vision that causes discomfort, annoyance, interference with vision, or eye fatigue. It is usually caused by uncontrolled light which strikes the eyes directly from a luminaire, but it can also be caused by the reflection of bright light by glossy or mirror-like surfaces. Luminaires should be selected with sufficient shielding, as shown in Fig. 1, and placed so glare can be held below an objectionable level. Luminaires should be mounted as far above the normal line of sight as possible and should be designed to limit both the brightness and the quantity of light emitted in the 45 to 85 deg zone since such light is likely to be well within the field of view and may interfere with vision. This precaution extends to the use of localized lighting equipment.

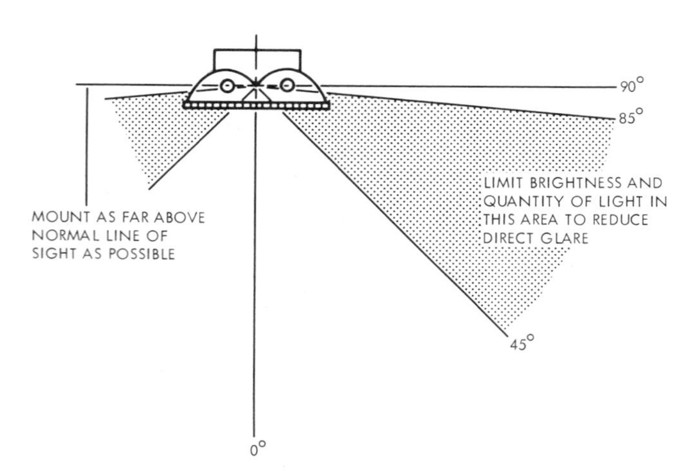

FIG. 1—DIRECT GLARE CONTROL FOR LUMINAIRES

2.2.3 Color. For performing some specialized seeing tasks, notably color discrimination processes and certain inspection work, light-source color may be an important factor of illumination quality. Color is also known to have certain psychological effects upon people and their emotions. These factors should be considered when selecting light sources to obtain quality lighting. For example, when grading or sorting eggs, fluorescent lamps are more effective than incandescent.

2.2.4 Environment. Certain environmental factors can greatly influence the quality of a lighting installation. Room surfaces should have high reflectance, matte finishes to help prevent excessive brightness ratios. The ceilings, walls, and floors can increase the utilization of light within a room by acting as a secondary large-area light source. Luminaires which direct some light upward toward a ceiling having a relatively high reflectance will help create a comfortable visual environment. Recommended reflectance values are presented in Table 4.

TABLE 4—RECOMMENDED MATTE REFLECTANCE VALUES FOR LIGHTING FARM BUILDINGS

Surface	Reflectance, percent
Ceilings	80 to 90
Walls	40 to 60
Desk and bench tops, machines & equipment	25 to 45
Floors	20 minimum

2.3 Lighting equipment

2.3.1 Lamps

2.3.1.1 Light sources available for farm lighting applications are incandescent, fluorescent, and high-intensity discharge (HID) (mercury, metal halide, high-pressure sodium, low-pressure sodium). The latter four sources are more efficient and generate 2 to 5 times as much light as the incandescent filament for the same amount of electrical energy. They also have a longer life. However, they require auxiliary electrical equipment, such as ballasts. Refer to Section 9—Industrial Lighting, in the 1981 Applications Volume of the IES Handbook.

2.3.1.2 HID lamps will normally be used outdoors, but are also suitable indoors where 4 m (13 ft) or more mounting height is available. HID sources require 3 to 7 min warm-up time and may not be suited to applications where they may be turned off and on in less than an hour or two. HID lamps differ in color rendering ability, in general being lower than fluorescent or incandescent.

2.3.1.3 Incandescent-filament lamps require no auxiliary electrical control equipment, and are available in many different types and sizes. They are a high brightness source and should be used in appropriate luminaires to minimize glare. Various types of reflectorized incandescent-filaments may be used to spotlight or floodlight specific areas.

2.3.1.4 Fluorescent lamps provide a large light source. Even though surface brightness of fluorescent lamps is relatively low, fluorescent lamps should always be used in a suitable luminaire that will minimize glare. Fluorescent lamps are available in different sizes, types, and colors, which can provide color rendering, cool or warm colors. But they are temperature sensitive, and light output drops when used in either very high or extremely low ambient temperatures. Fluorescent lamps used in cold areas require a special ballast designed for low temperature starting, and an enclosure to maintain lamp temperature. High humidity causes moisture condensation on lamps in open luminaires and may prohibit fluorescent lamps from starting. Refer to Section 8—Light Sources, in the 1981 Reference Volume of the IES Handbook.

2.3.2 Luminaires

2.3.2.1 The light from most lamps is emitted in many directions. The purpose of a luminaire is to control the direction at which light is emitted so glare will be reduced and light used more effectively on the objects to be seen. In outdoor applications, luminaires should direct light downward to minimize losses, but in indoor applications, it is desirable to have a portion of the light directed toward the ceiling. Some of the light striking a light-colored ceiling is reflected back to the seeing tasks and provides a well-balanced visual environment.

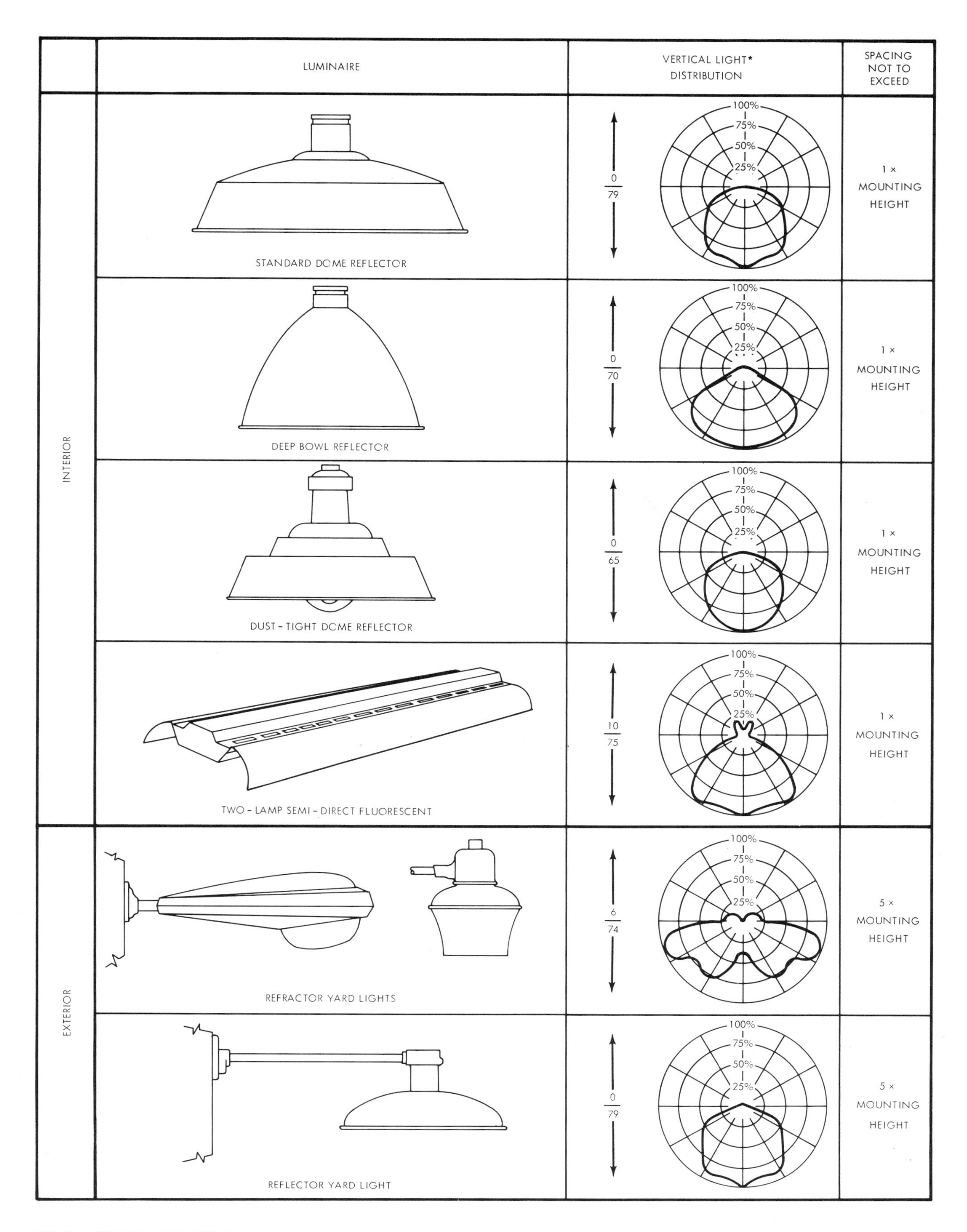

FIG. 2—TYPICAL LUMINAIRES

*The vertical light distribution curve for each luminaire is plotted on the basis of maximum candlepower as 100%. The figures to the left of the curves represent the percentage of the total lamp lumens directed above and below the luminaire. The sum of these two percentages is the efficiency of the luminaire.

2.3.2.2 Luminaires control light by reflectors (porcelain, enamel, aluminum, glass, or plastic), refractors (glass or plastic prismatic lenses), or by diffusers (glass or plastic diffusing shields). Often one or more of these principles are used in the same luminaire. Reflectors and refractors are used to direct the light for more efficient utilization. Refectors, refractors, and diffusing shields are used to prevent glare.

2.3.2.3 A luminaire prevents water from striking the hot lamps when they are used in damp or outdoor locations. There are some specifically constructed, reflectorized, incandescent filament lamps which can be exposed to the weather or water spray. See Fig. 2 for illustrations of typical luminaires which may be used.

2.3.3 Codes

2.3.3.1 The use of lighting equipment is governed by many federal, state, and local codes. Some public health codes specify minimum illumination levels required in processing plants, egg handling areas, milking and milk handling areas to maintain health standards. Public health lighting requirements are normally below those shown in Tables 1 and 2 since they are concerned only with sanitation. The levels recommended were selected for efficient performance of visual tasks but with concern for proper sanitation.

2.3.3.2 The modern farm includes many types of occupancies which may be wet, damp, corrosive, dirty, surrounded by combustible materials, or saturated with gasoline fumes. Therefore, it is important to follow the National Fire Protection Association Standard No. 70, National Electrical Code, and any local regulation which may be in effect when installing the lighting equipment. Assistance may be obtained by contacting a local electric power supplier or by referring to the Agricultural Wiring Handbook published by the National Food and Energy Council.

2.3.4 Maintenance

2.3.4.1 Lighting systems should be inspected and serviced regularly to maintain illumination levels for which they are designed. The elements of maintenance are light sources, luminaires, and room surfaces.

2.3.4.2 The lighting output of most lamps diminishes with use. In many cases, it is economical to replace lamps before they fail. Saving can sometimes be realized by replacing groups of lamps near the end of their useful life, rather than replacing individual lamps as they burn out.

2.3.4.3 The rate of depreciation of lamps and lighting equipment caused by dirt accumulation is dependent upon the atmospheric conditions and the type of lighting system employed. Losses up to 50 percent have resulted from poor maintenance.

2.3.4.4 Wall surfaces should be cleaned and painted regularly to improve reflective characteristics of the surfaces and to provide the proper visual environment. It is also important to keep windows clean in areas where they contribute to daytime illumination.

SECTION 3—GLOSSARY OF TERMS

3.1 **Illumination:** The quantity of light per unit volume on a surface when the time rate of flow of light is uniformly distributed.

3.2 **Visual task:** The seeing of all things that have to be seen at the given moment constitutes the visual part of any task, and the term visual task conventionally designates the sum total of all the things that have to be seen at the given moment.

3.3 **Work plane:** The plane at which work is done, and at which, illumination is specified and measured. Unless otherwise indicated, this is assumed to be a horizontal plane 0.75 m (2.5 ft) above the floor.

3.4 **Lumen:** The unit of the time rate of flow of light (lumenous energy) equal to the energy emitted through a unit solid angle (one steradian) from a uniform point source of one candela.

3.5 **Lux:** The unit of illumination when the meter is the unit of length. It is the illumination on a surface one square meter in area on which is uniformly distributed a flux of one lumen. It equals one lumen per square meter.

3.6 **Footcandle:** The unit of illumination when the foot is the unit of length. It is the illumination on a surface one square foot in area on which is uniformly distributed a flux of one lumem. It equals one lumen per square foot.

3.7 **Luminaire:** A complete lighting unit consisting of a lamp or lamps together with the parts designed to distribute the light, to position and protect the lamps, and to connect the lamps to the power supply.

3.8 **Uniformity ratio:** The ratio of maximum lux (footcandle) value to the minimum lux (footcandle) value over an area.

3.9 **Brightness ratio:** The ratio of the level of illumination between any relatively large areas within the visual field.

3.10 **Shielding angle:** The angle between a horizontal line through the light center and the line of sight at which the base source first becomes visible.

3.11 **Glare:** The effect of brightness or brightness differences within the visual field sufficiently high to cause annoyance, discomfort, or loss in visual performance.

3.12 **General lighting:** Lighting designed to provide a uniform level of illumination throughout the area involved.

3.13 **Supplemental lighting:** Lighting used to provide a specific amount or quality of illumination which cannot be readily obtained by the general lighting system, and which supplements the general lighting system.

3.14 **Local lighting:** Illumination provided over a relatively small area or confined space without any surrounding general lighting.

3.15 **Refractor:** A device used to redirect the illumination primarily by the process of refraction. (The bending of a ray of light as it passes obliquely from one medium to another in which its velocity is different.)

3.16 **Reflector:** A device used to redirect the light from a source primarily by the process of reflection. (A general term for the process by which part of the light leaves a surface or medium from the incident side.)

3.17 **Diffuser:** A device used to redirect the illumination by the process of diffuse transmission. (The process by which a portion of the illumination is re-emitted from a surface or medium on the non-incident side is a non-image forming [diffused] state).

3.18 **Louver:** An opaque or translucent member used to shield a source from direct view at certain angles or to absorb unwanted light. May also be a series of baffles.

References: Last printed in 1983 AGRICULTURAL ENGINEERS YEARBOOK; list available from ASAE Headquarters.

Cited Standards:

IES RP-5, Recommended Practice of Daylighting
NFPA No. 70, National Electrical Code

FLOOR AND SUSPENDED LOADS ON AGRICULTURAL STRUCTURES DUE TO USE

Developed by the ASAE Agricultural Structures Code and Loads Imposed on Structures Committees; approved by the ASAE Structures and Environment Division Standards Committee; adopted by ASAE December 1975; revised March 1977, December 1981, February 1987; revised editorially December 1987.

SECTION 1—PURPOSE AND SCOPE

1.1 This Engineering Practice presents probable floor and suspended loads due to building use and methods of applying the loads in building design.

E **1.2** This Engineering Practice includes recommended design loads resulting from livestock, suspended caged poultry, vehicles, and manure stored on a floor. It does not include loads on manure storages or wind and snow loads. Specifications for these applications are included in ASAE Engineering Practice EP288, Agricultural Building Snow and Wind Loads, and ASAE Engineering Practice EP393, Manure Storages, respectively.

SECTION 2—TERMINOLOGY

2.1 Terms used in this Engineering Practice are defined as follows:

2.1.1 Dead load: Gravitional force due to mass of all material used in the building construction.

2.1.2 Live load: Gravitional force due to mass of equipment, livestock, products, and other loads resulting from the use of the structure.

SECTION 3—FLOOR AND SUSPENDED LOADS

3.1 The recommended design live load due to the use of the area is listed in Table 1.

3.1.1 Bulk milk tanks. Design floors under bulk tanks in milk-rooms according to the mass of the tank plus contents and the support system of the tank.

3.1.2 Manure. Where manure can accumulate, increase the design load by 1040 kg/m² (65 lb/ft²) per meter (foot) of manure depth.

3.1.3 Cages. Base design loads for floor-supported cages on mass and support intervals.

3.1.4 Suspended poultry cages. Loads for suspended poultry cages are based on 4-row (double deck) or 6-row (triple deck) cages with two birds per 200 mm (8 in.) cage, or three birds per 300 mm (12 in.) cage, and 50 mm (2 in.) of manure accumulated on dropping boards under upper cages. For other configurations, consult the manufacturer.

3.1.5 Storage. Calculate design load for product storage on the basis of individual mass but no less than 490 kg/m² (100 lb/ft²).

3.1.6 Greenhouses. Bottom chord panel points of roof trusses shall be capable of safely supporting a minimum concentrated live load of 45 kg (100 lb) applied at any panel point.

3.1.7 Combined. Add equipment and vehicle loads to the animal loads in Table 1 where appropriate.

3.1.8 Heating, ventilating, and air conditioning (HVAC) and other equipment. Base design on equipment mass and support system specified.

3.1.9 Cranes and hoists. Procedures and data for structural loads from cranes and hoists are included in model building codes and Metal Building Systems Manual.

3.2 The recommended design loads on slotted floors in livestock pens are listed in Table 1.

3.2.1 An alternate requirement for slats, which may dictate design, is a single concentrated load of 115 kg (250 lb) located first for maximum moment, then for maximum shear, in slats and slat supports.

3.2.2 Design slotted floors in farrowing pens for a single concentrated load of 115 kg (250 lb) located first for maximum moment, then for maximum shear, in slats and slat supports.

TABLE 1—RECOMMENDED DESIGN FLOOR LIVE LOADS DUE TO USE

	Solid floors* and floor support		Slat per† unit length	
	kg/m²	lb/ft²	kg/m	lb/ft
Beef cattle				
Calves to 135 kg (300 lb)	245	50	225	150
Feeders, breeders	490	100	370	250
Dairy cattle (see 3.1.1)				
Calves to 135 kg (300 lb)	245	50	225	150
Mature	490	100	370	250
stall area	295	60	370	250
maternity or hospital pen	245	50	370	250
Swine (see 3.2.2)				
to 25 kg (50 lb)	170	35	75	50
90 kg (200 lb)	245	50	150	100
180 kg (400 lb)	315	65	225	150
225 kg (500 lb)	340	70	255	170
Sheep				
Feeders	195	40	150	100
Ewes, rams	245	50	180	120
Horses	490	100	370	250
Turkeys	145	30	37	25
Chickens (see 3.1.3, 3.1.4)				
Floor houses	100	20	22	15
Greenhouses	245	50	—	—
Manure (see 3.1.2)	1040‡	65§	—	—
Shops, storage, vehicles (see 3.3)				
	Suspended load			
	kg/m	lb/ft		
Chickens (see 3.1.4), suspended cages per length of cage row				
Full stair step (double deck, no dropping boards)	110	75		
Modified stair step (double deck, with dropping boards)	165	110		
Modified stair step (triple deck, with dropping boards)	225	150		
	kg/m²	lb/ft²		
Greenhouses (see 3.1.6)				
Crop	20	4		

*For floors that are outdoors, add suitable design snow loads on the ground. Increase solid floor live load 25% for floors supporting crowded animals (e.g., crowding pen, dairy holding pen with automatic gate, handling alleys near loading chute).

†Check slats for 115 kg (250 lb) concentrated load.

‡Per meter of depth.

§Per foot of depth.

3.2.3 Where slats are interconnected between supports so that three or more slats must deflect together, design each span of slat between support and interconnection, and between two interconnections, for the recommended load per unit length of slat. Design the full span of each slat for one-half the recommended load per unit length.

3.3 Vehicle loads. The following are for vehicles in storage or slow-moving vehicles, as in buildings, alleys, emptying manure storages, etc.

3.3.1 Uniformly distributed. The minimum design load on a floor area used for farm machinery with traffic limited to access and egress should be 730 kg/m² (150 lb/ft²). If the area will be occupied by either loaded farm trucks or large farm tractors [those with a mass exceeding 5900 kg (13,000 lb) including mounted equipment], the design load should be 975 kg/m² (200 lb/ft²).

3.3.2 Concentrated. In the absence of specific information, the minimum design loads due to probable concentrations of loads resulting from use of a floor area are:

For tractors and implements: 2270 kg (5000 lb) per wheel; tread print = 200 mm × 300 mm (8 in. × 12 in.).

For loaded trucks not exceeding 9075 kg (20,000 lb) gross vehicle weight (GVW): 3630 kg (8000 lb) per wheel; tread print = 100 mm × 510 mm (4 in. × 20 in.)

For loaded trucks exceeding 9075 kg (20,000 lb) GVW: 5440 kg (12,000 lb) per wheel; tread print = 150 mm × 560 mm (6 in. × 22 in.).

3.3.3 Loading and processing. In cases where the area (minimum traffic or driveway) is to serve as a place for loading, unloading, or processing, increase minimum design loads by 50% and allowable material stresses as permitted for impact, to allow for impact or vibrations of machinery or equipment.

3.3.4 On manure tank lids. For vehicle traffic on a manure tank lid, use 4540 kg (10,000 lb) axle load, or a loading equivalent of two 2270 kg (5000 lb) concentrated loads 1.2 m (4 ft) apart and oriented in any critical direction on the tank cover.

References:

1 MBMA. Metal building systems manual. Metal Building Manufacturers Assn., Cleveland, OH.

2 NGMA. Standard for design loads in greenhouse structures. National Greenhouse Manufacturers Assn., Pana, IL.

Cited Standards:

ASAE EP288, Agricultural Building Snow and Wind Loads
ASAE EP393, Manure Storages

E

ASAE Engineering Practice: ASAE EP379.1

CONTROL OF MANURE ODORS

Developed by the ASAE Agricultural Sanitation and Waste Management Committee; approved by the ASAE Structures and Environment Division Standards Committee; adopted by ASAE December 1975; reconfirmed December 1980, December 1985; revised December 1986.

SECTION 1—PURPOSE AND SCOPE

1.1 This Engineering Practice was prepared to assist engineers, pollution control officials, land use planners, and livestock and poultry producers in the location, planning, construction, and operation of animal holding and production enterprises so that potential societal conflicts caused by odors might be minimized.

1.2 Principles of odor generation, release, transport, and detection are the subject of continuing investigation. Many aspects of the relevant processes are only partially understood at this time. Decisions relating to odor control therefore represent judgements based upon limited knowledge.

1.3 Odor sensation is a personal response. Not all observers are equally sensitive, nor do they always agree as to the severity of an odor once it is detected. Though various instruments can be used to determine the identity and concentration of specific odorants, or to measure the intensity of odorant mixtures as perceived by humans, no instrument can yet reflect the qualitative appraisal of odorant mixtures necessary to establish its degree of unpleasantness. Odor intensity fluctuates with climatic conditions such as wind direction and atmospheric stability. Thus, precise documentation of the strength and nature of odor is generally unavailable.

SECTION 2—SOURCE AND IDENTIFICATION OF ODOROUS GASES AND VAPORS

2.1 Odors associated with animal enterprises are most frequently attributable to type of manure management. Odor that is emitted from the anatomy of livestock and poultry may also be a major odorous component of building ventilation air. Other potential odor sources are related to feed storage, processing and distribution.

2.2 Objectionable odors exist whenever odorous compounds are emitted and transported to an area in which their presence is undesirable to people. Emission of odorous vapors and gases is related to the volatility of the compound in question, the chemical composition of the media in which it is produced, temperature, and air movement.

2.3 Compounds released from decomposing manure which have been identified in the nearby atmosphere are listed in Table 1.

TABLE 1—COMPOUNDS RESULTING FROM THE ANAEROBIC DECOMPOSITION OF LIVESTOCK AND POULTRY MANURE

Alcohols	Amines
	Methylamine
Acids	Ethylamine
Butyric	Trimethylamine
Acetic	Diethylamine
Propionic	
Isobutyric	Esters
Isovaleric	
	Fixed gases
Carbonyls	Carbon dioxide (odorless)
	Methane (odorless)
Sulfur compounds	Ammonia
Hydrogen sulfide	
Dimethyl sulfide	Nitrogen heterocycles
Diethyl sulfide	Indole
Methylmercaptan	Skatole
Disulfides	

SECTION 3—TECHNIQUES TO MINIMIZE THE FORMATION OF ODOROUS GASES

3.1 Odorous gases emitted during collection, transport, storage, treatment, and land application of animal manures or manure slurries are principally products of anaerobic microbial metabolism. Treatment processes which inhibit this biological action will be effective in decreasing the production of some of these gases. When decomposition is completely stopped, odor release will cease after the volatile matter already decomposed have been released and dissipated.

3.1.1 Dehydration is one common technique for inhibiting anaerobic decomposition. When the moisture content of manure is lowered to 50 percent or less, the manure is sufficiently porous to permit air diffusion and to preclude anaerobic decomposition.

3.1.2 Disinfection can be used to inhibit anaerobic decomposition. Chlorine, lime, formaldehyde and other chemical disinfectants have been used, but costs may be unreasonable in the quantities required.

3.1.3 Aeration to maintain dissolved oxygen can be used to replace anaerobic decomposition with the nearly odorless aerobic process. Oxygenation capacity sufficient to satisfy the five-day biochemical oxygen demand, or the majority of the chemical oxygen demand, is generally required.

3.2 A variety of odor control chemicals, designed to aid in odor reduction have been marketed. The effectiveness of these products is difficult to measure, and an individual trial is suggested prior to purchase. (See paragraph 4.1.8).

3.2.1 Masking agents are mixtures of volatile oils which have a stronger odor than the manure, and are designed to cover-up the objectionable odor with a more acceptable odor.

3.2.2 Counteractants are based upon the characteristics of two properly selected odors which cancel one another so that the total intensity detected is less than that of the individual constitutents.

3.2.3 Deodorants are based upon the principle of eliminating or transforming the odorous constitutent so that it is not emitted. Deodorants may inhibit the biological activity or alter the digestive process by changing the enzyme balance.

3.3 Although chemical control and treatment process control may have useful applications, effective long-range solutions to odor problems are achieved by selecting manure collection, transport, storage, treatment and application procedures which are technically feasible and compatible with the total production system. Manure odor nuisance often results from poor housekeeping techniques as well as from problems inherent to manure management.

SECTION 4—MANAGING ANIMAL ENTERPRISES FOR ODOR CONTROL

4.1 Although neither a complete understanding of odor production nor fully adequate techniques for odor control are currently available, the following managerial procedures have proven helpful.

4.1.1 Locate a livestock operation at a reasonable distance from residential areas, places of employment, institutions, and other areas frequented by persons other than the operators of the animal enterprise. Although distances have not been established beyond which complaints are invalid, it is desirable to locate the livestock or poultry feeding facility 1600 m (1 mile) from housing developments and 400-800 m (1/4 to 1/2 mile) from neighboring residences. Wind direction and velocity, humidity, topography, temperature, and unique meteorological conditions (such as inversions) affect odor transport and detection.

4.1.2 Wind direction probability data should be considered in site selection to minimize the frequency of odor transport to housing developments or neighboring residences.

4.1.3 Feeding areas should be kept dry. Keeping manure-covered surfaces dry minimizes the primary source of odor from a livestock operation, that of anaerobic manure decomposition.

4.1.4 Manure-management systems should be designed and operated in a manner that prevents dirty, manure-covered animals. The warm body of an animal, when covered with wet manure, promotes accelerated bacterial growth and odor production. Once produced, the odorous by-products of manure decomposition are quickly vaporized by animal heat and emitted into the atmosphere. Use systems that separate the animal from its manure quickly, such as slotted floors for swine.

4.1.5 Appropriate selection of manure storage and treatment procedures can be helpful. Aerobic systems will in general reduce odor production. Other measures which inhibit anaerobic decomposition and reduce odor emission include dry manure storage and chlorine or lime addition. Digestion under controlled anaerobic conditions, where methane (biogas) is produced, will produce a residue free of offensive odors.

4.1.6 An orderly system for runoff collection and manure handling not only minimizes water pollution, but also promotes better drainage and reduces areas of odor production. It is important that accumulations of solids and polluted water in runoff control systems be expediently removed and applied to land to limit eventual odor emission as well as to preserve the functions of the facility.

4.1.7 A clean, orderly appearance of the livestock production unit, including landscaping, is effective in suggesting a non-offensive situation.

4.1.8 Dead animal disposal requires a definite plan to avoid odors, flies, and severe health risks. Removal from the site within 24 h is required in most areas. Pick-up by rendering workers is preferred where the service is available; otherwise, burial or incineration can be used.

4.1.9 Odor control chemicals have achieved limited use in animal enterprises. Because of their expense and the general lack of performance evaluation, use of odor control chemicals has been limited generally to short-term applications in particularly offensive areas, such as manure-storage pits immediately before hauling.

4.1.10 Land application is the primary method of animal waste management and is an integral part of nearly every manure handling system. Odors can be reduced by using the following land application procedures for liquid or solid manure:

4.1.10.1 Spread or apply manure within 4 days of excretion if possible to reduce time in anaerobic storage.

4.1.10.2 Avoid spreading when the wind would blow odors toward populated areas or nearby residences or businesses.

4.1.10.3 Avoid spreading or applying manure immediately before weekends and holidays when people are likely to be engaged in nearby outdoor and recreational activities.

4.1.10.4 Avoid spreading near heavily traveled highways.

4.1.10.5 Spread or apply manure in morning when air is warming and rising rather than in the late afternoon.

4.1.10.6 Use available weather information to best advantage. Turbulent breezes will dissipate and dilute odors. Rain will remove the odors from the atmosphere.

4.1.10.7 If possible, incorporate manure into the soil during or immediately after application. This can be done by 1) soil injection or 2) plowing or disking the soil during or after application. These practices not only minimize the spreading of odor but also preserve nutrients and reduce water pollution potential.

4.1.10.8 Apply manure uniformly and in a layer thin enough to insure drying in less than 5 days or less and to prevent fly propagation in warm weather.

ASAE Engineering Practice: ASAE EP381

SPECIFICATIONS FOR LIGHTNING PROTECTION

Developed by a subcommittee of the ASAE Structures and Environment Division Standards Committee; approved by the Structures and Environment Division Standards Committee; adopted by ASAE March 1976; reconfirmed December 1981, December 1982, December 1987.

SECTION 1—PURPOSE AND SCOPE

1.1 This Engineering Practice is intended as a guide for specifying farm lightning protection systems, and to check existing, new or proposed lightning protection systems against accepted standards of design, materials, and installation.

1.2 This Engineering Practice is applicable to protection of farm homes, barns, sheds, silos, slatted floors, fences, trees, barn equipment and other sizable bodies of conductance or inductance. Included are ordinary structures up to 23 m (75 ft) high.

1.3 Common assemblies on typical buildings are detailed to show current practices under normal conditions. Examples cannot cover the wide range of structural variations existing on farms. It is recommended that when a condition is not covered herein, the owner or his agent obtain engineering consultation by an engineer experienced in lightning protection principles, theories, and installation requirements, or from an established and recommended manufacturer or distributor of lightning protection materials and equipment.

1.4 In areas not covered by this Engineering Practice, refer to the National Fire Protection Association Standard No. 78, Lightning Protection Code; Underwriters' Laboratories Standard, Requirements for Master Label Lightning Protection; or Lightning Protection Institute Standard LPI-175, Installation Code.

SECTION 2—MATERIALS

2.1 Materials for lightning protection shall be inherently resistant to corrosion or properly protected against corrosion. No materials shall be combined which form an electrolytic coupling that accelerates corrosion.

2.2 Copper commonly required for commercial electrical work, with 98 percent conductivity when annealed, is the preferred material for farm lightning protection. Other acceptable materials are listed in paragraphs 2.3, 2.4 and 2.5. Material such as galvanized steel is not acceptable except as specified elsewhere in this Engineering Practice.

2.3 Alloyed metals used shall be substantially as resistant to corrosion and have the same conductivity as copper under similar conditions.

2.4 Copper-clad steel shall have a copper covering permanently and effectively welded to the steel core, in such proportions that conductance is not less than 30 percent of the conductance of an equivalent cross section of solid copper.

2.5 Aluminum is acceptable as a substitute for copper in lightning protection, with the stipulations that (a) aluminum shall not be used underground, in contact with ground or where air may be laden with corrosive elements; (b) when an aluminum system is joined with copper or copper-clad grounds, the union shall be made with approved bimetal connectors; (c) precautions be taken at connections with dissimilar metals; and (d) cable conductors be of electrical conductor grade aluminum.

2.6 Materials shall be used in those forms and sizes specified in the following Sections for air terminals, conductors, grounds, and other parts.

SECTION 3—AIR TERMINALS

3.1 Air terminals are the topmost elements of the lightning protection system and are designed to intercept a direct stroke. They must be so placed and spaced that it is assured lightning will strike one or more of the terminals—not a vulnerable part of the roof or a projection.

3.2 Air terminals shall be 9 mm (3/8 in.) minimum diameter for solid copper or 13 mm (1/2 in.) minimum for solid aluminum. Tubular air terminals shall have a wall thickness of 0.8 mm (0.032 in.) for copper and 1.63 mm (0.064 in.) for aluminum, with a minimum 16 mm (5/8 in.) outside diameter.

3.3 Air terminals shall extend above the protected object at least 254 mm (10 in.) but no more than 914 mm (36 in.). If over 610 mm (24 in.) high, air terminals shall be suitably braced.

3.4 Air terminals up to 610 mm (24 in.) high shall be spaced at intervals of 6.1 m (20 ft) or less; those 610 mm (24 in.) or higher shall be spaced at 7.6 m (25 ft) intervals or less. Terminals shall be placed on the ridges of gable, gambrel, and hip roofs of ordinary or high slope, and at the perimeters of flat or low-slope roofs. A shed roof with a high or normal slope shall be considered as half of a gable roof. There shall be an air terminal within 0.6 m (2 ft) of the end of each ridge or each corner of a flat or low-slope roof.

3.5 Air terminals shall be placed within 0.6 m (2 ft) of the edge along the perimeter of a flat or low-slope roof. A low-slope roof is one which is 12.2 m (40 ft) wide or less and has a pitch of 1/8 or less; or is over 12.2 m (40 ft) wide and has a pitch of 1/4 or less. The center of such a roof shall have intermediate air terminals at intervals not exceeding 15.2 m (50 ft).

3.6 Chimney air terminals may be anchored directly, or secured by an acceptable metal band around the chimney. No outside corner of a chimney shall be more than 0.6 m (2 ft) from an air terminal. Copper chimney air terminals and all related components shall be hot-dip lead coated to prevent corrosion. Aluminum points, cable and fittings on chimneys need not be lead coated.

3.7 All dormers and roof projections as high or higher than the main ridge shall be fully protected. Prominent dormer and roof extensions lower than the main roof require protection according to their size, within spacing requirements for air terminals. If a major dormer gives a roof a low-rise configuration, it shall be so treated. The lightning protection system shall be kept continuous through all levels.

3.8 Air terminals may be "dead ended" (connected by only one path to a main conductor) on dormers below the main ridge level if that conductor run is 4.9 m (16 ft) or less in length and is horizontal or downward coursing. Terminals may be "dead ended" on dormers or projections that are level with the main roof, provided the conductor run does not exceed 2.4 m (8 ft).

3.9 Air terminals shall be optional on metal roof projections, such as cupolas, ventilators, chimney extensions and caps, air conditioning units, etc., depending on conductivity. If the metal is of sufficient strength and conductivity to approximate that of a standard air terminal, it may be directly bonded as described in Section 7. Those not sufficiently conductive shall be protected with air terminals.

SECTION 4—CONDUCTORS

4.1 Main conductors are those used to (a) interconnect air terminals in a direct or a closed-loop roof system; (b) serve as downleads from the roof system to the ground system; (c) connect metal bodies of inductance or conductance to the main conducting system; and (d) serve as ground electrodes in some cases, or to connect ground rods in certain other cases. For conductor sizes, see Table 1.

4.2 Roof conductors shall be coursed along ridges, or at perimeters of flat or low-slope roofs to interconnect all air terminals, including those on chimneys, dormers and projections, and shall afford a two-way path to ground from the base of each terminal. (Exception: Conductor drops from a higher to a lower roof level are permitted without an extra downlead provided there are no more than two ridge or three flat roof air terminals on the lower level.)

4.3 Copper lightning protection materials shall not be installed on aluminum roofing or siding material or other aluminum surfaces; nor

TABLE 1—MINIMUM MATERIAL REQUIREMENTS FOR MAIN CONDUCTORS

Type of Conductor		Copper		Aluminum	
Cable	Min size ea strand	1.15 mm	(17 AWG)	1.63 mm	(14 AWG)
	Weight per 305 m (1000 ft)	85 kg	(187.5 lb)	43 kg	(95 lb)
	Cross sect area	0.30 cm^2	(59 500 CM)	0.499 cm^2	(98 500 CM)
Solid strip	Thickness	1.63 mm	(14 AWG)	2.05 mm	(12 AWG)
	Width	25.4 mm	(1.00 in.)	25.4 mm	(1.00 in.)
Solid bar	Weight per 305 m (1000 ft)	85 kg	(187.5 lb)	43 kg	(95 lb)
Tubular bar	Weight per 305 m (1000 ft)	85 kg	(187.5 lb)	43 kg	(95 lb)
	Min wall thickness	0.81 mm	(0.032 in.)	1.628 mm	(0.0641 in.)

should aluminum lightning protection materials be installed on copper surfaces.

4.4 When there is no alternative to coursing a conductor through the air, this may be done without support for a distance of 1 m (3 ft) or less, or with support of a 16 mm (5/8 in.) copperclad ground rod or its equivalent. A conductor may be coursed through the air for a distance up to 1.8 m (6 ft).

4.5 Roof conductors shall be coursed through or around obstructions in a horizontal plane with the main conductor. Conductors shall maintain a horizontal or downward course, free from "U" or "V" (down and up) pockets. No bend of a conductor shall form an angle of less than 90 deg, or have a radius of bend less than 203 mm (8 in.).

4.6 Metal roofing and siding, eave troughs, downspouts, and other metal parts are not acceptable as substitutes for lightning conductors. A lightning conductor system shall be applied to the metal siding of a metal-clad building in like manner as on buildings without such metal coverings.

4.7 Down conductors, which are continuations of roof conductors, shall be as widely separated as possible at diagonal corners of rectangular buildings, and diametrically opposite on cylindrical structures.

4.8 No less than two down conductors, with a proper ground for each, shall be provided on any structure. Buildings with perimeters exceeding 61 m (200 ft) shall have one additional down conductor for each 30 m (100 ft) or fraction. Irregular shaped buildings may require more down conductors in order to provide two-way paths to ground from air terminals on main ridges or on side wings.

4.9 Down conductors located in cattle yards, driveways or other vulnerable locations shall be guarded in such a manner as to prevent physical damage or displacement, to a distance not less than 1.8 m (6 ft) above ground level. If run through conducting pipe or tubing (of compatible metal), the conductor should be bonded to the conduit at top and bottom.

SECTION 5—GROUNDING

5.1 Proper grounds are critical to assure dissipation of a lightning discharge without damage. Extent of grounding will depend on character of the soil, ranging from two simple 3.0 m (10 ft) grounds for a small building located on deep conductive soil, to an elaborate network of cables and rods or plates buried in soil that is dry or rocky and of poor conductivity.

5.2 Minimum acceptable standard for each ground electrode shall be a copper-clad steel rod at least 13 mm (1/2 in.) in diameter and 3.0 m (10 ft) long. Rods of solid copper, 13 mm (1/2 in.) diameter, may be used in lieu of copper-clad steel. Stainless steel ground rods are also acceptable where acid soil conditions exist or other conditions warrant substitution of stainless steel.

5.3 Wherever practicable, connections to ground electrodes shall be made at points not less than 0.3 m (1 ft) below grade and 0.6 m (2 ft) out from the foundation. Grounds shall be distributed and placed at corners and other locations in a manner to direct the flow of current out from the building rather than under it. Placing of grounding under a building (as in extending a building) shall be kept at a minimum. There shall be a ground at each down conductor. An aluminum conductor shall not be attached to a surface coated with alkaline-base paint, embedded in concrete or masonry or installed in a location subject to excessive moisture.

5.4 In moist clay, the ground shall extend vertically not less than 3.0 m (10 ft) into the earth, and the earth shall be tamped along the full length of the ground.

5.5 In shallow top soil where bedrock is near the surface, the lightning conductor (extensions of the down conductors) shall be laid in trenches extending away from the building. Trenches shall be at least 3.7 m (12 ft) long and 0.3 m (1 ft) deep in clay soil, and at least 7.3 m (24 ft) long and 0.6 m (2 ft) deep in sandy or gravelly soil.

5.6 In moist sandy or gravelly soil of ordinary soil depth, two electrodes shall be driven at least 3.0 m (10 ft) deep at each ground. The conductor shall be extended out from the building in a trench at least 0.6 m (2 ft) of the wall. The two electrodes shall not be spaced more than 1.8 m (6 ft) apart.

5.7 Underground metallic water pipes or well casings shall be connected to the grounding system with main size conductors and special fittings with a minimum contact surface to the pipe 38 mm (1.5 in.) long and 13 cm^2 (2 $in.^2$) in area. Water pipe or well casing connections are in addition to the required number of regular grounds. If a metal water pipe or well casing enters a building, at least one down conductor of the lightning protection system shall be connected to it.

5.8 No connections shall be made to plastic or other non-conductive pipes. If such water pipes serve metal pipes within a structure, the metal pipes shall be treated as bodies of inductance.

5.9 Common grounding has been recognized as the most effective means of eliminating sideflashes between metal bodies of inductance. Therefore all grounding mediums are to be bonded to the lightning protection system, including electric and telephone service grounds, and piping systems including water service, gas piping, LP gas piping systems, underground conduits of metal, etc. Any sections of plastic pipe in piping systems shall be bridged with a length of main size conductor.

SECTION 6—CONNECTORS AND OTHER FITTINGS

6.1 More than 500 different lightning protection system parts are applicable to farm installations, thus they are covered here in only broad terms. One wrong part at a critical point may cause an electrical impedance with a resulting flash sufficient to ignite adjacent hay, straw, gas, fuel, etc.

6.2 Because of the above, it is extremely important to set guidelines assuring the integrity of all parts of a system. Therefore (1) all parts shall be of the same material as the conductor cable and other major

TABLE 2—MINIMUM MATERIAL REQUIREMENTS FOR SECONDARY CONDUCTORS

Type of Conductor		Copper		Aluminum	
Cable	Wire size	1.15 mm	(17 AWG)	1.63 mm	(14 AWG)
	Number of wires	14	(14)	10	(10)
Solid strip	Thickness	1.29 mm	(16 AWG)	1.29 mm	(16 AWG)
	Width	12.7 mm	(0.50 in.)	12.7 mm	(0.50 in.)
Solid rod	Wire size	4.12 mm	(6 AWG)	5.19 mm	(4 AWG)

components of the system, or proven to be compatible with such components by tests of electrical conductivity and electrolytic effects in combination; (2) all parts shall be guaranteed by the manufacturer as being manufactured for lightning protection purposes or as electrical components applicable to specific uses in lightning protection systems; (3) in any exception to the foregoing, the owner or his agent, such as architect or engineer, shall be responsible for obtaining certification that any alternate part used is equivalent in function, performance, and durability to the accepted standard part manufactured expressly for lightning protection, or an electrical part commonly used in lightning protection. An Underwriters' Laboratories Master Label, for example, constitutes such certification.

6.3 Ground rod clamps used to connect down conductors shall make contact with the ground rod for a minimum distance of 38 mm (1.5 in.) along the length of the rod and shall make contact with the cable itself for at least 38 mm (1.5 in.).

6.4 A disconnector shall be installed on all but one ground terminal of each building, except where exothermic connections are used at ground, the disconnector requirement is waived provided each connection is mechanically and electrically sound. Ground rod clamps or bimetal fittings used to connect aluminum conductors to copper conductors or to ground electrodes are acceptable disconnectors.

6.5 Fasteners shall be spaced not more than 0.9 m (3 ft) apart on all conductors. Nails, screws, or bolts shall be of the same material as the fasteners and shall be as resistant to corrosion as the fasteners. Galvanized or plated steel nails, screws or bolts are not acceptable.

6.6 Masonry fasteners shall have anchors with a diameter of not less than 6 mm (0.25 in.) and shall be anchored in the brick or stone, not in mortar joints. When set, the fit shall be tight against moisture and effects of frost and capable of withstanding a pull test of 445 N (100 lb).

6.7 Acceptable connector fittings shall be used on all lightning conductors at "end-to-end", "tee", or "Y" splices. They shall be attached to withstand a pull test of 890 N (200 lb).

6.8 Fittings for connection to metal tracks, gutters, downspouts, ventilators or other metal objects shall be of an acceptable type and shall be made tight by compression under bolt heads of the approved type.

6.9 "Crimp" and bolted clamps and splicers of stamped or cast metal of approved types are acceptable.

SECTION 7—BONDING OF METAL BODIES

7.1 Metal bodies may contribute to lightning hazard as either bodies of conductance on the roof or elevated and subject to direct lightning discharge, or as bodies of inductance, below the roof or inside a structure and subject to induced charges.

7.2 Metal bodies of conductance include ventilators, metal chimney extensions, flashings, valley and ridge rolls, metal roof railings, roof drains and soil vent pipes, etc.

7.3 Metal bodies of conductance within 1.8 m (6 ft) of main lightning conductor shall be bonded to that conductor with a size main conductor. Any body of conductance less than 1.8 (6 ft) from another body of conductance that has been bonded to the main conductor shall also be connected to the main conductor.

7.4 Cable leads shall, wherever possible, maintain a horizontal or downward path to the main conductor. A 203 mm (8 in.) radius for all bends shall be observed, and bends of more than 90 deg in any direction shall be avoided. All connections shall be made with bolt tension fittings of an acceptable type.

7.5 Metal stacks serving furnaces shall be bonded regardless of the distance from main conductors.

7.6 Metal bodies of inductance are those objects of appreciable size below the main roof or inside the buildings likely to become charged, at times, with a potential opposite to that of the ground or grounded lightning conductor system. Such metals within 1.8 m (6 ft) of the lightning conductor may induce flashes across the intervening gap.

7.7 Exterior bodies of inductance include door tracks, railings, low eave troughs, downspouts, metal door and window frames, hay tracks and elevators, manure tracks, guy wires, etc.

7.8 Interior bodies of inductance include isolated I-beams of extended length; steam, water or gas pipes; metal conduits; heavy machinery; metal ducts; metal doors; feed conveyors; feeders; ventilating fans; standing milking systems; cooling systems; etc. Metal bodies of inductance located within 1.8 (6 ft) of a lightning conductor shall be connected to the lightning conductor system. Metal bodies of inductance require interconnection even though they are over 1.8 m (6 ft) from a lightning conductor if they are within 1.8 m (6 ft) of a metal body already connected.

7.9 The interconnecting conductor for both exterior and interior bodies of inductance shall be an acceptable secondary conductor. The interconnecting conductor shall terminate in acceptable fittings. If a water supply system or milking system is present in the structure and the lightning conductor has been grounded to the water pipe, then the above metal bodies may be connected either to the water pipe system, the nearest lightning conductor, or to another metal body already connected to the system.

SECTION 8—ARRESTERS

8.1 Radio and television masts of metal, regardless of location on a building, shall be bonded to the main conductor of the lightning protection system with a main-size conductor and acceptable fittings.

8.2 To protect radio or TV equipment against surges, a lightning arrester shall be installed on the lead-in wire, tape, or cable and bonded to the lightning protection system directly or through common ground. Secondary service arresters shall be installed by the lightning protection contractor, electrical contractor, or the electric utility company. Such arresters shall be installed on both overhead and underground services at the electric service entrance, or at the interior service entrance box, depending on local regulations. Before installing a secondary service arrester, it should be determined that the neutral wire is adequately grounded preferably to a metal water pipe system that enters the ground.

SECTION 9—ALUMINUM SYSTEMS

9.1 Aluminum lightning protecton equipment shall not be installed on copper roofing materials or other copper surfaces.

9.2 Connection of aluminum conductors to ground equipment shall be made at a point not less than 0.3 m (1 ft) above grade level. Where downleads are concealed, the transition shall be at least 0.3 m (1 ft) above the lowest slab, floor or footing to be pierced and shall be accessible after installation for inspection purposes.

9.3 Aluminum conductors may be installed on a surface coated with alkaline paint or embedded in concrete or masonry (except in direct contact with earth) if run in watertight conduit (either metal or plastic). If metal conduit is used, the cable shall be securely bonded to it at both ends of any run.

SECTION 10—STRUCTURAL STEEL FRAMING

10.1 The structural steel framework of a building may be utilized as the main conductor of a lightning protection system. This type of installation is highly specialized and should not be undertaken by persons without full knowledge and training.

10.2 Air terminals shall be bonded directly to the steel framework or by conductors led through the ridge, or shall be connected together with a conductor on the exterior of the building that is itself connected to the steel framework.

10.3 If such a conductor is employed, it shall be bonded to the steel framework at intervals of not more than 18.3 m (60 ft). The minimum number of such connections required shall not be less than the number of groundings for the building.

10.4 Connections shall be made on cleaned areas of the steel framework by the use of bonding plates with bolt pressure cable connectors having a surface contact area of not less than 52 cm^2 (8 $in.^2$), bolted or welded securely to the steel or by welding or brazing a minimum of 38 mm (1.5 in.) of conductor to the steel. Properly accomplished exothermic welds are also acceptable for this type of connection.

10.5 Ground connections shall be made at approximately every other steel column, around the perimeter, and shall not average over 18.3 m (60 ft) apart. Ground terminals shall be attached to steel columns at the lowest available point with bonding plates having a surface contact area of not less than 52 cm^2 (8 $in.^2$), bolted or welded securely to cleaned areas of the structural steel, or by means of a sound exothermic connection.

10.6 If grounds are installed at dry locations, such as in sand, gravel, or rock, a counterpoise connected with each of the individual ground terminals shall be installed.

10.7 Grounding systems shall also comply with the requirements for grounding previously covered herein.

SECTION 11—SILO PROTECTION

11.1 Silo protection systems shall be installed in accordance with Sections 2-10, and in addition, shall meet the following requirements.

11.2 Silos standing alone more than 1.8 m (6 ft) away and not attached to the barn need not be protected as a part of barn protection. However, adjacent and attached silos shall be protected, and the systems bonded to the systems on the adjacent or attached structures.

11.3 Air terminals shall be equally spaced around the top edge of open or flat-roof silos at intervals not to exceed 6 m (20 ft), but in no case shall there be less than four air terminals provided. The air terminals shall be interconnected to a conductor coursed around the silo near the top edge, forming a closed loop.

11.4 Peak or dome-roof silos shall be provided with at least one air terminal.

11.5 Prefabricated all-metal silo tops or domes which are made electrically continuous before or after installation may be substituted for the air terminal. All-metal silos and hay keepers may be protected by two acceptable groundings attached to the all metal structure by bonding devices having a surface contact area of not less than 52 cm^2 (8 $in.^2$), provided such structures are constructed to be electrically continuous.

11.6 Down conductors shall be securely bonded to the lower edge of metal silo roofs by bolted joint fittings having a surface contact area not less than 19 cm^2 (3 $in.^2$).

11.7 If non-metal silos standing alone are to be protected, they shall be equipped with not less than two down conductors.

11.8 Silos adjacent to and attached to a barn shall be equipped with one down conductor if the distance from the base of a silo air terminal to the interconnection with the main conductor on the adjacent structure does not exceed 4.9 m (16 ft), and provided the top of the silo is lower than the main ridge of the adjacent structure. If this distance exceeds 4.9 m (16 ft), or if the top of the silo is higher than the main ridge of the adjacent structure, an additional down conductor is required.

11.9 The lightning protection system on adjacent or attached silos shall be interconnected, preferably underground, to the system installed on the adjacent or attached structure.

11.10 Twin silos with separate roofs shall be equipped with not less than three conductors, the center down conductor being common to both silos.

11.11 Metal parts of silos, such as bands, chutes, ladders, and lap-jointed roofs, shall not be used at substitutes for main conductors. Such metal parts are bodies of conductance and shall be bonded to the system.

SECTION 12—TREE PROTECTION

12.1 Trees do not afford protection from lightning to nearby buildings, and in many instances, they should be protected in view of their value. It is recommended that trees which have trunks located within 3 m (10 ft) of a protected building and branches extending above the building be equipped with lightning protection, not only to protect the tree but also to avoid the possibility of lightning striking the tree and sideflashing or grounding to the nearby structure.

12.2 One main cable should be coursed from the air terminal at the top of the main trunk or branch to the ground terminal. Acceptable secondary conductors should be coursed from miniature branch points, as far out on the main branches as possible, to the main conductor on the tree trunk.

12.3 To avoid possible injury to roots by locating depth grounding near trunks of trees, the conductor should be extended out and away from the base of a tree in a shallow trench to a distance of not less than 3.7 m (12 ft) or to the extremity of the overhanging branches. This conductor should terminate in a ground terminal. Depth groundings should be made outside the area of the root spread.

12.4 If the grounding on a protected building is within 7.6 m (25 ft) of a tree, the two systems should be interconnected. If the tree or grounding of the tree is within 7.6 m (25 ft) of a water pipe or a deep-well casing, a connection should be made between them.

12.5 Trees with trunks which exceed 1 m (3 ft) in diameter, and which have extra long branches, should have two down conductors. They should be led down opposite sides of the tree and connected to the two ground terminals. These two ground terminals may well be joined by a circular or semi-circular conductor or a counterpoise buried in a shallow trench.

12.6 If there are several trees in a row (all major trees), the ground terminals of the two trees not more than 24 m (80 ft) apart may be interconnected by a trench conductor coursed to the base of each intermediate tree. The down conductor of each intermediate tree may connect with the "trenched" interconnecting conductor. This practice avoids making independent groundings for each tree.

Cited Standards:

LPI-175, Installation Code
NFPA No.78, Lightning Protection Code
UL, Requirements for Master Label Lightning Protection

MANURE PRODUCTION AND CHARACTERISTICS

Developed by the Engineering Practices Subcommittee of the ASAE Agricultural Sanitation and Waste Management Committee; approved by the ASAE Structures and Environment Division Standards Committee; adopted by ASAE December 1976; reconfirmed December 1981, December 1982, December 1983, December 1984, December 1985, December 1986; reconfirmed for one year December 1987.

SECTION 1—PURPOSE AND SCOPE

1.1 Data on farm animal manure production and characteristics is presented to make readily available information for the planning, design and operation of structures and equipment for animal enterprises.

1.2 These data are combined from a wide base of published and unpublished information on animal waste production and characterization. Those making use of this information should recognize that these are median values and that actual values vary widely (by as much as 50%) due to differences in ration, animal age, and management practices.

TABLE 1—MANURE PRODUCTION AND CHARACTERISTICS PER 454 kg (1000 lb) LIVE WEIGHT*

Item	Units	Dairy		Beef		Swine		Sheep	Poultry		Horse
		Cow	Heifer	Yearling 182-318 kg (400-700 lb)	Feeder >318 kg (>700 lb)	Feeder	Breeder		Layer	Broiler	
Raw Waste (RW)	kg/day	37.2	38.6	40.8	27.2	29.5	22.7	18.1	24.0	32.2	20.4
	lb/day	82.0	85.0	90.0	60.0	65.0	50.0	40.0	53.0	71.0	45.0
Feces/Urine Ratio		2.2	1.2	1.8	2.4	1.2		1.0			4.0
Density	kg/m^3	1005.0	1005.0	1010.0	1010.0	1010.0	1010.0		1050.0	1050.0	
	lb/cu ft	62.7	62.7	63.0	63.0	63.0	63.0		65.5	65.5	
Total Solids (TS)	kg/day	4.7	4.2	5.2	3.1	2.7	1.9	4.5	6.1	7.7	4.3
	lb/day	10.4	9.2	11.5	6.9	6.0	4.3	10.0	13.4	17.1	9.4
	% of RW	12.7	10.8	12.8	11.6	9.2	8.6	25.0	25.2	25.2	20.5
Volatile Solids	kg/day	3.8			2.7	2.2	1.4	3.8	4.3	5.4	3.4
	lb/day	8.6			5.9	4.8	3.2	8.5	9.4	12.0	7.5
	% of TS	82.5			85.0	80.0	75.0	85.0	70.0	70.0	80.0
BOD_5†	% of TS	16.5			23.0	33.0	30.0	9.0	27.0		
COD‡	% of TS	88.1			95.0	95.0	90.0	118.0	90.0		
TKN§	% of TS	3.9	3.4	3.5	4.9	7.5		4.5	5.4	6.8	2.9
P‖	% of TS	0.7	3.9		1.6	2.5		0.66	2.1	1.5	0.49
K#	% of TS	2.6			3.6	4.9		3.2	2.3	2.1	1.8

* Numerical values for kg/day/1000 kg live weight are the same as those for lb/day/1000 lb live weight.

† 5-day biochemical oxygen demand.

‡ Chemical oxygen demand.

§ Total Kjeldahl nitrogen.

‖ Phosphorus as P.

Potassium as K.

ASAE Engineering Practice: ASAE EP388.1

DESIGN PROPERTIES OF ROUND, SAWN AND LAMINATED PRESERVATIVELY TREATED CONSTRUCTION POLES AND POSTS

Developed by the ASAE Materials of Construction Committee; approved by the Structures and Environment Division Standards Committee; adopted by ASAE December 1977; revised March 1983; reconfirmed for one year December 1987.

SECTION 1—PURPOSE AND SCOPE

1.1 The purpose of this Engineering Practice is to establish guidelines for the production and design of construction poles and posts that act as the main structural members in pole frame and post frame building.

SECTION 2—DEFINITIONS

2.1 Construction pole: A round, naturally tapered main structural member in a pole frame building. The pole is partially embedded and acts as a cantilever beam subjected to axial, bending, or combined loading.

2.2 Slabbed construction poles: Modified by slabbing before treatment to provide a continuous flat surface for attachment of sheathing and framing members, to permit more secure attachment of fasteners, and to facilitate the alignment and setting of intermediate wall and corner poles. The slabbing shall consist of a minimum cut to provide a single continuous flat face from groundline to top of intermediate wall poles, and two continuous flat faces at right angles to one another from groundline to top for corner and door poles. It should be recognized that preservative penetration is generally limited in heartwood, so that slabbing, particularly in the groundline area of poles with thin sapwood, may result in somewhat less preservative protection than that of an unslabbed pole.

2.3 Construction post: Sawn on four sides before treatment, is generally uniform in cross section along its length and serves the same function as a construction pole. The effect of the removal of sapwood on the retention of preservative is similar to slabbed construction poles. The construction post is the primary structural member in a post frame building.

2.4 Glued-laminated construction post: A structural timber manufactured in a timber laminating plant, consisting of assemblies of suitably selected and prepared wood laminations bonded together with adhesives. The laminated post serves the same function as a construction post in a post frame building.

2.5 Nailed-laminated construction post: A structural timber manufactured either partly or wholly in a plant or on the construction site, consisting of suitably selected wood laminations fabricated with nails or other mechanical fasteners. The laminations consist of preservative-treated wood below the groundline and untreated wood above the ground. The laminations are longitudinally spliced above the groundline with mechanical fasteners such as nails and metal plates.

SECTION 3—MATERIAL REQUIREMENTS

3.1 Poles. Seasoning requirements of paragraph 4.1.2 of American National Standard O5.1, Specifications and Dimensions for Wood Poles, shall apply. The grading requirements of paragraphs 4.1.3, 4.2, 4.3 and 4.4 of ANSI Standard O5.1, Specifications and Dimensions for Wood Poles, shall also apply.

3.2 Sawn posts. Sawn posts shall be graded by an approved agency in accordance with National Bureau of Standards PS20, American Softwood Lumber Standard, as either beams and stringers or posts and timbers.

3.3 Glued-laminated (glu-lam) posts. Glu-lam posts shall be manufactured using lumber grades in accordance with American Institute of Timber Construction Standard 117, Standard Specifications for Structural Glued-Laminated Timber of Softwood Species.

3.4 Nailed-laminated posts. Laminations shall be graded by an approved agency in accordance with NBS Standard PS20, American Softwood Lumber, as structural joists and planks.

3.5 Preservative treatment. All poles, solid and glued-laminated, and the embedded portion of nailed-laminated posts shall be pressure treated with a preservative in accordance with American Wood-Preservers' Association Standard C16, Wood Used on Farms—Preservative Treatment by Pressure Processes. Examples are given in Table 1. Treatment levels for fence posts are not recommended for the structural posts and poles of buildings.

TABLE 1—MINIMUM PRESERVATIVE RETENTIONS IN WEIGHT OF PRESERVATIVE PER VOLUME OF WOOD*

Material and usage	Creosote or creosote-coal tar	Creosote-petroleum	Pentachlorophenol	ACC	ACA or CCA	CZC	FCAP	Application AWPA standards//
				kg/m³ (lb/ft³)				
Poles† as round structural members:								
Southern pine, ponderosa pine	120.1 (7.5)	NR§	6.09 (0.38)	NR	9.61 (0.60)	NR	NR	C4
Red pine	168.2 (10.5)	NR	8.49 (0.53)	NR	9.61 (0.60)	NR	NR	C4
Coastal Douglas-fir	144.2 (9.0)	NR	7.21 (0.45)	NR	9.61 (0.60)	NR	NR	C4
Jack pine, lodgepole pine	192.2 (12.0)	NR	9.61 (0.60)	NR	9.61 (0.60)	NR	NR	C4
Western red cedar, western larch, intermountain Douglas-fir	256.3 (16.0)	NR	12.81 (0.80)	NR	9.61 (0.60)	NR	NR	C4
Posts, sawn four sides as structural members:								
All softwood species	192.2 (12.0)	192.2 (12.0)	9.61 (0.60)	NR	9.61 (0.60)	NR	NR	C2
Posts, fence:								
All softwood species								
Round, half-round, and quarter-round‡	128.1 (8.0)	128.1 (8.0)	6.41 (0.40)	8.01 (0.50)	6.41 (0.40)	9.93 (0.62)	4.97 (0.31)	C5
Sawn four sides	160.2 (10.0)	160.2 (10.0)	8.01 (0.50)	9.93 (0.62)	8.01 (0.50)	NR	NR	C2
Lumber:								
All softwood species								
In contact with soil	160.2 (10.0)	160.2 (10.0)	8.01 (0.50)	9.93 (0.62)	8.01 (0.50)	NR	NR	C2
Not in contact with soil	128.1 (8.0)	128.1 (8.0)	6.41 (0.40)	4.00 (0.25)	4.00 (0.25)	7.21 (0.45)	4.00 (0.25)	C2
Millwork:								
All softwood species	NR	NR	4.81 (0.30)	4.00 (0.25)	4.00 (0.25)	4.00 (0.25)	4.00 (0.25)	C2

NOTE: Water-borne preservatives or pentachlorophenol in suitable solvents should be used where dry surfaces are required or the materials are to be painted.
*From AWPA Standard C16, Wood Used on Farms—Preservative Treatment by Pressure Processes.
†Poles used as structural members will be assayed according to AWPA Standard C4, Poles—Preservative Treatment by Pressure Processes.
‡Creosote-coal tar is not recommended for Douglas-fir, western hemlock, western larch; these species are not recommended for half-round or quarter-round posts.
§NR—Not Recommended.
//See Cited Standards for titles.

SECTION 4—MANUFACTURING REQUIREMENTS

4.1 Poles. Requirements of Section 6 of ANSI Standard O5.1, Specifications and Dimensions for Wood Poles, shall apply.

4.2 Sawn posts. Requirements of NBS Standard PS20, American Softwood Lumber, shall apply.

4.3 Glu-lam posts. Wet-use adhesive shall be used and the posts shall be industrial appearance grade per AITC Standard 110, Standard Appearance Grades for Structural Glued-Laminated Timber. Manufacture shall conform to NBS Standard PS56, Structural Glued-Laminated Timber.

4.4 Nailed-laminated posts. Preservative-treated laminations shall be used on the embedded end of the post and the shortest treated lamination shall extend a minimum of 0.6 m (2 ft) above the exterior grade line. Untreated laminations and mechanical fasteners on longitudinal splices shall be protected from decay or corrosion by the building envelope or by other means.

SECTION 5—DESIGN STRESSES

5.1 Poles. Design stress shall be obtained following American Society for Testing and Materials Standards D2899, Establishing Design Stresses for Round Timber Piles, and D3200, Specifications and Methods for Establishing Recommended Design Stresses for Round Timber Construction Poles. Examples are given in Table 2, where species with similar properties have been combined.

5.2 Sawn posts. Design stresses for either beams and stringers or posts and timbers shall be obtained following procedures given in ASTM Standard D245, Establishing Structural Grades and Related Allowable Properties for Visually Graded Lumber. Such stress data are published for some species in National Forest Products Association Standard, National Design Specification for Wood Construction. Examples are given in Table 3.

5.3 Glu-lam posts. Design stresses shall be in accordance with methods given in ASTM Standard D3737, Establishing Stresses for Structural Glued-Laminated Timber (Glu-Lam) Manufactured from Visually Graded Lumber. Stress data for some species are given in AITC Standard 117, Standard Specifications for Structural Glued-Laminated Timber of Softwood Species. Examples are given in Table 4.

5.4 Nailed-laminated posts. The strength of a nailed-laminated post is, in general, limited by the bending strength of the joint between the treated and untreated lumber. Joints shall be designed from test data or by accepted methods of structural analysis.

SECTION 6—POLE DIMENSIONS

6.1 Length of pole or post shall be measured between its extreme ends and shall not be greater than 76 mm (3 in.) shorter or 152 mm (6 in.) longer than the nominal length.

TABLE 2—DESIGN STRESSES* FOR SELECTED SPECIES OF ROUND POLES

Species	Bending† F_b		Axial‡ compression F_c		Horizontal shear F_v		Compression perpendicular-to-grain $F_{c_\perp}$		Modulus of elasticity E	
	MPa	psi	MPa	psi	MPa	psi	MPa	psi	GPa	10^6 psi
Northern white cedar (EC)	8.27	1200	4.14	600	0.62	90	1.10	160	4.14	0.6
Western red cedar (WC)	10.34	1500	5.52	800	0.69	100	1.10	160	6.21	0.9
Ponderosa (WP) or lodgepole pine (LP)	9.65	1400	4.83	700	0.69	100	1.17	170	6.89	1.0
Red (NP) or jack (JP) pine	11.03	1600	5.52	800	0.69	100	1.17	170	7.58	1.1
Douglas fir§ (DF) or southern// (SP) pine	14.48	2100	6.89	1000	0.90	130	1.65	240	9.65	1.4
Western larch (WL)	15.86	2300	8.27	1200	0.90	130	1.86	270	10.34	1.5
Oak#	15.17	2200	6.89	1000	1.17	170	2.76	400	8.96	1.3

*Determined according to ASTM Standard D3200, Specifications and Methods for Establishing Recommended Design Stresses for Round Timber Construction Poles, for wet-use conditions.
†Safety factor = 1.3 per ASTM D2899, Establishing Design Stresses for Round Timber Piles.
‡Safety factor = 1.25 per ASTM D2899, Establishing Design Stresses for Round Timber Piles.
§Interior north or coast.
//Loblolly, longleaf, shortleaf, or slash.
#Includes white and northern red.

TABLE 3—DESIGN STRESSES* FOR SELECTED SPECIES AND GRADES OF SAWN POSTS

Species and grade	Bending F_b		Axial compression F_c		Horizontal shear F_v		Compression perpendicular-to-grain $F_{c_\perp}$		Modulus of elasticity E	
	MPa	psi	MPa	psi	MPa	psi	MPa	psi	GPa	10^6 psi
Ponderosa pine†										
Select structural	6.89	1000	5.00	725	0.45	65	1.72	250	7.58	1.1
No. 1	5.69	825	4.38	635	0.45	65	1.72	250	7.58	1.1
Northern pine										
Select structural	7.93	1150	5.65	820	0.45	65	2.00	290	8.96	1.3
No. 1	6.55	950	5.00	725	0.45	65	2.00	290	8.96	1.3
Douglas fir-larch										
Select structural	10.34	1500	7.21	1045	0.59	85	2.90	420	11.03	1.6
No. 1	8.27	1200	6.27	910	0.59	85	2.90	420	11.03	1.6
Southern pine‡										
No. 1 SR	9.31	1350	5.34	775	0.76	110	2.59	375	10.34	1.5
No. 2 SR	7.58	1100	4.31	625	0.65	95	2.59	375	9.65	1.4

*From National Fire Protection Association Standard, National Design Specification for Wood Construction, 1982, for wet-use conditions.
†Includes lodgepole and sugar pine.
‡Loblolly, longleaf, shortleaf, or slash.

TABLE 4—DESIGN STRESSES* FOR SELECTED VISUALLY-GRADED SPECIES AND COMBINATION OF GLUED-LAMINATED POSTS

Species and combination symbol-grade†	Bending‡ F_b MPa	Bending‡ F_b psi	Axial compression F_c MPa	Axial compression F_c psi	Horizontal shear F_v MPa	Horizontal shear F_v psi	Compression perpendicular-to-grain $F_{c_\perp}$ MPa	Compression perpendicular-to-grain $F_{c_\perp}$ psi	Modulus of elasticity E GPa	Modulus of elasticity E 10^6 psi
Douglas fir larch										
1	6.89	1000	6.03	875	0.83	120	1.76	255	8.27	1.2
2	8.96	1300	7.93	1150	0.83	120	1.76	255	9.65	1.4
3	10.34	1500	9.31	1350	0.83	120	2.07	300	10.34	1.5
Southern pine										
47	7.58	1100	5.86	850	1.00	145	1.76	255	8.27	1.2
48	8.96	1300	6.89	1000	1.00	145	2.07	300	9.65	1.4
49	9.65	1400	7.24	1050	1.00	145	1.76	255	9.65	1.4

*From AITC Standard 117, Standard Specifications for Structural Glued-Laminated Timber of Softwood Species, for wet-use conditions.
†Indicates combinations of grades of laminations from AITC Standard 117, Standard Specifications for Structural Glued-Laminated Timber of Softwood Species.
‡Values shown apply to 3 or more laminations.

6.2 Cross-section

6.2.1 Round poles are generally specified by class according to ANSI Standard O5.1, Specifications and Dimensions for Wood Poles, which gives minimum circumferences both at the top and 1.8 m (6 ft) from the butt for different species.

6.2.2 Sawn and glu-lam posts shall be specified by the actual cross-sectional dimensions at both ends of the posts. Unless otherwise specified, uniform taper will be assumed.

SECTION 7—STORAGE AND HANDLING

7.1 The handling and care of pressure-treated construction poles and posts shall be in accordance with AWPA Standard M4, Standard for the Care of Preservative-Treated Wood Products.

7.2 The requirements of this section shall include, but are not restricted to, the following items:

7.2.1 The use of handling tools and load-devices is not permitted in the groundline area (one foot above or two feet below specified groundline, as given in ANSI Standard O5.1, Specifications and Dimensions for Wood Poles.

7.2.2 Field fabrication in the groundline area is not permitted.

7.2.3 Poles and posts shall not be cut off from the butt end after treatment.

Cited Standards:

AITC 110, Standard Appearance Grades for Structural Glued-Laminated Timber
AITC 117, Standard Specifications for Structural Glued-Laminated Timber of Softwood Species
ANSI O5.1, Specifications and Dimensions for Wood Poles
ASTM D245, Establishing Structural Grades and Related Allowable Properties for Visually Graded Lumber
ASTM D2899, Establishing Design Stresses for Round Timber Piles
ASTM D3200, Specifications and Methods for Establishing Recommended Design Stresses for Round Timber Construction Poles
ASTM D3737, Establishing Stresses for Structural Glued-Laminated Timber (Glu-Lam) Manufactured from Visually Graded Lumber
AWPA C2, Lumber, Timbers, Bridge Ties and Mine Ties—Preservative Treatment by Pressure Processes
AWPA C4, Poles—Preservative Treatment by Pressure Processes
AWPA C5, Fence Posts—Preservative Treatment by Pressure Processes
AWPA C16, Wood Used on Farms—Preservative Treatment by Pressure Processes
AWPA M4, Standard for the Care of Preservative-Treated Wood Products
NBS PS20, American Softwood Lumber
NBS PS56, Structural Glued-Laminated Timber
NFPA, National Design Specification for Wood Construction

ASAE Engineering Practice: ASAE EP393.1

T

MANURE STORAGES

Developed by the Engineering Practices Subcommittee of the ASAE Agricultural Sanitation and Waste Management Committee; approved by the ASAE Structures and Environment Division Standards Committee; adopted by ASAE March 1978; reconfirmed December 1982, December 1983; removed reference to ASAE S288 and reconfirmed December 1984, December 1985, December 1986; revised March 1988.

SECTION 1—PURPOSE AND SCOPE

1.1 The objective of this Engineering Practice is to provide engineers with recommendations for the design and location of manure storage units. These storage units may or may not be an integral part of the livestock building.

1.2 Recommendations for filling, treating, or removing manure from storage are not included in this Engineering Practice.

SECTION 2—PRINCIPLES AND PRACTICES

2.1 Obtain approval from appropriate regulatory agencies during planning and prior to construction.

2.2 Storage volume requirements

2.2.1 Storage volume

$$S = \frac{N \times MW \times D}{MD} + BV + DV - SV$$

where

S = volume of storage required, m^3 (ft^3)

N = number of animal units

MW = weight of manure produced per animal unit per day, kg (lb) (see ASAE Data D384, Manure Production and Characteristics)

D = number of days of storage (see paragraph 2.2.1.1)

MD = manure density, kg/m^3 (lb/ft^3) (see ASAE Data D384, Manure Production and Characteristics)

BV = bedding volume = $VR \left(\frac{N \times B \times D}{BD} \right)$ (see paragraph 2.2.1.2)

where

VR = volume reduction factor (range, 0.3 to 0.5)

B = weight of bedding used per unit per day, kg (lb)

BD = loose bedding density, kg/m^3 (lb/ft^3)

DV = dilution volume, m^3 (ft^3) (see paragraph 2.2.1.3)

SV = shrinkage volume, m^3 (ft^3) (see paragraph 2.2.1.4)

2.2.1.1 Storage capacity for the period the ground is frozen, up to 180 days, is recommended in cold climates to avoid manure application on frozen ground. For warmer climates, storage capacity for at least 45 days is recommended. Cropping practices, distribution methods, and scheduling also affect selection of storage period.

2.2.1.2 The bedding volume depends on the quantity of bedding, and characteristics, type, void space, initial moisture content and water absorption capacity. Little or no bedding should be used with pump removal systems to avoid difficulty in handling.

2.2.1.3 The dilution volume is considered primarily in liquid and slurry manure storage. It includes milking center wastes, water, spillage, surface water and/or precipitation. For uncovered outdoor storage the quantity of precipitation added depends upon the watershed area, the storage period, and the climatic conditions. Dilution of 8 to 10% and 4 to 5% solids is recommended for tank wagon and irrigation removal, respectively.

2.2.1.4 The shrinkage volume is considered primarily in semisolid or solid manure storage but is difficult to estimate. The amount of shrinkage depends on the storage time, manure moisture content, manure temperature, bedding type, and manure depth.

2.2.2 Solid or semisolid manure storage

2.2.2.1 Liquid (precipitation and urine) from a storage unit may be drained to a detention pond or underground tank, held in the storage unit, or infiltrated in fields. It must not directly enter surface and groundwaters.

2.2.2.2 Detention ponds should hold at least twice the average precipitation falling onto the storage unit for the desired storage period or the amount specified by state and federal law. The factor of two provides sufficient storage for urine not retained in the bedding. Provide additional volume for rainfall runoff from the detention pond watershed.

2.2.2.3 If liquid is retained in the storage unit, the floor should be sloped to one side or end.

2.2.2.4 Liquids can be separated from solids using a floor drain or vertically slotted wall. For floor drains a 150 mm (6 in.) layer of corncobs, shredded bark or shredded cornstalks on the floor is necessary to attain satisfactory separation. Provide floor drains with removable grills. Slope floors slightly toward floor drain or vertically slotted wall. Install 150 mm (6 in.) or larger diameter noncorrosive pipe in the bottom of the drain to carry liquid to the detention pond.

2.2.2.5 Provide a means for emptying the detention pond, underground tank, or liquid in the storage unit.

2.2.3 Liquid or slurry manure storage

2.2.3.1 Tank

2.2.3.1.1 Tank depth is based on computed storage capacity plus freeboard above lowest inlet opening plus 200 mm (8 in.) of liquid remaining after emptying unless a sump is provided. The freeboard is to prevent overflow when agitating and to provide for ventilation system operation.

2.2.3.1.2 For tanks with slotted or solid covers, allow at least 300 mm (1 ft) of freeboard between the top of the manure and the bottom of the cover. Additional freeboard may be necessary to provide adequate ventilation space between the top of manure and the low member of the cover framing.

2.2.3.1.3 Provide access for agitating and emptying the tank.

2.2.3.1.4 Extra water must often be added to permit handling (see paragraph 2.2.1.3 for dilution). Fibrous materials, frozen manure and granular material may interfere with agitation or pumping.

2.2.3.2 Earthen storage basin

2.2.3.2.1 With a tank wagon removal system, minimize the basin watershed area. For irrigation removal system, a larger watershed area may be desirable to provide dilution of manure if not already sufficiently diluted with wastewater from the milking center or other sources (see paragraph 2.2.1.3). Reduce basin watershed area by diverting surface runoff away from the basin except when needed for dilution.

2.2.3.2.2 Provide a pumping platform, wall or sloping ramp access to agitate and/or empty the storage basin. Liquid removal by pumping and solid removal with wheel-mounted equipment is also possible. Larger units may require more than one agitation station.

2.3 Storage location

2.3.1 Distance between water wells and storage units should be at least 30 m (100 ft) or as required by state code.

2.3.2 Locate storage unit for year-round use, for easy access, for convenient filling and emptying, for pleasing appearance, and not in a flood plain.

2.3.3 Locate so surface water does not enter storage unit unless desired for dilution. Prevent storage unit runoff from draining toward buildings or directly to surface waters.

2.3.4 On sites with shallow soil over coarse sand and gravel, creviced limestone, or permeable bedrock, use construction procedures and materials that prevent groundwater pollution (clay liners, fabric liners, or concrete). Geologic conditions and treatments should be determined from data available for the area, performance of similar installations of geologic investigation when appropriate.

2.3.5 Between a milk house and storage unit, provide for at least 15 m (50 ft) or more if required by state code.

2.3.6 Avoid constructing a tank or earthen storage basin below the water table to prevent tank flotation or basin flooding. If no other site is suitable, design to avoid overland flooding and to resist flotation pressures.

2.4 Safety criteria

2.4.1 Enclose earthen storage basins, detention ponds, solid manure storages, and open top tanks with a woven wire or chain-link type fence, at least 1.5 m (5 ft) high, constructed to prevent humans, animals, or equipment from accidentally entering the storage unit.

2.4.2 Liquid or slurry manure storage

2.4.2.1 Avoid placing pumping ports inside a building where livestock are confined to avoid exposure to noxious gases during agitation and to prevent accidental entry of animals into the pit.

2.4.2.2 Protect tank openings with grills and/or covers to prevent humans, animals, and equipment from accidentally falling into a storage tank. Design removable covers and grills to prevent them from falling into the tank or being unintentionally removed.

2.4.2.3 Permanent ladders or steps are not recommended. If installed, they should be of noncorrosive material, and a warning sign must advise of the noxious gas hazard.

2.4.2.4 Vent tank to remove gases. Provide a gas trap or valve in the pipeline between a building and an outside tank.

2.4.2.5 Where a tank opens directly into a closed building (e.g., below slotted floors) continuously draw ventilating air across the surface and exhaust it outside the building.

2.4.2.6 If it is essential for a person to enter a tank, they should wear self-contained breathing equipment (chemical reaction filter masks are not sufficient protection) and have a safety rope around their chest. At least two persons should be outside the tank and capable of immediately pulling the person from the tank should they succumb to the noxious gases. Ventilate the tank with a fan for at least 30 min prior to entry and continuously while the person is in the tank.

2.4.2.7 Periodically inspect a tank for leaks and cracking and for deterioration of grills, covers and ladders; or other evidence of structural failure. Repair or replace the tank and components.

SECTION 3—MATERIALS AND STRUCTURAL DESIGN

3.1 Structural design

3.1.1 Structural design methods and criteria. Methods used for the structural analysis and design of manure storage units should be consistent with codes and standards developed by the engineering profession for the respective material.

3.1.2 Solid or semisolid manure manure storage

3.1.2.1 Vertical walls. (See paragraph 3.1.3 for wall design loadings.)

3.1.2.2 Sloping walls. Wall design loading as indicated in paragraph 3.1.3 shall be modified consistent with accepted engineering practice for sloping walls.

3.1.2.3 Earth walls. For permeable soils, an impermeable liner may be necessary to control groundwater pollution; minimum 150 mm (6 in.) compacted clay layer, plastic liner, or equivalent. Sidewall slopes should not be steeper than 2H:1V. The storage bottom should be above the high-water table and at least 1 m (3 ft) above creviced bedrock.

3.1.2.4 Floors and ramps. A concrete floor is recommended for convenient emptying of wastes and possible control of groundwater pollution. Earth or crushed rock may be used if a water pollution hazard does not exist, if the saturated soil will support the manure handling equipment, and if a structural floor is not needed.

3.1.2.4.1 On soils where water pollution or equipment support may be a problem, concrete floors and ramps are recommended. Design for equipment loads anticipated. For wheel loads up to about 2 300 kg (5 000 lb), 130 mm (5 in.) thick floors and 150 mm (6 in.) thick ramps over 150 mm (6 in.) granular fill are typically adequate. With no wheel loads, 100 mm (4 in.) thick floors are adequate.

3.1.2.4.2 On soils where water pollution will not be a problem but equipment support may be, 80 mm (3 in.) of wetted and compacted stone chips over 130 mm (5 in.) of crushed stone up to 40 mm (1.5 in.) diameter is typically adequate.

3.1.3 Liquid or slurry manure storage

3.1.3.1 Design loads for walls

3.1.3.1.1 Interior design load. Interior hydrostatic pressure is 9 400 Pa/m (60 lb/ft^2/ft) of depth. For that portion of walls below ground or backfill level, the exterior earth pressure may be assumed to resist 50% of the interior hydrostatic pressure from ground surface to a depth of 1.2 m (4 ft), and 100% of the interior hydrostatic pressure below the 1.2 m (4 ft) depth. Design partition walls for the full 9 400 Pa/m (60 lb/ft^2/ft) hydrostatic pressure on either side unless openings in the wall assure equal loads on both sides of the wall.

3.1.3.1.2 Exterior wall design load. External loads consist primarily of lateral earth pressures; hydrostatic pressures; wind loads; surcharge pressures; and floor or cover, building, and equipment loads.

3.1.3.1.2.1 Lateral earth pressures depend on the type and density of soil or backfill material, wall flexibility, structure shape, and water table level. Walls supported at both the top and bottom, cantilever walls having a base thickness to height ratio > 0.085, and walls of circular structures shall be considered rigid walls. Walls designed as a cantilever and having a base thickness to height ratio < 0.085 may be considered flexible walls. Table 1 provides lateral earth pressure values to be used.

3.1.3.1.2.2 Where the installation or operation permits the movement of heavy equipment within 1.5 m (5 ft) of the wall, the wall shall be designed for a minimum additional 4 800 Pa (100 lb/ft^2) surcharge load uniformly distributed over the wall height.

3.1.3.1.2.3 Walls projecting above the ground surface level or walls supporting a building shall be designed for appropriate wind loads. Refer to ASAE Engineering EP288, Agricultural Building Snow and Wind Loads.

3.1.3.1.2.4 Walls and columns subjected to loads from structure covers, buildings, and building floors shall be designed for the effects of such loads.

3.1.3.2 Design loads for footings. Footing load includes all dead and live loads above the footing level. Design soil bearing pressures should be determined from data on soil strength, performance of nearby structures, or soil testing and analysis when appropriate, along with engineering judgment. Footing designs shall consider differential lateral pressures resulting from variations in backfill height and the surcharge load of the stored material.

3.1.3.2.1 All footings shall be located at or below maximum frost depth. A compacted foundation of frost free material, such as drained granular material, extending to below frost depth may be used as an alternate to extending the structural footing.

3.1.3.3 Design loads for tank covers in addition to dead loads

3.1.3.3.1 If an outdoor tank cover is constructed at least 460 mm (18 in.) above ground level and not accessible to livestock or vehicle traffic, use a distributed live load of

TABLE 1—LATERAL EARTH PRESSURE VALUES*

Soil		Equivalent fluid pressure Pa/m (lb/ft^2/ft) of depth			
		Above seasonal high-water table†		Below seasonal high-water table‡	
Description	Classification#	Flexible wall	Rigid wall	Flexible wall	Rigid wall
Clean sand, gravel or sand gravel mixtures (maximum 5% fines)§	GP, GW, SP, SW	4 700 (30)	7 900 (50)	12 600 (80)	14 100 (90)
Well-graded sand, silt, and clay mixtures (less than 50% fines)	SC, SC-SW, GM, GM-GP, SM SM-SW, SM-SP, GC GM-GW, GC-GP, GC-GW	5 700 (35)	9 400 (60)	12 600 (80)	15 700 (100)
Low-plasticity silts and clays with significant sand and gravel or fine silty and clayey sands (more than 50% fines)	Gravelly sand CL and ML or fine SC or SM	7 100 (45)	11 800 (75)	14 100 (90)	16 500 (105)
Low- to medium-plasticity silts and clay lacking in sand and gravel (more than 50% fines)	CL, ML	10 200 (65)	13 400 (85)	14 900 (95)	17 300 (110)
High liquid limit silts and clays	CH, MH¶	15 700 (100)	15 700 (100)	18 100 (115)	18 100 (115)

*For lightly compacted soils (85 to 90% maximum standard density). Includes compaction by use of typical farm machinery.
†Also below seasonal high-water table if adequate drainage is provided.
‡Includes hydrostatic pressure.
§Generally, only washed materials are in this category.
¶Unsuitable for backfill.
#American Society for Testing and Materials Standard D2487, Classification of Soils and Engineering Purposes.

1 920 Pa (40 lb/ft^2) plus an appropriate snow load. Refer to ASAE Engineering Practice EP288, Agricultural Building Snow and Wind Loads.

3.1.3.3.2 If the tank cover is to support floor loads, use loads from ASAE Engineering Practice EP378, Floor and Suspended Loads on Agricultural Structures Due to Use.

3.1.3.4 Design loads for floors. Use loads from ASAE Engineering Practice EP378, Floor and Suspended Loads on Agricultural Structures Due to Use.

3.2 Materials

3.2.1 Reinforced concrete. Refer to the latest edition of American Concrete Institute Standard 318, Building Code Requirements for Reinforced Concrete.

3.2.1.1 Use air-entrained concrete with a minimum compressive strength of 24.1 MPa (3 500 lb/in^2).

3.2.1.2 Maximum aggregate sizes and recommended percentages of entrained air are as follows:

Maximum aggregate size	Percent air
38 - 64 mm (1.5 - 2.5 in.)	4 - 7
19 - 25 mm (0.75 - 1 in.)	5 - 8
10 - 13 mm (0.38 - 0.5 in.)	6 - 9

3.2.1.3 Water-cement ratio, by weight, shall not exceed 0.50.

3.2.2 For pressure-treated round poles, posts and square timbers or planks, refer to ASAE Engineering Practice EP388, Design Properties of Round, Sawn, and Laminated Preservatively Treated Construction Poles and Posts.

3.2.3 Identify all materials utilized in a storage structure such as steel, wood, concrete, and products made from them, plus coatings, linings, sealants, etc., with reference to the appropriate standards or specifications governing their quality.

References:

1. Bruns, E. G. and J. W. Crowley. 1971. Solid manure handling for livestock housing feeding and yard facilities in Wisconsin. Cooperative Extension Programs. Report No. A2418. University of Wisconsin-Extension.
2. Converse, J. C., C. O. Cramer, H. J. Larsen and R. F. Johannes. 1974. Storage lagoon vs. underfloor tank for dairy cattle manure. ASAE Paper No. 74-3028. ASAE, St. Joseph, MI 49085.
3. Sweeten, J. M., W. S. Allen and D. L. Reddell. 1973. Solid waste management for cattle feedlots. Regional Extension Project No. 1600. Agricultural Engineering Department, Texas A&M University.
4. USDA-SCS. 1975. Agricultural waste management field manual. U.S. Department of Agriculture-Soil Conservation Service, Washington, D.C.
5. USDA-SCS. 1977. National handbook of conservation practices. Practice Standard 313, Waste Storage Structure (1980). U.S. Department of Agriculture-Soil Conservation Service, Washington, D.C.

Cited Standards:

ACI 318, Building Code Requirements for Reinforced Concrete
ASAE EP288, Agricultural Building Snow and Wind Loads
ASAE EP378, Floor and Suspended Loads on Agricultural Structures Due to Use
ASAE D384, Manure Production and Characteristics
ASAE EP388, Design Properties of Round, Sawn, and Laminated Preservatively Treated Construction Poles and Posts
ASTM D2487, Classification of Soils for Engineering Purposes

USE OF THERMAL INSULATION IN AGRICULTURAL BUILDINGS

Developed by the ASAE Burning Characteristics of Thermal Insulation Subcommittee; approved by the ASAE Structures and Environment Division Standards Committee; adopted by ASAE December 1980; reconfirmed December 1985; revised February 1987.

SECTION 1—PURPOSE AND SCOPE

1.1 This Standard provides a method to evaluate and specify the type, amount, and manner of installation of thermal insulation in agricultural buildings. The need for this standard arises because thermal insulation in agricultural buildings may increase the fire risk by (1) contributing heat due to combustion, (2) providing a path for fire to spread in a building, (3) reducing the fire endurance of a fire-rated wall if the insulation is combustible, (4) contributing to the intensity of the fire by performing its natural function of reducing heat loss, and (5) generating smoke and harmful gases.

1.2 The scope includes consideration of burning characteristics, insulation values, and proper installation and protection of insulating materials.

SECTION 2—DEFINITIONS

2.1 An agricultural building is defined as a building primarily designed to house or store farm implements, hay, grain, poultry, livestock or other animal or plant products. Such a structure may be used part-time, temporarily, or seasonally for work involved with agricultural production. This structure is not to be considered a place of human habitation or one regularly used by the public.

2.2 Insulating material (insulation) is defined as any material installed for the primary purpose of reducing heat transmission.

SECTION 3—BURNING CHARACTERISTICS EVALUATION

3.1 Insulation which has passed the Factory Mutual Building Corner Test, as developed by Factory Mutual Research, Norwood, Massachusetts, and has a flame spread of 25 or less and smoke production of 450 or less when tested in accordance with ASTM Standard E84, Test Method for Surface Burning Characteristics of Building Materials, may be installed in an exposed manner to the interior of the building.

3.2 Insulation not meeting requirements of paragraph 3.1, but which has a flame spread of 75 or less and smoke production of 450 or less when tested in accordance with ASTM Standard E84, Test Method for Surface Burning Characteristics of Building Materials, shall be separated from the interior of the building by an ignition barrier capable of providing protection comparable to 13 mm (1/2 in.) cement plaster, 13 mm (1/2 in.) gypsum wallboard, 3 mm (1/8 in.) mineral board, 16 mm (5/8 in.) exterior type plywood*, 25 mm (1 in.) masonry or concrete or other material which performs similarly when tested in accordance with ASTM Standard E119, Method for Fire Tests of Building Construction and Materials.

SECTION 4—INSULATION VALUES

4.1 The insulation values of the various insulation materials shall be as established and published in the latest edition of American Society of Heating, Refrigerating and Air Conditioning Engineers, Inc., ASHRAE Handbook — Fundamentals Volume, or as determined by reputable testing laboratories using ASTM Standard C236, Test Method for Steady-State Thermal Performance of Building Assemblies by Means of a Guarded Hot Box. Laboratory test results should be made available upon request.

4.2 Installation of the insulation shall be specified according to manufacturers' recommendations and in a manner which will not allow degradation of the thermal properties of the insulation. If degradation is unavoidable, the loss in insulation value should be taken into account during the design.

4.3 The desirable thickness of insulation and thus the resistance to heat transfer is a question which must be decided after considering initial cost of material and installation, consequences of wear and condensation, operating costs for heating and/or cooling based on present and projected energy costs, and critical nature of building and contents being insulated. See Table 1 for a guide to insulation values in temperature controlled agricultural buildings.

TABLE 1—MINIMUM RECOMMENDED HEAT TRANSFER COEFFICIENTS FOR INSULATION*†

Winter design temperature‡		R_{SI}, $\frac{m^2 \cdot °C}{W}$		R, $\frac{h \cdot ft^2 \cdot °F}{Btu}$	
		Roof	Walls	Roof	Walls
- 4 °C	25 °F	1.4 §	1.1	8 §	6
-12 °C	10 °F	1.4 §	1.4	8 §	8
-21 °C	- 5 °F	2.1	1.4	12	8
-26 °C	-15 °F	2.8	2.1	16	12
-32 °C	-25 °F	3.5	2.8	20	16

*Total heat transfer coefficients include surface and air space conductances and insulation value of linings and sidings.
†The values shown do not represent the values necessary to provide a heat balance between heat produced by products or animals and the heat transferred through the building.
‡Design temperature may be exceeded 2 1/2% of the time.
§These values should be considered absolute minimum for summer radiant heat load. Where heat temperature and radiant heat load are severe, use an R_{SI} of 2.1 (R of 12).

SECTION 5—INSTALLATION

5.1 Insulation should be chosen and installed after due consideration is given to the temperature extremes which will be encountered. Some plastic insulations melt at elevated temperatures. Some insulations lose their fire retardation properties under long-term elevated temperature and humidity conditions. Some insulations may cause corrosive reactions with building components.

5.2 The insulation or covering materials should be durable, cleanable, moisture resistant, non-toxic to livestock and consistent with applicable Food and Drug Administration regulations. There should be no transfer of odor, taste, or toxicity to food and feed products stored in the building.

5.3 Insulation immediately adjacent to heaters, electrical panels and devices, and welding operations is more susceptible to fire and melting. Insulation in these and other similarly hazardous areas should be protected with ignition barriers.

5.4 Extreme care shall be taken when using a torch, welder, etc., around insulation during construction or renovation of a building. A person with a fire extinguisher shall maintain a fire watch during work and for at least two hours after work is complete.

5.5 Insulation should be installed in such a manner that a reasonably uniform insulation value exists over the entire insulated area.

5.6 Insulation should be installed so that it will not settle, or unacceptably sag. Insulation, if susceptible to deterioration from ultraviolet light, should be protected from the sun.

5.7 Insulation should be chosen and installed in a manner which will discourage the entrance and chewing by rodents, picking by birds, infestation by insects, and damage by livestock.

5.8 A vapor retarder should be chosen and installed with the insulation which provides a permanent resistance to the passage of water vapor of 14.3 ng/Pa·s·m^2 (0.25 perms) or less when installed. Vapor retarders shall be evaluated and installed in conformance with the latest edition of ASHRAE Handbook — Fundamentals Volume.

5.9 When there is a possibility of moisture migrating through the insulation, the wall or roof section shall be constructed so that the moisture can escape on the cold side.

SECTION 6—SPECIAL CONSIDERATIONS

6.1 Full-scale building fire tests conducted by insulation manufacturers on unoccupied wood frame buildings with various installation configurations, types, and amounts of insulation (all with 25 or less flame spread ratings according to ASTM Standard E84, Test Method for Surface Burning Characteristics of Building Materials) show that such buildings may burn rapidly when the wood frame is exposed on the interior of the building.

6.2 Because of the intense heat buildup possible in a well insulated building, consideration should be given to the use of automatic heat and smoke vents or burn-out panels and skylights in large, valuable buildings.

References:

1. ASHRAE Handbook — Fundamentals Volume. American Society of Heating, Refrigerating and Air-Conditioning Engineers, Atlanta, GA.

Cited Standards:

ASTM C236, Test Method for Steady-State Thermal Performance of Building Assemblies by Means of a Guarded Hot Box
ASTM E84, Test Method for Surface Burning Characteristics of Building Materials
ASTM E119, Method for Fire Tests of Building Construction and Materials

ASAE Engineering Practice: ASAE EP403

DESIGN OF ANAEROBIC LAGOONS FOR ANIMAL WASTE MANAGEMENT

Developed by the ASAE Agricultural Sanitation and Waste Management Committee; approved by the ASAE Structures and Environment Division Standards Committee; adopted by ASAE March 1981; reconfirmed December 1985, December 1986; reconfirmed for one year December 1987

SECTION 1—PURPOSE AND SCOPE

1.1 This Engineering Practice describes the minimum criteria for design and construction of anaerobic animal waste lagoons located in predominantly rural or agricultural areas.

SECTION 2—DEFINITION

2.1 Lagoons are impoundments made by constructing an excavated pit, dam, embankment, dike, levee, or by a combination of these procedures.

2.2 Anaerobic lagoons treat animal wastes by predominantly anaerobic biological action. Utilization of this treatment process applies where concentrated animal waste must be treated, handled, or retained to reduce sources of pollution, minimize health hazards, and maintain or improve the local environment.

SECTION 3—LAWS AND REGULATIONS

3.1 Federal regulations require that animal waste management system operations result in no point source of pollutants except as the result of catastrophic or chronic rainfall events. Anaerobic animal waste lagoons are designed so that residuals of the treatment or storage process may be returned to the soil for terminal disposition. All other federal, state, and local laws, rules, and regulations governing the use of animal waste management lagoons should be followed. Necessary approvals and permits for location, design, construction, and operation of the lagoon should be secured from accountable authorities.

SECTION 4—DESIGN CRITERIA

4.1 Characteristics of a successfully operating animal waste lagoon vary throughout the United States. Odor intensity is most often the criteria of success, but rates of sludge and supernatant accumulation are also important. This Engineering Practice provides recommended design values. Due to varying opinions of acceptable characteristics, other established design values may differ from these values. These criteria recommended in this section should be compared with locally accepted design criteria. Fig. 1 illustrates various construction details.

4.2 Location. For convenience the lagoon should be located adjacent to the source of waste or as near as is practical.

4.2.1 Odor. Anaerobic lagoons will produce undesirable odors at times, but these will be minimized by following recommended design and management practices. Odors are usually strongest during the changeover from winter to spring operating conditions. These odors are generally stronger in colder climates. Because of odor production, anaerobic lagoons are usually located in isolated areas. Additional information can be found in ASAE Engineering Practice EP379, Control of Manure Odors.

4.2.2 Residences. Lagoons should be located at least 90 m (300 ft) from residences or other places of occupancy or employment. This minimum separation may be extended to as much as 760 m (2500 ft) in northern states. A natural or constructed screen concealing a lagoon may be desirable.

4.2.3 Wind. Lagoons should be located so that prevailing winds disperse and transport lagoon odors away from residences. Spring winds are of greatest concern.

4.2.4 Water supply. Lagoons should be located so that water supply wells are not contaminated. A minimum distance of 90 m (300 ft) from wells is recommended; applicable state or local regulations should be followed.

4.2.5 Expansion. Space should be allowed for possible expansion of the animal enterprise and the lagoon facility consistent with long range plans.

4.3 Soil and Foundation. A site investigation should be made to determine the physical characteristics and suitability of each construction site. Lagoons should be located on soils of low permeability which seal quickly. Where the wetted soil surface should be sealed, an impermeable membrane such as polyethylene film, impermeable soil from a borrow area, incorporation of bentonite clay with the soil, or other material acceptable to the responsible government agency may be used.

4.4 Size. Anaerobic lagoons for treatment of animal waste are designed on the basis of waste load added per unit volume of capacity. Operating temperatures and loading procedures are major factors determining the rate of biodegradation.

4.4.1 Volume: Fig. 2 shows recommended maximum lagoon loading rates for the United States based on average monthly temperatures and corresponding biological activity rates. The maximum loading rate is expressed in terms of grams of volatile solids (VS) per day per cubic meter (lb VS/day/1000 ft^3) of lagoon volume. An adjusted loading rate allowing for animal species and ration is the product of the recommended maximum loading rate and the loading rate multiplier found in Table 1. Table 2 lists the daily VS production for various animal species. Related information can be found in ASAE Data D384, Manure Production and Characteristics.

4.4.1.1 Example. Problem: A central South Carolina swine producer wishes to build an anaerobic lagoon to treat the wastes from a 100-sow, farrow-to-finish, totally confined swine operation. What is the recommended volume of the lagoon?

Solution: Assume the following numbers and sizes of animals are housed:

100 sows, ave. wt. 160 kg. (350 lb.)
6 boars, ave. wt. 180 kg. (400 lb.)
10 replacements, ave. wt. 115 kg. (250 lb.)
750 pigs, ave. wt. 55 kg. (120 lb.)

Total animal weight is 59,480 kg. (129,900 lb.)

From Table 2, volatile solids (VS) production rate is 4.8 g per day per kg. animal weight (lb/day/1000 lb.)

Total daily VS production = 4.8 × 59,480
= 285,500 g (624 lb)

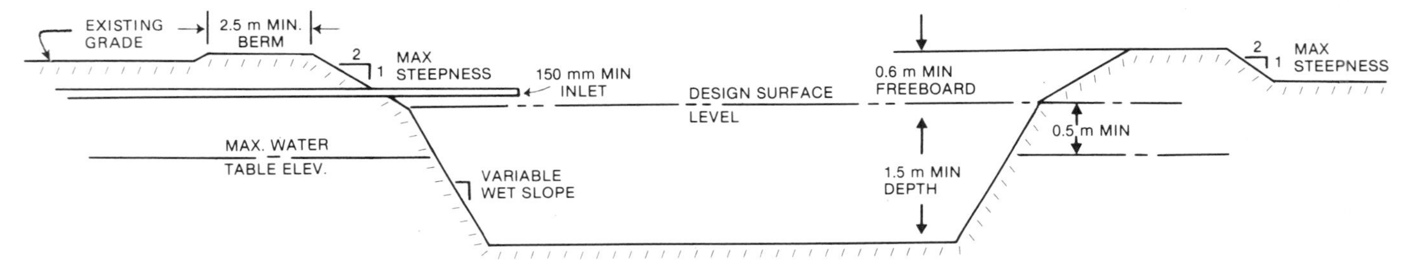

FIG. 1—ANAEROBIC LAGOON CONSTRUCTION DETAILS

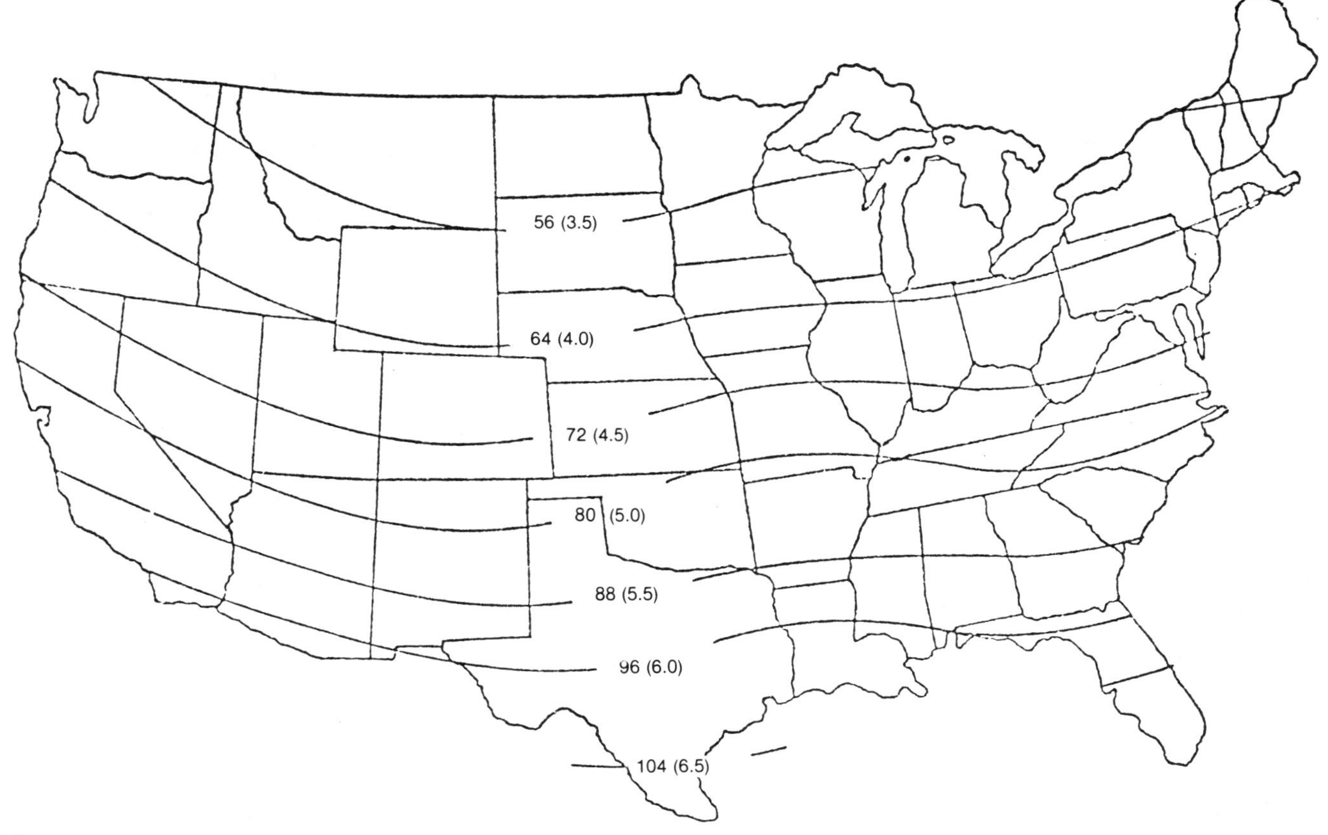

FIG. 2—RECOMMENDED MAXIMUM LOADING RATES FOR ANAEROBIC LAGOONS FOR ANIMAL WASTE IN g VS/day/m³ (lb VS/day/1000 ft³)

From Fig. 2, the recommended maximum loading rate for an anaerobic lagoon in central South Carolina is 84 g VS/day/m³ (5.25 lb VS/day/1000 ft³). Table 1 shows the loading rate multiplier for swine is 1.0.

$$\text{Design volume} = \frac{285{,}500}{84(1.0)}$$

$$= 3400 \text{ m}^3 \text{ (118,800 ft}^3\text{)}$$

Surface area and depth will be determined by site characteristics.

This example uses the maximum recommended loading rate. A lesser loading rate, which would result in a larger lagoon, may be preferable where odor reduction and greater storage capacity is desirable.

4.4.2 Depth. For installations of less than 300 m³, design depth should be at least 1.5 m (5 ft); greater depths are desirable (see paragraph 4.5.2). Where soil and groundwater conditions permit, depths of 6 m (20 ft) or more may be used. The design surface level should be at least 0.5 m (1.5 ft) above the highest level of the groundwater table.

4.4.3 Length-to-Width Ratio. Desirable length-to-width ratios vary according to detention time. Lagoons that are roughly round or square are favored for detention periods of 100 days or more or where lagoons are pumped down annually. Length-to-width ratios up to 5 to 1 may be used where detention periods are 30 days or less and where overflow to another lagoon cell or other treatment facility occurs daily.

4.4.4 Detention Time. Detention periods of several hundred days are common where influent is limited to animal waste and small amounts of water, where evaporation exceeds precipitation, and/or where recycling of supernatant is practiced. Long detention times can lead to buildup of dissolved salts such as Cl^-, NH_4^+, K^+, and Na^+, which inhibit biological action. When total salt concentration of the supernatant is in the range of 2000 to 3000 mg/l the lagoon should be pumped down, and if necessary, fresh water added (see paragraph 4.8).

4.4.4.1 Short detention periods may result where hydraulic loading rates are high and where large amounts of wash water enter the lagoon. Minimum detention periods of 15 days are recommended for single cell (single impoundment) lagoons.

4.4.5 Multiple Cell Lagoons. Multiple cell lagoons may be used when required or allowed by local conditions and/or regulation.

4.4.5.1 When cells operate in series, the initial cell receives the waste; the loading rate should not exceed 125% of the recommended maximum loading rate described in paragraph 4.4.1. Initial cells heavily overloaded in series operation result in severe odor problems.

TABLE 1—LOAD RATE MULTIPLIER FOR ANIMAL SPECIES AND RATION

Animal species	Ration	Loading rate multiplier
Swine	high concentrate	1.0
Dairy	high forage	0.75
Poultry	high concentrate	0.8
Beef	high concentrate	1.0
	high forage	0.75
Sheep	high concentrate	0.9
Horse	high forage	0.75

TABLE 2—DAILY VOLATILE SOLIDS PRODUCTION BY LIVESTOCK

Animal type	Volatile solids produced, per day per kg of animal (lb/day/1000 lb)
Swine	4.8
Dairy	8.6
Poultry	
Layer	9.4
Broiler	12.0
Beef	
Feeder	5.9
Sheep	8.5
Horse	7.5

4.4.5.2 Where multiple cells are operated in parallel, their volume should be determined according to paragraph 4.4.1.

4.5 Earth Embankment

4.5.1 Top Width. Minimum top width should be 2.5 m (8 ft.)

4.5.2 Side Slopes. Side slope on the dry side of the embankment should not be steeper than 2:1. Wet side slope should not be steeper than 2:1 above the design surface level. However, slopes to be moved should not be steeper than 3:1. Herbicides should be used on slopes steeper than 3:1. Below the design level, side slope should be as steep as possible without sacrificing soil stability. Deep lagoons and steep wet slopes reduce convection heat loss, enhance internal mixing, reduce odor emission, promote anaerobic conditions, minimize weed growth problems, reduce excavation cost, and reduce mosquito production.

4.5.3 Settlement. Embankment elevation should be increased at least 5 percent during construction to allow for settling.

4.5.4 Freeboard. The top of the settled embankment should be at least 0.6 m (2 ft) above the maximum design surface level. Where overflow is permitted, a spillway with a minimum capacity of 1.5 times the peak daily inflow rate may be installed, 0.3 m (1 ft) above the design level, to protect the embankment in the case of catastrophic or chronic rainfall events or failure of the overflow control system.

4.6 Inlet and Outlet

4.6.1 Inlet. Inlet devices should discharge the waste into the lagoon at a point beyond the cut slope of the embankment. Pipes or open channels may be used. The diameter or least dimension of the inlet should be at least 150 mm (6 in.) or at least 200 mm (8 in.) for dairy cattle waste. Inlets may be located above or below the lagoon surface. Submerged inlets may plug unless used at least twice daily. Trickle flow should be eliminated during freezing weather. Access to the inlet device should be provided for cleaning.

4.6.2 Outlet. An overflow device with a minimum capacity of 1.5 times the peak daily inflow rate may be installed at the lagoon surface level *only if* the overflow is to be contained in another lagoon cell or other treatment facility. Outlet devices should be installed in a manner that allows effluent to be taken at a level 150 - 450 mm (6 - 18 in) below the surface.

4.7 Effluent or Overflow

4.7.1 Discharge. Effluent from an anaerobic lagoon may not be discharged to streams, lakes, or other waterways, or be allowed to run off the owner's property. Lagoon overflow into a subsequent storage or treatment cell is permissible.

4.7.2 Handling. Lagoon supernatant and sludge are typically spread on land. Sprinkler or surface irrigation, tank spreaders, or other approved distribution procedures are often used. Supernatant should be applied to the land without producing surface runoff; in such amounts that undesirable levels of nutrients or toxic materials do not accumulate in the soil, plants, soil water, groundwater, or runoff; and with procedures that minimize odors. Continuous flow to a surface irrigation system should be avoided; biweekly or weekly dosing is preferable. Removal frequency may be dictated by availability of labor or a suitable application site, cropping schedules, rainfall or soil conditions.

4.8 Water Supply. Adequate water should be available to establish and maintain the minimum operating level of the lagoon at 60 percent of the design depth. Surface water may be diverted to the lagoon during the start-up period, but must be diverted away after the design depth has been established. The water supply may also be used to dilute lagoon contents and limit salt build-up.

4.9 Safety. A stock-tight fence should enclose the lagoon and warning signs posted to prevent children and others from using the lagoon for other purposes. Fences should be constructed to protect the embankment and to permit access for lagoon maintenance. See ASAE Engineering Practice EP250, Specifications for Farm Fence Construction.

4.10 Vegetation. Trees and shrubs should be cleared for a distance of at least 9 m (30 ft) surrounding the lagoon. This area and dry embankments should be maintained in grass or other low-growing vegetation for erosion prevention and appearance.

SECTION 5—OPERATION AND MAINTENANCE

5.1 Startup. The water level should be at least 60 percent of the design depth before waste is added. Such water may come from nearby lakes, streams, ponds, or wells, or directed surface runoff or roof drainage. Where possible, waste should be added slowly at first and increased over a period of 2 to 4 months to the design loading rate. Start-up in warm weather is more desirable than in cold weather, particularly in northern states.

5.2 Depth. A water depth of at least 60 percent of design depth should be maintained to minimize odor production. Daily loading is more satisfactory than less frequent loading. Manure solids should be covered by liquid at all times except as noted below.

5.3 Crust. Floating solids, especially where forage is fed, commonly cause crusting. The crust reduces odor emission and helps maintain anaerobic conditions and constant temperature.

5.4 Sludge Removal. Removal of lagoon sludge after many years of operation should be anticipated. Procedures and equipment for management of the sludge are beyond the scope of this document.

5.5 Inspection. Lagoons should be inspected periodically, particularly with regard to operating level and control of overflow. Maintenance is necessary to control rodent damage and growth of vegetation on embankments and surrounding areas.

Cited Standards:

ASAE EP250, Specifications for Farm Fence Construction
ASAE EP379, Control of Manure Odors
ASAE D384, Manure Production and Characteristics

HEATING, VENTILATING AND COOLING GREENHOUSES

Developed by the Greenhouse Engineering Practices Subcommittee of the ASAE Environment of Plant Structures Committee; approved by the ASAE Structures and Environment Division Standards Committee; adopted by ASAE April 1981; reconfirmed December 1985, December 1986; reconfirmed for one year December 1987.

SECTION 1—PURPOSE AND SCOPE

1.1 This Engineering Practice presents design information for heating, ventilating and cooling greenhouses. Both winter and summer ventilation problems are covered. Generally accepted methods of heating, ventilating and cooling are presented and the important design features of typical systems are indicated.

SECTION 2—DEFINITIONS

2.1 Heating: The addition of heat to the interior of the greenhouse from any energy source including the sun.

2.2 Ventilation: The process of exchanging air inside the greenhouse with outside air to control temperature, humidity, oxygen, and carbon dioxide levels.

2.2.1 Ventilation rate: The volume of air exchanged per unit of time per unit floor area. It is measured as cubic meters of air per second per square meter of greenhouse floor area because the heat load derives from solar radiation and is directly proportional to floor area. Ventilation rate is sometimes expressed as internal air volume change per unit of time.

2.3 Cooling: The removal of heat from the interior of the greenhouse.

2.4 Air circulation: The process of moving or mixing air within a greenhouse to control temperature, humidity and carbon dioxide distribution.

2.5 Infiltration: The air exchange which occurs through small, uncontrolled openings in the greenhouse due to air pressure and temperature differentials inside and outside the greenhouse.

2.6 Natural Ventilation: Desirable air exchange which occurs through controlled openings due to natural pressure variations inside and outside the greenhouse. Wind creates pressure gradients and solar energy creates vertical temperature gradients which facilitate natural ventilation.

2.7 Mechanical Ventilation: Desirable air exchange which occurs through controlled openings when moving air in and out of the greenhouse by means of fans. Fans may be located at either the air inlet (pressure ventilation) or the air exhaust (exhaust ventilation).

2.8 Horizontal Air Circulation: A circulation system utilizing propeller fans to generate a rotational air circulation pattern in large greenhouses.

2.9 Ridge and Furrow: A method of greenhouse construction where modular roof units are connected by gutters to cover large ground areas.

2.10 Aspirate: To circulate air continuously across an object such as a thermostat.

SECTION 3—PRINCIPLES OF HEATING AND HEAT LOSS DETERMINATION

3.1 Most undesired heat loss from a greenhouse occurs by conduction (Q_c) and infiltration (Q_i). The required size of a heating unit is determined by calculating the sum of Q_c plus Q_i.

3.2 The greenhouse heat loss by conduction, Q_c, (Watts) is determined by:

$$Q_c = UA_e\,(t_i\text{-}t_o)$$

where

U = overall heat transfer coefficient, W/m² °C (see Table 1)
A_e = exposed surface area, m²
t_i = inside temperature, °C
t_o = outside temperature, °C

TABLE 1—APPROXIMATIONS OF HEAT TRANSFER COEFFICIENTS FOR GREENHOUSE GLAZING METHODS AND MATERIALS

Greenhouse covering	Value of U W/m² °C	Value of U Btu/hr °F ft²
Single glass (sealed)	6.3	1.1
Single plastic	6.8	1.2
Single fiberglass	6.8	1.2
Double plastic, polyethylene	4.0	0.7
Rigid double wall acrylic	3.0	0.5
Double glass (sealed)	3.0	0.5
Double plastic over glass	3.0	0.5
Single glass and thermal blanket	3.0	0.5
Double plastic and thermal blanket	2.5	0.4
Double plastic, poly-pellets*	0.3	0.05

*A 130 mm thick layer of polystyrene pellets between a double plastic glazing.

3.3 Greenhouse heat loss by infiltration, Q_i, (Watts) can be estimated (assuming limited net moisture transfer) by:

$$Q_i = 0.5\ VN\ t_i\text{-}t_o)$$

where

Q_i = heat loss by infiltration, W
V = greenhouse internal volume, m³
N = number of air exchanges per hour (see Table 2)
t_i = inside temperature, °C
t_o = outside temperature, °C

3.4 The recommended heat transfer coefficient is nearly constant for greenhouses covered with a tight single glazing of glass, fiberglass or plastic (see Table 1).

3.5 The recommended heat transmission coefficient is nearly constant for greenhouses covered with any two glazings 20-100 mm (3/4-4 in.) apart (see Table 1).

3.6 Natural air exchanges can be very low during subfreezing conditions for all greenhouses since the leaks become sealed with frozen condensate.

TABLE 2—NATURAL AIR EXCHANGES FOR GREENHOUSES

Construction system	Air exchanges per hour*
New construction, glass or fiberglass	0.75-1.5
New construction, double layer plastic film	0.5 -1.0
Old construction, glass, good maintenance	1-2
Old construction, glass, poor condition	2-4

*Low wind or protection from wind reduces the air exchange rate. Values should be 0.5 or less for subfreezing outside temperatures since freezing condensate can seal small cracks.

3.7 Outside design temperatures for heating are given for various locations in the American Society of Heating, Refrigerating, and Air-Conditioning Engineers (ASHRAE) Handbook and Product Directory — Fundamentals volume. The lowest temperature for 99% of the time is recommended. A warmer temperature may be used for greenhouses intended for only early spring production such as bedding plant production. Consult local weather data where available.

3.8 Over half of all heat losses occur at night. Base maximum design heating load on 15°C (60°F) inside night temperature to meet the needs of most greenhouse plants, unless otherwise specified. Daytime thermostat settings are usually 15-10°C (60-70°F). The recommended daytime temperature is typically 5°C (10°F) higher on bright sunny days and 2.5°C (5°F) higher on semi-cloudy days. Solar radiation will usually provide the day temperature increase naturally.

3.9 Calculate heat loss through concrete or masonry walls by conventional means such as described in the ASHRAE Handbook and Product Directory—Fundamentals volume.

SECTION 4—HEATING AND AIR CIRCULATION SYSTEMS

4.1 Central heating

4.1.1 Central heating systems include standard black pipe containing hot water or steam, finned pipe containing hot water or steam, hot air distributed in sheet metal or plastic tubing, and electrical resistance strips.

4.1.2 The total length of radiation pipe is determined by the heating requirements, heating fluid, and pipe characteristics. Consult the ASHRAE Handbook and Product Directory—Fundamentals volume, chapter on pipe sizing, or manufacturer's data for radiation pipe design.

4.1.3 Greenhouses 9 m (30 ft) or less in width may be heated with pipes placed only along the sidewalls.

4.1.4 For single span houses wider than 9 m (30 ft), calculate radiation pipes along the side wall equivalent to the side wall heat loss and calculate the remaining piping on the basis of roof loss. Install pipes at least 300 mm (12 in.) above the top of crops grown on the ground. If benches are used in the house, size the pipe to provide a run of pipe under each bench.

4.1.5 Place piping in narrow-span ridge and furrow houses along the outside sidewalls and below the gutters between house sections.

4.1.6 Radiation piping may be placed below benches or directly above plant bed surfaces for root zone heating.

4.1.7 Naturally induced air circulation generally provides uniform temperatures at plant height. If air circulation fans are desired, they should be installed to aid the naturally induced circulation patterns. Overhead circulating fans which pull air upward and discharge it radially outward do not effectively improve temperature uniformity in houses with sidewall pipe heating systems.

4.1.8 All boilers must have appropriate valves, controls, and safety devices and must be installed in accordance with all applicable codes.

4.2 Spaced unit heaters

4.2.1 Downward discharge unit heaters placed overhead are not recommended because they do not maintain sufficiently uniform temperatures in the crop zone.

4.2.2 Horizontal discharge unit heaters designed specifically for greenhouse heating can be used without additional circulation fans for low growing crops on benches or in ground beds. There will be, however, significant temperature variations within the greenhouse.

4.2.3 Where more uniform temperatures are desired, transparent ducts can be used to distribute heat along the length of the greenhouse from horizontal discharge unit heaters (see paragraph 4.3).

4.2.4 If ducts are not used, circulation fans on unit heaters should be operated continuously for better temperature uniformity.

4.2.5 Two heaters with continuously operating circulation fans should be used for houses 20 m (66 ft) or less in length. Heaters should be located in opposite corners discharging parallel to the sidewalls.

4.2.6 For houses longer than 20 m (66 ft) and less than 40 m (130 ft) the heater fans may not be sufficient to develop a continuous horizontal air circulation pattern. In this case, it is recommended that two additional fans or heaters be provided at the middle of the greenhouse, one on each side of the greenhouse to aid in developing a horizontal air circulation pattern (see paragraph 4.4.)

4.2.7 Provide direct-fired unit-heaters with ducting to an outside source of combustion air. Design the duct to have a cross-section of 100 mm²/400 kJ (1 in.²/2500 Btu) heater capacity.

4.2.8 Vent all gas or oil fueled heaters outside according to manufacturer's directions and equip the stack with a weather cap.

4.2.9 Unvented gas or oil combustion units designed for carbon dioxide generation may serve as an emergency heat source. Do not use unvented direct-fired heating units for primary greenhouse heating as some byproducts of combustion are toxic to plants and people.

4.3 Perforated tube air distribution

4.3.1 Overhead perforated plastic tubes in combustion with unit heaters are used for greenhouse heating. Pairs of tube discharge holes are typically spaced 300-1000 mm (12-36 in.) apart depending on tube diameter and length.

4.3.2 Size the tubes to have an air entrance velocity of 5 m/s (17 ft/s) and an outlet velocity of 3 m/s (10 ft/s).

4.3.3 Arrange discharge holes so that air is not blown directly onto plants (see paragraph 4.4.5).

4.3.4 One tube is sufficient for greenhouses 9 m (30 ft) or less in width. Two or more tubes are necessary in wider greenhouses.

4.3.5 Circulate 1/4-1/3 of the air volume in the greenhouse through the tube each minute to obtain uniform heat distribution.

4.3.6 Circulation tubes should not exceed 50 m (160 ft) in length. Best distribution occurs when the length is 30 m (100 ft) or less.

4.3.7 Warm air discharge from unit heaters should be directed behind the plastic tube air circulation fan. The capacity of the air circulation fan must be equal to or greater than the heater fan capacity.

4.3.8 Run air circulation fans continuously during the winter heating season.

4.4 Horizontal air circulation

4.4.1 In ridge and furrow greenhouses, air can be moved down through one greenhouse and back through another. Use large diameter, low-horsepower propeller fans for most economical operation. Locate fans over the crop in each house.

4.4.2 In single greenhouses, mount propeller fans a distance of 1/4 the greenhouse width from the side walls. Air should move parallel to the side walls and be directed down along one side and back along the other side.

4.4.3 Mount the fans perpendicular to the ground and at least 0.5-1.0 m (2-3 ft) above the plants.

4.4.4 Place propeller fans along the direction of air movement at spacings of less than 30 times the fan diameter and 5 to 6 m (15-20 ft) from end walls.

4.4.5 The velocity across the plants should be no greater than 1.0 m/s (3.3 ft/s).

4.4.6 Select fans to provide airflow of 0.9-1.1 m³/min·m² (3-4 cfm/ft²) of floor area.

4.5 Floor heating

4.5.1 Floor heating is considered a good practice if plant containers can be set directly on the floor. It can provide up to 25% of the heat requirement for a double glazed structure.

4.5.2 Use either loose gravel or porous concrete (concrete without sand) for a 100 mm (4 in.) thick floor. If a water table exists within 2 m (6 ft) of the soil surface, insulate below the floor with 50 mm (2 in.) polystyrene board.

4.5.3 A typical installation circulates 35-40°C (95-105°F) water through 20 mm (3/4 in.) or larger diameter plastic pipe. Pipes are placed on the concrete or gravel.

4.5.4 For a greenhouse at 15°C (60°F) night temperature, 35°C (95°F) water circulated at 0.6-0.9 m/s (2-3 ft/s) through pipes spaced 300 mm (12 in.) on centers below a porous concrete floor, will provide 64 W/m² (20 Btu/hr·ft²) floor area, with no plants on the concrete.

4.5.5 Covering the floor with plants will decrease the floor heat transfer rate up to 25%.

4.6 Control selection and installation

4.6.1 Select controls capable of withstanding extremely humid and dusty conditions.

4.6.2 Control sensors should be continuously aspirated at 3 to 5 m/s (10-16 ft/s) to ensure air movement across the sensors to reduce temperature fluctuations and save heat. Aspirator fans should have totally enclosed motors.

4.6.3 The control sensors should be fully shaded from direct solar radiation. The shade should be made of a material having a low thermal conductivity and high reflectivity such as white painted wood. Wood should have a minimum thickness of 13 mm (0.5 in.).

4.6.4 Locate controls near the center of the greenhouse and near the plant growing area. The location should be representative of the plant requirements and should not be abnormally affected by heat ducts, ventilators, or sidewalls.

SECTION 5—NATURAL VENTILATING SYSTEMS

5.1 A pressure difference must be created by wind or temperature gradients for natural ventilation to occur. To take advantage of wind-created pressure differences, vent openings on both sides and ridge of a greenhouse are recommended. Greenhouses with only side vents depend on wind pressure to force air exchange and are usually ineffective.

5.2 Large vent openings will provide the most ventilation. The total vent area should be 15-25% of the floor area.

5.3 For single unit greenhouses, the combined sidewall vent area should be the same as the combined roof vent area.

5.4 Ridge vents should be top hinged and should run continuously the full length of the greenhouse.

5.5 Top hinged ridge vents should form a 60° angle with the roof when fully opened.

5.6 Automatic vent systems should be equipped with rain and high wind sensors to prevent crop and ventilator damage.

SECTION 6—FAN VENTILATING AND COOLING SYSTEMS

6.1 Basic considerations

6.1.1 The ventilation requirement of a greenhouse can be determined by considering the source of heating and cooling and writing an energy balance.

6.1.2 A maximum ventilation rate of 3/4-1 air change per minute is recommended for most greenhouses. Temperature rise from air inlet to exhaust is inversely proportional to airflow. During full sunlight, a 3/4 air change will result in a 5.5-6.5°C (10-12°F) temperature rise, whereas one air change will result in a 4.5-5.5°C (8-10°F) rise.

6.1.3 During winter, incoming air should be introduced into the house so that it mixes and is warmed by interior air before contacting the plants. Perforated distribution tubes are often used.

6.1.4 The maximum design ventilation rate is sometimes increased for elevations over 600 m (2000 ft) and light levels over 54 klx (5000 fc). Little information is available to substantiate the need for this modification.

6.1.5 Increase air velocities in greenhouses with air inlet to fan distances of less than 30 m (100 ft) by the factor:

$$F = \frac{5.5}{\sqrt{D}} \quad \left(F = \frac{10}{\sqrt{D}}\right)$$

where

D = inlet-to-fan distance in meters
D′ = inlet-to-fan distance in feet

6.1.6 Ventilation fans should deliver the required ventilation capacity at 0.03 kpa 0.12 in. H_2O) static pressure when all guards and louvers are in place unless specific design requirements require air delivery at higher static pressures.

6.1.7 Space fans no more than 7.5 m (25 ft) apart.

6.1.8 When possible, locate the fans on the downwind side or end of the greenhouse. If the fans must be located on the windward side, increase the design ventilation capacity at least 10%.

6.1.9 Maintain a clearance between the fan discharge and any obstruction of at least 1.5 times the fan diameter. Fans may be mounted in the roof if obstructions interfere with other mountings.

6.1.10 Exhaust fans should have freely operating pressure louvers on their exhaust side to prevent unwanted air exchange when fans are not operating.

6.1.11 Air intake louvers or shutters should open outward. They should be motorized and wired into the fan control circuit to open during fan operation.

6.1.12 Intake louver area should be at least 1.25 times the area of the fans.

6.1.13 Guard fans to prevent accidents. Use manufactured guards or install a woven wire mesh screen of at least 1.5 mm (16 gage) wire and 13 mm (0.5 in.) openings, for placement within 100 mm (4 in.) of moving parts. Guards more than 100 mm (4 in.) from moving parts may be made with woven wire mesh of 2.7 mm (12 gage) wire, and 50 mm (2 in.) openings.

6.1.14 Fans that have been tested and rated according to Air Movement and Control Association, Inc. (AMCA) Standard 210, Laboratory Methods of Testing Fans for Rating Purposes, and which bear the seal fo the AMCA Certified Rating Program, are recommended.

6.2 Pad and fan cooling

6.2.1 Size the ventilation system as recommended in paragraph 6.1.

6.2.2 The preferred pad-to-fan distance is 30-50 meters (100-150 ft). For very long houses, installing fans in the roof at the midpoint and pads at both ends will help to reduce air velocities across the plants, although a stagnant, hot spot will sometimes form directly under the fans.

6.2.3 The optimum air face velocity through the cooling pad is listed in Table 3 below.

TABLE 3—RECOMMENDED AIR VELOCITY THROUGH VARIOUS PAD MATERIALS

Type	Air face velocity through pad* m/s	ft/s
Aspen fiber mounted vertically 50-100 mm (2-4 in) thick	0.75	2.5
Aspen fiber mounted horizontally 50-100 mm (2-4 in) thick	1.0	3.3
Corrugated cellulose, 100 mm (4 in) thick	1.25	4.2
Corrugated cellulose, 150 mm (6 in) thick	1.75	5.8

*Velocity may be increased by 25% where construction is limiting.

6.2.4 Pads should cool the air to within 2°C (3.5°F) of the wet bulb temperature at a pressure loss not exceeding 0.015 kpa (0.06 in. H_2O).

6.2.5 The pad is normally run continuosuly along the side or end of the house opposite the ventilation fans. Vertical pad height should not exceed 2.5 m (8 ft) nor be less than 0.5 m (2 ft) for uniform water flow.

6.2.6 Aspen pads can be mounted vertically or horizontally. Vertical pads must be well secured to prevent sagging. Pads should be easy to install and replace.

6.2.7 Construct any air inlet so it may be readily covered without removing the pads.

6.2.8 When possible, locate pads on the prevailing wind side of the greenhouse. Pad location is not as important when the greenhouse is sheltered from prevailing winds by another building or greenhouse located within 7.5 m (25 ft).

6.2.9 Keep fans from exhausting directly into pads on an adjacent greenhouse, unless they are separated by at least 15 m (50 ft).

6.2.10 Offset adjacent greenhouse fans that face each other so the exahusting air from one fan will not blow directly against that of another fan. Offsetting is not necessary if facing fans are more than 4 fan diameters apart.

6.2.11 Cooled air tends to sink, so vertical baffles above the plant growing area are not necessary or recommended. Sometimes baffles covering the lower two-thirds of the area under benches are installed for better cooling at bench height.

6.2.12 A horizontal pad can be irrigated at a rate close to the cooling system evaporative requirements. The maximum recommended flow rate is 0.2 L/s·m² (0.3 gpm/ft²) of pad area. Lower rates can be achieved by intermittent operation of the pad irrigation system.

6.2.13 Recommended minimum water flow rates through vertically mounted pads and sump capacities are listed in Table 4.

6.2.14 Screen the water returned to the pump to filter out pad fibers and other debris. A 50 mesh inclined screen mounted below the return flow is effective. Cover the sump to protect from insects and other debris. Install removable caps or valves on the ends of the water distribution pipes to allow periodic flushing.

6.2.15 As water evaporates, the salt concentration is increased. In areas that have water with high mineral content a bleed-off system is necessary to prevent mineral precipitation in the pad. A continuous water bleed-off rate of 0.02 L/min·m³·s (0.05 gpm per 1000 cfm) of airflow will limit salts to about two concentrations.

6.2.16 If is preferable to protect the pad assembly by installing it inside any air inlet openings. The air inlet opening need not be continuous but should be uniformly distributed.

6.2.17 If the pad assembly is located outside an inlet vent, the opening should be continuous, have no large obstructions and be centered in relation to the pad. Maximum design air velocity through the vent is 1.8 m/s (5.9 ft/s).

6.2.18 When the height of the pad exceeds that of the air inlet, set the pad back from the vent a minimum distance of half the amount of the height difference.

6.3 Winter mechanical ventilation

6.3.1 The winter ventilation system must introduce small air quantities for necessary cooling without chilling any plants. Typically 10-20% of the summer ventilation requirement, or 0.005-0.01 m³/s·m² (1-2 cfm/ft²) of floor area is all that is necessary.

TABLE 4—RECOMMENDED WATER FLOW AND SUMP CAPACITY FOR VERTICALLY MOUNTED COOLING PAD MATERIALS

Pad type and thickness	Minimum water rate per lineal length of pad		Minimum sump capacity per unit pad area	
	L/min·m	gpm/ft	L/m²	gal/ft²
Aspen fiber 50-100 mm (2-4 in.)	4	0.3	20	0.5
Aspen fiber, desert conditions 50-100 mm (2-4 in.)	5	0.4	20	0.5
Corrugated cellulose, 100 mm (4 in.)	6	0.5	30	0.8
Corrugated cellulose, 150 mm (6 in.)	10	0.8	40	1.0

6.3.2 Winter ventilation is often combined with air circulation systems discussed in paragraphs 4.3 and 4.4.

6.3.3 Exhaust fans can be operated at low speed to bring outside air through a small shutter and into a perforated distribution tube.

6.4 Automatic controls

6.4.1 It is preferable to control ventilation volume in at least three stages. With fan ventilation, two speed motors can be used or increasing numbers of fans can be activated with increasing cooling requirements.

6.4.2 Set the first stage of ventilation 3°-6°C (5-10°F) above the heat setting to keep both systems from operating at the same time.

6.4.3 Humidistats are not recommended for greenhouse control because they require frequent calibration in high humidities. If humidistats are used, they should be aspirated at an air velocity of 3-5 m/s (10-16 ft/s).

6.4.4 Shield and locate thermostats and other measuring devices according to paragraph 4.6.

6.4.5 Air inlet vents should be controlled with the same thermostat that activates the fan system.

6.4.6 Line voltage controls normally are used on small installations with three or four exhaust fans. On large installations consider low-voltage controls for safety, improved sensitivity and the reduction of wiring costs.

6.4.7 Wire a manual control switch in parallel with each control stage to permit manual control when desired.

6.4.8 Install a safety disconnect switch near each fan and pump. Perform all wiring and electrical equipment installations according to applicable codes.

6.5 Cooling and ventilating small greenhouses

6.5.1 Small greenhouses [less than 30 m² (300 ft²) floor area] may be naturally ventilated by opening ridge vents and either doors or sidewall vents. If mechanical ventilation and evaporative cooling is installed, use the following design criteria:

Summer design ventilation rate per unit of floor area 0.06 m³/s·m² (12 cfm/ft²)

Package evaporative cooler fan capacity per unit of floor area 0.08 m³/s·m² (15 cfm/ft²)

6.5.2 Small greenhouses [less than 30 m² (300 ft²) floor area] can be cooled and ventilated with a package evaporative cooler for less cost and greater operating convenience than with a fan-and-pad system. Provide an exhaust opening, such as an automatic shutter or door on the opposite end of the house or a cracked-open ridge ventilator.

6.5.4 Maximum winter ventilation rate is about one-half the maximum summer ventilation requirement and will be greatest for lean-to houses attached to the east, south, or west wall of a building. Two speed fans or natural air flow are often used for ventilation.

Cited Standard:

AMCA 210, Laboratory Methods of Testing Fans for Rating Purposes

ASAE Engineering Practice: ASAE EP411.1

GUIDELINES FOR MEASURING AND REPORTING ENVIRONMENTAL PARAMETERS FOR PLANT EXPERIMENTS IN GROWTH CHAMBERS

Developed by the ASAE Environment of Plant Structures Committee; approved by the ASAE Structures and Environment Standards Committee; adopted by ASAE March 1982; revised March 1986.

SECTION 1—PURPOSE AND SCOPE

1.1 The purpose of this Engineering Practice is to set forth guidelines for the measurement of environmental parameters that characterize the aerial and root environment in a plant growth chamber.

1.2 This Engineering Practice establishes criteria that will promote a common basis for environmental measurements for the research community and the commercial plant producer.

1.3 This Engineering Practice promotes uniformity and accuracy in reporting data and results in the course of conducting plant experiments.

SECTION 2—INTRODUCTION

2.1 The aerial environment is characterized by the following parameters: air temperature, atmospheric composition including moisture and carbon dioxide concentration, air velocity, radiation, and the edge effects of wall/floor on these parameters.

2.2 The root environment is characterized by the following parameters: medium composition and quantity, nutrient concentrations, water content, temperature, pH, and electrical conductivity.

2.3 Measuring and reporting these various parameters will be covered in the sections that follow. The definitions of the parameters indicate the symbol and units, (symbol, units). Measurements should be made which accurately represent the mean and range of the environmental parameters to which the plants are exposed during the experimental period, to indicate the temporal variations, both cyclic and transient, and the spatial variations over the separate plants in the chamber.

2.4 The definitions, measurement techniques and reporting procedures provide criteria and promote uniformity in measuring and reporting environmental parameters, but these guidelines should not be used to select the environmental parameters applicable to a particular experiment. Other parameters may be applicable to special environments such as oxygen concentration in hydroponics, pollutant concentration in air quality research, and spectral quality ratios in photobiology.

2.5 When measurements are made, the chamber should be operating with containers and plants located in the chamber. Provision should be made to take all measurements with minimum disturbance to the operating environment.

SECTION 3—DEFINITIONS

3.1 Radiation: The emission and propagation of electromagnetic waves or particles through space or matter.

3.1.1 Radiant energy (Q_e, J): The transfer of energy by radiation.

3.1.2 Energy flow rate (ϕ_e, W): The rate of flow of energy, a fundamental radiometric unit; also called radiant power.

3.1.3 Spectral energy flow rate (ϕ_e,λ, $W \cdot nm^{-1}$): The radiant energy flow rate per unit wavelength interval at wavelength λ.

3.1.4 Energy flux (E_e, $W \cdot m^{-2}$): Radiant energy flow rate per unit plane surface area; also called irradiance and energy flux density, the latter to emphasize the unit area.

3.1.5 Spectral energy flux (E_e, λ, $W \cdot m^{-2} \cdot nm^{-1}$): The radiant energy flux per unit wavelength interval at wavelength λ.

3.1.6 Energy fluence (F_e, $J \cdot m^{-2}$): The radiant energy dose time integral per unit area based on the cross-sectional area of a spherical surface in totally diffuse radiation.

3.1.7 Spectral energy fluence (F_e,λ, $J \cdot m^{-2} \cdot nm^{-1}$): The energy fluence per unit wavelength interval at wavelength λ.

3.1.8 Energy fluence rate ($F_{e,t}$, $W \cdot m^{-2}$): The radiant energy fluence per unit time. The same as radiant energy flux (irradiance) for normal incidence radiation.

3.1.9 Spectral energy fluence rate ($F_{e,t},\lambda$ $W \cdot m^{-2} \cdot nm^{-1}$): The radiant energy fluence rate per unit wavelength interval at wavelength λ.

3.1.10 Photon (q): A quantum (the smallest, discrete particle) of electromagnetic energy with an energy of hc/λ (h = Planck's Constant; c = speed of light; λ = wavelength). Its energy is expressed in joules (J).

3.1.11 Photon flow rate (ϕ_p, $q \cdot s^{-1}$ or $mol \cdot s^{-1}$): The rate of flow of photons.

3.1.12 Photon flux (E_p, $q \cdot s^{-1} \cdot m^{-2}$ or $mol \cdot s^{-1} \cdot m^{-2}$): The photon flow rate per unit plane surface area; also called photon flux density to emphasize the unit area.

3.1.13 Spectral photon flux (E_p,λ, $q \cdot s^{-1} \cdot m^{-2} \cdot nm^{-1}$ or $mol \cdot s^{-1} \cdot m^{-2} \cdot nm^{-1}$): The photon flux per unit wavelength interval at wavelength λ.

3.1.14 Photon fluence (F_p, $q \cdot m^{-2}$ or $mol \cdot m^{-2}$): The photon dose time integral per unit area based on the cross-sectional area of a spherical surface in total diffuse radiation.

3.1.15 Photon fluence rate ($F_{p,t}$, $q \cdot m^{-2} \cdot s^{-1}$ or $mol \cdot m^{-2} \cdot s^{-1}$): The photon fluence per unit time. The same as photon flux for normal incidence radiation.

3.1.16 Spectral photon fluence rate ($F_{p,t},\lambda$, $q \cdot m^{-2} \cdot s^{-1} \cdot nm^{-1}$ or $mol \cdot m^{-2} \cdot s^{-1} \cdot nm^{-1}$): The photon fluence rate per unit wavelength interval at wavelength λ.

3.1.17 Spectral distribution: A functional or graphic expression of the relation between the spectral energy flux, spectral photon flux, or fluence rate per unit wavelength, and wavelength.

3.1.18 Photosynthetically active radiation (PAR, $q \cdot s^{-1} \cdot m^{-2}$ or $W \cdot m^{-2}$): The total radiation in the wavelength range of 400-700 nm contributing to photosynthetic productivity in relation to the relative photosynthetic quantum efficiency of the radiation. Measured as the photosynthetic photon flux (PPF) in average quanta$\cdot s^{-1} \cdot m^{-2}$ or μmoles of quanta$\cdot s^{-1} \cdot m^{-2}$, or photosynthetic irradiance (PI) in $W \cdot m^{-2}$ for the specified waveband, λ_1-λ_2 (400-700 nm).

3.1.19 Photomorphogenic radiation (PMR, $q \cdot s^{-1} \cdot m^{-2}$ or $W \cdot m^{-2}$): The radiation with wavelengths approximately ranging between 380-800 nm contributing to photomorphogenic responses (i.e., -flowering, reproduction, elongation, dormancy) in relation to the relative quantum efficiency of the spectral quality of the radiation in several discrete spectral regions. Measured as the photon flux in average quanta$\cdot s^{-1} \cdot m^{-2}$, μmoles of quanta$\cdot s^{-1} \cdot m^{-2}$ or in energy flux in $W \cdot m^{-2}$ for the specified waveband, λ_1-λ_2.

3.2 Light: Visually evaluated radiant energy, with wavelengths approximately ranging between 380 and 770 nm, based on the sensitivity of human eye.

3.2.1 **Illuminance (E_v, lx):** The luminous flux (light incident per unit area), a unit similar to the radiometric unit, energy flux.

NOTE: (a) The measurement of radiation with an instrument responding to light is not recommended for plant growth studies and if measured should be used only along with recommended radiation measurements for historical comparison. (b) Conversion factors from illuminance to radiation are spectrally sensitive and are applicable only to a specific source.

3.3 **Temperature:** The thermal state of matter with reference to its tendency to transfer heat. A measure of the mean molecular kinetic energy of that matter.

3.3.1 **Temperature, dry bulb (T, °C):** The temperature of a gas or mixture of gases indicated by an accurate thermometer protected from or corrected for incident radiation.

3.3.2 **Temperature, wet-bulb (T_w, °C):** The thermodynamic wet-bulb temperature is the temperature at which liquid or solid water, by evaporating into air, can bring the air to saturation adiabatically at the same temperature. Wet-bulb temperature (without qualification) is the temperature indicated by a wet-bulb psychrometer constructed and used according to specifications.

3.3.3 **Temperature, dewpoint (T_d, °C):** The temperature at which the condensation of water vapor in a space begins for a given state of humidity and pressure as the temperature of the vapor is reduced. Also, the temperature corresponding to saturation (100% relative humidity) for a given absolute humidity at constant pressure.

3.4 **Atmospheric moisture:** The water vapor component of the mixture of gases of the atmosphere.

3.4.1 **Water vapor density (p_r, $g{\cdot}m^{-3}$):** The ratio of the mass of water vapor associated with a cubic meter of air; also called absolute humidity.

3.4.2 **Relative humidity (H_r, percent):** The ratio of the mol fraction of water vapor present in the air to the mol fraction of water vapor present in saturated air at the same temperature and barometric pressure. It approximates the ratio of the partial pressure or density of the water vapor in the air to the saturation pressure or density, respectively, of water vapor at the same temperature.

3.5 **Air velocity (V, $m{\cdot}s^{-1}$):** The time rate of air motion along the major translational vector.

3.6 **Carbon dioxide concentration ([CO_2], $\mu mol{\cdot}m^{-3}$):** The carbon dioxide component of the mixture of gases of the atmosphere. Current expression of units of gas concentration are $\mu mol{\cdot}mol^{-1}$ or $\mu l{\cdot}1^{-1}$, but they do not express STP correction.

3.7 **Watering (volume, L):** The addition of water to the substrate specified as to the source, the times, the amount, and the distribution method.

3.8 **Substrate:** The media comprising the root environment specified as to type, amendments and its dimensions (container size).

3.9 **Nutrition:** The organic and inorganic nutrient salts necessary for plant growth and development. Formula and/or macro and micro nutrients are specified within the substrate as $mol{\cdot}m^{-3}$ or within liquid solution as $mol{\cdot}L^{-1}$.

3.10 **Hydrogen ion concentration (pH units):** The hydrogen ion concentration measured in the substrate or liquid media over a range of 0 to 14 pH units.

3.11 **Electrical conductivity (λ_c, $dS{\cdot}m^{-1}$):** The electrical conductivity within the solid or liquid media.

3.12 **Accuracy:** The extent to which the readings of a measurement approach the true values of measured quantities.

3.13 **Precision:** The ability of the instrument to reproduce measurements of measured quantities.

SECTION 4—INSTRUMENTATION

4.1 **Radiation.** Sensors should be cosine corrected and constructed of material of known stability, known response curve, and low temperature sensitivity. Such relationships should be specified and available for each sensor. By definition fluence measurements can only be taken with spherical sensors and cannot be derived from measurements taken with any plane surface sensors. The sensitivity and linearity over the spectral response and irradiance range should be specified by calibration or direct transfer from a calibrated instrument. Spectral measurements should be made with a bandwidth of 20 nm or less in the 300-800 nm waveband.

TABLE 1—EXPECTED INSTRUMENT PRECISION AND MEASUREMENT ACCURACY

Parameter	Instrument precision	Measurement accuracy of reading
Radiation		
Flux	±1%	±10%
Spectral flux	±1%	±5%
Light		
Illuminance	±10%	±10%
Temperature		
Air	±0.1 °C	±0.5 °C
Soil or liquid	±0.1 °C	±0.5 °C
Atmospheric moisture		
Relative humidity	±2%	±5%
Dewpoint temperature	±0.1 °C	±0.5 °C
Water vapor density	±0.1 $g{\cdot}m^{-3}$	±0.1 $g{\cdot}m^{-2}$
Air velocity	±2%	±5%
Carbon dioxide	±1%	±3%
pH		
H^+ concentration	±0.1 pH	±0.1 pH
Electrical conductivity		
Salt concentration	±5%	±5%

4.2 **Temperature.** Sensors should be shielded with reflective material and aspirated (> 3 $m{\cdot}s^{-1}$) for air measurements.

4.3 **Atmospheric moisture.** Measurement should be made by infrared analyzer, dewpoint sensor or psychrometer (shielded and aspirated at ≥ 3 $m{\cdot}s^{-1}$).

4.4 **Air velocity.** Sensor should have a range of 0.1 to 5.0 $m{\cdot}s^{-1}$.

4.5 **Carbon dioxide.** Measurement should be made by an infrared analyzer with a range of 0 to 45000 $\mu mol{\cdot}m^{-3}$ or greater.

4.6 **Hydrogen ion concentration.** Sensor should have a range of 3.0 to 10.0 pH units.

4.7 **Electrical conductivity.** Sensor should have a range of 10^{-2} to 10^{-4} $dS{\cdot}m^{-1}$ (10^2-10^4 ohms resistance).

4.8 Table 1 gives typical instrument precision and measurement accuracy expected. The percentages indicate full scale precision or accuracy. Further definition of these requirements can be found in reference 28.

SECTION 5—MEASUREMENT TECHNIQUE

5.1 **Photon and energy flux.** Measurements should be taken over the top of the plant canopy to obtain the average, maximum, and minimum readings, and at least at the start and end of each study and biweekly if studies extend beyond 14 days.

5.2 **Spectral photon or energy flux.** A measurement should be taken at the center of the growing area, at least at the start and end of each study.

5.3 **Air temperature.** Measurements should be made at the top of the plant canopy, obtaining average, maximum, and minimum readings at least daily, one hour or more after each light and dark period. Continuous measurements are recommended.

5.4 **Soil and liquid temperatures.** Measurements should be made at the center of the containers in the growing area, obtaining average, maximum, and minimum readings at the middle of the light and dark periods at the start of the experiment. Continuous measurements during the entire study are recommended.

5.5 **Atmospheric moisture.** Measurements should be made at the top of the plant canopy in the center of the growing area daily, one hour or more after each light and dark period. Continuous measurements are recommended.

5.6 **Air velocity.** Measurements should be taken at the top of the plant canopy, at the start and end of the studies. Obtain average, maximum, and minimum readings over the plants. If instantaneous devices are utilized, 10 consecutive readings should be taken at each location and averaged.

5.7 **Carbon dioxide.** Measurements should be taken at the top of the plant canopy continuously during the period of the study. A time sharing technique that provides a periodic measurement (at least hourly) in each chamber can be utilized.

5.8 Watering. The quantity of water added to each container or average per plant at each watering should be measured. Soil moisture should be measured to provide range between waterings.

5.9 Nutrition. Measurement of nutrients added to a volume of media or concentration of nutrients added in liquid culture should be obtained at each addition.

5.10 Hydrogen ion concentration. The pH of the liquid solutions in a nutrient culture system should be monitored daily and before each pH adjustment. The pH of the solution extracted from solid media should be measured at the start and end of studies and before and after each pH adjustment.

5.11 Electrical conductivity. Conductivity of the liquid solutions in a nutrient culture system should be monitored daily during the course of each study. Conductivity of the solution extracted from solid media should be measured at the start and end of each study.

SECTION 6—REPORTING

6.1 Photon or energy flux. Report the average and range over the containers at the start of the study and the decrease or fluctuations from the average over the course of the study. The source of radiation and the measuring instrument/sensor should be reported. Illuminance should not be reported except for historical comparison in conjunction with other radiation measurements.

6.2 Spectral photon or energy flux. Report the spectral distribution (graphical) and the integral (photon or energy flux) at the start of the study. The source of radiation and the measuring instruments should be reported.

6.3 Air temperature. Report the average daily readings with extremes over the growing area for the light and dark periods with the range of variations over the course of the study.

6.4 Soil and liquid temperatures. Report the average readings at the start of the study for the light and dark periods.

6.5 Atmospheric moisture. Report the daily average moisture level for both light and dark periods with the range over the course of the study.

6.6 Air velocity. Report the average and range over containers at the start and end of the study.

6.7 Carbon dioxide. Report the mean of hourly average concentrations and range of average readings over the period of the study.

6.8 Watering. Report the frequency of watering, source, and amount of water added daily to each container, and/or the range in soil moisture content between waterings.

6.9 Substrate. Report the type of soil and amendments, or components of soilless substrate, and container dimensions.

6.10 Nutrition. Report the nutrients added to solid media. Report the concentration of nutrients in liquid additions and in liquid culture solution along with the amount and frequency of all additions.

6.11 Hydrogen ion concentration. Report the mode and range in pH.

6.12 Electrical conductivity. Report the average and range in conductivity.

SECTION 7—SYNOPTIC TABLE

7.1 Table 2 is a synoptic table of the material presented in the previous section.

TABLE 2—GUIDELINES FOR MEASURING AND REPORTING ENVIRONMENTAL PARAMETERS FOR PLANT EXPERIMENTS IN GROWTH CHAMBERS*

Parameter	Units†	Measurements		
		Where to take	When to take	What to report
Radiation				
Photon flux†† $\lambda_1 - \lambda_2$ with cosine correction or Irradiance (energy flux) $\lambda_1 - \lambda_2$ with cosine correction	(quanta) μmol $s^{-1} \cdot m^{-2}$ $(\lambda_1 - \lambda_2)$ or $W \cdot m^{-2}$ $(\lambda_1 - \lambda_2)$	At top of plant canopy. Obtain maximum and minimum over plant growing area.	Minimum—at start and finish of each study and biweekly if studies extend beyond 14 days.	Average (± extremes) over containers at start of study. Percent decrease or fluctuation from average over course of the study. Source of radiation and instrument/sensor.
Spectral photon flux $\lambda_1 - \lambda_2$ in <20 nm bandwidths with cosine correction or Spectral irradiance (Spectral energy flux) $\lambda_1 - \lambda_2$ nm in <20 nm bandwidths with cosine correction	(quanta) μmol $s^{-1} \cdot m^{-2} \cdot nm^{-1}$ $(\lambda_1 - \lambda_2)$ or $W \cdot m^{-2}$ nm^{-1} $(\lambda_1 - \lambda_2$ nm)	At top of plant in center of growing area.	Minimum—at start and end of each study.	Spectral distribution of radiation with integral $(\lambda_1 - \lambda_2)$ at start of study. Source of radiation and instrument/sensor.
Temperature				
Air Shielded and aspirated ($\geq 3 m \cdot s^{-1}$) device	°C	At top of plant canopy. Obtain maximum and minimum over plant growing area.	Minimum—measure once daily during each light and dark period at least 1 h after light change. Desirable - continuous measurement.	Average of once daily readings (or hourly average values) for the light and dark periods of the study with ± extremes for the variation over the growing area.
Soil and liquid	°C	In center of container. Obtain maximum and minimum over plant growing area.	Minimum—measure at the middle of the light and dark period at the start of the study. Desirable - continuous measurement.	Light and dark period readings at the start of the study (or hourly average values for 24 h if taken).

TABLE 2— (continued)

Parameter	Units†	Measurements		
		Where to take	When to take	What to report
Atmospheric Moisture Shielded and aspirated ($3 m \cdot s^{-1}$) psychrometer, dewpoint sensor or infrared analyzer	Percent relative humidity, dewpoint temperature, or $g \cdot m^{-3}$	At top of plant canopy in center of plant growing area.	Minimum—once during each light and dark period at least 1 h after light changes. Desirable - continuous measurement.	Average of daily readings for both light and dark periods, with range of daily variation during studies.
Air Velocity	$m \cdot s^{-1}$	At top of plant canopy. Obtain maximum and minimum readings over growing area.	At start and end of studies. Take 10 successive readings at each location and age.	Average reading and range over containers at start and end of the study.
Carbon Dioxide	$\mu mol \cdot m^{-3}$	At top of plant canopy.	Minimum—hourly measurements. Desirable - continuous measurements.	Mean of hourly average concentrations and range of average concentration over the period of the study.
Watering	liter (L)	———	At times of water additions.	Frequency of watering. Amount of water added and/or range in soil moisture content between waterings.
Substrate	———	———	Water retention capacity.	Type of soil and amendments. Components of soilless substrate. Container dimensions.
Nutrition	Solid media $mol \cdot m^{-3}$ or $mol \cdot kg^{-1}$ Liquid culture $mol \cdot L^{-1}$	———	At times of nutrient additions.	Nutrients added to solid media. Concentration of nutrients in liquid additions and solutions culture. Amount and frequency of solution addition and renewal.
pH	pH units	In saturated media, extract from media or in solution of liquid culture.	Start and end of studies in solid media. Daily in liquid culture. Before each pH adjustment.	Mode and range during studies.
Electrical Conductivity	$dS \cdot m^{-1}$§ (decisiemens per meter)	In saturated media, extract from media or in solution of liquid.	Start and end of studies in solid media. Daily in liquid culture.	Average and range during studies.

*USDA North Central Regional (NCR 101) Committee on Growth Chamber Use June 1978; Revised by ASAE Environment of Plant Structures Committee Oct. 1978. Published in part in the following references: 3, 21, 22, 24, 29, 34, 35, 36; Revised NCR 101 Committee Mar. 1981, and Jun. 1985.

†Report in other subdivisions of indicated units if more convenient.

††Defined as photosynthetically active radiation (PAR) if $\lambda_1 - \lambda_2$ is 400-700 nm.

§$dS \cdot m^{-1}$ = mmho cm^{-1}.

References:

1. ASAE Engineering Practice EP285, Use of SI (Metric) Units. American Society of Agricultural Engineers, St. Joseph, MI 49085.
2. American Society for Horticultural Science Working Group on Growth Chambers and Controlled Environments. 1980. Guidelines for measuring and reporting the environment for plant studies. HortScience 15(6):719-720.
3. Bell, C. J. and D. A. Rose. 1981. Light measurement and the terminology of flow. Plant, Cell Environment 4:89-96.
4. Bickford, E. D. and S. Dunn. 1972. Lighting for plant growth. The Kent State University Press, Kent, OH.
5. Biggs, W. W. and M. C. Hansen. 1979. Instrumentation for biological and environmental sciences. LI-COR Inc., Lincoln, NE.
6. C. I. E. 1970. International lighting vocabulary. Commission Internationale de l'Eclairage, Publ. No. 17, Paris, France.
7. Downs, R. J. 1975. Controlled environments for plant research. Columbia University Press, New York, NY.
8. Geist, J. and E. Zalewski. 1973. Chinese nomenclature for radiometry. Applied Optics 12:435-436.
9. Holmes, M. G. and L. Fukshansky. 1979. Phytochrome photoequilibrium in green leaves under polychromatic radiation: a theoretical approach. Plant, Cell and Environment 2:59-65.
10. Holmes, M. G., W. H. Klein and J. C. Sager. 1985. Photons, flux, and some light on philology. HortScience 20(1):29-31.
11. IES Lighting Handbook. 1981. Fifth edition. Illuminating Engineering Society of North America, New York, NY.
12. Incoll, L. D., S. P. Long and M. R. Ashmore, 1977. SI units in publications in plant science. Current Advances in Plant Science 28:331-343.
13. Kerr, J. P., G. W. Thurtell and C. B. Tanner. 1967. An integrating pyranometer for climatological observer stations and meoscale networks. Journal of Applied Meteorology 6:688-694.
14. Kozlowski, T. T. (ed) 1968 and 1976. Water deficits and plant growth. Vol. 1, Development, control and measurements. Vol. 4, Soil water measurement, plant responses and breeding for drought resistance. Academic Press, New York, NY.
15. Krizek, D. T. 1982. Guidelines for measuring and reporting environmental conditions in controlled-environment studies. Physiol. Plant. 56-231-235.
16. Krizek, D. T. and J. C. McFarlane. 1983. Controlled-environment guidelines. HortScience 18(5):662-664 and Erratum 19(1):17.
17. Langhans, R. W. (ed.). 1978. A growth chamber manual. Cornell University Press, Ithaca, NY.
18. LI-COR. 1982. Radiation measurements and instrumentation, Publ. No. 8208-LM, LI-COR, Lincoln, NE.

19. McCree, K. J. 1972. Test of current definitions of photosynthetically active radiation against leaf photosynthesis data. Agricultural Meteorology 10:443-453.

20. McFarlane, J. C. 1981. Measurement and reporting guidelines for plant growth chamber environments. Plant Science Bulletin 27(2):9-11.

21. Mohr, H. and E. Schafer. 1979. Guest editorial-uniform terminology for radiation: A critical comment. Photochemistry and Photobiology 29:1061-1062.

22. Monteith, J. L. 1984. Consistency and convenience in the choice of units for agricultural science. Expl. Agric. 20(2):105-117.

23. NBS Tecnical Note 910-2. 1978. Self-study manual on optical radiation measurements, Part 1-Concepts. United States Government Printing Office, Washington, DC.

24. Norris, K. H. 1968. Evaluation of visible radiation for plant growth. Annual Review of Plant Physiology 19:490-499.

25. North Central Regional 101 Committee on Growth Chamber Use. 1984. Quality assurance procedures for accuracy in environmental monitoring-Draft proposal. Biotronics 13:43-46.

26. Percival Manufacturing Co. 1981. Guidelines: Measuring and reporting environment for plant studies. Available as a plastic card.

27. Rosenberg, N. J. 1974. Microclimate: The biological environment. John Wiley & Sons, New York, NY.

28. Rupert, C. S. and R. Latarjet. 1978. Toward a nomenclature and dosimetric scheme applicable to all radiations. Photochemistry and photobiology 28:3-5.

29. Sestak, Z., J. Catsky and P. G. Jarvis (eds). 1971. Plant photosynthetic production, manual of methods. Junk, The Hague, Netherlands.

30. Slayter, R. O. 1967. Plant-water relationships. Academic Press, New York, NY.

31. Spomer, L. A. 1980. Guidelines for measuring and reporting environmental factors in controlled environment facilities. Commun. Soil Science and Plant Analysis 11(12):1203-1208.

32. Spomer, L. A. 1981. Guidelines for measuring and reporting environmental factors in growth chambers. Agronomy Journal 73(2):376-378.

33. Thimijan, R. W. and R. D. Heins. 1983. Photometric, radiometric, and quantum light units of measure: a review of procedures for interconversion. HortScience 18(6):818-821.

34. Tibbitts, T. W. and T. T. Kozlowski (ed.) 1979. Controlled environment guidelines for plant research. Academic Press, New York, NY.

35. Tooming, K. G. 1977. Solar radiation and yield formation (Solnechnaya radiatsiya i formirovanio urozhaya) Gidrometeoizdat, Leningrad.

36. Zelitch, I. 1971. Photosynthesis, photorespiration and plant productivity. Academic Press, New York, NY.

PROCEDURE FOR ESTABLISHING VOLUMETRIC CAPACITY OF GRAIN BINS

Proposed by the Grain Bin Manufacturers Council of the Farm and Industrial Equipment Institute; approved by the ASAE Structures Group; approved by the ASAE Structures and Environment Division Standards Committee; adopted by ASAE March 1982; equation in paragraph 3.4.1 corrected September 1982; reconfirmed December 1986; reconfirmed for one year December 1987.

SECTION 1—PURPOSE AND SCOPE

1.1 The purpose of this Standard is to define the method for determining the volumetric capacity of cylindrical grain bins.

1.2 The purpose is also to establish an industry standard for compaction factors for all diameter bins.

SECTION 2—DEFINITIONS

2.1 Bin diameter: The diameter of the bin measured from the centerline or neutral axis of the corrugated sidewalls, or the inside diameter of a smooth-walled bin.

2.2 Open eave: The spaced relationship at the intersection of the bin sidewall and the bin roof which will allow free passage of air between the sidewall and the roof.

2.3 Tight eave: The sealed condition at the intersection of the bin sidewall and the bin roof. This seal, although not air tight, is sufficient to restrict the flow of air substantially and will prevent small grains from passing thru this space.

2.4 Eave height: The distance from the top of the permanent structural floor to the top of the bin sidewall. (Drying floors are not considered a permanent structural floor.)

2.5 Roof slope: The slope or inclination of the bin roof measured in degrees from horizontal.

2.6 Hopper slope: The slope or inclination of the hopper measured from horizontal.

2.7 Maximum angle of fill: The maximum angle that may be used for calculating the capacity of the roof area. This angle is measured from horizontal and shall be assumed to be 28 deg or the roof slope, whichever is less.

2.8 Standard U.S. bushel: Equivalent to 0.035 m^3 (1.25 ft^3).

2.9 Drying bin: A bin used in combination with an air heating/moving device for the purpose of removing excess moisture from the grain contained within the bin.

2.10 Storage bin: A bin used for the storage of dry grain. Air moving devices may be used in conjunction with these bins to help maintain the stored grain in good condition.

2.11 Hopper bin: A bin equipped with a conical hopper having a center discharge opening.

2.12 Drying bin capacity: Rated capacity of a bin equipped with a drying floor.

2.13 Level full storage capacity: Rated capacity of a bin used for storing dry grain, based on a uniform grain depth equal to the eave height of the bin, or 25 mm (1 in.) below the eave height in case of a bin with an open eave.

2.14 Peaked storage capacity: Rated capacity of a bin used for storage of dry grain, based on the appropriate level full storage capacity plus the capacity of the conical space within the roof area of the bin calculated using the maximum angle of fill (see paragraph 2.7).

2.15 Hopper bin storage capacity: Rated capacity of a hopper bin used for storage of dry grain, based on peaked storage capacity plus the capacity of the conical hopper.

SECTION 3—GRAIN BIN CAPACITIES

3.1 Capacities shall be based on volume only and shall be stated in terms of cubic meters (cubic feet) and/or standard U.S. bushel.

3.2 Drying bin capacity shall be calculated on the basis of level full storage capacity within 25 mm (1 in.) of the eave height of the bin. The space required for the installation of the manufacturer's standard drying floor shall also be deducted from the capacity of the bin. The capacity is determined by:

$$\text{Volume} = (\pi D^2/4)\,[(EH\text{-}K) - DFH]$$

where

D = bin diameter
EH = eave height of bin
K = constant = 25 mm (1 in.)
DFH = drying floor height (average height of floor if top of floor is not a flat horizontal surface)

3.3 Storage bin capacity may be expressed either in terms of level full storage capacity of peaked storage capacity.

3.3.1 Level full storage capacity reflects the volume contained within the cylindrical portion of the bin only.

3.3.2 Peaked storage capacity reflects the volume of the cylindrical portion of the bin plus the volume of the conical portion within the roof space. The volume of the conical portion shall be based on the maximum angle of fill (see paragraph 2.7).

3.3.3 Capacity of storage bins with open eaves shall be based upon filling within 25 mm (1 in.) of the eave height of the bin and shall be calculated as given below.

3.3.3.1 Level full capacity:

$$\text{Volume} = (\pi D^2/4)\,(EH - K)$$

where

D = bin diameter
EH = eave height of bin
K = constant = 25 mm (1 in.)

3.3.3.2 Peaked capacity:

$$\text{Volume} = (\pi D^2/4)\,(EH - K) + (\pi D^2/4)\,\{[(D/2)\tan\phi]/3\}$$

where

ϕ = maximum angle of fill

3.3.4 Capacity of storage bins with tight eaves shall be calculated as given below.

3.3.4.1 Level full capacity:

$$\text{Volume} = (\pi D^2/4)\,EH$$

where

D = bin diameter
EH = eave height of bin

3.3.4.2 Peaked capacity:

$$\text{Volume} = (\pi D^2/4)\,EH + (\pi D^2/4)\,\{[(D/2)\tan\phi]/3\}$$

where

ϕ = maximum angle of fill

3.4 Hopper bin capacity shall be calculated as given below.

3.4.1 Hopper bins with open eaves:

$$\text{Volume} = (\pi D^2/4)\,[EH - K] + (\pi D^2/4)\,\{[(D/2)\tan\phi]/3\} + (\pi D^2/4)\,\{[(D/2)\tan\gamma]/3\}$$

where

D = bin diameter
EH = eave height of bin
K = constant = 25 mm (1 in.)
ϕ = maximum angle of fill
γ = slope of hopper measured in degrees from horizontal

3.4.2 Hopper bins with tight eaves:

$$\text{Volume} = (\pi D^2/4)\,EH + (\pi D^2/4)\,\{[(D/2)\tan\phi]/3\} + (\pi D^2/4)\,\{[(D/2)\tan\gamma]/3\}$$

3.5 The following terminology should be used by manufacturers in their published grain bin capacity charts:

3.5.1 Drying bin capacity

3.5.2 Level full storage capacity

3.5.3 Peaked storage capacity

3.5.4 Hopper bin storage capacity

SECTION 4—INCREASE IN STORAGE CAPACITY DUE TO COMPACTION

4.1 Grain bulk densities in storage are generally greater than the bulk densities used for the purpose of commercial sales.

4.2 Rated capacities for all bins shall be determined on the basis of volumetric definitions cited earlier in this Standard.

4.3 Probable maximum storage capacity for any grain bin may be estimated by increasing grain bulk densities shown in ASAE Data D241, Density, Specific Gravity, and Weight-Moisture Relationships of Grain for Storage, by up to 6 percent. When these figures are used, they must be identified as estimated maximum capacities. Due to the number of uncontrollable variables which affect actual compaction within a bin, probable maximum storage capacity shall not be used unless accompanied by rated capacity values.

Cited Standard:

ASAE D241, Density, Specific Gravity, and Weight-Moisture Relationships of Grain for Storage

ENERGY EFFICIENCY TEST PROCEDURE FOR TOBACCO CURING STRUCTURES

Developed by the ASAE Special Crops Processing Committee; approved by the ASAE Electric Power and Processing Division Standards Committee; adopted by ASAE December 1982; reconfirmed December 1987.

SECTION 1—PURPOSE

1.1 The purpose of this Standard is to:

1.1.1 Promote uniformity and consistency in terms used to describe and evaluate the energy efficiency of forced-air tobacco curing structures.

1.1.2 Define a test procedure for a simulated six-day cure under external ambient conditions.

1.1.3 Provide a method for interpretation and presentation of test results such that the energy efficiency of different designs can be compared when operated under equivalent conditions.

1.1.4 Provide test data which can be used to improve the energy efficiency of tobacco curing structures.

1.2 This Standard does not define a set of functional requirements for tobacco curing structures nor give a procedure for determining if these requirements are met by a given design. It evaluates each design as presented.

SECTION 2—SCOPE

2.1 This standard is applicable to all batch-loaded forced-air tobacco curing structures, downdraft, updraft, and crossflow. It can be used to evaluate any curing structure where the specified site conditions are met.

2.2 The burner can be direct-fired or indirect-fired, and the fuel can be either gaseous or liquid. Provision is made for outside combustion air, if required.

2.3 The test conditions are based on the bright leaf tobacco curing schedule.

SECTION 3—TERMINOLOGY

3.1 Test: All events and data comprising a six-day simulated cure.

3.1.1 Test barn: The subject forced-air tobacco curing structure.

3.1.2 Test building: Structure in which the test is conducted.

3.2 Curing compartment: Space within the barn which is filled with tobacco.

3.2.1 Curing compartment area: Cross-sectional area perpendicular to the airflow.

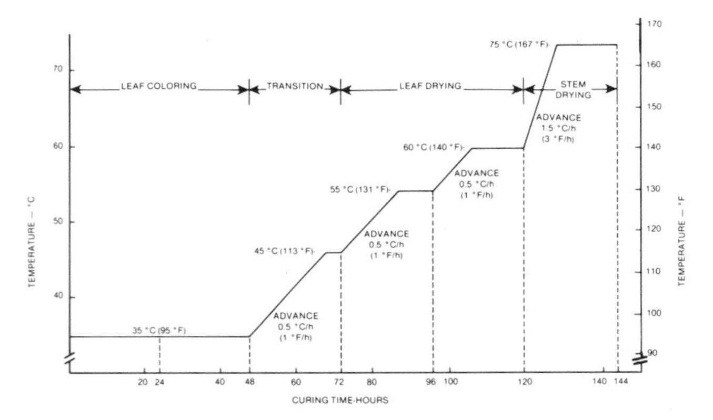

FIG. 1—STANDARD TEMPERATURE CYCLE FOR SIX-DAY CURE

TABLE 1—STATIC PRESSURE SETTING FOR SIMULATED SIX-DAY CURE

Time, h	Pressure adjustment	Static pressure Pa	in. H_2O
0		200	0.80
12	decrease 10	190	0.76
24	decrease 10	180	0.72
36	decrease 10	170	0.68
48	decrease 10	160	0.64
60	decrease 10	150	0.60
72	decrease 10	140	0.56
84	decrease 20	120	0.48
96	decrease 40	80	0.32
108	decrease 20	60	0.24
120	decrease 20	40	0.16
132	decrease 10	30	0.12
144		30	0.12

3.2.2 Loading end doors: Doors through which the curing compartment is filled.

3.2.3 Curing compartment volume: The curing compartment area times the distance through the tobacco. For updraft and downdraft barns, the distance through the tobacco is taken to be the height of the curing containers. For bulk racks, the distance through the tobacco is taken to be the distance from the drying floor to a point 15 cm (6 in.) above the top tier rail for an updraft barn, and for a downdraft barn it is the distance from the ceiling to a point 60 cm (24 in.) below the bottom tier rail. For crossflow barns the distance through the tobacco is the width of the curing container.

3.3 Thermostat schedule: Thermostat settings used for the simulated cure (Fig. 1).

3.4 Static pressure schedule: Static pressure settings used for the simulated cure (Table 1).

3.5 Inlet airflow schedule: Inlet airflow settings used for the simulated cure (Table 2).

3.6 Static pressure manifold: Device for obtaining a four-point average static pressure (Fig. 2).

3.7 Variable resistance plate: Device for restricting airflow through the curing compartment to simulate the resistance of the tobacco.

3.8 Inlet chamber: Chamber at the furnace end of the test barn used to monitor the flow of inlet air during the simulated cure.

3.8.1 Inlet fan: Variable speed fan used to set the inlet airflow given in Table 2.

3.8.2 Inlet duct: Duct used to measure the inlet airflow.

3.9 Furnace room wall: Wall between the curing compartment and the furnace.

TABLE 2—INLET AIRFLOW SETTINGS FOR SIMULATED SIX-DAY CURE

Time, h	Airflow adjustment	Air changes/h*
0		20
24	increase 10	30
48	increase 20	50
72	increase 10	60
96	decrease 30	30
120	decrease 10	20
144		20

*Based on curing compartment volume.

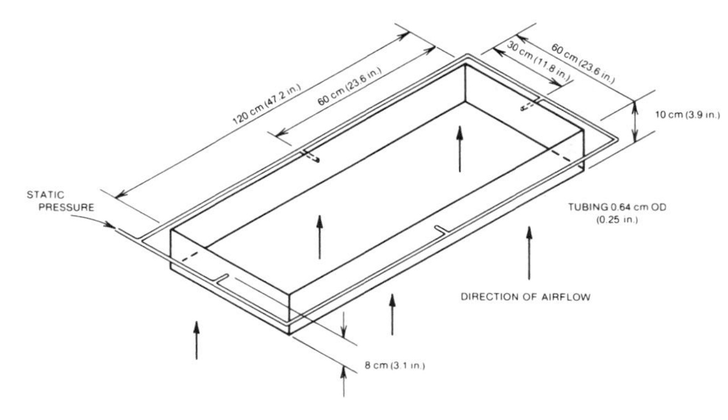

FIG. 2—STATIC PRESSURE MANIFOLD

3.10 Inlet vents: Vents on the test barn which are opened manually or automatically to introduce inlet air during the curing process.

3.11 Exhaust vents: Louvered vents on the test barn which open to exhaust air which has been passed through the tobacco.

3.12 Direct-fired furnace: Furnace which fires directly into the airstream which passes through the tobacco. The products of combustion go directly into the curing compartment.

3.13 Indirect-fired furnace: Furnace which fires into a combustion chamber that is vented outside the curing compartment.

3.14 Foundation insulation: Insulation around the foundation of the barn which may be applied by the manufacturer prior to shipment or may be installed at the site.

SECTION 4—SITE CONDITIONS

4.1 The test shall be conducted in a building which protects the test barn from direct solar radiation on any surface. A minimum of 3 m (10 ft) clearance must be allowed on each side and above the test barn. The building shall be naturally ventilated, but protect the test barn from direct incidence of ambient wind. If two or more barns are tested in the same building they must be at least 10 m (33 ft) apart.

4.2 The test barn shall be installed on a 10 cm (4 in.) thick concrete foundation pad insulated from the ground with 5 cm (2 in.) of board insulation having a thermal resistance of 1.4 $m^2 \cdot K/W$ (R = 8 $h \cdot ft^2 \cdot {}^\circ F/Btu$). The insulation shall be enclosed in a watertight envelope. The foundation shall have at least 5 cm (2 in.) of sand beneath the insulation.

4.3 Sealing of the foundation frame to the foundation pad shall be in accordance with specifications given in the manufacturer's literature. If foundation insulation is included in the pre-manufactured package, a description shall be included in the test report.

4.4 Fuel supplies and electrical connections, up to the point of connection with the test barn, shall comply with the following applicable standards of the National Fire Protection Association: NFPA No. 31, Standard for the Installation of Oil Burning Equipment; NFPA No. 54, Standard for the Installation of Gas Appliances and Gas Piping; NFPA No. 58, Standard for the Storage and Handling of Liquified Petroleum Gases; and NFPA No. 70, National Electrical Code.

4.4.1 LP gas furnaces shall be supplied with HDS grade propane.

4.4.2 Natural gas furnaces shall be supplied with natural gas.

4.4.3 Fuel oil furnaces shall be supplied with No. 2 fuel oil.

4.5 Indirect-fired units shall be vented to the outside of the test building.

4.6 An inlet chamber shall be installed at the furnace end of the test barn. It shall be the height and width of the barn. The length shall be equal to the width.

4.6.1 Connection of test barn to inlet chamber. The furnace room doors shall be removed. The resulting opening shall be sealed to a matching opening in the inlet chamber.

4.6.2 Inlet vents. The inlet vents on the test barn shall be opened to a setting equal to one-half the maximum setting used during a cure. If the barn has automatic vent control it shall be disconnected.

4.6.3 Outside combustion air. A duct shall be installed to supply outside combustion air to the burner for all indirect-fired furnaces. This duct shall be sized such that the pressure differential along its length does not affect burner operation.

4.6.4 Venting of combustion chamber. All indirect-fired furnaces shall be vented outside the test building with a vertical flue attached to the barn flue.

4.7 Air exhausted from the test barn exhaust vents shall be ducted outside the test building. This duct shall have a cross-sectional area at least twice the total area of the exhaust vents.

4.8 The variable resistance plate shall be installed such that the static pressure can be varied from outside the test barn. For barns with tier rails, it shall be installed at the first tier rail. For barns without tier rails, it shall be installed at a point one-third the distance from the delivery plenum to the return plenum. This plate shall have openings uniformly distributed over the entire curing compartment area. The open area shall be sufficient to obtain a minimum static pressure of 25 Pa (0.1 in. H_2O) measured at the static pressure manifold.

4.9 Test barn preparation

4.9.1 The loading end doors shall be closed and latched in the normal manner.

4.9.2 The test barn shall be inspected and adjusted as necessary by a manufacturer's representative after it is secured in the test configuration. Preliminary testing shall be done to insure satisfactory operation of the fan, burn, and all controls.

SECTION 5—TEST EQUIPMENT

5.1 Petroleum fuel

5.1.1 Consumption of gaseous fuel shall be measured with a totalizing positive displacement flow meter rated for a maximum flow of 500 MJ/h (500,000 Btu/h) with an accuracy of $\pm$ 0.5%. The meter shall be temperature compensated, and calibrated to operate at the burner line pressure recommended by the manufacturer.

5.1.2 Consumption of liquid fuel shall be measured with a totalizing flow meter rated for flows up to 0.25 L/min (4 gal/h) with an accuracy of $\pm$ 0.5%.

5.2 Electrical energy consumed by the test barn shall be recorded with an accuracy of $\pm$ 2%.

5.3 Temperature

5.3.1 Ambient temperature (T_a) shall be the average of five readings preferably made with a thermopile. One transducer each shall be mounted at the center of each side wall 2 m (6 ft) from the surface, and one shall be mounted at the center of the roof 2 m (6 ft) from the surface. One transducer shall be mounted at the center of the loading end 2 m (6 ft) from the surface, and one at the inlet to the inlet duct.

5.3.2 Delivery plenum temperature (T_d) shall be the average of temperatures measured at the level of the thermostat sensor in the curing compartment. The curing compartment area shall be divided into four quadrants and T_d shall be the average of the temperatures measured at the center of each quadrant.

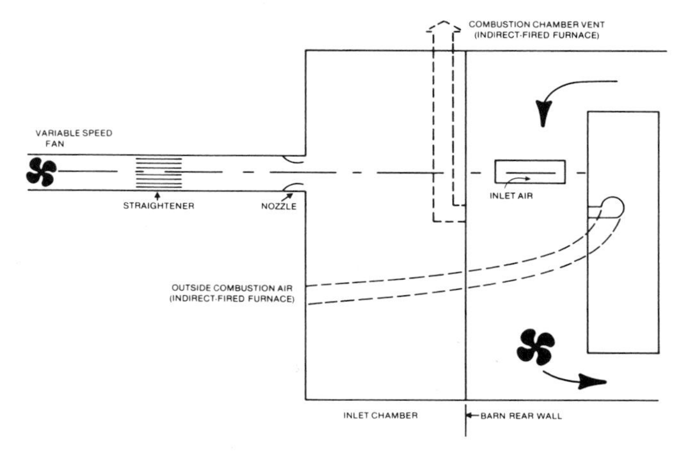

FIG. 3—POSITION OF INLET CHAMBER RELATIVE TO TEST BARN.

5.4 Static pressure shall be measured with a static pressure manifold (Fig. 2) placed at the center of the plane between the delivery plenum and the curing compartment.

5.5 The airflow delivered to the inlet chamber shall be measured in an inlet duct mounted such that the centerline coincides with the centerline of the inlet vents (Fig. 3). Flow rate in the inlet duct shall be measured in accordance with the American Society of Heating, Refrigerating and Air Conditioning Engineers Standard 51-75 (Air Moving and Conditioning Association Standard 210-74), Laboratory Methods of Testing Fans for Rating, using the outlet duct setup-nozzle on the end of the outlet duct shown in Fig. 8 of the Standard.

SECTION 6—TEST PROCEDURE

6.1 The test barn shall be started with the thermostat at an initial setting of 35 °C (95 °F), an initial static pressure of 200 Pa (0.8 in. H_2O), and an initial air exchange rate of 20 changes per hour. It shall be allowed at least ten hours to equilibrate prior to data collection at the zero hour.

6.2 Temperature shall be recorded hourly with an accuracy of ± 0.5 °C (± 1 °F). Humidity shall be recorded hourly with an accuracy of ± 2%.

6.3 The procedure during the one-half hour period before the end of each 12-hour interval shall be as follows:

6.3.1 Record the electrical energy consumption.

6.3.2 Set the static pressure in the delivery plenum in accordance with the static pressure schedule (Table 1), and record the new pressure with an accuracy of ± 2.5 Pa (± 0.01 in. H_2O).

6.3.3 Set the inlet airflow in accordance with the inlet airflow schedule (Table 2), and record the new value with an accuracy of ± 2%. Since this setting significantly effects energy consumption, it should be made as close to the hour as possible. The burner shall be turned off when this setting is made.

6.3.4 Record the fuel consumption.

6.3.5 Set the thermostat in accordance with the thermostat schedule (Fig. 1).

SECTION 7—DATA REDUCTION

7.1 Definitions

7.1.1 Thermostat index: A thermostat index (TI) shall be defined as:

$$TI = \sum_{i=1}^{n} (T_{ti} - T_{di})$$

where

T_{ti} = thermostat setting for the ith hour, °C (°F)

T_{di} = measured average temperature in the delivery plenum during the ith hour, °C (°F)

n = number of hours in test

The thermostat index shall be reported as degree-hours °C·h (°F·h), and as a plot of the hourly difference.

7.1.2 Integral ΔT: An integral ΔT for the test shall be defined as:

$$\Delta T = \sum_{i=1}^{n} (T_{di} - T_{ai})$$

where

T_{ai} = measured average ambient temperature for the ith hour, °C (°F)

7.1.3 Petroleum fuel index: A petroleum fuel index (PFI) shall be defined as units of energy in the total fuel consumed divided by the integral ΔT. It shall be reported in units of MJ/°C·h (Btu/°F·h). Conversion constants shall be obtained from the fuel supplier and shall be listed on the test report.

7.1.4 Inlet airflow: The inlet airflow (q) shall be defined as the air changes per hour times the curing compartment volume. It shall be reported in units of m^3/h (ft^3/h).

7.1.5 Psychrometric data: Definitions for specific volume of the inlet air (v) and enthalphy (h) shall be those given in ASAE Data D271, Psychrometric Data. The ambient humidity ratio (H) may be calculated from relative humidity or from a wet bulb measurement.

7.1.6 Exchanged air energy: Energy in the controlled exchanged airflow (XAE) shall be calculated as:

$$XAE = 0.001 \sum_{i=1}^{n} q_i (h_{di} - h_{ai}) / v_i$$

where

h_{di} = enthalpy of air in the delivery plenum based on T_{di} and H_i in kJ/kg (Btu/lb).

h_{ai} = enthalpy of ambient air based on T_{ai} and H_i in kJ/kg (Btu/lb).

XAE shall be reported in units of MJ (Btu).

7.1.7 Fuel efficiency index: The fuel efficiency index (FEI) shall be defined as the total measured exchanged air energy divided by the total fuel energy.

$$FEI = XAE/[(PFI)\Delta T]$$

7.1.8 Electrical energy: The electrical energy (EE) shall be defined as the total electrical energy required by the test barn to maintain the static pressure schedule and operate the controls. It shall be in units of kW·h.

7.1.9 Total efficiency index: The total efficiency index (TEI) shall be defined as:

$$TEI = \frac{XAE + 3.6\ EE}{(PFI)\Delta T + 3.6\ EE}$$

7.2 Computational procedure. The rectangular rule has been shown for all numerical integration.

SECTION 8—TEST REPORTING

8.1 The equipment being tested shall be described as:

8.1.1 The test barn shall be specified by manufacturer, model number, serial number, type, and configuration.

8.1.2 The fan shall be specified by manufacturer, model number, type, RPM, rating at 250 Pa (1 in. H_2O), and at 25 Pa (0.1 in. H_2O) discharge pressure.

8.1.3 The motor shall be specified by maximum rated hp, RPM, phase, volts, and amps at full load service factor.

8.1.4 The curing compartment shall be specified by the width (w), length (l), and depth (d).

8.2 In addition to information already mentioned, the test report shall include:

8.2.1 A copy of the data sheets.

8.2.2 A statement of the equipment performance during the test.

8.2.3 A plot of the temperature difference between the thermostat setting (T_t) and the delivery plenum temperature (T_d) for each hour of the test.

8.3 Results to be reported:

8.3.1 Thermostat Index (TI)

8.3.2 Petroleum Fuel Index (PFI)

8.3.3 Exchanged Air Energy (XAE)

8.3.4 Fuel Efficiency Index (FEI)

8.3.5 Electrical Energy (EE)

8.3.6 Total Efficiency Index (TEI)

Cited Standards:

ASAE D271, Psychrometric Data

ASHRAE 51-75, Laboratory Methods of Testing Fans for Rating

NFPA 31, Standard for the Installation of Oil Burning Equipment

NFPA 54, Standard for the Installation of Gas Appliances and Gas Piping

NFPA 58, Standard for the Storage and Handling of Liquified Petroleum Gases

NFPA 70, National Electrical Code

TERMINOLOGY AND RECOMMENDATIONS FOR FREE STALL DAIRY HOUSING FREE STALLS, FEED BUNKS, AND FEEDING FENCES

Developed by the ASAE Dairy Housing Committee; approved by the ASAE Structures and Environment Division Standards Committee; adopted by ASAE March 1984.

SECTION 1—PURPOSE AND SCOPE

1.1 The purpose of this Engineering Practice is to provide persons planning free stall dairy facilities with rationale, terminology and design recommendations. The recommendations cover free stalls, feed bunks, and feeding fences for mature cows.

1.2 This Engineering Practice is not intended to dictate specific designs to manufacturers or inhibit new design innovations. Therefore a range of good practice recommendations is presented. Designs outside the range may be desired in some specific cases. Reference is made to existing ASAE Standards and Engineering Practices wherever applicable.

SECTION 2—RESTING AREA

2.1 Free stall. A dairy cattle free stall is a defined area in which a cow may lie down. The cow is free to enter and leave the stall at will. The stall should be durable and designed with consideration of comfort, freedom from cow injury, stall cleanliness, bedding requirements, labor, and ease of removing downed cows. The free stall must be long enough to allow the cow to rest comfortably on the platform without injury, yet short enough that manure and urine fall into the alley. The free stall must be wide enough for the cow to lie comfortably, but narrow enough to keep the cow from turning around. Stalls that are too long can be altered by use of training rails.

2.1.1 The **stall partition** separates side-by-side free stalls from each other. Critical dimensions include length, height, space between bottom rail and stall surface, and open space within the partition. Strength of component parts and attachment methods should be selected considering abuse from large animals, equipment, and manure accumulations. Stall partitions may be (1) supported from the stall front assembly and a rear post installed in or near the curb; (2) suspended from the stall front assembly; or (3) supported by two free-standing posts.

2.1.1.1 The **bottom rail** must be high enough to reduce leg and ankle injury yet low enough to prevent cows from lying under the partition. A removable bottom rail is helpful for releasing a trapped cow but may be difficult to hold in place.

2.1.1.2 **Partitions** must be designed to take the load of 725 kg (1600 lb) animals leaning and rubbing on them. Free stall posts are most vulnerable at the curb line where manure and moist bedding hasten deterioration. Design, construction, and installation must consider corrosion.

2.1.2 The **stall front** prevents cows from moving too far into the stall and dunging within the stall. It often provides support for the front end of the stall partition. The stall front may be an exterior wall or special interior partition. Critical features include height and amount of open area. Some designs for facing stalls provide open space to allow cows to share the same head space. Strength of component parts and attachment methods should be selected considering abuse from large animals pushing, leaning, or rubbing on them. The partition and front must be high enough to discourage a cow from turning her head over the partition and walking further forward. A front that is too low has the same effect as making the stall longer.

2.1.3 The **stall curb** serves to separate the stall area from manure in the litter alley. It also holds the stall base within the stall area and may be used to space and anchor the rear post of the stall partition. Height and construction must allow for keeping manure and scraping equipment or flush water in the cow alley. The curb should be high enough to discourage a cow from lying part way in the stall. A curb that is too high may cause udder injury when cows enter and leave the stall. Unless long alleys are scraped frequently a higher curb is recommended to keep manure from overflowing into the stall during scraping.

2.1.4 A **stall base** of permanent or semi-permanent materials forms the "floor" of the free stall. The stall base is placed and maintained to provide desired elevation and slope for drainage. The stall base must support the cow and resist hollowing caused by hooves digging when a cow gets up. A stall base that slopes up to the front encourages the cow to lie with her head toward the front of the stall and provides drainage to the rear of the stall. Commonly used materials include clay, stone dust, concrete, and asphalt paving.

2.1.5 **Stall bedding** material is added on top of stall base to make the stalls more comfortable, reduce injuries, and absorb moisture and manure tracked into the stall. Common materials are straw, sawdust, wood chips, corn stalks, bark, peanut hulls, rice hulls, sand, and ground limestone. The use of short, fine materials reduces the amount dragging into the manure. Rubber mats on concrete may be used for cushioning.

2.1.6 **Bedding boards** may be embedded part way into the stall base directly under and parallel to the stall partition. The board is run from the stall front to the rear post or curb. This may reduce hollowing of soft stall bases and reduce the potential for an animal to get trapped under stall partitions.

2.1.7 A **bedding keeper** is a board or pipe along the rear of the stall to help retain bedding in the stall.

2.1.8 **Training (neck) rails** may be placed across the top rail of the free stall partition to keep the cows from standing too far forward in the stall. They may also be an integral support component for partitions with no rear posts. Cables can cause severe injury to the cow's neck and vertebrae and should be avoided.

2.2 Free stall dimensions. The top alley edge of the free stall curb is to be used as the primary reference line. (See Fig. 1)

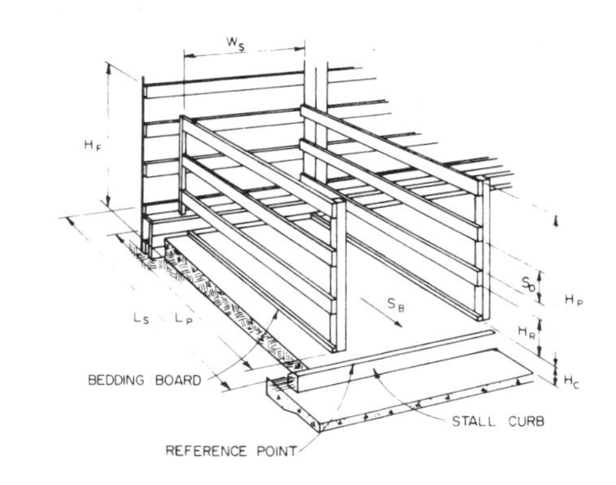

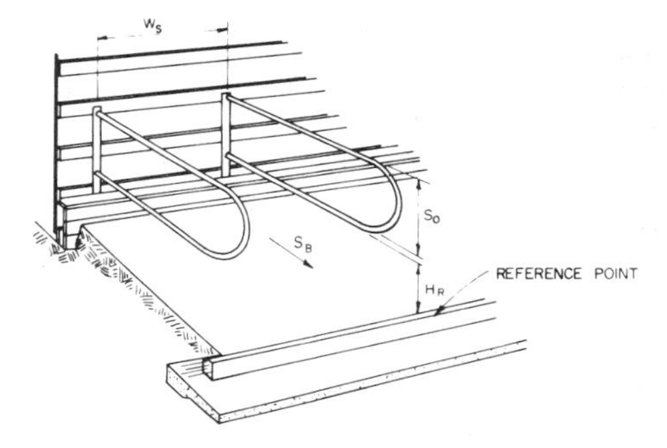

FIG. 1—FREE STALLS

2.2.1 **Bottom rail height (H_R)** is the vertical distance from the top of the curb to the lowest horizontal member of the partition.

2.2.2 **Partition open space (S_O)** is the spacing between rails, bar, or other openings in the stall partition and stall front.

2.2.3 **Stall base slope (S_B)** is the total slope of the stall base from the front of the stall to the top of the curb.

2.2.4 **Stall curb height (H_C)** is measured from the top of the alley floor to the top of the curb on the alley side.

2.2.5 **Stall front height (H_F)** is the vertical distance from the top of the curb to the top of the stall front.

2.2.6 **Stall length (L_S)** is the distance from the alley side of the curb to the stall front.

2.2.7 **Stall partition height (H_P)** is the vertical distance from the top of the curb to the top of the stall partition.

2.2.8 **Stall partition length (L_P)** is the distance from the extreme alley end of the partition to the front-most part of the partition.

2.2.9 **Stall width (W_S)** is the distance between the centerlines of partitions.

2.3 Free stall design data

H_C—stall curb height: 150 to 300 mm (6 to 12 in.)

H_F—stall front height: 1.2 to 1.6 m (4.0 to 5.3 ft)

H_P—stall partition height: 1.0 to 1.5 m (3.3 to 5.0 ft)

H_R—bottom rail height: 250 to 400 mm (10 to 16 in.)

L_P—stall partition length: (L_S − 0.3 m) to L_S (L_S − 1.0 ft to L_S)

L_S—stall length:
650 kg cows (1400 lb): 2.0 to 2.3 m (6.5 to 7.5 ft)
550 kg cows (1200 lb): 1.9 to 2.1 m (6.2 to 7.0 ft)

S_B—stall base slope: 40 to 60 mm (1.5 to 2.4 in.)

S_O—partition open space: 200 to 500 mm (8 to 20 in.)

W_S—stall width: 1.1 to 1.3 m (3.5 to 4.3 ft)

SECTION 3—FEEDING AREA

3.1 Feed bunks and fences. Cows are normally fed at a permanent location in or near the free stall barn, either from one or both sides of a feed bunk or along a fence line. The size of the feed bunk depends on cow measurements, amount and types of feed, feeding interval, and access time per cow. When separate ingredients are fed, sufficient space for all cows to eat must be provided. Use of chopped forages reduces space requirements. The minimum feeding space needed will result from providing continuous access to total mixed rations. Bunks must be sized to accomodate the volume of feed delivered (e.g., with a density of 320 to 400 kg/ma^3 (20 to 25 16/ft^3), approximately 0.1 m^3 (4 ft^3) will be needed per cow per day.). The feeding fence allows the cows to have access to feed, but separates them from it to prevent contamination. Feed bunks and fences also serve to restrain or confine the cows. Feeding areas must be arranged to allow access by feed handling equipment.

3.1.1 The **feed bunk length** provided is most affected by the feeding practice. If the amount of feed per cow is limited, enough bunk space is required for all cows to eat at the same time.

3.1.2 The **feed bunk width** is controlled by the reach of the cow. When the cows feed from both sides, the width can be twice as wide (see ASAE Data D321, Dimensions of Livestock and Poultry).

3.1.3 The **feed bunk height** depends on the toe-to-neck height of the cow. The bottom of the bunk can be 50 to 300 mm (2 to 12 in.) above the cow's front feet, depending on capacity desired see ASAE Data D321, Dimensions of Livestock and Poultry).

3.1.4 A **bunk step (sanitary step)** may be used at the base of the feed bunk to discourage the cow from backing next to the bunk, and thus help keep the bunk clean. The step also protects the bunk from damage during alley scraping.

3.1.5 **Control rail(s)** made of pipes, cables, or electric fence wires are often used to keep animals out of a feed bunk. These may be located directly above bunk side or offset in or out from it. They are located high enough to allow the animal to reach under them.

3.1.6 A **divider board** may be used to divide bunks accessible to animals from both sides. This is normally used in conjunction with a mechanical feeder that can be controlled to deliver different rations to either side of the bunk.

3.1.7 A paved **apron** may be used on a feed bunk located in an outside yard. The apron slopes away from the bunk to provide drainage and allows for easy removal of manure and snow and to prevent mud holes in dirt corrals. A rough, deep-grooved surface is recommended to prevent slippage.

3.1.8 A **roof or cover** may be provided for an outside bunk. This cover may protect feed and feeding equipment only, or be wide enough to protect feeding animals from precipitation or sunshine.

3.2 Feed bunk dimensions. The reference point is the juncture of the vertical portion of the feed bunk closest to the cow and the horizontal surface on which her front feet normally rest (step or platform) while feeding. Vertical dimensions are taken up or down from that point. Horizontal dimensions are taken away from the cow or towards the cow as she would stand facing the bunk. (See Fig. 2)

3.2.1 **Apron slope (S_A)** is the slope of the apron or alley expressed away from the bunk.

3.2.2 **Apron width (W_A)** is the horizontal dimension of the feed bunk apron or feed alley, including step.

3.2.3 **Back or divider board height (H_D)** is the vertical distance from reference point to the top of the back or divider board if present.

3.2.4 **Bottom height (H_F** is the vertical distance from the reference point to the bottom of the feed bunk floor.

3.2.5 **Bunk length (L_B) is the total outside length of the feed bunk.**

3.2.6 **Effective bunk length (L_E)** is the length of the bunk available for animal access. It equals bunk length (L_B) minus any major obstructions, such as waterers, building frames, or feeding equipment.

3.2.7 **Control rail(s) height(s) (H_C, H_{C_2}, etc)** is the vertical distance from the reference point to the control rails, the lowest subscript denotes the lowest rail when more than one are used.

3.2.8 **Control rail(s) offset(s) (O_{C_1}, O_{C_2})** is the horizontal offset of the control rail(s) from the reference point. A positive (+) sign indicates away from cow; a minus (−) sign indicates toward cow.

3.2.9 **Feed opening (O_F)** refers to the horizontal open portion of feeding fences or stanchions, clearance for animal heads.

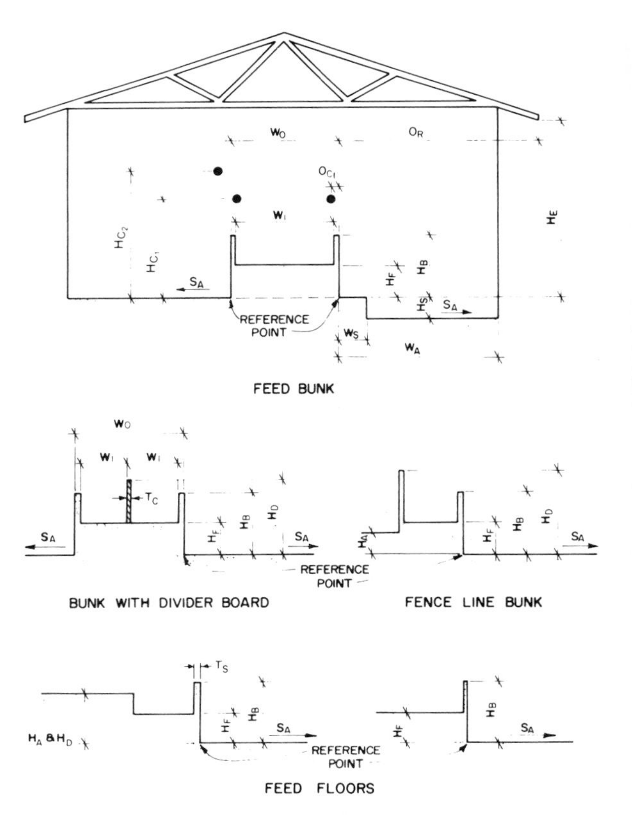

FIG. 2—FEED BUNKS AND FLOORS

3.2.10 Feeding alley height (H_A) is the vertical distance from the reference point to the feeding alley or driveway for fence line bunks. Required feed conveyor clearance equals divider board height (H_D) minus H_A.

3.2.11 Feeding space (S_F) refers to the lineal portion of the bunk available for single animals in bunks using vertical dividers or stanchions.

3.2.12 Inside width (W_I) is the horizontal dimension inside the bunk. It denotes the width of the space available for feed. When a divider board is used, W_I is the dimension from the inside of the divider board to the bunk side.

3.2.13 Outside width (W_O) is the total outside width or horizontal dimension of the feed bunk.

3.2.14 Post spacing (P_L) is the distance along the feed bunk between posts used for supporting feeders, fence lines, control lines, etc.

3.2.15 Roof height (H_E) is the vertical distance between the reference line and the lowest point on the bunk roof or cover.

3.2.16 Roof overhang (O_R) is the horizontal distance which the bunk roof or cover projects beyond the reference line.

3.2.17 Side height (H_B) is the vertical distance from the reference point to the top of the bunk side.

3.2.18 Side thickness (T_S) is the thickness of bunk side or fence line curb. For tapered sides this is the average thickness unless specified as top or bottom thickness.

3.2.19 Step height (H_S) is the vertical distance from the reference point to the juncture of the step face and the bunk apron.

3.2.20 Step width (W_S) is the horizontal dimension of the feed bunk step.

3.2.21 Center divider thickness (T_C) is the horizontal thickness of a divider in the bunk.

3.3 Feeding area design data

H_A—feeding alley height: $-H_s$ to 1000 mm ($-H_s$ to 20 in.)

H_B—side height: 600 to 700 mm (24 to 28 in.)

H_C—control rail height: H_B + (400 to 600 mm [H_B + (16 to 24 in.)]

H_D—divider board height: 500 to 1000 mm (20 to 40 in.)

H_E—roof height: 2000 to 4000 mm (6.5 to 13 ft)

H_F—bottom height: 100 to 300 mm (4 to 12 in.)

H_S—step height: 100 to 150 mm (4 to 6 in.)

$L_B$bunk length: See L_E for design data

L_E—effective bunk length
- feed not available continuously: 600 to 1000 mm/cow (24 to 40 in./cow)
- feed available continuously: 150 to 400 mm/cow (6 to 16 in./cow)

O_C—control rail offset: -300 to 300 mm (-12 to 12 in.)

O_F—feed opening: 200 to 400 mm (8 to 16 in.)

O_R—roof overhand: 1 to 3 m (3 to 10 ft)

P_L—post spacing: 1.2 to 5.0 m (4 to 16 ft) (depends on building design)

S_A—apron slope: 20 to 500 mm/m (0.2 to 0.5 in./ft)

S_F—feeding space: 400 to 1000 mm (16 to 40 in.)

T_S—side thickness: 40 to 200 mm (1.5 to 8 in.)

W_A—inside width: $W_Q - 2 (T_S) - T_C$

W_O—outside width
- Animals one side only: 750 to 800 mm (30 to 32 in.)
- Animals two sides: 1500 to 1650 mm (60 to 65 in.)

W_S—step width: 300 to 400 mm (12 to 16 in.)

T_C—center divider thickness: 25 to 300 mm (1 to 2 in.)

Cited Standard:

ASAE D321, Dimensions of Livestock and Poultry

LATERAL PRESSURE OF IRISH POTATOES STORED IN BULK

Developed by the Loads in Potato Storage Structures Subcommittee of the ASAE Structures Group; approved by the ASAE Structures and Environment Division Standards Committee; adopted by ASAE February 1985.

SECTION 1—PURPOSE AND SCOPE

1.1 These data provide information from which designers may calculate loads on walls, partitions, bin fronts, ducts, and appurtenances that are intended to resist lateral pressure of Irish potatoes stored in bulk.

1.2 These data encompass loads for both deep and shallow storage bins and include the effects of wet and dry potatoes.

1.3 The included data are for four different bins with one to three fillings of each bin.

SECTION 2—DEFINITIONS

2.1 Potato storage: A structure designed and constructed for storing potatoes in bulk where heat, light, moisture and ventilation needs of the potato are controlled.

2.2 Wet potatoes: Potatoes coated by water and/or chemical fungicide film, sprayed on during bin filling operations.

2.3 Dry potatoes: Potatoes put into the bin under normal harvesting moisture conditions without being sprayed.

2.4 Angle of internal friction (θ): The angle whose tangent equals the coefficient of friction between surfaces of the stored material. For potatoes it is approximated in practice as the filling angle of repose of the product (see Fig. 1) and may range from 28 deg for wet, round potatoes to 35 deg for dry, long potatoes.

2.5 Shallow bin: A storage bin in which $H/B < \tan(45 \text{ deg} + \theta/2)$; assumes $L > H$ and $L > B$ (see Fig. 1);
where

H = pile depth, m (ft)
B = bin width, m (ft)
L = bin length, m (ft)

2.6 Deep bin: A storage bin in which $H/B > \tan(45 \text{ deg} + \theta/2)$; assumes $L > H$ and $L > B$ (see Fig. 1).

2.7 Bulk density (W): The mass of the product per unit volume, in kg/m^3 (lb/ft^3). The bulk density of potatoes may range from 625 kg/m^3 (39 lb/ft^3) to 705 kg/m^3 (44 lb/ft^3). Large potatoes [greater than 89 mm (3.5 in.) diameter] with a low specific gravity of 1.050 may reach the smaller value. Small potatoes [less than 51 mm (2.0 in.) diameter] with a high specific gravity of 1.100 may reach the larger value. A bulk density of 673 kg/m^3 (42 lb/ft^3) is normally used for field-run potatoes.

2.8 Lateral pressure (P): The horizontal force per unit area of vertical wall surface in kPa (lbf/ft^2).

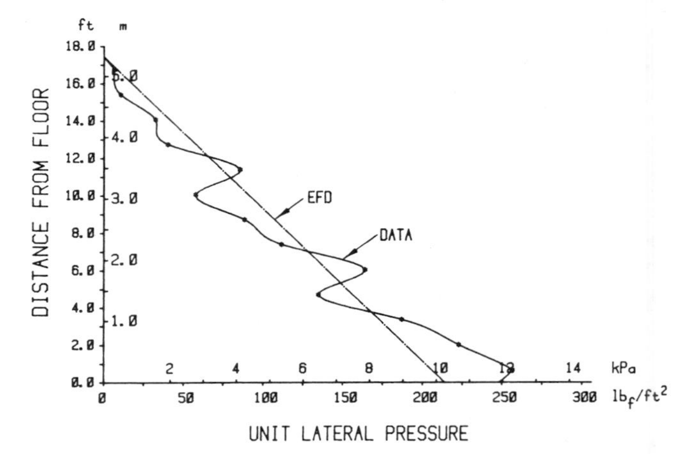

FIG. 2—TYPICAL MEASURED POTATO PRESSURE CURVE WITH CALCULATED EQUIVALENT FLUID DENSITY LINE FOR A SHALLOW BIN

2.9 Equivalent fluid density (EFD): Rankine's equation for active lateral pressure (P) of a cohesionless granular material with a level pile surface is given as

$$P = W(g/g_c)h \tan^2(45 \text{ deg} - \theta/2)$$

where

h = distance below top of pile, m (ft)
g = acceleration due to gravity, 9.8 m/s^2 (32.2 ft/s^2)
g_c = mass to force conversion factor. Equals 1 $kg \cdot m \cdot s^{-2}/N$ in SI system. (Equals 32.2 $lb \cdot ft \cdot s^{-2}/lbf$ in customary system.)
g/g_c = 9.8 N/kg (1 lbf/1 lb)

The terms W $\tan^2(45 \text{ deg} - \theta/2)$ are often grouped and called EFD. Units of EFD are kg/m^3 (lb/ft^3). Fig. 2 compares a typical measured pressure loading and a calculated EFD loading for a shallow bin.

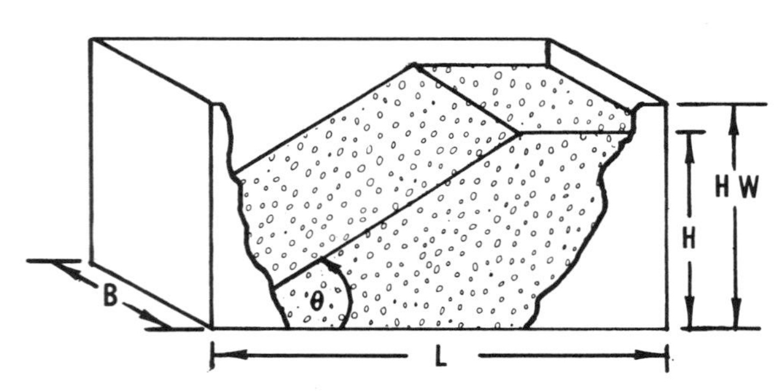

FIG. 1—BIN LAYOUT WITH SYMBOL REPRESENTATION SHOWING DIMENSIONS AND ANGLE OF REPOSE (See Section 2—Definitions, and paragraph 4.2, for symbols)

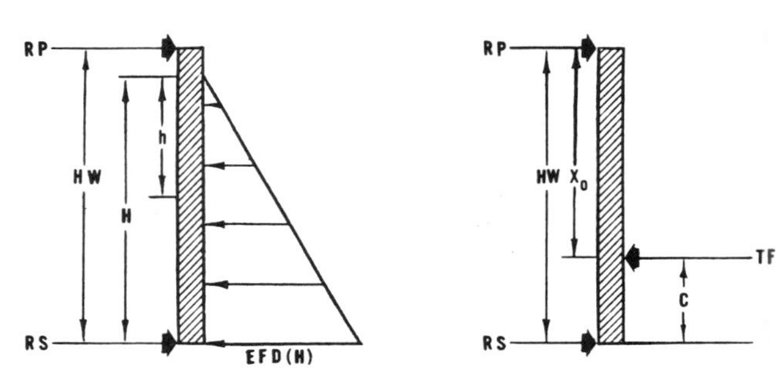

FIG. 3—FREE BODY DIAGRAMS OF BIN WALL (See Section 2—Definitions, and paragraph 4.2, for symbols)

TABLE 1—MEAN AND STANDARD DEVIATION (SD) OF ANNUAL MAXIMUM MEASURED TOTAL LATERAL FORCE (TF) AND CENTER OF FORCE (C) FOR POTATO BIN WALLS FOR SELECTED YEAR INTERVALS, IN SI UNITS

	Pile depth	Shallow bin			
		Dry potatoes (1974-76)		Wet potatoes (1977-78)	
		Mean	SD	Mean	SD
TF	3.5 m	7.1 kN/m	0.6 kN/m	9.6 kN/m	0.5 kN/m
	5.3 m	22.2 kN/m	2.2 kN/m	26.4 kN/m	1.1 kN/m
C	3.5 m	1.4 m	0.2 m	1.3 m	0.0 m
	5.3 m	1.7 m	0.1 m	1.7 m	0.1 m

	Pile depth	Deep bin	
		Dry potatoes (1950-52)	
		Mean	SD
TF	3.5 m	8.0 kN/m	1.4 kN/m
	5.3 m	15.9 kN/m	2.5 kN/m
C	3.5 m	1.3 m	0.1 m
	5.3 m	2.1 m	0.1 m

TABLE 2—MEAN AND STANDARD DEVIATION (SD) OF ANNUAL MAXIMUM MEASURED TOTAL LATERAL FORCE (TF) AND CENTER OF FORCE (C) FOR POTATO BIN WALLS FOR SELECTED YEAR INTERVALS, IN CUSTOMARY UNITS

	Pile depth	Shallow bin			
		Dry potatoes (1974-76)		Wet potatoes (1977-78)	
		Mean	SD	Mean	SD
TF	11.4 ft	500 lbf/ft	50 lbf/ft	650 lbf/ft	50 lbf/ft
	17.4 ft	1500 lbf/ft	150 lbf/ft	1800 lbf/ft	100 lbf/ft
C	11.4 ft	4.6 ft	0.6 ft	4.3 ft	0.0 ft
	17.4 ft	5.5 ft	0.4 ft	5.5 ft	0.2 ft

	Pile depth	Deep bin	
		Dry potatoes (1950-52)	
		Mean	SD
TF	11.4 ft	550 lbf/ft	100 lbf/ft
	17.4 ft	1100 lbf/ft	200 lbf/ft
C	11.4 ft	4.4 ft	0.3 ft
	17.4 ft	7.0 ft	0.3 ft

2.10 Total lateral force (TF): The summation of the lateral pressures for a unit of wall length, in kN/m (lbf/ft).

2.11 Center of force (C): The vertical distance measured from the bottom of the pile to the location at which the total lateral force (TF) acts, in m (ft). C is often expressed as a fraction of the pile height. The distance below the top of the wall to this point, where shear is zero and bending moment is maximum, is shown in Fig. 3 as X_o.

2.12 Percent error (E):

$$E = \frac{P_{AL} - P_{DATA}}{P_{DATA}} \; 100\%$$

where

P_{AL} = parameter calculated with an assumed load condition
P_{DATA} = parameter determined from actual data

TABLE 3—CALCULATED EQUIVALENT FLUID DENSITY (EFD) AND MEASURED CENTER OF FORCE (C) ESTIMATES FOR POTATO BIN WALLS, IN SI UNITS

Maximum calculated EFD

Pile depth m	Shallow bin: Center of force m	Shallow bin: Dry potatoes EFD kg/m³	Shallow bin: Wet potatoes EFD kg/m³	Deep bin: Center of force m	Deep bin: Dry potatoes EFD kg/m³
3.5	H/3.0	128 (1975)	169 (1978)	H/2.6	160 (1950)
5.3	H/3.5	179 (1976)	197 (1978)	H/2.6	127 (1951)

Mean and standard deviation (SD) of annual maximum calculated EFD

Pile depth m	Shallow bin					Deep bin		
	Center of force m	Dry potatoes (1974-76)		Wet potatoes (1977-78)		Center of force m	Dry potatoes (1950-52)	
		Mean kg/m³	SD kg/m³	Mean kg/m³	SD kg/m³		Mean kg/m³	SD kg/m³
3.5	H/3.0	120	10	163	9	H/2.6	145	26
5.3	H/3.5	161	16	191	8	H/2.6	114	18

Greatest mean EFD and standard deviation (SD) for all recordings for a single year

Pile depth m	Shallow bin					Deep bin		
	Center of force m	Dry potatoes (1975)		Wet potatoes (1978)		Center of force m	Dry potatoes (1950)	
		Mean kg/m³	SD kg/m³	Mean kg/m³	SD kg/m³		Mean kg/m³	SD kg/m³
3.5	H/3.0	115	10	124	15	H/2.6	139	19
5.3	H/3.5	147	24	158	11	H/2.6	116	7

TABLE 4—CALCULATED EQUIVALENT FLUID DENSITY (EFD) AND MEASURED CENTER OF FORCE (C) ESTIMATES FOR POTATO BIN WALLS, IN CUSTOMARY UNITS

Maximum calculated EFD

	Shallow bin			Deep bin	
Pile depth ft	Center of force ft	Dry potatoes EFD lb/ft^3	Wet potatoes EFD lb/ft^3	Center of force ft	Dry potatoes EFD lb/ft^3
11.4	H/3.0	8 (1975)	11 (1978)	H/2.6	10 (1950)
17.4	H/3.5	11 (1976)	12 (1978)	H/2.6	8 (1951)

Mean and standard deviation (SD) of annual maximum calculated EFD

	Shallow bin					Deep bin		
		Dry potatoes (1974-76)		Wet potatoes (1977-78)			Dry potatoes (1950-52)	
Pile depth ft	Center of force ft	Mean lb/ft^3	SD lb/ft^3	Mean lb/ft^3	SD lb/ft^3	Center of force ft	Mean lb/ft^3	SD lb/ft^3
11.4	H/3.0	7	0.6	10	0.6	H/2.6	9	1.6
17.4	H/3.5	10	1.0	12	0.5	H/2.6	7	1.1

Greatest mean EFD and standard deviation (SD) for all recordings for a single year

	Shallow bin					Deep bin		
		Dry potatoes (1975)		Wet potatoes (1978)			Dry potatoes (1950)	
Pile depth ft	Center of force ft	Mean lb/ft^3	SD lb/ft^3	Mean lb/ft^3	SD lb/ft^3	Center of force ft	Mean lb/ft^3	SD lb/ft^3
11.4	H/3.0	7	0.6	8	0.9	H/2.6	9	1.2
17.4	H/3.5	9	1.5	10	0.7	H/2.6	7	0.5

SECTION 3—MEASURED PRESSURE PATTERNS

3.1 The origin of the data included herein was a series of tests in which the following research procedure was used:

3.1.1 Shallow bin. Square pressure panels 305 mm (1 ft) by 305 mm (1 ft) with load cell transducers were installed in the wall from the surface of the pile to the bottom of the bin. Data for each year were obtained from weekly sampling of daily data logs over a 5-yr period.

3.1.2 Deep bin. Rectangular pressure panels 914 mm (3 ft) wide by 610 mm (2 ft) long, supported by calibrated steel bars, were installed in a continuous series. Data were recorded three to six times during each storage season over a 3-yr period.

3.2 Pressure pattern trend with increasing depth. A greater increase in lateral pressure (P) was noted in the lower 1.2 m (4 ft) to 1.8 m (6 ft) of the pile. Because of this pattern, data for actual pile depth, 5.3 m (17.4 ft), and a subset of these data for the top 3.5 m (11.4 ft), are shown.

SECTION 4—MEASURED DATA AND CALCULATED APPROXIMATIONS

4.1 Means and standard deviations of total lateral force (TF) per meter and per foot of wall length and center of force (C) for measured annual maximum pressure patterns during appropriate year intervals are tabulated in Tables 1 and 2 for both the top 3.5 m (11.4 ft) portion of the pile and the total pile.

4.2 Estimates of equivalent fluid density (EFD) and center of force (C) in Table 3 (Table 4) may be used to approximate pressure patterns, and **maximum bending moment of the wall (MB),** in N·m/m (ft·lbf/ft); **reaction at wall plate (RP)**, in kN/m (lbf/ft); and **reaction at wall sill (RS)**, in kN/m (lbf/ft). The estimates were calculated for a pinned-end, simple beam, where

HW = H + 0.5 m (HW = H + 1.6 ft)

where

HW = height of bin wall, m (ft)

4.3 The reported data present a skewed, non-normal distribution of undetermined type.

SECTION 5—ACCURACY OF APPROXIMATIONS

5.1 Design parameters for selected equivalent fluid density (EFD) and center of force (C) are compared in Table 5, by applying the error relationship from paragraph 2.12 to design parameters obtained by integration of measured maximum pressure patterns for a pinned-end, simple beam. A negative percent error (E) indicates that EFD predicts a design value lower than that obtained from the actual measured maximum pressures.

5.2 Since pressure trends were of such an irregular form, a curve did not give consistently better representation than a straight line approximation. As more results become available, smooth curves may be developed that are not now technically justified.

TABLE 5—PERCENT ERROR (E) OF CALCULATED DESIGN PARAMETERS USING SELECTED EQUIVALENT FLUID DENSITY (EFD) AND CENTER OF FORCE (C) WHEN COMPARED TO MAXIMUM DESIGN PARAMETERS CALCULATED FROM REPORTED MAXIMUM PRESSURE OCCURRENCES. SHALLOW BIN CONDITIONS. HW = 5.8 m (19 ft), H = 5.3 m (17.4 ft). SIMPLY SUPPORTED BEAM.

		Percent error (E)								
	Data source	Reaction sill			Reaction plate			Bending moment		
		C=H/3	C=H/3.3	C=H/3.5	C=H/3	C=H/3.3	C=H/3.5	C=H/3	C=H/3.3	C=H/3.5
EFD = 192 kg/m^3 (12 lb/ft^3)	Yaeger									
	Norchip, wet	-7	-3	0	9	-1	-9	3	-10	-20
	Norchip, dry	-4	0	3	46	32	21	25	10	-3
	Russet, wet	42	48	52	40	27	17	34	18	4
	Torabi									
	Russet, dry	20	25	29	76	60	47	14	0	-11
	Powell									
	Russet, dry	22	27	31	19	9	0	148	117	92
	Edgar									
	Pontiac, dry*	22	26	30	-5	-13	-21	6	-7	-17
EFD = 208 kg/m^3 (13 lb/ft^3)	Yaeger									
	Norchip, wet	1	5	9	18	8	-1	11	-2	-14
	Norchip, dry	4	8	12	58	43	31	36	19	5
	Russet, wet	54	60	65	52	38	26	46	28	13
	Torabi									
	Russet, dry	30	35	40	91	73	59	23	8	-4
	Powell									
	Russet, dry	32	37	42	29	18	8	168	135	108
	Edgar									
	Pontiac, dry*	32	37	41	3	-6	-14	15	1	-11

*Original data adjusted by $\sqrt{\frac{5.3\text{ m}}{3.0\text{ m}}}$ to approximate shallow bin conditions.

References: Last printed in 1985 STANDARDS; list available from ASAE Headquarters.

ASAE Standard: ASAE S466

NOMENCLATURE/TERMINOLOGY FOR LIVESTOCK WASTE/MANURE HANDLING EQUIPMENT

Proposed by the Farmstead Equipment Association of Farm and Industrial Equipment Institute; approved by ASAE Agricultural Sanitation and Waste Management Committee and ASAE Wagon Box, Forage Box, Manure Spreader and Farm Wagon Subcommittee; approved by ASAE Structures and Environment Division Standards Committee; adopted by ASAE March 1987.

SECTION 1—PURPOSE AND SCOPE

1.1 The purpose of this Standard is to define the terminology and establish uniformity within the field of livestock waste/manure handling equipment. See ASAE Standard S292, Uniform Terminology for Rural Waste Management, for definitions of the materials handled.

1.2 By these definitions, existing and newly developed equipment can be categorized, compared, and rated as to intended purpose and performance.

SECTION 2—DEFINITIONS

2.1 Transfer equipment/handling facilities

2.1.1 Manure transfer device: A device whose primary function is to move manure from a collection point to storage, and from storage to processing or utilization.

2.1.1.1 Displacement manure pumps: A positive displacement transfer device which utilizes a variety of methods to move primarily semi-solid manure through a pipeline.

2.1.1.1.1 Piston manure pump: A pump which uses a reciprocating piston and cylinder to move manure from a collection hopper into and through a pipeline. The pump may be mechanically or hydraulically powered. The cylinder can be of various shapes; round, square, etc.

2.1.1.1.1.1 Solid piston manure pump: A piston manure pump which uses a solid piston to move manure. The operation fully retracts the piston from the cylinder, charges the cylinder with manure, and then extends the piston into the cylinder to move the manure. Piston travel may be along a vertical, horizontal or inclined axis.

2.1.1.1.1.2 Hollow piston manure pump: A piston manure pump which uses a hollow piston equipped with a one-way gate which allows manure to flow through the piston on the retraction stroke and then moves it into the pipeline on the extension stroke. Piston travel is generally along an inclined axis.

2.1.1.1.2 Air pressure manure transfer system: A manure transfer system which utilizes a collection tank which is pressurized after filling, causing the collected manure to move into and through the pipeline.

2.1.1.1.3 Progressive cavity pump: A displacement pump where rotating components create a forward moving cavity that serves to transport manure. It can be used for agitation, transfer or irrigation and is typically powered hydraulically, electrically or by tractor PTO.

2.1.1.2 Centrifugal manure transfer pumps: A slurry and liquid (thin slurry) manure transfer pump which moves manure by pressure generated through a rotary centrifugal impeller and housing. Pumps are classified by their mechanical configuration and power source. Most pumps are equipped with a shredder or chopping device at the impeller inlet to aid in reducing large organic masses before they enter the pump housing.

2.1.1.2.1 Vertical shaft pump: A centrifugal transfer pump which utilizes a power source above the manure level and is connected to the centrifugal impeller by a vertical drive shaft. This pump may be permanently installed, usually in a reception pit, where it is powered by either tractor PTO or electric drive. It may also be used in a portable configuration where it is tractor PTO powered, and either 3-point hitch or 2-wheel trailer mounted. The transport mechanism may provide a means to rotate the pump to a vertical position within the storage structure.

2.1.1.2.2 Submersible pump: A centrifugal transfer pump which utilizes an electric or hydraulic motor attached directly to the impeller which is submerged in the manure slurry. The pump is typically supported by an attached cable and winch. The pump may be used as a stationary unit or as a portable unit in which case it is moved from reception pit to storage tank.

2.1.1.2.3 Horizontal shaft pump: A centrifugal pump with a horizontal shaft impeller housing where manure is drawn through an intake tube and through the impeller housing. It may be mounted directly to a tank wall, to a base or to a trailer. It is powered electrically or by tractor PTO.

2.1.2 Reception pit (holding unit): An in-ground tank (typically) whose purpose is to collect slurry or liquid manure for a short storage time (usually less than seven days) prior to transfer to manure storage. The tank may be round, rectangular or square in shape.

2.2 Agitation equipment: Slurry and thin slurry (liquid) manure handling equipment whose primary purpose is to agitate manure in storage and convert it into a homogeneous slurry. The equipment may also be used to transfer manure for final disposal or use.

2.2.1 Agitation pumps: Manure pumps which agitate by recirculating manure slurry, pumping it back into the manure storage either above or under the manure surface. These pumps are similar to transfer pumps, and also typically include a shredder chopping pump.

2.2.1.1 Vertical shaft pump: Similar to a vertical shaft centrifugal transfer pump. Generally tractor PTO powered and 3-point hitch or 2-wheel trailer mounted, it includes a mechanism to rotate the pump to a vertical position in the storage structure. It is used primarily with in-ground storage tanks. An axial flow prop may also be used along with the centrifugal impeller.

2.2.1.2 "Trail" or "lagoon" pump (inclined shaft pump): A centrifugal manure pump in which the power is supplied to the pump impeller from a source above the manure level by a drive shaft which assumes an inclined orientation. It is typically used in earthen storage basins. It may be 2-wheel trailer mounted, in which case the pump assembly does not rotate but the complete unit is backed over the basin's sloping sides and into the manure. It may also be 3-point hitch mounted where the pump and driveline assembly are lowered into the manure by pivoting at the tractor attaching frame.

2.2.1.3 Submersible agitator: (See submersible centrifugal manure transfer pump.)

2.2.1.4 Horizontal shaft pump: (See horizontal shaft centrifugal manure transfer pump.)

2.2.1.5 Progressive cavity pump: (See progressive cavity displacement manure transfer pump.)

2.2.2 Prop agitators: An axial flow propeller used for the purpose of agitation of manure slurry only.

2.2.2.1 "Trail" or "lagoon" prop agitator (inclined shaft prop): Similar to a trail or lagoon agitation pump where an axial flow prop is mounted in place of a centrifugal pump assembly and agitation nozzle. It is used most frequently in earthen storage basins.

2.2.2.2 **Submersible prop agitator:** Similar to a submersible transfer or agitation pump, but with an axial flow propeller rather than a centrifugal pump attached to the submerged power unit. It is typically used in a manure storage tank.

2.3 **Spreader:** A device to haul, unload and/or distribute manure. May be PTO or hydraulically driven.

2.3.1 **Box type:** A device with a horizontal floor, vertical side walls, and vertical or slanted front or rear wall. Manure is conveyed or pushed to the open end where a device, usually a beater, distributes or spreads the material being hauled. It is used primarily for solid and semi-solid manure.

2.3.2 **Open tank:** A device consisting of a half cylinder or modified half cylinder lying in a horizontal axis. A shaft runs the length of the cylinder and with its attachments, usually chains and hammers, is the sole means of shredding and unloading the manure being hauled. Delivery is either to the side or rear. It is used primarily for solid and semi-solid manure.

2.3.3 **"V" bottom:** An open or covered rectangular tank with a V-shaped bottom. A shaft runs the length of the tank at the bottom of the "V", and with its attachments (auger flighting or modified flighting) agitates and moves the manure to a second device which spreads the manure on the soil surface. The spreading device is generally a horizontal conveyor/paddle mechanism located at 90 deg to the center shaft. It is used primarily for semi-solid and slurry manure.

2.3.4 **Closed tank (tanker):** A completely closed structure, usually a horizontal cylinder or modified cylinder. The tank is filled either by creating a vacuum in the tank to pull manure into it, or by using an auxiliary pump. The tank may be emptied by gravity, by pressure discharge or by a mechanical pump or "slinger". The manure may be surface applied or injected below the soil surface. It is used primarily for slurry and thin slurry (liquid) manure.

2.4 **Manure separators:** A device or structure which brings about a partial separation of solid material from a liquid or slurry. The objective is to separate manures into solid and liquid fractions.

2.4.1 **Mechanical separators**

2.4.1.1 **Rotary strainers:** A slowly rotating, perforated cylinder mounted horizontally. Manure flows by gravity onto the cylinder at one end, where solids are scraped off the cylinder surface and moved to the exit end, and liquids pass through the screen where they are collected and removed.

2.4.1.2 **Vacuum filter:** A horizontally mounted, rotating perforated cylinder with a cloth fiber cover that uses vacuum to draw liquids out of manure. Manure flows onto the cylinder surface, liquids which pass through are collected and solids are scraped off the cloth at a separation point.

2.4.1.3 **Vibrating screen:** A circular or square shallow container with a replaceable screen bottom. The assembly is vibrated both vertically and horizontally. Manure flows into the container, where liquids pass through the screen and the solids are collected to the side of the container.

2.4.1.4 **Centrifugal separators:** A rapidly rotating device that uses centrifugal force to remove manure liquids from solids. One type, a relatively low speed design, uses a cylindrical or conical screen in which manure is fed into one end, solids are contained by the screen, scraped off and discharged from the opposite end while the liquid passes through. It may be installed vertically or horizontally. A second type, a higher speed decanter, uses a conical shaped bowl in which centrifugal force causes the denser solids to migrate to the bowl exterior where they are collected. Less dense liquids are forced to the center where they are collected.

2.4.1.5 **Belt pressure roller (belt press):** A roller and belt device whereby two concentrically running belts are used to squeeze the manure as it is deposited between the belts. The belts pass over a series of spring-loaded rollers where liquids are squeezed out or through the belt, and remaining solids are scraped off at a belt separation point.

2.4.1.6 **Perforated pressure roller (roll press):** One or more sets of parallel rollers between which manure passes. The upper roller is solid and may be of a compressible material. It acts to press liquids through openings in the lower perforated roller.

2.4.1.7 **Static inclined screen:** A screen, mounted on an incline, over which manure passes as it flows by gravity from a top head box. The liquid passes through the screen due to its flow momentum and surface tension, while solids continue over and flow off the end of the screen.

2.4.1.8 **Screw press:** A straight or tapered screw of fixed or varying pitch contained in a perforated or slotted cylinder. Liquids pass through the screen as the manure is conveyed along the cylinder, while solids are retained within the cylinder and are discharged out the end. The solids are retained by the screen and pushed out of the separator on the opposite side.

2.4.1.9 **Brushed screen/roller press:** A rectangular container with four vertical sides and a bottom consisting of two half cylindrical screens lying side by side which provide two stages of separation. Within each screen rotates a multiple brush and roller assembly which sweeps the manure across the screen. Manure is pumped into one side of the separator. The liquids are forced through the screen by the brush/roller while the solids are retained by the screen and pushed out of the separator on the opposite side.

2.4.2 **Gravity separation systems:** Structures which utilize gravity to collect more dense particulate solids by allowing them to settle out of highly liquid manure. The structure may be of any shape but with a relatively shallow depth.

2.4.2.1 **Settling tank:** A relatively short-term separation structure, smaller in size than a settling basin. The liquid is allowed to fully drain away for solids removal by mechanical means.

2.4.2.2 **Settling basins:** A relatively long-term separation structure, larger in size than a settling tank. Solids collection is by mechanical means once the liquids evaporate or have been drained away.

2.4.2.3 **Settling channels:** A continuous separation structure in which settling occurs over a defined distance in a relatively slow-moving manure flow. Baffles and porous dams may be used to aid separation by further slowing manure flow rates. Solids are removed mechanically once liquids are fully drained away.

2.5 **Aerators:** A device which brings about aeration of liquid and highly liquid manure for the purpose of accelerating aerobic decomposition.

2.5.1 **Surface aerator:** A partially submerged impeller whose action results in vigorous agitation and air entrainment. The impeller may be mounted on floats in a storage structure with varying liquid levels or fixed in a constant liquid level system. Power may be supplied by an electric or hydraulic motor coupled directly to the impeller.

2.5.2 **Turbine aerator:** A submerged axial flow pump in which the manure discharge is directed toward the surface where it breaks the surface and results in air entrainment. A variation of this aerator injects compressed air into the manure flow immediately downstream of the pump venturi. Power may be supplied by electric or hydraulic motor, or by a driveshaft from a surface power source.

2.5.3 **Air diffusers:** A submerged porous diffuser or air nozzle whose action results in direct air entrainment. Air is supplied via a surface compressor or blower.

Cited Standard:

ASAE S292, Uniform Terminology for Rural Waste Management

SOIL AND WATER RESOURCE MANAGEMENT

ASAE Notation:

The letter S preceding numerical designation indicates ASAE Standard; EP indicates Engineering Practice; D indicates Data. A decimal and numeral following the file number indicate the number of times a document has been revised. Thus ASAE S201.4 indicates Standard number 201, four times revised. The letter T after the designation indicates tentative status. Always refer to ASAE documents by complete designation to avoid confusion with standards of other organizations. For example: ASAE S201.4.

The symbol **T** preceding or in the margin adjacent to section headings, paragraph numbers, figure captions, or table headings indicates a technical change was incorporated in that area when this document was last revised. The symbol **T** preceding the title of a document indicates essentially the entire document has been revised. The symbol **E** used similarly indicates editorial changes or corrections have been made with no intended change in the technical meaning of the document.

ASAE Engineering Practice: ASAE EP260.4

DESIGN AND CONSTRUCTION OF SUBSURFACE DRAINS IN HUMID AREAS

Proposed by the Committee on Design and Construction of Tile Drains; approved by the Soil and Water Division Standards Committee; adopted by ASAE as a Tentative Recommendation 1952; revised 1959; reclassified as a full Recommendation 1962; revised December 1968; reconfirmed December 1973; revised and reclassified as an Engineering Practice December 1977; reconfirmed December 1982; revised December 1983; revised editorially February 1987.

SECTION 1—PURPOSE AND SCOPE

1.1 This Engineering Practice is intended as a guide to engineers in the design and construction of subsurface drains, particularly as a basis for writing detailed specifications for a specific drainage job. It is not designed to serve as a complete set of standards or specifications. This Engineering Practice is not directly applicable to drainage of irrigated lands in semi-arid and arid regions.

SECTION 2—MATERIALS

2.1 Clay and concrete drain tile

2.1.1 Quality. Clay and concrete drain tile should meet requirements specified by the American Society for Testing and Materials. The quality classes for clay and concrete tile are designated in American Society for Testing and Materials Standard C4, Specifications for Clay Drain Tile; ASTM Standard C412, Specifications for Concrete Drain Tile; and ASTM Standard C498, Specifications for Perforated Clay Drain Tile. Clay and concrete drain tile that meet ASTM specifications will generally give long and satisfactory service under most exposure conditions.

2.1.2 Tile exposed to acid and sulfate conditions. Concrete tile installed in acid soils or sulfate soils should be of the special-quality class described in ASTM Standard C412, Specifications for Concrete Drain Tile. Clay tile are acid resistant, and they are not affected by exposure to sulfate soils.

2.1.3 Tile exposed to freezing and thawing action

2.1.3.1 Clay or shale tile meeting the standard-quality class requirements described in ASTM Standard C4, Specifications for Clay Drain Tile, are frostresistant.

2.1.3.2 Concrete tile meeting standard-quality class requirements described in ASTM Standard C412, Specifications for Concrete Drain Tile, are frostresistant.

2.1.3.3 Drain tile should not be laid out or stacked in contact with wet ground or areas subject to flooding during the winter or periods of freezing and thawing.

2.1.4 Loading of drain tile. It is recommended that loads be determined for all the tile laid in deep and wide trenches, and that the quality-class or load requirements of the tile be specified to withstand the designed load.

2.2 Corrugated polyethylene drainage tubing

2.2.1 Quality. Corrugated polyethylene drainage tubing should meet requirements specified in ASTM Standard F405, Specification for Corrugated Polyethylene (PE) Tubing and Fittings, and ASTM Standard F667, Specification for 8, 10, 12, and 15-in. Corrugated Polyethylene Tubing. Quality classes are designated in this specification. Standard tubing is satisfactory for most agricultural drainage; heavy duty tubing is recommended where wide trenches are required, where side support is poor, or where rocky conditions are expected. Standard tubing can be specified for systems with prepared bedding (see paragraph 3.7.3) at depths no shallower than 0.6 m (2 ft) in soils not containing rock. Heavy duty tubing should be used in very narrow trenches (less than three pipe diameters wide) or very wide trenches [wider than o.d. + 0.3 m (1 ft)] and for difficult installation conditions such as where side support for the tubing is poor or non-existent, where depth of tubing is excessive [usually 3 m (10 ft)], or where rocky soil conditions are expected. (For information pertaining to installation instructions, see paragraph 4.8.2.)

2.2.2 Precautions during extremes in temperature

2.2.2.1 At colder temperatures, tubing stiffness increases and flexibility decreases thus care is necessary when rolls of tubing are uncoiled. Rapid uncoiling may cause the tubing to be stressed excessively and cracking may occur.

2.3 Other subsurface conduits. All other conduits regardless of material should be durable and perform their function as intended. Sewer pipe often can be used in subsurface drainage systems if joints or perforations provide for adequate water entry.

2.4 Envelopes and stabilizing materials

2.4.1 Most subsurface drains installed in stable soils will give satisfactory performance without the use of envelopes or other soil stabilizing materials. However, there are unstable soil situations (very fine sand or other cohesionless soil) where envelope materials are essential to ensure successful performance of the drainage system.

2.4.2 Conditions where envelopes or stabilizing materials are needed include (1) soils that easily fill a drain with sediment, such as coarse silt or very fine and medium sands in the size ranges of 0.05 to 0.5 mm; (2) soils that do not provide a stable foundation, such as saturated sands in a quick condition; and (3) soils that tend to seal or clog drain openings and limit water entry into the drain.

2.4.3 Envelope materials

2.4.3.1 Vegetative envelope or filter materials include straw, hay, ground corncobs, woodchips, sawdust and coconut fiber. These materials may be used with rigid conduits such as clay and concrete tile and pipe. Vegetative materials are not recommended for use with corrugated plastic tubing because the strength of the tubing depends on the development of lateral support. The minimum recommended thickness of packed vegetative materials around rigid conduits is 150 mm (6 in.). The effective life of vegetative materials depends on the type of material and the drainage condition. Local experience should be considered before vegetative materials are recommended.

2.4.3.2 Prefabricated envelope materials such as fiberglass sheets or mat (borosilicate type), spun-bonded nylon fabric, and other durable filter materials are satisfactory in some soils. Caution should be exercised in soils where iron or manganese oxide and other chemical deposits are probable because openings may become sealed, restricting water entry to the drain. Local experience should be considered before synthetic envelopes are recommended.

2.4.3.3 Mineral envelope materials include gravel, crushed stone, slag, and others. Design criteria for gravel envelopes have been developed and published in Section 16, Drainage of Agricultural Land, National Engineering Handbook, United States Department of Agriculture, Soil Conservation Service. Pit run sand and gravel may meet these criteria. Excessive fines, more than 5% finer than a No. 60 sieve, should be avoided. The minimum recommended thickness of mineral envelope is 75 mm (3 in.). The envelope material should completely encircle the conduits.

2.4.3.4 Stabilizing materials are used to provide a firm bedding condition. Stabilizing materials should have enough thickness and stiffness to maintain the conduit in proper alignment.

SECTION 3—DESIGN

3.1 Capacity and depth of existing outlet drains

3.1.1 Open ditches. Where the outlet is a stream or open ditch, it should provide free outlet for subsurface drains within a reasonable period of time after storm peak flow and should be large enough to remove the surface runoff from the watershed in time to prevent crop damage by overflow. Drainage runoff data, in accordance with ASAE Engineering Practice ASAE EP302, Design and Construction of Surface Drainage Systems on Agricultural Lands in Humid Areas, and/or ASAE Engineering Practice EP407, Agricultural Drainage Outlets—Open Channels, should be used in the design of new outlet ditches and in checking the capacity of existing ditches. The outlet ditch should be of adequate depth to ensure that the flow line (invert) of the main subsurface drain at the oulet should be at least 0.3 m (1 ft) above the normal low water flow in the ditch when the subsurface drains are laid at the adequate depth for the soil to be drained.

E

3.1.2 Existing subsurface drains. Where an existing drain is used as the outlet for a proposed system, it should be in good condition, free of failures, deep enough for adequate drainage and should have sufficient hydraulic capacity. If the existing subsurface drain does not have sufficient capacity, a new drain should be laid for the additional system, or a new drain installed large enough to handle both systems.

3.1.3 Pump outlets. When a suitable gravity outlet for the subsurface drainage system is not available, pumping should be considered. The pumping outlet should have sufficient capacity to remove water at a rate recommended in ASAE Engineering Practice EP369, Design of Agricultural Drainage Pumping Plants.

3.2 Rate of water removal (drainage coefficeint)

3.2.1 The drainage coefficeint is the rate of water removal expressed as the depth of water to be removed in 24 h. The frequency, intensity and duration of rainfall as well as the porosity and permeability of the soil, timeliness of field operation, and crop to be grown, should be considered in selecting the design drainage coefficient.

3.2.2 When the design rate of drop of the water table is known, the drainage coefficient should be computed from the drainable porosity.

3.2.3 If drainable porosity information is not available, the drainage coefficient may be selected from Tables 1 or 2. Fig. 1 provides solutions for design quantities of flow.

3.2.3.1 When the land to be drained has surface drainage provided either by natural topography or by artificial channels, land smoothing or diversions, the coefficients in Table 1 should be used.

3.2.3.2 Where it is necessary to admit surface water through screened or protected inlets into the system, the coefficients in Table 2 should be used.

TABLE 1—DRAINAGE COEFFICIENT
(No surface water admitted directly into the drain)

Soil	Field crops		Truck crops	
	mm/day	in./day	mm/day	in./day
Mineral	10 to 13	3/8 to 1/2	13 to 19	1/2 to 3/4
organic	13 to 19	1/2 to 3/4	19 to 38	3/4 to 1½

TABLE 2—DRAINAGE COEFFICIENT
(Surface water admitted directly into the drain)

Soil	Field Crops			
	Blind inlets		Open inlets	
	mm/day	in./day	mm/day	in./day
Mineral	13 to 19	1/2 to 3/4	13 to 25	1/2 to 1
organic	19 to 25	3/4 to 1	25 to 38	1 to 1½
	Truck Crops			
Mineral	19 to 25	3/4 to 1	25 to 38	1 to 1½
organic	38 to 51	1½ to 2	51 to 102	2 to 4

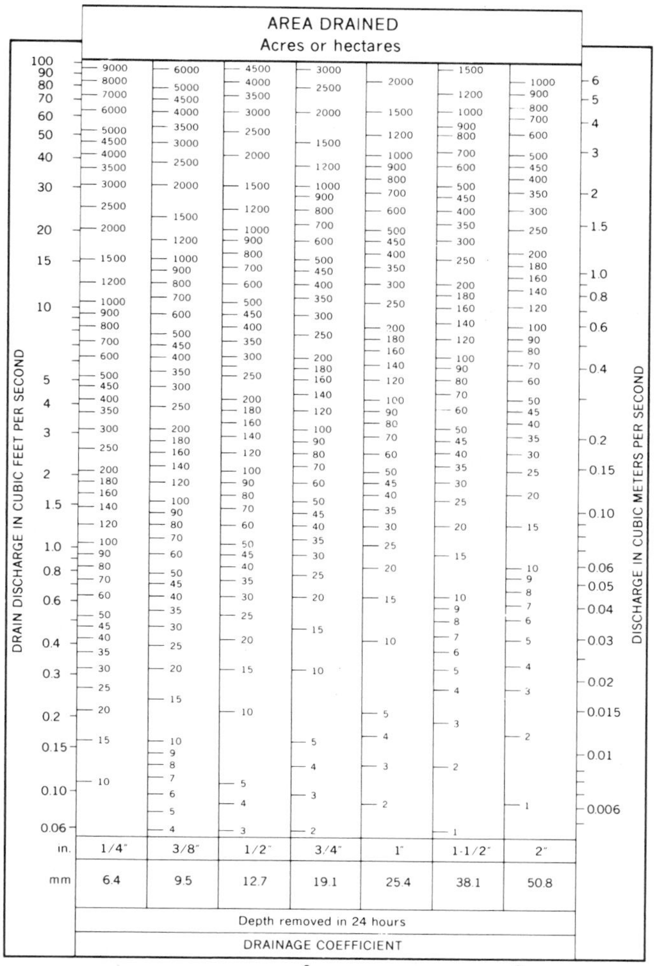

AREA DRAINED
Acres or hectares

DRAIN DISCHARGE IN CUBIC FEET PER SECOND: 100, 90, 80, 70, 60, 50, 40, 30, 20, 15, 10, 5, 4, 3, 2, 1.5, 1.0, 0.8, 0.6, 0.4, 0.3, 0.2, 0.15, 0.10, 0.06

1/4" (6.4 mm): 9000, 8000, 7000, 6000, 5000, 4500, 4000, 3500, 3000, 2500, 2000, 1500, 1200, 1000, 900, 800, 700, 600, 500, 450, 400, 350, 300, 250, 200, 180, 160, 140, 120, 100, 90, 80, 70, 60, 50, 45, 40, 35, 30, 25, 20, 15, 10

3/8" (9.5 mm): 6000, 5000, 4500, 4000, 3500, 3000, 2500, 2000, 1500, 1200, 1000, 900, 800, 700, 600, 500, 450, 400, 350, 300, 250, 200, 180, 160, 140, 120, 100, 90, 80, 70, 60, 50, 45, 40, 35, 30, 25, 20, 15, 10, 9, 8, 7, 6, 5, 4

1/2" (12.7 mm): 4500, 4000, 3500, 3000, 2500, 2000, 1500, 1200, 1000, 900, 800, 700, 600, 500, 450, 400, 350, 300, 250, 200, 180, 160, 140, 120, 100, 90, 80, 70, 60, 50, 45, 40, 35, 30, 25, 20, 15, 10, 5, 4, 3

3/4" (19.1 mm): 3000, 2500, 2000, 1500, 1200, 1000, 900, 800, 700, 600, 500, 450, 400, 350, 300, 250, 200, 180, 160, 140, 120, 100, 90, 80, 70, 60, 50, 45, 40, 35, 30, 25, 20, 15, 10, 5, 4, 3, 2

1" (25.4 mm): 2000, 1500, 1200, 1000, 900, 800, 700, 600, 500, 450, 400, 350, 300, 250, 200, 180, 160, 140, 120, 100, 90, 80, 70, 60, 50, 45, 40, 35, 30, 25, 20, 15, 10, 5, 4, 3, 2

1-1/2" (38.1 mm): 1500, 1200, 1000, 900, 800, 700, 600, 500, 450, 400, 350, 300, 250, 200, 180, 160, 140, 120, 100, 90, 80, 70, 60, 50, 45, 40, 35, 30, 25, 20, 15, 10, 9, 8, 7, 6, 5, 4, 3, 2, 1

2" (50.8 mm): 1000, 900, 800, 700, 600, 500, 450, 400, 350, 300, 250, 200, 180, 160, 140, 120, 100, 90, 80, 70, 60, 50, 45, 40, 35, 30, 25, 20, 15, 10, 9, 8, 7, 6, 5, 4, 3, 2, 1

DISCHARGE IN CUBIC METERS PER SECOND: 6, 5, 4, 3, 2, 1.5, 1.0, 0.8, 0.6, 0.4, 0.2, 0.15, 0.10, 0.06, 0.05, 0.04, 0.03, 0.02, 0.015, 0.01, 0.006

in.	1/4"	3/8"	1/2"	3/4"	1"	1-1/2"	2"
mm	6.4	9.5	12.7	19.1	25.4	38.1	50.8

Depth removed in 24 hours
DRAINAGE COEFFICIENT

Note: Use acres with ft^3/s and hectares with m^3/s

FIG. 1—CHART FOR DETERMINING DISCHARGE

3.2.4 When intercepting seepage or springs, the design rate of flow should be based on an estimate of the actual rate of flow as determined by field investigations. While subsurface drains are normally used for intercepting seepage, open channels should be used for severe conditions.

3.3 Capacity of mains and lateral drains

3.3.1 Drain capacity is the product of the nominal cross-sectional area of the drain and the velocity of flow. The Manning velocity equation is

$$V = \frac{1.49}{n} r^{2/3} s^{1/2}$$

where

V = velocity, m/s (ft/s)
r = hydraulic radius, m (ft)
s = slope, m/m (ft/ft)
n = Manning's roughness coefficient

Type of conduit	Nominal diameter, mm (in.)	Value of n
Clay or concrete drain tile	76 to 760 (3 to 30)	0.012 to 0.014
Corrugated plastic tubing	76 to 203 (3 to 8)	0.015
Corrugated plastic tubing	254 to 305 (10 to 12)	0.017
Corrugated plastic tubing	>305 (>12)	0.02
Corrugated metal pipe	76 to 760 (3 to 30)	0.025
Smooth wall pipe (tongue and groove or bell spigot joints)	76 to 760 (3 to 30)	0.012

3.3.2 Fig. 1 provides the discharge required for a given drainage area at the selected drainage coefficient. With this discharge and the grade of the drain the required size of clay concrete tile can be determined from Fig. 2. Fig. 3 provides the size for corrugated plastic tubing with "n" values varying for increasing diameter.

3.4 Drainage area. The selected drainage coefficient must be applied to the correct area as follows.

3.4.1 Where no surface water is admitted directly to the subsurface drainage system through inlets, the selected drainage coefficient should apply only to the soil area requiring drainage. An exception to this is where the runoff from an upland watershed spreads out over the area to be drained, thus adding an additional amount of infiltration over that produced by rainfall. If the runoff from this upland watershed cannot be diverted from or channeled through the area to be drained, the drainage coefficient should be applied to the entire watershed area.

3.4.2 When surface water is admitted directly to the subsurface drainage system through surface inlets, the entire watershed draining to the surface inlet should be used as the contributing area. An exception to this rule is where only a small amount of the runoff will be impounded at the location of the inlet and the remainder will flow away in a surface channel. The drain should then be designed of adequate size to prevent crop damage.

3.4.3 For seepage interceptor drains and for random drains the required size should be based on the estimated inflow rate.

3.5 Minimum drain size. The following recommendations serve as a guide in selecting the minimum size of subsurface drains.

3.5.1 The hydraulic capacity of the drain at the design grade should not be exceeded.

3.5.2 A 76 mm (3 in.) diameter drain may be installed where the minimum grade is as noted in paragraph 3.9.2.1; however 102 mm (4 in.) is generally the recommended size for flatter grades and for stable mineral soils. Subsurface drains installed in unstable mineral soils should use recommendations in paragraph 4.5.

3.5.3 A 127 mm (5 in.) minimum drain size is recommended in unstable organic soil.

3.5.4 Use of Figs. 1, 2, and 3: (1) Select drainage coefficient column in Fig. 1. (2) Find the number of acres or hectares to be drained. (3) Project a horizontal line to determine the discharge in either ft³/s or m³/s and then, using this discharge with Fig. 2 or 3, project a horizontal line until it intersects the vertical line representing the drain grade shown at the bottom of the chart. (4) The size shown in the space between diagonal solid lines where the point of intersection occurs is the required size to use. (5) The diagonal, partially dashed lines show the velocity of flow in the drain.

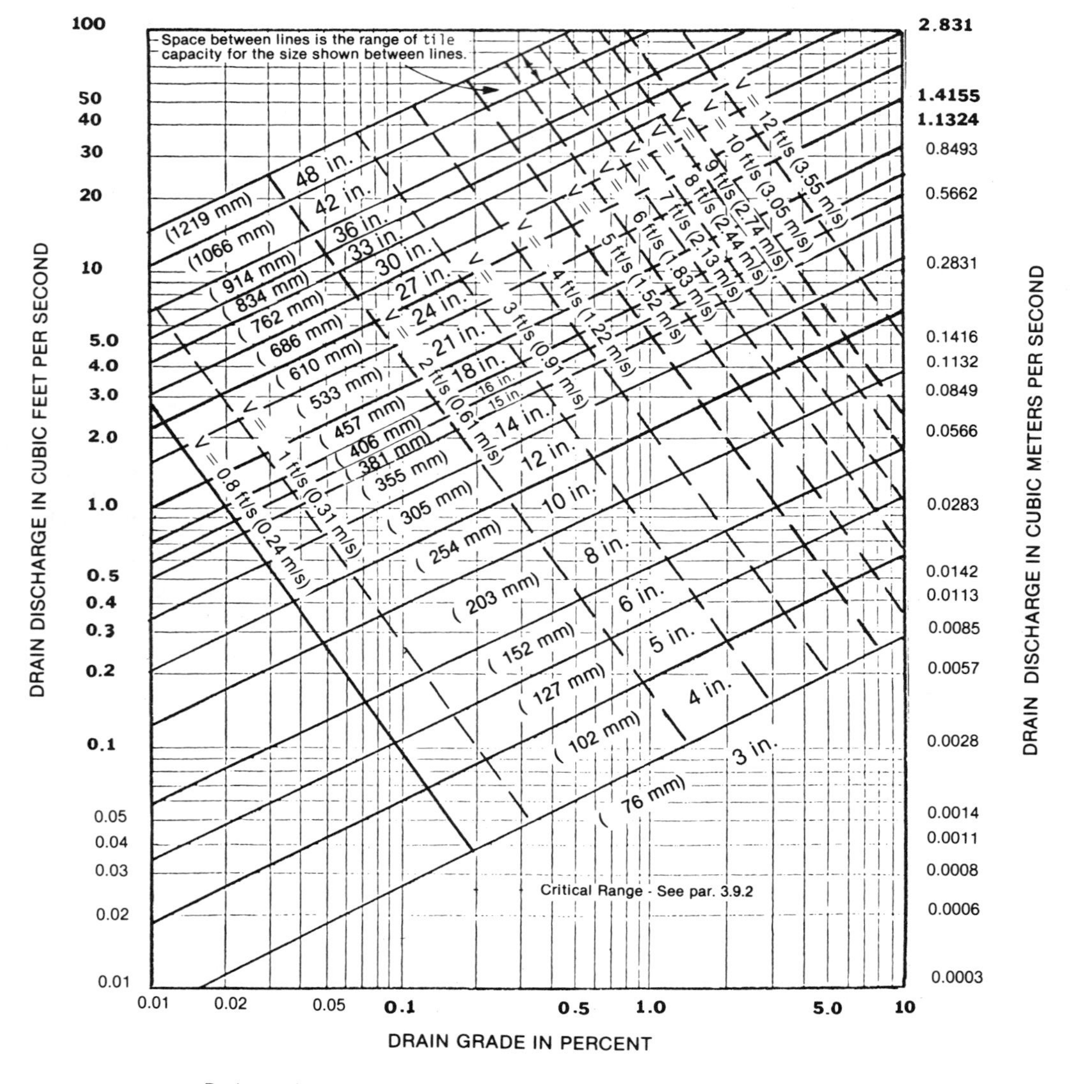

FIG. 2—GUIDE FOR DETERMINING THE REQUIRED SIZE OF CLAY AND CONCRETE DRAIN TILE (n = 0.013)

3.6 Minimum depth

3.6.1 Mineral soils

3.6.1.1 It is desirable for the depth of the laterals to be 0.9 to 1.2 m (2 to 4 ft). This may be reduced to 0.8 m (2.5 ft) under the following conditions provided the lateral spacing is adjusted (1) where slowly permeable soils exist; (2) where a layer of extremely tight soil, sand or large stones prohibits greater depth; (3) in depressional or impounded areas; and (4) where the outlet depth is limited.

3.6.1.2 Where 76 mm (3 in.) drains are used for combined drainage and subirrigation of shallow-rooted truck crops the minimum depth may be 0.5 m (1.5 ft) if heavy machinery is not used in the cropped area.

3.6.1.3 The minimum cover to protect a well-bedded subsurface drain from breakage by heavy machinery is 0.6 m (2 ft) for 76, 102, 127, and 152 mm (3, 4, 5, and 6 in.) diameters and 0.8 m (2.5 ft) for 203, 254, 305, 381 and 457 mm (8, 10, 12, 15 and 18 in.) diameters.

3.6.1.4 Organic soils should not be subsurface drained before initial subsidence has occurred. To produce initial subsidence temporary open drains should be constructed to carry off free water, and the area should be allowed to stand or be partially cultivated for a period of 3 to 5 yr before installing buried subsurface drains. Subsidence will likely continue after the initial stage and decrease with time to as little as 30 mm (1.2 in.) per year due to oxidation under cultivation, compaction, shrinkage and wind erosion. Following initial subsidence the soil strata at drain depth should be stable enough to maintain installed drain grades.

3.6.1.5 Controlled drainage should always be considered in organic soils. Experimental data show that the subsidence rate increases as the average depth to the water table during the crop season increases. If controlled drainage is not provided to hold subsidence to a minimum, the depth of cover should be increased to 0.9 m (3 ft).

3.6.2 Where it is impossible or uneconomical to secure the minimum cover specified above as under waterways and road ditches or at the outlet end of the main or near structures, the drain should be replaced with high-strength nonperforated pipe, and placed in a grand envelope, whenever possible.

3.7 Loading

3.7.1 Subsurface drains should be so installed that the load does not fracture the tile or cause excess deflection of flexible tubing.

3.7.2 Tile (concrete and clay). Loads on drain tile should be determined from formulas developed by Marston, Iowa, Experiment Station Bulletin 96, 1930. For live load calculations refer to Table 3.

3.7.3 Plastic tubing

3.7.3.1 Tubing should be so installed so it does not deflect more than 20% of its nominal diameter.

3.7.3.2 Loads on tubing installed under projecting conditions (wide trenches) and for live loads shall be computed as

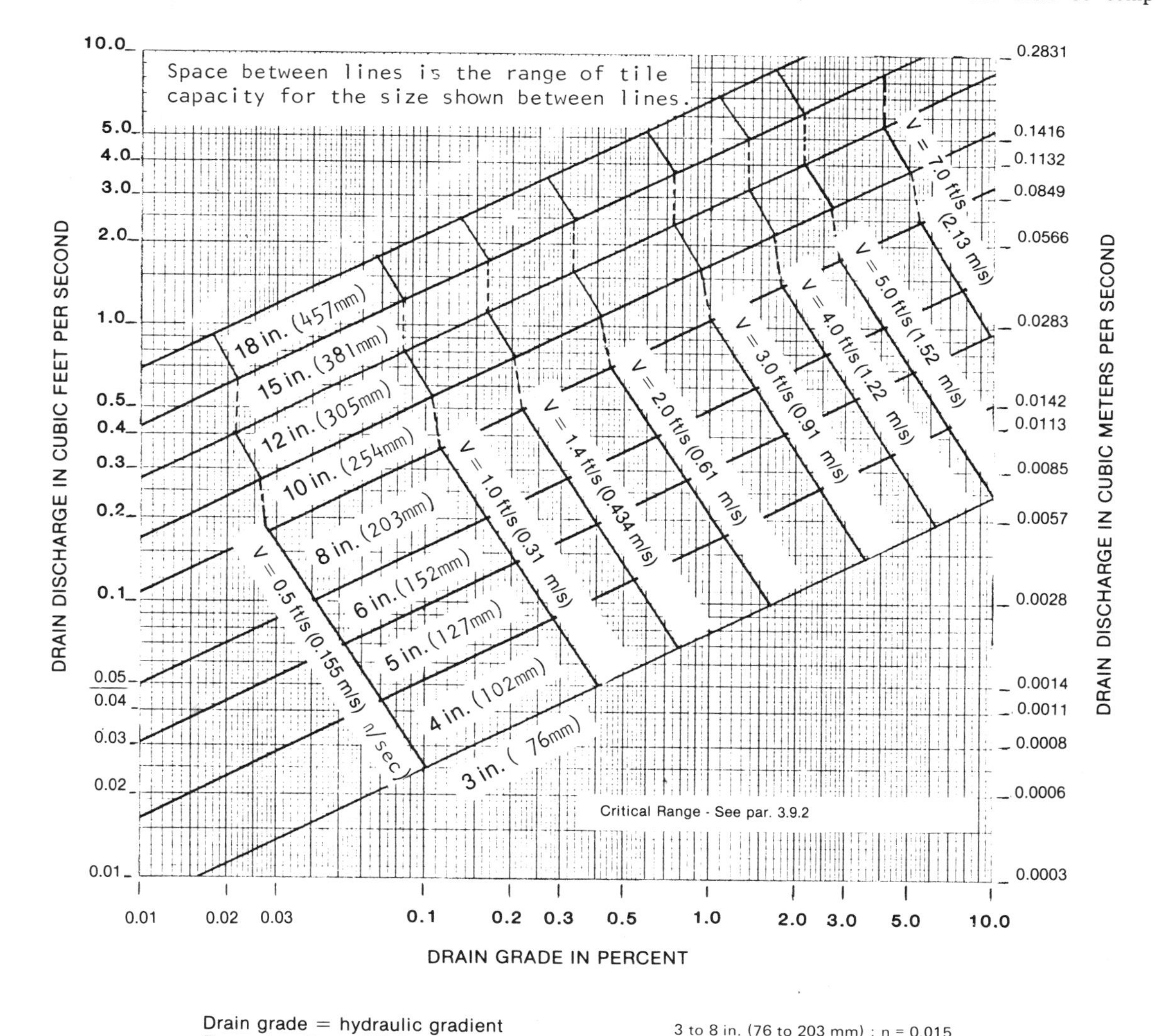

Drain grade = hydraulic gradient
Percent = ft per 100 ft, or m per 100 m
V = velocity in ft/s (m/s)

3 to 8 in. (76 to 203 mm) : n = 0.015
10, 12 in. (254, 305 mm) : n = 0.017
15, 18 in. (381, 457 mm) : n = 0.02

FIG. 3—GUIDE FOR DETERMINING THE REQUIRED SIZE OF CORRUGATED PLASTIC DRAINAGE TUBING

TABLE 3—PERCENT OF WHEEL LOADS TRANSMITTED TO UNDERGROUND CONDUIT

Depth of backfill over top of tile	Trench width at top of drain						
	0.3 m (1 ft)	0.6 m (2 ft)	0.9 m (3 ft)	1.2 m (4 ft)	1.5 m (5 ft)	1.8 m (6 ft)	2.1 m (7 ft)
0.3 m (1 ft)	17.0*	26.0*	28.6*	29.7*	29.9*	30.2*	30.3*
0.6 m (2 ft)	8.3	14.2	18.3	20.7	21.8	22.7	23.0
0.9 m (3 ft)	4.3	8.3	11.3	13.5	14.8	15.8	16.7
1.2 m (4 ft)	2.5	5.2	7.2	9.0	10.3	11.5	12.3
1.5 m (5 ft)	1.7	3.3	5.0	6.3	7.3	8.3	9.0
1.8 m (6 ft)	1.0	2.3	3.7	4.7	5.5	6.2	7.0

Live loads trandmitted are practically negligible below 1.8 m (6 ft).

*These percentages include both live load and impact transmitted to 0.30 m (1 lineal ft) of tile.

described in paragraph 3.7.2. For tubing installed under ditch conditions, the flexible load equation shall apply:

$$W_c = C_d w B_c B_d$$

where

W_c = the tubing load in kg/m
C_d = the load coefficient
w = soil weight in kg/m^3
B_c = the conduit width in m
B_d = the trench width in m (See Spangler, 1941, Iowa Engineering Experiment Station Bulletin 153 or American Society of Civil Engineers Manual No. 39, 1969, pp 185-228.)

The load carrying capacity of flexible tubing should be computed from the modified Iowa formula. (See Spangler, 1966, Soil Engineering.) The formula is

$$W_c = 93.16 \frac{Y}{D_1 K} (EI/r^3 + 0.061\ E')$$

where

W_c = supporting strength in kg/m length
y = allowable vertical deflection in m
D_1 = deflection lag factor (use 3.4)
K = bedding factor (use 0.096 for 90 deg groove angle or fitted circular groove)
EI/r^3 = ring stiffness in kN/m^2 from parallel plate test at 20% deflection
E′ = modulus of soil reaction or soil side support in kN/m^2.

Note: 1 kg/m^2 = 0.0098 kN/m^2 = 0.0098 kPa = 0.001422 lb/in./in. For backfill with loose granular soil, E′ = 345 kN/m^2 (50 psi) is suggested. (See Howard, 1977, J. Geotechnical Engineering Division, ASCE, 103:33-43, for other conditions.)

In rocky conditions, special care is needed to avoid local damage to the tubing that may lead to structural failure. The width of the trench should be at least 152 mm (6 in.) wider than the outside diameter of the tubing to permit placement of the supporting backfill along the lower sides of the tubing. Very narrow trenches may result in bridging of the backfill material above the tubing leaving the sides unsupported. Installation of corrugated plastic tubing in a narrow flat trench with uncompacted side support is not recommended. Maximum trench depths for a given set of conditions are given in Table 4.

3.7.4 Other loading requirements

3.7.4.1 To prevent overloading in deep and wide ditches, it is sometimes feasible to construct a subditch with a trenching machine, or by hand, in the bottom of a wide ditch which has been excavated by a bulldozer, dragline, power shovel, or backhoe. The width of the subditch measured at the top of the drain determines the load, and the width of the excavation above this point, is unimportant.

3.7.4.2 In deep trenches, it is good practice to backfill in several stages, thus allowing time for settlement between fillings. Live load calculations should be added to the soil load when determining class of drain and type of bedding. Table 3 shows the percent of wheel loads transmitted to the tile.

3.7.4.3 When the calculated load requirements exceed the crushing strengths for special quality concrete, or heavy duty clay drain tile or heavy duty tubing in applicable specifications, pipe capable of supporting the required loads should be specified.

3.8 Spacing

3.8.1 When a system of parallel laterals is used, the spacing between laterals should be based on soil type, soil permeability, drain depths, crops to be grown, applicable drainage coefficient and degree of surface drainage. Local drainage guides provide specific tile spacing recommendations.

TABLE 4—MAXIMUM TRENCH DEPTHS FOR TUBING BURIED IN LOOSE, FINE-TEXTURED SOILS, IN METERS (FEET)*

Nominal tubing diameter, mm (in.)	Tubing quality (ASTM)	Trench width at top of tubing, m (ft)			
		0.3 (1)	0.4 (1.3)	0.6 (2)	0.8 (2.6) or greater
102 (4)	Standard†	3.9 (12.8)	2.1 (6.9)	1.7 (5.6)	1.6 (5.2)
	Heavy-Duty ‡	§	3.0 (9.8)	2.1 (6.9)	1.9 (6.2)
152 (6)	Standard†	3.1 (10.2)	2.1 (6.9)	1.7 (5.6)	1.6 (5.2)
	Heavy-Duty ‡	§	2.9 (9.5)	2.0 (6.6)	1.9 (6.2)
203 (8)	Standard†	3.1 (10.2)	2.2 (7.2)	1.7 (5.6)	1.6 (5.2)
	Heavy-Duty ‡	§	3.0 (9.8)	2.1 (6.9)	1.9 (6.2)
254 (10)	‡	—	2.8 (9.2)	2.0 (6.6)	1.9. (6.2)
305 (12)	‡	—	2.7 (8.9)	2.0 (6.6)	1.9 (6.2)
381 (15)	‡	—	—	2.1 (6.9)	1.9 (6.2)
457 (18)	‡				

NOTE: These depths are based on limited research and should be used with CAUTION. Differences in commercial tubing from several manufacturers, including corrugation design and pipe stiffness and soil conditions, may change the assumptions; and, therefore, maximum depths may be more or less than stated above.

*Assumptions: E′ = 345 kN/m^2 (50 psi)
D_1 = 3.4
K = 0.096 (90 deg bedding angle)
w = 1750 kg/m^3 (109 lb/ft^3)
Y = 1.1x (x = horizontal deflection)

†Pipe stiffness 90 kN/m^2 (13 psi) for 20% deflection

‡Pipe stiffness 124 kN/m^2 (18 psi) for 20% deflection

§ Any depth is permissible for this or less width and for 0.2 m (0.67 ft) trench width for all sizes.

Revised table from Fenemor, A.D., B.R. Bevier, and G.O. Schwab, 1979. Prediction of deflection for corrugated plastic tubing. TRANSACTIONS of the ASAE 22(6):1338-1342

TABLE 5—TYPICAL DRAIN SPACINGS IN VARIOUS SOILS

Soil	Permeability	Spacing	
		m	ft
Clay and clay loam soils	Very slow	9 to 21	30 to 70
Silt and silty clay loam soils	Slow to moderately slow	18 to 30	60 to 100
Sandy loam soils	Moderate to rapid	30 to 91	100 to 300
Muck		15 to 61	50 to 200

3.8.2 When local drainage guides are not available, the general recommendations in Table 5 should be used for drains placed approximately 1.2 m (4 ft) deep. Detailed drain spacing equations and their limitations can be found in (1) Soil and Water Conservation Engineering, by Schwab, Frevert, Edminster and Barnes and (2) USDA, SCS National Engineering Handbook, Section 16, Drainage of Agricultural Land.

3.9 Grades

3.9.1 Grades of drains should be adequate to prevent reverse grades that might result from construction techniques or soil conditions. Subsurface drains with reverse grade or very little grade tend to fill with sediment. Drains on steep grades may have excessive velocities and be subject to pressure flow that results in an erosion hazard to the soils surrounding the drains.

3.9.2 Minimum grades

3.9.2.1 The grade should be as great as possible on flatlands, but adequate depth should not be sacrificed. The minimum grade of drains should be limited as shown in Table 6.

3.9.2.2 When grades flatter than the above are used for main lines the system should be designed to reduce the sedimentation as follows.

3.9.2.2.1 The system should have a free outlet so backwater conditions will not further reduce velocities.

3.9.2.2.2 Sediment traps and cleanout systems should be provided.

3.9.2.2.3 The entire system should be protected from sedimentation by the use of filters and envelopes suitable to prevent piping or erosion of the base material into the system.

3.9.2.2.4 Breathers and relief wells are needed for venting and assuring full hydraulic head operation.

3.9.3 Maximum grades

3.9.3.1 Short lateral drains not subject to pressure flow may be installed on grades up to 5%. On long laterals and main drains the maximum permissible velocity without protective measures should be limited as shown in Table 7.

3.9.3.2 Main drains that serve laterals and other drains with high peak capacities in excess of the design drainage coefficient may be subject to pressure flow. Under these flow conditions, the main line may be overdesigned for the excess capacity to prevent pressure flow and to reduce velocity as shown in Table 7, or special precaution should be taken as listed below.

TABLE 6—MINIMUM GRADE IN PERCENT

Inside diameter	Drains not subjected to fine sand or silt*		Where fine sand and silt may enter the drain†	
	tile	tubing	tile	tubing
76 mm (3 in.)	0.08	0.10	0.60	0.81
102 mm (4 in.)	0.05	0.07	0.41	0.55
127 mm (5 in.)	0.04	0.05	0.30	0.41
152 mm (6 in.)	0.03	0.04	0.24	0.32

*These grades provide for a minimum cleaning velocity of 0.15 m/s (0.5 ft/s).

†These grades provide for a minimum cleaning velocity of 0.42 m/s (1.4 ft/s).

TABLE 7—MAXIMUM VELOCITY WITHOUT PROTECTIVE MEASURES

Soil Texture	m/s	ft/s
Sand and sandy loam	1.1	3.5
Silt and silt loam	1.5	5.0
Silty clay loam	1.8	6.0
Clay and clay loam	2.1	7.0
Coarse sand or gravel	2.7	9.0

3.9.3.3 Protective measures for mains or laterals on steep grades or exceeding the criteria listed in Table 7 should include one or more of the following:

3.9.3.3.1 For clay or concrete tile

3.9.3.3.1.1 Use only tile uniform in size and shape with smooth ends.

3.9.3.3.1.2 Lay the tile to secure a tight fit with the inside section of one tile matching that of the adjoining sections.

3.9.3.3.1.3 Wrap open joints with tar-impregnated paper, burlap, or special filter material such as plastic sheets, fiberglass fabric or a properly graded sand and gravel.

3.9.3.3.1.4 Select least erodibile soil for blinding.

3.9.3.3.1.5 Tamp soil material carefully under and alongside the tile before backfilling.

3.9.3.3.1.6 Cement joints or use a drain with a watertight joint.

3.9.3.3.2 For corrugated plastic tubing or continuous pipe

3.9.3.3.2.1 For drains of this type with perforations, completely encase the drain with filter material, or use a properly graded sand and gravel filter.

3.9.3.3.2.2 Use nonperforated corrugated plastic tubing or continuous pipe with taped or leak-proof connections.

3.9.3.3.3 The necessity of a breather near the beginning of a steep grade and a relief well at the point where the steep grade changes to a flat grade should be considered. This will be determined by the velocities in the drain, the soil in which the drain is laid and the capacity of the drain below the steep grade with respect to that in the steep grade. If the capacity of the drain on the flat grade is such that the hydraulic gradeline will nearly reach to the surface of the ground for full flow, a relief well should be installed to prevent a blowout.

3.10 Alignment

3.10.1 Change in horizontal direction should be made in such a way that the specified grade is maintained. For tile lines the flow of water should not be impeded by excessive roughness and joint spacing on the outer side of the curve in accordance with the recommendations on tile openings in paragraph 4.5.

3.10.2 The change in horizontal direction should be made by one of the following methods.

3.10.2.1 A gradual curve which is compatible with the installation machine's ability to maintain grade.

3.10.2.2 The use of corrugated plastic tubing, manufactured bends or fittings.

3.10.2.3 The use of junction boxes and manholes.

3.11 Connections

3.11.1 For tile lines, manufactured connections or branches for joining two lines should be used when available. If connections are not available, the junction should be chipped, fitted, and sealed with cement mortar.

3.11.2 Mains should be laid deep enough so that the centerlines of laterals can be joined at or above the approximate centerline of the main.

3.11.3 For corrugated plastic tubing, manufactured couplers or fittings should be used at all joints and at all changes in direction where the centerline radius is less than three times the tubing diameter, at changes in diameter, and at the end of the line. All connections must be compatible with the tubing.

3.12 Outlet protection

3.12.1 When the main drain outlets into an open ditch, the end of the drain should always be protected against erosion and undermining. When the drain is corrugated plastic tubing, protection is also needed at the exposed portion against weather and animal damage, as well as crushing.

3.12.1.1 Where no surface water will enter the ditch at the location outlet, a minimum of 3.7 m (12 ft) of continuous rigid pipe should be used. At least two-thirds of its length should be embedded into the ditch bank with the overhanging length discharging at the toe of the ditch slope. When the placement of a pipe projecting into the ditch will cause a serious ice jam or be damaged by floating ice or debris, the pipe freeboard should either be reduced or the pipe recessed into the ditch bank. Corrugated thermoplastic tubing should not be used as an outlet pipe, unless the tubing is protected with a headwall.

3.12.1.2 Where surface water will enter the ditch at the location of the outlet, some type of structure should be used to lower the surface flow to the ditch. When there is no spoil bank, the straight-drop spillway is generally the best type of structure. If there is a spoil bank and sufficient temporary storage on the land is possible and permissible, a pipe drop-inlet structure will usually provide the best and most economical installation. Sometimes it may be possible to move the drain away from the waterway or divert the surface water to another location at least 18 to 23 m (60 to 75 ft) away and convey the surface flow into the ditch through a sodded chute.

3.12.2 Swinging gates, rods, screens or similar protection should be used on all outlets to exclude small animals unless the outlet is so located that it would be impossible for them to enter at the outlet end. Grating bars or attached screens should not be used on drain lines that have surface inlets since debris may enter through the inlets and collect on the gratings. Only gates should be used in these locations. Where plastic tubing is connected to old tile lines which may serve as animal runs, an animal guard should be installed within the line to restrict animal travel.

3.12.3 Outlets for lines that are affected by high tides or storm water flooding should be protected from backflow by means of automatic tidegates.

3.13 Junction boxes. Junction boxes should be used where three or more mains join. Junction boxes should be located in permanent fence rows or in noncultivated areas with the cover above ground to provide easy access for inspection. In cultivated fields, the box should be constructed so that the top is at least 0.5 m (1.5 ft) below the ground surface.

3.14 Relief well (stand pipe). Relief wells serve to relieve pressure in a line that might otherwise cause the line to blow out. A relief well can be constructed by placing a T connection in the line and fitting a pipe vertically into the T. The pipe should extend about 0.3 (1 ft) above the ground unless provisions have been made in the design for it to also serve as a surface inlet. The exposed end of the pipe should be covered with heavy wire mesh or grating. The size of the riser should be equal to or greater than the diameter of the line. Relief wells should be located at points where the line might become overloaded for a short period of time such as at the end of a steep section as noted in paragraph 3.9.2.2; and on lines that have surface inlets, particularly below large surface inlets.

3.15 Surface inlets

3.15.1 Surface inlets may be used where the construction of surface drains to remove the ponded water is not feasible or practical. Surface inlets may also be used to carry prolonged flows that might otherwise be carried in grass waterways. Surface inlets should be located on laterals or at least 2.4 m (8 ft) to one side of the main since inlets are a frequent source of trouble. All surface inlets should be provided with gratings to exclude coarse debris.

3.15.2 Surface inlets should be constructed to withstand crushing loads of agricultural machinery. High-strength walls and adequate concrete foundations below frost level are necessary to withstand frost heaving.

3.15.3 A sodded area around the inlet may reduce maintenance by trapping sediment.

3.16 Sediment traps

3.16.1 Sediment traps should only be used with open surface inlets, and as noted in paragraph 3.9.2.2.

3.16.2 In sandy or other unstable soils, protection against sediment entry should be provided by use of proper envelope materials (see paragraph 2.4).

3.17 Interception of hillside seepage. Hillside seepage should be intercepted and removed by placing the drain on or slightly above the impervious or slowly permeable layer and across the path of flow. Borings should be made to locate the restricting layers.

3.18 Trees. Roots of water-loving trees such as willow, cottonwood, elm and soft maple growing near subsurface drains may enter them and obstruct flow. The roots enter the drain to get water. Masses of roots sometimes grow until they completely fill the drain. Some shrubs, crops and grasses may cause trouble on shallow-depth lines. Where possible, water-loving trees should be removed from a distance of approximately 30 m (100 ft) on each side of the drain. A clearance of 15 m (50 ft) should be maintained from other species of trees, with the possible exception of fruit trees. If trees cannot be removed or the line rerouted, tile lines should be constructed with closed joints throughout the root-zone area of the trees. When the drain is of plastic tubing, nonperforated tubing should be used when passing through areas where root growth may create an obstruction in the line.

3.19 Buried cables, pipelines, highways, and other facilities. The crossing of buried cables, pipelines, and other facilities with drains should be avoided where possible. The number of crossings should be kept at a minimum by installing intercepting lines parallel to the facility. A section of watertight line should be used at each crossing. All crossings should meet the requirements of the owner of the facility.

3.20 Plan. A plan should be prepared for each subsurface installation for use by the contractor during construction and for the owner as a permanent record. If any changes are necessary during construction they must be noted on an as-built copy of the plan. It is recommended that this plan be filed with the abstract or deed of the property. The plan should include a map of the installation showing accurate locations and sizes of all lines and appurtenances together with a profile or copy of the staking notes on the main and submains. It is important that the plan be adequately referenced to existing physical features so that any segment can be located to facilitate maintenance and repair.

3.21 Iron ochre. Drains in some soils may be adversely affected by iron deposits (iron ochre) in the openings and inside the drain. The deposits usually associated with iron bacteria are filamentous and gelatinous and can clog drains and the drain envelope area. In sandy soils with high iron potential, it is important not to blind the drains with top soil (consisting of organic materials) and every effort should be made to keep spodic materials and top soil away from the envelope. The organic material will absorb iron sulfide flowing in the groundwater and thus contribute to a clogging problem. If a problem is known to occur, accessible maintenance openings should be provided for periodic jet cleaning.

SECTION 4—CONSTRUCTION

4.1 Controlling direction and grade

4.1.1 Automatic grade control may be used by utilizing grading devices capable of following the designated gradeline. Automatic grade control equipment is required for high-speed trenching equipment and plow installations over 10 m/min (30 ft/min).

4.1.2 Manual grade control with visual sight bars and targets can be used for trenching operations with installation rates less than 10 m/min (30 ft/min). To govern alignment, direction and grade stakes should be set 30 m (100 ft) or less apart on straight lines and 15 m (50 ft) or less on curves for all drain lines to be constructed. They should also be set at all intersections of mains and points of grade change. Grade-stake elevations should always be determined with an engineer's level or transit on grades less than 1%.

4.1.2.1 Targets are used as a guide to finish the trench to exact grade on all drain line construction. A minimum of four targets should be set along any given line of continuous grade to maintain accuracy. Each target should be set for depth of cut as indicated by its respective grade stake.

4.1.2.2 String lines and wires may be used in place of targets if care is exercised to keep them tight.

4.2 Trench width and bottom. The trench width should be the minimum required to permit installation and provide bedding conditions suitable to support the load on the drain but not less than 75 mm (3 in.) of clearance on both sides of the drain. The trench bottom should be relatively smooth and free of rock.

4.2.1 Tile. Tile should be bedded with ordinary care in an earth foundation shaped to fit the lower part of the tile. This can be accomplished with most trenching machines. When a backhoe is used to dig the trench, it will generally be necessary to hand grade and shape the trench bottom to fit the tile.

4.2.2 Corrugated plastic tubing

4.2.2.1 A specially shaped groove is required in the trench bottom where a gravel envelope is not specified. The groove provides side and bottom support to the lower part of the tubing and further provides a means of controlling alignment during installation. The groove may be semicircular, trapezoidal, or a 90 deg V. The 90 deg V-groove of sufficient depth will handle 76 to 152 mm (3 to 6 in.) tubing. The void under the tubing is advantageous for load bearing but may cause undermining by erosion on steep grades for larger sizes of pipe. Under steep grade conditions the bottom of the trench may be shaped to closely fit the tubing. The groove can be formed or cut in a number of ways. In all cases, some type of a forming tool is attached to the shoe of the trenching machine. One method is to install a forming tool on the bottom of the shoe and use compaction pressure to form the bedding groove. A better method is to install a device on the front of the finishing shoe and plow out the groove during the trenching operation. Special shaping cutters can also be attached to the trenching wheel. The latter two methods are preferred since they minimize soil compaction and permeability reduction (see ASTM Standard F449, Recommended Practice for Subsurface Installation of Corrugated Thermoplastic Tubing for Agricultural Drainage or Water Table Control).

4.2.2.2 Fig. 4 shows dimensions for a 90 deg V-groove. When the same 90 deg V-groove is used for 76 to 152 mm (3 to 6 in.) tubing, the depth that the tubing sets in the groove varies with the tubing size, therefore, this depth should be considered when setting the gradeline. For 203 mm (8 in.) diameter or larger tubing, a curved bottom which more nearly fits the tubing is recommended rather than the 90 deg V-groove.

4.3 Correction for overdigging grade. If the trench is excavated below the designed grade, it should be filled to grade with gravel or well-pulverized soil and tamped sufficiently to provide a firm foundation. The bottom of the trench must be shaped to grade.

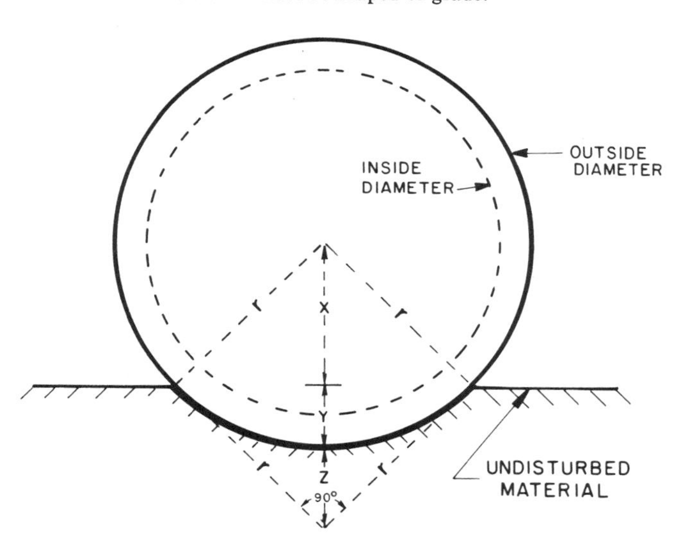

Tubing size mm (in.)	r (D/2) mm (in.)	x (0.707r) mm (in.)	y (0.293r) mm (in.)	z (0.414r) mm (in.)
76 (3)	46 (1.8)	33 (1.3)	13 (0.5)	18 (0.7)
102 (4)	61 (2.4)	43 (1.7)	18 (0.7)	25 (1.0)
127 (5)	76 (3.0)	53 (2.1)	23 (0.9)	30 (1.2)
152 (6)	91 (3.6)	64 (2.5)	28 (1.1)	38 (1.5)

Values are based on typical outside diameters which are assumed to be 20% greater than the inside diameters.

FIG. 4—DIMENSIONS FOR A 90 DEG-V SUPPORT GROOVE

4.4 Rock excavation. When the tubing is to be laid in a rock-cut, the trench should be overexcavated to a depth of 150 mm (6 in.) below grade level and this space filled with graded sand and gravel or well-pulverized soil and tamped sufficiently to provide a firm foundation. The bottom of the trench should then be shaped to grade. The trench should be filled with designed bedding or envelope material to the top of the rock-cut.

4.5 Construction features in noncohesive soils

4.5.1 When noncohesive soils are encountered, special construction features are required depending on the type and condition of the soil.

4.5.2 Unstable trench walls

4.5.2.1 Unstable or fluid soil conditions encountered in the trench wall, or base, such as may be found by excavation below a groundwater level or in a saturated sand, may cause tubing failure by caving of the trench sidewalls. This situation may also cause tile lines to fail due to tile misalignment. Where there are unstable trench sidewalls, a means must be provided to protect the tubing or tile from caving of the sidewalls until the drain has been properly laid and blinded. In some cases, the trencher shield behind the shoe may be lengthened to protect a greater length of the trench. It may be necessary to open the trench and allow the soil to drain and stabilize or initially incorporate a dewatering system to provide for temporary relief, enough to stabilize the soil so that open-trench excavations are possible.

4.5.2.2 The drain should be laid immediately after the shoe has passed. The trencher shield must provide enough time to surround the drain with blinding or envelope material which will provide some protection against caving. Provisions for safety during trenching operations shall be in compliance with the safety and health regulations for construction set forth in paragraph 4.16.2

4.5.3 Unstable trench bottom. Where an unstable trench bottom condition is encountered such as in fine sandy soils or in soils containing quicksand, extreme care must be taken to keep sediment from entering the drain and to provide a firm foundation for the drain.

4.5.3.1 The following factors should be observed when draining these soils.

4.5.3.1.1 Install the drain only when soil profile is in driest possible condition.

4.5.3.1.2 Use stabilizing envelope materials under the drain.

4.5.3.1.3 Use an envelope material to cover the remainder of the drain.

4.5.3.1.4 Use nonperforated tubing, self-sealing sewer or continuous rigid pipe where small pockets of noncohesive soils less than 30 m (100 ft) in length are encountered.

4.5.3.1.5 If the drain is tile, be sure to secure a snug fit of the joints.

4.5.3.1.6 If tubing is used, precautions must be taken to prevent it from floating.

4.5.3.2 If unstable soil material is encountered, it may be removed and replaced with a foundation and bedding of processed stone or processed gravel, suitably graded and acting as an impervious mat into which the unstable soil will not penetrate. The depth of this processed material used for foundation and bedding shall depend on the severity of the trench bottom soil condition. Install such special foundation and bedding material in maximum of 150 mm (6 in.) layers and compact each layer. If the foundation contains large particles to create a hazard to the drain, provide a cushion of acceptable bedding material between the foundation and the drain.

4.5.3.3 Where stabilizer materials do not furnish adequate support, the drain should be placed in a 90 deg rigid V-prefabricated foundation cradle with top of the V equal to the outside diameter of the drain. Each section of the cradle must provide rigidity and continuous support throughout the entire length of the cradle. Occasionally, it is necessary to place the cradle on pilings by driving pairs of posts along the edge of the cradle into solid material to provide the required support.

4.5.3.4 For drainage of unstable soils, a fast-moving trencher or drain plow that can maintain a continuous forward motion which disturbs the in-place material as little as possible should be used.

4.5.3.5 Sheet piling, pumping, freezing, and various other means have been used to install drain in unstable soils. Such methods are expensive and cannot be justified except where outlets serve large areas of high-priced lands.

4.5.4 Envelopes and envelope material

4.5.4.1 An envelope material installed around subsurface drains ensures proper bedding support and improves the flow of groundwater into the drain. When it is not feasible to form a bedding groove for plastic drain tubing, a granular envelope material can be used as an alternative bedding.

4.5.4.2 When a design gravel envelope material is specified, no special shaping or grooving of the trench bottom is required. The envelope material provides excellent drain support for all normal agricultural drain depths. The designed minimum envelope thickness may vary from 75 to 150 mm (3 to 6 in.) depending on the type of equipment used to install the material and the hydraulic conductivity and availability of the gravel material.

4.5.4.3 Figs. 5, 6, and 7 show the different placement of envelope materials used with tubing. Fig. 5 shows the envelope used where drain support is required for either tile or tubing. Installation as shown in Figs. 6 and 7 show the shape of the envelope used where fines are present in the soil. For all shapes when the trench is wider than the specified envelope width, the trench must be filled on both sides with bedding material or gravel envelope so there are no void spaces in the area between the drain and the walls of the trench.

4.5.4.4 The envelope material shall be placed to form an even, firm bedding under the drain, and the remainder of the material shall be carefully placed so as not to disturb the grade or alignment of the drain. Due care shall be exercised to ensure that no mud, excavated material, or foreign matter is mixed with the envelope material during the installation. No special compaction of gravel envelope is necessary except as required for ensuring a firm and even bedding for the drain.

4.5.5 **Filters and filter material.** A filter is used to restrict fine particles of silt and sand from entering the drain. A sand and gravel envelope designed as a filter may be used (see Figs. 5 and 6). Where this type of filter material is not available, artificial prefabricated filter material can be used. Protective filters are seldom required for agricultural drainage systems in humid areas because soils are generally stable. However, a filter is required for base materials such as uniform very fine to medium sands where high velocities can develop that will move the sands into the drain. Gravel envelopes are not normally designed as filters, but because they consist of a well-graded material related to the gradation of the base material they do act as partial filters. Most of the presently available artificial prefabricated filter material such as fiberglass, spun-bonded or knitted synthetic fabrics, and plastic filter cloth act as protective filters and with time may partially clog up and decrease the inflow into the drain. Where fiberglass filter material is used, it shall be manufactured from borosilicate-type glass and it shall be certified suitable for underground use. The fibers shall be of variable size, with some larger fibers intertwined in the mat in a random manner. During installation the material shall span all open joints and perforations without excessive stretch. Care shall be taken not to damage the material during installation. Any damaged areas shall be replaced before backfilling.

4.6 Construction features for mains on steep grades. Special design recommendations and construction features for drains on steep grades are outlined in paragraph 3.9.3.

4.7 Allowable variation from grade. The constructed grade must be such that the drain line as constructed will provide the capacity required for the drained area (see paragraph 3.2). A slight decrease in grade can be tolerated where the actual capacity of the drain exceeds the required capacity. A gradual variation of 30 mm (0.1 ft) from grade is allowable. No reverse grade shall be allowed.

4.8 Installing the drains

4.8.1 **Laying tile.** Each tile should be laid true to line and grade, except as provided in paragraph 4.7. Use of an automatic tile laying device is acceptable.

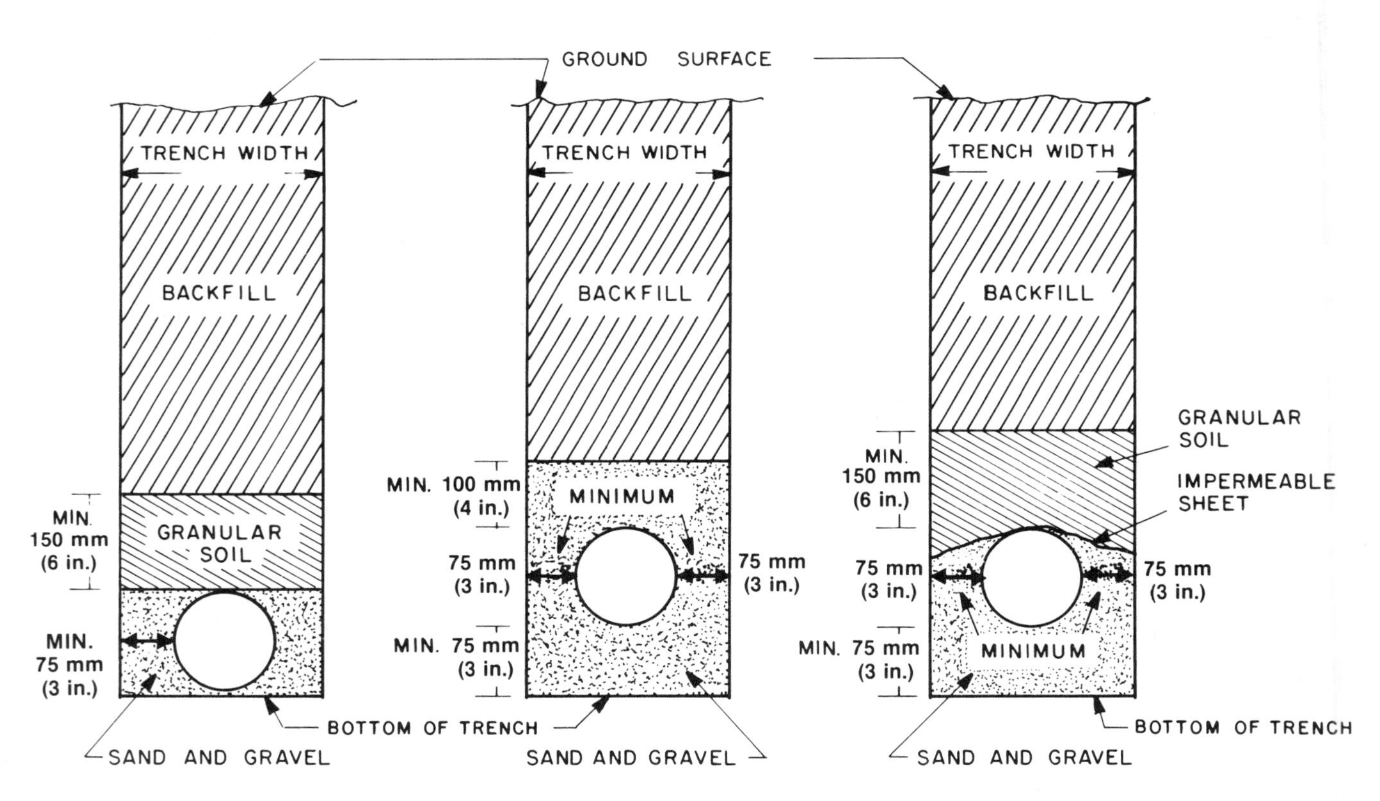

FIG. 5—TUBING ENCASED IN A SAND AND GRAVEL ENVELOPE FOR SUPPORT

FIG. 6—TUBING ENCASED IN A SAND AND GRAVEL ENVELOPE DESIGNED AS A FILTER

FIG. 7—TUBING ENCASED IN A SAND AND GRAVEL ENVELOPE DESIGNED AS A FILTER WITH AN IMPERMEABLE SHEET SUCH AS PLASTIC

4.8.1.1 Tile openings (joint spacing). In all soils, tile joint spacings should be large enough to permit entry of design flow and small enough to prevent entry of soil. The maximum joint spacing should be 3 mm (0.1 in.) except where special conditions indicate a wider spacing. Perforated tile may be used in place of plain tile.

4.8.1.2 Keeping tile clean. The tile should be kept clean while laying. All soil or debris in the tile should be removed before the next tile is laid.

4.8.2 Laying tubing. The tubing should be laid true to line and grade except as provided in paragraph 4.7.

4.8.2.1 Stretch during installations. Stretch that occurs during installation will cause some decrease in strength. Also, perforations may be pulled open wider than desirable. The stretch acquired during installation is influenced by the temperature of the tubing at the time it is being installed, the amount and duration of drag encountered when the tubing feeds through the installation equipment, and the stretch-resistance characteristics of the tubing. Tubing should not be stretched so as to reduce its stiffness to less than the allowable minimum pipe stiffness. Stretch is expressed as a percent increase in length and should not exceed 5%. The use of a power feeder is recommended for all sizes.

4.8.2.2 Weakening due to high temperature. The temperature of plastic tubing can reach 65 °C (150 °F) or more when laid out in a field on a hot bright day. The ability of corrugated polyethylene tubing to resist deflection is reduced about 40% when its temperature rises from 21 °C to 38 °C (70 °F to 100 °F) and by about 50% when its temperature rises from 21 °C to 49 °C (70 °F to 120 °F). Therefore, it is essential that the contractor takes precautions when these conditions exist so there is not a direct impact of sharp or heavy objects or excessive pull on the tubing during installation. The tubing will regain its strength when its temperature returns to that of the surrounding soil which is usually in 5 min or less after installation.

4.8.2.3 Brittleness due to low temperatures. Tubing stiffness increases and flexibility decreases as its temperature is lowered. If the tubing is rapidly uncoiled at low temperatures, it is stressed excessively and may crack. The tubing may have a tendency to coil, which presents difficulty in laying flat and level and necessitates extra care. Manufacturer's recommendations for handling the material under either hot or cold conditions should be checked.

4.8.2.4 Floating in water. Plastic tubing may float in water. Once it floats during installation, it is difficult to get backfill material around and over the tubing without getting the material underneath it, causing misalignment. Therefore, precautions must be taken to prevent floating. This can be accomplished by mechanically holding the tubing in place until blinding is completed.

4.9 Blinding

4.9.1 Blinding is the process of placing bedding material of loose stable-structured soil on the sides and over the top of the drain to a depth of 150 mm (6 in.). Except in those areas where chemical deposits in and around the drain area are a problem, the soil should be friable top soil or other porous soil. Fine sand should not be placed directly on or around the drain. In tight soils where blinding with soil is not adequate, a suitable envelope should be used. The bedding material must also permit water to easily reach the drain. The drain should be blinded immediately after installation. A number of different blinding methods have been proven acceptable. Some contractors place such importance on blinding that they hand-place selected material around and over the tubing. There are also a number of mechanical blinding devices which can be mounted on the trencher. These devices place selected material from near the top of the trench around and over the drain. Their main advantages are immediate blinding after laying the drain, selection of the most suitable blinding material that is readily available, and reduced labor requirements.

4.9.2 Blinding for tile

4.9.2.1 All tile should be blinded immediately to maintain alignment and to protect the tile from falling rocks, ditch cave-ins, and backfill operations.

4.9.2.2 Acceptable methods of automatic blinding and back-filling in one operation can be used provided the soil is suitable.

4.9.3 Blinding for tubing

4.9.3.1 Blinding will ensure proper alignment of the tubing in the groove and protect it when the remaining excavated material is placed in the trench. Careful soil placement on both sides of the tubing is necessary to provide good side support which will reduce deflection of the tubing. Tubing should be held in place in the trench until secured by blinding. This is especially important when the air temperature is below 7 °C (45 °F). Blinding quantities may be increased under these conditions. Blinding is not necessary where drains are placed in sand and gravel filters or envelopes.

4.9.3.2 No stones or other hard objects should be allowed in immediate contact with the tubing. Such objects apply point loads and may cause the tubing to fail. Blinding provides protection for the tubing during the backfilling operation when the impact of rocks and hard clods could damage it.

4.10 Inspection. All lines should be carefully inspected for grade alignment and other specifications prior to backfilling.

4.11 Backfilling

4.11.1 At the conclusion of each day's work, the uppermost end of the drain line should be stoppered and the trench backfilled to prevent sediment or debris from entering the line in the event of rain. Backfilling should be done if there is any danger of heavy rains or freezing temperatures. The upper end of each drain should be tightly covered with a manufactured plug or other equivalent material.

4.11.2 Various methods can be used to move the remaining excavated material back into the trench and mound it over the trench to allow for settling. This may be done by hand shovels, graders, bulldozers, augers and conveyor methods. The backfill material shall be placed in the trench in a manner that displacement of the drain will not occur. The direction of backfilling should be on an angle for the open trench so the material flows down the face of the backfill. Large stones, clods and heavy direct loads must be avoided in backfill operations.

4.11.3 When installing tubing on a hot day and the tubing feels warm to the touch [37 °C (98 °F)], delay the backfilling until the tubing reaches the soil temperature.

4.12 Early use damage. Tubing installed at shallow depths and without a gravel envelope may be crushed especially after a heavy rain, by livestock stepping on it, or by heavy wheeled traffic traveling over the trench. When the backfill is loose, livestock should not be permitted in fields where the tubing has been installed until proper settlement has occurred, which may be as long as one year.

4.13 Location of drains after construction. Clay and concrete tile lines can be located with a tile probe. Location of plastic tubing with a tile probe is more difficult. For tubing, the probe should not be larger than a standard 10 mm (0.4 in.) diameter rod with a 13 mm (0.5 in.) diameter sharpened tip. The probe may penetrate the tubing when the line is found. But penetration does not damage the tubing since the hole closes as the probe is withdrawn. If too large a probe is used, however, there is danger of collapsing the tubing.

4.14 Drain plow. Corrugated thermoplastic tubing is sometimes installed with a drain plow for humid-type agricultural subsurface drainage systems and for underground utility lines. Alignment, grade, and side-support provisions noted in preceding paragraphs are likewise applicable to drain plow installations. Special precautions are required with this method of installation to be sure other underground facilities are not damaged.

4.15 Storage

4.15.1 Storage requirements for clay and concrete tile are given in paragraph 2.1.3.3.

4.15.2 Corrugated thermoplastic tubing can be destroyed by fire. Therefore, precautions need to be taken to protect it from potential fire hazard. Care should also be provided during handling, to prevent unnecessary damage. Coils of tubing should be laid flat when stored for an extended period of time. Corrugated thermoplastic tubing can be damaged by rodents when stockpiled above ground. Where rodents are a potential problem, it is recommended that the ends of the tubing be capped to deter rodents from building nests inside the stored tubing. The tubing should also be checked before installing to be certain it is free of any plugs caused by birds, wasps, etc. Excessive exposure to ultraviolet rays from sunlight can be harmful to the tubing and therefore protection is required when tubing is subjected to long periods of storage in sunlight.

4.16 Safety

4.16.1 Safety standards for men and machines should be observed. Persons working in trenches should be protected from cave-ins by proper shoring and should not work alone. Moving parts of the machine should be protected by proper guards. Casual observers should not be permitted close to excaving operations.

4.16.2 Provisions for safety during trenching operations shall be in compliance with Occupational Safety and Health Administration, Code of Federal Regulations, Part 1926.652, Specific Trenching Requirements.

References: Last printed in 1985 STANDARDS; list available from ASAE Headquarters.

Cited Standards:

ASAE EP302, Design and Construction of Surface Drainage Systems on Agricultural Lands in Humid Areas
ASAE EP369, Design of Agricultural Drainage Pumping Plants
ASAE EP407, Agricultural Drainage Outlets—Open Channels
ASTM C4, Specification for Clay Drain Tile
ASTM C412, Specification for Concrete Drain Tile
ASTM C498, Specification for Perforated Clay Drain Tile
ASTM F405, Specification for Corrugated Polyethylene (PE) Tubing and Fittings
ASTM F449, Recommended Practice for Subsurface Installation of Corrugated Thermoplastic Tubing for Agricultural Drainage or Water Table Control
ASTM F667, Specification for 8, 10, 12, and 15-in. Corrugated Polyethylene Tubing
OSHA CFR Part 1926.652, Specific Trenching Requirements

DESIGN AND INSTALLATION OF NONREINFORCED CONCRETE IRRIGATION PIPE SYSTEMS

Developed by the Concrete Irrigation Pipe System Committee; approved by the Soil and Water Division Standards Committee; adopted by ASAE 1957; revised 1960, 1961, 1962, 1963, June 1968, May 1974, December 1978; reconfirmed December 1979, March 1981, March 1982, December 1982; revised December 1983; revised editorially February 1984.

SECTION 1—PURPOSE AND SCOPE

1.1 This Standard is intended as a guide to engineers in the design and installation of low or intermediate pressure nonreinforced concrete irrigation pipelines and for the preparation of detailed specifications for a particular installation. It is restricted to pipelines with vents or stands open to the atmosphere. It is not intended to serve as a complete set of design criteria and construction specifications.

1.2 The systems designed and/or installed under this Standard shall utilize pipe conforming to one or more of the following types of nonreinforced concrete irrigation pipe.

1.2.1 Pipelines with mortar joints. The pipe shall conform to the requirements of American Society for Testing and Materials Standard C118, Specifications for Concrete Pipe for Irrigation or Drainage.

1.2.2 Pipelines with rubber gasket joints. The pipe and gaskets shall conform to ASTM Standard C505, Specifications for Nonreinforced Concrete Irrigation Pipe with Rubber Gasket Joints.

1.2.3 Cast-in-place pipelines. The pipe shall conform to American Concrete Institute Standard 346, Specifications for Cast-in-Place Nonreinforced Concrete Pipe.

SECTION 2—DEFINITION OF TERMS

2.1 Appurtenances

2.1.1 Float Valve: A valve, actuated by a float in a stand, which automatically controls the flow of water into the stand.

2.1.2 Gate: A device used to control the flow of water to, from or in a pipeline. It may be opened and closed by screw action or by slide action; the latter is used only where pressures and velocities in the line are so low that sudden closure will not cause excessive water hammer. Types of gates, indicative of the place they will be used, are:

2.1.2.1 Line Gate: A hub-end screw-type gate which is mortared into the pipeline.

2.1.2.2 Stand gate: A gate in a stand which covers an inlet into a pipeline and which controls water flow into the pipeline. It may be either a screw or a slide type. A slide gate has a device to lock it in any desired position.

2.1.2.3 Stand gate pressure: A screw-type gate inside a stand which covers an outlet from a pipeline into the stand.

2.1.3 Inlet: An appurtenance to deliver water to a pipeline system.

2.1.3.1 Gravity inlet: A structure to control the flow of water from an open conduit into a pipeline. It may be combined with a baffle, gate, screen and/or a sand trap.

2.1.3.2 Pump stand: A structure where water enters a pipeline system from a pump or pressure system.

2.1.4 Outlet: An appurtenance to deliver water from a pipe system to the land or to any surface pipe system. An outlet may consist of a valve, riser pipe, and/or an outlet gate. Several types of outlets are defined as follows:

2.1.4.1 Alfalfa valve: An outlet valve attached to the top of a riser with an opening equal in diameter to the inside diameter of the riser pipe and an adjustable lid or cover to control water flow. A ring around the outside of the valve frame provides a seat and seal for a portable hydrant. Some alfalfa valves have a small air release valve in the cover to provide drainage following irrigation for mosquito abatement; it also provides supplemental air release during pipeline filling.

2.1.4.2 Modified alfalfa valve: This valve is similar to an alfalfa valve except the oustide ring is omitted. Only portable hydrants which fit directly over a riser pipe can be used with this outlet.

2.1.4.3 Orchard valve: An outlet valve installed inside a riser pipe with an adjustable cover or lid for flow control similar to an alfalfa valve. However, because the valve opening is smaller than the inside diamter of the riser, its flow capacity is less. The top of the riser may be (1) at or slightly below ground surface, (2) 6 to 12 in. (150 to 300 mm) aboveground surface with a notch in the side, or, (3) similarly above the ground with two or more outlet gates installed in the side of the riser.

2.1.4.4 Portable hydrant: An outlet used for connecting surface pipe to an alfalfa valve outlet.

2.1.4.5 Surface pipe outlet: An outlet for attaching surface pipe to a riser without using a portable hydrant.

2.1.4.6 Swivel-arm distributor: This outlet has a valve and two arms of gated pipe which swivel upward from the top of a riser (usually a steel pipe riser). When chained to a center post they are removed from the cultivation path and when dropped, the gates distribute the water into furrows.

2.1.4.7 Outlet gate: Usually a slide gate, or other type of gate, which controls the flow of water from an outlet.

2.1.4.8 Capped riser or pot: A riser extending above ground with a watertight cap over its top and outlet gates on its sides slightly above the ground surface (capped riser). To accommodate more outlet gates, a pot with a diameter larger than that of the riser pipe is sometimes installed on the top of the riser (capped pot). Die-cast screw-type valves are sometimes used on capped pots instead of outlet gates. Outlet gates must be placed on the outside of a capped riser or pot and tend to produce an erosive jet of water, which the die-cast valves eliminate.

2.1.4.9 Open pot: An outlet consisting of an orchard valve installed in the top of a riser with a section of larger diameter pipe mortared to the riser and extending above it. Two or more slide gates are mounted on the sides of the pot.

2.1.5 Stand: A structure formed from vertical sections of pipe or from cast-in-place concrete (box stand). It may service as a pump stand, gate stand, or float valve stand. It may also function as a vent or sand trap, or both. When gates are not required inside the stands, the stand may be capped and have a smaller vent pipe rising to a height of the hydraulic gradeline plus freeboard. Float valve stands are used on steep slopes and where the water supply rate can be varied to provide automatic control.

2.1.6 Vent: An appurtenance to the pipeline which permits the passage of air to or from the pipline.

2.2 Hydraulic terms: Hydraulic terms shall be as defined in the American Society of Civil Engineers Manual of Engineering Practice No. 11, except as noted below:

2.2.1 Freeboard: The vertical distance above the elevation of the hydraulic gradeline at working head to the tops of vents or stands.

2.2.2 Surge: That phenomenon wherein a rocking or oscillating motion of the water is set up, flow to be unsteady.

2.2.3 Water hammer: That phenomenon which occurs when water flowing in pipelines is rapidly decelerated or stopped, usually by a rapid or sudden gate or valve closure. The resulting pressure waves pass through the water at high velocities and can produce very high momentary pressures. Water hammer is not to be confused with surge in systems open to the atmosphere, although under certain conditions both may be initiated simultaneously.

2.2.4 Working head: The vertical distance that water will rise in a vent or stand above the centerline of the pipeline at design flow at any point in the system. On pipeline profiles the maximum working head shall be, at any point, the vertical distance from the centerline of the pipeline to a straight line drawn between the tops of consecutive vents and/or stands. It is, thus, the working head plus freeboard.

SECTION 3—DESIGN CRITERIA

3.1 Pipeline

3.1.1 Safety factors

3.1.1.1 External load limit. Although loads are generally light on this type of installation, where there are excessively high fills over the pipe, a safety factor of at least 1.25 shall be applied to the three-edge-bearing test in computing allowable heights of fill over precast nonreinforced concrete pipe. The loads shall be determined by the methods outlined in ASAE Engineering Practice EP260, Design and Construction of Subsurface Drains in Humid Areas.

3.1.1.2 Pressure. Maximum working head for cast-in-place pipelines shall be 15 ft (4.6 m) above the centerline of the pipe. Maximum working heads for precast nonreinforced concrete pipe shall not exceed one-fourth the certified hydrostatic test pressure as prescribed in ASTM Standards C118, Specifications for Concrete Pipe for Irrigation or Drainage, for mortar-jointed pipelines or one-third the certified hydrostatic test pressure as prescribed in ASTM Standard C505, Specifications for Nonreinforced Concrete Irrigation Pipe with Rubber Gasket Joints, for pipelines with rubber gasket joints.

3.1.1.3 Soil conditions

3.1.1.3.1 Concrete pipelines shall not be installed on sites where the sulfate salt concentration exceeds 1.0% as water soluble sulfate in soil samples, or 4000 parts per million sulfate in groundwater samples. Concrete pipe made with Type V cement or cement whose tricalcium aluminate content does not exceed 5% shall be used on sites where the water soluble sulfate content in soil samples is 0.20 to 1.0%, or where the sulfate content of groundwater samples ranges from 1000 to 4000 parts per million. Concrete pipe made with Type II cement or cement with a tricalcium aluminate content of not more than 8% shall be used on sites where the water soluble sulfate concentration in soil samples ranges from 0.10 to less than 0.20%, or the sulfate content in groundwater samples is from 150 to 1000 parts per million. Portland pozzolan cement, Types IP or IP-A, or Portland blast furnace slag cement, Types IS(MS) or IS-A(MS), ASTM Standard C595, Specifications for Blended Hydraulic Cements, also may be used for such exposures. There are no restrictions as to the type of cement used in concrete pipe for sites where the sulfate content is less than 0.10% in soil, or 150 parts per million in groundwater.

3.1.1.3.2 Cast-in-place pipe shall be used only in stable soils or soils that have been stabilized as in Section 3.3 of ACI Standard 346, Specifications for Cast-in-Place Nonreinforced Concrete Pipe, where the trench form conforms to the trench requirements per Chapter 3 of the above-referenced specification.

3.1.2 Friction loss. Friction loss for mortar-jointed or cast-in-place pipelines shall be computed using Scobey's concrete pipe equation with a coefficient of retardance $K_s = 0.310$ or Manning's equation with roughness coefficient $n = 0.013$. Similar coefficients should be used for pipe with rubber-gasket joints, except that for the smoothest makes of such pipe, the Scobey coefficient of retardance may range up to $K_s = 0.370$ and the Manning's roughness coefficient down to $n = 0.011$. Minor losses can be computed in accordance with current practices.

3.2 Stand requirements

3.2.1 Stands shall be placed at each inlet to a concrete irrigation pipe system and at such other points as required. All stands shall be installed on a base adequate to support the stand and to prevent undue movement or stress on the pipeline. All stands shall serve as vents in addition to their other functions as follows:

3.2.1.1 They shall avoid entrainment of air.

3.2.1.2 They shall provide 1 to 5 ft (0.3 to 1.5 m) of freeboard.

3.2.1.3 Stands constructed of concrete pipe having diameters greater than 24 in. (610 mm) shall use Class II Reinforced Concrete Pipe as specified in ASTM Standard C76, Specifications for Reinforced Concrete Culvert, Storm Drain, and Sewer Pipe.

3.2.1.4 Stands cast-in-place shall contain vertical steel reinforcing on not more than 1 ft (300 mm) centers and horizontal reinforcement to provide steel areas equal to or greater than the least values specified for Class II Reinforced Concrete Pipe in ASTM Standard C76, Specifications for Reinforced Concrete Culvert, Storm Drain, and Sewer Pipe.

3.2.1.5 The tops of all stands shall be at least 4 ft (1.2 m) above the ground surface. If visibility is not a factor, stands may be lower if covered or equipped with trash guards.

3.2.2 Pump stands. Pump stands shall be:

3.2.2.1 Concrete box stands with vertical sides suitably reinforced to withstand handling and installation stresses.

3.2.2.2 Nontapered stands of concrete pipe suitably reinforced to withstand handling and installation stresses.

3.2.2.3 Nontapered concrete pipe stands, capped and having a vent pipe to the height of the hydraulic gradeline plus freeboard.

3.2.2.4 Steel cylinder stands mortared to a single section of concrete pipe riser.

3.2.3 The centerline of the pump discharge pipe shall have a minimum vertical offset above the centerline of the outlet pipe equal to the sum of the diameters of the inlet and outlet pipes.

3.2.4 Check valves shall be used in the pump discharge line wherever the potential backflow from the pipeline would be sufficient to drain the pipeline or damage the pump.

3.2.5 Construction shall be such as to insure that the vibration from the pump discharge pipe is not carried to the stand.

3.2.6 Velocities in stands

3.2.6.1 Downward water velocities shall not exceed 2 ft/s (0.6 m/s). In no case shall such velocities exceed the average pipeline velocity.

3.2.6.2 If the size of the stand is decreased above the pump discharge pipe, the top vent portion shall be of such inside cross-sectional area that, if the entire flow of the pump were discharging through it, the average velocity would not exceed 10 ft/s (3.0 m/s).

3.2.7 Sand traps. Pump stands serving as sand traps shall have a minimum inside diameter of 30 in. (760 mm) and shall be constructed so that the bottom is at least 24 in. (610 mm) below the invert of the outlet pipeline. Suitable provisions for removing sand shall be provided.

3.2.8 Gate stands

3.2.8.1 Gate stands shall be cast-in-place or constructed from concrete pipe. Reinforcing requirements under paragraphs 3.2.1.3 and 3.2.1.4 apply.

3.2.8.2 Gate stand dimensions shall be sufficient to accommodate the gate or gates required.

3.2.8.3 Gate stands shall serve as vents.

3.2.8.4 Gate stands shall be of sufficient size that gates are accessible for repair.

3.2.9 Float valve stands. Float valve stands shall be of sufficient diameter to provide accessibility for maintenance and to dampen surge. (The wide-open friction loss for the valve approximates 2.4 velocity heads for the single disk type and 1.9 velocity heads for the double disk type.)

3.3 Vent requirements

3.3.1 Locations. Vents shall be placed:

3.3.1.1 At the downstream end of each lateral.

3.3.1.2 At a design point downstream from where there is opportunity for air entrainment and inadequate opportunity for escape of that air.

3.3.1.3 At high points wherever there are changes in grade downward in direction of flow of more than 10 deg.

3.3.1.4 At all turns of 90 deg or more with the exception of lines not more than 50 ft (15 m) in length.

3.3.2 The design point in 3.3.1.2 shall be determined by the equation

$$L = CVD$$

where

L = distance downstream from the air-entraining stand
C = 1.76 when L, D, and V are in feet and ft/s, respectively, and C = 5.77 when L, D, and V are in meters and m/s
V = maximum design velocity
D = inside diameter of the pipe

3.3.3 Any stand shall substitute for a vent.

3.3.4 There shall be considered opportunity for air entrainment at all gravity inlets and at pump stands where the pump might possibly pump air. When pumping from wells, if there is a downdraft of air into the well casing while the pump is in operation, the well shall be considered to pump air. In such case a vent shall be placed immediately downstream from the pump stand if the average downward velocity in the stand from the pump discharge to the pipeline exceeds 1 ft/s (0.3 m/s).

3.3.5 Size

3.3.5.1 The cross-sectional area of the vent shall be at least half the cross-sectional area of the pipeline (both inside measurements) for a distance of at least one pipeline diameter up from the centerline of the pipeline. Above this the vent may be reduced to 1/60 of the cross-sectional area of the pipeline, but not less than 2 in. (50 mm) diameter pipe shall be used.

3.3.5.2 Vents shall have a minimum freeboard of 1 ft (0.3 m) above the hydraulic gradeline. The maximum height shall not exceed the maximum working head of the pipe.

3.3.6 Air release valves. An air-vacuum release valve may be used in lieu of an open vent. The valve outlet shall have a 2 in. (50 mm) nominal minimum equivalent diameter. Two inch (50 mm) outlets shall be used for pipelines of 6 in. (150 mm) diameter or less, 3 in. (75 mm) outlets for pipelines of 7 to 10 in. (175 to 250 mm) diameter, and 4 in. (100 mm) outlets for pipelines of 12 in. (300 mm) and larger diameter. Air release valves used at summits in lieu of open vents shall continually release entrapped air.

3.4 Anchors and thrust blocks

3.4.1 Abrupt changes in pipeline grade or alignment require a stand of diameter greater than the pipeline or an anchor or thrust block to absorb any axial, side, or vertical thrust of the pipeline. An abrupt change shall be considered to be:

3.4.1.1 An angle of 45 deg or greater when the maximum working head is under 10 ft (3.0 m).

3.4.1.2 An angle of 30 deg or greater when the maximum working head is between 10 and 20 ft (3.0 and 6.1 m).

3.4.1.3 An angle of 15 deg or greater when the maximum working head is 20 ft (6.1 m) or more.

3.4.2 Anchors shall be used and designed as necessary to restrain any vertical thrust of the pipeline.

3.4.3 Thrust blocks shall be constructed of concrete placed to fill the space between the pipe and the undisturbed earth at the side of the trench on the outside of bends and tees as shown in Fig. 1, such that the block is in a direct line with the force resulting from the change in pipeline alignment. Plastic soil cement with at least one part cement to ten parts sandy or coarser texture soil similarly placed may be used.

3.4.4 The depth of the thrust blocks shall be equal to the full outside diameter of the pipe and shall have a minimum thickness of 6 in. (150 mm). The length normal to the direction of thrust is determined by the equation

$$L = C\frac{HD}{B}\sin\frac{a}{2}$$

where

L = length of thrust block
C = 98 when L, D, and B are in feet and lb/ft², respectively, and C = 15.4 when L, D, and B are in meter and kilopascals
H = maximum working head
D = inside diameter of the pipe
B = allowable passive pressure of the soil
a = deflection angle of the pipe bend

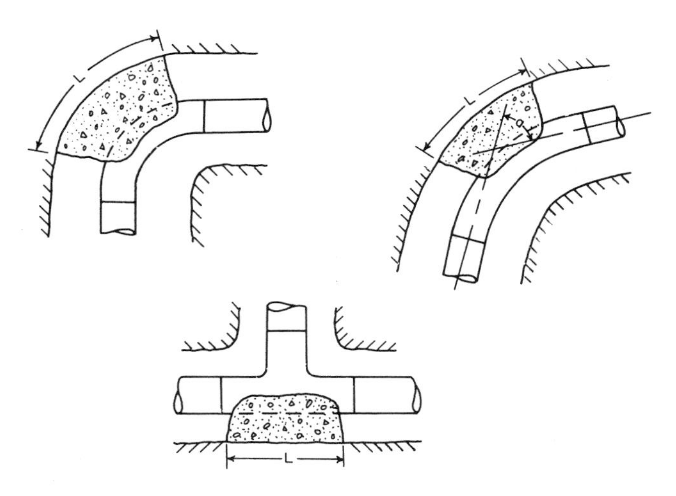

FIG. 1—THRUST BLOCKING FOR IRRIGATION PIPELINES.

3.4.5 The pipe shall be clean and wet when placing the thrust block to provide a good bond between concrete and pipe. Where adequate soil tests are not available, the allowable passive soil pressure shall be considered to be 500 lb/ft² (23.9 kPa).

SECTION 4—INSTALLATION

4.1. Size and location. The pipe and appurtenances shall conform to the standards specified and shall be located and constructed as shown on the design plans and in the construction specifications.

4.2 Placement

4.2.1 The pipelines shall be placed at sufficient depth below the land surface to permit covering the pipe a minimum of 2 ft (0.6 m) unless shallower covering is specified for rocky areas or other local conditions. If shallower covering is specified, provisions made shall be to protect the line from damage by vehicular traffic. Greater depths of cover shall be specified when the soil is subject to deep cracking or when other local conditions indicate a need.

4.2.2 Where trenches are excavated in soils containing rock or other hard materials, or in soils subject to appreciable swelling and shrinking on wetting or drying, or where the trench bottom is unstable, the trenches shall be overexcavated and backfilled with imported materials to sufficient depth to provide a suitable base. If water is in the trench, it shall be drained away or controlled in a manner to prevent damage to any joint mortar and to maintain a suitable base.

4.2.3 Provisions in paragraphs 4.2.1 and 4.2.2 apply to mortar-jointed and rubber-gasket pipe. Placement for cast-in-place pipe shall be as specified in ACI Standard 346, Specifications for Cast-in-Place Nonreinforced Concrete Pipe.

4.2.4 Rubber-gasket pipe shall not be placed with the joints rammed together so tight that longitudinal compression develops from wetting expansion of the pipe. If in doubt about any particular joint design, the end of the spigot shall be pulled back from the shoulder of the bell a slight distance not exceeding 0.06 in. (1.5 mm).

4.3 Joints and connections

4.3.1 Joints shall be mortar or rubber gasket, as specified and where required. All joints shall be constructed to leave the inside of the pipeline and appurtenances free of any obstruction which would reduce capacity below design standards.

4.3.2 Mortar for joints and bands shall consist of not less than one part Portland cement to three parts of clean, well-graded sand. The mortar should be plastic with a consistency such that it will readily adhere to the pipe and can be easily squeezed out of the joints. Mortar should be discarded if not used within 30 min after mixing with water.

4.3.3 The ends of pipe sections at all joints and connections shall be washed or brushed clean and thoroughly wetted immediately prior to placing the mortar. Care should be taken to assure that all joints and connections are filled with mortar before joining and that excess mortar is wiped from the inside of the pipe. The external bands shall be complete around the circumference of the pipe and no less than 0.4 in. (10 mm) thick at the joint.

4.3.4 Joints in stands and connections to appurtenances shall conform to the requirements of ASTM Standard C118, Specifications for Concrete Pipe for Irrigation or Drainage, or ASTM Standard C505, Specifications for Nonreinforced Concrete Irrigation Pipe with Rubber Gasket Joints.

4.3.5 Stoppage and horizontal joints for cast-in-place pipelines shall conform to the requirements of ACI Standard 346, Specifications for Cast-in-Place Nonreinforced Concrete Pipe. Connection joints shall be prepared by cleaning and freeing of loose or defective concrete, coatings, and foreign material. The contact faces of the pipe and fittings shall be wetted and the fitting mortared into place using bonding mortar as specified in the above-referenced specification.

4.4 Curing and backfilling

4.4.1 For mortar-jointed pipelines, the initial backfill of soil shall be placed around the pipe in lifts and to densities as specified in the project specifications and shall cover it to a depth of at least 6 in. (150 mm) for the full width of the trench and not more than seven sections of pipe behind the laying. If laying ceases for 2 h or more, the initial backfill shall be brought up to and cover the last completed joint. Nothing in this section shall prohibit complete backfilling while mortar bands are still plastic. If complete backfilling is not done at this time, the completion shall be delayed at least 20 h, but must be completed to the minimum specified cover or 2 ft (0.6 m), whichever is less, before water is put into the line.

4.4.2 Mortar joints are to be protected from drying out. If the soil in the initial backfill is not thoroughly moist, a suitable membrane shall be used to cover the mortar. Membranes consisting of one layer of kraft paper, or moistened paper cut from cement sacks, or membranes conforming to ASTM Standard C171, Specifications for Sheet Materials for Curing Concrete, or ASTM Standard C309, Specifications for Liquid Membrane-Forming Compounds for Curing Concrete, shall be considered suitable.

4.4.3 In areas where rips (longitudinal cracks or ruptures) have been known to occur in mortar-jointed pipelines, it is important that the soil used in the initial backfill be thoroughly moist and of a texture coarser than clay. (Soil will not ribbon in fingers when moist, or shall not have more than 30% by weight of material finer than 2 microns.) Generally, the required moisture content may be achieved by scheduling laying of the pipe no more than one day following trenching and by being sure the soil is moist when trenching is undertaken. This may indicate the need for deep irrigation along the trench line three days to a week before actual trenching.

4.4.4 All openings into non-rubber-gasket pipelines shall be covered or closed to prevent air circulation, except while work is actually in progress, and shall be kept closed until the pipeline is complete and is to be filled with water.

4.4.5 Curing and backfilling of cast-in-place pipelines shall be in accord with the provisions of ACI Standard 346, Specifications for Cast-in-Place Nonreinforced Concrete Pipe.

4.4.6 Backfilling over rubber-gasket nonreinforced concrete pipe shall be done in lifts and to densities specified in the project specifications.

4.5 Testing

4.5.1 All pipelines shall be tested to demonstrate that they function properly at the specified design capacity. At or below agreed design capacity there shall be no objectionable surge or water hammer. To be objectionable there shall be either:

4.5.1.1 Continuing, unsteady delivery of water.

4.5.1.2 Damage to the system.

4.5.1.3 Detrimental overflow from vents or strands.

4.5.2 Pipelines shall be tested for leaks by observing normal operation after a period of 2 wks of continuous wetting. However, unless mutually agreed upon, this leak test shall not be performed on cast-in-place pipe until after the pipe has had its initial 28-day curing period. All visible leaks shall be repaired. Losses shall not exceed 0.10, 0.05 or 0.02 ft^3/ft^2 (3.0, 1.5, or 0.6 cm^3/cm^2) of inside surface per 24 h for cast-in-place, mortar-jointed, and rubber-gasket pipelines, respectively. Water less than 50 °F (10 °C) shall not be used for testing mortar-jointed or cast-in-place pipelines.

4.6 Basis of acceptance. Unless otherwise specifically agreed to, the acceptability of the pipeline shall be determined by inspections to determine compliance with the provisions of this Standard with respect to the design of the pipeline and appurtenances, the materials used, workmanship, and the minimum installation requirements. Compliance shall not be binding for conditions related to earthquake or land settlement damage (except from improper trench bottom preparation or inadequate foundations for structures), external damage not caused by the parties involved, or leakage of mortar-jointed or cast-in-place pipe caused by water of temperature less than 50 °F (10 °C).

Cited Standards:

ASAE EP260, Design and Construction of Subsurface Drains in Humid Areas

ACI 346, Specifications for Cast-in-Place Nonreinforced Concrete Pipe

ASTM C76, Specifications for Reinforced Concrete Culvert, Storm Drain, and Sewer Pipe

ASTM C118, Specifications for Concrete Pipe for Irrigation or Drainage

ASTM C171, Specifications for Sheet Materials for Curing Concrete

ASTM C309, Specifications for Liquid Membrane-Forming Compounds for Curing Concrete

ASTM C505, Specifications for Nonreinforced Concrete Irrigation Pipe with Rubber Gasket Joints

ASTM C595, Specifications for Blended Hydraulic Cement

MINIMUM STANDARDS FOR ALUMINUM SPRINKLER IRRIGATION TUBING

Technical specifications were developed and approved by manufacturer's representatives, Technical Committee of the Sprinkler Irrigation Association, and the ASAE Sprinkler Irrigation Committee; adopted by ASAE January 1957; revised as a Tentative Standard March 1959; adopted as a full Standard 1962; reconfirmed by ASAE Soil and Water Division Standards Committee with editorial changes December 1967; reconfirmed December 1968, December 1969, December 1970; revised February 1972; revised editorially April 1976; reconfirmed December 1976, December 1977, December 1978, December 1979, March 1981, March 1982, December 1982, December 1983, December 1984, December 1985; revised February 1987.

SECTION 1—PURPOSE AND SCOPE

1.1 This Standard prescribes minimum requirements for design, manufacture, and test for "Class 150" irrigation tubing to be used in systems where the operating pressure will not exceed 1.0 MPa (145 psi).

1.2 This Standard has been formulated to afford reasonable protection and to provide a margin for deterioration in service so as to give a reasonably long period of usefulness to the users of aluminum irrigation tubing.

1.3 The interests of manufacturers have also been recognized by taking into consideration advancements in design and material and the evidence of experience. Specific sanction is given for the use of materials whose durability characteristics have been clearly demonstrated. Progress in the art has been anticipated by allowing the use of new materials and construction having safety characteristics equal or superior to those specified.

SECTION 2—GENERAL

2.1 Aluminum irrigation tubing shall be understood to include all aluminum tubing in irrigation systems to transport water at temperatures under 40 °C (104 °F). Other uses for which aluminum irrigation tube may be used, such as piping for gas, air, and other fluids, are not included in this Standard.

2.2 The specifications cited herein are minimum requirements. Dimensions and materials, unless otherwise specified, shall meet these specifications.

2.3 Dimensions

2.3.1 It is recommended that all aluminum tubing for lateral and main lines to be considered portable shall have outside diameters of 51, 76, 102, 127, 152, 178, 203, 229, 254, 279, 305, or 356 mm (2, 3, 4, 5, 6, 7, 8, 9, 10, 11, 12, or 14 in.). Fractional (in.) outside diameters, outside of allowable tolerances, are not recommended.

2.3.2 Variations from specified dimensions for portable irrigation tubing shall not exceed the amounts prescribed for extruded or for welded tube in American National Standard H35.2, Dimensional Tolerances for Aluminum Mill Products.

2.3.3 Permanent or buried irrigation lines that are not considered readily portable may have dimensions that are peculiar to the material used, unless restricted in other parts of this Standard.

2.4 Chemical composition. Tubing shall conform to the chemical composition requirements prescribed for extruded, welded, and drawn tubing in ASTM Standard B241, Specifications for Aluminum-Alloy Seamless Pipe and Seamless Extruded Tube; ASTM B313, Specifications for Aluminum-Alloy Round Welded Tubes; and ASTM B210, Specifications for Aluminum-Alloy Drawn Seamless Tubes.

2.5 Hydrostatic tests. On the basis of maximum operating pressure of 1.0 MPa (145 psi), all tubing must be capable of withstanding an internal hydrostatic test pressure of 3.0 MPa (435 psi) for 2 min without leaking.

2.6 Thickness of tubing. Minimum wall thickness is not a specification of this Standard. However, due to the relationship of wall thickness and diameter to mechanical characteristics of the tubing, the minimum wall thickness for each size of tubing and the specified yield strength of the material used must satisfy all of the following conditions:

TABLE 1—DENTING FACTORS FOR ALUMINUM IRRIGATION TUBING

Outside diameter		Minimum denting factor
mm	in.	N
51	2	280
76	3	280
102	4	280
127	5	300
152	6	370
178	7	450
203	8	580
229	9	750
254	10	980

2.6.1 To prevent excessive denting in handling or field use, the tubing must have a denting factor equal or superior to the denting factor in Table 1 for the pipe size under consideration.

The denting factor, D_f, is calculated by:

$$D_f = F_{ty} t^2$$

where

F_{ty} = minimum tensile yield strength of material in MPa (see Table 2)

t = specified wall thickness, mm

2.6.2 The tubing must be capable of spanning 9 m (29.5 ft) as a simple beam without permanent deflection or local buckling when filled with water at atmospheric pressure. The bending stresses (f) of aluminum alloy tubing that result shall not exceed the smaller of the two values found as follows:

f = 90% of the minimum tensile yield strength

or

$$f = 1.57\, F_{ty} - 2.47 \times 10^{-5}\, F_{ty}^2\, D/t$$

where

D = outside diameter, mm

t = wall thickness, mm

2.7 It is recommended that manufacturers of mechanical move systems use a safety factor of 2 in establishing their safety devices for protection due to torque loading.

2.7.1 The following formula for aluminum alloy tubing shall be used to compute the torque strength:

$$T = 79.3\, K\, D^{1/2} t^{5/2}$$

where

T = torque strength, N·m.

K = stiffening factor. (For extruded tubing, K = 1; for welded tubing, K = 1 or as specified from physical testing.)

t and D are as previously specified

TABLE 2—MECHANICAL PROPERTIES

Aluminum alloys and tempers ASTM No.	Minimum tensile Yield F_{ty}		Ultimate tensile strength, F_{tu}			
			Welded tubing‡		Seamless tubing	
	MPa	psi	MPa	psi	MPa	psi
3003-H14	117	17,000**	97	14,000	138	20,000**
3003-H16	145	21,000**	97	14,000	165	24,000**
3003-H18	165	24,000**	97	14,000	186	27,000**
3004-H32	145	21,000*	159	23,000	193	28,000*
3004-H34	172	25,000*	159	23,000	221	32,000*
5050-H34	138	20,000**	124	18,000	172	25,000**
5050-H36	152	22,000**	124	18,000	186	27,000**
5050-H38	165	24,000**	124	18,000	200	29,000**
5052-H32	159	23,000**	172	25,000	214	31,000**
5052-H34	179	26,000**	172	25,000	234	34,000**
5052-H36	200	29,000**	172	25,000	255	37,000**
5052-H38	214	31,000**	172	25,000	269	39,000**
5086-H32	193	28,000**	241	35,000	276	40,000**
5086-H34	234	34,000**	241	35,000	303	44,000**
5086-H36	262	38,000**	241	35,000	324	47,000**
5154-H32	179	26,000*	207	30,000	248	36,000*
5154-H34	200	29,000**	207	30,000	269	39,000**
5154-H36	221	32,000*	207	30,000	290	42,000*
5154-H38	234	34,000**	207	30,000	310	45,000**
6061-T6	241	35,000*†	165	24,000†(a)	262	38,000†(c)
6063-T6	172	25,000†(b)	117	17,000†(a)	207	30,000†(d)
6063-T31	193	28,000**			207	30,000

*ASTM B313, Specifications for Aluminum-Alloy Round Welded Tubes
**ASTM B210, Specifications for Aluminum-Alloy Drawn Seamless Tubes
†ASTM B241, Specifications for Aluminum-Alloy Seamless Pipe and Seamless Extruded Tube
‡ANSI/ASME Boiler and Pressure Vessel Code, Section IX, Welding and Brazing Qualifications, Par. QN-6(c)
(a) ANSI/ASME Boiler and Pressure Vessel Code, Section VIII, Pressure Vessels, Div. 1, Table UNF 23.1
(b) ASTM B210 allows 193 MPa (28,000 psi) for drawn tubing
(c) ASTM B210 allows 290 MPa (42,000 psi) for drawn tubing
(d) ASTM B210 allows 228 MPa (33,000 psi) for drawn tubing

2.8 Theoretical bursting pressure. The theoretical bursting pressure for the tubing shall be considered to be the pressure determined from the formula:

$$P = 2 F_{tu} t/D$$

where
P = bursting pressure, MPa
F_{tu} = ultimate tensile strength as given in Table 2 or those values guaranteed by the manufacturer
t and D are as previously specified

Cited Standards:

ANSI/ASME Boiler and Pressure Vessel Code
ANSI H35.2, Dimensional Tolerances for Aluminum Mill Products
ASTM B210, Specifications for Aluminum-Alloy Drawn Seamless Tubes
ASTM B241, Specifications for Aluminum-Alloy Seamless Extruded Tube
ASTM B313, Specifications for Aluminum-Alloy Round Welded Tubes

PRINCIPLES AND PRACTICES FOR PREVENTION OF MOSQUITO SOURCES ASSOCIATED WITH IRRIGATION

Initially prepared by the ASAE Soil and Water Division Committee on Irrigation System Design for Mosquito Control; finalized by the ASAE Committee on Surface Irrigation of the Irrigation Group; approved by the ASAE Soil and Water Division Steering Committee; adopted by ASAE as a Recommendation January 1958; revised December 1963, December 1968, March 1974, December 1974; reconfirmed and reclassified as an Engineering Practice December 1978; revised December 1979; reconfirmed December 1984; revised March 1985.

SECTION 1—PURPOSE AND SCOPE

1.1 The following principles and practices are recommended for prevention and elimination of man-made mosquito sources commonly associated with the engineering and agricultural phases of irrigation. While much concern is expressed over discomfort caused by large numbers of mosquitoes, human and animal diseases transmitted by these insects should receive primary attention. Cases of western equine encephalitis in both humans and horses are regularly reported from states with irrigation, and an association of this disease with specific irrigated areas has been demonstrated. The classic vector of western equine encephalitis, **Culex tarsalis**, is produced abundantly in irrigation wastewater. Experience has shown that man-made habitats in irrigated areas are more important mosquito sources than the natural mosquito-producing habitats. With mosquito insecticide resistance increasing and with limitations on insecticides available for mosquito control, emphasis should be placed on soil and water management as a control tool.

SECTION 2—BASIC PRINCIPLES

2.1 Irrigation developments often aggravate existing mosquito sources and create new ones unless appropriate preventive and control measures are provided. Mosquito production is augmented by creation of the free water conditions necessary for completion of the egg, larval and pupal stages of the mosquito life cycle. Examples of such conditions include reservoir construction and operation that encourage floating debris, emergent or floating vegetation, frequent fluctuations in water surface elevation, water standing for several days on irrigated fields and ponded water resulting from seepage or poor surface drainage. Incorporation of mosquito preventive and control measures should be made during the planning, construction, operation and maintenance of irrigation developments to reduce the magnitude of mosquito problems and eliminate the need for chemical control. Mosquito control planning and construction should be closely coordinated with other basic interests such as agriculture, soil and water conservation, flood control, hydroelectric power, wildlife management and recreation. The potential need for continuing funds to provide for ongoing surveillance and implementation of prevention and control measures should be considered as a part of pre-project planning.

2.2 Reduction or elimination of man-made mosquito sources can be achieved by proper water management and maintenance of facilities by personnel knowledgeable in both principles of irrigation and mosquito biology. Irrigation developers may obtain specifications and criteria for mosquito prevention and public health requirements from local, state and federal public health agencies during planning, construction and operational phases. It is desirable for public health and irrigation agencies to work closely together and with established agricultural education agencies in providing information on mosquito prevention and control to individuals.

SECTION 3—REQUIREMENTS

3.1 Storage reservoirs

3.1.1 Reservoir side slopes should be as steep as possible, consistent with soil characteristics and risk factors. Where steep side slopes are not feasible, the slopes should be lined with impervious material such as concrete to three feet below the waterline or mowed or treated periodically with herbicides to control vegetative growth.

3.1.2 Borrow areas located in the normal summer fluctuation zone or outside the reservoir basin should be made self-draining or diked and stocked with mosquito-eating fish. All depressions which will be flooded by the reservoir should be connected with the reservoir by drains to insure complete drainage or fluctuation of water within the depression.

3.1.3 The normal summer fluctuation zone should be completely cleared annually, preferably prior to spring reservoir refilling. Clearing should be to the maximum water level for the impoundment except for isolated trees and sparse vegetation along abrupt shore lines which will be exposed to wave action, or in other situations where no significant mosquito production is likely to occur. The upper clearing limit should not be lower than the elevation of the back-water curve at normal stream flow. Mechanical or approved chemical measures can be used to control vegetation.

3.1.4 Water level management should be employed to the maximum degree permitted by the primary functions of the reservoir to minimize conditions favorable for mosquito production. Continuously lowering reservoir levels during the mosquito breeding season is the principal water management measure to prevent breeding.

3.1.5 Seepage areas that develop below dams or behind dikes should be made to drain freely. Vegetation, debris and floatage should be removed from all mosquito-control drains to insure free flow.

3.2 Project conveyance and distribution systems

3.2.1 Lining, or other satisfactory seepage control measures, should be provided for all sections of canals and laterals located in porous material where excessive leakage would result in waterlogged areas, seeps or ponds.

3.2.2 Drainage should be provided to prevent any project ponding that would be favorable for mosquito production.

3.2.3 Provision should be made to prevent retention of ponded water by turnouts and other hydraulic structures when they are not in use.

3.2.4 Every effort should be made to establish delivery schedules which will provide farmers with adequate, but not excessive, water supplies to permit efficient irrigation. Delivery schedules and flow rates available to farmers should be as flexible as possible.

3.2.5 Vegetation and debris which prevent free flow should be removed from all conveyance channels, water control structures and drains.

3.3 Irrigated farms

3.3.1 All surface-irrigated fields should be properly leveled or graded to provide for efficient irrigation and for removal of excess water to prevent ponding that results in mosquito production. Sprinkler systems should be designed and operated to apply water at a rate equal to or less than the infiltration rate of the soil to prevent ponding or runoff. The life cycle of the mosquito will be disrupted if ponded water disappears before the larvae can complete their fourth instar or become pupae. Limiting pond duration to not more than three days will prevent adult mosquito emergence.

3.3.2 Application of irrigation water should be limited to that required to fill the crop root zone, plus reasonable losses and water needed to prevent a buildup of excessive salts in the crop root zone.

3.3.3 A conveyance system should be provided and maintained to prevent seepage from farm ditches and leakage from turnouts which would result in seeps or ponds favorable for mosquito production.

3.3.4 A drainage system should be provided for the timely removal of excess water from all portions of the farm into a drainage outlet or a properly designed return flow system with adequate capacity to efficiently utilize or remove the expected drainage waters. Use of return flow within the farm system should be encouraged, both from the standpoint of improving irrigation efficiency and reducing pollutant discharge (see ASAE Engineering Practice EP408, Design and Installation of Surface Irrigation Runoff Reuse Systems). Drainage and reuse systems should be constructed and maintained so that all channels drain freely and vegetation is controlled in both channels and water storage facilities.

3.4 Project drainage systems

3.4.1 Drainage outlet systems should be provided to expeditiously remove and dispose of waste irrigation water, natural runoff and seepage from both irrigated and non-irrigated lands affected by the distribution and use of irrigation water.

3.4.2 Drainage channels should be maintained to avoid ponding and to insure free flow at all time. Drainage outflows should be conducted to water bodies where mosquito production is not a problem.

3.4.3 Provisions should be made to prevent water from ponding behind spoil banks.

Cited Standard:

ASAE EP408, Design and Installation of Surface Irrigation Runoff Reuse Systems

ASAE Standard: ASAE S268.3

DESIGN, LAYOUT, CONSTRUCTION AND MAINTENANCE OF TERRACE SYSTEMS

Developed by ASAE Terrace and Related Slope Modification Systems Committee; approved by Soil and Water Division Steering Committee; adopted by ASAE as a Tentative Recommendation December 1962; revised and reclassified as a full Recommendation December 1963; reconfirmed December 1968, December 1969, December 1970; revised February 1972; reconfirmed December 1976, December 1977; revised and reclassified as a Standard December 1978; revised December 1983.

SECTION 1—PURPOSE AND SCOPE

1.1 This Standard is intended as a guide to engineers and technicians in the design, layout, construction and maintenance of terrace systems for erosion control, water conservation, and pollutant runoff control.

1.2 Terraces are earth embankments, channels or combinations of embankments and channels constructed across the slope at suitable spacings and with acceptable grades for one or more of the following purposes:

1.2.1 To reduce soil erosion

1.2.2 To provide for maximum retention of moisture for crop use

1.2.3 To remove surface runoff water at a nonerosive velocity

1.2.4 To reform land surface and improve farmability

1.2.5 To reduce sediment content in runoff water

1.2.6 To reduce peak runoff rates to installations downstream

1.2.7 To improve water quality

1.3 Terraces alone usually will not provide adequate control of erosion on sloping lands. Erosion control requires a complete water disposal system which may include waterways, underground outlets, water and sediment control basins, and drop structures. Other conservation practices usually recommended in conjunction with terrace systems are contour farming (parallel to terraces), crop rotations, contour stripcropping, conservation tillage, residue management, and good soil management. Each measure shall be planned for compatibility with modern farm equipment and shall be optimally beneficial for soil and water conservation. Natural features, permanent boundaries, and the location of auxiliary features such as fences and field roads shall be considered to enhance farmability during terrace system design and layout.

SECTION 2—CLASSIFICATION

2.1 All types of terrace systems can be classified according to alignment, cross section, grade and outlet.

2.1.1 Classification by alignment

2.1.1.1 Parallel terraces. To aid in farming operations, parallel terraces should be used whenever possible. Use long gentle curves when curves are necessary to make these terraces as straight as possible. Space the adjacent ridges equidistant whenever possible. On irregular topography divide terraces into parallel groups or sections with varying spacings for adjustment. Terrace spacings should be multiples of present and possible future equipment widths, especially for row crop equipment, to minimize point rows. Cut and fill along the terrace channel as necessary to establish satisfactory uniform or variable grades.

2.1.1.2 Nonparallel terraces. These terraces are designed with the terraces fitted to the contour of the land without regard to paralleling. In practice, adjustment is made to eliminate sharp turns and short point rows by installing additional outlets, using variable grade, and vegetated adjustment or turn areas.

2.1.2 Classification by cross section

2.1.2.1 Broadbase terrace. (Fig. 1) This terrace is constructed so that crops can be planted and machinery operated on the entire terrace. Ridge and channel slopes are constructed to fit the machinery being used.

2.1.2.1.1 Three-segment section. This terrace section consists of three-slope segments: the cut slope, front slope, and back slope. The width of this terrace section is usually 12 to 15 m (40 to 50 ft). This relatively narrow terrace section with steep slopes is suited only to tillage parallel to the terrace. Because of sharp angles at the channel bottom and ridge top it is important not to cross this terrace shape because of potential damage to terraces and equipment and operator safety.

2.1.2.1.2 Wide smooth segment. This terrace section is a much broader terrace section with smooth curves for the channel and ridge. This section may be safely crossed in tillage, planting, and harvesting operations and is recommended where the terrace will be crossed. The terrace section is usually over 18 m (60 ft) to as much as 25 m (80 ft) or more in width.

2.1.2.2 Conservation bench terrace. (Fig. 2) A broad level bench or flat channel is constructed below a sloping runoff contributing area. Conservation bench or flat channel terraces are designed for maximum conservation of water. Channel width will vary depending on the land slope, allowable depth of cut, precipitation, machine width, and contributing runoff area.

2.1.2.3 Steep-backslope terrace. (Fig. 3) This terrace is constructed with a very steep backslope which is usually seeded to permanent grass. This terrace is used to reduce slopes between terraces and improve farmability. When field slopes are uniform, a balanced cross section can be used. When field slopes are irregular and fills are necessary along the terrace length, transitions may be required from excess borrow section to excess fill section.

2.1.2.3.1 Balanced cross section. This terrace is normally constructed entirely from the lower side. If exposed subsoil is not productive, the top soil may be moved downhill first and then moved back over the excavated area.

2.1.2.3.2 Excess fill section. Where cuts below the terrace are undesirable, the terrace ridge may be built entirely of soil from excess borrow sections of the terrace or from a planned borrow area. A fill section is desirable when cuts below the terrace would result in increased row grades below the terrace or where terraces are built on shallow soils.

2.1.2.3.3 Excess borrow section. Where deep cuts are required to improve terrace alignment, both the backslope and/or channel may be excavated. A ridge crossed by a terrace may be cut down during construction to develop acceptable terrace channel grades, provided the subsoil responds to fertilizers and the channel would not erode. In this case, the cross section may resemble the ridgeless terrace channel defined in paragraph 2.1.2.6.

2.1.2.4 Narrow base terrace. (Fig. 4) The ridge of this terrace is constructed with both sides of the ridge steep, permanently vegetated, and not farmed. They are economical, eliminate the hazard of tractor or harvest tip over, eliminate the problem of fitting ridge slope to machinery width, and permit placing the inlet to underground outlets where they do not interfere with machinery.

2.1.2.5 Bench terrace. (Fig. 5) This terrace is constructed with a wide level bench. The ridge has steep side slopes which are seeded to a permanent vegetation. The terrace is used mainly to provide more efficient distribution of water under both irrigated and dryland production.

2.1.2.6 Ridgeless channel terrace. (Fig. 6) This terrace is constructed on nearly flat to gently sloping land to remove excess surface runoff water at nonerosive velocities. The cut material from the channel is used to smooth the interterrace interval.

2.1.3 Classification by grade

2.1.3.1 Level terraces. These terraces are constructed where water conservation is the primary objective with erosion control as an accompanying benefit. They are usually used in low-rainfall regions to retain runoff for storage in the soil and may be used in high-rainfall regions on very permeable soils. The ends of the channel may be blocked or left open.

2.1.3.2 Graded terraces. These terraces are constructed to reduce field slope lengths and remove runoff, with tolerable soil loss as the primary objective. Water conservation may be an accompanying benefit. The grade to the outlet may be uniform or variable.

2.1.4 Classification by outlet. Terraces may be classified as blocked outlets (all water infiltrates in the terrace channel), permanently vegetated outlets (grassed waterway or a vegetated area) or underground outlets (water is removed from the terrace channel by underground conduits to reduce erosion and remove less land from production). Combinations of these may be used to meet topography and soil conditions.

SECTION 3—DESIGN CRITERIA

3.1 Spacing

3.1.1 The interval used in spacing terraces is measured from the terrace ridge to the next lower terrace channel, but excluding all permanently vegetated slopes. It is sometimes called the eroding slope length. This dimension is called the horizontal interval. The distance from the center of the channel of one terrace to the center of the channel of the next terrace is defined as terrace spacing. Terraces to control water erosion should be spaced by one of the following methods:

3.1.1.1 Where data are available for application of the universal soil loss equation, the horizontal spacing of terraces should not exceed the slope length determined for contour cultivation by using the allowable soil loss, the most intensive use expected for the land, and the expected level of residue and soil management. On gently sloping land, anticipated runoff and the economical terrace cross section size may determine spacing, rather than soil loss between terraces.

3.1.1.2 The equation $VI = XS + Y$ or $HI = (XS + Y)100/s$ may be used for determining the interval,

where

VI = vertical interval in meters
HI = horizontal interval in meters
X = a variable with values from 0.12 to 0.24. Values of X for different zones are shown in Fig. 7
S = the weighted average land slope of the land draining into the terrace in meters per 100 m
Y = a variable with values of 0.3, 0.6, 0.9 or 1.2

Values of Y are influenced by soil erodibility, cropping systems, and crop management practices. The low value is applicable for very erodible soils with conventional tillage methods, where little to no residue is left on the surface. The high value is applicable to erosion resistant soils where no-till planting methods are used and large amounts of residue are left [a minimum of 3.3 metric tons per hectare (1.5 tons per acre) of straw or its equivalent] on the soil surface.

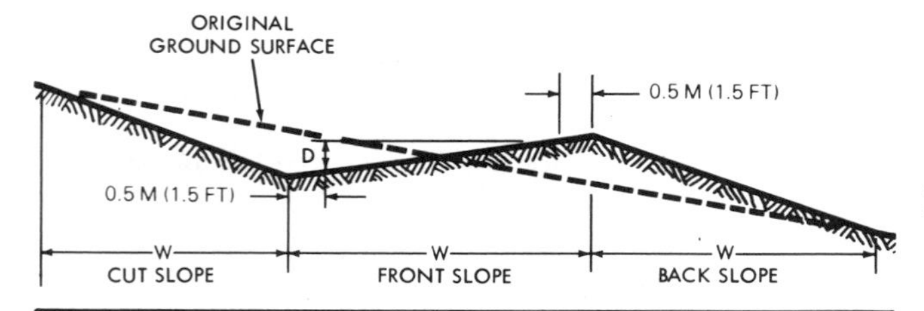

FIG. 1—BROADBASE TERRACE CROSS SECTION†

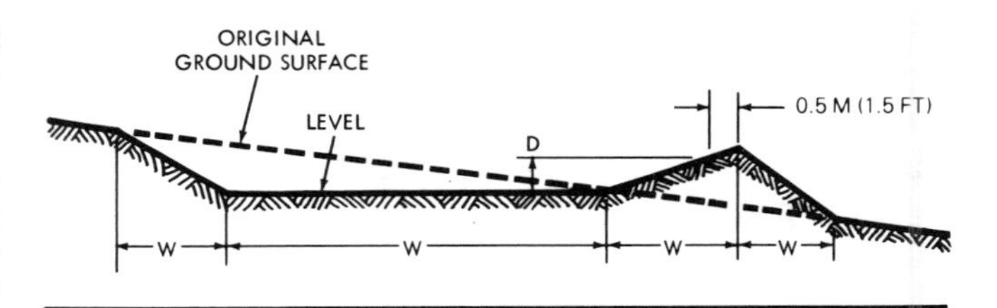

FIG. 2—CONSERVATION BENCH TERRACE CROSS SECTION†

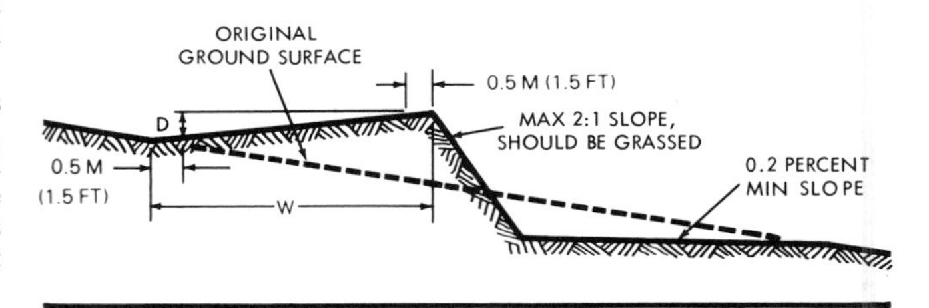

FIG. 3—STEEP-BACKSLOPE TERRACE CROSS SECTION†

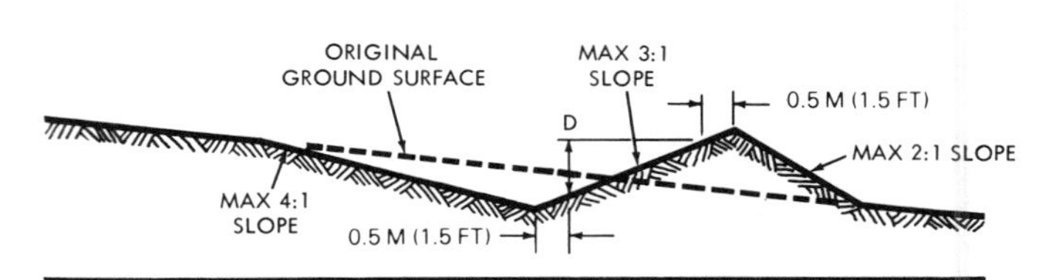

FIG. 4—NARROW-BASE TERRACE CROSS SECTION†.

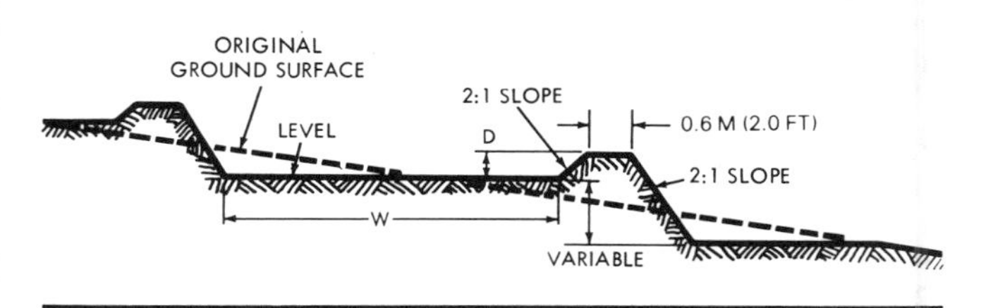

FIG. 5—BENCH TERRACE CROSS SECTION†

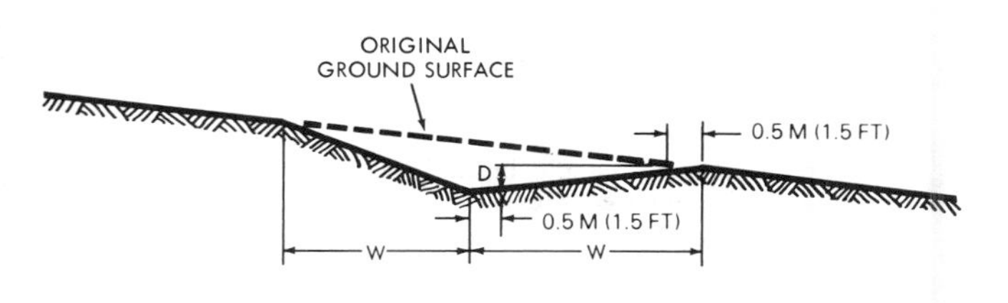

FIG. 6—RIDGELESS CHANNEL TERRACE CROSS SECTION†

†W is multiple of machinery width used to farm the land.
D is the design depth of terrace channel.

The maximum constructed slope allowed is 6:1 unless local conditions permit modification of the design criteria, or the drawings show a steeper maximum slope.

In each area, local conservation agency engineering standards and specifications designate recommended channel depths, cross sectional areas and terrace slopes for all locations and terrace configurations.

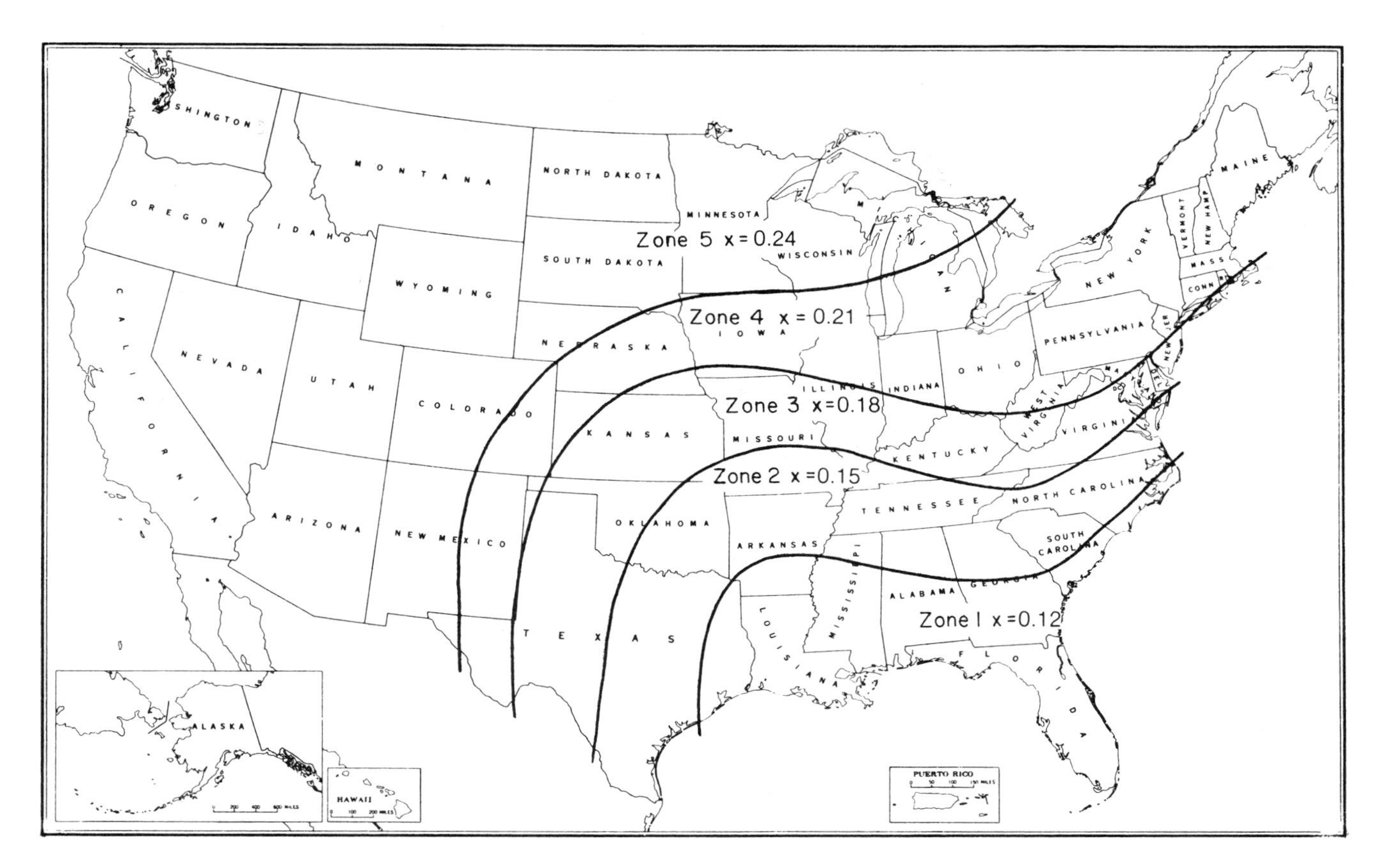

FIG. 7—VALUES OF X IN EQUATION VI = XS + Y or HI = (XS + Y)100/s (See Paragraph 3.1.1.2)

3.1.2 Spacings determined by the above methods may be increased as much as 0.15 m (0.5 ft) or 10%, whichever is greater, to provide better alignment or location, to miss obstacles, to adjust for agricultural machinery, or to reach a satisfactory outlet. Spacings may be increased an additional 10% for terraces with underground or blocked outlets.

3.1.3 Level terraces for water conservation and erosion control should also be spaced using the procedure outlined in paragraph 3.1.1.

3.1.4 The spacing is adjusted to match present and expected future equipment widths, especially for row crop equipment. Care should be taken to use spacings which match many widths. Multiples of 36, 18, 9 m (120, 60, and 30 ft) fit the most row crop machinery spacing combinations. Thus multiples of these spacings should be used whenever possible. Ideally, terrace spacings should allow an even number of passes of each piece of equipment used.

3.1.5 Because of the difficulty of farming steep slopes, steep backslope and narrow base terraces are often used when slopes exceed 6 to 8%. Benching of land between terraces by erosion occurs naturally and is desired to reduce slopes and improve farmabilty. The wider the spacing, the greater the amount of earth movement required to bench the land and thus the wider the spacing, the greater will be the time required for land to become benched by the erosion process.

3.2 Alignment. To make farming operations easier, terraces should be made parallel with long gentle curves where practical. Smooth curves are extremely important in modern agriculture where the trend is to narrow rows and wide equipment. If terraces traverse natural ridges or natural drainage depressions, alignment and acceptable curvature become more difficult. Improved alignment, to make farming easier and to avoid point rows, is possible in most terrace layouts. Seven methods are available to improve alignment, reduce curvature and decrease the area of point rows in a field. These methods may be used singly or in combination to improve terrace layouts.

3.2.1 Use turn strips on long, flat ridges which are of lesser slope than the rest of the field.

3.2.2 Vary grade along the terrace channel within limits specified in paragraph 3.5.

3.2.3 Vary depth of cut along the terrace line.

3.2.4 Shape or smooth land prior to terrace construction.

3.2.5 Vary the amount of excavation or fill along the terrace channel or ridge for steep backslope terraces. With steep backslope terraces much greater leeway is provided for improvement in terrace alignment.

3.2.6 Use channel blocks with level terraces where it is desirable to adjust the terrace channel elevation along the same terrace line.

3.2.7 Use multiple outlets and shorter terrace drainage lengths. Shortened terrace lengths allow greater variation in channel grade and more opportunities for paralleling. Underground outlets can shorten the terrace drain length and do not shorten the field length as would a waterway.

3.2.8 Use water and sediment control basins in conjunction with terraces in particularly difficult situations, to allow greater flexibility.

3.3 Capacity. Graded terraces should have enough discharge capacity to handle the peak rate of runoff expected from a 24-h, 10-yr frequency storm. In computing the capacity use an n value of 0.06 in the Manning formula.* Level terraces and graded terraces with underground outlets that impound runoff water on the field should be designed to handle the runoff volume from a 24-h, 10-yr frequency storm.

3.4 Cross section. The terrace cross section shall be proportioned to fit the ground slope, the crops grown, and the farm machinery used. Additional ridge height should be added to provide for settlement, channel sediment deposits, ridge erosion, the effect of normal tillage

---

*The Manning Formula: $V = \dfrac{r^{2/3}\, s^{1/2}}{n}$

where
- V = velocity, m/s
- r = hydraulic radius—the area divided by wetted perimeter, m
- s = slope of channel, m/m
- n = roughness coefficient

Rate of Discharge: Q = AV

where
- Q = capacity, m^3/s
- A = area, m^2
- V = velocity, m/s

operations, and a safety factor. The ridge and channel should each have a minimum width of 0.9 m (3 ft) at the design elevation.

3.5 Channel grade

3.5.1 Minimum grades should be such that ponding in the channel will not cause serious delay in field operations.

3.5.1.1 Soils with slow internal drainage—0.2%

3.5.1.2 Soils with good internal drainage—0.0%

3.5.2 Maximum grades will vary with the type of terrace and the purpose of the terrace. The critical condition will occur when the terraces have been recently cultivated, and there is no vegetative cover. When terraces use an underground outlet, sediment and water will be stored in the channel, thus reducing velocities and allowing steeper grades near the outlet.

3.5.3 Maximum velocities should be 0.8 m/s (2.5 ft/s) for soils with high organic content, 0.6 m/s (2.0 ft/s) for most soils, and 0.5 m/s (1.5 ft/s) for extremely erodible soils. Velocities are to be computed by Manning's formula* using an n value of 0.035.

3.5.4 Channels may be permanently vegetated if necessary to control erosion and if appropriate design *n* values are used for the vegetation grown.

3.6 Outlets. All terraces must have adequate outlets. The outlet requirements vary with the type of terraces.

3.6.1 Permanently vegetated outlets. Vegetated outlets may be used for graded or open-end level terraces. The outlet must convey runoff from the terrace and rows to a point where no erosion damage will occur. It should be established in advance of terrace construction, if necessary, to insure vegetative cover. Vegetated outlets may be shaped waterways, grassed areas or natural wooded draws.

3.6.2 Underground outlets. Underground outlets may be used with graded or level terraces. The outlet consists of an intake riser, the underground conduit, and the outlet pipe. The outlet is designed to remove the stored runoff water durng the design drain period within a maximum of 48 h. Shorter drain periods are desirable for soil with poor internal drainage and crops that do not tolerate inundation or wet soil.

3.6.3 Soil infiltration. Level closed-end terraces normally use water infiltration into the soil in the channel as the outlet. Soil infiltration should permit draining the terrace channel within 48 h. The channel should be level along the majority of the length between terrace blocks in order to provide the maximum soil profile storage area. The end blocks on level terraces should be designed so water will flow over the end block before overtopping the terrace ridge.

3.6.4 Combinations of soil infiltration, permanently vegetated outlets, and underground outlets may be used within a terracing system to minimize interference with farming operations and to maximize the conservation of soil and water.

SECTION 4—PLANNING AND LAYOUT

4.1 Planning with owner/operator. The process of selecting a terrace and outlet system for a field should be a joint effort between the technician or engineer doing the layout and design and the land owner/operator. The system should meet the owner/operator's goals of farmability as well as the goal of controlling erosion. In conjunction with other conservation measures, such as residue management, contouring, crop rotations, and conservation tillage, the erosion losses should average less than the tolerable limit.

4.2 Farmabilty. In general, planning should begin with the best layout or design. If the first choice will not work or is unacceptable, substitutions should be made and the process repeated until an acceptable plan is developed.

4.2.1 Systems parallel to permanent boundaries. (Fig. 8) In this system permanent boundaries (such as ridges or major drainage divides, public roads, major drainage patterns, rock outcrops, and property lines) serve as the paralleling guide. This system may require extensive use of water and sediment control basins in addition to other aids to improve alignment discussed in paragraph 3.2. Residue management and conservation tillage will normally be needed.

4.2.2 Parallel terrace groups system. (Fig. 9) When complete parallel terrace systems are not possible, groups of two or more terraces may be made parallel. A correction then occurs between the next group of parallel terraces. In this system, varying channel grades, channel slope direction, and some underground outlets are used to improve the design.

4.2.3 Improved alignment contour system. (Fig. 10) This system follows the general contour of the land with terraces not parallel. Minor cuts and fills and variations in grade are commonly used to improve alignment.

4.3 Owner/operator selection of system type. It should be understood by the owner/operator that several terrace systems with different levels of farmability are possible for most fields and that initial costs and maintenance requirements will differ for these systems. Once the owner/operator has a clear understanding of the differences and their ramifications, he/she is ready to make a choice.

4.4 Planning. Parallel terrace systems are commonly planned with the aid of topographic maps. A map with a contour interval of 0.6 m (2 ft) maximum can be used satisfactorily. Closer contours may be more desirable on flatter slopes. First, permanent boundaries (i.e., ridges, public roads, major drainage ways, rock outcrops, etc.) should be located. Next, field features such as outlets, turnrows, and field roads should be located. The design terrace interval should be determined and the top and bottom terraces located using the multiples of the design or adjusted interval.

4.4.1 The second or third terrace from the top is often used as a key for parallel systems. The selection of a key terrace is dependent upon such physical factors as the location of outlets, the location of ridges, and the uniformity of the slope. Methods used to parallel terraces include: (1) items listed in paragraph 3.2; (2) divide terraces into short segments; (3) use underground outlets to parallel terraces through depressional locations; and (4) any combination of these methods. As the layout continues downhill, the selection of additional key terraces may be necessary with correctional areas provided as needed. Terrace curves should have a radius of 30 m (100 ft) or greater for farming with wide, high speed machinery. Grassed turn areas should be planned when sharp terrace curves are necessary. Sharp terrace curves can be reduced by varying the depth of cut in the terrace channels. Several tentative systems may be laid out on the topographic map for comparison. The best plan is selected and staked on the field for construction.

4.5 Field-trial method. The same criteria and steps used to plan a terrace system on a topographic may are also used for planning in the field. However, the terraces are staked in the field as the system is planned. Several trials may be necessary before the best system can be selected.

4.6 Construction drawings and specifications should show the location, cut and fill requirements, channel grade, constructed cross section, and location of each terrace outlet. Sufficient copies of the drawings and specifications shall be provided to the landowner/operator for themselves and for the contractor.

SECTION 5—CONSTRUCTION

5.1 Many different machines are successfully used to build terraces. These include elevating or belt terracers, whirlwind terracers, moldboard or disk plows, oneways, graders, and scrapers all powered by farm tractors. Heavy construction machinery includes the grader, elevating grader, bulldozer, and scraper. The haul distance and amount of soil to be moved per unit length are factors to consider in selecting construction equipment.

SECTION 6—MAINTENANCE

6.1 Maintenance is essential to retain adequate ridge height and channel capacity. Structures must be maintained at or near design specifications to insure adequate capacity so design events do not overload and damage the system. Erosion by water, wind, and tillage operations tends to reduce ridge height and fill the channel. A regular and adequate maintenance program restores the ridge height and channel capacity and keeps the terrace shape. Inadequate maintenance results in an overtopped terrace during a large storm. This failed terrace often causes serious field erosion and possible failure of downstream terraces.

6.2 Maintenance procedures. Terrace maintenance should be done in a logical sequence beginning with a measurement of ridge height. When the height meets the design specification, maintenance has been adequate and past maintenance practices should be continued. When the ridge height does not meet the design specification, maintenance has been inadequate and must be improved. The second step is to

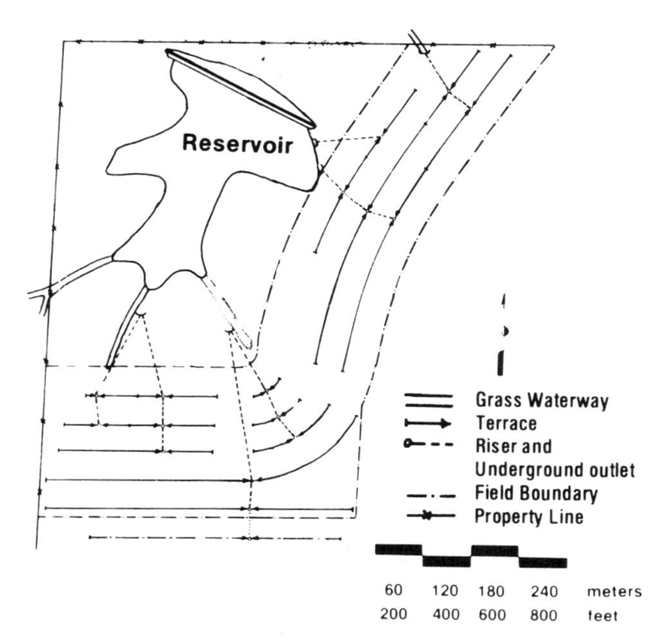

FIG. 8—SYSTEM PARALLEL TO PERMANENT BOUNDARY. (This system of parallel terraces combines water and sediment control structures to maintain alignment parallel with field boundaries. The pasture boundary adjacent to the pond has been relocated to be parallel with terraces. Farmability of this system would be excellent; it may acutally improve compared with no terraces.)

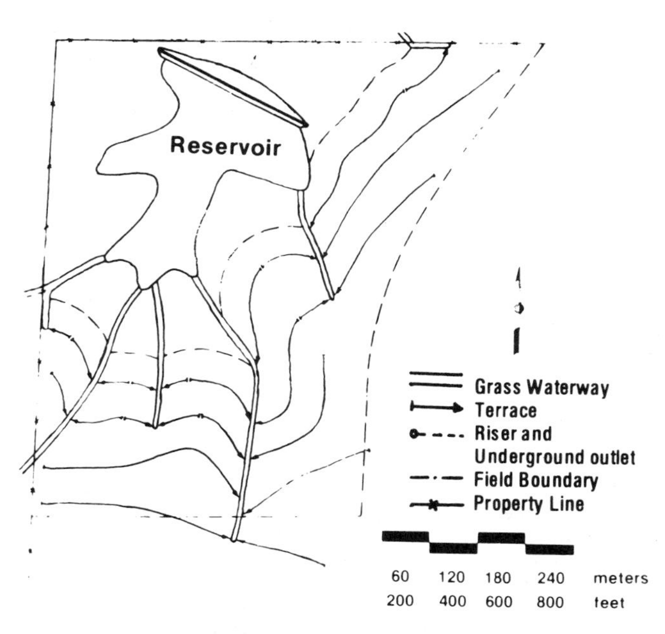

FIG. 10—IMPROVED ALIGNMENT CONTOUR SYSTEM. (Many waterways within the field divide the field into 5 small fields, greatly decreasing farmability. Many point row areas between each terrace result in poor farmability.)

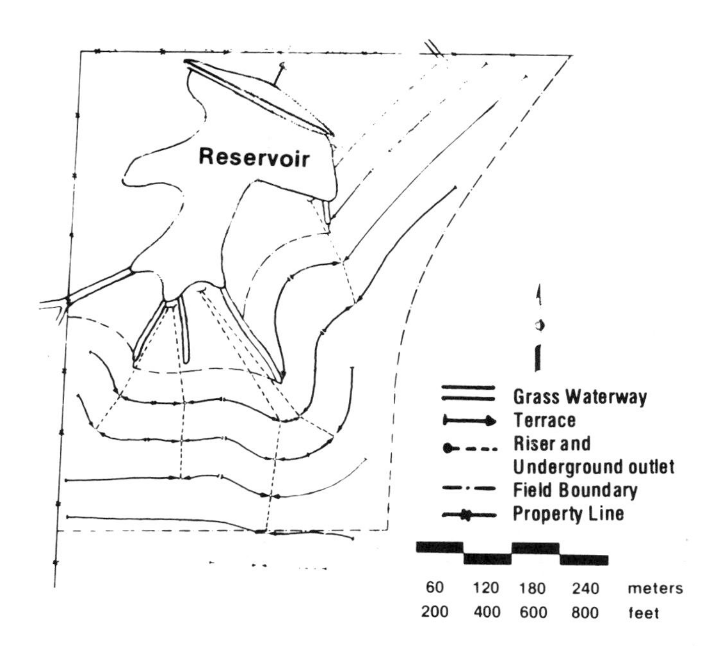

FIG. 9—PARALLEL TERRACE GROUPS SYSTEM. (This system utilizes underground outlets to improve alignment at drainageways and avoid dividing field into many small fields. The farmability of this system would be very good but would still have many point row areas.)

repair damages. The third step is to carry out the normal practice of working up the terrace. The fourth and final step is to insure that the transition to the terrace outlet is adequate.

6.2.1 Check for needed repairs. The terrace system may be damaged by erosion, siltation, machinery, and settling following construction. A good time to check for needed repairs is after a heavy runoff event when erosion, sedimentation and uneveness in channel and ridge elevations are easy to see. Specific items to be repaired are low or narrow spots in the terrace ridge, ponding in the terrace channel, transition to the outlet, and the terrace outlet itself.

6.2.2 Terraces may be rebuilt by any equipment which moves soil from one point to another. Common farm equipment used for maintenace includes moldboard and disk plows, oneways, three-point rear blades and bottomless scraper blades, front-end blades, and front-end loaders. Construction equipment including graders, bulldozers, belt terracers, and scrapers may also be used for terrace maintenance.

6.2.3 Terrace outlet. The terrace system must have an adequate outlet with free unrestricted drainage. Transition to the vegetated oulet has often been a maintenance problem. When the terrace oulet occurs at the edge of the field, a common problem is inadequate ridge height and channel capacity because of tillage over terraces adjacent to the field boundary. A small scraper, front blade, rear blade or front-end loader have all proven useful for moving soil from the channel to the ridge top. The ends of closed-end level terraces must be adequately closed at all times in order to contain the design runoff event.

6.2.3.1 Trash accumulation, sediment deposits, and soil banked up by tillage have sometimes been a problem with underground outlets. Adequate outlet capacity requires keeping inlet risers of underground lines free and open and maintaining adequate runoff and sediment storage capacity adjacent to the outlet.

6.2.4 Maintenance in arid and semi-arid regions. In arid and semi-arid areas, terrace maintenance should be performed during the season when wind erosion is least likely and the establishment of plant cover most likely.

SECTION 7—SAFETY

7.1 Caution should be used when operating equipment on the terrace. Slopes steeper than 4 to 1 should not be farmed except with special equipment. Safety devices such as protective frames or protective canopies may be required to protect equipment operators.

7.2 Structures should be designed with safety in mind. How the structure will be constructed and used should be visualized. There should be information in the final design on the limitations and safety features of the structure. If hazards do exist, it should be so stated to the person or agency responsible for construction and maintenance of the structure. A record of this information should be kept on file.

ASAE Standard: ASAE S289.1

CONCRETE SLIP-FORM CANAL LININGS

Proposed by the ASAE Soil and Water Division Concrete Slip-Form Canal Linings Committee; approved by the Soil and Water Division Steering Committee; adopted by ASAE June 1965; reconfirmed with editorial changes February 1970; reconfirmed December 1974; revised editorially February 1978; reconfirmed December 1979, March 1981, March 1982, December 1983, December 1984; revised April 1986.

SECTION 1—PURPOSE AND SCOPE

1.1 The growing demand for water in the world increases the need for conserving the available water supply. In open conveyance systems, conservation of irrigation water and reduced maintenance costs through control of weeds and seepage losses can best be accomplished by providing a dependable impervious lining. Non-reinforced concrete lining, which has a high hydraulic efficiency, is a dependable means of water control and seepage reduction in canals. The purpose of this Standard is to present recommended standards for concrete slip-form canal linings in the interest of reducing the cost of this type of lining.

1.2 The use of continuous excavating, trimming, and lining equipment has proved to be the most economical method of constructing concrete-lined, trapezoidal canals. Available equipment permits construction to practical tolerances in alignment, grade, and concrete thickness. The universal acceptance and use of standard trapezoidal canal sections will enable manufacturers to standardize excavating and lining equipment. This will permit a reduction in special engineering and equipment manufacturing costs, will make replacement and service parts readily available, and will result in a net reduction in the cost of completed linings.

1.3 These standards are restricted to irrigation canals that have a bottom width not greater than 1.8 m (72 in.), and a total depth of lined section not greater than 2.1 m (84 in.). Linings for larger canals usually involve more complex engineering and economic considerations; and they may require special criteria and construction requirements for the canal section, as well as for the lining.

1.4 The wide variance in design requirements by those involved in designing irrigation works prohibits developing criteria for establishing freeboard or foundation treatment. The scope is limited to establishing standard dimensions and shape of the canal sections.

SECTION 2—BASIC CONCEPTS

2.1 Minimum required thickness for concrete linings as shown in Table 1 are considered adequate to meet the requirements of canals included in this Standard. These are based on the assumptions that the concrete will be of high quality (see Appendix A), that it will be placed on a firm, unyielding subgrade, and will be protected against such destructive forces as external hydrostatic pressures, uplift caused by expansive clays, and frost heave. Provisions for underdrainage to relieve uplift and to minimize frost damage, treatment of foundations involving expansive clays, and determining the design of the concrete mix are design considerations not included in this Standard.

2.2 The following tolerances for lining thickness, alignment, and grade are considered adequate to assure quality installations:

Item	Tolerance
Departure from established alignment	50 mm (2 in.) on tangents 100 mm (4 in.) on curves
Departure from established profile grade	0.03 m (0.1 ft)
Reduction in lining thickness	10% of specified thickness

Available excavating and lining equipment will permit construction of linings within the above tolerances. Abrupt deviations from design grade or horizontal alignment shall not be permitted.

TABLE 1—MINIMUM REQUIRED NOMINAL THICKNESS FOR SLIP-FORM CONCRETE LININGS#

DESIGN VELOCITY*	Climatic area†			
	Warm		Cold	
	Minimum thickness			
m/s (ft/s)	mm	in.	mm	in.
Less than 2.7 (9.0)	38	1.5	50	2.0
2.7-3.6 (9.0-12.0)	50	2.0	63	2.5
3.6-4.5 (12.0-15.0).......	63	2.5	76	3.0

*Velocities in short chute sections shall not be considered design velocity.
†Climatic area:
Warm - Average midwinter temperature is 4.4 °C (40 °F) and above.
Cold - Average midwinter temperature is less than 4.4 °C (40 °F).
#For canals that have a bottom width not greater than 1.8 m (72 in.), design capacity not greater than 2.8 m^3/s (100 ft^3/s), and a maximum velocity of 4.5 m/s (15 ft/s).

2.3 Standard subgrade dimensions shall be maintained, and any deviations from standard lining thickness will be made by modifying the inside dimensions of the lining. Modifications made in this manner can be more economically accomplished than by changing subgrade dimensions.

2.4 To control cracking in the lining caused by shrinkage and temperature change, longitudinal and transverse grooves should be provided as shown in Fig. 1. The groove dimensions specified will provide adequate space for placing common mastic sealing compounds.

2.5 Contraction or control joints. Contraction joints shall be provided to control cracking caused by shrinkage and temperature change. These joints are to be made by providing grooves in accordance with detail shown in Fig. 1. Transverse grooves shall be uniformly spaced at intervals of 2.5 to 4.5 m (8 to 15 ft) based on canal size and lining thickness. Where depth of section exceeds 1.2 m (4 ft), provide longitudinal grooves on each side as shown in Fig. 1.

2.6 Construction joints. Construction joints shall be butt-type, formed square with the lining surface and at right angles to the canal.

SECTION 3—STANDARD SECTION

3.1 The recommended standard cross sections consist of three sections with 1:1 side slopes, three sections with 1.25:1 side slopes, and five sections with 1.5:1 side slopes. The dimensional details of these standard sections are shown in Fig. 1, and Table 2.

3.2 The eleven standard sections provide a desirable overlap in carrying capacity which permits adequate flexibility in selecting a section for optimum economy in materials and other construction costs.

3.3 Standard canal sections. Standard canal sections are based on subgrade dimensions to provide standardized excavation and slip-form equipment to permit economical construction.

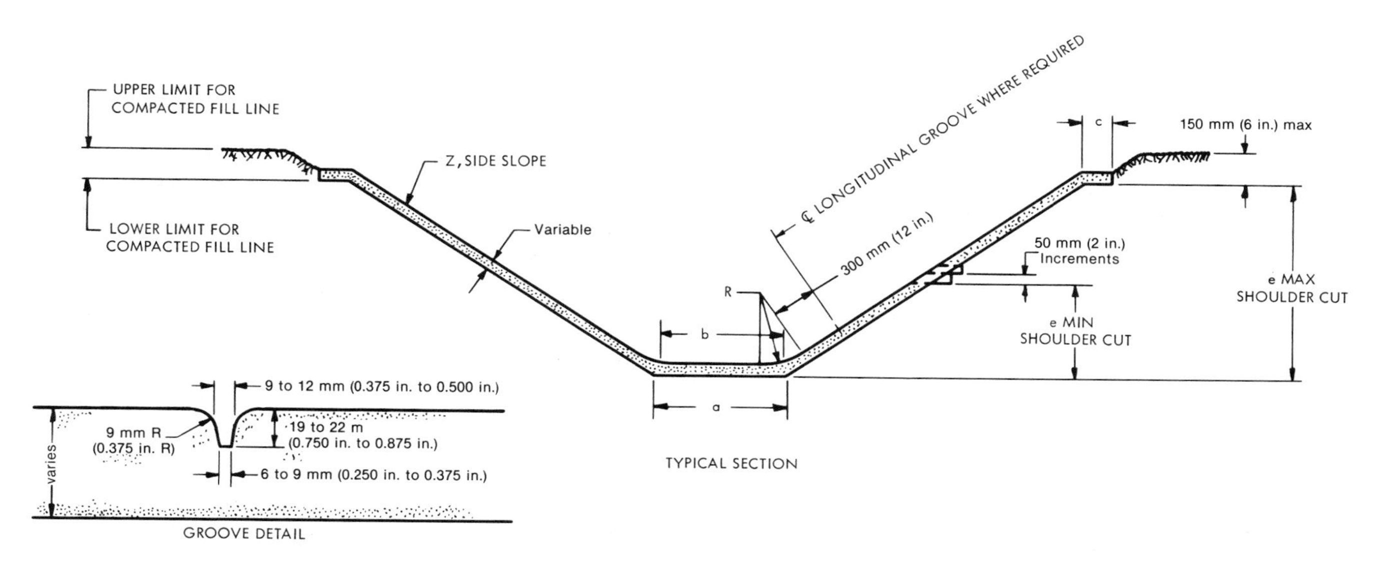

FIG. 1—STANDARD TRAPEZOIDAL CANAL SECTIONS

TABLE 2—DIMENSIONS OF STANDARD SLIP-FORM TRAPEZOIDAL CANAL SECTIONS

Section	Z	a		b		c		eMin		eMax		R	
		mm	in.	mm	in.	mm	in.	mm	in.	mm	in.	mm	in.
A-1	1:1	350	14.1	300	12	100	4	400	15	750	30	200	9
A-1.5	1:1	500	20.1	450	18	100	4	400	15	750	30	200	9
A-2	1:1	650	26.1	600	24	100	4	400	15	750	30	200	9
B-1	1.25:1	350	13.8	300	12	100	4	450	18	900	36	300	12
B-1.5	1.25:1	500	19.8	450	18	100	4	450	18	900	36	300	12
B-2	1.25:1	650	25.8	600	24	100	4	450	18	900	36	300	12
C-2	1.5:1	650	25.5	600	24	150	6	600	24	1200	48	450	18
C-3	1.5:1	950	37.5	900	36	150	6	700	27	1400	54	450	18
C-4	1.5:1	1250	49.5	1200	48	150	6	800	33	1700	66	450	18
C-5	1.5:1	1550	61.5	1500	60	150	6	900	36	1800	72	450	18
C-6	1.5:1	1850	73.5	1800	72	150	6	1000	42	2100	84	450	18

APPENDIX A—CONCRETE GUIDE SPECIFICATIONS FOR CONCRETE SLIP-FORM CANAL LININGS

A1—SCOPE

A1.1 This specification covers materials used, and the manufacture, delivery, inspection, testing, and curing of concrete for use in slip-form lining of irrigation canals.

A2—APPLICABLE SPECIFICATIONS AND REFERENCES

A2.1 The latest issue of the cited standards and references listed at the last page of this Standard shall govern except where superseded by particular requirements of this specification.

A3—CONCRETE MIX

A3.1 The cement used shall be Portland cement, Type I, IA, II, IIA, or V as specified for the job. On sites where the sulfate concentration is more than 0.1%, special sulfate-resistant cements shall be used as follows:

Sulfate Concentration	Cement Type
Over 0.1%	II, IIA, or V
Over 0.3%	V

A3.2 Aggregates shall conform to American Society for Testing and Materials Standard C33, Specifications for Concrete Aggregates, except that locally available aggregate materials may be used if they are well-graded, clean and durable. Maximum size of aggregates shall not exceed one-third of the specified lining thickness.

A3.3 Mixing water for concrete shall be clean and free of acids, alkalis, oils, sulfates, and other harmful materials. A good rule for all jobs is that the water used should be fit to drink.

A3.4 Each classification of concrete shall be furnished in accordance with Table A1.

A3.5 When using locally available aggregate materials or materials other than the quality specified in ASTM Standard C33, Specifications for Concrete Aggregates, a minimum cement content of 8.5 bags of cement per m^3 (6.5 bags of cement per yd^3) of concrete shall be used. Concrete containing locally available aggregate materials shall also meet a minimum compressive strength at 28 d of 20.7 MPa (3,000 $lbf/in.^2$).

A4—VERIFICATION OF MIX DESIGN

A4.1 Any concrete mix being used on a canal lining job shall be subject to sampling and testing to determine compliance with the 28-d strength requirement prescribed in this specification. If the test shows the average 28-d compressive strength of three specimens was less than required, the job shall not be accepted as meeting this specification.

TABLE A1—CONCRETE MIX REQUIREMENTS

Class of concrete	Compressive strength at 28 d		Aggregate size (per ASTM Standard C33 Specifications for Concrete Aggregates) max		Cement content* (Slump—permit slip-form operation)	
	MPa	$lbf/in.^2$	mm	in.	$bags/m^3$	$bags/yd^3$
Normal						
For mild exposure	17.3	2,500	19	0.75	6.5	5.0
avg 20 or less	20.7	3,000	19	0.75	7.0	5.5
freeze-thaw cycles/yr	24.2	3,500	19	0.75	8.0	6.0
Air-entrained						
For avg exposure avg 20 to 80 freeze-thaw cycles/yr	20.7	3,000	19	0.75	7.5	5.75
For severe exposure avg 80 or more freeze-thaw cycles/yr	24.2	3,500	19	0.75	8.5	6.5

*For concrete containing Type V, sulfate-resistant portland cement, increase the cement requirements 20%.

A5—TOLERANCE IN SLUMP

A5.1 Concrete used in canal linings shall be proportioned so it can be consolidated thoroughly and still be of low enough slump to permit the material to stay in place on the side slopes.

A6—AIR-ENTRAINED CONCRETE

A6.1 Air-entrained concrete shall contain 4 to 7.0% air, by volume. In areas of the country with fine aggregate, it will be necessary to increase the air content of the concrete. This entrained air may be obtained by using an air-entraining portland cement meeting specifications of ASTM Standard C150, Specifications for Portland Cement.

A7—ADMIXTURES

A7.1 When air-entrained concrete is specified and an admixture is used to secure the desired air content, the admixture shall conform to specifications of ASTM Standard C260, Specifications for Air-Entraining Admixtures for Concrete.

A8—MEASURING MATERIALS

A8.1 When ready-mixed concrete is used, the cement, aggregates, water, and admixtures shall be measured and combined strictly in accordance with specifications of ASTM Standard C94, Specifications for Ready-Mixed Concrete.

A9—MIXING AND DELIVERY

A9.1 Ready-mixed concrete shall be mixed and delivered to the point designated by the means and standards set forth by specifications of ASTM Standard C94, Specifications for Ready-Mixed Concrete.

A9.2 No water shall be added on the job unless authorized by the responsible technician or engineer. The amount of water added shall be recorded on all copies of the delivery ticket. If water is permitted to be added to mixed concrete upon arrival at the job, an additional mixing of 20 revolutions of the drum shall be required.

A9.3 Concrete delivered in outdoor temperatures lower than 4.4 °C (40 °F) shall arrive at the site of the work having a temperature not less than 15.6 °C (60 °F), and not greater than 32.2 °C (90 °F) unless otherwise specified or permitted by the technician or engineer.

A10—TESTS

A10.1 Cylinders for strength tests shall be made in accordance with ASTM Standard C31, Making and Curing Concrete Test Specimens in the Field. During the first 24 h all test specimens shall be covered and kept at air temperatures between 15.6 °C (60 °F) and 26.7 °C (80 °F). At the end of 24 h, specimens shall be carefully transported to the testing laboratory where molds shall be removed, and cylinders shall be cured in a moist condition at 23 ° ± 1 °C (73.4 ° ± 3 °F) until time of test.

A10.2 A strength test for any class of concrete shall consist of four standard cylinders made from a composite sample secured from a single load of concrete in accordance with ASTM Standard C172, Sampling Fresh Concrete, with one cylinder tested at 7 d and three cylinders tested at 28 d. The test results at 28 d shall be the average of the strength of three specimens determined in accordance with ASTM Standard C39, Test for Compression Strength of Cylindrical Concrete Specimens, except that if one specimen shows manifest evidence of improper sampling, molding or testing, it shall be disregarded.

A10.3 Strength tests shall be made for each of the following conditions: Each day's pour; each class of concrete; each change of supplies or source; and for each 76 m^3 (100 yd^3) of concrete or fraction thereof.

A10.4 To conform to the requirements of this specification, the average of all of the strength tests representing each class of concrete shall be equal to or greater than the specified strength and not more than one test in 10 (if less than 10 tests made per job, substitute "per job" for "in 10") shall have an average value of less than 85% of specified strength. No one test shall be less than 70% of design strength.

A10.5 A record shall be made of the delivery ticket number for the particular load of concrete tested, and the exact location in the work at which each load represented by a strength test is deposited.

A11—DELIVERY TICKETS

A11.1 Duplicate delivery tickets shall be furnished with each load of concrete delivered to the job: One for the contractor, and one for the responsible technician or engineer. Delivery tickets shall provide the following information:

A11.1.1 Date

A11.1.2 Name of ready-mix concrete plant

A11.1.3 Job location

A11.1.4 Contractor

A11.1.5 Type (standard or air-entrained) and brand name of cement

A11.1.6 Class and specified cement content in bags per m^3 (yd^3) of concrete

A11.1.7 Truck number

A11.1.8 Time dispatched

A11.1.9 Amount of concrete in load in m^3 (yd^3)

A11.1.10 Admixtures in concrete, if any

A11.1.11 Maximum size of aggregate

A11.1.12 Water added at job, if any

A11.1.13 Weather conditions: Approximate temperature, relative humidity, wind velocity

A12—INSPECTION AND TESTING

A12.1 All required sampling, preparing specimens, and testing shall be performed by an independent laboratory or a person acceptable to the engineer. The cost of all tests shall be paid for as mutually agreed.

A13—CONTROL JOINTS

A13.1 Contraction and construction joints shall be provided at the locations and to the dimensions as shown in Fig. 1.

A13.2 "A cold applied sealer conforming to ASTM Standard D1850 or a hot-poured sealer meeting Federal Specification SS-S-1401B, Sealing Compound, Hot Applied for Concrete and Asphalt Pavements, or other suitable materials approved for this application shall be used. CAUTION: Do not overheat hot, pour-type sealers."

A14—CURING

A14.1 A pigmented concrete curing compound shall be applied to the concrete surface within 20 min after placing or finishing the concrete. The sealing compound shall be applied in a manner that will ensure a continuous, uniform membrane over the surface. Coverage shall not exceed 3.7 m^2/l (150 ft^2/gal).

A14.2 Curing compounds shall conform to specifications of ASTM Standard C309, Specifications for Liquid Membrane-Forming Compounds for Curing Concrete.

A14.3 In lieu of applying a curing compound, water-curing by ponding water in the lined ditch for a minimum of 5 d shall be acceptable. Care shall be taken to prevent damage to fresh concrete.

References:

ACI 211.2, Practice for Selecting Proportions for Structural Lightweight Concrete
ACI 305, Practice for Hot Weather Concreting
ACI 306, Practice for Cold Weather Concreting
ASTM C70, Method of Test for Surface Moisture in Fine Aggregate
ASTM C138, Test for Unit Weight, Yield and Air Content (Gravimetric) of Concrete
ASTM C143, Test for Slump of Portland Cement Concrete
ASTM C173, Test for Air Content of Freshly Mixed Concrete by the Volumetric Method
ASTM C231, Test for Air Content of Freshly Mixed Concrete by the Pressure Method
ASTM C566, Method of Test for Total Moisture Content of Aggregate by Drying
ASTM D1190, Specifications for Concrete Joint Sealer, Hot-Poured Elastic Type

Cited Standards:

ASTM C31, Making and Curing Concrete Test Specimens in the Field
ASTM C33, Specifications for Concrete Aggregates
ASTM C39, Test for Compression Strength of Cylindrical Concrete Specimens
ASTM C94, Specifications for Ready-Mixed Concrete
ASTM C150, Specifications for Portland Cement
ASTM C172, Sampling Fresh Concrete
ASTM C260, Specifications for Air-Entraining Admixtures for Concrete
ASTM C309, Specifications for Liquid Membrane-Forming Compounds for Curing Concrete
ASTM D1850, Specifications for Concrete Joint Sealer, Cold-Application Type
FED. SPEC. SS-S-1401B, Sealing Compound, Hot Applied for Concrete and Asphalt Pavements

ASAE Engineering Practice: ASAE EP302.3

DESIGN AND CONSTRUCTION OF SURFACE DRAINAGE SYSTEMS ON AGRICULTURAL LANDS IN HUMID AREAS

Developed by ASAE Surface Drainage Committee; approved by ASAE Soil and Water Division Steering Committee; adopted by ASAE as a Recommendation December 1966; revised March 1972, March 1973; reconfirmed December 1977; reconfirmed and reclassified as an Engineering Practice December 1978; reconfirmed December 1983, December 1984, December 1985; revised December 1986.

SECTION 1—PURPOSE

1.1 This Engineering Practice is intended to improve the design, construction and maintenance of surface drainage systems which are adapted to modern farm mechanization. It is limited to agricultural or farm-size areas, 259 ha (640 ac) or less, in the humid region of the eastern United States.

1.2 Surface drainage is normally required for efficient crop production on slowly permeable soils with restrictive topography. It is not required when excess water is removed naturally. Typical problem areas are glaciated areas, coastal plains, bottomlands, deltas, and old lake beds. Surface drainage may eliminate the need for subsurface drains under certain conditions. Surface drains also apply to farm mains used to collect water from field drains and subsurface drains.

SECTION 2—DEFINITIONS AND TERMINOLOGY

2.1 Definitions and terminology are intended to relate only to this Engineering Practice and its application.

2.2 Drainage system classification

2.2.1 Bedding: A surface drainage method accomplished by plowing land to form a series of low narrow ridges separated by parallel dead furrows. The ridges are oriented in the direction of the greatest land slope.

2.2.2 Crowning: The process of forming the surface of land into a series of broad low ridges separated by paralleled field laterals.

2.2.3 Land grading: The process of forming the surface of land to predetermined grades so each row or surface slopes to a drain.

2.2.4 Land smoothing: Smoothing the land surface with a land plane or land leveler to eliminate minor depressions and irregularities without changing the general topography.

2.2.5 Parallel system: A system of parallel laterals with most of the row drains or field drains normal to the laterals.

2.2.6 Random system: A system of meandering row drains, field drains and/or field laterals that are located in and drain depressions in a field.

2.2.7 Water leveling: A method of land grading wherein fields are divided into segments, separated by permanent contour levees, and leveled to a series of planes with zero grades by flooding then scraping the highs between levees and allowing the resulting slurry to settle.

2.3 Grades or slopes

2.3.1 Cross slope: The slope perpendicular to crop rows.

2.3.2 Reverse grade: A grade on a field surface, or in a crop row, or in a drainage channel, that slopes in the direction opposite to the prevalent grade.

2.3.3 Row grade: The grade of crop rows.

2.4 Surface drainage: The diversion or orderly removal of excess water from the surface of land by means of improved natural or constructed channels, supplemented when necessary by shaping and grading of land surfaces to such channels.

2.5 Surface drains or ditches

2.5.1 Diversion: A channel constructed across the slope to intercept surface runoff and conduct it to a safe outlet.

2.5.2 Farm main: An outlet ditch serving an individual farm.

2.5.3 Field drain: A shallow graded channel, usually having relatively flat side slopes, that collects water within a field.

2.5.4 Field lateral: The principal ditch for adjacent fields or areas on a farm. Field laterals receive water from row drains, field drains, and, in some areas from field surfaces, and carry it to the farm mains.

2.5.5 Interceptor drain: A channel located across the flow of groundwater and installed to collect subsurface flow before it resurfaces. Surface water is also collected and removed.

2.5.6 Row drain: A small drain constructed with a plow or similar implement which cuts across rows or through low areas to provide drainage into field drains or field laterals. Also a minimum field drain. Also known as plow drain, quarter drain, header ditch or annual drain.

2.6 Other terms

2.6.1 Berm: The undisturbed area between the edge of a ditch and the edge of the spoil bank.

2.6.2 Pipe drop: A pipe, with or without headwalls, used as an erosion control structure at a transition to drop water into a deeper drain.

2.6.3 Transition: The outlet section of a drain that discharges into a deeper drain. Transitions may be vegetated or non-vegetated sections, or erosion control structures.

SECTION 3—PRINCIPLES

3.1 The objectives of surface drainage are:

3.1.1 To prevent water from ponding on land surfaces or in surface drains that are crossed by farm equipment.

3.1.2 To remove excess water in time to prevent damage to crops.

3.1.3 To accomplish the above without excess erosion.

3.2 Complete drainage systems shall be planned and designed with all aspects considered. The collection system, consisting of the land surface and shallow drains, removes excess surface water from individual areas within fields. The disposal system, consisting of somewhat larger ditches, receives drainage water from the collection system and removes it from the land. Transitions, including structures, that permit drains to discharge into larger or deeper drains must provide for adequate erosion control without holding water on field surfaces for excessive periods of time. The system shall be designed as an integral part of the soil and water facilities, the farm layout, and cropping plan. Farm roads shall be planned to complement the drainage system and to facilitate farming operations.

3.3 There are several methods of draining agricultural land. Land surfaces that pocket water can be drained by cutting and filling. Depressions can be drained by row drains, field drains, and field laterals. The methods can be combined, thereby reducing the number of depressions and the number of drains. A brief discussion of the principal surface drainage methods follows.

3.3.1 Land smoothing eliminates minor depressions and irregularities in a field, thereby improving surface drainage and increasing farm machinery efficiency. Frequently this is an interim measure that will be replaced by land grading when economic conditions permit.

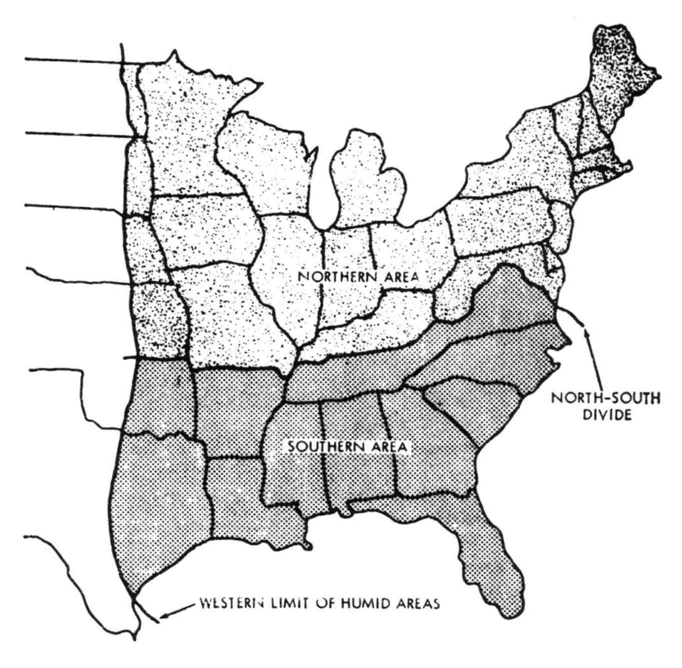

FIG. 1—KEY MAP FOR DRAINAGE CURVES
(Humid areas of eastern U.S. where surface drainage may be needed.)

3.3.2 Land grading, by carefully designed cutting and filling operations, provides excellent surface drainage.

3.3.3 Water leveling was developed for fields which are to be flooded and provides adequate drainage for rice and other selected, close growing, water tolerant crops.

3.3.4 Random drainage systems provide for drainage of the low areas by row drains, field drains and field laterals. These systems are relatively low in cost and give a high return on the investment. They require frequent maintenance and are obstacles to farm machinery, particularly if drains are closely spaced. On severely undulating topography, random drainage may be the only economical system.

3.3.5 Parallel drainage systems facilitate mechanization by eliminating short rows and point rows. Crossable field drains or row drains are constructed between parallel field laterals. This permits farm equipment to travel greater distances without turning. Field surfaces between field laterals should be improved by land smoothing or land grading for maximum benefits.

3.3.6 Bedding, under most conditions, is not considered an acceptable drainage practice for row crops because rows adjacent to the dead furrows will not drain satisfactorily. It is an acceptable practice for hay and pasture crops in some areas. Normally there is some crop loss in and adjacent to the dead furrows.

3.3.7 Crowning was developed for sugar cane production in areas that formerly had ample hand labor, is well established in some areas and provides excellent drainage. Crowning requires more maintenance than most of the other systems. The large number of field laterals take land out of production, and they are a source of erosion and sedimentation, and weed and grass infestation.

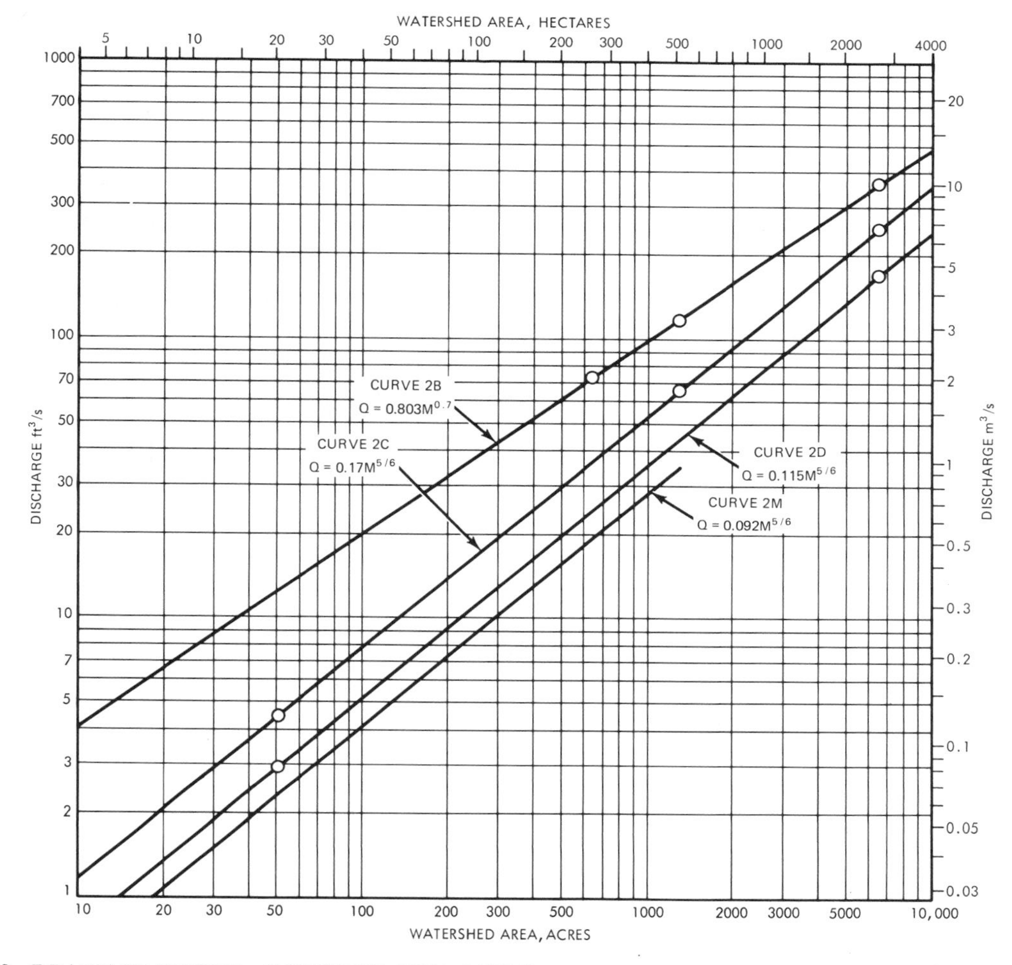

FIG. 2—DRAINAGE CURVES—NORTHERN AREA OF U.S.
(See Fig. 1 for area) Based on USDA Soil Conservation Service (SCS) National Engineering Handbook, sect. 16, chap. 6.

Crowning with crossable field laterals provides excellent drainage for the pasture crops commonly grown in the sugar cane areas.

3.3.8 Diversions intercept upland runoff and prevent it from overflowing bottomlands. This simplifies installation of drainage systems on bottom lands.

3.3.9 Interceptor drains intercept and remove subsurface water. They are used on long slopes with one percent or steeper grades and shallow, permeable surface soils overlying relatively impermeable subsoils.

SECTION 4—DESIGN

4.1 Surface drainage system design criteria are based on the assumption that all lands to be drained will be suitable for agricultural use after drainage. Design shall consider construction and maintenance needs and irrigation requirements where applicable. The rate of water removal, in terms of depth per unit of time, to be provided by the drainage system depends on several interrelated factors such as rainfall characteristics, soil properties and cropping patterns. For most row crops, surface drainage systems shall complete removal of excess water from the soil surface within 24 h after rainfall ceases. More rapid removal may be necessary for high-value truck crops. A longer time is permissible for grasslands and woodlands.

4.2 The intensity of drainage desired is expressed in terms of drainage curves (Figs. 1, 2, and 3). These empirical curves were developed from a large number of field measurements of drainage flow rates and observations of the adequacy of drainage. The curves will not provide for peak flows from large storms. Excess runoff will be discharged as overland flow temporarily flooding adjacent lands. Curves with the higher coefficient values represent a higher intensity drainage.

4.2.1 The drainage curves are applicable to drainage areas having average slopes of less than 4.7 m/km (25 ft/mile).

4.2.1.1 The basic equations used to describe the drainage curves of Figs. 2 and 3 are of the form:

$$Q = K_1CM^{5/6}$$

One exception is curve 2B which is:

$$Q = K_2CM^{0.7}$$

where

Q = design discharge, m^3/s (ft^3/s)

K^1 = unit conversion factor equal to 0.06 if in SI metric units and 1 if in English units

K^2 = unit conversion factor equal to 0.0534 if in SI metric units and 1 if English units

C = coefficient relating to desired level of drainage Numerical values for the coefficient are shown on Figs. 2 and 3

M = drainage area, ha (ac)

4.2.2 Fig. 1 shows the extent of the humid area of the eastern United States where surface drainage may be needed. Regional areas are delineated to show coverage of Figs. 2 and 3.

4.2.3 Fig. 2 is for the northern regional area:

Curve 2B for excellent farm drainage.
Curve 2C for good farm drainage (basic curve for grain crops).
Curve 2D for fair drainage (basic curve for improved pastures).
Curve 2M for agricultural drainage in the Red River Valley in

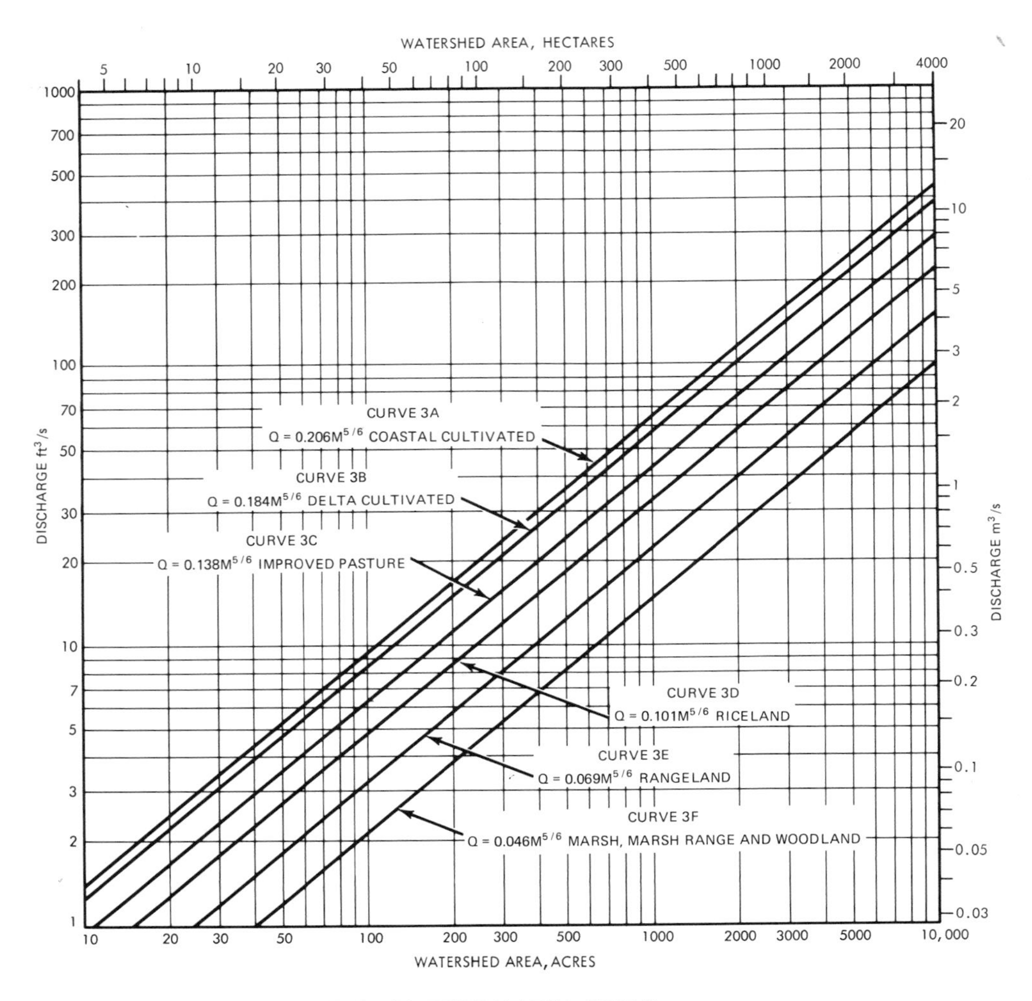

FIG. 3—DRAINAGE CURVES—SOUTHERN AREA OF U.S.
(See Fig. 1 for area) Based on USDA SCS National Engineering Handbook, sect. 16, chap. 6.

ROW DRAIN

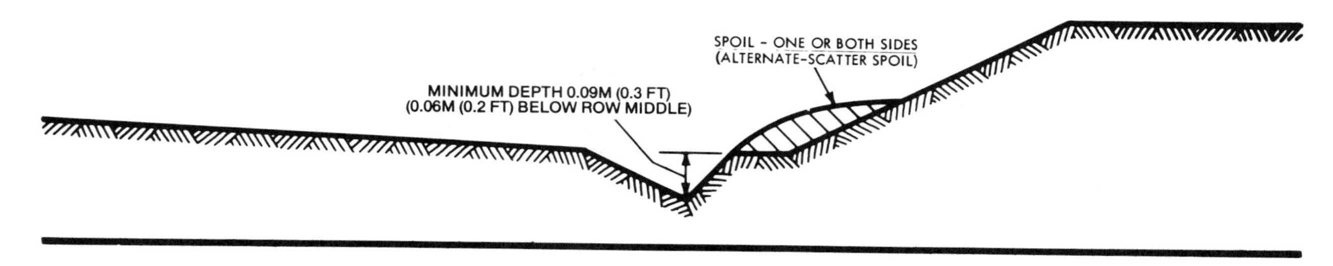

VEE FIELD DRAIN

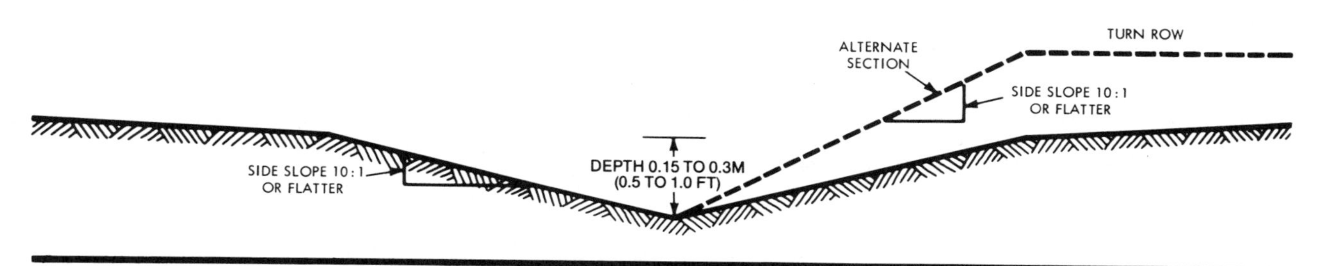

½ -VEE FIELD DRAIN

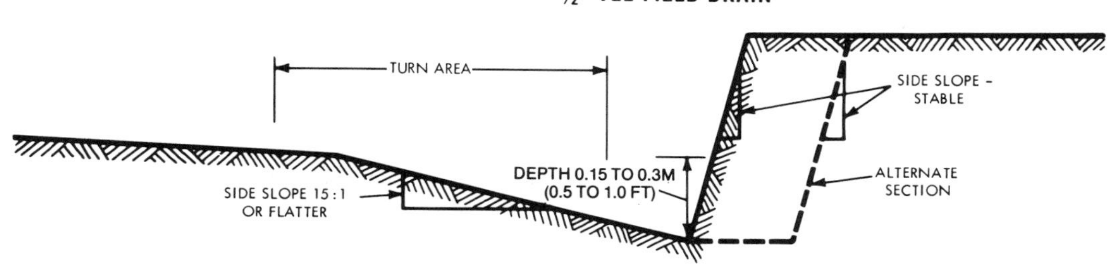

TRAPEZOIDAL FIELD DRAIN

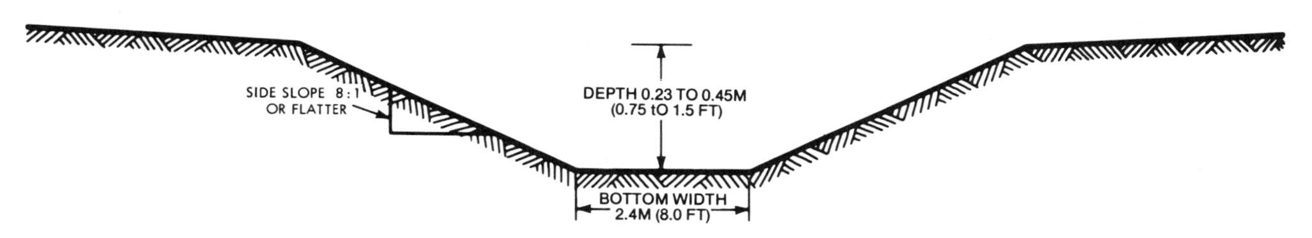

FIG. 4—FIELD DRAINS

Minnesota, North Dakota, and South Dakota.

(NOTE: No recommendation is made for woodland drainage.)

4.2.4 Fig. 3 is for the southern regional area:
Curve 3A for row-crop drainage in the gulf coastal and coastal plains area.
Curve 3B for row-crop drainage in the deltas and bottomlands of the major rivers.
Curve 3C for improved pasture drainage.
Curve 3D for irrigated ricelands.
Curve 3E for upland range and unimproved pasture drainage.
Curve 3F for gulf coast marsh and marsh range, and woodland drainage.

4.2.5 The value for M should be determined for each separate drainage components. For each row drain, this would be that amount of land drained by it. For each collector component, such as field drain, field lateral and farm main, the value for M would be the total area drained by the system.

TABLE 1—RECOMMENDED ROW DRAIN AND FIELD DRAIN DIMENSIONS

Type	Depth M	Depth Ft	Bottom Width M	Bottom Width Ft	Side Slope
Row drain	0.09*	0.3*	0	0	—
Vee	0.15 to 0.3	0.5 to 1.0	0	0	10:1 or flatter†
One-half vee	0.15 to 0.3	0.5 to 1.0	—	—	15:1 or flatter
Trapezoidal	0.23 to 0.45	0.75 to 1.5	2.4	8	8:1 or flatter

*Minimum, 0.06 m (0.2 ft) below row middles for row crops.
†10 horizontal to one vertical.

TABLE 2—RECOMMENDED FIELD LATERAL SIDE SLOPES

Cross Section Type	Depth M	Depth Ft	Recommended Side Slope	Minimum Side Slope
Vee	0.3 to 0.6	1.0 to 2.0	6:1	3:1
Vee	0.61 and over	2.1 and over	4:1	3:1
Trapezoidal	0.3 to 0.9	1.0 to 3.0	4:1	2:1
Trapezoidal	0.91 and over	3.1 and over	2.5:1	1:1

4.3 Ditch design

4.3.1 Field drains (Table 1 and Fig. 4) are located at the lower ends of field rows, through surface depressions, above barriers that trap runoff, and where required to divert runoff from lower areas. Drains located at or near soil changes provide locations for field roads, permitting the soils to be managed differently during farming operations. Field drains need not be designed to contain the quantities of flow indicated on the drainage curves, as their primary purpose is to remove residual surface water after volume runoff has passed out of the field. Recommended design grades are from 0.10 to 0.30%, and should never be less than 0.05%. Grades may be uniform or may increase or decrease. Abrupt changes to flatter grades may result in excessive sedimentation.

4.3.1.1 Row drains are acceptable field drains provided they have adequate grades; topography permits the disposal of spoil without restricting drainage into the row drains; and adequate maintenance is provided.

4.3.1.2 Vee field drains with side slopes steeper than 10:1 are easily constructed and maintained with blade equipment. These drains are difficult to cross with farm machinery, fill rapidly with sediment, and must be re-established yearly. They are not recommended under normal conditions.

4.3.1.3 Vee field drains that are 0.15 to 0.30 m (0.5 to 1.0 ft) deep with 10:1, or flatter, side slopes are excellent when crop rows on one or both sides terminate at or near the centers of the drains. Field roads adjacent to these field drains will not interfere with mechanized operations if field road side slopes are not steeper than 10:1.

4.3.1.4 One-half Vee field drains are not crossed by farm machinery. Machinery turns within the cross sections without crossing the bottoms of the drains. Turn areas must be smoothed and any obstructions at the bottoms of the drains removed. There are no restrictions on the side slope opposite the field other than that it must be stable.

4.3.1.5 Trapezoidal field drains are satisfactory in some areas. Where soils are not erosive, trapezoidal field drains can be constructed to grade without uniform depths and can be expected to have long lives with little maintenance. Where soils are erosive and field drains are not vegetated, trapezoidal field drains with the recommended depths will fill quickly with eroded material.

4.3.1.6 Double field drains are recommended only on ungraded land for use in wide, shallow depressions where runoff enters from both sides and the excavated soil must be placed between the twin drains. Recommended minimum distances between the centerlines of the twin field drains is 15 m (50 ft) for drains 0.15 m (0.5 ft) deep, plus 3.0 m (10 ft) for each additional 0.03 m (0.10 ft) of depth. Lesser distances may be necessary at the outlet ends because of construction requirements.

4.3.2 Subsurface drains are located adjacent to field drains in some areas where subsurface drainage is effective. The installation of subsurface drains lowers water tables, aids tillage, and reduces maintenance requirements.

4.3.3 Field laterals (Table 2 and Fig. 5) shall be designed to discharge the rates of flow indicated by the appropriate drainage curves. Field laterals occupy productive land and are costly to construct and maintain. They shall be spaced as widely apart as field conditions permit.

4.3.3.1 The design cross sections of field laterals shall meet the combined requirements for capacities, erosion control, depths, side slopes, maintenance and, if needed, allowance for sedimentation. Minimum field lateral depths, regardless of design capacities, shall be 0.3 m (1.0 ft).

4.3.3.2 The water surface elevations used in the design of field laterals shall be as near natural ground surfaces as practical to hold ditch depths to a minimum at row drain and field drain entrances. Flat side slopes and, for trapezoidal sections, wide bottom widths can be used to reduce depths.

4.3.3.3 The spoil from field laterals should, in most cases, be spread or removed. When spoil is removed, tillable low levees prevent water entry except at planned inlets. Spoil banks are usually recommended only as an undesirable alternate but may be preferred in some locations.

4.3.3.4 Shaped spoil banks should have side slopes on the field sides of 6:1 or flatter to permit operation of farm equipment. The area which contributes runoff to the channel side slopes should be kept to a minimum for erosion control purposes. For this reason side slopes on the channel sides can be as steep as the angle of repose of the spoil material and should be no flatter than the channel side slopes. Shaped spoil banks ordinarily should not exceed 1.0 m (3 ft) in height.

4.3.3.5 When spoil is shaped, berms may not be required. When used, berms should be adequate to satisfy the intended purpose and, in no case, should be less than 0.6 m (2 ft) wide.

4.3.3.6 When spoil is dumped and left unshaped, the recommended berm widths are shown in Table 3.

4.3.3.7 The excavated material from field laterals should be placed and shaped to minimize overbank wash, provide access for maintenance equipment, prevent excavated material from washing or rolling back into the channel and prevent sloughing of channel banks caused by heavy loads near the channel.

4.3.3.8 Field laterals and mains with their spoils can be hazardous, therefore, side slopes, berm widths, etc., should be chosen to provide maximum safety in construction and maintenance operations. There should be information in the final design on the limitations and safety features. If hazards do exist, it should be so stated to the persons or agency responsible for construction and maintenance. A record of this information should be kept on file.

4.3.4 Farm mains shall be designed to carry the volumes of flow indicated on the drainage curves. Where discharge into swamps, marshes or lakes is contemplated, high water marks shall be checked to determine whether gravity flow will be adequate. Pump drainage may be necessary. Specifications for the design and construction of farm mains usually require freeboard added to the design depths but are generally the same as for field laterals.

TABLE 3—RECOMMENDED BERM WIDTHS FOR UNSHAPED SPOIL

Average Depth of Ditch M	Average Depth of Ditch Ft	Minimum Berm Width* M	Minimum Berm Width* Ft
0.6 to 1.2	2.0 to 4.0	Depth of ditch	Depth of ditch
1.3 to 1.8	4.1 to 6.0	2.4	8
1.9 to 2.4	6.1 to 8.0	3.0	10
Over 2.4	Over 8.0	4.5	15

*In locations where soils are not stable, wider berm widths may be necessary.

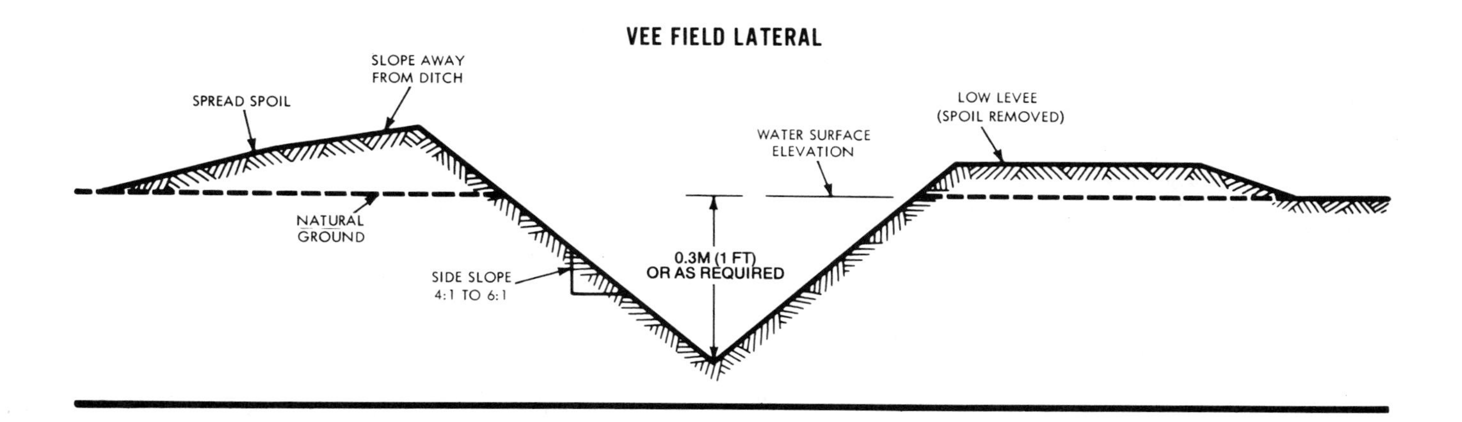

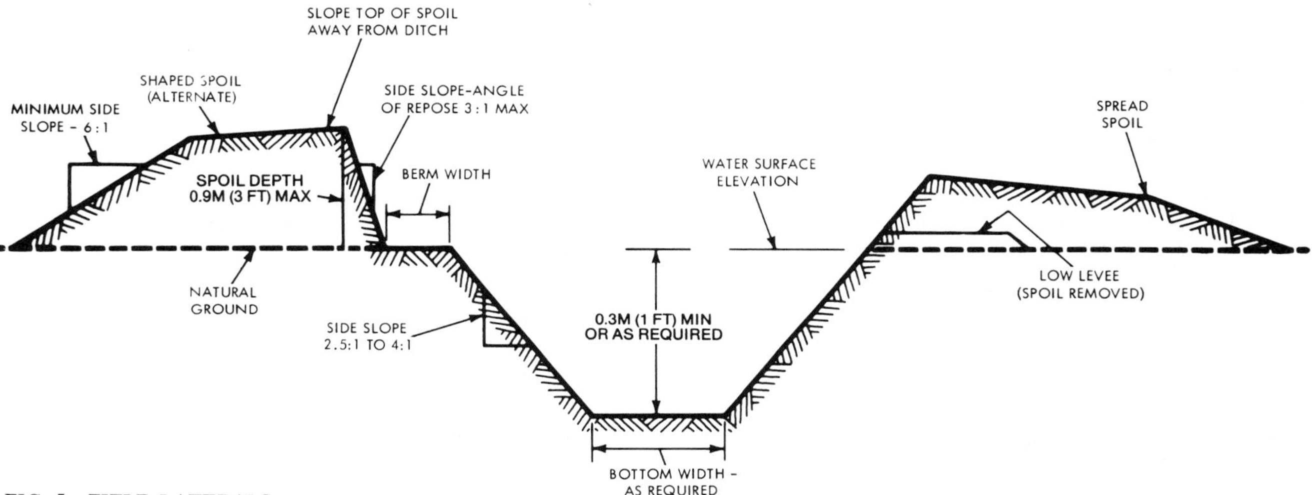

FIG. 5—FIELD LATERALS

4.4 Design for land surface modification

4.4.1 Land smoothing to eliminate minor depressions usually may be directed in the field without detailed surveys or plans. However, surveys may be required in critical portions of some fields, when visual observations do not provide the accuracy required.

4.4.2 Land grading for drainage may be planned with row grades and cross slopes that are uniform or that vary within limits imposed by soils, topography and economics. Areas to be graded shall be planned for a minimum number of field drains with the drains located, where possible, normal to field laterals and crop rows.

4.4.2.1 Surface drainage will be adequate if all reverse row grades that form depressions are eliminated. Minimum grade limits should include a tolerance in construction that will permit the elimination of all depression either in original construction or by post construction touchup. Reverse row grades can be eliminated with relative ease on fields designed with 0.20% minimum grades. Unusual precision in construction is required with 0.10% and flatter grades.

4.4.2.2 Land is frequently graded for both irrigation and surface drainage. In these cases, design limits to meet irrigation requirements may take precedence over those from drainage.

4.4.2.3 Economic considerations and the requirements for efficient mechanized operations may justify exceeding the recommended maximum row grade and cross slope limits. When the recommended limits are exceeded, provisions should be made to control erosion and sedimentation.

4.4.2.4 Recommended row grades range from 0.10 to 0.50%. Grades may be uniform, or increase or decrease.

4.4.2.5 Cross slopes normally should not exceed 0.5%. Reverse cross slopes are satisfactory providing field drains have the minimum required grades and drainage outlets are adequate.

4.4.2.6 Lengths of rows shall be determined by soil changes, soil depths, topography, irrigation and erosion control requirements, and grading costs.

4.4.2.7 The maximum allowable depth of cut depends upon soils and costs. Soils data shall be used for determining maximum allowable depths of cut.

4.4.3 Water leveling shall be planned to increase the size and improve the shape of the areas between contour levees. The conventional vertical spacing between levees shall be increased and levee alignment shall be improved. Straight, parallel levees are preferred.

4.4.4 Bedding shall be designed so the beds are parallel to the direction of the greatest land slope. Field laterals oriented normal to the dead furrows should be spaced at intervals that will provide adequate outlets.

4.4.5 Crowning shall be designed as precision drainage systems. Crowns may slope from the center to both sides or in only one direction. Minimum cross slopes shall be 0.3%. Row drains shall be installed at all points where there are reverse row grades. Field drains or row drains are required at the ends of the rows. Field laterals shall be located parallel to the crowns to provide outlets for the row drains, and also normal to the crowns at intervals as needed to conduct flow to farm mains.

4.4.6 Diversions and interceptor drains are specialized drainage structures. See ASAE Engineering Practice EP268, Design, Layout, Construction and Maintenance of Terrace Systems, for diversion design recommendations. Interceptor drains shall be designed to meet the field condition as disclosed by soil borings and topography.

SECTION 5—CONSTRUCTION

5.1 Good quality construction is essential for adequate drainage. Qualified engineers should be responsible for surveys, designs, layouts and checking to guarantee that construction meets specifications. Landowners and contractors shall be provided with information necessary for good quality construction.

5.1.1 Land smoothing will provide adequate drainage only if depressions are eliminated by smoothing the land surface. Row drains, field drains, and/or laterals shall be installed in the remaining low areas.

5.1.2 Land grading is hampered by trash and vegetation which shall be destroyed or removed prior to construction and kept under control while the work is being done. Plowing trash under the land surface is not recommended.

5.1.2.1 Fields shall be scarified prior to construction if there are hardpans. Fields shall not be graded when wet, because this impairs the physical condition of the soil.

5.1.2.2 Grading shall be checked prior to and following finishing operations. A tolerance of 30 mm (0.1 ft) may be allowed in checking, providing there are no reverse grades in row direction.

5.1.2.3 Land smoothers shall be passed over the field surface at least three times as a finishing operation. The first two passes shall be on opposite diagonals and the last pass shall be in the row direction.

5.1.2.4 Fields shall be scarified either before or after the final land smoothing operations to loosen the cut surfaces and to blend the fill material with the underlying soil.

5.1.3 Water leveling shall be constructed with adequate permanent levees and water control devices. After partial flooding, "highs" between levees shall be cut to sufficient depths and the soil-water slurries agitated and allowed to settle so that, upon draining and drying, level surfaces are formed.

5.1.4 Bedding shall be constructed so all beds are full-bodied and free of pockets or depressions. The parallel dead furrows between the beds shall be free of water-holding pockets.

5.1.5 Crowning shall be constructed so all crowns are free of pockets or depressions. Depths of cuts and fills shall be established at key points by surveys. Experienced operators can then construct crowns with only visual checking between the key points. Crowns shall be land smoothed after rough construction.

5.1.6 Field drains must be constructed without reverse grades if they are to give satisfactory performance.

5.1.6.1 Field drains having irregular depths and cross sections are subject to differential deposition of sediment, which creates ponding and grade maintenance problems. Therefore, when feasible, the land surfaces adjacent to and above the drains should be graded to have the same grades (cross slopes) as the drains.

5.1.6.2 Spoil material excavated from Vee drains located in depressions should be removed from the site so water can enter the drains freely from both sides. Where this is not possible, spoil material shall be shaped or spread and provision made for entrance of side-water by use of constructed inlets and row drains. Sometimes this may be accomplished by feathering the spoil material with blade equipment so water enters freely.

5.1.7 Field laterals shall be constructed with uniform side slopes for ease of maintenance. Productive spoil excavated from field laterals shall be shaped to facilitate cropping and vegetative control, or spread so that row crops can be planted parallel and adjacent to the laterals, or removed and used in land grading operations. Spread spoil shall be deposited so water enters the laterals only at planned inlets. Shaped spoil banks shall be constructed with smooth crowns and sloped to provide drainage away from the excavated channels. Nonproductive spoil shall be dumped to minimize areas of productive land taken out of production.

5.1.8 Transitions that do not require special design for erosion control shall be constructed without restricting drainage. Specially graded channel transitions for erosion control shall be precisely constructed, because poor workmanship may easily reduce or eliminate their erosion-resistant qualities. Local experience may indicate that graded transitions will become stable after minor erosion has occurred; otherwise they require establishment of grass or other erosion resistant material for protection.

SECTION 6—CONSTRUCTION EQUIPMENT

6.1 Proper construction equipment and skilled operators are required for good quality work. Proper grade control must be assured for satisfactory drainage performance. Grade control can be obtained by either laser controlled leveling or by the use of grade stakes.

6.1.1 Tractor and scraper combinations are used for excavating, for hauling excavated material to fill areas, and for spreading excavated material. Tractors and scrapers are used for land grading, and to construct and maintain field drains and diversions. Under favorable conditions, they are used to construct and maintain field laterals with flat side slopes. Track-type tractors and large pan scrapers have been used for many years. In recent years, larger wheel-type farm tractors have been used with small pan scrapers pulled in tandem and with self-loading scrapers. Smaller wheel-type farm tractors have been used with single small pan-type scrapers and have become very popular in some areas. The tractor-scraper combination is the most versatile type of earth moving equipment commonly used in surface drainage work.

6.1.2 Water leveling blades are specially designed, tractor-mounted blades that loosen soil by scraping adjacent to or under water reducing the soil and water to a fluid slurry.

6.1.3 Land smoothers, commonly known as land levelers or land planes, are used for land smoothing or to finish land grading. They will automatically fill small low areas and remove small high areas when traversing a field. They cannot accomplish the major grading operations done by tractors and scrapers. Land smoothers are used periodically to maintain the surface of land that has previously been graded or smoothed.

6.1.4 Motor graders and blade-type graders pulled by tractors are used to excavate relatively small channels, such as field drains or field laterals, where the excavated material can be spread or shaped along one or both sides of the channels. Graders are used to construct diversions and to crown land. They are used extensively to maintain field drains and field laterals. A grader is an efficient tool where firm foundation conditions exist and the work area is free of stumps.

6.1.5 Draglines and backhoes are used to construct and maintain field laterals, farm mains, and interceptor drains. Occasionally they are used to construct field drains under wet conditions. Draglines are adapted for work under wet conditions, for channels having steep side slopes, and where excavated spoil material can be deposited alongside the excavated channel. Excavated spoil ordinarily must be shaped or spread by some other type of equipment.

6.1.6 Bulldozers are used to crown land, to construct diversions, to spread spoil, and for maintenance. Occasionally, they are used to construct and maintain field drains and field laterals with flat side slopes. Bulldozers are suitable for use where excavated material is moved short distances.

6.1.7 Farm plows may be used for bedding and to construct row drains. They are used to maintain beds and crowns.

SECTION 7—EROSION CONTROL

7.1 Erosion of deep productive soils in fields that require surface drainage may be of little direct economic importance. On shallow soils, erosion may over a period of time result in irreparable losses in productivity. Sedimentation that results from erosion obstructs nearby channels. The justifiable intensity of erosion control measures on field surfaces is, therefore, dependent on the depths of productive soils and the relative costs of preventing erosion and maintaining channels.

7.1.1 Erosion control on land surfaces is normally required only where land grades exceed 1.0% or where there are reverse row grades. By concentrating water and causing it to overflow row beds, reverse row grades are a major contributing cause of erosion across rows. Graded rows, precisely constructed so that there are no reverse row grades and so that each row carries its own water, are an effective erosion control practice. Bench leveling is usually not recommended for surface drainage because construction and maintenance costs are not justified by benefits.

7.1.2 Field drain side slopes and the row middles immediately above the field drains are subject to rill erosion with resultant sedimentation in the field drains. To control this erosion, portions of field drains that are 0.15 m (0.5 ft) below field level or below row middles shall be vegetated except vegetation is not necessary for field drains re-established yearly.

7.1.3 Field drains with grades of about 0.30% and steeper shall be vegetated to prevent channel erosion. The initial erosive grades vary somewhat with soils.

7.1.4 Field laterals are subject to serious erosion in some areas where they are located on erodible soils and steeper grades. These laterals should be vegetated. Where there is a doubt, the maximum permissible grade for unvegetated channels should be checked. Permissible grades are dependent on the scour conditions which are likely to occur in the channels.

7.1.5 Sedimentation in field laterals is usually expected. In areas where sedimentation is anticipated, field laterals should be overcut 0.1 m (0.3 ft) or more to provide for sedimentation.

7.1.6 Inlets to field laterals for surface water shall be constructed only at planned points. Continuous low levees, spread spoil or spoil banks shall prevent entry except at planned inlets.

7.1.7 The outlets of field drains and field laterals shall be checked for erosiveness. In some cases the depths of channels can be controlled by design to reduce or eliminate erosion. In other cases the water level in the larger channels may reduce velocities sufficiently to effectively control erosion. Where additional erosion control measures are indicated, graded channel transitions shall be considered. Such transitions may be grass waterways or non-vegetated channels, depending on the erosion potential. When graded transitions are not adequate for erosion control, pipe drops, chutes, drop structures or other suitable erosion control structures shall be used.

7.1.8 Pipe drops are relatively inexpensive, usually have long lives, and provide crossings for field roads. Pipe drops shall have design capacities of 1.25 to 1.50 times the volumes of flow indicated on the drainage curves. Small diameter pipe drops are subject to blockage by debris. Pipe drop inverts shall be 0.03 to 0.09 m (0.1 to 0.3 ft) below design grades at the inlet ends of the pipes. Insofar as practical pipe drops should outlet at least 0.6 m (2 ft) above the ditch bottoms at the outlet ends of the pipes.

7.1.9 All uncultivated field lateral side slopes and spoil banks shall be vegetated as soon as practical after construction. Vegetive plants shall be selected for their erosion resistance, wildlife habitat and beautification values.

SECTION 8—MAINTENANCE

8.1 Timely maintenance is essential for continued adequate performance of a drainage system. Maintenance operations shall be planned when the system is designed. Inspection of the entire drainage system should be made following heavy rainfall, and at least once each year. Tillage operations should retain the shape of field surfaces as initially constructed. The two-way plow is desirable for primary tillage because dead furrows and back furrows can be eliminated or located to meet the needs of the drainage system. Where the one-way plow is used, dead furrows and back furrows shall be kept to a minimum, and shall be located so they will not restrict water flow. Unless dead and back furrows are parts of the drainage systems as in bedding, their locations shall be alternated annually.

8.1.1 Row drains and field drains may require cleaning following tillage operations. When the field drains are not cultivated, weed growth shall be kept under control by mowing, spraying, or burning.

8.1.2 Field laterals require periodic maintenance. Suitable chemical weed and bush sprays are effective in controlling cattails, willows, and other undesirable vegetation. Mowing is effective where side slopes are not steeper than 3:1. Controlled grazing of livestock is a good method. Laterals require cleanout where excess deposition has occurred.

8.1.3 Land grading maintenance is critical during the first year or two after construction. Settlement of the fill areas may make several annual land smoothing operations necessary. In some cases, particularly where deep fills have been made, it may be necessary to cut and fill again with tractors and scrapers to eliminate depressions and reverse row grades. It is recommended that the land planing operation be substituted for one of the secondary tillage operations. The final land maintenance operations shall be in the direction of the crop rows.

8.1.4 Transitions shall be checked frequently for restriction of drainage. Removal of deposition downstream from the transition may be the only maintenance required.

8.1.5 Pipe drops shall be checked and maintained to see that the flow is not restricted, that overtopping of the fills is not causing excessive erosion and that piping is not occurring along the conduits.

Cited Standard:

ASAE EP268, Design, Layout, Construction and Maintenance of Terrace Systems

ASAE Standard: ASAE S313.2

SOIL CONE PENETROMETER

Proposed by the ASAE Cultural Practices Equipment Committee; approved by the ASAE Power and Machinery Division Technical Committee; adopted by ASAE as a Recommendation December 1968; revised February 1971; reconfirmed December 1975; reclassified as a Standard December 1978; reconfirmed December 1980; revised February 1985.

SECTION 1—PURPOSE AND SCOPE

1.1 The soil cone penetrometer (Fig. 1) is recommended as a measuring device to provide a standard uniform method of characterizing the penetration resistance of soils. The force required to press the 30° circular cone through the soil, expressed in kilopascals, is an index of soil strength called the "cone index."

1.2 The Standard is intended for the following purposes:

1.2.1 To provide a common method of expressing general soil mechanical conditions and facilitate the reporting and interpretation of soil data by different research workers.

1.2.2 To assist those who work with different soils and soil conditions and who need a measure of soil mechanical properties for comparative purposes.

1.2.3 To provide a common system of characterizing soil properties from which it may be possible to develop performance and prediction relationships.

1.3 The Standard serves only as a means of measuring, describing, and reporting the composite soil resistance to penetration, and does not provide specific soil values such as cohesion, angle of friction or coefficient of soil-metal friction. Since the measured results are influenced by the method of use, a procedural method is included.

SECTION 2—DEFINITIONS

2.1 Cone penetrometer: A 30° circular stainless steel cone with driving shaft per Fig. 1.

2.2 Base area: The cross-sectional area at the base of the cone expressed in mm^2 (in.^{2}).

2.3 Cone index: The force per unit base area required to push the penetrometer through a specified very small increment of soil. Values may be reported as X kPa at Y mm depth or X kPa average in the Y to Z mm depth, or CI_Y or CI_{Y-Z}. Cone base size should be stated.

2.4 Depth profile: The measured depth-penetration resistance relation of a specific soil condition and location.

SECTION 3—TEST APPARATUS AND PROCEDURE

3.1 The hand-operated soil cone penetrometer shown in Fig. 1 has a cone and a graduated driving shaft. Two cone base sizes are recommended: 323 mm^2, 20.27 mm diameter (0.5 in.^{2}, 0.798 in. diameter) with 15.88 mm (0.625 in.) diameter shaft for soft soils (Fig. 1a); and 130 mm^2, 12.83 mm diameter (0.2 in.^{2}, 0.505 in. diameter) with 9.53 mm (0.375 in.) diameter shaft for hard soils (Fig. 1b). American Iron and Steel Institute, AISI 416 stainless steel, machined to a smooth finish 1.6 μm (63 microinches) maximum is recommended as a suitable initial cone finish. The 129 mm^2 (0.2 in.^{2}) base area cone is suitable for hand-operated instruments on hard soil if the shaft does not exceed a length of 457 mm (18 in.). On extremely hard soils, a mechanically driven 323 mm^2 (0.5 in.^{2}) base area cone is suitable because of the stronger driving shaft. The overall form of the penetrometer equipment is not part of this Standard. Graduations on the driving shaft are 25.4 mm (1.0 in.) apart and are used to identify depth on hand-operated devices. The measuring device of hand-operated units should have a cone index capacity of approximately 2000 kPa (290 psi) for the 320 mm^2 (0.5 in.^{2}) base area penetrometers and not exceed 5000 kPa (725 psi) for the 129 mm^2 (0.2 in.^{2}) base area penetrometers in order to be suitable for most agricultural soil conditions. A direct readout of cone index is desirable. This Standard specifies two cone sizes. Nonstandard data should be reported but should be accompanied by a description of the nonstandard penetrometer and its method of use. In very soft soils when operating at great depths, care should be taken so the shaft does not drag. A smaller diameter shaft may be necessary in some cases. Penetrometer cones should be replaced when the base diameter wear exceeds 3% and therefore affects the cone index by 5%. A 1.5 mm (0.06 in.) shoulder on the base of the cone provides longer cone life without materially affecting accuracy of readings. Good judgment should be used to discard cones for other factors such as excessive bluntness, or abnormal wear. Soil conditions rapidly modify the cone shape and finish, but change in diameter is suggested as the criterion for discard of the cone.

3.2 Recording and mechanized penetrometers may have different overall forms, but the cone, size of driving shaft, and operating speed should conform with this Standard.

3.3 The cone is pushed into the soil at a uniform rate of approximately 30 mm/s (72 in./min). The surface reading is measured at the instant the base of the cone is flush with the soil surface. Subsequent readings should be made continuously or as frequently as possible while maintaining a 30 mm/s (72 in./min) penetration rate. Manual penetrometer readings should be made at depth increments of 50 mm (2 in.) or less. The depths are indicated by markings on the shaft of the cone penetrometer. Should it be necessary to stop the penetrometer at some depth (as would probably be the case where only one man was performing the cone index test), the penetration and measurement may be resumed without introducing errors. In very hard soils it may not be possible to achieve a rate as high as 30 mm/s (72 in./min), but somewhat slower rates will not result in significant errors.

3.4 Manually operated penetrometers may require an instrument operator and an assistant to record data.

3.5 Five to seven readings should be taken to establish the cone index and to verify the presence of unique layers in the soil profile.

SECTION 4—GUIDES FOR THE INTERPRETATION AND REPORTING OF DATA

4.1 The cone index may be expressed as a function of depth. Average cone index values may be expressed for specific increments of the profile or in complete depth-penetration relations. When increments of the profile are characterized, the size and location of the increments should be reported for the corresponding cone index values. Area profiles in any plane may be graphically constructed by connecting points of equal cone index for specific areas of interest. Available data on soil moisture, density, or classification should be included with penetrometer data to enhance interpretation of the reported results.

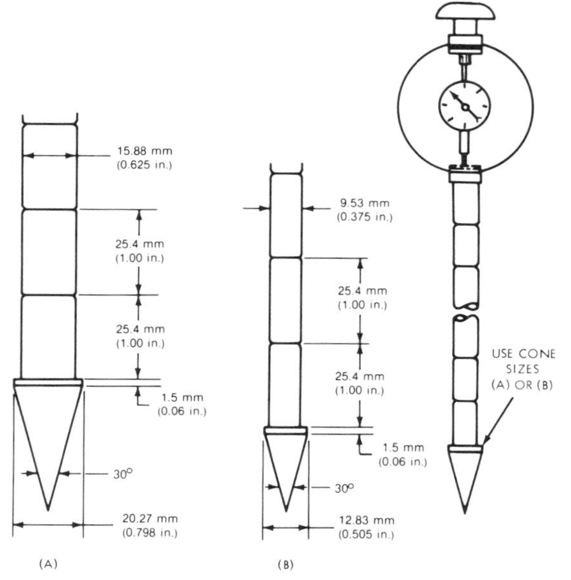

FIG. 1—SOIL CONE PENETROMETER

ASAE Standard: ASAE S330.1

PROCEDURE FOR SPRINKLER DISTRIBUTION TESTING FOR RESEARCH PURPOSES

Developed by the ASAE Sprinkler Irrigation Committee; approved by ASAE Soil and Water Division Standards Committee; adopted by ASAE as a Recommendation December 1969; reconfirmed December 1974; reclassified as a Standard March 1979; reconfirmed March 1981, March 1982, December 1982, December 1983, December 1984; revised March 1985.

SECTION 1—PURPOSE AND SCOPE

1.1 This Standard has the following two purposes:

1.1.1 To provide a basis for the accumulation of data on the distribution characteristics of sprinklers.

1.1.2 To provide a uniform method for the presentation of the data described in paragraph 1.1.1.

1.2 The data collected are to be of such extent and accuracy as to assist sprinkler system designers in making decisions regarding the water distribution pattern of sprinklers.

1.3 This Standard describes the types and methods of obtaining and recording pertinent climatic data. There must be a sufficient amount of data so that apparent conflicts between results of different investigators can be resolved.

1.4 No attempt is made here to define analysis procedures.

1.5 This Standard applies primarily to rotating sprinklers, and is not intended for testing nonrotating sprinklers such as spray nozzles.

SECTION 2—SPRINKLER DESCRIPTION AND SELECTION

2.1 Number of sprinklers. Single sprinkler tests only are covered in these procedural recommendations. It is generally desirable to perform more than one test under ostensibly the same sprinkler operating and climatic conditions. Each test shall be reported separately.

2.2 Selection of sprinklers. Any sprinkler used in these tests shall be chosen from normal production runs and shall be representative of all such sprinklers being produced.

2.3 Description of sprinkler. The sprinkler shall be described in such a way that a completely unambiguous reference can be made to it at a future date. This description shall include, but not necessarily be limited to, the following:

Manufacturer
Model name and number
Serial number or other identifying mark
Nozzle diameter(s) and description(s)
Entrance fitting description (size, type, etc.)
Type of bearing
Other identifying information (e.g., straightening vanes, type of drive, nozzle angle(s), etc.)

SECTION 3—TESTING INSTALLATION

3.1 Sprinkler location and installation

3.1.1 The sprinkler shall be located in an area which has either a bare surface or less than 80 mm (3 in.) of vegetative growth. The slope of the test area shall not exceed 1.5%. The site shall be located such that there is a minimum clear distance upwind of the pattern area of 6 heights of any windbreak for each 0.45 m/s (1 mile/h) of wind speed up to a maximum of 30 heights for winds of 2.24 m/s (5 mile/h) or greater, and a minimum clear distance downwind of the pattern area equal to 5 heights of any downwind windbreak. A map showing location and height of windbreaks shall be included on the Standard Data Presentation Form. Tests shall not be run when these conditions are not satisfied.

3.1.2 The height of the main nozzle shall be 0.6 m (2 ft) above the average elevation of the tops of the four nearest collectors. Other heights may be used for special purposes, but in all cases the nozzle height shall be clearly shown on the data.

3.1.3 The sprinkler riser shall be vertical within 1 deg. The riser shall be restrained to prevent vibration.

3.2 Collector description and location

3.2.1 All collectors used to measure distribution shall be the same. They shall be designed such that the water does not splash in or out and such that evaporation is kept to a minimum. The collector opening should be circular with a minimum diameter of 80 mm (3.1 in.). The collector shall be completely described on the data sheet. If an evaporation suppressant is used, its type and method of application shall be reported.

3.2.2 A square grid pattern of collectors shall be used. The sprinkler shall be located in the center of a grid square (midway between four adjacent collectors). A minimum of 80 collectors shall be maintained within the wetted area such that the entrance portion is horizontal, as estimated by visual means.

3.2.3 The average height of the tops of the four collectors nearest the sprinkler shall be no greater than 0.3 m (1 ft) above the ground or top of the vegetation. This distance shall be reported as collector height. The collectors shall be in a plane parallel to the average land slope.

3.3 Climatic measuring equipment and location

3.3.1 The wind movement during the test period shall be measured with a rotating-cup totalizing anemometer with an accuracy of ± 3%, or a device of equal or better accuracy. Floating ball type devices are not satisfactory. The wind direction shall be measured with a wind vane on the basis of eight points of the compass.

3.3.2 Wind measuring equipment shall be located within the clear area as described in paragraph 3.1, but outside the sprinkler pattern and at a height of 2 m (6.5 ft).

3.3.3 Dry and wet bulb temperature measurements shall be made at a location where the microclimate is essentially unaffected by the operation of the sprinkler. This will normally be upwind of the pattern area.

SECTION 4—MEASUREMENTS

4.1 Sprinkler pressure. One or both of the following pressure measurements may be used:

4.1.1 The nozzle pressure is defined as the pitot-static pressure at the vena-contracta of the jet from the main (largest) nozzle. It shall be measured prior to a test with a pitot tube and a pressure-indicating device accurate to within ± 2% (at the sprinkler pressure).

4.1.2 The base pressure is defined as the pressure head at the sprinkler. This shall be measured at a point on the riser at least 10 riser diameters downstream from any change of direction of flow or change in pipe cross-sectional area. Position and pressure shall be recorded. The base pressure shall vary no more than ± 3% during a test.

4.2 Sprinkler flow. The flow through the sprinkler shall be measured and reported at the test pressure. Accuracy of the measuring technique shall be ± 3%. A calibrated water meter or weight-volume-time methods may be used.

4.3 Sprinkler rotation. The rate of rotation of the sprinkler shall be measured and reported. The uniformity of rotation through the four quadrants shall be measured and reported.

4.4 Climatic data

4.4.1 Wind measurements shall be taken at intervals of 15 min maximum. Velocity shall be recorded to the nearest 0.04 m/s (0.1 mile/h) and directions to the nearest octant. Direction shall be keyed to one of the principal axes of the Standard Data Presentation Form.

4.4.2 Wet and dry bulb temperatures shall be measured at intervals of 15 min maximum.

4.5 Depth of application. The depth of application in each collector shall be determined to an accuracy ± 2% of the average application depth, and reported either in application depth or rate units in a table showing the location of the collector relative to the sprinkler (see Standard Data Presentation Form).

SECTION 5—TEST DURATION

5.1 The preferable test duration is 1 h. Other test durations may be used, but the circumstances and time must be clearly stated on the test sheet. Sprinklers shall be started and stopped at the same position and true total time recorded.

SECTION 6—REPORTING

6.1 Information to be recorded. The data outlined in Section 2—Sprinkler Description and Selection, Section 3—Testing Installation, Section 4—Measurements, and Section 5—Test Duration, of this Standard shall be recorded on forms similar to the Standard Data Presentation Form shown in Figs. 1 and 2. A separate set of sheets shall be prepared for each sprinkler test.

6.2 Deviations from recommended procedure. Deviations from the recommended procedure shall be indicated on the Standard Data Presentation Form.

6.3 Additional data. Additional data on the conduct of a test should be included if it will benefit the system designer or help explain variations in results.

Test Conditions

Testing Agency ____________ Date ____________ Test No. ________

Sprinkler Specifications (Par. 2.3) ____________

Test Location ____________ Weather ____________

1. Vertical distance from riser gage to nozzle, mm (in.) (Par. 4.1.2) ____________
2. Inside diameter of riser, mm (in.) ____________
3. Collector height, m (ft) (Par. 3.2.3) ____________
4. Collector entrance diameter, mm (in.) (Par. 3.2.1) ____________
5. Sprinkler height, m (ft) (Par. 3.1.2) ____________
6. Flow rate, L/min (gpm) (Par. 4.2) ____________
7. Description of collector (Par. 3.2.1) ____________

8. Evaporation suppressant used ____________ Volume/collector ____________
9. Data during test:

Time	Wind		Temp., °C (°F)		R.H.%	Rotation rate, seconds per full revolution					Pres., kPa (psi)	
	m/s (mile/h)	Dir.	DB	WB		Full	1st Q	2nd Q	3rd Q	4th Q	Noz	Base

10. Map of test area. Give the following:
    a. Location of sprinkler.
    b. Location of climatic measurement equipment.
    c. Wind direction during test period.
    d. Distance from sprinkler to all windbreaks (upwind, downwind, and to side).
    e. Heights of all windbreaks.

This test ________ does ________ does not meet the criteria for sprinkler testing set forth in ASAE S330, Procedures for Sprinkler Distribution Testing for Research Purposes.

FIG. 1—STANDARD DATA PRESENTATION FORM, Test Conditions

Testing Agency ________________ Date ________________ Test. No. ________

Sprinkler Specifications ____

____

Test Location ________________ Weather ________________

Above data is in the following units: ____________. Mark locations of sprinkler with plus (+) sign. Indicate prevailing wind direction by an arrow and give its least angle of deviation from a line parallel to one of the principal axes of this sheet. Grid spacing is ______ m (ft).

FIG. 2—STANDARD DATA PRESENTATION FORM, Map

ASAE Standard: ASAE S339

UNIFORM CLASSIFICATION FOR WATER HARDNESS

Developed by the ASAE Water Treatment and Use Committee; approved by the ASAE Soil and Water Division Standards Committee; adopted by ASAE as a Recommendation December 1970; reconfirmed December 1975; reconfirmed and reclassified as a Standard April 1977; reconfirmed March 1982, December 1986; reconfirmed for one year December 1987.

SECTION 1—PURPOSE AND SCOPE

1.1 The purpose of the classification herein is to establish uniformity of qualitative description of water hardness. This classification was adopted from classifications used by various governmental agencies and industry groups. It is intended that this Standard will serve as a guide in the development of useful new terms and definitions.

1.2 This classification is appropriate for classifying the overall hardness of all potable water supplies and other supplies which are to be softened by the ion exchange process.

1.3 Standard procedure for the determination of water hardness can be found in "Standard Methods," American Public Health Association, New York, N.Y.

SECTION 2—CLASSIFICATION

2.1 Classification for water hardness: (hardness as $CaCO_3$)

Term	Milligrams per Liter	Grains per Gallon*
Soft	0-60	0-3.5
Moderate	61-120	3.5-7
Hard	121-180	7-10.5
Very hard	over 180	over 10.5

*1 grain per gallon = 17.1 milligrams per liter

INSTALLATION OF FLEXIBLE MEMBRANE LININGS

Developed by the Flexible Membrane Linings Committee; approved by the Soil and Water Division Standards Committee; adopted by ASAE as a Recommendation December 1970; reconfirmed December 1975; revised and reclassified as an Engineering Practice April 1977; reconfirmed March 1982, December 1982, December 1983; revised March 1986.

SECTION 1—PURPOSE AND SCOPE

1.1 Flexible membranes are used extensively as seepage barriers for ponds, reservoirs, lagoons, landfills, waste ponds, and canals. These membranes, when properly installed, may be expected to render satisfactory service in controlling liquid loss and pollution.

1.2 Their use should be considered wherever liquid loss due to seepage is of such proportion as to prevent the facility from economically fulfilling its planned function; where the conservation of soil or liquid resources is being impaired; or where liquid leakage is damaging land or creating pollution and health hazards.

1.3 The purpose of this Engineering Practice is to provide the designer of water or specialty liquid, not hazardous waste, facilities with guidelines for the proper installation of membrane materials. It is not intended to serve as a complete set of specifications. Neither is it intended to replace the judgment of personnel who are intimately familiar with site conditions or the overriding circumstances of the project.

SECTION 2—GENERAL REQUIREMENTS

2.1 Since flexible membranes may be subject to mechanical damage during installation, it is important to consider beforehand the field procedure which will be utilized in the lining operation. The manufacturer or his local representative may provide assistance, especially in selecting the proper prefabricated membrane width and method of packaging or delivery.

2.2 It is desirable that soil subgrades be firm enough to support the personnel or equipment to be used during installation. All facilities should be designed so that they will not fail structurally without a lining since there is always the possibility of lining damage. The surface of the soil subgrade should be finished smoothly and as projection-free as possible to minimize the possibility of puncture or damage to the membrane when placed on the subgrade.

2.2.1 All soil clods, brush, roots, rocks, sod, or foreign material which might puncture the lining material shall be eliminated from the area. Rolling or surface compaction of the subgrade is encouraged to provide an extra measure of safety. Backfilled depressions, including areas where stumps, trees or large rocks have been removed, are particularly subject to localized settling and should always be compacted with care. A representative of the contracting officer or representative of the owner shall approve the surface on which the membrane is to be installed before commencing work.

2.2.2 If the subgrade is coarse-textured and open after preparing and compacting, or in rocky soils, a one to two inch cushion layer of well-graded sands and gravelly sands (soil group SW or finer, as defined in American Society for Testing and Materials Standard D2487, Classification of Soils for Engineering Purposes) should be applied. Geotextiles may also be used to provide a suitable surface.

2.2.3 After subgrade preparation, the soil surface under the lining should be treated with a high quality, nonselective soil sterilant, if weed growth is considered to be a likely problem. Treatment of the subgrade with a sterilant may, however, pose a problem with shallow groundwater aquifers. The use of sterilants under these conditions should be evaluated in concert with state permitting authorities. Care should be taken not to apply sterilants outside the lined area if landscaping or grasses are to be established.

2.3 When a flexible membrane lining is installed over deteriorated rigid (concrete) or semirigid (asphalt) linings, the same precautions regarding foreign materials or sharp projections which could puncture the membrane apply. The deteriorated structure should be inspected carefully and all such hazards smoothed. In addition, all visible cracks should be filled with mortar or other suitable material provided that material is compatible with the membrane.

2.3.1 While properly installed membrane linings are durable and watertight, they should not be relied upon for structural support. The structural stability of badly deteriorated and cracked rigid linings shall be ascertained prior to relining with a flexible membrane.

2.4 Membrane thickness. Selection of the membrane thickness should be based on the soil texture, liquid characteristics and the susceptibility of the lining to damage during or after installation. The minimum thickness of rubber sheeting, either fabric reinforced or unsupported, is 0.75 mm (30 mil where 1 mil = 0.001 in.) regardless of the soil subgrade. The minimum thickness of plastic sheeting is 0.25 mm (10 mil) for sands (greater than 50% of the soil particles pass No. 4 sieve which includes soil groups SM, SP, and SW as defined in ASTM Standard D2487, Classification of Soils for Engineering Purposes) and 0.50 mm (20 mil) for gravels (greater than 50% of the soil particles retained on No. 4 sieve which includes soil groups GC, GM, GP, and GW). The National Sanitation Foundation Standard NSF No. 54, Flexible Membrane Liners, can also be used as a guide when selecting minimum membrane thickness for specific applications.

2.5 Sizing. Membrane linings should be supplied in sections as large as practical. However, availability of equipment to handle large rolls or packages may limit the section size specified. Table 1 indicates the approximate unit weight of typical rubber and plastic sheeting.

2.5.1 Factory seams. Joints are constructed in a controlled factory environment to fabricate large panels in order to minimize field seaming. Types of seam joints include: adhesive, bodied solvent, dielectric, extrusion welded, solvent, thermal, and vulcanized. A seam joint should provide adequate initial seam strength to meet installation and service requirements. Factory seams shall meet the physical property requirements for the specific flexible membrane material as outlined in the material property tables of NSF Standard No. 54, Flexible Membrane Liners.

2.5.2 Field seams. Joints are constructed on site to complete the fabrication of the membrane from factory panels. Field seams are constructed in an uncontrolled outside environment. Field seaming should only be performed when weather conditions are favorable. The contact surfaces of the materials should be clean of dirt, dust, moisture, or other foreign materials. The materials to be field seamed shall lay flat against one another, shall be aligned with sufficient overlap, and shall be bonded in accordance with the suppliers recommended procedures. Wrinkles should be smoothed prior to seaming. Seams should be made so there are no loose edges. All seams on the slopes should be oriented perpendicular to the liquid surface whenever practical (with the slope) to reduce stress on the joint. Any seam which parallels the liquid surface should have the top sheet overlap the bottom sheet.

TABLE 1—APPROXIMATE UNIT WEIGHT OF SHEETING TYPES

Membrane type	Gram per square meter per millimeter thickness	Pound per square yard per mil thickness
Rubber	0.625	0.065
Plastic (Depends on Type)	0.385 to 0.625	0.040 to 0.065

2.5.3 Seam testing. Both factory and field seams shall be tested. Seams should be inspected by both nondestructive and destructive testing techniques to verify seam integrity. For destructive testing, seam samples should be taken from the panels at regular intervals. In addition, on-site seam samples can be made with identical liner material, adhesive, and technique which need not be taken from the actual field seam.

2.6 Placement. Flexible membrane linings should not be stretched during installation and should be installed in a relaxed state or with slight slack in both directions to compensate for thermal shrinkage, settling of the sub-base, etc.

2.7 Placement of cover material. The use of cover material (earth, sand, gunite, concrete) is important for most membrane linings to reduce mechanical damage as well as reduce the seepage rate through holes. The sequence for placement of cover material must be determined by job and site peculiarities. In general, cover material on all slopes shall be placed beginning at the bottom and proceeding toward the top. Cover soil placement shall be conducted in such a manner that the lining will not be displaced or damaged by equipment or overburden. Sliding cover material over the lining should be avoided. The cover material shall not be placed when the temperature is below 5 °C (41 °F), unless the condition of the cover soil is such that its placement will not rupture the lining (relatively dry and free flowing). Care must be taken in the cover material placement when the air temperature is over 40 °C (104 °F), as the puncture resistance of plastic membranes diminishes as the temperature increases.

SECTION 3—RESERVOIR AND POND LINING

3.1 Design criteria. Agricultural or conservation reservoirs and ponds to be lined shall be designed to meet the general requirements of United States Department of Agriculture Soil Conservation Service Standards and Specifications, SCS 359, Waste Treatment Lagoon; SCS 378, Pond; SCS 425, Waste Storage Pond; SCS 436, Irrigation Storage Reservoirs; SCS 521, Pond Sealing or Lining; SCS 552, Irrigation Pit or Regulating Reservoirs; and SCS 648, Wildlife Watering Facilities.

3.2 Membrane type selection. Whether a buried or exposed flexible membrane liner is used, is mainly an economic decision depending on material cost, the particular application requirements, cost and availability of cover material, and effective life of the lining material.

3.2.1 Buried liners. Many liner materials are covered for one or more reasons: protection from solar radiation, particularly ultraviolet; protection from mechanical damage by maintenance machines and workers, by animals, or by vandals; reduction of temperature variation; and protection from lifting by wind. By burying the liner long-term watertight integrity is generally improved.

3.2.1.1 Where a membrane is to be buried in the structure for seepage control, the overexcavation of the earthen structure must be considered in the design. Soil cover should be a minimum of 0.3 m (1 ft). The bottom 0.15 m (0.5 ft) next to the membrane should not be coarser than silty sand (soil group SM). Where a flexible membrane is to be buried for protection against mechanical damage, a minimum of 0.45 m (1.5 ft) of cover is required, with additional thickness desirable when the cover material is of high plasticity (see Fig. 1).

3.2.1.2 Cover material shall be sufficiently stable to minimize erosion caused by wave action and liquid scouring on the sloping sides as the liquid level in the facility is varied. Likewise, all cover material must be sufficiently stable to resist sliding on the side slopes because of the weak shear plane created by the membrane. Side slopes for reservoirs utilizing buried linings should not be steeper than 3 to 1. In some cases side slope can be increased if a rigid cover of reinforced mortar or masonry is used provided the slope does not exceed the angle of repose for the embankment and the membrane is properly anchored before the rigid cover is applied.

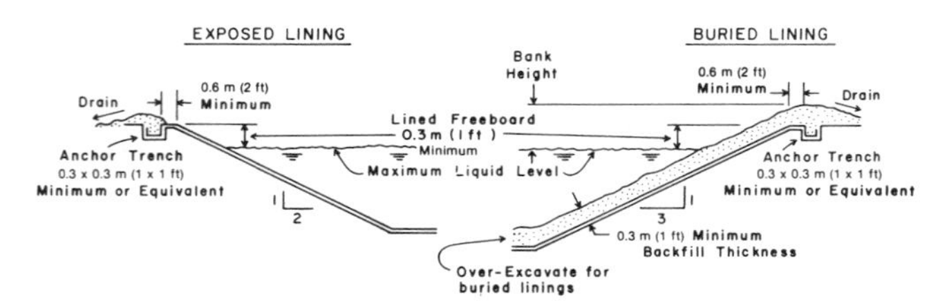

FIG. 1—RESERVOIR ANCHORING DETAILS

3.2.2 Exposed liners. Materials which are exposed are generally resistant to deterioration from weathering, various kinds of mechanical damage, etc. For reservoirs, mechanical damage can often be minimized by prevention of public access and fencing to control animal access.

3.2.2.1 Where exposed membrane linings are to be utilized in earthen structures, the side slope of the reservoir may be increased to as much as 2 to 1, depending on the material offered, provided this does not exceed the angle of repose for the embankment material.

3.2.2.2 Exposed membranes may be used to line deteriorated rigid reservoir structures having side walls or abutments steeper than 1 to 1. The membrane may require anchoring depending on the material selected. This may be done either mechanically or with suitable adhesives recommended by the membrane manufacturer. Redwood nailing strips, rolled galvanized strips and aluminum bar stock are often used above the liquid level to hold the vertical membrane in position. Generally, only reinforced membranes should be used when slopes are steeper than 2 to 1.

3.2.3 The minimum lined freeboard should be at least 0.3 m (1 ft) above the maximum level of the liquid surface in reservoirs of less than 0.4 ha (1 acre) surface. In larger reservoirs, additional bank height should be provided to prevent overtopping by wind-generated waves. Generally, add 0.3 m (1 ft) of lined bank height for every 240 m (800 ft) of additional wind reach, above 60 m (200 ft). The lined freeboard should be equivalent to the minimum permissible bank height to prevent overtopping of the reservoir liquid (see Fig. 1).

3.2.4 Provision must be made to prevent surface flooding from run-off around the reservoir perimeter due to poor drainage. The reservoir design should not permit surface run-off, or other incident liquid, from eroding the soil bank structure or seeping between the subgrade and the membrane.

3.2.5 The flexible membrane lining shall be anchored about its perimeter at the top of the reservoir embankment. In earthen structures this may be accomplished by means of an anchor trench at least 0.3 m (1 ft) wide and 0.3 m (1 ft) deep or in accordance with the manufacturers recommendation. The membrane should extend across the bottom and up the far side at the trench to form a "U" shape as shown in Fig. 1 or in accordance with the manufacturers recommendation. The trench may be located either on the side slope of the embankment or on the bench level if one is established. When located on the bench level, the trench should not be closer than 0.6 m (2 ft) to the top of the side slope. Anchor trenches shall be carefully backfilled and compacted after the membrane is in place.

3.2.6 Fill pipes, drains, overflow structures, gauge supports, or other fixtures that penetrate the membrane lining shall be suitably flashed and sealed to prevent leakage. The ground-surrounding structures must be properly compacted to minimize differential settlement which could result in damage to the membrane.

3.2.7 In lined reservoirs where the flexible membrane is buried, the liquid entrance-way should be protected with an additional heavy covering of large gravel or riprap to protect the earthen covering from erosion. A de-energizing structure may also be needed to reduce turbulence of the entering liquid, particularly if the intake structure is a pipe or flume. Likewise, drainage outlets should utilize vortex breakers to minimize disturbance of the protective earthen cover. In lined reservoirs where an exposed membrane is utilized, similar precautions are recommended to insure lining stability and positioning.

3.2.8 Where groundwater tables intercept the reservoir invert, where it is anticipated that gas or liquid/water may accumulate under the lining, or where exposed liners may be damaged by wind, a pressure relief system should be provided. Various methods can be used to control the problems depending on their nature and severity. These include weighting the liner with a protective cover and/or various pressure relief systems. Systems useful in relieving either gas or water pressures include (a) perforated pipe wrapped in a geotextile placed in trenches under the liner and vented to the atmosphere and (b) open granular filled trenches with vents to the atmosphere and possibly drain outlets if excessive groundwater is

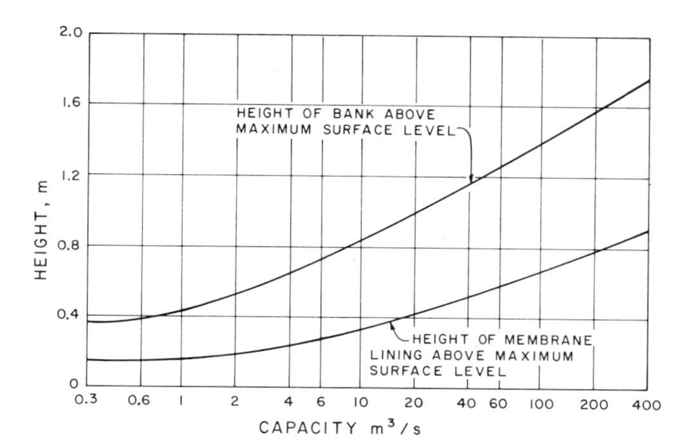

FIG. 2—BANK HEIGHT AND FREEBOARD FOR MEMBRANE LINED CANALS

anticipated. Systems useful in relieving gas pressure include (a) open granular fill or geotextile drainage fabric over the subgrade, (b) sloped bottom and sides, (c) three-sided gas units, and (d) wind cowels. Consideration should be given to using a soil cover as a complement to the pressure relief system.

SECTION 4—CANAL LINING

4.1 Design criteria. Irrigation canals and ditches to be lined shall be designed to meet the general requirements of existing USDA Soil Conservation Service Standards and Specifications, or existing United States Department of Interior Bureau of Reclamation Standards where applicable. In this case, however, a flexible membrane lining is to be utilized as the impervious material to control seepage.

4.1.1 A lined irrigation canal shall have enough capacity to function as designed without danger of overtopping.

4.2 Flexible membrane linings should be buried using suitable cover to protect them from damage, exposure to the elements, and injury by turbulent water, stock, plant growth, and maintenance equipment. The depth of cover depends on cover material, size of canal, water velocity, and canal slopes. The minimum recommended cover depth is 0.25 m (0.83 ft) plus 25 mm (1 in.) for each 0.3 m (1 ft) of water depth. To minimize costs, approved excavated material should be used for up to one-half the cover requirement, Fig. 3. The upper layer should be sand and gravel cover which is well graded. If possible, the sand and gravel material should be obtained from an approved borrow area. Using material that requires no blending lowers cost.

4.2.1 Canals with buried membrane linings must be constructed so that side slopes will be stable. Slope requirements will vary with different types of cover soil, but shall not be steeper than 2 to 1.

4.2.2 Stream velocities in canals lined with buried flexible membranes shall not exceed the nonerosive velocity for the earth cover material. Local information on velocity limits for specific soils should be used when available, however design stream velocity generally should not exceed 1 m/s (3 ft/s).

4.2.3 The canal cross section shall be sized for sufficient capacity to carry the required flow under maximum retardance conditions. In new construction, the prism should reflect an economical shape based on hydraulic design. A Manning roughness coefficient, n, of 0.025 is usually used for canals with capacities less than 2.8 m^3/s (100 ft^3/s) and 0.025 to 0.0225 for larger capacities.

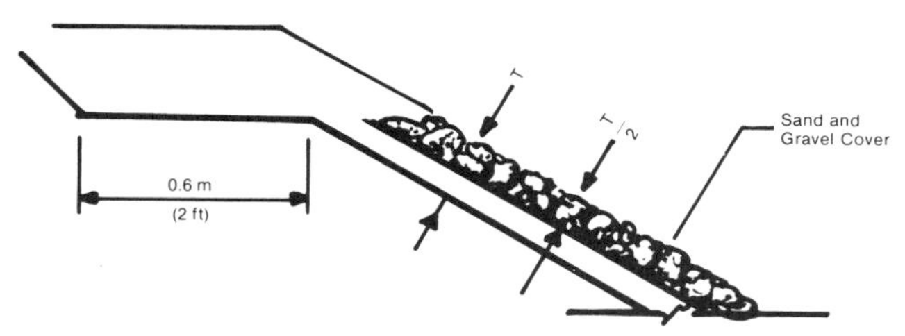

FIG. 3—CANAL ANCHORING DETAILS (MINIMUM BACKFILL THICKNESS, T)

4.2.4 Water surface elevations shall be established to provide adequate hydraulic head for successful operation of all ditches or other conveyance structures diverting from the canal. A minimum head of 0.15 m (0.5 ft) shall be provided, with 0.3 m (1 ft) preferred.

4.2.5 The lined freeboard allowed in canals should not be less than 0.15 m (0.5 ft) above maximum design operating level, increasing to 1 m (3 ft) as the canal capacity increases from 0.3 to 400 m^3/s (110 to 14,000 ft^3/s). The design bank height above the maximum surface level should be twice the permissible lined freeboard (see Fig. 2). In addition, provision should be made to prevent soil bank erosion from surface run-off around the canal perimeter caused by poor drainage.

4.2.6 The buried membrane lining should be anchored on each side of the canal along the berm so that the specified freeboard will be maintained. The membrane should be anchored as shown in Fig. 3. The upstream and downstream ends of the membrane sections should be overlapped in the downstream direction a minimum of 1 m (3 ft). The placement of the protective cover material will secure the lap joint. Since the cover material will be utilized to stabilize the membrane, care should be taken to insure full cover depth over the upstream end of the first section of lining.

4.2.7 In some instances, notably the rehabilitation of deteriorated rigid canal linings (cracked concrete or asphalt), it may be desirable and economical to install a flexible membrane lining inside the structure over the old lining. This should be considered only after the possibility of damage to the exposed lining from cleaning equipment, vandals or other causes has been carefully weighed. The specifications for this type of construction should be left up to the designer who should consider the climate, age of concrete, type of cracking, and the structure size.

SECTION 5—MEMBRANE MATERIALS

5.1 The sheet, including factory seams if present, shall be watertight and visually free of pinholes, particles of foreign matter, undispersed raw material and other manufacturing defects that might affect serviceability. If irregularities in the form of pock-marks appear excessive in a sheet or portion thereof, its rejection should be negotiated between the manufacturer and the owner. Rubber and plastic materials supplied for installation as flexible linings shall be compounded of high quality ingredients to produce durable, watertight membranes. Ingredients shall be thoroughly mixed to insure complete dispersion throughout the compound prior to processing into sheeting. The finished membrane material shall be free from defects which would affect its long-term serviceability as an impermeable lining.

Note: Pock-marks are oblong depressions, cavities or craters on the surface of the sheet which may have dimensions of 3 mm (0.12 in.) by 1.5 mm (0.062 in.) and may have a depth approaching half the sheet thickness.

5.1.1 Dimensional tolerance. Where membrane material is purchased under job specifications which detail the physical size of sheeting pieces, actual material dimensions shall not be smaller than specified. The measurements shall be made at ambient temperature in a relaxed state, or under light tension to remove wrinkles. Thickness tolerances shall be as outlined in the materials property tables of NSF Standard No. 54, Flexible Membrane Liners.

5.1.2 Certification. Purchasing authorities vary in their requirements as to material sampling and certification. It is not the intention of these guidelines to restrict this authority. At the very minimum, the purchasing agency should insist on contractor or material supplier certification that the material supplied for the project meets or exceeds the physical and performance properties referenced in paragraph 5.2. Such certification shall include identification of material, quantity represented, and certified physical property test results covering all of the specification test requirements.

5.1.3 Packaging. Unless otherwise specified the flexible membrane lining shall be packaged for shipment in rolls or accordion folded. The packages shall be suitably protected so that the material will not be damaged by commercial carrier shipment. The flexible membrane lining shall be properly protected, as specified by the manufacturer, against weather elements during transportation and storage.

5.1.4 The details of roll or package marking shall be outlined in the job specifications. As a minimum, each roll or container of lining material shall be marked or identified with the name of the material, the quantity contained therein (area and thickness), the name of the supplier, and the contract or project identification number.

5.2 Material specification. Minimum physical requirements for some commercially available materials which may be used as flexible membrane linings are outlined in material property tables on the NSF Standard No. 54, Flexible Membrane Liners. The following membrane materials are included in the NSF publication:

Polyvinyl chloride (PVC)
Oil resistant polyvinyl chloride (PVC-OR)
Chlorinated polyethylene (CPE)
Butyl rubber (IIR)
Polychloroprene (CR)
High density polyethylene (HPDE)
Ethylene-propylene diene terpolymer (EPDM)
Epichlorohydrin polymers (CO)
Polyethylene ethylene propylene alloy (PE-EP-A)
High density polyethylene elastomeric alloy (HDPE-A)
Chlorosulfonated polyethylene (CSPE)
Chlorosulfonated polyethylene low water absorption (CSPE-LW)
Thermoplastic nitrile - PVC (TN-PVC)
Thermoplastic EPDM (T-EPDM)
Ethylene interpolymer alloy (EIA)
Chlorinated polyethylene alloy (CPE-A)

Other materials may be used for a particular application and may be so specified. The physical requirements listed are intended to ensure good workmanship and quality, but are not necessarily adequate for design purposes depending on specific performance requirements. Minimum physical requirements for low density polyethylene can be found in ASTM Standard D3020, Specifications for Polyethylene and Ethylene Copolymer Plastic Sheeting for Pond, Canal, and Reservoir Lining. Some materials are unsupported while others are fabric reinforced. Materials differ significantly in their chemical resistance and should be carefully selected for compatability with the liquid to be stored or conveyed.

SECTION 6—OPERATION AND MAINTENANCE

6.1 The owner should be provided with guidelines for operation and maintenance of the liner system, which should include recommendations on subjects such as:

Frequency of documentation of inspection
Testing and repair of liner
Monitoring of waste characteristics
Monitoring of observation wells
Animal and plant control
Erosion control
Security and safety
Unacceptable practices

Cited Standards:

ASTM D2487, Classification of Soils for Engineering Purposes
ASTM D3020, Specifications for Polyethylene and Ethylene Copolymer Plastic Sheeting for Pond, Canal, and Reservoir Lining
NSF 54, Flexible Membrane Liners
SCS 359, Waste Treatment Lagoon
SCS 378, Pond
SCS 425, Waste Storage Pond
SCS 436, Irrigation Storage Reservoirs
SCS 521, Pond Sealing or Lining (A, Flexible Membranes)
SCS 552, Irrigation Pit or Regulating Reservoirs (A, Irrigation Pit and B, Regulating Reservoirs)
SCS 648, Wildlife Water Facilities

ASAE Engineering Practice: ASAE EP369.1

T # DESIGN OF AGRICULTURAL DRAINAGE PUMPING PLANTS

Developed by the ASAE Pump Drainage Committee; approved by the ASAE Soil and Water Division Standards Committee; adopted by ASAE May 1974; reconfirmed December 1978, December 1979, March 1981, March 1982, December 1986; revised December 1987.

SECTION 1—PURPOSE AND SCOPE

1.1 This Engineering Practice sets forth principles and practices useful to engineers in the planning and design of pumping plants for drainage of agricultural land. It is not intended as complete specifications and does not include pumping plants for deep well drainage.

SECTION 2—CLASSIFICATION OF PUMPING PLANTS

2.1 For purposes of planning and design, agricultural pumping plants are classified as:

2.1.1 Pumping plants for surface drainage. These have open sumps for collecting and temporarily storing both surface and groundwater. Ditches and adjacent lowland may be used to provide added storage away from the sump area.

2.1.2 Pumping plants for subsurface drainage. These have closed sumps for storage of groundwater only.

SECTION 3—PLANNING OF AREA TO BE PUMP DRAINED

3.1 The drainage system of the area served by the pumps should be planned to meet both drainage needs of the area and efficient operation of the pumps.

3.1.1 Runoff from high ground, when removable by gravity flow to a suitable outlet, should be excluded from the pumped area by diversion around the area or by channeling through the area.

3.1.2 Protection of the pumped area against overflow or backwater from the outlet should be provided by perimeter dikes designed against overtopping, wave action, erosion, and instability of high water stages. Disposal of interior drainage water should be provided through the dike in conduits protected by gates that prevent backflow during high exterior water stages and permit outflow during low water stages.

3.1.3 The drainage system of the pumped area should provide:

3.1.3.1 An optimum sump water stage commensurate with the design hydraulic gradient of the drainage system.

3.1.3.2 Optimum use of undeveloped swamp or wooded lowland near or adjoining the pumping plant to supplement sump storage for reduction of the pumping rate.

3.1.3.3 Drainage channels to the pumps with ample depth and capacity to avoid excessive water drawdown and channel scour.

3.1.4 Pumping plant location should be at or near the point of lowest elevation of the pumped area and accessible to an adjacent outlet.

3.1.4.1 Site adjustments should be made as necessary to provide foundations without excessive use of piling, freedom from flooding of operating equipment, access to fuel or power supply, and protection from vandalism.

3.1.4.2 Very lowland within a pumped area can often be drained more economically and satisfactorily with a second pump rather than designing the entire system for the lowest land.

3.2 The outlet channel from the pumped area shall have capacity for the discharge and shall comply with applicable drainage laws and local codes.

SECTION 4—PUMPING PLANT CAPACITY

4.1 The pumping rate should be determined from the drainage requirement, plus an allowance for seepage (when this is significant), less temporary storage.

4.1.1 The drainage requirement will vary with climate, topography, soils, land use, and types of crops. The drainage requirement may be expressed as a coefficient or quantity of water to be removed per unit of area per unit of time. Drainage coefficients are established for specific areas and crops and are available in state and some local drainage guides.

4.1.2 Pumping plant capacity for surface drainage should be based on surface drainage design coefficients. In addition to state and local guides, specific coefficients can be found in ASAE Engi neering Practice EP302, Design and Construction of Surface Drainage Systems on Agricultural Lands in Humid Areas.

4.1.3 Pumping plant capacity for subsurface drainage should be based on subsurface drainage design coefficients. In addition to state and local guides, specific coefficients can be found in ASAE Engineering Practice EP260, Design and Construction of Subsurface Drains in Humid Areas.

4.2 Pumping plant capacity of some areas has been established directly from experience with existing installations and local conditions. This may provide the best basis of design for plants near such sites.

SECTION 5—PUMP SELECTION

5.1 Pump selection involves determination of type, characteristics, capacity, head, size, and number of pumps in conjunction with kind and source of power, shape and size of sump, housing, and method of plant operation. Actual selection of the drainage pump will usually be based on maximum, minimum and rated or average discharge; maximum, minimum and rated or average total head; and the type of power to be supplied.

5.2 The centrifugal pump is the most common type used in agricultural drainage. There are three kinds of centrifugal pumps; radial flow, axial flow, and mixed flow. The discharge section may be a volute (radial or mixed flow), turbine (radial or mixed flow), or a straight continuation of the suction or inlet section (axial or mixed flow).

5.2.1 Radial flow pumps operate through partial conversion of velocity head, imparted by the impeller to the water, into pressure head. The conversion is accomplished by the gradual change in cross-sectional area within the impeller casing for the volute, and in a similar manner by fixed vanes for the turbine. Radial flow pumps are commonly referred to as centrifugal pumps.

5.2.2 Axial flow pumps develop head mostly by the propelling or lifting action of the vanes on water. Axial flow pumps are commonly referred to as propeller pumps. Propeller pumps are especially suited to low head and high volume requirements of most drainage sites.

5.2.3 Mixed flow pumps, with modified blades and casing features common to both radial and axial flow pumps, apply both axial and radial thrust in moving water through the pump.

5.2.4 Fig. 1 is an allocation chart useful in selecting the applicable type of pump when approximate head and discharge requirements have been determined. The crosshatched band indicates an overlap in application or choice of selection.

5.3 Pump performance varies with head, speed, discharge, and power relationships. The effect of these factors on efficiency of the pumping operation may be obtained from pump manufacturer's performance curves or pump characteristics for each kind and size of pump made.

5.3.1 Efficiency characteristics of drainage pumps on a percentage basis are shown in Fig. 2. For capacity variation (see Fig. 2a), the radial flow pump covers a wide span, while for head variation (see Fig. 2b), which is most important for drainage purposes, the propeller (axial flow) pump has a wider span. Generally, efficiency increases with the size of pumps. Fig. 3 shows the general shape of a head-discharge curve in which abscissa and

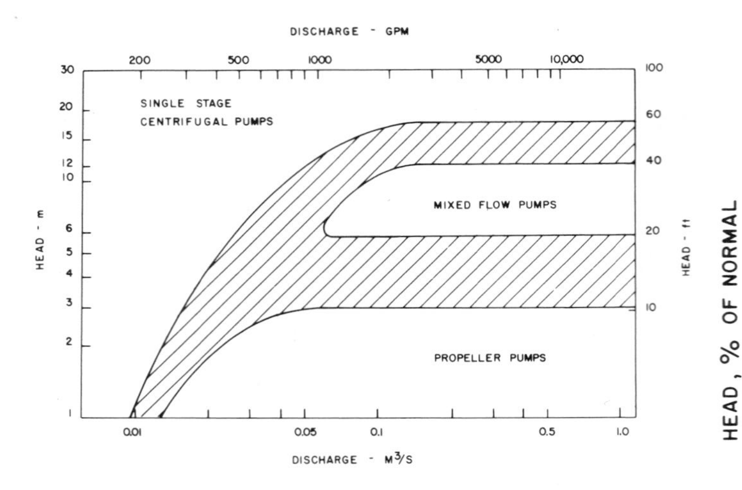

FIG. 1—PUMP TYPE SELECTION CHART. Crosshatched band shows the overlap in application or choice of selection.

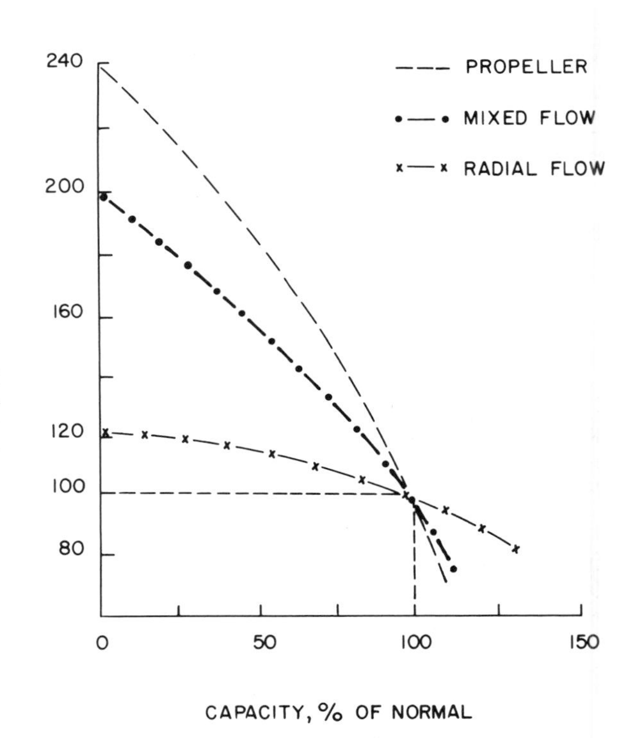

FIG. 3—PUMP SELECTION BASED ON PERCENT OF DESIGN VALUE AT MAXIMUM EFFICIENCY POINT

(a)

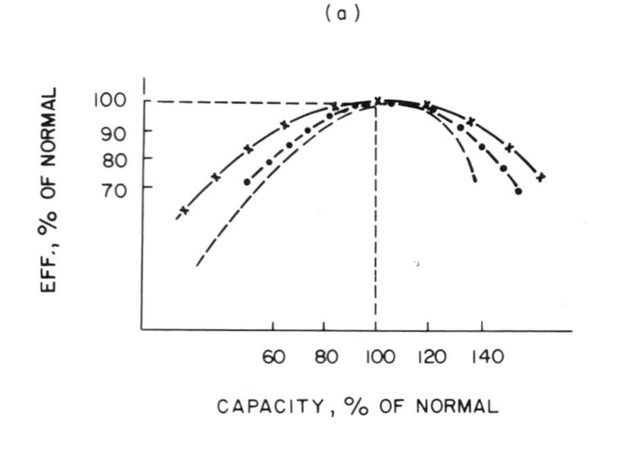

--- PROPELLER

•—• MIXED FLOW

x—x RADIAL FLOW

(b)

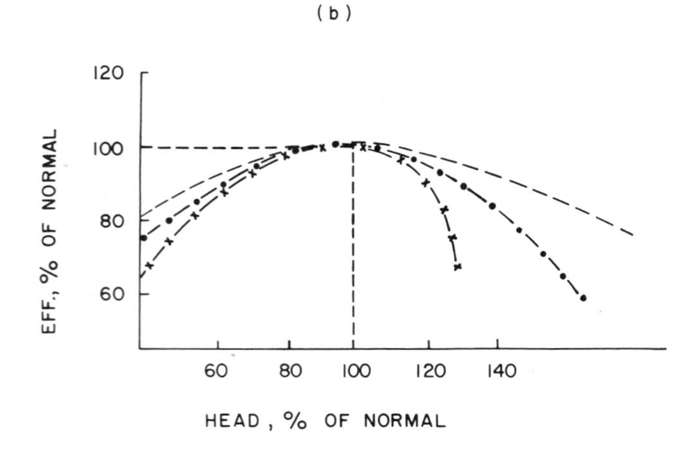

FIG. 2—EFFICIENCY CHARACTERISTICS OF DRAINAGE PUMPS

ordinate are expressed in percent of design value at maximum efficiency point and in terms of total dynamic head. The steep head-discharge curve for propeller pumps shows that this type should not be used at lower than design capacity because of increasing power requirements.

5.3.2 Cavitation, the partial vacuum that can form in water passing through the pump impeller, causes vibration and noise from water hammer, reduction in discharge, and pump deterioration. Performance curves are useful in preventing cavitation through proper selection and operation of the pump which avoids:

5.3.2.1 Heads much higher than that at optimum efficiency of the pump.

5.3.2.2 Capacities much lower than capacity at peak efficiency of the pump.

5.3.2.3 Suction lifts abnormally higher, or net positive suction heads (NPSH) lower than that rated for the pump.

5.3.2.4 Speeds higher than those recommended for the pump.

5.4 Pumping head, H_t, is the dynamic or total energy head that the pump must impart at design discharge.

$$H_t = H + H_v + H_f \quad \text{[1]}$$

where

H = static head
H_v = velocity head
H_f = friction head

5.4.1 Static head, or lift, is the vertical distance between the free water surface at the suction side of the pump and the free water surface at the discharge side of the pump when discharge is submerged, or at the center of the discharge pipe when discharge is not submerged. Design head is measure from the lowest water level on the suction side to the highest water level on the discharge side.

5.4.2 Velocity head, H_v in m (ft) is calculated

$$H_v = \frac{V^2}{2g} = C_2\frac{Q^2}{D^4} \quad \text{[2]}$$

where

C_2 = 0.083 (0.002 6)
V = column velocity at discharge, m/s (ft/s)
Q = pump capacity, m^3/s (gpm)
D = column diameter, m (in.)
g = acceleration of gravity, 9.81 m/s^2 (32.2 ft/s^2)

5.4.3 Friction head, H_f, is the headloss between the entrance and discharge sides of the pump installation. Friction head losses are comprised of entrance losses; internal losses as shear in sealing rings, water impact on the impeller, and friction in wearing rings and seals; and friction and exit losses in the discharge pipe. Entrance and internal losses in well designed pumps are minimal and are usually compensated for in the specifications and selection tables provided. Thus discharge pipe friction loss may be the only friction head value to be accounted for in the usual pumping plant design. Losses for various sizes of pipe and discharge rates can be calculated for smooth pipe from

$$H_f = \frac{C_3 L Q^{1.9}}{D^{4.9}} \qquad [3]$$

where

H_f = total friction loss in line, m (ft)
C_3 = coefficient = 2.62×10^{-9} (4.64×10^{-9})
L = length of pipe, m (ft)
Q = total discharge, L/s (gpm)
D = inside diameter of pipe, m (ft)

Losses are for straight pipe however bends up to 45 deg can be included without substantial error. Sharp bends should not be used.

5.4.4 Loss of head through automatic drainage gates (flap gates) is small and is often neglected in total head calculations. The loss of head can be calculated from

$$L = C_4 V^2 \times e^{\frac{C_4' V}{\sqrt{D}}} \qquad [4]$$

where

L = head loss, m (ft)
V = velocity through the gate, m/s (ft/s)
D = diameter of outlet, m (ft)
C_4 = coefficient = 0.408 (0.125)
C_4' = coefficient = -2.1 (-1.15)

5.5 Size of pumps depends upon total head and quantity of water pumped. The rated size is usually designated by the pipe column diameter at the discharge end of the pump.

5.5.1 The design column velocity of a propeller pump may range from 2 to 4 m/s (7 to 13 ft/s) with highest efficiency usually occurring at values of 2.4 to 3.0 m/s (8 to 10 ft/s). Three m/s (10 ft/s) is commonly used as an initial estimate in developing the pumping plant design.

5.5.2 Pump size, D, can be calculated from

$$D = C_5 \left(\frac{Q}{V}\right)^{1/2} \qquad [5]$$

where

D = size of the pump, m (in.)
C_5 = coefficient = 1.13 (0.64)
Q = pump capacity, m^3/s (gpm)
V = column velocity, m/s (ft/s)

5.6 The discharge pipe should be watertight and of adequate capacity.

5.6.1 The discharge pipe may be gradually enlarged away from the pump to decrease velocity head and thereby decrease the power cost.

5.6.2 Submerged disharges are recommended to keep heads as low as possible. An automatic drainage gate (flap gate) should be used at the end of the discharge pipe to prevent reverse flow through the pump.

5.6.2.1 In cold areas ice may prevent the automatic drainage gate from opening. On large pumps a dual discharge may be provided. On small pumps a riser permanently above high water can be installed behind the automatic drainage gate. Discharge through this alternate usually destroys ice around the main automatic drainage gate.

5.6.2.2 Erosion protection should be considered when discharging water into unprotected channels.

5.6.3 A siphon may be used when drainage water is pumped over a dike for reduction in both total head and power cost.

5.6.3.1 Priming head must be greater than operating head. The pump must be capable of delivering a full cross section of water over the crest at a velocity greater than 1.5 m/s (5 ft/s). However, a limit on maximum lift of the siphon and a maximum velocity of flow at the crest must be maintained to prevent reducing inside pressure at the crest enough to cause cavitation. The top of the discharge pipe should be installed at an elevation below the minimum design water level in the outlet channel to reduce pumping head. The pipe section at the crest must have sufficient strength to resist external atmospheric pressure.

5.6.3.2 A siphon breaker must be installed at the high point in the discharge pipe to prevent reverse flow during a power interruption. The siphon breaker should actuate when power is interrupted or when flow reverses. A flap gate at the end of the discharge pipe may serve in lieu of a siphon breaker for small pumping plants.

5.6.3.3 An air release valve is required to insure smooth priming of the siphon. This is preferably installed at the outlet end of the high part in the discharge pipe.

5.7 Size and number of pumps are determined chiefly by quantity of water to be pumped. Pumping from small areas or from groundwater only is usually handled by one pump. When pumping from large areas and when high crop values or farm improvements require some flood protection, several pumps will provide more efficient pumping over a wider range of pumping rates and will also provide continued pumping should one pump fail. The most flexible load distribution is achieved with two pumps where one is about twice the capacity of the other; and for three pumps where each has equal capacity. At least one pump should be selected to operate efficiently over long periods. Optimum efficiency is not so essential when a pump is used for short periods at peak stages or discharges.

5.8 Typical drainage pump installations and head relationships that should be considered in design are shown in Fig. 4.

5.8.1 Submersible, sewage type pumps can be used effectively for capacities up to 50 L/s (800 gpm).

5.8.2 In-line pumps may be used in large diameter outlet drain pipe where it is difficult to provide a satisfactory sump.

SECTION 6—STORAGE

6.1 The volume of drainage runoff to be stored in the sump and auxiliary areas will vary from a maximum of the total runoff of the design storm to a minimum set by the pumping rate that will prevent excessive numbers of pump starts and stops by the power source.

6.2 For manual operation of the pumps, pump starts may be limited to 2 per day for convenience of the operator. The volume of storage for 2 starts per day may be calculated from

$$S = C_6 Q \qquad [6]$$

where

S = volume of storage, m^3 (ft^3)
C_6 = coefficient = 10.84 (24)
Q = pump capacity, l/s (gpm)

6.3 For automatic operation of pumps, cycles of operation should be limited to about 10 per hour. A cycle of operation includes running and starting time. Running time should not be less than 3 min.

6.3.1 The minimum storage for automatic operation may be calculated from

$$S = \frac{C_7 Q}{n} \qquad [7]$$

where

S = volume of storage, m^3 (ft^3)
C_7 = coefficient = 0.90 (2)
Q = pump capacity, L/s (gpm)
n = number of cycles per hour

6.4 Sump dimensions for automatic operation can be varied to fit field conditions and to be in line with economical installation. Generally the sump should not be deep but should be large and

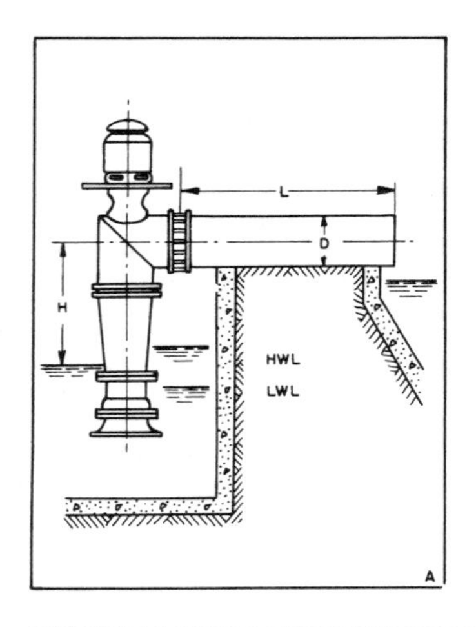

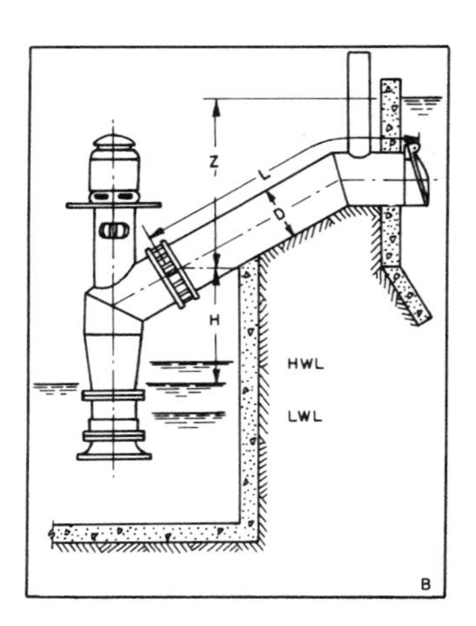

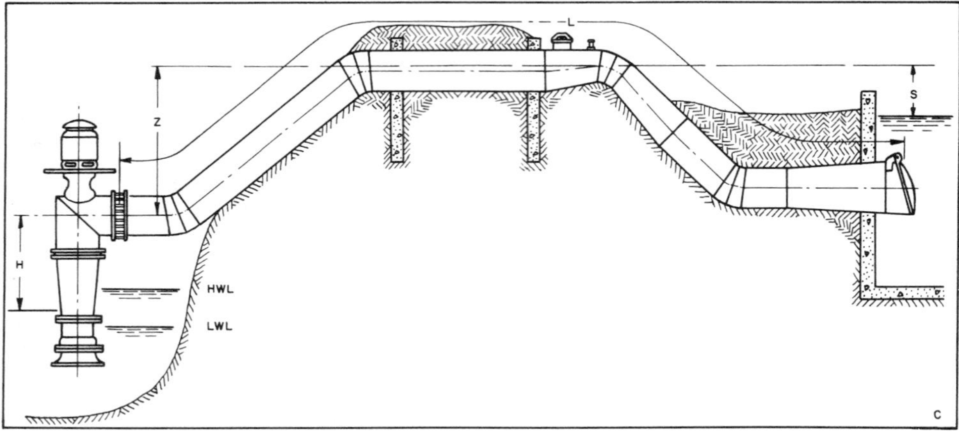

FIG. 4—TYPICAL DRAINAGE PUMP INSTALLATIONS. HWL & LWL = High and low water level: L = length, H = head to centerline of pipes, Z = discharge head, D = diameter of pipe, and S = average syphon head.

shallow. For efficient automatic operation, a storage depth of 0.6 m (2 ft) for closed sumps and 0.3 m (1 ft) for open sumps is recommended. Greater sump storage depth usually adds to both installation and operation cost of the pumping plant. In some soils, wide and frequent fluctuations in water depth cause serious bank sloughing and channel erosion.

SECTION 7—SUMP DESIGN

7.1 The drainage sump is a pit, tank, section of ditch or low area which serves as a collection point from which the drain waters are pumped. The shape, size, and position of the sump with respect to the pump affect the efficiency of the pumping operation.

7.1.1 Ditches and low elevations in the watershed provide added storage when large volumes of surface water must be handled or where manual operation is used.

7.1.2 Corrugated metal, steel or concrete pipe and other suitable material of large diameter may be used as a closed sump for subsurface drainage systems. A pipe of adequate strength and size to support the pump should be set vertically in an excavated trench to the designed elevation. A concrete floor should then be poured. Similar adjoining structures can be built and connected to provide added storage. Figs. 5 and 6 show typical closed sump installation using corrugated metal pipe. Concrete ballast can be added to the floor and walls for necessary mass in overcoming uplift, or the base slab can be extended outside the vertical walls of the sump.

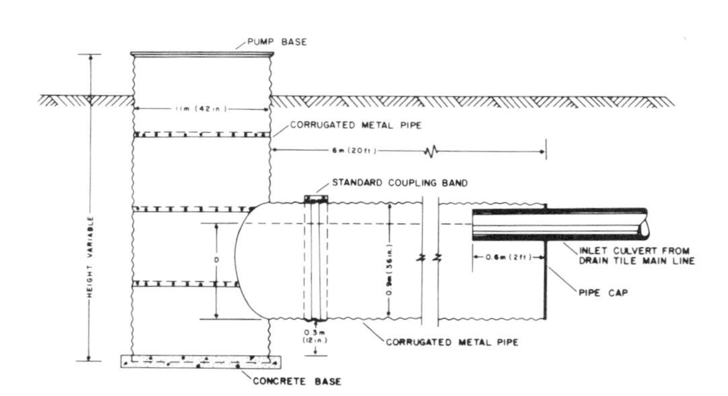

FIG. 5—TYPICAL PUMP DRAINAGE SUMP

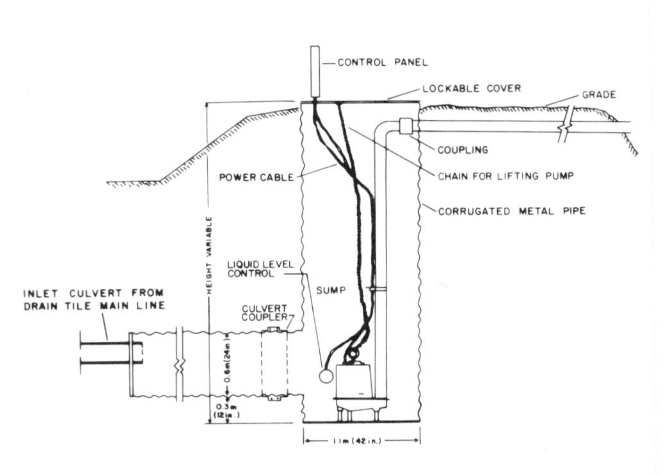

FIG. 6—TYPICAL SUBMERSIBLE PUMP DRAINAGE SUMP

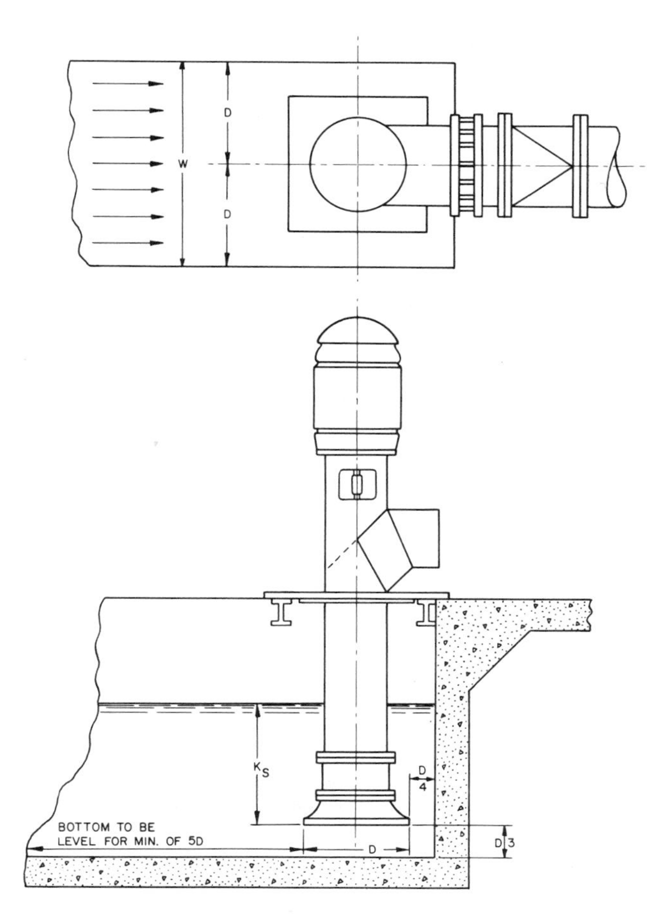

FIG. 7—RECOMMENDED CLEARANCES FOR DRAINAGE SUMPS. (See Equation 8 for K_s calculation)

7.1.3 Other materials for constructing closed sumps include cast-in-place concrete, prefabricated concrete silo staves, and prefabricated septic tanks. Cast-in-place concrete construction is often difficult because of soil conditions. Concrete silo staves make inexpensive sump walls when stable foundations can be provided for laying a floor or ring footing. Backfill against the staves must be uniformly and carefully placed to maintain the circular shape. Prefabricated septic tanks may be connected in series to provide the necessary storage volume.

7.2 Efficient pump operation requires special sump dimensions and pump clearances for smooth, free flow of water into the suction intake pipe. Essential dimensions and clearances are shown in Fig. 7.

7.2.1 Sump water level must be maintained above a minimum submergence of the pump suction intake and is determined by the greater of the following independent conditions:

7.2.1.1 Water cover over the propeller sufficient to keep the sump self-priming.

7.2.1.2 Net positive suction head (NPSH) as determined by the pump manufacturer to overcome cavitation.

7.2.1.3 Submergence above the suction bell, as usually recommended in manufacturers' literature for vortex suppression, can be approximated from

$$K_s = \frac{0.64\,Q}{D^2} \quad \text{[8]}$$

where

K_s = submergence, m (ft)
Q = pumping rate, m^3/s (ft^3/s)
D = pipe diameter, m (ft)

7.2.1.4 Water storage capacity should be sufficient to prevent too frequent pump starts and stops.

7.2.2 Shape of the pump bay of an open sump should be square or rectangular since a circular sump will tend to accentuate water rotation sufficiently to interfere with pumping operation. Use of a baffle between wall and pump, from floor to minimum static water level, or placement of the pump near one side of the sump will counteract rotation.

7.2.3 Floor of the approach channel and sump should be level for a distance of 5 diameters of the suction bell measured from the centerline of the pump. Approach velocity to the pump should be within 0.3 to 0.6 m/s (1 to 2 ft/s) unless the sump configuration has been checked for eddy currents by the manufacturer. In large pumps, approach velocity may reach 0.9 m/s (3 ft/s). Trash rack openings below minimum static water level should permit water passage at design velocity.

7.2.4 Sidewall clearance of pumps should not be less than 2 times the diameter of the suction bell.

7.2.5 Minimum clearance between adjacent pump centerlines should be the sum of the diameters of the suction bells.

7.2.6 Back wall clearance of pumps may be reduced to that necessary for maintaining the pumps.

7.2.7 Where ditch storage is provided, the culvert inlet passing water to the sump should be of sufficient size to handle the maximum pump discharge at a velocity not exceeding 0.46 m/s (1.5 ft/s).

7.2.8 Subsurface drain outlets discharging into a sump should be located as far from the pump as possible to minimize the effects of turbulence and air entrainment.

7.2.9 The maximum water level in the sump should be about the invert elevation of the drain discharging into the sump.

7.2.10 Additional information on sump dimensions and multiple pits may be found in the centrifugal pump section of Hydraulic Institute Standards for Centrifugal, Rotary and Reciprocal Pumps.

SECTION 8—POWER SOURCE

8.1 Driving units for pumps may be electric motors or gasoline, diesel or gas fueled engines. Electric motors provide an economical installation when an adequate and reliable source of electric power is available close to the site at reasonable cost. Internal combustion engines should be used when the source of electric power is not reliable or is too costly.

8.2 The required power of the driving unit can be calculated from

$$P = \frac{C_9\,Q\,H_t}{N_p N_g N_m} \quad \text{[9]}$$

where

p = power, W (brake horsepower)
Q = pump discharge, m^3/s (gpm)
H_t = total dynamic head, m (ft)
N_p = overall efficiency of the pump
N_g = overall efficiency of gear drive (if used)
N_m = overall efficiency of motor or engine
C_9 = coefficient = 9 800 (0.000 25)

8.3 Electric power is convenient for pump drainage operation and easily adapted to automatic operation.

8.3.1 Vertical, squirrel cage induction type or synchronous type constant speed motors are normally used for pump drives. Since propeller pumps require a high starting torque, the available line voltage must be checked with the electric power supplier. If the voltage is sustained under 90% or over 110% of nameplate voltage, special motors are required.

8.3.2 Induction motors will usually have adequate torque for average conditions. Synchronous motors should always be checked for starting conditions. Requirements for synchronous motors are full voltage starting and suitability for across the line starting service.

8.3.3 Choice between motor types should be made on the basis of lowest (present value) cost, including the installation cost. Motors and controls should conform to standards of the National Electrical Manufacturers Association.

8.3.4 Three-phase power is desirable and usually necessary. Rotary of static phase converters may be used to generate three-phase power from single phase power lines to operate a drainage pump in the range 10 to 50 kW (13 to 67 hp).

8.4 Gasoline, propane or natural gas, and diesel engines can be adapted for variation in speed over a broad range and are limited only by the range of critical speeds which produce dangerous torsional vibrations. Ability to throttle speed and reduce pump discharge at low lifts is an advantage. A float-operated throttle may be used to reduce speed with decreased inflow rates.

SECTION 9—DRIVE HEADS

9.1 Drive heads are required to connect the power source to the pump. Loss of efficiency through these units ranges from none in direct connections to about 5% in gear connections, and up to 10% in multiple belt connections. Proper direction of pump rotation must be maintained in all drive applications.

9.1.1 A right-angle gear drive should be used to connect a standard electric motor to vertical pumps. Such drives are available in sizes up to 225 kW (300 hp) and with stepdown gear ratios to match higher speed motors to slower speed pumps. Direct-connected, hollow shaft electric motors should be used in a vertical position.

9.1.2 Tractors may be used to drive right-angle gear drives at 540 or 1,000 rpm. This saves the user the investment and maintenance of a separate power unit for each pump.

9.1.3 Multiple V-belt heads and flat pully drives should be used on small pump installations.

9.1.4 Safety devices for drive systems should conform to ASAE Standards S203 Rear Power Take-Off for Agricultural Tractors, S207 Operating Requirements for Tractors and Power Take-Off Driven Implements, and S318 Safety for Agricultural Equipment.

SECTION 10—AUTOMATIC OPERATION

10.1 Automatic operation is used primarily with electric power sources. Considerations in applying automatic equipment are:

10.1.1 Added installation and maintenance costs.

10.1.2 Possible lack of interest on the part of owner or operator if operation is too highly automated.

10.1.3 Compatibility of controls compatible with local electrical codes.

10.1.4 Possibility of moisture getting into or submergence of controls.

10.1.5 Where freezing occurs, use of a 500W waterproof electric heating cable to enclose the automatic water level control system so ice does not form.

10.1.6 Greater requirement for operator vigilance and maintenance, including lubrication and trash rack cleaning.

10.2 Electric telemetering systems for remote control of pumps have three basic elements:

10.2.1 A detecting element which senses the change in water level for making power starts and stops. This is usually a float or adjustable immersion electrode. The float should be enclosed in a corrosion resistant well. Either type of element is satisfactory at sites where freezing does not occur.

10.2.2 An electrical transmitter which converts the sensed signal into switch action. Pressure of the float against adjustable stops on a vertical float rod activates the switch on float systems, whereas im mersion electrodes respond to water contact at preset vertical posi tions to activate a switch.

10.2.3 A relay control unit on the pump which energizes the starter on the pump.

10.3 Automatic controls are available for internal combustion engines such as an electric starting system for air-cooled engines. Such systems are not readily adaptable to most pump installations.

10.4 Protective controls must be supplied for all installations.

10.4.1 Automatic or electric units should have safety shutoff against overload, low voltage, and excessive heating.

10.4.2 Internal combustion engines should have controls which protect against overheating, low oil pressure, and excessive speed.

10.4.3 Auxiliary warning devices such as signal lights on the pumphouse which light when pumps are operating, should be provided for attention of the owner or operator.

SECTION 11—HOUSING

11.1 Type of housing for pumping plants depends upon importance and size of the plant, type of power used, and plant location.

11.1.1 Factors influencing need for housing are air temperature; wind; moisture from humidity, precipitation and flooding; fuel storage; safety; vandalism; and plant appearance.

11.1.2 Materials used for housing should be fire resistant, water proof, durable, and easily maintained.

11.1.3 Size of housing need only be sufficient for the containment and servicing of the equipment. Provision should be made for removing the pumps when required.

11.1.4 When a submersible pump is used in a closed pump, a protective cover should be placed over the sump to prevent accidents and entry of trash and debris into the sump. All sumps should be covered, or a safety rail should be placed around the sump to prevent people and animals from falling into them.

SECTION 12—TRASH RACKS

12.1 Trash racks must be provided to prevent entry of floating debris into sumps where damage to pumps might occur.

12.1.1 Trash racks should be set not less than 2½ times the suction bell diameter in front of the centerline of the pump.

12.1.2 Velocity of flow through the trash rack should not exceed 0.6 m/s (2 ft/s).

12.1.3 Recommended trash rack bar spacing in millimeters (inches) is as follows:

Pump diameter	Bar spacing
406 (16)	19 (0.75)
457-610 (18-24)	25-38 (1-1.5)
762-1067 (30-42)	51 (2)
1067 (42)	64-76 (2.5-3)

12.1.4 Trash racks should be sloped so cleaning by hand is easily done, or they must be equipped with mechanical cleaners.

12.1.5 Where trash racks are not feasible, an independent strainer cage with vertical bars should surround the intake. If excessive trash is expected, a secondary fence screen should be placed across the ditch to catch weeds and floating debris.

12.1.6 Basket strainers are not practical with electric drives because they are easily plugged and difficult to clean. Electric motors may overload due to the increasing NPSH where a strainer is plugged.

Cited Standards:

ASAE S203, Rear Power Take-Off for Agricultural Tractors
ASAE S207, Operating Requirements for Tractors and Power Take-Off Driven Implements
ASAE EP260, Design and Construction of Subsurface Drains in Humid Areas
ASAE EP302, Design and Construction of Surface Drainage Systems on Agricultural Lands in Humid Areas
ASAE S318, Safety for Agricultural Equipment
Hydraulic Institute Standards for Centrifugal, Rotary and Reciprocal Pumps

DESIGN, INSTALLATION AND PERFORMANCE OF UNDERGROUND, THERMOPLASTIC IRRIGATION PIPELINES

Developed by the ASAE Irrigation Water Supply and Conveyance Committee; approved by the ASAE Soil and Water Division Standards Committee; adopted by ASAE April 1975; reconfirmed December 1979, February 1981; revised April 1982; reconfirmed December 1986; reconfirmed for one year December 1987.

SECTION 1—PURPOSE AND SCOPE

1.1 Purpose. Thermoplastic pipe is manufactured in several size classifications from different materials of various grades, types and formulations involving many different specifications. It is used for applications other than irrigation where certain requirements often apply to pipe used for a specific purpose. This Standard pertains to thermoplastic pipe used underground for irrigation and is intended to:

1.1.1 Provide minimum guidelines for engineers and others in planning, designing and specifying thermoplastic pipe commonly used for irrigation. It is not intended as a complete specification nor to replace the judgment of personnel familiar with site conditions or other controlling factors.

1.1.2 Consolidate applicable reference information and technical data in readily available form.

1.1.3 Establish uniform standards for materials used in the manufacture of thermoplastic irrigation pipe and to promote uniformity in classifying, pressure rating, testing and marking the pipe.

1.1.4 Establish minimum requirements for the design, installation and testing of pipelines which are necessary for the satisfactory performance and safe operation of the irrigation system and to prevent damage to the system.

1.2 Scope. This Standard applies to underground, thermoplastic pipelines used in the conveyance of irrigation water to the point of distribution and may or may not apply to potable water systems.

1.2.1 High pressure pipelines. This term applies to underground pipelines constructed of thermoplastic pipe from 21 to 710 mm (1/2 to 27 in.) nominal diameter that are closed to the atmosphere, and subject to internal pressures, including surge pressures, from 550 to 2170 kPa (80 to 315 psi).

NOTE: Nominal pipe size in millimeters is the actual outside pipe diameter to the nearest millimeter for OD controlled pipe and the actual inside diameter to the nearest millimeter for ID controlled pipe.

1.2.2 Low pressure pipelines. This term applies to underground thermoplastic pipelines 114 to 630 mm (4 to 24 in.) nominal diameter that are used in systems subject to pressures of 545 kPa (79 psi) or less.

SECTION 2—DEFINITIONS

2.1 Design area: The specific land area in which pipelines are planned and located to serve as integral parts of an irrigation water distribution or conveyance system, designed to facilitate conservation, use and management of water and soil resources, and which the supplier or designer and purchaser mutually understand to be irrigated.

2.2 Irrigation system: All equipment required to apply water to the design area.

2.3 Irrigation pipelines: Includes the underground, thermoplastic pipelines and appurtenances installed in an irrigation system.

2.4 Outlets: Appurtenances required to deliver water from the pipeline to an individual sprinkler or to a lateral of sprinklers, to surface pipe located on the ground, to distribution pipe or laterals containing surface or subsurface emitters or tricklers, to surface valves, or to open ditches.

2.5 Hydrostatic design stress: The estimated maximum tensile stress in the wall of the pipe in the circumferential orientation, due to internal hydrostatic water pressure, that can be applied continuously with a high degree of certainty that failure of the pipe will not occur.

2.6 Pressure rating (PR): The estimated maximum pressure that water in the pipe can exert continuously with a high degree of certainty that failure of the pipe will not occur.

2.7 Dimension ratio (DR): The ratio of pipe diameter to wall thickness.

2.7.1 For outside diameter (OD) based pipe, which includes polyvinyl chloride (PVC), acrylonitrile-butadiene-styrene (ABS) pipe and some polyethylene (PE) pipe, the ratio is calculated by dividing the pipe's average outside diameter by the pipe's minimum wall thickness. The minimum wall thickness shall not be less than 1.52 mm (0.060 in.). Certain DR values have been selected as standard and given the designation Standard Dimension Ratio (SDR). The SDR and DR values for PVC and ABS are rounded to the nearest 0.5.

2.7.2 For inside diameter (ID) based pipe, which includes some PE pipe, the ratio is calculated by dividing the average inside diameter of the pipe by the pipe's minimum wall thickness. The minimum wall thickness shall not be less than 1.52 mm (0.060 in.). The SDR values shall be rounded to the nearest 0.1.

2.8 Relation between standard dimension ratio, hydrostatic design stress and pressure rating: The following expression, commonly known as the ISO equation (from International Organization for Standardization Standard ISO 161/1-1978, Thermoplastic Pipes for the Transport of Fluids—Nominal Outside Diameters and Nominal Pressures—Part 1: Metric Series), is used to relate standard dimension ratio, hydrostatic design stress, and pressure rating:

2.8.1 For OD based pipe:

$$2\,S/P = R - 1$$

or

$$2\,S/P = (D_o/t) - 1$$

where

S = hydrostatic design stress, kPa (psi)
P = pressure rating, kPa (psi)
D_o = average outside diameter, mm (in.)
t = minimum wall thickness, mm (in.)
R = dimension ratio, DR (equals D_o/t for PVC, ABS, and other OD based pipe)

2.8.2 For ID based pipe:

$$2\,S/P = R + 1$$

or

$$2\,S/P = (D_i/t) + 1$$

where

R = dimension ratio, DR (equals D_i/t for ID based pipe such as some PE pipe
D_i = average inside diameter, mm (in.)

SECTION 3—DESIGN CRITERIA

3.1 Working pressure

3.1.1 General. The pipeline shall have a pressure class rating (see Table 1) greater than the static or working pressure plus surge at any point in the system. Surge pressures should not exceed 28 percent of the pipe's pressure class rating; therefore, if surge is not

TABLE 1—PRESSURE RATINGS (PR) FOR NONTHREADED THERMOPLASTIC PIPE*†

SDR‡		PVC materials (all pipes OD based)								PE materials (pipes made to both OD & ID basis)						ABS materials (all pipes OD based)					
		PVC 1120 PVC 1220 PVC 2120		PVC 2116		PVC 2112		PVC 2110		PE 3408		PE 3406 PE 3306 PE 2306		PE 2305		ABS 1316		ABS 2112		ABS 1210	
OD based pipe	ID based pipe	psi	kPa§	psi	kPa	psi	kPa	psi	kPa	psi	kPa	psi	kPa	psi	kPa	psi	kPa	psi	kPa	psi	kPa
	5.3									250	1725	200	1380	160	1105						
	7.0									200	1380	160	1105	125	860						
11.0	9.0									160	1105	125	860	100	690						
13.5	11.5	315	2170	250	1725	200	1380	160	1105							250	1725	200	1380	160	1105
17.0	15.0	250	1725	200	1380	160	1105	125	860	100	690	80	550	63	435	200	1380	160	1105	125	860
21.0		200	1380	160	1105	125	860	100	690	80	550	64	440			160	1105	125	860	100	690
26.0		160	1105	125	860	100	690	80	550	64	440	50	345			125	860	100	690	80	550
32.5		125	860	100	690	80	550	63	435	50	345	40	275			100	690	80	550	64	440
41.0		100	690	80	550	63	435	50	345	40	275	31	215			80	550	64	440	50	345
51.0		80	550	63	435	50	345	40	275							64	440	50	345	40	275
64.0		63	435	50	345	40	275	30	205												
81.0		50	345	40	275	30	205	25	170							40	275	30	205	25	170
93.5//		43	295																		
50 ft head		22	150																		

*For water at 23 °C (73.4 °F).
†Pressure ratings are determined by the ISO equation as shown in paragraph 2.8 using the hydrostatic design stress values shown in Table 5.
‡SDR = Standard Dimension Ratio determined as shown in paragraph 2.7.
§kPa = kilopascals, kN/m^2
//The dimension ratio 93.5 is non standard and is referred to as DR (Dimension Ratio)

TABLE 2—MAXIMUM ALLOWABLE PRESSURE FOR NONTHREADED THERMOPLASTIC PIPES WHEN SURGE PRESSURES ARE NOT KNOWN*†

SDR		PVC materials (all pipes OD based)								PE materials (pipes made to both OD & ID basis)						ABS materials (all pipes OD based)					
		PVC 1120 PVC 1220 PVC 2120		PVC 2116		PVC 2112		PVC 2110		PE 3408		PE 3406 PE 3306 PE 2306		PE 2305		ABS 1316		ABS 1212		ABS 1210	
OD based pipe	ID based pipe	psi	kPa	psi	kPa	psi	kPa	psi	kPa	psi	kPa	psi	kPa	psi	kPa	psi	kPa	psi	kPa	psi	kPa
	5.3									180	1240	144	995	115	795						
	7.0									144	995	115	795	90	620						
11.0	9.0									115	795	90	620	72	495						
13.5	11.5	227	1565	180	1240	144	995	115	795							180	1240	144	995	115	795
17.0	15.0	180	1240	144	995	115	795	90	620	72	495	58	400	45	310	144	995	115	795	90	620
21.0		144	995	115	795	90	620	72	495	58	400	45	310			115	795	90	620	72	495
26.0		115	795	90	620	72	495	58	400	45	310	36	250			90	620	72	495	58	400
32.5		90	620	72	495	58	400	45	310	36	250	29	200			72	495	58	400	46	315
41.0		72	495	58	400	45	310	36	250	29	200	22	150			58	400	46	315	36	250
51.0		58	400	45	310	36	250	29	200							46	315	36	250	29	200
64.0		45	310	36	250	29	200	22	150												
81.0		36	250	29	200	22	150	18	125							29	200	22	150	18	125
93.5		31	215																		
50 ft head		21	145																		

*Maximum allowable working pressure = pressure rating (PR) x 0.72 for SDR and DR pipe.
†For water at 23°C (73.4°F).

TABLE 3—MAXIMUM, OR CRITICAL, SURGE PRESSURE FOR THERMOPLASTIC PIPE

Pipe SDR (or DR)		Surge pressure* per ft/s (0.3 m/s) of sudden change in flow velocity					
OD based	ID based	For pipe material of 400,000 psi (2800 MPa) modulus (includes most PVC)		For pipe material of 300,000 psi (2100 MPa) modulus (includes most ABS)		For pipe material of 100,000 psi (700 MPa) modulus (includes most PE)	
		psi	kPa	psi	kPa	psi	kPa
	5.3	28.1	195	24.3	170	14.0	95
	7.0	25.1	175	21.7	150	12.5	85
11.0	9.0	22.5	155	19.5	135	11.2	75
13.5	11.5	20.3	140	17.6	120	10.2	70
17.0	15.0	18.0	125	15.6	110	9.0	60
21.0		16.1	110	13.9	95	8.0	55
26.0		14.4	100	12.5	85	7.2	50
32.5		12.9	90	11.2	75	6.4	45
41.0		11.4	80	9.9	70	5.7	40
51.0		10.2	70	8.8	60	5.1	35
64.0		9.1	65	7.9	55	4.5	30
81.0		8.1	55	7.0	50	4.0	30
93.5		7.5	50	6.5	45	3.2	20

$$* \quad P = V \left(\frac{3960\ ET}{et + 300{,}000D} \right)^{1/2}$$

where

P = surge pressure, psi
V = sudden change in velocity, ft/sec
E = modulus of elasticity of pipe material, psi
t = pipe wall thickness, inch
D = pipe inside diameter (ID), inch

See also: Seipt, W.R. 1974. Water hammer considerations for PVC pipeline in irrigation systems. TRANSACTIONS of the ASAE 17(3): 417-423.

TABLE 4—PRESSURE RATING SERVICE FACTORS FOR TEMPERATURES FROM 23 TO 60 °C (73.4 TO 140 °F) FOR PVC AND PE PIPES*†

Temperature °C	°F	PVC factor	PE factor
23	73.4	1.00	1.00
26.7	80	0.88	0.92
32.2	90	0.75	0.81
37.8	100	0.62	0.70
43.3	110	0.50	
48.9	120	0.40	
54.4	130	0.30	
60.0	140	0.22	

*To obtain the pressure rating for a temperature above 23 °C (73.4 °F), multiply the pressure rating at 23 °C (73.4 °F) as given in Table 1 or Table 2 as appropriate by the corresponding service factor. For PE pipe having improved strength retention with an increase in temperature, and PE pipe used at temperatures exceeding 38 °C (100 °F), the manufacturer should be consulted for recommended service factors.

†For ABS pipe used at temperatures above 23 °C (73.4 °F) service factors recommended by the manufacturer should be used.

known, the working pressure shall not exceed the maximum allowable working pressure as given in Table 2 for the particular pipe and SDR or DR used. Maximum or critical pressure as a function of pipe SDR or DR is shown in Table 3 for thermoplastic pipe having different moduli of elasticity.

3.1.2 Service factor. All pressure ratings are determined in a water environment of 23 ± 2 °C (73.4 ± 3.6 °F). As the temperature of the environment or fluid increases, the pipe becomes more ductile. Therefore, the pressure rating must be decreased for use at higher temperatures to allow for safe operation of the pipe. The service factors for PVC and PE are shown in Table 4. For PE pipe having improved strength retention with an increase in temperature and PE pipe used at temperatures exceeding 38 °C (100 °F), the manufacturer should be consulted for recommended service factors. For ABS pipe, service factors recommended by the manufacturer should be used.

3.2 System capacity. The design capacity of the pipeline shall be sufficient to provide an adequate flow of water for all methods of irrigation planned.

3.3 Friction losses. For design purposes, friction head losses shall be no less than those computed by the Hazen-Williams equation using a flow coefficient (C) equal to 150.

3.4 Flow velocity. The design water velocity in a pipeline when operating at system capacity should not exceed 1.5 m/s (5 ft/s) unless special considerations are given to the control of surge or water hammer and adequate protection from these pressures is provided (see paragraph 3.1.1 and Table 3). Adequate pressure and/or air relief valves shall be used with all velocities.

3.5 Outlets. Outlets shall have adequate capacity at the pipeline working pressure to deliver the design flow to the distribution system at the design operating pressure of the respective systems, i.e., sprinklers, surface pipe, emitters, tricklers, etc.

3.6 Check valves. A check valve shall be installed between the pump discharge and the pipeline where detrimental back flow may occur. It shall be designed to close, without slamming shut, at the point of zero velocity before damaging reversal of flow can occur.

3.7 Pressure relief valves. These shall be installed between the pump discharge and the pipeline when excessive pressures can develop by operating with all valves closed. Pressure relief valves or surge chambers shall be installed on the discharge side of the check valve where back flow may occur and at the end of the pipeline when needed to relieve surge.

3.7.1 Low pressure systems. Pressure relief valves may be used as alternatives to serve the pressure relief functions of vents and stands open to the atmosphere. They do not function as air release valves and should not be substituted for such valves where release of entrapped air is required.

3.7.1.1 Pressure relief valves shall be large enough to pass the full pump discharge with a pipeline pressure no greater than 50 percent above the permissible working head of the pipe.

3.7.1.2 Pressure relief valves shall be marked with the pressure at which the valve starts to open. Adjustable valves shall be installed in such a manner to prevent changing of the adjustment marked on the valve.

3.7.2 High pressure systems. The ratio of nominal size pressure relief valves to pipeline diameter shall be no less than 0.25. Pressure relief valves shall be set to open at a pressure no greater than 34.5 kPa (5 psi) above the pressure rating of the pipe or the lowest pressure rated component in the system.

3.8 Air release and vacuum relief valves. Air release and vacuum relief valves shall be installed at all summits, at the ends, and at the entrance of pipelines to provide for air escape and air entrance. Combination air-vacuum release valves which provide both functions may be used.

3.8.1 Air flow capacity. Valves having large orifices to exhaust large quantities of air from pipelines when filling and to allow air to enter to prevent a vacuum when draining are required at the end and entrance of all pipelines. Valves intended to release entrapped air only may have smaller orifices and are required at all summits.

3.8.2 Low pressure systems (not open to the atmosphere).

3.8.2.1 Air-vacuum release valves shall be provided at each of the locations described in paragraph 4.5.3.

3.8.2.2 The size of valve outlet for low pressure systems shall be as specified in paragraph 4.6.2.

3.8.3 High pressure systems. The ratio of air release valve diameter to pipe diameter for valves intended to release air when filling the pipe should not be less than 0.1. However, smaller diameter valves may be used as a means of limiting water hammer pressures by controlling air release where filling velocities cannot be controlled. Equivalent valve outlet diameters of less than 0.1 are permitted for continuously acting air release valves. Adequate vacuum relief must still be provided. It is not only very important to select the correct air release or vacuum breaker valve, but also to select the right size and to locate valves properly at all places where needed. Air vacuum release valves shall be used as follows (all valve diameters refer to the total cross-sectional flow area of the vent or port outlet).

Pipe diameter mm (in.)	Minimum air-vacuum release valve outlet diameter mm (in.)
102 (4) or less	13 (0.5)
127-203 (5-8)	25 (1)
254-500 (10-20)	51 (2)
530 (21) or larger	0.1 pipe diameter

3.9 Draining. Provisions shall be made for draining the pipeline completely where a hazard is imposed by freezing temperatures, drainage is recommended by the manufacturer of the pipe, or drainage of the line is specified for any reason. Where provisions for drainage are required, drainage outlets shall be located at all low places in the line. The outlets may drain into dry wells or to points of low elevation. If drainage cannot be provided by gravity, provisions shall be made to empty the line by pumping.

3.10 Flushing. Where provision is needed to flush the line free of sediment, a suitable valve shall be installed at the distal end of the pipeline.

3.11 Gate stands and float valve stands. When these are used in low pressure pipelines not open to the atmosphere, refer to the criteria in paragraphs 4.4.1 and 4.4.2.

SECTION 4—SPECIAL DESIGN CRITERIA FOR LOW PRESSURE PIPELINE SYSTEMS OPEN TO THE ATMOSPHERE

4.1 Stands, general. Stands shall be used wherever water enters the pipeline to avoid entrapment of air, to prevent surge pressures, to avoid collapse due to negative pressures, and to prevent pressure from exceeding the head class of the pipe. Stands shall be supported on a base adequate to support the stand and prevent movement or undue stress on the pipeline. Stands shall be designed:

4.1.1 To allow at least 0.3 m (1 ft) of freeboard above design working head. The stand height above the centerline of the pipeline shall be such that neither the static head nor the design working head plus freeboard shall exceed the head class of the pipe.

4.1.2 With the top of each stand at least 1.2 m (4 ft) above the ground surface, except for surface gravity inlets, which shall be equipped with trash racks and covers.

4.1.3 With downward water velocities in stands not to exceed 0.6 m/s (2 ft/s). The inside diameter of the stand shall not be less than the inside diameter of the pipeline.

4.2 Pump stands. When the water velocity of an inlet exceeds three times the velocity of the outlet, the centerline of the inlet shall have a minimum vertical offset from the centerline of the outlet at least equal to the sum of the diameters of the inlet and outlet pipes. The cross-sectional area of the stands may be reduced above a point 0.3 m (1 ft) above the top of the upper inlet, but in no case shall the reduced cross section be such that it would produce an average velocity of more than 3 m/s (10 ft/s) if the entire flow was discharging through it.

4.2.1 Types. Pump stands shall be one of the following types:

4.2.1.1 Steel cylinder stands.

4.2.1.2 Concrete box stands with vertical sides, suitably reinforced.

4.2.1.3 Nontapered stands of concrete pipe, suitably reinforced.

4.2.1.4 Nontapered stands of concrete pipe, capped and having a vent pipe of a height exceeding the hydraulic gradeline plus freeboard.

4.2.2 Vibration control. Construction shall insure that the vibration from the pump discharge is not transmitted to the stand. Vibration control also applies to low-head pipelines not open to the atmosphere when pump stands are used.

4.3 Sand traps. Sand traps, when combined with a stand, shall have a minimum inside dimension of 762 mm (30 in.) and shall be constructed so that the bottom is at least 610 mm (24 in.) below the invert of the outlet pipeline. The downward velocity of flow of the water in a sand trap shall not exceed 0.08 m/s (0.25 ft/s). Suitable provision for cleaning sand traps shall be provided.

4.4 Gate stands and float valve stands

4.4.1 Gate stands. Gate stands shall be of sufficient dimension to accommodate the gate or gates, and shall be large enough to make the gates accessible for repair.

4.4.2 Float valve stands. Float valve stands shall be large enough to provide accessibility for maintenance and to dampen surge.

4.5 Vent requirements. Vents shall be designed into the system to provide for the removal of air and protection from surge.

4.5.1 Vents shall have a minimum freeboard of 0.3 m (1 ft) above the hydraulic gradeline. The maximum height of the vent above the centerline of the pipeline must not exceed the working head class of the pipe.

4.5.2 Vents shall have a cross-sectional area of at least one-half the cross-sectional area of the pipeline (both inside measurements) for a distance of at least one pipeline diameter up from the centerline of the pipeline. Above this elevation the vent may be reduced to 51 mm (2 in.) in diameter.

4.5.3 Vents shall be located as follows:

4.5.3.1 At the downstream end of each lateral.

4.5.3.2 At all summits of the line.

4.5.3.3 At points where there are changes in grade of more than 10 deg (18 percent) in a downward direction of flow.

4.5.3.4 Immediately below any stand if the downward velocity in the stands exceeds 0.6 m/s (2 ft/s).

4.6 Air-vacuum release valves

4.6.1 An air-vacuum release valve may be used in lieu of an open vent, but either a vent or an air-vacuum release valve shall be provided at each of the locations listed in paragraph 4.5.3.

4.6.2 Air-vacuum release valve outlets shall have a 51 mm (2 in.) minimum diameter. The valves shall be used as follows:

Pipe diameter mm (in.)	Minimum air-vacuum release valve outlet diameter mm (in.)
152 (6) or less	51 (2)
178-254 (7-10)	76 (3)
305 (12) or larger	102 (4)

NOTE: Air-vacuum release valves shall not replace the open stand required in paragraph 4.1.

SECTION 5—PIPE MATERIALS

5.1 Compounds. This Standard covers pipe made from the compounds that are listed and identified in this section by code classification and that are further defined and identified by hydrostatic design stress rating. The respective pipe compound shall have an established long term hydrostatic design stress rating as given in Table 5 when tested in accordance with paragraph 5.1.1. The compound shall meet the short term test requirement denoted by its code classification and defined in the relevant American Society for Testing and Materials Standards referenced in paragraph 5.2.

5.1.1 Sustained pressure. The pipe shall not fail, balloon, burst, or weep as defined in Section 4 of ASTM Standard D1598, Test for Time-to-Failure of Plastic Pipe Under Long-Term Hydrostatic Pressure. The pipe shall be treated in accordance with the following section of the applicable ASTM Standard:

5.1.1.1 PVC: Section 7.5 of ASTM Standard D2241, Specifications for PVC Plastic Pipe, at the appropriate test pressure given in Table 3 of that specification or Table 6 of this Standard.

TABLE 5 — MAXIMUM HYDROSTATIC DESIGN STRESS FOR THERMOPLASTIC PIPE

Compound	ASTM code classification	Type, grade	Standard code designation*	Hydrostatic design stress† psi	Hydrostatic design stress† MPa
PVC	12454-B	I, 1	PVC 1120	2000	13.8
PVC	12454-C	I, 2	PVC 1220	2000	13.8
PVC	14333-D	II, 1	PVC 2120	2000	13.8
PVC	14333-D	II, 1	PVC 2116	1600	11.0
PVC	14333-D	II, 1	PVC 2112	1250	8.6
PVC	14333-D	II, 1	PVC 2110	1000	6.9
PE	IVC-P34	III, 4	PE 3408	800	5.5
PE	IVC-P34	III, 4	PE 3406	630	4.3
PE	IIIC-P33	III, 3	PE 3306	630	4.3
PE	IIC-P23	II, 3	PE 2306	630	4.3
PE	IIC-P23	II, 3	PE 2305	500	3.4
ABS	3-5-5	I, 3	ABS 1316	1600	11.0
ABS	4-4-5	II, 1	ABS 2112	1250	8.6
ABS	5-2-2	I, 2	ABS 1210	1000	6.9

*Applies to compounds for pressure pipe

†Hydrostatic design stress = $\frac{\text{long-term hydrostatic strength}^{\ddagger}}{2.0}$

‡Long-term hydrostatic strength is determined by ASTM Standard D1598, Test for Time-to-Failure of Plastic Pipe Under Long-Term Hydrostatic Pressure, and ASTM Standard D2837, Obtaining Hydrostatic Design Basis for Thermoplastic Pipe Materials.

NOTE: Recommended design stress values are issued by the Plastics Pipe Institute, New York, NY and are reissued periodically. Design stress values were issued in Technical Report TR-4, 1982.

TABLE 6—SUSTAINED PRESSURE TEST CONDITIONS FOR PVC PLASTIC PIPE*†

SDR (DR)	Pressure required for test‡							
	PVC 1120 PVC 1220 PVC 2120		PVC 2116		PVC 2112		PVC 2110	
	psi	kPa	psi	kPa	psi	kPa	psi	kPa
51	170	1170	135	930	115	795	90	620
64	135	930	105	725	90	620	75	515
81	105	725	85	585	70	485	60	415
93.5	90	620						
50 ft head	83	570						

*Requirements in addition to those listed in ASTM Standard D2241, Specification for Poly (Vinyl Chloride) (PVC) and Chlorinated Poly (Vinyl Chloride) (CPVC) Plastic Pipe (SDR-PR), for SDR rated PVC plastic pipe.
†With water at 23 °C (73.4 °F).
‡The fiber stresses used to derive the test pressures are as follows:

	psi	MPa
PVC 1120, PVC 1220, PVC 2120	4200	29.0
PVC 2116	3360	23.2
PVC 2112	2800	19.3
PVC 2110	2300	15.9

5.1.1.2 PE: Section 7.7 of ASTM Standard D2239, Specifications for Polyethylene Plastic Pipe, at the appropriate test pressures given in Table 3 of that specification.

5.1.1.3 ABS: Section 7.4 of ASTM Standard D2282, Specifications for ABS Plastic Pipe, at the appropriate test pressures given in Table 3 of that specification.

NOTE: Tests of pipe made with different diameters and wall thicknesses but with the same material shall not be required to reestablish long-term hydrostatic design rating since this is a compound qualifying test.

5.2 Compound code classification

5.2.1 PVC: ASTM Standard D1784-12454 B (Type I, Grade 1)
12454 C (Type I, Grade 2)
14333 D (Type II, Grade 1)

5.2.2 PE: ASTM Standard D1248-II C-P23 (Type II, Grade 3, Class C)
III C-P33 (Type III, Grade 3, Class C)
IV C-P34 (Type III, Grade 4, Class C)

5.2.3 ABS: ASTM Standard D1788-5-2-2 (Type I, Grade 2)
3-5-5 (Type I, Grade 3)
4-4-5 (Type II, Grade 1)

5.3 Rework materials. Clean rework material generated from the manufacturer's own pipe production may be used by the same manufacturer, as long as the pipe produced meets all the requirements of this Standard.

5.4 Physical requirements

5.4.1 Workmanship. The pipe shall be homogeneous throughout and free from visible cracks, holes, foreign inclusions, or other defects. The pipe shall be as uniform as commercially practicable in color, opacity, density and other physical properties.

5.4.2 Dimensions and tolerances

5.4.2.1 Wall thickness. The wall thickness and tolerances shall be determined in accordance with the appropriate sections of ASTM Standard D2122, Determining Dimensions of Thermoplastic Pipe and Fittings, and shall be as shown in Tables 7, 8 and 9 of this Standard.

5.4.2.2 Diameters. The outside diameter or inside diameter of the pipe shall be determined in accordance with the appropriate sections of ASTM Standard D2122, Determining Dimensions of Thermoplastic Pipe and Fittings, and shall be as shown in Tables 10 and 11 of this Standard.

5.5 PVC pipe requirements

5.5.1 Burst pressure. The minimum burst pressure shall be determined in accordance with Section 7.5 of ASTM Standard D2241, Specifications for PVC Plastic Pipe, and as given in Table 4 of ASTM Standard D2241 or Table 12 of this Standard.

5.5.2 Flattening. There shall be no evidence of splitting, cracking, or breaking when the pipe is tested in accordance with Section 7.6 of ASTM Standard D2241, Specification for PVC Plastic Pipe.

TABLE 7—WALL THICKNESS AND TOLERANCE IN MILLIMETERS (INCHES) FOR PVC AND ABS PIPE: OD CONTROLLED

Nominal pipe size in.	mm *		50 ft head	DR 93.5	SDR 81	SDR 64	SDR 51	SDR 41	SDR 32.5	SDR 26	SDR 21	SDR 17	SDR 13.5
1/2	21	IPS											1.57 + 0.51 (0.062 + 0.020)
3/4	27	IPS										1.57 + 0.51 (0.062 + 0.020)	1.98 + 0.51 (0.078 + 0.020)
1	33	IPS									1.60 + 0.51 (0.063 + 0.020)	1.96 + 0.51 (0.077 + 0.020)	2.46 + 0.51 (0.097 + 0.020)
1 1/4	42	IPS								1.63 + 0.51 (0.064 + 0.020)	2.00 + 0.51 (0.079 + 0.020)	2.49 + 0.51 (0.098 + 0.020)	3.12 + 0.51 (0.123 + 0.020)
1 1/2	48	IPS								1.85 + 0.51 (0.073 + 0.020)	2.29 + 0.51 (0.090 + 0.020)	2.84 + 0.51 (0.112 + 0.020)	3.58 + 0.51 (0.141 + 0.020)
2	60	IPS							1.85 + 0.51 (0.073 + 0.020)	2.31 + 0.51 (0.091 + 0.020)	2.87 + 0.51 (0.113 + 0.020)	3.56 + 0.51 (0.140 + 0.020)	4.47 + 0.53 (0.176 + 0.021)
2 1/2	73	IPS							2.11 + 0.51 (0.083 + 0.020)	2.79 + 0.51 (0.110 + 0.020)	3.48 + 0.51 (0.137 + 0.020)	4.29 + 0.51 (0.169 + 0.020)	5.41 + 0.66 (0.213 + 0.026)
3	89	IPS							2.74 + 0.51 (0.108 + 0.020)	3.43 + 0.51 (0.135 + 0.020)	4.24 + 0.51 (0.167 + 0.020)	5.23 + 0.64 (0.206 + 0.025)	6.58 + 0.79 (0.259 + 0.031)
3 1/2	102	IPS							3.12 + 0.51 (0.123 + 0.020)	3.91 + 0.51 (0.154 + 0.020)	4.83 + 0.58 (0.190 + 0.023)	5.97 + 0.71 (0.235 + 0.028)	7.52 + 0.91 (0.296 + 0.036)
4	114	IPS				1.79 + 0.51 (0.070 + 0.020)		2.79 + 0.51 (0.110 + 0.020)	3.51 + 0.51 (0.138 + 0.020)	4.39 + 0.53 (0.173 + 0.021)	5.44 + 0.66 (0.214 + 0.026)	6.73 + 0.81 (0.265 + 0.032)	8.46 + 1.02 (0.333 + 0.040)
	105	PIP	1.65 + 0.51 (0.065 + 0.020)		1.65 + 0.51 (0.065 + 0.020)		2.06 + 0.51 (0.081 + 0.020)	2.57 + 0.51 (0.101 + 0.020)					
5	141	IPS						3.45 + 0.51 (0.136 + 0.020)	4.34 + 0.53 (0.171 + 0.021)	5.44 + 0.69 (0.214 + 0.021)	6.73 + 0.81 (0.265 + 0.032)	8.31 + 0.99 (0.327 + 0.039)	10.46 + 1.24 (0.412 + 0.049)
6	168	IPS				2.64 + 0.51 (0.104 + 0.020)		4.11 + 0.51 (0.162 + 0.020)	5.18 + 0.61 (0.204 + 0.024)	6.48 + 0.79 (0.255 + 0.031)	8.03 + 0.97 (0.316 + 0.038)	9.91 + 1.19 (0.390 + 0.047)	12.47 + 1.50 (0.491 + 0.059)
	156	PIP	1.78 + 0.51 (0.070 + 0.020)	1.78 + 0.51 (0.070 + 0.020)	1.93 + 0.51 (0.076 + 0.020)		3.05 + 0.51 (0.120 + 0.020)	3.81 + 0.51 (0.150 + 0.020)					
8	219	IPS				3.43 + 0.51 (0.135 + 0.020)		5.33 + 0.64 (0.210 + 0.025)	6.73 + 0.81 (0.265 + 0.032)	8.43 + 1.02 (0.332 + 0.040)	10.41 + 1.24 (0.410 + 0.049)	12.90 + 1.55 (0.508 + 0.061)	
	207	PIP	2.03 + 0.51 (0.080 + 0.020)	2.21 + 0.51 (0.087 + 0.020)	2.57 + 0.51 (0.101 + 0.020)		4.06 + 0.51 (0.160 + 0.020)	5.05 + 0.61 (0.199 + 0.024)					
10	273	IPS				4.27 + 0.51 (0.168 + 0.020)		6.65 + 0.79 (0.262 + 0.031)	8.41 + 1.02 (0.331 + 0.040)	10.49 + 1.27 (0.413 + 0.050)	12.98 + 1.55 (0.511 + 0.061)	16.05 + 1.93 (0.632 + 0.076)	
	259	PIP	2.54 + 0.51 (0.100 + 0.020)	2.77 + 0.51 (0.109 + 0.020)	3.20 + 0.51 (0.126 + 0.020)		5.08 + 0.61 (0.200 + 0.024)	6.32 + 0.76 (0.249 + 0.030)					
12	324	IPS				5.05 + 0.61 (0.199 + 0.024)		7.90 + 0.94 (0.311 + 0.037)	9.96 + 1.19 (0.392 + 0.047)	12.45 + 1.50 (0.490 + 0.059)	15.39 + 1.85 (0.606 + 0.073)	19.05 + 2.29 (0.750 + 0.090)	
	311	PIP	3.05 + 0.51 (0.120 + 0.020)	3.33 + 0.51 (0.131 + 0.020)	3.84 + 0.51 (0.151 + 0.020)		6.10 + 0.74 (0.240 + 0.029)	7.59 + 0.91 (0.299 + 0.036)					
14	363	PIP	3.56 + 0.51 (0.140 + 0.020)	3.89 + 0.51 (0.153 + 0.020)	4.50 + 0.53 (0.177 + 0.021)		7.11 + 0.86 (0.280 + 0.034)	8.89 + 1.02 (0.350 + 0.040)	11.20 + 1.27 (0.441 + 0.050)	13.97 + 1.78 (0.550 + 0.070)	17.30 + 1.90 (0.681 + 0.075)	21.36 + 2.41 (0.841 + 0.095)	26.92 + 3.18 (1.060 + 0.125)
15	389	PIP	3.81 + 0.51 (0.150 + 0.020)	4.17 + 0.51 (0.164 + 0.020)	4.80 + 0.58 (0.189 + 0.023)		7.62 + 0.91 (0.300 + 0.036)	9.47 + 1.14 (0.373 + 0.045)					
16	406	IPS	4.06 + 0.51 (0.160 + 0.020)	4.34 + 0.52 (0.171 + 0.021)	5.03 + 0.61 (0.198 + 0.024)		7.98 + 0.96 (0.314 + 0.038)	9.91 + 1.19 (0.390 + 0.047)	12.50 + 1.50 (0.492 + 0.059)	15.62 + 1.87 (0.615 + 0.074)			
18	466	IP		5.00 + 0.61 (0.197 + 0.024)	5.77 + 0.69 (0.227 + 0.027)		9.14 + 1.09 (0.360 + 0.043)	11.38 + 1.37 (0.448 + 0.054)	14.35 + 1.72 (0.565 + 0.068)				
18	475	PIP		5.08 + 0.61 (0.200 + 0.024)	5.87 + 0.71 (0.231 + 0.028)		9.32 + 1.12 (0.367 + 0.044)	11.58 + 3.23 (0.456 + 0.127)	14.61 + 1.75 (0.575 + 0.069)				
20	518	IP		5.54 + 0.66 (0.218 + 0.026)	6.40 + 0.76 (0.252 + 0.030)		10.16 + 1.22 (0.400 + 0.048)	12.65 + 1.52 (0.498 + 0.060)	15.95 + 1.91 (0.628 + 0.075)				
21	560	PIP		5.99 + 0.71 (0.236 + 0.028)	6.91 + 0.84 (0.272 + 0.033)		10.97 + 1.32 (0.432 + 0.052)	13.66 + 3.81 (0.538 + 0.150)	17.22 + 2.06 (0.678 + 0.081)				
24	610	IPS		6.53 + 0.79 (0.257 + 0.031)	7.52 + 0.90 (0.296 + 0.036)		11.96 + 1.44 (0.471 + 0.057)	14.86 + 1.78 (0.585 + 0.070)	18.75 + 2.25 (0.738 + 0.089)				
24	630	PIP		6.76 + 0.81 (0.266 + 0.032)	7.77 + 0.94 (0.306 + 0.037)		12.34 + 1.47 (0.486 + 0.058)	15.37 + 4.29 (0.605 + 0.169)	19.38 + 2.34 (0.763 + 0.092)				
27	710	PIP					13.92 + 1.68 (0.548 + 0.066)	17.32 + 4.83 (0.682 + 0.190)	21.84 + 2.62 (0.860 + 0.103)				

*Pipe is not currently available in metric sizes in the United States except for 475, 560, 630 and 710 mm sizes.

TABLE 8—WALL THICKNESS AND TOLERANCE IN MILLIMETERS (INCHES) FOR PE PIPE: ID CONTROLLED

Nominal pipe size in	mm	SDR 15	SDR 11.5	SDR 9	SDR 7	SDR 5.3
½	16	1.52 + 0.51 (0.060 + 0.020)	1.52 + 0.51 (0.060 + 0.020)	1.75 + 0.51 (0.069 + 0.020)	2.26 + 0.51 (0.089 + 0.020)	2.97 + 0.51 (0.117 + 0.020)
¾	21	1.52 + 0.51 (0.060 + 0.020)	1.83 + 0.51 (0.072 + 0.020)	2.34 + 0.51 (0.092 + 0.020)	3.00 + 0.51 (0.118 + 0.020)	3.94 + 0.51 (0.155 + 0.020)
1	27	1.78 + 0.51 (0.070 + 0.020)	2.31 + 0.51 (0.091 + 0.020)	2.97 + 0.51 (0.117 + 0.020)	3.81 + 0.51 (0.150 + 0.020)	5.03 + 0.61 (0.198 + 0.024)
1¼	35	2.34 + 0.51 (0.092 + 0.020)	3.05 + 0.51 (0.120 + 0.020)	3.89 + 0.51 (0.153 + 0.020)	5.00 + 0.61 (0.197 + 0.024)	6.60 + 0.79 (0.260 + 0.031)
1½	41	2.72 + 0.51 (0.107 + 0.020)	3.56 + 0.51 (0.140 + 0.020)	4.55 + 0.51 (0.179 + 0.020)	5.84 + 0.71 (0.230 + 0.028)	7.72 + 0.91 (0.304 + 0.036)
2	52	3.50 + 0.51 (0.138 + 0.020)	4.57 + 0.56 (0.180 + 0.022)	5.84 + 0.71 (0.230 + 0.028)	7.49 + 0.89 (0.295 + 0.035)	0.91 + 1.19 (0.390 + 0.047)
2½	63	4.19 + 0.51 (1.65 + 0.020)	5.46 + 0.64 (0.215 + 0.025)			
3	78	5.21 + 0.51 (0.205 + 0.020)	6.78 + 0.81 (0.267 + 0.032)			
4	102	6.81 + 0.81 (0.268 + 0.032)	8.89 + 0.107 (0.350 + 0.042)			
6	154	10.26 + 1.22 (0.404 + 0.048)	13.38 + 1.60 (0.527 + 0.063)			

TABLE 9—DIAMETERS, WALL THICKNESSES AND TOLERANCES FOR PE PIPE IN IPS SIZING SYSTEM, OD CONTROLLED.*

Nominal	Pipe outside diameter: Actual outside diameter		Tolerance (±) on average OD		Minimum wall thickness†: SDR 32.5		SDR 26		SDR 21		SDR 17		SDR 11	
in.	in.	mm	in.	mm	in.	mm	in.	mm	in.	mm	in.	mm	in.	mm
3	3.500	88.9	0.016	0.41	0.108	2.7	0.135	3.4	0.167	4.2	0.206	5.2	0.318	8.1
4	4.500	114.3	0.020	0.51	0.138	3.5	0.173	4.4	0.214	5.4	0.265	6.7	0.409	10.4
5	5.563	141.3	0.025	0.64	0.171	4.3	0.214	5.4	0.265	6.7	0.327	8.3	0.506	12.8
6	6.625	168.3	0.030	0.76	0.204	5.2	0.255	6.5	0.315	8.0	0.390	9.9	0.602	15.3
8	8.625	219.1	0.039	0.99	0.265	6.7	0.332	8.4	0.411	10.4	0.507	12.9	0.784	19.9
10	10.750	273.1	0.048	1.22	0.331	8.4	0.413	10.5	0.512	13.0	0.632	16.0	0.977	24.8
12	12.750	323.8	0.057	1.45	0.392	10.0	0.490	12.4	0.607	15.4	0.750	19.0	1.159	29.5
14	14.000	355.6	0.063	1.60	0.431	10.9	0.538	13.7	0.667	16.9	0.824	20.9	1.273	32.3
16	16.000	406.4	0.072	1.83	0.492	12.5	0.615	15.6	0.762	19.4	0.941	23.9	1.455	37.0
18	18.000	457.2	0.081	2.06	0.554	14.2	0.692	17.6	0.857	21.8	1.059	26.9	1.636	41.6
20	20.000	508.0	0.090	2.27	0.615	15.6	0.769	19.5	0.952	24.2	1.176	29.9	—	—
22	22.000	558.8	0.099	2.51	0.677	17.2	0.846	21.5	1.048	26.6	1.294	32.9	—	—
24	24.000	609.6	0.108	2.74	0.738	18.7	0.923	23.4	1.143	29.0	1.412	35.9	—	—

*OD based PE pipe is also made in the metric sizing system, based on ISO 161/1-1978, Thermoplastics Pipes for the Transport of Fluids — Nominal Outside Diameters and Nominal Pressures — Part 1: Metric Series. Specifications for this pipe can be obtained from the Plastic Pipe Institute, New York, NY.

†Wall thickness variability in any diametrical cross section shall not exceed 12% when calculated in accordance with ASTM Standard D2122, Determining Dimensions for Thermoplastic Pipe and Fittings.

TABLE 10—OUTSIDE DIAMETER AND TOLERANCE FOR PVC AND ABS PIPE

					TOLERANCES(±): For average OD		For maximum & minimum (out-of-round): 50 ft head DR 93.5 SDR 81 SDR 64 SDR 51		SDR 41 SDR 32.5 SDR 26 SDR 21 / SDR 17 SDR 13.5	
Nominal in.	Pipe size mm		Average OD in.	mm	in.	mm	in.	mm	in.	mm
½	21	IPS	0.840	21.34	0.004	0.10	0.015	0.38	0.008	0.20
¾	27	IPS	1.050	26.67	0.004	0.10	0.015	0.38	0.010	0.25
1	33	IPS	1.315	33.40	0.005	0.13	0.015	0.38	0.010	0.25
1¼	42	IPS	1.660	42.16	0.005	0.13	0.015	0.38	0.012	0.30
1½	48	IPS	1.900	48.26	0.006	0.15	0.030	0.76	0.012	0.30
2	60	IPS	2.375	60.32	0.006	0.15	0.030	0.76	0.012	0.30
2½	73	IPS	2.875	73.02	0.007	0.18	0.030	0.76	0.015	0.38
3	89	IPS	3.500	88.90	0.008	0.20	0.030	0.76	0.015	0.38
3½	102	IPS	4.000	101.60	0.008	0.20	0.050	1.27	0.015	0.38
4	114	IPS	4.500	114.30	0.009	0.23	0.050	1.27	0.015	0.38
	105	PIP	4.130	104.90	0.009	0.23	0.050	1.27	0.015	0.38
5	141	IPS	5.563	141.30	0.010	0.25	0.050	1.27	0.030	0.76
6	168	IPS	6.625	168.28	0.011	0.28	0.050	1.27	0.035	0.89
	156	PIP	6.140	155.96	0.011	0.28	0.050	1.27	0.030	0.76
8	219	IPS	8.625	219.08	0.015	0.38	0.075	1.90	0.045	1.14
	207	PIP	8.160	207.26	0.015	0.38	0.075	1.78	0.042	1.07
10	273	IPS	10.750	273.05	0.015	0.38	0.075	1.90	0.050	1.27
	259	PIP	10.200	259.08	0.015	0.38	0.075	1.90	0.050	1.27
12	324	IPS	12.750	323.85	0.015	0.38	0.075	1.90	0.060	1.52
	311	PIP	12.240	310.90	0.018	0.46	0.075	1.90	0.060	1.52
14	363	PIP	14.280	362.71	0.021	0.53	0.075	1.90	0.070	1.78
15	389	PIP	15.300	388.62	0.023	0.58	0.075	1.90	0.075	1.90
16	406	IPS	16.000	406.40	0.024	0.61	0.075	1.90		
18	466	IP	18.360	466.34	0.027	0.69	0.100	2.50		
18	475	PIP	18.701	475.00	0.028	0.71	0.100	2.50		
20	518	IP	20.400	518.16	0.030	0.76	0.100	2.50		
21	560	PIP	22.047	560.00	0.033	0.84	0.100	2.50		
24	610	IPS	24.000	609.6	0.036	0.91	0.125	2.50		
24	630	PIP	24.803	630.00	0.037	0.94	0.125	2.50		
27	710	PIP	27.953	710.00	0.047	1.19	0.125	2.50		

TABLE 11—INSIDE DIAMETER AND TOLERANCE FOR PE PIPE

Nominal pipe size in.	mm		ID in.	mm	Tolerance in.	mm
½	16	IPS	0.622	15.80	±0.010	0.25
¾	21	IPS	0.824	20.93	+0.010	0.25
					−0.015	0.38
1	27	IPS	1.049	26.64	+0.010	0.25
					−0.020	0.51
1¼	35	IPS	1.380	35.05	+0.010	0.25
					−0.020	0.51
1½	41	IPS	1.610	40.89	+0.010	0.25
					−0.020	0.51
2	52	IPS	2.067	52.50	+0.015	0.38
					−0.020	0.51
2½	63	IPS	2.469	62.71	+0.015	0.38
					−0.025	0.64
3	78	IPS	3.068	77.93	+0.015	0.38
					−0.030	0.76
4	102	IPS	4.026	102.26	+0.015	0.38
					−0.035	0.89
	102	PIP	4.000	101.6	+0.020	0.51
					−0.020	0.51
6	154	IPS	6.065	154.05	+0.020	0.51
					−0.035	0.89
	152	PIP	6.000	152.4	+0.025	0.64
					−0.025	0.64
8	203	PIP	8.000	203.2	±0.040	1.02
10	254	PIP	10.000	254.0	±0.040	1.02
12	305	PIP	12.000	304.8	±0.040	1.02
15	381	PIP	15.000	381.0	±0.040	1.02

TABLE 12—BURST PRESSURE REQUIREMENTS FOR PVC PLASTIC PIPE*†

SDR (DR)	Minimum burst pressure‡			
	PVC 1120 PVC 1220 PVC 2120		PVC 2116 PVC 2112 PVC 2110	
	psi	kPa	psi	kPa
51	260	1790	200	1380
64	200	1380	160	1105
81	160	1105	125	860
93.5	140	965		
50 ft head	127	875		

*Requirements in addition to those listed in ASTM Standard D-2241, Specification for Poly(Vinyl Chloride)(PVC) and Chlorinated Poly(Vinyl Chloride) (CPVC) Plastic Pipe (SDR-PR), for SDR rated PVC plastic pipe.
†With water at 23 °C (73.4 °F).
‡The fiber stresses used to derive the test pressures are as follows:

	psi	MPa
PVC 1120, PVC 1220, PVC 2120	6400	44.1
PVC 2116, PVC 2112, PVC 2110	5400	34.5

5.5.3 Extrusion quality. The pipe shall not flake or disintegrate when tested in accordance with ASTM Standard D2152, Test for Quality of Extruded PVC Pipe by Acetone Immersion.

5.5.4 Impact resistance. The pipe shall be tested in accordance with ASTM Standard D2444, Test for Impact Resistance of Thermoplastic Pipe and Fittings by Means of a Tup (Falling Weight), using a 89 N (20 lb), Type B tup with a flat plate at 23 ± 2 °C (73.4 ± 3.6 °F) and shall meet the test levels shown in Table 13 of this Standard. The impact test shall be made on new production pipe at the time of manufacture.

5.6 PE pipe requirements

5.6.1 Thickness of outer layer. For pipe produced by simultaneous multiple extrusion, that is, pipe containing two or more concentric layers, the outer layer shall be at least 0.51 mm (0.020 in.) thick.

5.6.2 Bond. For pipe produced by simultaneous multiple extrusion, the bond between the layers shall be strong and uniform. It shall not be possible to separate any two layers with a probe or a point of a knife blade so that the layers separate cleanly at any point.

5.6.3 Carbon black. The pipe extrusion compound shall contain at least 2 percent carbon black when tested in accordance with Section 7.5 of ASTM Standard D2239, Specifications for Polyethylene (PE) Plastic Pipe. For pipe produced by simultaneous multiple extrusion, this requirement shall apply to the outer layer.

5.6.4 Density. The polyethylene base resin (uncolored PE) in the pipe compound shall have a density in the range from 0.926 to 0.940 Mg/m^3 for pipe made from Grade P23 and 0.941 to 0.965 Mg/m^3 for pipe made from Grade P33 and Grade P34 of ASTM Standard D1248, Specifications for Polyethylene Plastic Molding and Extrusion Materials, when determined in accordance with Section 7.6 of ASTM Standard D2239, Specifications for Polyethylene Plastic Pipe.

5.6.5 Burst pressure. The minimum burst pressure for PE plastic pipe shall be determined in accordance with Section 7.8 and Table 4 of ASTM Standard D2239, Specifications for Polyethylene Plastic Pipe.

TABLE 13—IMPACT REQUIREMENTS FOR PVC AND ABS PIPE*

Nominal pipe size in.	mm		50 ft. head ft·lbf	Nm	DR 93.4 ft·lbf	Nm	SDR 81 ft·lbf	Nm	SDR 64 ft·lbf	Nm	SDR 51 ft·lbf	Nm	SDR 41 ft·lbf	Nm	SDR 32.5 ft·lbf	Nm	SDR 26 ft·lbf	Nm	SDR 21 ft·lbf	Nm	SDR 17 ft·lbf	Nm	SDR 13.5 ft·lbf	Nm
½	21	IPS																					25	35
¾	27	IPS																			36	50	36	50
1	33	IPS																	38	50	38	50	38	50
1¼	42	IPS															50	70	50	70	50	70	50	70
1½	48	IPS													50	70	50	70	50	70	50	70	50	70
2	60	IPS													63	85	63	85	63	85	63	85	63	85
2½	73	IPS													63	85	63	85	63	85	63	85	63	85
3	89	IPS													75	100	75	100	75	100	75	100	75	100
3½	102	IPS													75	100	75	100	75	100	75	100	75	100
4	114	IPS							50	70					80	110	100	135	100	135	100	135	100	135
	105	PIP	30	40			30	40			70	95	70	95										
5	141	IPS													90	120	110	150	110	150	110	150	110	150
6	168	IPS							60	80					100	135	120	165	120	165	120	165	120	165
		PIP	30	40	30	40	30	40			80	110	80	110										
8	219	IPS							70	95					100	135	130	175	130	175	130	175	130	175
	207	PIP	30	40	30	40	30	40			90	120	90	120										
10	273	IPS							80	110					110	150	140	190						
	259	PIP	60	80	60	80	60	80			100	135	100	135										
12	324	IPS							100	135					110	150	150	205						
	311	PIP	100	135	100	135	100	135			110	150	110	150										
14	363	PIP	110	150	100	135	110	150			120	165	120	165	120	165	150	205	150	205	150	205	150	205
15	389	PIP	110	150	110	150	110	150			120	165	120	165										
16	406	IPS	110	150	110	150	110	150			120	165	150	205	120	165	150	205						
18	466,475	IP,PIP			110	150	110	150			120	165	150	205	150	205								
20	518	IP			110	150	110	150			120	165	150	205	150	205								
21	560	PIP											150	205										
24	610,630	IPS,PIP			110	150	110	150			120	165	150	205	150	205								
27	710	PIP											150	205										

*When tested in accordance with ASTM Standard D2444, Test for Impact Resistance of Thermoplastic Pipe and Fittings by Means of a Tup (Falling Weight, using an 89 N (20 lb) type B tup with a flat plate at 23 °C (73.4 °F) on new production pipe at the time of manufacture.

5.6.6 **Environmental stress cracking.** There shall be no loss of pressure in the pipe when tested in accordance with Section 7.9 of ASTM Standard D2239, Specifications for Polyethylene Plastic Pipe.

5.7 **ABS pipe requirements**

5.7.1 **Burst pressure.** The minimum burst pressure shall be determined in accordance with Section 7.6 and Table 4 of ASTM Standard D2282, Specifications for ABS Plastic Pipe.

5.7.2 **Impact resistance.** The pipe shall be tested in accordance with ASTM Standard D2444, Test for Impact Resistance of Thermoplastic Pipe and Fittings by Means of a Tup (Falling Weight), using a 89 N (20 lb) type B tup with a flat plate at 23 ± 2 °C (73.4 ± 3.6 °F) and shall meet the test levels shown in Table 13 of this Standard. The impact test shall be made on new production pipe at the time of manufacture.

5.8 **Joints**

5.8.1 **General.** All joints shall be constructed to withstand the design maximum working pressures for the pipeline without leakage, and without internal obstruction which could reduce its capacity below design requirements, except that insert fittings for joining PE pipe are permitted. Manufacturer's recommendations for joining pipe shall be used when not in conflict with requirements of paragraph 5.8.

5.8.2 **Sockets and couplings.** The integral bell or separate coupling shall meet the same strength requirements as the pipe. When joint assembly requires use of separate couplings, one such coupling of the same class and size shall be furnished with each length of pipe.

5.8.3 **Solvent cements.** Solvent cements only for use with PVC pipe and fittings shall meet the requirements of ASTM Standard D2564, Specifications for Solvent Cements for PVC Plastic Pipe and Fittings. Solvent cements only for ABS pipe and fittings shall meet the requirements of ASTM Standard D2235, Specifications for Solvent Cement for ABS Plastic Pipe and Fittings. The pipe manufacturer should be consulted for the type of cement recommended for joining large diameter pipes. Safe handling of solvent cements shall conform to ASTM Standard F402, Recommended Practice for Safe Handling of Solvent Cements Used for Joining Thermoplastic Pipe and Fittings.

5.8.4 **Rubber gasket joints.** Rubber gasket joints shall conform to ASTM Standard D3139, Specifications for Joints for Plastic Pressure Pipes Using Flexible Elastometric Seals.

5.8.5 **Plastic risers.** Plastic risers shall have at least the same strength as the pipe, including risers with use limited to subsurface attachment.

5.9 **Fittings**

5.9.1 **General.** All fittings, such as couplings, reducers, bends, tees and crosses shall be made of material that is recommended for use with the pipe and shall be installed in accordance with the recommendations of the manufacturer. Where fittings made of steel or other materials subject to corrosion are used in the line, they shall be adequately protected by wrapping with plastic tape or by coating with high quality corrosion preventatives. Where plastic tape is used, all surfaces to be wrapped shall be thoroughly cleaned and then coated with primer compatible with the tape prior to wrapping.

5.9.2 **Requirements.** Fittings for IPS sized pipe shall meet all the dimensional and quality requirements given in the following ASTM Standards:

- ASTM Standard D2468, ABS Plastic Pipe Fittings, Socket-Type, Schedule 40
- ASTM Standard D2469, ABS Plastic Pipe Fittings, Socket-Type, Schedule 80
- ASTM Standard D2609, PE Plastic Insert Fittings
- ASTM Standard D3261, Butt Heat Fusion for PE Plastic Pipe and Tubing
- ASTM Standard D2672, Bell-End PVC Plastic Pipe
- ASTM Standard D3036, PVC Plastic Line Couplings, Socket-Type
- ASTM Standard D2466, PVC Plastic Pipe Fittings, Socket-Type, Schedule 40
- ASTM Standard D2467, PVC Plastic Pipe Fittings, Socket-Type, Schedule 80

SECTION 6—MARKING

6.1 **General.** The pipe shall be marked at intervals of not more than 5 ft (1.5 m). (The metric marking shall not be required until it becomes a national standard. In this section customary units are shown first.)

Marking shall include the following:

6.1.1 The nominal pipe size, e.g. 4 in. (114 mm).

6.1.2 The pipe OD sizing system when applicable (IPS, IP, or PIP), e.g. PIP.

6.1.3 The ASTM Standard numeric designation for sizing systems other than PIP, e.g. ASTM Standard D2241.

6.1.4 The type of plastic pipe material in accordance with the designation code, e.g. PVC 1120.

6.1.5 **Pressure rating**

6.1.5.1 **Low pressure pipe.** The pressure rating shall be shown in psi and/or in feet of head; e.g. 22 psi (152 kPa) 50 ft (15.2 m) head.

6.1.5.2 **SDR pipe.** The pressure class rating in psi for water at 73.4 °F (23 °C); e.g. 200 psi (1379 kPa) @ 73.4 (23), or the standard dimension ratio as calculated in paragraph 2.8; e.g. SDR 21, or both: e.g. 200 psi (1379 kPa) @ 73.4 (23) SDR 21.

6.1.6 The manufacturer's name or trademark and code.

6.1.7 Pipe intended for the conveyance of potable water shall also include the seal or mark of the laboratory making the evaluation for this purpose, spaced at intervals specified by the laboratory.

SECTION 7—INSTALLATION REQUIREMENTS

7.1 **General.** The thermoplastic pipe shall be installed in accordance with the manufacturer's recommendations. If these are not available, then for pipe 152 mm (6 in.) diameter or less, ASTM Standard D2774, Recommended Practice for Underground Installation of Thermoplastic Pressure Piping, or this Standard shall be followed. Recommendations in ASTM Standard D2321, Underground Installation of Flexible Thermoplastic Sewer Pipe, may also be followed.

7.2 **Trench construction**

7.2.1 **Trench bottom.** The trench bottom should be continuous, firm, relatively smooth and free of rocks or other hard objects larger than 13 mm (0.5 in.) in size. Where ledge rock, hard pan or boulders are encountered, the trench bottom shall be undercut and filled with bedding material, using sand or compacted fine-grained soils to provide a minimum depth of bed between the pipe and rock of 100 mm (4 in.). Where unstable trench bottom conditions are encountered, stabilizing methods and materials to provide adequate and permanent support shall be used.

7.2.2 **Trench width.** The width of the trench at any point below the top of the pipe should not be greater than necessary to provide adequate room for joining the pipe and compacting the initial backfill. The trench width should be sufficient to provide adequate room for joining the pipe in the trench, if this is necessary; filling and compacting the side fills; and snaking the pipe from side-to-side along the bottom of the ditch, if recommended by the pipe manufacturer. Trench widths above the top of the pipe should not be greater than 0.6 m (2 ft) wider than the pipe diameter, except that in unstable soils where sloughing or caving may occur or where required by regulations or local conditions, the sidewalls above the top of the pipe may be sloped.

7.2.2.1 **Low pressure pipe.** Maximum and minimum trench widths below the top of the pipe for low pressure pipe shall be as follows:

Pipe size		Approximate trench width minimum		maximum	
in.	mm	in.	mm	in.	mm
4	102	16	400	30	760
6	152	18	450	30	760
8	203	20	510	30	760
10	254	22	560	30	760
12	305	24	610	30	760
14	356	26	660	30	760
15	381	27	690	30	760
18	457-475	30	760	36	910
20	508	32	810	36	910
24	610-630	36	910	42	1070
27	710	40	1020	46	1170

7.2.3 **Trench depth.** The trench depth should be determined with consideration given to requirements imposed by trench bottom, pipe size and cover conditions (see paragraphs 7.2.1 and 7.7). The depth shall be sufficient to ensure placement of the top of the pipe 0.25 m (10 in.) below the frost line unless the requirements of paragraph 3.9 are satisfied.

7.2.4 **Safety.** Provisions shall be made to insure safe working conditions where unstable soil, trench depth or other conditions impose a safety hazard to personnel working in the trench.

7.3 Placement

7.3.1 **General.** Special handling and an awareness of temperature effects on thermoplastic pipe are needed to prevent permanent distortion and pipe damage when handling during unusually warm or cold weather. Prior to any backfilling beyond light backfill for shading, and prior to connecting to other facilities, the pipe shall be allowed to come to within a few degrees of the temperature it will reach after complete covering. The pipeline shall be installed to provide protection from hazards imposed by traffic crossing, farming operations, freezing temperatures, or soil cracking. If the pipe is assembled above ground, it should be lowered into the trench with care to prevent dropping or damaging the pipe or its joints. Treatment such as dragging or excessive bending which could cause excessive joint stressing, displacement or pull-out should be avoided.

7.3.2 **Deflection and bending.** The pipe shall be installed in a manner to ensure that excessive deflection in elastomeric seal joints and excessive bending of the pipe do not occur during installation. Bending stresses should be avoided and at no time should the pipe be blocked or braced to hold a bend. The pipe manufacturer should be consulted for maximum permissible deflection limits and minimum pipe bending radii.

7.3.3 **Connection to a rigid structure.** Where differential settlement could create a concentrated loading on a pipe or joint, as at the connection of a buried pipe to a rigid structure such as a stand, extra care should be taken to compact the foundation and bedding adjoining the structure. A supporting structure beneath the joint and the pipe or a flexible joint also may be used.

7.3.4 **Bell holes for rubber gasket joints.** When the pipe being installed is provided with rubber gasket joints, bell holes shall be excavated in the bedding material to allow for the unobstructed assembly of the joint. Care should be taken that the bell hole is no larger than necessary to accomplish proper joint assembly. When the joint has been made, the bell hole should be carefully filled with initial backfill material to provide adequate support of the pipe throughout its entire length.

7.4 Thrust blocking

7.4.1 **General.** Thrust blocking prevents the line from moving and is required primarily with rubber gasket joints. Unequal forces due to water pressure at changes in pipeline alignment result in thrust loads. The thrust block transfers this load from the pipe to a wider load bearing surface. Thrust blocks are required at the following locations:

7.4.1.1 Where the pipe changes the direction of the water (i.e., ties, elbows, crosses, wyes and tees).

7.4.1.2 Where the pipe size changes (i.e., reducers, reducing tees and crosses).

7.4.1.3 At the end of the pipeline (i.e., caps and plugs).

7.4.1.4 Where there is an in-line valve.

7.4.2 **Placement.** The thrust block must be formed against a solid trench wall that has been excavated by hand. Damage to the bearing surface of the trench wall may result from excavation by mechanical equipment. The size and type of thrust block depends on pipe size, line pressure, type of fitting, degree of bend and type of soil. Thrust block size can be calculated by the procedures shown in Table 14.

7.4.3 **Side thrust on curves.** An outward pressure exists on all deflections from a straight line. Generally, good soil properly tamped in sufficient to hold side thrust. If the soil is unstable, blocking should be placed against the pipe on the outside radius on each side of a gasketed coupling. Do not thrust block the coupling itself.

7.4.4 **Construction of thrust blocks.** Thrust blocks are anchors placed between the pipe or fittings and the solid trench wall. The recommended blocking is concrete having a calculated compressive strength of at least 13.8 MPa (2000 psi). The concrete mixture is one part cement, two parts washed sand and four parts gravel. Thrust blocks should be constructed so the bearing surface is in direct line with the major force created by the pipe or fitting (see Table 14). The earth bearing surface should be undisturbed with only the simplest of forms required.

7.5 Initial backfill

7.5.1 **General.** The pipe should be uniformly and continuously supported over its entire length on firm stable material. Blocking should not be used to change pipe grade or to intermittently support pipe across excavated sections.

7.5.2 **Special considerations.** Special consideration must be given to soils, backfilling, and bedding procedures for 457 mm (18 in.) diameter and larger low pressure pipe to ensure protection of the pipe under the maximum loading conditions to which it may be subjected. Special engineering design and soils analysis may be needed to determine the supportive strength of the soils intended for use as backfill.

7.5.3 All low pressure pipelines shall be water-strutted or filled with water prior to backfilling. The backfill must be compacted to the required or on adequate density for all low pressure pipe. Either the water packing method or hand or mechanical backfilling methods may be used for backfill consolidation.

7.5.3.1 **Water packing.** When water packing is used, the pipeline must first be filled with water, all air removed, and the pipe kept full during the backfill operation. The initial backfill material shall be as specified in paragraph 7.5.3.2. The backfill, before wetting, shall be 300 to 450 mm (12 to 18 in.) deep over the top of the pipe. Water packing is accomplished by adding water in such quantity as to thoroughly saturate the initial backfill. While saturated, rods, shovels, concrete vibrators or other means may be used to help consolidate the backfill around the pipe, taking care not to float the pipe. After saturation, the pipeline shall remain full until after final backfill is made. The wetted fill shall be allowed to dry until firm enough to walk on before final backfill is begun.

7.5.3.2 **Hand or mechanical backfilling.** The initial backfill in contact with the pipe and immediately surrounding it shall be of fine-grained material free from rocks, stones, or clods greater than approximately 19 mm (0.75 in.) diameter and earth clods greater than approximately 50 mm (2 in.) diameter. The backfill shall be tamped in layers not to exceed 150 mm (6 in.) lift and compacted firmly around the pipe and up to at least 152 mm (6 in.) above the top of the pipe. The backfill material shall be sufficiently damp to permit thorough compaction under and on each side of the pipe to provide support free from voids. Care should be taken to avoid deforming, displacing, or damaging the pipe during this phase of the operation.

7.6 Final backfill

7.6.1 After pipeline testing, final backfill shall be placed and spread in approximately uniform layers in such a manner as to fill the trench completely so that there will be no unfilled spaces under or around rocks or lumps of earth in the backfill. Final backfill shall be free of large rocks, frozen clods and other debris greater than 75 mm (3 in.) in diameter.

7.6.2 Rolling equipment or heavy tampers should *not* be used to consolidate the final backfill until *after* the minimum depth of cover has been placed and then only with pipe having wall thicknesses greater than that of SDR-41.

7.7 Minimum depth of cover

7.7.1 **General.** At low places on the ground surface, extra fill may be placed over the pipeline to provide the minimum depth of cover. In such cases, the top width of the fill shall be no less than 3 m (10 ft) and the side slope no steeper than 4 horizontal to 1 vertical. The minimum depth shall be as follows:

Pipe size mm (in.)	Minimum depth of cover mm (in.)
13-64 (0.5-2.5)	460 (18)
76-102 (3-4)	610 (24)
>102 (>4)	760 (30)

7.7.2 **Minimum cover for load applications.** At least 760 mm (30 in.) cover over the top of the pipe shall be provided before the trench is wheel-loaded for both low pressure and high pressure pipe.

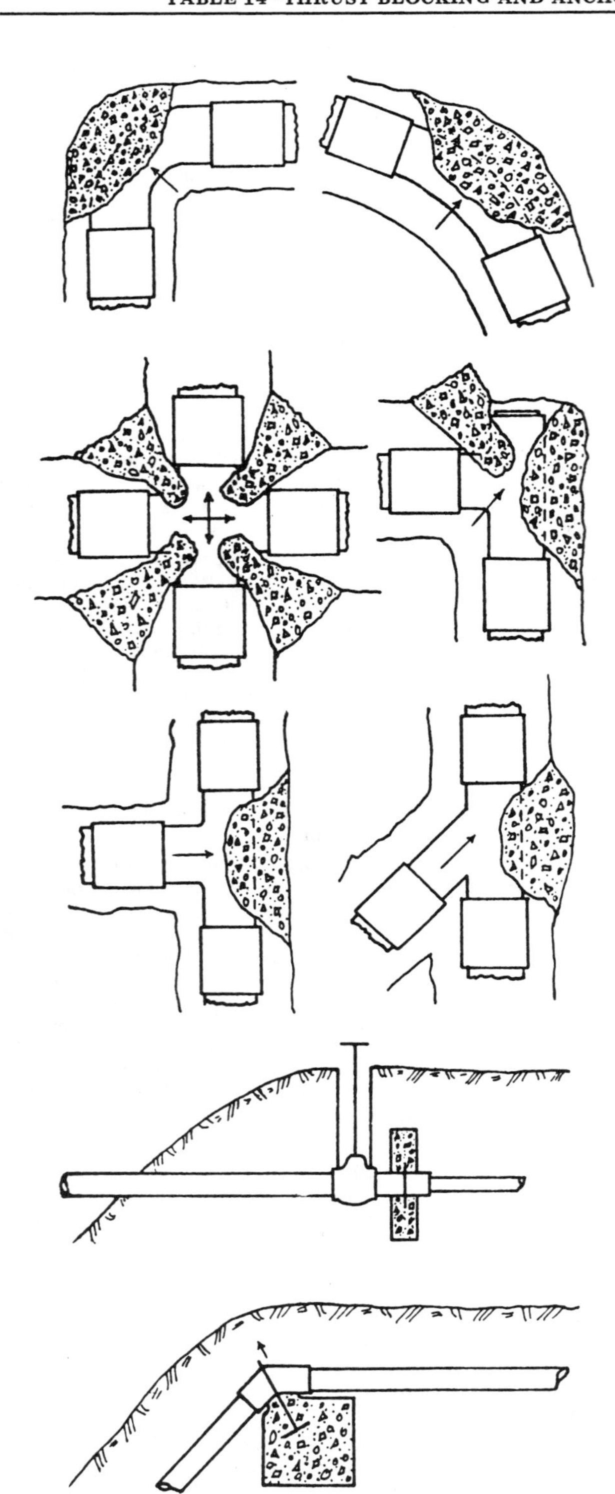

Step 1. Multiply the working pressure by the appropriate value shown in the following table to obtain total thrust in N (lb):

PIPELINE THRUST FACTORS* †

Pipe size in.	mm	Dead end or tee	90° Elbow	45° Elbow	22 ½° Elbow
1½	48	2.94	4.16	2.25	1.15
2	60	4.56	6.45	3.50	1.78
2½	73	6.65	9.40	5.10	2.60
3	89	9.80	13.9	7.51	3.82
3½	102	12.8	18.1	9.81	4.99
4	114	16.2	23.0	12.4	6.31
5	141	24.7	35.0	18.9	9.63
6	168	34.8	49.2	26.7	13.6
8	219	59.0	83.5	45.2	23.0
10	273	91.5	130.0	70.0	35.8
12	324	129.0	182.0	98.5	50.3
14	363	160.2	226.5	122.6	62.6
15	389	183.9	260.0	140.7	71.9
16	406	201.1	284.4	153.8	78.6
18	475	274.7	388.4	210.1	107.4
20	518	326.9	462.2	250.1	127.8
21	560	381.8	539.9	292.1	149.3
24	630	483.2	683.2	369.6	188.9
27	710	613.7	867.8	469.5	239.9

*Based on thrust per kPa (psi) pressure
†Blocking for cross may not be needed with long branch lines.

Step 2. Determine the bearing strength of the soil from the table below:

BEARING STRENGTH OF SOILS

Soils and safe bearing loads	lb/ft^2	kPa
Sound shale	10 000	500
Cemented gravel and sand difficult to pick	4 000	200
Coarse and fine compact sand	3 000	150
Medium clay-can be spaded	2 000	100
Soft clay	1 000	50
Muck	0	0

Step 3. Divide the total thrust obtained in Step 1 by the bearing strength of the soil to get the area needed, m^2(ft^2).

SIDE THRUST ALTERNATIVE PROCEDURE

Pipe size in.	mm	Side thrust-per degree* lb	N
1½	48	5.1	22.7
2	60	7.9	35.1
2½	73	11.6	51.6
3	89	17.1	76.1
3½	102	22.4	99.6
4	114	28.3	125.9
5	141	43.1	191.7
6	168	60.8	270.5
8	219	103.0	458.2
10	273	160.0	711.7
12	324	225.0	1000.8
14	363	278.2	1237.4
15	389	319.6	1421.6
16	406	349.3	1553.7
18	475	477.3	2123.0
20	518	568.0	2526.5
21	560	663.6	2951.7
24	630	839.6	3734.5
27	710	1066.2	4742.5

*Based on side thrust per 689 kPa (100 psi) pressure per degree of deflection.

NOTE: Multiply side thrust from table by degrees of deflection times kPa (psi) divided by 100 to obtain total side thrust in N (lb).

7.8 Maximum depth of cover. The maximum depth of cover for low pressure pipe shall be 1.2 m (4 ft). For other classes of pipe, the pipe manufacturer should be consulted for maximum depths of cover greater than approximately 1.2 m (4 ft).

SECTION 8—TESTING

8.1 General. Low pressure pipelines shall be thoroughly and completely tested for pressure strength and leakage before backfill operations are undertaken. If it is necessary to partially backfill the line before testing to hold the line in place, the partial backfill shall be undertaken as specified in paragraph 7.5. Only the body of the pipe sections shall be covered with all joints and connections left uncovered for inspection. High pressure pipelines may be tested after backfilling.

8.2 Filling. The line shall be slowly filled with water. The velocity of the water input shall not exceed 0.2 m/s (1 ft/s). Adequate provision shall be made for air release while filling, taking care to bleed all entrapped air in the process. The pressure shall be slowly built up to the maximum design working pressure. Pressurizing should take at least ten minutes for pipelines 102 mm (4 in.) and smaller in diameter and having a test pressure of 690 kPa (100 psi) in a test section of 305 m (1000 ft). For larger diameters, longer lines and higher pressures, proportionately longer build-up times shall be used.

8.3 Inspection. The pipeline shall be inspected in its entirety while the maximum working pressure is maintained. Where leaks are discovered, they shall be promptly repaired and the line retested.

8.4 Flow capacity. It shall be demonstrated by testing that the pipeline will function properly at design capacity. At or below design capacity, there shall be no objectionable surge or water hammer.

8.5 Objectional flow conditions. Objectional flow during testing conditions shall include continuing unsteady delivery of water, damage to the pipeline, detrimental overflow from vents or stands, or sudden or rapid changes in flow velocity at either start-up or shutdown including emergency shut-off, particularly in lines appreciably longer than 305 m (1000 ft).

SECTION 9—BASIS OF ACCEPTANCE

9.1 Requirements. The acceptability of the pipeline shall be determined by inspections to check compliance with all the provisions of this Standard with respect to the design of the line, the pipe and appurtenances used, and the minimum installation requirements.

SECTION 10—CERTIFICATION

10.1 General. All materials shall conform to these minimum requirements and to the tests prescribed in the applicable ASTM Standards.

10.2 Certification. When required, the pipe shall be certified by a qualified testing laboratory for compliance with the requirements set out in this Standard.

SECTION 11—PLANS AND SPECIFICATIONS

11.1 General. Plans and specifications for construction of underground thermoplastic irrigation pipelines shall be in keeping with this Standard and shall describe the requirements for application of the practice to achieve its intended purpose.

Cited Standards:

ASTM D1248, Specification for Polyethylene Plastics Molding and Extrusion Materials

ASTM D1598, Test for Time-to-Failure of Plastic Pipe Under Constant Internal Pressure

ASTM D1784, Specification for Rigid Poly (Vinyl Chloride) Compounds and Chlorinated Poly (Vinyl Chloride) Compounds

ASTM D1788, Specification for Rigid Acrylonitrile-Butadiene-Styrene (ABS) Plastics

ASTM D2122, Determining Dimensions of Thermoplastic Pipe and Fittings

ASTM D2152, Test for Quality of Extruded PVC Pipe by Acetone Immersion

ASTM D2235, Specification for Solvent Cement for Acrylonitrile-Butadiene-Styrene (ABS) Plastic Pipe and Fittings

ASTM D2239, Specification for Polyethylene (PE) Plastic Pipe (SDR-PR)

ASTM D2241, Specification for Poly (Vinyl Chloride) (PVC) and Chlorinated Poly (Vinyl Chloride) (CPVC) Plastic Pipe (SDR-PR)

ASTM D2282, Specification for Acrylonitrile-Butadiene-Styrene (ABS) Plastic Pipe (SDR-PR)

ASTM D2321, Underground Installation of Flexible Thermoplastic Sewer Pipe

ASTM D2444, Test for Impact Resistance of Thermoplastic Pipe and Fittings by Means of a Tup (Falling Weight)

ASTM D2466, Specification for Socket-Type Poly (Vinyl Chloride) (PVC) and Chlorinated Poly (Vinyl Chloride) (CPVC) Plastic Pipe Fittings, Schedule 40

ASTM D2467, Specification for Socket-Type Poly (Vinyl Chloride) (PVC) and Chlorinated Poly (Vinyl Chloride) (CPVC) Plastic Pipe Fittings, Schedule 80

ASTM D2468, Specification for Socket-Type Acrylonitrile-Butadiene-Styrene (ABS) Plastic Pipe Fittings, Schedule 40

ASTM D2469, Specification for Socket-Type Acrylonitrile-Butadiene-Styrene (ABS) Plastic Pipe Fittings, Schedule 80

ASTM D2564, Specification for Solvent Cements for Poly (Vinyl Chloride) (PVC) Plastic Pipe and Fittings

ASTM D2609, Specification for Plastic Insert Fittings for Polyethylene (PE) Plastic Pipe

ASTM D2672, Specification for Bell-End Poly (Vinyl Chloride) (PVC) Pipe

ASTM D2774, Recommended Practice for Underground Installation of Thermoplastic Pressure Piping

ASTM D2837, Obtaining Hydrostatic Design Basis for Thermoplastic Pipe Materials

ASTM D2855, Standard Recommended Practice for Making Solvent Cemented Joints with Polyvinyl Chloride Pipe and Fittings

ASTM D3036, Specification for Poly (Vinyl Chloride) (PVC) Plastic Line Couplings, Socket-Type

ASTM D3139, Specification for Joints for Plastic Pressure Pipes using Flexible Elastomeric Seals

ASTM D3261, Butt Head Fusion for PE Plastic Fittings for PE Plastic Pipe and Tubing

ASTM F402, Recommended Practice for Safe Handling of Solvent Cements Used for Joining Thermoplastic Pipe and Fittings

ASTM F412, Definitions of Terms Relating to Plastic Piping Systems

ASTM F477, Specification for Elastomeric Seals (Gaskets for Joining Plastic Pipe)

ISO 161/1-1978, Thermoplastic Pipes for the Transport of Fluids—Nominal Outside Diameters and Nominal Pressures—Part 1: Metric Series

ASAE Standard: ASAE S394

SPECIFICATIONS FOR IRRIGATION HOSE AND COUPLINGS USED WITH SELF-PROPELLED, HOSE-DRAG AGRICULTURAL IRRIGATION SYSTEMS

Developed jointly by the ASAE Sprinkler Irrigation Committee and the Traveler Division of The Irrigation Association; approved by the Soil and Water Division Standards Committee; adopted by ASAE December 1978; reconfirmed December 1983.

SECTION 1—PURPOSE AND SCOPE

1.1 This Standard establishes minimum performance levels, functional properties and physical characteristics for new irrigation hose and couplings used with self-propelled, hose-drag agricultural irrigation systems (travelers).

SECTION 2—DESCRIPTIONS AND DEFINITIONS

2.1 Irrigation hose as defined for purposes of this Standard shall be:

2.1.1 Used with a hose-drag (traveler) system.

2.1.2 Used in a double (side-by-side) configuration with one end attached to a water source, and the other moveable end attached to the traveler.

2.1.3 Pulled across a field by the traveler attached to its moveable end.

2.1.4 Laid on the ground during irrigation.

2.1.5 Flexible when subjected to normal operating pressure and collapsible to a flat cross section when purged of water.

2.2. **Elongation:** An increase in the hose length caused by pressurizing the hose.

2.3 **Snaking:** The deviation in location of the hose from the original straight line, laid out position, caused by elongation.

2.4 **Kinking:** The vertical transverse folding from the normally round configuration of a pressurized hose.

2.5 **Pressure:** The internal pressure measured in kilopascals (psi) at the inlet end of the hose or as otherwise designated.

SECTION 3—BURST PRESSURE

3.1 Irrigation hose shall have a minimum burst pressure of 2000 kPa (290 psi), determined in accordance with American Society for Testing and Materials Standard D380, Testing Rubber Hose.

SECTION 4—LONGITUDINAL STRENGTH

4.1 Hose shall meet the following minimum longitudinal strength: $UTS = KLD^2$.

where

	Metric (SI)	Customary
K = constant	0.29	1.26
L = hose length	meters	feet
D = hose diameter	centimeters	inches
UTS = ultimate longitudinal strength at failure to loss of service	kilograms	pounds

4.2 The procedure for testing hose for longitudinal strength shall be ASTM Standard D378, Testing Rubber Belting, Flat Type, but modified in the preparation of the sample. The test sample of hose shall be a longitudinal strip cut from the hose. This strip shall be long enough to leave exposed a minimum of 300 mm (12 in.) between the jaws of the test machine. The width of the sample shall be wide enough so that a minimum of 10 percent of the total design number of warp yarns in the hose circumference remain intact throughout the length of the sample.

4.3 The working load as measured under actual field conditions shall not exceed 35 percent of the longitudinal strength.

SECTION 5—COUPLINGS

5.1 The pull coupling assembly when attached to the hose shall withstand a minimum load of 1.5 times the working load of the hose, as set forth in paragraph 4.3.

5.2 The hose and coupler assembly used to test for the above performance shall be a minimum of 1 m (3 ft) in length. The coupler end of the assembly shall be gradually brought up to the required tension and held constant for 5 min.

5.3 The assembly shall not fail by breakage or slippage of the coupler or hose or in any other manner while exposed to the required tension.

SECTION 6—ELONGATION

6.1 Maximum allowable elongation for hose shall be 3 percent of its length when pressurized to 700 kPa (100 psi).

6.2 Testing for elongation shall be done with a sample length. Two marks shall be made on a relaxed sample, at least 3 m (10 ft) apart, and the distance between the marks (L_1) shall be accurately measured along a straight hose. The hose shall be pressurized to 700 kPa (100 psi) and a second distance measurement (L_2) taken. L_2/L_1 shall not exceed 1.03.

SECTION 7—KINKING

7.1 The minimum internal pressures at which kinking does not occur vary with nominal hose diameters. These minimum internal pressures shall be:

Nom. Dia.	Min. Pressure
5 in.	550 kPa (80 psi)
4 & 4 1/2 in.	500 kPa (72 psi)
3 & 3 1/2 in.	400 kPa (58 psi)
2 1/2 in.	350 kPa (50 psi)

7.2 The test procedure for kinking of irrigation hose shall be as follows:

7.2.1 The test hose, a minimum of 60 m (180 ft) in length, shall be laid out straight on a smooth, level soil surface with one end connected to the source of water pressure.

7.2.2 The hose shall be pressurized with water to a static internal pressure of 700 kPa (100 psi), and all air shall be bled from the hose.

7.2.3 The test shall be performed with no water flowing.

7.2.4 The free hose end shall then be returned toward the fixed hose end, a minimum of 20 m (60 ft) measured from the free hose end, in a "J" pattern simulating field operating conditions. The two legs of the hose shall be parallel and have a maximum lateral separation of 2 m (6 ft).

7.2.5 The hose shall be towed by its moveable end, at a constant rate of 0.01 m/s (2 ft/min), for a minimum of 6 m (20 ft) and observed for kinking.

7.2.6 If no kinking is observed, then paragraphs 7.2.1 through 7.2.5 shall be repeated at reduced pressure intervals of 50 kPa (7 psi) until kinking is observed in the loop of the hose. This pressure shall be considered the kinking pressure of the hose.

SECTION 8—ADHESION OF COVER TO FABRIC

8.1 The mechanical or chemical adhesion of the hose cover to the fabric of the hose shall be adequate to withstand the intended operating environment for the hose.

8.2 The procedure for testing adhesion of the hose cover to the fabric of the hose shall be the circumferential 25 mm (1 in.) strip method rated at a minimum of 4.5 kg (10 lb), per ASTM Standard D413, Tests for Rubber Property—Adhesion to Flexible Substrate.

SECTION 9—DIAMETER

9.1 The hose shall be nominally rated for internal diameter.

9.2 The hose shall be no more than 1.6 mm (1/16 in.) under the rated nominal diameter at the inside of the hose when the hose is without pressure, measured in accordance with ASTM Standard D380, Testing Rubber Hose, except that the plug gauge method shall be used.

SECTION 10—HOSE IDENTIFICATION

10.1 Each hose shall have its own serial number for identification. This marking shall be permanently attached to the hose in at least two places, 3 to 12 m (10 to 40 ft) from each end.

SECTION 11—OZONE RESISTANCE

11.1 The testing procedure for ozone resistance shall be ASTM Standard D518, Test for Rubber Deterioration—Surface Cracking.

11.1.1 The minimum level of ozone for testing purposes shall be as specified in ASTM Standard D1149, Test for Rubber Deterioration—Surface Ozone Cracking in a Chamber (Flat Specimens).

11.1.2 No cracks shall be visible under seven power magnification after 48 h of testing.

SECTION 12—HOSE MAINTENANCE

12.1 Operators instructions shall be provided with each hose outlining safety precautions, care of hose in the field, storage and repairs.

SECTION 13—HOSE DATA

13.1 Irrigation hose descriptive material shall include, but not be limited to, diameter, wall thickness, mass per unit length, length and working load.

Cited Standards:

ASTM D378, Testing Rubber Belting, Flat Type
ASTM D380, Testing Rubber Hose
ASTM D413, Tests for Rubber Property—Adhesion to Flexible Substrate
ASTM D518, Test for Rubber Deterioration—Surface Cracking
ASTM D1149, Test for Rubber Deterioration—Surface Ozone Cracking in a Chamber (Flat Specimens)

ASAE Standard: ASAE S395

SAFETY FOR SELF-PROPELLED, HOSE-DRAG AGRICULTURAL IRRIGATION SYSTEMS

Developed jointly by the ASAE Sprinkler Irrigation Committee and the Traveler Division of The Irrigation Association; approved by the Soil and Water Division Standards Committee; adopted by ASAE March 1979; reconfirmed December 1983.

SECTION 1—PURPOSE AND SCOPE

1.1 This Standard is intended to improve the degree of personal safety for operators and others during the normal application, operation and service of self-propelled, hose-drag agricultural irrigation systems.

1.2 This Standard is not intended to cover irrigation systems other than those with components as described in Section 2.

SECTION 2—DEFINITIONS

2.1 Self-propelled, hose-drag agricultural irrigation systems consist of:

2.1.1 A cable-drawn chassis that carries a sprinkler and includes a method of propulsion designed to gradually move the chassis and drag a feeder hose across a field.

2.1.2 A full or part circle sprinkler head that is sized and rated commensurate with the system design.

2.1.3 A drag-hose that is designed to feed water to the chassis and sprinkler. The hose shall be sized by internal diameter and length, and rated by working load commensurate with system design.

2.1.4 Hose handling equipment that facilitates the location, relocation and storage of the drag-hose. This equipment may be a separate device or may be mounted integrally to the chassis.

SECTION 3—SERVICE

3.1 Operator's manuals shall be provided including, but not limited to, suggested cable anchorage, safe operation, servicing, adjustments and storage.

3.2 Service shall be performed in accordance with the operator's manual.

3.3 Servicing or adjusting shall not be done with the system in operation, unless specific instructions on servicing or adjusting with the system in operation are provided in the operator's manual.

3.4 Ready access shall be provided to the areas requiring regular servicing, adjusting or inspecting.

3.5 Access shall be provided to areas that may require repair.

3.6 All other operation and servicing shall be in accordance with provisions of ASAE Standard, S318, Safety for Agricultural Equipment.

SECTION 4—OPERATION CONTROLS

4.1 Controls shall be identified and be readily visible at their points of operation. If symbols are used, they shall be in accordance with ASAE Standard S304, Symbols for Operator Controls on Agricultural Equipment.

4.1.1 Operating instructions for each control shall be on the machine whenever possible, and in the operator's manual.

4.2 All other provisions on operator controls shall be in accordance with ASAE Standard S318, Safety for Agricultural Equipment.

SECTION 5—POWER TAKE-OFF DRIVEN SHAFTS

5.1 Both towed and integral-type power take-off driven equipment shall be equipped with shielding for that portion of the power shaft that is furnished as a part of the driven equipment to prevent the operator from coming in contact with positively-driven rotating members of the power shaft. The shield for the power shaft between the driven equipment and tractor rear power take-off shall be integral with and journaled on the rotating member. Where integral power take-off driven equipment is of a design requiring removal of the tractor master shield, such driven equipment shall also include protection for that portion of the tractor power shaft which protrudes from the tractor.

5.2 A sign or decal stating plainly "Danger—Remove From Tractor When Not In Use" shall be firmly affixed to power take-off driven shafts supplied with systems.

5.3 All other power take-off equipment shall be in accordance with provisions of ASAE Standard S318, Safety for Agricultural Equipment.

SECTION 6—SHIELDS AND GUARDS

6.1 Guards shall be firmly affixed to prevent hand, foot or clothing contact with any moving parts, either fast or slow, rotating or reciprocal, that may be contacted during the normal course of operation, adjusting or servicing.

6.2 Power shaft drives shall be shielded if operating personnel are exposed to them while the drives are in motion.

6.3 Access doors and shields, which present the risk of personal injury from functional components when not in place, shall not be readily detached from the machine.

6.4 All other provisions pertaining to shields and guards shall be in accordance with ASAE Standard S318, Safety for Agricultural Equipment.

SECTION 7—TRAVEL ON HIGHWAYS

7.1 All road-towable system components shall be identified in accordance with ASAE Standard S276, Slow-Moving Vehicle Identification Emblem, and applicable state and federal laws and regulations.

7.2 Hitch pins and other hitching devices shall be provided with a retainer to prevent accidential unhitching.

7.3 Appurtenances shall have means for securing them to the chassis.

SECTION 8—SAFETY SIGNS

8.1 System equipment shall have appropriately displayed safety signs in accordance with ASAE Standard S318, Safety for Agricultural Equipment, when necessary to alert the operator and others to the risk of personal injury in normal operation and servicing.

8.1.1 Following are some of the equipment and operator practices that may be hazardous:

- Riser pipes and sprinklers that may contact electrical lines
- Rapid movement of machine due to component failure
- Sudden movement of machine when cable is being reeled out
- Fast reverse action of sprinkler
- Cable anchor failure

SECTION 9—SYSTEM LOAD RATINGS

9.1 Systems shall be load-rated for capacity of pull. This information shall be in the operator's manual and displayed on the machine.

9.2 Ratings for system load capacity shall have a minimum 2:1 safety factor, except for the hose.

9.2.1 The hose shall have a working load rating as outlined in ASAE Standard S394, Specifications for Irrigation Hose and Couplings Used with Self-Propelled, Hose-Drag Agricultural Irrigation Systems.

9.2.2 The coupler-hose mating connection shall have a load-rating in accordance with ASAE Standard S394, Specifications for Irrigation Hose and Couplings Used with Self-Propelled, Hose-Drag Agricultural Irrigation Sysytems.

Cited Standards:

ASAE S276, Slow-Moving Vehicle Identification Emblem
ASAE S304, Symbols for Operator Controls on Agricultural Equipment
ASAE S318, Safety for Agricultural Equipment
ASAE S394, Specifications for Irrigation Hose and Couplings Used with Self-Propelled, Hose-Drag Agricultural Irrigation Systems

ASAE Standard: ASAE S398.1

PROCEDURE FOR SPRINKLER TESTING AND PERFORMANCE REPORTING

Proposed by The Irrigation Association, Sprinkler Manufacturer Division, Sprinkler Testing Subcommittee; approved by the ASAE Soil and Water Division Standards Committee; adopted by ASAE as a Tentative Standard December 1979; reconfirmed March 1982, December 1982, December 1983, December 1984; revised and reclassified as a full Standard March 1985.

SECTION 1—PURPOSE

1.1 This Standard has the following three purposes:

1.1.1 To define a common test procedure for the collection of sprinkler test data such as pressure, flow rate, and radius of throw, which may be used for the purpose of publishing performance specifications for sprinklers whose areas of coverage have uniform radii.

1.1.2 To provide methods for the interpretation of test data for sprinkler performance specifications, as derived from paragraph 1.1.1.

1.1.3 To provide a method to readily distinguish which performance specifications have been developed using this procedure.

1.2 The sprinkler performance specifications presented are to be of such extent and accuracy as to assist irrigation system designers when comparing the basic performance of various types and makes of sprinklers.

1.3 This Standard describes the types and methods of obtaining and recording pertinent test data. There must be a sufficient amount of data recorded and retained to support published performance specifications so apparent conflicts can be resolvd.

SECTION 2—SCOPE

2.1 This procedure is used only to determine the radius of throw. No attempt is made here to define product use, design or application procedures, nor to determine uniformity of distribution, or the uniformity of the radius of throw throughout the area of coverage.

2.2 Single sprinkler tests only are covered in this Standard.

SECTION 3—SPRINKLER SELECTION AND DESCRIPTION

3.1 Number of sprinklers. To establish typical test data, it is recommended that a sufficient quantity of sprinklers of the same model number and nozzle size be tested under ostensibly the same operating and climatic conditions. Each test shall be recorded separately, but shall be combined with others for the development of performance specifications.

3.2 Selection of sprinklers. Sprinklers used for securing data shall be chosen at random from normal production runs.

3.3 Description of sprinklers. Each sprinkler shall be described in such a way that a specific reference can be made to it at a later date. This description shall include, but not necessarily be limited to the following:

Name of manufacturer
Approximate date of manufacture
Model name, number and nozzle size (Complete customer ordering identification)

SECTION 4—SITE CONDITIONS AND TEST EQUIPMENT

4.1 Sprinkler site

4.1.1 The sprinkler shall be located in an area where the surface is smooth or where vegetative growth is less than 150 mm (6 in.) in height. The surface grade shall not exceed 2% within the wetted area of sprinkler under test.

4.2 Sprinkler mounting

4.2.1 The sprinkler nozzle height above the nearest collector(s) for test purposes is defined in Table 1.

TABLE 1—SPRINKLER NOZZLE HEIGHT ABOVE COLLECTORS

Sprinkler Type	Sprinkler Inlet Size (Nom. Pipe Dia.)	Maximum Nozzle Height Above Collector mm	in.
a) Riser mounted; rotating	1 1/4 in. or smaller	915	36
b) Riser mounted; rotating	1 1/2 in. or larger	1830	72
c) Riser mounted; non-rotating	All	460	18
d) Grade mounted; all	All	Sprinkler lid level with the collector in the non-operating position.	
e) Hose end, base mounted; all	All	Bottom of sprinkler base to be level with the collector inlet.	

4.2.2 The sprinkler shall remain vertical (within 2 deg) throughout the duration of the test.

4.2.3 The sprinkler riser shall be made from schedule 40 steel pipe. The riser nominal pipe size shall be the same size as the sprinkler inlet connection. The base pressure measurement location shall be defined as a point a distance of at least five times the nominal sprinkler inlet diameter from the last upstream direction change or change in pipe cross-sectional area. The pressure tap shall be perpendicular to the riser and shall not extend into the inside diameter of the riser. Riser stream straightening vanes may be used when data are collected if such vanes are supplied as standard equipment with the sprinkler.

4.3 Collector description and location

4.3.1 All collectors used for any one test shall be identical. They shall be such that the water does not splash in or out. The type of collector shall be identified and recorded on the data sheet. If an evaporation suppressant is used, its type and method of application shall be identified and recorded on the data sheet.

4.3.2 Spacing of collectors for radius of throw determination is given in Table 2.

TABLE 2—SPACING OF COLLECTORS

Sprinkler Radius of Throw, m (ft)	Maximum Collector Spacing Center to Center, m (ft)
0.3–3 (1–10)	0.30 (1.0)
3–6 (10–20)	0.60 (2.0)
6–12) (20–39)	0.75 (2.5)
> 12 (> 39)	1.50 (5.0)

4.3.3 The position of all collectors shall be maintained such that the entrance portion is level.

4.3.4 The above ground height of the top of any collector shall be a maximum 0.9 m (3 ft) above the ground. Collectors shall be placed such that the vertical change in height between successive collectors shall not exceed a grade of 1%.

4.4 Wind measuring equipment and location for outdoor tests

4.4.1 Wind movement during the test period shall be determined with a rotating cup anemometer or device of equal or better accuracy. The wind direction shall be determined with a wind vane indicating at least 8 points of the compass.

4.4.2 Wind velocity sensing equipment shall be located at a minimum height of 2.0 m (6.5 ft). For sprinklers with a trajectory height of more than 2.0 m (6.5 ft), the sensor height shall be equal to the highest point of the main stream ± 10%.

4.4.3 The wind sensing equipment shall be located outside the wetted area and at a location that is representative of the wind conditions at the sprinkler location. The maximum distance of the sensor location shall not exceed 45 m (150 ft) from the wetted area of the sprinkler under test.

SECTION 5—MEASUREMENTS

5.1 Sprinkler pressure. The sprinkler base pressure shall not vary more than ± 3% during the test period. Pressure shall be measured with pressure measuring devices accurate within ± 3% of the sprinkler test pressure, and recorded.

5.2 Sprinkler flow. The flow through the sprinkler shall be measured to an accuracy of ± 3% of the sprinkler flow rate, and recorded.

5.3 Sprinkler radius of throw

5.3.1 The radius for rotating sprinklers shall be defined as the distance measured from the sprinkler centerline to the farthest point at which the sprinkler deposits water at the minimum rate of 0.26 mm/h (0.01 in./h) over the inlet surface area of the collector. The sprinkler shall be operating from its own drive mechanism.

5.3.2 The radius for non-rotating sprinklers shall be defined as the farthest distance measured from the sprinkler centerline to the point at which the sprinkler deposits water at the minimum rate of 0.26 mm/h (0.01 in./h) typically measured at any arc of coverage except at the arc extremes of part circle sprinklers.

5.4 Sprinkler rotation. The sprinkler's speed of rotation shall be measured only while the sprinkler is rotating from its own drive mechanism and shall be recorded.

5.5 Collector readings. The amount of water in each collector shall be accurately determined and recorded showing the location of the collectors relative to the sprinkler. For multi-leg tests, the readings for each leg shall be recorded independently.

5.6 Wind data

5.6.1 It is recommended that wind velocity be recorded continuously during the test period on a chart recorder. If continuous recording equipment is not available, wind velocity measurements shall be taken at the beginning and end of the test and at intervals not to exceed 10% of the test period or at 3 min intervals, whichever is greater. Wind velocities above 0.44 m/s (1.5 ft/s) shall be recorded to the nearest 0.22 m/s (0.75 ft/s) of movement. Wind direction shall be recorded to the nearest octant and shall be keyed to one of the principle axes of the data presentation form.

5.6.2 For single leg tests, the speed of rotation, collector precipitation data and radius data shall be measured only when the maximum wind velocity is 0.44 m/s (1.5 ft/s) or less.

5.6.3 For multiple leg tests, the speed of rotation, collector precipitation rate and radius data shall be measured only when the average wind velocity is less than 1.3 m/s (4.4 ft/s). In no case, during the measurement intervals, shall wind velocity exceed 2.2 m/s (7.3 ft/s).

5.7 Test period. Any test period may be used as long as all sprinkler streams pass over all collector legs an equal number of times. The actual or equivalent (if accelerated) test period shall be recorded.

5.8 Test date. The calendar date and time of day shall be recorded.

SECTION 6—TEST RECORDS AND DATA REDUCTION

6.1 Information to be recorded. The data outlined in Sections 3, 4 and 5 shall be recorded on appropriate forms. Supplemental data describing the conduct of the test may be included on the form.

6.2 Radius determination

6.2.1 In a single leg test, if the last collector measuring any water receives water at a rate greater than 0.26 mm/h (0.01 in./h), the radius of coverage shall be the distance from the sprinkler centerline to the centerline of that collector plus 0.3 m (1 ft). For rotary sprinklers with a flow rate of less than 0.126 L/s (2.0 gpm), the above specification shall be modified to 0.13 mm/h (0.005 in./h). The radius for all nozzle sizes of that specific model may be determined with this precipitation rate.

6.2.2 For multiple leg tests the radius shall be the average of all of the legs tested as defined in paragraphs 5.3 and 6.2.1.

6.3 Interpolation of data. When interpolation is used to determine the published radius of throw performance specifications, the specification shall be derived from the data secured for the nearest published nozzle size, at equal nozzle size increments, above and below the nozzle size to be interpolated. A minimum of four pressure-radius points from the larger and smaller nozzle sizes shall be used for interpolation. The points shall include the maximum and minimum published pressure points and be at equal increments not to exceed 138 kPa (20 psi). The published performance specification of the interpolated nozzle size shall be the average radius at the same pressure point of the larger and smaller nozzles.

SECTION 7—PUBLISHED DATA

7.1 The year of test will be published with data obtained and reported by the methods previously outlined.

7.1.1 Pressure. Data shall be listed in kilopascals (kPa) or pounds per square inch (psi).

7.1.2 Flow rate. Flow rates listed in the performance specifications shall be as defined in Sections 4 and 5 of this procedure. Data shall be listed in liters per second (L/s) and/or gallons per minute (gpm).

7.1.2.1 Data shall be listed to at least the nearest 0.006 L/s (0.1 gpm) for flow rates up to 1.58 L/s (25 gpm). Flow rates greater than 1.58 L/s (25 gpm) shall be listed to at least the nearest 0.06 L/s (1 gpm).

7.1.3 Radius of throw. The radius of throw for both full and part circle sprinklers shall be listed to the nearest 0.3 m (1.0 ft).

7.2 Supplemental data

7.2.1 Sprinkler spacing. Methods for the determination of specific sprinkler spacings are not a part of this procedure, and when listed in published performance specifications shall be clearly labeled "manufacturers recommended spacing" or "manufacturers maximum recommended spacing". Spacing data shall not be referenced to this ASAE Standard.

7.2.2 Other supplemental data. Other supplemental data may be published in the performance specifications. However, such data shall not be referenced to this Standard.

SECTION 8—CERTIFICATION

8.1 Published performance specifications should be revised if product design changes alter the performance.

8.1.1 Certification statement. "Company name" certifies that pressure, flow rate, and radius data for this product were determined and listed in accordance with ASAE Standard S398.1, Procedure for Sprinkler Testing and Performance Reporting, and are representative of performance of production sprinklers at the time of publication. Actual product performance may differ from the published specifications due to normal manufacturing variations and sample selection. All other specifications are solely the recommendation of "Company name".

8.1.2 Data certification. The statement in paragraph 8.1.1 shall appear on each specification page referencing the ASAE Standard. Each model complying with this procedure shall be appropriately referenced to this statement.

ASAE Engineering Practice: ASAE EP400.1

DESIGNING AND CONSTRUCTING IRRIGATION WELLS

Developed by the ASAE Ground Water Management Committee; approved by the ASAE Soil and Water Division Standards Committee; adopted by ASAE September 1980; reconfirmed December 1985; revised February 1987.

SECTION 1—PURPOSE AND SCOPE

1.1 This Engineering Practice is intended as a guide for preparing specifications for irrigation well construction, the objective being to obtain economical wells of high efficiency which are relatively sand free with a long projected life. In addition to this Engineering Practice, well design and construction should conform to all applicable health, safety, and other governmental regulations.

1.2 Many of the details presented herein also are suitable for domestic, municipal, and industrial wells. The scope of the Engineering Practice, however, is directed to wells constructed to obtain ground water for irrigation purposes.

1.3 The pump, power unit, and irrigation systems are not part of this guide. However, because they are all interdependent, certain related factors are included.

1.4 A detailed presentation of well standards is given in the National Water Well Association-Environmental Protection Agency Standard EP-570/9-75-001, Manual of Water Well Construction Practices, and in the American Water Works Association Standard A100-84, Deep Wells. Reference should be made to these two publications for comprehensive industrial and municipal well standards.

SECTION 2—DEFINITIONS

2.1 D_i size: The particle size diameter at which i percent of a granular material by weight is smaller, commonly referred to as the i percent passing size.

2.2 Drawdown: The difference between the pumping water level at a given discharge rate and the static water level.

2.3 Filter pack: Pack material carefully selected to increase permeability near the well inlet and to control movement of aquifer material into the well.

2.4 Pack material: Selected gravel or sand which is introduced to fill the annulus between the screen and/or casing and the borehole wall.

2.5 Perforated casing: Well inlet with machine-made openings usually formed by punching, pressing, or sawing.

2.6 Pump column: A pipe inside the well casing through which water is conveyed from the pump to the surface.

2.7 Screen: A manufactured well inlet with precisely dimensioned and shaped openings.

2.8 Specific capacity: Well discharge divided by the water level drawdown after a specified pumping duration.

2.9 Stabilizer pack: Pack material used only as formation and borehole stabilizer. Placed in a well where the character of the aquifer does not require a filter pack.

2.10 Test hole: A hole, usually of small diameter, e.g., 100 mm (4 in.), drilled through underground formations to obtain data necessary to evaluate the aquifer(s) and the site potential for locating a production well and to obtain formation-material samples for well design.

2.11 Uniformity coefficient: The ratio of the D_{60} size to the D_{10} size of a granular material.

2.12 Well casing: A pipe installed within the borehole to prevent collapse of sidewall material, receive and protect pump and pump column, and transport water from well inlet to pump intake.

2.13 Well development: The process of removing fine materials from the well intake zone which are either natural formation materials or materials introduced during well construction for the purpose of stabilizing the well intake zone and the pack material and increasing permeability of the intake zone and the pack material.

2.14 Well efficiency: Ratio of theoretical drawdown to actual or measured drawdown. Theoretical drawdown is best estimated from adjacent observation well data obtained during well test.

2.15 Well inlet: That part of the well which has openings in either a screen or perforated casing through which water enters.

2.16 Well intake zone: The portion of the well surrounding the well inlet which is modified by the well construction and development processes. This includes the space between the well inlet and the undisturbed aquifer.

2.17 Well test: Determination of well yield vs. drawdown relationship with time. Test may also provide information on water quality and suspended solids content.

2.18 Well yield: A discharge rate that can be sustained for some specified period of time.

SECTION 3—GENERAL RECOMMENDATIONS

3.1 Several factors are pertinent to irrigation well construction but are separate from the immediate design and construction details. The following factors should be included in the written contract(s).

3.1.1 Location and scope of work. The location of the proposed irrigation well work should be expressed as accurately as possible and practical. The extent of the work is to be clearly described with the understanding that the exact conditions under the ground surface are not completely known prior to commencement of the work. It may be advisable to have separate contracts for test hole drilling and for the irrigation well itself since irrigation well specifications and contract should not be completed until test hole data has been analyzed.

3.1.2 Permits, licenses, and registration. All applicable governmental regulations regarding permits, licenses, and registration should be satisfied. A determination of who has the responsibility to furnish necessary information, submit the required forms to the proper agencies, and secure the appropriate documents should be made.

3.1.3 Insurance, health, and safety requirements. Adequate liability, personal and property insurance for drilling operations and a healthy and safe working environment should be maintained.

3.1.4 Access to the well construction site. After the final decision on exact location of site(s) is made, access to this land with sufficient working space for the drilling operations should be granted.

3.1.5 Obstructions. The site should be investigated prior to start of drilling and the owner informed of any hazards or obstructions, such as overhead power lines, buried cables, and buried pipes, which may affect the location of the test or production well.Any known underground obstructions not readily visible should be identified.

3.1.6 Water supply for drilling. The source of the water for drilling and who is responsible for it should be clearly understood. The water should be taken from the cleanest source available.

3.1.7 Records. A copy of the completed well construction details, well log, development record, and well test data should be supplied to the owner.

3.1.8 Water-level measurement access. When the permanent pump is installed in the well after completion, a sealed access to the well casing should be provided to allow future water level measurements. It is recommended that a permanent air line be installed at the time the pump is set. Refer to paragraph 8.3.1 for details.

SECTION 4—DESIGN

4.1 Test holes. Most aquifers are variable in areal extent, thickness, and composition. One or more test holes may be required to locate a suitable aquifer(s) and obtain formation and water samples.

4.1.1 Number and location of test holes. The number of test holes required to properly select a production well location usually cannot be precisely determined before test drilling begins. Therefore, as the work proceeds, the number of test holes should be mutually agreed upon, often in consultation with an engineer or hydrologist.

4.1.2 Test hole diameter. Test holes should be drilled sufficiently large to obtain accurate samples of the formations encountered and to allow proper testing, e.g., geophysical logging. The holes should be at least 75 mm (3 in.) and preferably 100 mm (4 in.) or more in diameter.

4.1.3 Test hole depth. Knowledge of the geology of the area and inspection of logs of other wells or test holes are useful in estimating the approximate depth to which test holes should be drilled. Each test hole should be drilled to a sufficient depth for determining the desirable or undesirable characteristics of the formations. Normally, this will be through the entire water-bearing formation. However, where thick formations are encountered and high capacity wells are not required, the drilling may not penetrate the entire formation.

4.1.4 Test hole casing. Test holes may require casing to prevent caving and/or to maintain the hole for extended periods. All materials required shall be furnished, but may be recovered after the work is completed unless purchased in place.

4.1.5 Test hole log. A complete and accurate log of each test hole shall be kept and a copy furnished to the owner along with the location of each test hole. The log should include: (a) sample description at intervals of depth, (b) drilling time, (c) drilling action, and (d) fluid losses. Samples of the water-bearing material should be taken at each lithologic change in formation and as often as necessary to accurately log the formations penetrated. This is normally every 1.5 m (5 ft) but should be reduced to a shorter interval in thin, highly stratified aquifers. Each sample should contain at least 0.5 kg (1 lb) of the formation material.

4.1.6 Sampling procedure. Accurate samples are important to successful well design. Sampling procedures vary with drilling method and type of sampling equipment used. Detailed procedures should be included in the contract.

4.1.7 Geophysical logging. Additional information for determination of superior water producing aquifers (for well inlet location) and location of aquifers producing poor quality water (for sealing) may be obtained by geophysical logging equipment, particularly in a test hole selected near the location for constructing the production well. The most common geophysical logs for irrigation wells are natural gamma ray, electrical resistance, and spontaneous potential logs.

4.1.8 Test hole abandonment. Test holes shall be sealed as soon as they have served their function. Sealing is accomplished by filling the hole with material less permeable than that which was removed. If a test hole penetrates more than one aquifer, a definite seal between the water-bearing formations to prevent commingling of water must be provided where a difference in water quality may exist.

4.2 Well intake zone. The well intake zone consists of: (1) The modified aquifer region surrounding the well inlet, or (2) filter pack and modified region of aquifer, or (3) stabilizer pack and modified region of aquifer.

4.2.1 Stabilizer pack. The pack material should be clean and at least as permeable as the natural formation. Uniformity and particle size distribution are not critical. Minimum particle size is the width of openings in well inlet (see paragraph 4.3).

4.2.2 Filter pack. When pack material is necessary to both stabilize the well bore and serve as a filter, the aggregate material should be uniformly graded and be of particular specific sizes. A uniformity coefficient of less than 2.5 is preferable. A filter pack is especially useful with aquifers composed of fine material of relatively uniform particle size when, without the pack, the proper width of the openings in the well inlet would be very small, e.g., less than 1.0 mm (0.04 in.), restricting the flow. Also a filter pack is used when the formation is extensively stratified.

4.2.2.1 Thickness and length of filter pack. The thickness of the filter pack should normally be between 50 and 150 mm (2 and 6 in.). A 50 mm pack is capable of stopping all sand and can be placed from the surface if the filter pack is uniform and is placed in small increments. Four centering guides are recommended for thin packs and should be placed at closely spaced intervals along the casing or screen. The pack material should extend through the entire aquifer exposed to the well bore. If water is desired from more than one aquifer and it is not necessary to maintain separation of the waters, the pack length may be almost the entire depth of the well.

4.2.2.2 Size of filter pack material. To obtain stable filtering action and high well efficiency, the D_{50} size of the pack material is normally 4 to 6 times larger than the D_{50} size of the aquifer if a very uniform pack is available. This could be greater if the aquifer and/or the available pack is very non-uniform.

4.2.2.3 Quality of filter pack. Clean, well-rounded, smooth, uniform, stable, water-insoluble, and non-acid soluble particles are to be used.

4.2.2.4 Surface Sealing. Surface sealing by cementation should be required. A gravel pipe (capped 50 to 100 mm diameter pipe) should be installed through the seal to facilitate replacement of the pack if necessary.

4.3 Well inlet. Provision must be made to permit the movement of water from the aquifer through the pack, if one is present, into the casing. In consolidated aquifers an open borehole through the water producing zone may be satisfactory. In unconsolidated aquifers, screen or certain perforated sections must be used to allow water to enter the well with a minimum of head loss while keeping the surrounding material out of the casing. Thus, the bottom of the well must be sealed to prevent any unconsolidated material from entering. To minimize head loss through the inlet, sufficient inlet open area is necessary to limit maximum velocities to 0.03 m/s (0.1 ft/s).

4.3.1 Opening size. The smallest dimension of the well inlet opening is considered the opening size. Preferably the openings should be tapered with the smaller openings on the outside of the inlet.

4.3.1.1 Opening size for a naturally developed well. When a well is to be completed without a pack, the inlet opening size should normally be about equal to the D_{60} size of the aquifer. This will allow 60% of the aquifer material to pass through into the well and be removed during well development.

4.3.1.2 Opening size for filter packed wells. When a well is to be completed with a pack, the inlet opening size should normally be equal to the D_{10} size of the pack material.

4.3.2 Types of inlet openings. Two types of openings are acceptable for irrigation wells. These are commercial screens and machine-made perforations. Opening a portion of the casing for an inlet with chisels, picks, cutting torches, etc., should not be allowed. Characteristics which should be considered in screen selection include: effect of screen on well development, mechanical strength, open area, opening configuration, and the desired life as affected by the anticipated underground environment.

4.3.2.1 Screen. A well screen is a fabricated cylindrical device with closely sized openings designed to permit passage of water with minimum resistance while retaining naturally occurring or artificially placed materials.

4.3.2.2 Perforation. A perforated well inlet is a machine-made product with relatively smooth openings typically formed by either milling, sawing or punching (for large openings).

4.3.3 Inlet length. The length of the inlet is determined by the desired capacity of the well, the aquifer thickness, and the type of aquifer (artesian or non-artesian). The well inlet openings should be located adjacent to the most permeable areas of the aquifer(s). Blank casing may be used between inlet areas where major aquifers are separated by less permeable formations if these areas are accurately identified and located by reliable test hole information which includes geophysical logs.

4.3.3.1 Artesian conditions (confined aquifer). The inlet length may be up to 70 or 80% of the aquifer thickness when it is an artesian aquifer of similar material throughout. The possibility of the aquifer changing in the future from artesian conditions to water table conditions should be considered.

4.3.3.2 **Water table conditions (unconfined aquifer).** When the water table is in the major aquifer, the well inlet should preferably be in the bottom one-third to one-half of the formation. When the water table is considerably above the top of the major aquifer, the well inlet may extend throughout the entire thickness of the major aquifer.

4.3.4 **Inlet diameter.** Where possible, it is desirable to locate the pump above the well inlet to avoid cascading water into the well, air entrainment, and air-caused incrustation of inlet openings. However, this practice may not be possible in some thin aquifers. Inlet diameter is normally the same as that of the casing. Casing and inlet diameter is normally chosen to: 1) allow proper clearance for installation and efficient operation of pump and, 2) to assure good hydraulic efficiency of the well.

4.3.5 **Inlet material.** Numerous materials may be used for the well inlet. A particular material is selected for a specific situation on the basis of the quality of the groundwater, strength requirements, and cost.

4.3.5.1 **Water quality.** A chemical analysis of the groundwater will indicate whether or not it has corroding or incrusting tendencies. An acid treatment management program can control incrustation, but the well inlet material must be able to endure the acid treatment, which may require heavier material. If incrustation is expected, the size of the inlet openings should be slightly larger than normal. Corrosion-resistant material and slightly smaller than normal openings should be used, if corroding tendencies of ferrous materials are detected.

4.3.5.2 **Strength requirements.** The well inlet material must be strong enough to withstand usual handling during installation, column load if supporting casing, collapse pressure from sidewall load, and maximum open area configuration for efficiency.

4.3.5.3 **Types of material.** Steel, bronze, stainless steel, galvanized steel, and polyvinyl chloride are typical materials used for well inlets. Other alloys are available for very corrosive waters and other specialized conditions. Dissimilar metals should not be used in the well inlet or between the inlet and casing because of galvanic corrosion.

4.4 **Well casing.** The well casing maintains the borehole at a desired size by preventing unstable formation materials from collapsing and closing the hole, serves as an enclosure for the pump, and is a vertical conduit for the flow of water from the well inlet to the level of the pump inlet.

4.4.1 **Casing size.** The inside diameter of the well casing should be 50 to 100 mm (2 to 4 in.) larger than the maximum outside diameter of the pump and pump column that are likely to be installed to allow the pump to hang freely in the well. This annular space is also useful to allow the lowering of an electrical line or air line to monitor water levels for future well management. If a submersible pump is to be used, adequate annular space as specified by the pump manufacturer must be provided. Also, the maximum allowable vertical water velocity in the well casing is 1.5 m/s (5 ft/s).

4.4.2 **Casing material.** Several materials may be used for the well casing. A particular material is selected for a specific situation on the basis of the quality of the groundwater, strength requirements, and cost.

4.4.2.1 **Water quality.** A chemical analysis of the groundwater will indicate whether or not it has corroding or incrusting tendencies. Incrustation is not of major concern for well casing but the casing material must be of sufficient thickness and characteristics to endure acid treatment of the well inlet. If corroding tendencies are detected, the casing material must be of sufficient thickness and characteristics to resist deterioration and failure.

4.4.2.2 **Strength requirements.** Well casing wall thickness should be sufficient to withstand anticipated formation and hydrostatic pressures imposed on the casing during its installation, well development, and use.

4.4.2.3 **Types of material.** Typical materials used for well casing include mild steel, black steel, asbestos-cement, fiberglass, polyvinyl chloride, stainless steel, copper alloys and galvanized steel. Dissimilar metals should not be joined because of galvanic corrosion.

SECTION 5—CONSTRUCTION

5.1 Irrigation well construction consists of the drilling operation, installing the well inlet, installing the casing, installing any pack material, and protecting the quality of the aquifer.

5.2 **Placement of equipment and materials.** All equipment and materials should be placed in the well in a manner which meets design specifications, unless some good and sufficient reason is encountered during construction to deviate from design specifications. Changes should be mutually agreed upon and a written record of them maintained.

5.2.1 **Borehole.** The borehole should be round, plumb, straight and of adequate diameter to permit satisfactory and proper installation of the well casing and inlet.

5.2.2 **Inlet.** The screen or perforated sections should be handled carefully and placed at the correct depths to match the desired formations. When the well is to be packed the entire length of the inlet should be centered in the borehole by the placement of centering guides at vertical intervals of about 6 m (20 ft) to insure that the pack material will fill the entire intake area uniformly. The centering guides should be placed 90 deg apart circumferentially.

5.2.3 **Casing.** The casing should be relatively plumb and straight to permit installation and operation of the pump. The usual requirement for plumbness of casing and screen is that it not deviate from the vertical by more than two-thirds the casing diameter per 30 m (100 ft) of depth as determined by gaging. The deviations must be reasonably consistent regarding direction. A normal standard for straightness requires that a 12 m (40 ft) long bucket (bailing bucket or pipe) with a diameter of 25 mm (1 in.) less than the casing be lowered freely to the total depth of the well or to the deepest possible pump setting.

5.2.4 **Pack material.** Pack material should be placed carefully to prevent separation and bridging. The use of a tremie (a temporary pipe which extends almost to the well bottom that is withdrawn as pack material is fed through it) will help prevent layering of material caused by different rates of settling, bridging of pack materials, and non-uniform placement of pack material in the intake area. Thin packs that are uniform in size may be added from the surface if done very slowly.

5.2.5 **Drilling fluid additives.** It is often necessary to seal unconsolidated formations in order to prevent excessive loss of drilling fluids. Natural clays or organic materials may be placed in the drilling fluid to temporarily seal the pores in the borehole walls. However, these introduced materials must be removed during the development process.

5.2.6 **Aquifer protection.** The upper 6 m (20 ft) of the annular space between the borehole and the well casing must be carefully filled with material which does not contain potentially harmful bacteria or chemicals and that is of a texture which is at least as fine as the original material which was removed. Often a cement grout is used in this space to prevent surface water entering the aquifer through the area around the well casing. If the natural water level in the ground is less than 6 m (20 ft) deep, the depth of finer textured material around the casing should be down to the static water level. Surface casing is appropriate for additional aquifer protection. State regulations should be consulted for additional requirements regarding aquifer protection.

5.3 **Construction certification.** A certification that the well was constructed in such a manner as to satisfy all original design specifications shall be given unless the specifications were modified in writing by prior and mutual agreement. Any modifications of the original specifications shall be included in the certification.

SECTION 6—WELL DEVELOPMENT

6.1 It is necessary to develop the well to obtain its maximum capacity for a given drawdown. The developing process removes fine material from the formation near the well screen, thereby opening the passages, so that the water can enter the well more freely. The process is accomplished by forcing water in and out of the screen openings. The well should be developed as soon as possible after construction. All fine material pulled into the well during the developing operation should be removed from the well before installing the test pump.

6.2 **Well development methods.** Often more than one method is used to successfully develop a well.

6.2.1 Development by surge block or swab. After the casing and filter pack (if required) have been installed, the well may be thoroughly surged or swabbed. The surge block should be alternated with the use of a bailer to keep the screen and casing clean.

6.2.2 Development by bailing. Although not a highly recommended method, the bailer alone may be used in a similar manner as that of a surge block in wells with diameters greater than 200 mm (8 in.). The bailing unit should have a line speed of at least 2.5 m/s (500 ft/min) during hoisting (5 m/s or 1000 ft/min is recommended) and free-fall during lowering of the bailer. The bailer shall fit closely inside the casing being no smaller than 50 mm (2 in.) than the well inlet. The bailing unit shall be capable of removing at least 2L/s (30 gal/min) continuously for one hour from 200 mm (8 in.) diameter well to 8L/s (125 gal/min) from 400 mm (16 in.) diameter wells.

6.2.3 Development by pumping. Development by pumping may be employed as an additional or final development step when used in conjunction with other methods. It is not recommended as a singular method of development. A pump must be supplied for this method and may be the same pump used for the well test. The suggested procedure for pump development is to begin pumping at about one-fourth or one-fifth of the desired yield of the well and to continue pumping until the water becomes clear. Then the well should be surged by turning off the pump, allowing the water to run back into the well, and then pumping again. This process of alternating pumping and not pumping continues until no sand is pumped at the first discharge rate. The process is repeated at 50, 75, and 100 and if possible 125% of the desired well yield.

6.2.4 Development with explosives. In formations of semiconsolidated or consolidated material, explosives may sometimes increase production. Upon authorization from the owner, the part of the formation indicated as most promising by the drillers log and/or by the electric log may be blasted or fractured. Explosives such as dynamite, nitroglycerin, and water gel, may be used. The size and spacing of the shots should be agreed upon. Broken rock and fine materials dislodged by the stimulation process shall be removed. Such loose material may be removed by drilling, bailing, air-lifting, or a combination of these methods.

6.2.5 Development with chemicals. Numerous chemicals are used to aid in well development. Several polyphosphates are available to disperse clays and mud cakes. Oxidizing agents may be necessary to break down organic type drilling fluids. Acid may be used in limestone formations to enlarge openings and improve production.

6.2.6 Development by jetting. Development by high-pressure hydraulic jetting may be required to remove the foreign material and fines that accumulate in the formation during the drilling process. If practical, the well should be pumped during the jetting process to remove the fine particles. The jetting tool may also be utilized for precise and uniform injection of chemicals (see paragraph 6.2.5) used in the development process.

6.2.7 Development by air pumping. Air pumping is accomplished by injecting a high volume of compressed air in the well. The uplift action provides a surging effect. A more effective air pumping method utilizes a double-packer tool to selectively develop the well inlet.

6.3 Disinfection. If well testing is not required, the well must be disinfected with a chlorine solution to destroy any bacteria introduced during the construction and development processes. The well should then be properly capped for protection.

SECTION 7—WELL TESTING

7.1 The well shall be adequately tested for discharge and drawdown characteristics with a test pump after the well is developed and the discharge is free of sand.

7.2 Static water level. The static water level in the well is measured with an air pressure device, wetted tape, or a standard electrical depth device prior to test pumping. In order to obtain a realistic static water level, the measurement should be made at least 12 h after any work has been done on the well which may have disturbed the water level.

7.3 Pumping water level. For test pumping purposes the water level may be lowered substantially below the desired or anticipated production pumping level. The lowest level normally attempted is that which results from a rate equal to 125% of the desired pumping rate after equilibrium is reached.

7.4 Pumping rates. Several patterns of pumping rates may be used in pump testing a well. The pattern selected depends on the type of information desired, e.g., aquifer constants, degree of well development, maximum discharge within permissible drawdown, and acceptable discharge for pump selection. It is important that the pumping test provide adequate data for efficient pump selection.

7.4.1 Maximum discharge. If problems in the continuity of the water supply are not anticipated, the one flow rate which gives the lowest allowable pumping water level as described in paragraph 7.3 is the maximum discharge rate. This flow rate should be maintained until the drawdown remains stable for a minimum of 1 h and the entire test should never be less than 8 h. However, test periods of 24 h or more are preferred, especially in low permeability water table aquifers.

7.4.2 Decreasing step-drawdown test. This pattern of pumping rates is suitable for determining the range of acceptable discharges for selecting the permanent pump. The first pumping rate and pumping time is described in paragraph 7.4.1. At the end of the first pumping period the rate is reduced to approximately 80% of the initial pumping rate. This rate is continued until the discharge and pumping water level in the well remain constant for at least 30 min. Then the process is repeated for pumping rates of 60, 40, and 20% of the initial pumping rate. Each pumping rate should continue until the discharge and pumping water level remain constant for at least 30 min.

7.4.3 Constant discharge test. A constant rate test may be specified to determine certain aquifer characteristic coefficients. The well is pumped at the established rate (see paragraph 7.4.1) or other specified rate criterion without substantial variations in rate for the time period required. Periodic measurement of pumping rate should be made and noted during the test period. Pumping water levels may be measured with either of the following time schedules.

7.4.3.1 Regular intervals. Water level measurements are made every 5 min for the first 30 min of pumping, every 10 min for the next hour, every 30 min for the next 4.5 h, then every hour for the duration of the test period. At the conclusion of pumping, the recovery of water level is measured in the same time sequence for at least 50% of the elapsed pumping time.

7.4.3.2 Logarithmic intervals. Pumping water levels should be measured with sufficient frequency so that at least 10 data points are located throughout each logarithmic graph cycle. For example, depth is measured at approximately 1, 1.2, 1.5, 2, 2.5, 3, 4, 5, 6, 7, and 8 min after the start of the test. Then measurements are continued at all succeeding decimal multiples of these intervals to the end of the test, e.g., 10, 12, 15, 20, 25, 30, 40, 50, 60, 70, 80, 100, 120, 150 min.

7.5 Continuity. The pump should not be stopped during the test pumping period. If the pump is stopped for less than 1 h, the test may be resumed but measurements should not be recorded until the water level in the well has remained steady for 1 h. If a constant rate test is being conducted the test should not be resumed until a sufficient recovery period has allowed the static water table to return to its original level and then restart the complete test.

7.6 Data. For each pumping rate test the flow rates (discharge) and pumping water levels must be measured accurately and recorded. The rate is measured with a standard orifice, flow meter, or other measuring device with an error of less than 5%. Water level measurements should be accurate to 30 mm (0.1 ft), if possible.

7.7 Aquifer protection. During and after all testing work, the well should be properly protected to prevent the entry of foreign material and contaminated water from any source.

SECTION 8—PUMPING EQUIPMENT AND IRRIGATION SYSTEMS

8.1 The pump, power unit, and irrigation systems are not part of these irrigation well recommendations. However, they are all intimately related.

8.2 Irrigation system. The final design of the irrigation system must await the results of the well testing. The design discharge of the water supply is an important component of the irrigation system design.

8.3 Pump. The permanent pump for the irrigation well cannot be selected until after the well has been tested and the design discharge determined. Also the pump for the irrigation well must depend on the irrigation system design for the total head that must be produced.

8.3.1 Air line. When the irrigation pump is installed an air line may be attached to the pump column for water level determination. The air line is a 6 to 9 mm (1/4 to 3/8 in.) copper, polyethylene, or galvanized tubing which extends from the ground surface to below the lowest water level to be measured, but terminating at least 1 m (3 ft) above the pump intake. The vertical distance from the center of the pressure gauge to the bottom of the air line must be carefully recorded to allow accurate water level determinations to be made.

8.3.2 Water meters. A water meter should be installed at the pump discharge to observe the pumping rate and account for total water used.

8.3.3 Disinfecting. After the pump has been installed the entire well should be disinfected with a chlorine solution to destroy any bacterial accumulations.

Cited Standards:

AWWA A100-84, Deep Wells
NWWA-EPA-570/9-75-001, Manual of Water Well Construction Practices

T DESIGN AND INSTALLATION OF MICROIRRIGATION SYSTEMS

Developed by the ASAE Subsurface and Trickle Irrigation Committee; approved by the ASAE Soil and Water Division Standards Committee; adopted by ASAE December 1980; reconfirmed December 1985, December 1986; revised April 1988.

SECTION 1—PURPOSE AND SCOPE

1.1 The purpose of this Engineering Practice is to establish minimum recommendations for the design, installation and performance of microirrigation systems; including trickle, drip, subsurface, bubbler and spray irrigation systems. This Engineering Practice should encourage sound system design and operation and enhance communication among involved personnel.

1.2 Provisions of this Engineering Practice are primarily those that affect the adequacy and uniformity of water application, filtration requirements, water treatment, and water amendments.

SECTION 2—DEFINITIONS

2.1 Chemical water treatment: Chemical treatment of the water to make it acceptable for use in microirrigation systems. This may include acids, fungicides and bactericides used to prevent emitter clogging or used for pH adjustment.

2.2 Control station: The control station may include facilities for water measurement, filtration, treatment, addition of amendments, flow and pressure control, timing of application and backflow prevention.

2.3 Crop area: The field surface area allocated to each plant. In tree crops the tree crop area is the spacing multiplied by the row spacing.

2.4 Design area: The specific land area which is to be irrigated by the microirrigation system.

2.5 Design emission uniformity: An estimate of the uniformity of emitter discharge rates throughout the system, as described by the equation in paragraph 3.5.2.

2.6 Emitters: The devices used to control the discharge from the lateral lines at discrete or continuous points.

2.6.1 Emission point: Point where the water is discharged from an emitter.

2.6.2 Line-source emitters: Water is discharged from closely spaced perforations, emitters or a porous wall along the lateral line.

2.6.3 Point-source emitters: Water is discharged from emission points that are individually and relatively widely spaced, usually over 1 m (3.3 ft). Multiple-outlet emitters discharge water at two or more emission points.

2.7 Emitter discharge rate: The discharge rate at a given operating pressure from an individual point-source emitter expressed as a volume per unit time or from a unit length of line-source emitter expressed as a volume per unit length per unit time.

2.8 Emitter operating pressure: The average operating pressure of the emitters within any simultaneously operated portion of the system.

2.9 Evapotranspiration: The combined effects of evaporation from the soil and plant surfaces and transpiration from plants. Peak evapotranspiration is the maximum rate of daily evapotranspiration.

2.10 Filtration system: The assembly of independently controlled physical components used to remove suspended solids from irrigation water. This may include both pressure and gravity-type devices and such specific units as settling basins or reservoirs, screens, media beds and centrifugal force units.

2.11 Lateral: The water delivery pipeline that supplies water to the emitters from the manifold pipelines.

2.12 Main and submain: The water delivery pipelines that supply water from the control station to the manifolds.

2.13 Manifold: The water delivery pipeline that supplies water from the submain or main to the laterals.

2.14 Manufacturer's coefficient of variation (C_v): This is a measure of the variability of discharge of a random sample of a given make, model and size of emitter, as produced by the manufacturer and before any field operation or aging has taken place.

$$C_v = \frac{s}{\bar{x}}$$

where
$\bar{x}$ = the mean discharge of emitters in the sample
s = the standard deviation of the discharge of the emitters in the sample

$$s = \left[\frac{\Sigma_{i=1}^{n} (x_i - \bar{x})^2}{n-1} \right]^{1/2}$$

where
x_i = the discharge of an emitter
n = the number of emitters in the sample

If a line-source emitter is used, the individual discharges from holes on a one-meter or other specified length of emitter tape are used. This term can also be used to describe the variability in the downstream pressure from pressure control valves or the variability in discharge from flow control valves or orifices.

2.15 Microirrigation: The frequent application of small quantities of water on or below the soil surface as drops, tiny streams or miniature spray through emitters or applicators placed along a water delivery line. Microirrigation encompasses a number of methods or concepts; such as bubbler, drip, trickle, mist or spray and subsurface irrigation.

2.15.1 Bubbler irrigation: The application of water to the soil surface as a small stream or fountain, where the discharge rates for point-source bubbler emitters are greater than for drip or subsurface emitters but generally less than 225 L/h (60 gal/h). Because the emitter discharge rate normally exceeds the infiltration rate of the soil, a small basin is usually required to contain or control the water.

2.15.2 Drip and trickle irrigation: The application of water to the soil surface as drops or tiny streams through emitters. Often the terms drip and trickle irrigation are considered synonymous. For trickle and drip irrigation, discharge rates for point-source emitters are generally less than 8 L/h (2 gal/h) for single-outlet emitters, and discharge rates for line-source emitters are generally less than 12 L/h per meter (1 gal/h per foot) of lateral.

2.15.3 Spray irrigation: The application of water by a small spray or mist to the soil surface, where travel through the air becomes instrumental in the distribution of water. Discharge rates for point-source spray emitters are generally lower than 175 L/h (45 gal/h).

2.15.4 Subsurface irrigation: The application of water below the soil surface through emitters, with discharge rates generally in the same range as drip irrigation. This method of water application is different from and not to be confused with the method where the root zone is irrigated by water table control, herein referred to as subirrigation.

2.16 Microirrigation systems: The physical components required to apply water by microirrigation. System components that may be required include the emitters, lateral lines, manifold lines, main and submain lines, filter, chemical injectors, flow control station and other necessary items.

2.17 Peak daily irrigation water requirement: The net quantity of water needed to meet the peak daily evapotranspiration rate occurring during the growing season expressed in mm/day (in./day).

2.18 Percent area wetted: The area wetted as a percentage of the total crop area.

2.19 Pumping station: The pump or pumps that provide water and pressure to the system, together with all necessary appurtenances such as base, sump, screens, valves, motor controls, motor protection devices, fences and shelters.

2.20 Subunit: The main manifold and lateral pipelines which operate simultaneously and have independent flow control.

2.21 System operating pressure: The average operating pressure downstream from the pumping and control station where the main lines begin.

2.22 Water amendment: The fertilizer, herbicide, insecticide or other material added to the water for the enhancement of crop production or as a chemical water treatment to reduce emitter clogging.

2.23 Wetted area: The average irrigated soil area in a horizontal plane located at or below the emitter.

SECTION 3—DESIGN, INSTALLATION, AND PERFORMANCE

3.1 System capacity. Microirrigation systems shall have a design capacity adequate to satisfy the peak irrigation water requirement as described in paragraphs 3.1.1 and 3.1.2 of each and all crops to be irrigated within the design area. The capacity shall include an allowance for water losses (evaporation, runoff, deep percolation) that may occur during application periods. The system shall have the capacity to apply a stated amount of water to the design area in a specified net operation period. The system should have a minimum design capacity sufficient to deliver the peak daily irrigation water requirements in about 90% of the time available or not more than 22 h of operation per day. If a system is designed with a capacity less than the peak daily irrigation water requirement, the design capacity shall be stated in writing.

3.1.1 Design according to peak irrigation requirement. Where irrigation provides all or part of the water to the crop, the system shall have the capacity to meet the peak daily irrigation requirements of all crops irrigated within the design area. Unless field research with microirrigation systems is available, peak daily irrigation water requirements for crops determined with conventional irrigation systems should be used to determine system capacity.

3.1.2 Special cases. If specified by the user (e.g., economic considerations, especially in areas of frequent rainfall) and/or for special uses, the system may be designed with a capacity to apply a required volume of water, which is less than peak, to a design area in a specified net operating period.

3.2 Emitter discharge rate. The following conditions shall be met:

3.2.1 For drip, subsurface, and spray irrigation, the emitter discharge rate should not create runoff within the immediate application area. Small depressional ponds may develop beneath or above an emitter, but channelization to a furrow or other nearby low-lying area should be avoided. In fields with varying soil types, this criterion shall apply to the soil with the lowest infiltration rate unless it is less than 15% of the area irrigated.

3.2.2 For bubbler irrigation, a basin beneath the plant canopy will be required for water control. Applications shall generally be confined to the basin area.

3.2.3 Where natural precipitation and/or stored soil water is not sufficient for germination, special provisions shall be made for germination, or the microirrigation system shall apply water at a rate sufficient to adequately wet the soil to germinate seeds or establish transplants. The depth of a subsurface system for use on annual crops shall be limited by the ability of the system to germinate the seeds, unless it is stated in writing that other provisions will be required for this function.

3.2.4 Proper emitter discharge rate shall be determined and specified. Infiltration rates for different types of local, bare soils may be obtained from responsible agricultural technicians. In the absence of such advice, the proper emitter discharge rate may be estimated on the basis of past experience with similar soil types. In new areas field tests are recommended.

TABLE 1—RECOMMENDED CLASSIFICATION OF MANUFACTURER'S COEFFICIENT OF VARIATION (C_v)

Emitter type	C_v range	Classification
Point-source	<0.05	excellent
	0.05 to 0.07	average
	0.07 to 0.11	marginal
	0.11 to 0.15	poor
	>0.15	unacceptable
Line-source	<0.10	good
	0.10 to 0.20	average
	>0.20	marginal to unacceptable

3.3 Number and spacing of emitters. The number and spacing of emitters along the lateral line depend upon the emitter discharge rate, system capacity, soil water-holding capacity, lateral spread of water from the emission points, crop being grown, depth of irrigated root zone, desired water application efficiency and emitter discharge variability. Information on soil water-holding capacity and effective crop rooting depth can usually be obtained from responsible agricultural technicians, or may be estimated on the basis of past experience with similar crops and soil types. The lateral spread of water may also be estimated from past experience, but in new areas field tests are recommended. The area wetted as a percent of the total crop area may range from a low of 20% for widely spaced crops, such as trees for irrigation in high rainfall regions, to a high of over 75% for row crops in low rainfall regions.

3.4 Operating pressure. The design operating pressure shall be in accordance with the recommendations of the manufacturers. The system operating pressure must compensate for pressure losses through system components and field elevation effects.

3.5 Water application uniformity. The water application uniformity (for nonpressure compensating emitters) is affected by the operating pressure, emitter spacing, land slope, pipeline size, emitter discharge rate and emitter discharge variability. The emitter discharge variability is due to pressure and temperature changes, manufacturing variability, aging and clogging.

3.5.1 Emitter manufacturing variability. The expected manufacturer's coefficient of variation (C_v) should be available for new emitters operated at a constant temperature and near the design emitter operating pressure. A general guide for classifying C_v values is shown in Table 1.

3.5.2 Design emission uniformity. To estimate design emission uniformity in terms of C_v and pressure variations at the emitter, the following equation is suggested:

$$EU = 100 \left[1.0 - \frac{1.27\, C_v}{\sqrt{n}}\right] \frac{q_m}{q_a}$$

where

EU = the design emission uniformity, %

n = for a point-source emitter on a perennial crop, the number of emitters per plant; for a line-source emitter on an annual or perennial row crop, either the lateral rooting diameter of the plants divided by the same unit length of lateral line used to calculate C_v or 1, which is greater

C_v = the manufacturer's coefficient of variation for point or line-source emitters

q_m = the minimum emitter discharge rate for the minimum pressure in the subunit, L/h (gal/h)

q_a = the average or design emitter discharge rate for the subunit, L/h (gal/h)

Table 2 shows recommended ranges of EU values.

3.5.3 Allowable pressure variations. The following recommendations are made to reduce pressure loss and minimize pipeline sizes in microsystems:

3.5.3.1 Pressure differences at the emitters throughout the system (or block or subunit) should be maintained in a range such that the desired design emission uniformity (EU) is obtained. For example, from the equation in paragraph 3.5.2, with an EU of 80%, a C_v of 0.10, and one emitter per plant, the ratio between the minimum and average emitter discharge rate should be no less than 0.92. Since the allowable pressure loss

TABLE 2—RECOMMENDED RANGES OF DESIGN EMISSION UNIFORMITY (EU)

Emitter type	Spacing (m)	Topography	Slope, %	EU range, %
Point source on perennial crops	>4	uniform steep or undulating	<2 >2	90 to 95 85 to 90
Point source on perennial or semipermanent crops	<4	uniform steep or undulating	<2 >2	85 to 90 80 to 90
Line source on annual or perennial crops	All	uniform steep or undulating	<2 >2	80 to 90 70 to 85

corresponding to the minimum emitter discharge rate will differ depending on the emitter characteristics, the allowable pressure variation should be stated in writing for the specific emitter type and C_v specified.

3.5.3.2 Field shape and slope frequently dictate the most economical lateral direction. Whenever possible, laterals should be laid downslope for slopes of less than 5% if lateral size reduction can be attained. For steeper terrain, lateral lines should be laid along the field contour and pressure compensating emitters should be specified or pressure control devices used along downsloped laterals.

3.5.3.3 Excessive main or submain pressure differences can result in widely varying manifold or lateral takeoff pressures. In some instances, these excessive variations cannot be controlled by main or submain size alone. The only practical alternative is to design for adequate pressure at the lateral lines and properly regulate the pressures at the manifold or lateral lines. This pressure regulation may be accomplished by using automatic presesure regulators, fixed orifice or flow control pipe restrictions, or manually set valves.

3.5.3.4 Pipe sizes for mains and submains should be chosen after considering pipe costs and power costs, while keeping flow velocities within recommended limits for surge control and accounting for the effect on design emission uniformity from the resulting pressure variations. Additional information can be found in ASAE Standard S376, Design, Installation and Performance of Underground, Thermoplastic Irrigation Pipelines.

3.6 Filtration systems. A general design recommendation for the water filtration system should include location, size, specification of allowable suspended material sizes, types of filter or filters, and maintenance requirements.

3.6.1 Location. A primary filter shall be located after the pump and chemical injection point to remove both large and fine particles from the flow. Secondary filters may be used downstream from the primary filter to remove any particles which may pass through the primary filter during normal or cleaning operations. When secondary filters are used, the size of the openings is usually larger than that of the primary filter to minimize needed attention. Water meters, solenoid-operated valves and final pressure regulators should follow the primary and secondary filters. Lateral-line or in-line filters can be used as additional protection.

3.6.2 Size. Filter flow openings shall be sufficiently small to prevent the passage of unwanted particles into the system. When available, recommendations of the emitter manufacturer shall be used to select the size of the filtration system. In the absence of manufacturer's recommendations, the filter size should be based on the diameter of the emitter opening or the type and size of contaminants to be filtered. The capacity of the filter should be sufficiently large to permit the rated flow without frequent cleaning. Filters that are to be cleaned by hand should not require more than daily maintenance. The maximum permissible head loss across the filter shall be 70 kPa (10 psi) before filter cleaning is required.

3.6.3 Types. Filtration may be accomplished through the use of pressure filters (screen and media) and gravity filters (centrifugal separators, gravity screen filters and settling basins).

3.6.3.1 Pressure screen filters. This filter consists of a screen made of metal, plastic, or synthetic cloth enclosed in a special housing used to limit maximum particle size. The presence of algae in irrigation water tends to cause screen blockage and can considerably reduce filtering capacity. Screens are classified according to the number of openings per inch with standard wire size for each screen size. Most manufacturers recommend 150 to 75 micron (100 to 200 mesh) screens for emitters, but some recommend screens as coarse as 600 microns (30 mesh). Screen flow capacity should not exceed 135 L/s per square meter (200 gpm per square foot) of screen opening.

3.6.3.2 Media filters. Media filters consist of fine gravel and sand of selected sizes placed in pressurized tanks. Media can lose effectiveness with time (due to rounding, etc.) and should be replaced after extended usage. Media filters are not easily plugged by algae and can remove relatively large amounts of suspended solids before cleaning is needed. Cleaning is accomplished by forcing water backwards through the filter (backflushing). Media filters in current use will retain particle sizes in the range of 25 to 200 microns. In general, water flow rates through the filters should be between 10 and 18 L/s per square meter (14 and 26 gpm/per square foot) of filtration surface area. Media filters should be followed by secondary screen filter or a rinse cycle valve to prevent carryover of contaminants following the backwashing process.

3.6.3.3 Centrifugal separators. Sand separators, hydrocyclones or centrifugal filters remove suspended particles that have a specific gravity greater than water. These filters are ineffective in removing most organic solids. A sand separator can effectively remove a large number of sand particles and may be installed on the suction side of the pump as a prefilter to reduce pump wear.

3.6.3.4 Settling basins. Settling basins, ponds or reservoirs can be used as a form of pre-filtration treatment, but unless covered, the water is exposed to wind-blown contaminants and algae growth. Open reservoirs can be treated with commercially available algicides; however, care must be taken to avoid potential environmental hazards. When used, basins should be sized to limit turbulence and permit a minimum of 15 min for water to travel from the basin inlet to the pumping system intake.

3.6.3.5 Gravity screen filters. Gravity screen filters rely upon gravity instead of water pressure to move water through the screen. Pressure losses across gravity screen filters rarely exceed 7 kPa (1 psi) consequently they can be used in systems where pressure losses must be minimized. They are also effective in removing organic (i.e., algae) as well as inorganic contaminants.

3.7 Flushing system: To assist in keeping sediment buildup at a minimum, automatic or hand flushing of all microirrigation pipelines is recommended on a regular time schedule. Filtration should be effective enough so that flushing of the system is needed no more frequently than once per week.

3.7.1 Location. Valves shall be provided at the ends of mains and submains and provisions made for flushing of lateral lines. All connections and pipeline fittings shall be large enough in diameter to facilitate flushing.

3.7.2 Capacity. A minimum flow velocity of 0.3 m/s (1 ft/s) is needed for flushing of lateral lines. Because only a few lateral lines can be flushed at one time, the flushing system should be adequately valved so that subunits can be flushed independently.

3.8 Chemical water treatment: The need for chemical water treatment depends primarily on the type of microirrigation system used and the composition of the water. Acids and bactericides are both used for prevention of emitter clogging and renovation after clogging occurs.

3.8.1 Acids. The least expensive acid available can be used, usually at a concentration which is sufficient to offset calcium, magnesium or iron carbonate and bicarbonate precipitation. Another method for dealing with water high in bicarbonates is to aerate the water and hold it in a reservoir until equilibrium is reached and the precipitates have settled out.

NOTE: The reduction of water pH may cause harmful effects on crop production if the soil pH is low.

3.8.2 Bactericides. Calcium hypochlorite, sodium hypochlorite, chlorine gas, or other algicides and bactericides can be added continuously or periodically at the control station to inhibit bacterial growth. Registration of these chemicals for use in microirrigation systems may be required. For high pH waters, acids can be injected to adjust pH to increase the bacteria-killing property of the hypochlorites. If hypochlorite and acids are used simultaneously, they shall be injected from separate sources to avoid the possibility of generating lethal chlorine gas. Gas chlorination or hypochlorination is not recommended for water containing more than 0.4 mg/L (0.4 ppm) dissolved iron, since it can lead to the precipitation of iron that may not be filterable and may deposit in pipelines and emitters. The amount of chlorine to be added depends on the chlorine demand and the potential for bacterial emitter clogging. Where bacterial control is needed, enough chlorine should be added to have some measurable amount of free residual chlorine (at least 0.1 mg/L [0.1 ppm]) present at the ends of all lateral lines after a chemical injection period. When continuous or frequent monitoring for chlorine is not possible, it may be prudent to target for 1.0 mg/L (1.0 ppm) free chlorine at the ends of laterals to provide a sufficient safety factor to cover fluctuations in chlorine demand between monitoring times.

3.9 Fertilization system. Microirrigation systems provide a convenient method of supplying nutrient materials to the crop; however, the effects of the nutrient on the system should be considered.

3.9.1 Nitrogen. Ammonium sulfate, ammonium nitrate and urea have been used at low concentrations with no harmful effects on the water or irrigation system. Anhydrous ammonia, aqueous ammonia or ammophos increase pH and can cause chemical precipitation which can clog emitters, particularly with high pH water.

3.9.2 Phosphorus. Commonly used phosphate fertilizers tend to precipitate in irrigation waters with high hydroxyl, calcium and manganese contents. Phosphorous fertilizers can become immobilized in the soil. Phosphoric acid and water-soluble organic phosphorous compounds have been successfully used in microsystems; however, common practice is to apply the phosphorous separately and not through the irrigation system.

3.9.3 Potassium. There are no problems associated with potassium application through microirrigation systems.

3.9.4 Micronutrients. Manganese, zinc, iron, copper, etc., may be applied as soluble salts through the irrigation system. These should each be injected separately and apart from other fertilizers and chemicals to avoid chemical interaction and precipitation in emitters. The effects of improper application amounts on the nutrient balance in the soil should be considered when applying micronutrients.

3.10 Injection system. ASAE Engineering Practice EP409, Safety Devices for Applying Liquid Chemicals Through Irrigation Systems, specifies required safety devices when injecting chemicals into irrigation systems. The following are some of the more important considerations that shall be considered in the design of a chemical or fertilizer injection system:

3.10.1 Injection method and rate. Fertilizers may be injected by a differential pressure system, venturi injector or by pumping under pressure (pressure injected) into the irrigation water. Other chemicals such as acids, bactericides and chlorine which require a constant concentration rate entering the irrigation water shall be injected by constant rate injection devices only. Chemicals shall be injected into the system before the primary filter. The required rate of chemical injection depends on the initial concentration of chemical and the desired concentration of chemical to be applied during the irrigation. An injector that will operate within a range of injection rates for a number of chemicals may be desired.

3.10.2 Concentration. The concentration of chemicals to be injected in the irrigation water is normally very low, with ranges of 4-100 mg/L (4-100 ppm) for fertilizers and chemicals and 0.5-10 mg/L (0.5-10 ppm) for bactericides. Chemical concentrations should be routinely measured after filters and before main pipelines, and occasionally at the end of the last lateral line, to check if the entire microirrigation system is being treated.

3.10.3 Storage tank capacity. Large, low cost tanks constructed from epoxy-coated metal, plastic or fiberglass are usually practical when injection pumps are used. For a pressure differential injection system, the high pressure rated chemical tank should have enough capacity for a complete application.

3.10.4 Contamination of water supply. When the water supply or pump fails, it may be possible to have reverse flow from the injection or irrigation system. A pressure switch or other means should be used to turn off the injector pump or close a solenoid-operated valve on a differential pressure system so that chemicals cannot be injected when the irrigation pump is not operating. A vacuum-breaker and check valve should be placed between the water supply pump and the injection point to prevent backflow from the pipeline to the water supply. In addition, municipalities may require other backflow prevention valves at connections to municipal water lines to prevent reverse flow, particularly when water is being treated with chemicals or amendments.

3.10.5 Resistance to chemicals. All hardware shall be resistant to reaction with the chemicals being injected.

3.11 Flow monitoring: A flow measuring device shall be installed as part of the control station to aid in scheduling irrigations and monitoring system performance. Flow measurements can indicate if pressure regulation is malfunctioning, excessive leaks exist, emitters are clogging or emitter opening enlargement is occurring. Some flow measuring devices give both the accumulated flow and instantaneous flow rate. Accumulated flow over a specified period of time can be used to determine flow rate if instantaneous readings are not available.

3.12 Safety

3.12.1 The design or operation of the irrigation systems shall prevent the leaking or spraying of water on electrical lines or power units.

3.12.2 Protective devices on chemical injection equipment shall be provided to prevent contamination of the water supply or unplanned discharge of chemicals. A water source should be provided near the chemical tank for washing off chemicals if skin contact occurs. Protective clothing is advisable when handling chemicals.

3.12.3 Pumps and power units shall be set on a firm base and kept in proper alignment. The installation shall comply with ASAE Standard S318, Safety for Agricultural Equipment.

3.12.4 Wiring and starting equipment for electrically-operated plants shall comply with overload and low voltage protection requirements, and any applicable electrical codes. Electrical installation shall comply with ASAE Standard S397, Electrical Service and Equipment for Irrigation.

3.12.5 Pumps and power units shall be provided with protective devices. Thermostats shall be provided which stop the power units when engine or motor temperatures exceed safety points. If failure of the water supply might cause the pump to lose its prime, the motor or engine shall be properly protected. Chemical and fertilizer injection equipment shall be automatically turned off when the irrigation pump is not operating. Detailed discussions of irrigation safety devices are found in ASAE Engineering Practice EP409, Safety Devices for Applying Liquid Chemicals Through Irrigation Systems.

Cited Standards:

ASAE S318, Safety for Agricultural Equipment

ASAE S376, Design, Installation and Performance of Underground, Thermoplastic Irrigation Pipelines

ASAE S397, Electrical Service and Equipment for Irrigation

ASAE EP409, Safety Devices for Applying Liquid Chemicals Through Irrigation Systems

ASAE Engineering Practice: ASAE EP407

AGRICULTURAL DRAINAGE OUTLETS—OPEN CHANNELS

Developed by the ASAE Surface Drainage Committee; approved by the ASAE Soil and Water Division Standards Committee; adopted by ASAE January 1982; reconfirmed December 1986; reconfirmed for one year December 1987.

SECTION 1—PURPOSE AND SCOPE

1.1 The purpose of this Engineering Practice is to provide planners and engineers with planning, design, construction and maintenance information, and criteria for agricultural drainage outlets by means of open channels, in order to provide adequate drainage for agricultural production and protection of environmental values. This Engineering Practice is compatible with ASAE Engineering Practices ASAE EP302, Design and Construction of Surface Drainage Systems on Agricultural Lands in Humid Areas, and ASAE EP260, Design and Construction of Subsurface Drains in Humid Areas.

1.2 This Engineering Practice covers all engineering phases of open-channel planning, design, construction, and maintenance for projects within the following limits:

1.2.1 Watersheds that are larger than 2.6 km^2 (1 $mile^2$) and smaller than 260 km^2 (100 $mile^2$) with average slopes generally less than 4.7 m/km (25 ft/mile).

1.2.2 Channels that are excavated in earthen material to convey design flows within banks designed to be stable and maintainable.

1.2.3 Channel work that does not have a serious impact on significant environmental values or that has adverse effects minimized.

1.2.4 Project development that is physically and economically sound, environmentally and socially acceptable and compatible with long-term use of natural resources.

1.2.5 Project development that uses safe construction and maintenance practices.

SECTION 2—RESOURCE DEVELOPMENT

2.1 The removal of excess water from land used for agricultural production is essential to obtain optimum crop qualities and yields.

2.2 Land resources have a wide variety of values that satisfy many objectives and needs of people. Use of these resources for agricultural production may have impact on other values. To reduce this impact, drainage development should focus primarily on existing pasture and cropland or land suitable for agricultural production. Objectives should be evaluated for maximizing economic returns, while minimizing adverse environmental impacts, and providing maximum beneficial use of water, land and natural resources.

SECTION 3—SITE INVESTIGATIONS

3.1 Reconnaissance and preliminary investigations should consist of the following:

3.1.1 A field reconnaissance, by an interdisciplinary team, to establish the scope and objectives and probable conflicts of resource usages. The watersheds above and downstream from the problem area should be included to determine possible detrimental effects. Discussions with landowners and others will provide information on local support for the project, types of improvement needed, and an indication of their priorities.

3.1.2 An investigation to establish physical and economic feasibility. This investigation is an extension of the field reconnaissance survey. It should include (a) a working map of the project and watershed showing the pertinent physical features; (b) a generalized soil and land use map; (c) an environmental assessment; (d) a preliminary plan for each objective; (e) an estimate of project cost and benefits; and (f) an assessment of impacts. Use should be made of existing data such as aerial photographs, topographic maps, property ownership, local surveys, detailed soil maps, and legal requirements.

3.2 Adequacy of an outlet for a proposed channel should be examined to insure that it will safely carry the discharge from the project.

3.3 Legal requirements of all state and federal laws should be strictly observed.

3.4 Engineering field survey data should have an established system of horizontal and vertical controls which adequately define channel alignment, channel bank and bottom profiles, topography of adjacent land and sufficient cross sections to accurately identify channel excavation and hydraulic conditions. Whenever possible, mean sea level datum using third order vertical control or better should be used. Drawings and maps should be prepared displaying survey data, topography, drainage patterns, soils, land use, land capability, drainage area, benefited area, and ownership.

3.5 Subsurface investigations should be made to obtain information on the soils, geology, and water table elevations. Samples for laboratory analysis should be obtained as needed. Piezometers are useful for determining groundwater elevations and movement.

3.5.1 Information from each boring should be located on a plan view and plotted on the profile. Logs should be plotted using the Unified Soil Classification System.

3.5.2 Laboratory tests should be made for mechanical analysis, plasticity indices, and dispersion. In addition, sheer strength, hydraulic conductivity, moisture aeration curves, and drainable pore space data may be required for final design.

SECTION 4—DESIGN DISCHARGES

4.1 The design discharge of the channel or system of channels should be the discharge that provides for the most economical and feasible level of agricultural production.

4.1.1 The design discharge should remove excess water from the watershed in 24 h for most agricultural crops. A longer time period may be allowed for woodland, pasture and some crops.

4.1.2 Floodwater retarding structures should be considered when selecting a design discharge.

4.2 Drainage coefficients are defined as that discharge capacity which removes a specified depth (volume) of water from the watershed in a 24 h period. The drainage coefficients include the capacity for direct runoff and base flow from the watershed. In some instances, especially large watersheds fed by many subsurface drains or seeps and springs and in irrigation areas, separation of flow sources may be desirable for design purposes. The accretion discharge from subsurface drains or seeps and springs may be a significant portion of the flow in channels in humid areas. In arid regions the flow from irrigated lands may be the only inflow to some channel systems.

4.2.1 Where applicable, ASAE Engineering Practice ASAE EP260, Design and Construction of Subsurface Drains in Humid Areas, should be used to determine subsurface drainage discharge for channel design capacities. Seepage and discharge from springs should be considered and measured if of significant magnitude.

4.2.2 For arid areas, drainage and seepage discharge from irrigated lands may be a large portion of the required collection channel capacities. Experience in similar watersheds will provide the best estimate of discharges. Measured discharges have ranged from 0.0018 to 0.70 $m^3/s{\cdot}km$ (0.10 to 40.0 $ft^3/s{\cdot}mile$) of channel. The discharge may also be rationalized from the amount of irrigation water that is applied.

$$q = i\frac{\frac{P + C}{100}}{T}$$

where

q = drainage coefficient, mm/24 h (in./24 h)
P = deep percolation from irrigation and leaching, percent
C = field canal losses, percent
i = irrigation application, mm (in.)
T = time between irrigation, days

4.2.3 Drainage coefficients for collection channel systems for irrigated areas are usually in the range of 3 to 6 mm/24 h (0.12 to 0.24 in./24 h). The disposal system for several irrigated tracts usually has a smaller combined coefficient than the sum of the collection system components. This flow may be in the range of 0.15 to 0.30 $m^3/s{\cdot}km^2$ (2 to 4 $ft^3/s{\cdot}mile^2$). Deep percolation, consumptive use by trees and phreatophyes and land temporarily not irrigated along with cultural practices account for the reduction in combined flow to the disposal system.

4.3 Drainage curves provide the best values of surface drainage discharge in humid regions. The curves apply to watersheds with average land slopes of 1 percent or less and are applicable to watersheds with minor areas of steeper land.

4.3.1 The basic formula historically used to develop drainage curves is in customary units:

$$Q = CM^{5/6}$$

where

Q = design discharge, ft^3/s
C = coefficient relating to watershed and storm characteristics and the level of drainage to be provided for the crops grown
M = drainage area, $mile^2$

4.3.2 The basic formula converted to SI units is

$$Q = 0.013\ CM^{5/6}$$

where

Q = design discharge, m^3/s
0.013 = conversion factor to enable use of SI units in the original formula.
C = coefficient having the same definition and magnitude as in paragraph 4.3.1.
M = drainage area, km^2

4.3.3 The coefficient C and the 5/6 power integrate the effect of all parameters influencing the discharge. Judicious evaluation of the coefficient is essential for successful drainage project designs.

4.3.4 Values of C and associated drainage curves that have provided adequate design discharges for numerous drainage projects are shown in Figs. 1 and 2. Fig. 1 is in SI units while Fig. 2 displays the same data in customary units. These curves can be used with confidence for design. Gaged runoff data, if available, can be used to evaluate C.

4.3.5 Historical application of drainage curves used locally should be reviewed before selecting final design discharges. These curves provide for out-of-bank flow and storage of water on the land for a duration of at least 24 h.

4.4 Adjustments of drainage curve values are necessary to approximate actual discharges from nontypical watersheds in humid areas.

4.4.1 Curve values for discharges from watersheds with noncontributing areas such as swamps and lakes should be adjusted. These adjustments may be made by deducting the volume of runoff withheld or by adjusting drainage areas before entering the design curves.

4.4.2 Controlled overflow from outside the natural watersheds may necessitate increased design discharges.

4.4.3 Design discharges from relatively steep watersheds should be derived from hydrographs for selected storms or by using C values adjusted upward to provide higher levels of protection against flooding.

4.4.4 The design discharge below a junction depends on the ratio of the drainage areas. An emperical procedure called the 20-40 rule is a recommended method for determining the design discharge below a junction.

4.4.4.1 When the watershed of one of the ditches is from 40 to 50 percent of the total watershed, the design discharge of the channel below the junction should be the sum of the design discharges of the ditches above the junction.

4.4.4.2 When the watershed of a lateral is less than 20 percent

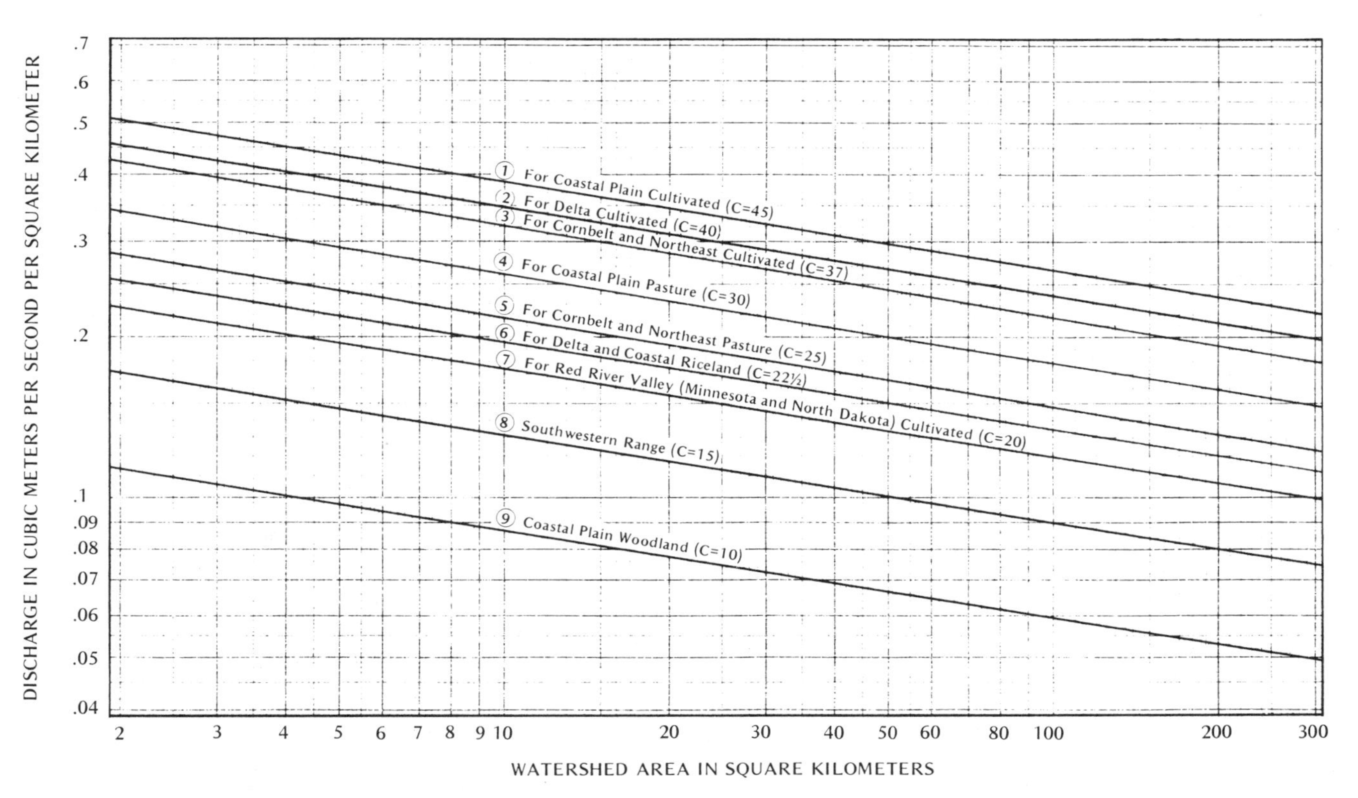

FIG. 1—TYPICAL DRAINAGE CURVES SUITABLE FOR DESIGN IN TYPICAL DRAINAGE WATERSHED IN HUMID AREAS, SI UNITS

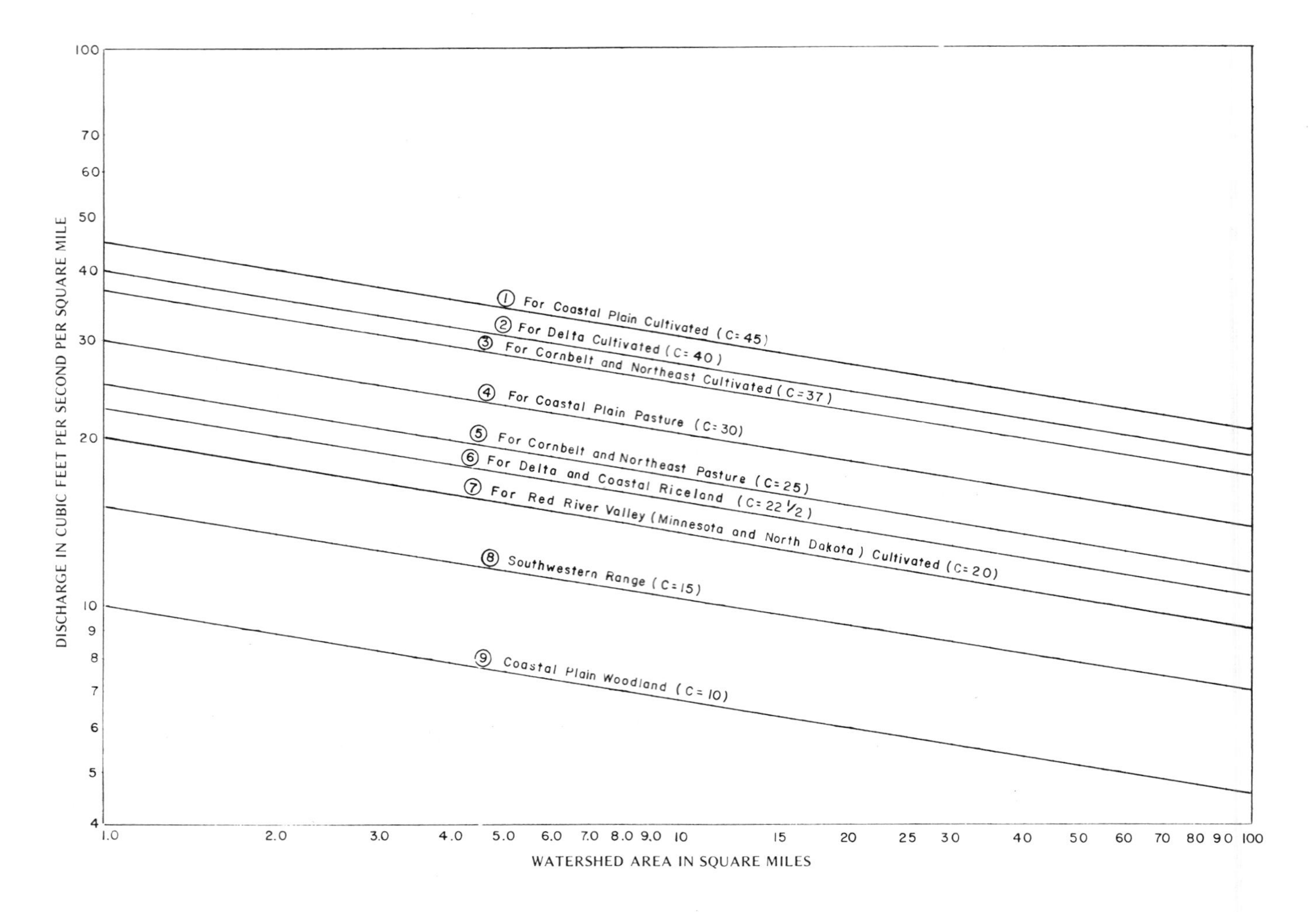

FIG. 2—TYPICAL DRAINAGE CURVES SUITABLE FOR DESIGN IN TYPICAL DRAINAGE WATERSHED IN HUMID AREAS, CUSTOMARY UNITS

of the total watershed, the design discharge of the channel below the junction should be the design discharge from the total watershed treated as a unit.

4.4.4.3 When the watershed of a lateral is in the range of 20 to 40 percent of the total watershed, the design discharge of the channel below the junction should be interpolated between the values obtained using the methods outlined in paragraphs 4.4.4.1 and 4.4.4.2.

4.5 In arid regions where drainage curves are not available, peak discharge calculations should determine design discharges.

4.6 Stream gage data including peaks and volumes should be used for development of design discharges whenever feasible. The gage data will give credence to values obtained using drainage coefficients, drainage curves or peak discharge calculations.

SECTION 5—CHANNEL DESIGN

5.1 Channel locations should be determined by considering topography, existing drainage patterns, and groundwater movement.

5.1.1 Channels should be located to provide adequate outlets for on-farm drainage systems, subsurface drainage, interception channels and irrigation systems. The smaller laterals, when feasible, should be located along farm boundaries or parallel to field and other boundaries. The larger main channels should be located to serve the laterals. These channels should generally follow existing drainage patterns.

5.1.2 The location of channels in flatland areas requires careful study of microrelief and watershed boundaries. Overland flooding between watersheds is a common problem. Diversion or interception channels or dikes may be required to provide positive watershed boundary control.

5.1.3 Channels in the narrow flood plains of rolling or hilly land usually follow the natural channels. Many of these channels have natural levees built by sedimentation processes with drainage parallel to the channels.

5.1.4 Meandering in existing channels is an indication of instability. Straightening these channels may increase instability.

5.1.5 The locations of some channels are fixed by farm boundaries or rights-of-way owned or controlled by districts or similar organizations.

5.1.6 The locations of the channels should consider the need for easements and rights-of-way for construction, inspection, operation and maintenance.

5.1.7 The locations of channels should be planned to make use of existing bridges and structures. Care should be taken to minimize the isolation of fields or parts of fields from the rest of a farm.

5.2 The design capacity of channel systems should be based on the design discharges of their watersheds.

5.2.1 Design discharges should be determined and tabulated for all locations that may require changes in channel design. Typical locations are at junctions, crossings, changes in drainage area size and at changes in channel grades.

5.3 Design elevations for the channel hydraulic gradeline and bottom should be calculated beginning at the outlet.

5.3.1 The design of a tributary junctioning with an outlet requires establishing the stage on the outlet at design discharge.

5.3.1.1 When the drainage area of the outlet and tributary are approximately the same size, the design elevations should be determined using the same discharge frequencies.

5.3.1.2 An estimate of outlet hydraulic gradeline elevation may be calculated by developing a water surface profile for the outlet. The profile should start downstream from the point of intersection and extend upstream past the point of intersection.

5.3.2 A water surface control line should be established on profile sheets, using control elevations obtained from field surveys. These may include low ground, channel banks, levees, bridge and structure elevations, and the control elevations of laterals.

5.3.3 The hydraulic gradeline should approximate the water control line. The preliminary gradeline should be developed using uniform flow conditions. This should be checked by water surface profile calculations.

5.3.4 Energy gradients in most channels are not important, but may be required for design of in-channel structures.

5.3.5 The distance between the channel bottom and the hydraulic gradeline is the depth of the channel. The bottom elevation may be controlled by the need for subsurface drainage, existing channel bottoms, soil material, water tables, depth of laterals, outlet elevation, depth for stability design, or other considerations.

5.4 Channels should be designed using as uniform a cross section as possible. A trapezoidal cross section is most common.

5.4.1 Soil materials should be carefully examined and analyzed to determine the minimum acceptable side slope. A review of existing channels in similar material should be made to verify the stability of the selected side slope. Consideration should be given to establishing vegetation, maintenance, and safety in determining final side slopes.

5.4.2 The design capacity of a channel should be for an aged condition using Manning's equation and a value of Manning's n that can be reasonably maintained. Manning's n for the channel may be determined using several graphs or charts.

5.4.3. Channels should be designed to minimize sedimentation, maintenance, erosion, and bank failure. A minimum velocity of approximately 0.6 m/s (2 ft/s) should be used to prevent excessive sedimentation and maintenance problems. Analytical procedures for stability analysis are described by USDA Soil Conservation Service, Technical Release No. 25—Design of Open Channels (1977).

5.5 Excavated material should be placed and shaped in a manner which will minimize overbank wash, provide access for maintenance equipment, prevent excavated material from washing or rolling back into the channel, and prevent sloughing of channel banks caused by heavy loads too near the edge of the channel. Caution should be used when operating equipment on the shaped spoil. Slopes steeper than 4:1 should be farmed with special equipment. Safety devices, such as protective frames or canopies, may be required to protect equipment operators. Fig. 3 shows dimensions for a typical channel system.

5.5.1 As a general rule, spoil material should be equally divided and placed along each side of the channel so that the locations of overbank flow into the channel are reduced to a minimum. An exception is when construction is from one side only. In these cases, the spoil should be placed on the construction side and the offside left undisturbed.

5.5.2 Spoil material should be disposed of in a manner to enhance the aesthetics of the site to the extent feasible.

5.6 Berms provide travelways for maintenance and for work areas to facilitate spoil spreading. Berms may be required to prevent sloughing of channel banks caused by heavy soil loads near the edge of the channel.

5.6.1 The minimum berm width should be 3 m (10 ft) or twice the channel depth with a maximum berm width of 6 m (20 ft).

5.6.2 If the spoil banks are to be leveled to blend into cultivated land, berms are not required.

5.7 Channel alignment should consider geologic and soil conditions. In many cases, the alignment of the existing channel may be satisfactory.

5.7.1 The radius of curvature influences velocity and the stability of side slopes. Recommended design criteria for curvature are given in Table 1.

TABLE 1—RECOMMENDED DESIGN CRITERIA FOR CURVATURE

Channel top width		Slope	Minimum radius of curvature		Approximate degree of curve
m	ft	m/m (ft/ft)	m	ft	deg
4.6	15	0.0006	91	300	19
4.6	15	0.0006-0.0012	122	400	14
5-11	15-35	0.0006	152	500	11
5-11	15-35	0.0006-0.0012	183	600	10
11	35	0.0006	183	600	10
11	35	0.0006-0.0012	244	800	7

5.7.2 Final channel alignment should provide adequate access for on-farm drainage systems.

5.8 Where feasible, vegetation should be planned for channel side slopes, berms, and spoil areas. Vegetation stabilizes the soil, reduces damage from sediment and runoff to downstream areas, provides wildlife habitat, and enhances natural beauty. The side slopes and berms should be seeded within 24 h after excavation. The spoil should be seeded and fertilized immediately after spreading or shaping. Temporary seedings or mulching should be used during seasons when permanent seeding is not practical. The types and species of vegetation should be governed by site locations and local conditions.

SECTION 6—IN-CHANNEL STRUCTURES

6.1 In-channel structures may be installed to control water levels, channel grades, or to provide crossings. Other structures may include water gates, deflectors, and low weirs to form riffles and pools.

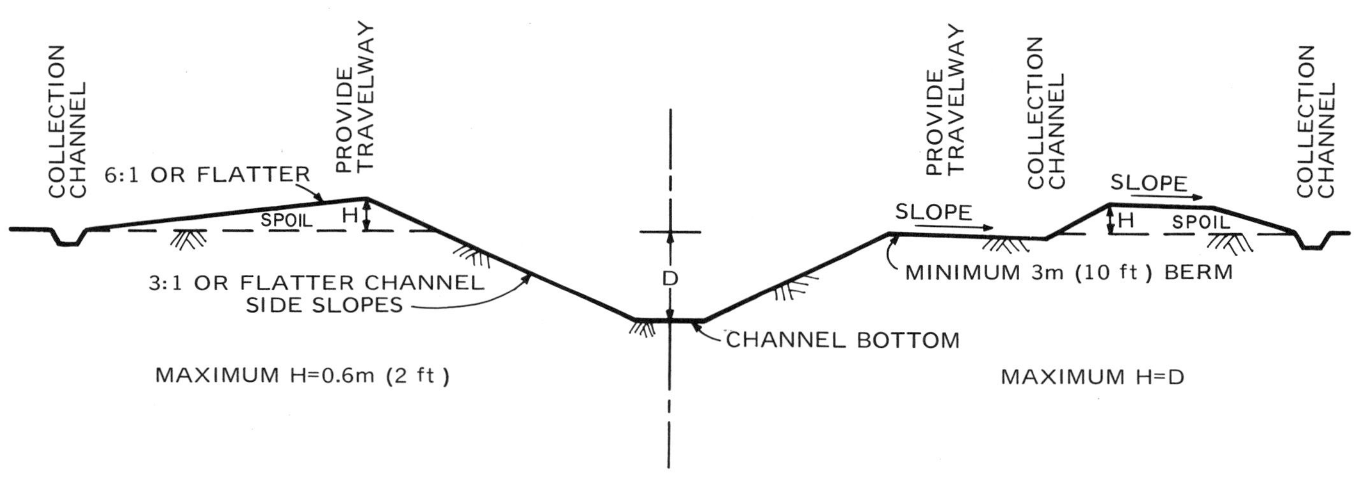

FIG. 3—TYPICAL METHODS OF HANDLING SPOIL AND LAYOUTS OF CHANNEL DIMENSIONS

6.1.1 Structures should be designed commensurate with their purposes and the probable damages associated with failures. In no case should the design capacity be less than the design discharge of the channel.

6.1.2 Channel crossings for highways should meet the design criteria of appropriate government agencies. They should normally be designed to pass at least the 25-yr flood without severe damage. Farm road crossings, bridges, and culverts should pass at least the design discharge. Stream fords or ramps properly designed are generally useful crossings for channels where the restrictive flows are not of more than 72 h duration. The top of the stone or concrete should be approximately at the channel bottom.

6.1.3 Flood flows greater than the design discharges should normally be bypassed. A spillway area should be placed upstream and downstream at least 15 m (50 ft) from the structure. The structure should be designed with elevated wingwalls that are tied into short reaches of spoil, so that at flood flows the excess water is safely bypassed around the structure, and the structure appears like an island. The associated fill or shaped spoil should be vegetated to improve stability.

6.2 Side slope and bottom protection should be considered in conjunction with structures, curves, and erosive soil materials.

6.2.1 Protection can be provided in several ways, such as riprap blankets to protect soils, and drainage rock toes to control seepage. Blankets of rock riprap may be used to reduce erosion at channel junctions, pipe outlets, overland flow returns to channels, and overflow sections.

6.2.2 Rock riprap design should be based on velocity or tractive stress. The need for a filter or base material should be considered.

SECTION 7—SIDE INLET CHANNEL STRUCTURES

7.1 Control of overbank flow into the channel is essential. Spoil material plus pipe inlets should be used to control water from field surfaces and small drains entering the channel. The hydrologic design criteria for these structures should be based on the discharge expected from a storm commensurate with the probable damage associated with structure failure. For most structures, discharges from drainage curves should be used. The design should take into account the wide fluctuation in water surface stages in the channel that may exist at time of operation.

7.2 Three different types of structures are normally used to lower large volumes of water from smaller channels into larger channels. All earthwork associated with these structures should be adequately compacted and promptly seeded or otherwise protected from erosion. If debris from the watershed is expected, then full flow structures should be considered.

7.2.1 Pipe structures may be used. Drop inlets should be attached to large volume, high head pipe drops. If ice or floating debris is a problem, the outlet end of the pipe should be placed flush with the channel side slope or recessed into the slope.

7.2.2 Chute structures can be designed for a wide range of capacities and heads. They can be constructed from many different materials, such as concrete, rock or sod.

7.2.3 Straight drop spillways should be used to lower water from large drainage areas.

7.3 Seepage is the main stability problem associated with side inlet structures. Foundation investigations may be a few soil borings for a culvert type pipe or a thorough geologic investigation for a large structure. Anti-seep collars should be used on pipe structures when the head exceeds 1.5 m (5 ft). Inlets should be protected from erosion by riprap and/or flared end sections.

7.4 Concrete structures should be protected from piping. Headwall extensions should be used to prevent piping around the structures. Drainage systems may be needed to control uplift and piping.

SECTION 8—CONSTRUCTION

8.1 Detail plans and specifications should be prepared for the structural improvements. These plans should establish the lines, grades and technical requirements for the proposed work and should be satisfactory for contracting. Recommended organization is: Cover Sheet with General Location Map; Ownership Map; Plan and Profile Sheets; Typical Sections; Structural Details; and Vegetative Treatment and Construction Specifications. A design report should be developed. Recommended major headings of the report are:

I. CORRESPONDENCE
II. GENERAL
  A. DESIGN RESUME
  B. SUMMARY OF DESIGN DECISIONS
III. HYDRAULICS
IV. GEOLOGY
V. SOIL TESTING
VI. EMBANKMENT AND FOUNDATION DESIGN OR STABILITY ANALYSIS
VII. STRUCTURAL
VIII. QUANTITIES AND PERFORMANCE TIME
IX. SPECIFICATIONS
X. DRAWINGS
XI. MAINTENANCE

8.2 Construction installation should be by formal contract with payments based on measurable items of work. A job showing to prospective bidders followed by a preconstruction conference with the successful bidder is recommended. Safety requirements and satisfactory control during construction should be stressed.

8.3 The construction should be checked and documented as needed to satisfy all concerned that the job is progressing satisfactorily.

8.4 As-built plans showing all changes made during construction should be completed for a permanent record of construction.

SECTION 9—OPERATION AND MAINTENANCE

9.1 Channels are constructed to carry the design discharge after aging. Channel aging normally includes the growth of grasses or similar vegetation on the side slopes and berms if they carry part of the design flow. Vegetative growth in excess of that considered in design should be removed to maintain channel capacities. Sediment bars and slides that impinge on the channel cross section should be removed. Structures may require periodic repair or replacement.

9.2 An operation and maintenance plan should be developed. An organization adequately financed should be established with responsibility for carrying out operation and maintenance. Channels should be inspected annually and after major storms. Emergency work should be done as needed. The design report should provide criteria for determining when repair, sediment removal or reconstruction is needed.

References: Last printed in 1982 AGRICULTURAL ENGINEERS YEARBOOK; list available from ASAE Headquarters.

Cited Standards:

ASAE EP260, Design and Construction of Subsurface Drains in Humid Areas
ASAE EP302, Design and Construction of Surface Drainage Systems on Agricultural Lands in Humid Areas

DESIGN AND INSTALLATION OF SURFACE IRRIGATION RUNOFF REUSE SYSTEMS

Developed by the ASAE Surface Irrigation Committee; approved by the ASAE Soil and Water Division Standards Committee; adopted by ASAE November 1981; revised February 1987.

SECTION 1—PURPOSE AND SCOPE

1.1 This Engineering Practice is intended as a guide to engineers and technicians in the design and installation of systems collecting runoff from surface irrigated fields for subsequent use on the same or another field. It is not intended to serve as a complete set of design criteria and specifications for construction.

SECTION 2—DEFINITIONS

2.1 Application duration: The elapsed time from the beginning of water application to an irrigation set to the time at which the inlet flow is terminated for that set.

2.2 Cutback irrigation: The reduction of the inflow stream after the advance phase of irrigation.

2.3 Cycling system: A system which recycles runoff as soon as a small reservoir accumulates enough water to provide proper pump operation.

2.4 Field application duration: The elapsed time from the beginning of water application to the first irrigation set to the time at which water application is terminated on the last irrigation set of a field.

2.5 Final infiltration rate: The soil infiltration rate at the maximum infiltration opportunity time for an irrigation.

2.6 Irrigation set: Refers to the area irrigated at one time within a field.

2.7 Pumping plant: In a reuse system the pumping plant consists of a trash screen, pump, power unit, and system operational controls.

2.8 Pump system: A system that utilizes a reservoir to store runoff water until a sufficient quantity of water is stored for efficient reuse. The reuse water is returned to the irrigation system with a pump.

2.9 Reservoir inflow rate: The rate at which runoff enters the storage reservoir.

2.10 Runoff: Surface water leaving the end of a field.

2.11 Reuse system: A system designed to collect and reuse runoff from a surface irrigated field for irrigation.

2.12 Runoff duration: The elapsed time between the beginning and end of a runoff event.

2.13 Runoff rate: The rate at which water flows off the end of a field.

2.14 Storage reservoir: A facility which accumulates runoff from irrigation or rainfall for later or immediate reuse. Capacity may range from small reservoirs, which will require intermediate pumping, to large reservoirs to store runoff from multiple irrigations.

2.15 Time of cutback: The elapsed time from the beginning of an irrigation set to the time at which cutback is begun.

SECTION 3—REUSE SYSTEM COMPONENTS

3.1 All systems should be designed in conformance with local regulations (reservoir construction, safety precautions, etc.). Nevertheless, there are certain features of design which depend directly on the operation of the reuse system.

3.2 Runoff from irrigated areas is intercepted by a system of drainage channels and carried to the storage reservoir. Capacity of the channels should be sufficient to carry runoff from irrigation, as well as runoff from precipitation.

3.3 Where channels transport suspended solids, a desilting basin should be built to settle suspended solids before the runoff reaches the storage reservoir.

3.4 A storage reservoir is needed to store the runoff until it is redistributed in the farm irrigation system. The desired control of water at the point where the runoff is returned to the irrigation system shall be considered in determining the capacity of the reservoir. Small reservoirs with frequently cycling pumping plants may be used where the runoff is returned to an irrigation regulation reservoir or a pipeline with the flow controlled by a float valve. However, if the irrigation distribution system does not include facilities for regulating fluctuating flows, the reservoir shall be made large enough to provide the regulation needed to permit efficient use of the water.

3.5 Water should enter the storage reservoir from drainage channels through a cantilevered pipe inlet, a drop structure, a chute, or other type of structure designed to prevent erosion.

3.6 A trash screen, pump, power unit, and system operational controls are the key components in a reuse system pumping plant.

3.6.1 Trash in the runoff water which might interfere with pump operation and redistribution of the irrigation water must be removed to prevent its entry into the reuse system. Basket screens are normally installed on the pump inlet pipe. The screened surface area should be approximately 14 times the cross-sectional area of the intake pipe. Depth of water over the basket screen must be sufficient to prevent the pump from drawing air. Screen restrictions do not apply to pumps fitted with self-cleaning screens.

3.6.2 If the reuse pump delivers water to the same pipeline as the irrigation source, the two should be properly matched. Flow from two sources must have equal operating pressures in a parallel system. If operating characteristics are mismatched, efficiency of one or both units may be adversely affected, possibly to a point of reverse flow in one of the sources.

3.6.3 Turbine, axial flow, horizontal centrifugal, and submerged centrifugal pumps can be appropriate for reuse systems.

3.6.3.1 Turbine, axial flow and submerged centrifugal pumps can be mounted on a structure built over the water, on a floating raft, on a trailer, or on the bank with a channel or short length of pipe to direct water to the intake pipe. These pumps are self-priming and the impeller must be submerged.

3.6.3.2 Horizontal centrifugal pumps are usually mounted on the bank with a suction line extending into the water. A foot valve at the submerged inlet and a discharge valve at the pump allow the pump to be primed by a hand pump or a vacuum line from the manifold of an internal combustion engine. Self-priming centrifugal pumps may also be used for low lift conditions.

3.6.4 Pumps can be powered by internal combustion engines or electric motors, either directly connected or driven by V-belts or right angle gear drives. All equipment should meet the safety requirements of ASAE Standard S318, Safety for Agricultural Equipment.

3.6.5 Reuse systems, especially those powered with electric motors, are adaptable to automatic controls. Automatic controls are of three types: (a) water level controls in the storage reservoir, (b) time controls, and (c) safety controls. If the pumping plant is to be started automatically, fail-safe priming is necessary.

3.6.5.1 Air-cell gate switches, float-operated switches, and electrode sensors are used to turn pumping plants on and off at preset water levels.

3.6.5.2 Clock-operated time controls are sometimes used on large storage reservoirs to turn the pumping plant on and off.

3.6.5.3 Electric motors should have safety shutoff switches in the event of overload, low voltage, ground fault, or excessive heating (see ASAE Standard S397, Electrical Service and Equipment for Irrigation). Internal combustion engines should have safety shutoff controls in case of overheating, low oil pressure, or excessive speed. All units should shut off automatically in case of loss of water pressure.

3.7 The intake to either pipeline or open channel gravity reuse systems should be fitted with a screen to prevent trash that might cause plugging or hamper irrigation operation from entering the system. The screen should have sufficient open area and the depth of water should be great enough so that flow is not controlled by entrance conditions.

3.8 Reuse systems require a pipeline or open ditch to convey runoff from the storage reservoir to the primary distribution system.

3.8.1 Sizes of pipelines or ditches will depend on the capacity of the reuse pump and slope.

3.8.2 Aboveground pipelines can be of aluminum, steel, or ultraviolet resistant plastic. Plastic, concrete, asbestos cement, steel, and plastic-covered aluminum can be buried underground.

3.8.3 Air relief, pressure relief, and vacuum relief valves should be provided to protect pipelines (see ASAE Standard S376, Design, Installation and Performance of Underground, Thermoplastic Irrigation Pipelines).

3.8.4 If the reuse system is connected directly to an irrigation well by a pipeline, a check valve or other suitable backflow prevention device in accordance with applicable statutes is required at the pump on the irrigation well to prevent runoff from entering the well and contaminating the groundwater. Also, a check valve should be installed at the reuse pump to prevent pumping from the main well directly into the storage reservoir.

SECTION 4—SYSTEM MANAGEMENT

4.1 Runoff should be applied to a succeeding irrigation set or another field. Recirculating runoff to the same irrigation set that is generating runoff, without an equivalent reduction of the primary inflow, results only in temporarily storing water on the field surface. This will not significantly increase the infiltration of water, but will increase the rate of runoff and may increase erosion.If additional furrows are started as runoff increases, application durations will be different in various parts of the field and labor will be increased.

4.2 Improvement in application uniformity and potentially on-farm irrigation efficiency can result if stored runoff is used to achieve an initially larger furrow stream. In this design, runoff is pumped from a reservoir to increase stream size only during the furrow advance period. Pumping is stopped after runoff begins or the supply is used up. As a result, runoff from one set is used to supplement inflow at the beginning of the next set. This reduces percolation and runoff so that a minimum amount of water must be reused.

4.3 Reservoir storage can affect labor, water and energy if runoff is placed in a reservoir of large enough size to retain it all for later use. With such a reservoir, cutback streams are usually not needed, thereby saving labor. Moderate size initial streams can provide reasonably uniform water distribution along furrows and more runoff is produced than if a cutback system is used. However, this runoff can be retained on the farm by returning it to a storage reservoir. Erosion may increase with some soils, which would then require periodic reservoir cleaning at an increased cost.

4.4 The reuse of irrigation water increases the potential for low water quality, i.e., pesticides and fertilizer. Caution should be exercised to avoid contamination of another field by unwanted pesticides. With erosive soils, stream size must be kept small enough to keep erosion to a minimum.

SECTION 5—DETERMINATION OF RUNOFF

5.1 Volume and rate of runoff are important considerations in the design of surface irrigation reuse systems. Runoff can be determined by direct measurement or estimated from a knowledge of the soil being irrigated and the irrigation management practices followed.

5.2 Direct measurement

5.2.1 Direct measurement of runoff can be made in drainage channels which collect and transport runoff from irrigated fields. Flow rate measuring devices include V-notch, rectangular, broad-crested and trapezoidal (Cipolletti) weirs; and Parshall, cutthroat, long-throated, trapezoidal, and H-type flumes.

5.2.2 Clock-operated water stage recorders, electronic data logging devices and direct flow rate integrators can be used to obtain a hydrograph of a runoff event. If a recorder isn't used, the beginning time of runoff should be noted and the runoff rate measured a minimum of three times during the rising portion of the hydrograph to define the runoff curve. As maximum runoff rate occurs soon after application of irrigation water ceases, assuming there has been no cutback in flow, one measurement should be made at this time. The rapidly decreasing recession flow following the runoff peak constitutes a small part of the total runoff on most soils and one or two measurements are adequate during the declining portion of the hydrograph.

5.2.3 Runoff volume is determined from a summation of the product of time interval and average flow rate per interval as determined from the runoff hydrograph. Where possible, runoff should be measured from different irrigation events that supply the storage reservoir throughout the irrigation season, especially late in the season. The management of irrigation water when runoff measurements are made should be representative of irrigator practice.

5.2.4 The volume of runoff can vary considerably among sets in a given field. Factors affecting runoff volume include: infiltration rate, furrow roughness, slope, variability of inflow and length of run. The ratio of runoff volume to volume of water applied for individual sets generally follows a log-probability relationship. Systems designed to handle runoff at a high confidence level would need either a large enough storage capacity to handle the greater than average runoff volume or a pump with more capacity than the average runoff rate. Due to the variability in runoff volume among sets, twice the average runoff volume is a good estimate of the runoff volume to be stored.

5.3 Estimates based on soil type and management

5.3.1 With optimum irrigation management, runoff duration on slowly to moderately permeable soils does not usually exceed one-half the application duration. Runoff duration of one-half the application duration with a maximum runoff rate of two-thirds the initial application rate would result in a runoff volume of 17 to 33% of the application volume.

5.3.2 On more permeable soils, rapid advance rates and shorter application durations or lengths of run are necessary to minimize deep percolation losses. On these soils, runoff duration will be approximately 75% of application duration and actual runoff volume will be approximately 35% of application volume.

5.3.3 Since many variables can influence water infiltration during irrigation, and application rates usually are not accurately controlled, runoff volumes can vary considerably from the average runoff representative of good water management. A design runoff volume that is twice the average runoff volume will normally be exceeded less than 10% of the time if good water management is practiced. Therefore, the actual design runoff volume should not exceed 50% of application volume on slowly to moderately permeable soils and 60% on permeable soils where larger stream sizes are used to reduce advance time and minimize deep percolation. Where the larger stream sizes and rapid advance times are not used on permeable soils (greater deep percolation losses are allowed), the design runoff may be the same as for slowly and moderately permeable soils.

5.3.4 Decreased net water applications lead to increased irrigation application efficiencies because runoff losses are reduced.

5.3.5 By purposely using large streams to achieve uniform distribution and purposely having a large amount of runoff, where soils are nonerosive, a reuse system may be used to reduce labor for managing and adjusting water flow during an irrigation. This practice results in increased runoff duration and volume which can result in an initially adequately designed system, anticipating cutback streams, to have inadequate capacity. Any need to accommodate increased runoff should be anticipated and considered in determining the storage capacity and reuse pumping rate.

5.3.6 Highly efficient water management results in substantially less runoff duration and volume than normal. Conditions under which runoff can be reduced even further are flat slopes, little or no lower end leaching requirements for salinity control, appreciable seasonal rainfall, and permissible lower end soil water deficits where drought tolerant crops are grown. The design runoff based on the expected use of such management practices should normally not be less than one-half the design runoff for normal management practices.

TABLE 1—MINIMUM STORAGE RESERVOIR DIMENSIONS FOR CYCLING REUSE SYSTEMS

Runoff inflow	Inside diameter of circular storage reservoir, m			
	Depth in storage reservoir between on and off levels, m			
L/s	0.6	1.2	1.8	2.4
3.0	0.6*	0.4*	0.4*	0.3*
6.0	0.9	0.6*	0.5*	0.4*
9.0	1.1	0.8	0.6*	0.5*
12.0	1.2	0.9	0.7	0.6*
15.0	1.4	1.0	0.8	0.7
18.0	1.5	1.1	0.9	0.8
21.0	1.6	1.2	0.9	0.8
24.0	1.7	1.2	1.0	0.9
27.0	1.8	1.3	1.1	0.9
30.0	1.9	1.4	1.1	1.0
33.0	2.0	1.4	1.2	1.0
36.0	2.1	1.5	1.2	1.1

*Pipe diameter too small to allow for servicing system.

SECTION 6—STORAGE RESERVOIR SIZING

6.1 The most important items in the design of reuse systems are the selection of storage reservoir volume and pump capacity. Both are dependent on the type of reuse system planned.

6.2 Cycling systems have a small storage reservoir with an electric motor-powered pump automatically controlled by water level-actuated switches.

6.2.1 Cycling systems must not cycle more than 15 times per hour to conform with pump manufacturer's recommendations regarding motor overheating.

6.2.2 For maximum cycle rate, 15 times per hour, the minimum storage reservoir and pump size are related by:

$$S = 0.06P = 0.06I_d$$

where
S = capacity of the storage reservoir, m^3
P = pumping rate, L/s
I_d = design inflow rate to the storage reservoir, L/s

If all runoff is to be reused (no loss):

$$I_d = I_{max}$$

where
I_{max} = maximum runoff rate, L/s

6.2.3 Storage reservoir dimensions should conform to those given in Table 1. Additional information on storage reservoir design is given in ASAE Engineering Practice EP369, Design of Agricultural Drainage Pumping Plants.

6.2.4 Fluctuations in pressure and flow rate make water deliveries from a cycling system difficult to manage. Such systems are usually used to deliver the runoff water to a regulating reservoir or a major supply reservoir.

6.3 Pump systems utilize a storage reservoir with sufficient capacity to collect the runoff from one or two irrigation sets.

6.3.1 When using a design considering irrigation runoff only, the size of the storage reservoir from which accumulated runoff will be pumped continuously is:

$$S = V_d \left(1 - \frac{\bar{I}_d}{P}\right)$$

where
V_d = design volume of runoff, m^3
$\bar{I}_d$ = average runoff inflow rate corresponding to the design volume of inflow, L/s

If all the runoff is to be utilized:

$$V_d = V$$

where
V = total runoff, m^3

Otherwise V_d is limited by some other criterion (e.g., 90% of runoff will be recycled, the remainder will be released).

6.3.2 The average inflow rate, based on the design volume of runoff and the time over which runoff occurs, is:

$$\bar{I}_d = \frac{V_d}{0.06T_t}$$

where
T_t = duration of runoff, min

6.3.3 The pumping duration, T_p, is:

$$T_p = \frac{\bar{I}_d\, T_t}{P}$$

6.3.4 Runoff can be reapplied to the same field from which it is collected. In this case, the first set is irrigated by the primary supply stream only, and the last set (n + 1) is irrigated by the water stored from runoff. The total number of sets irrigated by the supply stream is n. Intermediate sets are irrigated from both sources. The number of furrows irrigated in sets 2 through n is constant, if field and soil characteristics are uniform. The number of furrows irrigated in set 1 and set n + 1 is less than in the others.

6.3.4.1 The total number of sets, n, irrigated by the supply stream is:

$$n = \frac{T}{t}$$

where
T = field application duration, min
t = set application duration (constant, regardless of source of water), min

The time, T, is determined from previous knowledge of the gross water to be applied and the inflow rate of the supply stream, Q_s. The time, t, is the application duration of each set, which is determined by the designer.

6.3.4.2 The number of furrows irrigated by the supply stream, f_s, is:

$$f_s = \frac{Q_s}{q}$$

where
Q_s = inflow rate of the supply stream, L/s
q = inflow to each furrow, L/s

6.3.4.3 The number of furrows irrigated by the runoff, f_p, is:

$$f_p = \frac{P}{q}$$

6.3.4.4 The total number of furrows to be irrigated, F, is:

$$F = n(f_s + f_p)$$

Since F is determined from the field size and furrow spacing, it is obvious that there must be a balance between the chosen pumpback rate, the application duration, and the supply stream rate.

6.3.5 The volume of water in storage in the reservoir at the end of any given set, except for set n + 1 when water is pumped from the reservoir only, is:

$$V_i = 0.06\ t\ R_f\ [Q_s i + P(i-1)] - 0.06\ t\ P\ (i-1)$$

where
V_i = volume of runoff at the end of set i, m^3
R_f = total runoff volume expressed as a fraction of the applied volume

The runoff fraction can be determined by one of the methods outlined in Section 5—Determination of Runoff. The storage reservoir should be large enough to hold the largest runoff volume calculated by the above formula for sets 1 through n.

6.4 Other reuse systems utilize runoff reservoirs which have sufficient storage volume to hold the runoff from several days of irrigation or the runoff from an entire field.

6.4.1 The most flexible systems are those which store all the runoff for the field application duration. These allow use of pumps of any convenient size (to deliver cutback or main streams), but they require large storage reservoirs. The storage capacity of such reservoirs is calculated by:

$$S = 0.06\ \bar{I}_d T_t$$

No cycling is required during the emptying of the reservoir, and pumping can commence at the convenience of the irrigator.

6.5 Runoff from expected rainfall should be considered when designing any resue system.

References: Last printed in 1982 AGRICULTURAL ENGINEERS YEARBOOK; list available from ASAE Headquarters.

Cited Standards:

ASAE S318, Safety for Agricultural Equipment
ASAE EP369, Design of Agricultural Drainage Pumping Plants
ASAE S376, Design, Installation and Performance of Underground, Thermoplastic Irrigation Pipelines
ASAE S397, Electrical Service and Equipment for Irrigation

SAFETY DEVICES FOR APPLYING LIQUID CHEMICALS THROUGH IRRIGATION SYSTEMS

Developed by the ASAE Irrigation Management Committee; approved by the ASAE Soil and Water Division Standards Committee; adopted by ASAE January 1982; reconfirmed for one year December 1986.

SECTION 1—PURPOSE AND SCOPE

1.1 This Engineering Practice specifies required safety devices when injecting chemicals into irrigation systems. Many irrigators apply water-soluble or liquid fertilizers, herbicides, insecticides, and other chemicals through their irrigation systems. With approved chemicals for this method of application and with properly engineered irrigation and injection systems, these chemicals can be easily and uniformly applied along with irrigation water.

1.2 A water supply pollution hazard may exist if proper safety devices are not installed on the chemical injection equipment. Two specific hazards are: (1) the irrigation pumping plant may shut down from mechanical or electrical failure while the injection equipment continues to operate, causing a mixture of water and chemicals to backflow into the irrigation well or other water source, or possibly causing chemicals to empty unnecessarily into the irrigation system, and (2) the chemical injection system may stop while the irrigation pump continues to operate, causing water to backflow through the chemical supply tank and overflow onto the ground.

1.3 The scope of this Engineering Practice is to describe the safety devices needed to prevent the two hazards described in Section 1.2. Additional equipment required for mixing or storing the chemicals is not covered by this Engineering Practice.

SECTION 2—DEFINITIONS

2.1 Backflow prevention devices: Safety devices used to prevent pollution or contamination by flow of a mixture of water and other chemicals back into the irrigation well or potable water supply. The term backsiphonage is often used to describe the backflow.

2.2 Interlock injection devices: Safety equipment used to insure that if the irrigation pumping plant stops, the chemical injection pump will also stop. This will prevent the possibility of emptying the entire chemical mixture from the supply tank into the irrigation pipeline. Simultaneously, a safety device is needed to prevent water from flowing back through the injection pump and overflowing the chemical supply tank, whenever the injection pump is turned off and the irrigation pump is still operating.

SECTION 3—BACKFLOW PREVENTION DEVICES

3.1 Selection of the proper type of device depends upon the fluid that can backflow, i.e., whether it is toxic or nontoxic, and whether there can be backpressure or backsiphonage. State and local regulations and codes must be followed in selecting these devices.

3.2 One nonmechanical (air gap) and four different types of mechanical backflow prevention devices exist.

3.2.1 Air gap. An air gap is a physical separation between the free flowing discharge end of a water pipeline and an open or nonpressurized receiving vessel. To have an acceptable air gap, the end of the discharge pipe must be at least twice the diameter of the pipe above the topmost rim of the receiving vessel. In no case can this distance be less than 2.5 cm (1 in.). This is a simple and effective type of protection. However, an additional pump is required downstream of the receiving vessel to repressurize the water before entering the irrigation system.

3.2.2 Atmospheric vacuum breaker. An atmospheric vacuum breaker has a movable element or plunger, which prevents water from spilling from the device during flow and drops down to provide a vent opening following cessation of flow. This device cannot be installed where backpressure persists for a long duration and can be used only to prevent backsiphonage. An atmospheric unit should not be used with shutoff valves downstream and must be installed at least 15 cm (6 in.) above the highest outlet or the topmost overflow rim of a nonpressure tank. These units are installed primarily in lawn and turf irrigation systems that are connected to potable water supplies.

3.2.3 Pressure vacuum breaker. The pressure vacuum breaker contains, within a single body, a spring-loaded check valve and a spring-loaded, air-opening valve which opens to admit air whenever the pressure within the body approaches atmospheric. Like the atmospheric vacuum breaker, the pressure vacuum breaker cannot be installed where there can be backpressure, only where there can be backsiphonage. The pressure vacuum breaker can have shutoff valves downstream of the device. It must be installed at least 30 cm (12 in.) above the highest downstream outlet.

3.2.4 Double check valve. The double check valve assembly is composed of two single, independently acting check valves. It can handle both backsiphonage and backpressure, but is not recommended for toxic chemicals. A minimum of 30 cm (12 in.) above the ground level or grade is suggested.

3.2.5 Reduced pressure principle device. This device consists of two independently acting check valves, together with a pressure differential relief valve that is located between the two check valves. It can also be used for backsiphonage and backpressure control and can handle most toxic chemicals. A minimum clearance of 30 cm (12 in.) above the ground level or grade is suggested to ensure an air gap between the relief valve and any water that might puddle beneath the device.

3.3 Backflow prevention devices should be installed between the water supply or pump and the chemical injection line (see Figs. 1 and 2).

SECTION 4—INTERLOCK INJECTION DEVICES

4.1 For an irrigation pump driven by an internal combustion engine, an easy and adequate interlocking mechanism is to have the chemical

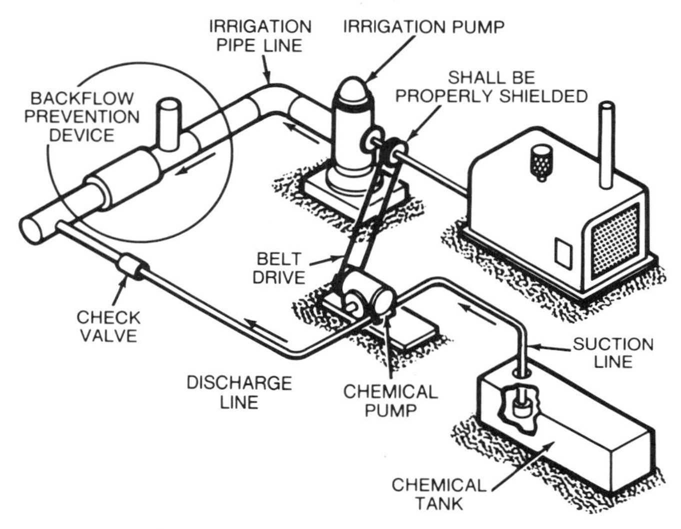

FIG. 1—SAFETY DEVICES FOR INJECTION OF CHEMICALS IN IRRIGATION SYSTEMS USING ENGINE POWER UNITS

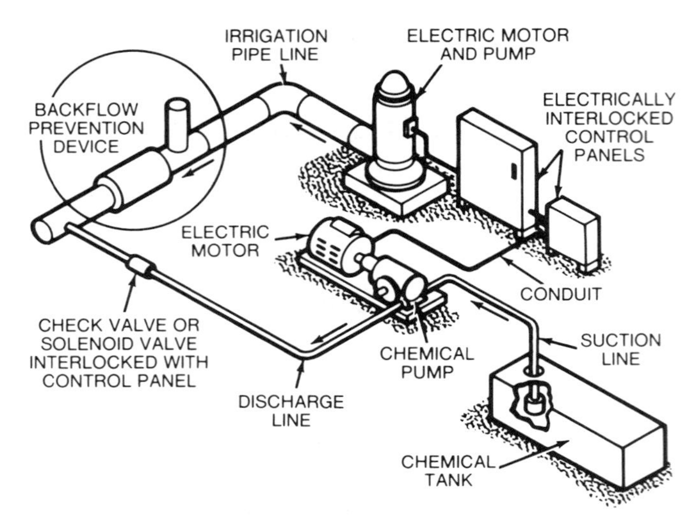

FIG. 2—SAFETY DEVICES FOR INJECTION OF CHEMICALS IN IRRIGATION SYSTEMS USING ELECTRIC MOTORS

injection pump belted to the drive shaft or an accessory pulley of the engine. This will ensure that both the irrigation pump and chemical injection pump will be turned off at the same time. Also, a mechanical, normally-closed check valve is required in the chemical injection line to stop backflow from the irrigation system into the chemical supply tank when the injection pump is stopped. This will prevent the chemical supply tank from overflowing. Interlocking is shown in Fig. 1 for an internal combustion engine.

4.2 For an electrical motor driven irrigation pump, a separate electric motor is usually needed to power the chemical injection pump. The electric controls for the two electric motors should be interlocked so that the injection motor will stop when the electric motor on the irrigation pump stops. A check valve, vacuum breaker (atmospheric vacuum breaker, pressure vacuum breaker, or similar antisiphon device), or normally-closed solenoid valve can be placed in the injection line after the injection pump. A solenoid valve will provide a positive shutoff on the chemical injection line so that neither the chemical nor the water can flow in either direction if the chemical injection pump stops. An electric solenoid valve must be electrically connected with the motor driving the electrical injection pump and electrically interlocked with the irrigation pump. The interlocking procedure using electric motors is shown in Fig. 2.

SECTION 5—ADDITIONAL SAFETY PRECAUTIONS

5.1 In addition to providing devices for the injection of liquid chemicals, safety precautions should be provided to protect workers against accidental discharge or spillage of chemicals.

5.2 A water source should be provided near the chemical supply tank and injection pump for washing off any chemicals that contact the skin. Protective goggles, face shields, and clothing should also be worn when making chemical dilutions. Concentrated chemicals should always be added to the water in preparing dilutions in a chemical supply tank. Water should not be added directly to a small volume or quantity of concentrated chemicals under any circumstances.

5.3 Bulk chemicals, in particular, and diluted solutions should be stored in secured areas. Information, such as label directions, on chemical storage, temperature, life, handling, etc., should be posted and followed. Also, local safety regulations and codes for worker protection should be adhered to under all situations.

EVALUATION OF FURROW IRRIGATION SYSTEMS

Developed by the ASAE Surface Irrigation Committee; approved by the ASAE Soil and Water Division Standards Committee; adopted by ASAE February 1987.

SECTION 1—PURPOSE AND SCOPE

1.1 The purpose of this Engineering Practice is to establish procedures for evaluating the performance of furrow irrigation systems. This practice should encourage standardization of the terminology and procedures used in furrow irrigation evaluation.

1.2 This Engineering Practice outlines a thorough evaluation process. Depending upon the objectives of a particular evaluation and the resources available, some of the procedures could be simplified or omitted.

SECTION 2—DEFINITIONS

2.1 Application efficiency (AE): The ratio of the average depth of irrigation water infiltrated and stored in the root zone to the average depth of irrigation water applied, expressed as a percent.

2.2 Application efficiency of low quarter (AELQ): The ratio of the average low-quarter depth of irrigation water infiltrated and stored in the root zone to the average depth of irrigation water applied, expressed as a percent.

2.3 Christiansen's uniformity coefficient (CU): The average depth of irrigation water infiltrated minus the average absolute deviation from this depth, all divided by the average depth infiltrated.

2.4 Deep percolation percentage (DPP): The ratio of the average depth of irrigation water infiltrated and drained out of the root zone to the average depth of irrigation water applied, expressed as a percent.

2.5 Distribution uniformity (DU): The ratio of the average low-quarter depth of irrigation water infiltrated to the average depth of irrigation water infiltrated, expressed as a percent.

2.6 Irrigation efficiency (IE): The ratio of the average depth of irrigation water which is beneficially used to the average depth of irrigation water applied, expressed as a percent. Beneficial uses include satisfying the soil moisture deficit and any leaching required to remove salts from the root zone.

2.7 Management allowed deficit (MAD): The desired soil moisture deficit at the time of irrigation.

2.8 Runoff percentage (RP): The ratio of the equivalent depth of irrigation water running off the field to the depth of irrigation water applied, expressed as a percent.

2.9 Soil moisture deficit (SMD): The depth of water required to bring a specific depth of soil to field capacity at a particular time.

2.10 Storage percentage (SP): The ratio of the average depth of irrigation water infiltrated and stored in the root zone to the soil moisture deficit, expressed as a percent.

SECTION 3—EQUIPMENT

3.1 General use
- **3.1.1** Data forms
- **3.1.2** Clipboards
- **3.1.3** Stop watches or other watches capable of measuring seconds
- **3.1.4** Shovels

3.2 Station layout
- **3.2.1** Measuring tape
- **3.2.2** Stakes, laths or flags
- **3.2.3** Hatchet or hammer
- **3.2.4** Marking crayon or paint

3.3 Furrow characteristics
- **3.3.1** Measuring tape
- **3.3.2** Level or transit, tripod and rod
- **3.3.3** Profilometer or other device for measuring furrow geometry

3.4 Soil sampling and soil moisture measurements
- **3.4.1** Soil auger or probe
- **3.4.2** Bulk density equipment
- **3.4.3 Gravimetric method**
  - **3.4.3.1** Soil cans or other watertight containers
  - **3.4.3.2** Scale and oven (laboratory)
- **3.4.4 Neutron scattering method**
  - **3.4.4.1** Access tubes
  - **3.4.4.2** Neutron source and detector
- **3.4.5 Water budget method**
  - **3.4.5.1** Weather instruments

3.5 Infiltration measurements
- **3.5.1 Blocked furrow infiltrometer method**
  - **3.5.1.1** Metal plates
  - **3.5.1.2** Sledgehammer and driving plate or board
  - **3.5.1.3** Supply reservoir and staff gauge
  - **3.5.1.4** Hook or point gauge
  - **3.5.1.5** Buckets or water supply line
- **3.5.2 Inflow-outflow method**
  - **3.5.2.1** Flow measuring devices
  - **3.5.2.2** Equipment for installing the devices (levels, shovels, etc.)

3.6 Inflow and runoff measurements
- **3.6.1** Flow measuring devices
- **3.6.2** Equipment for installing the devices (levels, shovels, etc.)

3.7 Wetted cross-sectional measurements
- **3.7.1** Carpenter's tape or other measuring device

SECTION 4—FIELD PROCEDURES

4.1 Selection of test furrows. The furrows chosen for study should be representative of the irrigated area. Factors such as alternate furrow irrigation (see paragraph 5.9.2.4) and soil compaction differences should be considered. Three or more test furrows should be selected. If the test furrows are in the same proximity, data collection will be facilitated. The use of guard or buffer furrows between the test furrows is recommended. For on-farm evaluations, it may be desirable to measure more furrows less intensively.

4.2 Furrow characteristics. Without disturbing the test furrows, the physical characteristics of the furrows should be noted and recorded.

4.2.1 Length. The length of the furrows should be measured using a tape or other suitable means.

4.2.2 Slope. The furrow slope can be measured using surveying instruments. Any irregularities or discontinuities in slope should be noted.

4.2.3 Spacing. The distance between the centerlines of adjacent furrows should be measured. If water is not introduced into each furrow during a given irrigation, then the distance between wetted furrows should also be recorded.

4.2.4 Geometry. The geometry of the furrow cross section should be described. This may be done in general terms by noting the furrow shape (i.e., vee, trapezoidal, parabolic, or retangular) and the approximate depth and top width. More quantitatively, a profile measuring device may be used. One such device, called a profilometer, uses vertical rods to indicate relative ground elevations. The furrow cross section can then be mathematically described using the profilometer data.

4.2.5 Soil type and condition. If the soil type is not already known, several representative soil samples should be taken for textural analysis. Additionally, the physical condition of the soil should be qualitatively evaluated as to roughness, cracking, amount of residue, etc. Tillage practices and the soil compaction history are also important pieces of information.

4.3 Soil moisture deficit. Several methods are available for determining the moisture status of the soil prior to irrigation. One or more of these methods may be used in the evaluation. In the process of making soil moisture determinations, the test furrows should not be disturbed. Guard furrows can be used for this purpose. Following are some of the methods for determining soil moisture deficit.

4.3.1 Soil appearance and feel. An experienced evaluator can estimate the soil moisture deficit from the appearance and feel of soil samples taken at various depths. Charts are available as an aid in making these determinations [Merriam and Keller (1978), p.5; Merriam et al. (1980), p.759].

4.3.2 Soil moisture measurement. The soil moisture status can be determined from direct measurements. The measurements should be made at several depths and at several locations along the furrows. In order to determine the soil moisture deficit, the moisture content corresponding to field capacity must also be known.

4.3.2.1 Gravimetric method. The gravimetric method involves weighing soil samples both before and after oven drying. To obtain the moisture content on a volumetric basis, the moisture content on a dry weight basis is multiplied by the ratio of the soil bulk density to the density of water.

4.3.2.2 Neutron scattering method. The neutron scattering method, when calibrated for the particular soil type, measures water content on a volumetric basis. The access tubes for the neutron probe enable repeated measurements to be made at the some location and depth.

4.3.3 Water budget. The soil moisture deficit can be estimated using a water budget approach. Starting with the soil at a known moisture content, water additions to the soil root zone (precipitation, irrigation) and water losses (evapotranspiration, drainage) are monitored. Some of the components of this water balance may need to be estimated. Periodic checks against a measured soil moisture content are desirable.

4.4 Infiltration. The infiltration characteristics of the soil may be measured prior to or during the irrigation event. Since infiltration rates vary depending on local conditions, such as soil moisture content, the tests should be conducted as close to the time of irrigation as possible and under representative conditions. Preferably, the measurements should be made in guard furrows rather than test furrows. If a system is to be evaluated for the entire season, infiltration characteristics should be evaluated during at least the first, second, and one other irrigation event.

4.4.1 Blocked furrow infiltrometer. Metal plates driven into the soil are used to block off a furrow segment, typically up to five meters in length. Water is then ponded in this furrow segment to a depth approximating that observed during irrigation events. The water in the infiltrometer is kept at a relatively constant level as indicated by a hook or point gauge. Water is periodically added to maintain this level, and the volume of water added is recorded with time. The test should be continued until an essentially constant infiltration rate has been reached, or until the elapsed time exceeds the expected duration of the irrigation event.

4.4.2 Recirculating furrow infiltrometer. A modification of the infiltrometer described above, in which water is continuously recycled over the furrow segment, may be used in an attempt to better simulate the flow conditions of an actual irrigation. One recycling infiltrometer design incorporates inflow and outflow sumps, two small pumps, and a water supply reservoir. The decline in reservoir water volume with time is a measure of the infiltration rate in the furrow segment.

4.4.3 Inflow-outflow measurement. Infiltration rates along a furrow segment can be determined by taking the difference in measured flow rates at the beginning and end of the segment. Flumes or other suitable flow measuring devices may be used. The length of the test segment can be selected based on the soil infiltration characteristics. As the length of the test segment decreases, the infiltration opportunity time along the segment becomes more uniform, but the measured difference in flow rates becomes smaller and the accuracy of flow readings has greater relative importance. The volume of water going into storage in the furrow channel may also have to be considered in the analysis. Care should be taken in selecting and using flow measuring devices. Factors such as measurement accuracy and backwater effects should be considered. Individual calibration of the devices is highly desirable.

4.5 Inflow. Accurate determination of inflow rates is one of the most important aspects of furrow irrigation evaluation. If the objective of the evaluation process is to analyze current irrigation practices, then "normal" inflow stream sizes should be used. If the objective is to study system performance under a variety of inflow conditions, then a range of stream sizes should be applied to the test furrows. Inflow rates may be measured with flumes, orifices, or weirs, or by measuring the time required to fill a container of known volume. Except on relatively flat slopes where ponding may become a problem, flumes may be the most practical of the furrow flow measuring devices. These devices should be located as close to the beginning of the furrow as possible. When automatic recorders are not used, flow readings should be taken periodically throughout the irrigation event and the time of the reading recorded. Frequent observations may be required when the inflow rate does not remain constant with time. It is usually desirable to have a constant inflow rate.

4.6 Advance. The advance of the water front in the test furrows should be monitored. This can be accomplished by recording the time at which the water front reaches each station. The stations should be marked with laths, stakes or flags prior to the irrigation event. Station 0 + 00 is the beginning of the furrow. A convenient interval between stations is 20 or 25 m.

4.7 Runoff. After advance is complete on freely draining furrows, the rate of runoff from each furrow should be recorded with time. Again, flumes are probably the most practical devices for measuring runoff rates. A further discussion on determining runoff rates may be found in ASAE Engineering Practice EP408, Design and Installation of Surface Irrigation Runoff Reuse Systems.

4.8 Wetted cross-section. As time permits during the evaluation process, the flow depth and top width along the furrow should be measured and recorded. The station location for each measurement should be noted.

4.9 Recession. The time of cutoff, when inflow to the furrow ceases, should be noted. Data should also be collected on the recession of water from each furrow. As a minimum, the recession times at the head and tail ends of the furrow should be recorded. Determining when recession occurs is a subjective judgment, due to the micro-relief of furrow bottoms. One possible criterion is to use the time at which longitudinal water movement ceases.

4.10 Post-irrigation soil moisture. The moisture status of the soil following irrigation may be determined with any of the methods described in paragraph 4.3. Depending upon the soil texture and the method of moisture measurement, these determinations should be made one to three days after the irrigation event. Soil probing can indicate the location of the wetting front in some soils. It may be of interest to know whether or not the wetting fronts from adjacent furrows have met.

SECTION 5—ANALYSES

5.1 In order to quantitatively evaluate the performance of a furrow irrigation system, the distribution of infiltrated water along the furrow must be known. This distribution can be estimated from a mathematical function describing infiltration and from data on infiltration opportunity times along the furrow. Two methods of estimating the infiltration function will be presented (see paragraph 5.6), but certain intermediate data analyses are required for either or both methods.

5.2 Infiltration tests

5.2.1 Units. Furrow infiltration rates are typically described with units of depth per unit time. Correspondingly, cumulative infiltration has units of depth. This infiltrated depth is commonly taken to be the infiltrated volume divided by the product of the furrow length and the spacing between adjacent wetted furrows.

The result is actually an equivalent infiltrated depth dependent on geometric as well as soil factors.

5.2.2 Plots. Using data collected with either the blocked furrow method (see paragraph 4.4.1), the recirculating infiltrometer method (see paragraph 4.4.2), or the inflow-outflow method (see paragraph 4.4.3), plots can be made of infiltration rate and cumulative infiltration versus infiltration opportunity time.

5.2.3 Equations. In addition to graphical representation, a mathematical equation can be fitted to the infiltration test data. In furrow irrigation studies, one of the following infiltration formulas is commonly used.

5.2.3.1 Kostiakov

$$z = k\tau^a$$

where

z = infiltrated depth (L)
τ = infiltration opportunity time (T)
k, a = empirical fitting parameters

5.2.3.2 Modified, or extended, Kostiakov

$$z = k\tau^a + c\tau$$

where

c = the final infiltration rate (LT^{-1})

The modified form of the Kostiakov equation is preferred for relatively large infiltration opportunity times, when the infiltration rate reaches its final, or steady state value.

5.3 Inflow and runoff

5.3.1 Volumes. The inflow and runoff data yield hydrographs which can be integrated to determine the total inflow and runoff volumes, respectively. The difference in these two values is a measure of the total volume of water that has infiltrated into the soil. These volumes can also be expressed as an equivalent depth over an area defined by the furrow length and the spacing between wetted furrows.

5.3.2 Final infiltration rate. For long irrigation events in which the runoff rate becomes relatively constant, the difference between the measured inflow and runoff rates can be used in estimating the parameter, c, the furrow's final infiltration rate.

5.4 Advance and recession

5.4.1 Plots. The advance and recession data together define the infiltration opportunity time at each station. These data can be plotted, preferably as the advance (or recession) time versus the distance along the furrow. Then the abscissa denotes horizontal distance and the vertical distance between the advance and recession curves is the infiltration opportunity time (see Fig. 1). When recession is fairly rapid (or a small proportion of the total irrigation event) and extensive recession data are not collected, the recession curve is often assumed to be linear between the first and last stations, or assumed to be a horizontal line corresponding to the time of cutoff. If there are significant changes in field slope or in infiltration characteristics, recession should be measured where these changes occur.

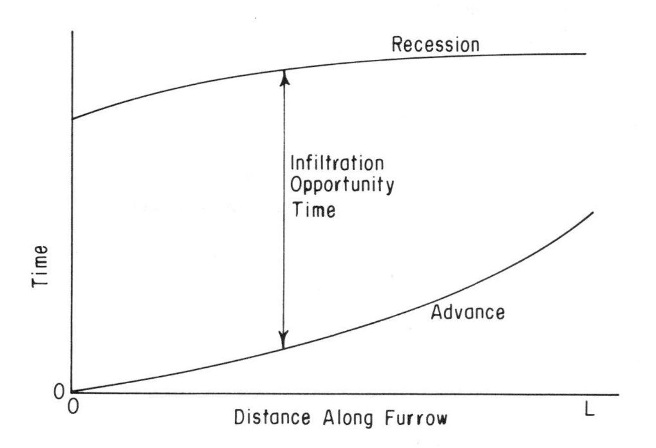

FIG. 1—TYPICAL CURVES DESCRIBING THE ADVANCE AND RECESSION OF THE WATER FRONT

5.4.2 Equations. Mathematical equations can be fitted to the advance data to provide a continuous function relating advance times and distances. One equation that may be appropriate is a simple power function of the form

$$t_x = bx^h$$

where

x = advance distance (L)
t_x = time of advance to x (T)
b,h = empirical fitting parameters

If desired, an equation can also be fitted to the recession data.

5.5 Surface storage. The amount of water in surface storage can be an important factor in volume balance calculations. If sufficient wetted cross-section measurements are made (see paragraph 4.8), it is possible to estimate surface storage volumes from these measurements. More often, the inlet cross-sectional flow area is multiplied by the advance distance and by a shape factor to arrive at a surface storage volume. The shape factor is normally assumed to have a constant value between 0.70 and 0.80 during advance. The inlet flow area can be measured in the field or can be approximated using a uniform flow formula (e.g., the Manning equation). In the latter case, the selection of a value for the roughness coefficient may present some difficulties. Roughness is known to vary depending upon the surface condition of the furrow. USDA Soil Conservation Service procedures assume that the Manning roughness coefficient for furrows is equal to 0.04.

5.6 Determining the infiltration function parameters

5.6.1 Approach 1 - adjusted infiltration test results. The infiltration equation developed from infiltrometer or inflow-outflow tests, together with observed infiltration opportunity times, can be used to estimate the cumulative depth of infiltration at each station. By integrating along the furrow using the trapezoidal rule and multiplying by the furrow spacing, an estimate of the furrow's total infiltrated volume can be obtained. This value is then compared to the difference between measured inflow and runoff volumes. If they are not in close agreement, the parameters of the infiltration function should be adjusted accordingly. Usually, this adjustment is made by increasing or decreasing the value of the coefficient, k, in the infiltration equation.

5.6.2 Approach 2 - volume balance during advance. The infiltration function may also be deduced from volume balance calculations, in which the cumulative infiltrated volume at some time during advance is equated to the difference between the cumulative inflow volume and the surface storage volume at that time. For the case of a power advance equation and the modified Kostiakov infiltration equation, this volume balance relationship may be expressed as:

$$(\sigma_z k t_x^a + \frac{hct_x}{1+h})Wx = V_{in} - \sigma_y A_o x$$

where

W = spacing between wetted furrows (L)
V_{in} = cumulative inflow volume to the furrow (L^3)
A_o = cross-sectional flow area at the inlet (L^2)
σ_y = surface storage shape factor (usually defined as a constant between 0.70 and 0.80)
σ_z = subsurface shape factor defined as:

$$\sigma_z = \frac{h + a(h-1) + 1}{(1+a)(1+h)}$$

When the available data include inflow (see paragraph 5.3), advance (see paragraph 5.4), and flow area at the inlet (see paragraph 5.5), the unknowns in the volume balance equation are the three infiltration parameters. If the basic infiltration rate, c, is estimated with the method described in paragraph 5.3 (or with some other method), only k and a remain as unknowns. The values of these two parameters can be determined by direct solution of the volume balance equation at two points during advance, or by using several points during advance and a "best fit" procedure. If runoff data are available, it is advisable to perform the overall volume balance check described in paragraph 5.6.1, and adjust the parameter values if necessary. The volume balance relationship given above is sometimes simplified by assuming the constant term in the infiltration function is negligible (especially for short infiltration opportunity times).

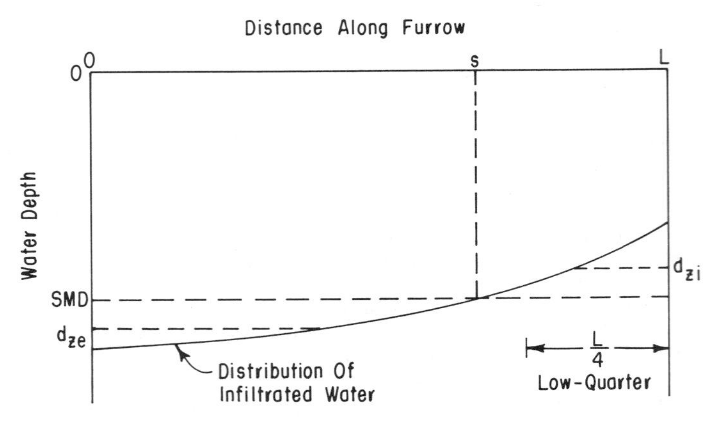

FIG. 2—TYPICAL DISTRIBUTION OF INFILTRATED WATER FOR THE CASE OF "UNDERIRRIGATION"

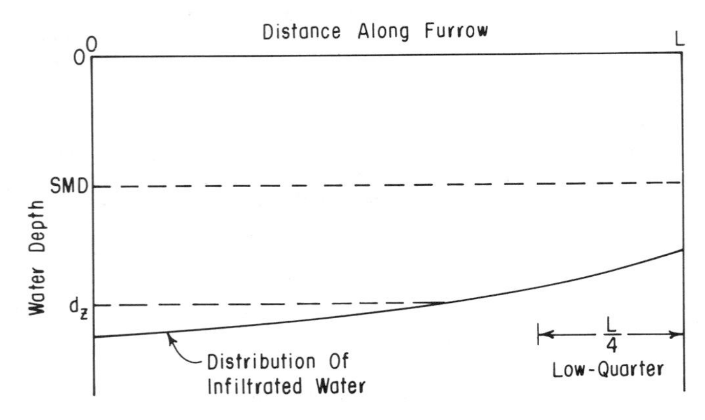

FIG. 3—TYPICAL DISTRIBUTION OF INFILTRATED WATER FOR THE CASE OF "COMPLETE IRRIGATION"

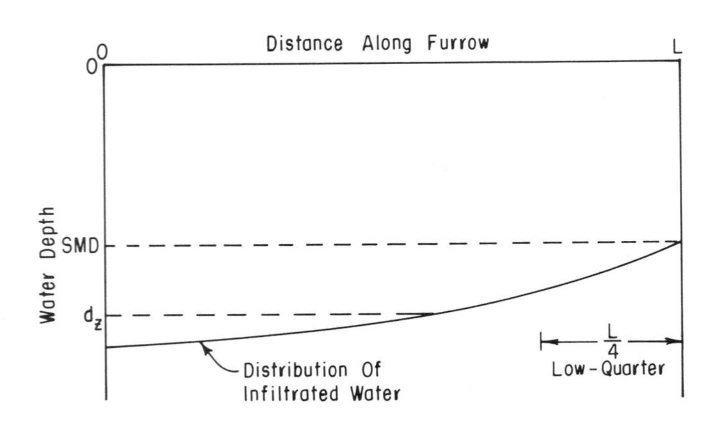

FIG. 4—TYPICAL DISTRIBUTION OF INFILTRATED WATER FOR THE CASE OF "OVERIRRIGATION"

5.7 Distribution of infiltrated water. Using infiltration opportunity time data and the derived infiltration function, the distribution of infiltrated water along the furrow can be determined. One shortcoming of this technique for determining the distribution of infiltrated water is that it ignores the effect of wetted perimeter on infiltration. In Figs. 2 - 4, typical distributions are presented for the cases of "underirrigation", "complete irrigation", and "overirrigation", respectively. With the complete irrigation case, deep percolation losses can be significant. In humid and semi-humid regions, it may be desirable to intentionally underirrigate, i.e., to not completely refill the crop root zone during the irrigation. Underirrigation can result in more efficient use of rainfall without sacrificing crop yields. It will also help reduce deep percolation if the uniformity of water application is relatively high. Intentional overirrigation may be of beneficial use in the leaching of salts.

5.8 Performance parameters. The performance parameter calculations may be made using either water depths or water volumes. The furrow spacing and length are used to convert from one to the other. The following equations are written for water depths. Figs. 2 - 4 serve to illustrate the concepts involved. It should be noted that the application efficiency, deep percolation percentage, and runoff percentage sum to 100%.

5.8.1 Underirrigation case (Fig. 2)

5.8.1.1 Application efficiency

$$AE = \frac{SMD \cdot s + d_{zi}(L-s)}{L \cdot d_a} \times 100$$

where

AE	=	application efficiency, percent
SMD	=	soil moisture deficit
L	=	furrow length
s	=	distance down the furrow at which the depth of water infiltrated is equal to the soil moisture deficit
d_{zi}	=	average infiltrated water depth over the inadequately irrigated area
d_a	=	average depth of water applied to the furrow (if a runoff recovery system is in use, include the total quantity of water applied minus the quantity recovered)

5.8.1.2 Application efficiency of low quarter

$$AELQ = \frac{d_{sq}}{d_a} \times 100$$

where

AELQ	=	application efficiency of low quarter, percent
d_{sq}	=	average low quarter depth of water infiltrated and stored in the root zone

5.8.1.3 Irrigation efficiency

$$IE = \frac{d_b}{d_a} \times 100$$

where

IE	=	irrigation efficiency, percent
d_b	=	average depth of irrigation water beneficially used

5.8.1.4 Christiansen's uniformity coefficient

$$CU = 1 - \frac{\sum_{i=1}^{n} |d_i - d_z|}{d_z \cdot n}$$

where

CU	=	Christiansen's uniformity coefficient
d_z	=	average depth of water infiltrated

d_i = depths of water infiltrated at discrete points along the furrow
n = number of discrete points

5.8.1.5 Distribution uniformity

$$DU = \frac{d_{zq}}{d_z} \times 100$$

where
DU = distribution uniformity, percent
d_{zq} = average depth of water infiltrated in the quarter of the area infiltrating the least water

5.8.1.6 Storage percentage

$$SP = \frac{SMD \cdot s + d_{zi}(L-s)}{L \cdot SMD} \times 100$$

where
SP = storage percentage

5.8.1.7 Deep percolation percentage

$$DPP = \frac{s(d_{ze} - SMD)}{L \cdot d_a} \times 100$$

where
DPP = deep percolation percentage
d_{ze} = average infiltrated water depth over the area receiving more than the soil moisture deficit

5.8.1.8 Runoff percentage

$$RP = \frac{d_r}{d_a} \times 100$$

where
RP = runoff percentage
d_r = equivalent depth of irrigation water running off the field

5.8.2 Complete and overirrigation cases (Figs. 3 and 4)

5.8.2.1 Application efficiency

$$AE = \frac{SMD}{d_a} \times 100$$

5.8.2.2 Application efficiency of low quarter

$$AELQ = \frac{SMD}{d_a} \times 100$$

5.8.2.3 Irrigation efficiency

$$IE = \frac{d_b}{d_a} \times 100$$

5.8.2.4 Christiansen's uniformity coefficient

$$CU = 1 - \frac{\sum_{i=1}^{n} |d_i - d_z|}{d_z \cdot n}$$

5.8.2.5 Distribution uniformity

$$DU = \frac{d_{zq}}{d_z} \times 100$$

5.8.2.6 Storage percentage

$$SP = 100\%$$

5.8.2.7 Deep percolation percentage

$$DPP = \frac{d_z - SMD}{d_a} \times 100$$

5.8.2.8 Runoff percentage

$$RP = \frac{d_r}{d_a} \times 100$$

5.8.2 When several furrows are being evaluated, the performance parameters should be calculated for each individual furrow. If desired, these results can then be averaged to describe the field system.

5.8.3 Since so many factors are involved, it is impossible to make a universal statement as to acceptable values for the performance parameters. Rather, furrow irrigation systems must be evaluated on a case by case basis.

5.9 Improving system performance. The results of the evaluation process may reveal shortcomings in system performance. Some of the possible problems are: (1) application depths which are too large or too small in relation to the soil moisture deficit, (2) excessive runoff, (3) excessive deep percolation, and (4) excessive soil erosion. Available are a variety of physical and operational modifications which can be implemented to improve system performance.

5.9.1 Physical changes

5.9.1.1 Furrow length. Where feasible, the furrow length may be changed (usually shortened).

5.9.1.2 Furrow shape and spacing. Furrow shape and spacing can be altered to affect the hydraulic and infiltration characteristics of the furrow.

5.9.1.3 Topography. Through land shaping, the furrow slope may be changed or the uniformity of slope may be improved.

5.9.1.4 Furrow direction. Changing the direction of furrows in the field or using contour furrows may provide a more desirable furrow length or slope.

5.9.1.5 Soil condition. Through tillage or compaction operations, the physical condition of the soil surface can be modified, thereby affecting the soil infiltration rate and the surface roughness.

5.9.1.6 Runoff recovery. Water lost to runoff can be recovered for reuse by installing a runoff recovery system. (See ASAE Engineering Practice EP 408, Design and Installation of Surface Irrigation Runoff Reuse Systems)

5.9.2 Operational changes

5.9.2.1 Inflow rate. The inflow rate to the furrow may be increased or decreased as required. "Cutback" irrigation could be practiced, in which the inflow rate is reduced during the runoff phase.

5.9.2.2 Time of cutoff. The time of cutoff of inflow is an important operational parameter which can be varied.

5.9.2.3 Irrigation frequency. Irrigation frequency is related to the management allowed deficit, which may need to be adjusted based on soil infiltration characteristics. If excessive infiltration is a problem and the soil surface tends to crack between irrigations, lighter and more frequent irrigations may be helpful. Conversely, on soils with low infiltration rates, the time interval between irrigations may have to be lengthened to increase infiltration by inducing cracking.

5.9.2.4 Alternate furrow irrigation. Another management option is the practice of irrigating every other furrow and leaving the intervening ones dry. In some situations, this practice could improve system performance by reducing deep percolation and/or runoff losses.

References:

1 Kruse, E. G., Chmn. 1978. Describing irrigation efficiency and uniformity. Journal of the Irrigation and Drainage Division. ASCE, 104(IR1):35-41.

2 Merriam, J. L. and J. Keller. 1978. Farm irrigation system evaluation: a guide for management. Utah State Univ., Logan.

3 Merriam, J. L., M. N. Shearer, and C. M. Burt. 1980. Evaluating irrigation systems and practices. Chapter 17 **in** M. E. Jensen (Ed.), Design and Operation of Farm Irrigation Systems. ASAE, St. Joseph, MI.

Cited Standard:

ASAE EP408, Design and Installation of Surface Irrigation Runoff Reuse Systems

DRIP/TRICKLE POLYETHYLENE PIPE USED FOR IRRIGATION LATERALS

Developed by the Drip Irrigation Tubing Standards Committee of The Irrigation Association and the ASAE Subsurface and Trickle Irrigation Committee; approved by ASAE Soil and Water Division Standards Committee; adopted by ASAE December 1985.

SECTION 1—PURPOSE AND SCOPE

1.1 This Standard covers requirements and methods for testing of polyethylene (PE) pipe made in standard dimensions for drip irrigation. Included are criteria for classifying PE plastic materials and PE pipe, a system of nomenclature for PE plastic pipe, and requirements and methods of test for materials, workmanship, dimensions, sustained pressure, burst pressure, and environmental stress crack resistance. Methods of marking are also given.

NOTE: This Standard is presently written for PE material. Other thermoplastic materials may be suitable for this specification and if presented will be tested in accordance with this Standard.

SECTION 2—DEFINITIONS

2.1 General: Definitions are in accordance with both American Society for Testing and Materials Standard D883-80c, Definitions of Terms Relating to Plastics, and ASTM Standard F412-81a, Standard Definitions and Terms Relating to Plastic Piping Systems. Abbreviations are in accordance with ASTM Standard D1600-81, Abbreviations of Terms Relating to Plastics, unless otherwise indicated. The abbreviation for polyethylene is PE.

2.2 Hoop stress: The tensile stress in the wall of the pipe in the circumferential orientation due to internal hydrostatic pressure.

2.3 Relation between stress and pressure:

$$P = \frac{2St}{d+t} \quad \text{(for inside diameter controlled pipe)}$$

where

S = stress, kPa (psi)
P = pressure, kPa (psi)
d = average inside diameter, mm (in.)
t = minimum wall thickness, mm (in.)

2.4 Standard thermoplastic pipe internal dimension ratio (d/t) (SIDR): The standard thermoplastic pipe internal dimension ratio (SIDR) is the ratio of the pipe internal diameter to wall thickness. For drip irrigation pipe it is calculated by dividing the average inside diameter of pipe in millimeters or in inches by the minimum wall thickness in millimeters or in inches. The SIDR values should be rounded off to the nearest tenth (0.1).

2.5 Failure: Bursting, cracking, splitting, or weeping (seepage of liquid) of the pipe during test.

2.6 Long-term hydrostatic strength (LTHS): The estimated tensile stress in the wall of the pipe in the circumferential orientation that, when applied continuously, will cause failure of the pipe at 100,000 h. This is the intercept of the stress vs. time regression line with the 100,000-h coordinate.

2.7 Hydrostatic design stress, basic (HDSB): The 20,000-h stress value for a compound obtained from the stress vs. time regression line used to establish LTHS.

2.8 Hydrostatic design stress (HDS): The estimated maximum tensile stress in the wall of the pipe in the circumferential orientation due to internal hydrostatic pressure that can be sustained continuously with a high degree of certainty that failure of the pipe will not occur.

2.9 Service (safety) factor: The number 0.5 (which takes into consideration all the appropriate variables and degree of safety involved in the use of thermoplastic pressure piping installation) which is multiplied by HDSB to give the HDS.

2.10 Pressure rating (PR): The estimated maximum internal pressure that water in the pipe can exert continuously with a high degree of certainty that failure of the pipe will not occur within 20,000 h of continuous pressurization. This is equivalent to approximately 10 years of irrigation lateral service life with a 100% safety factor. The assumption is that each lateral will be pressurized approximately 1/4 of the elapsed time.

SECTION 3—MATERIALS

3.1 Grades. This specification covers PE pipe made from grade P14, P23, P33, and P34 PE plastics as defined by ASTM Standard D1248-81a, Specifications for Polyethylene Plastics Molding Extrusion Materials, in which these requirements are based on short-term tests.

3.2 Extrusion compound. The PE plastic extrusion compound shall meet the requirements of grade P14, P23, P33, or P34; class C as described in ASTM Standard D1248-81a, Specifications for Polyethylene Plastics Molding Extrusion Materials.

3.3 Rework plastic. Clean, rework material, generated from the manufacturer's own production, may be used by the same manufacturer as long as the tubing produced meets all the requirements of this specification.

3.4 General. Polyethylene plastics used to make pipe meeting the requirements of this specification are categorized by means of two criteria; (1) short-term strength tests and (2) environmental stress crack resistance (ESCR).

3.5 Long-term hydrostatic design stresses. This specification covers PE pipe made from PE plastics as defined by the hydrostatic design stresses developed on the basis of long-term tests (LTHS, ASTM Standard D2837-76, Obtaining Hydrostatic Design Basis for Thermoplastic Pipe Materials) as modified for a 20,000-h service life as follows:

on a log/log graph, plot the straight line relationship between short-term hydrostatic failure stress [ASTM Standard D1599-82, Test for Short-Time Hydraulic Failure Pressure of Plastic Pipe, Tubing, and Fittings] times 0.5 and 100,000-h hydrostatic design stress times 0.5. By interpolation of the graph, determine the 20,000-h design stress for rating PE irrigation pipe and for use in the formula in paragraph 2.3.

3.6 Elevated temperature hydrostatic design stresses. Hydrostatic design stresses at elevated temperatures, above 23 °C (73 °F), shall be determined by running a series of short-time hydraulic failure tests per paragraph 5.9, at standard temperature and at the anticipated design temperature. The 20,000-h hydrostatic design stress value shall be adjusted by direct ratio as determined from the short-time hydraulic failure pressure tests to reflect the design temperature.

SECTION 4—PIPE IDENTIFICATION

4.1 General. This specification covers the identification of PE pipe manufactured for drip irrigation. PE pipe manufactured to this specification shall be labeled at intervals not to exceed 2 m (5 ft) or each roll tagged with the following minimum information

(1) manufacturer's name or logo,
(2) inside diameter/minimum wall thickness expressed to 0.025 mm (0.001 in.),
(3) resin grade, and
(4) ASTM Standard D1248-81a, Specifications for Polyethylene Plastics Molding Extrusion Materials.

SECTION 5—MANUFACTURING REQUIREMENTS AND METHODS OF TESTING

5.1 Workmanship. The pipe shall be homogeneous throughout and free from visible cracks, holes, foreign inclusions, or other defects. The pipe shall be uniform in color, opacity, density, and other physical properties.

5.2 Sampling. The selection of the sample or samples of pipe shall be as agreed upon by the purchaser and the seller. In case of no prior agreement, any sample selected by the testing laboratory shall be deemed adequate.

5.2.1 Test specimens. Not less than 50% of the test specimens required for any pressure test shall have at least a part of the marking in their central sections. The central section is that portion of pipe which is at least one pipe diameter away from an end closure.

5.3 Conditioning. Condition the test specimens at 23 ± 2 °C (73.4 ± 3.6 °F) and 50 ± 5% relative humidity for not less than 40 h prior to test in accordance with Procedure A of ASTM Standard D618-61, Conditioning Plastics and Electrical Insulating Materials for Testing, for those tests where conditioning is required.

5.4 Test conditions. Conduct the tests in the standard laboratory atmosphere of 23 ± 2 °C (73.4 ± 3.6 °F) and 50 ± 5% relative humidity, unless otherwise specified in the test methods or in this specification.

5.5 Dimensions and tolerances. Any length of pipe may be used to determine the dimensions. Measure in accordance with ASTM Standard D2122-81, Determining Dimensions of Thermoplastic Pipe and Fittings.

5.5.1 Inside diameter. Measure the inside diameter of the pipe with a tapered plug gauge in accordance with Section 7 of ASTM Standard D2122-81, Determining Dimensions of Thermoplastic Pipe and Fittings. Tolerance on ID shall be −0.00 + 0.15 mm (0.006 in.); i.e., ID shall be no smaller than the specified/stated value and no greater than the specified value plus 0.15 mm (0.006 in.).

5.5.2 Wall thickness. Make micrometer measurements of the wall thickness in accordance with Section 6 of ASTM Standard D2122-81, Determining Dimensions of Thermoplastic Pipe and Fittings, to determine the maximum and minimum values. Measure the wall thickness at both ends of the tubing to the nearest 0.025 mm (0.001 in.). The minimum wall thickness shall be as established by stress calculation. Tolerance on wall thickness shall be −0.00 + 0.15 mm (0.006 in.).

5.5.3 Maximum outside diameter. The final actual OD of the pipe shall not exceed the sum of the specified ID plus two specified minimum wall thicknesses by more than 0.25 mm (0.010 in.). The maximum OD shall be calculated as

OD (Maximum) = Specified ID + Specified Minimum Wall Thickness + 0.25 mm (0.010 in.)

5.6 Carbon black. Determine the carbon black content of the tubing in accordance with ASTM Standard D1603-76, Test for Carbon Black in Ethylene Plastics. Use a minimum of two test specimens. The tubing shall contain at least 2% carbon black content (ASTM Standard D1248-81a, Specifications for Polyethylene Plastics Molding Extrusion Materials) when tested.

5.7 Resin density. Determine the density of the pipe compound in accordance with ASTM Standard D1505-68, Test for Density of Plastics by the Density-Gradient Technique, or ASTM Standard D792-66, Tests for Specific Gravity and Density of Plastics by Displacement, using three specimens. Determine the percentage of carbon black by weight in accordance with paragraph 5.6. Calculate the density of the PE base resin (uncolored PE) in the pipe compound as follows

$$Dr = Dp - 0.0044C$$

where

Dr = density of resin, mg/mm^3
Dp = density of pipe compound, mg/mm^3
C = percentage by weight of carbon black

The PE base resin (uncolored PE) in the pipe compound shall have a density (ASTM Standard D1248-81a, Specifications for Polyethylene Plastics Molding Extrusion Materials) in the following ranges when tested

0.910 to 0.925 mg/mm^3 for pipe made from grade P14
0.926 to 0.940 mg/mm^3 for pipe made from grade P23
0.941 to 0.965 mg/mm^3 for pipe made from grade P33
0.041 to 0.965 mg/mm^3 for pipe made from grade P34

5.8 Short-time hydraulic failure pressure. Determine the minimum failure pressure with at least five specimens in accordance with ASTM Standard D1599-82, Test for Short-Time Hydraulic Failure Pressure of Plastic Pipe, Tubing, and Fittings. The time of testing of each specimen shall be between 60 and 70 s. The minimum failure pressure for the pipe when tested is calculated by the following formula using stress values shown below

$$P = \frac{2S}{SIDR+1}$$

where

P = minimum failure pressure, kPa (psi)
S = test stress, kPa (psi)
SIDR = ratio of the pipe internal diameter to wall thickness (standard thermoplastic pipe internal dimension ratio)

TEST STRESS (S) WITH WATER AT 23 °C (73.4 °F)

Grade designation	Test stress, kPa (psi)
PE 1404	8600 (1250)
PE 2305	13800 (2000)
PE 2306, 3306, 3406	17380 (2520)

5.9 Environmental stress crack resistance (ESCR). Tag six randomly selected 250 mm (10 in.) long specimens for this test. Insert in one end a barbed fitting which has a shape as defined in Fig. 1 and which will elongate the pipe circumferentially by 30%. Bend the other end of the specimens back 180 deg as shown in Fig. 2 and retain in that position. Submerge specimens in a 10% by volume nonylphenoxy poly (ethyleneoxy) ethanol ("Igepal C - 630") solution at a temperature of 76.7 ± 2.8 °C (170 ± 5 °F) for a 48-h period. A crock pot or oven with temperature control may be used; however, the Igepal solution should be stirred to maintain a 10% solution in the bath. After 48 h, examine the specimens at the bend and also at the insert end for signs of stress cracking. Record any sign of cracking as a failure. There shall be no surface cracking, cracks, or splits on pipe when tested.

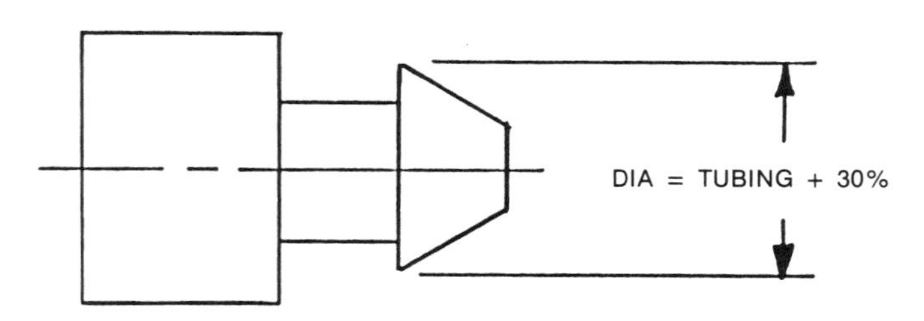

FIG. 1—BARBED FITTING FOR ESCR TEST

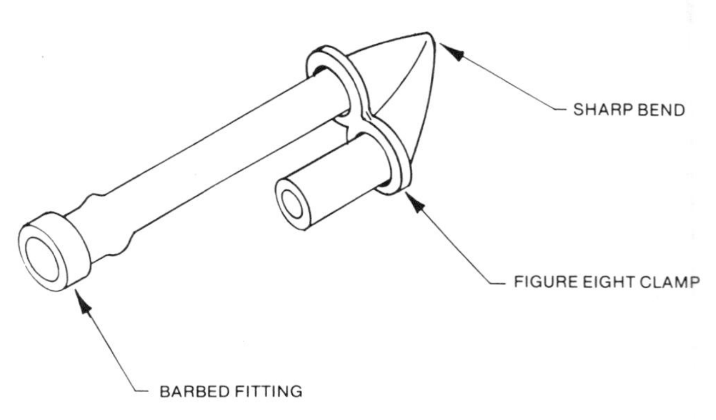

FIG. 2—BEND FOR ESCR TEST

SECTION 6—RETEST AND REJECTION

6.1 If any failure occurs, the pipe may be retested to establish conformity. If failure occurs on the retested specimens, the pipe shall be deemed not in conformance with the Standard.

Cited Standards:

ASTM D543-67, Resistance of Plastics to Chemical Re-agents
ASTM D618-61, Conditioning Plastics and Electrical Insulating Materials for Testing
ASTM D792-66, Tests for Specific Gravity and Density of Plastics by Displacement
ASTM D883-80c, Definitions of Terms Relating to Plastics
ASTM D1248-81a, Specifications for Polyethylene Plastics Molding Extrusion Materials
ASTM D1505-68, Test for Density of Plastics by the Density-Gradient Technique
ASTM D1599-82, Test for Short-Time Hydraulic Failure Pressure of Plastic Pipe, Tubing, and Fittings
ASTM D1600-81, Abbreviations of Terms Relating to Plastics
ASTM D1603-76, Test for Carbon Black in Ethylene Plastics
ASTM D2122-81, Determining Dimensions of Thermoplastic Pipe and Fittings
ASTM D2837-76, Obtaining Hydrostatic Design Basis for Thermoplastic Pipe Materials
ASTM F412-82a, Standard Definitions and Terms Relating to Plastic Piping Systems

ASAE Standard: ASAE S436

TEST PROCEDURE FOR DETERMINING THE UNIFORMITY OF WATER DISTRIBUTION OF CENTER PIVOT, CORNER PIVOT, AND MOVING LATERAL IRRIGATION MACHINES EQUIPPED WITH SPRAY OR SPRINKLER NOZZLES

Developed jointly by The Irrigation Association and the ASAE Sprinkler Irrigation Committee; approved by the ASAE Soil and Water Division Standards Committee; adopted by ASAE June 1983.

SECTION 1—PURPOSE AND SCOPE

1.1 The purpose of this Standard is to establish a uniform method of collecting water distribution data in the field and calculating the coefficient of uniformity from the data.

1.2 This Standard covers evaluation of water distribution on center pivots, corner pivots, and moving lateral irrigation machines.

SECTION 2—TERMS AND DEFINITIONS

2.1 Center pivot: An automated irrigation machine consisting of a sprinkler line rotating about a pivot point and supported by a number of self-propelled towers. The water is supplied at the pivot point and flows outward through the line supplying each of the individual sprinkler outlets. The entire unit irrigates a basically circular shape.

2.2 Corner pivot: An automated irrigation machine consisting of a center pivot with an additional span or other equipment attached to the end of the machine that allows the overall center pivot radius to increase or decrease in relation to the field boundaries. The water is supplied as in a center pivot machine.

2.3 Moving lateral: An automated irrigation machine consisting of a sprinkler line supported by a number of self-propelled towers. The entire unit moves in a generally straight or slightly curved path and irrigates a basically rectangular area. The water can be supplied to the machine from a variety of locations along the pipeline and is then distributed to the individual sprinkler outlets. For the purpose of this test procedure, the names "linear", "lateral" and "moving lateral" are synonymous.

2.4 Sprinkler package: A collection of spray or sprinkler distribution hardware such as sprinklers, nozzles, pressure or flow control devices, and supporting plumbing designed for a specific machine and set of operating parameters.

SECTION 3—TEST CONDITIONS AND EQUIPMENT

3.1 All collectors used for any one test must be identical and shaped such that water does not splash in or out. The lip of the collector shall be symmetrical and without depressions. The diameter of the opening must be at least 80 mm (3.2 in.) and the height at least 100 mm (3.9 in.).

3.2 The collectors shall be uniformly spaced and located in a straight line with a maximum collector spacing of 30% of the average sprinkler-wetted diameter. The wetted diameter of the individual sprinkler is to be determined based upon the vertical distance from the sprinkler nozzle to the collector described in paragraph 3.5. With the height of the nozzle known, the wetted diameter shall be defined as the wetted dimension perpendicular to the pipeline while the machine is stationary. In no case shall the collector spacing exceed 4.5 m (14.8 ft).

3.2.1 At least two (2) lines of collectors must be used for any evaluation.

3.3 The collectors must be located so that no interference from crop growth occurs. Crop canopies or other obstructions higher than the elevation of the collector, but below the nozzle height, shall be at a distance from the collector of at least twice the height of the crop canopy or obstruction on both sides of the collector rows. For nozzles operating below the crop canopy height, the collectors shall be located so that an unobstructed distance of 1.25 times the sprinkler-wetted radius is maintained on either side of the collector rows.

3.4 The entrance portion of the collectors must be level and located between 300 mm (11.8 in.) and 1.25 m (4.1 ft) above the ground. The discharge height of the nozzle must be at least 1.5 m (4.9 ft) above the elevation of the collector. Nozzle and collector heights shall be recorded.

3.5 Wind velocity during the test period shall be measured with a rotating anemometer or equivalent device.

3.5.1 The wind direction shall be determined with a vane indicating at least eight points of the compass. Wind direction passing over the rows of collectors shall be recorded.

3.5.2 The wind velocity measuring equipment shall be located at a minimum height of 2.0 m (6.6 ft) and within 200 m of the test site, in a location that is representative of the wind conditions at the test site.

3.5.3 The rotating anemometer shall have a threshold velocity not to exceed 0.3 m/s (0.7 mile/h) and be capable of measuring the actual velocity within ±10%.

3.5.4 The entire test shall be conducted with a wind velocity of less than 1.0 m/s (2.2 mile/h).

3.6 Tests shall be run during periods that will minimize the effect of evaporation. It is recommended that the test be run at night.

3.6.1 All data collection shall be taken as soon as possible after the machine has passed over the collector.

3.6.2 Alternate procedures for minimizing evaporation are to use an evaporation suppressant or a specially designed collector. Record type and method used.

3.6.3 If an evaporation adjustment is made on the collected data, a minimum of three control collectors containing the anticipated catch shall be placed at the test site and monitored to determine the rate of evaporation.

3.7 The test must be conducted in an area which has elevation changes which are within the design specifications of the sprinkler package.

SECTION 4—TEST AND PROCEDURE

4.1 It shall be verified that the sprinkler package has been installed and adjusted according to the sprinkler design specifications. If not installed or adjusted properly, the package shall be corrected before the proceeding.

4.2 The system pressure shall be adjusted to the design specifications using a pressure measuring device capable of measuring accurately within ±5% of the sprinkler package pressure setting.

4.3 The machine shall be operated at a speed which will deliver an average depth of not less than 12.5 mm (0.5 in.).

4.4 The distribution data shall be recorded by measuring the volume of water caught in the collectors. The measuring device shall be accurate to ±3% of the average catch.

4.5 The machine shall be started and passed over the lines of collectors either tangentially or perpendicularly, depending on whether the machine is a pivot or lateral. (See Section 5—Center Pivot, Section 6—Corner Pivot and Section 7—Lateral Move.)

4.6 "Extreme deviating observations" or data points which are unreasonable high or low may be eliminated from the analysis. Under no circumstances can the number of eliminated "extreme observations" exceed 0.5% of the total number of data points.

4.7 In addition to those points outlined in paragraph 4.6, any obviously incorrect data points caused by such occurrences as leaking or tipped containers, or other explainable variances from the test conditions can be automatically excluded from the data.

4.8 Data collection points located on the outer edge of the wetted area may be eliminated from the analysis whenever the readings fall below 70% of the average catch.

4.9 If the sprinkler package is designed with an endgun, the test is to be performed with the endgun in its normal operational mode.

SECTION 5—CENTER PIVOT

5.1 The collectors shall be located along lines extending radially from the pivot point. The outer ends of the radial lines shall be no more than 50 m (165 ft) apart.

5.2 The collectors on the inner 20% of the total length may be eliminated from the analysis.

5.3 All other conditions as outlined in Section 3—Test Conditions and Equipment, and Section 4—Test and Procedure, must be followed.

SECTION 6-CORNER PIVOT

6.1 The collectors shall be located along lines extending radially from the pivot point.

6.2 Data collections shall be made with the corner pivot operating at its maximum radius and with the corner pivot operating at its minimum radius.

6.3 All other conditions as outlined in Section 3—Test Conditions and Equipment, and Section 4—Test and Procedure, must be followed.

SECTION 7—LATERAL MOVE

7.1 The collectors shall be located along lines parallel to the pipeline, and the collector rows shall be no more than 4.5 m (15 ft) apart.

7.2 All other conditions as outlined in Section 3—Test Conditions and Equipment, and Section 4—Test and Procedure, must be followed.

SECTION 8—CALCULATIONS

8.1 The center pivot coefficient of uniformity is to be calculated from the Heermann and Hein modified equation.

$$C_u = 100 \left[1.0 - \left(\frac{\sum\limits_{\eta} S_s \left| D_s - \dfrac{\sum\limits_{\eta} D_s S_s}{\sum\limits_{\eta} S_s} \right|}{\sum\limits_{\eta} D_s S_s} \right) \right]$$

where

C_u = Heermann and Hein uniformity coefficient
D_s = Total depth of application at a distance S from the center (collection depth)
S = Distance from the center to the collector
s = Subscript denoting a point at a distance S
η = Number of catch containers

8.2 Corner pivots are to be evaluated in the same manner as outlined in paragraph 8.1. If total field coefficient of uniformity is required, additional data must be collected and additional weighting factors applied.

8.3 Lateral move uniformity is to be calculated based on the Christiansen equation:

$$C_u = 100 \left(1 - \frac{\sum\limits_{\eta} \left| D_s - \overline{D} \right|}{\sum\limits_{\eta} D_s} \right)$$

where

C_u = Christiansen uniformity coefficient
$\overline{D}$ = Mean of catch container volumes

8.4 Separate coefficient of uniformity calculations shall be made for each line of collectors as well as a composite coefficient of uniformity value for all data points.

SECTION 9—EVALUATIONS

9.1 The results from such analysis may not reflect all the variations in depth of application that may occur as the machine operates in any given field, but allows a standard by which the performance of various machines may be compared.

9.2 The calculated coefficient of uniformity based on field data is subject to test condition variations, and as such, shall be used only as an indication of actual sprinkler package performance. For instance, the coefficient of uniformity of a particular machine can vary from one irrigation to another, primarily due to the effects of wind velocity and/or evaporative demand.

9.3 The coefficient of uniformity is only one factor in evaluating total system performance. Application rates, runoff, amount of water applied, pump performance, and overall system management can greatly affect the total level of performance.

References: Last printed in 1985 ASAE STANDARDS; list available from ASAE Headquarters.

ASAE Standard: ASAE S442

WATER AND SEDIMENT CONTROL BASINS

Developed by the ASAE Conservation Systems Committee; approved by the ASAE Soil and Water Division Standards Committee; adopted by ASAE October 1986.

SECTION 1—PURPOSE AND SCOPE

1.1 This Standard is intended to guide engineers and technicians in the planning, design, layout, construction and maintenance of water and sediment control basins. These basins consist of an earth embankment with a level top section constructed across a watercourse or gulley to form a sediment trap and a water detention basin.

1.2 This Standard applies where gully erosion, water management and/or sediment is a problem; where soil and site conditions are suitable to build earth embankments; where topography is suitable to provide adequate detention storage; and where water will infiltrate or can be discharged through underground outlets in time to prevent damage to crops or other vegetation.

1.3 Water and sediment control basins differ from terraces in that they do not necessarily follow the contour of the land, and they do not necessarily reduce slope length as defined in the Universal Soil Loss Equation. Therefore reduction in sheet and rill erosion may be slight. However, they may be used in conjunction with terrace systems.

SECTION 2—GENERAL RECOMMENDATIONS

2.1 Water and sediment control basins may be used to reduce gully erosion, trap sediment, manage surface runoff, improve downstream water quality, protect terrace systems, and reduce downstream flooding.

2.2 They are not intended to control sheet and rill erosion. In combination with terraces, contouring, conservation cropping systems, conservation tillage, crop residue management, and/or permanent vegetation, they can be part of a resource management system needed to protect the soil and water resource base.

2.3 Water and sediment control basins should be used on cultivated land where sheet and rill erosion is controlled using other conservation practices, or they may be used on permanently vegetated land.

SECTION 3—TERMINOLOGY

3.1 Water and sediment control basins are classified according to the embankment cross section.

3.1.1 Broadbase: Constructed so agricultural machinery may be operated safely on the entire cross section, usually not steeper than 5:1 (horizontal to vertical). Embankment slopes should be constructed to fit machinery widths.

3.1.2 Steep-backslope: Constructed with a stable steep backslope (1.5:1 or flatter) which is kept in permanent vegetation, usually grass. The front slope and basin are constructed to be farmed and to fit machinery widths.

3.1.3 Narrow-base: The back and front slopes are constructed with steep but stable slopes (1.5:1 or flatter) and kept in permanent vegetation, usually grass (not recommended where cattle have access).

3.2 The outlets for water and sediment control basins are classified by their function.

3.2.1 Underground outlet: The outlet consists of an inlet riser and underground pipe that discharges water through or under the embankment to an adequate underground or surface outlet.

3.2.2 Infiltration outlet: Water collected in the basin is discharged by infiltration into the soil.

3.3 A watercourse is a channel or low area where water concentrates and flows during a runoff event.

SECTION 4—DESIGN

4.1 Spacing is the horizontal distance, perpendicular to the embankment centerline, from one basin to the next. Spacing of water and sediment control basins should generally be equivalent to terrace intervals (see ASAE Standard S268, Design, Layout, Construction and Maintenance of Terrace Systems).

4.1.1 When terrace spacing is not used, spacing should limit the drainage area between basins so that the peak design runoff will not cause erosive velocities in the watercourse between basins. This velocity is computed by using Manning's formula and an n value of 0.03. Where permanent vegetation is not present, maximum velocities should be no more than 0.8 m/s (2.5 ft/s) for soils with high organic content, 0.6 m/s (2.0 ft/s) for most soils, and 0.5 m/s (1.5 ft/s) for extremely erodible soils. If permanently vegetated, the watercourse should be evaluated as a grassed waterway.

4.1.2 The system of basins shall be parallel where possible and spaced to accommodate agricultural machinery making complete trips around the field. When determining spacing, consideration shall be given to embankment slope lengths, top width, inlet location, and crops.

4.1.3 The drainage area of each basin shall be limited so flooding, infiltration, or seepage will not damage crops or create other water problems. For safety reasons, the drainage area should normally be limited to a size which produces no more than 12,500 m^3 (10 acre·ft) of runoff from the design storm.

4.2 The embankment orientation and crop row direction shall be approximately parallel to the field contours to permit areas of contour farming. Permanent features such as field boundaries, watercourses, ridges, terraces, and rock outcroppings should also be considered when determining the location of the basins.

4.3 Cross sections of the embankment may vary (see paragraph 3.1), but generally shall be patterned after terrace cross sections (see ASAE Standard S268, Design, Layout, Construction and Maintenance of Terrace Systems).

4.3.1 The minimum effective top width for various heights shall be:

Fill height		Effective top width	
m	ft	m	ft
0 — 1.5	0 — 5	1.0	3
1.5 — 3.0	5 — 10	2.0	6
3.0 — 4.6	10 — 15	2.5	8

4.3.2 The constructed height of the embankment shall be at least 5% greater than the design height to allow for settlement.

4.3.3 The maximum settled height shall be 4.6 m (15 ft) measured from the natural ground at the centerline of the embankment.

4.3.4 To maintain traction and support for passage of center pivot or sideroll irrigation systems which travel across the embankment centerline, construct elevated roadways to provide a crossing for the irrigation system.

4.4 Capacity of the basin shall be sufficient to control the runoff from a 10-yr frequency, 24-h duration storm without overtopping. The capacity of basins designed to provide flood protection or to function with other structures may need to be increased and shall be adequate to control a storm of a frequency and duration consistent with the potential hazard. The detention basin size may be determined by flood routing the design storm through the structure. It is desirable, at a minimum, to provide temporary storage for all of the runoff from a 2-yr frequency, 24-h duration storm.

4.5 A maximum of 0.3 m (1 ft) freeboard may be added to the design height leaving a low (design height) area to provide for an emergency spillway. The emergency spillway shall not contribute runoff to a lower basin in series that does not also have an emergency spillway. The emergency spillway should be located near an end of the embankment where the erosion hazard is minimized.

4.6 The elevation of the settled embankment should be at least 5 cm (0.2 ft) higher in the center than at the ends to allow overflow at locations where a breach will result in a minimum amount of damage.

4.7 The basin capacity should include storage for the anticipated 10-yr sediment accumulation (using the Universal Soil Loss Equation and a delivery ratio of 100%), unless provisions are made for periodic sediment removal from the basin to maintain the design capacity.

4.8 Outlets shall be either a riser inlet to an underground outlet or the soil shall be porous enough to allow infiltration. Underground outlets shall meet requirements in ASAE Engineering Practice EP260, Design and Construction of Subsurface Drains in Humid Areas.

4.8.1 A single underground pipe with adequate capacity may be used as the underground outlet for several basins.

4.8.2 Outlets shall have adequate capacity to discharge water in time to prevent severe crop damage, seepage or embankment stability problems. Usually, discharge in less than 24 h is recommended. However, 48 h may be allowed where crops have some tolerance to water, e.g., corn, and where soils have good internal drainage.

4.8.3 When used in conjunction with terraces, the outlets shall meet the requirements for an adequate terrace outlet.

4.8.4 If underground outlets are used, the system shall be hydraulically designed so the emptying time of the basins will be as uniform as possible and so the water does not flow from one basin to another through the outlet. If orifices are not used to control the flow into the underground pipe, the system shall be designed as a full flow system with pipe sizes controlling discharge and emptying time.

4.8.5 Inlets shall be located to minimize their interference with farm equipment. Normally space is allowed for one pass of equipment between the inlet and the ridge unless the front slope is not farmed.

4.9 If a water and sediment control basin diverts water across a slope, this portion of the basin shall meet terrace criteria (see ASAE Standard S268, Design, Layout, Construction and Maintenance of Terrace Systems).

4.10 Broadbase cross sections should normally be used only where average land slopes are less than 6%. Steep-backslope or narrow-base cross sections should be used on steeper land.

4.11 Location for water and sediment control basins and their outlets is best determined from topographic maps. Maps with a 0.6 m (2 ft) contour interval should provide sufficient detail to locate all minor watercourses and determine the drainage areas.

SECTION 5—CONSTRUCTION

5.1 Several methods can be used to build water and sediment control basin embankments.

5.1.1 Soil depth and type determine construction methods and allowable cuts or fills.

5.1.2 Compaction adequate to prevent seepage and failure of the embankment should be accomplished by passing construction equipment uniformly over 15-30 cm (6-12 in.) layers of fill. In arid climates or other dry conditions, water should be added to insure adequate compaction.

5.1.3 Where basins are built in conjunction with terraces, it may be desirable to move soil from the terrace line cuts to make the basin fills. Topographic improvement should be considered and fills should be made in low areas, normally at watercourses, and borrow should be from the ridges. Consideration may be given to the use of one borrow area for an entire system on fragile soils. Topsoil should be saved and placed on the surface of cuts and fills.

5.2 It is preferable to install the underground outlet one construction season in advance of the embankment construction to allow settlement of the backfill over the pipe. Otherwise, the pipe trench under the terrace base should be backfilled by constructing the sides of the trench with a 1:1 slope down to the top of the pipe and then compacting the backfill, being careful not to crush the pipe. In arid areas, water should be added to facilitate adequate compaction. Noncohesive backfill may be water packed.

5.3 Pipe and inlet materials should meet the requirements specified in ASAE Engineering Practice EP260, Design and Construction of Subsurface Drains in Humid Areas.

5.4 Slopes and disturbed areas that are not to be farmed should be permanently vegetated with erosion resistant vegetation. If soil or climatic conditions preclude use of vegetation and protection is needed, coverings of organic or gravel mulches may be used, considering the potential for flotation and removal by water erosion. Environmental factors and wildlife food and habitat shall be considered in selecting the vegetation species.

SECTION 6—MAINTENANCE

6.1 Maintenance needs can be determined by inspections of the embankment, basin area, erosion resistant cover, inlets, and underground outlets.

6.2 Inspections should be made after each major runoff event, but at least annually. Repairs shall be made as soon as feasible to prevent further damage to the systems.

6.2.1 Repair piping failures, rodent damage, livestock damage, slides, and erosion of the embankment. Replace damaged, crushed or failed inlets and underground outlets.

6.2.2 Remove excess sediment that encroaches seriously on the detention capacity. Sediment removal operations should leave the inlet in the lowest area of the basin so it will drain properly. Sediment can be used to enlarge the embankment resulting in a bench leveling effect, or it can be redistributed over the field.

6.2.3 Fill eroded areas. Reseed or mulch bare areas.

6.2.4 Remove and clean trash from inlets after each significant runoff producing storm.

6.2.5 Remove sediment and trash from outlets and repair rodent guards.

SECTION 7—SAFETY

7.1 The plan layout should be designed with safety in mind. Potential hazards should be recognized and identified on the plans and pointed out to the owner/operator responsible for construction, farming, and maintenance. Plans should delineate farmable and nonfarmable areas.

7.2 The embankment slopes and basin configuration should be designed considering machinery limitations and safety. Slopes steeper than 5 to 1 are difficult to farm. Machinery operating on the embankments near the steep backslopes are in danger of sliding or breaking down the slope. Inlets are a hazard to farm equipment; wire, steel or wood posts, and metal risers can easily damage some equipment. Flags or other warning devices should be attached to inlets so that they can be avoided during farming operations.

7.3 Caution should be used when operating equipment on embankment slopes. Slopes steeper than 4 to 1 should not be farmed except with special equipment. Safety devices such as protective frames or protective canopies may be required to protect equipment operators. Inexperienced equipment operators should be trained to farm water and sediment control basin systems.

Cited Standards:

EP260, Design and Construction of Subsurface Drains in Humid Areas
S268, Design, Layout, Construction and Maintenance of Terrace Systems

ASAE Engineering Practice: ASAE EP463

DESIGN, CONSTRUCTION AND MAINTENANCE OF SUBSURFACE DRAINS IN ARID AND SEMIARID AREAS

Developed by the ASAE Drainage of Irrigated Lands Committee; approved by the ASAE Soil and Water Division Standards Committee; adopted by ASAE February 1985.

SECTION 1—PURPOSE AND SCOPE

1.1 This Engineering Practice is intended as a guide to engineers in the design and construction of subsurface drains in arid and semiarid regions where irrigation is often used to provide adequate water for crops. It is not designed to serve as a complete set of specifications or standards. This Engineering Practice is intended to complement ASAE Engineering Practice ASAE EP260.4, Design and Construction of Subsurface Drains in Humid Areas, which is used as a primary reference for this Engineering Practice.

SECTION 2—INVESTIGATIONS

2.1 Introduction

2.1.1 There are many types of drainage problems and the investigation of each must be varied as required to solve the particular problem. Before beginning, a clear understanding should be obtained of the purpose of the investigation. Plans should be formulated to determine the minimum amount of data required for the specific project or report and the best way of obtaining such data.

2.1.2 In irrigated areas, there is a close relationship between irrigation and drainage. The amount of drainage water to be removed by subsurface drains is related to the irrigation water applied and the irrigation management in the area. Data related to irrigation include infiltration, consumptive use, water quality and salinity.

2.2 Pre-reconnaissance investigations

2.2.1 Topographic maps of the area are usually available from the U.S. Geological Survey, the U.S. Bureau of Reclamation, the USDA Soil Conservation Service, or irrigation districts who have done work in the area. The maps should cover the area of interest and also surrounding areas which may be contributive to the excess water problem. Topographic maps often contain information relative to seep areas in nonirrigated lands, or areas with a continuing drainage problem.

2.2.2 Soil surveys are important to drainage investigations. Most irrigated areas have detailed standard soil surveys prepared by the USDA Soil Conservation Service or the U.S. Bureau of Reclamation. Soils information may be available from the same agencies that have topographic maps. In addition to the general information contained in the standard soil surveys, additional investigation may be required. Soil borings to approximately one-and-one-half times the estimated depth of drains will be needed to determine depth and thickness of different soil strata. Deeper holes may be necessary to locate and identify artesian aquifers or deep barriers. Location of soil layers of low and high hydraulic conductivity is important. Layers below drain levels may include clay pans, shale, sandstone, gravel, or quicksand. Existence of these materials will modify drainage design and the design of drain envelopes, and may influence construction practices.

2.2.3 Geologic reports for the area should be examined to identify sources of groundwater that might influence drainage requirements.

2.2.4 Aerial photographs are a good source of information and will indicate seep areas and possible outlets as well as property lines. Infrared photos may provide additional information concerning the extent of poorly drained areas, especially if taken at the proper time of year.

2.2.5 Well logs for borings in the area will provide valuable information about deep barriers and possible artesian pressures in areas to be drained. Copies of pertinent well logs should be obtained where possible.

2.2.6 Climatic records are important to determine the pattern of rainfall and the response of fluctuating water tables to natural precipitation.

2.2.7 Local interviews with people cognizant of drainage problems are important. Landowners or farmers in the area may be able to give valuable information relative to seasonal variation of drainage problems and the response of the water table and crops to water management practices in the area.

2.3 Field reconnaissance. After all reasonably available recorded information has been obtained, it is important to make a field reconnaissance of the area of interest to evaluate the condition of the area, the problems of access, and the locations of possible outlets.

2.3.1 Outlets for subsurface drainage systems in irrigated areas may be difficult to obtain. Special attention should be paid to the possible location of outlets for planned drainage systems as well as to the capacity and flow characteristics of these outlets. A determination should be made as to whether discharge permits will be needed for these outlets.

2.3.2 Existing drainage systems in the area should be noted and marked on the topographic maps as possible outlets and as features to be considered in the design of the new drainage system.

2.3.3 Soils needing drainage, which are apparent from a field reconnaissance, should also be noted on the maps. Some of these areas do not show on topographic maps or aerial photographs.

2.4 Field investigations. Some types of information are common to all types of drainage investigation, e.g., information on soils, land use and cropping patterns; and information related to precipitation, evapotranspiration, runoff and stream flow. Whether drains are open or closed, it is necessary to make profile, cross-section and soil profile investigations along the routes where outlet drains are to be installed. For subsurface drainage, additional information is needed on groundwater, perched water tables, salinity, hydraulic conductivity, and conditions below the upper soil horizons, in addition to information that would normally be required for surface drainage.

2.4.1 Water table levels in the soil are of particular importance and must be carefully investigated. Observations can be made of water levels in borrow pits, in wells, and in excavations made for other purposes in the area. In addition, a number of water table observation wells should be established. These may be open auger holes which are drilled at various locations for a preliminary investigation, or they may be permanent water table observation wells which have been established for collecting information over a longer period of time. Soil profiles should be logged when holes for the water table observation wells are drilled. Observations should be made of free water, mottling of the soil, and estimated hydraulic conductivity of various strata. The diameter of the water table observation well is relatively unimportant except in fine-textured soils. Large diameter holes which may be a danger to animals in the area should be covered. After the wells are installed, time should be allowed for water levels to come to equilibrium. Permanent water table observation wells can be constructed by drilling a 102 mm (4 in.) diameter hole, putting a 25 or 51 mm (1 or 2 in.) diameter perforated plastic or steel pipe in the center of the hole, backfilling the hole to within 0.3 m (1 ft) of the surface with coarse sand, and then packing the upper part of the hole with finer soil material to prevent the entrance of surface water. Bottoms of pipes should be bedded in sand, or closed to prevent the entry of unstable soil materials. If the water table observation well is part of a large network of pipes designed to collect information over a long period of time, elevations should be obtained for the pipe tops so that direction of groundwater movement can be determined.

2.4.2 Piezometers should be installed in any area where significant upward or downward movement of water is suspected. Piezometers can be installed by jetting, or driving, using standard techniques. At least two piezometers should be installed at different depths in any location. Piezometers should be installed in fence lines or in protected areas where they will not be dangerous to farm equipment, tires, or animals. Tests of the sensitivity of the piezometers should be made periodically by filling them with water or by extracting some water and observing the rate of return to equilibrium to determine whether they are functioning properly.

2.4.3 Groundwater sources that contribute to the drainage problem should be identified from observations and data obtained using water table observation wells and piezometers. Groundwater contours, depths to water table, and water table fluctuations with time should be plotted on appropriate maps. Such information is important when determining whether inceceptor or relief type drains should be installed.

2.4.4 Water quality may need consideration if U.S. Environmental Protection Agency, local or state agency discharge permits are required. Environmental impact statements may be necessary depending on local laws and conditions. In some areas, a permit may be required to discharge drainage water regardless of its quality.

2.4.5 Hydraulic conductivity should be carefully determined, since the functioning of any drainage system and the drain spacing is directly dependent on the hydraulic conductivity of the soil. In-place bail-out auger hole tests should be made where a high water table exists. Pump-in auger hole tests should be made if a high water table does not exist. Procedures for determining hydraulic conductivities can be found in *Drainage Manual* (U.S. Bureau of Reclamation), *Drainage of Agricultural Lands* (USDA Soil Conservation Service), *Drainage Engineering* (J. N. Luthin), and *Drainage for Agriculture — Agronomy Monograph No. 17* (American Society of Agronomy).

2.4.5.1 Since drain spacing and performance is also dependent on the relative location of the barrier with respect to the depth of the drains, reasonable efforts should be made to identify subsurface barrier conditions. All auger holes made for the investigation of water table position, soil profile characteristics, and hydraulic conductivities should be evaluated at least to the depths that the drains will be installed. One hole in ten should extend to the barrier, three holes in ten should extend to a deeper depth than the drain, and six holes in ten should be at least 3 m (10 ft) deep. A barrier layer can be considered as any layer whose hydraulic conductivity is one-fifth or less of the hydraulic conductivity of the layer above it. Where soils are layered, horizontal hydraulic conductivities should be determined for each layer by progressively deepening the auger hole and making a determination for each layer. Procedures for determining the weighted hydraulic conductivity of the saturated profile are given in ASAE Special Publication 903CO466, *Guide for Investigation of Subsurface Drainage Problems on Irrigated Lands.*

2.4.6 Soil gradation at the depth that the drains are to be installed should be determined for use in the design of drain envelopes. In most irrigated areas, soil structural stability is limited and drain envelopes are required. After the soil gradation, or mechanical analysis, has been made, a suitable gravel envelope gradation can be selected to protect the drains. Specifications for gravel envelope design can be found in *Drainage of Agricultural Lands* (USDA Soil Conservation Service) and ASAE Special Publication 903CO466, *Guide for Investigation of Subsurface Drainage Problems in Irrigated Lands.* See paragraph 3.9 for envelope materials other than gravel.

2.4.7 Recharge rate, or drainage coefficient, should be determined. Local experience in drain design should be considered. Drainage coefficients may already be available from local government agencies. Estimates of recharge rates can be made from data on irrigation efficiency, evapotranspiration, and irrigation interval by observing water table responses of existing systems to irrigation and rainfall events.

SECTION 3—DESIGN

3.1 Outlets for buried pipe drains

3.1.1 Open channels. If the proposed outlet is to be a stream or open ditch, it should provide a free flowing outlet for subsurface drains within a reasonable period of time after peak storm or other short duration flows, and should be large enough to remove normal surface runoff and operational waste from irrigated lands without submerging the pipe drain outlet. ASAE Engineering Practice ASAE EP407, Agricultural Drainage Outlets—Open Channels, provides planning, design, construction and maintenance information for open channels.

3.1.1.1 The outlet channel should be of adequate depth to insure that the minimum flow line (invert) of the main subsurface drain at the outlet shall be a minimum of 152 mm (6 in.) above the normal water surface. In determining the flow capacity and the normal water surface in the outlet channel, consideration should be given to the possibility of discharge from additional drains that may be constructed upstream of the drains currently being designed.

3.1.1.2 Where an open channel is either nonexistent or unsatisfactory for use as an outlet, a new channel may have to be constructed. Design of a new channel should satisfy the criteria mentioned above. Some general guidelines for design of open drains can be found in *Drainage Manual* (U.S. Bureau of Reclamation), *Drainage of Agricultural Lands* (USDA Soil Conservation Service), *Drainage for Agriculture — Agronomy Monograph No. 17* (American Society of Agronomy), or *TR-25, Planning and Design of Open Channels* (USDA Soil Conservation Service), as well as many other books on hydraulics and open channel flow.

3.1.2 Existing buried pipe drains. Where an existing buried pipe drain is to serve as an outlet for a proposed system, the existing drain should be carefully examined. It should be in good structural condition, free of roots and/or sediment that would impair flow, deep enough for adequate drainage, and large enough to carry the added drainage water. If the existing subsurface drain does not have sufficient capacity, a separate main outlet should be installed for the additional drains, or the existing main should be enlarged. The resultant outlet must be of sufficient size to handle both systems.

3.1.3 Pump outlets. When a suitable gravity outlet for the subsurface drainage system is not available, pumping should be considered. The pumping outlet should have sufficient capacity to remove the maximum flow expected from the area to be drained. In determining the size of the pumping plant, possible future additions to the drainage system should be considered. Refer to ASAE Engineering Practice ASAE EP369, Design of Agricultural Drainage Pumping Plants, for information on drainage pumping plants.

3.2 Field investigations preliminary to design. A properly designed drainage system depends upon accurate and adequate field data. Details of field investigation procedures can be found in ASAE Special Publication 903CO466, *Guide for Investigation of Subsurface Drainage Problems on Irrigated Lands.* Basic field data are necessary for the proper design of a subsurface drainage system irrespective of the procedure used to determine the system layout (see Section 2—Investigations).

3.2.1 Soil texture. Sufficient subsurface investigations should be conducted to determine the texture of the soil materials from the ground surface to the drainage barrier throughout the area proposed for drainage (see paragraphs 2.2.2 and 2.4.6).

3.2.2 Barrier. The barrier is any layer that is impermeable or whose horizontal hydraulic conductivity is one-fifth of the weighted hydraulic conductivity of the soil layers above. Although this is a somewhat arbitrary definition or standard, it has worked out satisfactorily in practice. The depth to barrier throughout the area proposed for drainage must be determined (see paragraph 2.4.5).

3.2.3 Hydraulic conductivity. The horizontal hydraulic conductivity is required for the various soil textures throughout the area requiring drainage. The weighted hydraulic conductivity for the soil from the barrier to the maximum allowable water table height should be used in the drain spacing formula (see paragraph 2.4.5).

3.2.4 Water table. Permanent or temporary water table elevations and piezometric heads need to be determined before beginning design (see paragraphs 2.4.1 and 2.4.2).

3.3 Drain spacing. Drain spacing can be computed by a variety of methods. Basically the methods can be divided into two categories, the steady-state method such as Donnan's formula and the transient flow method used by the U.S. Bureau of Reclamation. Strong consideration should be given to spacing methods which have proved successful in the area of the proposed project. In addition to design parameter values determined by field investigations, values for the following additional parameters must be selected by the design engineer for use in the drain spacing computations.

3.3.1 Root zone depth. The root zone depth is the minimum design depth from the ground surface to the water table. A value of 1.2 m (4 ft) is normally used as a design value for most crops in arid regions. A value of 1 m (3.3 ft) can sometimes be used in semiarid regions.

3.3.2 Drain depth. Drain depth is dependent on root zone depth, soil properties, availability of a suitable outlet, and economics. Normally, the minimum drain depth is about 2 m (6.5 ft) in arid regions and about 1.5 m (4.9 ft) in supplementally irrigated lands in semiarid regions. The type of drain installation equipment available in the area will have a definite impact on drain depths and drain construction costs.

3.3.3 Deep percolation. Deep percolation from any source causes buildup of the water table. The transient flow method for calculating drain spacing requires that deep percolation and buildup of the water table from each source of recharge (rainfall, snowmelt or irrigation) be estimated and accounted for. The deep percolation rate becomes the recharge rate for steady-state spacing computations (see paragraph 2.4.7).

3.3.3.1 In the planning stage of new surface irrigation projects or in areas where values are not available for deep percolation, Table 1 may be used to arrive at an approximate value. For sprinkler, trickle, or other specialized methods of irrigation, deep percolation values may be lower than those shown in Table 1.

3.3.4 Specific yield. Specific yield is defined as the volume of water released from the soil under the force of gravity for a unit drop of the water table. It is expressed as a percent of the unit volume of soil.

$$\text{Specific yield} = \frac{\text{Volume of water drained}}{\text{Unit volume of drained soil}} \times 100$$

TABLE 1—APPROXIMATE DEEP PERCOLATION FROM SURFACE IRRIGATION (Percent of net input into soil)

Soil texture	Percent
LS	30
SL	26
L	22
SiL	18
SCL	14
CL	10
SiCL	6
SC	6
C	6

Estimated values for specific yield may be obtained from Fig. 1, developed by the U.S. Bureau of Reclamation. The specific yield value is required in the U.S. Bureau of Reclamation's transient flow procedure.

3.3.5 Water table buildup. The water table buildup used in the transient flow equation is determined by dividing the deep percolation from a recharge event by the specific yield.

3.3.6 Recharge rate. The recharge rate (Q_d) used in the steady-state calculations is determined by dividing the unit depth of deep percolation by the shortest time interval between deep percolation events, usually the time between irrigations at the peak of the season.

3.3.7 Drain spacing equations. Numerous methods for calculating drain spacing exist, but the Donnan steady-state equation and the U.S. Bureau of Reclamation transient flow method are in common use.

3.3.7.1 Steady-state method. Donnan's steady-state equation can be expressed as:

$$L^2 = \frac{4K(b^2 - a^2)}{Q_d}$$

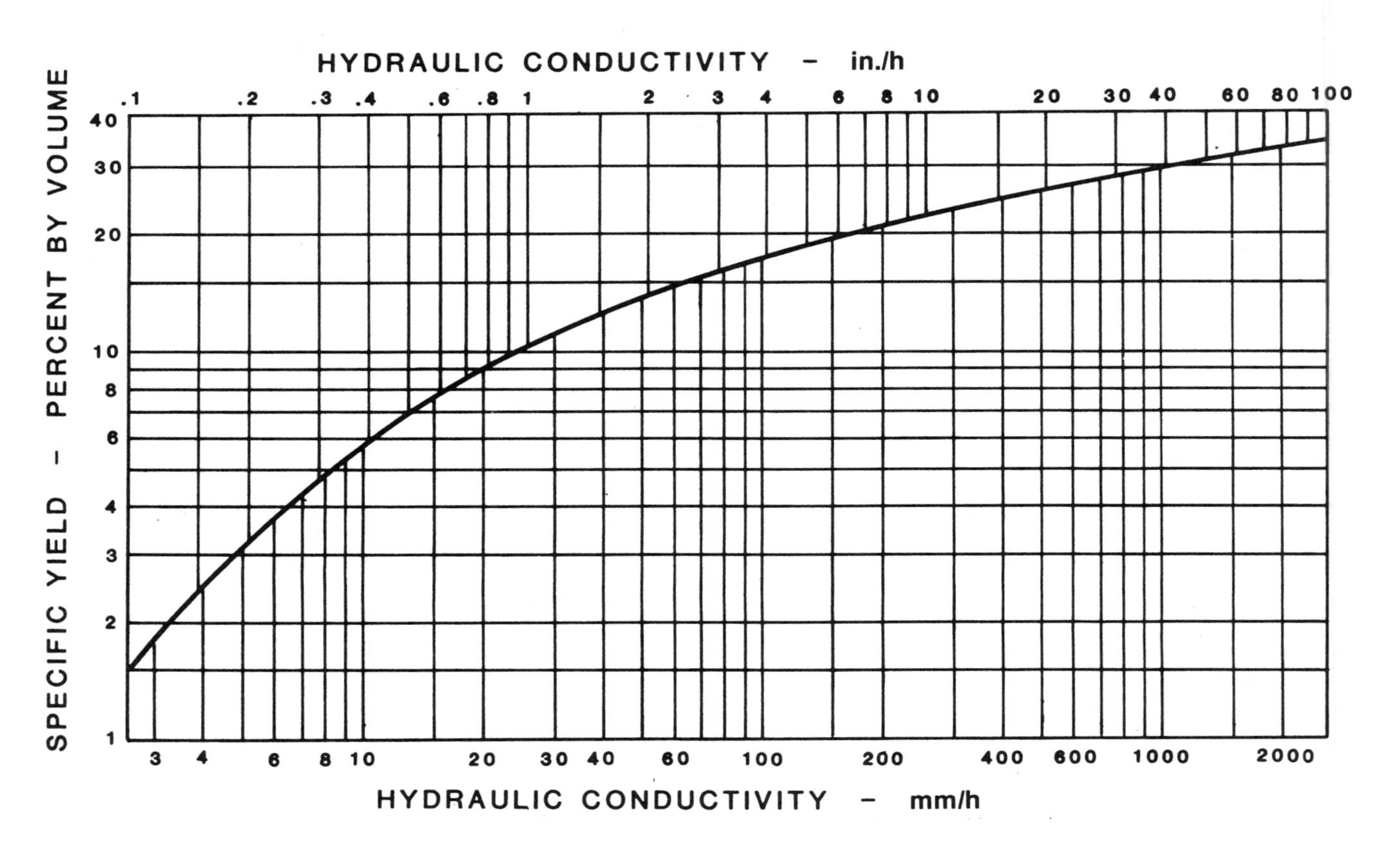

FIG. 1—CURVE SHOWING GENERAL RELATIONSHIP BETWEEN SPECIFIC YIELD AND HYDRAULIC CONDUCTIVITY

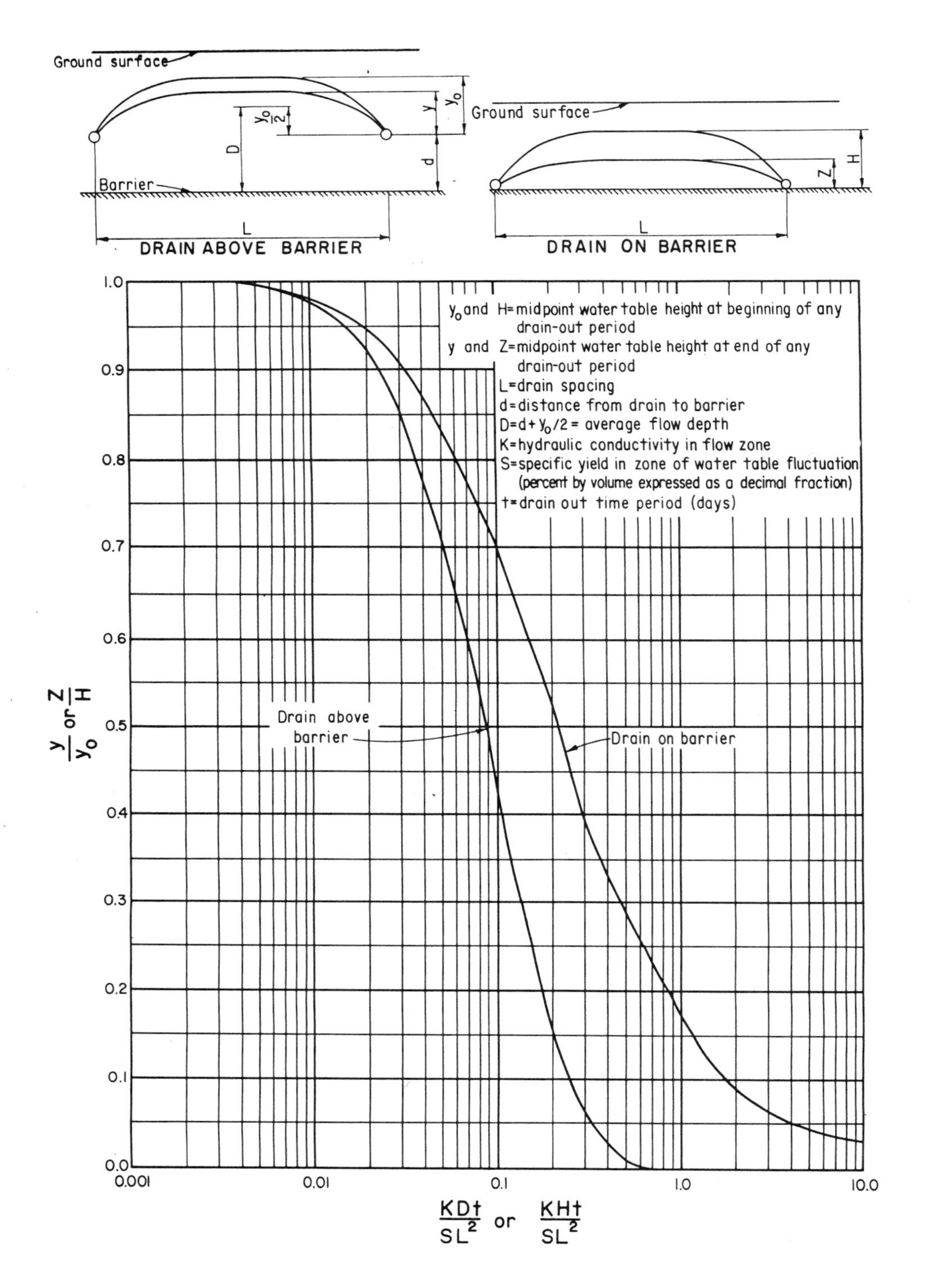

FIG. 2—CURVES SHOWING RELATIONSHIP OF PARAMETERS FOR DRAIN SPACING USING TRANSIENT FLOW THEORY. THESE CURVES CAN BE USED FOR CALCULATIONS IN EITHER SI OR CUSTOMARY UNITS AS LONG AS DIMENSIONAL UNIFORMITY IS MAINTAINED

where

L = drain spacing, m (ft)
K = hydraulic conductivity, m/day (ft/day)
a = distance from drain depth to barrier, m (ft)
b = distance from maximum allowable water table height to the barrier, m (ft)
Q_d = recharge rate, m/day (ft/day) (see paragraph 3.3.6)

When the drain is to be placed on or near the barrier, it has been found through field experience by the U.S. Bureau of Reclamation that drain spacing can be calculated by the following equation:

$$L^2 = \frac{8Kb^2}{Q_d}$$

K = 3.05 m/day
S = 18%
Depth to barrier = 9.14 m
Drain depth = 2.44 m
d = 6.70 m
L = 442 m

	Deep percolation per irrigation	Buildup
Snowmelt	25.4 mm	0.140 m
	25.4 mm	0.140 m

1	2	3	4	5	6	7	8
Irrigation no.	Time period, t, days	Buildup per irrigation, meters	y_o, meters	D, meters	$\frac{KDt}{SL^2}$	$\frac{y}{y_o}$	y, meters
6							
	117		1.22	7.31	0.0742	0.575	0.701
	116		0.701	7.05	0.0709	0.590	0.414
SM		0.140					
	45		0.554	6.98	0.0272	0.870	0.482
1		0.140					
	25		0.622	7.01	0.0152	0.958	0.596
2		0.140					
	20		0.736	7.07	0.0123	0.978	0.720
3		0.140					
	14		0.860	7.13	0.0087	0.985	0.847
4		0.140					
	14		0.987	7.19	0.0087	0.985	0.972
5		0.140					
	14		1.112	7.26	0.0088	0.985	1.095
6		0.140					
			1.235				

K = 10 ft/day
S = 18%
Depth to barrier = 30 ft
Drain depth = 8 ft
d = 22 ft
L = 1450 ft

	Deep percolation per irrigation	Buildup
Snowmelt	1.0 in.	0.46 ft
	1.0 in.	0.46 ft

1	2	3	4	5	6	7	8
Irrigation no.	Time period, t, days	Buildup per irrigation, feet	y_o, feet	D, feet	$\frac{KDt}{SL^2}$	$\frac{y}{y_o}$	y, feet
6							
	117		4.00	24.00	0.0742	0.575	2.30
	116		2.30	23.15	0.0710	0.590	1.35
SM		0.46					
	45		1.82	22.91	0.0272	0.870	1.58
1		0.46					
	25		2.04	23.02	0.0152	0.958	1.95
2		0.46					
	20		2.41	23.20	0.0123	0.978	2.36
3		0.46					
	14		2.81	23.41	0.0087	0.985	2.77
4		0.46					
	14		3.22	23.61	0.0087	0.985	3.17
5		0.46					
	14		3.63	23.82	0.0088	0.985	3.58
6		0.46					
			4.04				

FIG. 3—COMPUTATION OF WATER TABLE FLUCTUATION WITH A DRAIN ABOVE THE BARRIER AND AN ASSUMED DRAIN SPACING OF 442 M OR 1450 FT.

Example: Steady-state with drain above the barrier:

K = 3.05 m/day (10 ft/day)
a = 6.71 m (22 ft)
b = 7.92 m (26 ft)
deep percolation/irrigation = 25.4 mm (1 in.)
time between irrigations at peak of season = 14 days

$$Q_d = \frac{25.4 \text{ mm}}{14 \text{ days } (1000 \text{ mm/m})} = 0.00181 \text{ m/day}$$

$$L^2 = \frac{4(3.05)(7.92^2 - 6.71^2)}{0.00181}$$

$$L = 345 \text{ m}$$

$$Q_d = \frac{1 \text{ in.}}{14 \text{ days } (12 \text{ in./ft})} = 0.00595 \text{ ft/day}$$

$$L^2 = \frac{4(10)(26^2 - 22^2)}{0.00595}$$

$$L = 1{,}136 \text{ ft}$$

3.3.7.2 Transient flow method. The U.S. Bureau of Reclamation's transient flow method for determining drain spacing takes into account the transient nature of groundwater recharge and discharge. This method gives drain spacings which produce dynamic equilibrium of the true fluctuating water table while filling the requirement of keeping the water table below a specified depth. Fig. 2 shows graphically the relationship between the dimensionless parameters Y/Y_o vs. KDt/SL^2 and Z/H vs. KHt/SL^2 based on the transient flow theory. This figure shows the relationship for the water height at the midpoint between drains for cases where drains are located above a barrier or on a barrier. Fig. 2 can be used to find the proper drain spacing for dynamic equilibrium of the water table. However, the computations should be accomplished for the entire year to account for water removal by the drains during non-irrigated periods. Fig. 3 shows an example of computed water table fluctuations with a drain spacing of 442 m (1,450 ft) and with the drain above a barrier, using the transient flow method.

In Fig. 3,
K = hydraulic conductivity, m/day (ft/day)
S = specific yield (from Fig. 1)
For definition of other parameters see Fig. 2

Explanation of each column in Fig. 3:

Column (1): number of each successive increment of recharge such as snowmelt (SM), rain, or irrigation.

Column (2): length of drain-out period, i.e., time between successive increments of recharge or length of incremental drain-out period.

Column (3): instantaneous buildup from each recharge increment (deep percolation divided by specific yield).

Column (4): water table height above drains at midpoint between drains immediately after each buildup or at beginning of incremental time periods during the non-irrigation season (Col. 8 of preceding period plus Col. 3 of current period).

Column (5): average depth of flow, $D = d + Y_o/2$ (d should be limited to L/4).

Column (6): a calculated value representing the flow conditions during any particular drainout period.

Column (7): value taken from the curve in Fig. 2.

Column (8): midpoint water table height above drain at end of each drainout period (Col. 4 x Col. 7).

Fig. 3 shows a final y_o = 1.23 m (4.04 ft), which is approximately equal to the maximum allowable water table height of 1.22 m (4.00 ft). Therefore, the spacing of 442 m (1,450 ft) results in dynamic equilibrium. However, the above calculations do not account for head loss due to convergence. As a result, the computed drain spacing is too large and must be corrected as described in paragraph 3.3.8 below.

3.3.8 Correction for convergence. When groundwater flows toward a drain, the flow converges near the drain. This convergence causes a head loss in the groundwater flow system and must be accounted for in the drain spacing computations. Several methods of correcting for convergence have been developed.

3.3.8.1 One method of correcting for convergence uses a single formula which gives the correction to be subtracted from the calculated spacing. The formula, which can be used when the barrier depth below the drain is less than 10% of the drain spacing, is:

$$\text{Correction} = D \ln \frac{D}{4r}$$

where
D = average depth of flow m (ft), i.e., maximum value from Col. 5 of Fig. 3 for transient flow calculations, or (a+b)/2 for the steady-state formula.
r = outside radius of pipe plus the radial thickness of any gravel envelope, m (ft).

For the barrier and drain depths used in the above calculations and assuming 102 mm (4 in.) pipe with 152 mm (6 in.) gravel envelope to obtain an effective drain radius of about 203 mm (8 in.):

Correction = 7.31 ln (7.31/0.81)
= 7.31 ln 9.02
= 16.1 m which must be subtracted from the computed drain spacing

Correction = 24 ln (24/2.7)
= 24 ln 8.89
= 52.4 ft which must be subtracted from the computed drain spacing

From the examples:

Steady-state:	L = 345 - 16 = 329 m	Steady-state:	L = 1136 - 52 = 1084 ft
Transient flow:	L = 442 - 16 = 426 m	Transient flow:	L = 1450 - 52 = 1398 ft

3.3.8.2 Another method includes the convergence loss of head near the drain directly in the primary spacing calculation. This method accounts for this head loss by using an equivalent depth to barrier to replace the measured value. Two equations allow computation of equivalent depth to a barrier (a_e or d_e) for use in place of a or d in the Donnan or the transient method, respectively. The equivalent depth equation requires use of the initially unknown drain spacing, L. This means that a trial and error method must be used, starting with an estimated drain spacing. The advantage of using the equivalent depth method is that the correct drain spacing can be obtained directly without making the empirical correction at the end. The equivalent depth equations are:

when:

$$0 < \frac{d}{L} < 0.31$$

use:

$$d_e = a_e = \frac{d}{1 + \frac{d}{L}\,(2.55 \ln\frac{d}{r} - c)}$$

$$c = 3.55 - 1.6\frac{d}{L} + 2(\frac{d}{L})^2$$

and when:

$$\frac{d}{L} > 0.31$$

use:

$$d_e = a_e = \frac{L}{2.55\,(\ln\frac{L}{r} - 1.15)}$$

If the example drain spacings are computed using equivalent depth to barrier, the respective drain spacings are:

Steady-state: L = 333 m (1091 ft)
Transient flow: L = 357 m (1172 ft)

3.3.8.3 Since the steady state drain spacing computed in this example is less than the spacing computed by the transient method, the steady-state spacing is slightly conservative. For the same hydrologic conditions, i.e., dynamic equilibrium over the period of a year, the water table between the steady-state designed drains would only rise to a maximum height of 1.1 m (3.6 ft) above the drains compared to a rise of 1.2 m (4.0 ft) for the transient case design. Depending on the local hydrology, calculated steady-state drain spacings may be greater or less than transient flow drain spacings. Computation of drain spacings by the dynamic equilibrium transient flow method is best. Further explanation of this procedure and associated graphs can be found in the U.S. Bureau of Reclamation's *Drainage Manual* and other drainage publications.

3.4 Interceptor drains. Interceptor drains are installed across the slope to intercept groundwater moving downslope from some source. Interceptor drains should be placed as close to the barrier as practical and are usually placed just upslope from where the surface of the barrier converges with the ground surface. This usually means the drain is located at a break in slope to control the water table on the lower slope. An interceptor drain installed on the barrier will effectively intercept most of the water moving downhill. Specific conditions will determine the need for additional drains either upslope or downslope from the initial interceptor.

3.5 Drain discharge

3.5.1 The discharge per unit length of equally spaced drains can be computed using the following formulas:

For drains above a barrier: $$q = \frac{2\pi K\, Y_o\, D}{86{,}400\ L}$$

For drains on a barrier: $$q = \frac{4KH^2}{86{,}400\ L}$$

where

q = discharge per unit length of drain, m^3/s (ft^3/sec)
y_o or H = maximum height of water table above drain invert, m (ft), i.e., the maximum value from Col. 4 of Fig. 3 of transient flow calculation or (b−a) from steady-state calculation
K = weighted average horizontal hydraulic conductivity of the soil profile between maximum water table height and barrier, m/day (ft/day)
D = average flow depth ($D = d + Y_o/2$), i.e., the maximum value from Col. 5 of Fig. 3 of transient flow computation or (a+b)/2 for steady-state calculation
L = drain spacing, m (ft)

3.5.2 The discharge of an interceptor drain located on a barrier can be estimated by using the following basic equation:

$$q = KiA$$

where

q = discharge per unit length of drain, m^3/s (ft^3/sec)
K = weighted average horizontal hydraulic conductivity of the saturated soil profile above the barrier, m/s (ft/sec)
i = slope of the water table measured normal to the groundwater contours upslope from the interceptor
A = area through which flow occurs, m^2 (ft^2), i.e., a vertical plane one unit of length wide from the water surface to the barrier where it is measured

3.5.2.1 This value of q is the total amount of water moving within the saturated profile above the barrier. For practical purposes, however, the drain can be expected to intercept only that portion of the saturated profile above the water surface in the drain.

3.5.3 When the drain is above the barrier, the equation for discharge from a unit length of drain becomes:

$$q = KiA\,\frac{y}{y + d}$$

where

y = height of maximum water surface above the invert of the drain, m (ft)
d = distance from drain invert to barrier, m (ft)

3.5.4 The discharge (Q) from a drain is equal to discharge per unit length (q) times length (ℓ), i.e., $Q = q\ell$.

3.6 Grade and alignment

3.6.1 The proper installation and functioning of pipe drains require rigid control of grade and alignment. The minimum grade for a closed drain should be 0.001; however, steeper grades are more desirable. With steeper grades, the control required during construction can be less exacting and there is also less change of drain clogging.

3.6.2 In determining the grade of the proposed drain, a slope should be chosen that is easy to work with in the field. For example, it is easier for the contractor to establish, and for the inspector to check, grade if a slope of 0.002 is used instead of 0.00213.

3.7 Minimum drain size. The following recommendations serve as a guide in selecting the minimum size of subsurface drains.

3.7.1 The hydraulic capacity of the drain at the design slope should not be exceeded. Common practice in some areas is to design buried pipe drains to flow at less than maximum capacity because the volume of water to be drained can only be estimated.

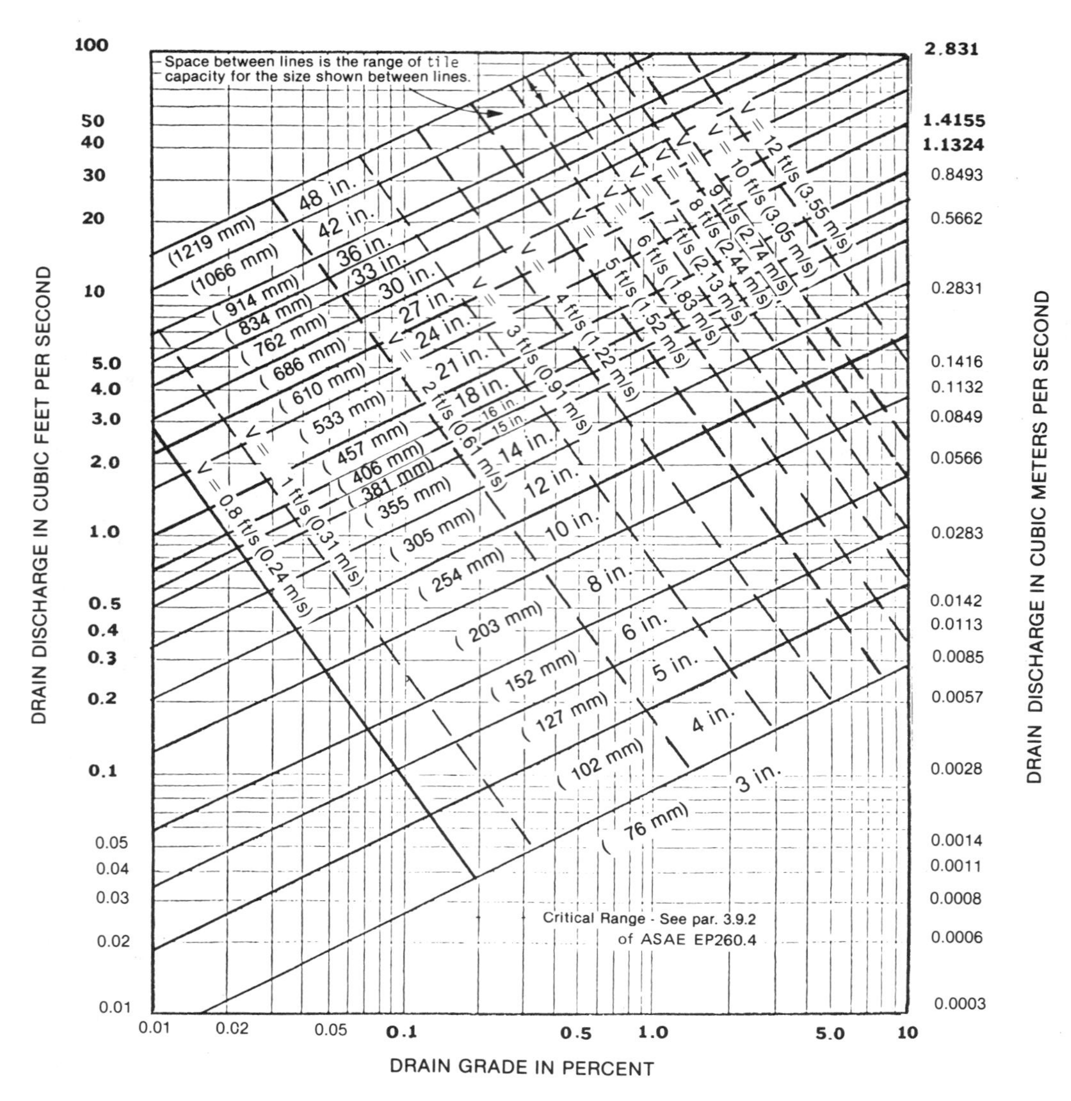

Drain grade = hydraulic gradient
(Percent = ft per 100 ft, or m per 100 m)

V = velocity in f/s (m/s)

FIG. 4—GUIDE FOR DETERMINING THE REQUIRED SIZE OF CLAY AND CONCRETE DRAIN TILE (n=0.013)

3.7.2 A 76 mm (3 in.) diameter drain may be installed in locations where the grade is 0.2% or more for clay or concrete pipe and where the grade is 0.3% for corrugated plastic tubing. For flatter grades, however, 102 mm (4 in.) inside diameter pipe is generally the recommended minimum size.

3.7.3 Figs. 4 and 5 may be used to determine the minimum pipe size required. When designing for the pipe to run less than full, a larger pipe than the pipe size determined from Figs. 4 and 5 may be required.

3.8 Drain layout. For maximum effectiveness, drains should be laid out perpendicular to the direction of groundwater movement. Using calculated drain spacing as a guide, drains are laid out to fit the topography of the area to be drained. Since the drain system should be the most economical one without sacrificing performance, it is usually necessary to adjust the alignment to avoid existing head ditches, buildings or other improvements. Good judgment must be exercised in determining how far the practical layout can deviate from the calculated spacings and still assure a satisfactory level of groundwater control.

3.9 Envelopes and filters

3.9.1 Since buried pipe drains are installed in all kinds of soils, it is usually necessary to lay the pipe in a suitable envelope. Gravel is the most acceptable envelope material for drains in problem soils. Design criteria for gravel envelopes have been developed and published in *Drainage of Agricultural Land, Section 16 of National Engineering Handbook* (USDA Soil Conservation Service) and in *Drainage Manual* (U.S. Bureau of Reclamation). Additional criteria and discussion of envelope and filter materials is contained in ASAE Engineering Practice ASAE EP260.4, Design and Construction of Subsurface Drains in Humid Areas.

3.9.2 Organic envelope or filter materials should be carefully evaluated to insure that they will not encourage ochre formation in the drain. Organic envelopes should not be used with corrugated plastic tubing because these materials usually do not provide adequate lateral support.

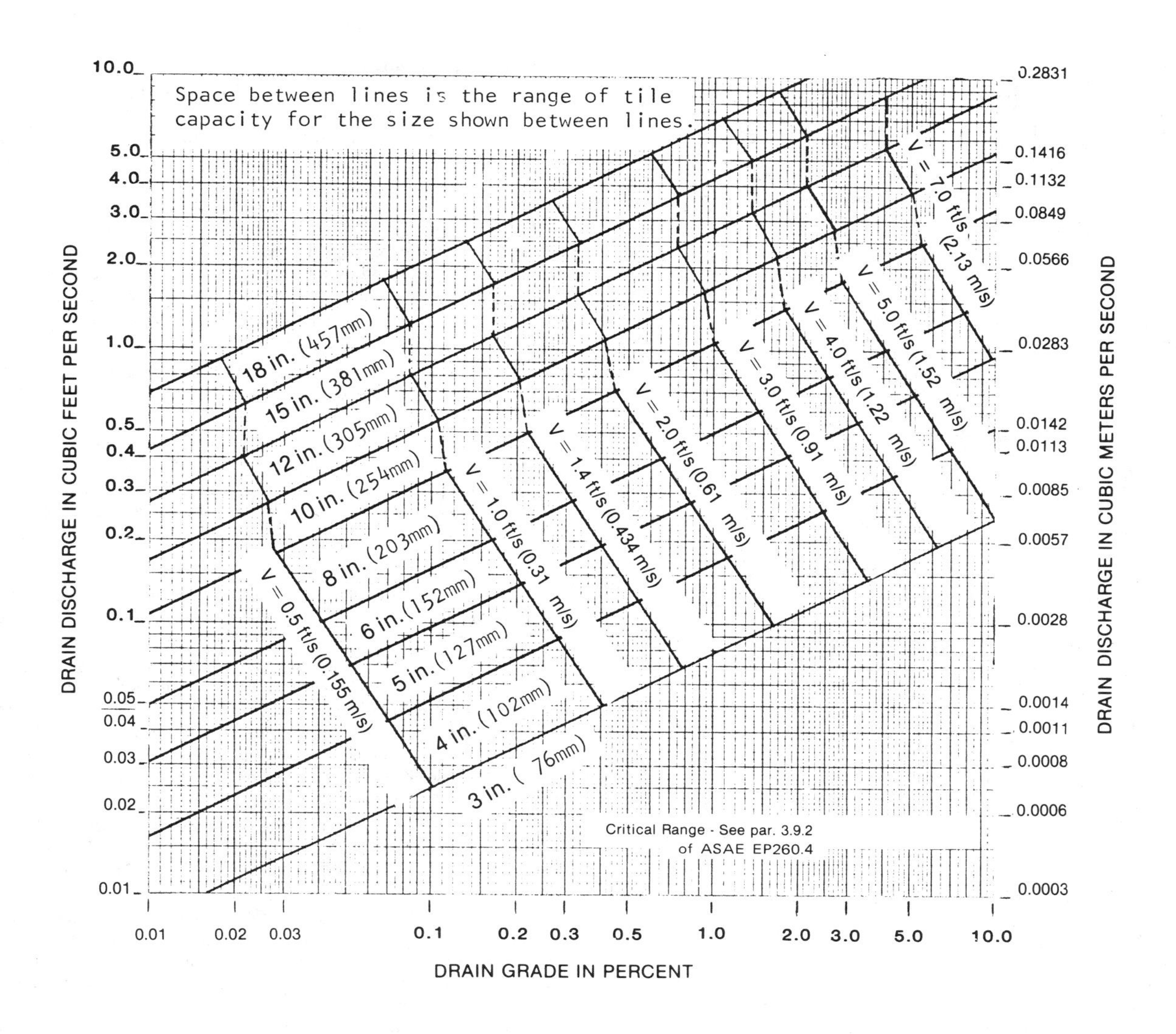

FIG. 5—GUIDE FOR DETERMINING THE REQUIRED SIZE OF CORRUGATED PLASTIC DRAINAGE TUBING

3.10 Connections

3.10.1 For tile lines, manufactured connections or branches for joining two lines should be used when available. If manufactured connections are not available, the junction should be chipped, fitted, and sealed with cement mortar.

3.10.2 For corrugated plastic tubing lines, manufactured couplers should be used at all joints, at all changes in direction where the centerline radius is less than three times the tubing diameter, at changes in diameter, and at the end of the line. All connections must be compatible with the tubing and should not restrict the normal flow cross-section of the tubing.

3.10.3 A 90 deg angle is as satisfactory as a 30 deg or 45 deg angle where a lateral line joins a main line. Laterals should be joined at the most convenient angle.

3.11 Manholes

3.11.1 Manholes may be located in pipe drains to serve as junction boxes, silt and sand traps, observation wells, discharge measurement facilities, or entrances to the drain for maintenance. These structures generally extend above the ground surface and provide for easy location of drain lines. There are no set criteria for the spacing of manholes. In general, a manhole should be used at important junction points on a drain or at major changes in alignment on collector and suboutlet drains. Manholes are not required at every junction of closely spaced (less than 213 m [700 ft]) relief, interceptor or collector drains. Manholes are usually not required at grade changes if the grade becomes steeper. Special effort should be made to locate manholes where they will not interfere with farming operations. The functional limits of drain cleaning equipment in common use, such as the pressurized water jet cleaner, should be considered in determining the manhole spacing.

3.11.1.1 If a manhole cannot be justified for the purposes described above, a simple Y-section, T-section, or holes made in the collector pipe can be used to connect the relief or interceptor drains to the collector drain. Changes in pipe diameter should be made at a manhole, if convenient.

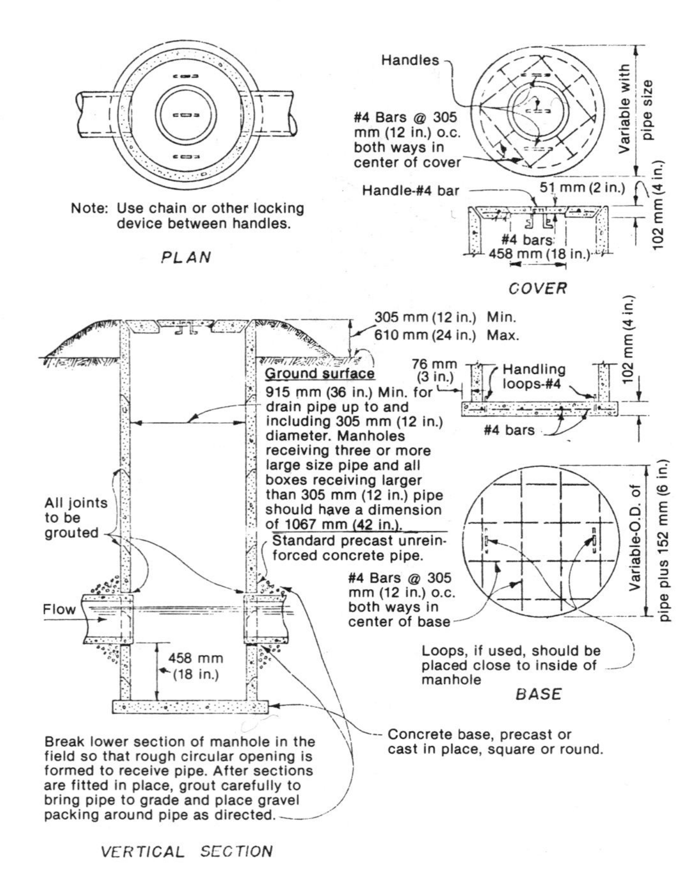

FIG. 6—TYPICAL MANHOLE DESIGN FOR A CLOSED DRAIN

3.11.1.2 Manholes should extend a minimum of 0.3 m (1 ft) and a maximum of 0.6 m (2 ft) above the natural ground surface to make them easy to recognize. They should be placed in fence rows or at other out-of-the-way locations if at all possible. Neither a manhole nor a cleanout is required at the upper end of a line, but this end of the line must be plugged. The location of the plugged end should be recorded both in field books and on as-built drawings. When clean-out risers are used at the end of the line, they should be constructed with durable materials and placed at a sufficient angle to permit the insertion of cleaning equipment.

3.11.2 To compensate for the head losses within a manhole, the general practice has been to provide a drop at the invert elevation between the inlet and outlet pipes. This practice is satisfactory but not absolutely necessary and sometimes creates problems on level land where the grade-lines have to be greater than the gradients of the land surface. For this condition, the top of the inlet and outlet pipes can be placed at the same elevation. If design data show the inlet pipe to be at capacity at the manhole, the outlet pipe size will be increased and the necessary drop will be available in the larger pipe. If a size change is not required at the manhole, neither pipe will be at capacity and the slight head loss required will be available in the unused capacity of the pipes.

3.11.2.1 The base of the manhole should be about 0.46 m (1.5 ft) below the bottom of the effluent pipe to form a trap that will catch any debris that may enter the line. Upon completion of a new drain, all traps should be cleaned out and the manhole covers set. Traps should also be cleaned periodically as a maintenance item.

3.11.2.2 Fig. 6 shows a standard design for a manhole. Manholes may also be constructed of asphalt-dipped or asbestos-bonded corrugated metal pipe (CMP) where salinity of the soil and water is low and where stability is a problem for heavy concrete pipe.

3.12 Outlet structures

3.12.1 The outlet end of a buried pipe drain, if not properly protected, will be undercut by the action of the discharging water. This undercutting will cause the drain to shift out of proper grade and alignment, often resulting in the blockage of the outlet and costly maintenance.

3.12.2 To prevent misalignment, 3.7 to 4.6 m (12 to 15 ft) of heavy gage, asphalt-dipped or asbestos-bonded corrugated-metal pipe or asbestos-cement pipe should be placed at the outlet end of buried pipe drains. The end of the outlet pipe should extend a minimum of 0.3 m (1 ft) from the soil surface into the outlet channel.

3.12.3 Some drain outlets require flap valves to keep high flows in the open drain from entering the pipe drain.

3.12.4 In areas where rodents may be a problem, drain outlets should have a rodent screen or guard installed over the end of the pipe.

3.12.5 Riprap or coarse gravel protection should be installed around the pipe drain outlet and extended at least to the toe of the slope below the outlet. Where the drain is to discharge onto materials such as caliche or basalt, the extra protection may be eliminated.

3.13 Surface water

3.13.1 The use of combination drains which attempt to handle both surface runoff and subsurface water in the same pipe is not recommended.

3.13.2 In some instances it may be necessary to dispose of small amounts of surface water through a subsurface drainage system. Special precautions should be taken to remove weeds, trash and silt from the surface water. The minimum precaution should be to install a self-cleaning trash rack which would prevent entry of debris. A desilting pond should be provided if the water contains significant amounts of sediment.

3.14 Trees. Roots of water-loving trees such as willow, cottonwood, poplar and other phreatophyes may enter and clog the drains. All such vegetation within 15 m (50 ft) on each side of the drain should be removed.

3.15 Buried cables, pipelines, highways and other facilities. Where buried cables, pipelines and other facilities are crossed, the backfill in the trench should be compacted to minimize future damage from settlement. All crossings should meet the requirements of the owner of the facility being crossed.

3.16 Plan. A plan should be prepared for each subsurface installation for use by the contractor during construction and for the owner as a permanent record. If any changes are necessary during construction they must be noted on an as-built copy of the plan. It is recommended that this plan be filed with the abstract or deed of the property. The plan should include a map of the installation showing accurate locations and sizes of all lines and appurtenances together with a profile or a copy of the staking notes for the main and submains. It is important that the plan be adequately referenced to existing physical features so that any segment can be located to facilitate maintenance and repair.

SECTION 4—CONSTRUCTION AND MATERIALS

4.1 General

4.1.1 Purpose. Construction requirements and limitations as well as material specifications should be clearly and adequately documented for the protection of both the contractor and his client. The following specifications provide minimum standards for construction and materials for subsurface drain construction in irrigated areas.

4.1.2 Description of work. A sufficiently detailed description of the work to be performed, including engineering drawings and supplemental specifications, shall be prepared to assure that the contractor will submit a valid bid. The description shall include job location, available transportation facilities, access routes, and general site conditions.

4.1.3 Division of responsibility. Invitations to bid shall clearly define areas of responsibility divided between the contractor and his client.

4.1.4 **Line and grade for pipe drains.** With the low flows that occur at various times in many pipe drains, any departure from established grade may result in sediment collecting in the low areas of the pipe which may eventually clog the drain. The maximum allowable departure from grade should not exceed 10% of the inside diameter of the drain pipe and in no case should the departure exceed 30 mm (0.1 ft). Where departures occur, the rate of return to established grade should not exceed 2% of the pipe diameter per joint for concrete or clay pipe or per 0.9 m (3 ft) of length for plastic tubing. The maximum allowable horizontal departure from alignment should not exceed 20% of the inside diameter of the drain pipe with a rate of return to the established line not to exceed 5% per joint for concrete or clay pipe or per 0.9 m (3 ft) of length for plastic tubing. Construction right-of-way boundaries shall be located and marked in such a way that all construction crews can easily identify the limits for operations specified in paragraphs 4.2 and 4.3.

4.1.5 **Tests and inspections**

4.1.5.1 **Materials.** Unless otherwise specified, all materials used in construction shall be certified as meeting requirements of the specifications. All materials used for pipe drains shall be visually inspected at the jobsite prior to installation. All defective material shall be rejected and removed from the jobsite.

4.1.5.2 **Tests for completed drains.** Completed drain lines can be tested using an inflated ball 25 mm (1 in.) in diameter less than the diameter of the pipe being tested. The ball shall float freely through the full length of the pipe. An alternative method is to draw a solid, torpedo-shaped "pig" 25 mm (1 in.) in diameter less than the diameter of the pipe through the line using a hand line attached to the "pig" and pulled by one man. Any obstruction in the line shall be removed, and all damaged or collapsed pipe and tubing shall be replaced.

4.2 Environment

4.2.1 **General.** Care shall be exercised to preserve the natural environment. Movement of crews and equipment shall be within the specified right-of-way and over routes provided for access to the worksite.

4.2.2 **Preservation of trees.** Special care shall be exercised where trees or shrubs outside the right-of-way are exposed to injuries by construction operations and chemicals used during construction. "Injury" includes bruising, scarring, tearing, and breaking of roots, trunk, or branches. Injured trees and shrubs shall be repaired or replaced as nearly as practical to original conditions.

4.2.3 **Preservation of landscape.** Before borrow areas and gravel pits are abandoned, the sides shall be brought to stable slopes. Slope intersections shall be rounded to provide a natural appearance. Waste banks and piles shall be leveled and trimmed to regular lines and shaped to provide a natural appearance.

4.2.4 **Air and water pollution.** Applicable laws, orders, and regulations related to air and water pollution shall be followed. Activities shall be performed in a manner that will prevent discharge of solid matter, contaminants, wastes, and other pollutants into the air, streams, watercourses, lakes, or groundwater bodies in excess of allowable rates.

4.2.5 **Cleanup and disposal of wastes**

4.2.5.1 **Cleanup.** Construction and storage areas shall be kept free from accumulations of waste materials and rubbish. Before a completed job site is vacated, all temporary facilities and materials shall be removed. All work areas shall be left in a neat condition conforming to the natural appearance, as provided in paragraph 4.2.3.

4.2.5.2 **Disposal of wastes.** Wastes shall be disposed of in accordance with state and local laws. Necessary burning or disposal permits shall be secured from state and local authorities. Materials to be disposed of by dumping shall be hauled to an approved dump. Sanitation requirements as given in *Subpart D — Occupational Health and Environmental Controls, Part 1926 — Safety and Health Regulations for Construction, Occupational Safety and Health Standards and Interpretations* (Occupation Safety and Health Administration, U.S. Department of Labor) shall be followed.

4.2.5.3 **Disposal of unsuitable backfill materials.** Excavated materials not suitable for backfill shall be disposed of in waste banks. Waste banks shall be left with even and uniform surfaces as provided in paragraph 4.2.3. Disposal areas should be designated in bid specifications.

4.2.5.4 **Disposal of excess material suitable for backfill.** Excavated materials suitable for backfill shall be used for backfill or other required earthwork. Excess material from trench excavation, remaining after backfilling according to paragraph 4.3.3, shall be spread uniformly over the construction right-of-way or placed in areas designated in the plans and specifications. Smoothing of the construction right-of-way shall be adequate to re-establish slope and irrigation characteristics equal to those existing prior to construction or to the satisfaction of the client.

4.3 Earthwork

4.3.1 **Clearing.** The portion of the right-of-way required for access and for construction work shall be cleared of all trees, brush, rubbish, and other objectionable matter. All willow, cottonwood, poplar, and other phreatophyte growth within the right-of-way for buried pipe drains shall be removed by dozing, pulling, or other means to remove as much of the root system as practical without grubbing. Cleared materials shall be disposed of in accordance with paragraph 4.2.5.2.

4.3.2 **Classification of excavation.** Where necessary, the following classifications for excavation of earth material shall be used for payment.

4.3.2.1 **Rock excavation.** Boulders or detached pieces of solid rock more than 0.76 m^3 (1 yd^3) in volume shall be classed as rock excavation. Rock is defined as sound solid masses, layers, or ledges of mineral matter of such hardness that one of the two following situations applies:

4.3.2.1.1 It cannot be effectively loosened or broken by ripping in a single pass with a tractor-mounted hydraulic ripper with one digging point adequately sized for use with a crawler-type tractor rated from 157 to 179 flywheel kW (210 to 240 flywheel hp) operating in low gear.

4.3.2.1.2 It cannot be loosened or broken down by a 2.7 kg (6 lb) drifting pick.

4.3.2.2 **Common excavation.** Common excavation includes all material other than rock excavation as defined in paragraph 4.3.2.1.

4.3.3 **Excavation and backfill of drain trenches**

4.3.3.1 **General.** Trench excavation and backfill shall include trenches for both open and closed joint pipe, and trenches for perforated and unperforated plastic tubing. Trenches shall not remain open longer than necessary. In areas under irrigation, water service interrupted by trenching operations should be restored within 7 days or less depending upon crop requirements.

4.3.3.2 **Safety.** Drain trenches are often greater than 1.8 m (6 ft) deep. All trench excavation safety regulations in *Part 1926 — Safety and Health Regulations for Construction, Occupational Safety and Health Standards and Interpretations* (Occupational Safety and Health Administration, U.S. Department of Labor), as well as any state or local regulations that may apply, shall be followed (also see ASAE Engineering Practice ASAE EP260.4, Design and Construction of Subsurface Drains in Humid Areas).

4.3.3.3 **Excavation**

4.3.3.3.1 Where separation of material is required to assure replacement of suitable topsoil, excavated material from the lower part of the trench shall be prevented from mixing with either excavated or in-place topsoil on either side of the trench.

4.3.3.3.2 Where shielded excavation is required, a gravel envelope, if used, shall be placed in a manner that assures the envelope will be in contact with uncompacted and otherwise undisturbed earth materials at the sides and in the bottom of the trench. Shielded excavation is specified to assure proper functioning of the completed drains and is not to be construed as a safety measure. The excavation shall be performed with a trench excavator with an attached shield, or by other methods which produce equivalent results.

Puddling of soil immediately around the drain during installation shall not be permitted (also see ASAE Engineering Practice ASAE EP260.4, Design and Construction of Subsurface Drains in Humid Areas).

4.3.3.3.3 Where materials in the bottom of the trench are too unstable to assure uniform grade of the completed drain, the trench shall be over-excavated as necessary and backfilled to bottom grade of the gravel envelope with suitable material, such as a coarse gravel, with the following gradations:

Retained on 127 mm (5 in.) screen	— None
Retained on 102 mm (4 in.) screen	— 0% to 20%
Retained on 76 mm (3 in.) screen	— 0% to 30%
Retained on 51 mm (2 in.) screen	— 20% to 50%
Retained on 19 mm (3/4 in.) screen	— 20% to 50%
Passing No. 4 screen	— Less than 5%

(Also see ASAE Engineering Practice ASAE EP260.4, Design and Construction of Subsurface Drains in Humid Areas.) As much care as practical shall be taken to prevent puddling the bottom of the trench.

4.3.3.4 Backfill. Backfill material shall be obtained from the trench excavation. Where insufficient material is available, borrow sources may be used provided the material is comparable to that excavated from the trench. Backfill placed within 0.3 m (1 ft) of the drain that is installed without a gravel envelope shall not contain rocks greater than 76 mm (3 in.) in diameter or frozen material greater than 203 mm (8 in.) in any dimension. Loads of backfill shall not be dropped directly onto a completed drain. Backfill operations shall be subject to the following provisions:

4.3.3.4.1 Where separation of excavation is required, the top portion of the trench shall be backfilled with suitable topsoil material to the depth of the existing topsoil material or 0.6 m (2 ft), whichever is less. Excavation material from the lower portion of the trench which contains gravel, rocks or other materials detrimental to farming operations shall be excluded from the top 0.6 m (2 ft) of the trench or less as directed.

4.3.3.4.2 Backfilling of trenches shall proceed as soon as practical to 0.3 m (1 ft) below the top of the trench. A 1.2 m (4 ft) minimum depth of backfill shall be placed on the drain as close behind the drain laying operation as possible. The trench should be backfilled to allow for settlement to normal ground surface (also see ASAE Engineering Practice ASAE EP260.4, Design and Construction of Subsurface Drains in Humid Areas).

4.3.3.4.3 Pipe with closed joints and non-perforated plastic tubing may be surrounded with compacted earth materials. Backfill placed with 0.1 m (0.33 ft) of the pipe or tubing shall be free of stone or rock greater than 44 mm (1.75 in.) in any dimension.

4.3.3.4.4 All suitable material excavated from drain trenches shall be used as backfill where possible. Any excess material suitable for backfill shall be spread uniformly over the construction right-of-way or placed in other designated areas in accordance with paragraph 4.2.5.4. All excess material unsuited for use as backfill shall be disposed of in accordance with paragraph 4.2.5.3.

4.3.3.5 Gravel envelope placement. Graded envelope material conforming to paragraph 4.4.3 should be placed around open-joint pipe and corrugated plastic tubing. The gravel envelope shall be handled and placed in a manner that prevents separation or segregation of materials. Introduction of fine soils by stockpiling envelope material on bare ground and scooping under the stockpile for placement shall not be permitted (also see ASAE Engineering Practice ASAE EP260.4, Design and Construction of Subsurface Drains in Humid Areas).

4.3.3.5.1 The envelope material shall be placed to form an even, firm bedding under the pipe. The remainder of the material shall be carefully placed around the pipe, avoiding movement of the pipe.

4.3.3.5.2 Envelope material must be placed in such a manner that it contacts undisturbed earth material on the bottom and sides of the trench. Where the trench has been over-excavated to provide stability, the envelope must contact the coarse stabilizing material. Puddling the envelope or placing the envelope on puddled soils shall not be permitted.

4.3.3.6 Compacting backfill at crossings. Backfill placed above the gravel envelope in trenches crossing main roads, waterways, and other public facilities shall be compacted for the full width of the trench. Backfill shall be compacted around closed-joint pipe and unperforated tubing where gravel envelope is not used. Backfill material placed within 0.3 m (1 ft) of the pipe or tubing shall not contain rocks larger than 76 mm (3 in.) in any dimension. All materials shall be compacted to 95% of laboratory standard for soil density (dry) in accordance with American Society for Testing and Materials Standard D 698, Test Methods for Moisture-Density Relations of Soils and Soil-Aggreggate Mixtures, Using 5.5-lb (2.5-kg) Rammer and 12-in. (304.8-mm) Drop.

4.3.3.7 Saturating backfill in drain trenches. This construction procedure is optional. After backfill material has been placed to within 0.3 m (1 ft) of the top of the trench, water shall be ponded over the full width of the trench. Dikes shall be constructed in the trench at intervals of 61 m (200 ft) or less and the backfill shall be saturated. The entire backfilled surface shall be ponded, and all of the ponded water shall be allowed to seep into the soil. Surface draining of the water from ponded areas shall not be permitted. The trench shall be patrolled during ponding and any sink holes or areas of excessive settlement shall be filled immediately. After saturation, backfill shall be completed in accordance with paragraphs 4.3.3.4.2 and 4.3.3.4.4.

4.4 Materials

4.4.1 General. Materials shall be certified to meet specifications. Materials specified by reference to Federal Specifications, Federal Standards, American National Standards Institute, American Society for Testing and Materials, American Society of Civil Engineers, or other standard specifications or codes, shall comply with the latest editions or revisions thereof in effect on the date that response is made to the bid invitation. Where types, grades, or other options offered in the reference specifications are not specified, the materials furnished will be acceptable if they conform to any one of the reference specifications.

4.4.2 Pipe and tubing for drains. In addition to the following paragraphs, see ASAE Engineering Practice ASAE EP260.4, Design and Construction of Subsurface Drains in Humid Areas.

4.4.2.1 Concrete drainage pipe. Unreinforced concrete drainage pipe shall be Class C manufactured and tested according to U.S. Bureau of Reclamation Standard Specifications for Unreinforced Concrete Drainage Pipe, February 15, 1971; ASTM Standard C 14, Specification for Concrete Sewer, Storm Drain, and Culvert Pipe; ASTM Standard C 412, Specification for Concrete Drain Tile; ASTM Standard C 118, Specification for Concrete Pipe for Irrigation or Drainage; or ASTM Standard C 444, Specification for Perforated Concrete Pipe. In addition, the following requirements must be met:

4.4.2.1.1 Low-alkali cement is required for drain pipe unless aggregates to be used are known to be essentially inert.

4.4.2.1.2 All pipe shall be steam cured for 48 h between 38 and 60 °C (100 and 140 °F) or moist cured for 7 days. Ambient temperature rise within the steam curing enclosure shall not exceed 17 °C/h (30 °F/h). Temperatures shall not be allowed below 4 °C (40 °F) before or during the curing process.

4.4.2.1.3 Absorption shall not exceed 6.5% in accordance with ASTM Standard C 14, Specification for Concrete Sewer, Storm Drain, and Culvert Pipe.

4.4.2.1.4 Pipe shall be air dried 30 days before placement.

4.4.2.1.5 Calcium chloride shall not be used in cement for concrete pipe.

TABLE 2—GRADATION RELATIONSHIP BETWEEN BASE MATERIAL AND DIAMETERS OF GRADED ENVELOPE MATERIAL

Base material 60% passing (diameter of particles, mm)	Gradation limitations for envelope (diameter of particles, mm)											
	Lower limits, percent passing						Upper limits, percent passing					
	100	60	30	10	5	0	100	60	30	10	5	0
0.02-0.05	9.52	2.0	0.81	0.33	0.3	0.074	38.1	10.0	8.7	2.4	—	0.59
0.05-0.10	9.52	3.0	1.07	0.38	0.3	0.074	38.1	12.0	10.4	3.0	—	0.59
0.10-0.25	9.52	4.0	1.30	0.40	0.3	0.074	38.1	15.0	13.1	3.8	—	0.59
0.25-1.00	9.52	5.0	1.45	0.42	0.3	0.074	38.1	20.0	17.3	5.0	—	0.59

4.4.2.1.6 When sulfate concentrations in the soil exceeds 0.2% or 1,000 mg/L in ground water, Type V cement shall be used in manufacture of concrete pipe.

4.4.2.2 Clay pipe. Clay drainage pipe shall be manufactured and tested according to the following specifications:

4.4.2.2.1 Both standard strength clay pipe and extra strength clay pipe shall conform to Federal Specification SS-P-361E, Pipe, Clay, Sewer (Vitrified Fittings and Perforated Pipe), or ASTM Standard C 700, Specification for Vitrified Clay Pipe, Extra Strength, Standard Strength, and Perforated.

4.4.2.2.2 Pipe sections for 102 to 457 mm (4 to 18 in.) diameter pipe shall not be longer than 914 mm (36 in.).

4.4.2.3 Corrugated plastic drain tubing. Plastic drain tubing shall be certified by the manufacturer as meeting one or more of the following specifications listed below. Also, tag ends and poorly cut slots or holes shall not be permitted.

4.4.2.3.1 Heavy duty tubing shall conform to ASTM Standard F 405, Specification for Corrugated Polyethylene (PE) Tubing and Fittings; ASTM Standard F 667, Specification for 10, 12, and 15-inch Corrugated Polyethylene Tubing; or ASTM Standard F 800, Specification for Corrugated Poly(vinyl chloride) Tubing and Compatible Fittings.

4.4.2.3.2 USDA Soil Conservation Service Specifications 606, Subsurface Drain, or 606R, Manufacturing Corrugated Plastic Tubing Using Reprocessed Polyethylene Resin.

4.4.2.3.3 U.S. Bureau of Reclamation Standard Specifications for Polyethylene or Polyvinyl-Chloride Corrugated Plastic Drainage Tubing.

4.4.2.4 Closed-joint drainage pipe and unperforated plastic tubing. Where closed-joint pipe is specified, the option may be given of using concrete and clay pipe or unperforated corrugated plastic tubing. Plastic tubing shall meet the requirements of paragraph 4.4.2.3 except that the perforations shall be omitted. Type F rubber gasket joints conforming to ASTM Standard C 443, Specification for Joints for Circular Concrete Sewer and Culvert Pipe, Using Rubber Gaskets, shall be used for concrete and clay pipe. Concrete and clay pipe shall conform to requirements in paragraphs 4.4.2.1 and 4.4.2.2.

4.4.3 Gravel envelope. Envelope materials are often used in the construction of subsurface drain lines. The following represents one criterion for a gravel envelope. The envelope shall have a minimum thickness of 102 mm (4 in.) around the pipe. All materials passing the No. 200 sieve, and material larger than 38 mm (1.5 in.), shall be removed from the envelope material (see paragraphs 2.4.6 and 3.9.1).

4.4.3.1 Table 2 shows acceptable gradations for typical irrigated soils. Only sound, nonreactive aggregate shall be used as envelope materials. Limestone and lime shall be removed. All organic and other deleterious substances that may change over time shall be removed. Special care shall be taken to avoid introducing fine soil particles into the envelope when transferring envelope material from any storage site or stockpile and when placing material around the drain pipe or tubing.

4.4.3.2 A presentation of another criterion for a gravel envelope and other envelope materials is given in ASAE Engineering Practice ASAE EP260.4, Design and Construction of Subsurface Drains in Humid Areas.

4.5 Drain Line Installation

4.5.1 ASAE Engineering Practice ASAE EP260.4, Design and Construction of Subsurface Drains in Humid Areas, presents drain line installation guidelines.

4.6 Structures

4.6.1 Manholes and junction boxes. Where specified, manholes shall be placed on a level, monolithically cast concrete base. Openings for installing drainpipe or tubing shall not exceed a radius 25 mm (1 in.) greater than the outside diameter of the drain pipe or tubing.

4.6.2 Pipe outlets to open drains. Corrugated-metal or asbestos-cement pipe can be used for outlets to open drains. Galvanized pipe and couplings shall conform to Federal Specification WW-P-405B, Pipe, Corrugated (Iron or Steel, Zinc Coated). Metal pipe shall be galvanized or bituminous coated and shall be handled in a manner that avoids damage to the coating. Any damage to the bituminous or galvanized coating shall be repaired by application of galvanizing compound and bituminous coating.

4.6.3 Existing works. Where existing works must be removed or destroyed in the process of construction, the replaced works shall be of equal or better quality and condition. Interruption of utilities, water deliveries, and traffic on roads that must be crossed by a drain line shall be minimized. Appropriate permits, licenses, and access to right-of-way shall be obtained for crossing, removal, and replacement of existing works. There should be a firm understanding between the contractor and the landowner concerning the responsibility of each.

4.7 Drain pipe storage. ASAE Engineering Practice ASAE EP260.4, Design and Construction of Subsurface Drains in Humid Areas, paragraph 4.15, presents storage guidelines.

SECTION 5—MAINTENANCE

5.1 Minimizing maintenance needs

5.1.1 Design. The maintenance needs for a subsurface drainage system can be minimized with good design. Every system needs a free flowing outlet. This can be either a gravity flow outlet or a pumped outlet. Drain lines must be designed for adequate capacity based on expected drainage flow. Drains must be designed for specified minimum flow velocities. The envelope material should be specified on the basis of soil characteristics. Design features such as inspection pipes and inlets for flushing can make maintenance easier.

5.1.2 Materials. All commonly used drain tube materials can be used to make satisfactory drains when properly fabricated, stored, transported and installed. Some plastic materials will deteriorate when exposed to sunlight for extended periods of time.

5.1.3 Installation. The drain system must be installed according to design specifications to minimize maintenance requirements. Examples of poor installation practices that can result in excessive maintenance needs are deformed or broken tubing, poor joints between drain lines that allow sediment to enter drains, poor grade control, and poor placement of envelope material.

5.2 Indicators of maintenance need. Indicators of maintenance needs should be easily observed and simple to measure, and should indicate the need for maintenance early enough that corrective action can be made before serious crop damage or damage to the drain system occurs. Reduced crop yields are evidence that a salinity or waterlogging problem has developed and that drain maintenance is overdue.

5.2.1 **Reduced outflow rate.** One of the first indications of the need for maintenance of a subsurface drainage system is a reduced outflow rate. Changes in outflow rate can be easily monitored by using a ruler or depth gage to determine the depth of flow at the outlet.

5.2.2 **Wet spots in field.** A clogged or partially obstructed drain will result in poor drainage which often will cause persistent wet spots in the field.

5.2.3 **Soil salinity.** A major purpose of drainage in most irrigated areas is soil salinity control. An increase in soil salinity indicates the need for maintenance of the subsurface drainage system or that the drain sytem is inadequate.

5.2.4 **Water table levels.** The water table should not invade the root zone. Water table levels can be measured in temporary auger holes dug midway between the drains in the field or in observation wells that are left in the field for extended periods of time.

5.2.5 **Aerial photography.** Black-and-white, colored and infrared photographs can be used to indicate areas of poor drainage and crop stress in a field. Areas of poor drainage usually can be visually identified from a low-flying aircraft.

5.3 **Evaluation of indicators.** Indicators of poor drainage can be caused by factors other than the need for maintenance. Reduced outflow rate can be caused by inadequate irrigation to provide a water table above the drain lines. Wet spots can result because either a natural or developed hard pan limits vertical movement of water in the soil or because the drain capacity is inadequate. Crop damage and reduced yields can be caused by poor irrigation scheduling, inadequate fertility, machine damage, insect damage and other causes.

5.4 **Field investigation**

5.4.1 **Map and design data.** A map and design data for the system should be available when making a detailed field investigation of a drainage system. A subsurface drainage map should include location of drain lines and adequate field boundary and road locations for identification of the area. The design data should include the name of the installer and such data as description of the drain tube used, depth and spacing of drains, envelope material and size of drains (also see paragraph 3.16).

5.4.2 **Outlet inspection.** A free flowing outlet is necessary for the proper functioning of a subsurface drainage system. Any obstructions of the outlet should be removed before further investigation of the drainage problem is made.

5.4.3 **Excavation.** Malfunctions in the drain lines are located by systematically excavating portions of the lines across the field or in the vicinity of poor drainage areas. Excavations are continued until the cause of the problem is located. Backfill must be placed in a manner that will not cause mechanical damage to the drain or permit soil to enter the drain.

5.5 **Maintenance methods.** The outlet should be free flowing before performing additional drain maintenance.

5.5.1 **High-pressure jetting.** A satisfactory high pressure jetting operation should remove all silt, roots, iron and manganese deposits, and most carbonates and silicates from the drains without damaging the drain. High-pressure jetting can be used with plastic drains as well as concrete and clay drains if the nozzle is continually moved during operation.

5.5.2 **Chemical cleaning.** A mixture of SO_2 and water can be used to dissolve and flush iron and manganese deposits that clog subsurface drains. Chemical removal of roots is not recommended.

5.5.3 **Drain replacement.** In cases when drains are filled with calcium and silicate deposits, high-pressure jetting may not remove the material, so replacement of the drains may be necessary. If the drain tubing had insufficient strength and has collapsed, it must be replaced.

5.5.4 **Tree removal.** Water-loving trees such as willow, cottonwood, poplar and other phreatophytes whose roots may enter and clog the drains should not be allowed within 15 m (50 ft) of any buried pipe drain. Any such growth should be eliminated by spraying or some other method.

5.6 **Pollution control.** Subsurface drain flow usually contains few contaminants. Maintenance operations on subsurface drains can cause considerable amounts of contaminants to flow from the drains in a relatively short period of time. These contaminants will consist primarily of silt, clay, organic material, iron and manganese oxides, and a few silicates and carbonates. Permits may be necessary to dump these contaminants into drainage channels. In some locations special arrangements may be necessary to provide for disposal of the contaminants.

References: Last printed in 1985 STANDARDS; list available from ASAE Headquarters.

Cited Standards:

ASAE EP260, Design and Construction of Subsurface Drains in Humid Areas
ASAE EP369, Design of Agricultural Drainage Pumping Plants
ASAE EP407, Agricultural Drainage Outlets — Open Channels
ASTM C 14, Specification for Concrete Sewer, Storm Drain, and Culvert Pipe
ASTM C 118, Specification for Concrete Pipe for Irrigation or Drainage
ASTM C 412, Specification for Concrete Drain Tile
ASTM C 443, Specification for Joints for Circular Concrete Sewer and Culvert Pipe, Using Rubber Gaskets
ASTM C 444, Specification or Perforated Concrete Pipe
ASTM C 700, Specification for Vitrified Clay Pipe, Extra Strength, Standard Strength and Perforated
ASTM D 698, Tests for Moisture-Density Relations of Soils and Soil-Aggregate Mixtures, Using 5.5-lb (2.5-kg) Rammer and 12-inch (304.8-mm) Drop
ASTM F 405, Specification for Corrugated Polyethylene (PE) Tubing and Fittings
ASTM F 667, Specification for 10, 12, and 15-inch Corrugated Polyethylene Tubing
ASTM F 800, Specification for Corrugated Poly (vinyl chloride) Tubing and Compatible Fittings
Federal Specification SS-P-361E, Pipe, Clay, Sewer (Vitrified Fittings and Perforated Pipe) (July 6, 1973)
Federal Specification WW-P-405B, Pipe, Corrugated (Iron or Steel, Zinc Coated) (June 14, 1974)
OSHA CFR Part 1926, Subpart D — Occupational Health and Environmental Controls
USBR Standard Specifications for Unreinforced Concrete Drainage Pipe (February 15, 1971)
USBR Standard Specifications for Polyethylene or Polyvinyl-chloride Plastic Corrugated Drainage Tubing
USDA-SCS Specification 606, Subsurface Drain
USDA-SCS Specification 606R, Manufacturing Corrugated Plastic Tubing Using Reprocessed Polyethylene Resin

The ASAE COOPERATIVE STANDARDS PROGRAM

STANDARDS—ESSENTIAL TO AGRIBUSINESS AROUND THE WORLD

ASAE Standards, Engineering Practices, and Data, developed and maintained through the Cooperative Standards Program, provide:

- **INTERCHANGEABILITY** among interfunctional products manufactured by two or more organizations.
- **REDUCTION IN VARIETY OF COMPONENTS** required to serve an industry, thus improving availability and economy.
- **IMPROVED DEGREE OF HUMAN SAFETY** in operation and application of products, materials and equipment.
- **PERFORMANCE CRITERIA** for products, materials, or systems.
- **A COMMON BASIS FOR TESTING, DESCRIBING, OR INFORMING** regarding performance and characteristics of products, methods, materials or systems.
- **INCREASED EFFICIENCY** of engineering effort in design and production.
- **DESIGN DATA** in readily available form.
- **A SOUND BASIS FOR CODES, LEGISLATION, EDUCATION** related to the agricultural industry. Promote uniformity of practice.
- **INPUT FOR ISO DRAFT PROPOSALS** (International Organization for Standardization).

STANDARDIZATION PROCEDURES

TABLE OF CONTENTS

INTRODUCTION

ASAE has long been active in the development of standards as evidenced by the appointment of a committee on Standardization of Farm Machinery, Farm Tractors, and Their Tests during the 3rd Annual Meeting of ASAE at Ames, Iowa, December, 1909 and publication of ASAE Recommendation, Conventional Signs for Agricultural Engineers, in 1912.

This document includes development and approval procedures of the Cooperative Standards Program of the American Society of Agricultural Engineers. It outlines relationships of ASAE committees, ASAE members and other related groups and interested parties in the standardization procedures. It presents official ASAE policies including use of the metric system, and relationships with other organizations; it outlines procedures for developing ASAE Standards, Engineering Practices, and Data; and it presents the format in which they should be prepared for efficient, effective action and/or disposition.

Article B2 of the ASAE Bylaws and Rules states in part that "The objectives of the Society (as stated in the ASAE Constitution) shall be accomplished by advancing the theory and practice of engineering in agriculture and the allied sciences and arts by developing and promulgating standards, codes, formulas and recommended practices."

Revisions and additions. Recommendations for revisions and additions to "ASAE Standardization Procedures" shall be referred to the ASAE Committee on Standards. A copy of such recommendations should be submitted to ASAE Headquarters.

Revised April 1985

PART I
PURPOSE, PRINCIPLES, AND POLICIES

SECTION 1—PURPOSE OF STANDARDIZATION

1.1 Standards, Engineering Practices, and Data (hereafter referred to collectively as standards) are normally generated for one or more of the following reasons:

1.1.1 To provide interchangeability between similarly functional products manufactured by two or more organizations, thus improving mechanical compatibility, safety and performance for users.

1.1.2 To reduce the variety of components required to serve an industry, thus improving availability and economy for manufacturer and user.

1.1.3 To improve the degree of personal safety during operation and application of products and materials.

1.1.4 To establish performance criteria for products, materials, or systems.

1.1.5 To provide a common basis for testing, describing, or informing regarding the performance and characteristics of products, methods, materials or systems.

1.1.6 To provide design data in readily available form.

1.1.7 To develop a sound basis for codes, education and legislation related to the agricultural industry; and promote uniformity of practice among states.

1.1.8 To provide a voice in international standardization.

1.1.9 To increase efficiency of engineering effort.

1.2 Standards are developed and adopted because of a need for action on a common problem. Their effectiveness is dependent upon voluntary compliance with the standards adopted. It is, therefore, essential that affected groups be invited to participate in the development of and conformance with such standards.

SECTION 2—PRINCIPLES OF STANDARDIZATION

2.1 Standards are basically engineering specifications prepared to define materials, products, processes, tests, testing procedures and performance criteria in an effort to achieve certain specified purposes. Therefore, standards must accurately and specifically define the properties required without unnecessary, restrictive specifications which thwart originality or progress.

2.2 The following basic principles shall be considered in determining the acceptability of proposed standards:

2.2.1 The standard shall have a clearly defined *purpose* and *scope.*

2.2.2 Only those specifications, essential to achieving the stated purpose within the stated scope, should be included. Unessential, descriptive material should be excluded.

2.2.3 Standards shall not assign responsibility for conformance to or application of its provisions and specifications.

2.2.4 The words "shall" and "must" are to designate requirements for the conformance to that specific standard. They define the specifications that are necessary to achieve the stated "purpose." The words "should" and "may" are to designate options or advisory specifications which are desirable but which are not essential for achieving the stated "purpose."

2.2.5 Standards concerned with safety and/or health of personnel shall not state or imply that conformance with the standard automatically makes the product or material completely safe or completely healthful. They shall not state or imply that products or materials which do not comply with the standard are basically unsafe or unhealthy.

2.2.6 The following statement should be considered for use in the scope statement of each standard where appropriate:

> "In addition to the design and configuration of equipment, hazard control and accident prevention are dependent upon the awareness, concern, and prudence of personnel involved in the operation, transport, maintenance, and storage of equipment or in the use and maintenance of facilities."

2.3 Many standards are of international significance and importance. ASAE standards should be harmonized with other national and international standards where appropriate.

SECTION 3—GENERAL POLICY ON STANDARDS DEVELOPMENT

3.1 ASAE assumes no responsibility for results attributable to the application of ASAE Standards, Engineering Practices, and Data. They are informational and advisory only. Their use by anyone engaged in industry and trade is entirely voluntary. Conformity does not ensure compliance with applicable ordinances, laws and regu lations. Prospective users are responsible for protecting themselves against liability for infringement of patents.

3.2 There shall be three basic categories of standardization documents recognized by the ASAE, these being defined as follows:

3.2.1 ASAE Standard: A definite terminology, specification, performance criteria, or procedure providing interchangeability; enhancing quality, safety, economy or compatibility; viewed as a proper and adequate model or example. Standards may include: (1) definitions, terminology, graphic symbols and abbreviations; (2) performance criteria for materials, products or systems; (3) testing procedures; (4) specifications or ratings regarding size, weight, volume, etc.

3.2.1.1 ASAE Tentative Standard: Classification of "tentative" may be assigned to any new standard, if sufficient justification for approval exists pending clarification or introduction of minor changes. The tentative classification shall be used when provisions of the standard are based on new technology which may be beyond or lead the current state-of-the-art, or when "hardware" required for conformance with the standard is not readily available.

3.2.2 ASAE Engineering Practice: A practice, procedure or guide accepted as appropriate, proper and desirable for general use in design, installation or utilization of systems or system components and based upon current knowledge and the state-of-the-art.

3.2.3 ASAE Data: Numerical values, including statistics, and relationships either mathematical or graphical, organized, codified and uniquely applicable to engineering in agriculture. Data need not be free of variation.

3.3 Where conformance with a proposed standard would require the use of a patented item, ASAE shall obtain assurance that a license shall be made available without compensation to any applicant who wishes to conform to the standard prior to approval of the proposal.

3.4 In discharging their responsibilities, the members of all division standards and technical committees, or other groups organized to carry on standardization activities, shall function as individuals and not as agents or representatives of their employers. Members are appointed to ASAE technical committees and subcommittees on the basis of their personal qualifications and ability to contribute to the work of the committee.

3.5 All interested groups and individuals shall be provided the opportunity to participate in the development of proposed standards.

3.6 Proposed, unapproved voluntary standards should include the statement, "Draft: For Review Only—Not for Distribution."

3.7 Use of SI units

3.7.1 Specifications shall be expressed in SI units or in both SI and customary units (see Section 11). ASAE considers it desirable and necessary for the profession to use the International System of Units, SI (Systeme International d'Unites), and thereby make its standards more understandable and useful to all readers.

3.7.2 ASAE Engineering Practice EP285, Use of SI (Metric) Units, includes a list of preferred SI units. These SI units are in agreement with ISO 1000, SI Units and Recommendations for the Use of Their Multiples and of Certain Other Units. (International Organization for Standardization)

3.7.3 ASAE recognizes a genuine and increasing requirement for improving the compatibility of specifications and products produced according to SI and customary units. Technical committees are, therefore urged to work toward simplification of measurement units in the best interests of industry and the profession. This effort may include active participation with other organizations whose goals are to seek a solution to this problem on a national as well as an international scope.

SECTION 4—POLICY ON ORGANIZATIONAL RELATIONSHIPS

4.1 ASAE relationships with other organizations are encouraged. In the interest of ASAE these statements of policy should be considered.

4.1.1 All bodies, such as divisions, committees and sections, officially recognized within the ASAE structure are urged to cooperate with, and give technical advice to, government, civic and industry groups working on problems in agricultural engineering. Official ASAE groups are free to consult with corresponding groups of other technical or professional organizations in areas of overlapping interest. This policy applies at the local, regional, and national levels of ASAE.

4.1.2 City, county or state groups will usually find it most expedient and effective to cooperate with ASAE bodies at the section level. Federal, national or international groups can normally be best served at the national level of ASAE.

4.1.3 In establishing or maintaining a specific ASAE group relationship with other organizations, those policies apply:

4.1.3.1 It shall be the responsibility of official ASAE groups who work with other organizations, private and governmental, to acquaint those organizations with existing ASAE standards wherever applicable, and to assist them in interpreting such information.

4.1.3.2 If no current ASAE standard is applicable to the problem at hand, the ASAE group shall cooperate with the other organization involved in preparing an initial proposal for review. When agreement of the combined group is reached, the proposal shall be referred to the appropriate ASAE division standards committee for action. The standards committee shall promptly review the material and make a disposition as follows:

4.1.3.2.1 The division standards committee may consider the material to be of national significance, and process it through regular development procedures used for ASAE standards, giving reasonable notice of this ongoing activity to potentially interested parties. After review, revision and final approval by regular voting procedure, the material may be published and used as an ASAE standard, as applicable.

4.1.3.2.2 The division standards committee may consider the material to be technically sound, but not of sufficient national significance to justify processing it as an ASAE standard. In this case, the originating group shall be so advised. Such material may then be published for local use only as having been developed with the knowledge and advice of ASAE; but it shall not be represented as an ASAE standard.

4.1.3.2.3 The division standards committee may reject the material as being deficient to the extent that it shall not be given further consideration by an ASAE group. In this case, the outside organization shall be so advised and given specific reasons for the decision.

4.1.4 As individual ASAE members frequently serve in various capacities with other organizations or groups, each member is urged to encourage use of established ASAE standards and policies. They are also urged to advise ASAE of areas where the Society may render service in the national interest. Individual ASAE members shall avoid giving the impression that they represent ASAE, *unless* they have been officially appointed to do so.

SECTION 5—COOPERATIVE STANDARDS DEVELOPMENT (ASAE/SAE) POLICY OF ASSIGNED AREAS OF STANDARDIZATION

5.1 This definitive policy, adopted in June 1964, regarding development and maintenance of standards, delineates the fields of endeavor or areas of responsibility between the American Society of Agricultural Engineers and the Society of Automotive Engineers. The best interests of industry, government, the engineering profession, and society will be served if the following points are considered:

5.1.1 Conflicting standards should be avoided.

5.1.2 This division of endeavor should not apply to society activities outside the area of developing standards.

5.2 Areas of responsibility defined

5.2.1 Cooperative standardization effort relative to self-propelled machines and agricultural implements and implement components used essentially and primarily in agricultural machinery should be the responsibility of ASAE.

5.2.2 Cooperative standardization effort relative to agricultural tractors and tractor components should be the responsibility of SAE.

5.2.3 Cooperative standardization efforts involving the relationships of agricultural tractors and/or mounted, towed or self-propelled machines with the soil should be the responsibility of ASAE.

5.2.4 Where the subject matter influences the implement and tractor combination (especially in the field of safety, providing for each make of tractor to successfully operate each make of implement, or improving the overall operation of the implement and tractor combination, etc.), it is of vital interest to both ASAE and SAE and should, therefore, be processed and recognized by both Societies.

5.2.5 In cases where the material in question will benefit the user by receiving approval from both Societies, such material should be processed and recognized by both Societies.

5.2.6 Conversely, joint ASAE-SAE standards should be avoided unless they can be properly classified per paragraphs 5.2.4 or 5.2.5, above.

5.3 Organizational responsibilities

5.3.1 The desired results should be achieved if the following organizational provisions in SAE and ASAE are made effective:

5.3.1.1 The technical committees in ASAE and SAE which process material that is expected to be approved by both groups as standards should appoint or designate an individual to serve as a liaison member on the two Society technical committees involved. It will be his specific responsibility to:

5.3.1.1.1 Keep each Society technical committee informed as to the views, activities and actions of the other Society's technical group.

5.3.1.1.2 Follow the material through the various steps of voting procedure in each Society with the aim of securing simultaneous publication of a compatible document by each Society in the same year.

5.3.1.1.3 Check the joint material as published by each Society, and if the material does not correspond, to instigate the necessary action so as to make the material correspond.

5.3.1.2 It is recommended that ASAE designate liaison representatives from the ASAE Power and Machinery Division Standards Committee to:

5.3.1.2.1 The SAE Agricultural Tractor Test Code Subcommittee

5.3.1.2.2 The SAE Agricultural Tractor Tire Subcommittee

5.3.1.2.3 The SAE Agricultural Tractor Technical Committee. (This member will be responsible for those items not covered by paragraphs 5.3.1.2.1 and 5.3.1.2.2.)

5.3.1.3 It is recommended that ASAE and SAE designate the same individuals as their official liaison representatives on the respective SAE committees.

PART II
PROCEDURES

SECTION 6—PROPOSAL FOR AN ASAE STANDARD

6.1 Any individual, committee or organization, whether associated with ASAE or not, may propose standardization material, or may express the need for development of standards by ASAE. This includes proposals for new standards and revisions to existing documents.

6.1.1 Before an individual or group expends time and effort developing a proposal, it *shall* be referred to the appropriate ASAE division standards committee. A written, specific outline of the standard needed should be directed to the appropriate division standards committee chairman, with a copy to ASAE Headquarters. Proposals should avoid duplication of content covered in any existing ASAE standards. Individuals or groups proposing standards shall suggest individuals or groups best qualified to assist in developing the proposal, with recognition of need to make pending activities known to affected, interested parties.

6.1.2 If it is not apparent which ASAE division standards committee should review the proposal, it should be referred to the ASAE Director of Standards. The ASAE Committee on Standards shall, where necessary, interpret ASAE policy and provide proper interdivisional coordination on such matters.

SECTION 7—DEVELOPMENTAL GUIDELINES

7.1 Any division standards committee receiving a standardization proposal shall:

7.1.1 Determine whether the proposal is within the division's area of interest and whether it should be developed in light of existing standards.

7.1.2 Assign responsibility for developing the proposal in writing with a copy to the ASAE Director of Standards.

7.1.2.1 The division standards committee may assign the task of developing the complete proposal to an existing ASAE committee; refer it to existing committees in other organizations; or create a subcommittee for the purpose. Appointments for such subcommittee activities are made by the division standards committee chairman and require no additional approval. When a special subcommittee is appointed, it is recommended that the chairman and a majority of committee members serve until the assigned task has been completed. When an existing committee is assigned a task, it should retain sufficient membership for continuity, throughout the development of the standard. Subcommittees shall follow these ASAE standardization procedures regarding opportunity for input from interested parties.

7.1.2.2 The division standards committee may request technically qualified individuals or existing groups outside ASAE to contribute to the standardization effort.

SECTION 8—STANDARDS WRITING COMMITTEE RESPONSIBILITIES

8.1 Committee structure and balance

8.1.1 Committees generating a needed standard shall keep their respective division standards committee advised of status or progress and provide a list of groups (within or outside ASAE) that may have a vital interest.

8.1.2 Committees developing standards proposals shall strive to obtain broad representation and shall provide an opportunity for qualified individuals from substantially interested producer, consumer and general interest groups to participate and vote. With respect to technical background and experience, a substantial balance of voting members shall be maintained. There shall be no opportunity for domination by any single interest.

8.2 The *purpose* and the *scope* of the standard should first be agreed upon by the working group, and then established in clearly defined terms. The scope should delineate anticipated coverage. It is essential that the purpose and scope be appropriate for cooperative effort within the policies, as well as the Constitution, Bylaws and Rules of the Society.

8.3 Only specifications that are essential to achieving the goals defined in the purpose and scope are to be included in the standard.

8.4 As soon as action is taken to develop a standard it shall be brought to the attention of the Director of Standards so that same can be brought to attention of interested parties through appropriate ASAE publications.

8.5 The research of material, interchange of ideas and information between all interested groups, with expedient and concise development in accord with format and procedures is the responsibility of the originating committee.

8.6 The writing committee is responsible for the technical accuracy of the material and must have the latest information bearing on the particular area of standardization.

8.7 Committees that originate new standards formally approved by the Society should develop a news release in cooperation with the Director of Standards for use in announcing availability of the standard.

8.8 Standards writing committee records. A committee or subcommittee, assigned by a division standards committee to develop standards shall keep a record of its membership (name, title and address), of all drafts and of revisions of proposals. Each draft shall be appropriately dated. Copies of these records shall be submitted to the responsible division standards committee and to ASAE Headquarters. The committee may be requested to meet with the division standards committee, as a group or by representation, to discuss development of proposals.

8.8.1 Committees (or subcommittees) developing

proposals shall attempt to reconcile differences of opinion before a vote is taken among committee members. The basis for differing opinions and/or rationale for reconcilliation or disposition of dissenting views shall be recorded for review by the division standards committee(s) and ASAE Committee on Standards.

8.9 Standards approvals and reports

8.9.1 Decisions in ASAE voluntary standards writing activities shall reflect a substantial agreement by all parties at interest with an attempt to resolve all negative comments.

8.9.2 After an assigned technical committee or subcommittee approves a proposal by a required 3/4 favorable vote of the committee, after deducting waive votes, 40 copies of the proposal shall be forwarded to the Director of Standards with an information copy directly to the division standards committee chairman. (See Section 14, also)

8.9.3 The division standards committee is responsible for verifying the technical accuracy of standards developed by writing committees with technical expertise related to that division. When accuracy, committee balance, format, etc., are assured, the proposed document is approved when substantial agreement and consensus of all those concerned have been assured but in no case with less than with a 3/4 favorable vote. It is then forwarded to the Director of Standards, with appropriate transmittals of actions, for consideration by the Committee on Standards (T-1). (See paragraph 14.2)

8.9.4 The Committee on Standards (T-1) shall be responsible for the approval of standards from the viewpoint of policy and interdivisional coordination. To be approved, a standard must have the *unanimous* approval of committee members representing the technical divisions. (See paragraph 14.3)

SECTION 9—DIVISION STANDARDS COMMITTEE RESPONSIBILITIES

9.1 The division standards committee has responsibility for approval of technical writing committee proposals as noted in paragraph 8.9.3.

9.2 The Standards Committee of each ASAE division (EES-03, FPE-03, PM-03, SW-03, SE-03) is responsible for coordinating all standardization efforts within its division.

9.3 All proposed ASAE standards *shall* be processed through one or more of the division standards committees. When a standardization proposal involves more than one division, the ASAE Committee on Standards (T-1) will determine which division shall be ultimately responsible.

9.4 The division representative or alternate to the Committee on Standards (T-1) shall provide liaison between the division standards committee and the Committee on Standards.

9.5 Subcommittees may be established to prepare the technical material and arrange it in acceptable format.

SECTION 10—COMMITTEE ON STANDARDS (T-1) RESPONSIBILITIES

10.1 The Committee on Standards (T-1) has responsibility for approval of all proposed standards as noted in paragraph 8.9.4.

10.2 The Committee on Standards shall obtain assurance that known interests have had opportunity to participate in or be represented in the development of proposed standards. The Committee on Standards shall seek assurance from appropriate division standards committees that proposed standards are technically accurate.

10.3 The Committee on Standards shall keep informed regarding national and international standardization activities relating to ASAE standardization policy and procedure, and shall share pertinent information with division standards committees.

10.4 The Committee on Standards shall make a semi-annual report to the ASAE Board of Directors through the Technical Council.

SECTION 11—PROPOSED STANDARD FORMAT AND CONTENT

11.1 General

11.1.1 Each proposal should include a statement of purpose. Each proposal should include a scope statement indicating areas, classes of equipment, etc., for which the proposal is and is not applicable. (See Section 2)

11.1.2 Each ASAE technical committee shall specify the units and the actual values to be shown. It is essential that functional interchangeability between parts produced according to SI and customary units be maintained.

11.1.3 Decimals rather than common fractions shall be used except where common fractions are used in customary units as nominal designations and are necessary to convey proper meaning (i.e., 3/4 in. pipe).

11.1.4 Implied measurement tolerances are to be avoided. When specifying measurements in either customary or SI units which require tolerances, such tolerances shall be included.

11.2 Format for standards proposals/submittals

11.2.1 Committees that propose ASAE standards should submit 40 copies *single-spaced* with lines numbered on either margin according to the following general format:

"DRAFT FOR REVIEW ONLY—
NOT FOR DISTRIBUTION"

1 Proposed ASAE Standard
2 TITLE OF STANDARD IN CAPS date ____________
3 Proposed by the (Committee or organization)
4 for consideration by ASAE ____________
5 ____________ Committee
6
7 SECTION 1—PURPOSE AND SCOPE
8 1.1 Xxxxx ....
9 ....
10 1.2 Xxxxx ....
11 ....
12 SECTION 2—(BRIEF TITLE
13 DESCRIBING MATERIAL IN SECTION)
14 2.1 Heading or lead in, if appropriate, should be underscored but not
15 capitalized, except initial capital, unless it involves a phrase that would
16 normally be capitalized if it were in the body of the text.
17 2.2 Number and indent paragraphs. This will identify paragraphs for
18 correspondence and review, and it will organize subject matter in our
19 line form
20 2.2.1 Xxxxx ....
21 2.2.2 Xxxxx ....
22 ....
23 2.2.2.1 Xxxxx ....
24 ....
25 2.2.2.2 Xxxxx ....

11.3 Terms and definitions, alphabetizing

11.3.1 Scope. Terminology and definitions shall be limited to those actually used within the text of a standard or those otherwise recommended for use.

11.3.2 Alphabetizing. When listing terms and definitions, alphabetize by listing the significant words first, and the modifiers second. Example:

Aeration: The movement of air . . .
Dryer, batch: Any dryer wherein the product . . .
Valve, excess-flow: A check valve which . . .
Valve, quick-acting: A manually operated . . .

11.4 Formulas. The formula should be indented in the column, and the legend should be shown below the formula. *Where* should begin at the left-hand margin. Example:

$$P = 2St/d$$

where
P = bursting pressure, Pa(psi)
t = specified wall thickness, mm(in.)
D = specified outside dia, mm(in.)
S = allowable working stress, N/m^2(psi)

11.5 Dimensioning

11.5.1 Dimensioning drawings. ANSI Y14.5, Dimensioning and Tolerancing, should be used as a guide in dimensioning drawings.

11.5.1.1 Abbreviations. Certain abbreviations are commonly used on drawings, and are universally understood. Examples are dia, max. Abbreviations should be in accordance with existing industry and military standards. It is recommended that less common abbreviations be used as little as practicable, because misinterpretations are less likely if words are spelled out.

11.5.1.2 Identification of linear units. The abbreviation in. is used for inch. The abbreviation mm is used for millimeter. On drawings where all dimensions are in inches or millimeters the abbreviations are omitted. Where there is any chance of confusion, the drawing format should contain a note "Unless otherwise specified, all dimensions are in 'inches' or 'millimeters,' as applicable." Where customary equivalents are shown, the customary dimensions shall be enclosed in parentheses and identified by the abbreviation, for example, (12 in.). Where the equivalents are used extensively, the abbreviations may be omitted provided a general note on the drawing states that "Dimensions in parentheses are in inches."

11.5.2 Dual dimensioning. When both customary and SI units are to be used in the text of a standard, the SI units should be printed in decimal form, followed by customary equivalents in parentheses. Example:

"... extending 104 mm (4.1 in.) above and 216 mm (8.5 in.) below." For dual dimensioning in tables, customary equivalents should be placed in a column adjacent to the SI units as shown in Table 1, paragraph 11.7.

11.6 Drawings, figures, graphs. The illustrations of the following standards in AGRICULTURAL ENGINEERS YEARBOOK may be used to determine appropriate style: ASAE S316, S203, EP340. ASAE illustrations are usually drawn twice finished size desired, and then reduced 50 percent. Each illustration should be identified by a figure number and brief caption in capital letters.

11.7 Formal tables: Formal tables should be presented as shown below. To avoid crowding, and to permit revisions and editing, it is often best to place tables lengthwise on the paper. Customary and SI dimensions can be placed in adjacent columns as in the example below. Example:

TABLE 1—TITLE IN CAPS*

Stub heading	Column caption	Column caption†		Caption for both SI and customary dimensions	
		Subcaption	Subcaption	mm	in.
Line heading					
Subheading	xxx	xxx	xxx‡	25.4	1.00
Subheading	xxx	xxx	xxx	50.8	2.00
Line heading					
Subheading	xxx	xxx	xxx	etc.	etc.
Subheading	xxx	xxx	xxx	etc.	etc.

*Table footnotes.
†Table footnotes.
‡Table footnotes.

11.8 Informal tables. Informal tables are used for simple tabulation that is easily incorporated into the sentence structure. In most instances, there should not be more than a stub and one column. Tabulations of more than one column should usually be made into formal tables. Informal tables should not be numbered because they are part of the sentence structure and are not referred to more than once in the text. The following illustration shows correct use of an informal table:

Dimensions comprising the standard specification are divided into three categories:

Category	Maximum drawbar power
I	20 to 45
II	40 to 100
III	80 and over

11.9 Footnotes for tables. Footnotes should be placed directly beneath the table. They should be identified consecutively beginning with the title, proceeding through column captions from left to right, and then moving down the line headings. Asterisks, daggers, etc., should be used as footnote symbols:

*, †, ‡, §, ||, #, then double symbols, e.g. **, ††.

SECTION 12—CITATIONS AND REFERENCES

12.1 Cross referencing ASAE standards. When citing an ASAE standard, provide complete title and numerical identification each time mentioned. Example:

" . . . as specified in ASAE Standard S209, Agricultural Tractor Test Code."

Refer to the sections and paragraphs by title and number. Example:

"Section 1—Test Conditions, paragraph 1.1.3."

12.2 Cross referencing other standards. Cite responsible organization, title and complete numerical designation the first time mentioned. Example:

"... in accordance with American Society for Testing and Materials Standard E380, Metric Practice."

Use organization acronym thereafter. Example:

"... in accordance with ASTM Standard E380, Metric Practice."

12.3 Date of adoption or revision. When the date of a cited standard is not given, reference to the latest revision of that standard is intended. Example:

ASTM Standard E380   ASAE Standard S201

When a standardization date is given, only the standard so dated is indicated for reference. Example:

ASTM Standard E380-76   ASAE Standard S201.4

12.4 Credit lines. Credit lines for tables, curves and illustrations may be listed.

12.5 Bibliographical reference

12.5.1 When considered necessary by the responsible ASAE division standards committee, bibliographical references may be included at the end of ASAE standards the first time published in the book of ASAE STANDARDS. After the first year published, extensive references will not be listed, but their availability from ASAE Headquarters will be noted at the end of the document.

12.5.2 References to articles in periodicals and books should include the following information in the order indicated:

References to Articles in Periodicals	Reference to Books
Last name of author	Last name of author
First name or initials	First name or initials
Year of publication	Year of publication
Title of article	Title of book
Title of periodical	Edition number
Volume number	Name of publisher
Issue number or date of issue	Place of publication
First and last pages of article	

SECTION 13—SUBMITTING PROPOSED STANDARDS FOR APPROVAL

13.1 Number of copies. Individuals, committees or other organizations submitting proposals for consideration by a division standards committee should submit single-spaced typed copies, with lines numbered as shown in paragraph 11.2.1. Forty (40) copies should be forwarded to the ASAE Standards Department. Should a division standards committee request an advisory review by another committee or organization, additional copies may be requested. Proposals should be prepared as prescribed by these Procedures.

13.2 Letter of transmittal. Above distribution should be accompanied by an equal number of copies of a letter of transmittal that:

13.2.1 Outlines the need for the standard;

13.2.2 Indicates approval by the committee preparing the proposal;

13.2.3 Lists names, titles and business affiliations of committee members who developed the proposal;

13.2.4 Lists the individuals, committees or other organizations who were consulted during preparation of the proposal with a summary of the response obtained.

13.3 The information provided in the letter of transmittal will advise the ASAE division standards committees and the Committee on Standards of the extent to which those affected by the proposal have participated in its development, as well as attitude toward same.

SECTION 14—STANDARDS VOTING PROCEDURES

14.1 Subcommittees and originating committees. Committees and subcommittees responsible for developing or revising standards shall approve proposals by a 3/4 vote (after deducting waive votes) before submitting them to a division standards committee. Committee members offering no documented response to a ballot proposal during the 30-day ballot period or during a subsequent 15-day reminder period will be counted as having cast *waive* votes if at least 80% of the committee membership voted. Standards listed on any published agenda may also be approved by 3/4 of committee members present, after deducting waive votes, at an official meeting of the committee. (See paragraph 8.9.3)

14.2 Division standards committees

14.2.1 Division standards committees and the ASAE Committee on Standards may vote in committee or by letter ballot. The committee members should receive proposals in writing so they may study them at their respective places of business before casting their votes. The ASAE Director of Standards shall conduct letter ballots for division standards committees and the Committee on Standards. He shall maintain a permanent record of the results and provide committees with a summary of results and comments.

14.2.2 Letter ballots will be due within 30 days of distribution. The vote may be *approval, disapproval,* or *waive. Disapproval* votes on division standards committee ballots must be accompanied by a statement of reason for the negative vote and suggestions for improvement. Comments and negative vote statements shall be reviewed by the division standards committee. All comments returned with letter ballots shall be returned to the committee who submitted the proposal for future consideration. *Approval* votes may also be accompanied by suggestions for improving the proposal in a subsequent revision. *Approval* votes shall *not* be cast if it is felt that technical commentary should be incorporated prior to publication. In such a case a disapproval vote should be cast. *Waive* votes may be cast by reason of not being qualified to vote on the proposal. Committee members offering no documented response to a ballot proposal during the 30-day ballot period or during a subsequent 15-day reminder period will be counted as having cast *waive* votes if at least 80% of the committee membership voted. Approval of letter ballots by 3/4 of division standards committee membership, after deducting waive votes, is required for all proposals affecting the technical sense of a standard. Objections raised by letter ballot may be resolved by discussion and voice vote or subsequent letter ballot as directed by the committee chairman. Standards listed on a published agenda may also be approved by 3/4 of committee members present after deducting waive votes, at any official meeting of the committee.

14.2.3 Liaison representatives to division committees are appointed to function as representatives of their respective organizations to report on activities, offer comments or suggestions, and to carry back instructions or information to their respective groups. Since they are not members of the committee, they shall not be entitled to vote.

14.3 Committee on Standards. Technical division

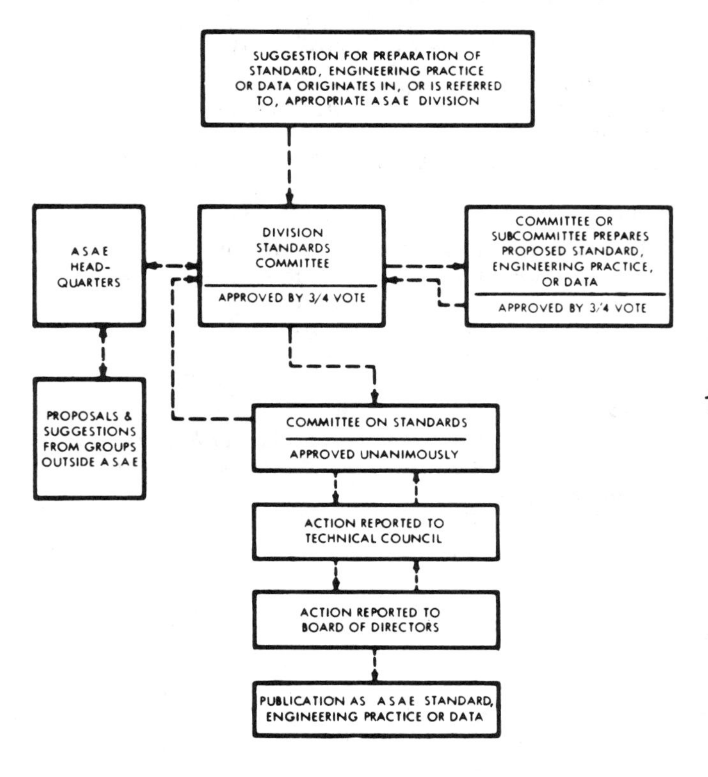

representatives and their alternates are expected to attend meetings and vote on all proposals in committee and by letter ballot. Letter ballots will be due within 30 days. Votes on letter ballots may be for *approval* or *disapproval.* A *waive* vote is not permitted in this committee. *Disapproval* votes must be accompanied by suggestions for action that would enable the member to approve the proposal. Unanimous approval by division representatives is required. One approval vote cast by the representative from all divisions represented in the Committee on Standards constitutes unanimous approval. Ballots of the *representatives* will be first tabulated for committee decisions. If a *representative's* vote is not received by the end of the 30-day voting period, his *alternate's* vote will be counted to represent the division.

14.4 Technical Council. The Technical Council, and ultimately ASAE Board of Directors, may accept or reject specific standards action as reported by the Committee on Standards.

SECTION 15—NORMAL CHRONOLOGICAL SCHEDULES

15.1 Proposals may be forwarded at any time to the ASAE Director of Standards for consideration by division standards committees. Since major standards review and approval activity occur during the two national meetings of ASAE, the following timetable is suggested:

March 20 and September 20—Committees, subcommittees and non-ASAE organizations should submit proposals to the ASAE Director of Standards. The number of copies requested in Section 11 should be provided.

March 30 and September 30—ASAE Director of Standards should ballot division standards committees

May 1 and October 25—ASAE Director of Standards should ballot Committee on Standards.

June 20 and December 10—During Summer and Winter ASAE meetings, division standards committees and the Committee on Standards review results of ballots and resolve objections or comments. The Committee on Standards reports to the Technical Council of the ASAE Board of Directors.

SECTION 16—STANDARDS IDENTIFICATION/ ASSIGNMENTS

16.1 ASAE standards shall be identified as follows:

16.1.1 Appropriate identification shall be assigned by ASAE Headquarters.

16.1.2 Numerical identification will be assigned from a continuing series of numbers beginning with number 201.

16.1.3 The Letter *S, EP,* or *D* will precede the numerical identification of a standard, engineering practice or data. The numerical identification of a given standardization item will be retained by that item even though its classification as standard, engineering practice or data may change. Example:

If standardization item number 201 is classified as an *Engineering Practice,* it shall be identified EP201. If it is reclassified as a *Standard,* it shall be identified S201.

16.1.4 The letters ASAE shall precede the assigned identification. Example:

ASAE S201 (Note: When citing in text, use "ASAE Standard S201, . . .")

16.1.5 Technical revisions, which affect performance, interchangeability, or application of a standard shall be indicated by adding to the assigned identification a decimal and numeral denoting the number of times the item has been revised. Example:

ASAE S201.4 identifies an ASAE standard which has undergone its fourth technical revision.

A revision numeral shall be assigned or changed only if the revision is a result of regular ASAE voting procedure.

16.1.6 An ASAE standard having tentative status shall be identified with the letter T. Example:

ASAE S360T

SECTION 17—STANDARDIZATION APPEALS PROCESS

17.1 Any person submitting an unresolved dissenting view on a standards proposal shall have the right to appeal approval of the proposal to the appropriate division standards committee. If the dissenting view cannot be satisfactorily resolved by the division standards committee at a scheduled meeting, the appeal may be submitted within 30 days to the ASAE Committee on Standards for appropriate disposition. A further appeal may be made within 30 days after Committee on Standards action to the Technical Council of the ASAE Board of Directors. Appeals shall be submitted in writing, and may be supported by oral argument. Appeals shall be heard at the next scheduled meeting of the appropriate appeals body, not to exceed six (6) months from the date of receipt of said written appeal.

SECTION 18—MAINTAINING PUBLISHED STANDARDS

18.1 Review. Each division standards committee is responsible for initiating a review of standards placed in its charge at intervals of 5 years or less, or as requested by the Committee on Standards. Reviews shall normally be conducted by the committee or subcommittee that developed the standard. The division standards committee, through regular voting procedures, will:

18.1.1 Revise. Revise, or arrange for revision of, the material to coincide with current needs, or

18.1.2 Reconfirm. Reconfirm the material without change if it is current, accurate, and is of value*, or

18.1.3 Withdraw. Recommend withdrawal if it no longer serves a useful purpose. Titles of withdrawn standards shall be listed in subsequent editions of the book of ASAE STANDARDS for 5 years.

18.2 The dates of revision or reconfirmation shall be recorded in the printed heading of the standard affected.

18.3 The ASAE Director of Standards shall notify division standards committees and the Committee on Standards when standardization items are due for 5-year review and shall initiate follow-up when necessary.

18.4 Technical revisions. Technical changes, deletions, or additions that alter the technical sense of a standard must be approved by the appropriate ASAE division standards committee and by the ASAE Committee on Standards through regular voting procedures. Proposed technical revisions should be submitted according to procedures outlined in Section 9.

18.5 Editorial revisions. Editorial revisions are those which do not alter the technical sense. Anyone may propose editorial revisions to the appropriate standards committee, or to the ASAE Director of Standards who will consult the responsible division standards committee chairman when necessary to ascertain whether the proposed changes may alter the technical sense of the standard. All committees should be alert to the need for editorial revisions in standards for which they are responsible.

18.6 Revising existing standards. When submitting proposed revisions of existing standards, it is important that reviewing committees be completely aware of each proposed change they are asked to approve. When possible type, pen or paste up the additions, and line through proposed deletions, on the most recent version of the existing standard; then make machine copies for balloting. If proposed revisions are extensive, requiring a completely retyped draft, the changes should be *identified* as *changes,* insofar as possible. Additions should be underscored. Deletions should be typed in the new draft and then lined through. Completely new paragraphs should be identified parenthetically (proposed addition). A cover letter or statement should list the changes, additions, and deletions and where appropriate give reasons for same.

18.7 Committee records. A committee assigned to review or revise a standard shall maintain records of individuals and organizations participating and of all drafts and revisions in the same manner as in developing a new proposal. (See paragraph 8.8)

*Action by a division standards committee to reconfirm standards without change does not require approval by the Committee on Standards, but must be reported to it as a matter of record.

SECTION 19—SUBMITTING ASAE STANDARDS FOR CONSIDERATION BY ANSI AS AMERICAN NATIONAL STANDARDS

19.1 Purpose. The purpose of this section is to establish criteria for determining which ASAE standards should be referred by ASAE to ANSI for consideration as an American National Standard, and to establish requirements for officially approving the submission of ASAE standards to ANSI.

19.2 Criteria. It is preferable, but not necessary, that an ASAE standard, to be submitted to ANSI, be judged sufficiently refined and representative of established technology or the state-of-the-art that frequent revisions or corrections are unlikely. The standard should, however, meet one or more of the following criteria:

19.2.1 The standard defines products or processes, engineering practices or includes data of interest to groups not served by ASAE. Adoption as an American National Standard will eliminate the need for development of similar standards by these groups.

19.2.2 The community served by ASAE will benefit from recognition of the ASAE standard and the subsequent compliance by groups in other fields.

19.2.3 Adoption as an American National Standard will enhance use of and conformance with the ASAE standard in certain geographical or trade areas, especially when such recognition cannot be achieved otherwise. The additional public review accompanying approval as an American National Standard enhances the stature of that standard.

19.2.4 Adoption as an American National Standard will establish the U.S. position in international trade or standards activities. All U.S. counterpart standards to international standards proposals should be submitted for recognition as American National Standards so that they may be recognized by other countries as the U.S. position. American National Standards may be used directly as a basis for international standards.

19.3 Approval. A standard shall be submitted to ANSI for consideration as an American National Standard after such submittal has been approved by:

19.3.1 The originating ASAE committee, in accordance with paragraph 14.1. (If the standard was initially proposed by a committee or organization outside ASAE, the division standards committee (03) shall inform the committee or organization that the standard is being considered for submission to ANSI.)

19.3.2 The division standards committee, in accordance with paragraph 14.2.

19.3.3 The Committee on Standards, in accordance with paragraph 14.3.

19.3.4 The question of submission of an ASAE standard to ANSI shall be considered on a ballot separate from ballots required and used for normal approval of proposed ASAE standards.

19.3.5 Submission of revisions to ASAE standards that have previously been adopted as American National Standards require no further balloting within ASAE. The revision shall automatically be submitted to ANSI by the Director of Standards.

SECTION 20—FORMAT FOR INDICATING HARMONIZATION OF ASAE STANDARDS WITH INTERNATIONAL (ISO) STANDARDS

20.1 When an ASAE standard corresponds to an ISO standard (i.e., covers the same subject), the following information shall be printed on the first page of the ASAE standard.

20.1.1 The corresponding ISO standard number and year of publication, along with a footnote symbol, shall be added following the ASAE standard number.

Example: ASAE Standard: S366.1 (*SAE J1036, + ISO 5675-1981)

20.1.2 A footnote explaining the relationship between the ASAE standard and the ISO standard shall be added immediately preceding the history paragraph located at the top of every ASAE standard. The complete ISO number, year of publication, and title shall be used. Terms to be used to describe this relationship are:

(1) "equivalent" (if identical)
(2) "equivalent with editorial changes"
(3) "equivalent with minor technical deviations" (A minor technical deviation is a technical deviation that does **not** render **unacceptable** under the terms of the ASAE standard anything that **is acceptable** under the terms of the ISO standard or vice versa.)
(4) "corresponds to" or "is based upon" (if not equivalent due to major technical deviations. A major technical deviation is a technical deviation that renders **unacceptable** under the terms of the ASAE standard anything that **is acceptable** under the terms of the ISO standard or vice versa.)
(5) "includes" (if the entire ISO standard is included and reprinted within the larger ASAE standard)
(6) "references" (if the entire ISO standard is included by reference, but not reprinted, within the larger ASAE standard)

20.2 In all cases editorial changes or technical deviations should be explained or otherwise indicated so that the user can determine where the ASAE and ISO standards differ.

Example: + ASAE S366.1 is equivalent with editorial changes to International Standard ISO 5675-1981, Agricultural Tractors and Machinery—Hydraulic Couplers for General Purposes—Specifications. The editorial changes consist of a complete redraft without technical deviation.

INDEX

TECHNICAL COMMITTEES

(Other committees of ASAE are not included. Contact ASAE Headquarters for a complete roster of Society committees.)

Appointments are for the Society year (close of June 1988 meeting in Rapid City, South Dakota through close of June 1989 meeting in Quebec City, Quebec) except as noted. Appointments are subject to change as deemed appropriate by appointees, the ASAE President, or the appointing ASAE Vice President.

TECHNICAL DIVISIONS/INSTITUTES

TECHNICAL COUNCIL of the BOARD OF DIRECTORS

Leroy K. Pickett, Vice President (89)

Glenn E. Hall, Electrical and Electronic Systems (90)
David R. Thompson, Emerging Technologies (89)
Arthur A. Teixeira, Food and Process Engineering (90)
Jack W. Crane, Power and Machinery (89)
Donald K. McCool, Soil and Water (90)
William H. Brown, Structures and Environment (89)

NONDIVISIONAL TECHNICAL COMMITTEES

T-1 COMMITTEE ON STANDARDS

Rollin D. Schnieder, chairman
Harmon L. Towne, vice chairman
Russell H. Hahn, staff secretary

EES Representatives—James L. Steele (89); F. Spencer Givens III(90), alternate

FPE Representatives—Jay S. Marks (90); Rakesh K. Singh (91), alternate

PM Representatives—Earl A. Hudson (91); Rollin D. Schnieder (89), alternate

SW Representatives—E. Gordon Kruse (89); Gylan L. Dickey (91), alternate

SE Representatives—Harmon L. Towne (89); James A. Lindley (90), alternate

American National Standards Institute:
Organizational Member Council—Russell H. Hahn, staff representative

T-2 AGRICULTURAL SAFETY COMMITTEE

Glen H. Hetzel, chairman
L. Dale Baker, vice chairman
Robert A. Aherin, secretary
Karl C. Anderson, past chairman

Terms expiring June 1989—
Robert A. Aherin, Wesley F. Buchele, Charles F. Brundage, Thomas G. Carpenter, Robert D. Davis, Coy W. Doty, Glen H. Hetzel, Edwin J. Matthews, Ronald L. McAllister, Rollin D. Schnieder, Herbert D. Sullivan, Wendell M. Van Syoc, Billy B. Willis

Terms expiring June 1990—
L. Dale Baker, Dwight A. Benninga, Martin A. Berk, William R. Borghoff, Rodney W. Carpenter, John H. Crowley, John E. Hirzel, Murray Madsen, J. Edson McCanse, Larry R. Piercy, Mark A. Purschwitz, B. Bennett Reak, William A. Wathen

Terms expiring June 1991—
Karl C. Anderson, William J. Becker, Rodney W. Carpenter, David W. Cayton, Howard J. Doss, William E. Field, Wayne E. Hartman, David E. Knorr, Wallace McDougall, Peter G. Neilson, John B. Sevart, Gary L. Smith, Ivan L. Winsett

Liaison Representatives to Division Standards Committees—
EES-03 — Ivan L. Winsett
FPE-03 — To be appointed
PM-03 — Thomas G. Carpenter
SW-03 — Coy W. Doty
SE-03 — Rodney W. Carpenter

T-2/1 Safety Goals and Plans Subcommittee

Rollin D. Schnieder, chairman
Karl C. Anderson, past chairman
Glen H. Hetzel (T-2 chm.)
L. Dale Baker (T-2 vice chm.)
William E. Field (T-2 past chm.)

T-4 BIOENGINEERING COMMITTEE

See BI-02

T-6 AGRICULTURAL SYSTEMS ANALYSIS COMMITTEE

See KS-02

T-7 AQUACULTURAL ENGINEERING COMMITTEE

See AQ-02

T-8 MECHANIZATION MANAGEMENT COMMITTEE

See P-101

T-9 ENVIRONMENTAL QUALITY COORDINATING COMMITTEE

Donald L. Basinger, chairman
John L. Goodenough, vice chairman
William L. Magette, secretary
Lawson M. Safley, Jr., past chairman

EES Representatives—Gordon F. Williams, Jr. (89), John L. Goodenough (90), William D. Mayfield (91)

FPE Representatives—William F. Ritter (89), Louis A. Licht (90), Charles C. Ross (91)

PM Representatives—Eugene P. Columbus (89), Alan D. Brashears (90), Herschel Elliott (91)

SW Representatives—Arthur B. Holland (89), Richard K. White (90), John M. Sweeten (91)

SE Representatives—James A. Moore (89), James A. Lindley (90), Larry D. Jacobson (91)

Regional Representatives—

Illinois-Wisconsin—Donald L. Day (90, SE)
Gary D. Bubenzer (91, SW)

Mid-Central—Dale H. Vanderholm (89, SW),
James L. Baker (90, SW), Delynn R. Hay (91, SW)

North Atlantic—William L. Magette (89, SW),
Donald L. Basinger (90, SW)

North Central—Terrence E. Huntrods (89, SW),
Charles J. Clanton (90, SE), Charles H. Ullery (91, SW)

Pacific Northwest—David C. Moffit (89, SW),
Mark F. Madison (90, SW), John J. R. Feddes (91, SE)

Pacific—David J. Hills (89, SE), Mark E. Grismer (90, SW)
Durham K. Giles (91, EPP)

Rocky Mountain—Ronald C. Sims (89, SW),
Robert S. Freeburg (90, SE)

Southeast—Lawson M. Safley, Jr. (89, SE),
Philip W. Westerman (90, SW), Adelbert B. Bottcher (91, SW)

Southwest—Clifford B. Fedler (89, SW), James L. Revel (90, SE)
Carl L. Griffis (91, EES/FPE)

Tri-State—Karen M. Mancl (89, SW), Theodore L. Loudon (90, SW),
Andrew D. Ward (91, SW)

T-9/2 Nominating Subcommittee

Lawson M. Safley, Jr., chairman
Arthur B. Holland

T-10 METRIC COORDINATING COMMITTEE

Donnell R. Hunt, chairman
Alvin C. Bailey, vice chairman and secretary

Terms expiring June 1989—
Page L. Bellinger, Richard E. Linhardt (ER), Richard L. Phillips

Terms expiring June 1990—
Alvin C. Bailey, Donnell R. Hunt (PM), William E. Hart (SW)

Terms expiring June 1991—
Myron G. Britton (SE), Harvey J. Hirning (EES), John H. Pedersen, John E. Turnbull (Canadian liaison)

ASAE Publications Dept. Executive Committee
(P-501)—Donnell R. Hunt, liaison representative

T-11 ENERGY COMMITTEE

Stanley J. Clark, chairman
Charles L. Peterson, vice chairman
Wayne A. LePori, secretary
Earle E. Gavett, past chairman

Terms expiring June 1989—
Hugh J. Hansen, D. Houston Luttrell, Kenneth L. McFate, W. Ralph Nave,
Bill A. Stout

Terms expiring June 1990—
Zane R. Helsel, Bryan M. Jenkins, Blaine F. Parker, Charles L. Peterson,
Richard A. Peterson, James H. Ruff

Terms expiring June 1991—
Phillip C. Badger, R. Nolan Clark, Earle E. Gavett, Wayne A. LePori,
Charles C. Ross, Norman Smith

Terms expiring June 1992—
James L. Butler, Stanley J. Clark, Robert G. Curley, Albert Garcia III,
Dennis L. Larson, Douglas E. Pond

ELECTRICAL AND ELECTRONIC SYSTEMS DIVISION

Glenn E. Hall, Director (90)

EES-01 EXECUTIVE COMMITTEE

Bailey W. Mitchell (90), chairman
Carl E. Bohman (91), vice chairman
Joe M. Bunn (92), secretary
Joe A. Gregory (89), past chairman
Glenn E. Hall (90), director

EES-02 STEERING COMMITTEE

Bailey W. Mitchell (90), chairman
Carl E. Bohman (91), vice chairman
Joe M. Bunn (92), secretary
Joe A. Gregory (89), past chairman
Glenn E. Hall, director (88)

Committee Chairmen—Barry Bauman, Gary W. Buller, Suhas R. Ghate,
Marvin J. Pitts, Richard A. Peterson, Terry J. Siebenmorgan, David G. Sokol,
Gregory L. Stark, Joseph Tecza, Luther R. Wilhelm

Terms expiring June 1989—
F. Spencer Givens III, J. Wayne Mishoe

Terms expiring June 1990—
L. Allen Butler, Stephen W. Searcy

Terms expiring June 1991—
Robert J. Gustafson, James L. Steele

Other Representatives reporting to the Steering Committee

CIGR, International Commission of Agricultural Engineering,
Sect. IV: Rural Electrification—Fred W. Bakker-Arkema, technical correspondent

ASAE Environmental Quality Coordinating Committee (T-9)—Gordon F. Williams, Jr. (89), John L. Goodenough (90), William D. Mayfield (91), liaison representatives

ASAE Metric Coordinating Committee (T-10)—Harvey J. Hirning (91), liaison representative

ASAE Paper Awards Committee (M-116)—Gary M. Hyde (91), liaison representative

ASAE Research Committee (P-303)—Robert J. Gustafson (89), Stuart O. Nelson (90), Glenn E. Hall (91), Zachary A. Henry (91), liaison representatives

ASAE Cooperative Standards Program Finance Committee (E-22)—Stanley A. Weeks (89), Douglas H. Aeilts (90), liaison representatives

ASAE Publications Executive Committee (P-501)—Clifford A. Flood, Jr. (90), liaison representative

ASAE Refereed Publications Committee (P-511)—J. Wayne Mishoe (91), liaison representative

ASAE Monographs Committee (P-512)—Luther R. Wilhelm (89), Hugh J. Hansen (91), liaison representatives

ASAE Textbook Committee (P-514)—Zachary A. Henry (90), liason representative

ASAE Special Publications Committee (P-515)—Hugh J. Hansen (89), liaison representative

EES-02/2 EES DIVISION STRUCTURE AND VITALITY SUBCOMMITTEE

Clifford A. Flood, Jr., chairman
Gary W. Buller, Glenn E. Hall, J. Wayne Mishoe

EES-03 STANDARDS COMMITTEE

Ivan L. Winsett, chairman
Donavon L. Bakker, vice chairman
Russell H. Hahn, staff secretary
Joe M. Bunn, past chairman

Terms expiring June 1989—
F. Spencer Givens III, Joe A. Gregory, Bruce W. Moechnig, Ronnie G. Morgan

Terms expiring June 1990—
Donavon L. Bakker, Richard W. Guest, Cecil B. Medders, Richard A. Peterson

Terms expiring June 1991—
Joe M. Bunn, Gary L. Downey, Manjit K. Misra, James L. Steele, Ivan L. Winsett

Voting Representatives of Technical Committees reporting to EES-03
H. David Currence, Agricultural Wiring and Utilization Committee (EES-33)
Barry Bauman, Electrical Code for Agricultural Committee (EES-35)
David C. Ludington, Milk Handling Equipment Committee (EES-41)
Gary W. Buller, Electrical Controls for Farmstead Equipment Committee (EES-42)
Richard A. Peterson, Electric Utilization Research Committee (EES-44)
Joe M. Bunn, Radiation Committee (EES-48)
Marvin L. Stone, Instrumentation and Controls Committee (EES-53)
Luther R. Wilhelm, Computers Committee (EES-54)

ASAE Committee on Standards (T-1)—James L. Steele,
representative; F. Spencer Givens, III, alternate

ASAE Agricultural Safety Committee (T-2)—
Ivan L. Winsett, liaison representative

Institute of Electrical and Electronic Engineers—
J. M. van Name, liaison representative

EES-03/1 Overhead Clearance of Power Lines Subcommittee

Paul Proctor, chairman
EES: Lyle E. Stephens, Ivan L. Winsett
PM: Rollin D. Schnieder, Walter L. Scott
SW: Bill E. Berry, Donald R. Sisson

EES-03/2 Early Identification of Needs for Standards Subcommittee

Barry Bauman, chairman
Gary L. Downey, Richard A. Peterson

EES-04 PAPER AWARDS COMMITTEE

Gary I. Wallin (89), chairman
Robert S. Brashear (90), John L. Goodenough (91), John W. Goodrum (92)

EES-05 PUBLICATIONS REVIEW COMMITTEE

Marvin J. Pitts, chairman and division editor

Terms expiring June 1989—
Robert J. Gustafson, Bailey W. Mitchell, Marvin R. Paulsen,

Terms expiring June 1990—
Michael J. Delwiche, B. Derrell McLendon, Marvin J. Pitts, James L. Steele

Terms expiring 1991—
J. Wayne Mishoe, Robert S. Sowell, Harold A. Hughes

EES-06 PROGRAM COMMITTEE

Carl E. Bohman, chairman
Joe M. Bunn, vice chairman

Technical Committee Program Chairmen—Donavon L. Bakker, H. David Currence, Ivan L. Winsett, Thomas A. Bon, Gary W. Buller, Sundaram Gunasekaran, Michael J. Delwiche, Marvin L. Stone

EES-33 AGRICULTURAL WIRING AND UTILIZATION COMMITTEE

Gregory L. Stark, chairman
H. David Currence, vice chairman
Robert J. Gustafson, secretary
Richard A. Spray, past chairman

Terms expiring June 1989—
Robert H. Brown, Gary W. Buller, William K. Dick, Gary L. Downey,
Joe A. Gregory, Hugh J. Hansen, Richard S. Hiatt, Kenneth L. McFate,
Richard A. Peterson, Richard A. Spray, Gregory L. Stark, Ivan L. Winsett

Terms expiring June 1990—
Douglas H. Aeilts, Robert E. Callies, Micheal D. Clement, H. David Currence,
F. Spencer Givens III, Elwyn S. Holmes, David E. Maki, Cecil B. Medders,
Randolph J. Morgan, Mark Thomas

Terms expiring June 1991—
Donavon L. Bakker, Keith B. Coffman, Lowell J. Endahl, John P. Hall,
Lorin E. Krueger, Marvin Nabben, Paul J. Shea, John D. Turrel, Elijah J. Tyson

Terms expiring June 1992—
Barry Baumann, H. Cecil Beggs, Paul W. Benson, Robert J. Gustafson,
Sherwood D. Skinner, LaVerne E. Stetson, Truman C. Surbrook

EES-35 ELECTRICAL CODE FOR AGRICULTURE COMMITTEE

Barry Bauman, chairman
Ivan L. Winsett, vice chairman
LaVerne E. Stetson, secretary
Leo H. Soderholm, past chairman

Terms expiring June 1989—
Gary W. Buller, Kenneth L. McFate, Marvin Nabben

Terms expiring June 1990—
Micheal D. Clement, H. David Currence, Richard S. Hiatt, Cecil B. Medders, Edgar H. Smith, Mark Thomas

Terms expiring June 1991—
Douglas H. Aeilts, Paul M. Anderson, Robert J. Gustafson, Ivan L. Winsett

Terms expiring 1992—
Barry Baumann, H. Cecil Beggs, Gary L. Downey, Gregory L. Stark, LaVerne E. Stetson, Truman C. Surbrook

National Fire Protection Association,
Electrical Code Committee:
Panel 19—Barry Bauman, representative;
Truman C. Surbrook, alternate

EES-41 MILK HANDLING EQUIPMENT COMMITTEE

Joseph Tecza, chairman
Todd Williams, vice chairman
Daniel J. Aneshansley, secretary
David C. Ludington, past chairman

Terms expiring June 1989—
Daniel J. Aneshansley, James A. Carrano, Richard W. Guest, Duncan M. Thompson, Todd Williams

Terms expiring June 1990—
Gerald R. Bodman, David C. Ludington, Sybren Y. Reitsma

Terms expiring 1991—
Mofazzal H. Chowdhury, Paul R. George, Scott A. Sanford, T. Wyatt Smith, Joseph Tecza

EES-42 ELECTRICAL CONTROLS FOR FARMSTEAD EQUIPMENT COMMITTEE

Gary W. Buller, chairman
Thomas A. Bon, vice chairman
Ivan L. Winsett, secretary
Norman D. Reese, past chairman

Terms expiring June 1989—
Donavon L. Bakker, H. David Currence, Gary L. Downey, Zachary A. Henry, Marvin Nabben, Terry J. Siebennmorgan, Paul D. Thompson, Fred C. Vosper, Earl H. Williams, Richard L. Witz

Terms expiring June 1990—
John G. Greiner, Gary M. Hyde, Gerard Muegerl, Norman D. Reese, Ivan L. Winsett

Terms expiring June 1991—
Thomas A. Bon, Nelson L. Buck, Gary W. Buller, Mark C. Leen, David C. Ludington, Paul Proctor, Edgar H. Smith, Harold R. Wakefield

EES-42/1 Protection of Electronic Controls Subcommittee
Norman D. Reese, chairman
Gary W. Buller, H. David Currence, Harold R. Wakefield

EES-42/2 Electric Fence Controller Subcommittee
Harold R. Wakefield, chairman
James L. Baker, Jack Bishop, Greg Burton, Gerard Muegerl, Ed Sheldon, Wilhelm Weinreich, Kirk Wolfgram

EES-44 ELECTRIC UTILIZATION RESEARCH COMMITTEE

Richard A. Peterson, chairman
Gary W. Buller, vice chairman
Gregory L. Stark, secretary
Kenneth L. McFate, past chairman

Terms expiring June 1989—
William T. Cox, Hugh J. Hansen, Zachary A. Henry, Robert J. Gustafson, Richard A. Peterson

Terms expiring June 1990—
Robert H. Brown, Lowell J. Endahl, Olin W. Ginn, Sherwood D. Skinner

Terms expiring June 1991—
Barry Bauman, Micheal D. Clement, Joe A. Gregory, Truman C. Surbrook

Terms expiring June 1992—
Ray L. Bollaert, Gary W. Buller, Kenneth L. McFate, Cecil B. Medders, Paul Proctor, Gregory L. Stark

Terms expiring June 1993—
Douglas H. Aeilts, Richard A. Hiatt, Richard A. Spray

EES-48 RADIATION COMMITTEE

Terry J. Siebenmorgan, chairman
Sundaram Gunasekaran, vice chairman
Joe M. Bunn, past chairman

Terms expiring June 1989—
L. Allen Butler, Jerome J. Gaffney, James G. Hartsock, Zachary A. Henry, Michael Krones, William F. McClure, Christos C. Mpelkas, Steven G. Schmitt, Terry J. Siebenmorgan, John S. Smith, Jr., James M. Stanley, Raymond A. Stermer, J. C. Webb, Jerry W. White

Terms expiring June 1990—
Albert Garcia, III, Sundaram Gunasekaran, David R. Massie, Marvin R. Paulsen, John C. Sager, Lalit R. Verma, Donald F. Wanjura

Terms expiring 1991—
Gerald S. Birth, Joe M. Bunn, John L. Goodenough, Glenn A. Kranzler, Stuart O. Nelson

EES-53 INSTRUMENTATION AND CONTROLS COMMITTEE

David G. Sokol, chairman (89)
Michael J. Delwiche, vice chairman (89)
Ralph P. Cavalieri, secretary (90)
Carl E. Bohman, Instrument News editor (89)
Marvin L. Stone, past chairman

Terms expiring June 1989—
Carl E. Bohman, Richard K. Byler, Glenn A. Kranzler, William F. McClure, Marvin J. Pitts, David G. Sokol, Marvin L. Stone

Terms expiring June 1990—
Ralph P. Cavalieri, James M. Ebeling, Zachary A. Henry, Gary Kah, Bruce L. Upchurch, John Wilkinson, Gerald C. Zoerb

Terms expiring June 1991—
Michael J. Delwiche, Roy C. Harrell, Gordon R. Hjertaas, Derrell B. McLendon, Stephen W. Searcy, James L. Steele, Luther R. Wilhelm

Terms expiring June 1992—
Henry A. Affeldt, Jr., Young J. Han, Digvar S. Jayas, Charles T. Morrow

Advisors:
Theodore E. Divine, Gains E. Miles, Bailey W. Mitchell, Ronald T. Schuler, Raymond A. Stermer

EES-54 COMPUTERS COMMITTEE

Luther R. Wilhelm, chairman
Marvin L. Stone, vice chairman
James M. McKinion, secretary
Shahab Sokhansanj, computer news editor
Kenneth A. Sudduth, past chairman

Terms expiring June 1989—
Max P. Gassman, George E. Meyer, Stephen W. Searcy, Luther R. Wilhelm

Terms expiring June 1990—
Carl E. Bohman, Glyn R. Boone, Richard K. Byler, James M. Ebeling, Dale E. Marshall, J. Wayne Mishoe, Bailey W. Mitchell, Marvin J. Pitts, David J. Wolak

Terms expiring June 1991—
Roy C. Harrell, Ward R. Simonton, Shahab Sokhansanj, Dennis G. Watson

Terms expiring June 1992—
Gary W. Buller, Richard C. Coddington, Young J. Han, Jerry R. Lambert, James M. McKinion, Charles T. Morrow, Marvin L. Stone, Kenneth A. Sudduth

JOINT COMMITTEES

(These committees are administered by other technical divisions, but incorporate technical interests of the Electrical and Electronic Systems Division.)

Rural Water Supplies Committee, SW-214
ASHRAE Liaison Committee, SE-300
Environmental Physiology Committee, SE-301
Environment of Animal Structures Committee, SE-302
Environment of Plant Structures Committee, SE-303
Environment of Stored Products Committee, SE-304
Solar Energy Committee, SE-308

EMERGING TECHNOLOGIES DIVISION

David R. Thompson, Director (89)

ET-01 EXECUTIVE COMMITTEE

Robert M. Peart, chairman
John A. Sturos, vice chairman
Robert B. Fridley, secretary
David R. Thompson, director (89)

ET-02 STEERING COMMITTEE

Robert M. Peart, chairman
John A. Sturos, vice chairman
Robert B. Fridley, secretary
David R. Thompson, director (89)
Gaines E. Miles, chairman ET-04
To be selected, chairman ET-07

Chairmen and elected Representatives of each Emerging Area as follows:
John R. Barrett, John A. Collier, James H. Dooley, Awatif E. Hassan, James McKinnon, Penn A. Peters, Raul Piedrahita, Norman R. Scott

ET-02/2 NOMINATING COMMITTEE
Robert M. Peart, chairman
James H. Dooley, Awatif E. Hassan, James M. McKinion, Raul Piedrahita

ET-04 PUBLICATIONS REVIEW AND PAPER AWARD COMMITTEE

Gaines E. Miles, chairman & division editor

Associate Editors — R. Kenneth Matthes, Gaines E. Miles, Jaw-Kai Wang, Roy E. Young

ET-06 PROGRAM COMMITTEE

John A. Sturos, chairman
Robert H. Brock, James H. Dooley, Thomas M. Losordo, Thomas L. Thompson

ET-07 DEVELOPMENT AND INTERSOCIETY COMMITTEE

To be selected, chairman
To be selected, vice chairman
To be selected, Secretary

Secretaries and Development Officers of Each Emerging Area as follows:
David E. Brune, Thomas Cathcart, James L. Fridley, Gerald W. Isaacs, Arthur T. Johnson, A. Dale Whittaker, Harold T. Wiedemann, Roy E. Young

AQ-20—AQUACULTURAL ENGINEERING AREA

AQ-01 AQUACULTURAL ENGINEERING EXECUTIVE COMMITTEE

John A. Collier, chairman
Thomas M. Losordo, vice chairman
Thmas Cathcart, secretary
Raul Piedrahita, ET-02 rep
David E. Brune, development officer

AQ-02 AQUACULTURAL ENGINEERING ADVISORY COMMITTEE (formerly T-7)

John A. Collier, chairman
Thomas M. Losordo, vice chairman
Thomas Cathcart, secretary
Raul Piedrahita, ET-02 rep
David E. Brune, development officer
Jaw-Kai Wang, associate editor

Terms expiring June 1989—
John A. Collier, Robert A. Fridley, Thomas M. Losordo, Carol S. Stwalley, Michael R. Williamson, John W. Zahradnik

Terms expiring June 1990—
Larry O. Bagnell, John P. Bolte, David E. Brune, Thomas Cathcart, Gary L. Rogers, Jaw-Kai Wang, Frederick W. Wheaton

Terms expiring June 1991—
Hjalmar D. Bruhn, Marshall J. English, Roger E. Garrett, Barney K. Huang, Thomas B. Lawson III, Raul Piedrahita, Robert S. Pile

BI-30—BIOENGINEERING AREA

BI-01 BIOENGINEERING EXECUTIVE COMMITTEE

Norman R. Scott, chairman
James H. Dooley, vice chairman
Arthur T. Johnson, secretary
James H. Dooley, ET-02 rep
Roy E. Young, development officer

BI-02 BIOENGINEERING ADVISORY COMMITTEE (formerly T-4)

Norman R. Scott, chairman
James H. Dooley, vice chairman
Arthur T. Johnson, secretary
James H. Dooley, ET-02 rep
Roy E. Young, development officer
Roy E. Young, associate editor

Terms expiring June 1989—
Michael J. Delwiche, Eva A. Lissick, George E. Meyer, Arne Mollner, Ronald E. Pitt, Larry W. Turner, Dennis P. Stombaugh, Garrett L. Van Wicklen

Terms expiring June 1990—
J. Robert Cooke, James H. Dooley, William R. Fox, Roger E. Garrett, Conly L. Hansen, Jean B. Hunter, J. Wayne Mishoe, Norman R. Scott, Roy E. Young

Terms expiring June 1991—
Frank Abrams, David P. Chynoweth, R. Bruce Curry, Arthur T. Johnson, George E. Merva, Jonathan W. Pote, Ward Simonton

FE-40—FOREST ENGINEERING AREA

FE-01 FOREST ENGINEERING EXECUTIVE COMMITTEE

Penn A. Peters, chairman
Robert H. Brock, vice chairman
Harold T. Wiedeman, secretary
Awatif E. Hassan, ET-02 rep
James L. Fridley, development officer

FE-02 FOREST ENGINEERING ADVISORY COMMITTEE (formerly T-12 and PM-55)

Penn A. Peters, chairman
Robert H. Brock, vice chairman
Harold T. Wiedeman, secretary
Awatif E. Hassan, ET-02 rep
James L. Fridley, development officer
R. Kenneth Matthes, associate editor

Terms expiring June 1989—
Colin S. Ashmore, Cleveland J. Biller, Charles M. Crowell, Michael R. Duncan, Brian C. Horsfield, Jerry L. Kroger, Charles N. Lee, R. Kenneth Matthes, Richard J. McClimans, Harold T. Wiedemann

Terms expiring June 1990—
Farnum A. Burbank, James O. Burrows, Nels S. Christopherson, Samuel J. Coughran, James Earnest, Jerry L. Edwards, James L. Fridley, Harry G. Gibson, Bruce Hartsough, Dan W. McKenzie, Charles N. Mann, Al Menzel, Walter L. Moden, Jr., Norman G. Sears, Norman Smith, Bryce J. Stokes, James G. Storms, Michael A. Thompson, William P. Tully, Clyde G. Vidrine

Terms expiring June 1991—
William H. Aldred, Perry H. Ballek, Robert H. Brock, Clayton L. Enix, Awatif E. Hassan, John B. Holtman, Michael R. Lambert, John A. Miles, John V. Perumpral, Penn A. Peters, John A. Sturos
Non-member advisors:
David F. Gibson, W. Dale Greene, Ben D. Jackson, Leonard R. Johnson, John Mann, Thomas C. Meisel, Edwin S. Miyata, Roberto Munila, Thomas W. Reisinger, Don L. Sirois, John Walters, William F. Watson

ANSI Committee Z249: Safety Requirements for Forest Management and Silvicultural Operations—
Samuel J. Coughran, representative (Also reports to PM-03)

Society of Automotive Engineers, Subcommittee XXVIII, Forestry and Logging Equipment—Don L. Sirois, liaison representative

U.S. Technical Advisory Committee for ISO/TC 23/SC 15, Machinery for Forestry—Don L. Sirois, liaison representative

KS-50—KNOWLEDGE SYSTEMS AREA

KS-01 KNOWLEDGE SYSTEMS EXECUTIVE COMMITTEE

John R. Barrett, chairman
Thomas L. Thompson, vice chairman
A. Dale Whittaker, secretary
James M. McKinion, ET-02 rep
Gerald W. Isaacs, development officer

KS-02 KNOWLEDGE SYSTEMS ADVISORY COMMITTEE (formerly T-6)

John R. Barrett, chairman
Thomas L. Thompson, vice chairman
A. Dale Whittaker, secretary
James M. McKinion, ET-02 rep
Gerald W. Isaacs, development officer
Gaines E. Miles, associate editor
Jerry R. Lambert, program co-chairman
James W. Jones, program co-chairman

Area Committee Chairmen:
R. Bruce Curry, W. David Shoup, Fred E. Sistler, Robert S. Sowell, Ronald H. Thieme

KS-11 AGRICULTURAL OPERATIONS MANAGEMENT

W. David Shoup, chairman
Tony V. Harrison, vice chairman
Erdal H. Ozkan, secretary
Brian K. Cardin, James C. Frisby, Douglas J. Glunz, W. Fred Odom, Stan E. Prussia

KS-12 ROBOTICS, MACHINE VISION, IMAGE PROCESSING AND PATTERN RECOGNITION

Fred E. Sistler, chairman
Lawrence J. Kutz, vice chairman
Roy C. Harrell, Gaines E. Miles, Roy E. Young

Non-member Advisor—
Ken Jordan

KS-13 SYSTEMS OPTIMIZATION

Robert S. Sowell, chairman
A. Wayne Anderson, vice chairman/secretary
Donald A. Bender, Heber D. Bouland, Lung-Hua Chen, Maurice R. Gebhardt, J. Ben Holtman, James W. Jones, Dennis L. Larson, Ronald W. McClendon, Jaw-Kai Wang, Robert C. Ward

KS-14 KNOWLEDGE ENGINEERING

Ronald H. Thieme, chairman
A. Dale Whittaker, vice chairman/secretary
John R. Barrett, John P. Bolte, Bernie A. Engel, Peter R. Flynn, Conrad D. Heatwole, Don D. Jones, Pierce H. Jones, D. Earl Kline, Jerry R. Lambert, James M. McKinion, Robert M. Peart, Kenneth H. Solomon, Robert S. Sowell, Ngoc Chi Thai, Kuan-Cheng Ting, Braham P. Verma

Non-member advisor:
Karen Wisiol

KS-15 SIMULATION

R. Bruce Curry, chairman
Ronald W. McClendon, vice chairman/secretary
Carl E. Anderson, Donald A. Bender, Thomas C. Bridges, James W. Jones, Jerry R. Lambrt, J. Wayne Mishoe, George E. Meyer, Robert E. Muller, John F. Reid, C. Alan Rotz, James H. Young

Non-member advisor:
David Elwell

FOOD AND PROCESS ENGINEERING INSTITUTE

Arthur A. Teixeira, Director (90)

FPE-01 EXECUTIVE COMMITTEE

Gerald H. Brusewitz, chairman
Martin R. Okos, vice chairman and program chairman
Fred A. Payne, secretary and technical chairman
John R. Rosenau, past chairman
Milford A. Hanna, standards chairman
Manjeet S. Chinnan, publications chairman
Kenneth C. Diehl, Jr., education chairman
Arthur A. Teixeira, director (90)

FPE-02 STEERING COMMITTEE

Gerald H. Brusewitz, chairman
Martin R. Okos, vice chairman
Fred A. Payne, secretary
John R. Rosenau, past chairman
Arthur A. Teixeira, director (90)

Chairs of all Committees and Groups as Follows:
Dennis E. Bilton, David P. Bresnanhan, L. Allen Butler, Manjeet S. Chinnan, Charles J. Clanton, Jerome J. Gaffney, Albert Garcia III, Milford A. Hanna, Conly L. Hansen, Dennis R. Heldman, Michael J. Lichtensteiger, Fred A. Payne, S. Paul Singh, Richard L. Stroshine

FPE-02/2 Nominating Subcommittee

John R. Rosenau, chairman
Dwayne A. Suter, Vincent E. Sweat, Arthur A. Teixeira, David R. Thompson

FPE-03 STANDARDS GROUP

Milford A. Hanna, chairman
Marvin R. Paulsen, vice chairman
Evelyn E. Rosentreter, staff secretary
Jay S. Marks, past chairman

Terms expiring June 1989—
Brad N. Borgman, Manjeet S. Chinnan, William A. Dean, Jr., Eugene W. Ford, Marvin O. Marmorine, Ronnie G. Morgan, Michael O'Brien, David R. Thompson

Terms expiring June 1990—
Ronald K. Allen, Clayton F. Brasington, Pamela K. Hardt-English, Jeffrey M. Jenniges, Charles T. Morrow, Rakesh K. Singh

Terms expiring June 1991—
Maynard E. Anderson, Thomas V. DeBrock, Milford A. Hanna, Jay S. Marks, Marvin R. Paulsen, Kenneth R. Swartzel

ASAE Committee on Standards (T-1)—
Jay S. Marks, representative; Rakesh K. Singh, alternate

ASAE Agricultural Safety Committee (T-2)—
Representative to be appointed

ASAE Cooperative Standards Program Committee (A-311)—
Jeffrey M. Jenniges, representative

Cooperative Standards Activities with ASAE Representatives Reporting to FE-03

American Society of Mechanical Engineers, F2.1: Food, Drug and Beverage Equipment Committee—
Marvin O. Marmorine

FPE-04 PUBLICATIONS GROUP

Manjeet S. Chinnan, chairman and rep. to A-501
Luther R. Wilhelm, vice chairman
James F. Steffe, past chairman

FPE-041 REFEREED PUBLICATIONS COMMITTEE

Manjeet S. Chinnan, chairman
Luther R. Wilhelm, vice chairman
Dwayne A. Suter, editor

Associate Editors—
Fred W. Bakker-Arkema, Kenneth C. Diehl, Jr., Clifford A. Flood, Milford A. Hanna, Marvin O. Marmorine, Gerald E. Rehkugler, Kenneth R. Swartzel, Vincent E. Sweat, Luther R. Wilhelm

Terms expiring June 1989—
Gerald H. Brusewitz, Manjeet S. Chinnan, Stanley E. Prussia, Dwayne A. Suter

Terms expiring June 1990—
Robert Y. Ofoli, R. Paul Singh, Vincent E. Sweat, Luther R. Wilhelm, Carlos A. Zuritz

Terms expiring June 1991—
Khe V. Chau, Yen-Con Hung, Digvir S. Jayas, Arthur A. Teixeira, John H. Wells

FPE-042 FPEI NEWS COMMITTEE

Khe V. Chau, chairman
Wayne A. Maley (ex officio), staff

Terms expiring June 1989—
Manjeet S. Chinnan, Abdel Ghaly, Randy A. Hartwig, C. Gene Haugh, Kenneth J. Hellevang, William M. Miller, Jeffrey C. Nash, Yahya I. Sharaf-Eldeen

Terms expiring June 1990—
Leslie F. Backer, Richard H. Barton, Lino R. Correia, Digvir S. Jayas, Terry M. Midden, Stanley E. Prussia, Ralph E. Smith, Arthur A. Teixeira, B. A. Twigg

Terms expiring June 1991—
Terry Burford, Khe V. Chau, Mark Casada, G. A. Dodge, J. Bruce Litchfield, R. Paul Singh, Jeff Soldner, Brahm P. Verma, Pie Yi Wang

FPE-043 PAPER AWARDS AND SPECIAL PUBLICATONS COMMITTEE

Yen-Cen Hung, chairman

FPE-06 PROGRAM GROUP

Martin R. Okos, chairman

FPE-061 GENERAL PROGRAM COMMITTEE

Martin R. Okos, chairman
Sundaram Gunasekaran, FPE-701
Terry J. Sokhansanj, FPE-702
Digvir S. Jayas, FPE-703
John S. Cundiff, FPE-704
Robert A. Romero, FPE-705
Michael J. Delwiche, FPE-706
Charles C. Ross, FPE-707
Larry R. Johnson, FPE-708
Richard P. Egg, FPE-709
Khe V. Chau, FPE-710
Michael Richmond, FPE-711

FPE-062 SPEAKER EVALUATIONS COMMITTEE

Yen-Con Hung, chairman

FPE-063 SPECIAL PROGRAMS COMMITTEE

John M. Hyde, chairman

FPE-70 TECHNICAL GROUP

Fred A. Payne, chairman
Conly L. Hansen, past chairman
L. Allen Butler, FPE-701
Richard L. Stroshine, FPE-702
David P. Bresnahan, FPE-703
Fred A. Payne, FPE-704
Dennis E. Bilton, FPE-705
Bruce L. Upchurch, FPE-706
Charles J. Clanton, FPE-707
L. Allen Butler, FPE-708
Albert Garcia III, FPE-709
Jerome J. Gaffney, FPE-710
S. Paul Singh, FPE-711

FPE-701 PHYSICAL PROPERTIES OF AGRICULTURAL PRODUCTS COMMITTEE

(Joint with Power and Machinery Div.)

L. Allen Butler, chairman
Sundaram Gunasekaran, vice chairman
Charles R. Hurburgh, secretary
Larry D. Swetnam, past chairman

Terms expiring June 1989—
Gerald H. Brusewitz, Sundaram Gunasekaran, Robert J. Gustafson, Charles R. Hurburgh, Virendra M. Puri, Marvin R. Paulsen, Ajit K. Srivastava, Larry D. Swetnam

Terms expiring June 1990—
L. Allen Butler, Ashim K. Datta, Kamyar Haghigi, Digvir S. Jayas, Shahab Sokhansanj, Richard L. Stroshine, Vincent E. Sweat

Terms expiring June 1991—
Paul Chen, Michael J. Delwiche, Clifford B. Fedler, Palaniappa Krishnan, J. Bruce Litchfield, G.S.V. Raghavan, Bruce L. Upchurch

FPE-702 GRAIN AND FEED PROCESSING AND STORAGE COMMITTEE

(Joint with Structures and Environment Div.)

Richard L. Stroshine, chairman
Shahab Sokhansanj, vice chairman and secretary
Terry J. Siebenmorgan, program chairman (87-89)
Roger C. Brook, past chairman

Leslie F. Backer, Carl J. Bern, Thomas C. Bridges, Donald B. Brooker, L. Allen Butler, Cheng S. Chang, Khe V. Chau, Do Sup Chung, Clifford B. Fedler, Clifford A. Flood, Jr., George H. Foster, Robert J. Gustafson, Glenn E. Hall, Ekramul Haque, Joseph P. Harner, Floyd L. Herum, Mark H. Huss, Digvir S. Jayas, Darrell E. Lischynski, Manjit K. Misra, Bruce W. Moechnig, R. Vance Morey, David W. Morrison, Vu Thai Nguyen, Ronald T. Noyes, Kulbir S. Pannu, L. Jeffrey Patton, Marvin R. Paulsen, Richard O. Pierce, Terry J. Siebenmorgan, Charles E. Sukup, Thomas L. Thompson, Lalit R. Verma, Gerald M. White, William F. Wilcke, David L. Williams, Michael E. Wolff, James H. Young

FPE-703 FOOD PROCESSING COMMITTEE

David P. Bresnahan, chairman
Digvir S. Jayas, vice chairman
Pamela K. Hardt-English, past chairman

Terms expiring June 1989—
R. C. Anantheswaran, David P. Bresnahan, Ralph P. Cavalieri, Dilip Changara, Khe V. Chau, Manjeet S. Chinnan, David P. Gorby, Steven G. Grall, Pamela K. Hardt-English, Ronnie G. Morgan, Robert Y. Ofoli, Fred A. Payne, John H. Wells

Terms expiring June 1990—
Ashim K. Datta, Sundaram Gunasekaran, Yen-Con Hung, Digvir S. Jayas, Jeffrey M. Jenniges, John W. Larkin, J. Bruce Litchfield, Alfredo C. Rodriguez, Rakesh K. Singh, Arthur A. Teixeira, Carlos A. Zuritz

Terms expiring June 1991—
Maynard E. Anderson, Jay S. Marks, Jeff Nash, Pie Yi Wang

FPE-704 SPECIAL CROPS PROCESSING COMMITTEE

Fred A. Payne, chairman
John D. Cundiff, vice chairman
Mark E. Casada, secretary
Stanley E. Prussia, past chairman

Terms expiring June 1989—
John G. Alphin, Arnold G. Berlage, Mark E. Casada, Lalit R. Verma, Linus R. Walton

Terms expiring June 1990—
Zachary A. Henry, Ivan W. Kirk, Palaniappa Krishnan, Fred A. Payne, Thomas R. Rumsey

Terms expiring June 1991—
William S. Anthony, Fred W. Bakker-Arkema, Gerald D. Christenbury, Stanley E. Prussia, Larry D. Swetnam, William F. Wilcke

Terms expiring 1992—
John S. Cundiff, Barry C. Frey, Gino J. Mangialardi, Jr., Jonathan C. Popp, James H. Young

FPE-705 TRANSPORTATION, HANDLING, AND WAREHOUSING OF AGRICULTURAL PRODUCTS COMMITTEE

Dennis E. Bilton, chairman
Robert A. Romero, vice chairman
Michael T. Talbot, secretary
Khe V. Chau, past chairman

Terms expiring in June 1989—
Maynard E. Anderson, Dennis E. Bilton, Darrel E. Campbell, Khe V. Chau, William M. Miller, Paul H. Orr, Robert A. Romero, Raymond A. Stermer, Fredrick W. Wheaton

Terms expiring June 1990—
Richard A. Cavaletto, Manjeet S. Chinnan, Jerome J. Gaffney, Suhas R. Ghate, Temple G. Grandin, Sundaram Gunasekaran, Martin L. Hellickson, Chi Ngoc Thai

Terms expiring June 1991—
B. Hunt Ashby, Paul Chen, Stanley E. Prussia, Steven A. Sargent, Michael T. Talbot

Non-member advisory:
S. J. Parker

American Society of Mechanical Engineers,
MH-1: Standardization of Pallets—B. Hunt Ashby, representative; David W. Cayton, alternate

International Materials Management Society,
MH-10: Packaging Dimensions—Representative to be appointed

FPE-706 FRUIT AND VEGETABLE PACKINGHOUSE OPERATIONS COMMITTEE

Bruce L. Upchurch, chairman
Michael J. Delwiche, vice chairman
William M. Miller, secretary
Suhas R. Ghate, past chairman

Terms expiring June 1989—
Dennis E. Bilton, Ralph P. Cavalieri, Manjeet S. Chinnan, Michael J. Delwiche, Stephen R. Delwiche, Jerome J. Gaffney, Daniel E. Hammett, Michael O'Brien, Robert A. Romero, Bruce L. Upchurch

Terms expiring June 1990—
Galen K. Brown, Richard A. Cavaletto, Suhas R. Ghate, William M. Miller, Steven A. Sargent, James F. Thompson

Terms expiring June 1991—
Brad N. Borgman, Khe V. Chau, Stanley E. Prussia, Roger P. Rohrbach, Thomas R. Rumsey, David C. Slaughter

FPE-707 FOOD PROCESSING WASTE MANAGEMENT AND UTILIZATION COMMITTEE

Charles J. Clanton, chairman
Charles C. Ross, vice chairman
Edward Valentine, secretary
Frank J. Humenik, past chairman

Terms expiring June 1989—
James C. Converse, Conly L. Hansen, Louis A. Licht, Charles C. Ross, Rakesh K. Singh, John M. Sweeten, Jr.

Terms expiring June 1990—
Charles J. Clanton, David R. Nelson, Abolghasen A. Shahbazi, James L. Walsh, Robert R. Zall

Terms expiring June 1991—
James C. Barker, Abdel Ghaly, Philip R. Goodrich, Frank J. Humenik, Jeffrey Nash, William F. Ritter, Michael Thomas

FPE-708 FARM MATERIALS HANDLING COMMITTEE
(Joint with Power and Machinery, and Structures and Environment Divs.)

L. Allen Butler, chairman
Larry R. Johnson, vice chairman
Joseph P. Harner, secretary
Dwight A. Benninga, past chairman

Terms expiring June 1989—
Larry R. Johnson, Bruce A. McKenzie, Manjit K. Misra, Ronald T. Noyes, Kent L. Rysdon, Franklin J. Wade, John W. Worley

Terms expiring June 1990—
L. Allen Butler, Jack W. Crane, Joseph P. Harner, Kenneth J. Hellevang

Terms expiring June 1991—
Dwight A. Benninga, Lawrence H. Ellebracht, Naftali Galili, Gino J. Mangialardi, Calvin B. Parnell, Jr., Lawrence L. Wickstrom, David L. Williams

FPE-709 ENERGY AND ALTERNATE PRODUCTS FROM BIOMASS COMMITTEE
(Joint with Power and Machinery Div.)

Albert Garcia III, chairman
Richard P. Egg, vice chairman and program chairman
Samuel G. McNeill, secretary
Bryan M. Jenkins, past chairman

Terms expiring June 1989—
Robert S. Brashear, Barney R. Eiland, James M. Ebeling, Albert Garcia III, Phillip R. Goodrich, John W. Goodrum, Milford A. Hanna, David T. Hill, Bryan M. Jenkins, Wayne A. LePori, Samuel G. McNeill, Fred A. Payne, Charles L. Peterson, Daniel B. Waddle, Larry P. Walker, Douglas W. Williams

Terms expiring June 1990—
Marvin O. Bagby, Don A. Bender, Robert G. Curley, Cady R. Engler, William C. Fairbank, Clifford B. Fedler, Abdel Ghaly, Harry G. Gibson, Andrew G. Hashimoto, Harold M. Keener, John M. Krochta, Frederick W. Landers, Dennis L. Larson, Bruce W. Moechnig, Rollin D. Schnieder, Dennis D. Schulte, Abolghasem A. Shabazi

Terms expiring June 1991—
Phillip C. Badger, Ramadan Ben-Hassan, Jacqueline D. Broder, William L. Bryan, Charlie G. Coble, Jack W. Crane, Harry W. Downs, Richard P. Egg, Gene L. Iannotti, Charles M. Jones, James Linden, Charles E. Sukup, David H. Vaughan, David P. Chynoweth

FPE-709/1 Biological Conversion Subcommittee
David P. Chynoweth, chairman
Philip C. Badger, William L. Bryan, Harry W. Downs, Abdel Ghaly, Cady R. Engler, Clifford B. Fedler, James R. Fischer, Abdel Ghaly, Andrew G. Hashimoto, Ted Landers, Bruce W. Moechnig, Sangha Oh, Bradley K. Rein, Rollin W. Schnieder, Dennis D. Schulte

FPE-709/2 Thermochemical Biomass Conversion Subcommittee
Samuel G. McNeill, chairman
John R. Barrett, Jr., Avtar Binning, Abdel Ghaly, Harold M. Keener, Gerald L. Kline, Tom R. Reed, Clarence B. Richey, David H. Vaughan

FPE-709/4 Production, Harvest, Handling and Storage Subcommittee
John S. Cundiff, chairman
Charlie G. Coble, John S. Cundiff, Richard P. Egg, Barney R. Eiland, Harold R. Sumner

FPE-710 HEAT AND MASS TRANSFER COMMITTEE

Jerome J. Gaffney, chairman
Khe V. Chau, vice chairman
Ashim K. Datta, secretary

Terms expiring June 1989—
Mrinal Bhattacharya, Yen C. Hung, Harold M. Keener, Martin R. Okos, Robert Y. Ofoli, Benjamin Medine, Alfredo C. Rodriguez, R. Paul Singh, Shahab Sokhansanj, Arthur A. Teixeira

Terms expiring June 1990—
Khe V. Chau, Ashim K. Datta, Clifford A. Flood, Jerome J. Gaffney, Sundaran Gunasekaran, James L. Julson, John W. Larkin, Bruce Litchfield, Vincent E. Sweat, Linus R. Walton

Terms expiring June 1991—
C. Direlle Baird, Dilip Chandarana, John D. Floros, Kamyar Haghighi, M. Anandha Rao, Sudhir K. Sastry, Rakesh K. Singh, Marvin L. Stone, Lalit R. Verma, Carlos A. Zuritz

FPE-711 PACKAGING COMMITTEE

S. Paul Singh, chairman
Michael Richmond, vice chairman
Ed Church, secretary
Khe V. Chau, Manjeet S. Chinnan, Wayne LePori, Chester J. Mackson, M. Ananda Rao, R. Paul Singh, James F. Thompson, Dennis Young

FPE-80 EDUCATION AND PRIORITIES GROUP

Michael J. Lichtensteiger, chairman
Robert Y. Ofoli, vice chairman

FPE-811 EDUCATION COMMITTEE

Michael J. Lichtensteiger, chairman
Robert Y. Ofoli, vice chairman
Bruce J. Litchfiilip C. Badger, William L. Bryan, Harry W. Downs, Abdel Ghaly, Cady R. Engler, Clifford B. Fedler, James R. Fischer, Abdel Ghaly, Andrew G. Hashimoto, Ted Landers, Bruce W. Moechnig, Sangha Oh, Bradley K. Rein, Rollin W. Schnieder, Dennis D. Schulte

FPE-812 PRIORITIES: RESEARCH AND EXTENSION COMMITTEE

Dennis R. Heldman, chairman
R. Paul Singh, Vincent E. Sweat, David R. Thompson

JOINT COMMITTEES

(These committees are administered by other technical divisions, but incorporate technical interests of the Food and Process Engineering Institute.)

Environment of Stored Products Committee, SE-304

POWER AND MACHINERY DIVISION

Jack W. Crane, Director (89)

PM-01 EXECUTIVE COMMITTEE

James H. Ruff, chairman
Cletus E. Schertz, vice chairman
Roy E. Young, past chairman
Jack W. Crane, director (89)

PM-02 STEERING COMMITTEE

James H. Ruff, chairman
Cletus E. Schertz, vice chairman
Roy E. Young, past chairman
Jack W. Crane, director (89)

Plus all elected officers of PM-03, PM-04 and PM-07 and all voting members of PM-06 including all chairs of Technical Committees as follows:
Michael F. Broder (41), H. Erdal Ozkan (42), Donald A. Bender (43), C. Alan Rotz (44), Lyle E. Stephens (45), James W. Ramsey (46), Leonard L. Bashford (47), Lawrence N. Shaw (48), Roy V. Baker, Jr. (50), J. K. Schueller (51), David L. Apple (52), John K. Schueller (53), Gary L. Barker (54), Gordon R. Hjertaas (58), George A. Duncan (59), Charles R. Frank (60), Darrel L. Roberts (61), Walter N. Nawrocki (62)

Division Editor—Donald C. Erbach

Other Committees or Activities Reporting to Steering Committee

CIGR, International Commission of Agricultural Engineering, Sect. III: Agricultural Machinery—Lawrence H. Skromme, technical correspondent

ASAE Committee on Standards (T-1)—Earl A. Hudson (91), representative; Rollin D. Schnieder (89), alternate

ASAE Environmental Quality Coordinating Committee (T-9)—Eugene P. Columbus (89), Alan D. Brashears (90), Herschel Elliott (91), liaison representatives

ASAE Metric Coordinating Committee (T-10)—Alvin C. Bailey (90), liaison representative

ASAE Paper Awards Committee (A-116)—Malcolm E. Wright, (89), liaison representative

ASAE Research Committee (A-209)—Robert B. Fridley (89), Marshall F. Finner (89), Paul K. Turnquist (90), Jack W. Craig (90), liaison representatives

ASAE Continuing Education Committee (A-416)—Max P. Gassman (88), L. S. "Pete" Simonson (90), liaison representatives

ASAE Refereed Publications Committee (A-511)—David W. Smith (91), liaison representative

ASAE Monographs Committee (A-512)—Larry F. Stikeleather (91), Sverker Persson (89), liaison representatives

ASAE Textbook Committee (A-514)—Roger P. Rohrbach (91), liaison representative

ASAE Special Publications Committee (A-515)—J. Lyle Shaver, liaison representative

ASAE CIGR Liaison Subcommittee (A-611/2)—Robert H. Wilkinson, liaison representative

PM-03 STANDARDS COMMITTEE

Earl A. Hudson, chairman
Thomas D. Ogle, vice chairman
Russell H. Hahn, staff secretary
Cletus E. Schertz, past chairman

Terms expiring June 1989—
Thomas F. Crusinberry, Hugh F. Grow, Thomas D. Ogle, Jack P. Palmer, Clark E. Renner, Cletus E. Schertz, Arthur S. Tobiassen

Terms expiring June 1990—
James A. Koch, J. Edson McCanse, Rollin D. Schnieder, Walter L. Scott, Barrie L. Smith, Edmund J. Zeglen

Terms expiring June 1991—
John P. Beal, Thomas G. Carpenter, Robert D. Reed, Robert B. Skromme, Earl H. Williams

Terms expiring June 1992—
David J. Gustafson, Wallace McDougall, Richard G. Moe, Mark H. Sickman

Terms expiring June 1993—
B. Jack Butler, Thomas S. Colvin, Earl A. Hudson, Arnold B. Skromme, Richard J. Straub, Paul E. Young, Arnold Zimmerman

ASAE Committee on Standards (T-1)—Earl A. Hudson, representative; Rollin D. Schnieder, alternate

ASAE Agricultural Safety Committee (T-2)—Thomas G. Carpenter, liaison representative

Cooperative Standards Activities with ASAE Representatives Reporting to PM-03

American National Standards Institute Committees:
- B6: Standardization of Gears—Carroll K. Reece
- B71: Safety Standards for Lawn Mowers, Snow Throwers, Power Edgers and Trimmers, Garden Tractors and Related Equipment and Attachments—John J. Slazas (PM-52)
  - B71.4: Safety Specifications for Commercial Turf Care Equipment—John J. Slazas (PM-52)
- B92: Splines and Splined Shafts—Carroll K. Reece
- B93: Fluid Power Systems and Components—David L. Newcom (PM-51)
- K61: Storage and Handling of Anhydrous Ammonia—Mansel M. Mayeux
- Z249: Safety Requirements for Forest Management and Silvicultural Operations—Samuel J. Coughran (PM-55)
- U.S. Technical Advisory Groups for International Standardization:
  - ISO/TC 23, Tractors and Machinery for Agriculture and Forestry—Russell H. Hahn
    - ISO/TC 23/SC 5, Equipment for Working the Soil—Roy G. Brandt (PM-03/12)
    - ISO/TC 23/SC 7, Equipment for Harvesting and Conservation—Barrie L. Smith (PM-03/13)
    - ISO/TC 23/SC 9, Equipment for Sowing, Planting and Distributing Fertilizers—John W. Hummel (PM-03/14)
    - ISO/TC 23/SC 13, Powered Lawn and Garden Equipment—Norman F. Weir (PM-52)
  - ISO/TC 31/SC 5, Tires, Rims and Valves—Thomas W. Freiburger (PM-03/8)
  - ISO/TC 131, Fluid Power Systems and Components—Ken K. Koch (PM-51)

American Society for Testing and Materials:
- D2: Petroleum Products and Lubricants
  - Tech B, Automotive Lubricants—Orville J. Nelson (PM-47)
- E35: Committee on Pesticides—B. Jack Butler (PM-41)

American Society of Mechanical Engineers:
- B15: Committee on Mechanical Power Transmission Apparatus—Wallace W. McDougall
- B18: Standardization of Bolts, Nuts, Rivets, Screws, and Similar Fasteners—E. R. Friesth
- B29: Chains, Attachments, and Sprockets for Power Transmission and Conveying—Gail G. Worsley

Society of Automotive Engineers:
- Agricultural Tractor Technical Committee—Robert D. Reed
- Agricultural Tractor Test Code Subcommittee—Robert D. Reed
- Agricultural Tractor Tire Subcommittee—Robert D. Reed

Farm and Industrial Equipment Institute, Farm Equipment Division Engineering Committee—J. Lyle Shaver

PM-03/1 Early Identification of Needs for Standards Subcommittee

Ralph A. Gerhardt, chairman

L. Dale Baker, Ralph E. Baumheckel, Frank E. Buckingham, Wilbur M. Davis, David P. Fritz, Harold W. George, Robert C. Lanphier, Cletus E. Schertz, Rollin D. Schnieder, Joe Shiver, Arthur S. Tobiassen, Kenneth L. Von Bargen, Franklin C. Walters

PM-03/3 Wagon Box, Forage Box, Manure Spreader and Farm Wagon Subcommittee

David R. Scheffler, chairman

Paul C. Gordon, Nicholas Hamm, Lauri Heikenen, Earl A. Hudson, Orvill J. Nelson, Robert B. Skromme

PM-03/4 Agricultural Loader Subcommittee

Kurt I. Cook, chairman

Thomas L. Bailey, John G. Christopher, Stanley R. Clark, David A. Horrman, J. Ross McMillan, Harold A. Ralston, David Schneider, Norman F. Weir, Keith A. Wheeler

PM-03/6 Auger Flight Subcommittee

Dathan R. Kerber, chairman

Dwight A. Benninga, James W. Bos, Garry W. Busboom, Richard L. Day, Wayne Hoefer, Earle C. Morton, T. William Waldrop

PM-03/8 Implement Tire Subcommittee

Richard P. Bernhardt, chairman

John J. Freer, Thomas W. Freiburger, Dathan R. Kerber, Earle C. Morton, Glenn E. Riffe, James H. Taylor, B. Bob Willis, Malcolm E. Wright

U.S. TAG for ISO/TC 31, Tires, Rims and Valves—Thomas W. Freiburger, representative (Also reports to PM-03)

PM-03/9 SMV Emblem Subcommittee

Robert A. Aherin, chairman
Walter L. Scott, vice chairman
Lee Forrester, Wayne E. Hartman, John Mone, James P. Murphy, Robert Nelson, Fred Ranck, William Stalker, Charles E. Trussell

PM-03/10 Terminology Subcommittee

Gordon P. Barrington, Max R. North, co-chairmen
Wallace McDougall, Robert B. Skromme

PM-03/11 Forward Planning for Standards Subcommittee

Arnold B. Skromme, chairman
Thomas G. Carpenter, David J. Gustafson, Jack P. Palmer, Cletus E. Schertz, Rollin D. Schnieder, Arnold Zimmerman

PM-03/12 U.S. TAG for ISO/TC 23/SC 5, Equipment for Working the Soil

Roy G. Brandt, chairman

Clarence E. Johnson, Howard L. Lewison, J. David Long, Robert L. Schafer, Robert J. Szucs, Julius R. Williford

PM-03/13 U.S. TAG for ISO/TC 23/SC 7, Equipment for Harvesting and Conservation

Barrie L. Smith, chairman
Larry W. Gutekunst, vice chairman
Orlin W. Johnson, secretary

Richard P. Bernhardt, Richard A. DePauw, David H. Diebold, Marshall F. Finner, Larry W. Gutekunst, Raymond P. Hoemsen, Orlin W. Johnson, John T. King, Gary L. Kunz, Ronald McAllister, Allan E. Neal, E. William Rowland-Hill, Robert A. Stelzer, Kenneth L. Von Bargen

PM-03/14 U.S. TAG for ISO/TC 23/SC 9, Equipment for Sowing, Planting and Distributing Fertilizers

John W. Hummel, chairman

David M. Anderson, Loren E. Bode, Louis F. Bouse, Maurice R. Gebhardt, Richard W. Hook, Clarence E. Johnson, David M. McLeod, Harry H. Takata, Jodie D. Whitney, Julius R. Williford

PM-04 PUBLICATIONS REVIEW COMMITTEE

Donald C. Erbach, chairman and division editor
Carroll E. Goering, vice chairman

Terms expiring June 1989—
Thomas G. Carpenter, Richard E. Muck, Charles L. Peterson, Donald L. Peterson, Charles W. Suggs

Terms expiring June 1990—
Norman B. Akesson, Ervin G. Humphries, David A. Pacey, Randall C. Reeder, William E. Rowland-Hill, Ernest W. Tollner, Lalit R. Verma, Joel T. Walker

Terms expiring June 1991—
W. Stanley Anthony, Alvin C. Bailey, Donald C. Erbach, R. Kenneth Matthes

PM-05 NOMINATING COMMITTEE

Roy E. Young, chairman

Terms expiring June 1989—
Donald C. Erbach, Earl A. Hudson

PM-06 TECHNICAL COORDINATING COMMITTEE

Charlie G. Coble, chairman

Terms expiring June 1989—
Charlie G. Coble

Terms expiring June 1990—
Joseph F. Gerling

Terms expiring June 1991—
Elbert C. Dickey

Plus Chairmen of all PM Technical Committees PM-41 through PM-62 as follows:

Michael F. Broder (41), H. Erdal Ozkan (42), Donald A. Bender (43), C. Alan Rotz (44), Lyle E. Stephens (45), James W. Rumsey (46), Leonard L. Bashford (47), Lawrence N. Shaw (48), Roy V. Baker, Jr. (50), John K. Schueller (51), David L. Apple (52), John K. Schueller (53), Gary L. Barker (54), Gordon R. Hjertaas (58), George A. Duncan (59), Charles R. Frank (60), Darrel L. Roberts (61), Walter T. Nawrocki (62)

PM-07 PROGRAM COMMITTEE

Lester J. Thompson, chairman
Maurice R. Gebhardt, vice chairman

Terms expiring June 1989—
Eddie C. Burt, Lester J. Thompson, Jodie D. Whitney, John E. Wood

Terms expiring June 1990—
Maurice R. Gebhardt, Joseph F. Gerling, Randall C. Reeder

Terms expiring June 1991—
Loren E. Bode, Edward J. Bortner, Thomas S. Colvin

Plus a Nonvoting Representative from PM-03 and from PM Technical Committees PM-41 through PM-62.

PM-41 AGRICULTURAL PEST CONTROL AND FERTILIZER APPLICATION COMMITTEE

Michael F. Broder, chairman
Steven C. Young, vice chairman and program chairman
P. Krishnan, secretary
John B. Solie, past chairman

Terms expiring June 1989—
Loren E. Bode, Henry D. Bowen, Michael F. Broder, James B. Carlton, H. Willard Downs, J. D. Fish, Robert D. Fox, Dennis K. Kuhlman, Masoud Salyani, John B. Solie, Richard W. Whitney, M. Herbert Willcutt, Wesley E. Yates, Steven C. Young

Terms expiring June 1990—
Norman B. Akesson, Donald B. Churchill, Joseph F. Gerling, John L. Goodenough, P. Krishnan, S. Edward Law, Alton O. Leedahl, Louis A. Liljedahl, H. Erdal Ozkan, Thomas F. Reed, R. Barry Rogers, Daniel E. Roush, Jodie D. Whitney, Edgar A. Wood

Terms expiring June 1991—
Louis F. Bouse, David L. Cochran, Durham K. Giles, Robert D. Grisso, William M. Lyle, David M. McLeod, Donald L. Reichard, Larry D. Sands, David B. Smith, Harold R. Sumner, Ray Treichler, Harold Wiedemann, Thomas H. Williams

Non-member advisor:
R. W. Tate

ASTM E35: Committee on Pesticides—B. Jack Butler, representative (Also reports to PM-03)

Entomological Society of America—David B. Smith, liaison representative

Weed Science Society of America—
Loren E. Bode, liaison representative

PM-41/1 Liquid Materials Application Subcommittee
Robert D. Fox, chairman

PM-41/2 Aviation Subcommittee
Dennis K. Kuhlman, chairman

PM-41/3 Dry Materials Application Subcommittee
J. D. Fish, chairman

PM-41/5 Chemigation Subcommittee
David M. McLeod, Chairman

PM-41/6 Pest Control Systems Modelling Subcommittee
John L. Goodenough, chairman

PM-42 CULTURAL PRACTICES EQUIPMENT COMMITTEE
H. Erdal Ozkan, chairman
Lorrin H. Schwartz, vice chairman
Glenn M. Olson, secretary
J. David Long, past chairman

Terms expiring June 1989—
J. David Long, Gordon E. Monroe, Roy W. Morling, Glenn M. Olson, H. Erdal Ozkan, Richard L. Parish, Lorrin H. Schwartz, Ernest W. Tollner, Dale E. Wilkins, J. Ray Williford

Terms expiring June 1990—
J. Gill Christopher, Harry C. Deckler, Donald C. Erbach, James K. Fornstrom, Dawson Hastings, John W. Hummel, John E. Morrison, Joseph F. Prem, David P. Shelton

Terms expiring June 1991—
Dan Augsburger, Maurice R. Gebhardt, Dean A. Knoblock, Oliver M. Kruse, Harold J. Luth, Randall C. Reeder, Harold J. Schramm, John C. Siemens

PM-42/1 Program Subcommittee
Ernest W. Tollner, chairman

PM-42/2 Spring Seminar Subcommittee
Lorrin H. Schwartz, chairman

PM-42/4 Standards Liaison Subcommittee
Earl H. Williams, chairman

PM-43 FARM MACHINERY MANAGEMENT COMMITTEE
Donald A. Bender, chairman
Harold J. Schramm, vice chairman and program chairman
John A. Smith, (NE), secretary
Donald R. Daum, past chairman

Terms expiring June 1989—
Ross K. Brown, Curtis J. Crothers, Karl B. Emerson, Raymond P. Hoemsen, Dave A. Mowitz, James W. Rumsey

Terms expiring June 1990—
Thomas S. Colvin, Vernon L. Hoffman, Palaniappa Krishnan, David A. Pacey, Samuel D. Parsons, W. David Shoup, John A. Smith, (NE), Lyle E. Stephens, Michael B. Thurmond

Terms expiring June 1991—
Donald A. Bender, Frank E. Buckingham, Thomas G. Carpenter, Lung-Hua Chen, Donald R. Daum, Loren D. Gautz, Joe A. Gliem, Thomas D. Ogle, Harold J. Schramm, Ronald T. Schuler, John C. Siemens, Frank M. Zoz

PM-44 FORAGE HARVESTING AND UTILIZATION COMMITTEE
C. Alan Rotz, chairman
Donald H. Pettengill, vice chairman
Kevin J. Shinners, secretary
Glenn A. Musser, past chairman

Terms expiring June 1989—
Walter K. Bilanski, James L. Butler, Richard P. Cromwell, Donald R. Landphair, C. Alan Rotz, Robert A. Stelzer, Fred D. Tompkins

Terms expiring June 1990—
Kurt I. Cook, H. Willard Downs, Philip F. Fleming, Richard G. Moe, Lalit R. Verma

Terms expiring June 1991—
Paul M. Anderson, George A. Duncan, Donald H. Pettengill, Ronald E. Pitt, Joseph Schriver, David Schleffler, Kevin J. Shinners, Kenneth L. Von Bargen

American Forage and Grassland Council—
C. Alan Rotz, representative

PM-44/2 Forage Length-of-Cut Standard Subcommittee
Marshall F. Finner, chairman
John D. Anderson, James L. Butler, David H. Diebold, Glenn A. Musser, Roger L. Villers, Kenneth L. Von Bargen

PM-44/3 Forage Harvesting Standards Subcommittee
David H. Diebold, chairman
John Bocksnick, Kurt I. Cook, Marshall F. Finner, Glenn A. Musser, Sverker Persson, Kevin J. Shinners

PM-44/4 Round Baler Standards Subcommittee
Roger W. Frimml, cochairman
Mark D. Schrock, cochairman
Paul M. Anderson, Bobby L. Bledsoe, James L. Butler, Roger W. Frimml, Rodney S. Horn, Lorne C. Heslop, Edwin O. Margerum, Richard G. Moe, Mark D. Schrock

PM-45 SOIL DYNAMICS RESEARCH COMMITTEE
Lyle E. Stephens, chairman
Randy L. Raper, vice chairman and program chairman
Henry D. Bowen, secretary
Alvin C. Bailey, past chairman

Terms expiring June 1989—
Lyle M. Carter, Stanley J. Clark, Lyle E. Stephens, James D. Summers, E. Dale Threadgill, Shrinivas K. Upadhyaya

Terms expiring June 1990—
Alvin C. Bailey, Henry D. Bowen, Eddie C. Burt, Thomas G. Carpenter, Wayne Coates, Floyd Dowell, Clarence E. Johnson, Radhey L. Kushwaha, Randy L. Raper, Philip E. Risser, Wm. Chuck Sahm, Chi N. Thai

Terms expiring June 1991—
Paul D. Ayers, Nelson L. Buck, Donald C. Erbach, Robert D. Grisso, Robert G. Holmes, John V. Perumpral, G. S. Vijaya Raghavan, Ernest W. Tollner, Dale E. Wilkins, Randall K. Wood, Ligun Chi, Mark Evans, Gerald E. Thierstein, Julius R. Williford, Alvin R. Womac

Advisory members:
W. E. Larsen (MN), R. N. Yong

PM-45/2 Controlled Traffic for Crop Production Subcommittee
Julius R. Williford, chairman
Wesley F. Buchele, Lyle M. Carter, William T. Dumas, Jr., Donald C. Erbach, Gary W. Krutz, Wayne A. LePori, Edwin J. Matthews, J. Edson McCanse, Roy W. Morling, John E. Morrison, Jr., G. S. Vijaya Raghavan, Randy L. Raper, David R. Smith, James H. Taylor, Albert C. Trouse, Jr., John A. True, Ward B. Voorhees, Dale E. Wilkins, Arnold Zimmerman

PM-45/3 Soil Classification and Measurement Subcommittee
Stanley J. Clark, chairman
Alvin C. Bailey, Eddie C. Burt, Donald C. Erbach, Clarence E. Johnson John V. Perumpral, G. S. Vijaya Raghaven

PM-45/5 Soil-Plant Dynamics Subcommittee
Paul D. Ayers, chairman
Donald C. Erbach, Megh R. Goyal, Roy W. Morling, G. S. Vijaya Raghavan, Ernest W. Tollner, Larry G. Wells, Dale E. Wilkins

PM-46 TRACTIVE AND TRANSPORT EFFICIENCY COMMITTEE
James W. Rumsey, chairman
Robert D. Reed, vice chairman and program chairman
Costas Kozabassis, secretary
Randall K. Wood, past chairman

Terms expiring June 1989—
William W. Brixius, Eddie C. Burt, Leonard B. Dell-Moretta, Raymond P. Hoemsen, Henry L. Kucera, Robert R. Reed, James W. Rumsey, James H. Taylor, Gerald W. Turnage, Shrinivasa K. Upadhyaya

Terms expiring June 1990—
Richard M. Beeghly, Henry D. Bowen, Thomas G. Carpenter, William J. Chancellor, William R. Evans, Awatif E. Hassan, Radhey L. Kushwaha, William L. Schubert, Kelvin P. Self, Betty J. Wells

Terms expiring June 1991—
S. Colin Ashmore, Richard P. Bernhardt, Lyle F. Bohnert, Rex L. Clark, Kenneth W. Domier, Costas Kotzabassis, Loran C. Lopp, H. Erdal Ozkan, Dan Salm, Lon R. Shell, Andrew Shorter, Roc Tyson, Randall K. Wood

Non-member advisor:
R. N. Yong

PM-47 TRACTOR COMMITTEE
Leonard L. Bashford, chairman
Jonathan M. Chapman, vice chairman and program chairman
Carroll E. Goering, secretary
Edward J. Bortner, past chairman

Terms expiring June 1989—
Leonard L. Bashford, Edward J. Bortner, William W. Brixius, Jonathan Chaplin, Stanley R. Clark, Lester F. Larsen, Louis I. Leviticus, Donald G. Painter, Lester J. Thompson

Terms expiring June 1990—
Charles W. Anderson, Kenneth W. Domier, Carroll E. Goering, William D. Hanford, Robert E. Kyle, Orville J. Nelson, David A. Pacey, Robert D. Reed

Terms expiring June 1991—
Wayne A. LePori, Paul E. Lockie, Theodore D. Mathewson, Charles E. Sheets, Paul K. Turnquist, James M. Woodward

ASTM Committee D2: Petroleum Products and Lubricants, Tech. B, Automotive Lubricants—Orville J. Nelson, representative (Also reports to PM-03)

PM-47/1 Lecture Series Subcommittee
David C. Shropshire, chairman
Roger D. Mayhew, secretary
Donald L. Henderson, past chairman

Terms expiring June 1989—
William J. Howard, John V. Perumpral

Terms expiring June 1990—
Donald L. Henderson, Roger D. Mayhew

Terms expiring June 1991—
Carroll E. Goering, Gerald R. Mortensen

Terms expiring June 1992—
Charles E. Sheets, Jr., David C. Shropshire

PM-48 FRUIT AND VEGETABLE HARVESTING COMMITTEE

Lawrence N. Shaw, chairman
Galen K. Brown, vice chairman
Jodie D. Whitney, secretary
Henry A. Affeldt, Jr., program chairman
Donald L. Peterson, past chairman

Terms expiring June 1989—
David W. Cayton, Ervin G. Humphries, Donald L. Peterson, Bruce E. Plumb, John H. Posselius, Bernard R. Tennes, Gary R. Van Ee, Robert E. Williamson

Terms expiring June 1990—
Galen K. Brown, Charlie G. Coble, James L. Halderson, Gary M. Hyde, Roger P. Rohrbach, Dennis R. Schultz, Lawrence N. Shaw, Henry E. Studer

Terms expiring June 1991—
Henry A. Affeldt, Lung-hua Chen, Gerald D. Christenbury, Wayne Coates, Dale E. Marshall, Hayden M. Soule, Roger Theriault, Jodie D. Whitney, John Wilhort

American Society for Horticultural Science, Working Group on Mechanization—Dale E. Marshall, liaison representative

PM-50 COTTON HARVESTING AND MECHANIZATION COMMITTEE

Roy V. Baker, Jr., chairman
William F. Lalor, vice chairman
Donald W. Van Doorn, secretary
Jefferson D. Bargeron III, past chairman

Terms expiring June 1989—
William S. Anthony, Jefferson D. Bargeron III, Ivan W. Kirk, Weldon Laird, William J. Naarding, Bill M. Norman, Glenn M. Olson, Donald W. Van Doorn, Lambert H. Wilkes

Terms expiring June 1990—
Roy V. Baker, Jr., Alan D. Brashears, Robert G. Curley, Clifford B. Fedler, S. Ed Hughs, Samuel T. Rayburn, Ervin J. West, J. Ray Williford

Terms expiring June 1991—
Russell D. Copley, Perry T. Isbell, Andrew G. Jordan, William F. Lalor, William D. Mayfield, Calvin B. Parnell, Jr., Fredrick M. Shofner, G. Neil Thedford

PM-51 TRACTOR AND IMPLEMENT HYDRAULIC COMMITTEE

John K. Schueller, chairman
Michael Mailander, vice chairman and program chairman
Jerome A. Moore, secretary
Gary W. Krutz, past chairman

Terms expiring June 1989—
Stanley R. Clark, David Hansen, Larry M. Delfs, Ken K. Koch, Jerome A. Moore, Phillip Flor, Arthur B. Zimmerman

Terms expiring June 1990—
John G. Christopher, George N. Clark, Larry D. Gaultney, Michael Mailander, Orville J. Nelson, Robert E. Schott, John K. Schueller

Terms expiring June 1991—
Loren L. Alderson, Gary W. Krutz, Louis I. Leviticus, Joe Mayo, Frank J. Simak

ANSI Committee B93: Fluid Power Systems and Components—David L. Newcom (Also reports to PM-03)

U.S. TAG for ISO/TC 131, Fluid Power Systems and Components—Ken K. Koch (Also reports to PM-03)

PM-52 SMALL TRACTOR AND POWER EQUIPMENT COMMITTEE

Norman F. Weir, chairman
David L. Apple, vice chairman and chairman-elect
Larry Gruenberger, secretary
John P. Nelson, past chairman

Terms expiring June 1989—
David L. Apple, Wesley F. Buchele, Ron Cuba, Edward J. Hengen, Lamont Meinen, Anthony A. Saiia, John J. Slazas, Dan Whalen

Terms expiring June 1990—
Martin A. Berk, Larry Gruenberger, Jerry Johnson, Edson J. McCanse, Roger D. Mayhew, Russell V. Rouse, Dale H. Schairer, Norman F. Weir, James L. Wirsbinski

Terms expiring June 1991—
Hugh F. Grow, Marvin L. Joray, Guy R. Midtbo, Orville B. Olson, Edward W. Puffer, David S. Sassman, David K. Schirer, Owen Schumacher, Lawrence N. Shaw, Neill C. Woelffer

ANSI Committee B71: Safety Standards for Lawn Mowers, Snow Throwers, Power Edgers and Trimmers, Garden Tractors and Related Equipment and Attachments—John J. Slazas, representative (Also reports to PM-03)

ANSI Subcommittee B71.4: Safety Specifications for Commercial Turf Care Equipment—John J. Slazas, representative (Also reports to PM-03)

Outdoor Power Equipment Institute and Society of Automotive Engineers small tractor and power equipment activities—John J. Slazas, liaison representative

U.S. TAG for ISO/TC 23/SC 13, Powered Lawn and Garden Equipment—Norman F. Weir, representative (Also reports to PM-03)

PM-52/1 Definitions Subcommittee
Chairman to be appointed

PM-52/2 Implement and Attachment Subcommittee
Roger D. Mayhew, chairman

PM-52/3 Program Subcommittee
Orville B. Olson, chairman

PM-52/4 Safety Subcommittee
David S. Sassaman, chairman

PM-52/5 Standards Subcommittee
Hugh F. Grow, chairman
(Also reports to PM-03)

PM-53 GRAIN HARVESTING COMMITTEE

John K. Schueller, chairman
Richard E. McMillan, vice chairman
Mark D. Schrock, secretary
Mike Huhman, program chairman
Orlin W. Johnson, past chairman

Terms expiring June 1989—
Robert L. Capstick, Thomas B. Haar, Michael Huhman, Orlin W. Johnson, Ramesh Kumar, Ron F. McNeil, Earle C. Morton, Ajit K. Srivastava

Terms expiring June 1990—
Richard P. Bernhardt, Michael A. Buresch, Richard A. DePauw, Masahiro Iwashita, Rodney L. Kushwaha, Allan E. Neal, Barrie L. Smith, Neil L. West

Terms expiring June 1991—
Walter K. Bilanski, H. Willard Downs, Lawrence H. Ellebracht, Leslie G. Hill, Raymond P. Hoemsen, Gary L. Kunz, Richard E. McMillen, E. William Rowland-Hill, Cletus E. Schertz, Mark D. Schrock, John K. Schueller

PM-53/3 Testing and Terminology Standards Subcommittee
, chairman
Geoffrey F. Cooper, Richard A. DePauw, Lawrence H. Ellebracht, Richard E. McMillen, Barrie L. Smith

PM-53/5 Chaff and Straw Characteristics Subcommittee
Neil L. West, chairman
Lawrence H. Ellebracht, Rodney L. Kushwaha, Earle C. Morton, Cletus E. Schertz

PM-54 MECHANIZATION OF AGRICULTURAL RESEARCH COMMITTEE

Gary L. Barker, chairman
Julius R. Williford, vice chairman and program chairman
Robert S. Freeland, secretary
Lowrey A. Smith, past chairman

Terms expiring June 1989—
Ronald T. Schuler, William P. Simpson, Joel T. Walker, Dale E. Wilkins, Julius R. Williford

Terms expiring June 1990—
Gary L. Barker, Wesley F. Buchele, Donald C. Erbach, Robert S. Freeland, Charles F. Garman, John W. Hummel, Lowrey A. Smith, Richard A. Wesley

Terms expiring June 1991—
Alan D. Brashears, Rex L. Clark, Barney R. Eiland, Thomas H. Garner, Gordon E. Monroe, Fred D. Tompkins

International Association on Mechanization of Field Experiments, Executive Committee—Wesley F. Buchele, representative

PM-58 AGRICULTURAL EQUIPMENT AUTOMATION COMMITTEE

Gordon R. Hjertaas, chairman
Marvin J. Pitts, vice chairman
Charles A. Parrish, secretary
Stephen W. Searcy, past chairman

Terms expiring June 1989—
Frederick W. Nelson, Marvin J. Pitts, Jeffrey L. Ruckman, Stephen W. Searcy, Lowrey A. Smith, Ronald E. Squires, Marvin L. Stone

Terms expiring June 1990—
Carl E. Bohman, Roy B. Dodd, Willard R. Ellis, James R. Fries, Gerhard Jahns, Larry J. Kutz, David G. Sokol, John E. Wood, Steven C. Young

Terms expiring June 1991—
Rex L. Clark, Gordon R. Hjertaas, Gary M. Hyde, Neil B. McLaughlin, Charles A. Parrish, Terence D. Pickett, John F. Reid, John K. Schueller, James N. Wilson

Non-member advisors:
T. G. Kirk, John V. Stafford

PM-59 NURSERY AND GREENHOUSE MECHANIZATION COMMITTEE

George A. Duncan, chairman
William H. Aldred, vice chairman
Gregory L. Branch, secretary
Bryan M. Maw, past chairman

Terms expiring June 1989—
John W. Bartok, Farnum M. Burbank, George A. Duncan, Richard F. Dudley, Gene A. Giacomelli, Robert G. Holmes, David S. Ross, Roger Theriault, Earl C. Yaeger

Terms expiring June 1990—
William H. Aldred, Michael D. Clement, James H. Dooley, Karl R. Huber, James A. Mullins, Serker Persson, Ted H. Short, Giles O. Van Duyne, Roy E. Young

Terms expiring June 1991—
Gregoray L. Branch, Harold L. Brewer, Gene A. Giacomelli, Awatif E. Hassan, Richard L. Parish, Lawrence N. Shaw, Braham P. Verma, Daniel H. Willits, Earl C. Yeager

PM-60 TESTING AND RELIABILITY COMMITTEE

Charles R. Frank, chairman
Michael B. Thurmond, vice chairman
David A. Skinner, secretary

Terms expiring June 1989—
Douglas A. Bargiel, John E. Langdon, Robert D. Davis, Frank E. Woeste

Terms expiring June 1990—
Ray A. Brandt, David A. Skinner, Lester J. Thompson

Terms expiring June 1991—
Lin H. Bowen, Charles R. Frank, Michael B. Thurmond

PM-61 HUMAN FACTORS COMMITTEE

Darrell L. Roberts, chairman
William J. Becker, vice chairman and secretary
Richard A. Cavaletto, program chairman
Stanley E. Prussia, past chairman

Terms expiring June 1989—
David C. Beppler, Alan L. Dorris, Patrick C. Conlin, Jerry R. Duncan, Thomas P. Hillstrom, John B. Meyers, Robert B Rummer, Robert H. Wilkinson

Terms expiring June 1990—
William S. Anthony, L. Dale Baker, William J. Becker, M. Stephen Kaminaka, Darrell L. Roberts, W. David Shoup, Charles W. Suggs, Ronald F. Zitko

Terms expiring June 1991—
Robert A. Aherin, Richard A. Cavaletto, Edwin J. Matthews, Dennis J. Murphy, Frank D. Perkinson, Stanley E. Prussia, Herbert D. Sullivan, Roger Tormoehlen

PM-62 AGRI-INDUSTRY SUPPLIERS COMMITTEE

Walter T. Nawrocki, chairman
Dieter Bloecks, vice chairman
Roy G. Brandt, secretary

Terms expiring June 1989—
Marilyn M. Everest, Lanny L. Leppo, Mars Paterson

Terms expiring June 1990—
Walter E. Hull, James F. Lamb, Walter T. Nawrocki

Terms expiring June 1991—
Dieter Bloecks, Roy G. Brandt, E. Joel Martin

JOINT COMMITTEES

(These committees are administered by other technical divisions, but incorporate technical interests of the Power and Machinery Division.)

Physical Properties of Agricultural Products Committee, FPE-701
Farm Materials Handling Committee, FPE-708
Energy and Alternate Products from Biomass Committee, FPE-709

SOIL AND WATER DIVISION

Donald K. McCool, Director (90)

SW-01 EXECUTIVE COMMITTEE

Larry A. Kramer, chairman (88-89)
James M. Steichen, vice chairman (88-89)
Keith E. Saxton, secretary (88-89)
Terry A. Howell, past chairman (88-89)
Donald K. McCool, director (88-90)

SW-02 STEERING COMMITTEE

Larry A. Kramer, chairman (88-89)
James A. Steichen, vice chairman (88-89)
Keith E. Saxton, secretary (88-89)
Terry A. Howell, past chairman (88-89)
Donald K. McCool, director (88-90)
William E. Hart, chairman SW-03 (88-89)
Gary D. Bubenzer, chairman SW-05 (88-89)
Norman R. Fausey, chairman SW-06 (88-89)
R. Nolan Clark, chairman SW-07 (88-89)
Charles H. Ullery, chairman SW-08 (88-89)

Plus the officers of SW-21, SW-22, SW-23, SW-24, SW-25 and SW-26 shown below.

Other Committees or Activities Reporting to Steering Committee

American Institute of Hydrology, Intersociety Committee—Charles T. Haan, representative

American Society of Agronomy—Harry B. Pionke, liaison representative

CIGR, International Commission of Agricultural Engineering, Sect. I: Soil and Water Sciences—David B. Palmer, technical correspondent

FIEI Tillage and Crop Production Equipment Council—Lee P. Herndon, liaison representative

Soil Conservation Society of America—Lloyd N. Mielke, liaison representative

ASAE Committee on Standards (T-1)—E. Gordon Kruse (89) representative; Gylan L. Dickey (91) alternate

ASAE Ad Hoc Technical Advisory Group to ISO/TC23/SC18, Irrigation and Drainage Equipment—Allen R. Dedrick, liaison representative

ASAE Agricultural Safety Committee (T-2)—Coy W. Doty, liaison representative

ASAE Environmental Quality Coordinating Committee (T-9)—Arthur B. Holland (89), Richard K. White (90), John M. Sweeten (91), liaison representatives

ASAE Energy Committee (T-11)—R. Nolan Clark, liaison representative

ASAE Forest Engineering Advisory Committee (FR-402)—John L. Nieber, Donald E. McCandless, Jr., Robert A. Wiles, liaison representatives

ASAE Paper Awards Committee (A-116)—J. Kent Mitchell (90), liaison representative

ASAE Historic Commemoration Committee (A-123)—E. Gordon Kruse, liaison representative

ASAE Research Committee (A-209)—L. Donald Meyer (90), R. Nolan Clark (90), Edward A. Hiler (91), Dale H. Vanderholm (92), liaison representatives

ASAE Continuing Education Committee (A-416)—Saied Mostaghimi, liaison representative

ASAE Publications Executive Committee (A-501)—John M. Laflen, liaison representative

ASAE Refereed Publications Committee (A-511)—Gary D. Bubenzer, Harold R. Duke (88), liaison representatives

ASAE Monographs Committee (A-512)—John L. Nieber (89), liaison representatives

ASAE Agricultural Engineering Editorial Board (A-513)—Calvin D. Mutchler, liaison representative

ASAE Textbook Committee (A-514)—Glenn O. Schwab (89), liaison representative

ASAE Special Publications Committee (A-515)—Edward A. Hiler, liaison representative

SW-03 STANDARDS COMMITTEE

William E. Hart, chairman (88-91)
G. Morgan Powell, vice chairman (88-91)
Evelyn E. Rosentreter, staff secretary

Terms expiring June 1989—
Vincent F. Bralts, Robert C. Sears

Terms expiring June 1990—
John L. Brewer, Dale A. Christensen, Gylan L. Dickey, Coy W. Doty, Donald M. Edwards, William R. Morrison

Terms expiring June 1991—
Allen R. Dedrick, E. Gordon Kruse, Leonard R. Massie, Roger C. Moe, Matt Plotkin, Kenneth H. Solomon, Jay D. Stradinger

ASAE Committee on Standards (T-1)—E. Gordon Kruse, representative; Gylan L. Dickey, alternate

ASAE Agricultural Safety Committee (T-2)—Coy W. Doty, liaison representative

Technical Advisory Group to ISO/TC 23/SC 18, Irrigation and Drainage Equipment
(Appointed by Irrigation Association, the U.S. Administrator for SC 18)

Allen R. Dedrick, chairman
Kenneth H. Solomon, vice chairman

Jim Anshutz, Brice E. Boesch, Verne Bray, John Brewer, John A. Chapman, Jerry Hockert, Allan S. Humpherys, Donald W. Jensen, R. D. Johnston, Ruben A. Koch, E. Gordon Kruse, Eugene R. Lindemann, George H. Lockwood, Theda Lockwood, James W. Mason, Charles Meis, Austin J. Miller, David W. Miller, Claude H. Pair, F. Frank Penkava, Matt Plotkin, Robert C. Sears, Ronald E. Sneed, LaVerne E. Stetson, Jay D. Stradinger, Glen E. Stringham, Glenn O. Tribe, Charles V. Trivette, Jr., F. Derek Wood, Thomas H. Young

William E. Hart (AZ) (chm. SW-03),
G. Morgan Powell (vice chm. SW-03)
Robert D. von Bernuth (chm. SW-241),
Walter L. Trimmer (vice chm. SW-241)
Jerry L. Chesness (chm. SW-245),
John M. Langa (vice chm. SW-245)

Cooperative Standards Activities with ASAE Representatives Reporting to SW-03

American Society for Testing and Materials,
C13: Concrete Pipe—
C-15: Manufactured Masonry Units—
C-27: Precast Concrete Products—
D-18: Soil and Rock—
D-19: Water—
F-17: Plastic Piping Systems—
Swayne F. Scott, liaison representative to above committees

SW-05 PUBLICATIONS REVIEW COMMITTEE

Gary D. Bubenzer, division editor (87-89)

Associate Editors (1988-90)—Richard L. Bengtson, Carl R. Camp, Norman A. Fausey, James L. Fouss, Donald W. Fryrear, Terry A. Howell, Byron H. Nolte, Earl C. Stegman

Associate Editors (1987-89)—Elias Bloom, Elbert C. Dickey, Ronald L. Elliott, James R. Gilley, John M. Laflen, Walter J. Rawls, Michael D. Smolen, John M. Sweeten, Jr.

SW-06 PAPER AWARDS COMMITTEE

Norman R. Fausey, chairman (88-89)

Terms expiring June 1989—
Karen M. Mancl, G. Morgan Powell

Terms expiring June 1990—
Jales K. Koelliker, Allan S. Humpherys, Dennis L. Hurtz

SW-07 NOMENCLATURE COMMITTEE

R. Nolan Clark, chairman

Terms expiring June 1989—
Elias Bloom, Elbert C. Dickey, Curtis L. Larson, John A. Replogle, Glenn O. Schwab

SW-08 SOIL AND WATER RESOURCES LAW COMMITTEE

Charles H. Ullery, chairman (88-89)
David R. Nelson, vice chairman (88-89)

Terms expiring June 1989—
Herman N. McGill, Donald L. Pfost, Swayne F. Scott, Ronald E. Sneed

Terms expiring June 1990—
Lloyd B. Baldwin, Donald J. Brosz, William R. Folsche, David Miller, David R. Nelson, Richard C. Peralta

Terms expiring June 1991—
Elias Bloom, Carl E. Bouchard, Dorrell C. Larsen, Gerald L. Westesen, Alan D. Wood, Charles H. Ullery, James N. Krider, Fletcher C. Armstrong

Liaison Members, Vice Chairmen of SW Groups—
Karen M. Mancl, G. Morgan Powell, James K. Koelliker, William E. Altermatt, Allan S. Humpherys, Dennis L. Hurtz

SW-09 FORWARD PLANNING COMMITTEE

Terry A. Howell, chairman (88-89)
James R. Gilley, Walter D. Lembke, David B. Palmer, Clarence W. Richardson

SW-21—HYDROLOGY GROUP

Kenneth L. Campbell, chairman (88-90)
James K. Koelliker, vice chairman (88-90)

SW-211 POROUS MEDIA FLOW COMMITTEE

Adrian W. Thomas, chairman (87-89)
Mary L. Wolfe, vice chairman (87-89)

Terms expiring June 1989—
Donald L. Brakensiek, Lowell A. Disrud, Eric Flaig, Albert R. Jarrett, John L. Nieber

Terms expiring June 1990—
Timothy J. Gish, Walter J. Rawls

Terms expiring June 1991—
Otto J. Helweg, James R. Hoover, Rameshwan S. Kanwar, Ashok Katyal, Dale A. Lehman, Avinash Patwardhan, Shiv Prasher, Ramesh P. Rudra, Adel Shirmohammadi, Roger Stillwater, Kim A. Tan, Shu-Tung Chu, Tammo S. Steenhuis

SW-212 PRECIPITATION AND RUNOFF COMMITTEE

Adrian W. Thomas, chairman (88-90)
Loyd K. Ewing, vice chairman (88-90)

Terms expiring June 1989—
Mitchell L. Griffin, Donald K. McCool, Keith C. McGregor, Allen Thompson

Terms expiring June 1990—
Carol Drungil, Loyd K. Ewing, Michael C. Hirschi, W. Carlisle Mills, Robert A. Young

Terms expiring June 1991—
J. Hari Krishna, Chandra Madramootoo, Walter J. Rawls, Donald E. Woodward

SW-213 EVAPOTRANSPIRATION COMMITTEE

Derrell L. Martin, chairman (88-90)
Sun-Fu Shih, vice chairman (88-90)

Terms expiring June 1989—
Robert D. Burman, Carl R. Camp, Robert J. Edling, Carroll A. Hackbart, Cristu-Silviu V. Papadopol, Wesley Rosenthal, Sun-Fu Shih, Maluneh Yitayew

Terms expiring June 1990—
Philip L. Barnes, Walter C. Bausch, Norman L. Klocke

Terms expiring June 1991—
Christopher L. Butts, Steve Thompson, Bryan W. Maw, Steven E. Hinkle, Claudio Stockle, Christopher Neal, Clyde R. Bogle, Thomas Scherer, Donald C. Slack

SW-214 RURAL WATER SUPPLIES COMMITTEE

(Joint with Electrical and Electronic Engineering Systems and Structures and Environment Divs.)

Thomas D. Glanville, chairman (88-90)
Fred N. Swader, vice chairman (88-90)
Charles H. Ullery, secretary (88-90)

Terms expiring June 1989—
Derrell B. McLendon, Lawton E. Samples

Terms expiring June 1990—
Delynn R. Hay, Charles H. Ullery, Dennis M. Sievers, Lu Cole, John J. Kolega, Richard L. Witz

Terms expiring June 1991—
Dorota Haman, Mark Herriott, Roger E. Machmeier, Chris Wilker, Kelly Downing

SW-215 HYDROLOGIC SYSTEMS COMMITTEE

W. Carlisle Mills, chairman (88-90)
Andrew D. Ward, vice chairman (88-90)

Terms expiring June 1989—
Donald L. Brakensiek, Eric Flaig, Mitchell L. Griffin, Conrad D. Heatwole, John L. Nieber, James D. Nelson, Harvey H. Richardson

Terms expiring June 1990—
Carol Drungil, Richard C. Peralta, Sun-Fu Shih, Ian D. Moore, Andrew D. Ward

Terms expiring June 1991—
John W. Chenoweth, Gerald N. Flerchinger, J. Hari Krishna, Paul I. Welle

SW-217 HYDRAULICS AND TRANSPORT PROCESSES COMMITTEE

John E. Gilley, chairman (87-89)
Theo A. Dillaha III, vice chairman (87-89)

Terms expiring June 1989—
David B. Beasley, Clarence W. Robinson, Michael D. Smolen

Terms expiring June 1990—
Dennis C. Flanagan, George R. Foster, Jeff McBurnie, L. Donald Meyer, David C. Ralston, Bruce N. Wilson

Terms expiring June 1991—
Adelbert B. Bottcher, Maureen E. Guck, James R. Hoover, Albert R. Jarrett, Daniel E. Line, William L. Magette, Marilyn J. Mroz, Mark A. Nearing, W. Howard Neibling, Kenneth G. Renard, Ramesh P. Rudra

SW-22—EROSION CONTROL GROUP

Ronald R. Allen, chairman (87-89)
G. Morgan Powell, vice chairman (87-89)

SW-223 EROSION CONTROL RESEARCH COMMITTEE

George R. Foster, chairman (88-90)
Larry C. Brown, vice chairman (88-90)

Terms expiring June 1989—
David B. Beasley, John B. Borreli, Andrew J. Bowie, Thomas G. Franti, Donald K. McCool, Keith C. McGregor, Gregory McIssac, Oscar Perez, Curtis H. Shelton, Allen L. Thompson

Terms expiring June 1990—
Elbert C. Dickey, Gary E. Formanek, Donald W. Fryrear, William Kranz, Thomas R. McCarty, Ian D. Moore, Kyung H. Yoo

Terms expiring June 1991—
Bily J. Barfield, Richard L. Bengtson, Marlon A. Breve, James M. Gregory, Lawrence J. Hagan, Albert R. Jarrett, Ann Kenimer, John M. Laflen, Daniel E. Line, Saied Mostaghimi, Ramesh P. Rudra, Ted M. Zobeck

SW-224 POLLUTION BY SEDIMENT COMMITTEE

Ernest W. Tollner, chairman (87-89)
Saied Mostaghimi, vice chairman (87-89)

Terms expiring June 1989—
Joseph R. DelVecchio, John C. Hayes, Flint Holbreck, Ramesh P. Rudra, Mark A. Smith, Sally A. Stokes, Richard C. Warner, Bruce N. Wilson, Ted M. Zobeck

Terms expiring June 1990—
Lloyd N. Mielke, J. Kent Mitchell, Carl E. Murphree, Jr., W. Howard Neibling, David R. Nelson, Frank R. Schiebe

Terms expiring June 1991—
Wayne R. Anderson, Andrew J. Bowie, John Brach, Larry C. Brown, James M. Hamlett, Michael D. Smolen, Guye H. Willis

SW-225 CONSERVATION SYSTEMS COMMITTEE

Lee F. Herndon, chairman (87-89)
Lloyd N. Mielke, vice chairman (87-89)

Terms expiring June 1989—
Arthur M. Brate, Emeron P. Christensen, Gary E. Formanek, James M. Gregory, Michael Herschi, John D. Jelinski, Clarence E. Johnson, Joseph H. Marter, Gene L. Nimmer, Donald L. Pfost, John R. Ramsden, Walter K. Twitty, Jr., Ross L. Ulmer

Terms expiring June 1990—
Clifford J. Baumer, Gregory L. Brenneman, John M. Brumett, Dennis K. Carman, Thomas S. Colvin, Bobby G. Moore, David R. Nelson, Oscar Perez, Douglas C. Seibel, David P. Shelton, Leon Wendte, Kyung H. Yoo, Robert A. Young

Terms expiring June 1991—
Leland O. Anderson, William P. Annable, Daniel J. Baumert, Hugh A. Curry, Daniel E. Line, Donald E. McCandless, Jr., Roger C. Moe, Carl E. Murphree, Jr., Dale E. Wilkins

SW-23—DRAINAGE GROUP

Norman R. Fausey, chairman (87-89)
William E. Altermatt, vice chairman (87-89)

SW-231 DRAINAGE RESEARCH COMMITTEE

Tammo S. Steenhuis, chairman (88-91)
Andrew J. Ward, vice chairman (88-91)

Terms expiring June 1989—
Joseph Bornstein, Cade E. Carter, Sie-Tan Chieng, Deanna S. Durnford, Forrest T. Izuno, William R. Johnston, Edwin J. Monke, Avinash Patwardhan, David L. Rausch

Terms expiring June 1990—
Shiv O. Prasher, Richard D. Wenberg

Terms expiring June 1991—
Robert O. Evan, Mark Grismer, James S. Rogers, Daniel L. Thomas, Chandra A. Madramootoo, Todd Troolen, Charles A. Neumann, Rameshwar S. Kanwar, Eric Desmond

SW-232 SUBSURFACE DRAINAGE IN HUMID AREAS COMMITTEE

John R. Johnston, chairman (86-89)
Larry D. Geohring, vice chairman (86-89)

Terms expiring June 1989—
Philip Brink, Coy W. Doty, Carroll J. W. Drablos, Fredrick H. Galehouse, Ross W. Irwin, Forrest T. Izuno, Carl W. Marsee, John L. Nieber, Richard D. Wenberg

Terms expiring June 1990—
William E. Altermatt, John E. Burnham, Dana C. Chapman, Mike Cook, Harry G. Hirth, James Jacobs, Stanley L. Seevers, Daniel L. Thomas

Terms expiring June 1991—
Harold W. Belcher, Robert O. Evans, James L. Fouss, K. D. (Ken) Konyha, Stewart W. Melvin, John F. Rice, Adel Shirmohammadi

SW-233 SURFACE DRAINAGE COMMITTEE

Fredrick H. Galehouse, chairman (87-90)
John F. Rice, vice chairman (87-90)

Terms expiring June 1989—
James Lee Evans, Byron H. Nolte

Terms expiring June 1990—
Dennis K. Carman, Hugh A. Curry, James Jacobs, Paul E. Lucas, Wayne Wood

Terms expiring June 1991—
Gary L. Montgomery, Richard T. Smith, Eric Desmond, Arthur M. Brate

SW-234 DRAINAGE OF IRRIGATED LANDS COMMITTEE

James E. Ayars, chairman (87-90)
Blaine R. Hanson, vice chairman (87-90)

Terms expiring June 1989—
Bishay G. Bishay, Glenn J. Hoffman, Robert J. Pofahl, Tom Spofford, Robert J. Toedter

Terms expiring June 1990—
Jimmy L. Gartung, Deane M. Manbeck, Philip M. Myers, Charles A. Neumann, Rodney G. Tekrony

Terms expiring June 1991—
William E. Altermatt, Alan R. Bender, Keith A. Campbell, Darrel W. DeBoer, Frank C. Stambach, Terence H. Podmore, Elwin A. Ross, Lyman S. Willardson, Al Qziz Eddebbach

SW-236 BENEFITS OF DRAINAGE COMMITTEE

Tom L. Zimmerman, chairman (86-89)
Harold W. Belcher, vice chairman (86-89)

Terms expiring June 1989—
Cade E. Carter, Carroll J. W. Drablos, Deanna S. Durnford, Norman R. Fausey, Leonard R. Massie, Edwin J. Monke, James S. Rogers

Terms expiring June 1990—
Richard L. Bengtson, Blaine R. Hanson, Shiv O. Prasher, Rodney G. Tekrony

Terms expiring 1991—
Sie-Tan Chieng,Seyed A. Madani, Kieth A. Campbell, Andrew C. Chang, Walter K. Twitty

SW-24—IRRIGATION GROUP

Darrell W. DeBoer, chairman (88-90)
Allan S. Humphreys, vice chairman (88-90)

SW-241 SPRINKLER IRRIGATION COMMITTEE

Robert D. von Bernuth, chairman (87-90)
Walter L. Trimmer, vice chairman (87-90)

Terms expiring June 1989—
Donald J. Brosz, Carl R. Camp, Dale A. Christensen, Robert J. Edling, Carroll A. Hackbart, Steven E. Hinkle, William M. Lyle, Kurt J. Maloney, Robert C. Sears, Ronald E. Sneed

Terms expiring June 1990—
Gerald Buchleiter, Freddie Lamm, Dennis C. Kincaid, Mark A. Locke, Derrell L. Martin, Len J. Ring, Eugene W. Rochester, Christopher J. Striby, E. Dale Threadgill, Wesley W. Wallender, Hal D. Werner

Terms expiring June 1991—
Jerry Gerdes, Jacob L. LaRue, Eugene R. Lindemann, David Miller, Arland D. Schneider

SW-242 SURFACE IRRIGATION COMMITTEE

Ronald L. Elliott, chairman (86-89)
Thomas Trout, vice chairman (86-89)

Terms expiring June 1989—
Day L. Bassett, Richard D. Black, Dean E. Eisenhauer, Yu-Si Fok, Peter Livingston, John L. Merriam

Terms expiring June 1990—
Allie W. Blair, Gylan L. Dickey, Robert G. Evans, E. Gordon Kruse, Donald L. Reddell, Wesley W. Wallender, C. Dean Yonts

Terms expiring June 1991—
Albert J. Clemmens, Allen R. Dedrick, William E. Hart, Joseph C. Henggeler, Duane Hickerson, Thomas W. Ley, Eugene R. Lindemann, Harry L. Manges, Wesley W. Wallender

SW-243 IRRIGATION WATER SUPPLY AND CONVEYANCE COMMITTEE

Jerry A. Wright, chairman (87-90)
Allen R. Dedrick, vice chairman (87-90)

Terms expiring June 1989—
James D. Burks, Charles M. Burt, Coy W. Doty, Svat Jonas, Larry D. King

Terms expiring June 1990—
William E. Hart, Allan S. Humpherys, Eugene R. Lindemann, Peter Livingston, Walter L. Trimmer

Terms expiring June 1991—
Wayne Clyma, Debbie Davies, Harry J. Gibson, Gary Kah, Fadi Kamand, Joel D. Palmer, John A. Replogie

Advisory member:
George H. Abernathy, Robert Novick

SW-244 IRRIGATION MANAGEMENT COMMITTEE

Dale A. Bucks, chairman (86-89)
Glenn J. Hoffman, vice chairman (86-89)

Terms expiring June 1989—
Richard D. Black, Vincent F. Bralts, Robert G. Evans, Dorata Haman, Douglas J. Hunsaker, Darnell R. Lundstrom, William M. Lyle, Christopher Neale, Edward J. Sadler, Hal D. Werner

Terms expiring June 1990—
John Brewer, Gerald Buchleiter, Gary W. Buttermore, Carl R. Camp, Ronald L. Elliott, Dale F. Heermann, Shrikant Jaytap, Freddie Lamm, Saied Mostaghimi, Claude J. Phene, John A. Replogle, Danny H. Rogers, Ronald E. Sneed, Robert D. von Bernuth

Terms expiring June 1991—
Walter C. Bausch, John R. Busch, Gerald D. Christenbury, Dale A. Christensen, Dean E. Eisenhauer, Carroll A. Hackbart, Otto J. Helweg, Jacob L. LaRue, Jerry Lambert, Theodore L. Loudon, Harry Manges, Robert W. Schottman, Donald C. Slack, Kenneth H. Solomon, James R. Stansel, Todd P. Trooien, Jerry A. Wright, F. Scott Wright, E. Dale Threadgill

SW-245 MICROIRRIGATION COMMITTEE

John M. Langa, chairman (88-91)
Vincent F. Bralts, vice chairman (88-91)

Terms expiring June 1989—
Dale A. Bucks, Robert J. Edling, Glenn J. Hoffman, Harry L. Manges, Claude J. Phene, Gajendra Singh, Christopher J. Striby, E. Dale Threadgill, Mulunah Yitayew

Terms expiring June 1990—
Gary A. Clark, Robert G. Evans, Jerry Gerdes, Kenneth M. Lomax, Marshall J. McFarland, Donald J. Pitts, I-Pai Wu

Terms expiring June 1991—
Carl R. Camp, Jerry L. Chesness, Megh R. Goyal, Dorota Haman, David J. Hills, David M. McLeod, Jackie W. D. Robbins, Allan G. Smajstrla, Thomas H. Williams

SW-25—WATER RESOURCE STRUCTURES GROUP

Lloyd E. Thomas, chairman (88-90)
Dennis L. Hurtz, vice chairman (88-90)

SW-251 PLANNING COMMITTEE

Allen B. Colwick, chairman (86-89)
Joseph R. DelVecchio, vice chairman (86-89)

Terms expiring June 1989—
Coy W. Doty, Kyle L. Moran, Lynn R. Shuyler

Terms expiring June 1990—
Harry G. Gibson, Larry D. Hasty, Robert C. Robison, James W. Stingel

Terms expiring June 1991—
Robert E. Dunn, Otto J. Helweg, Karl R. Klingelhofer, Lee A. Mulkey, Ross L. Ulmer, Paul I. Welle

SW-252 DESIGN COMMITTEE

Alan D. Wood, chairman (87-89)
Darrel M. Temple, vice chairman (87-89)

Terms expiring June 1989—
Elias Bloom, Charles L. Hahn, Peter Livingstone, Hershel R. Read

Terms expiring June 1990—
David L. Camper, Donald L. Newton, David C. Ralston, Peter E. Wright

Terms expiring June 1991—
Ronald G. Scheffler, Philip H. Smith

SW-253 CONSTRUCTION AND MAINTENANCE COMMITTEE

Ronald W. Hayes, chairman (88-90)
James L. Evans, vice chairman (88-90)

Terms expiring June 1989—
William J. Bowers, Sheldon L. Dynes, James Lee Evans, Mervin R. Ice, Gary L. Montgomery

Terms expiring June 1990—
Micheal E. Andreas, John C. Brach, Wayne C. Foust, Ernest U. Gingrich, Jerry W. Leonard, Philip M. Myers, John F. Ourada, Richard C. Purcell

Terms expiring June 1991—
Robert A. Bird, Harry G. Hirth, Marvin L. Knabach, Edwin L. Minnick, Fred Scheutz, James L. Sell, Robert A. Wiles

SW-26—COUNTRYSIDE ENGINEERING GROUP

William F. Ritter, chairman (87-89)
Karen M. Mancl, vice chairman (87-89)

SW-262 HOME SEWAGE DISPOSAL COMMITTEE

Theodore L. Loudon, chairman (88-89)
Albert R. Rubin, vice chairman (88-89)
J. Ross Harris, secretary (88-89)

Terms expiring June 1989—
Marie E. Davis, Roger C. Machmeier, Fred N. Swader, Richard L. Witz

Terms expiring June 1990—
Frank J. Humenik, Charles G. McKiel, Cristu-Silvin V. Papadopol

Terms expiring June 1991—
Eldridge R. Collins, Art J. Gold, Kenneth M. Lomas, James D. Nelson

ASTM Committee C-27: Precast Concrete Products—
Swayne J. Scott, representative
(Also reports to SW-03)

SW-263 LAND APPLICATION OF WASTE COMMITTEE

(Joint with Structures and Environment Div.)

Joseph L. Taraba, chairman (88-89)
Clifford B. Fedler, vice chairman (88-89)
Jame C. Barker, secretary (88-89)

Terms expiring June 1989—
Frank J. Humenik, James A. Moore, Albert R. Rubin, Tamim M. Younos

Terms expiring June 1990—
James C. Converse, Herschel A. Elliott, Richard J. McClimans, John N. Schneider, Stephen B. Smith, Richard K. White

Terms expiring June 1991—
Douglas W. Hamilton, Victor W. E. Payne, John H. Schreider

SW-264 RECLAMATION OF DISTURBED LANDS COMMITTEE

Richard C. Warner, chairman (88-89)
Tamim M. Younos, vice chairman (88-89)
Jackie W. D. Robbins, secretary (88-89)

Terms expiring June 1989—
Larry W. Caldwell, Joseph H. Harrington, John L. Roll, Lloyd E. Thomas, Robert A. Wiles, Samuel E. Young

Terms expiring June 1990—
Herschel A. Elliott, Richard J. McClimans, Daniel Storm

Terms expiring June 1991—
Gary Felton, Peter Forsythe, Mervin R. Ice

SW-265 RURAL LANDSCAPE COMMITTEE

Ronald W. Tuttle, chairman (88-89)
Robert A. Wiles, vice chairman (87-89)
Fred N. Swader, secretary (87-89)

Terms expiring June 1989—
Elias Bloom, Victor W. E. Payne, Richard M. Rovang

Terms expiring June 1990—
Leland O. Anderson, Neil F. Bogner, Robert L. Burris, Duane Coen, Thomas A. Keep, Gregory McIsaac, Joan I. Nassauer, Allen D. Wood

Terms expiring June 1991—
Janna Marie Coen, Andy Hall, Arthur B. Holland, Dexter W. Johnson, Donald W. Lake, John H. Pedersen, Marilyn J. Mroz, Lewis L. Studer, Ross L. Ulmer

JOINT COMMITTEES

(These committees are administered by the Structures and Environment Division, but incorporate technical interests of the Soil and Water Division.)

Environmental Physiology Committee, SE-301
Climate and Meteorology Committee, SE-307
Agricultural Sanitation and Waste Management Committee, SE-412

STRUCTURES AND ENVIRONMENT DIVISION

William H. Brown, Director (89)

SE-01 EXECUTIVE COMMITTEE

Dennis P. Stombaugh, chairman
Harvey B. Manbeck, 1st vice chairman
Daniel L. Wambeke, 2nd vice chairman
Bruce A. Martin, secretary
Mylo A. Hellickson, past chairman
William H. Brown, director (89)

SE-02 STEERING COMMITTEE

Dennis P. Stombaugh, chairman
Harvey B. Manbeck, 1st vice chairman
Daniel L. Wambeke, 2nd vice chairman
Bruce A. Martin, secretary
Mylo A. Hellickson, past chairman
William H. Brown, director (89)

Members-at-Large—Sidney A. Thompson (92), James E. Kirchhofer (89), Martin L. Hellickson (90), John A. Nienaber (91)

Group Chairmen—Rodney W. Carpenter, Lawson M. Safley, Jr., Paul N. Walker

Group Vice Chairmen—Kifle G. Gebremedhin, Brian J. Holmes, Garrett L. Van Wicklen

Group 2nd Vice Chairmen—Kenneth J. Guffey, Kevin A. Janni, Douglas G. Overhults

Division Publications Editor—Lawson M. Safley, Jr. (89)

Other Committees or Activities Reporting to Steering Committee

CIGR, International Commission of Agricultural Engineering Sect. II: Rural Construction and Related Equipment—Arthur J. Muehling, technical correspondent

ASAE Committee on Standards (T-1)—Harmon L. Towne (89), representative; James A. Lindley (90), alternate

ASAE Agricultural Safety Committee (T-2)—Rodney W. Carpenter (90), liaison representative

ASAE Instrumentation and Controls Committee (EES-53)—Bailey W. Mitchell (89), liaison representative

ASAE Bioengineering Advisory Committee (BI-302)—R. Bruce Curry (89), liaison representative

ASAE Environmental Quality Coordinating Committee (T-9)—James A. Moore (89), Bruce A. Martin (90), Larry D. Jacobson (91), liaison representatives

ASAE Metric Coordinating Committee (T-10)—Myron G. Britton (89), liaison representative

ASAE Energy Committee (T-11)—Larry P. Walker (89), liaison representative

ASAE Paper Awards Committee (A-116)—Roger A. Nordstedt (90), liaison representative

ASAE Research Committee (A-209)—Martin L. Hellickson (91), James R. Fischer (89), Norman R. Scott (90), liaison representatives

ASAE Cooperative Standards Program Committee (A-311)—Curtis D. Fankhauser (89), Bruce A. Martin (90), liaison representatives

ASAE Engineering Registration Committee (A-414)—Dwaine S. Bundy (89), liaison representative

ASAE Continuing Education Committee (A-416)—Calvin O. Cramer (89), liaison representative

ASAE Refereed Publications Committee (A-511)—Lawson M. Safley, Jr. (89), liaison representative; James A. Lindley (89), member-at-large

ASAE Monographs Committee (A-512)—Raymond L. Huhnke (89), liaison representative

ASAE Textbook Committee (A-514)—Leslie L. Christianson (89), liaison representative

ASAE Special Publications Committee (A-515)—Leslie L. Christianson (90), liaison representative

ASAE CIGR Liaison Subcommittee (A-611/2)—Arthur J. Muehling (89), liaison representative

SE-03 STANDARDS COMMITTEE

Dennis D. Schulte, chairman
Lewis A. Schaper, vice chairman
Evelyn E. Rosentreter, staff secretary
Daniel L. Wambeke, past chairman

Terms expiring June 1989—
Michael F. Brugger, Lawrence A. Donoghue, James E. Gordon, Charles M. Milne, James P. Murphy, Ronald L. Sutton

Terms expiring June 1990—
Harvey B. Manbeck, Lewis A. Schaper, John M. Sweeten, Harmon L. Towne, Daniel L. Wambeke, Larry T. Wyatt

Terms expiring June 1991—
Rodney W. Carpenter, William A. Cook, James A. Lindley, Bruce A. Martin, Vernon M. Meyer, Dennis D. Schulte

ASAE Committee on Standards (T-1)—Harmon L. Towne, representative; James A. Lindley, alternate

ASAE Agricultural Safety Committee (T-2)—Rodney W. Carpenter, liaison representative

Cooperative Standards Activities with ASAE Representatives Reporting to SE-03

ANSI/ASCE Standards Committee A58, Building Code Requirements for Minimum Design Loads in Buildings and Other Structures—Robert A. Parsons

SE-04 PAPER AWARDS COMMITTEE

Michael B. Timmons (89), chairman
Richard Gates (89), Don A. Bender (90)

SE-05 PUBLICATIONS REVIEW COMMITTEE

Lawson M. Safley, Jr., division editor

Terms expiring June 1989—
W. Harold Allen, Leslie F. Backer, Norman E. Collins, Jr., Harvey B. Manbeck, John H. Pedersen, Lawson M. Safley, Jr., Glen H. Smerage, Michael B. Timmons

Terms expiring June 1990—
Louis D. Albright, Robert W. Bottcher, Michael F. Brugger, Charles J. Clanton, Daniel H. Willits

Terms expiring June 1991—
David T. Hill, C. Roland Mote, John A. Neinaber, Dennis M. Sievers, John D. Simmons, Sidney A. Thompson, Garrett L. Van Wicklen

SE-06 PROGRAM COMMITTEE

Harvey B. Manbeck, chairman
Daniel L. Wambeke, vice chairman

Terms expiring June 1989—
Kifle G. Gebremedin, Brian J. Holmes, Harvey B. Manbeck, Garrett L. Van Wicklen, Daniel L. Wambeke

SE-07 TECHNICAL ISSUES AND AWARENESS COMMITTEE

Leslie L. Christianson, chairman
Larry A. Donoghue, vice chairman
Dwaine S. Bundy, past chairman

Terms expiring June 1989—
Dwaine S. Bundy, Leslie L. Christianson

Terms expiring June 1990—
James A. Lindley, John C. Sager, Frank E. Woeste

Terms expiring June 1991—
Dennis E. Buffington, Lawrence A. Donoghue, John A. Nienaber

SE-07/1 Animal Welfare/Care Subcommittee

, chairman
Duane Sellner, vice chairman
Joseph M. Zulovich, secretary
Robert E. Graves, past chairman

Terms expiring June 1989—
Marcus J. Milanuk, Duane Sellner, Joseph M. Zulovich

Terms expiring June 1990—
Leslie L. Christianson, Duane C. Crisp, John E. Dixon, Robert E. Graves, Kevin A. Janni, Herman F. Mayes, Peter A. Phillips, Larry T. Wyatt

Terms expiring June 1991—
William A. Bailey, James A. DeShazer, Carl W. Van Gilst, G. Leroy Hahn, James E. Kirchhofer, Joe L. Koon, Donald G. Stevens

SE-20—STRUCTURES GROUP

Rodney W. Carpenter, chairman
Kifle G. Gebremedhin, 1st vice chairman
Kenneth J. Guffey, 2nd vice chairman
Sidney A. Thompson, past chairman

Terms expiring June 1989—
Donald A. Bender, William H. Bokhoven, Myron G. Britton, Calvin O. Cramer, Harvey B. Manbeck, Charles M. Milne, Virendra M. Puri, Daniel L. Wambeke, Ray W. Wilson (IN), Frank E. Woeste

Terms expiring June 1990—
Robert A. Aldrich, Ed L. Bahler, Ray Bucklin, William H. Friday, Phillip C. Hammar, George W. A. Mahoney, Ira J. Ross, Jay A. Runestad, Calvin E. Siegel, Ronald L. Sutton, Sidney A. Thompson

Terms expiring June 1991—
Dwaine S. Bundy, Rodney W. Carpenter, Lawrence A. Donoghue, Curtis D. Fankhauser, Kifle G. Gebremedhin, Kenneth J. Guffey, Michael C. Momb, Eric B. Moysey, John H. Pedersen, Gerald Riskowski, John L. Smith

SE-20/1 Grain Bin Ladders, Cages and Walkways Subcommittee

Daniel L. Wambeke, chairman

William H. Bokhoven, Rodney W. Carpenter, Curtis D. Fankhauser, Phillip C. Hammar, Neil F. Meador

SE-20/4 Loads Due to Bulk Grains and Fertilizers Subcommittee

Ray Bucklin, chairman

William H. Bokhoven, Myron G. Britton, Rodney W. Carpenter, Curtis D. Fankhauser, Phillip C. Hammar, Harvey B. Manbeck, Dwight W. Moore, Eric B. Moysey, Virendra M. Puri, Ira J. Ross, Charles V. Schwab, Sidney A. Thompson, Daniel L. Wambeke

SE-20/12 Diaphragm Design of Metal Clad Post-Frame Rectangular Buildings Subcommittee

Harvey B. Manbeck, chairman

Gary Anderson, Ed L. Bahler, Dwaine S. Bundy, Kifle G. Gebremedhin, Kenneth J. Guffey, Michael C. Momb, Calvin E. Siegel, John Turnbull, Frank E. Woeste

SE-20/13 Lateral Foundation Resistance Subcommittee

William H. Friday, chairman

Gary Anderson, Ed L. Bahler, Donald A. Bender, William H. Bokhoven Phillip C. Hammar, John L. Smith, Ronald L. Sutton, John H. Pedersen Gerald Riskowski, Leo Shirek

SE-20/14 Review of EP 388.1

William H. Friday, chairman

Ed L. Bahler, Donald A. Bender, Calvin O. Cramer, Harvey B. Manbeck, Gerald Riskowski, Frank E. Woeste

SE-30—ENVIRONMENT GROUP

Paul N. Walker (89), chairman
Brian J. Holmes (90), 1st vice chairman
Kevin A. Janni (91), 2nd vice chairman
Dwaine S. Bundy, past chairman

SE-300 ASHRAE LIAISON COMMITTEE

(Joint with Electrical and Electronic Systems Div.)

Herschel H. Klueter, chairman
Kevin A. Janni, vice chairman
Leslie L. Christianson, secretary
Kifle G. Gebremedhin, past chairman

Terms expiring June 1989—
Kifle G. Gebremedhin, Joe L. Koon, Robert A. Parsons, Richard J. Roberts

Terms expiring June 1990—
Louis D. Albright, Leslie L. Christianson, Donald G. Colliver, Jerome J. Gaffney, Abdelkader E. Ghaly, Kenneth J. Hellevang, Kevin A. Janni, Sudhir K. Sastry

Terms expiring June 1991—
Lowell E. Campbell, Herschel H. Klueter, David C. Ludington, Frank Wiersma

Chairmen of SE-30 Group Committees—
Robert A. Aldrich, Herschel H. Klueter, James L. Halderson, Brian W. Maw, George E. Meyer, Douglas G. Overhults, William H. Peterson, Gordon F. Williams, Jr.

SE-301 ENVIRONMENTAL PHYSIOLOGY COMMITTEE

(Joint with Electrical and Electronic Systems, and Soil and Water Divs.)

Brian W. Maw, chairman
Larry W. Turner, vice chairman
Thomas A. Costello, secretary
Ronald W. McClendon, past chairman

Terms expiring June 1989—
Thomas C. Bridges, Joe L. Koon, Brian W. Maw, Ronald W. McClendon, Gary Myers, John A. Nienaber

Terms expiring June 1990—
Kenneth R. Beerwinkle, Thomas A. Costello, R. Bruce Curry, Richard S. Gates, James W. Jones, Pierce H. Jones, Larry W. Turner

Terms expiring June 1991—
John R. Barrett, Gary L. Barker, Lung Hua Chen, Timothy McDonald, James M. McKinion, Wesley Rosenthal, Dennis P. Stombaugh, James Usry, Donald Wanjura, Joseph Zulovich

SE-302 ENVIRONMENT OF ANIMAL STRUCTURES COMMITTEE

(Joint with Electrical and Electronic Systems Div.)

Douglas G. Overhults, chairman
Garrett L. Van Wicklen, vice chairman
John A. Nienaber, secretary
Larry D. Jacobson, past chairman

Terms expiring June 1989—
Robert W. Bottcher, Donell P. Froehlich, Raymond L. Huhnke, Larry D. Jacobson, Kevin A. Janni, Gerard Muegerl

Terms expiring June 1990—
Larry E. Christenson, Leslie L. Christianson, Mark Doyle, Robert L. Fehr, John A. Nienaber, Douglas G. Overhults, Jerald Plessing

Terms expiring June 1991—
Eldridge R. Collins Jr., Brian J. Holmes, Bruce A. Martin, James P Murphy, Garrett L. Van Wicklen, Frank Wiersma

SE-303 ENVIRONMENT OF PLANT STRUCTURES COMMITTEE

(Joint with Electrical and Electronic Systems Div.)

Robert A. Aldrich, chairman
Robert J. Downs, vice chairman
Eldon Muller, secretary
Gene A. Giacomelli, past chairman

Terms expiring June 1989—
Robert A. Aldrich, Harold E. Gray, Michael Krones, William A. Bailey, Roy E. Young, Paul N. Walker, Lora J. Sandholm

Terms expiring June 1990—
Louis D. Albright, John W. Bartok, Michael F. Brugger, Robert J. Downs, Gene A. Giacomelli, David S. Ross, Robert Rynk, K. C. Ting

Terms expiring June 1991—
Maynard Bates, James H. Dooley, Eldon R. Muller, W. J. Roberts, J. C. Sager, T. H. Short, J. F. Thompson, D. H. Willits

Advisory members:
Donald T. Krizek, T. W. Tibbits

SE-304 ENVIRONMENT OF STORED PRODUCTS COMMITTEE

(Joint with Electrical and Electronic Systems Div. and Food and Process Engineering Inst.)

James L. Halderson, chairman
James H. Hunter, vice chairman
Secretary to be selected
Lewis A. Schaper, past chairman

Terms expiring June 1989—
John C. Anderson, Denny C. Davis, Nathan H. Gellert, Suhas R. Ghate, James H. Hunter, David N. Moutner, Errol D. Rodda

Terms expiring June 1990—
Richard W. Allen, Roger C. Brook, Martin L. Hellickson, Sudhir K. Sastry, James F. Thompson, Henry Waelti

Terms expiring June 1991—
Jon M. Carson, James L. Halderson, Kenneth J. Hellevang, Lewis A. Schaper, Douglas Small

SE-307 CLIMATE AND METEOROLOGY COMMITTEE

(Joint with Soil and Water Div.)

George E. Meyer, chairman
James M. McKinion, vice chairman
Secretary to be appointed
G. Morgan Powell, past chairman

Terms expiring June 1989—
Dale E. Linvill, George E. Meyer, Fred V. Nurnberger, John F. Reid

Terms expiring June 1990—
Galen Campbell, Robert F. Cullum, R. Bruce Curry, Joyce A. Fox, Dale E. Linvill, James M. McKinion, John W. Mishoe, Kevin Robbins

Terms expiring June 1991—
Robert J. Edling, Robert D. Fox, James W. Jones, Marshall J. McFarland, Myron P. Molnau, G. Morgan Powell

Non-member advisors:
H. Christie, A. Richardson, B. D. Tanner

American Meteorological Society—Dale E. Linvill, liaison representative

SE-308 SOLAR ENERGY COMMITTEE

(Joint with Electrical and Electronic Systems Div.)

William H. Peterson, chairman
Shahab Sokhansanj, vice chairman
William Wilcke, secretary
Fred C. Vosper, past chairman

Terms expiring June 1989—
Marcus J. Milanuk, Gerard Muegerl, William H. Peterson, Virendra M. Puri, Thomas R. Rumsey

Terms expiring June 1990—
Edgar J. Carnegie, Dennis L. Carson, R. Nolan Clark, Jerry O. Newman, Blaine F. Parker, Fred C. Vosper, William F. Wilcke

Terms expiring June 1991—
George Abernathy, Kamyar Enshayan, R. Peter Fynn, K. C. Ping, Shahab Sokhanfanj

SE-309 ENVIRONMENTAL AIR QUALITY COMMITTEE

Gordon F. Williams, Jr., chairman
Eugene P. Columbus, vice chairman
Robert E. Harrison, secretary
John M. Sweeten, Jr., past chairman

Terms expiring June 1989—
Jefferson D. Bargeron III, Glenn E. Hall, Robert E. Harrison, John M. Sweeten, Jr., Gordon F. Williams, Jr.

Terms expiring June 1990—
Dwaine S. Bundy, Donald G. Colliver, Eugene P. Columbus, Albert J. Heber, Charles R. Martin, J. Ronald Miner, Calvin B. Parnell, Jr.

Terms expiring June 1991—
Roy V. Baker, N. Ross Bulley, Peter E. Hillman, S. Ed Hughes, Andrew G. Jordan,

SE-309/1 Cotton Dust Subcommittee

S. Ed Hughs, chairman
Roy V. Baker, Jr., Robert Bethea, Andrew G. Jordan, Ivan W. Kirk, Gino J. Mangialardi, William D. Mayfield, J. H. (Tony) Price, Gordon F. Williams, Jr.

SE-309/2 Grain Dust Subcommittee

Gordon F. Williams, Jr., chairman
David F. Aldis, Glenn E. Hall, Robert Jacko, Fang S. Lai, Charles R. Martin, Bruce W. Moechnig, Calvin B. Parnell, Jr., Gary I. Wallin

SE-309/3 Air Quality for Structures Subcommittee

John M. Sweeten, Jr., chairman
Dwaine S. Bundy, Donald G. Colliver, Albert J. Heber, Peter E. Hillman, Kevin A. Janni, J. Ronald Miner

SE-40—SYSTEMS GROUP

Lawson M. Safley, Jr., chairman
Garrett L. Van Wicklen, 1st vice chairman
Douglas G. Overhults, 2nd vice chairman
Bruce A. Martin, past chairman

SE-401 HUMAN HOUSING COMMITTEE

Mary Ellen Bodman, chairman
Thomas H. Greiner, vice chairman
Larry R. Piercy, secretary
Jacob Pos, past chairman

Terms expiring June 1989—
Donald G. Colliver, Homer T. Hurst, Larry R. Piercy, Jacob Pos

Terms expiring June 1990—
Harry J. Braud, Jr. Axel R. Carlson, Donald L. Day, Thomas H. Greiner, Donald G. Jedele, James A. Lindley, David R. Mears, Jerry O. Newman, Jerome R. Smith

Terms expiring June 1991—
Mary Ellen Bodman, Dennis E. Buffington, Ben N. Cox, John J. Kolega, George W. A. Mahoney, Roger A. Miller, John D. Simmons, Barbara J. Wiersma

SE-402 BEEF HOUSING COMMITTEE

Charles J. Clanton, chairman
Howard L. Person, vice chairman
Joseph C. Garner, secretary
James P. Murphy, past chairman

Terms expiring June 1989—
Ray Bucklin, Charles J. Clanton, Joseph C. Garner, John George (KS), Conrad B. Gilbertson, Don D. Jones

Terms expiring June 1990—
Dexter W. Johnson, James P. Murphy, Howard L. Person, Mark A. Wiwi

Terms expiring June 1991—
William H. Collins, Robert L. Mensch, John M. Sweeten, Jr., Larry W. Turner

SE-402/1 Beef Dimensions Subcommittee

Mark A. Wiwi, chairman

SE-402/2 Horses Subcommittee

Dexter W. Johnson, chairman

SE-403 DAIRY HOUSING COMMITTEE

Brian J. Holmes, chairman
Larry W. Turner, vice chairman
Ray Bucklin, secretary
Steven A. Larson, past chairman

Terms expiring June 1989—
Donald W. Bates, Ray Bucklin, Robert E. Graves, Dexter W. Johnson, Wilmot W. Irish, Nevin W. Wagner, Harold R. Wakefield

Terms expiring June 1990—
Phillip D. A. Johnson, Rodney O. Martin, Steven E. Taylor, Larry W. Turner, Jerry L. Wille, Larry T. Wyatt

Terms expiring June 1991—
William G. Bickert, Michael F. Brugger, William H. Collins, Lynn D. George, Brian J. Holmes, Steven A. Larson

SE-404 SWINE HOUSING COMMITTEE

Marcus J. Milanuk, chairman
Arthur J. Muehling, vice chairman
Robert M. Butler, secretary
Douglas G. Overhults, past chairman

Terms expiring June 1989—
L. Bynum Driggers, Robert L. Fehr, Warren D. Goetsch, Vernon M. Meyer, John A. Nienaber, Douglas G. Overhults

Terms expiring June 1990—
Larry E. Christenson, Larry D. Jacobson, Marcus J. Milanuk, Arthur J. Muehling, Peter A. Phillips, Steve H. Pohl, Gerald Riskowski

Terms expiring June 1991—
Robert M. Butler, Ray H. Crammond, Jay Harmon, Daniel J. Meyer, Michael A. Veenhuizen, Nevin W. Wagner

SE-405 POULTRY HOUSING SYSTEMS COMMITTEE

Robert W. Bottcher, chairman
Olin L. Vanderslice, vice chairman
Joe L. Koon, secretary
Michael B. Timmons, past chairman

Terms expiring June 1989—
Robert W. Bottcher, Dennis E. Buffington, Joe L. Koon, John P. H. Mason, Ralph P. Prince, Charles C. Ross

Terms expiring June 1990—
Gerald R. Baughman, Neil E. Blackwell, Jon M. Carson, Michael Czarick, III, Richard S. Gates, Bruce A. Martin, John D. Simmons, Michael B. Timmons, Joseph M. Zulovich

Terms expiring June 1991—
James M. Allison, Ivan L. Berry, Lewis E. Carr, Daniel G. Hansen, Olin L. Vanderslice, Garrett L. Van Wicklin

SE-412 AGRICULTURAL SANITATION AND WASTE MANAGEMENT COMMITTEE

(Joint with Soil and Water Div.)

David C. Moffitt, chairman
David T. Hill, vice chairman
David Friederick, secretary
James A. Moore, past chairman

Terms expiring June 1989—
Charles D. Fulhage, Abdelkader E. Ghaly, David J. Hills, Marvin L. Knabach, David C. Moffitt, Charles C. Ross, Lawson M. Safley, Jr., Dennis D. Schulte, Philip W. Westerman

Terms expiring June 1990—
Ten Hong Chen, Charles J. Clanton, Clifford B. Fedler, Larry R. Johnson, James N. Krider, Jack D. Messner, Orville J. Nelson, Victor W. E. Payne

Terms expiring June 1991—
James C. Converse, Robert F. Cullum, Donald Joe Gribble, J. David Jelinski, Joseph L. Taraba, Stanley A. Weeks, Richard K. White, Ping-Yi Yang

Terms expiring June 1992—
John P. Burt, Donald L. Day, David Friederick, David T. Hill, James A. Moore, Dennis M. Sievers, John M. Sweeten Jr.

Poultry Science Association, Intersociety Committee on Environmental Quality—Donald L. Day, representative

SE-412/3 Energy and Nutrient Recovery Subcommittee

Dennis D. Schulte, chairman

SE-412/5 Engineering Practices Subcommittee

Lawson M. Safley, Jr., chairman

JOINT COMMITTEES

(These committees are administered by other technical divisions, but incorporate technical interests of the Structures and Environment Division.)

Farm Materials Handling Committee, FPE-708
Grain and Feed Processing and Storage Committee, FPE-702
Rural Water Supplies Committee, SW-214
Land Application of Waste Committee, SW-263

CURRENT ASAE STANDARDS PROJECTS

(January 1988)

The following projects to develop new ASAE standards and projects to revise existing ASAE standards are being undertaken by various ASAE committees shown below. Persons interested in these projects should contact the appropriate committee (see roster of Technical Committees in this book) or ASAE Standards Department. Projects to revise existing documents are identified by an asterisk.

GENERAL ENGINEERING FOR AGRICULTURE

A-205, Agricultural Mechanics Education Committee

- *X415 Safety Color Code for Educational and Training Laboratories

FPE-702, Grain and Feed Processing and Storage Committee

- *X271.2 Psychrometric Data

AGRICULTURAL EQUIPMENT (Field and Farmstead)

T-2, Agricultural Safety Committee

- *X279.8 Lighting and Marking of Agricultural Equipment on Highways
- *X338.1 Safety Chain for Towed Equipment
- *X365.1T Brake Test Procedures and Brake Performance Criteria for Agricultural Equipment
- X493 Guarding for Agricultural Equipment

EES-41, Milk Handling Equipment Committee

- *X300.1 Terminology for Milking Machine Systems
- X445 Test Equipment and Its Application for Measuring Milking Machine Operating Characteristics

FPE-702, Grain and Feed Processing and Storage Committee

- *X303.2 Test Procedures for Solids-Mixing Equipment for Animal Feeds

FPE-708, Farm Materials Handling Committee

- *X361.2 Safety for Agricultural Auger Conveying Equipment

PM-03, Power and Machinery Division Standards Committee

- X482 Drawbars - Agricultural Wheel Tractors
- X485 Screw Type Swivel Mounted Jacks

PM-03/3, Wagon Box, Forage Box, Manure Spreader and Farm Wagon Subcommittee

- *X317 Improving Safety on Enclosed Mobile Tanks for Transporting and Spreading Agricultural Liquids and Slurry
- *X373 Safety for Self-Unloading Forage Boxes

PM-03/4, Agricultural Loader Subcommittee

- *X301.2 Front-End Agricultural Loader Ratings
- *X355.1 Safety for Agricultural Front End Loaders

PM-03/8, Implement Tire Subcommittee

- *X385.3 Combine Harvester Tire Loading and Inflation Pressures
- X430 Agricultural Equipment Tire Loadings and Inflation Pressures

PM-03/9, SMV Emblem Subcommittee

- *X276.3 Slow-Moving Vehicle Identification Emblem

PM-41, Agricultural Pest Control and Fertilizer Application Committee

- *X327.1 Terminology and Definitions for Agricultural Chemical Application
- X469 Sprayer Tank Agitation
- X471 Sprayer Nozzle Wear

PM-41/1, Liquid Materials Application Subcommittee

- *X387.1 Test Procedures for Measuring Deposits and Airborne Spray from Ground Swath Sprayers

PM-42, Cultural Practices Equipment Committee

- *X236 Guide for Planning and Reporting Tillage Experiments
- *X291.1 Terminology and Definitions for Soil Tillage and Soil-Tool Relationships
- *X414 Terminology and Definitions for Agricultural Tillage Implements
- X477 Terminology for Soil-Engaging Components for Planters, Drills, and Seeders

PM-43, Farm Machinery Management Committee

- *X495 Uniform Terminology for Agricultural Machinery Management (formerly S322)
- *X496 Agricultural Machinery Management (formerly S391)
- *X497 Agricultural Machinery Management Data (formerly D230)

PM-44, Forage Harvesting and Utilization Committee

- *X328.1 Dimensions for Compatible Operation of Forage Harvesters, Forage Wagons and Forage Blowers

---

*Project to revise an existing ASAE document.

PM-44/2, Forage Length-of-Cut Standard Subcommittee

*X424 Method of Determining and Expressing Particle Size of Chopped Forage Materials by Screening

PM-44/3, Forage Harvesting Standards Subcommittee

*X472 Terminology for Forage Harvesters and Forage Harvesting

PM-44/4, Round Baler Standards Subcommittee

X498 Terminology for Round Balers

PM-45, Soil Dynamics Research Committee

*X313.2 Soil Cone Penetrometer

PM-51, Tractor and Implement Hydraulic Committee

X489 Hydraulic Pressure Available for Remote Use on Agricultural Tractors

PM-58, Agricultural Equipment Automation Committee

X434 Aperture for Entry of Electrical Wiring Into Agricultural Tractor Cabs

PM-61, Human Factors Committee

*X304.5 Symbols for Operator Controls on Agricultural Equipment

POWERED LAWN AND GARDEN EQUIPMENT

PM-52, Small Tractor and Power Equipment Committee

X432 Towed Loads and Braking for Lawn and Garden Tractors

X478 Operator Protection Systems for Lawn and Garden Ride-On Tractors

PM-52/2, Implement and Attachment Subcommittee

X431 2000 RPM Front and Midmount PTO for Lawn and Garden Tractors

ELECTRICAL AND ELECTRONIC SYSTEMS

EES-33, Agricultural Wiring and Utilization Committee

X473 Equipotential Planes for Livestock Farm Areas

EES-42, Electrical Controls for Farmstead Equipment Committee

*X417.1 Specifications for Alarm Systems Utilized in Agricultural Structures

X500 Performance Standard for Electric Fence Controllers

EES-42/1, Protection of Electronic Controls Subcommittee

X499 Electrical Surge and Transient Protection of Stationary Electronic Controls and Equipment

PM-58, Agricultural Equipment Automation Committee

X455 Environmental Considerations in Development of Mobile Agricultural Electrical/Electronic Components

*Project to revise an existing ASAE document.

FOOD AND PROCESS ENGINEERING

EES-41, Milk Handling Equipment Committee

*X254.3 Uniform Terminology for Bulk Milk Handling

FPE-701, Physical Properties of Agricultural Products Committee

*X245.4 Moisture Relationships of Grain

*X368.1 Compression Test of Food Materials for Convex Shape

FPE-702, Grain and Feed Processing and Storage Committee

*X248.3 Construction and Rating of Equipment for Drying Farm Crops

*X274 Flow of Wheat Through Orifices

X448 Thin Layer Drying

FPE-703, Food Processing Committee

X451 Steam Flow Measurement

X453 Dehydration During Freezing and Blanching

X454 Filler Efficiency

FPE-704, Special Crops Processing Committee

X488 Energy Efficiency of Peanut Drying Equipment

STRUCTURES, LIVESTOCK, AND ENVIRONMENT

SE-20, Structures Group

*X250.2 Specifications for Farm Fence Construction

*X252.1 Tower Silos: Unit Weight for Silage and Silo Capacities

*X378.3 Floor and Suspended Loads on Agricultural Structures Due to Use

*X388.1 Design Properties of Round, Sawn and Laminated Preservatively Treated Construction Poles and Posts

*X401.1 Use of Thermal Insulation in Agricultural Buildings

*X412T Grain Bin Ladders, Cages and Walkways

X486 Structural Foundation Embedment

SE-20/4, Loads Due to Bulk Grains and Fertilizers Subcommittee

*X413 Procedure for Establishing Volumetric Capacity of Grain Bins

X433 Loads Exerted by Free Flowing Grains on Bins

SE-20/12, Diaphragm Design of Roofing and Cladding on Farm Buildings Subcommittee

X484 Diaphragm Design of Metal Clad Post Frame Rectangular Buildings

SE-302, Environment of Animal Structures Committee

*X270.5 Design of Ventilation Systems for Poultry and Livestock Shelters

*X282.1 Design Values for Livestock Fallout Shelters

SE-303, Environment of Plant Structures Committee

*X406 Heating, Ventilating and Cooling Greenhouses

X460 Greenhouse Construction

SE-304, Environment of Stored Products Committee

X475 Commodity Storage Environments

SE-308, Solar Energy Committee

X423 Testing Solar Heating Equipment (Agricultural & Residential)

SE-403, Dairy Housing Committee

*X444 Terminology and Recommendation for Free Stall Dairy Housing Free Stalls, Feed Bunks, and Feeding Fences

SE-404, Swine Housing Committee

X470 Manure Storage Safety

SE-412, Agricultural Sanitation and Waste Management Committee

X461 Feedlot Runoff Controls

X462 Guidelines for Designing and Managing Systems to Land Apply Livestock Manure

SE-412/5, Engineering Practices Subcommittee

*X384 Manure Production and Characteristics

*X393 Solid and Liquid Manure Storages

*X403 Design of Anaerobic Lagoons for Animal Waste Management

SOIL AND WATER RESOURCE MANAGEMENT

SW-214, Rural Water Supplies Committee

*X339 Uniform Classification for Water Hardness

SW-225, Conservation Systems Committee

X425 Underground Outlets for Conservation Practices

X464 Grassed Waterways for Agricultural Runoff

X492 Design, Construction and Maintenance of Diversions

*Project to revise an existing ASAE document.

SW-232, Subsurface Drainage in Humid Areas Committee

*X260.4 Design and Construction of Subsurface Drains in Humid Areas

X479 Design, Installation and Operation of Subirrigation Systems

X480 Design of Subsurface Drains in Humid Areas

X481 Construction of Subsurface Drains in Humid Areas

SW-233, Surface Drainage Committee

*X407 Agricultural Drainage Outlets—Open Channels

SW-24, Irrigation Group

X491 Graphic Symbols for Irrigation Design

SW-241, Sprinkler Irrigation Committee

X439 Procedure for Testing and Reporting Low Pressure Spray Distribution Device Performance as Used on Mechanical Move Irrigation Equipment

X494 Traveler Irrigation Machines—Distribution Uniformity Test Method

SW-242, Surface Irrigation Committee

X490 Terminology for Irrigation Scheduling and Irrigation Efficiency

SW-243, Irrigation Water Supply and Conveyance Committee

*X376.1 Design, Installation and Performance of Underground, Thermoplastic Irrigation Pipelines

X437 PVC Aboveground Irrigation Pipe

X438 Minimum Requirements for Irrigation Pumping Plants

SW-244, Irrigation Management Committee

*X409 Safety Devices for Applying Liquid Chemicals Through Irrigation Systems

SW-245, Microirrigation Committee

*X405 Design, Installation and Performance of Trickle Irrigation Systems

X458 Field Evaluation of Trickle Irrigation Emitters and Systems

X467 Irrigation Equipment Emitting-Pipe Systems

X468 Irrigation Equipment Emitters Specifications and Test Methods

Subject Index to ASAE Standards, Engineering Practices, Data

See also numerical index of standards, engineering practices, and data on page 622.

ASAE Notation:

The letter S preceding numerical designation indicates ASAE Standard; EP indicates Engineering Practice; D indicates Data. A decimal and numeral following the file number indicate the number of times a document has been revised. Thus ASAE S201.4 indicates Standard number 201, four times revised. The letter T after the designation indicates tentative status. Always refer to ASAE documents by complete designation to avoid confusion with standards of other organizations. For example: ASAE S201.4.

The symbol **T** preceding or in the margin adjacent to section headings, paragraph numbers, figure captions, or table headings indicates a technical change was incorporated in that area when this document was last revised. The symbol **T** preceding the title of a document indicates essentially the entire document has been revised. The symbol **E** used similarly indicates editorial changes or corrections have been made with no intended change in the technical meaning of the document.

A

*New standards since last publication.
†Revised or reclassified standards since last publication.
‡Approved by ANSI as an American National Standard.
§Earlier revision approved by ANSI as an American National Standard.
||Reconfirmed since last publication.

B

*New standards since last publication.
†Revised or reclassified standards since last publication.
‡Approved by ANSI as an American National Standard.
§Earlier revision approved by ANSI as an American National Standard.
|| Reconfirmed since last publication.

D

*New standards since last publication.
†Revised or reclassified standards since last publication.
‡Approved by ANSI as an American National Standard.
§Earlier revision approved by ANSI as an American National Standard.
||Reconfirmed since last publication.

E

---

*New standards since last publication.
†Revised or reclassified standards since last publication.
‡Approved by ANSI as an American National Standard.
§Earlier revision approved by ANSI as an American National Standard.
||Reconfirmed since last publication.

*New standards since last publication.
†Revised or reclassified standards since last publication.
‡Approved by ANSI as an American National Standard.
§Earlier revision approved by ANSI as an American National Standard.
||Reconfirmed since last publication.

*New standards since last publication.
†Revised or reclassified standards since last publication.
‡Approved by ANSI as an American National Standard.
§Earlier revision approved by ANSI as an American National Standard.
||Reconfirmed since last publication.

I

L

*New standards since last publication.
†Revised or reclassified standards since last publication.
‡Approved by ANSI as an American National Standard.
§Earlier revision approved by ANSI as an American National Standard.
||Reconfirmed since last publication.

M

O

*New standards since last publication.
†Revised or reclassified standards since last publication.
‡Approved by ANSI as an American National Standard.
§Earlier revision approved by ANSI as an American National Standard.
||Reconfirmed since last publication.

P

Q

R

*New standards since last publication.
†Revised or reclassified standards since last publication.
‡Approved by ANSI as an American National Standard.
§Earlier revision approved by ANSI as an American National Standard.
||Reconfirmed since last publication.

S

*New standards since last publication.
†Revised or reclassified standards since last publication.
‡Approved by ANSI as an American National Standard.
§Earlier revision approved by ANSI as an American National Standard.
||Reconfirmed since last publication.

T

*New standards since last publication.
†Revised or reclassified standards since last publication.
‡Approved by ANSI as an American National Standard.
§Earlier revision approved by ANSI as an American National Standard.
||Reconfirmed since last publication.

*New standards since last publication.
†Revised or reclassified standards since last publication.
‡Approved by ANSI as an American National Standard.
§Earlier revision approved by ANSI as an American National Standard.
||Reconfirmed since last publication.

U

V

W

---

*New standards since last publication.
†Revised or reclassified standards since last publication.
‡Approved by ANSI as an American National Standard.
§Earlier revision approved by ANSI as an American National Standard.
||Reconfirmed since last publication.

Numerical Index to ASAE Standards, Engineering Practices, Data

ASAE Notation:

The letter S preceding numerical designation indicates ASAE Standard; EP indicates Engineering Practice; D indicates Data. A decimal and numeral following the file number indicate the number of times a document has been revised. Thus ASAE S201.4 indicates Standard number 201, four times revised. The letter T after the designation indicates tentative status. Always refer to ASAE documents by complete designation to avoid confusion with standards of other organizations. For example: ASAE S201.4.

The symbol **T** preceding or in the margin adjacent to section headings, paragraph numbers, figure captions, or table headings indicates a technical change was incorporated in that area when this document was last revised. The symbol **T** preceding the title of a document indicates essentially the entire document has been revised. The symbol **E** used similarly indicates editorial changes or corrections have been made with no intended change in the technical meaning of the document.

*New standards since last publication.
†Revised or reclassified standards since last publication.
‡Approved by ANSI as an American National Standard.
§Earlier revision approved by ANSI as an American National Standard.
‖Reconfirmed since last publication.

*New standards since last publication.
†Revised or reclassified standards since last publication.
‡Approved by ANSI as an American National Standard.
§Earlier revision approved by ANSI as an American National Standard.
||Reconfirmed since last publication.

*New standards since last publication.
†Revised or reclassified standards since last publication.
‡Approved by ANSI as an American National Standard.
§Earlier revision approved by ANSI as an American National Standard.
||Reconfirmed since last publication.

*New standards since last publication.
†Revised or reclassified standards since last publication.
‡Approved by ANSI as an American National Standard.
§Earlier revision approved by ANSI as an American National Standard.
||Reconfirmed since last publication.

DOCUMENTS WITHDRAWN FROM PUBLICATION AS ASAE STANDARDS WITHIN THE LAST FIVE YEARS:

Published last in 1987:

S412T Grain Bin Ladders, Cages and Walkways

Published last in 1983:

S297.1T Enclosure-Type Shielding of Forward Universal Joint and Coupling Means of Agricultural Implement Power Drive Lines (Superseded by S318.7)

INTERNATIONAL STANDARDS

These International Standards, relating to ASAE's scope of technical interest, were developed within the International Organization for Standardization, ISO, by Technical Committee ISO/TC23, Tractors and Machinery for Agriculture and Forestry.

Copies are available from the American National Standards Institute, the United States member body of ISO, at prices ranging from $13 to $43 US each. Copies are also available from the Standards Council of Canada and other national member bodies of ISO.

ANSI	SCC
1430 Broadway	2000 Argentia Rd, Suite 2-401
New York, NY 10018	Mississauga, Ontario L5N 1V8

ISO 500-1979	Agricultural tractors - Power take-off and drawbar - Specification
ISO 563-1981	Equipment for harvesting - Knife sections for agricultural cutter bars
ISO 730/1-1977	Agricultural wheeled tractors - Three-point linkage - Part 1: Categories 1, 2 and 3
ISO 730/2-1979	Agricultural wheeled tractors - Three-point linkage - Part 2: Category 1 N (Narrow hitch)
ISO 730/3-1982	Agricultural wheeled tractors - Three-point linkage - Part 3: Category 4
ISO 789/1-1981	Agricultural tractors - Test procedures - Part 1: Power tests
ISO 789/2-1983	Agricultural tractors - Test procedures - Part 2: Hydraulic power and lifting capacity
ISO 789/3-1982	Agricultural tractors - Test procedures - Part 3: Turning and clearance diameters
ISO 789/4-1986	Agricultural tractors - Test procedures - Part 4: Measurement of exhaust smoke
ISO 789/5-1983	Agricultural tractors - Test procedures - Part 5: Partial power PTO - Non-mechanically transmitted power
ISO 789/6-1982	Agricultural tractors - Test procedures - Part 6: Centre of gravity
ISO 2057-1981	Agricultural tractors - Remote control hydraulic cylinders for trailed implements
ISO 2288-1979	Agricultural tractors and machines - Engine test code (bench test) - Net power
	Amendement 1-1983
ISO 2332-1983	Agricultural tractors and machinery - Connections - Clearance zone for the three-point linkage of implements
ISO 3339/0-1986	Tractors and machinery for agriculture and forestry - Classification and terminology - Part 0: Classification system and classification Bilingual edition
ISO 3462-1980	Tractors and machinery for agriculture and forestry - Seat reference point - Method of determination
ISO 3463-1984	Agricultural and forestry wheeled tractors - Protective structures - Dynamic test method and acceptance conditons
ISO 3600-1981	Tractors and machinery for agriculture and forestry - Operator manuals and technical publications - Presentation
ISO 3737-1976	Agricultural tractors and self-propelled machines - Test method for enclosure pressurization systems
ISO 3767/1-1982	Tractors, machinery for agriculture and forestry, powered lawn and garden equipment - Symbols for operator controls and other displays - Part 1: Common symbols
	Addendum 1-1985
ISO 3767/2-1982	Tractors, machinery for agriculture and forestry, powered lawn and garden equipment - Symbols for operator controls and other displays - Part 2: Symbols for agricultural tractors and machinery
	Addendum 1-1985
ISO 3776-1976	Agricultural tractors - Anchorages for seat belts
ISO/TR 3778-1987*	Agricultural tractors - Maximum actuating forces required to operate controls
ISO 3789/1-1982	Tractors, machinery for agriculture and forestry, powered lawn and garden equipment - Location and method of operation of operator controls - Part 1: Common controls
ISO 3789/2-1982	Tractors, machinery for agriculture and forestry, powered lawn and garden equipment - Location and method of operation of operator controls - Part 2: Controls for agricultural tractors and machinery
ISO 3789/3-1982	Tractors, machinery for agriculture and forestry, powered lawn and garden equipment - Location and method of operation of operator controls - Part 3: Controls for powered lawn and garden equipment
ISO 3835/1-1976	Equipment for vine cultivation and wine making - Vocabulary - Part 1 Bilingual edition
ISO 3835/2-1977	Equipment for vine cultivation and wine making - Vocabulary - Part 2 Bilingual edition
ISO 3835/3-1980	Equipment for vine cultivation and wine making - Vocabulary - Part 3 Bilingual edition
ISO 3835/4-1981	Equipment for vine cultivation and wine making - Vocabulary - Part 4 Bilingual edition
ISO 3835/5-1982	Equipment for vine cultivation and wine making - Vocabulary - Part 5 Bilingual edition

*ISO/TR, technical reports, provide either interim progress reports or factual information/data different from that normally incorporated in an International Standard.

ISO 3918-1977	Milking machine installations - Vocabulary Bilingual edition
ISO 3965-1977	Agricultural wheeled tractors - Determination of maximum travel speed
ISO 3971-1977	Rice milling - Symbols and equivalent terms Bilingual edition
ISO 4002/1-1979	Equipment for sowing and planting - Part 1: Concave disks type D1 - Dimensions
ISO 4002/2-1977	Equipment for sowing and planting - Disks - Part II: Flat disks type D2 with single bevel - Dimensions
ISO 4004-1983	Agricultural tractors and machinery - Track widths
ISO 4102-1984	Equipment for crop protection - Sprayers - Connection threading
ISO/TR 4122-1977*	Equipment for working the soil - Dimensions of flat disks - Type A
ISO 4197-1981	Equipment for working the soil - Hoe blades - Fixing dimensions
ISO 4252-1983	Agricultural tractors - Access, exit and the operator's workplace - Dimensions
ISO 4253-1977	Agricultural tractors - Operator's seating accommodation - Dimensions
ISO 4254/1-1985	Tractors and machinery for agriculture and forestry - Technical means for ensuring safety - Part 1: General
ISO 4254/2-1986	Tractors and machinery for agriculture and forestry - Technical means for providing safety - Part 2: Anhydrous ammonia applicators
ISO/TR 5007-1980*	Agricultural wheeled tractors - Operator seat - Measurement of transmitted vibration
ISO 5008-1979	Agricultural wheeled tractors and field machinery - Measurement of whole-body vibration of the operator
ISO 5395/1-1984	Power lawn mowers, lawn tractors, and lawn and garden tractors with mowing attachments - Safety requirements and test procedures - Part 1: Definitions Bilingual edition
ISO 5395/2-1981	Power lawn mowers, lawn tractors, and lawn and garden tractors with mowing attachments - Safety requirements and test procedures - Part 2: Basic requirements
ISO 5395/3-1985	Power lawn mowers, lawn tractors, and lawn and garden tractors with mowing attachments - Safety requirements and test procedures - Part 3: Requirements for rotary mowers Addendum 1-1986
ISO 5395/4-1983	Power lawn mowers, lawn tractors, and lawn and garden tractors with mowing attachments - Safety requirements and test procedures - Part 4: Requirements for cylinder (reel) mowers
ISO 5669-1982	Agricultural trailers and trailed equipment - Braking cylinders - Specifications
ISO 5670-1984	Agricultural trailers - Single-acting telescopic tipping cylinders - 25 MPa (250 bar) series - Types 1, 2 and 3 - Interchangeability dimensions
ISO 5673-1980	Agricultural tractors - Power take-off drive shafts for machines and implements
ISO 5674-1982	Tractors and machinery for agriculture and forestry - Guards for power take-off drive shafts - Test methods
ISO 5675-1981	Agricultural tractors and machinery - Hydraulic couplers for general purposes - Specifications
ISO 5676-1983	Tractors and machinery for agriculture and forestry - Hydraulic coupling - Braking circuit
ISO 5678-1979	Equipment for working the soil - Clearance zones and main dimensions for S-Type cultivator tines
ISO 5679-1979	Equipment for working the soil - Disks - Classification, main fixing dimensions and specifications
ISO 5680-1979	Equipment for working soil - Tines and shovels for cultivators - Main fixing dimensions
ISO 5681-1981	Equipment for crop protection - Vocabulary
ISO 5682/1-1981	Equipment for crop protection - Spraying equipment - Part 1: Test methods of sprayer nozzles
ISO 5682/2-1986	Equipment for crop protection - Spraying equipment - Part 2: Test methods for agricultural sprayers
ISO 5687-1981	Equipment for harvesting - Combine harvester - Determination and designation of grain tank capacity and unloading device performance
ISO 5690/1-1985	Equipment for distributing fertilizers - Test methods - Part 1: Full width fertilizer distributors
ISO 5690/2-1984	Equipment for distributing fertilizers - Test methods - Part 2: Fertilizer distributors in lines
ISO 5691-1981	Equipment for planting - Potato planters - Method of testing
ISO 5692-1979	Agricultural vehicles - Mechanical connections on towed vehicles - Hitch-rings - Specifications
ISO 5696-1984	Trailed agricultural vehicles - Brakes and braking devices - Laboratory test method
ISO 5697-1982	Agricultural and forestry vehicles - Determination of braking performance
ISO 5698-1979	Agricultural machinery - Hoppers - Manual loading height
ISO 5699-1979	Agricultural machines, implements and equipment - Dimensions for mechanical loading with bulk goods
ISO 5700-1984	Agricultural and forestry wheeled tractors - Protective structures - Static test method and acceptance conditions
ISO 5702 1983	Equipment for harvesting - Combine harvester component parts - Equivalent terms Trilingual edition
ISO 5703-1979	Equipment for vine cultivation and wine making - Grape presses - Methods of test
ISO 5704-1980	Equipment for vine cultivation and wine making - Grape-harvesting machinery - Test methods
ISO 5707-1983	Milking machine installations - Construction and performance
ISO 5708-1983	Refrigerated bulk milk tanks

*ISO/TR, technical reports, provide either interim progress reports or factual information/data different from that normally incorporated in an International Standard.

ISO 5709-1981	Equipment for internal farm work and husbandry - Metal grids for cattle stalls
ISO 5710-1980	Equipment for internal farm work and husbandry - Continuous scraper conveyors for stalls
ISO 5715-1983	Equipment for harvesting - Dimensional compatibility of forage harvesting machinery
ISO 5717-1981	Equipment for harvesting - Combine harvester - Sickle guards of fingers - Dimensions
ISO 5721-1981	Agricultural tractors - Operator's field of vision
ISO 6094-1981	Agricultural and forestry tractors - Accidental overturning report form
ISO 6097-1983	Agricultural tractors and self-propelled machines - Performance of heating and ventilation systems in closed cabs - Method of test
ISO 6489/1-1980	Agricultural vehicles - Mechanical connections on towing vehicles - Part 1: Hook type - Dimensions
ISO 6489/2-1980	Agricultural vehicles - Mechanical connections on towing vehicles - Part 2: Clevis type - Dimensions
ISO 6531-1982	Machinery for forestry - Portable chain saws - Vocabulary Trilingual edition
ISO 6532-1982	Machinery for forestry - Portable chain saws - Technical data
ISO 6533-1983	Forestry machinery - Portable chain saws - Front hand guard - Dimensions
ISO 6534-1985	Forestry machinery - Portable chain-saws - Front hand-guard - Determination of strength
ISO 6535-1983	Forestry machinery - Portable chain saws - Chain brake - Performance
ISO 6686-1981	Equipment for crop protection - Anti-drip devices - Determination of reduction of nozzle flow rate
ISO 6687-1982	Machinery for foresty - Winches - Performance requirements
ISO 6688-1982	Machinery for forestry - Disc trenches - Disc-to-hub flange fixing dimensions
ISO 6689-1981	Equipment for harvesting - Combines and functional components - Definitions, characteristics and performance
ISO 6690-1983	Milking machine installations - Mechanical tests
ISO 6720-1981	Agricultural machinery - Equipment for sowing, planting, distributing fertilizers and spraying - Recommended working widths
ISO 6814-1983	Machinery for forestry - Mobile and self-propelled machinery - Identification vocabulary
ISO 6815-1983	Machinery for forestry - Hitches - Dimensions
ISO 6816-1984	Machinery for forestry - Winches - Classification and nomenclature
ISO 6880-1983	Machinery for agriculture - Trailed units of shallow tillage equipment - Main dimensions and attachment points
ISO 7071-1981	Agricultural tractors - Mounting for front ballast weights
ISO 7072-1982	Agricultural wheeled tractors - Three-point-linkage - Linch pins - Dimensions
ISO 7112-1982	Machinery for forestry - Portable brush saws - Vocabulary Trilingual edition
ISO 7113-1982	Machinery for forestry - Brush saws - Saw blades - Dimensions and peripheral speed
ISO 7182-1984	Acoustics - Measurement at the operator's position of airborne noise emitted by chain saws (*see also T 43)*
ISO 7224-1983	Equipment for vine cultivation and wine making - Mash pumps - Methods of test
ISO 7256/1-1984	Sowing equipment - Test methods - Part 1 - Single seed drills (precision drills)
ISO 7256/2-1984	Sowing equipment - Test methods - Part 2: Seed drills for sowing in lines
ISO 7293-1983	Forestry machinery - Portable chain saws - Engine performance and fuel consumption
ISO 7505-1986	Forestry machinery - Chain saws - Measurement of hand-transmitted vibration
ISO 7714-1985	Irrigation equipment - Volumetric valves - General requirements and test methods
ISO 7749/1-1986	Irrigation equipment - Rotating sprinklers - Part 1: Design and operational requirements
ISO 7914-1986	Forestry machinery - Portable chain saws - Minimum handle clearance and sizes
ISO 7915-1985	Forestry machinery - Portable chain saws - Handles - Determination of strength
ISO 7918-1985	Forestry machinery - Portable brush-saws - Circular saw-blade guard - Dimensions
ISO 8016-1985	Machinery for agriculture - Wheels with integral hub
ISO 8026-1985	Irrigation equipment - Irrigation sprayers - General requirements and test methods
ISO/TR 8059-1986*	Irrigation equipment - Automatic irrigation systems - Hydraulic control
ISO 8169-1984	Equipment for crop protection - Sprayers - Connecting dimensions for nozzles and manometers
ISO 8224/1-1985	Traveller irrigation machines - Part 1: Laboratory and field test methods
ISO 8334-1985	Forestry machinery - Portable chain-saws - Determination of balance
ISO 8380-1985	Forestry machinery - Portable brush-saws - Circular saw-blade guard - Strength
ISO 8524-1986	Equipment for distributing granulated pesticides or herbicides - Test method
ISO 8759/1-1985	Agricultural wheeled tractors - Front-mounted linkage and power take-off - Part 1: Power take-off
ISO 8759/2-1985	Agricultural wheeled tractors - Front-mounted linkage and power take-off - Part 2: Front linkage
ISO 8910-1987	Machinery and equipment for working the soil - Mouldboard ploughs - Working elements - Vocabulary Trilingual edition
ISO 8912-1986	Equipment for working the soil - Roller sections - Coupling and section width

*ISO/TR, technical reports, provide either interim progress reports or factual information/data different from that normally incorporated in an International Standard.

ORDERING INFORMATION

Standards Books

- Additional copies of this volume are available at $65.00 list; $27.00 to members of ASAE.*
- ASAE Standards books for most years, 1954 to the present, are also available at above prices.

Individual Standards

- Copies of individual Standards, either current or superseded, are available at $10.00 each; $5.00 to members of ASAE.* A 10% discount applies to orders of 10 or more copies. Quantity prices are quoted on request.

Microfilm

ASAE Standards on microfilm are available from Information Handling Services, P. O. Box 1154, Englewood, CO 80150 (303) 790-0600.

---

*U.S. Dollars. Prices subject to change.

Contact:

ASAE Order Department
2950 Niles Road
St. Joseph, MI 49085-9659 USA
Phone: 616-429-0300 (8 am to 5 pm Eastern Time Zone)
After hours leave message on recorder.
FAX: 616-429-3852